MECHANICS OF SOLIDS

The Rodney Hill
60th Anniversary Volume

RODNEY HILL

MECHANICS OF SOLIDS

The Rodney Hill 60th Anniversary Volume

Edited by

H. G. HOPKINS and M. J. SEWELL

PERGAMON PRESS

OXFORD · NEW YORK · TORONTO · SYDNEY · PARIS · FRANKFURT

U.K.	Pergamon Press Ltd., Headington Hill Hall, Oxford OX3 0BW, England
U.S.A.	Pergamon Press Inc., Maxwell House, Fairview Park, Elmsford, New York 10523, U.S.A.
CANADA	Pergamon Press Canada Ltd., Suite 104, 150 Consumers Road, Willowdale, Ontario M2J 1P9, Canada
AUSTRALIA	Pergamon Press (Aust.) Pty. Ltd., P.O. Box 544, Potts Point, N.S.W. 2011, Australia
FRANCE	Pergamon Press SARL, 24 rue des Ecoles, 75240 Paris, Cedex 05, France
FEDERAL REPUBLIC OF GERMANY	Pergamon Press GmbH, 6242 Kronberg-Taunus, Hammerweg 6, Federal Republic of Germany

First edition 1982

British Library Cataloguing in Publication Data

Mechanics of solids.
1. Mechanics – Addresses, essays, lectures
I. Hopkins, H G
II. Sewell, M J
531 QC127.6 80–40995

ISBN 0–08–025443–8

Printed in Great Britain by A. Wheaton & Co. Ltd., Exeter
Phototypeset by Macmillan India Ltd., Bangalore

Editors' Preface

Rodney Hill celebrates his sixtieth birthday on the 11th June 1981. This Volume is presented to him by the contributors, as their glad tribute to a man whose dedication of purpose and high level of achievement have permanently influenced developments in large and basic areas of the mechanics of solids, especially, although not exclusively, in the theory and applications of plasticity in metals.

We are confident that the Volume will be accompanied by good wishes to him, on this anniversary and for the future, from numerous scientists of international standing in many countries. As Editors, we decided at the outset that the occasion called for the planning of a substantial Volume having some balanced structure, in the hope that it might achieve a degree of permanence in the literature of the subject. The eighteen articles contained herein therefore provide a mosaic of selected subject areas. The choice was made from topics which Rodney Hill himself has sought to illuminate at one time or another, or from certain topics immediately adjacent thereto in either physical or mathematical or engineering terms. This policy of a restricted number of substantial articles implied, regretfully, that we were not able to invite papers from all who we are sure would have wished to contribute.

We regard it as a privilege to have had the opportunity to edit the Volume. It is a pleasure to record the alacrity with which contributors gave their support, and the ready co-operation from many people which has helped the project to move ahead smoothly. We make no attempt here to give a technical assessment of the work of Rodney Hill, which happily is still continuing strongly. However, we have felt that the inclusion of a general biographical sketch, and also of a complete current bibliography, would be widely appreciated. In the preparation of these we have been greatly assisted by a number of people. We express our thanks to Rodney himself, for his ready response to our requests for autobiographical source material. We are also grateful to several associates from the different periods of Rodney's career for being willing to exercise their memories for us and provide recollections of events sometimes many years ago. The photograph was taken by Edward Leigh of King's Parade, Cambridge.

Particular thanks are due to the Publishers for their most helpful collaboration with us, and for their agreement to publish the Volume given when it was yet unwritten, but existed simply as a proposal in the form of a list of authors and subject areas; and also to Mr. Robert Maxwell himself for his encouraging personal interest.

H. G. Hopkins, *Manchester* M. J. Sewell, *Reading*
30 *April* 1980

Editors' Preface

Contents

Rodney Hill

Biographical Note

RODNEY HILL was born on the 11th June 1921 at Stourton, near Leeds, in Yorkshire. He comes from a family with deep roots in the practical and cultural traditions of the West Riding, although with no known mathematical ability in an earlier generation. Rodney's father, Harold Harrison Hill, had been an only child and he was educated at the University of Leeds, gaining an M.A. for postgraduate work in history. He also took an external London degree in economics. After wartime service in the Royal Navy he became a schoolmaster, and was eventually Senior History Master at Leeds Boys' Modern School. Rodney's mother had been a student at Leeds School of Art. Rodney himself was also an only child, in an immediate home background which encouraged scholarship and self-sufficiency.

Rodney entered Leeds Grammar School with a scholarship in 1932, and there gave regular prize-winning evidence of all-round intellectual ability not only in mathematics, but equally in art, English literature, and other Arts subjects. During this period he taught himself to play the piano, and became proficient at chess in which he was later to represent Cambridge University and town. Thus were developing the powers of accurate observation and analysis to be brought to bear on the mathematics and physics which became his formal specialism from the age of 15. The customary large-team games did not attract him at school, but Rodney enjoyed the one-to-one sports of squash, fencing, and golf. He left school as Head of House, and in December 1938 he was awarded an Open Major Scholarship at Pembroke College, Cambridge. However, it needed the State and County Scholarships gained in the preceding summer to make a financially independent undergraduate.

Hill went up to Cambridge to read Mathematics in October 1939, against a background of external events which must have seemed the least auspicious since the very founding of the University. Major Scholars were expected to take Part II of the Tripos in two years instead of three by omitting all first-year courses. This imposed a heavy workload, to be carried under spartan conditions created by wartime restrictions such as blackout and rationing combined with antique College plumbing. For example, there was no running hot water, the nearest bath was two courts away, and the winter allocation of one sack of coal per week fuelled a fire in one's room only in the evenings. Hill was not deflected by the adverse general situation from his aim of a first-class honours degree, and he became a Wrangler in June 1941. This entitled him to take Part III of the Mathematical Tripos, in the applied mathematical part of which quantum mechanics figured prominently at that time. However, he felt obliged to volunteer for war-work, and so lost the opportunity for advanced training which those lecture courses would have provided.

The Mathematical Laboratory at Cambridge University had become established by the beginning of the war, with Professor J. E. Lennard-Jones, F.R.S., as its Director, and it soon became closely allied to the External Ballistics Department of the Ordnance Board under the control of the Ministry of Supply. Hill was directed to work in the Laboratory in the late summer of 1941. He was assigned the task of calculating the trajectories of shells for the compilation of range tables, by small-arc iteration on desk machines such as the Brunsviga. After some months in that activity Hill began helping to operate the Bush differential analyser for solving the internal ballistics equations ("internal" indicates that the shell is still travelling down the gun barrel). In this context he was able to show the existence of connected families of solutions, so reducing the need for separate computations, and thus wrote his first research paper.

The developing war effort in 1942 saw the appointment of Lennard-Jones as Chief Superintendent of Armament Research, with a staff of ultimately over 3000 under his authority in various parts of the country, in an Armament Research Department reconstituted from the old Research Department at Woolwich Arsenal. Lennard-Jones made his headquarters at Fort Halstead, an isolated but attractive location on a spur of the North Downs near Sevenoaks in Kent. Among his innovations was the setting up of new research groups headed by various scientists of distinction brought in for the purpose. Professor Nevill Mott, F.R.S., became the first superintendent of a group for theoretical research in armaments, which became a permanent feature at Fort Halstead. In the spring of 1943 Mott assembled a team of able young scientists from Cambridge, Woolwich, and elsewhere, many of whom subsequently went on to influential and distinguished careers in universities or government service. They included J. W. Maccoll (Mott's deputy), J. Corner, D. C. Pack, A. F. Devonshire, L. Howarth, E. H. Lee, I. N. Sneddon, C. K. Thornhill, and J. H. Wilkinson. Rodney Hill joined the team in May 1943 and he stayed for three years.

It can be seen that this was a pivotal date in the career not only of Hill, but also of many other now well-known scientists. Their attention was brought at a formative age to a range of practical and urgent problems, and their contribution deserves to be widely recognized. The atmosphere in the Theoretical Research Branch was stimulating, with Mott frequently sketching the line of approach to a problem almost on the back of a postcard, and expecting his recruits to fill in the details. The Branch was responsible for advising military designers about both basic theory and day-to-day problems concerning the performance of explosives, fragmentation bombs, internal and external and terminal ballistics, attack and defence of tanks and other armoured vehicles, and so on. For example, they were involved in reconstructing the German V-2 rocket from intelligence reports and estimating its range.

The physical background to this activity was utterly different from Cambridge. Fort Halstead was a vast establishment not only of civilian but also of military personnel, with war equipment much in evidence. A balloon barrage was strung out overhead along the North Downs. V-1 flying bombs and V-2 rockets descended all around and in the areas like Petts Wood nearer London where staff were billeted in private houses. In fact Fort Halstead lay within what became known as "flying bomb alley", and one flying bomb landed in the Establishment, blasting windows in the Theoretical Research Branch. Fortunately this was at night, otherwise the Branch could have suffered heavy casualties.

Problems brought to the Theoretical Research Branch were distributed initially according to the specialisms of the more senior members, some of whom had acquired relevant experience at Woolwich Arsenal. Those problems which were quite new in context tended to go to the young inexperienced graduates newly arrived from university. This was indeed a baptism of fire for them, but it was a test which was to reveal Hill's true métier. One of his initial assignments was the deep penetration of very thick armour by Munroe jets and by high-velocity shells with tungsten–carbide cores. This required a mechanics of plastic deformation with unlimited magnitude, and thus was aroused Hill's interest in the field in which he later became perhaps the foremost world authority. At this stage, however, he had no prior knowledge of the physics and metallurgy of plasticity, and little of stress, strain or the tensors which the mathematics would eventually require. There was no useful textbook, but G. I. Taylor had written one or two helpful reports on shaped charges and Munroe jets. Nevertheless, working at first with Mott and Pack, Hill was soon able to show, for example, that penetration by a tungsten–carbide core with pure ogival head would be seriously degraded if too much of the tip were ground conical (the British practice for manufacturing convenience). The demonstration was achieved not only theoretically, but also in field trials planned by Hill in collaboration with an experimental group under Dr. Charles Sykes, F.R.S.

In such ways Hill acquired a training in pertinent science in the prime of his youth, and the influence of it may be recognized repeatedly in his later work. The problems at Fort Halstead called for simple but effective mathematics guided by physical intuition and a willingness to communicate with others, including non-mathematicians and experimentalists. There was not time for complicated mathematics, there were no electronic computers to assist it, and the experimental data were usually too crude to warrant it anyway. He acquired a lasting taste for a pragmatic blend of rigour, elegance, and simple realism in the application of mathematics.

The sense of purpose discovered at this time was noticed by colleagues as a cheerful and sparkling earnestness. Popular relaxations among the group at Cambridge had included music, books, and lightning chess. At Fort Halstead ballroom dancing was added as a consuming passion for some, and Hill was not slow to find that he had medal-winning ability in this new enthusiasm. He met his future wife, Jeanne Wickens, early in 1945. She had been transferrred to work at Fort Halstead from the bombing range at Shoeburyness. Previously she had trained as a dancer and teacher of ballet, but the war cut short a promising career. They were married in Cambridge in 1946, and they have one daughter, born in 1955. The strength of his wife's support could already be detected in the Preface to Hill's first book, and the passage of years has happily reinforced this bond.

In April 1946 Hill was seconded by the Ministry of Supply to work with a group of metal physicists under Dr. E. Orowan, F.R.S., at the Cavendish Laboratory in Cambridge. By this time the applied mechanics of both solids and fluids was being forced to push the boat out onto a sea of nonlinear problems, and away from the haven of linearity in which much pre-war work had lingered. The trend was evident not only in England, of course, but in other countries too. Hill found himself in demand as the sole adviser on continuum plasticity in England, not only concerning problems arising from the interests at Fort Halstead, but also for new theories of metal-working processes needed by engineers in the steel industry. He obtained a Cambridge Ph.D. in 1948 for a Thesis entitled "Theoretical studies of the plastic deformation of metals". In February 1949 he moved to Sheffield, having accepted an invitation to head a new Section in the Metal Flow Research Laboratory of the British Iron and Steel Research Association.

From the Ph.D. Thesis grew a much more extensive monograph on *The Mathematical Theory of Plasticity*, published at the Clarendon Press, Oxford, in 1950. This very rapidly established Hill as an international authority. The final draft was written in his spare time, i.e. in the evenings and weekends left to him after discharging his responsibilities at B.I.S.R.A. He was then still only in his 28th year, and it is timely to recall a remark from the review of the book in *Engineering*: "The author has done his work so well that it is difficult to see how it could be bettered. The book should rank for many years as an authoritative source of reference." This prognostication was fully borne out. The book was in print at Oxford for 21 years, Japanese and Russian translations have been made, and total sales currently approach 13 000.

The work in Sheffield produced new opportunities for collaboration with practical laboratory work, for example on rolling and wire drawing. Once again the physical environment was very different from Cambridge. This was before the days of Clean Air Acts, and the Laboratory was situated in a smoky valley, one of the grimiest and most derelict areas of the city. The theoreticians themselves were housed in a temporary hut while main buildings were being constructed. In compensation, however, Hill enjoyed walking with his wife on the moors outside the city at weekends. At the suggestion of Professor H. W. Swift, Hill gave courses on slipline field analysis in the Mechanical Engineering Department at Sheffield University. Swift was very helpful in promoting these new ideas within the Iron and Steel Institute and in the industry. At this time, too, Hill was joined by A. P. Green and shortly afterwards by J. F. W. Bishop, who thus became the first of a later sequence of postgraduate students wishing to learn the new subject of mathematical plasticity from him at first hand.

The next move was to Bristol University in September 1950. A three-year Research Fellowship was created for Hill, jointly in the Departments of Physics and Theoretical Mechanics. As his list of publications reveals, joint papers were written here from the standpoints both of physics (relating single crystal behaviour to that of a polycrystalline aggregate—with Bishop who had moved with him) and of engineering (bars and tubes under combined loading—with M. P. L. Siebel and B. Crossland). For the latter, advantage was taken of the laboratory facilities in the Mechanical Engineering Department, with the encouragement of Professor J. L. M. Morrison. There was also the opportunity for more teaching and related reading. A new family relaxation taken up at this time was the study of field botany and mycology. Ever since, and in many parts of the English countryside, this has proved an absorbing diversion for the mind from the exigencies of a sustained research programme.

The *Journal of Mechanics and Physics of Solids* was launched with the encouragement of the infant Pergamon Press in 1952. Hill suggested the title and the general aim of a forum for effective applied mathematics, linked with experimentation, in engineering science. Any new journal must begin energetically if it is to stamp out an identity for itself in the international community, and Hill took the lead by contributing 10 articles (some jointly) to the total of 55 which appeared in the first two Volumes. From then onwards the *Journal* has been regarded as among the foremost in its general field, and unique in flavour. Hill served as Editor-in-Chief until handing over in 1968 to H. G. Hopkins.

The University of Nottingham had received its Charter, and independence from London, in 1948, and was shortly to embark on two decades of substantial expansion. Professor H. R. Pitt was appointed in 1950 to head the existing Mathematics Department, and he was soon instrumental in securing the creation of a new Chair of Applied Mathematics. Rodney Hill applied, and was offered the post in 1953 while still only 31. It was his responsibility to modernize the teaching of applied mathematics. Hill took over some existing courses himself, and instigated new ones with the aim of encouraging research students. His undergraduate lectures were characterized by conciseness and a tendency for brevity. He would never exceed the time limit. But those students who took the trouble to write down what he said, in addition to what was written on the blackboard, found after reflection that they had a first-class and substantial set of notes.

Funds gradually became available for essential new staff in applied mathematics, and J. E. Adkins was appointed. He was followed later by two of Hill's Ph.D. students,

R. J. Knops and M. J. Sewell. Another major innovation which he sought was the creation, in 1960, of a Department of Theoretical Mechanics to teach and do research within the Faculty of Engineering. This was headed first by Adkins as Professor until his untimely death in 1964, and since by A. J. M. Spencer, and is known as a flourishing and influential Department. Hill was also consulted by the Mining Engineering Department on subsidence problems. These were evident in many parts of the city, which is built over part of the Nottinghamshire coalfield. He proposed a theory of ground movement based on the idea that extracting a seam of coal by the longwall method resulted in an edge dislocation of the rock strata. The idea was developed by D. S. Berry, who led a stress-analysis section recruited for the purpose in the Mining Department.

During this period Hill undertook the administrative tasks customarily associated with an established Chair, and without previous experience of such work. If the quantity of his personal research decreased because of this, there was also a change to work of a broader and more general scope. It may only have been a coincidence that the emergence of interest in the so-called rational continuum mechanics was taking place in some American and British universities at this time. Hill's writings demonstrate an independent view of these developments, and no taste at all for axiomatics. He was beginning to lay down the basis of general studies of non-uniqueness and instability in continua which were to prove highly influential over the next two decades, and which in due course brought further students and able collaborators.

The University of Cambridge conferred the degree of Sc.D. upon Rodney Hill in 1959. The highest honour to which any British scientist aspires followed in 1961, when he was elected a Fellow of the Royal Society. This gave much pleasure to his colleagues at Nottingham and to his friends elsewhere.

By this time Nottingham University was widely recognized as a centre of activity in applied mathematics, and in the mechanics of solids in particular, and Professor Hill could justifiably feel that his various efforts to that end had borne fruit. He resigned from his Chair as from October 1962, and the University conferred on him an Honorary Professorship while he remained in Nottingham for the ensuing academic year. During this time Hill found considerable satisfaction in the writing of an undergraduate textbook on *Principles of Dynamics*. This was the product of 10 years spent re-thinking and teaching classical dynamics. It is a subject which, as the book's Preface shows, Hill regards as an exemplar of scientific theory and method, which is indispensable for students of structural engineering, applied mathematics and theoretical physics alike. The book does not shirk conceptual difficulties, and it leads the reader to think deeply and afresh about the subject.

In 1963 Hill was elected to a Berkeley Bye-Fellowship (for research) at Gonville and Caius College, Cambridge. This he held for 6 years until the University conferred a personal Readership in the Mechanics of Solids. Thus he became a member of the teaching staff of the Department of Applied Mathematics and Theoretical Physics, and in 1972 a personal Professorship was conferred. During this Cambridge period the foundations laid at Nottingham for the unambiguous formulation of incremental boundary value problems at large strain were built upon in various directions. Properties of heterogeneous media (including fibre composites), single crystals, continuum plasticity, and an independent reformulation of rubber elasticity were explored. A sequence of new research students found fruitful employment in these or older fields. They included I. F. Collins, D. J. F. Ewing, J. P. Miles, R. W. Ogden, G. P. Parry, L. J. Walpole, and N. J. B. Young. During the seventies Hill was gratified to be able to collaborate with several distinguished scholars from abroad, who chose to come to Cambridge for their sabbatical leave in order to work with him.

This is not the place to offer an assessment of Rodney Hill's impact on the theoretical and applied mechanics of solids. His list of scientific articles speaks for itself. Currently (February 1980) there are 2 books and 132 articles, 44 of the latter being written jointly with a total of 20 collaborators. He has himself contributed to *Festschriften* for G. I. Taylor, N. I. Muskhelishvili, W. Prager, F. K. G. Odqvist, V. V. Novozhilov, and A. A. Ilyushin. His style of writing is spare and concise. Every word is weighed, and must be weighed again by the reader. Mathematics has frequently to be read with paper and pencil to hand—they are essential when reading Hill. He does not offer an easy ride—he would not think one worth while. A comment that this or that paper was not easy to read may draw the swift reply that it was not easy to write. But the reader who persists finds a rich vein of understanding, and fresh thinking at every turn.

His standards of scholarship and intellectual honesty are the highest. He is ready in his appreciation of the good work of others; and he has been sharp in candid criticism of misguided thinking or slack presentation (especially by those mature enough to know better) if he thought the subject-matter would be best served thereby—as some celebrated footnotes and book reviews testify.

The outward character of the man is not unlike his papers: physically tall and slim, with the long fingers of a pianist, and having a quiet but compelling presence. His unusually deep reserve has meant that casual social gatherings and conferences have held less interest and been less rewarding for him than for others. Some lack of robustness in health has militated against unnecessary travel. Thus, his General Lectures given on such occasions as the 9th International Congress of Theoretical and Applied Mechanics at Brussels in 1956, or the 3rd British Theoretical Mechanics Colloquium at Newcastle upon Tyne in 1961, have been rare events.

On the other hand, in circumstances where his contribution could be telling his effort has been unstinting. His students testify to the very thorough and considerate way in which he supervised their research work, with a just and effective balance of encouragement and the measured push, whichever was currently deserved. His colleagues, if they would have preferred to have known him better, have respected his desire for privacy and understood his wish to optimize the unique contribution which he has made.

The honorary degree of Doctor of Science was conferred upon Rodney Hill by the University of Manchester in 1976, and by the University of Bath in 1978. Also in 1978 he was awarded the Theodore von Kármán Medal, instituted in 1960 for annual award by the American Society of Civil Engineers for "distinguished achievement in engineering mechanics". The only previous British recipient of this medal was G. I. Taylor.

In 1979 he resigned his Chair, but was elected to another Fellowship at Gonville and Caius College, Cambridge. His flow of papers shows no sign of diminishing, and his friends will hope that that short legend "by R. Hill" will appear in the scientific literature many more times yet.

GEOFFREY HOPKINS — MICHAEL SEWELL

List of Publications by Rodney Hill

A. Scientific Papers

1943a (With J. Crank.) Families of solutions of the internal ballistics equation, U.K. Ministry of Defence (Procurement Executive), unpublished report.

b (With J. Crank.) Effect of various types of band resistance on gun ballistics, U.K. Ministry of Defence (Procurement Executive), unpublished report.

1944a (With N. F. Mott and D. C. Pack.) Penetration by Munroe jets, U.K. Ministry of Defence (Procurement Executive), unpublished report.

b (With D. C. Pack and N. F. Mott.) Penetration of armour by high-velocity projectiles and Munroe jets, U.K. Ministry of Defence (Procurement Executive), unpublished report.

c (With D. C. Pack.) Penetration into wood, U.K. Ministry of Defence (Procurement Executive), unpublished report.

d Cavitation phenomena in ductile materials and the dynamic term in the resistance to penetration, U.K. Ministry of Defence (Procurement Executive), unpublished report.

1945a The analysis of projectile penetration of non-ferrous ductile materials, U.K. Ministry of Defence (Procurement Executive), unpublished report.

b The theory of armour penetration by tungsten carbide shot and the effect of headshape on performance at high velocities, U.K. Ministry of Defence (Procurement Executive), unpublished report.

c (With E. H. Lee.) The weakening of a gun barrel due to the presence of a hole normal to the axis, U.K. Ministry of Defence (Procurement Executive), unpublished report.

d (With N. F. Mott and R. F. Bishop.) The theory of indentation and hardness tests. *Proc. Phys. Soc.* **57,** 147–159.

1947a Critical survey of soil mechanics with suggestions for future research, U.K. Ministry of Defence (Procurement Executive), unpublished report.

b (With E. H. Lee and S. J. Tupper.) The theory of wedge indentation of ductile materials, *Proc. R. Soc. London* A **188,** 273–289.

c (With E. H. Lee and S. J. Tupper.) The theory of combined plastic and elastic deformation with particular reference to a thick tube under internal pressure, *Proc. R. Soc. London* A **191,** 278–303.

d (With D. C. Pack.) An investigation, by the method of characteristics, of the lateral expansion of gases behind a detonating slab of explosive, *Proc. R. Soc. London* A **191,** 524–541.

e Comments on a paper by E. Siebel, *J. Iron Steel Inst.* **156,** 513.

1948a A theory of earth movement near a deep underground explosion, U.K. Ministry of Defence (Procurement Executive), unpublished report.
b Remarks on the ORG theory of tank sinkage, U.K. Ministry of Defence (Procurement Executive), unpublished report.
c An analysis of RRL experiments on cavitation in mild steel, U.K. Ministry of Defence (Procurement Executive), unpublished report.
d Theoretical analysis of the stresses and strains in extrusion and piercing, *J. Iron Steel Inst.* **158,** 177–185.
e A theory of yielding and plastic flow of anisotropic metals, *Proc. R. Soc. London* A **193,** 281–297.
f A variational principle of maximum plastic work in classical plasticity, *Q. Jl Mech. Appl. Math.* **1,** 18–28.
g (With S. J. Tupper.) Theory of plastic deformation in wire drawing, *J. Iron Steel Inst.* **159,** 353–359.
h Special problems of indentation and compression, *Proceedings of the Seventh International Congress for Applied Mechanics, London, 5–11 September 1948,* **1,** 365–377, published by the Organizing Committee.
1949a The theory of plane plastic strain for anisotropic metals, *Proc. R. Soc. London* A **198,** 428–437.
b The calculation of stresses in the ironing of metal cups, *J. Iron Steel Inst.* **161,** 41–44.
c General features of plastic–elastic problems as exemplified by some particular solutions, *J. Appl. Mech.* **16,** 295–300.
d Plastic distortion of non-uniform sheets, *Phil. Mag.* **40,** 971–983.
e Plastic yielding of notched bars, *Q. Jl Mech. Appl. Math.* **2,** 40–52.
1950a A comparative study of some variational principles in the theory of plasticity, *J. Appl. Mech.* **17,** 64–66.
b Relations between roll-force, torque, and applied tensions in strip rolling, *Proc. Instn Mech. Engrs* **163,** 135–140.
c (With L. Bourne.) On the correlation of the directional properties of rolled sheet in tension and cupping tests, *Phil. Mag.* **41,** 671–681.
d Inhomogeneous deformation of a plastic lamina in a compression test, *Phil. Mag.* **41,** 733–744.
e A theoretical investigation of the effect of specimen size in the measurement of hardness, *Phil. Mag.* **41,** 745–753.
f A theory of the plastic bulging of a metal diaphragm by lateral pressure, *Phil. Mag.* **41,** 1133–1142.
1951a (With M. P. L. Siebel.) On combined bending and twisting of thin tubes in the plastic range, *Phil. Mag.* **42,** 722–733.
b (With E. H. Lee and S. J. Tupper.) A method of numerical analysis of plastic flow in plane strain and its application to the compression of a ductile material between rough plates, *J. Appl. Mech.* **18,** 46–52.
c (With I. M. Longman.) Cold rolling of very thin strip, *Sheet Metal Ind.* 705–706.
d On the state of stress in a plastic–rigid body at the yield point, *Phil. Mag.* **42,** 868–875.
e (With J. F. W. Bishop.) A theory of the plastic distortion of a polycrystalline aggregate under combined stresses, *Phil. Mag.* **42,** 414–427.
f (With J. F. W. Bishop.) A theoretical derivation of the plastic properties of a polycrystalline face-centred metal, *Phil. Mag.* **42,** 1298–1307.

1952a (With R. F. Bishop and N. F. Mott.) The theory of indentation and hardness tests, *Selected Government Research Reports*, Vol. 6, *Strength and Testing of Materials*, Part I, *Theoretical Papers on Strength and Deformation*, pp. 1–16, Her Majesty's Stationery Office, London.

b (With E. H. Lee and S. J. Tupper.) The theory of wedge indentation of ductile materials, *ibid.*, pp. 17–35.

c (With E. H. Lee.) The theory of wedge penetration at oblique incidence and its application to the calculation of forces on a yawed shot impacting on armour plate at any angle, *ibid.*, pp. 36–46.

d (With E. H. Lee and S. J. Tupper.) The theory of combined plastic and elastic deformation with particular reference to the deformation of a thick tube under internal pressure, *ibid.*, pp. 47–76.

e (With E. H. Lee and S. J. Tupper.) The theory of the autofrettage of a closed-end tube, *ibid.*, pp. 77–87.

f (With E. H. Lee and S. J. Tupper.) A method of numerical solution of the general plane plastic problem, and its application to the compression of a ductile material between rough plates, *ibid.*, pp. 88–104.

g The two-dimensional theory of plastic yielding of deeply-notched bars in tension, *ibid.*, pp. 177–190.

h The theory of the extrusion of metals, *ibid.*, pp. 191–204.

i (With S. J. Tupper.) A theoretical analysis of the stresses and strains in wire-drawing, *ibid.*, pp. 205–224.

j General methods of solution of the axially symmetric problem in plasticity, *ibid.*, pp. 225–246.

k A theory of the yielding and plastic flow of anisotropic metals, *ibid.*, pp. 247–260.

l The elastic behaviour of a crystalline aggregate, *Proc. Phys. Soc.* A **65,** 349–354.

m A note on estimating yield-point loads in a plastic–rigid body, *Phil. Mag.* **43,** 353–355.

n On discontinuous plastic states, with special reference to localized necking in thin sheets, *J. Mech. Phys. Solids* **1,** 19–30.

o (With A. P. Green.) Calculations on the influence of friction and die geometry in sheet drawing, *J. Mech. Phys. Solids* **1,** 31–36.

p (With E. H. Lee and S. J. Tupper.) Plastic flow in a closed ended tube with internal pressure, *Proceedings of the First U.S. National Congress of Applied Mechanics, Chicago, 11–16 June 1951*, edited by E. Sternberg, 561–567, American Society of Mechanical Engineers, New York.

1953a Some reflections on the hollow charge problem, U.K. Ministry of Defence (Procurement Executive), unpublished report.

b Further reflections on the hollow charge problem, U.K. Ministry of Defence (Procurement Executive), unpublished report.

c A note on the back-pull factor in strip drawing, *J. Mech. Phys. Solids* **1,** 142–145.

d (With M. P. L. Siebel.) On the plastic distortion of solid bars by combined bending and twisting, *J. Mech. Phys. Solids* **1,** 207–214.

e On the mechanics of cutting metal strips with knife-edged tools, *J. Mech. Phys. Solids* **1,** 265–270.

f A new method for determining the yield criterion and plastic potential of ductile metals, *J. Mech. Phys. Solids* **1,** 271–276.

1953g (With B. Crossland.) On the plastic behaviour of thick tubes under combined torsion and internal pressure, *J. Mech. Phys. Solids* **2,** 27–38.

1954a The plastic torsion of anisotropic bars, *J. Mech. Phys. Solids* **2,** 87–91.

b On Inoue's hydrodynamical analogy for the state of stress in a plastic solid, *J. Mech. Phys. Solids* **2,** 110–116.

c On the limits set by plastic yielding to the intensity of singularities of stress, *J. Mech. Phys. Solids* **2,** 278–285.

d The mechanics of machining: a new approach, *J. Mech. Phys. Solids* **3,** 47–53.

1955a Towards a rationale of eccentricities of hollow charges, U.K. Ministry of Defence (Procurement Executive), unpublished report.

b On related pairs of plane elastic states, *J. Mech. Phys. Solids* **4,** 1–9.

1956a The mechanics of quasi-static plastic deformation in metals, *Surveys in Mechanics—The G. I. Taylor 70th Anniversary Volume*, edited by G. K. Batchelor and R. M. Davies, 7–31, Cambridge University Press.

b On the problem of uniqueness in the theory of a rigid–plastic solid—I, *J. Mech. Phys. Solids* **4,** 247–255.

c (With A. P. Green and J. F. W. Bishop.) A note on the deformable region in a rigid–plastic body, *J. Mech. Phys. Solids* **4,** 256–258.

d On the problem of uniqueness in the theory of a rigid–plastic solid—II, *J. Mech. Phys. Solids* **5,** 1–8.

e New horizons in the mechanics of solids, *J. Mech. Phys. Solids* **5,** 66–74.

f (With G. Power.) Extremum principles for slow viscous flow and the approximate calculation of drag, *Q. Jl Mech. Appl. Math.* **9,** 313–319.

1957a On the problem of uniqueness in the theory of a rigid–plastic solid—III, *J. Mech. Phys. Solids* **5,** 153–161.

b On uniqueness and stability in the theory of finite elastic strain, *J. Mech. Phys. Solids* **5,** 229–241.

c The problem of uniqueness in the theory of rigid–plastic solids—IV, *J. Mech. Phys. Solids* **5,** 302–307.

d Stability of rigid–plastic solids, *J. Mech. Phys. Solids* **6,** 1–8.

1958 A general theory of uniqueness and stability in elastic–plastic solids, *J. Mech. Phys. Solids* **6,** 236–249.

1959 Some basic principles in the mechanics of solids without a natural time, *J. Mech. Phys. Solids* **7,** 209–225.

1960a (With M. J. Sewell.) A general theory of inelastic column failure—I, *J. Mech. Phys. Solids* **8,** 105–111.

b (With M. J. Sewell.) A general theory of inelastic column failure—II, *J. Mech. Phys. Solids* **8,** 112–118.

1961a Bifurcation and uniqueness in non-linear mechanics of continua, *Problems of Continuum Mechanics* (Contributions in Honor of the Seventieth Birthday of Academician N. I. Muskhelishvili, 16th February 1961), edited by M. A. Lavrent'ev *et al.* (English edition edited by J. R. M. Radok), 155–164, Society for Industrial and Applied Mathematics, Philadelphia.

b Discontinuity relations in mechanics of solids, *Progress in Solid Mechanics*, edited by I. N. Sneddon and R. Hill, **2,** 247–276, North-Holland, Amsterdam.

c Uniqueness in general boundary-value problems for elastic or inelastic solids, *J. Mech. Phys. Solids* **9,** 114–130.

1962a Theories of elastic behaviour of reinforced solids, British Iron and Steel Research Association, Sheffield, Report P/19.
b Acceleration waves in solids, *J. Mech. Phys. Solids* **10,** 1–16.
c Constitutive laws and waves in rigid/plastic solids, *J. Mech. Phys. Solids* **10,** 89–98.
d Uniqueness criteria and extremum principles in self-adjoint boundary-value problems of continuum mechanics, *J. Mech. Phys. Solids* **10,** 185–194.
e (With M. J. Sewell.) A general theory of inelastic column failure—III, *J. Mech. Phys. Solids* **10,** 285–300.
1963a New derivations of some elastic extremum principles, *Progress in Applied Mechanics—The Prager 60th Anniversary Volume*, 99–106, Macmillan, New York.
b A general method of analysis for metal-working processes, *J. Mech. Phys. Solids* **11,** 305–326.
c Elastic properties of reinforced solids: some theoretical principles, *J. Mech. Phys. Solids* **11,** 357–372.
1964a Theory of mechanical properties of fibre-strengthened materials—I. Elastic behaviour, *J. Mech. Phys. Solids* **12,** 199–212.
b Theory of mechanical properties of fibre-strengthened materials—II, Inelastic behaviour, *J. Mech. Phys. Solids* **12,** 213–218.
1965a Continuum micro-mechanics of elastoplastic polycrystals, *J. Mech. Phys. Solids* **13,** 89–101.
b Theory of mechanical properties of fibre-strengthened materials—III, Self-consistent model, *J. Mech. Phys. Solids* **13,** 189–198.
c A self-consistent mechanics of composite materials, *J. Mech. Phys. Solids* **13,** 213–222.
1966a Generalized constitutive relations for incremental deformation of metal crystals by multi-slip, *J. Mech. Phys. Solids* **14,** 95–102.
b The extremal stress-field concept, *J. Mech. Phys. Solids* **14,** 239–243.
c A remark on diagonal streaming in plane plastic strain, *J. Mech. Phys. Solids* **14,** 245–248.
1967a On the classical constitutive laws for elastic/plastic solids, *Recent Progress in Applied Mechanics –The Folke Odqvist Volume*, edited by B. Broberg, J. Hult, and F. Niordson, 241–249, Almqvist & Wiksell, Stockholm; Wiley, New York.
b The essential structure of constitutive laws for metal composites and polycrystals, *J. Mech. Phys. Solids* **15,** 79–95.
c (With D. J. F. Ewing.) The plastic constraint of V-notched tension bars, *J. Mech. Phys. Solids* **15,** 115–124.
d Ideal forming operations for perfectly plastic solids, *J. Mech. Phys. Solids* **15,** 223–227.
e On the vectorial superposition of Hencky–Prandtl nets, *J. Mech. Phys. Solids* **15,** 255–262.
f Eigenmodal deformations in elastic/plastic continua, *J. Mech. Phys. Solids* **15,** 371–386.
1968a On constitutive inequalities for simple materials—I, *J. Mech. Phys. Solids* **16,** 229–242.
b On constitutive inequalities for simple materials—II, *J. Mech. Phys. Solids* **16,** 315–322.

1969 Some aspects of the incremental behaviour of isotropic elastic solids under finite strain, *Problems in Mechanics: Deformation of Solid Bodies–V. V. Novozhilov 60th Anniversary Volume*, edited by L. I. Sedov (in Russian), 459–466, Academy of Sciences, SSSR; University of Leningrad.

1970 Constitutive inequalities for isotropic elastic solids under finite strain, *Proc. R. Soc. London* A **314,** 457–472.

1971 On macroscopic measures of plastic work and deformation in microheterogeneous media (in Russian), *Prikl. Mat. Mekh.* **35,** No. 1, 31–39; Special Issue in honour of the 60th birthday of A. A. Ilyushin (in English), *J. Appl. Math. Mech.* **35,** 11–17.

1972a On constitutive macro-variables for heterogeneous solids at finite strain, *Proc. R. Soc. London* A **326,** 131–147.

b An invariant treatment of interfacial discontinuities in elastic composites, *Continuum Mechanics and Related Problems of Analysis–N. I. Muskhelishvili 80th Anniversary Volume*, edited by L. I. Sedov, 597–604, Academy of Sciences, SSSR; National Committee for Theoretical and Applied Mechanics, SSSR.

c (With J. R. Rice.) Constitutive analysis of elastic–plastic crystals at arbitrary strain, *J. Mech. Phys. Solids* **20,** 401–413.

1973 (With J. R. Rice.) Elastic potentials and the structure of inelastic constitutive laws, *SIAM J. Appl. Math.* **25,** No. 3, 448–461; Special Issue dedicated to William Prager on his 70th birthday and retirement from Brown University, edited by A. C. Pipkin.

1974 (With D. W. Kim.) Some theoretical aspects of hydrostatic extrusion and allied processes, *J. Mech. Phys. Solids* **22,** 73–84.

1975a On the elasticity and stability of perfect crystals at finite strain, *Math. Proc. Camb. Phil. Soc.* **75,** 225–240.

b (With J. W. Hutchinson.) Bifurcation phenomena in the plane tension test, *J. Mech. Phys. Solids* **23,** 239–264.

1976 Plastic analysis of pressurized cylinders under axial load, *Int. J. Mech. Sci.* **18,** 145–148.

1977a (With F. Milstein.) Principles of stability analysis of ideal crystals, *Phys. Rev.* B **15,** 3087–3096.

b (With F. Milstein.) Theoretical properties of cubic crystals at arbitrary pressure—I, Density and bulk modulus, *J. Mech. Phys. Solids* **25,** 457–477.

1978a Aspects of invariance in solid mechanics, *Adv. Appl. Mech.* **18,** 1–75.

b (With F. Milstein.) Theoretical properties of cubic crystals at arbitrary pressure—II, Shear moduli, *J. Mech. Phys. Solids* **26,** 213–239.

1979a Theoretical plasticity of textured aggregates, *Math. Proc. Camb. Phil. Soc.* **85,** 179–191.

b On the theory of plane strain in finitely deformed compressible materials, *Math. Proc. Camb. Phil. Soc.* **86,** 161–178.

c (With F. Milstein.) Theoretical properties of cubic crystals at arbitrary pressure—III, Stability, *J. Mech. Phys. Solids* **27,** 255–279.

d (With F. Milstein.) Divergences among the Born and classical stability criteria for cubic crystals under hydrostatic loading, *Phys. Rev. Lett.* **43,** 1411–1413.

1980a (With F. Milstein and K. Huang.) Theory of the response of cubic crystals to [111] loading, *Phys. Rev.* B **21,** 4282–4291.

1980b (With B. Storåkers.) Plasticity and creep of pressurized membranes: a new look at the small-deflection theory, *J. Mech. Phys. Solids* **28,** 27–48.

c Basic stress analysis of hyperbolic regimes in plastic media, *Math. Proc. Camb. Phil. Soc.* **88,** 359–369.

d Cavitation and the influence of headshape in attack of thick targets by non-deforming projectiles, *J. Mech. Phys. Solids* **28,** 249–263.

1981 Invariance relations in thermoelasticity with generalized variables, *Math. Proc. Camb. Phil. Soc.*, to appear.

B. Scientific Books

1950 *The Mathematical Theory of Plasticity*, ix + 355 pp., The Oxford Engineering Science Series, Clarendon Press, Oxford; reprinted 1956, 1960, 1964, 1967, 1971; now o.p. Japanese translation, 1954. Russian translation, 1956.

1964 *Principles of Dynamics*, vii + 191 pp., Pergamon Press, Oxford.

Contributors

Dr. J. F. W. Bishop, Risley Nuclear Power Development Establishment, United Kingdom Atomic Energy Authority, Risley, Warrington, Cheshire WA3 6AT, England.

Professor B. Budiansky, Division of Applied Sciences, Harvard University, Pierce Hall, Cambridge, Massachusetts 02138, U.S.A.

Professor P. Chadwick, School of Mathematics and Physics, University of East Anglia, Norwich NR4 7TJ, England.

Dr. D. R. J. Chillingworth, Department of Mathematics, University of Southampton, Southampton S09 5NH, England.

Professor I. F. Collins, Department of Theoretical and Applied Mechanics, University of Auckland, Private Bag, Auckland, New Zealand.

Professor J. D. Eshelby, Department of the Theory of Materials, University of Sheffield, Sheffield S10 2TN, England.

Dr. J. H. Gittus, Springfields Nuclear Power Development Laboratories, United Kingdom Atomic Energy Authority, Springfields, Salwick, Preston PR4 0RR, England.

Professor K. S. Havner, Department of Civil Engineering, North Carolina State University, Raleigh, North Carolina 27650, U.S.A.

Professor H. G. Hopkins, Department of Mathematics, University of Manchester Institute of Science and Technology, P.O. Box No. 88, Manchester M60 1QD, England.

Professor J. W. Hutchinson, Division of Applied Sciences, Harvard University, Pierce Hall, Cambridge, Massachusetts 02138, U.S.A.

Professor W. Johnson, University Engineering Department, Trumpington Street, Cambridge CB2 1PZ, England.

Professor R. J. Knops, Department of Mathematics, Heriot–Watt University, Riccarton, Currie, Edinburgh EH14 4AS, Scotland.

Professor P. B. Mellor, Department of Mechanical Engineering, University of Bradford, Bradford, West Yorkshire BD7 1DP, England.

Professor F. Milstein, Department of Mechanical and Environmental Engineering, University of California, Santa Barbara, California 93106, U.S.A.

Professor A. Needleman, Division of Engineering, Brown University, Providence, Rhode Island 02912, U.S.A.

Professor R. W. Ogden, Department of Mathematics, Brunel University, Uxbridge, Middlesex UB8 3PH, England.

Professor J. R. Rice, Division of Applied Sciences, Harvard University, Pierce Hall, Cambridge, Massachusetts 02138, U.S.A.

Professor M. J. Sewell, Department of Mathematics, University of Reading, Whiteknights, Reading RG6 2AX, England.

Miss S. Slutsky, Division of Applied Sciences, Harvard University, Pierce Hall, Cambridge, Massachusetts 02138, U.S.A.

Dr. G. D. Smith, School of Mathematics and Physics, University of East Anglia, Norwich NR4 7TJ, England.

Professor A. J. M. Spencer, Department of Theoretical Mechanics, University of Nottingham, University Park, Nottingham NG7 2RD, England.

Dr. V. Tvergaard, Department of Solid Mechanics, Building 404, Technical University of Denmark, 2800 Lyngby, Denmark.

Professor J. R. Willis, School of Mathematics, University of Bath, Claverton Down, Bath BA2 7AY, England.

Distortions and Interactions in a Fast Reactor Core

by J. F. W. Bishop

Risley Nuclear Power Development Establishment,
United Kingdom Atomic Energy Authority

Summary

Design styles employed to support fast reactor cores are outlined. The phenomena of swelling by void formation, and irradiation creep due to intense neutron bombardment of the metal components in the core of a fast reactor, are described and quantified. The distortions which result from these effects in the fuel elements of a fast reactor are outlined. Interaction effects between sub-assemblies are described and the relevant material deformation processes noted. Calculation methods developed in the U.K. for dealing with the interactions are outlined. Examples of the application of the most developed code to the design of a Commercial Fast Reactor and to the operation of an existing Large Fast Reactor are given.

1. Introduction

In 1964, two researchers at the Dounreay Experimental Reactor Establishment, C. Cawthorne and E. J. Fulton, were examining a rare piece of material, a piece of stainless steel which had been irradiated in the core of the Dounreay Fast Reactor. They showed (1966) that this material now contained some microscopic holes, a few hundred Angström units in diameter, which were essentially voids not containing gas at a significant pressure. It has transpired that this was the first sighting of what is the major materials phenomenon at the centre of the Fast Reactor core design task.

This review paper describes the story for dealing with core distortions as it has unfolded in the U.K., and reference has been made to some of those involved. The author wishes to acknowledge at the outset that parallel endeavours are being undertaken throughout the world fast reactor community, and contributions and discussions from those sources and other U.K. sources have conditioned progress in the U.K.

2. Core Design Styles

In a fast reactor a fuel material, which is typically the mixed oxides of plutonium and uranium, is sealed into stainless steel tubes of some 6 mm diameter and 2500 mm length: the length occupied by the fuel is about 1000 mm. An array of such fuel "pins" is suitably supported inside an hexagonal metal duct or wrapper, and when equipped with appropriate endfittings and other internal furniture, it constitutes a "sub-assembly". A typical reactor contains a central group of these core sub-assemblies, surrounded by an annular zone of radial blanket sub-assemblies (fuelled by natural or depleted uranium dioxide), which in turn is surrounded by a further annular region of reflector elements which limit the escape of neutrons from the core. This article will largely use the Prototype Fast Reactor, also located at Dounreay, for illustration: that reactor of 600 MW thermal power (250 MW electric) has a core of approximately 75 sub-assemblies surrounded by a radial blanket zone of 45 sub-assemblies and a reflector of 105 units outwardly similar to sub-assemblies.

All of these fuel and reflector units are supported at the bottom end in an inlet plenum which, in the case of the sodium-cooled reactor, contains liquid sodium at some 400° C and a pressure of about 6 atm. The sodium velocity through the pin bundles is up to *ca.* 8 m/sec. Three support styles for the sub-assemblies are commonly considered: these are illustrated schematically in Fig. 2.1. The style can be "free-standing" in which the sub-assemblies are vertically cantilevered from a support spike at the lower end. Or it can be "passively restrained" with support at the lower end and contact between sub-assemblies at one or more levels of contact pads somewhere along the length of the wrapper, and with a restraint structure surrounding the total group of sub-assemblies. Or it can be "actively restrained", in which case the restraint structure is tightened down onto the total core aggregation before power is raised.

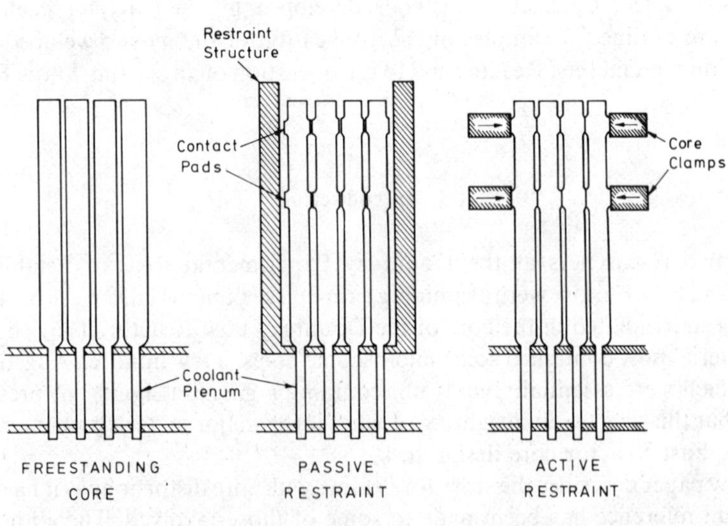

FIG. 2.1. Core support styles for fast reactors.

Actual cores may depart in some detail from these illustrations, e.g. passive restraint may have various degrees of stiffness. Of the existing fast reactors, the U.K. PFR and the Russian BN 350 and BN 600 use freestanding cores; the French Phénix, Japanese Joyo, German SNR 300, and American FFTF use passive restraint. The Russian experimental reactor BOR 60 appears to use an active restraint. PFR is equipped with an adjustable restraint ring which is not currently being used and would probably be used in the passive mode if activated.

3. Neutron-induced Void Swelling and Irradiation Creep

In the Introduction reference was made to the production of neutron-induced voids by high-energy neutron bombardment of metals. There is a second materials phenomenon which needs to be taken into account in the analysis of fast reactor cores. This "irradiation creep" is analogous to thermal creep, but occurs at all reactor temperatures including those at which thermal creep is of negligible magnitude.

The energy required to displace a metal atom from its lattice site is about 25 eV. The neutrons which result from uranium and plutonium fission, both of which processes occur in a fast reactor core, are emitted over a broad energy band up to over 10 MeV. When account is taken of the relative masses of metal atoms and neutrons, it can be shown that not only do initial displacements of clad atoms result from neutron bombardment, but cascades of displacements can follow the initial collision. In the centre of a core, each metal cladding atom will be knocked out of its lattice site approximately every 4 days during the 400 or so days in which it resides in the core, giving a total of about 100 displacements per atom (dpa). The overwhelming majority of the displaced atoms will relocate in a vacant lattice site created by the displacement process. However, a fraction of the vacancies will aggregate together on a suitable nucleus and in turn form void embryos. The early stages of irradiation establish a distribution of void embryos throughout the grains of the metal up to a threshold dose at which the embryos begin to grow more rapidly by aggregating more vacancies. There are many variations of detail on this picture, but a perspective can be established by thinking in terms of the threshold period occupying up to *ca.* 100 days, with final void diameters up to 1000 Å units ($\equiv 10^{-4}$ mm) in cladding from spent fuel which has been irradiated for some 400 days. Figure 3.1, published by Bramman *et al.* (1977), is a transmission electron

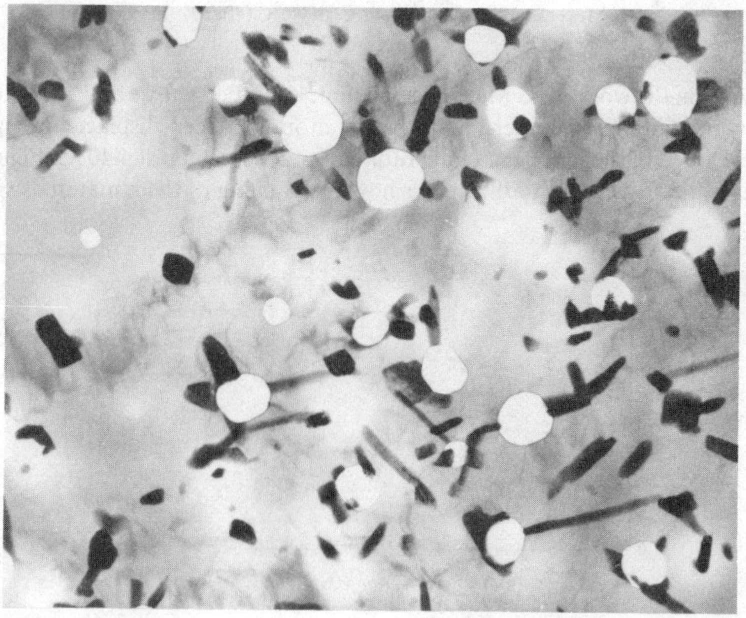

FIG. 3.1. Voids in cold-worked M316 stainless steel at 608°C, 33 dpa ($N/2$) (after Bramman *et al.*, 1977).

microscope photograph of voids in 20% cold-worked 316 type stainless steel in an area 4 × 3.2 microns. Large voids are here seen to be collocated with large carbide precipitates, whilst the voids remote from precipitates are generally much smaller. This material was irradiated at *ca.* 608°C to 33 dpa (*N*/2) calculated on a modified Nelson model, showing 2.4% volume swelling.

As with much of engineering practice, a satisfactory *description* of the bulk behaviour of the material does not require the microscopic detail to be known: an understanding of the phenomenon and the development of materials of desired properties does, of course, require study at the microscopic level.

Phenomenological studies using fast reactor irradiation have shown that void swelling is a complex property dependent on irradiation temperature, neutron fluence (neutrons/cm^2), material, and fabrication history. It is sensitive to minor alloying elements and impurities and the dispersion of such constituents.

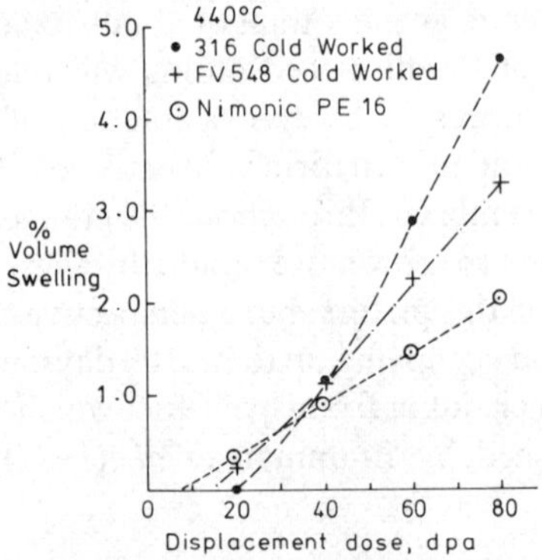

FIG. 3.2. Void swelling in three fuel element structural alloys at 440°C (after K. M. Swanson and J. K. Butler).

Figure 3.2, from the work of K. M. Swanson and J. K. Butler, shows how the void swelling magnitude varies with the number of times each atom has been displaced in engineering materials developed in the U.K. at a temperature of practical interest, 440°C. Figure 3.3, by J. I. Bramman and C. Brown (1979), shows how the swelling of these materials varies with

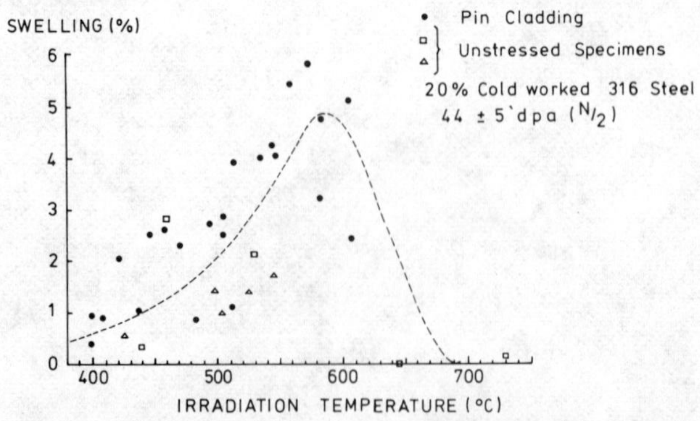

FIG. 3.3. Temperature dependence of void swelling in cold-worked 316 steel irradiated to 44 dpa (*N*/2) (after Bramman and Brown, 1979).

temperature at a particular dose level. Design data have been produced which enable the engineer to deal with the range of conditions encountered in fast reactors. It will be seen that dimensional changes need to be taken into account which are at least an order of magnitude larger than those met in conventional structural analysis due to stressing and thermal expansion. It should be noted that Figs. 3.2 and 3.3 quote average properties. The scatter band encompasses considerably larger magnitudes. Finally, it should be noted that it is currently assumed that the stressing which occurs in fuel applications does not materially affect the void swelling magnitude, in line with the available experimental data.

Irradiation creep has been ascribed to the stress-induced preferential absorption of point defects (created by neutron bombardment) by lattice dislocations at temperatures up to *ca.* 400°C, the SIPA mechanism, augmented by the preferential movement of vacancies thermally emitted from vacancy loops occurring within cascades of displacements at higher temperatures: D. R. Harries (1976) has reviewed relevant mechanisms. Experimental work, such as that of D. Mosedale and G. W. Lewthwaite (1972), has shown that irradiation creep may exhibit an incubation region, within which limited creep occurs, followed by a region where the creep rate is linearly proportional to stress and to displacement rate. The temperature dependence of irradiation creep is not a well-characterized property, and experimental data vary from showing no temperature dependence in the core temperature range to a marked increase of creep rate with increasing temperature. Designers have access to equations for relevant materials obtained from irradiations in fast reactors. Figure 3.4 shows a typical dose dependence of the parameter shear strain γ divided by the shear stress τ for stainless steels reported by Mosedale *et al.* (1977): nickel-base alloys can behave substantially differently as can be seen from Fig. 2 of that paper.

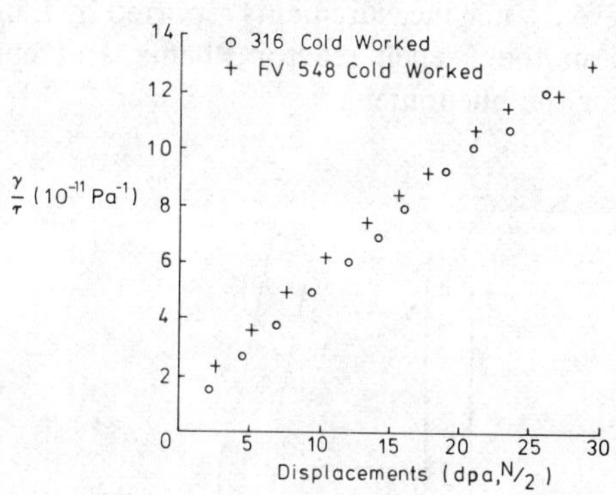

FIG. 3.4. Dose dependence of irradiation creep magnitude in two stainless steels irradiated at 280°C (after Mosedale *et al.*, 1977).

4. Sub-assembly Distortions

The neutron flux in a fast reactor is of highest intensity at the centre of the core and it decreases radially outwards, and vertically both upwards and downwards. It can adequately be envisaged for the purposes of this article as falling in both directions as a cosine dependency although the real situation is considerably more complex. The flux in the core centre is many times that in the breeder and reflector regions. It will be readily apparent from a consideration of Fig. 3.2 that since swelling magnitude is a monotonic function of displacement dose (which is directly related to flux), irradiated sub-assemblies will swell in a non-uniform fashion and in particular the side nearer to the core centre will swell more than the side remote from it. In consequence, the sub-assembly will bow with a curvature convex to the core centre. From a determination of the radial and axial flux and temperature gradients in the sub-assembly, an effective estimate can be made of the bowed shape of the sub-assembly as a function of dwell time in the reactor.

The second major distortion mode occurring in a sub-assembly arises from the fact that the wrapper or duct housing the fuel pins is connected to the pressurized inlet coolant plenum at the bottom end, and discharges into a (free) pool at the top. The gaps between the sub-assemblies are connected to the pool and therefore there is a net outwards pressure on the wrapper which is largest at the bottom, falling to zero at the top. The initial wrapper hexagonal cross-section will therefore increase in size due to voidage swelling and will distort in the direction of circularity due to irradiation creep under the net outwards pressure: the distortion in the wrapper cross-section due to the two mechanisms is termed "dilatation".

Sub-assembly bowing and dilatation (as described above) are the two processes currently taken into account in considering the behaviour of the entire fuel–blanket–reflector assembly. It will be appreciated that this is a simplification of the true situation. In practice the wrapper may also be loaded from within by mechanical interactions with the fuel pins and a grid support structure, but at present these effects are probably of greater importance to pin behaviour than to sub-assembly structural performance. Figure 4.1 schematically indicates the distortions which would occur in a freestanding sub-assembly by the two mechanisms stated. In Fig. 4.2 some measurements reported by Languille *et al.* (1979) from sub-assemblies irradiated in the French reactor Phénix are reproduced to establish a quantitative perspective for the phenomena.

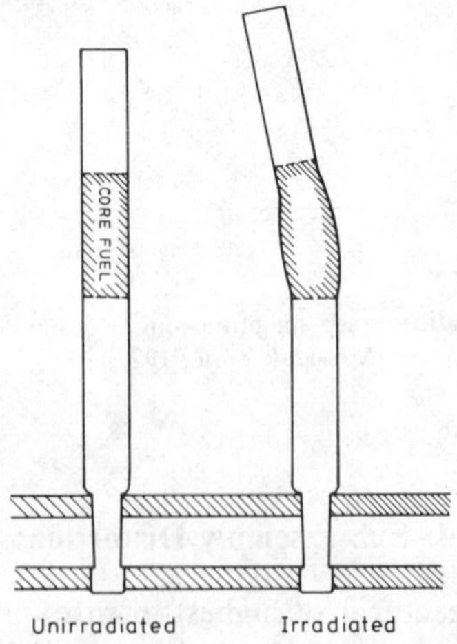

FIG. 4.1. Schematic representation of sub-assembly bowing, dilatation, and length increase resulting from irradiation.

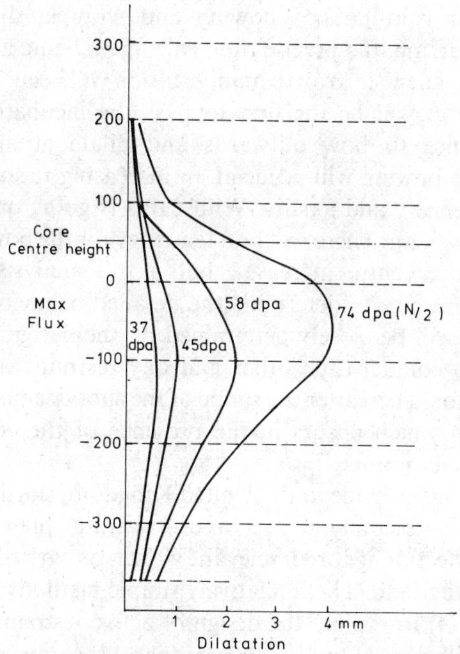

FIG. 4.2. Dilatation in four Phénix wrappers as a function of displacement dose (after Languille *et al.*, 1979).

5. Core Interactions

Consider initially a freestanding core style in which the sub-assemblies are cantilevered vertically from the bottom support grid with gaps of (say) 5 mm between all the hexagon faces of neighbouring sub-assemblies. Because of the neutron flux gradients noted in Section 4, there will be a corresponding radial power gradient which in turn will result in a radial temperature gradient. This will lead to differential thermal expansion and the sub-assembly will undergo thermal bow, generally convex to the core centre: the whole core will tend to a wheatsheaf configuration. If the freedom for this wheatsheafing is limited, as in a restrained core, then a set of inter-sub-assembly reactions are set up leading (generally) to elastic bending of the sub-assemblies. Reactor designers need to take careful note of the distorted configuration because the fissile material in the core is now in a different location compared with the zero power state and it will have different neutronic characteristics, i.e. it will have a different reactivity. Since the change in configuration develops over the short period that the reactor is raised to power, it has a direct bearing on the control characteristics of the reactor through the power coefficient of reactivity.

At some subsequent time, depending on the detail of the irradiation creep laws for the materials involved, any inter-sub-assembly loads developed due to thermal bow will commence to relax, with the elastic strain converted to permanent strain. If then the reactor is shut down, reversed thermal bowing will occur and the core assembly will not return to its original configuration.

The two processes of void-induced bowing and wrapper dilatation are meanwhile nucleating and then occurring, the precise timescale again being a function of the detailed materials laws operating. The void growth manifestation will occur firstly in the centre of the core because this material will be the first to pass the incubation threshold. The sub-assemblies will commence to bow outwards and dilate at around core mid-height. Progressively in time the bowing will occur at an increasing radius in the reactor and the dilatation will spread radially and axially. Whilst this is going on, sub-assemblies will be elongating due to void swelling. There will be a tendency for the bowing to be more severe as the radius from the core centre increases, but a full analysis of flux gradients and displacement rates is necessary to determine the detailed behaviour of a real core.

At some time, which will be closely determined by the design details of the core, sub-assemblies will begin to contact each other and any restraint structure which has been engineered into the reactor. Thereafter the shape of the sub-assemblies is further determined by the irradiation creep which occurs in the presence of the contact loads and by the frictional sliding at contact points.

The early theoretical studies made in the United Kingdom, starting in 1971, examined the restraint of a bowing sub-assembly at 2, 3, and 4 restraint planes with irradiation creep occurring, but elastic deflection ignored: relevant work was carried out by R. G. Anderson, I. R. Collinge, and J. Litherland. These relatively simple methods were used by N. J. Davis and J. A. G. Holmes (1974) to explore the design of active restraint systems. This led to the realization that for a coefficient of friction greater than $1/\sqrt{3}$, an active system might not be able to close the inter-sub-assembly gaps, a fact which was found independently in the United States.

The first U.K. code SABOW which analysed an array of sub-assemblies was produced by R. E. Haigh. In the version which was used to arrive at the U.K. choice of passive restraint for large (1300 MW (e)) cores, the code took account of voidage bowing and dilatation (increase in size of wrapper cross-section due to swelling), and the reversal of bowing due to irradiation creep: it did not include an irradiation creep term in the distortion of wrapper cross-section. At the restraint planes, frictional forces were allowed and the wrapper cross-section was assumed to behave according to a load-deflection modulus. Two versions of the code were developed, one a "spoke" model, dealing with all the sub-assemblies lying on a diametral traverse of the core, the other a whole-core version. The method of attack was to relax out the errors in loads and displacements at all wrapper faces where contact occurred at each increment of a marching solution, the basic requirements being force equilibrium, matching displacements at mating faces, and the satisfaction of the deformability relation for the wrapper cross-section at pad levels. Historically this code has been central in U.K. fast reactor design. It converged most rapidly, as is common in relaxation solutions, in the hands of a skilled user, and this has been a significant fact in its being superseded by a later code, CRAMP, produced by R. C. Perrin (1979).

The code CRAMP in its simplest terms finds the equilibrium elastic configuration of N interacting sub-assemblies which have known (free) shapes and have displacement boundary conditions round the perimeter of the whole assembly. This is achieved by minimizing the strain energy of the system, giving a unique displacement and interaction load solution. Since the distorted shape is conditioned by irradiation creep under the sub-assembly interactions (as well as by the void swelling distortions), the sequence of the development of contact points and the resulting loads must be followed. A solution method for achieving this using linear programming theory was contributed by J. K. Reid. Friction at the contact points can be taken into account by allowing tangential forces to occur up to the limiting friction case at which point sliding may occur: the extent of the sliding (or even a cascade of sub-assembly sliding increments) is determined both by the details of the array and whether a lower value of dynamic friction is used compared with the static coefficient of friction. The elastic and creep distortion of the sub-assembly due to bending and inter-facially coupled loading at the constraint planes is incorporated in a (Miller) matrix, arising from the work of J. P. Ellington, J. Barnes, and A. G. Miller.

The calculation code CRAMP has been developed to the stage where it can deal with a 30° sector (incorporating 62 sub-assemblies) of a large core, appropriate to a Central Power Station and to full core calculations for the Dounreay Prototype Reactor with 235 sub-assemblies with five possible (axial) planes of inter-sub-assembly contact. Further development is continuing to refine the core descriptive input and to enable real core operations, such as replacement of sub-assemblies and control rods, to be modelled.

6. Indicative Results

There are two principal users of the calculation methods outlined, namely, reactor designers and reactor operators. The designers are faced with the task of selecting the design detail and associated materials such that the core power and temperatures can be satisfactorily controlled in all normal and fault conditions (including such events as earthquakes). They require that the loading on the sub-assemblies and the restraint structure will lead to acceptable deformations, and that objects can be removed from the core and from the reactor and replaced within the load limits of the handling equipment and without suffering damage.

As an aid to this task, they calculate the history of the core loads and displacements for on- and off-power conditions and taking into account the planned sequence of refuelling operations for replacing spent fuel. Extensive calculations of this kind have been undertaken by J. A. Gatley, V. Beckett, S. F. Loveday, A. R. Lunt, and their colleagues in the Nuclear Power Company for possible core designs and materials for a Commercial Demonstration Fast Reactor. Figure 6.1 is from work by D. Barnes and shows the shapes of sub-assemblies located on a diameter of a core after 200 days' irradiation with frictional contact between sub-assemblies and with two above-core contact planes at 3746 mm and 2596 mm above the datum (located at the support face for the sub-assembly). It will be seen that as one proceeds radially outwards from sub-assembly S/A 1, the bow becomes more pronounced up to the core boundary at S/A 31: the three radial breeder rows 37, 43, 50 also have a substantial bow. The core centre height is at 2056 mm, so it will be seen that compared with the new off-power core, the core has compacted (and therefore increased in reactivity) at 200 days. The characteristics of the two above-core restraint planes can be seen where the tops move outwards and the lower plane compacts towards the reactor centre. Other output from the CRAMP program gives the sub-assembly discharge loads, axial elongation, contact points, etc.

It can be inferred from the above paragraphs, which touch on the designers' task, that a current central issue in the management of a large operating fast reactor is how the core is distorting as burn-up proceeds and at what stage it would be prudent to remove the fuel to keep it within the design parameters for operation and fuel handling. The use of CRAMP to predict the behaviour of the PFR core and other internals has now been established as a

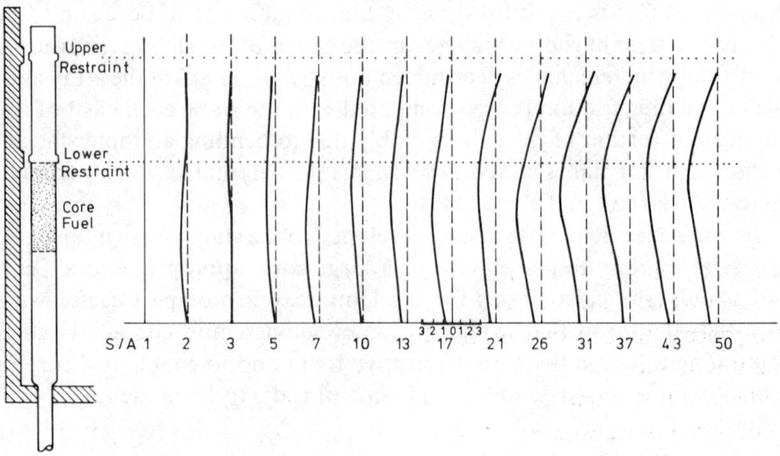

FIG. 6.1. Calculated distortions in Commercial Demonstration Fast Reactor wrappers after 200 full power days' irradiation (after Barnes, 1980, private communication).

routine through the work of J. A. Dodd, A. J. Brook, H. Addison, A. T. D. Butland, and their colleagues. The results are presented as sub-assembly length increase, bowing magnitude and wrapper across-flats increase, sub-assembly extraction loads, and stresses in parts of the support structure. The results are given for a range of possible material properties following the reactor power history.

From a critical inspection of this information, the operators decide on monitoring operations, both in the reactor and in the post-irradiation examination facilities to get reasonably precise data on how the structural materials or specific sub-assemblies are behaving, followed by undertaking sub-assembly rotations or discharges in appropriate cases.

In Fig. 6.2 a segment of the print-out for the predicted magnitudes and directions of head bows of sub-assemblies after a further 60-day run of the reactor from the present time (January 1980), and based on what are considered to be best estimate materials properties, is shown: the whole print-out contains 175 hexagons. Table 6.1 shows the predicted current head bow in millimetres, on a best estimate and on a pessimistic basis, both for the present moment and after one more run, for the sub-assembly marked in the figure.

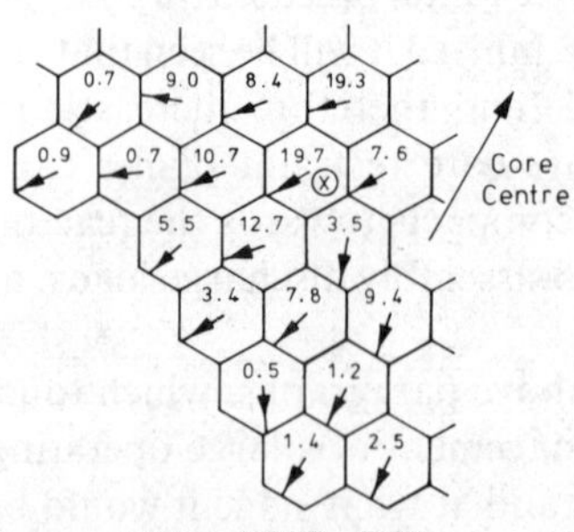

FIG. 6.2. Calculated head bows of PFR sub-assemblies (after Brook and Addison, 1980, private communication).

TABLE 6.1. Predicted head bow in a PFR sub-assembly (mm)

	Best estimate	Pessimistic
Current	6	17
Next run	20	37

The operators judged that the 37 mm bow could be excessive, and therefore a determination was made of the current head bow. It was close to 9 mm. It was decided therefore to proceed to the next run without removing or rotating by 180° the sub-assembly in question. It is of interest to note that the predicted extraction load for the sub-assembly on best estimate data if it is removed after one more run is 0.4 kN. On the pessimistic data it would be 10.2 kN, just above the normal operating limit of the discharge machine. It should, perhaps, be stated that other examples would show that the best estimate data do not apply to other sub-assemblies, and that a different course of action would be required.

7. Conclusion

This article has outlined the phenomena currently taken into account in predicting the distortions in fast reactor cores. The routines are guiding the management of the PFR core and enabling higher endurances to be obtained than could otherwise be undertaken. The calculation models are capable of further refinement and it remains to be seen to what extent that is required. Separate routines have been evolved to deal with the temperature changes caused by the distortions and these are also employed to consider at what point fuel must be discharged.

Acknowledgement

It will be clear to the reader that this article has drawn on work over a wide field of expertise. It has not been appropriate to name all those colleagues who have been involved, and the author takes this opportunity warmly to acknowledge the work which they have initiated and undertaken.

References

Bramman, J. I., Brown, C., Watkin, J. S., Cawthorne, C., Fulton, E. J., Barton, P. J., and Little, E. A. (1977) Void swelling and microstructural changes in fuel pin cladding and unstressed specimens irradiated in DFR, *Proc. Int. Conf. on Radiation Effects in Breeder Reactor Structural Materials, Scottsdale*, 479–507, edited by M. L. Bleiberg and J. W. Bennett, Metallurgical Society of A.I.M.E., New York.

Bramman, J. I. and Brown, C. (1979) The temperature dependence of void swelling of fast reactor irradiated 316 stainless steel, *Proc. Int. Conf. on Irradiation Behaviour of Metallic Materials for Fast Reactor Core Components, Ajaccio*, 155–160, edited by J. Poirier and J. M. Dupuoy, C.E.A., France.

Cawthorne, C. and Fulton, E. J. (1966) Voids in irradiated stainless steel, *Nature* **216,** 575–576.

Harries, D. R. (1976) Irradiation creep in non-fissile metals and alloys, AERE-R8422.

Holmes, J. A. G. (1974) Support of PFR sub-assemblies and associated developments, *Conference on Fast Reactor Power Stations*, 307–318, B.N.E.S., London.

Languille, A., Lambert, G., Marbach, G., Menière, J. M., Mikailoff, H., Blanchard, P., Millet, P., Peray, R., and Ratier, J. L. (1979) Examination results for spent Phénix and Rapsodie fuel elements, *European Nuclear Conference, Hamburg*, edited by R. Farmakes, European and American Nuclear Societies.

Mosedale, D. and Lewthwaite, G. W. (1972) Irradiation creep in some austenitic stainless steels, Nimonic PE16 alloy and nickel, *Proc. Conf. on Creep Strength in Steel and High Temperature Alloys*, 169–188, Iron and Steel Inst., Sheffield.

Mosedale, D., Harries, D. R., Hudson, J. A., Lewthwaite, G. W., and McElroy, R. J. (1977) Irradiation creep in fast reactor core component materials, *Proc. Int. Conf. on Radiation Effects in Breeder Reactor Structural Materials, Scottsdale*, 209–228, edited by M. L. Bleiberg and J. W. Bennett, Metallurgical Society of A.I.M.E., New York.

Perrin, R. C. (1979) The restrained core modelling program, CRAMP, Lecture at 5-SMIRT Conference, Berlin.

Void Growth and Collapse in Viscous Solids

by B. Budiansky, J. W. Hutchinson, and S. Slutsky

Division of Applied Sciences, Harvard University

Summary

Deformation of an isolated void in an infinite block of linearly or nonlinearly viscous material is studied for remote axisymmetric stressing. The material is isotropic and incompressible with a power-law dependence of strain-rate on stress. Included in the family of materials are a linearly viscous solid and a rigid–perfectly plastic solid at the extreme limit of nonlinearity. Evolution of the void shape and size is analyzed in detail for voids starting as spheres in the linearly viscous material under all possible combinations of remote axisymmetric stressing. Asymptotic, or self-similar, shapes towards which the voids evolve are exhibited as are the associated rates of growth or collapse. Under conditions of high remote triaxial stressing the growth-rate of the asymptotic void is found to be identical to that of a spherical void with the same volume. The influence of nonlinearity on the growth-rate and deformation of a spherical void is investigated. Under high triaxiality conditions the behavior of a void in the nonlinearly viscous or rigid–perfectly plastic material is qualitatively different from that of a void in a linearly viscous material. In particular, a void in a nonlinear block undergoing tensile straining with sufficiently large superimposed remote hydrostatic tension grows more rapidly in directions perpendicular to the straining direction than along it and becomes significantly oblate. Asymptotic growth-rates are estimated numerically for the nonlinear materials. The results for the isolated void are used in a simple, approximate way to investigate the roles of material nonlinearity and stressing conditions on the strain required for ductile failure by void growth and coalescence.

1. Introduction

The roles of material nonlinearity and stress state on void growth or collapse in viscous solids are studied in this article. The materials considered are isotropic and incompressible with power-law dependence of strain-rate on stress. Under simple tension σ, the associated strain-rate $\dot{\varepsilon}$ is

$$\dot{\varepsilon} = \dot{\varepsilon}_0(\sigma/\sigma_0)^n \tag{1.1}$$

where $\dot{\varepsilon}_0$ and σ_0 are reference strain-rate and stress quantities. The hardening exponent n ranges from unity to infinity. For multiaxial stress states σ_{ij}, (1.1) is generalized to

$$\dot{\varepsilon}_{ij} = \frac{3}{2}\dot{\varepsilon}_0(\sigma_e/\sigma_0)^{n-1}s_{ij}/\sigma_0, \tag{1.2}$$

where s_{ij} is the stress deviator, $\sigma_e = (3s_{ij}s_{ij}/2)^{1/2}$ is the effective stress, and $\dot{\varepsilon}_{ij}$ is the strain-rate. For convenience, (1.2) will be used throughout in the more compact form

$$\dot{\varepsilon}_{ij} = \frac{1}{2\eta}\sigma_e^{n-1}s_{ij} \tag{1.3}$$

where η is a viscosity-like parameter defined by

$$\eta = \sigma_0^n/(3\dot{\varepsilon}_0). \tag{1.4}$$

Included in this family of materials is a linear Newtonian viscous solid with $n = 1$ and a rigid–perfectly plastic solid with $n \to \infty$. Hill (1956) singled out the family of materials (1.3) as being particularly appropriate for the investigation of a wide range of material behaviors—a viewpoint which is adopted here.

The study presented below is restricted to an investigation of the growth or collapse of an isolated void in an infinite block of material. Furthermore, the void is axisymmetric with respect to the x_3-axis and the stresses remote from the void are also axisymmetric with respect to this axis. The following notation for the nonzero remote stresses will be used throughout:

$$\sigma_{11}^{\infty} = \sigma_{22}^{\infty} = T, \qquad \sigma_{33}^{\infty} = S. \tag{1.5}$$

In addition, it is useful to introduce the notation

$$\sigma_m = \frac{1}{3}\sigma_{ii}^{\infty} = \frac{1}{3}(S+2T) \tag{1.6}$$

for the remote mean stress, and to let

$$\sigma = S - T \tag{1.7}$$

so that $\sigma_e^{\infty} = |\sigma|$. The nonzero remote strain-rates are

$$\dot{\varepsilon}_{33}^{\infty} = -2\dot{\varepsilon}_{11}^{\infty} = -2\dot{\varepsilon}_{22}^{\infty} \equiv \dot{\varepsilon}, \tag{1.8}$$

where

$$\dot{\varepsilon} = \frac{1}{3\eta}|\sigma|^{n-1}\sigma. \tag{1.9}$$

Figure 1.1 displays this notation.

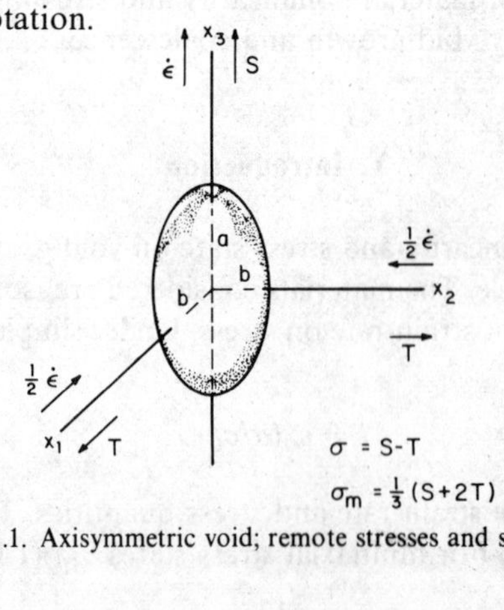

FIG. 1.1. Axisymmetric void; remote stresses and strains.

Sections 2 and 3 are devoted to the evolution of a void in the Newtonian viscous solid ($n = 1$) under all possible combinations of S and T. At the beginning of the deformation history the void is taken to be spherical. Depending on the combination of S and T imposed on the block, the void grows or collapses and its shape evolves towards an asymptotic shape which can be characterized as a needle, cylinder, spheroid, or crack. These asymptotic shapes are determined, as are their growth-rates.

Void behavior in the linearly viscous solid is used as a guide in carrying out the analysis of void growth in the nonlinearly viscous solid. Voids with needle-like or cylinder-like shapes are analyzed in Section 4, and the ranges of S and T, for which these shapes are those approached asymptotically by initially spherical voids, are determined. The results of McClintock (1968) for the rigid–perfectly plastic solid are reproduced. Contact is also made with Tracey's (1971) results for cylindrical voids in power-hardening materials.

A detailed analysis of the growth-rate of a spherical void as it depends on n and on S and T is presented in Section 5. The results presented required extensive numerical calculations, but a simple formula derived in Section 6 gives a reasonably accurate approximation for the growth-rate of the void when the remote mean stress exceeds the remote effective stress. In the limit of a rigid–perfectly plastic solid this formula reduces to the Rice and Tracey (1969) formula for high triaxiality conditions.

Some surprising predictions for the deformation of the spherical void are discovered. At high n with $S > T$, the void grows more rapidly perpendicular to the remote straining direction than along it when σ_m is sufficiently large compared to σ. A similar tendency was found by Andersson (1977) for an initially spherical void in a finite cylindrical block with boundary conditions chosen to model high triaxiality conditions at a crack tip. The paper concludes in Section 7 with a brief study of ductile fracture by void growth based on the solutions of Sections 2–6. Estimates of the influence of material nonlinearity and remote stress state on the fracture strain are obtained.

2. Governing Equations for a Void in a Linearly Viscous Solid

An isolated ellipsoidal cavity in an infinite, linearly viscous solid will always remain an ellipsoid under the action of a uniform stress state (constant or variable) at infinity, although it may change its size and shape, and rotate. This fact follows from Eshelby's (1957) paper, and holds for all homogeneous linear relations

$$\sigma_{ij} = L_{ijkl}\dot{\varepsilon}_{kl} \tag{2.1}$$

between strain-rate and stress, including anisotropic ones. (The Eshelby result was for linear elasticity and small strains, but the extension to linearly viscous solids and arbitrary strains is immediate, as long as $\dot{\varepsilon}_{ij}$ is considered to be the current velocity strain.) These results persist for the two-dimensional problems of elliptical cavities in linearly viscous solids analyzed by Berg (1962) and McClintock *et al.* (1966) in their early studies of void growth and coalescence.

Here we shall consider spheroids (prolate or oblate) of the form

$$(x_1/b)^2 + (x_2/b)^2 + (x_3/a)^2 = 1 \tag{2.2}$$

and study their evolution under the action of the axial and transverse stresses (1.5) lined up with the axes of the spheroid. We shall ultimately adopt the incompressible, isotropic relation

$$\dot{\varepsilon}_{ij} = \frac{1}{2\eta} s_{ij} \tag{2.3}$$

that follows from (1.3) for $n = 1$. It will be useful, initially, to use the compressible relationship

$$\sigma_{ij} = 2\eta\left(\dot{\varepsilon}_{ij} + \frac{v}{1-2v}\dot{\varepsilon}_{kk}\delta_{ij}\right) \tag{2.4}$$

and then let the viscous Poisson's ratio v approach $\frac{1}{2}$ at an appropriate point in the analysis. Symmetry and isotropy imply that the cavity will not rotate, and so we address the problem of calculating its shape and size change. In particular, we shall calculate the strain-rate $\dot{\varepsilon}^I$ *in* the cavity in terms of S, T, and η, and then, with

$$\dot{a}/a = \dot{\varepsilon}^I_{33} \quad \text{and} \quad \dot{b}/b = \dot{\varepsilon}^I_{11} = \dot{\varepsilon}^I_{22}, \tag{2.5}$$

we shall be able to study the evolution of the void geometry.

Hill (1965) defines the tensor $\mathbf{Q}$ relating the cavity strain-rate to the remote stresses $\boldsymbol{\sigma}^\infty$ according to

$$\sigma^\infty_{ij} = Q_{ijmn}\dot{\varepsilon}^I_{mn} \tag{2.6}$$

and shows that

$$Q_{ijmn} = L_{ijmn} - L_{ijkl}S_{klmn} \tag{2.7}$$

where $\mathbf{S}$ is the Eshelby (1957) tensor. (Eshelby relates a "stress-free" transformation strain $\boldsymbol{\varepsilon}^T$ in an ellipsoidal region of a homogeneous linear elastic material to the actual "constrained" strain $\boldsymbol{\varepsilon}^c$ by $\boldsymbol{\varepsilon}^c = \mathbf{S}\boldsymbol{\varepsilon}^T$.) In the present spheroidal cavity problem, with $\mathbf{L}$ defined by (2.4) and $\boldsymbol{\sigma}^\infty$ by (1.5), we find

$$\frac{\sigma_m}{\eta} = \frac{2(1+v)}{3(1-2v)}\left[(1 - S_{3333} - 2S_{1133})\dot{\varepsilon}^I_{33} + 2(1 - S_{3311} - S_{1111} - S_{1122})\dot{\varepsilon}^I_{11}\right], \tag{2.8}$$

$$\frac{\sigma}{3\eta} = \frac{2}{3}\left[(1 - S_{3333} + S_{1133})\dot{\varepsilon}^I_{33} - (1 + 2S_{3311} - S_{1111} - S_{1122})\dot{\varepsilon}^I_{11}\right]. \tag{2.9}$$

From Eshelby (1957),

$$\begin{aligned}
8\pi(1-v)S_{3333} &= 3a^2 I_{aa} + (1-2v)I_a,\\
8\pi(1-v)S_{3311} &= 3b^2 I_{ab} - (1-2v)I_a,\\
8\pi(1-v)S_{1133} &= 3a^2 I_{ab} - (1-2v)I_b,\\
8\pi(1-v)S_{1111} &= 3b^2 I_{bb} + (1-2v)I_b,\\
8\pi(1-v)S_{1122} &= 3b^2 I_{bc} - (1-2v)I_b,
\end{aligned} \tag{2.10}$$

where, with $\lambda = a/b$,

$$\begin{aligned}
\frac{I_b}{2\pi} \equiv \beta &= \lambda(1-\lambda^2)^{-3/2}\{\cos^{-1}\lambda - \lambda(1-\lambda^2)^{1/2}\}, \quad \lambda < 1,\\
&= \lambda(\lambda^2-1)^{-3/2}\{\lambda(\lambda^2-1)^{1/2} - \cosh^{-1}\lambda\}, \quad \lambda > 1,
\end{aligned} \tag{2.11}$$

and

$$\begin{aligned} I_a &= 4\pi - 2I_b, \\ I_{ab} &= [3(a^2-b^2)]^{-1}(I_b - I_a), \\ I_{aa} &= 4\pi(3a^2)^{-1} - 2I_{ab}, \\ I_{bb} &= 3I_{bc} = \pi b^{-2} - 3I_{ab}/4. \end{aligned} \tag{2.12}$$

Elimination of the components of S in (2.8) using (2.10) and (2.12) gives

$$S + 2T = \frac{\eta\,(1+\nu)}{\pi\,(1-\nu)}\left[I_b\dot{\varepsilon}^l_{33} + (4\pi - I_b)\dot{\varepsilon}^l_{11} \right]. \tag{2.13}$$

For the incompressible material ($\nu = 1/2$) this becomes

$$S + 2T = 3\eta\left[2\beta\dot{\varepsilon}^l_{33} + (4-2\beta)\dot{\varepsilon}^l_{11} \right], \tag{2.14}$$

where β is defined in (2.11). Similarly, (2.9) reduces to

$$S - T = 3\eta\left[\alpha\dot{\varepsilon}^l_{33} - \alpha\lambda^{-2}\dot{\varepsilon}^l_{11} \right] \tag{2.15}$$

in the incompressible limit where for all $\lambda = a/b$

$$\alpha = \lambda^2(\lambda^2-1)^{-1}(3\beta-2). \tag{2.16}$$

Inverting (2.14) and (2.15) and noting (2.5), one finds

$$\eta\frac{\dot{a}}{a} = \frac{\lambda^2[(6-5\beta+\alpha)S + 2(\alpha-2\beta)T]}{6\alpha[2\lambda^2+\beta(1-\lambda^2)]} \tag{2.17}$$

and

$$\eta\frac{\dot{b}}{b} = \frac{\lambda^2[(\alpha-2\beta)S + 2(\alpha+\beta)T]}{6\alpha[2\lambda^2+\beta(1-\lambda^2)]}. \tag{2.18}$$

For a given history of S and T, (2.17) and (2.18) govern the evolution of the size and shape of the void. The evolution of the void can also be expressed in terms of the rate of change of its aspect ratio $\lambda = a/b$ and the rate of change of its volume $V = 4\pi ab^2/3$. Since

$$\frac{\dot{\lambda}}{\lambda} = \frac{\dot{a}}{a} - \frac{\dot{b}}{b} \tag{2.19}$$

it follows from (2.17) and (2.18) that

$$\eta\dot{\lambda} = \frac{\lambda^3[(2-\beta)S - 2\beta T]}{2\alpha[2\lambda^2+\beta(1-\lambda^2)]}. \tag{2.20}$$

With starting aspect ratio $\lambda_0 = a_0/b_0$ at time $t = 0$, (2.20) provides the time as a function of λ as

$$\frac{t}{\eta} = \int_{\lambda_0}^{\lambda} \frac{2\alpha[2\lambda^2+(1-\lambda^2)\beta]}{\lambda^3[(2-\beta)S - 2\beta T]}\,d\lambda \tag{2.21}$$

assuming S and T do not vary with time. From

$$\frac{\dot{V}}{V} = \frac{\dot{a}}{a} + 2\frac{\dot{b}}{b} \tag{2.22}$$

it also follows that

$$\eta\frac{\dot{V}}{V} = \frac{\lambda^2[(2+\alpha-3\beta)S+2\alpha T]}{2\alpha[2\lambda^2+\beta(1-\lambda^2)]}. \tag{2.23}$$

Elimination of time in (2.23) in favor of λ as the dependent variable using (2.20) gives

$$\ln\left[\frac{V}{V_0}\right] = \int_{\lambda_0}^{\lambda} \frac{(2+\alpha-3\beta)S+2\alpha T}{\lambda[(2-\beta)S-2\beta T]}\,d\lambda, \tag{2.24}$$

where $V_0 = 4\pi a_0 b_0^2/3$ is the starting volume.

If the void is momentarily a sphere of radius $a = b$, then (2.11) reduces to $\beta = 2/3$, and (2.16) has the limit $\alpha = 2/5$; (2.20) and (2.23) give

$$\eta\dot{\lambda} = \frac{5}{6}(S-T) = \frac{5}{6}\sigma \tag{2.25}$$

and

$$\eta\frac{\dot{V}}{V} = \frac{1}{4}(S+2T) = \frac{3}{4}\sigma_m. \tag{2.26}$$

These results can be restated as

$$\dot{\lambda} = \frac{5}{2}\dot{\varepsilon} \tag{2.27}$$

and

$$\frac{1}{\dot{\varepsilon}}\frac{\dot{V}}{V} = \frac{3}{4}\left[\frac{S+2T}{S-T}\right] \tag{2.28}$$

in terms of the remote strain-rate $\dot{\varepsilon} = \sigma/(3\eta)$ as given by (1.9).

3. Evolution of an Initially Spherical Void under Various S/T in a Linearly Viscous Material

Equation (2.24) relates the evolving aspect ratio and the volume of the void, independently of time and viscosity. This relation holds for time-varying S and T as long as S/T is fixed. In turn, the aspect ratio is related to time and viscosity through (2.20) or (2.21). A differential relation between the aspect ratio and the remote logarithmic strain ε may be obtained immediately from (2.20) and $\dot{\varepsilon} = (S-T)/(3\eta)$ as

$$\frac{d\lambda}{d\varepsilon} = \frac{3\lambda^3[(2-\beta)S-2\beta T]}{2\alpha(S-T)[2\lambda^2+\beta(1-\lambda^2)]}, \tag{3.1}$$

and this relation also holds for time-varying S and T if S/T is fixed. Volume histories as functions of remote strain ε may now be calculated parametrically in terms of λ, with V/V_0 obtained from (2.24) and ε found from integration of (3.1).

Figures 3.1(a) through 3.1(g) illustrate what happens to initially spherical voids for all representative combinations of S and T. The chart in Fig. 3.1(a) summarizes the asymptotic shapes and values of V/V_0, λ, a, and b towards which the void evolves. In Figs. 3.1(b) through 3.1(g) the void volume normalized by its starting volume is plotted as a function of the remote stretch in the 3-direction (or its inverse), which is related to the remote logarithmic strain by

$$(L/L_0)_\infty = e^{\varepsilon}. \tag{3.2}$$

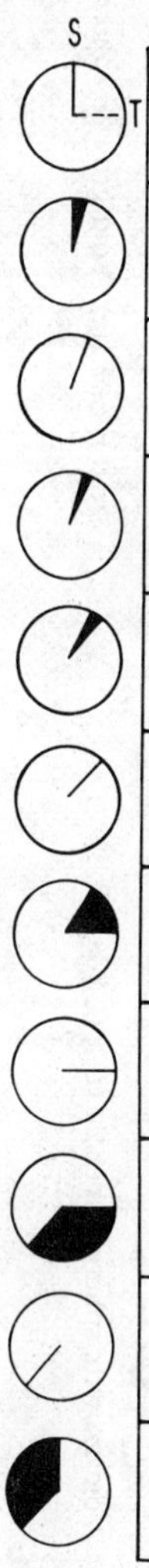

	ASYMPTOTIC VALUES V/V_0	$\lambda = a/b$	a/a_0	b/b_0	ASYMPTOTIC SHAPES
S = 1, T = 0	1.264	∞	∞	0	NEEDLE
S > 0, S/T > 4	∞	∞	∞	0	NEEDLES
S > 0, S/T = 4	∞	∞	∞	.7942	CYLINDER
S > 0, 4 > S/T ≥ 2	∞	∞	∞	∞	CYLINDERS
S > 0, 2 > S/T > 1	∞	λ > 1	∞	∞	PROLATE SPHEROIDS
S > 0, S/T = 1	∞	1	∞	∞	SPHERE
T > 0, 1 > S/T > 0	∞	0 < λ < 1	∞	∞	OBLATE SPHEROIDS
S = 0, T > 0	2.356	0	0	∞	CRACK
S < 0, S < T	0	0	0	finite	CRACKS
S < 0, S/T = 1	0	1	0	0	POINT
T < 0, S > T	0	∞	∞	0	NEEDLES

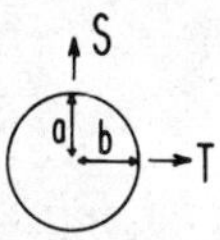

FIG. 3.1(a). Asymptotic geometries of initially spherical voids.

Since the material is incompressible, $(L/L_0)_\infty$ is also the inverse of the remote area reduction. The curves in Figs. 3.1(b)–(g) are independent of the time variation of S and T as long as S/T does not change in a given history.

The plots of Figs. 3.1(b)–(g) will be discussed in turn, starting at 12 o'clock in the T-S plane and proceeding clockwise. In Fig. 3.1(b) for $0 \leqslant T/S \leqslant 1/4$ with $S > 0$, the void asymptotes

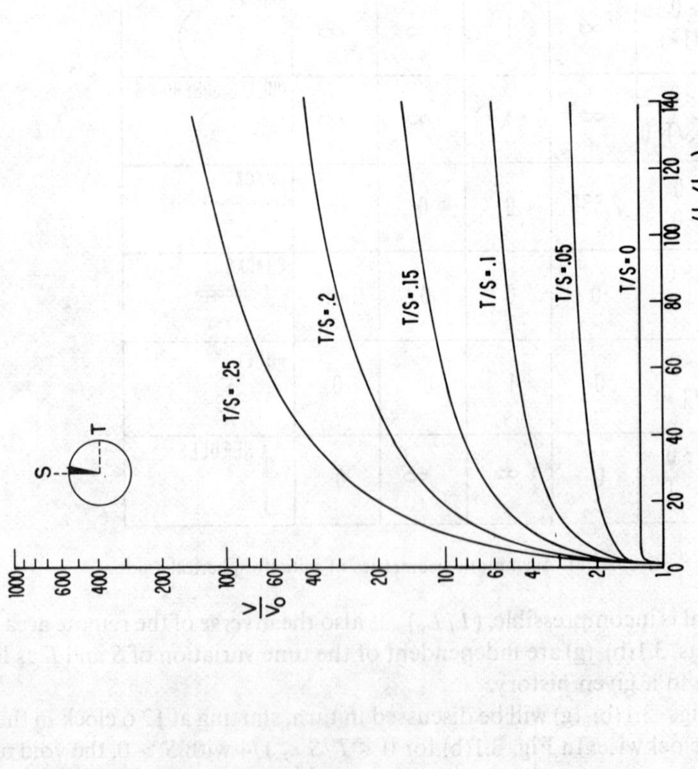

FIG. 3.1(b). Void growth: sphere to needle ($T \geq 0$, $S > 4T$).

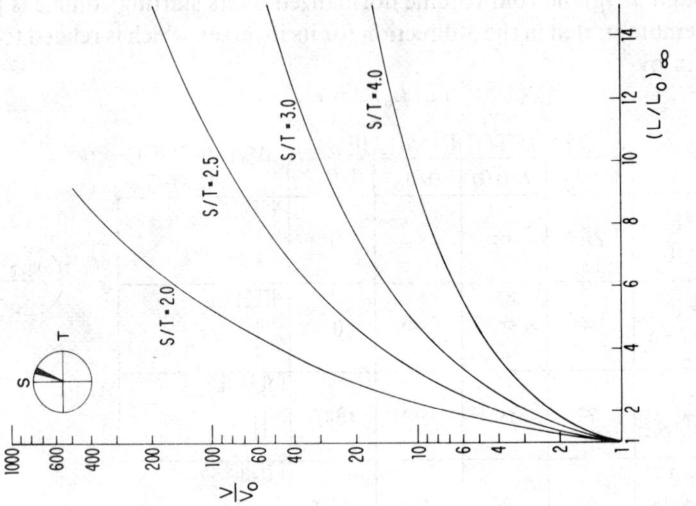

FIG. 3.1(c). Void growth: sphere to cylinder.

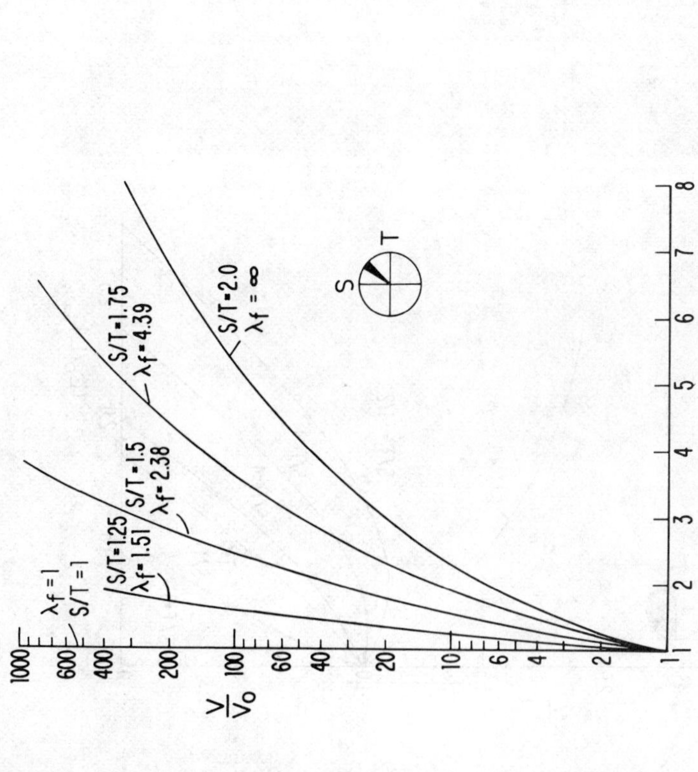

FIG. 3.1(d). Void growth: sphere to prolate spheroid.

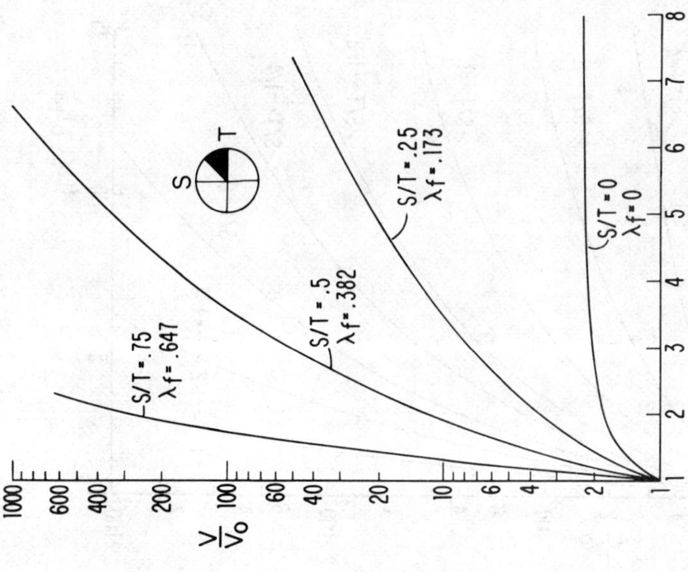

FIG. 3.1(e). Void growth: sphere to oblate spheroid.

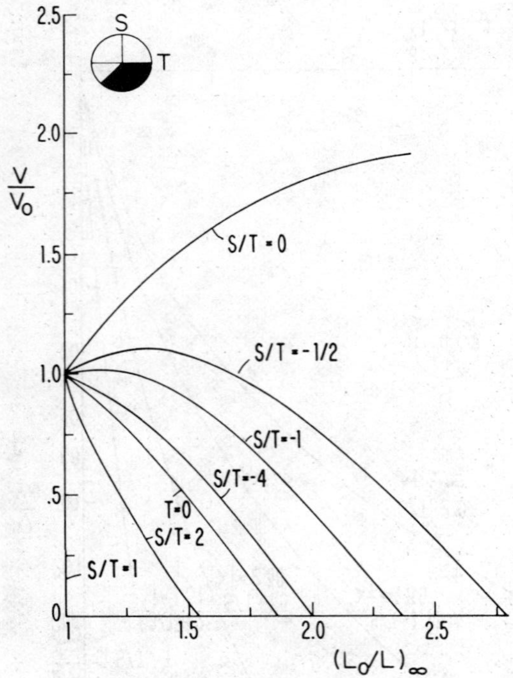

FIG. 3.1(f). Void collapse: sphere to crack ($S < 0, T > S$).

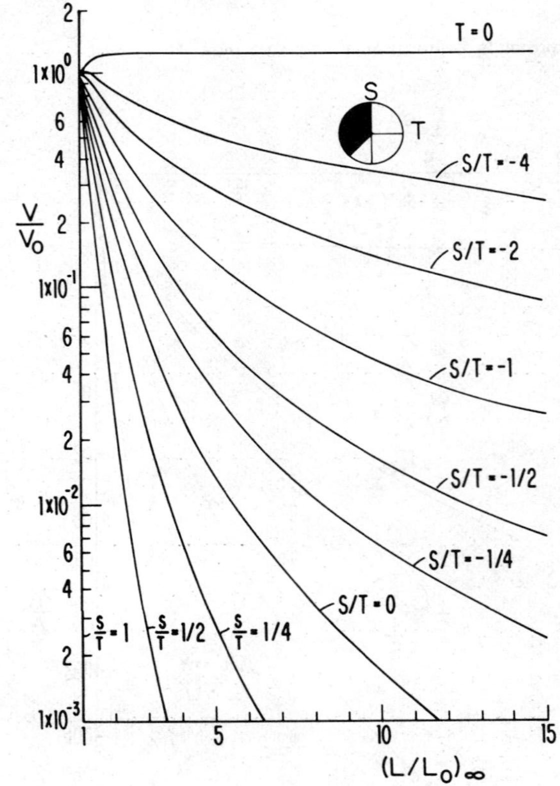

FIG. 3.1(g). Void collapse: sphere to needle ($T < 0, S > T$).

to a needle-like shape of infinite length and zero radius but with infinite volume. Exceptionally for remote uniaxial tension ($S > 0, T = 0$), the volume asymptotes to a value which is only 26.4% above the starting volume. For $1/4 \leqslant T/S \leqslant 1/2$ with $S > 0$, the void evolves towards a cylinder-like shape of infinite length and radius, except for $T/S = 1/4$, where $b/b_0 \to 0.794$.

For any segment of an infinite cylinder with the x_3-axis as its axis of revolution and subject to remote stresses S and T,

$$\eta \frac{\dot{V}}{V} = \eta \left(\dot{\varepsilon}^I_{33} + 2\dot{\varepsilon}^I_{11} \right) = T \tag{3.3}$$

independent of S. This result, easily found by elementary considerations, agrees with the limit of (2.23) for $a \to \infty$. Equation (3.3) provides the asymptotic value of $\dot{V}/V$ for the range of S and T in which the void asymptotes to a needle or a cylinder, i.e.

$$\eta \frac{\dot{V}}{V} \to \eta \left[\frac{\dot{V}}{V} \right]_f = T \quad (S > 0,\ 0 \leqslant T/S < 1/2), \tag{3.4}$$

where here and in the sequel a subscript f denotes values associated with an asymptotic shape. As will be seen later, this asymptote also applies in the range $T < 0$ with $S > T$. The bottom two curves in Fig. 3.2 for $S/T = 3$ and $S/T = 4$ show $\dot{V}/V$ as a function of V/V_0 dropping from the starting value (2.26) when the void is spherical to essentially the asymptotic value (3.4) in only a doubling of V over V_0. Thus although the void is still far from a cylinder in appearance, its volume-rate of change is well approximated by its asymptote.

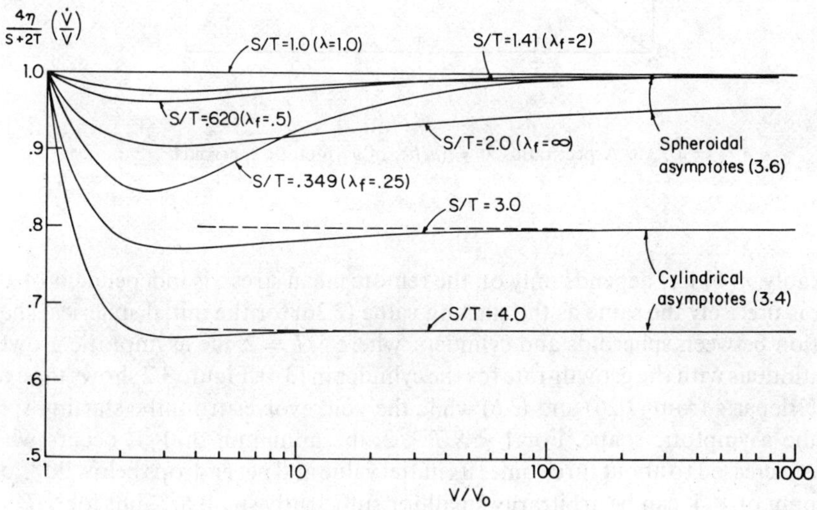

FIG. 3.2. Evolution of void growth rates.

In the range $1 < S/T < 2$ with $T > 0$ the void asymptotes to a prolate spheroid ($\lambda \to \lambda_f > 1$); when $S/T = 1$ the void remains spherical; and for $0 < S/T < 1$ with $T > 0$ the void asymptotes to an oblate spheroid ($\lambda \to \lambda_f < 1$). To discover the asymptotic aspect ratio λ_f and the associated rate of volume change in the range $0 < S/T < 2$, set $\dot{\lambda} = 0$ in (2.20) and note

$$\beta \to \beta_f = 2S/(S+2T). \tag{3.5}$$

The relation between λ_f and S/T plotted in Fig. 3.3 is obtained from (3.5) and (2.11). The asymptotic shape satisfies $\dot{\varepsilon}^l_{11} = \dot{\varepsilon}^l_{33}$ and, thus, combining (2.14) and (2.22), gives

$$\eta\left[\frac{\dot{V}}{V}\right]_f = \frac{3}{4}\sigma_m = \frac{1}{4}(S+2T) \quad (0 < S/T < 2). \tag{3.6}$$

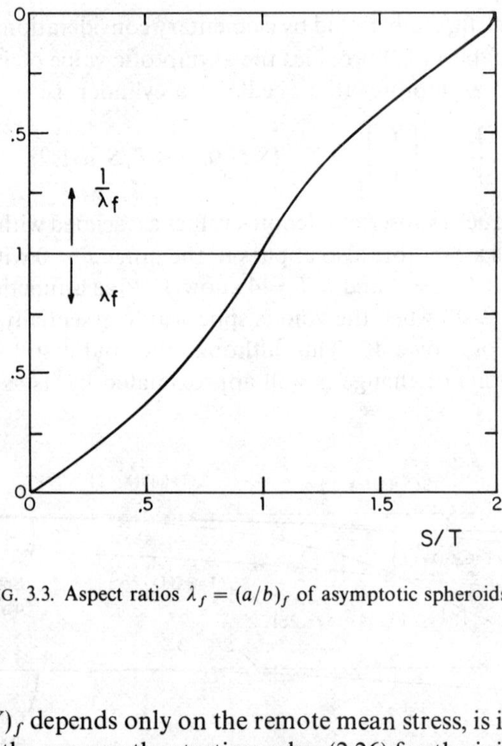

FIG. 3.3. Aspect ratios $\lambda_f = (a/b)_f$ of asymptotic spheroids.

Remarkably, $\eta(\dot{V}/V)_f$ depends only on the remote mean stress, is independent of λ_f, and, moreover, is precisely the same as the starting value (2.26) for the initial spherical shape. At the transition between spheroids and cylinders, where $S/T = 2$, the asymptotic growth-rate (3.6) is continuous with the growth-rate for the cylinder in (3.4). Figure 3.2 shows the extent to which $\dot{V}/V$ departs from (2.26) and (3.6) while the void evolves from the starting spherical shape to the asymptotic shape. For $1 < S/T < 2$, the minimum of $\dot{V}/V$ occurs when the volume has increased to about three times its initial value and never drops below 90 % of (3.6). The minimum of $\dot{V}/V$ can be arbitrarily small for sufficiently small S/T, but for $S/T > 0.3$ it does not fall below 80 % of (3.6).

When $S < 0$ with $S < T$ in Fig. 3.1(f), the void is crushed to a penny-shaped crack ($a_f = 0$) in the x_1–x_2 plane. With the exception of the case $S = 0$ for which $b/b_0 \to \infty$ and $V/V_0 \to 2.356$, the volume goes to zero and b approaches a finite value at a finite remote strain, as can be determined from the intercept with the abscissa. In uniaxial compression ($S < 0$, $T = 0$) a remote length reduction of about 50 % is needed to completely flatten out a spherical void; the final radius of the crack is 1.41 times the radius of the starting sphere.

Finally, for $T < 0$ with $S > T$ in Fig. 3.1(g), the void is crushed to an infinitely long needle ($a \to \infty, b \to 0$) with zero volume. Complete collapse is not attained at finite remote strain but is only approached asymptotically with (3.4) as the asymptote for $\dot{V}/V$. Under transverse pressure alone ($S = 0, T < 0$), an overall area reduction ratio of 12:1 is needed to squeeze the void into a needle-like cavity having 0.1 % of its original spherical volume.

Under all-around hydrostatic pressure ($S = T < 0$) the void remains spherical and shrinks to a point. In this case, however, the spherical shape is unstable in that any axisymmetric departure from sphericity will drive the void to either a penny-shaped crack with finite radius or a needle with finite length, both with zero volume. Similarly, the asymptotic shapes found in the tensile range, $0 < S/T < 2$, are also possible solutions when the signs of both S and T are reversed. In fact, if the void started with the asymptotic aspect ratio λ_f rather than as a sphere, the aspect ratio would remain unchanged as the void collapsed to a point. However, it is easy to show that these shapes are unstable in compression, just as in the case of the spherical void when $S = T < 0$.

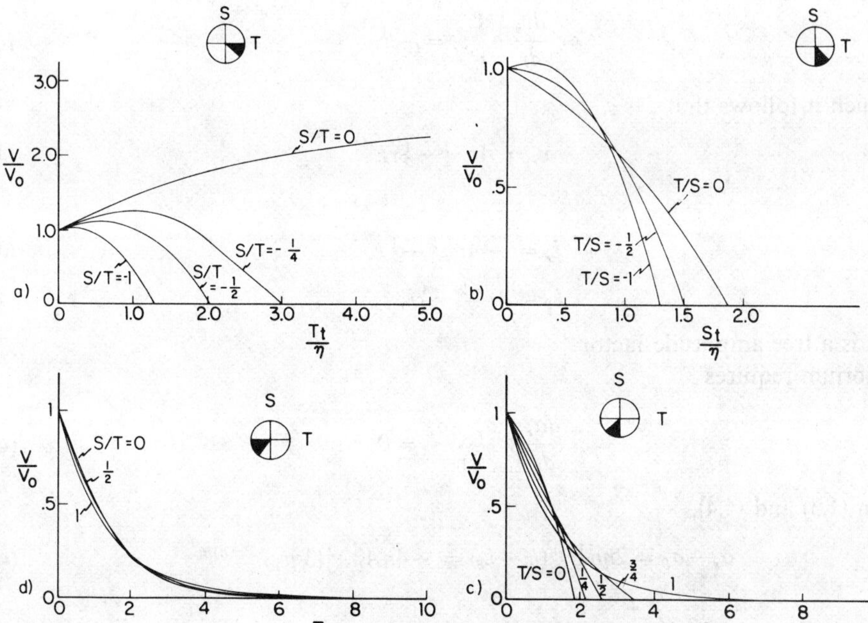

FIG. 3.4. Void-volume time histories, $S \leqslant 0$.

Time histories of the volume for $S < 0$ are shown in Fig. 3.4. Here use is made of (2.21) and it is now assumed that S and T are time-independent. For $S < 0$ and $T > S$ (Figs. 3.4(a) through 3.4(c)) $V \to 0$ at finite time; for $S < 0$ and $T \leqslant S$, $V \to 0$ as $t \to \infty$. The nearly identical responses in Fig. 3.4(d) for $0 \leqslant S/T \leqslant 1$ reflect the fact that the void very quickly assumes the asymptotic growth-rate (3.4) for the cylinder or needle.

4. Deformation of an Infinite Cylindrical Void in a Nonlinearly Viscous Material

Let the axis of an infinite cylindrical void of radius b coincide with the x_3-axis. The surface of the void is traction-free, while far from the void the stresses are specified by S and T as in (1.5). Surrounding the void is the incompressible, nonlinearly viscous material (1.3). Solutions to this problem have been given by McClintock (1968) and Rice and Tracey (1969) for the rigid–plastic limit ($n = \infty$) and by Tracey (1971) for arbitrary values of n. The solution presented here is in a form which is particularly simple and appropriate for the void growth problem. In addition, the solution will be used to infer the ranges of S/T corresponding to needles or cylinders as asymptotic shapes.

In a cylindrical coordinate system (r, θ, z) the nonzero in-plane velocity component is $v_r(r)$ and the nonzero strain-rate components are

$$\dot{\varepsilon}_r = \frac{dv_r}{dr}, \quad \dot{\varepsilon}_\theta = \frac{1}{r}v_r, \quad \dot{\varepsilon}_z = \dot{\varepsilon}, \tag{4.1}$$

where $\dot{\varepsilon}$ is given by (1.9). Incompressibility requires

$$\frac{dv_r}{dr} + \frac{1}{r}v_r = -\dot{\varepsilon} \tag{4.2}$$

from which it follows that

$$v_r = Ar^{-1} - \tfrac{1}{2}r\dot{\varepsilon} \tag{4.3}$$

and

$$\dot{\varepsilon}_r = -Ar^{-2} - \tfrac{1}{2}\dot{\varepsilon}, \tag{4.4}$$

$$\dot{\varepsilon}_\theta = Ar^{-2} - \tfrac{1}{2}\dot{\varepsilon}, \tag{4.5}$$

where A is a free amplitude factor.

Equilibrium requires

$$\frac{d\sigma_r}{dr} + \frac{\sigma_r - \sigma_\theta}{r} = 0. \tag{4.6}$$

But from (1.3) and (4.4),

$$\sigma_r - \sigma_\theta = 2\eta\sigma_e^{1-n}(\dot{\varepsilon}_r - \dot{\varepsilon}_\theta) = -4\eta A r^{-2}(3\eta\dot{\varepsilon}_e)^{(1-n)/n} \tag{4.7}$$

where

$$\dot{\varepsilon}_e \equiv (2\dot{\varepsilon}_{ij}\dot{\varepsilon}_{ij}/3)^{1/2} = (4A^2r^{-4}/3 + \dot{\varepsilon}^2)^{1/2}. \tag{4.8}$$

Substitution of (4.7) into (4.6) and integration of (4.6) from $r = b$, where $\sigma_r = 0$, to $r = \infty$, where $\sigma_r = T$, gives

$$T = (3\eta)^{1/n}(4A/3)\int_b^\infty (4A^2r^{-4}/3 + \dot{\varepsilon}^2)^{(1-n)/2n} r^{-3}\,dr. \tag{4.9}$$

By (1.9), this can be rewritten as

$$\frac{T}{S-T} = \frac{4A}{3\dot{\varepsilon}}\int_b^\infty \left[1 + \left(\frac{2A}{\sqrt{3}\dot{\varepsilon}}\right)^2 r^{-4}\right]^{(1-n)/2n} r^{-3}\,dr. \tag{4.10}$$

Lastly, with the change of integration variable

$$x = \omega\left(\frac{b}{r}\right)^2, \quad \text{where} \quad \omega = \frac{2A}{\sqrt{3}\dot{\varepsilon}b^2}, \tag{4.11}$$

(4.10) gives the final result

$$\frac{S}{T} = 1 + \left[\frac{1}{\sqrt{3}}\int_0^{\omega} (1+x^2)^{(1-n)/2n} dx\right]^{-1}. \tag{4.12}$$

The rate of volume change of a segment of the cylindrical void with current length L is

$$\dot{V} = 2\pi Lb\dot{b} + \pi b^2 \dot{L} = 2\pi LA, \tag{4.13}$$

where the last equality follows from (4.4) and $\dot{\varepsilon}_\theta = \dot{b}/b$, $\dot{\varepsilon} = \dot{L}/L$ at the cavity. Therefore

$$\frac{1}{\dot{\varepsilon}}\frac{\dot{V}}{V} = \frac{2A}{\dot{\varepsilon}b^2} = \sqrt{3}\omega \tag{4.14}$$

and consequently (4.12) with (4.14) provides the desired relation between S/T and $\dot{V}/(\dot{\varepsilon}V)$.

For the linearly viscous material ($n = 1$), (4.12) integrates to

$$\frac{1}{\dot{\varepsilon}}\frac{\dot{V}}{V} = \frac{3T}{S-T} \tag{4.15}$$

in agreement with the result obtained by combining (3.4) and (1.9). For a rigid–perfectly plastic material ($n = \infty$), (4.12) can be integrated to give McClintock's (1968) formula

$$\frac{1}{\dot{\varepsilon}}\frac{\dot{V}}{V} = \sqrt{3}\sinh\left(\frac{\sqrt{3}T}{S-T}\right). \tag{4.16}$$

This formula brings out the large amplification of the volume growth-rate when triaxiality is high, i.e. when $T \gg S-T$ or, equivalently, when $\sigma_m \gg \sigma$. The formula is valid for arbitrary signs of S and T, and high hydrostatic pressure in combination with remote straining $\dot{\varepsilon}$ has a correspondingly large effect on the rate of void collapse.

The solution (4.12) can be used to infer whether a void which starts as a sphere, say, will evolve to an asymptotic shape which is a cylinder or needle. First, suppose S and T are in the range in which the void asymptotes to a cylindrical void, which as previously defined requires that $a/b \to \infty$ with $b \geqslant 0$. If the void is to asymptote to a cylindrical void, the cylinder solution must for consistency grow in radius, or at least not decrease in radius, but in such a way that its elongation per unit length is even more rapid. That is, for S and T in the range such that the cylindrical void solution satisfies

$$0 \leqslant \frac{1}{\dot{\varepsilon}}\frac{\dot{b}}{b} < 1 \quad \text{with} \quad \dot{\varepsilon} > 0, \tag{4.17}$$

the asymptotic shape is expected to be a cylinder. By a similar argument the asymptotic shape will be a needle ($b \to 0$, $a \to \infty$) when

$$\frac{1}{\dot{\varepsilon}}\frac{\dot{b}}{b} < 0 \quad \text{with} \quad \dot{\varepsilon} > 0. \tag{4.18}$$

From (4.13) and (4.14),

$$\frac{1}{\dot{\varepsilon}}\frac{\dot{b}}{b} = \frac{1}{2}\left(\frac{1}{\dot{\varepsilon}}\frac{\dot{V}}{V} - 1\right) = \frac{1}{2}(\sqrt{3}\omega - 1). \tag{4.19}$$

The transition of the asymptotic shapes from needles to cylinders occurs when

$$\frac{\dot{b}}{\dot{\varepsilon}b} = 0 \quad \text{or} \quad \frac{\dot{V}}{\dot{\varepsilon}V} = \sqrt{3}\omega = 1, \tag{4.20}$$

which from (4.12) corresponds to

$$S/T = 1 + \sqrt{3}\left[\int_0^{1/\sqrt{3}} (1+x^2)^{(1-n)/2n} dx\right]^{-1}. \tag{4.21}$$

For $n = 1$ (4.21) gives $S/T = 4$ and, for $n = \infty$, $S/T = 4.153$. A plot of the transition values of S/T as a function of n is shown in Fig. 4.1. The transition between asymptotic shapes which are cylinders and those which are pseudo-spheroids occurs when

$$\frac{\dot{b}}{\dot{\varepsilon}b} = 1 \quad (\dot{\varepsilon} > 0) \quad \text{or} \quad \frac{\dot{V}}{\dot{\varepsilon}V} = \sqrt{3}\omega = 3, \tag{4.22}$$

whence

$$S/T = 1 + \sqrt{3}\left[\int_0^{\sqrt{3}} (1+x^2)^{(1-n)/2n} dx\right]^{-1}. \tag{4.23}$$

For $n = 1$, (4.23) gives $S/T = 2$, and, for $n = \infty$, $S/T = 2.315$; the results for intermediate values of n are shown in Fig. 4.1.

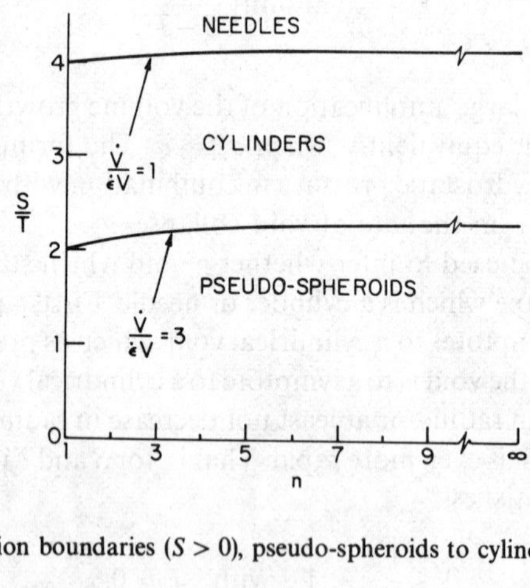

FIG. 4.1. Transition boundaries ($S > 0$), pseudo-spheroids to cylinders to needles.

5. Deformation-rate of a Spherical Void in a Nonlinearly Viscous Solid

The initial rates of deformation and dilatation of a spherical void in an infinite block of the incompressible, nonlinearly viscous solid characterized by (1.3) is studied in this Section. As in the linear problem, the block is subject to the remote stresses (1.5), but now attention will be directed mainly to the effects of axial and transverse stresses S and T that are either both positive or both negative.

The minimum principle for velocities stated by Hill (1956) will be generalized to a form appropriate for infinite domains, and then used to find approximate Rayleigh–Ritz solutions. Consider a traction-free void of arbitrary shape inside a finite spherical region whose radius R will subsequently be permitted to become unbounded. The outer spherical surface is denoted by S_R, its outward unit normal by $\mathbf{n}$, and tractions $T_i = \sigma_{ij}^{\infty} n_j$ consistent with the uniform stress state $\boldsymbol{\sigma}^{\infty}$ given by (1.5) are imposed on S_R. The material region surrounding the void and contained within S_R is denoted by V_R. Then (Hill, 1956) the velocities v_i minimize the functional

$$\Phi = \int_{V_R} W(\dot{\varepsilon})\, dV - \int_{S_R} \sigma_{ij}^{\infty} n_i v_j\, dS, \tag{5.1}$$

where the usual admissibility conditions

$$v_{i,i} = 0, \quad \dot{\varepsilon}_{ij} = \tfrac{1}{2}(v_{i,j} + v_{j,i}), \tag{5.2}$$

are imposed, and

$$W(\dot{\varepsilon}) = \int_0^{\dot{\varepsilon}} \sigma_{ij}\, d\dot{\varepsilon}_{ij} = (3\eta)^{1/n} \left(\frac{n}{n+1}\right) (\dot{\varepsilon}_e)^{(n+1)/n}, \tag{5.3}$$

where

$$\dot{\varepsilon}_e = (2\dot{\varepsilon}_{ij}\dot{\varepsilon}_{ij}/3)^{1/2}. \tag{5.4}$$

Here $\dot{\varepsilon}_e$ is an effective strain-rate which, from (1.3), is related to σ_e according to the simple tension formula (1.1).

Now let

$$v_i = v_i^{\infty} + \tilde{v}_i \quad \text{and} \quad \dot{\varepsilon}_{ij} = \dot{\varepsilon}_{ij}^{\infty} + \dot{\tilde{\varepsilon}}_{ij}, \tag{5.5}$$

where $\mathbf{v}^{\infty}$ and $\dot{\boldsymbol{\varepsilon}}^{\infty}$ are the velocity and uniform strain-rates due to $\boldsymbol{\sigma}^{\infty}$ in the absence of the void and where

$$\dot{\tilde{\varepsilon}}_{ij} = \tfrac{1}{2}(\tilde{v}_{i,j} + \tilde{v}_{j,i}), \quad \tilde{v}_{k,k} = 0. \tag{5.6}$$

The minimum principle remains valid if the Φ in (5.1) is replaced by

$$\Phi = \int_{V_R} [W(\dot{\varepsilon}) - W(\dot{\varepsilon}^{\infty})]\, dV - \int_{S_R} \sigma_{ij}^{\infty} n_j \tilde{v}_i\, dS \tag{5.7}$$

and this new Φ is minimized with respect to the additional velocity field $\tilde{\mathbf{v}}$. Next, by the principle of virtual work,

$$\int_{S_R} \sigma_{ij}^{\infty} n_j \tilde{v}_i\, dS = \int_{V_R} \sigma_{ij}^{\infty} \dot{\tilde{\varepsilon}}_{ij}\, dV + \int_S \sigma_{ij}^{\infty} n_j \tilde{v}_i\, dS, \tag{5.8}$$

where $\mathbf{n}$ is the unit normal to the void surface S pointing into V_R. Thus, (5.7) can be written as

$$\Phi = \int_{V_R} [W(\dot{\varepsilon}) - W(\dot{\varepsilon}^{\infty}) - \sigma_{ij}^{\infty} \dot{\tilde{\varepsilon}}_{ij}]\, dV - \int_S \sigma_{ij}^{\infty} n_j \tilde{v}_i\, dS. \tag{5.9}$$

It is now possible to let the outer radius R become infinite. For $x \to \infty, \dot{\bar{\varepsilon}} \to 0$, and

$$W(\dot{\varepsilon}) - W(\dot{\varepsilon}^{\infty}) - \sigma_{ij}^{\infty}\dot{\bar{\varepsilon}}_{ij} = O(\dot{\bar{\varepsilon}}_{ij}\dot{\bar{\varepsilon}}_{ij}). \tag{5.10}$$

Thus the volume integral in (5.9) is well defined for $R \to \infty$ for all additional strains which satisfy

$$\dot{\bar{\varepsilon}} = o(r^{-s}) \quad \text{as} \quad r \equiv (x_i x_i)^{1/2} \to \infty, \quad \text{where} \quad s > 3/2. \tag{5.11}$$

In the linear problem ($n = 1$) for the spherical void, $\dot{\bar{\varepsilon}} \sim r^{-3}$ for large r. For the spherically symmetric solution with $S = T$, $\dot{\varepsilon} \sim r^{-3}$ for large r for all n and (5.9) is well defined even though $\dot{\varepsilon}^{\infty} = 0$.

In the rigid–perfectly plastic limit, (5.9) continues to apply, but σ^{∞} must satisfy $s_{ij}^{\infty} s_{ij}^{\infty} = 2\sigma_0^2/3$. In this limit, the minimum principle reduces to the one used by Rice and Tracey (1969) in their rigid–plastic analysis. For finite n a principle suitable for infinite domains that is essentially the same as the one derived here was communicated to one of us (B.B.) by J. R. Rice in 1967.

5.1. *Spherically symmetric solution* ($S = T$)

We digress briefly to give the solution for the case in which the remote stress conditions are equivalent to hydrostatic pressure or tension ($\sigma_m = S = T$). We shall also give the solution for the spherical void in a finite spherical region with outer radius R since this simple solution brings out the strong influence of the void volume fraction when the material is nonlinear.

The strain-rate components of a spherically symmetric solution in a spherical coordinate system (r, θ, ϕ) are related to the radial velocity $v_r(r)$ by

$$\dot{\varepsilon}_r = v_{r,r}, \quad \dot{\varepsilon}_\theta = \dot{\varepsilon}_\phi = r^{-1} v_r. \tag{5.12}$$

Incompressibility requires $v_{r,r} + 2r^{-1}v_r = 0$, which implies that the most general spherically symmetric velocity is

$$v_r = Ar^{-2}. \tag{5.13}$$

With the aid of (1.3) and (5.12), the radial equilibrium equation can be integrated subject to $\sigma_r = 0$ at $r = a$ and $\sigma_r = \sigma_m$ at $r = R$. The equation for A which results is

$$A = \operatorname{sign}(\sigma_m)\frac{a^3}{6\eta}\left[\frac{3}{2n}|\sigma_m|\right]^n\left[1 - \rho^{1/n}\right]^{-n}, \tag{5.14}$$

where $\rho = (a/R)^3$ is the ratio of void volume to total volume. The dilatation rate of the void is $\dot{V}/V = 3v_r(a)/a$, and so we have

$$\eta\frac{\dot{V}}{V} = \tfrac{1}{2}\operatorname{sign}(\sigma_m)\left[\frac{3}{2n}|\sigma_m|\right]^n\left[1 - \rho^{1/n}\right]^{-n}. \tag{5.15}$$

The influence of void volume fraction on the dilatation rate is seen by expanding (5.15) in terms of small ρ:

$$\eta\frac{\dot{V}}{V} = \frac{1}{2}\operatorname{sign}(\sigma_m)\left[\frac{3}{2n}|\sigma_m|\right]^n\left[1 + n\rho^{1/n} + \ldots\right]. \tag{5.16}$$

Note that the lowest order influence of the nonzero void volume fraction enters as $\rho^{1/n}$. A void volume fraction of only one-tenth of 1% ($\rho = 10^{-3}$) already increases $\dot{V}$ by almost 30%

above the prediction for the infinite region ($\rho = 0$) when $n = 3$. Interaction effects between voids are therefore expected to be significant at relatively low volume fractions for the nonlinear materials, and this must be borne in mind in any attempts to use solutions for the isolated void in an infinite block to make predictions for porous media. Conversely, numerical calculations carried out for finite blocks of material cannot be expected to reflect accurately the behavior of a void in an infinite block unless the void volume fraction is extremely small.

5.2. *Numerical solution* ($S \neq T$)

For $\sigma = S - T \neq 0$, a Rayleigh–Ritz method based on the minimum principle was used to generate approximate solutions in the nonlinearly viscous range, including the rigid–perfectly plastic limit. Attention will be focused on the results for the rates of volume change of the voids and for the character of their changes from an initially spherical shape.

A general representation of the incompressible, axisymmetric additional velocity field can be written in terms of a stream function $\chi(r, \theta)$ as

$$\tilde{v}_r = -r^{-2}(\sin\theta)^{-1}(\chi\sin\theta)_{,\theta}, \quad \tilde{v}_\theta = r^{-1}\chi_{,r}. \tag{5.17}$$

For a field which is symmetric with respect to $\theta = \pi/2$ ($x_3 = 0$), χ can be expressed as

$$\chi(r,\theta) = A\cot\theta + \sum_{k=2,4,\ldots} P_{k,\theta}(\cos\theta)f_k(r), \tag{5.18}$$

where $P_k(\cos\theta)$ is the Legendre polynomial of degree k. The first term in (5.18) generates the spherically symmetric contribution (5.13) to the field. The amplitude A and the unknown functions $f_k(r)$ must be chosen to minimize Φ in (5.9). From (5.17)

$$\tilde{v}_r = Ar^{-2} + \sum_{k=2,4,\ldots} n(n+1)P_k(\cos\theta)r^{-2}f_k(r) \tag{5.19}$$

and

$$\tilde{v}_\theta = \sum_{k=2,4,\ldots} P_{k,\theta}(\cos\theta)r^{-1}f_{k,r}(r). \tag{5.20}$$

The rate of change of void volume is given by

$$\dot{V} = \int_S v_r dS = 4\pi A \quad \text{or} \quad \dot{V}/V = 3Aa^{-3} \tag{5.21}$$

and is thus uniquely tied to the spherically symmetric field.

The functions $f_k(r)$ were approximated by a finite number of terms according to

$$f_k(r) = \sum_{i=1} A_k^{(i)}r^{2-i} = A_k^{(1)}r + A_k^{(2)} + A_k^{(3)}r^{-1} + A_k^{(4)}r^{-2} + \ldots. \tag{5.22}$$

Expressions for the associated additional strain-rates are readily derived from

$$\begin{gathered}\dot{\tilde{\varepsilon}}_r = \tilde{v}_{r,r}, \quad \dot{\tilde{\varepsilon}}_\theta = r^{-1}\tilde{v}_{\theta,\theta} + r^{-1}\tilde{v}_r, \quad \dot{\tilde{\varepsilon}}_\phi = -\dot{\tilde{\varepsilon}}_r - \dot{\tilde{\varepsilon}}_\theta, \\ \dot{\tilde{\varepsilon}}_{r\theta} = \tfrac{1}{2}(r^{-1}\tilde{v}_{r,\theta} - r^{-1}\tilde{v}_\theta + \tilde{v}_{\theta,r}).\end{gathered} \tag{5.23}$$

For convenience, the amplitude factors A and $A_k^{(i)}$ ($k = 2, 4, \ldots; i = 1, 2, \ldots$) are denoted collectively by $\{A_j\}$, and these are the variables with respect to which Φ must be minimized. For prescribed S and T, Φ in (5.9) is

$$\Phi(\{A_j\}) = 4\pi \int_a^\infty r^2\,dr \int_0^{\pi/2} \sin\theta [W(\dot{\varepsilon}) - W(\varepsilon^\infty) - \sigma_{ij}^\infty \dot{\varepsilon}_{ij}]d\theta - 4\pi a^2 \int_0^{\pi/2} \sigma_{ij}^\infty n_j \dot{v}_i \sin\theta\, d\theta \tag{5.24}$$

and the last integral in (5.24) is just

$$4\pi a^2 \int_0^{\pi/2} \sigma_{ij}^\infty n_j \dot{v}_i \sin\theta\, d\theta = 4\pi\sigma_m A + \frac{8\pi}{5}\sigma \sum_i A_2^{(i)} a^{2-i}(4-i). \tag{5.25}$$

The double integral in (5.24) cannot be evaluated analytically in terms of the amplitude factors except in the spherically symmetric case when A is the only nonzero factor or when $n = 1$. Minimization of Φ with respect to these amplitude factors was achieved by a numerically implemented Newton–Raphson method. By the change of variable $\mu = a/r$, the r-integration in the double integral was converted to an integration over the range $0 \leqslant \mu \leqslant 1$. At each of ten Gaussian integration stations of μ, the inner integration with respect to θ was carried out (for assigned values of the A_j's) with a ten-point Gaussian quadrature formula. These values were then used in the ten-point formula for the outer integration with respect to μ. The accuracy of the scheme was checked against the exact evaluation of Φ for $n = 1$ and for the spherically symmetrical case for general n. In all cases checked, the numerical integration gave Φ to five or six significant figures.

For a given set of N amplitude factors the minimum condition is

$$\partial\Phi/\partial A_j = 0, \quad j = 1, \ldots, N. \tag{5.26}$$

With $\{A_j\}$ as an estimate of the solution to (5.26), the improved Newton–Raphson estimate $\{A_j + \Delta A_j\}$ is obtained from

$$\sum_{m=1}^{N} \Delta A_m \partial^2\Phi/\partial A_m \partial A_j = -\partial\Phi/\partial A_j, \tag{5.27}$$

where the partial derivatives are evaluated at $\{A_j\}$. All partial derivatives were found numerically from finite difference formulas such as

$$\frac{\partial\Phi}{\partial A_j} = \frac{\Phi(A_1, \ldots, A_j + \delta, \ldots) - \Phi(A_1, \ldots, A_j - \delta, \ldots)}{2\delta}. \tag{5.28}$$

The scheme proved highly effective and displayed the quadratic convergence characteristic of this method in the vicinity of the minimum.

The choice of functions used for the representation of $\chi(r, \theta)$ in (5.18) and (5.22) was guided by the results of trial calculations aimed at reaching the minimum of Φ most efficiently. In addition to the spherically symmetric term, the P_2 terms are the most important, and in some of our exploratory calculations as many as six free amplitude factors ($A_2^{(i)}, i = 1, \ldots, 6$) were used to represent $f_2(r)$ in (5.22). Of these six, $A_2^{(2)}$, $A_2^{(3)}$ and $A_2^{(4)}$ were found to be more important than any other set of three and led to results which were not appreciably different from those obtained with the full set of six. Our final choice of χ, which was used in determining most of the numerical results presented below, was

$$\chi(r,\theta) = A\cot\theta + P_{2,\theta}(\cos\theta)\left[A_2^{(2)} + A_2^{(3)}r^{-1} + A_2^{(4)}r^{-2}\right] + P_{4,\theta}(\cos\theta)\left[A_4^{(2)} + A_4^{(3)}r^{-1} + A_4^{(4)}r^{-2}\right], \tag{5.29}$$

with a total of seven free amplitude factors.

The numerical minimization procedure gave results for the linear case ($n = 1$) which were accurate to six significant figures. However, this accuracy reflects mainly on the numerical evaluation of Φ since (5.29) contains the exact solution when $n = 1$, i.e.

$$A = \frac{1}{4\eta}a^3\sigma_m, \quad A_2^{(2)} = -\frac{5}{3}A_2^{(4)}a^2 = \frac{5}{54\eta}a^3\sigma, \tag{5.30}$$

with the other amplitudes zero. For $n > 1$, the terms containing P_4 have a minor quantitative influence on the shape change but are not crucial as far as qualitative predictions are concerned. A form of parameter tracking was used to generate numerical results. For a given S and T, the solution for $n = 1$ was used as the first guess of the solution at a somewhat larger n and when that solution had been found it, in turn, was used as the first guess for a yet higher n. Calculations for $n = \infty$ were carried out separately.

As a check on our numerical method and to make contact with the analysis of Rice and Tracey (1969), we specialized our scheme to the rigid–perfectly plastic limit ($n = \infty$) and studied the results of choosing the same approximating functions used by Rice and Tracey. Their analyses only involved two degrees of freedom: the amplitude A of the spherically symmetric field and the amplitude of a P_2 term with a fixed, associated r-variation $f_2(r)$. They did, however, repeat their calculations for six choices of $f_2(r)$. Our specialization to the choices of Rice and Tracey did reproduce their numerical results in all cases checked. While $\dot{V}$ was only slightly affected by the choice of $f_2(r)$, Rice and Tracey found substantial variation in the corresponding shape changes. For convenience, let us denote the rate of change of the vertical axis of the spherical void by $\dot{a}$ and let $\dot{b}$ be the rate of change of the equatorial radius. For $\sigma_m > \sigma > 0$, Rice and Tracey found that for four of their r-variations, the shape became prolate ($\dot{a} > \dot{b}$), while for the other two $f_2(r)$ choices it became oblate ($\dot{a} < \dot{b}$). Rice and Tracey discounted the possibility of evolution towards an oblate shape in the presence of a positive axial strain-rate at infinity and attributed this anomalous prediction to poor choices for $f_2(r)$. They evidently did not note, however, that the two "poor" choices gave lower values for the minimum of Φ than did some of the other four choices. It will be seen that for $n > 1$ evolution towards an oblate shape is the rule rather than the exception when $\sigma > 0$ and the triaxiality ratio σ_m/σ is high.

Numerical results obtained by minimizing Φ with respect to the seven amplitudes in (5.29) will now be presented. Figures 5.1(a) and 5.1(b) give plots of $\dot{V}/(\dot{\varepsilon}V)$ as a function of σ_m/σ for the case when both σ_m and σ are positive (i.e. $S + 2T > 0$ and $S > T$) or are both negative. The solid line curves are the numerical results obtained from (5.29). As already mentioned, $\dot{V}$ is only weakly dependent on a wide variety of choices in approximating functions when $\sigma_m/\sigma > 1$. In fact, for sufficiently large σ_m/σ, accurate results for $\dot{V}$ are obtained by using just the spherically symmetric term, and the dashed line curves in Fig. 5.1(b) are based on an approximate formula derived under this assumption in Section 6. The dashed line curve for the rigid–perfectly plastic solid ($n = \infty$) coincides with the corresponding high triaxiality approximation of Rice and Tracey (1969).

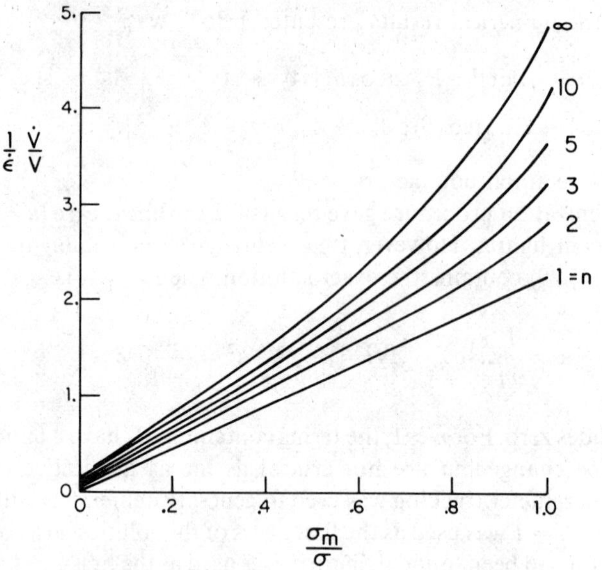

FIG. 5.1(a). Dilatation rate of spherical void, low triaxiality.

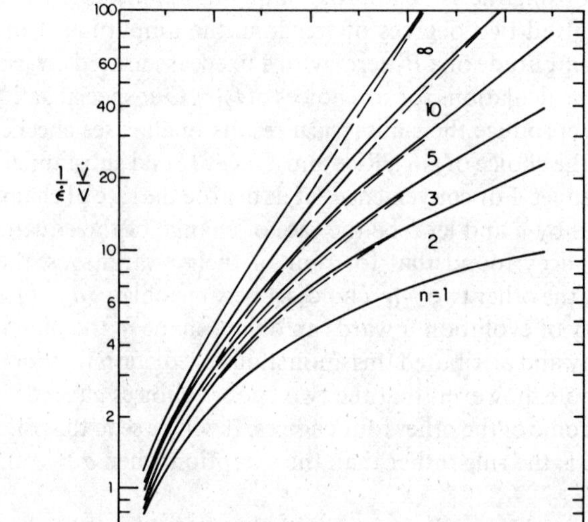

FIG. 5.1(b). Dilatation rate of spherical void, moderate to high triaxiality.

We note from Fig. 5.1(b) that nonlinearity plays an exceptionally strong role in the expansion (or contraction) of the volume growth-rate of the void, as measured by $\dot{V}/(\dot{\varepsilon}V)$, in the high triaxiality range where $\sigma_m/\sigma > 1$. A similar conclusion was drawn by Hellan (1975) on the basis of a high triaxiality approximation not unlike that which will be discussed in Section 6.

Numerical results related to the shape change of the spherical void are shown in Fig. 5.2, where curves of $\dot{a}/\dot{b}$ are plotted against S/T for the case in which S and T are both positive (or both negative). For the linearly viscous solid ($n = 1$),

$$\frac{\dot{a}}{\dot{b}} = \frac{9\sigma_m + 20\sigma}{9\sigma_m - 10\sigma} = 1 + \frac{30(S-T)}{16T - 7S}, \tag{5.31}$$

and the void grows towards a prolate shape when $0 < T < S$ and towards an oblate shape when $0 < S < T$, as has already been discussed in Section 3. In contrast, a spherical void in a rigid–perfectly plastic solid ($n = \infty$) grows towards an oblate shape when $0 < T < S < 1.85T$, and $\dot{b}$ can be more than twice $\dot{a}$. This seemingly anomalous effect is biggest at high n and essentially disappears for n less than 2. For $0 < S < T$ the situation reverses with the void becoming prolate in the nonlinear material for S/T sufficiently close to unity.

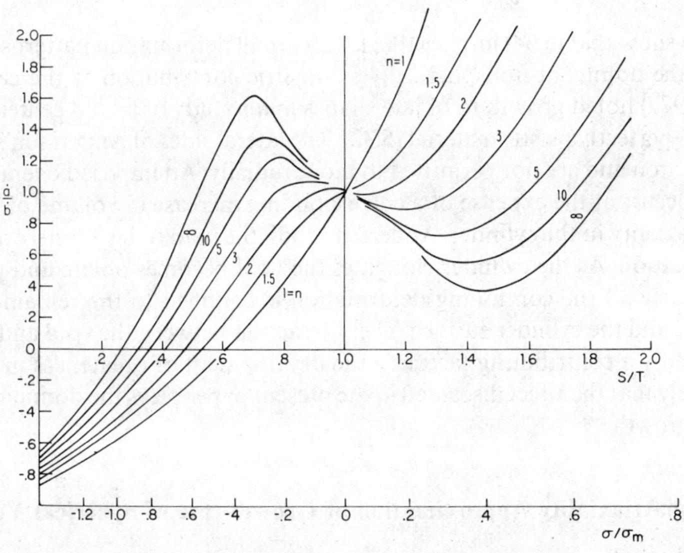

FIG. 5.2. Initial shape change of spherical void.

Void oblateness is obviously significant in void interaction and coalescence under tensile straining, since the cross-sectional area of the void perpendicular to the straining axis is a more important variable in void coalescence than is its volume. A qualitative explanation of the origin of this counter-intuitive effect will now be given.

The deformation of the void in the linearly viscous solid is in accord with geometrical intuition in that one expects a remote tensile strain-rate $\dot{\varepsilon}$ to elongate the void in the direction of straining and detract from growth of its equator. The converse behavior for the void in the nonlinear material under sufficiently high triaxiality is a consequence of two factors—the nonlinear coupling between $\dot{\varepsilon}$ and the spherically symmetric field, and the nature of the flow law (1.3) at high n. Equation (1.3) can be written as

$$\dot{\varepsilon}_{ij} = \tfrac{3}{2}\dot{\varepsilon}_e(s_{ij}/\sigma_e), \quad \text{where} \quad \dot{\varepsilon}_e = \dot{\varepsilon}_0(\sigma_e/\sigma_0)^n. \tag{5.32}$$

In the rigid–perfectly plastic limit of (5.32), $\sigma_e \equiv \sigma_0$ wherever yield occurs. The spherically symmetric solution ($\sigma = 0$, $\sigma_m > 0$) gives rise to strain-rates given by

$$\dot{\varepsilon}_r = -\dot{\varepsilon}_e, \quad \dot{\varepsilon}_\theta = \dot{\varepsilon}_\phi = \tfrac{1}{2}\dot{\varepsilon}_e. \tag{5.33}$$

Imposition of a remote tensile strain-rate $\dot{\varepsilon} > 0$ induces a large spherically symmetric contribution to the void deformation when $\sigma_m \gg \sigma > 0$ at high n, as has already been discussed. Thus the stresses at the void are determined largely by the spherically symmetric term at high n when triaxiality is high and, from (5.32), this means that the strain-rates are still given approximately by (5.33) at the void. But imposition of a positive $\dot{\varepsilon}$ gives rise to a variation of $\dot{\varepsilon}_e$ at the cavity which is largest near the equator and falls to a lower value at the poles. (This observation, which is in accord with intuition, holds for all n, including the linear material.) It follows from this variation and from (5.33) that at high n the void equator lengthens more rapidly than the meridians, i.e.

$$2\int_0^\pi \dot{\varepsilon}_\theta(a,\theta)a\,d\theta < 2\pi a\dot{\varepsilon}_\phi(a,\pi/2). \tag{5.34}$$

It is then easy to show that (5.34) implies that $\dot{a} < \dot{b}$ for all deformation patterns in which the P_2 variation is the dominant nonspherically symmetric contribution at the cavity.

Andersson (1977) noted growth to oblate shapes in his study of a void centered in a finite cylinder of rigid–perfectly plastic material (5.32). The lateral sides of Andersson's cylinder are free of shear traction and are not permitted to move radially. An imposed overall elongation of the cylinder occurs at the expense of a corresponding increase in volume of the void and induces high triaxiality in the cylinder. Andersson took the void to be a sphere at the start of the overall elongation. As the cylinder elongates the void becomes oblate and grows in this fashion until nearly all the continuing deformation is confined to the remaining ligament between the void and the cylinder surface. The interaction between the void and the cylinder wall is undoubtedly a contributing factor to the development of oblateness in Andersson's void, but it is likely that the effect discussed in the present paper plays the dominant role in the early stages of growth.

6. High-triaxiality Approximation of Growth-rate of Spherical Void

A good approximation to $\dot{V}$ when $|\sigma_m/\sigma|$ is larger than about unity can be obtained by neglecting all but the spherically symmetric contribution to the velocity $\tilde{\mathbf{v}}$ in the minimization procedure discussed in Section 5. The end result, after some further approximation, is a simple formula for the dilatation-rate of the spherical void. The analysis generalizes a similar one for the rigid–perfectly plastic case given by Rice and Tracey (1969), and the formula derived for $(\dot{V}/\dot{\varepsilon}V)$ agrees with their results in the limit $n \to \infty$.

With only the radially symmetric contributions

$$\tilde{v}_r = Ar^{-2}, \quad \tilde{\varepsilon}_r = -2\tilde{\varepsilon}_\theta = -2\tilde{\varepsilon}_\phi = -2Ar^{-3}, \tag{6.1}$$

used in the calculation of Φ in (5.24), we find directly

$$\Phi = 2\pi\int_0^\pi \sin\theta\,d\theta \int_a^\infty \left\{(3\eta)^{1/n}\left(\frac{n}{n+1}\right)[\dot{\varepsilon}^2 - \dot{\varepsilon}Ah(\theta)r^{-3} + 4A^2r^{-6}]^{(n+1)/2n} - (3\eta)^{1/n}\left(\frac{n}{n+1}\right)|\dot{\varepsilon}|^{(n+1)/n} - \sigma Ah(\theta)r^{-3}\right\}r^2\,dr - 4\pi\sigma_m Aa^3 \tag{6.2}$$

in terms of A and $\dot{\varepsilon}$, with $\sigma = S - T = (3\eta)^{1/n}|\dot{\varepsilon}|^{(1-n)/n}\dot{\varepsilon}$, $\sigma_m = (S+2T)/3$ and $h(\theta) = 1 + 3\cos 2\theta$.

Introduction of the key variable

$$\Lambda = \frac{\dot{V}}{\dot{\varepsilon}\, V} = \frac{3A}{\dot{\varepsilon} a^3} \tag{6.3}$$

and the substitution $z = \frac{2}{3}|\Lambda|\,(a/r)^3$ reduces (6.2) to $\Phi = (4\pi a^3/3)\,(n/(n+1))\,(\sigma\dot{\varepsilon})\,F$, with

$$F = \frac{|\Lambda|}{3}\int_0^{\pi} \sin\theta\, d\theta \int_0^{\frac{2}{3}|\Lambda|}\left\{\left[1 - \frac{mzh}{2} + z^2\right]^{(n+1)/2n} - 1 + \left(\frac{n+1}{4n}\right) mzh\right\}\frac{dz}{z^2} - \left(\frac{n+1}{n}\right)\frac{\Lambda\sigma_m}{\sigma}, \tag{6.4}$$

where $m = \operatorname{sign}(\Lambda)$.

The quantity Λ may replace A as the single free variable in the minimization of Φ, and so $dF/d\Lambda = 0$ gives

$$\frac{m}{3}\int_0^{\pi} \sin\theta\, d\theta \int_0^{\frac{2}{3}|\Lambda|}\left\{\left[1 - \frac{mzh}{2} + z^2\right]^{(n+1)/2n} - 1 + \left(\frac{n+1}{4n}\right) mzh\right\}\frac{dz}{z^2}$$

$$+ \frac{m}{2|\Lambda|}\int_0^{\pi}\left\{1 - \frac{m|\Lambda|h}{3} + \frac{4}{9}|\Lambda|^2\right\}^{(n+1)/2n} \sin\theta\, d\theta - \frac{m}{|\Lambda|} = \left(\frac{n+1}{n}\right)\left(\frac{\sigma_m}{\sigma}\right). \tag{6.5}$$

This can be used, via numerical integration, to provide the desired estimate of the relation between σ_m/σ and $\dot{V}/(\dot{\varepsilon} V)$. However, we can obtain a useful analytical approximation to this result. It is seen that for large $|\Lambda|$ the dominant contribution of the integrals in (6.5) is

$$\frac{2m}{3}(n+1)\left|\frac{2\Lambda}{3}\right|^{1/n}. \tag{6.6}$$

This implies that $m \equiv \operatorname{sign}(\Lambda) = \operatorname{sign}(\sigma_m/\sigma)$ for sufficiently small $|\sigma|$, and also provides the asymptotic approximation

$$\Lambda = \frac{3m}{2}\left|\frac{3\sigma_m}{2n}\right|^n. \tag{6.7}$$

Except for $n = 1$, where it is exact, this estimate for $\dot{V}/(\dot{\varepsilon} V)$ is too crude since it is invalid for $n = \infty$ and has a very small range of validity for large n. The next term in the asymptotic expansion of the left-hand side of (6.5) for large Λ is a constant which is of magnitude comparable to (6.6) as $n \to \infty$ and must be included if an estimate of Λ that is uniformly valid for all n is to be obtained.

To get this better estimate, we write the left-hand side of (6.5) as

$$\frac{2m}{3}(n+1)\left[\left|\frac{2\Lambda}{3}\right|^{1/n} + G(n,m) + \dots\right], \tag{6.8}$$

where, by subtracting (6.6) from (6.5), we can get

$$G(n,m) = -1 + \frac{1}{2(n+1)}\int_0^{\pi}\sin\theta\, d\theta \int_0^{1}\left\{\left[1 - \frac{mzh}{2} + z^2\right]^{(n+1)/2n} - 1 + \left(\frac{n+1}{4n}\right) mzh\right\}\frac{dz}{z^2}$$

$$+ \frac{1}{2(n+1)}\int_0^{\pi}\sin\theta\, d\theta \int_1^{\infty}\left\{\left[1 - \frac{mzh}{2} + z^2\right]^{(n+1)/2n} - z^{(n+1)/n}\right\}\frac{dz}{z^2}. \tag{6.9}$$

The terms omitted in (6.8) vanish for large $|\Lambda|$. The improved estimate for Λ becomes

$$\frac{\dot{V}}{\dot{\varepsilon}V} = \frac{3m}{2}\left\{\frac{3}{2n}\left|\frac{\sigma_m}{\sigma}\right| - G(m, n)\right\}^n \tag{6.10}$$

and this is a uniformly valid asymptotic approximation, for large $\dot{V}/(\dot{\varepsilon}V)$, for all n.

The formula still requires numerical evaluation of the integrals in (6.9), but now we expand $G(n, m)$ into

$$G(n, m) = -\left[1 + \frac{g(m)-1}{n} + \dots\right] \tag{6.11}$$

for large n; then

$$g(m) = \lim_{\varepsilon \to 0}\left[\ln \varepsilon - \tfrac{1}{2}\int_0^{\pi} \sin\theta \, d\theta \int_{\varepsilon}^{\infty}\left\{\left[1 - \frac{mzh}{2} + z^2\right] - 1 - z\right\}\frac{dz}{z^2}\right], \tag{6.12}$$

which can be evaluated analytically to give

$$\begin{aligned} g(1) &= \ln 3 - \frac{2}{3} \cong 0.4319, \\ g(-1) &= \frac{2\pi}{9\sqrt{3}} \cong 0.4031. \end{aligned} \tag{6.13}$$

Finally, we use the approximation

$$-G(n, m) \cong \frac{(n-1)[n+g(m)]}{n^2}, \tag{6.14}$$

which is correct to order $1/n$ at $n = \infty$, and correctly vanishes at $n = 1$. A comparison between this approximation to $G(n, m)$ and the numerical evaluation of $G(n, m)$ in (6.9) is shown in Fig. 6.1 for $m = 1$; the comparison for $m = -1$ is equally good. Used in (6.10), this gives our final approximate formula

$$\frac{\dot{V}}{\dot{\varepsilon}V} = \frac{3m}{2}\left\{\frac{3}{2n}\left|\frac{\sigma_m}{\sigma}\right| + \frac{(n-1)[n+g(m)]}{n^2}\right\}^n. \tag{6.15}$$

In the limit $n \to \infty$,

$$\begin{aligned} \frac{\dot{V}}{\dot{\varepsilon}V} &= \frac{3m}{2}\exp[g(m)-1]\exp\left[\frac{3}{2}\left|\frac{\sigma_m}{\sigma}\right|\right] \\ &= 0.850\exp\left[\frac{3}{2}\left|\frac{\sigma_m}{\sigma}\right|\right] \quad \text{for} \quad \frac{\sigma_m}{\sigma} > 0, \\ &= -0.826\exp\left[\frac{3}{2}\left|\frac{\sigma_m}{\sigma}\right|\right] \quad \text{for} \quad \frac{\sigma_m}{\sigma} < 0, \end{aligned} \tag{6.16}$$

in agreement with the high triaxiality formulas derived by Rice and Tracey (1969).

A comparison of (6.15) with the unapproximated high triaxiality relation (6.5) is made in Fig. 6.2, where the ratio of the two estimates of $\dot{V}/(\dot{\varepsilon}V)$ is plotted against $|\sigma/\sigma_m|$ for various n-values for $m = 1$. Results from (6.5) were obtained by numerical integration. The comparison for $n = 1$ is not shown since both results are exact in this case. The simple formula (6.15) is

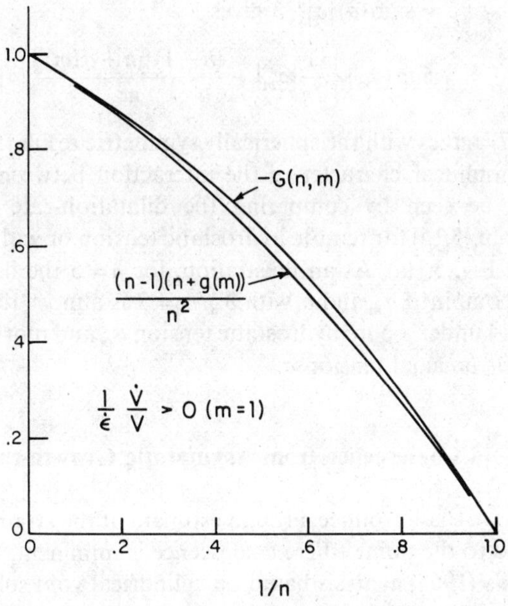

FIG. 6.1. Accuracy of approximation to $G(n, m)$.

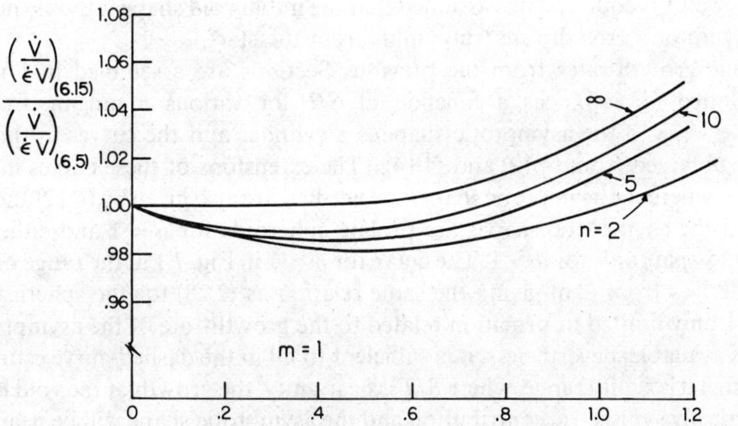

FIG. 6.2. Accuracy of analytic approximation for dilatation-rate in high triaxiality range.

within 2% of the unapproximated result (6.5) for $|\sigma/\sigma_m| < 1$, which is the range of applicability of the high triaxiality approximation. Similar claims hold for the case $m = -1$. Results from the simple formula (6.15) are shown as dashed line curves in Fig. 5.1(b), where they are compared to the predictions obtained from the more extensive calculations of Section 5. For $\sigma_m/\sigma > 2$, (6.15) clearly provides an excellent approximation; and even for σ_m/σ as small as unity, (6.15) underestimates the more accurate results by no more than 20%.

Multiplying (6.15) by $\eta\dot{\varepsilon} = \text{sign}(\sigma)|\sigma|^n/3$ gives

$$\eta\frac{\dot{V}}{V} = \tfrac{1}{2}\,\text{sign}(\sigma_m)\left\{\frac{3}{2n}|\sigma_m| + \frac{(n-1)[n+g(m)]}{n^2}|\sigma|\right\}^n \tag{6.17}$$

and it is seen that (6.17) agrees with the spherically symmetric result (5.15) (with $\rho = 0$) when $\sigma = 0$. The strongly nonlinear character of the interaction between σ and σ_m in the high triaxiality range can be seen by comparing the dilatation-rate from (6.17) with the corresponding rate from (5.15) for remote hydrostatic tension or with the rate under remote uniaxial tension from Fig. 5.1(a). As an illustration, for $n=5$ the dilatation-rate $\dot{V}/V$ of a spherical void under combined σ_m and σ with $\sigma_m/\sigma = 2$ is almost 100 times as great as the dilatation-rate of a void under a pure hydrostatic tension σ_m and more than 10 times as great as that of a void under uniaxial tension σ.

7. Strain to Void Coalescence from Asymptotic Growth-rates ($S > T > 0$)

In this final Section we shall produce a rough estimate of the strain needed to enlarge the voids in a porous solid to the point where coalescence is imminent, following an approach similar to McClintock's (1968) analysis based on cylindrical void solutions. We shall make use of the growth-rate of the asymptotic voids for this purpose since this greatly simplifies the calculations. In Section 3 it was seen that the growth-rate of a void of initially spherical shape in the linearly viscous solid closely approaches the asymptotic value after only a doubling of the initial volume. It seems reasonable to expect similar behavior of voids in the nonlinear materials as well. Of course, if it is assumed that the initial void shape *is* the asymptotic shape then the asymptotic growth-rate truly holds from the start.

Asymptotic growth-rates from the previous Sections are assembled in Fig. 7.1, where we have plotted $(\dot{V}/\dot{\varepsilon}V)_f^{-1}$ as a function of S/T for various n ranging from 1 to ∞. For $1 \leqslant (\dot{V}/\dot{\varepsilon}V)_f \leqslant 3$ the asymptotic shape is a cylinder and the curves in this portion of Fig. 7.1 are obtained from (4.12) and (4.14). (The extensions of these curves into the range $(\dot{V}/\dot{\varepsilon}V)_f < 1$, where the asymptotic shapes are needles, are also given by (4.12) and (4.14).) For $(\dot{V}/\dot{\varepsilon}V)_f > 3$ the asymptotic shapes are prolate spheroids for $n = 1$ and either prolate or oblate pseudo-spheroids for $n > 1$. The curve for $n = 1$ in Fig. 7.1 in the range of asymptotic spheroids derives from (3.6) giving the same relation as (2.28) for the spherical void.

For $n > 1$ only limited information related to the growth-rate of the asymptotic pseudo-spheroids is available; nevertheless, it is sufficient to fill in the dashed-curve estimates in Fig. 7.1. In the high triaxiality range, where S/T is near unity, the growth of the void is dominated by the spherically symmetric contribution and the asymptotic shape will be nearly spherical, although slightly oblate for $n > 1$. The growth-rate for the spherical void should provide a good approximation to the asymptotic growth-rate in this range, and (6.15) has been used to plot the curves in Fig. 7.1 near $S/T = 1$. For increasing S/T the oblateness of the asymptotic shape first increases and then decreases to zero ($a = b$). For still larger S/T the asymptotic shape is increasingly prolate until it becomes cylindrical at $(\dot{V}/\dot{\varepsilon}V)_f = 3$. The results for the shape change of the spherical void in Fig. 5.2 can be used to estimate the transition of the asymptotic void from an oblate to a prolate shape. We assume that, when $\dot{a}/\dot{b} = 1$ for the sphere, the asymptotic shape will be a pseudo-spheroid with $a = b$ (approximately) and that the growth-rate for the sphere should be a good approximation to the asymptotic growth-rate associated with this transitional shape. The circles in Fig. 7.1 are estimates of these

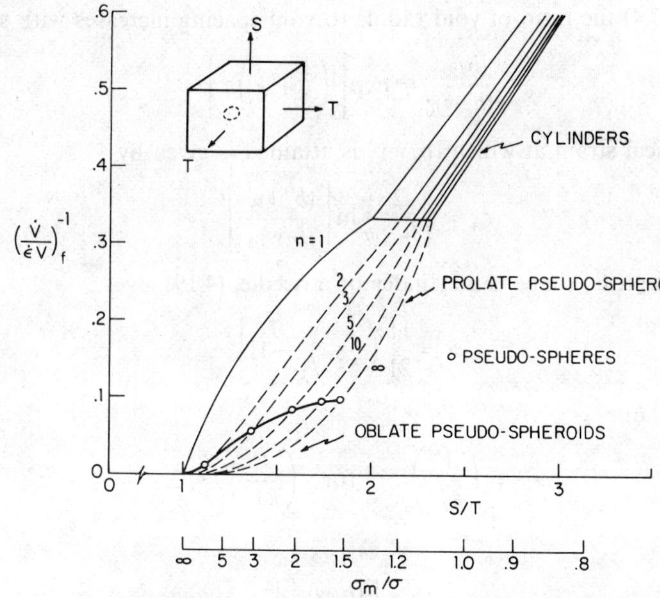

FIG. 7.1. Asymptotic dilatation rates; analytic (sclid lines) and estimated (dashed lines).

transitional points so obtained. The remainder of the dashed curve for each n is faired in to connect smoothly the high triaxiality results, the transition point, and the cylindrical limit of the prolate shapes at $(\dot{V}/\dot{\varepsilon}V)_f = 3$. (Subsequent work addressing the accurate calculation of the asymptotic void shapes and their associated growth-rates, published in the *Proceedings of the 15th Congress of Theoretical and Applied Mechanics at Toronto*, reveals that the estimates in Fig. 7.1 are quite reasonable.)

Now we can turn to the problem of estimating the strain at which voids coalesce in a porous block subject to overall stresses S and T in (1.5) with $S > T > 0$. We neglect interaction between the voids and treat each void as if it is isolated in an infinite block, but we relate the growing equatorial radius b of the void in the plane perpendicular to the straining direction to the shrinking distance w between voids in the same plane. Coalescence becomes imminent when b/w grows to some sufficiently large value, $(b/w)_{cr}$.

Whether the shape is a pseudo-spheroid, a cylinder, or a needle,

$$\psi = \dot{b}/(\dot{\varepsilon}b) \tag{7.1}$$

is constant for an asymptotic void growing under fixed S/T. With $b = b_0$ as the initial equatorial radius of the void at the start of straining, (7.1) gives

$$b/b_0 = \exp(\psi\varepsilon). \tag{7.2}$$

The spacing between the voids in the plane normal to S contracts according to

$$\dot{w}/w = -\dot{\varepsilon}/2, \tag{7.3}$$

so that, if w_0 is the initial spacing,

$$w/w_0 = -\exp(\varepsilon/2). \tag{7.4}$$

From (7.2) and (7.4) the ratio of void radius to void spacing increases with strain as

$$\frac{(b/w)}{(b/w)_0} = \exp\left[\left(\frac{1}{2}+\psi\right)\varepsilon\right] \tag{7.5}$$

and thus the critical strain at which $(b/w)_{cr}$ is attained is given by

$$\varepsilon_{cr} = \frac{2}{1+2\psi}\ln\left[\frac{(b/w)_{cr}}{(b/w)_0}\right]. \tag{7.6}$$

When the asymptotic shape is a cylinder or a needle, (4.19) gives

$$\psi = \frac{1}{2}\left[\left(\frac{\dot{V}}{\dot{\varepsilon}V}\right)_f - 1\right] \tag{7.7}$$

and thus, from (7.6),

$$\frac{\varepsilon_{cr}}{c} = \left(\frac{\dot{V}}{\dot{\varepsilon}V}\right)_f^{-1} \quad \text{for} \quad \left(\frac{\dot{V}}{\dot{\varepsilon}V}\right)_f \leqslant 3, \tag{7.8}$$

where

$$c = \ln\left[\frac{(b/w)_{cr}}{(b/w)_0}\right]^2. \tag{7.9}$$

Asymptotic spheroids or pseudo-spheroids satisfy

$$\psi = \frac{1}{3}\left(\frac{\dot{V}}{\dot{\varepsilon}V}\right)_f \tag{7.10}$$

whence

$$\frac{\varepsilon_{cr}}{c} = 1\Big/\left[1+\frac{2}{3}\left(\frac{\dot{V}}{\dot{\varepsilon}V}\right)_f\right] \quad \text{for} \quad \left(\frac{\dot{V}}{\dot{\varepsilon}V}\right)_f \geqslant 3, \tag{7.11}$$

which equals (7.8) when $(\dot{V}/\dot{\varepsilon}V)_f = 3$. Curves of σ_m/σ as a function of ε_{cr}/c in the range in which the asymptotic voids are cylinders or needles are shown in Fig. 7.2(a) and are obtained from (7.8) and (4.12). Similar curves on an expanded scale for ε_{cr}/c are shown in Fig. 7.2(b) for the range in which the asymptote voids are pseudo-spheroids. (McClintock's 1968 results for cylinders are essentially equivalent to those in Fig. 7.2(a) but he extrapolates them into the pseudo-spheroid range of Fig. 7.2(b).) The curve for $n = 1$ is obtained from (7.11) and the exact result for $(V/\varepsilon V)_f$ from (3.6), while the curves for $n > 1$ are obtained from the estimates of $(V/\varepsilon V)_f$ in Fig. 7.1.

Interaction between voids is likely to accelerate the growth of the voids and, therefore, to reduce the strain to coalescence below what has been predicted here. Steps to analyze interaction have been taken by Tracey (1971), who models interaction effects for cylindrical voids using a thick cylindrical shell configuration, and by Andersson (1977) with his void in a cylindrical cell which was discussed earlier. More elaborate finite element calculations for planar arrays of voids in plane strain have been carried out by Burke and Nix (1979), Needleman (1972), and Nemat-Nasser and Taya (1976). While it is clear that interaction effects are important in the coalescence phenomenon, it is less clear at what stage (i.e. at what b/w) it becomes essential to allow for interaction. It may be that simple results such as those in Figs. 7.2(a) and (b) which ignore interaction have approximate validity if a judicious choice

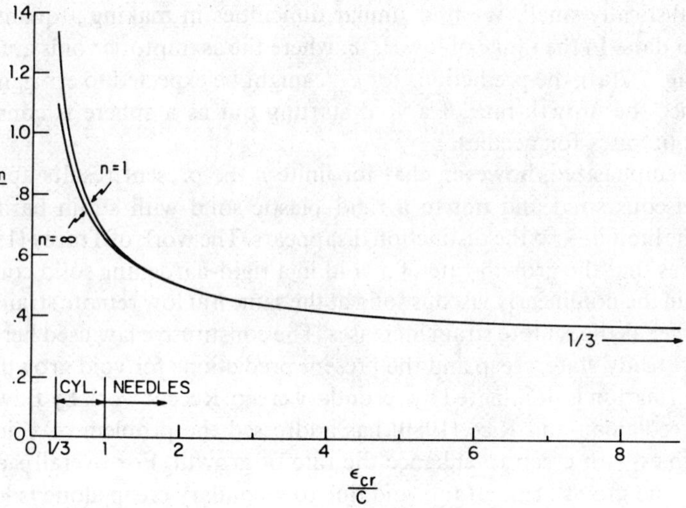

FIG. 7.2(a). Critical strain for void coalescence, low triaxiality range.

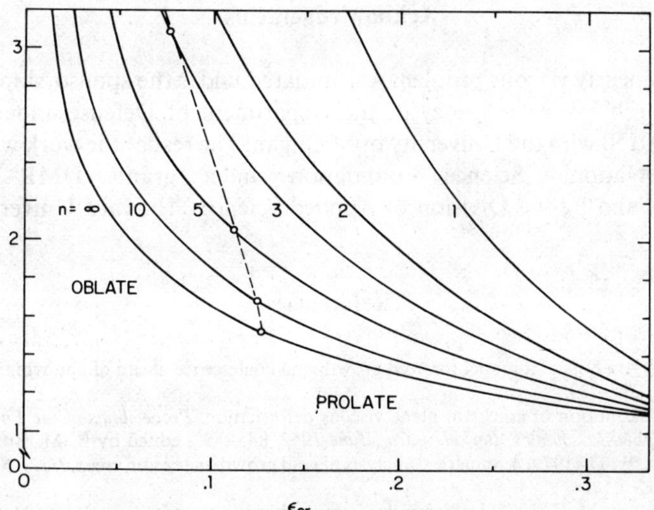

FIG. 7.2(b). Critical strain for void coalescence, moderate to high triaxially range.

of $(b/w)_{cr}$ is made with the aid of more accurate calculations or with experimental data. Hancock and Mackenzie (1976) and Mackenzie *et al.* (1977) have carried out a major study of the effect of triaxiality on the ductile fracture of several structural steels. They obtained experimental data in the form of σ_m/σ against ε_{cr} for σ_m/σ in the range 1/3 to 1.4, and they compared their data with curves similar to those in Fig. 7.2(a) based on McClintock's (1968) analysis. While the qualitative trend of σ_m/σ against ε_{cr} was found to be correct, the value of $(b/w)_{cr}/(b/w)_0$ needed to bring the theoretical curve into juxtaposition with the data was usually unrealistically small. We find similar difficulties in making a quantitative correlation with their data. In the range of low σ_m/σ, where the asymptotic voids are needles (i.e. for $\varepsilon_{cr}/c > 1$ in Fig. 7.2(a)), the predictions for ε_{cr}/c might be expected to err significantly on the high side, since the growth-rate of a void starting out as a sphere is considerably larger than asymptotic rates for needles.

It must be emphasized, however, that for finite n the present results apply strictly to a nonlinearly viscous solid and not to a rigid–plastic solid with strain hardening index n, although in the limit $n \to \infty$ the distinction disappears. The work of Tracey (1971) and Hellan (1975) indicates that the growth-rate of a void in a rigid-hardening solid coincides with the rate of a void in the nonlinearly viscous solid at the same n at low remote strain ε and becomes somewhat larger as the remote strain increases. The constitutive law used here also describes secondary, or steady-state, creep and the present predictions for void growth apply directly when the deformation is dominated by secondary creep. Recent work by Edward and Ashby (1979) and Needleman and Rice (1980) has addressed the problem of void growth when diffusion couples with creep to enhance the rate of growth. For overall states of uniaxial tension, when the growth-rate of the void due to secondary creep alone is low, diffusion is found to make a substantial increase in the growth-rate of the void over an important range of the parameters characterizing diffusion and creep in common metals.

Acknowledgements

Work on the linearly viscous problem was initiated under the sponsorship of the Defense Advanced Research Projects Agency of the Department of Defense under Contract No. MDA903–76C–0250 with the University of Michigan. The rest of the work was supported in part by the National Science Foundation under grants DMR–77–24295 and ENG–78–10756, and by the Division of Applied Sciences, Harvard University.

References

Andersson, H. (1977) Analysis of a model for void growth and coalescence ahead of a moving crack tip, *J. Mech. Phys. Solids* **25,** 217–233.

Berg, C. A. (1962) The motion of cracks in plane viscous deformation, *Proceedings of the Fourth U.S. National Congress of Applied Mechanics, Berkeley, Ca., June 1962,* 885–892, edited by R. M. Rosenberg. ASME.

Burke, M. A. and Nix, W. D. (1979) A numerical analysis of void growth in tension creep, *Int. J. Solids Structures* **15,** 55–71.

Edward, G. H. and Ashby, M. F. (1979) Intergranular fracture during power-law creep, *Acta Metall.* **27,** 1505–1518.

Eshelby, J. D. (1957) The determination of the elastic field of an ellipsoidal inclusion, and related problems, *Proc. R. Soc. Lond.* A **241,** 376–396.

Hancock, J. W. and Mackenzie, A. C. (1976) On the mechanism of ductile failure in high-strength steels subjected to multi-axial stress-states, *J. Mech. Phys. Solids* **24,** 147–169.

Hellan, K. (1975) An approximate study of void expansion by ductility or creep, *Int. J. Mech. Sci.* **17,** 369–374.

Hill, R. (1956) New horizons in the mechanics of solids, *J. Mech. Phys. Solids* **5,** 66–74.

Hill, R. (1965) A self-consistent mechanics of composite materials, *J. Mech. Phys. Solids* **13,** 213–222.

Mackenzie, A. C., Hancock, J. W., and Brown, D. K. (1977) On the influence of state of stress on ductile failure initiation in high strength steels, *Engr. Fracture Mech.* **9,** 167–188.

McClintock, F. A., Kaplan, S. M., and Berg, C. A. (1966) Ductile fracture by hole growth in shear bands, *Int. J. Fracture Mech.* **2,** 614–627.

McClintock, F. A. (1968) A criterion for ductile fracture by growth of holes, *Trans. ASME,* Series E, *J. Appl. Mech.* **35,** 363–371.

Needleman, A. (1972) Void growth in an elastic–plastic medium, *Trans. ASME,* Series E, *J. Appl. Mech.* **39,** 964–969.

Needleman, A. and Rice, J. R. (1980) Plastic creep flow effects in the diffusive cavitation of grain boundaries, *Acta Metall.* **28,** 1315–1332.

Nemat-Nasser, S. and Taya, M. (1976) Model studies of ductile fracture–I Formulation, *J. Franklin Inst.*, special issue *Basis of the Finite Element Method* **302,** 463–472, edited by K. Washizu.

Rice, J. R. and Tracey, D. M. (1969) On the ductile enlargement of voids in triaxial stress fields, *J. Mech. Phys. Solids* **17,** 201–217.

Tracey, D. M. (1971) Strain-hardening and interaction effects on the growth of voids in ductile fracture, *Engr. Fracture Mech.* **3,** 301–316.

Surface Waves in Cubic Elastic Materials

by P. Chadwick and G. D. Smith

School of Mathematics and Physics, University of East Anglia, Norwich

Summary

In Part I of this article a programme is formulated for investigating the existence of surface waves in a semi-infinite body with traction-free boundary composed of an arbitrary anisotropic elastic material. Part II describes in detail the realization of this programme for materials with cubic symmetry, and the results are applied in Part III, first to the general existence problem, then to three particular configurations of the boundary and surface wave normals which display features of especial interest and have previously been the subject of numerical studies. Although intended primarily as a contribution to the theory of elastic surface waves, the article gives extensive consideration to body waves and provides, *inter alia*, a more complete account of slowness surfaces and acoustic polarization fields for cubic media than has been available hitherto.

1. Introduction

In 1977 the present writers published a survey of recent work on surface waves in anisotropic elastic media, giving prominence to the major advances which have stemmed from the analogy between surface waves and uniformly moving line dislocations first recognized by Stroh (1962). Implicit in this review (Chadwick and Smith, 1977; cited henceforth as CS) is a programme for studying the existence of surface waves in elastic materials belonging to any given symmetry class, but applications of the general concepts and methods were restricted to partial results for hexagonal crystals (subsequently extended by Lothe and Al'shits (1977) and Al'shits and Lothe (1978)). Our aim in the present article is to make the programme explicit and to implement it as fully as possible for elastic materials with cubic symmetry.

The contents are divided into three parts, entitled Formulation, Realization, and Application. The fundamental result in the theory of surface waves, as it now stands, is the existence theorem of Barnett and Lothe (1974). Part I states and explains this theorem and articulates the considerations involving body waves which enter into its application. The procedure which we adopt has two main steps, requiring, first, a sufficiently detailed understanding of the slowness surface $\mathscr{S}$ of the transmitting material, and, second, the characterization of a certain subset of the body waves associated with the outer sheet of $\mathscr{S}$

which generate no traction on a set of parallel planes. (Body waves leaving a family of parallel planes traction-free are said to be *exceptional*, and the term *exceptional limiting* is applied to the members of the subset.) Part I is wholly concerned with anisotropic elastic materials of unspecified symmetry and is thus, in spirit, a continuation of CS.

In Parts II and III attention is confined to cubic elastic materials. Part II begins with a classification of cubic crystals based at first on relations among the three elastic constants, but ultimately connected to the structure of $\mathscr{S}$ and the acoustic polarization field. (A somewhat similar approach to the classification of orthorhombic elastic materials has been developed by Musgrave (1981).) The programme formulated in Part I is then put in hand. A synoptic description of $\mathscr{S}$ is evolved by a combination of analysis and numerical calculation, and a systematic search for exceptional limiting waves is prosecuted. The second stage is extended beyond the limits of strict necessity by an investigation of polarization fields for cubic media and their relationship to exceptional waves, but we believe this enlargement of the frame of reference to be justified by the new results and broadened perspectives which it affords.

Part III brings the findings of Part II to bear on the basic question of the existence of surface waves in cubic elastic materials. It is known, on general grounds, that the propagation of a unique surface wave represents the usual state of affairs, so the purpose of our initial survey is to identify placements of the boundary and surface wave normals relative to the material symmetry axes which are exceptional in this regard. We single out for detailed analysis three particular cases for which numerical results are available (in Farnell, 1970, Section V), and which serve also to exemplify the classes of exceptional limiting waves defined at the end of Part II. The conclusions reached in Part III are entirely consistent with Farnell's calculations, and they considerably strengthen the theoretical basis for his and other computational studies of the behaviour of surface waves in cubic elastic media.

Being, in a sense, a sequel to CS, this article inevitably draws heavily on its predecessor. Frequent references to CS occur in the early Sections and, indeed, a fair acquaintance with either CS or the papers of Barnett *et al.* (1973) and Barnett and Lothe (1973, 1974) is a prerequisite for Part I. Parts II and III are substantially self-contained.

Our usage of vector and tensor algebra follows the book by Chadwick (1976a). When no statement is made to the contrary, vectors and tensors are real and their components relate to a fixed orthonormal basis $e = \{\mathbf{e}_1, \mathbf{e}_2, \mathbf{e}_3\}$. The rectangular Cartesian coordinates associated with e are denoted by x_1, x_2, x_3, and the comma notation is used for partial derivatives with respect to these variables. Latin subscripts take the values 1, 2, 3 throughout. The summation convention applies on repetitions of the suffixes p, q, r, and s, but not to other repeated subscripts.

I. FORMULATION

2. Elastic Body Waves

We bring together here the basic results in the theory of body waves in an anisotropic elastic body which are required in the later developments.

2.1. *Constitutive and field equations*

We are concerned with a homogeneous elastic body which is anisotropic in relation to a natural (i.e. stress-free) reference configuration N. In Part I no restriction is placed on the symmetry of the material composing the body, but the existence of a strain-energy function is presumed throughout. The components of **B**, the fourth-order elasticity tensor in N, thus have the symmetry properties

$$B_{jikl} = B_{klij} = B_{ijkl}. \tag{2.1}$$

We assume that **B** also satisfies the strong convexity condition

$$B_{pqrs}S_{pq}S_{rs} > 0 \quad \forall \text{ non-zero real symmetric tensors } \mathbf{S}, \tag{2.2}$$

this requirement being equivalent to positive definiteness of the strain-energy function.

The constitutive equations describing the mechanical response of the material in small deformations from N are

$$\sigma_{ij} = B_{ijrs}\, u_{s,r}, \tag{2.3}$$

where u_i and σ_{ij} are the components of the displacement vector **u** and the stress tensor $\boldsymbol{\sigma}$. It follows from (2.3) that the components of the traction **t**(**k**) on a surface segment with outward unit normal **k** are given by

$$t_i(\mathbf{k}) = \sigma_{pi}k_p = B_{pirs}u_{s,r}k_p. \tag{2.4}$$

In the absence of body forces the equations governing the displacement components in a small amplitude motion of the body are

$$B_{pirs}u_{s,pr} = \rho\partial^2 u_i/\partial t^2, \tag{2.5}$$

t being the time and ρ the uniform density of the material in the state N.

2.2. *Acoustical tensors*

The *acoustical tensor* **Q**(.) relative to the natural configuration N and *associated acoustical tensors* **P**(. , .) and **R**(. , .), also referred to N, are defined by the component relations

$$\begin{aligned} Q_{ij}(\mathbf{n}) &= B_{pirj}n_p n_r \quad && \forall\, \mathbf{n} \in \mathscr{U}, \\ P_{ij}(\mathbf{m}, \mathbf{n}) &= B_{ijrs}m_r n_s \quad && \forall\, \mathbf{m}, \mathbf{n} \in \mathscr{U}, \\ R_{ij}(\mathbf{m}, \mathbf{n}) &= B_{pirj}m_p n_r \quad && \forall \text{ orthogonal } \mathbf{m}, \mathbf{n} \in \mathscr{U}, \end{aligned} \tag{2.6}$$

where $\mathscr{U}$ is the set of all unit vectors. We observe that **Q**(.) is an even function and, in view of (2.2) and (2.1), positive definite, while **P**(. , .) is odd in each of its arguments. We also take note of the following relations, consequent upon the symmetries (2.1):

$$\begin{aligned} \mathbf{Q}^T(\mathbf{n}) = \mathbf{Q}(\mathbf{n}), \quad \mathbf{m}.\{\mathbf{Q}(\mathbf{n})\mathbf{m}\} = \mathbf{n}.\{\mathbf{Q}(\mathbf{m})\mathbf{n}\}, \\ \mathbf{P}^T(\mathbf{m}, \mathbf{n}) = \mathbf{P}(\mathbf{n}, \mathbf{m}) = \mathbf{P}(\mathbf{m}, \mathbf{n}), \\ \mathbf{Q}(\mathbf{n})\mathbf{m} = \mathbf{P}(\mathbf{m}, \mathbf{n})\mathbf{n}, \quad \mathbf{P}(\mathbf{m}, \mathbf{n})\mathbf{p} = \mathbf{R}(\mathbf{p}, \mathbf{n})\mathbf{m}. \end{aligned} \tag{2.7}$$

2.3. *Body waves: basic results*

In the remainder of this Section the configuration N is taken to be of infinite extent and we consider solutions of the equations of motion (2.5) of the form

$$u_i = f(x_q n_q - ct)p_i. \tag{2.8}$$

This displacement field represents a progressive plane body wave advancing with constant speed c in the direction defined by the *wave normal* $\mathbf{n} \in \mathscr{U}$ and linearly polarized in the direction defined by $\mathbf{p} \in \mathscr{U}$. From (2.3), (2.4), and $(2.6)_2$, the stress and the traction excited by this wave have components

$$\sigma_{ij} = f'(x_q n_q - ct)P_{ij}(\mathbf{n}, \mathbf{p}), \quad t_i(\mathbf{k}) = f'(x_q n_q - ct)P_{pi}(\mathbf{n}, \mathbf{p})k_p. \tag{2.9}$$

On substituting the expression (2.8) into (2.5) and cancelling the second derivative of the waveform $f(.)$, we obtain a vector relation which, with the aid of the definitions $(2.6)_{1,2}$, can be expressed in direct notation in the alternative forms

$$\mathbf{Q}(\mathbf{n})\mathbf{p} = \rho c^2 \mathbf{p} = \mathbf{P}(\mathbf{n}, \mathbf{p})\mathbf{n}. \tag{2.10}$$

This, the *propagation condition*, is the central result in the theory of elastic body waves. It exhibits the polarization $\mathbf{p}$ as an eigenvector of the acoustical tensor $\mathbf{Q}(\mathbf{n})$ with accompanying eigenvalue ρc^2, necessarily positive. With an arbitrary wave normal $\mathbf{n}$ there are thus associated three body waves whose speeds of propagation $c_i(\mathbf{n})$ (assumed positive) are determined by the *secular equation*

$$\det\{\mathbf{Q}(\mathbf{n}) - \rho c^2 \mathbf{I}\} = 0, \tag{2.11}$$

and whose polarizations $\mathbf{p}_i(\mathbf{n})$ form an orthonormal set. We adhere to the conventions

$$c_1(\mathbf{n}) \leqslant c_2(\mathbf{n}) \leqslant c_3(\mathbf{n}), \quad [\mathbf{p}_1(\mathbf{n}), \mathbf{p}_2(\mathbf{n}), \mathbf{p}_3(\mathbf{n})] = 1, \tag{2.12}$$

and refer to the set of all waves with speed $c_i(\mathbf{n})$ as a *branch* of body wave propagation.

Since, for a given branch, the wave speed depends upon the wave normal, constructive interference can take place between the constituents of a packet of sinusoidal waves propagating within a cone of small angle. It is accordingly possible to define a *group* (or *ray*) *velocity* $\mathbf{C}$ for this branch which can also be interpreted as the velocity with which the energy content of the body wave is propagated. The group velocity is given by

$$\mathbf{C} = (\rho c)^{-1}\mathbf{Q}(\mathbf{p})\mathbf{n} = (\rho c)^{-1}\mathbf{P}(\mathbf{n}, \mathbf{p})\mathbf{p} \tag{2.13}$$

(see, for example, Schouten, 1954, p. 173).

2.4. *The slowness surface*

Crucial to the extensive geometrical considerations in this article is the *slowness surface* of the material, constructed in the following manner. On the radius vector from O (the origin of the coordinates x_i) defined by $\mathbf{n}$, mark off the three points whose distances from O are $c_i^{-1}(\mathbf{n})$. Then, as $\mathbf{n}$ ranges over the unit sphere $\mathscr{U}$, the marked points sweep out a three-sheeted surface. The sheets are denoted by $\mathscr{S}_i$ and their union is the slowness surface $\mathscr{S}$.

The position vector $\mathbf{s}$ relative to O of a representative point of $\mathscr{S}_i$ is called the *slowness vector* and is given by

$$\mathbf{s} = c_i^{-1}(\mathbf{n})\mathbf{n}, \tag{2.14}$$

where $c_i(\mathbf{n})$ is a root of the secular equation (2.11). Equations (2.11) and (2.14) together completely specify $\mathscr{S}$, and we deduce from the even property of the acoustical tensor that each of the sheets $\mathscr{S}_i$ is centrosymmetric with respect to O. It also follows from (2.11) that $\mathscr{S}$ is a surface of degree six, which means that no straight line can intersect it in more than six points. According to the convention $(2.12)_1$, the branches of body wave propagation with speeds $c_1(\mathbf{n})$, $c_2(\mathbf{n})$, and $c_3(\mathbf{n})$ correspond in turn to the outer, medial, and inner sheets of $\mathscr{S}$.

When occasion demands we therefore use the adjectives outer, medial, and inner to distinguish the three branches.

If one or both of the signs of equality holds in $(2.12)_1$, two or all three sheets of $\mathscr{S}$ intersect on the radius vector defined by **n**, which is then called an *acoustic axis* (see Fedorov, 1968, p. 94). A point of intersection of two (or three) sheets is referred to as a *singular point* of $\mathscr{S}$: such singularities are the only points at which the slowness surface may not possess a unique normal. At an ordinary point of $\mathscr{S}_i$ the normal direction is aligned with the group velocity of the associated branch of body wave propagation, given by (2.13) (Schouten, 1954, p. 179). Two types of singularities are common: *tangential points*, at which two sheets share a tangent plane (and therefore remain smooth); and *conical points*, at which the intersecting sheets are asymptotic to the complementary halves of a circular cone.

3. Elastic Surface Waves

In the theory of surface waves we deal with an anisotropic elastic body for which the natural configuration N is of semi-infinite extent. We denote by **k** the unit vector directed along the outward normal to the plane boundary of the body. A surface wave may be adequately characterized as a small perturbation of N giving rise to a displacement field of the form

$$\mathbf{u} = \sum_{\alpha=1}^{3} \exp\{i\kappa(\mathbf{x}\,.\,\mathbf{j} + p_\alpha \mathbf{x}\,.\,\mathbf{k} - v_s t)\}\mathbf{a}_\alpha \quad (\mathbf{j}\,.\,\mathbf{k} = 0). \tag{3.1}$$

This expression describes an inhomogeneous wave with (real) wave number κ propagating with speed v_s in the direction defined by the unit vector **j**. The complex scalars p_α and the complex vectors $\mathbf{a}_\alpha$ in (3.1) must be chosen so that the boundary is traction-free and $\mathbf{u} \to \mathbf{0}$ as $\mathbf{x}\,.\,\mathbf{k} \to -\infty$, and this entails the solution of an eigenvalue problem (CS, Section VII, A).

We now outline the considerations bearing on the existence of surface waves in elastic media of unrestricted symmetry.

3.1. *The transonic state*

Inspection of (3.1) shows that the distribution of displacement associated with a surface wave is identical in each plane orthogonal to the vector $\mathbf{j} \wedge \mathbf{k}$. We therefore describe the plane R spanned by **j** and **k** and containing O as the *reference plane*. The intersection of R with the slowness surface $\mathscr{S}$ is called the *slowness section* and denoted by S. It follows from the properties of $\mathscr{S}$ noted in Section 2.4 that S is a sextic curve, consisting of three closed branches $\mathsf{S}_i = \mathscr{S}_i \cap \mathsf{R}$, each centrosymmetric relative to O.

Suppose now that a straight line L in R, codirectional with the unit vector **k**, approaches S in the direction $-\mathbf{j}$ (see Fig. 3.1(a)). When L first encounters the outer branch S_1, it may touch this curve at one, two, or three points, and at a point of contact L may be tangent to one, two, or all three of S_i. Bearing in mind that a straight line in R can intersect S in at most six points (not necessarily distinct), there are six possibilities altogether. The perpendicular distance of L from O is the reciprocal of a speed v which increases as L moves in the indicated direction. When L touches S_1, v reaches a critical value $\hat{v}$, known as the *limiting speed*, and we say that the *transonic state* has been attained. A discussion of the mathematical and physical

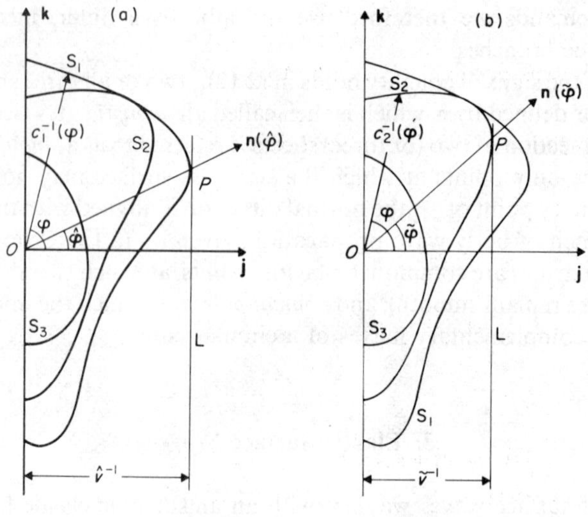

FIG. 3.1. Illustration of transonic and second transonic states.

implications of the transonic state is given in CS, Section VI, B, C, together with descriptions and illustrations of the six types of transonic states which may occur.

Figure 3.1(a) depicts a transonic state of Type 1, the simplest and most common variety. Here L touches S_1 at a single point P, corresponding to a body wave travelling in the direction OP with slowness

$$c_1^{-1}(\hat{\varphi}) = \hat{v}^{-1} \sec \hat{\varphi}. \tag{3.2}$$

The term *limiting body wave* is applied to this wave, and the inclination $\hat{\varphi}$ of OP to the surface wave normal **j** is referred to as the *critical angle*. Since the normal to S_1 at P is aligned with **j**, the group velocity of the limiting wave has no component normal to the boundary of the body. When the transonic state is not of Type 1, each limiting body wave has this property, the number L of such waves being two for states of Types 2 and 4, and three for states of Types 3, 5, and 6 (CS, Fig. 2).

The transonic state is described as *normal* when there is at least one limiting body wave which does not satisfy the condition of zero traction on the boundary of the body. If, on the other hand, each limiting body wave meets this requirement, the transonic state is said to be *exceptional*.

3.2. *The tensors* $\mathbf{S}_A(v)$

Two approaches are available to the investigation of surface wave solutions of (2.5) of the form (3.1). One is based on the algebraic eigenvalue problem controlling the quantities p_α and $\mathbf{a}_\alpha$, and the other, developed mainly by Barnett and Lothe (1973), makes use of integral representations of certain combinations of the eigenvalues and eigenvectors. In studies of the existence and uniqueness of surface waves the second method has clear advantages, and we direct the reader to CS, Sections IV to IX for a full exposition.

The integral representations in the Barnett–Lothe theory give rise to three real tensors $\mathbf{S}_A(v)(A = 1, 2, 3)$ which assume decisive rôles in the determination of surface wave properties. These tensors are defined in the interval $\hat{\imath}^+ = \{v : 0 \leqslant v < \hat{v}\}$ by

$$\mathbf{S}_1(v) = -\pi^{-1}\int_{-\frac{1}{2}\pi}^{\frac{1}{2}\pi} \mathbf{Q}^{-1}(\varphi, v)\mathbf{R}^T(\varphi, v)d\varphi, \tag{3.3}$$

$$\mathbf{S}_2(v) = \pi^{-1}\int_{-\frac{1}{2}\pi}^{\frac{1}{2}\pi} \mathbf{Q}^{-1}(\varphi, v)d\varphi,$$

$$\mathbf{S}_3(v) = \pi^{-1}\int_{-\frac{1}{2}\pi}^{\frac{1}{2}\pi} \{\mathbf{R}(\varphi, v)\mathbf{Q}^{-1}(\varphi, v)\mathbf{R}^T(\varphi, v) - \mathbf{Q}(\varphi, v)\}\, d\varphi,$$

where

$$\begin{aligned} \mathbf{Q}(\varphi, v) &= \mathbf{Q}(\cos\varphi\mathbf{j} + \sin\varphi\mathbf{k}) - \rho v^2\cos^2\varphi\mathbf{I}, \\ \mathbf{R}(\varphi, v) &= \mathbf{R}(\sin\varphi\,\mathbf{j} - \cos\varphi\,\mathbf{k}, \cos\varphi\,\mathbf{j} + \sin\varphi\,\mathbf{k}) - \rho v^2\sin\varphi\cos\varphi\,\mathbf{I}, \end{aligned} \tag{3.4}$$

and the acoustical tensors $\mathbf{Q}(.)$ and $\mathbf{R}(.,.)$ are given by $(2.6)_{1,3}$. The strong convexity condition (2.2) ensures that $\mathbf{Q}(\varphi, v)$ is invertible for all φ and $v \in \hat{\imath}^+$, whence, keeping in mind $(2.7)_1$, $\mathbf{S}_2(v)$ is symmetric and positive definite in $\hat{\imath}^+$. The tensor $\mathbf{S}_3(v)$ is also symmetric; its derivative $\mathbf{S}_3'(v)$ is positive definite in $\hat{\imath}^+$, and, subject to (2.2), $\mathbf{S}_3(0)$ is negative definite. $\mathbf{S}_1(v)$ has the properties

$$\det \mathbf{S}_1(v) = 0, \quad \operatorname{tr} \mathbf{S}_1(v) = 0, \tag{3.5}$$

from which it may be deduced, by writing down the characteristic equation, that the eigenvalues of $\mathbf{S}_1(v)$ are 0, $\pm\{\frac{1}{2}\operatorname{tr}\mathbf{S}_1^2(v)\}^{\frac{1}{2}}$. Lastly, we note that the tensors $\mathbf{S}_A(v)$ are not independent, but satisfy the relation

$$\mathbf{S}_3(v) = -\mathbf{S}_2^{-1}(v)\{\mathbf{I} + \mathbf{S}_1^2(v)\}, \tag{3.6}$$

among others (CS, Section IV, D). (Proofs of the quoted definiteness results are given in CS, Section VI, D. Equations (3.5) are not established in CS, but follow directly from relations given there. To prove $(3.5)_1$, form the determinant of each side of equation (4.37) of CS, and to reach $(3.5)_2$ take the trace of (4.30) in CS and use the fact, apparent from (4.29), that the sum of the eigenvalues of the 6-tensor S is zero.)

The tensor with the most immediate bearing on the behaviour of surface waves is $\mathbf{S}_3(v)$, a necessary and sufficient condition for the semi-infinite anisotropic elastic body under consideration to transmit a unique surface wave in the direction $\mathbf{j}$ being the existence of a speed v_s in the subsonic interval $\hat{\imath} = \hat{\imath}^+ - \{0\}$ at which

$$\operatorname{rank} \mathbf{S}_3(v) = 1 \tag{3.7}$$

(CS, Section VII, B). In view of the connection (3.6), the positive definiteness in $\hat{\imath}$ of $\mathbf{S}_2(v)$, and the observation about the eigenvalues of $\mathbf{S}_1(v)$ made in the last paragraph, an alternative form of the *secular equation* (3.7) is

$$F(v) \equiv 1 + \tfrac{1}{2}\operatorname{tr}\mathbf{S}_1^2(v) = 0. \tag{3.8}$$

For reasons given in Section 3.4, (3.8) is preferred to (3.7) in the present article.

3.3. *The Barnett–Lothe existence theorem*

The manner in which the existence of a surface wave is linked to the nature of the transonic state is revealed by the following theorem (Barnett and Lothe, 1974; CS, Section IX).

(i) If the transonic state is normal, a unique surface wave exists.
(ii) If the transonic state is exceptional and not of Type 1, no surface wave exists.
(iii) If the transonic state is exceptional and of Type 1, a unique surface wave exists if $\operatorname{tr} \mathbf{S}_3(\hat{v}) > 0$. Otherewise, there is no surface wave.

It is important to remember that, in these statements, the term surface wave alludes to a disturbance of the type (3.1) *propagating with a speed* v_s *in the subsonic interval* $\hat{\imath}$.

In general terms, the Barnett–Lothe theorem affirms that the existence of a unique surface wave is normally in prospect and that such a wave fails to exist only if all the limiting body waves are, in effect, degenerate surface waves in which the displacement field does not decay with increasing distance from the traction-free boundary. It follows that, in situations where a surface wave does not exist, the orientations of the normals $\mathbf{j}$ and $\mathbf{k}$ relative to the crystallographic axes of the material can be expected to display a high degree of symmetry. Once existence is seen to represent normality, interest turns to cases of non-existence, and in this connection a symmetrical interrelationship of geometrical and material characteristics, of the kind just anticipated, is highly expedient mathematically.

The proof of the existence theorem hinges on the behaviour in the transonic limit $v \uparrow \hat{v}$ of the eigenvalues $\lambda_i(v)$ of $\mathbf{S}_3(v)$ (CS, Section IX, B). It is found that one eigenvalue tends to infinity when the transonic state is normal, while all three eigenvalues approach finite limits in the case of an exceptional transonic state, L of these limits being zero where L is the number of limiting waves. When the transonic state is exceptional and of Type 1, $L = 1$ and two eigenvalues are finite and non-zero at $\hat{v}$. If these values are positive, $\operatorname{tr} \mathbf{S}_3(\hat{v}) > 0$ and a unique surface wave exists in accordance with part (iii) of the theorem. If they are negative, however, $\det \mathbf{S}_3(v) < 0$ in $\hat{\imath}$ and no surface wave exists. Nevertheless, in the latter case, if the two eigenvalues which are non-zero at $\hat{v}$ can be meaningfully continued into an interval i containing $\hat{\imath}$, they may vanish together at a speed exceeding $\hat{v}$. We pursue this possibility in Section 3.4.

3.4. *Supersonic surface waves*

It is known, from the study of special cases, that disturbances possessing all the attributes of a surface wave but travelling with a speed exceeding the critical value $\hat{v}$ can propagate in certain directions in anisotropic elastic media. Detailed computations carried out by Farnell (1970) indicate that such *supersonic surface waves* belong to a continuous branch of solutions, the other members being pseudo-surface waves, i.e. disturbances of the form (3.1) in which one of the partial waves transmits energy away from the boundary of the body and thus violates the requirement that $\mathbf{u} \to \mathbf{0}$ as $\mathbf{x} \cdot \mathbf{k} \to -\infty$. Also present is a branch of genuine surface wave solutions, degenerating, as the special direction is approached, into a body wave (sometimes referred to as a *bulk surface wave*). These findings have been supported by theoretical analysis, the most complete study, by Lothe and Al'shits (1977) and Al'shits and Lothe (1978), having been made for waves propagating in a zonal plane of a hexagonal material at right angles to the symmetry axis and in nearby directions.

In order to bring supersonic surface waves within the ambit of the theory outlined in Sections 3.1 to 3.3, we must examine the possibility of continuing two of the eigenvalues $\lambda_i(v)$ of $\mathbf{S}_3(v)$ beyond the limiting speed $\hat{v}$ when the transonic state is exceptional and of Type 1. (If the transonic state is exceptional and not of Type 1, no supersonic surface wave can exist since $\lambda_i(\hat{v}) = 0$ for at least two values of i.) In this case (exceptional and of Type 1) the tensors $\mathbf{S}_1(v)$ and $\mathbf{S}_3(v)$ are bounded at $\hat{v}$, but $\mathbf{S}_2(v)$ is unbounded on account of the leading approximation for small values of $\hat{v}-v$ containing the term $(\hat{v}^2-v^2)^{-\frac{1}{2}}$ (CS, Section IX, B, 2). The corresponding approximation to $\mathbf{S}_3(v)$ includes the term $(\hat{v}^2-v^2)^{\frac{1}{2}}$ which, while non-divergent, has no real extension to supersonic speeds. It is not possible, therefore, to continue $\mathbf{S}_2(v)$ and $\mathbf{S}_3(v)$, but this difficulty does not apply to $\mathbf{S}_1(v)$ and we have, in (3.8), a form of the secular equation for surface waves which is valid in some interval $\tilde{i} = \{v : 0 < v < \tilde{v}\}$, with $\tilde{v} > \hat{v}$, when the transonic state is exceptional and of Type 1.

The information about the tensors $\mathbf{S}_A(v)$ given in Section 3.1 tells us that the eigenvalues of $\mathbf{I}+\mathbf{S}_1^2(v)$ are 1 and $F(v)$ (twice), whence, from (3.6),

$$F^2(v) = -\det \mathbf{S}_2(v) \det \mathbf{S}_3(v).$$

Moreover, $\mathbf{S}_2(v)$ is positive definite in $\hat{i}$ and $\mathbf{S}_3(v)$ is negative definite in $(0, v_s)$ or $\hat{i}$ according as a (subsonic) surface wave does or does not exist. We can thus take $F(v)$ to be positive in $(0, v_s)$ or $\hat{i}$ and replace the condition $\operatorname{tr} \mathbf{S}_3(\hat{v}) > 0$ by $F(\hat{v}) < 0$ in part (iii) of the Barnett–Lothe theorem. The existence of a supersonic surface wave is assured if it can be shown that $F(\hat{v}) > 0$ and $F(v) < 0$ for some $v > \hat{v}$, and *a fortiori* if there is a *second limiting speed* $\tilde{v}$ such that $F(v) \to -\infty$ as $v \uparrow \tilde{v}$. The special cases considered in CS (Sections VI, E; VIII, A, 2, C, 2; IX, E) and in Part III below lend strong support to the hypothesis that $\tilde{v}$ is defined by the *second transonic state*, arising when the line L, continuing its motion in the direction $-\mathbf{j}$, is again tangent to the slowness section, and that $F(v) \to -\infty$ as $v \uparrow \tilde{v}$ if this state is normal. The second transonic state is illustrated in Fig. 3.1(b) where the contact is between L and S_2, and is of Type 1.

4. A Programme for Investigating the Existence of Surface Waves

In Sections 2 and 3 we have assembled the basic ingredients required for studying the existence of surface waves, both subsonic and supersonic, in an anisotropic elastic body. In principle, the purport of the Barnett–Lothe theorem and the additional considerations presented in Section 3.4 can be systematically explored for materials with arbitrary symmetry and we now specify a suitable procedure. There are two main stages, the first directed to the slowness surface and the second to the subset of body waves possessing the properties of limiting waves in exceptional transonic states. We give sharper definition to this scheme in the following subsections.

4.1. *The nature of $\mathscr{S}$*

The purpose of this part of the programme is to comprehend the form of the slowness surface in sufficient detail for the circumstances in which the various types of transonic states occur to be discernible. The essential components of this understanding are, first, a complete classification of the singular points of $\mathscr{S}$ by location and type; second, explicit approximations giving the form of the sheets $\mathscr{S}_i$ in neighbourhoods of singularities and symmetry axes; and, third, a global visualization of $\mathscr{S}$ incorporating the foregoing features. Normally, the first two objectives can be reached analytically, but in most cases the construction of a complete slowness surface calls for numerical computation.

4.2. *Exceptional limiting body waves*

The second part of the programme is aimed at categorizing all elastic body waves which serve as limiting waves in exceptional transonic states. We apply the term *exceptional limiting wave* to such a body wave and draw from Fig. 3.1 (a) and the discussion in Section 3.1 the following conditions which it must satisfy.

A. It must leave free of traction a family of parallel planes.
B. Its representative point on S must belong to the outer branch S_1.
C. If $\mathbf{k} \in \mathscr{U}$ is normal to the traction-free planes, the tangent to S_1 at the representative point must be parallel to $\mathbf{k}$.
D. If the tangent in condition C meets S_1 again, it must do so tangentially.

In connection with item A, we need only remark, on the basis of $(2.9)_2$, that if the stress field generated by a body wave gives zero traction on a plane surface, the traction vanishes on every plane parallel to it. The association of a limiting wave with S_1, the codirectionality of $\mathbf{k}$ and the line L in Fig. 3.1 (a), and the fact that each point of intersection of L with S is a tangent point account for the other requirements.

4.3. *Choice of materials*

In Part II we seek to fulfil the aims set out above in respect of elastic materials with cubic symmetry. The reasons for this choice are twofold: a wide variety of physically significant and practically important crystals belong to the cubic system; and the mathematical developments are at once amenable to analysis and of sufficient scope to admit a diversity of special cases. It can not be claimed that the programme formulated in Part I is realized with total finality, but the results obtained are reasonably complete, especially for the restricted type of three-constant elasticity to which all known materials with cubic symmetry conform.

5. Exceptional Waves

Although the set of all exceptional limiting waves is the relevant object of study from the standpoint of the theory of surface waves, considerable interest attaches to elastic body waves meeting the first of the conditions listed in Section 4.2, but not necessarily the others. In the later Sections we therefore use this larger class of body waves, termed *exceptional waves*, as an avenue of approach to exceptional limiting waves, and, by way of preparation, we conclude Part I with an account of what is presently known about exceptional waves in a general anisotropic elastic medium.

5.1. *Basic properties*

It follows from $(2.9)_2$ that an elastic body wave with wave normal $\mathbf{n}$ and polarization $\mathbf{p}$ is an exceptional wave if and only if there exists a unit vector $\mathbf{k}$ such that

$$\mathbf{P}(\mathbf{n}, \mathbf{p})\mathbf{k} = \mathbf{0}. \tag{5.1}$$

The vector **k** defines the direction normal to the family of parallel planes which remain free of traction during the passage of the wave.

By appeal to (5.1) and $(2.7)_{3,4}$ and the positivity of ρc^2, we deduce from $(2.10)_2$ and $(2.13)_2$ that

$$\mathbf{k}.\mathbf{p} = 0, \quad \mathbf{k}.\mathbf{C} = 0. \tag{5.2}$$

Provided that **p** and **C** are linearly independent, as is usually the case, these relations yield the useful formula

$$\mathbf{k} = \{\mathbf{C}.\mathbf{C} - (\mathbf{p}.\mathbf{C})^2\}^{-\frac{1}{2}}\mathbf{p}\wedge\mathbf{C} \tag{5.3}$$

for **k**. It may be noted that **k** and **n** are linearly independent since $(2.7)_5$ and the strong convexity condition (2.2) prohibit **n** from being a null vector of $\mathbf{P}(\mathbf{n}, \mathbf{p})$.

If the condition (5.1) holds, $\mathbf{P}(\mathbf{n}, \mathbf{p})$ is a singular tensor, and, conversely, if $\mathbf{P}(\mathbf{n}, \mathbf{p})$ is singular it possesses a unit null vector. A necessary and sufficient condition for the body wave given by (2.8) to be exceptional is thus

$$D(\mathbf{n}, \mathbf{p}) \equiv \det \mathbf{P}(\mathbf{n}, \mathbf{p}) = 0. \tag{5.4}$$

5.2. *An alternative existence criterion*

A second condition, necessary and sufficient for a body wave to be exceptional and, for some purposes, more convenient than (5.4), has been obtained by Al'shits and Lothe (1979c). Let **q** and **r** be the polarization vectors of the other body waves with wave normal **n**. Since **p**, **q**, and **r** are orthonormal, we see from $(5.2)_1$ that **k** is expressible as a linear combination of **q** and **r**, and we then infer from (5.1) that the vectors

$$\mathbf{P}(\mathbf{n}, \mathbf{p})\mathbf{q} \quad \text{and} \quad \mathbf{P}(\mathbf{n}, \mathbf{p})\mathbf{r} \tag{5.5}$$

are linearly dependent. Hence

$$\mathbf{0} = \{\mathbf{P}(\mathbf{n}, \mathbf{p})\mathbf{q}\} \wedge \{\mathbf{P}(\mathbf{n}, \mathbf{p})\mathbf{r}\} = \mathbf{P}^*(\mathbf{n}, \mathbf{p})(\mathbf{q}\wedge\mathbf{r}) = \pm\mathbf{P}^*(\mathbf{n}, \mathbf{p})\mathbf{p}, \tag{5.6}$$

$\mathbf{P}^*(.,.)$ being the adjugate of $\mathbf{P}(.,.)$. In particular,

$$\mathbf{n}\cdot\{\mathbf{P}^*(\mathbf{n}, \mathbf{p})\mathbf{p}\} = 0. \tag{5.7}$$

(Al'shits and Lothe (1979c) write this condition as

$$[\mathbf{n}, \mathbf{R}(\mathbf{q}, \mathbf{p})\mathbf{n}, \mathbf{R}(\mathbf{r}, \mathbf{p})\mathbf{n}] = 0.$$

To derive this, use $(5.6)_1$ and the connection $(2.7)_6$.) Equation (5.7) is evidently a necessary condition for the body wave under consideration to be exceptional. It is also sufficient. For (5.7) and (5.6) imply that **n** is coplanar with the vectors (5.5), and we see from the propagation condition $(2.10)_2$ that each of these vectors is orthogonal to **n**. Hence the vectors (5.5) are linearly dependent and it follows that $\mathbf{P}(\mathbf{n}, \mathbf{p})$ has a unit null vector.

For an arbitrary tensor **A**,

$$\mathbf{A}^* = (\mathbf{A}^T)^2 - (\operatorname{tr}\mathbf{A})\mathbf{A}^T + \tfrac{1}{2}\{(\operatorname{tr}\mathbf{A})^2 - \operatorname{tr}\mathbf{A}^2\}\mathbf{I}.$$

With the aid of this identity and the relation

$$\mathbf{P}(\mathbf{n}, \mathbf{p})\mathbf{n} = [\mathbf{n}\cdot\{\mathbf{P}(\mathbf{n}, \mathbf{p})\mathbf{p}\}]\mathbf{p},$$

supplied by the propagation condition $(2.10)_2$, we can recast the condition (5.7) as

$$E(\mathbf{n}, \mathbf{p}) \equiv [\{\operatorname{tr} \mathbf{P}(\mathbf{n}, \mathbf{p})\}^2 - \operatorname{tr} \mathbf{P}^2(\mathbf{n}, \mathbf{p})](\mathbf{n} \cdot \mathbf{p}) + 2[\mathbf{p} \cdot \{\mathbf{P}(\mathbf{n}, \mathbf{p})\mathbf{p}\} - \operatorname{tr} \mathbf{P}(\mathbf{n}, \mathbf{p})][\mathbf{n} \cdot \{\mathbf{P}(\mathbf{n}, \mathbf{p})\mathbf{p}\}] = 0. \tag{5.8}$$

5.3. *The polarization field*

The existence criteria (5.4) and (5.8) focus attention on the polarization field of elastic body waves in the chosen material. If, for a particular branch, the distribution of the polarization vector $\mathbf{p}(\mathbf{n})$ on the unit sphere $\mathscr{U}$ is known, the equation

$$D(\mathbf{n}, \mathbf{p}(\mathbf{n})) = 0, \quad \text{or} \quad E(\mathbf{n}, \mathbf{p}(\mathbf{n})) = 0,$$

represents a curve on $\mathscr{U}$ whose points define the wave normals of body waves in the branch which are exceptional. Plainly, then, a knowledge of the polarization field is pre-essential to the search for exceptional waves. Consistently with the terminology introduced in Section 2.4, we refer to $\mathbf{p}_1(\mathbf{n})$, $\mathbf{p}_2(\mathbf{n})$, and $\mathbf{p}_3(\mathbf{n})$ as the *outer*, *medial*, and *inner branches* of the polarization field, respectively.

The propagation condition (2.10) is invariant under the replacement of $\mathbf{n}$ by $-\mathbf{n}$ and homogeneous of degree 1 in $\mathbf{p}$. It is therefore possible for a branch of the polarization field to be either even or odd (i.e. $\mathbf{p}(-\mathbf{n}) = \mathbf{p}(\mathbf{n})$ or $\mathbf{p}(-\mathbf{n}) = -\mathbf{p}(\mathbf{n}) \, \forall \, \mathbf{n} \in \mathscr{U}$), and the convention $(2.12)_2$ then implies that either all three branches are even or two branches are odd and the third branch even (Al'shits and Lothe, 1979a).

At a *double point* (common to two sheets of $\mathscr{S}$), two of the polarizations can be any orthogonal pair of unit vectors in the two-dimensional eigenspace of $\mathbf{Q}(\mathbf{n})$ associated with the repeated eigenvalue, and at a *triple point* (belonging to all three sheets) the three polarizations can be any orthonormal triad. This indicates that discontinuities are to be expected in the polarization field, and an independent line of argument leads to the same conclusion. A well-known topological theorem asserts that if a field of unit vectors on a sphere in three-dimensional Euclidean space is continuous, then at least one of its members is directed radially to the sphere (see, for example, Chinn and Steenrod, 1966, p. 126). It follows that a branch of the polarization field containing no longitudinal wave is not continuous on $\mathscr{U}$, and hence that a continuous branch containing no transverse wave is accompanied by two discontinuous branches (cf. Al'shits and Lothe, 1979b).

Fundamental studies of polarization fields in crystal acoustics have been made by Al'shits and Lothe (1979a, b) and Holm and Lothe (1980), and it emerges from this work that discontinuities are due primarily to the presence of acoustic axes associated with conical points. If two conical points belong to the same pair of sheets of $\mathscr{S}$, each of the corresponding branches of the polarization field contains a singular curve, connecting the points of intersection of the acoustic axes with $\mathscr{U}$, on which the polarization suffers a change of direction. Because of the ambiguity in the sign of $\mathbf{p}$ permitted by the propagation condition (2.10), the configuration of such curves, which we refer to as *cuts*, is open to choice. (Al'shits and Lothe (1979a, b, c) employ the term *line of antisign*.)

It has been proved by Holm (see Al'shits and Lothe, 1979b) that when acoustic axes are wholly absent, all three branches of the polarization field are even. In the vast majority of crystals, including all those with cubic symmetry, acoustic axes occur, and we are then concerned with the alternative case in which two branches are odd and the other one even. We describe in Section 9 the construction of complete polarization fields for cubic elastic materials within the broad framework adumbrated here.

5.4. *Existence theorems*

We close this general discussion of exceptional waves by establishing three existence theorems, due to Al'shits and Lothe (1979c), which collectively demonstrate that such waves are a regular feature of body wave fields in crystalline media. The proofs, which differ from those given in the original paper, rest on the observation that the function $D(.\,,.)$ defined in (5.4) (and likewise $E(.\,,.)$ defined in (5.8)) is an odd function of each of its arguments.

1. *In a branch of the polarization field which is even and continuous, every great circle on* $\mathscr{U}$ *contains at least one pair of antipodal points representing wave normals of exceptional waves.* The stated properties imply that, for the branch in question, $d_1(\mathbf{n}) \equiv D(\mathbf{n}, \mathbf{p}(\mathbf{n}))$ is an odd function of $\mathbf{n}$, continuous on $\mathscr{U}$. In a continuous map from a circle to the real line the extremities of at least one diameter have the same image (Chinn and Steenrod, 1966, pp. 62–63). On an arbitrary great circle of $\mathscr{U}$ there thus exist directions $\pm\mathbf{n}'$ such that $d_1(\mathbf{n}') = d_1(-\mathbf{n}')$, whence, owing to the oddness of $d_1(.)$, $d_1(\mathbf{n}') = 0$. The result follows on appeal to the criterion (5.4).

2. *In a branch of the polarization field containing a cut, terminating at a conical point C, there exists a curve of exceptional wave solutions radiating from C.* On describing a circuit about C from one edge of the cut to the other, as indicated in Fig. 5.1, we change the sign of $\mathbf{p}$ while returning $\mathbf{n}$ to its initial value. The sign of $D(\mathbf{n}, \mathbf{p})$ is therefore reversed, which means that this function has a zero on the circuit. Since the choice of circuit is unrestricted and $d_1(.)$ is continuous away from the cut, we conclude that a curve of exceptional wave solutions emanates from C. It may be noted that this argument is unaffected if C is an end-point of more than one cut: specifically, if n cuts terminate at C, n curves of exceptional wave solutions start from C. The incidence of exceptional waves associated with C is, of course, qualitatively the same in each of the branches of body wave propagation to which this singularity belongs.

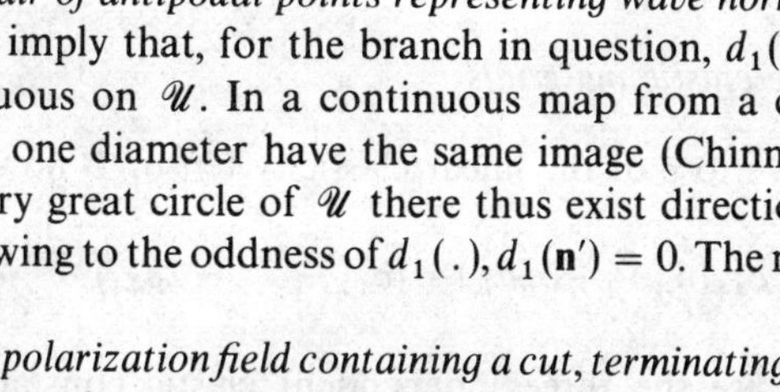

FIG. 5.1. Exceptional wave solutions associated with a conical point.

3. *At least one exceptional wave can propagate along an acoustic axis.* When $\mathbf{n}$ is directed along an acoustic axis it is possible to transmit in this direction a body wave with polarization $\cos\zeta\mathbf{p}' + \sin\zeta\mathbf{p}''$, where $\mathbf{p}', \mathbf{p}''$ is any pair of orthogonal unit vectors in the two-dimensional eigenspace of $\mathbf{Q}(\mathbf{n})$ and ζ is arbitrary. Now the function $d_2(\zeta) \equiv D(\mathbf{n}, \cos\zeta\mathbf{p}' + \sin\zeta\,\mathbf{p}'')$ maps the circle $0 \leqslant \zeta < 2\pi$ into the reals and has the odd property $d_2(\zeta + \pi) = -d_2(\zeta)$. The reasoning used in the first proof thus affirms the existence of an angle ζ' such that $d_2(\zeta') = 0$ and, from (5.4), the body wave with polarization $\cos\zeta'\mathbf{p}' + \sin\zeta'\mathbf{p}''$ is exceptional.

II. REALIZATION

6. Cubic Symmetry

We turn at this stage to the case in which the elastic material at hand has cubic symmetry. There then exist three mutually orthogonal fourfold axes of material symmetry and it is natural to align with them the orthonormal basis e defining the frame of reference. The term *face* is applied to a symmetry plane (spanned by two base vectors), and a plane bisecting two faces is called a *face bisector*. Half-lines situated in two faces, in one face and one face bisector, and in three face bisectors are referred to in turn as *axes*, *face diagonals*, and *body diagonals*. There are, in all, 3 faces, 6 face bisectors, 6 axes, 12 face diagonals, and 8 body diagonals, the members of each set being equivalent under the prevailing symmetry.

6.1. *Classification of cubic elastic materials*

The components relative to e of the linear elasticity tensor $\mathbf{B}$ now take the form

$$B_{ijkl} = c_{12}\delta_{ij}\delta_{kl} + c_{44}(\delta_{ik}\delta_{jl} + \delta_{il}\delta_{jk}) + (c_{11} - c_{12} - 2c_{44}) \sum_{p=1}^{3} \delta_{ip}\delta_{jp}\delta_{kp}\delta_{lp}, \tag{6.1}$$

where c_{11}, c_{12}, and c_{44} are the three independent elastic constants characterizing the mechanical response of the material. It may be readily verified that necessary and sufficient conditions for the elasticities (6.1) to satisfy the strong convexity condition (2.2) are

$$c_{11} + 2c_{12} > 0, \quad c_{11} - c_{12} > 0, \quad c_{44} > 0. \tag{6.2}$$

It is convenient henceforward to work with the quantities

$$a = c_{11} - c_{44}, \quad b = c_{12} + c_{44}, \quad c = c_{44}, \tag{6.3}$$

rather than c_{11}, c_{12}, c_{44}, and to think of a particular cubic material as represented by a point with rectangular Cartesian coordinates (a, b, c). The inequalities

$$\max\{0, \tfrac{1}{2}(b-a)\} < c < a + 2b \tag{6.4}$$

are equivalent to (6.2) and they imply that

$$a + b > 0, \quad a + 2b > 0. \tag{6.5}$$

The subset of cubic materials exhibiting physically realistic behaviour thus corresponds to open segments, defined by (6.4), of lines parallel to the c-axis passing through the points of the sector W of the (a, b)-plane specified by (6.5).

We divide W into the four subsectors

$$\begin{aligned} C_1 &= \{(a, b)\colon\ -\tfrac{1}{2}a < b < 0\}, & C_2 &= \{(a, b)\colon\ 0 < b < a\}, \\ C_3 &= \{(a, b)\colon\ 0 < a < b\}, & C_4 &= \{(a, b)\colon\ -b < a < 0\}, \end{aligned} \tag{6.6}$$

and the two half-lines

$$D_a = \{(a, b)\colon\ a = 0,\ b > 0\}, \quad D_b = \{(a, b)\colon\ b = 0,\ a > 0\}. \tag{6.7}$$

(There remains the half-line $a = b > 0$. The points (a, a, c), with $0 < c < 3a$, correspond to isotropic elastic materials. In the present context, however, isotropy is a degenerate case of no interest and we assume from now on that $a \neq b$.) This partitioning of W is illustrated in

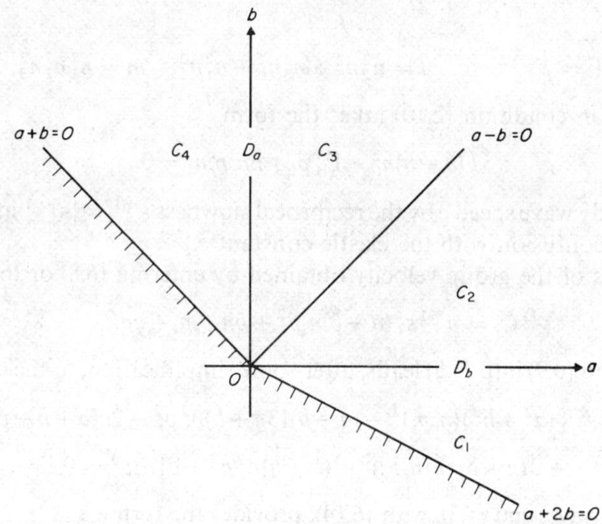

FIG. 6.1. Classification of elastic materials with cubic symmetry.

Fig. 6.1. The classification of cubic materials effected by the definitions (6.6) and (6.7) is related in Sections 8 and 9 to the structure of the slowness surface and the polarization field. The pronounced differences, in respect of wave propagation characteristics, between cubic media for which $a > b$ and those for which $a < b$ are well known (Farnell, 1970, Section V, B–E), but the finer subdivision given here seems not to have been adopted in previous work. On the evidence of available compilations of data (Huntington, 1958, pp. 62, 64; Musgrave, 1970, pp. 278–279; Auld, 1973, pp. 368–374) the (a, b) pairs of all known crystals with cubic symmetry belong to either C_2 or C_3. Nevertheless, a complete account of wave propagation in cubic elastic media must embrace the full range of elastic constants compatible with the strong convexity condition (2.2).

For brevity we refer subsequently to a cubic material whose elasticities define a point in, say, C_2 as a *C_2 material.*

6.2. *Acoustical tensors and basic relationships*

On combining (6.1) with the definitions (2.6) and utilizing the connections (6.3), we arrive at the following expressions for the components, relative to the natural basis *e*, of the acoustical tensors of a cubic elastic material. (We remind the reader at this point of the summation convention stated in the Introduction.)

$$Q_{ij}(\mathbf{n}) = \{(a-b)n_i^2 + c\}\delta_{ij} + bn_i n_j, \tag{6.8}$$

$$P_{ij}(\mathbf{m}, \mathbf{n}) = \{(a-b)m_i n_i + (b-c)m_p n_p\}\delta_{ij} + c(m_i n_j + m_j n_i), \tag{6.9}$$

$$R_{ij}(\mathbf{m}, \mathbf{n}) = (a-b)m_i n_i \delta_{ij} + (b-c)m_i n_j + cm_j n_i. \tag{6.10}$$

With the use of (6.8) the secular equation (2.11) becomes

$$\Gamma^3 - a\Gamma^2 + (a^2 - b^2)l\Gamma - (a-b)^2(a+2b)m = 0, \tag{6.11}$$

with

$$\Gamma = \rho s^{-2} - c, \quad l = n_2^2 n_3^2 + n_3^2 n_1^2 + n_1^2 n_2^2, \quad m = n_1^2 n_2^2 n_3^2, \tag{6.12}$$

and the propagation condition (2.10) takes the form

$$\{(a-b)n_i^2 - \Gamma\}p_i + bn_r p_r n_i = 0. \tag{6.13}$$

(We replace the body wave speed c by the reciprocal slowness $s^{-1} = |\mathbf{s}|^{-1}$ in $(6.12)_1$ and (6.14) in order to avoid confusion with the elastic constant c.)

The components of the group velocity obtained by entering (6.8) or (6.9) into (2.13) are

$$C_i = \rho^{-1}s\{(a-b)n_i p_i^2 + bn_r p_r p_i + cn_i\}. \tag{6.14}$$

Substitution from (6.9) into (5.8) leads, after some simplification, to the existence criterion

$$E(\mathbf{n}, \mathbf{p}) \equiv \{(a^2+b^2)(n_r p_r)^2 - (a-b)(3a+b)n_r^2 p_r^2 - 2c(a+b-c)\}n_s p_s$$
$$+ 2(a-b)\{b(n_r p_r)^2 + (a-b)n_r^2 p_r^2 + c\}n_s p_s^3 = 0 \tag{6.15}$$

for an exceptional wave, and (5.3), with (6.14), provides the formula

$$k_i = K\{(a-b)p_j p_k(n_j p_j - n_k p_k) + c(n_j p_k - n_k p_j)\} \tag{6.16}$$

for the components of a unit vector normal to the traction-free planes. In (6.16), K is a normalization factor and $\{i, j, k\}$ a cyclic permutation of $\{1, 2, 3\}$.

6.3. *Elementary segments*

The secular equation (6.11), with (6.12), is seen to be invariant under the transformations $n_i \to -n_i$ and $(n_i, n_j) \to (n_j, n_i)$, which implies that the slowness surface $\mathscr{S}$ of a cubic elastic material has reflexional symmetry about each face and each face bisector. In consequence of

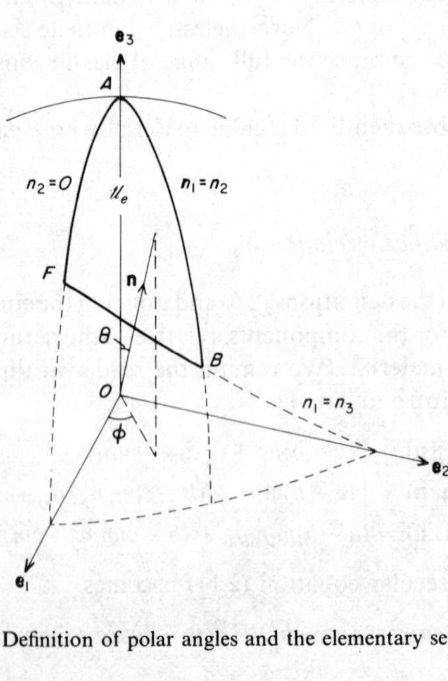

FIG. 6.2. Definition of polar angles and the elementary segment $\mathscr{U}_e$.

these symmetries, each of the sheets $\mathscr{S}_i$ consists of 48 identical quasi-triangular segments, one vertex of the typical segment lying on an axis, one on a face diagonal and the third on a body diagonal (Miller and Musgrave, 1956).

The calculations presented in the rest of Part II refer in the first place to the portion $\mathscr{U}_e$ of the unit sphere $\mathscr{U}$ shown in Fig. 6.2. The vertices A, F, and B have position vectors $\mathbf{e}_3, 2^{-\frac{1}{2}}(\mathbf{e}_1+\mathbf{e}_3)$, and $3^{-\frac{1}{2}}(\mathbf{e}_1+\mathbf{e}_2+\mathbf{e}_3)$, and are thus situated on an axis, a face diagonal, and a body diagonal respectively. In terms of the polar angles θ and ϕ defined in Fig. 6.2, the edges of $\mathscr{U}_e$ are given by

$$\begin{aligned} AF &= \{(\theta,\phi)\colon 0 \leqslant \theta \leqslant \tfrac{1}{4}\pi,\ \phi = 0\}, \\ AB &= \{(\theta,\phi)\colon 0 \leqslant \theta \leqslant \theta_c,\ \phi = \tfrac{1}{4}\pi\}, \\ FB &= \{(\theta,\phi)\colon \theta = \cot^{-1}(\cos\phi),\ \tfrac{1}{4}\pi \leqslant \theta \leqslant \theta_c,\ 0 \leqslant \phi \leqslant \tfrac{1}{4}\pi\}, \end{aligned}$$

where $\theta_c = \tan^{-1} 2^{\frac{1}{2}} = 54.74°$.

In future we refer to $\mathscr{U}_e$ and the subsets of $\mathscr{S}_i$ and $\mathbf{p}_i(\mathbf{n})$ with domain $\mathscr{U}_e$ as *elementary segments.*

7. Properties of Body Waves

In this Section we collect the results concerning the propagation of body waves in cubic elastic media which are needed in the remainder of the article. Accounts of parts of the material covered here are given in the books by Musgrave (1970, pp. 108–117), Auld (1973, pp. 213–219, 385–388), and Dieulesaint and Royer (1974, pp. 149–155).

7.1. *Longitudinal and transverse waves*

Multiplying each term of the propagation condition (6.13) by p_i and summing from $i = 1$ to 3 gives

$$\Gamma = (a-b)n_r^2 p_r^2 + b(n_r p_r)^2, \tag{7.1}$$

and on using this result to eliminate Γ from (6.13) we obtain

$$(a-b)(n_i^2 - n_r^2 p_r^2)p_i = b n_s p_s (n_r p_r p_i - n_i). \tag{7.2}$$

If the body wave under consideration is either longitudinal (with $p_i = n_i$) or transverse ($n_s p_s = 0$), the right side of (7.2) is zero. Since $a \neq b$, we then arrive at the reduced propagation conditions

$$(n_i^2 - n_r^2 n_r^2)n_i = 0, \quad \text{with} \quad \Gamma = (a-b)n_r^2 n_r^2 + b, \tag{7.3}$$

for longitudinal and

$$(n_i^2 - n_r^2 p_r^2)p_i = 0, \quad \text{with} \quad \Gamma = (a-b)n_r^2 p_r^2, \tag{7.4}$$

for transverse waves.

It follows from $(7.3)_1$ that either $n_i = 0$ or $n_i^2 = n_r^2 n_r^2$ for each i. There are three possibilities:

(i) $n_i^2 = 1, \quad n_j = n_k = 0, \quad$ with $\quad \Gamma = a$;
(ii) $n_i = 0, \quad n_j^2 = n_k^2 = \tfrac{1}{2}, \quad$ with $\quad \Gamma = \tfrac{1}{2}(a+b)$;
(iii) $n_i^2 = n_j^2 = n_k^2 = \tfrac{1}{3}, \quad$ with $\quad \Gamma = \tfrac{1}{3}(a+2b)$;

$\{i, j, k\}$ being a permutation of $\{1, 2, 3\}$. The only directions in which a longitudinal wave can travel are thus along an axis (case (i)), a face diagonal (case (ii)), and a body diagonal (case (iii)).

Equation $(7.4)_1$ presents the alternatives $p_i = 0$ and $n_i^2 = n_r^2 p_r^2$ for each i. If the wave normal does not lie in a face or a face bisector, n_1^2, n_2^2, and n_3^2 are distinct and non-zero, whence two of the components p_i are zero. But then, since $n_s p_s = 0$, one of the n_i's must be zero, which is untrue. Transverse waves in a cubic material are thus confined to directions of propagation in the faces and face bisectors.

Assured that our findings will incorporate all the longitudinal and transverse waves that exist, we proceed to examine in further detail body wave propagation in a face and a face bisector. On account of the symmetry properties described in Section 6.3, attention can be confined to wave normals belonging to the perimeter AFB of the elementary segment $\mathscr{U}_e$.

7.2. *Propagation of body waves in a face: the edge AF*

For a typical wave normal in AF, we have, with reference to Fig. 6.2,

$$n_1 = \sin\theta, \quad n_2 = 0, \quad n_3 = \cos\theta; \quad l = \tfrac{1}{4}\sin^2 2\theta, \quad m = 0.$$

The roots of the secular equation (6.11) are hence Γ_1, Γ_2, and 0 where

$$\Gamma_1, \Gamma_2 = \tfrac{1}{2}a \pm \tfrac{1}{2}(a^2\cos^2 2\theta + b^2\sin^2 2\theta)^{\frac{1}{2}}, \tag{7.5}$$

the positive square root being implied. We assume for the present that $0 < \theta \leqslant \frac{1}{4}\pi$, consideration of the vertex A being postponed to the end of this subsection.

In view of the definition $(6.12)_1$ the relative magnitudes of the body wave speeds follow the algebraic order of the roots $\Gamma_1, \Gamma_2, 0$, which, from (7.5), is determined by the signs of a, b, and $a-b$. The first four columns of Table 7.1 identify the branches of body wave propagation to which the roots of the secular equation correspond for each of the six subclasses of cubic elastic materials defined in (6.6) and (6.7). The associated branches of the polarization field are derived from the propagation condition (6.13) and the results, in direct notation, occupy the last three columns of Table 7.1. The angle α and the unit vector $\mathbf{a}_i(.)$ appearing there are defined by

$$\alpha = \begin{cases} \tfrac{1}{2}\tan^{-1}(|b/a|\tan 2\theta) & (0 \leqslant \alpha \leqslant \tfrac{1}{4}\pi), \quad a \neq 0, \quad b \neq 0, \\ \tfrac{1}{4}\pi, & a = 0, \\ 0, & b = 0, \end{cases} \tag{7.6}$$

$$\mathbf{a}_i(\zeta) = \cos\zeta\,\mathbf{e}_j + \sin\zeta\,\mathbf{e}_k,$$

$\{i, j, k\}$ being a cyclic permutation of $\{1, 2, 3\}$. The variation of α with θ is represented graphically in Fig. 7.1. As pointed out in Section 5.3, the propagation condition allows any polarization vector to be replaced by its negative. The construction of complete polarization fields is undertaken in Section 9 and the signs of polarization vectors calculated in the meantime are chosen so as to meet the requirement $(2.12)_2$ and fit the global pattern which eventually emerges.

Inspection of Table 7.1 and Fig. 7.1 shows that, for each subclass and every wave normal in $AF - \{A, F\}$, precisely one branch of body wave propagation is transverse, the polarization, $\mathbf{e}_2$, being normal to the face, and the branch in question being either the outer or the medial. The polarizations of the other branches, which are neither longitudinal nor transverse, lie in

TABLE 7.1. Body wave speeds and polarization vectors for wave normals on the edge $AF - \{A\}$. $(\langle x \rangle = \{(c+x)/\rho\}^{\frac{1}{2}})$

Subclass	Wave speed			Polarization		
	Outer	Medial	Inner	Outer	Medial	Inner
$C_1 \cup D_b$	$\langle 0 \rangle$	$\langle \Gamma_2 \rangle$	$\langle \Gamma_1 \rangle$	$\mathbf{e}_2$	$\mathbf{a}_2(\frac{1}{2}\pi - \alpha)$	$\mathbf{a}_2(\pi - \alpha)$
$C_2 \cup D_b$	$\langle 0 \rangle$	$\langle \Gamma_2 \rangle$	$\langle \Gamma_1 \rangle$	$\mathbf{e}_2$	$\mathbf{a}_2(\frac{1}{2}\pi + \alpha)$	$-\mathbf{a}_2(\alpha)$
$C_3 \cup D_a$	$\langle \Gamma_2 \rangle$	$\langle 0 \rangle$	$\langle \Gamma_1 \rangle$	$\mathbf{a}_2(\frac{1}{2}\pi + \alpha)$	$\mathbf{e}_2$	$\mathbf{a}_2(\alpha)$
C_4	$\langle \Gamma_2 \rangle$	$\langle 0 \rangle$	$\langle \Gamma_1 \rangle$	$\mathbf{a}_2(\pi - \alpha)$	$\mathbf{e}_2$	$\mathbf{a}_2(\frac{1}{2}\pi - \alpha)$

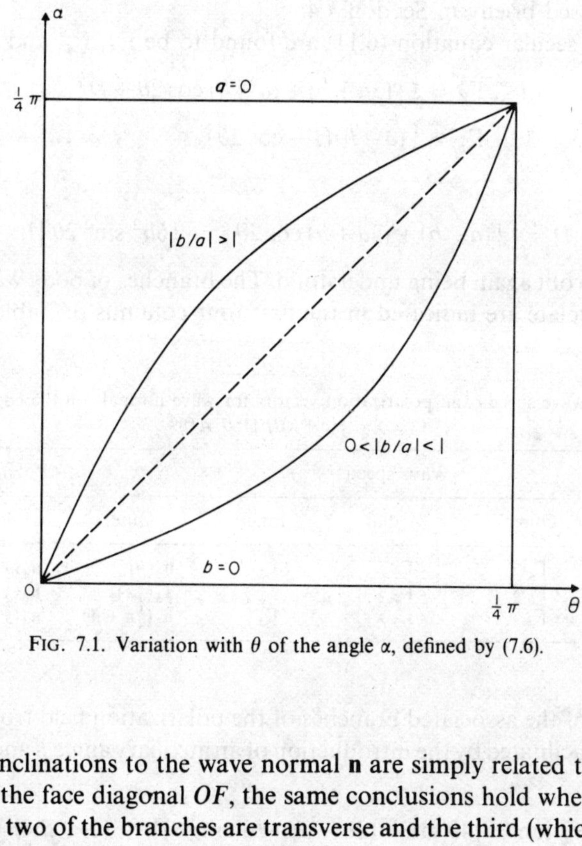

FIG. 7.1. Variation with θ of the angle α, defined by (7.6).

the face and their inclinations to the wave normal $\mathbf{n}$ are simply related to the angle α. For propagation along the face diagonal OF, the same conclusions hold when $b = 0$, but when $b \neq 0$ (and $\alpha = \frac{1}{4}\pi$) two of the branches are transverse and the third (which is the medial for C_1 materials, elsewhere the inner) is longitudinal.

For all wave normals in $AF - \{A\}$, the roots of the secular equation (and hence the wave speeds) are distinct, so no acoustic axis is present. When $\mathbf{n}$ acts along OA, however, $\theta = 0$ and, from (7.5), $\Gamma_2 = 0$, $\Gamma_1 = 0$, or $\Gamma_1 = \Gamma_2 = 0$ according as a is positive, negative, or zero. The symmetry axes are thus acoustic axes. When $a \neq 0$, the confluent branches are transverse and circularly polarized and the other branch, which is the outer for C_4 materials and otherwise the inner, is longitudinal. When $a = 0$, the three wave speeds are equal on the axes and the branches coalesce in a spherically polarized wave.

7.3. *Propagation of body waves in a face bisector: the edges AB and FB*

Since the edges AB and FB of $\mathscr{U}_e$ are both situated in face bisectors, results obtained for one of them extend in a straightforward manner to the other. We therefore give detailed consideration only to AB. The components of a typical wave normal and the values of l and m are, in this case,

$$n_1 = n_2 = 2^{-\frac{1}{2}}\sin\theta, \quad n_3 = \cos\theta,$$
$$l = \sin^2\theta(\tfrac{1}{4}\sin^2\theta + \cos^2\theta), \quad m = \tfrac{1}{4}\sin^4\theta\cos^2\theta.$$

We assume initially that $0 < \theta < \theta_c$, thereby excluding the vertices A and B, and also that $b \neq 0$. Wave propagation along the axis OA has already been dealt with and we defer to the end of this subsection the situation in which $\mathbf{n}$ is directed along the body diagonal OB. The case $b = 0$ is discussed briefly in Section 7.4.

The roots of the secular equation (6.11) are found to be Γ_3, Γ_4, and Γ_5 where

$$\left.\begin{aligned}\Gamma_3, \Gamma_4 &= \tfrac{1}{8}\{(3a+b)+(a-b)\cos 2\theta \pm D\},\\ \Gamma_5 &= \tfrac{1}{4}(a-b)(1-\cos 2\theta),\end{aligned}\right\} \tag{7.7}$$

and

$$D = [\{(a-b)+(3a+b)\cos 2\theta\}^2 + 16b^2\sin^2 2\theta]^{\frac{1}{2}}, \tag{7.8}$$

the positive square root again being understood. The branches of body wave propagation to which these roots relate are indicated in the first four columns of Table 7.2.

TABLE 7.2. Body wave speeds and polarization vectors for wave normals on the edge $AB-\{A, B\}$. $(\langle x\rangle = \{(c+x)/\rho\}^{\frac{1}{2}}, b \neq 0)$

Subclass	Wave speed			Polarization		
	Outer	Medial	Inner	Outer	Medial	Inner
C_1	$\langle\Gamma_4\rangle$	$\langle\Gamma_5\rangle$	$\langle\Gamma_3\rangle$	$\mathbf{b}_3(\beta)$	$\mathbf{a}_3(-\frac{1}{4}\pi)$	$\mathbf{b}_3(\frac{1}{2}\pi+\beta)$
C_2	$\langle\Gamma_5\rangle$	$\langle\Gamma_4\rangle$	$\langle\Gamma_3\rangle$	$\mathbf{a}_3(\frac{3}{4}\pi)$	$\mathbf{b}_3(\frac{1}{2}\pi+\beta)$	$-\mathbf{b}_3(\beta)$
$C_3\cup D_a\cup C_4$	$\langle\Gamma_4\rangle$	$\langle\Gamma_5\rangle$	$\langle\Gamma_3\rangle$	$\mathbf{b}_3(\frac{1}{2}\pi+\beta)$	$\mathbf{a}_3(\frac{3}{4}\pi)$	$\mathbf{b}_3(\beta)$

The calculation of the associated branches of the polarization field from the propagation condition (6.13) is facilitated by the introduction of an auxiliary angle β and a unit vector $\mathbf{b}_i(.)$ defined by

$$\beta = \tfrac{1}{2}\tan^{-1}\{4b\tan\theta/(2a-(a+b)\tan^2\theta)\} \quad (0 \leqslant \beta \leqslant \tfrac{1}{2}\pi), \tag{7.9}$$
$$\mathbf{b}_i(\zeta) = \cos\zeta\,\mathbf{e}_i + 2^{-\frac{1}{2}}\sin\zeta(\mathbf{e}_j+\mathbf{e}_k),$$

$\{i, j, k\}$ again being a permutation of $\{1, 2, 3\}$. Curves illustrating the dependence of β on θ are shown in Fig. 7.2. The values taken on $AB-\{A, B\}$ by the branches of the polarization field are set out in the final three columns of Table 7.2. We see from these entries that, again, exactly one branch is transverse for each wave normal in the interior of the chosen edge, the branch concerned being either the outer or the medial. The polarization of the transverse wave, $\pm 2^{-\frac{1}{2}}(\mathbf{e}_1-\mathbf{e}_2)$, is normal to the face bisector containing AB. The other branches are neither

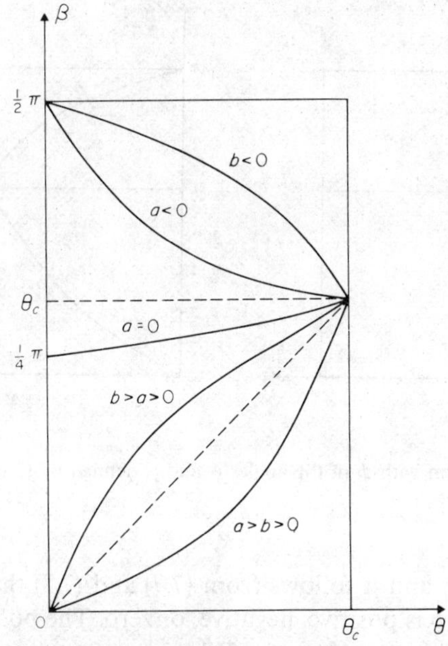

FIG. 7.2. Variation with θ of the angle β, defined by (7.9).

longitudinal nor transverse and the inclinations to the wave normal of their polarizations, which are confined to the face bisector, are determined by β.

In order to adapt the results just obtained for $AB-\{A, B\}$ to the third edge $FB-\{B\}$, we must interchange n_2 and n_3 and replace θ by

$$\psi = \tan^{-1}(2^{\frac{1}{2}}\cot\phi) \quad (\theta_c \leqslant \psi \leqslant \tfrac{1}{2}\pi). \tag{7.10}$$

In place of Table 7.2 we then have Table 7.3 in which $\Gamma'_\Lambda(\Lambda = 3, 4, 5)$ is derived from Γ_Λ, as defined in (7.7), by substituting ψ for θ, and the angle γ is given by

$$\gamma = \tfrac{1}{2}\tan^{-1}\{2^{\frac{3}{2}} b \tan\phi/(a\tan^2\phi - (a+b))\} \quad (0 \leqslant \gamma \leqslant \tfrac{1}{2}\pi). \tag{7.11}$$

The variations of ψ and γ with φ are shown in Fig. 7.3.

When $b \neq 0$, the roots Γ_Λ are distinct for all wave normals in $AB-\{A, B\}$ and the same is true of Γ'_Λ in $FB-\{B\}$. Acoustic axes are therefore absent from these edges. When $\mathbf{n}$

TABLE 7.3. Body wave speeds and polarization vectors for wave normals on the edge $FB-\{B\}$. $(\langle x\rangle = \{(c+x)/\rho\}^{\frac{1}{2}}, b \neq 0)$

Subclass	Wave speed			Polarization		
	Outer	Medial	Inner	Outer	Medial	Inner
C_1	$\langle\Gamma'_4\rangle$	$\langle\Gamma'_3\rangle$	$\langle\Gamma'_5\rangle$	$\mathbf{b}_2(\gamma)$	$\mathbf{b}_2(\frac{1}{2}\pi+\gamma)$	$\mathbf{a}_2(\frac{3}{4}\pi)$
C_2	$\langle\Gamma'_4\rangle$	$\langle\Gamma'_5\rangle$	$\langle\Gamma'_3\rangle$	$-\mathbf{b}_2(\frac{1}{2}\pi+\gamma)$	$\mathbf{a}_2(\frac{3}{4}\pi)$	$-\mathbf{b}_2(\gamma)$
$C_3 \cup D_a \cup C_4$	$\langle\Gamma'_5\rangle$	$\langle\Gamma'_4\rangle$	$\langle\Gamma'_3\rangle$	$\mathbf{a}_2(\frac{3}{4}\pi)$	$-\mathbf{b}_2(\frac{1}{2}\pi+\gamma)$	$\mathbf{b}_2(\gamma)$

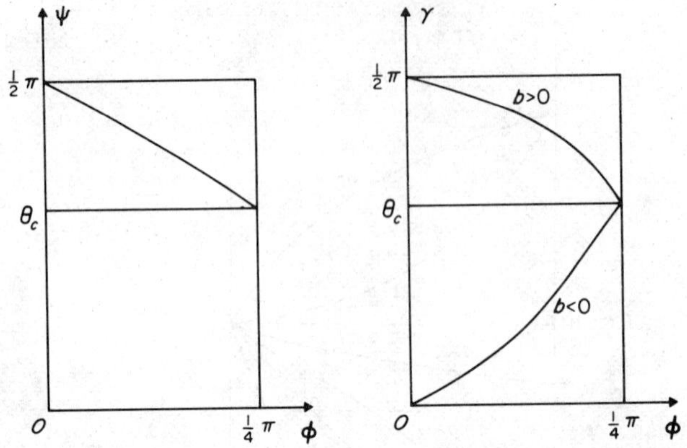

FIG. 7.3. Variation with ϕ of the angles ψ and γ, defined by (7.10) and (7.11).

is directed along OB, $\theta = \theta_c$ and it follows from (7.7) and (7.8) that $\Gamma_4 = \Gamma_5$, $\Gamma_3 = \Gamma_5$, or $\Gamma_3 = \Gamma_4 = \Gamma_5$ according as b is positive, negative, or zero. The body diagonals, in common with the axes, are therefore acoustic axes. When $b \neq 0$, the intersecting branches are transverse and circularly polarized and the remaining branch, which is the outer for C_1 materials and otherwise the inner, is longitudinal. When $b = 0$, the three wave speeds are equal on the body diagonals, implying the existence of a spherically polarized wave.

7.4. *The degenerate case b = 0*

The subclass D_b is exceptional in being the only one for which the secular equation and the propagation condition can be solved explicitly for all wave normals.

When $b = 0$, the roots of (6.11) are an_1^2, an_2^2, an_3^2, and when the components of $\mathbf{n}$ are distinct the accompanying polarization vectors, furnished by (6.13), are $\mathbf{e}_1$, $\mathbf{e}_2$, $\mathbf{e}_3$, respectively. Throughout the interior of the elementary segment $\mathscr{U}_e$ the ordering of the n_i's is $n_2 < n_1 < n_3$, and these inequalities also hold in the interior of the edge AF. For all wave normals in this domain the outer, medial, and inner branches of body wave propagation thus have speeds $\{(an_i^2 + c)/\rho\}^{\frac{1}{2}}$ with $i = 2, 1$, and 3, respectively, and the related branches of the polarization field have the constant values $\mathbf{e}_2$, $\mathbf{e}_1$, and $\mathbf{e}_3$. No acoustic axes are present and no body wave is either longitudinal or transverse.

Each point of the edges $AB - \{B\}$ and $FB - \{B\}$ is situated in a face bisector and the associated directions are therefore acoustic axes. For each wave normal in this set two branches of body wave propagation combine to form a wave circularly polarized in the face orthogonal to the face bisector containing $\mathbf{n}$, and the polarization of the third branch, which is the inner for wave normals in $AB - \{B\}$ and the outer for wave normals in $FB - \{B\}$, is normal to this face. Only when $\mathbf{n}$ points along OA and OF are these waves longitudinal and transverse. As remarked at the end of Section 7.3, the three branches merge in a spherically polarized wave when $\mathbf{n}$ acts along OB.

No additional longitudinal and transverse waves appear, then, when $b = 0$, but every direction in the face bisectors, and not merely the axes and face diagonals, is an acoustic axis.

8. The Slowness Surface

We are now in a position to carry out the first part of the programme drawn up in Section 4, involving the detailed examination of $\mathscr{S}$. An extensive study of the wave and slowness surfaces of cubic elastic materials has been made by Miller and Musgrave (1956), and some additional details have been supplied by Duff (1960). The emphasis is different here, being placed on cubic elasticity in its full generality rather than particular crystals of technical importance. We deal in turn with the steps enumerated in Section 4.1.

8.1. *Location of singular points*

The facts about acoustic axes established in Section 7 show that the slowness surface of a cubic elastic material always has singular points on the axes and body diagonals. When $b = 0$, every point situated on just one face bisector is singular, and it is known that these additional points complete the set of singularities in this case. In fact, when $b \neq 0$ the singular points on the axes and body diagonals are the only ones and we now sketch a direct proof of this result.

The argument is based on the bijection between the elementary segment $\mathscr{U}_e$ and a region V in the first quadrant of the (l, m)-plane. The numbers l and m are defined in (6.12) as elementary symmetric polynomials in the squares of the components of the wave normal $\mathbf{n}$, and since the sum of these squares is unity, the n_i^2's are the roots of the cubic equation

$$x^3 - x^2 + lx - m = 0. \tag{8.1}$$

The mappings $\mathscr{U}_e \to V$ and $V \to \mathscr{U}_e$ are defined by $(6.12)_{2,3}$ and (8.1) respectively. The cubic (8.1) yields a real unit vector $\mathbf{n}$ if and only if it has non-negative real roots, and the boundary ∂V of V separates the pairs (l, m) which meet this requirement from those which do not. At a point of ∂V, (8.1) therefore has a repeated root and we infer that ∂V is defined by the vanishing of the discriminant of (8.1). This condition can be put as

$$m = \tfrac{1}{27}\{9l - 2 \pm 2(1-3l)^{\frac{3}{2}}\}, \tag{8.2}$$

from which we perceive that ∂V is a cusped curve closed by a segment of the positive l-axis. We thus arrive at the representation of V shown in Fig. 8.1, the origin A', the cusp B' $(l = \frac{1}{3}$, $m = \frac{1}{27})$, and the point $F'(\frac{1}{4}, 0)$ corresponding to the vertices A, B, and F of $\mathscr{U}_e$.

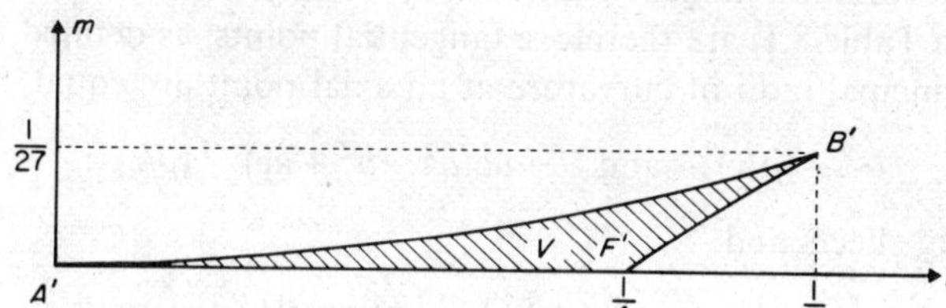

FIG. 8.1. The region V of the (l, m)-plane.

We dispose first of triple points, common to all three sheets of $\mathscr{S}$. At such a point the left side of the secular equation (6.11) must be a perfect cube which requires that

$$l = \tfrac{1}{3}\{1-(b/a)^2\}^{-1}, \quad m = \tfrac{1}{27}\{1-(b/a)\}^{-2}\{1+2(b/a)\}^{-1}.$$

These relations are compatible with the inequalities $0 \leqslant l \leqslant \frac{1}{3}$, $0 \leqslant m \leqslant \frac{1}{27}$, apparent from Fig. 8.1, only if either $a = 0$ or $b = 0$. We conclude, therefore, that the triple points brought to light in Sections 7.2 and 7.3 (situated on the axes when $a = 0$ and the body diagonals when $b = 0$) are the only ones.

At a double point, belonging to two sheets of $\mathscr{S}$, (6.11) has a repeated root, implying that its discriminant is zero. The result of solving this condition for m is

$$m = \tfrac{1}{27}(a-b)^{-2}(a+2b)^{-1}[9a(a^2-b^2)l - 2a^3 \pm 2\{a^2 - 3(a^2-b^2)l\}^{\frac{3}{2}}]. \tag{8.3}$$

Equation (8.3), like (8.2), represents a cusped curve in the (l, m)-plane. The two curves intersect at A' and B' and coincide when $b = 0$, thus providing independent confirmation of the locations of double points noted at the beginning of this subsection. The task of proving that $\mathscr{S}$ has no other singularities now reduces to verifying that when $b \neq 0$ the curve given by (8.3) meets the region V only at A' and B'. This is a matter of simple inequalities and we omit the details; a full account has been given by Smith (1978, pp. 31–43).

The sheets of $\mathscr{S}$ to which the singular points belong are specified in the final paragraphs of Sections 7.2 and 7.3, and this information is summarized in Table 8.1 below. The additional singular arcs which are present when $b = 0$ link $\mathscr{S}_1$ and $\mathscr{S}_2$, or $\mathscr{S}_2$ and $\mathscr{S}_3$, according as the end-points lie on an axis and a body diagonal, or on face and body diagonals.

8.2. *Local approximations*

The shapes of the sheets $\mathscr{S}_i$ near the axes and the face and body diagonals may be investigated by perturbation methods. The first step in the procedure is to expand l and m as Taylor series in a small parameter ε measuring the deviation of the wave normal from the nominated direction, $\mathbf{n}_0$ say. The resulting expressions are next entered into the secular equation (6.11) alongside a suitable development of Γ in powers of ε. The leading term of the expansion of Γ is the root of (6.11) corresponding to $\mathbf{n}_0$ and the chosen sheet $\mathscr{S}_i$, and is therefore known explicitly. Finally, equating coefficients of like powers of ε yields an approximation to Γ which can be used to determine the principal radii of curvature at $\mathbf{n}_0$ if $\mathscr{S}$ is smooth there, or the envelope of tangent planes if it is not. We now describe the outcome of such calculations for the three sets of special directions.

When $a \neq 0$, the sheets $\mathscr{S}_i$ are smooth on the axes and the two sheets which intersect there accordingly have a common tangent plane. Except for D_a materials, the singularities on the axes (labelled T in Table 8.1) are therefore tangential points, as defined in Section 2.4. For each sheet, the principal radii of curvature at an axial point are equal, their values being

$$-(\rho/c)^{\frac{1}{2}} \quad \text{and} \quad -ac(a^2-b^2+ac)^{-1}(\rho/c)^{\frac{1}{2}} \tag{8.4}$$

for the intersecting sheets and

$$-a(ac+b^2)^{-1}\{\rho(a+c)\}^{\frac{1}{2}} \tag{8.5}$$

for the remaining sheet. It is clear from (8.4) that one of the confluent sheets is always convex and, with reference to the definitions (6.6) and (6.7), we find that the other member of the pair is convex for C_1, D_b, C_2, and C_4 materials. For a C_3 material this sheet is convex if

$$a^2 - b^2 + ac > 0 \tag{8.6}$$

and concave if the left side of (8.6) is negative. With regard to (8.5), the conditions (6.4) imply that $ac + b^2 > 0$, so the separate sheet is convex for C_1, D_b, C_2, and C_3 materials and concave for materials in C_4.

When $a = 0$, $\mathscr{S}$ has triple points on the axes. The medial sheet is smooth and convex in this case, and $\mathscr{S}_1$ and $\mathscr{S}_3$ approximate locally to the complementary halves of a circular cone of semi-angle $\tan^{-1}(2c/b)$ $(> \frac{1}{4}\pi)$.

When $b \neq 0$, the sheets $\mathscr{S}_i$ are smooth and disjoint on the face diagonals. The principal curvatures are no longer equal, however, the principal axes of curvature being directed orthogonally to the face diagonal in the face and the face bisector in which it lies. It is found that $\mathscr{S}_1$ and $\mathscr{S}_3$ are invariably convex and that $\mathscr{S}_2$ has a saddle point on the face diagonals for C_1, D_a, and C_4 materials. For materials in C_2 and C_3 satisfying the respective conditions

$$-(a-b)(2a+b)+2bc < 0 \qquad \text{for} \quad a > b > 0, \tag{8.7}$$

$$-(b-a)(a+2b)+(a+b)c < 0 \qquad \text{for} \quad b > a > 0, \tag{8.8}$$

$\mathscr{S}_2$ again has a saddle point, but when the reverse inequalities hold $\mathscr{S}_2$ is convex on the face diagonals. In the case of a saddle point, the sections of $\mathscr{S}_2$ in the face and the face bisector on which the point lies are locally concave and convex, respectively.

The results for the case $b = 0$ given in Section 7.4 provide a complete representation of $\mathscr{S}$ in analytic form. This is discussed in Section 8.3.

The two sheets of $\mathscr{S}$ which meet on the body diagonals when $b \neq 0$ are not smooth at the point of intersection, but approximate to matching half-cones with semi-angle

$$\tan^{-1}\left(2^{\frac{1}{2}}\frac{a-b+3c}{|a-b|}\right)$$

(de Klerk and Musgrave, 1955). The singularities labelled C in Table 8.1 are therefore conical points (see Section 2.4). The third sheet is smooth on the body diagonals with both its principal radii of curvature equal to

$$-3b(4a^2+7ab-2b^2+9bc)^{-1}\{3\rho(a+2b+3c)\}^{\frac{1}{2}}. \tag{8.9}$$

The third factor in (8.9) is positive, by virtue of the inequalities (6.4), so the separate sheet is convex on the body diagonals for materials in C_2, C_3, D_a, and C_4 and concave for C_1 materials.

The facts about the slowness surface elicited in this subsection are summarized in Table 8.1.

TABLE 8.1. Properties of the sheets of $\mathscr{S}$ on the axes, face diagonals, and body diagonals (C = conical point, T = tangential point, Cx = convex, Cv = concave, SP = saddle point)

Subclass	Axis			Face diagonal			Body diagonal		
	$\mathscr{S}_1$	$\mathscr{S}_2$	$\mathscr{S}_3$	$\mathscr{S}_1$	$\mathscr{S}_2$	$\mathscr{S}_3$	$\mathscr{S}_1$	$\mathscr{S}_2$	$\mathscr{S}_3$
C_1	T, Cx	T, Cx	Cx	Cx	SP	Cx	Cv	C	C
C_2	T, Cx	T, Cx	Cx	Cx	SP or Cx	Cx	C	C	Cx
C_3	T Cx or Cv	T, Cx	Cx	Cx	Cx or SP	Cx	C	C	Cx
D_a	C	Cx	C	Cx	SP	Cx	C	C	Cx
C_4	Cv	T, Cx	T, Cx	Cx	SP	Cx	C	C	Cx

8.3. *Global visualization of $\mathscr{S}$*

In order to arrive at a general view of the slowness surface for cubic elastic materials, we have solved (6.11), with $(6.12)_1$ and (2.14), numerically to yield the three values of the slowness vector **s** corresponding to specified directions **n**. The results are presented in the form of quarter-sections of $\mathscr{S}$ in planes through the axis defined by $\mathbf{e}_3$. The points on such a section are generated by fixing the polar angle ϕ in Fig. 6.2 successively at the values $\frac{1}{32}n\pi$ ($n = 0, 1, \ldots, 8$) and increasing the second polar angle θ from 0 to $\frac{1}{2}\pi$ in steps of 2°.

Results for materials belonging to C_1, C_2, C_3, D_a, and C_4 are displayed graphically in Figs. 8.2 to 8.6 respectively. The subclasses C_2 and C_3 are represented in these calculations by actual materials, KCl and Ni, but elsewhere it has been necessary to use contrived constants

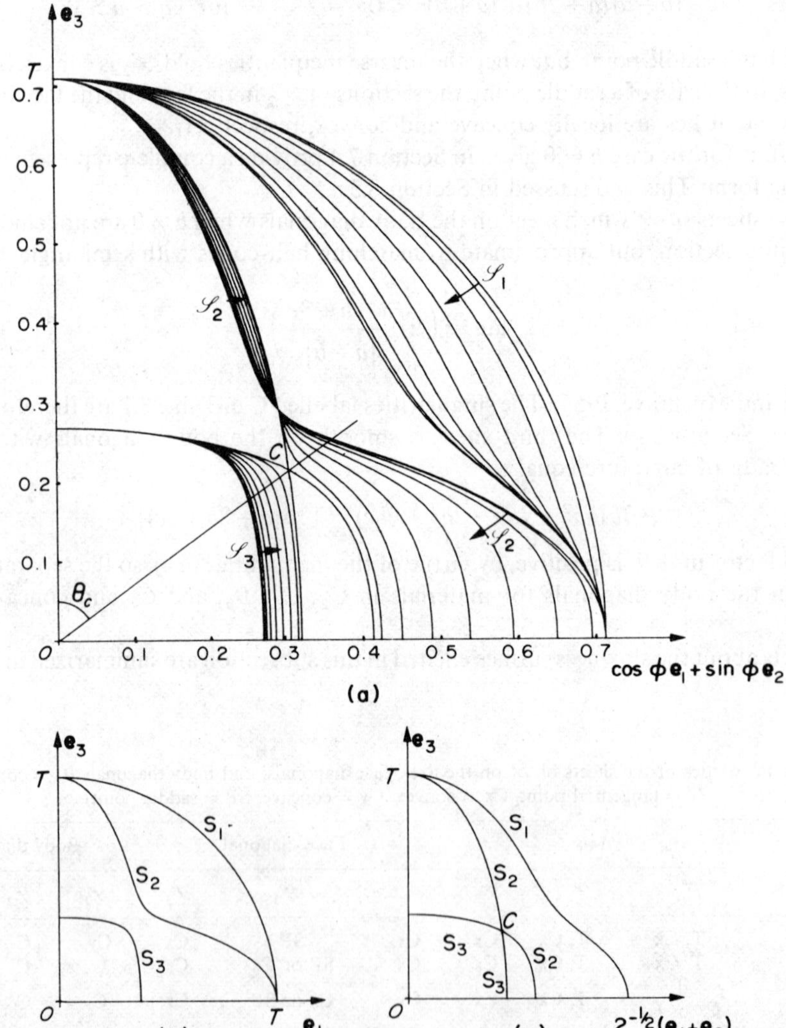

FIG. 8.2. Quarter-sections of $\mathscr{S}$ for a C_1 material ($a = 6$, $b = -2$, $c = 1$ ($\times 10^{10}$ Pa), $\rho = 5000\ \mathrm{kg\,m^{-3}}$).

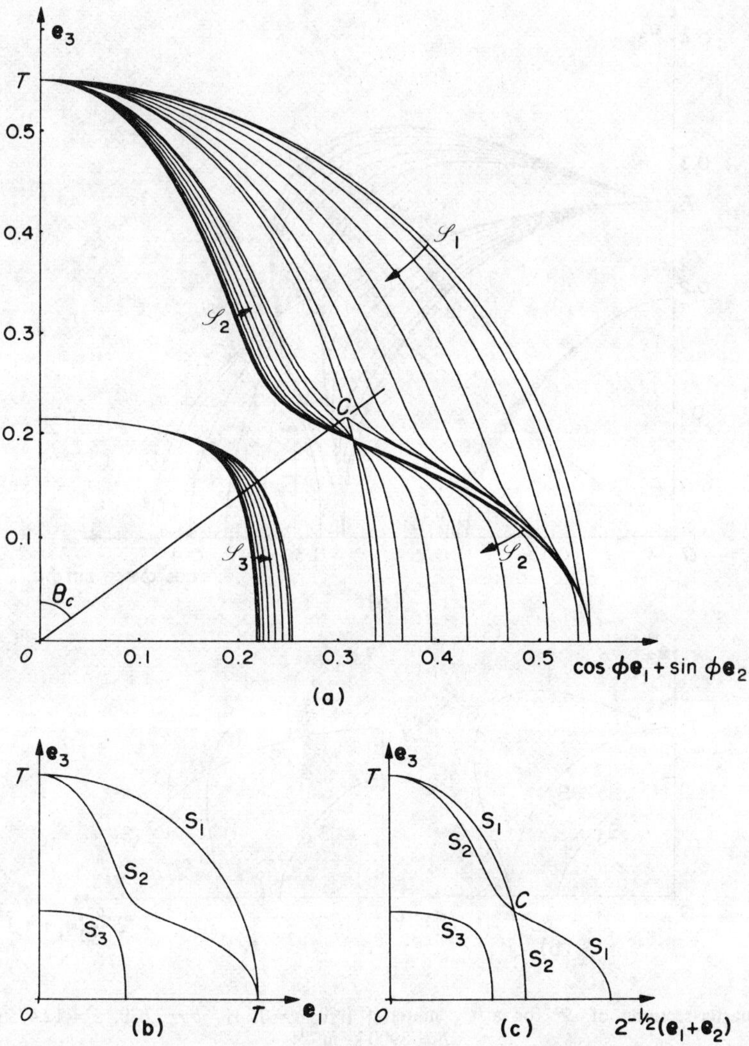

FIG. 8.3. Quarter-sections of $\mathscr{S}$ for a C_2 material (KCl: $a = 3.40$, $b = 1.29$, $c = 0.63$ ($\times 10^{10}$ Pa), $\rho = 1980\,\mathrm{kg\,m^{-3}}$).

(recorded in the captions) since no known cubic crystal qualifies for membership of C_1, D_a, or C_4. In each of Figs. 8.2(a) to 8.6(a) the nine computed quarter-sections of $\mathscr{S}$ are superimposed. The arrows in these diagrams indicate the trend of the sections of a given sheet as ϕ increases from 0 to $\frac{1}{4}\pi$, the curve bearing the arrow being situated in the face $\phi = 0$ and the curve farthest from it in the face bisector $\phi = \frac{1}{4}\pi$. For ease of reference the extreme quarter-sections are reproduced separately in parts (b) and (c) of Figs. 8.2 to 8.6.

An inspection of these figures confirms that the features set out in Table 8.1 are present in due order. Further comment on the nature of $\mathscr{S}$ is made in Section 8.4, but a remark is in order here on a property put clearly in evidence by the adopted mode of presentation of the numerical results. This is the occurrence of a point common to all the sections of either $\mathscr{S}_1$

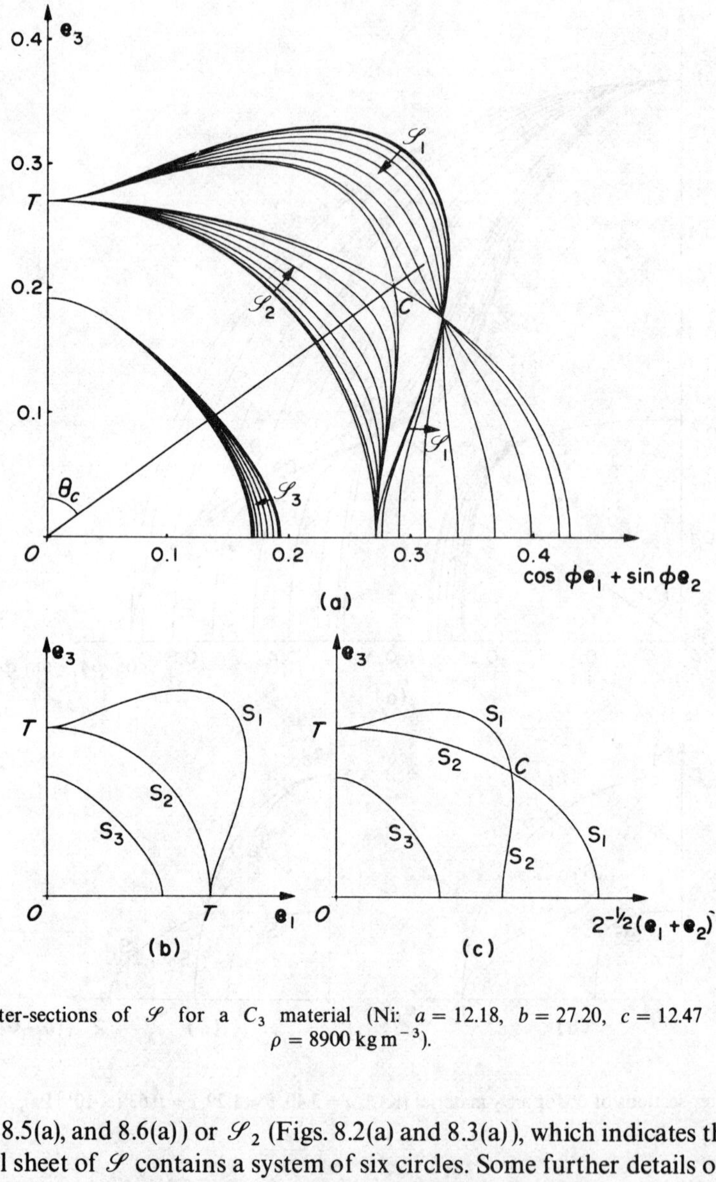

FIG. 8.4. Quarter-sections of $\mathscr{S}$ for a C_3 material (Ni: $a = 12.18$, $b = 27.20$, $c = 12.47$ ($\times 10^{10}$ Pa), $\rho = 8900\ \mathrm{kg\,m^{-3}}$).

(Figs. 8.4(a), 8.5(a), and 8.6(a)) or $\mathscr{S}_2$ (Figs. 8.2(a) and 8.3(a)), which indicates that the outer or the medial sheet of $\mathscr{S}$ contains a system of six circles. Some further details of this system have been given in a separate note (Chadwick and Smith, 1979).

No numerical calculations for D_b materials have been included in the preceding survey since in this case a complete visualization of $\mathscr{S}$ can be secured analytically. We know from Section 7.4 that when $b = 0$ the roots of the secular equation (6.11) are an_i^2 and thus, via the definitions (2.14) and $(6.12)_1$, the left side of (6.11) can be expressed as the product of the three factors $as_i^2 + cs_p s_p - \rho$ where s_i are the components of the slowness vector relative to the natural basis e. It follows that $\mathscr{S}$ is the union of three identical oblate spheroids given by

$$(a+c)s_i^2 + c(s_j^2 + s_k^2) = \rho, \tag{8.10}$$

with $\{i, j, k\}$ ranging over the permutations of $\{1, 2, 3\}$. The intersections of the sheets $\mathscr{S}_i$

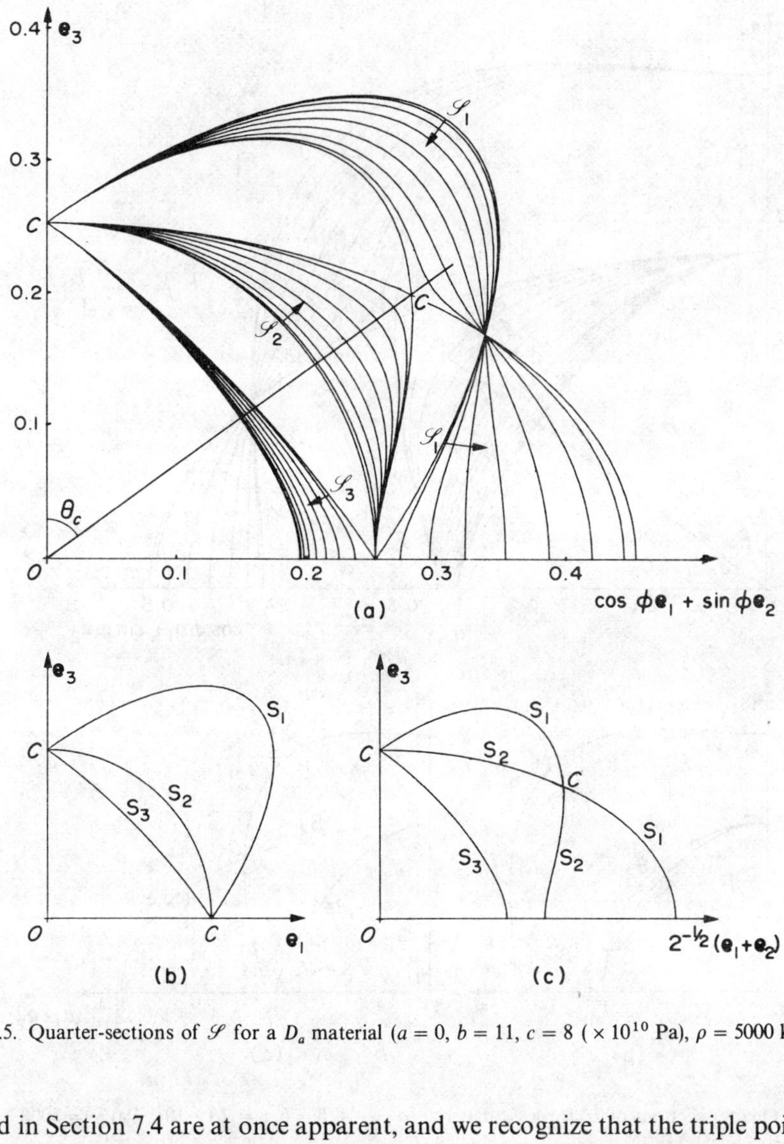

FIG. 8.5. Quarter-sections of $\mathscr{S}$ for a D_a material ($a = 0$, $b = 11$, $c = 8$ ($\times 10^{10}$ Pa), $\rho = 5000\ \mathrm{kg\,m^{-3}}$).

described in Section 7.4 are at once apparent, and we recognize that the triple points on the body diagonals are not conical points since they lie on only three tangent planes. We refer to these singularities as *tetrahedral points.*

8.4. *Incidence of the various types of transonic states*

On the basis of the results obtained in Sections 8.1 to 8.3, we can now assess the extent to which the various types of transonic states (mentioned in Section 3.1) can occur in cubic elastic materials. The general conclusions which emerge are that states of Types 1, 2, and 4 are universal in media of this kind, but that states of Types 3, 5, and 6 are totally absent.

The first fact to be noted is that no line through a singular point on $\mathscr{S}_1$ which is conical or tetrahedral in character can intersect $\mathscr{S}_1$ only tangentially, since we have shown that the

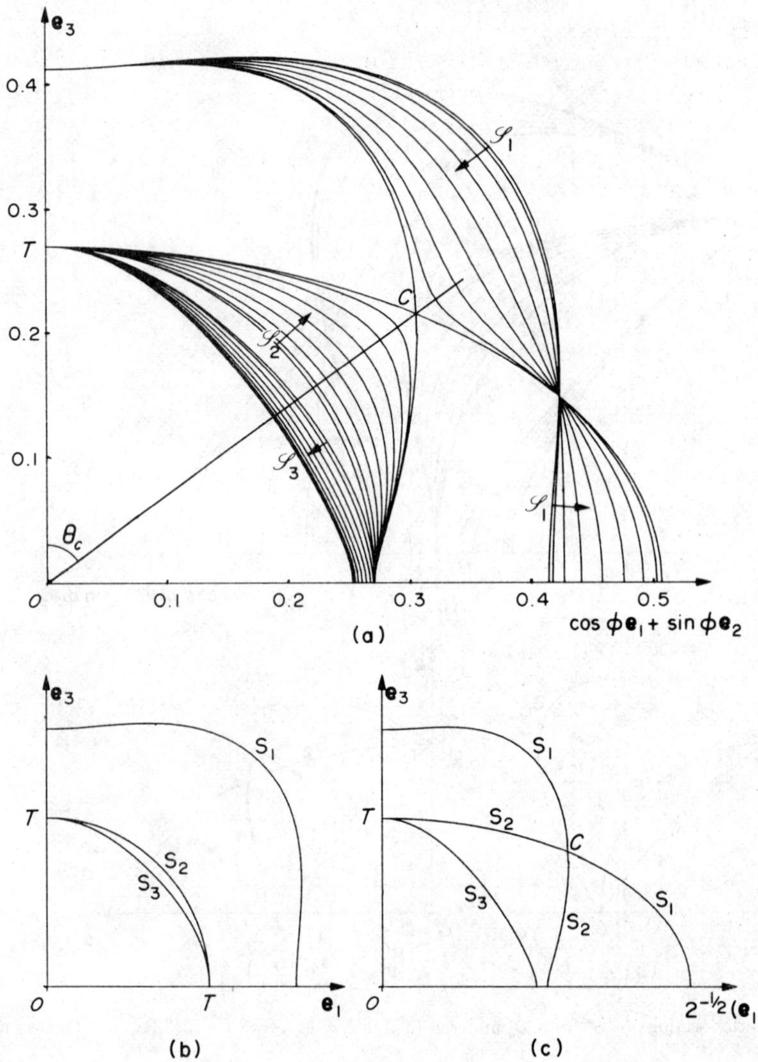

FIG. 8.6. Quarter-sections of $\mathscr{S}$ for a C_4 material ($a = -4$, $b = 6$, $c = 7$ ($\times 10^{10}$ Pa), $\rho = 5000$ kg m^{-3}).

outer sheet is always concave near such a singularity. We also observe that transonic states of Type 1 are the most commonly encountered variety in cubic materials and, indeed, in all genuinely anisotropic elastic media.

A transonic state of Type 2 can only arise in association with a tangential point linking $\mathscr{S}_1$ and $\mathscr{S}_2$ at which a tangent line can be drawn having no other point in common with $\mathscr{S}$. This means that $\mathscr{S}_1$ must be convex near the double point, and we see from Table 8.1, bearing in mind the description of $\mathscr{S}$ in the case $b = 0$ given at the end of Section 8.3, that the possibilities are limited to materials in C_1, D_b, C_2, and the portion of C_3 distinguished by the condition (8.6). Since the tangential points are located on the axes, the implied direction of surface wave propagation is along an axis with the traction-free boundary occupying any plane containing this half-line (recall Fig. 3.1).

It is evident from the last column but two of Table 8.1 and the form of $\mathscr{S}$ for D_b materials that $\mathscr{S}_1$ is never a convex surface. Transonic states of Type 4 are always possible, therefore, in cubic media. As an example, we may cite the case of surface wave propagation along an axis when $\mathscr{S}_1$ is concave or conical at the axial points. Table 8.1 shows this to be so for materials in C_4, D_a, and the part of C_3 defined by the inequality (8.6) reversed. The free boundary can be any plane containing the direction of propagation.

Transonic states of the other types do not occur. A prerequisite for a state of Type 3 is a tangent to $\mathscr{S}_1$ at a triple point and since, in cubic materials, singularities of this sort are conical or tetrahedral, no such tangent can be drawn. For a state of Type 5 to arise, a tangent to $\mathscr{S}_1$ at a tangential singularity must touch $\mathscr{S}_1$ at one other point and this is ruled out on symmetry grounds alone, the restriction of the tangential points to the axes ensuring that a tangent at such a singularity could have only an even number of additional intersections with $\mathscr{S}_1$. Finally, a state of Type 6 calls for the existence of a line tangent to $\mathscr{S}_1$ at three distinct points, but this too is incompatible with the overriding cubic symmetry.

9. The Polarization Field

The remaining Sections of Part II are directed towards the second part of the programme laid down in Section 4, namely the determination of the set of exceptional limiting wave solutions for cubic elastic materials. In the present Section we utilize the information obtained in Sections 7 and 8 in constructing complete polarization fields.

9.1. *Elementary segments*

It has been shown in Sections 7 and 8 that, in cubic elastic materials, longitudinal and transverse waves and acoustic axes are confined to the faces and face bisectors. Since these are precisely the sets of directions for which explicit solutions of the secular equation and propagation condition are available, we are in the fortunate position, with this class of media, of being able to treat analytically the features of body wave propagation of greatest interest.

As the first step in the further investigation of the polarization field, we return briefly to the results in the last three columns of Tables 7.1 to 7.3, which give the restrictions to the spherical triangle AFB of the elementary segments of the branches. Examination of these entries confirms that each branch is continuous at the vertex F, corresponding to propagation along a face diagonal. At the other vertices, two of the branches are connected by acoustic axes. These branches are discontinuous, the mismatch being permitted by the indeterminacy of the polarizations associated with an acoustic axis. The third branch, which is longitudinal, is continuous.

On account of the boundary of $\mathscr{U}_e$ containing all the acoustic axes, each branch of the polarization field can be made continuous in the interior of $\mathscr{U}_e$. The interior fields may be computed by solving numerically (6.11) to (6.13), and the simple selection rules stated in Table 9.1 ensure that the results are consistent with the boundary fields given in Tables 7.1 to 7.3.

TABLE 9.1. Selection rules for the polarization field in the interior of $\mathcal{U}_e$

Subclass	Branch		
	Outer	Medial	Inner
$C_1 \cup D_b \cup C_2$	$p_2 > 0$	$p_1 > 0$	$p_3 < 0$
$C_3 \cup D_a$	$p_1 > 0$	$p_2 > 0$	$p_3 > 0$
C_4	$p_3 < 0$	$p_2 > 0$	$p_1 > 0$

9.2. *Complete fields*

The elementary segments of the branches of the polarization field are now fully determined and, by appropriately replicating these units, we can proceed to assemble complete fields. The discussion which follows refers to materials for which $b \neq 0$: the polarization field for D_b materials has peculiar features which are described at the end of this subsection.

It is known, from the general considerations outlined in Section 5.3, that the presence of conical points on the body diagonals entails the insertion of cuts in the relevant branches of the polarization field, and it follows from the construction of the elementary segments, described in Section 9.1, that these cuts must be restricted to the face bisectors. Each point of intersection of $\mathcal{U}$ with a body diagonal is an end-point of three edges like AB, connecting the body diagonal to the adjacent axes, and three edges like FB whose other ends are on face diagonals. In pursuance of a first basic rule that continuity be preserved as far as possible, we place cuts on at most one of these triples. The choice of cuts is thus non-unique, as pointed out in Section 5.3, and additional non-uniqueness arises from the possibility of reversing any pair of entire branches, allowed by the orientation convention $(2.12)_2$.

Reference to Tables 8.1, 7.2, and 7.3 shows that the conical points are associated with the medial and inner branches for C_1 materials and with the outer and medial branches for materials in C_2, C_3, D_a, and C_4, and that, in each case, one branch is transverse on AB, but not on FB, and the other branch transverse on FB, but not on AB. This latter property affords a systematic basis for the assignment of cuts, and we adopt as our second basic rule that cuts in branches with conical points shall run along the edges like AB and FB on which the branch is non-transverse. Up to reversals of pairs of branches, the polarization field is now completely specified.

For cubic elastic materials with $b \neq 0$, the branches of the polarization field, composed by the procedure just described, are of five types, labelled E_1, E_2 and O_1, O_2, O_3, the E-types being even and the O-types odd vector fields on $\mathcal{U}$. The five fields are shown in diagrammatic form in Fig. 9.1, and the types of the outer, medial, and inner branches for the various subdivisions of the cubic system are set out in Table 9.2. We see that in each instance two branches are odd and one even, consistently with $(2.12)_2$.

In explanation of Fig. 9.1, the following points should be noted. Each diagram represents 24 copies of the elementary segment, covering a hemisphere. Cuts are shown in heavy line and the arrows indicate the direction of the projection of $\mathbf{p}$ on the tangent plane of $\mathcal{U}$. Longitudinal waves are not marked, though their presence is apparent on the body diagonals in E_1, where they are directed outward to $\mathcal{U}$, along the face diagonals in O_1, where they reverse direction on the cut, and along the face and body diagonals in O_3. Information about

TABLE 9.2. Types of the branches of the polarization field. The letters in brackets specify the nature of the field on the axes: L + and L − indicate outwardly and inwardly directed longitudinal waves, T a circularly polarized wave, and S a spherically polarized wave

Subclass	Branch		
	Outer	Medial	Inner
C_1	E_1(T)	O_1(T)	O_2(L −)
C_2	E_2(T)	O_2(T)	O_3(L −)
C_3	O_2(T)	E_2(T)	O_3(L +)
D_a	O_2(S)	E_2(S)	O_3(S)
C_4	O_2(L −)	E_2(T)	O_3(T)

the remaining longitudinal waves, which occur on the axes, is given in Table 9.2. Whenever O_3 is longitudinal along the axes it is a continuous field, in conformity with the theorem given in Section 5.3, $\mathbf{p} \cdot \mathbf{n}$ being of one sign on $\mathscr{U}$. Otherwise, O_3 fails to be continuous, in common with the other four vector fields shown in Fig. 9.1.

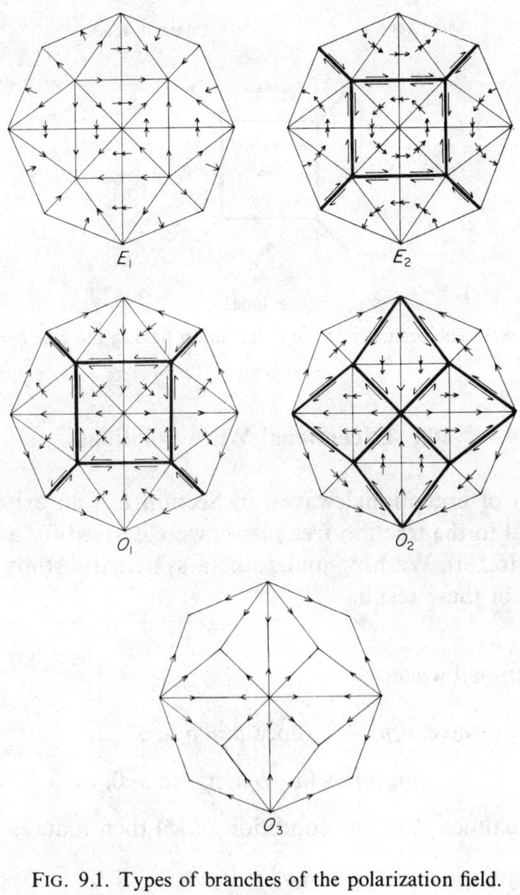

FIG. 9.1. Types of branches of the polarization field.

We come finally to the construction of a complete polarization field for a D_b material. Since, in this case, the singularities on the body diagonals are triple points, cuts must be placed in each branch. The structure of the field is determined by the observation, tacit in Section 7.4, that if $n_i^2 < n_j^2 < n_k^2$, then the polarizations of the outer, medial, and inner branches are $\mathbf{e}_i$, $\mathbf{e}_j$, $\mathbf{e}_k$, respectively. It remains only to select the direction of $\mathbf{p}$ and in Fig. 9.2 this is done in such a way that the outer branch is even and the other branches odd (cf. Table 9.2). While, in the fields shown in Fig. 9.1, the discontinuities on cuts are simply reversals of the polarization, the fields in Fig. 9.2 turn through a right angle when a cut is crossed. This difference arises from the fact that when $b = 0$ the cuts consist entirely of double points at which two of the polarizations are incompletely determined.

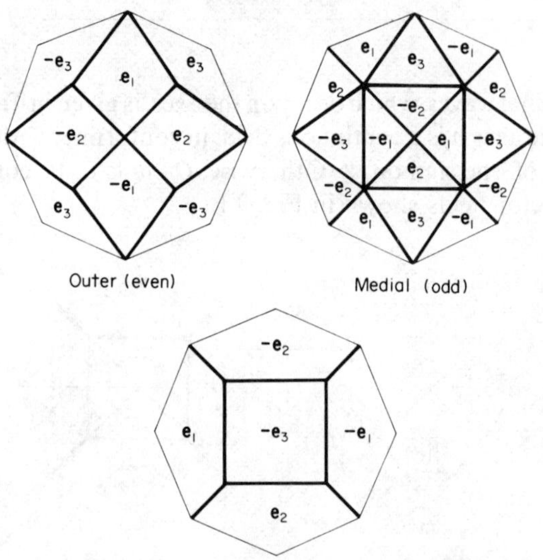

FIG. 9.2. The branches of the polarization field in the case $b = 0$.

10. Exceptional Wave Solutions

At the last mention of exceptional waves, in Section 6.2, an existence criterion and a formula for the normal to the traction-free planes were derived for a cubic elastic material (equations (6.15) and (6.16)). We now undertake a systematic study of exceptional wave solutions on the basis of these results.

10.1. *Transverse exceptional waves*

For a transverse body wave, $n_s p_s = 0$. Since $a \neq b$ and

$$b(n_r p_r)^2 + (a-b)n_r^2 p_r^2 + c > 0,$$

as a result of the inequalities (6.4), the condition (6.15) then reduces to

$$n_s p_s^3 = 0. \tag{10.1}$$

The fact that (10.1) is a purely geometrical relation makes it appear likely that transverse waves will normally be exceptional. On the other hand, it is plain from (6.15) that a non-transverse wave can be exceptional only if a connection holds among the material constants a, b, c and this is to be expected only in rather special circumstances.

It has been shown in Section 7.1 that the directions of propagation of transverse body waves in cubic elastic media are confined to the faces and face bisectors, and it is a simple matter to submit to the requirement (10.1) each of the six possibilities that arise. With reference to the elementary segment $\mathscr{U}_e$ (see Fig. 6.2), the cases to be considered correspond to wave normals in (i) $AF - \{A, F\}$, (ii) $AB - \{A, B\}$, (iii) $FB - \{F, B\}$, and at (iv) A, (v) F, (vi) B. It turns out that the transverse waves in categories (i) to (v) are exceptional, but that under the sixth heading only three waves have this property. For the circularly polarized transverse wave propagating along OB we have

$$\begin{gathered} n_1 = n_2 = n_3 = 3^{-\frac{1}{2}}, \\ p_1 = 2^{-\frac{1}{2}} \cos \zeta + 6^{-\frac{1}{2}} \sin \zeta, \quad p_2 = -2^{-\frac{1}{2}} \cos \zeta + 6^{-\frac{1}{2}} \sin \zeta, \quad p_3 = -(\tfrac{2}{3})^{\frac{1}{2}} \sin \zeta, \end{gathered} \tag{10.2}$$

where ζ is arbitrary. Hence

$$n_s p_s^3 = -(2^{\frac{3}{2}}/3) \sin \zeta \sin (\zeta - \tfrac{2}{3}\pi) \sin (\zeta - \tfrac{4}{3}\pi),$$

proving that the wave is exceptional only if ζ takes one of the values 0, $\frac{2}{3}\pi$, $\frac{4}{3}\pi$.

The information supplied by (6.16) on the orientation of the traction-free planes is as follows. For all transverse waves travelling in a face (categories (i), (iv), and (v)), the vectors $\mathbf{n}$, $\mathbf{p}$, and $\mathbf{k}$ are orthonormal so that, without loss of generality, we can take $\mathbf{k} = \mathbf{n} \wedge \mathbf{p}$. For transverse waves travelling in a face bisector, but not along a face or body diagonal, we find that

$$\mathbf{k} = -2^{-\frac{1}{2}} \cos \delta (\mathbf{e}_1 + \mathbf{e}_2) + \sin \delta \mathbf{e}_3 = \cos (\delta - \theta) \mathbf{n} \wedge \mathbf{p} + \sin (\delta - \theta) \mathbf{n}, \tag{10.3}$$

with

$$\delta = \tan^{-1} \left(\frac{a - b + 2c}{2c} \tan \theta \right) \quad (0 < \delta < \tfrac{1}{2}\pi), \tag{10.4}$$

for category (ii). For category (iii), $\mathbf{e}_2$ and $\mathbf{e}_3$ must be interchanged and θ replaced by the angle ψ defined in (7.10). Lastly, for the three exceptional transverse waves in category (vi), $(10.3)_2$ and (10.4) apply with $\theta = \theta_c (= \tan^{-1} 2^{\frac{1}{2}})$.

We notice that whenever two transverse exceptional waves travel in the same direction (categories (iv) to (vi)), the respective families of parallel planes which are rendered traction-free do not coincide, the normals to these families being orthogonal in cases (iv) and (v) and at an angle $\cos^{-1} \{1 - \frac{3}{2} \cos^2 (\delta(\theta_c) - \theta_c)\}$ to one another in case (vi). This implies, in particular, that a transonic state of Type 2 in a cubic elastic material is necessarily normal. (In a general anisotropic elastic body it is not possible for two transverse waves with a common wave normal $\mathbf{n}$ to be exceptional in relation to the same family of parallel planes. For if this possibility did exist, it would follow from $(5.2)_1$ that $\mathbf{P}(\mathbf{n}, \mathbf{p}_\alpha)\mathbf{n} = \mathbf{0}$ $(\alpha = 1, 2)$, $\mathbf{p}_1$ and $\mathbf{p}_2$ being the polarizations of the transverse waves. It has been noted in Section 5.1, however, that these relations are incompatible with the strong convexity condition (2.2). *A transonic state of Type 2 with transverse limiting body waves is therefore normal.*)

The preceding results, in conjunction with Tables 7.1 to 7.3, 8.1, and 9.2, enable us to mark on diagrams of the kind shown in Fig. 9.1 the wave normals of transverse exceptional waves, with the results shown in Fig. 10.1. The exceptional wave solutions form the curves shown in heavy line which fall into three sets: the three great circles in the faces; twelve arcs of great circles joining pairs of body diagonals and passing, in pairs, through the axial points; and the twelve complementary arcs of the great circles in the face bisectors. These sets occur in the branches of types E_1, O_1, and O_2, respectively, and the branch of type E_2 contains the union of the first two sets. The dots in the plot of O_2 in Fig. 10.1 represent isolated transverse exceptional waves propagating along the axes. For C_4 materials these solutions are absent, but for this subclass and D_a they occur in the branch O_3. Otherwise, O_3 is free from transverse exceptional wave solutions, which accounts for its non-appearance in Fig. 10.1.

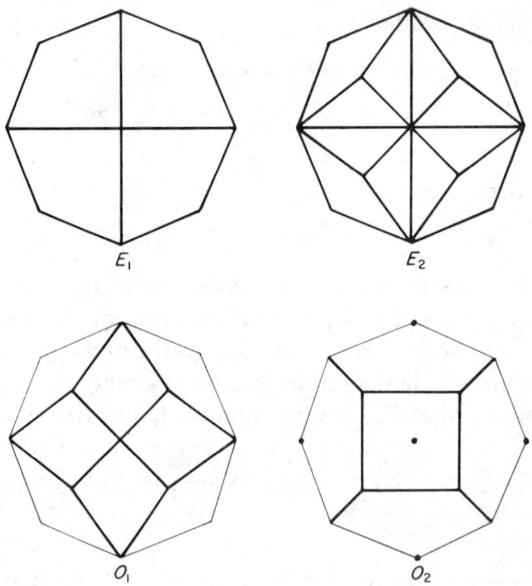

FIG. 10.1. The distribution of transverse exceptional waves in branches of the polarization field of types E_1, E_2, O_1, and O_2.

The diagrams in Figs. 9.1 and 10.1 bearing the same label are found to have properties fully in accord with the existence theorems proved in Section 5.4. The field E_1, which is even and continuous, apart from tangential points on the axes, has at least three pairs of exceptional wave solutions on each great circle, in conformity with Theorem 1. In the discontinuous branches E_2, O_1 and O_2, there is a curve of exceptional wave solutions associated with each of the three cuts terminating at a conical point, as required by Theorem 2 and depicted in Fig. 5.1. In fact, in each of these branches the union of the cuts and the curves of transverse exceptional wave solutions includes the six great circles in the face bisectors. When the pairs of branches joined by conical points are juxtaposed, the solution curves may be regarded as complete circles, passing from one branch to the other at a conical point in the manner of a path through a branch point on a Riemann surface. With regard to Theorem 3, it is easy to verify, with the aid of Tables 8.1 and 9.2, that on every acoustic axis there are situated, in each of the branches which it connects, exceptional wave solutions.

10.2. *The degenerate case* $b = 0$

An initial approach to the question of whether or not there exist non-transverse exceptional waves can usefully be made via the case $b = 0$ for which the polarization field is known explicitly.

Away from the face bisectors the components of the polarization vector are of the form $p_i = \delta_{ij}$ (see Fig. 9.2), and (6.15) then gives $E = 2c^2 n_j$. A body wave not propagating in a face bisector is therefore exceptional if and only if it is transverse, and the wave normal then lies in a face. To examine the case of wave propagation in a face bisector, we suppose that $\mathbf{n}$ lies on the edge $AB - \{B\}$ of $\mathscr{U}_e$. Then either $\mathbf{p} = \mathbf{e}_3$ or $\mathbf{p} = \cos\zeta\,\mathbf{e}_1 + \sin\zeta\,\mathbf{e}_2$, with ζ arbitrary. When $\mathbf{p} = \mathbf{e}_3$, we have $E = 2c^2 n_3$, and since $n_3 \neq 0$ on AB, this wave is not exceptional. In the second case, we find from (6.15) that

$$E = 2^{\frac{1}{2}} c n_1 \sin(\zeta + \tfrac{1}{4}\pi)(2c - a \sin 2\zeta).$$

This wave is therefore exceptional if (a) $n_1 = 0$, corresponding to a transverse wave travelling along OA, (b) $\zeta = \frac{3}{4}\pi$ or $\frac{7}{4}\pi$, again representing a transverse wave, and (c) $\zeta = \frac{1}{2}\sin^{-1}(2c/a)$. Possibility (c), which exists only if $2c \leqslant a$, does not relate to a transverse wave. Evidently, then, not all exceptional waves in cubic elastic media are transverse.

10.3. *Non-transverse exceptional waves*

The analytical results presented in Sections 10.1 and 10.2 have been supplemented by numerical evaluations of the scalar function $E(\mathbf{n}, \mathbf{p})$ defined in (6.15) at the nodes of a fine mesh covering the interior of the elementary segment $\mathscr{U}_e$. This computation, which effectively scans the three branches of the polarization field for exceptional waves, has been carried out for 50 cubic crystals with elastic constants belonging to the subclasses C_2 and C_3, and also for a range of hypothetical materials in C_1, D_a, and C_4. (In C_2: AgBr, AgCl, $Ba(NO_3)_2$, CaF_2, CsBr, CsCl, CsI, FeS_2, KBr, KCl, KF, KI, Mo, NaBr, $NaBrO_3$, NaCl, $NaClO_3$, NaF, NaI, NH_4Br, NH_4Cl, PbS, TlBr, TlCl, W. In C_3: Ag, Al, Au, BaF_2, C (diamond), Cr_2FeO_4, Cu, Cu_3Au, CuZn, Fe, Fe_3O_4, GaAs, GaSb, Ge, InSb, K, Li, LiF, MgO, Na, Ni, Pb, Si, Th, ZnS. Sources of physical data for these materials have been indicated in Section 6.1.)

For all but two of the C_2 and C_3 materials examined, no non-transverse exceptional waves were found, the anomalous substances being FeS_2 and Na. These are the crystals whose representative points in the (a, b)-plane are closest to the half-lines D_a and D_b respectively (see Fig. 6.1): for FeS_2, $a/b = 4.3$ and for Na, $a/b = 0.016$. In Na, the non-transverse exceptional waves are confined to the inner branch of the polarization field and the associated wave normals form 14 closed curves on $\mathscr{U}$ surrounding the axes and the body diagonals. In FeS_2, the medial and inner branches contain non-transverse exceptional waves, the solution curves encircling the axes in the medial and the axes and body diagonals in the inner branch.

The branches of the polarization field which admit non-transverse exceptional waves in the aforementioned cases are of the odd types O_2 and O_3 (see Table 9.2), and so long as the solution curves are simple circuits about the axes and body diagonals, the general considerations explained in Section 5.4 imply that they will be limited to branches of odd parity. The numerical scans which have been carried out for materials in C_1, D_a, and C_4 bear out this connection and provide instances of non-transverse exceptional waves occurring in branches of type O_1 as well as in the other odd types. The feature of greatest interest in these results, to which we return in Section 11.1, is the presence of curves of non-transverse solutions surrounding the body diagonals in the outer branch of the polarization field for some materials in C_4.

11. Exceptional Limiting Waves

Our aim in this, the concluding Section of Part II, is to characterize the subset of the exceptional wave solutions found in Section 10 which meet the extra requirements B, C, and D formulated in Section 4.2.

11.1. *Examination of outer branches*

The effect of condition B is to restrict exceptional limiting waves to $\mathscr{S}_1$ and hence to the outer branch of the polarization field. It follows, therefore, from Table 9.2, that we are concerned here with the patterns of transverse exceptional wave solutions of types E_1, E_2, and O_2 in Fig. 10.1, appropriate to materials in C_1, C_2, and $C_3 \cup D_a \cup C_4$, respectively, and with the eight closed curves of non-transverse solutions which may appear in the outer branch, of type O_2, for C_4 materials. (Non-transverse exceptional wave solutions may also occur in the outer branch of the polarization field for D_b materials, as shown in Section 10.2. However, the representative points of these waves on $\mathscr{S}_1$ lie in the "grooves" formed by the intersection of pairs of the ellipsoids (8.10), so there is no possibility of meeting the constraints C and D.)

We deal first with transverse exceptional waves advancing along the axes and body diagonals. For propagation along an axis, it has been noted in Section 10.1 that the polarization **p** can lie in any direction orthogonal to **n** and that the normal to the traction-free planes is $\mathbf{k} = \mathbf{n} \wedge \mathbf{p}$. The reference plane R (spanned by **k** and **n**, as shown in Fig. 3.1 (a)) can therefore be any plane through the axis, and the wave can be limiting exceptional only if $\mathscr{S}_1$ is convex on the axes. But then, from Table 8.1, the transonic state is of Type 2 and, as pointed out in Section 10.1, normal. The possibility of a transverse wave travelling along a body diagonal having the exceptional limiting property can be ruled out since, with reference to Table 8.1, $\mathscr{S}_1$ is either concave on the body diagonals or has a conical point there, and in neither case could conditions C and D be satisfied. We recognize, then, that further consideration of the parts of Fig. 10.1 specified above can be confined to waves traversing neither an axis nor a body diagonal.

All such waves are polarized normal to the face or face bisector in which **n** lies, and this property, in conjunction with (5.2) and the fact, noted in Section 2.4, that the group velocity of a branch of body wave propagation is directed normally to the cognate sheet of $\mathscr{S}$, leads to two important conclusions. First, the face or face bisector is the reference plane R since it contains the normals **n** and **k**, and the traction-free planes are perpendicular to R. The normal to $\mathscr{S}_1$ at a point in a face lies in that face and the same is true for a face bisector away from acoustic axes. Secondly, therefore, the group velocity **C** is in R and orthogonal to both the tangent line L and the normal **k**, which means that the exceptional wave meets condition C.

In each of E_1, E_2, and the outer branch in Fig. 9.2 there is a transverse exceptional wave in every direction in a face. We see from Figs. 8.2(b) and 8.3(b) that S_1, the outer branch of the slowness section, is a circle, and it is at once apparent that condition D is satisfied. For materials in $C_1 \cup D_b \cup C_2$ (i.e. when $a > b$), a transverse body wave propagating in a face, but not along an axis, is therefore an exceptional limiting wave. The transonic state is of Type

1 and since $\mathbf{k} = \mathbf{n} \wedge \mathbf{p}$, the traction-free planes contain the wave normal $\mathbf{n}$. The situation regarding condition D for the transverse exceptional waves travelling in a face bisector is illustrated in Fig. 11.1, where the cases $a > b > 0$ (appropriate to E_2) and $b > a$ (to O_2) are treated separately. Bearing in mind Fig. 3.1(a) and the results established in the last paragraph, we deduce that the subsets of these waves which qualify as exceptional limiting waves are governed by the restrictions $0 < \theta < \theta_-$ and $\theta_+ < \theta < \frac{1}{2}\pi$ on the wave normal, applying when $a > b > 0$ and $b > a$, respectively. Here, θ_- is the smaller and θ_+ the larger of the values of θ at the points of contact of L with S_1 in the unique transonic state of Type 4 which can eventuate when R is a face bisector and the direction of surface wave propagation is non-axial. This Type 4 state is of course normal, since the limiting wave represented by the point of contact of L with the non-elliptical portion of S_1 is not exceptional in relation to the planes with unit normal $\mathbf{k}$. By the same token, normality is a general attribute of transonic states of Type 4 in cubic elastic materials.

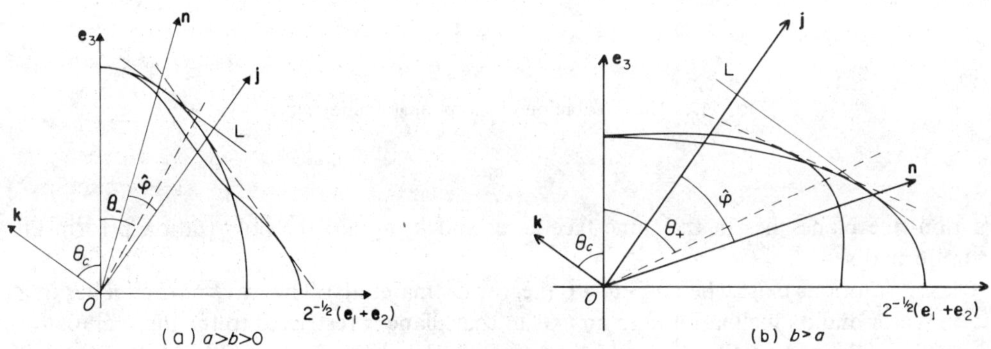

FIG. 11.1. Definition of the angles θ_- and θ_+, and illustration of the transonic state in Case 3.

There remain the non-transverse exceptional waves mentioned in the opening paragraph. For C_4 materials, $\mathscr{S}_1$ has concavities near the axes and body diagonals (see Fig. 8.6), and it appears inevitable, from the numerical calculations which we have performed for cubic solids in this subclass, that an appreciable part of the closed solution curve encompassing a body diagonal will be "shielded" by $\mathscr{S}_1$ and thus put at odds with condition D. For the exposed" portion, condition C necessitates a connection between the tangent space of $\mathscr{S}_1$ and the outer branch of the polarization field which could only hold fortuitously. Thus, while the existence of non-transverse exceptional limiting waves can not be altogether excluded, the fact that such solutions are confined to the "non-physical" subdivision C_4, and require, moreover, a precise tuning of the material properties, prompts us to regard them as pathological.

11.2. *Classification of solutions*

The transverse exceptional limiting wave solutions arrived at in the last subsection fall into three classes.

Class 1 solutions exist when $a > b$ (i.e. for C_1, D_b, and C_2 materials). The wave normal $\mathbf{n}$ can have any non-axial orientation in a face and the wave speed is $(c/\rho)^{\frac{1}{2}}$. The unit normal $\mathbf{k}$ to the

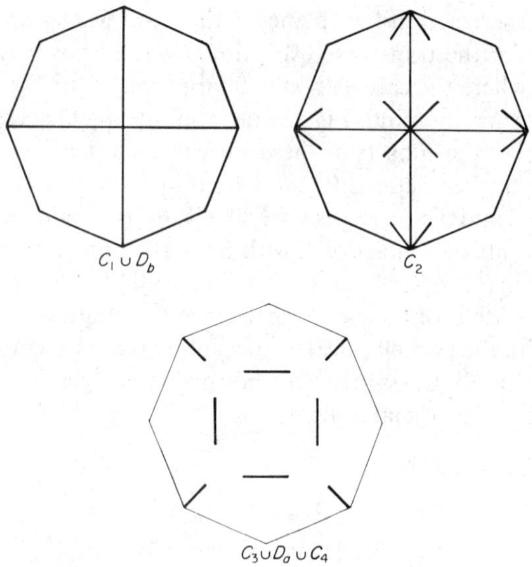

FIG. 11.2. The distribution of exceptional limiting waves.

traction-free planes lies in the same face as **n**; and **k**, **n**, and the polarization **p** form an orthonormal set.

Class 2 solutions exist when $a > b > 0$ (i.e. for C_2 materials). The wave normal **n** lies in a face bisector and its inclination θ to an axis in that plane is restricted to the interval $(0, \theta_-)$ where θ_- is defined in Fig. 11.1(a). The wave speed is $[\{\frac{1}{2}(a-b)\sin^2\theta + c\}/\rho]^{\frac{1}{2}}$. The unit normal **k** to the traction-free planes is situated in the same face bisector as **n**: its inclination to the axis is $\frac{1}{2}\pi - \delta$ and the angle between **k** and **n** is $\frac{1}{2}\pi - (\delta - \theta)$, δ being defined by (10.4). The polarization **p** is normal to the face bisector.

Class 3 solutions exist when $a < b$ (i.e. for C_3, D_a, and C_4 materials). The orientations of **n**, **k**, and **p** are the same as for exceptional limiting waves in Class 2 except that the interval to which the angle θ is restricted is now $(\theta_+, \frac{1}{2}\pi]$, θ_+ being defined in Fig. 11.1(b).

For all the solutions the transonic state is of Type 1 and the surface wave normal **j** is orthogonal to **k** and coplanar with **k** and **n**.

The diagrams in Fig. 11.2 display the distribution on $\mathscr{U}$ of the directions in which transverse exceptional limiting waves can propagate for each subclass of cubic elastic materials.

III. APPLICATION

12. The Existence of Surface Waves in Cubic Elastic Media

The relationship of the matters discussed in Part II to the theory of surface waves has been described in Section 4. We come now to the detailed application of these results to the existence of surface waves, both subsonic and supersonic, in elastic materials with cubic symmetry.

12.1. *General survey*

We are concerned hereafter with a semi-infinite body composed of a cubic elastic material. The plane traction-free surface of the body has outward unit normal **k**, and we are to examine the possibility of a surface wave, of the form (3.1), propagating in a given direction, normal to **k**, specified by the unit vector **j**.

As explained in Section 3.1, the vectors **j** and **k** define the reference plane R which intersects the slowness surface $\mathscr{S}$ of the material in the slowness section S. The tangency to the outer branch S_1 of a line in R codirectional with **k** determines the transonic state. It has been shown in Section 8.4 that only transonic states of Types 1, 2, and 4 can occur in cubic media, and we know from Sections 10.1 and 11.1 that states of Types 2 and 4 are always normal. The Barnett–Lothe existence theorem, stated in Section 3.3 and amended in Section 3.4, thus tells us that a unique (subsonic) surface wave can travel in the direction **j** if the transonic state is not both of Type 1 and exceptional, and that the same conclusion holds when the transonic state is of Type 1 and exceptional if $F(\hat{v}) < 0$, $\hat{v}$ being the limiting speed and $F(.)$ the function defined in (3.8).

In cubic media a subsonic surface wave can therefore fail to exist only if the transonic state is of Type 1 and exceptional and $F(\hat{v}) > 0$. There then arises the possibility that a supersonic surface wave can travel in the direction **j** and, as pointed out in Section 3.4, such a wave indeed exists if $F(\hat{v}) > 0$ and $F(v) < 0$ for some $v > \hat{v}$. The fact, established in Section 11.1, that an exceptional transonic state of Type 1 can arise only when R is a face or a face bisector delimits the incidence of supersonic surface waves in cubic media, and the expectation, raised in Section 3.3, that the normals **j** and **k** will be symmetrically orientated relative to the directions of material symmetry is evidently fulfilled.

12.2. *Particular cases of exceptional transonic states of Type* 1

As a framework for his numerical investigation of surface waves in cubic elastic media, Farnell (1970, Section V) chose three placements of the normal **k** relative to the axes of symmetry of the transmitting material. In the selected cases **k** is directed along an axis, a face diagonal, and a body diagonal, and the free surface of the body is thus situated in turn in a face, a face bisector, and an octahedral plane. Surface wave propagation in a face had previously been studied by Stoneley (1955), Gazis *et al.* (1960), and Buchwald and Davis (1963), and the latter authors also considered directions of propagation in a face bisector. In the remainder of the present article we take up Farnell's cases and show how an improved theoretical understanding of some salient features of his results may be obtained. We find that in each case there is just one direction (or, more precisely, a set of directions equivalent under cubic symmetry) in which a supersonic surface wave can travel, and these possibilities are fully analysed on the basis of (3.8).

(a) *Propagation of surface waves in a face.* Consulting the classification given in Section 11.2, we see that the only exceptional limiting body wave which leaves a face traction-free is the member of Class 3 for which $\theta = \delta = \frac{1}{2}\pi$. The wave normal **n** is therefore along a face diagonal. We deduce from the specialized form of the Barnett–Lothe theorem stated in Section 12.1 that a unique (subsonic) surface wave can propagate in every direction in a face when $a > b$ and in every direction except, possibly, the face diagonals when $a < b$. The typical exceptional direction, which is designated *Case* 1, is discussed in Section 14.

The foregoing conclusion is wholly consistent with Farnell's results (see Figs. 4, 8, 18, and

21 of his article). For each of the C_2 materials included in the computations there is a unique surface wave for all **j** situated in the face. For all the C_3 materials except C (diamond) the behaviour described at the beginning of Section 3.4 is encountered near the face diagonals, a unique surface wave being obtained elsewhere in the face. This suggests that $F(\hat{v}) < 0$ for some C_3 materials (including C), but that, for the remainder, $F(\hat{v}) > 0$ and $F(v) < 0$ for some $v > \hat{v}$. A criterion for the existence of a supersonic surface wave in Case 1 is derived in due course (see (14.13) below).

(b) *Propagation of surface waves in a face bisector.* On referring again to Section 11.2, we find that there is only one exceptional limiting wave for which the traction-free planes include a face bisector. This is the wave in Class 1 for which **n** points along a face diagonal. The existence theorem then asserts that a unique (subsonic) surface wave can progress in every direction in a face bisector when $a < b$ and in every direction except, possibly, the face diagonals when $a > b$. The typical exceptional direction, termed *Case* 2, is the subject of Section 15.

Agreement with Farnell's results is apparent from Figs. 11, 13, and 19 of his article. For all the C_3 materials examined, a unique surface wave is recorded for all directions of propagation in the face bisector, while the C_2 materials fall into two groups: one crystal (W) exhibits qualitatively the same behaviour as C_3 materials, and for all the others the regular branch of surface waves degenerates and a supersonic surface wave appears on the face diagonals. The condition governing this subdivision of C_2 is obtained later (see (15.6)).

(c) *Propagation of surface waves in an octahedral plane.* It is clear, from Section 11.2, that we must look to Classes 2 and 3 for an exceptional limiting wave which produces no traction on an octahedral plane. The inclination of **k** to the axis $\theta = 0$ is then θ_c and since, according to the definitions of Classes 2 and 3, this angle equals $\frac{1}{2}\pi - \delta$, with δ given by (10.4), we have

$$\theta = \tan^{-1}\{2^{\frac{1}{2}}c/(a-b+2c)\}. \tag{12.1}$$

There is a body wave in Class 2 or Class 3 with the desired property according as $0 < \theta < \theta_-$ or $\theta_+ < \theta < \frac{1}{2}\pi$. As in Case 1, the reference plane $\overline{R}$ is a face bisector, and since each octahedral plane is intersected orthogonally by three face bisectors, there are, in this instance, six equivalent directions of surface wave propagation (referred to henceforth as *octahedral diagonals*) for which an exceptional transonic state of Type 1 may materialize. We deduce from the existence theorem that for all other directions in an octahedral plane a unique (subsonic) surface wave exists. The representative exceptional direction is treated in Section 16 under the heading *Case* 3.

Farnell's article presents numerical results for 16 crystals in C_2 and C_3, and his Fig. 20 shows that in none of them do supersonic surface waves propagate in an octahedral plane. The analysis developed in Section 16 provides a means of deciding whether or not, in a specified cubic solid, supersonic surface waves exist, and examples of such a wave are given. Finally, we note that, for the exceptional limiting wave associated with surface wave propagation along the octahedral diagonal given by $\theta = \frac{1}{2}\pi - \theta_c$, $\phi = \frac{1}{4}\pi$, (10.3) and (10.4), with $\delta = \frac{1}{2}\pi - \theta_c$, give

$$\mathbf{k}\,.\,\mathbf{n} = 3^{-\frac{1}{2}}(a-b)\{(a-b+2c)^2+2c^2\}^{-\frac{1}{2}}.$$

Since $a \neq b$, the wave normal is hence inclined to the traction-free boundary, being directed upward to it if $a > b$ and downward if $a < b$. This property and the fact that the group velocity of the limiting body wave is orthogonal to **k** have been commented on by Farnell (1970, Section V, D; see, in particular, Fig. 17).

13. Preliminary Consideration of Particular Cases

Our main objective now is to determine $F(v)$ from the definition (3.8) for Cases 1, 2, and 3. To this end the tensor $\mathbf{S}_1(v)$ must be calculated from (3.3) and (3.4), and we proceed, not by referring $\mathbf{S}_1(v)$ to the natural basis e, but by introducing a second basis in which the surface wave and boundary normals, **j** and **k**, are base vectors. In order to keep the algebra reasonably concise we utilize the following combinations of elastic constants:

$$d=\tfrac{1}{2}(a+b)+c,\quad e=\tfrac{1}{2}(a-b)+c,\quad f=\tfrac{1}{2}(a+b)-c,$$
$$g=\tfrac{1}{3}(a-b),\quad h=\tfrac{1}{3}(a+2b)+c. \tag{13.1}$$

We also use the notation

$$x'=x-\rho v^2,$$

where x stands for any of the constants $a, b, \ldots, h$.

13.1. *Components relative to a rotated basis*

Let $k=\{\mathbf{k}_1,\mathbf{k}_2,\mathbf{k}_3\}$ be the orthonormal basis in which $\mathbf{k}_1=\mathbf{j}$, $\mathbf{k}_2=\mathbf{k}$ and $\mathbf{k}_3=\pm\mathbf{j}\wedge\mathbf{k}$, the sign being chosen so that k is congruent to the basis e defined at the beginning of Section 6. There is associated with the bases k and e a proper orthogonal tensor $\mathbf{K}=\mathbf{k}_r\otimes\mathbf{e}_r$ with components $K_{ij}=\mathbf{e}_i.\mathbf{k}_j$ in relation to e.

From the definitions (3.4) and the formulae (6.8) and (6.10), the components relative to e of the modified acoustical tensors $\mathbf{Q}(\varphi,v)$ and $\mathbf{R}(\varphi,v)$ for a cubic elastic material are

$$Q_{ij}(\varphi,v)=\{(c-\rho v^2\cos^2\varphi)+(a-b)m_i^2\}\delta_{ij}+bm_im_j,$$
$$R_{ij}(\varphi,v)=\{-\rho v^2\sin\varphi\cos\varphi+(a-b)l_im_i\}\delta_{ij}+(b-c)l_im_j+cl_jm_i,$$

where

$$\mathbf{l}=\sin\varphi\,\mathbf{k}_1-\cos\varphi\,\mathbf{k}_2,\quad \mathbf{m}=\cos\varphi\,\mathbf{k}_1+\sin\varphi\,\mathbf{k}_2$$

are orthogonal unit vectors spanning the reference plane R. The components of these tensors relative to **k** are

$$\begin{aligned}\tilde{Q}_{ij}(\varphi,v)&=K_{pi}K_{qj}Q_{pq}(\varphi,v)\\&=(c-\rho v^2\cos^2\varphi)\delta_{ij}+b\tilde{m}_i\tilde{m}_j+(a-b)\sum_{r=1}^{3}m_r^2K_{ri}K_{rj},\end{aligned} \tag{13.2}$$

$$\begin{aligned}\tilde{R}_{ij}(\varphi,v)&=K_{pi}K_{qj}R_{pq}(\varphi,v)\\&=-\rho v^2\sin\varphi\cos\varphi\,\delta_{ij}+(b-c)\tilde{l}_i\tilde{m}_j+c\tilde{l}_j\tilde{m}_i\\&\quad+(a-b)\sum_{r=1}^{3}l_rm_rK_{ri}K_{rj},\end{aligned} \tag{13.3}$$

where

$$\tilde{l}_i=K_{pi}l_p=\sin\varphi\,\delta_{i1}-\cos\varphi\,\delta_{i2},\quad \tilde{m}_i=K_{pi}m_p=\cos\varphi\,\delta_{i1}+\sin\varphi\,\delta_{i2}. \tag{13.4}$$

We observe that, in the right sides of (13.2) and (13.3), the 23, 32, 31, and 13 components of each term except, possibly, the last are zero. When the indicated components of the final terms also vanish, the advantage of working with the rotated basis k is evident. For the particular cases to be considered we take:

$$\text{Case 1.}\quad \mathbf{k}_1 = 2^{-\frac{1}{2}}(\mathbf{e}_1+\mathbf{e}_3),\quad \mathbf{k}_2=\mathbf{e}_2,\quad \mathbf{k}_3 = 2^{-\frac{1}{2}}(-\mathbf{e}_1+\mathbf{e}_3); \tag{13.5}$$

$$\text{Case 2.}\quad \mathbf{k}_1 = 2^{-\frac{1}{2}}(\mathbf{e}_1+\mathbf{e}_2),\quad \mathbf{k}_2 = 2^{-\frac{1}{2}}(-\mathbf{e}_1+\mathbf{e}_2),\quad \mathbf{k}_3=\mathbf{e}_3; \tag{13.6}$$

$$\text{Case 3.}\quad \mathbf{k}_1 = 6^{-\frac{1}{2}}(\mathbf{e}_1+\mathbf{e}_2+2\mathbf{e}_3),\quad \mathbf{k}_2 = 3^{-\frac{1}{2}}(-\mathbf{e}_1-\mathbf{e}_2+\mathbf{e}_3),\quad \mathbf{k}_3 = 2^{-\frac{1}{2}}(\mathbf{e}_1-\mathbf{e}_2); \tag{13.7}$$

and in each instance it is duly found, from (13.2) to (13.4), that $\tilde{Q}_{ij}(\varphi, v)$ and $\tilde{R}_{ij}(\varphi, v)$ are zero when $i = 3 \neq j$ or $j = 3 \neq i$. We say that the tensors $\mathbf{Q}(\varphi, v)$ and $\mathbf{R}(\varphi, v)$ are *plane* relative to the base vector $\mathbf{k}_3$.

Inspection of the definition (3.3) shows that $\mathbf{S}_1(v)$ is also plane relative to $\mathbf{k}_3$ which implies that

$$\tilde{S}^{(1)}_{33}(v) = -\pi^{-1}\int_{-\frac{1}{2}\pi}^{\frac{1}{2}\pi}\{\tilde{R}_{33}(\varphi, v)/\tilde{Q}_{33}(\varphi, v)\}d\varphi$$

is an eigenvalue. We find from (13.2) to (13.7) that, in each of Cases 1, 2, and 3,

$$\tilde{R}_{33}(\varphi, v) = -\tfrac{1}{2}(\partial/\partial\varphi)\tilde{Q}_{33}(\varphi, v),\quad \tilde{Q}_{33}(\tfrac{1}{2}\pi, v) = \tilde{Q}_{33}(-\tfrac{1}{2}\pi, v).$$

The 33 component of $\mathbf{S}_1(v)$ is therefore the zero eigenvalue implied by $(3.5)_1$, and since $\mathbf{S}_1(v)$ is traceless the definition $(3.8)_1$ takes the form

$$F(v) = 1+\{\tilde{S}^{(1)}_{11}(v)\}^2+\tilde{S}^{(1)}_{12}(v)\tilde{S}^{(1)}_{21}(v) \tag{13.8}$$

when referred to k.

The integral representations of the components of $\mathbf{S}_1(v)$ appearing in (13.8) are given by (3.3) as

$$\tilde{S}^{(1)}_{\alpha\beta}(v) = -\pi^{-1}\int_{-\frac{1}{2}\pi}^{\frac{1}{2}\pi}\Delta^{-1}(\varphi, v)s_{\alpha\beta}(\varphi, v)d\varphi, \tag{13.9}$$

with

$$\Delta = \tilde{Q}_{11}\tilde{Q}_{22}-\tilde{Q}^2_{12} \tag{13.10}$$

and

$$s_{11} = \tilde{Q}_{22}\tilde{R}_{11}-\tilde{Q}_{12}\tilde{R}_{12},\quad s_{12} = \tilde{Q}_{22}\tilde{R}_{21}-\tilde{Q}_{12}\tilde{R}_{22},\quad s_{21} = \tilde{Q}_{11}\tilde{R}_{12}-\tilde{Q}_{12}\tilde{R}_{11}, \tag{13.11}$$

the arguments φ, v being suppressed in (13.10) and (13.11). The calculation of Δ and the $s_{\alpha\beta}$'s from (13.2) to (13.7) is straightforward and we pass directly to the results.

Case 1

$$\begin{aligned}\Delta &= (a+c)c\sin^4\varphi-\{b^2-cc'-(a+c)d'\}\sin^2\varphi\cos^2\varphi+c'd'\cos^4\varphi,\\ s_{12} &= (a+c)c\sin^2\varphi-(b-c)c'\cos^2\varphi,\\ s_{21} &= c\{(b-c)\sin^2\varphi-d'\cos^2\varphi\}.\end{aligned} \tag{13.12}$$

Case 2

$$\begin{aligned}\Delta &= de\sin^4\varphi-(a^2-dd'-ee')\sin^2\varphi\cos^2\varphi+d'e'\cos^4\varphi,\\ s_{12} &= de\sin^2\varphi-fe'\cos^2\varphi,\\ s_{21} &= e(f\sin^2\varphi-d'\cos^2\varphi).\end{aligned} \tag{13.13}$$

Case 3

$$\begin{aligned}
\Delta &= (c+g)h\sin^4\varphi + 2^{\frac{1}{2}}gh\sin^3\varphi\cos\varphi \\
&\quad + \{(c+g)(c+g') - (b+2g)^2 + hd'\}\sin^2\varphi\cos^2\varphi \\
&\quad - 2^{\frac{1}{2}}g(b+g-c')\sin\varphi\cos^3\varphi + \{(c+g')d' - \tfrac{1}{2}g^2\}\cos^4\varphi, \\
s_{11} &= 2^{-\frac{3}{2}}gh\sin^2\varphi + \{2c(b+g) + \tfrac{1}{4}gh - \tfrac{1}{2}b\rho v^2\}\sin\varphi\cos\varphi \\
&\quad + 2^{-\frac{3}{2}}g(2c-h')\cos^2\varphi - \tfrac{1}{4}\partial\Delta/\partial\varphi, \\
s_{12} &= (c+g)h\sin^2\varphi + 2^{-\frac{1}{2}}gh\sin\varphi\cos\varphi + (2c-h)(c+g')\cos^2\varphi, \\
s_{21} &= -(c+g)(2c-h)\sin^2\varphi - 2^{-\frac{1}{2}}g(2c-h)\sin\varphi\cos\varphi \\
&\quad - \{(c+g)d' - \tfrac{1}{2}g^2\}\cos^2\varphi.
\end{aligned} \tag{13.14}$$

The expressions for s_{11} omitted from (13.12) and (13.13) are odd functions of φ about $\varphi = 0$, whence, due to the evenness of Δ in Cases 1 and 2, $\tilde{S}^{(1)}_{11}(v) = 0$ and (13.8) simplifies to

$$F(v) = 1 + \tilde{S}^{(1)}_{12}(v)\tilde{S}^{(1)}_{21}(v). \tag{13.15}$$

It is plain that Cases 1 and 2 are mathematically similar and much simpler than Case 3.

13.2. *The second transonic state*

It follows from (2.11) and $(3.4)_1$ that

$$\det \mathbf{Q}(\varphi, v) = \rho^3 \prod_{r=1}^{3} \{c_r^2(\varphi) - v^2\cos^2\varphi\}, \tag{13.16}$$

where $c_i(\varphi)$ are the speeds of the body waves with wave normal $\cos\varphi\mathbf{j} + \sin\varphi\mathbf{k}$. When $\mathbf{Q}(\varphi, v)$ is plane relative to $\mathbf{k}_3$ we also have

$$\det \mathbf{Q}(\varphi, v) = \Delta(\varphi, v)\tilde{Q}_{33}(\varphi, v), \tag{13.17}$$

with $\Delta(\varphi, v)$ given by (13.10). It is verified below that in each of Cases 1, 2, and 3 there is an interval Φ of the angle φ, including the critical value $\hat{\varphi}$, in which

$$\tilde{Q}_{33}(\varphi, v) = \rho\{c_1^2(\varphi) - v^2\cos^2\varphi\}, \tag{13.18}$$

and we recall in this connection the convention $(2.12)_1$. We deduce from (13.16) to (13.18) that

$$\Delta(\varphi, v) = \rho^2\{c_2^2(\varphi) - v^2\cos^2\varphi\}\{c_3^2(\varphi) - v^2\cos^2\varphi\} \quad \forall\, \varphi \in \Phi. \tag{13.19}$$

In the second transonic state, defined in Section 3.4 and illustrated in Fig. 3.1(b), the projection on the surface wave normal $\mathbf{j}$ of the slowness vector of the medial branch of body wave propagation attains a maximum. The critical values of v and φ in this state thus satisfy the relations

$$\tilde{v} = c_2(\tilde{\varphi})\sec\tilde{\varphi}, \quad (d/d\varphi)\{c_2(\varphi)\sec\varphi\}\big|_{\varphi=\hat{\varphi}} = 0 \tag{13.20}$$

(cf. (3.2) and CS, equations (6.14), $(6.15)_1$). Differentiation of (13.19) yields, after rearrangement,

$$\begin{aligned}
\tfrac{1}{2}\rho^{-2}(\partial/\partial\varphi)\Delta(\varphi, v) &= c_2(\varphi)\cos\varphi\,[(d/d\varphi)\{c_2(\varphi)\sec\varphi\} \\
&\quad - c_2^{-1}(\varphi)\sin\varphi\{c_2^2(\varphi)\sec^2\varphi - v^2\}][c_3^2(\varphi) - v^2\cos^2\varphi] \\
&\quad + \cos^2\varphi\{c_2^2(\varphi)\sec^2\varphi - v^2\}\{c_3(\varphi)c_3'(\varphi) \\
&\quad + v^2\sin\varphi\cos\varphi\},
\end{aligned}$$

and in view of (13.20) we arrive at the conditions

$$\Delta = 0, \quad \partial\Delta/\partial\varphi = 0 \quad \text{at} \quad \varphi = \tilde{\varphi}, v = \tilde{v}. \tag{13.21}$$

14. Case 1. The Surface Wave along a Face Diagonal in a Face

The further analysis of Case 1 given in this Section centres on the calculation of $F(.)$, and in order to interpret the behaviour of this function we also examine the second transonic state. It is shown that when a subsonic surface wave does not exist, there is a unique supersonic surface wave, a conclusion implicit in earlier work (Chadwick, 1976b), but supported there only by *ad hoc* arguments.

14.1. *Transonic states*

In Case 1 the reference plane is a face bisector; the normals $\mathbf{j}(=\mathbf{k}_1)$ and $\mathbf{k}(=\mathbf{k}_2)$ are directed along a face diagonal and an axis respectively, and we are concerned with cubic materials in C_3, D_a, and C_4. Equations (13.2), (13.4), and (13.5) yield

$$\tilde{Q}_{33}(\varphi, v) = c \sin^2 \varphi + e' \cos^2 \varphi,$$

and we gather from Figs. 8.4(c) to 8.6(c), Table 7.3, and equations $(7.7)_2$ and $(13.1)_2$ (keeping in mind Fig. 3.1(a) also) that

$$\rho c_1^2(\varphi) = c \sin^2 \varphi + e \cos^2 \varphi \quad \text{for} \quad -\tfrac{1}{2}\pi + \theta_c < \varphi < \tfrac{1}{2}\pi - \theta_c. \tag{14.1}$$

The relation (13.18) is therefore verified and the interval Φ is, in this case, $(-\frac{1}{2}\pi + \theta_c, \frac{1}{2}\pi - \theta_c)$. The critical angle and the limiting speed are, from (14.1) and CS, equations (6.13) and (6.14),

$$\hat{\varphi} = 0, \quad \hat{v} = (e/\rho)^{\frac{1}{2}}. \tag{14.2}$$

The discussion given in Section 8.2, in conjunction with Figs. 3.1(b) and 8.4(c) to 8.6(c), shows that the second transonic state is of Type 1 for C_3 materials violating the condition (8.8) and is otherwise of Type 4. In the former case,

$$\tilde{\varphi} = 0, \quad \tilde{v} = (c/\rho)^{\frac{1}{2}}, \tag{14.3}$$

and in both cases $\tilde{\varphi} \in \Phi$. We learn from Table 9.2 that the medial branch of the polarization field is of type E_2 when $a < b$, and, as Fig. 10.1 makes clear, the interior of the edge FB of $\mathscr{U}_e$ (corresponding to the positive angles in Φ) contains no exceptional wave solutions. On the face diagonal defined by $\mathbf{j}$, $\mathbf{p} = \mathbf{k}$ and we infer from $(5.2)_1$ that the planes relative to which this transverse wave is exceptional do not include the boundary of the body. We conclude that, irrespective of its type, the second transonic state is normal.

14.2. *Determination of the function* $F(.)$

In order to determine $\tilde{S}_{12}^{(1)}(v)$ and $\tilde{S}_{21}^{(1)}(v)$ from (13.9) and (13.12), we employ the functions $\alpha(.)$ and $\beta(.)$ defined by

$$\{\alpha^{-1}(v), \beta^{-1}(v)\} = \pi^{-1} \int_{-\frac{1}{2}\pi}^{\frac{1}{2}\pi} \Delta^{-1}(\varphi, v)\{\sin^2 \varphi, \cos^2 \varphi\} d\varphi. \tag{14.4}$$

Then, omitting arguments from now on,

$$\tilde{S}^{(1)}_{12}(v) = -(a+c)c\alpha^{-1} + (b-c)c'\beta^{-1}, \quad \tilde{S}^{(1)}_{21}(v) = -c\{(b-c)\alpha^{-1} - d'\beta^{-1}\}. \tag{14.5}$$

The reasoning developed in Section 13.2 assures us that $\Delta > 0$ for all φ and for all v in $\bar{i}$, so we can set

$$\Delta = (p_1^2 \sin^2\varphi + q_1^2 \cos^2\varphi)(p_2^2 \sin^2\varphi + q_2^2 \cos^2\varphi), \tag{14.6}$$

where p_1, p_2, q_1, and q_2 are real. With the use of the formula

$$\pi^{-1}\int_{-\frac{1}{2}\pi}^{\frac{1}{2}\pi} (p^2 \sin^2\varphi + q^2\cos^2\varphi)^{-1} d\varphi = (pq)^{-1} \quad (p>0,\ q>0),$$

we derive from (14.4) and (14.6) the relations

$$p_1^2\alpha^{-1} + q_1^2\beta^{-1} = (p_2q_2)^{-1}, \quad p_2^2\alpha^{-1} + q_2^2\beta^{-1} = (p_1q_1)^{-1},$$

leading to

$$\alpha = p_1p_2r, \quad \beta = q_1q_2r \quad \text{with} \quad r = p_1q_2 + p_2q_1. \tag{14.7}$$

A comparison of $(13.12)_1$ and (14.6) provides the connections

$$\begin{aligned} p_1p_2 &= \{(a+c)c\}^{\frac{1}{2}}, \quad q_1q_2 = (c'd')^{\frac{1}{2}}, \\ r^2(v) &= -b^2 + cc' + (a+c)d' + 2\{(a+c)cc'd'\}^{\frac{1}{2}}, \end{aligned} \tag{14.8}$$

which, with (14.5) and (14.7), enable us to write down explicit formulae for $\tilde{S}^{(1)}_{12}(v)$ and $\tilde{S}^{(1)}_{21}(v)$.

On entering these expressions into (13.15) we find, after some simplification, that

$$F(v) = \{(a+c)c'd'\}^{-\frac{1}{2}}r^{-2}(v)[(cc')^{\frac{1}{2}} + \{(a+c)d'\}^{\frac{1}{2}}]G(v), \tag{14.9}$$

with

$$G(v) = c'^{\frac{1}{2}}\{(a+c)d' - (b-c)^2\} - \rho v^2\{(a+c)cd'\}^{\frac{1}{2}}. \tag{14.10}$$

The secular equation for a surface wave propagating along a face diagonal in a face is thus $G(v) = 0$, a result first obtained, by an entirely different method, by Stoneley (1955, equation (22)). It may be readily proved that, subject to the inequalities (6.4), $G(v)$ is positive at $v = 0$, negative at $v = (c/\rho)^{\frac{1}{2}}$, and zero at exactly one speed between these values. Equations (14.9) and (14.10) hold for all cubic materials. When $a > b$, $\hat{v} = (c/\rho)^{\frac{1}{2}}$, and there exists a unique subsonic surface wave, as predicted on general grounds in Section 12.1.

14.3. *Further discussion of Case* 1

When $a < b$, regard must be had to the singularities of $F(.)$. As v increases from the limiting speed given by $(14.2)_2$, $F(v)$ is first singular at either $v = (c/\rho)^{\frac{1}{2}}$ or at a zero of $r(v)$, and in view of the considerations set out in Section 14.1, these possibilities might be expected to correspond to the second transonic state being of Type 1 or of Type 4.

Noting first, by appeal to $(14.8)_3$ and $(13.1)_1$, that the inequality (8.8) may be restated as $r^2((c/\rho)^{\frac{1}{2}}) < 0$, we turn for further clarification to the conditions (13.21). Substituting for Δ from (14.6), putting $P = p_1p_2$, $Q = q_1q_2$, and using the definition $(14.7)_3$, we arrive at the relations

$$\begin{aligned} (P\sin^2\tilde{\varphi} - Q\cos^2\tilde{\varphi})^2 + r^2\sin^2\tilde{\varphi}\cos^2\tilde{\varphi} &= 0, \\ 2\sin\tilde{\varphi}\cos\tilde{\varphi}\{2(P+Q)(P\sin^2\tilde{\varphi} - Q\cos^2\tilde{\varphi}) + r^2(\cos^2\tilde{\varphi} - \sin^2\tilde{\varphi})\} &= 0. \end{aligned} \tag{14.11}$$

There are two solutions, the first given by (14.3), the second by

$$r(\tilde{v}) = 0, \quad \tilde{\varphi} = \tan^{-1}(Q/P)^{\frac{1}{2}} = \tan^{-1}\{c'd'/(a+c)c\}^{\frac{1}{4}}, \tag{14.12}$$

and the anticipated interpretation of the singularities of $F(.)$ is thus confirmed.

A unique surface wave exists in Case 1 which is subsonic or supersonic according as the zero of $G(v)$ in $(0, (c/\rho)^{\frac{1}{2}})$ is less than or greater than $\hat{v}$. From (14.10) and $(14.2)_2$, the criterion for the existence of a supersonic wave is therefore

$$\{2(b-a)\}^{\frac{1}{2}}\{(a+c)b-(b-c)^2\} - (a-b+2c)\{(a+c)bc\}^{\frac{1}{2}} > 0. \tag{14.13}$$

The interval in which $F(.)$ is defined as a real-valued function is $[0, (c/\rho)^{\frac{1}{2}})$, from which must be deleted the smaller zero of r if $\tilde{v}$ is given by $(14.12)_1$.

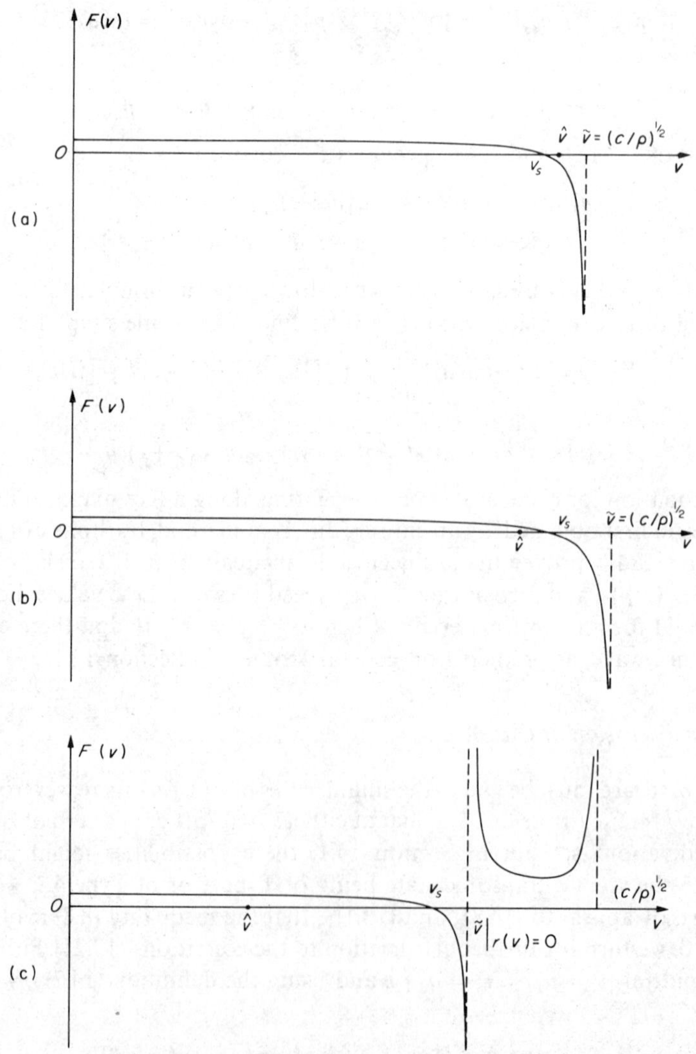

FIG. 14.1. Possible forms of the variation of $F(v)$ with v in Case 1.

The variation of $F(v)$ with v has been computed for each of the 25 C_3 materials listed in Section 10.3, and the results fall into the three classes revealed by the above analysis and illustrated by the diagrams in Fig. 14.1. Materials displaying the behaviour shown in Fig. 14.1(a) fail to meet the condition (14.13) and transmit a subsonic surface wave. When (14.13) is satisfied, the surface wave is supersonic and the form of $F(.)$ depends on the relative magnitudes of $(c/\rho)^{\frac{1}{2}}$ and the smaller zero of $r(v)$. The two possibilities, depicted in parts (b) and (c) of Fig. 14.1, reflect the occurrence of c'^{-1} and r^{-1} on the right side of (14.9) as fractional and integer powers, and they correspond respectively to the second transonic state being of Type 1 or of Type 4. Complete consistency is found with the results of Farnell (1970, Section V, B, E). Of the materials studied, only BaF_2, C, and Fe_3O_4 admit a subsonic surface wave, and only in Al, Cr_2FeO_4, and MgO is the second transonic state of Type 1. The state of affairs represented in Fig. 14.1(c) therefore appears to be most typical of Case 1.

15. Case 2. The Surface Wave along a Face Diagonal in a Face Bisector

Because of the mathematical resemblance of Case 2 to Case 1, this Section follows much the same course as its predecessor and arrives at broadly similar results. We concentrate on the essential features of the situation and omit much of the analytical detail.

15.1. *Transonic states*

Case 2 refers to cubic elastic materials in C_1, D_b, and C_2. The reference plane is a face and the normals $\mathbf{j}(=\mathbf{k}_1)$ and $\mathbf{k}(=\mathbf{k}_2)$ are directed along face diagonals. Figures 8.2(b) and 8.3(b) and Table 7.1 show that S_1 is a circle of radius $(\rho/c)^{\frac{1}{2}}$, and on calculating $Q_{33}(\varphi, v)$ from (13.2), (13.4), and (13.6) we corroborate the relation (13.18). The interval Φ is $(-\frac{1}{2}\pi, \frac{1}{2}\pi)$ and

$$\hat{\varphi} = 0, \quad \hat{v} = (c/\rho)^{\frac{1}{2}}. \tag{15.1}$$

The second transonic state may be of Type 1 or Type 4. For C_2 materials disobeying the condition (8.7), L touches S_2 at a single point and

$$\tilde{\varphi} = 0, \quad \tilde{v} = (e/\rho)^{\frac{1}{2}}. \tag{15.2}$$

Otherwise, there are two points of contact. In each case, $\tilde{\varphi} \in \Phi$. The following considerations, based on Table 9.2 and Fig. 10.1, ensure that the second transonic state is always normal. When $a > b$, the medial branch of the polarization field is of type O_1 or O_2, and the only exceptional wave normals in a face are directed along the axes and face diagonals. The axial directions are not relevant to Case 2, however, and the traction-free planes associated with the exceptional wave in O_2 with wave normal $\mathbf{k}_1$ have unit normal $\mathbf{k}_3$, not $\mathbf{k}_2$.

15.2. *The function F(.)*

To determine $F(v)$ we proceed exactly as in Section 14.2, calculating $\tilde{S}^{(1)}_{12}(v)$ and $\tilde{S}^{(1)}_{21}(v)$ from (13.9) and (13.13), carrying out the integrations with the aid of the functions defined in (14.4), and substituting the eventual expressions into (13.15). The result finally obtained is

$$F(v) = (dd'e')^{-\frac{1}{2}}r^{-2}(v)\{(dd')^{\frac{1}{2}} + (ee')^{\frac{1}{2}}\}G(v), \tag{15.3}$$

with

$$G(v) = e'^{\frac{1}{2}}(dd' - f^2) - \rho v^2 (ded')^{\frac{1}{2}} \tag{15.4}$$

and

$$r^2(v) = -a^2 + dd' + ee' + 2(ded'e')^{\frac{1}{2}}. \tag{15.5}$$

Equations (15.3) to (15.5) are valid for all cubic media. When $a < b$, the definitions $(13.1)_{1,2}$ give $e < c < d$, and $F(.)$, $G(.)$, and $r(.)$ are thus defined as real-valued functions in the interval $[0, (e/\rho)^{\frac{1}{2}})$ which, *pace* $(15.1)_2$, does not contain $\hat{v}$. Since, moreover, $G(v)$ changes sign just once between 0 and $(e/\rho)^{\frac{1}{2}}$, a unique subsonic surface wave exists, in agreement with the general discussion in Section 12.1.

Returning to Case 2, we deduce from $(13.1)_{1,2}$ that $c < d < e$, $c < d = e$, and $c < e < d$ for materials in C_1, D_b, and C_2, respectively. The domain in $v \geqslant 0$ of $G(.)$ and $r(.)$ is hence $[0, (d/\rho)^{\frac{1}{2}})$ for C_1 and D_b, and $[0, (e/\rho)^{\frac{1}{2}})$ for C_2 materials. In each case, $\hat{v}$ and precisely one zero of $G(v)$ are included. The interpretation of the second transonic state offered by (13.21) applies to Case 2 as well as Case 1. Equations (15.2) hold when this state is of Type 1 and (14.12), with the final member of $(14.12)_2$ replaced by $\tan^{-1}(d'e'/de)^{\frac{1}{4}}$, when it is of Type 4. It is easy to verify, from (15.5) and $(13.1)_{1,2}$, that the inequality (8.7) is equivalent to $r^2((e/\rho)^{\frac{1}{2}}) < 0$. When the second transonic state is of Type 1, $F(.)$ has the same domain as $G(.)$ and $r(.)$, and $F(v) \to -\infty$ as v approaches the smaller of $(d/\rho)^{\frac{1}{2}}$ and $(e/\rho)^{\frac{1}{2}}$. When this state is of Type 4, the domain of $F(.)$ is obtained by deleting from the common domain of $G(.)$ and $r(.)$ the smaller zero of $r(v)$. Regardless of the type, $F(v)$ has exactly one zero in its domain and there is accordingly a unique surface wave, subsonic or supersonic according as this zero is less than or greater than $\hat{v}$. The condition for the existence of a supersonic surface wave is thus, from (15.4), $(15.1)_2$, and $(13.1)_{1\text{-}3}$,

$$(a-b)^{\frac{1}{2}}(3a+3b-2c) - \{(a+b)(a+b+2c)(a-b+2c)\}^{\frac{1}{2}} > 0. \tag{15.6}$$

Numerical evaluation of $F(v)$ over its domain has been carried out for the 25 C_2 materials specified in Section 10.3, and the results indicate that the behaviour depicted in Fig. 14.1(c) (with $(e/\rho)^{\frac{1}{2}}$ replacing $(c/\rho)^{\frac{1}{2}}$) is typical of Case 2. Indeed, only one of these media (W, which is very nearly isotropic) transmits a subsonic surface wave, the second transonic state then being of Type 1, as in Fig. 14.1(a). In each of the other solids a supersonic surface wave occurs in conjunction with a second transonic state of Type 4. As in Case 1, our findings agree in full with the results of Farnell (1970, Section V, C, E).

16. Case 3. The Surface Wave along an Octahedral Diagonal in an Octahedral Plane

We have remarked in Section 13.1 that Case 3 is mathematically more complex than the other cases and, in essence, this is because the direction of surface wave propagation no longer coincides with a line of symmetry of the slowness section. The results obtained in this Section are consequently less explicit than their counterparts for Cases 1 and 2, but they constitute, nevertheless, a tolerably complete analysis of Case 3.

16.1. *Transonic states*

Case 3 can arise when $a > b > 0$ (i.e. for C_2 materials) and when $a < b$ (corresponding to C_3, D_a, and C_4 materials). The reference plane R is a face bisector with $\mathbf{j}(=\mathbf{k}_1)$ acting along the octahedral diagonal in which R intersects the octahedral plane with normal $\mathbf{k}(=\mathbf{k}_2)$. The arc of S_1 on which the transonic state occurs is elliptical, and the orientations in R of **j**, **k**, and **n**, the wave normal of the exceptional limiting wave, are shown in Fig. 11.1, parts a and b referring to the possibilities $a > b > 0$ and $a < b$, respectively. From these diagrams and the definitions of Classes 2 and 3 in Section 11.2 we derive the relation

$$\rho c_1^2(\varphi) = c\cos^2\theta + e\sin^2\theta, \quad \varphi = \tfrac{1}{2}\pi - \theta_c - \theta, \tag{16.1}$$

use being made of $(13.1)_2$. When $\tilde{Q}_{33}(\varphi, v)$ is calculated from (13.2), (13.4), and (13.7), the connection (13.18) is verified, and inspection of Fig. 11.1 shows the interval Φ to be $(\tfrac{1}{2}\pi - 2\theta_c, \tfrac{1}{2}\pi)$ or $(-\tfrac{1}{2}\pi, \tfrac{1}{2}\pi - 2\theta_c)$ according as $a > b > 0$ or $a < b$. When $\varphi = \hat{\varphi}$, (12.1) gives $\theta = \tan^{-1}(2^{-\frac{1}{2}}c/e)$, and we deduce from (16.1) and (3.2) that

$$\hat{v} = \left\{\frac{3ce}{\rho(c+2e)}\right\}^{\frac{1}{2}}, \quad \hat{\varphi} = \tan^{-1}\left(2^{\frac{1}{2}}\frac{e-c}{c+2e}\right).$$

When $a > b > 0$, it is evident from Fig. 11.1(a) that the second transonic state is of Type 1 and $\hat{\varphi} \in \Phi$. The medial branch of the polarization field is then, *vide* Table 9.2, of type O_2 and a glance at Fig. 10.1 confirms that the transonic state is normal. When $a < b$, $\hat{\varphi} \notin \Phi$ and the second encounter of L with S falls, like the first, on S_1 (see Fig. 11.1(b)).

16.2. *Determination of the function F(.)*

On substituting the expressions $(13.14)_{2\text{–}4}$ into (13.9) we obtain

$$\begin{aligned}\tilde{S}_{11}^{(1)} &= -2^{-\frac{3}{2}}gh\lambda - \{2c(b+g) + \tfrac{1}{4}gh - \tfrac{1}{2}b\rho v^2\}v - 2^{-\frac{3}{2}}g(2c-h')\mu,\\ \tilde{S}_{12}^{(1)} &= -(c+g)h\lambda - 2^{-\frac{1}{2}}ghv - (2c-h)(c+g')\mu,\\ \tilde{S}_{21}^{(1)} &= (c+g)(2c-h)\lambda + 2^{-\frac{1}{2}}g(2c-h)v + \{(c+g)d' - \tfrac{1}{2}g^2\}\mu,\end{aligned} \tag{16.2}$$

where

$$\{\lambda(v), v(v), \mu(v)\} = \pi^{-1}\int_{-\frac{1}{2}\pi}^{\frac{1}{2}\pi} \Delta^{-1}(\varphi, v)\{\sin^2\varphi, \sin\varphi\cos\varphi, \cos^2\varphi\}\,d\varphi, \tag{16.3}$$

and arguments have been omitted from (16.2). Since $\Delta(\tfrac{1}{2}\pi, v) = \Delta(-\tfrac{1}{2}\pi, v)$, from $(13.14)_1$, the term $-\tfrac{1}{4}\partial\Delta/\partial\varphi$ in s_{11} makes no contribution to $\tilde{S}_{11}^{(1)}$. As in Cases 1 and 2, $\Delta > 0$ for all φ and for all $v \in \tilde{i}$. We can therefore express $(13.14)_1$ in the form

$$\Delta = (c+g)h\prod_{\alpha=1}^{2}(\sin^2\varphi + 2A_\alpha\sin\varphi\cos\varphi + B_\alpha\cos^2\varphi), \tag{16.4}$$

where A_1, A_2, B_1, and B_2 are unique and

$$R_\alpha^2 = B_\alpha - A_\alpha^2 > 0 \quad (\alpha = 1, 2). \tag{16.5}$$

The integrals (16.3) may be evaluated by making the change of variable $\xi = \tan\varphi$, decomposing the integrands into partial fractions and using the formula

$$\pi^{-1}\int_{-\infty}^{\infty}\frac{L\xi + M}{\xi^2 + 2A\xi + B}\,d\xi = \frac{M - AL}{(B - A^2)^{\frac{1}{2}}} \quad (B > A^2).$$

We find that

$$\begin{aligned}\lambda &= G^{-1}\{(2CA_1+BB_1)R_1^{-1}-(2CA_2+BB_2)R_2^{-1}\},\\ \mu &= G^{-1}\{(2AA_1-B)R_1^{-1}-(2AA_2-B)R_2^{-1}\},\\ \nu &= -G^{-1}\{(AB_1+C)R_1^{-1}-(AB_2+C)R_2^{-1}\},\end{aligned} \tag{16.6}$$

with

$$A = A_1 - A_2, \quad B = B_1 - B_2, \quad C = A_1B_2 - A_2B_1, \quad G = (c+g)h(B^2+4AC). \tag{16.7}$$

It remains to calculate the coefficients A_1, A_2, B_1, and B_2 from the four relations found by identifying corresponding terms in $(13.14)_1$ and (16.4). We define

$$\begin{aligned}k &= \{(c+g)h\}^{-\frac{1}{4}}\{(c+g')d'-\tfrac{1}{2}g^2\}^{\frac{1}{4}}, \quad l = 2^{-\frac{1}{2}}g\{(c+g)k\}^{-1},\\ m &= \{(c+g)hk^2\}^{-1}\{(c+g)(c+g')-(b+2g)^2+hd'\},\\ n &= -2^{-\frac{1}{2}}g\{(c+g)hk^3\}^{-1}(b+g-c'),\end{aligned} \tag{16.8}$$

and set

$$A_1, A_2 = \tfrac{1}{2}kl(1 \pm y), \quad B_1, B_2 = k^2\{x \pm (x^2-1)^{\frac{1}{2}}\}. \tag{16.9}$$

Then two of the relations are satisfied and the other two impose the requirements

$$l^2(1-y^2)+2x = m, \quad l\{x - y(x^2-1)^{\frac{1}{2}}\} = n. \tag{16.10}$$

The result of eliminating y between (16.10) is the cubic equation

$$2x^3 - mx^2 + 2(ln-1)x - (l^2+n^2-m) = 0 \tag{16.11}$$

for x. We see that (16.11) has a real root exceeding unity, and the uniqueness of the quadratic factors in (16.4) guarantees that no other root has this property. It should also be noted that $k(v)$, defined by $(16.8)_1$, is real in the subsonic interval $\hat{i}$ when $a > b > 0$, and also when $a < b$ and

$$\tfrac{2}{27}(b-a)^2(a+2b) - \tfrac{1}{6}(b-a)(a+5b)c + bc^2 < 0. \tag{16.12}$$

We have here an indication that all C_2 materials admit the transonic state illustrated in Fig. 11.1(a), but that only in the C_3, D_a, and C_4 materials meeting the restriction (16.12) can the state shown in Fig. 11.1(b) arise.

When the appropriate root of (16.11) has been extracted, y is given by $(16.10)_2$ and A_1, A_2, B_1, and B_2 by (16.9). By using successively (16.7), (16.6), (16.2), and (13.8), we can then calculate $F(v)$.

16.3. *Further discussion of Case* 3

The analysis given in Section 16.2 shows that Case 3 is universal in C_2 materials, but unusual when $a < b$. When $a > b > 0$, further information about the second transonic state can be obtained by subjecting the expression (16.4) for Δ to the conditions (13.21). There are three solutions in this case: $\check{v}$ is the smallest of the zeros of $R_1(v)$ and $R_2(v)$, and $\check{\varphi} = \tan^{-1}(-A_\alpha(\check{v}))$, where $R_\alpha(\check{v}) = 0$. This result does not hold when $a < b$ since $\check{\varphi}$ no longer belongs to the interval of validity of (13.18).

The variation of $F(v)$ with v has been determined numerically for each of the C_2 materials recorded in Section 10.3. In every case the graphical output is similar to Fig. 14.1(a) and the surface wave is subsonic. Of the 25 C_3 crystals listed there, only three, CuZn (β-brass), Li, and Na, satisfy the condition (16.12). In Na the surface wave is again subsonic, but CuZn and Li transmit a supersonic surface wave in Case 3 and the $F(v) - v$ curves for these materials have the features shown in Fig. 14.1(b). Our conclusions once more agree with the computations of Farnell (1970, Section V, D, E) which extend to both C_2 and C_3 materials, including Na, but not CuZn and Li.

Acknowledgement

We are grateful to Professor J. Lothe for showing us, in advance of publication, the results on exceptional waves embodied in Section 5.

References

Al'shits, V. I. and Lothe, J. (1978) Surface waves in hexagonal crystals, *Sov. Phys. Crystallogr.* **23**, 509–515.
Al'shits, V. I. and Lothe, J. (1979a) Elastic waves in triclinic crystals. I. General theory and the degeneracy problem, *Sov. Phys. Crystallogr.* **24**, 387–392.
Al'shits, V. I. and Lothe, J. (1979b) Elastic waves in triclinic crystals. II. Topology of polarization fields and some general theorems, *Sov. Phys. Crystallogr.* **24**, 393–398.
Al'shits, V. I. and Lothe, J. (1979c) Elastic waves in triclinic crystals. III. The problem of existence of exceptional surface waves and some of their general properties, *Sov. Phys. Crystallogr.* **24**, 644–648.
Auld, B. A. (1973) *Acoustic Fields and Waves in Solids*, Vol. 1, Wiley, New York.
Barnett, D. M. and Lothe, J. (1973) Synthesis of the sextic and integral formalism for dislocations, Green's functions, and surface waves in anisotropic elastic solids. *Phys. Norv.* **8**, 13–22.
Barnett, D. M. and Lothe, J. (1974) Consideration of the existence of surface wave (Rayleigh wave) solutions in anisotropic elastic crystals, *J. Phys. F* **4**, 671–686.
Barnett, D. M., Lothe, J., Nishioka, K., and Asaro, R. J. (1973) Elastic surface waves in anisotropic crystals: A simplified method for calculating Rayleigh velocities using dislocation theory, *J. Phys. F* **3**, 1083–1096.
Buchwald, V. T. and Davis, A. (1963) Surface waves in elastic media with cubic symmetry, *Q. Jl. Mech. Appl. Math.* **16**, 283–294.
Chadwick, P. (1976a) *Continuum Mechanics. Concise Theory and Problems*, Allen & Unwin, London.
Chadwick, P. (1976b) The existence of pure surface modes in elastic materials with orthorhombic symmetry, *J. Sound Vib.* **47**, 39–52.
Chadwick, P. and Smith, G. D. (1977) Foundations of the theory of surface waves in anisotropic elastic materials, *Advances in Applied Mechanics* **17**, 303–376, Academic Press, New York.
Chadwick, P. and Smith, G. D. (1979) Circles on the slowness surface of a cubic elastic material, *J. Elasticity* **9**, 393–396.
Chinn, W. G. and Steenrod, N. E. (1966) *First Concepts of Topology*, Singer, New York.
de Klerk, J. and Musgrave, M. J. P. (1955) Internal conical refraction of transverse elastic waves in a cubic crystal, *Proc. Phys. Soc.* B **68**, 81–88.
Dieulesaint, E. and Royer, D. (1974) *Ondes Élastiques dans les Solides*, Masson et Cie, Paris.
Duff, G. F. D. (1960) The Cauchy problem for elastic waves in an anisotropic medium, *Phil. Trans. R. Soc. Lond.* A **252**, 249–273.
Farnell, G. W. (1970) Properties of elastic surface waves, *Physical Acoustics* **VI**, 109–166, edited by W. P. Mason and R. N. Thurston, Academic Press, New York and London.
Fedorov, F. I. (1968) *The Theory of Elastic Waves in Crystals*, Plenum Press, New York.
Gazis, D. C., Herman, R., and Wallis, R. F. (1960) Surface elastic waves in cubic crystals, *Phys. Rev.* **119**, 533–544.
Holm, P. and Lothe, J. (1980) The topological nature of the polarization field for body waves in anisotropic elastic media, *Proc. R. Soc. Lond.* A **370**, 331–350.
Huntington, H. B. (1958) *The Elastic Constants of Crystals*, Academic Press, New York and London.
Lothe, J. and Al'shits, V. I. (1977) Existence criterion for quasibulk surface waves, *Sov. Phys. Crystallogr.* **22**, 519–525.
Miller, G. F. and Musgrave, M. J. P. (1956) On the propagation of elastic waves in aeolotropic media. III. Media of cubic symmetry, *Proc. R. Soc. Lond.* A **236**, 352–383.
Musgrave, M. J. P. (1970) *Crystal Acoustics*, Holden-Day, San Francisco.

Musgrave, M. J. P. (1981) On an elastodynamic classification of orthorhombic media, *Proc. R. Soc. Lond.* A **374**, 401–429.
Schouten, J. A. (1954) *Tensor Analysis for Physicists* (2nd edn.), Clarendon Press, Oxford.
Smith, G. D. (1978) Surface waves in cubic elastic media, Ph.D. thesis, University of East Anglia, Norwich.
Stoneley, R. (1955) The propagation of surface elastic waves in a cubic crystal, *Proc. R. Soc. Lond.* A **232**, 447–458.
Stroh, A. N. (1962) Steady state problems in anisotropic elasticity, *J. Math. Phys.* **41**, 77–103.

Universal Bifurcation Problems

by D. R. J. CHILLINGWORTH

Department of Mathematics, University of Southampton

Summary

Recent advances in singularity theory have led to new insights in perturbed bifurcation theory. We outline some of the main ideas, showing how the natural geometric notion of transversality leads to computable criteria for rigorously transforming multi-parameter bifurcation problems into tractable form by relating them to certain universal problems. This has implications also for imperfection-sensitivity and structural stability.

1. Introduction and Formulation of the Problem

Singularity theory is about local behaviour of differentiable nonlinear mappings. Bifurcation theory involves understanding the local structure of the zero sets of nonlinear mappings (operators), and so it is to be expected that singularity theory should provide useful insights into bifurcation problems. However, during its flowering throughout the last decade, singularity theory has seemingly failed to address itself to the problems that concern applied mathematicians grappling with nonlinear equations varying with one or many parameters. The theory of universal unfoldings of singularities has not answered the right questions for bifurcation theory.

Recently, though, the scene has changed, notably with the elegant work of Golubitsky and Schaeffer (1979a, b). They have shown how the techniques of singularity theory can be adapted to meet precisely the main requirements of perturbed bifurcation theory, with or without constraints of symmetry. This article is an introduction to their methods, describing the underlying geometric ideas and algebraic tools in the hope that their theory may become somewhat more approachable by workers in continuum mechanics or other applied fields who may wish to use it. There are no essentially new results, although in Section 5 there are observations on families of linear operators that generalize theorems of Shearer (1978).

For an accessible treatment of singularity theory the book by Gibson (1979) is earnestly recommended: it would provide a valuable companion to Golubitsky and Schaeffer's papers.

The article by Callahan (1974) makes enjoyable reading; see also the excellent survey by Arnol'd (1968). The books by Golubitsky and Guillemin (1974) and Lu (1976) contain a large amount of information. Singularities of scalar-valued mappings (functions), important for variational problems, are the subject of many surveys and texts, such as Wassermann (1974), Bröcker and Lander (1975), Sussmann (1975), Zeeman and Trotman (1976), Zeeman (1977), and Golubitsky (1978). In this context also the copiously illustrated and entertaining book of Poston and Stewart (1978) should not be missed.

The bifurcation problems that concern us in this article consist of describing the solutions to

$$G(\lambda, x) = 0 \tag{1.1}$$

as an equation in x as the real parameter λ is varied. Here x belongs to some finite or infinite-dimensional real linear space X, and G is a nonlinear operator taking values in some finite or infinite-dimensional real linear space Y. Write $G(\lambda, x) = G_\lambda(x)$ and let Σ_λ be the set of zeros of G_λ; then the problem is formally posed as:

Question 1. How does Σ_λ vary as λ varies?

The usual approach is to look in (λ, x)-space for the zero set Σ of G itself, and then to read off from Σ a description of Σ_λ for each λ. In Fig. 1.1 we give some familiar and elementary

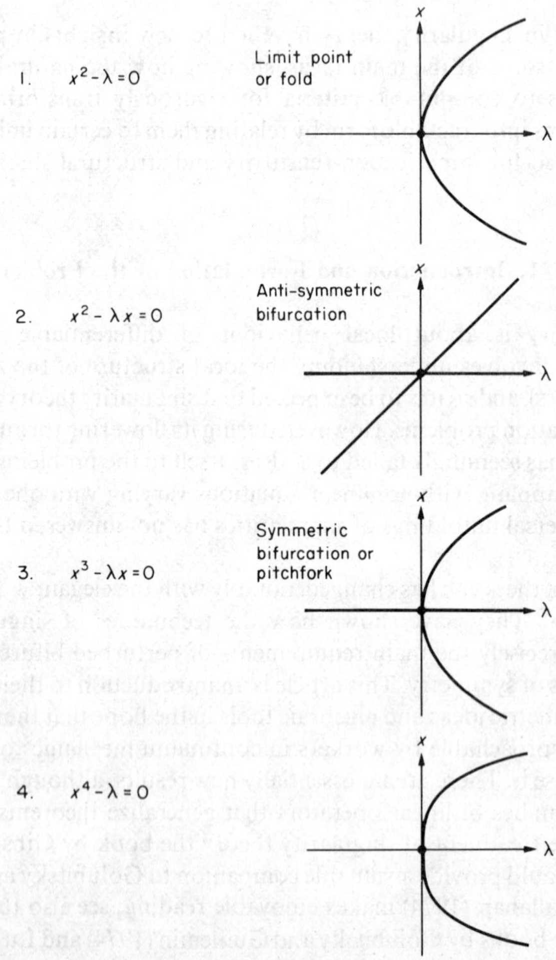

FIG. 1.1. Elementary examples of bifurcations.

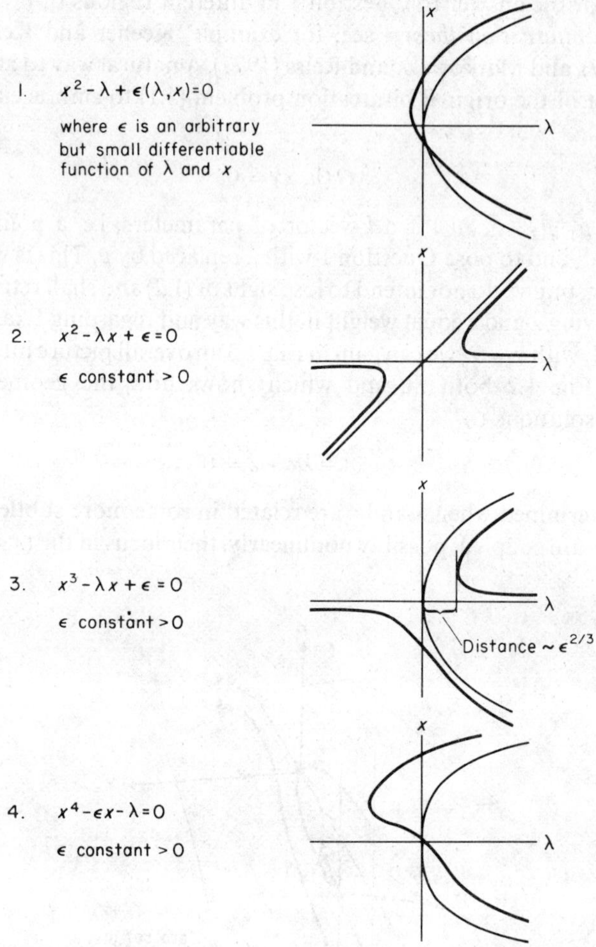

FIG. 1.2. Examples in Fig. 1.1 after the inclusion of imperfections.

examples in the case when X and Y are each one-dimensional, introduced here as they illustrate general points to be made later on.

These examples are not only familiar in themselves, but it is also well known how they respond to perturbations in the original problem (1.1) (Fig. 1.2). This is of crucial importance in modelling real phenomena, since data for G are determined only to within a margin of experimental or computational error. For example, the $\frac{2}{3}$-power law for *imperfection-sensitivity* in Example 3 in Fig. 1.2 is very relevant to simple buckling problems: see Koiter (1945), Sewell (1966), and Thompson and Hunt (1973). To understand a bifurcation problem fully we must therefore also answer:

Question 2. How is the answer to Question 1 affected if G is perturbed slightly?

To make this more tractable it is sensible to put in explicit perturbations of G controlled by parameters $\alpha_1, \alpha_2, \ldots, \alpha_r$ (perhaps $r = 1$) to produce a new problem

$$G(\lambda, \alpha_1, \alpha_2, \ldots, \alpha_r, x) = 0, \tag{1.2}$$

and then to ask for the answer to Question 1 in different regions of $(\alpha_1, \alpha_2, \ldots, \alpha_r)$-space. This is *perturbed bifurcation theory*: see, for example, Keener and Keller (1973), Keener (1974), Hale (1977), and Matkowsky and Reiss (1977). A natural way to attack (1.2) is then to widen the concept of the original bifurcation problem (1.1) to embrace any problem of the form

$$G(\mu, x) = 0, \tag{1.3}$$

where now $\mu = (\mu_1, \mu_2, \ldots, \mu_k)$ is a k-vector of parameters, i.e. a point in k-dimensional euclidean space $\mathbb{R}^k$, and to pose Question 1 with λ replaced by μ. This is what we shall do for most of this article, but we do not intend to lose sight of (1.2) and shall return to it in Section 9.

Note that by giving λ and ε equal weight in this way and regarding Example 3 in Fig. 1.2 as an instance of (1.3) with $k = 2$, we can gain in Fig. 1.3 an overall picture into which Examples 3 of Fig. 1.1 and Fig. 1.2 both fit, and which shows how the geometrical form of the configuration of solutions to

$$x^3 - \lambda x + \varepsilon = 0 \tag{1.4}$$

can readily be determined when λ and ε are related in some more subtle way than simply ε = constant. If λ, ε are coupled, possibly nonlinearly, their locus in the (λ, ε)-plane is a curve Γ

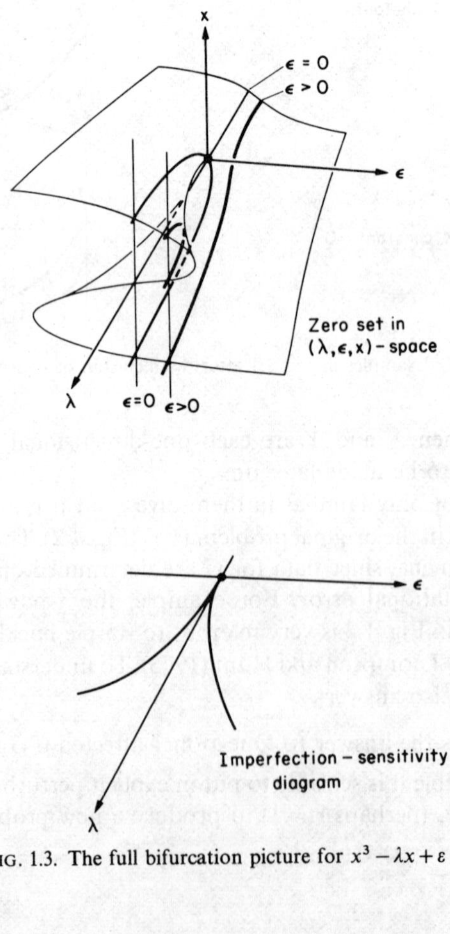

FIG. 1.3. The full bifurcation picture for $x^3 - \lambda x + \varepsilon = 0$.

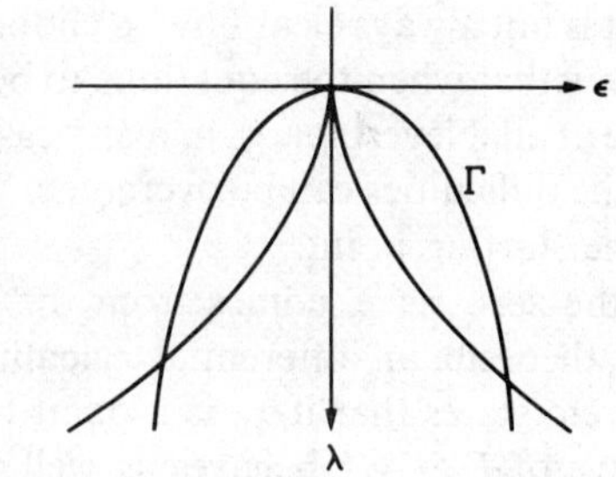

(a) Curve Γ in (λ, ϵ) – space

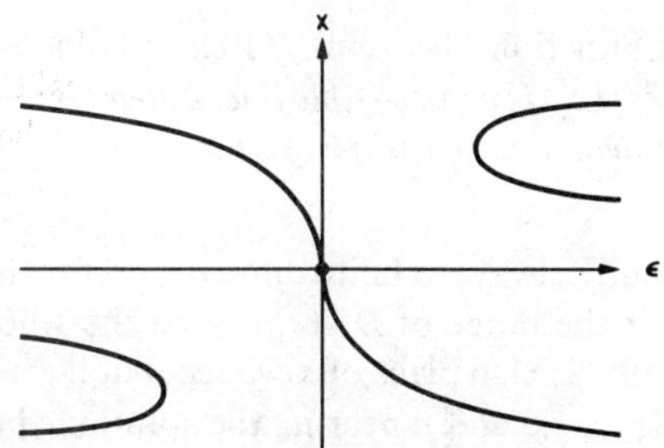

(b) Bifurcation diagram along Γ

FIG. 1.4. A one-parameter bifurcation problem as a curve through a two-parameter problem.

and so the relevant zero set Σ is just that part of the solution locus to (1.4) which lies above Γ. For example, if $\varepsilon = \lambda^{1/2}$ then Γ is a parabola as in Fig. 1.4(a) and the solution locus over Γ, considered as parameterized by ε, is as in Fig. 1.4(b). Whether or not this gives any extra insight is a matter of taste.

Of crucial significance in this example, however, is the fact that *no further small imperfections will materially affect the bifurcation behaviour* of the two-parameter problem $G(\lambda, \varepsilon, x) = 0$, however ingeniously these extra imperfections may be inserted. This is doubtless well known in solid mechanics as an empirical truth, but without the aid of singularity theory a rigorous proof would be hard to find. A formal proof of course requires a formal statement of the assertion to be proved, and this will be given in Section 6. A bifurcation problem which is robust against further small imperfections is said to be *structurally stable*, and the aim in what follows is to outline a theory and method for classification of structurally stable k-parameter bifurcation problems.

2. Preliminaries

The techniques in this article are all from differential calculus, so we assume G is differentiable (in the Fréchet sense in infinite-dimensional spaces) as many times as necessary for our purposes. In fact to simplify matters we suppose G is infinitely differentiable. This can present serious difficulties, as it is not always clear how to choose function space appropriate to a given problem in such a way that when the equations to be solved are expressed as (1.1) the resulting G is differentiable at all. Nevertheless, in many cases including elastostatics (see Marsden and Hughes, 1978), the difficulties can be overcome. We shall not go into this here, but take differentiability as the starting point.

An immediate fact about the zero set Σ comes from the Implicit Function Theorem, perhaps the most important theorem in differential calculus. Let F be a differentiable nonlinear operator. The theorem states that if x_0 is a point for which the linear operator $L = DF(x_0)$ (the first derivative of F at x_0) behaves as well as possible, then near x_0 the nonlinear operator F also behaves as well as possible. In infinite dimensions we suppose the spaces are Banach spaces (complete normed linear spaces) and the linear operators are bounded (continuous).

2.1. THEOREM (Implicit Function Theorem). *If x can be split as $x = (v, z)$ in such a way that the partial derivative $(\partial F/\partial z)(x_0)$ is an invertible linear operator, then for each c near $F(x_0)$ the set of solutions to $F(x) = c$ near x_0 is the graph of a function $z = f_c(v)$ differentiable with respect to v and c.*

In many cases, and certainly always in finite dimensions, for the hypothesis of the theorem to hold at x_0 it suffices for the range of $DF(x_0)$ to be the whole of Y.

By applying this to G with (μ, x) in place of x we see that if $x = (v, z)$ and $(\partial G/\partial z)(\mu_0, x_0)$ is invertible then for $c_0 = G(\mu_0, x_0)$ and μ near μ_0 the solutions in x to $G(\mu, x) = c_0$ form the graph of z as a differentiable function of v, and this varies differentiably with μ. Consequently the only points (μ_0, x_0) in Σ around which interesting changes in Σ_μ can occur are those for which this form of the Implicit Function Theorem fails. We call these *singular points* of Σ; the places where the theorem holds are *regular* points.

Let (μ_0, x_0) be a singular point of Σ, and now suppose that there is a linear projection P from Y to a "smaller" space such that (μ_0, x_0) becomes a regular point for the projected problem

$$PG(\mu, x) = 0. \tag{2.1}$$

This will always be possible in finite dimensions, and in many infinite dimensional cases also. Then the zero set Σ' for PG is locally the graph of z as a differentiable function g of μ and v, and Σ' is precisely the locus of those (μ, x) which are taken by G into the zero set or null space N_P for P. Since $(\mu, x) = 0$ implies $PG(\mu, x) = 0$ it follows that the Σ we seek is contained entirely in Σ'. In fact Σ is the part of the graph of g lying above the zero set Σ'' for the problem

$$\overline{G}(\mu, v) \equiv G[\mu, v, g(\mu, v)] = 0 \quad \text{in} \quad N_P \tag{2.2}$$

(Fig. 2.1). The force of this observation is that in cases when v and $\overline{G}(\mu, v)$ belong to finite-dimensional spaces, which can be arranged, for example, when $(\partial G/\partial x)(\mu_0, x_0)$ is a Fredholm operator (see Section 5) the original bifurcation problem (1.3) has been converted into a finite-dimensional one (2.2). This reduction technique is a slightly generalized version of the *method of Liapunov–Schmidt*: see Cesari (1976) and Hale (1977). It is by this method that we are enabled to treat many infinite-dimensional problems rigorously as finite dimensional ones.

The hypothesis of the Implicit Function Theorem can be expressed as: $DF(x_0)$ has a linear *right inverse*, together with assumptions about the splitting of coordinates (automatically

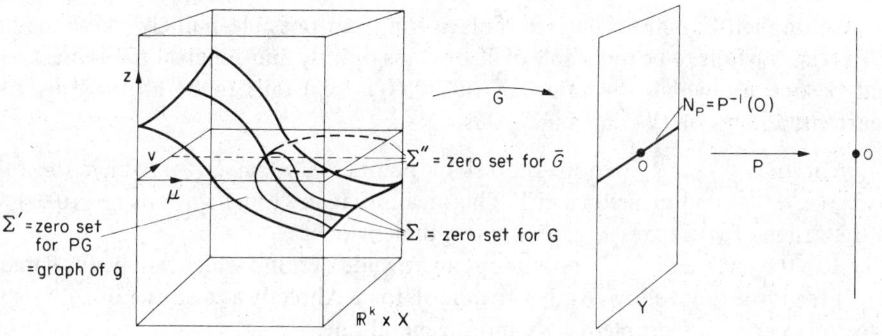

FIG. 2.1. Illustration of Liapunov–Schmidt reduction.

satisfied in finite dimensions). A right inverse for any operator H from X to Y is an operator M from Y to X such that HM is the identity operator I_Y on Y. In our case a linear right inverse for $DF(x_0)$ would be the inverse of $(\partial F/\partial z)(x_0)$.

2.2. COROLLARY. *If the Implicit Function Theorem holds for the (nonlinear) operator F at x_0, then F has a differentiable (nonlinear) right inverse N defined on some neighbourhood of $c_0 = F(x_0)$, satisfying $N(c_0) = x_0$.*

Proof. Take, for example, $N(c) \equiv f_c(v_0)$ in the notation of Theorem 2.1, where $x_0 = (v_0, z_0)$.

In the case when $DF(x_0)$ is itself invertible (i.e. $z = x$) it follows that N is a genuine inverse for F locally, a result usually called the *Inverse Function Theorem*. An operator or mapping such as F which, on some neighbourhood of a given point x_0, is one-to-one, differentiable, and has inverse also differentiable is called a *local diffeomorphism* in the jargon of differential topology. We think of it as an invertible regular *nonlinear local change of coordinates*.

A further note about terminology. A nonlinear operator, when translated from analysis into topology, becomes a (nonlinear) *mapping*. This is the word we shall prefer to use as the article proceeds. It is also synonymous with *function* although we usually reserve that for scalar-valued mappings. Geometrically, the zero set for a mapping F from X to Y is pictured as the set of points in X that land on zero in Y when F is applied to them. To study bifurcation problems $G(\mu, x) = 0$ is to study mappings from (μ, x)-space $\mathbb{R}^k \times X$ to the space Y. If we are emphasizing x as the variable and μ as a controlling parameter, then we call G a *k-parameter family of mappings* from X to Y, or often just a *k-parameter family*. When local bifurcation behaviour near some given (μ_0, x_0) is all that concerns us, we think of $G(\mu, x)$ as a k-parameter *perturbation* of $G(\mu_0, x) \equiv G_{\mu_0}(x)$ near $x = x_0$. In singularity theory the term is k-parameter *deformation*, but we shall avoid this because of its possible confusion with the use of the same word in solid mechanics. Another evocative term from singularity theory is k-parameter *unfolding*, conveying the idea of the bud at x_0 for G_{μ_0} blooming into flower as μ is varied, although it is not officially strictly synonymous with deformation (cf. Gibson, 1979). We shall nevertheless use it in this way in Section 8.

3. Understanding the Global Picture

For the next step in analysing the structure of the zero set Σ for G we need to find the singular points in Σ. The most well-behaved type of singular point (μ_0, x_0) would be one at which the Implicit Function Theorem fails as simply as possible, namely where the range of $(\partial G/\partial x)(\mu_0, x_0)$ fails to be the whole of Y, but fails by only one dimension. The next best type would be one for which the range of $(\partial G/\partial x)(\mu_0, x_0)$ fails to be all of Y by just two dimensions, and so on. We formalize this.

3.1. *Notation.* Let R_q denote the set of points (μ, x) in Σ for which the range of $(\partial G/\partial x)(\mu, x)$ has codimension q in Y. This means that any y in Y splits as $y = (u, w)$ where u is in the range of $(\partial G/\partial x)(\mu, x)$ and w is q-dimensional.

To grasp the nature of R_q we now adopt an attitude of mind which will be fundamental to many of the ideas that follow. Rather than look for R_q directly as a subset of (μ, x)-space we replace this particular problem by a more general one.

Let $\mathscr{L}(X, Y)$ be the space of *all* (continuous) linear operators from X to Y. Now pose the problem of *understanding the structure of the subset* S_q *of* $\mathscr{L}(X, Y)$ *consisting of all operators of corank q, i.e. whose range has codimension q in Y*. The assignment of the linear operator $(\partial G/\partial x)(\mu, x)$ to the point (μ, x) defines a mapping G_1 from (μ, x)-space to $\mathscr{L}(X, Y)$, and we immediately see by definition:

3.2. PROPOSITION. *The set* R_q *is the intersection of* Σ *with* $G_1^{-1}(S_q)$, *i.e. those points* (μ, x) *taken to elements of* S_q *by the application of* G_1.

Pictorially we visualize something like Fig. 3.1. This allows us to decompose the problem of finding R_q into two parts as follows:

1. Find S_q in $\mathscr{L}(X, Y)$.
2. Investigate how G_1 maps (μ, x)-space into $\mathscr{L}(X, Y)$, and how it relates Σ to the configuration of S_q.

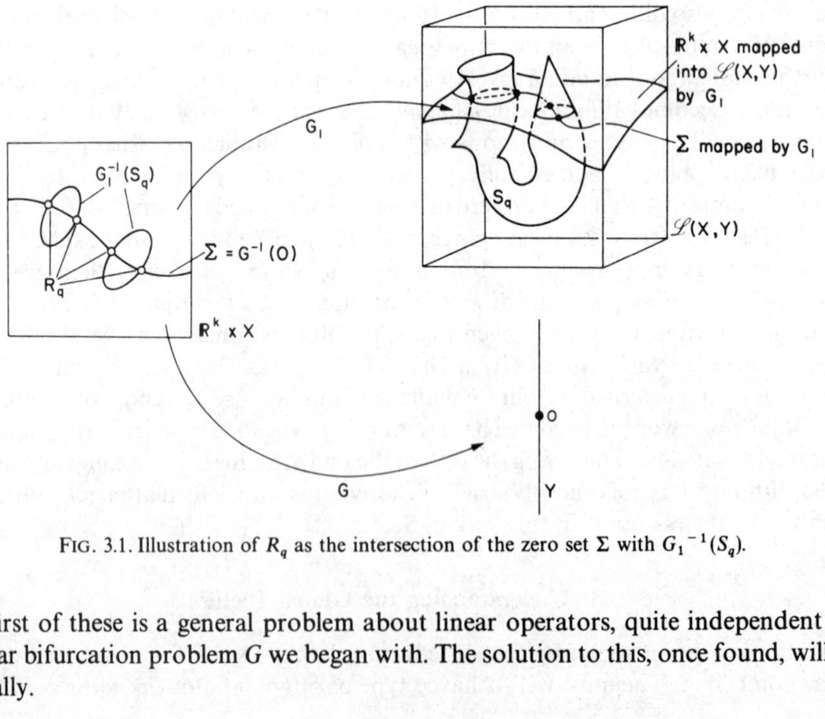

FIG. 3.1. Illustration of R_q as the intersection of the zero set Σ with $G_1^{-1}(S_q)$.

The first of these is a general problem about linear operators, quite independent of the particular bifurcation problem G we began with. The solution to this, once found, will apply universally.

The second depends on G. Even without knowing much about G, however, we can still make optimistic assertions of a general nature. If G_1 hits S_q in *general position* in some sense we might hope to read off a description of R_q, at least locally, by "pulling" the local description of S_q through G_1 back into (μ, x)-space. It turns out that this can indeed be done in rather a precise way. In order to do it we have first to clarify a few elementary ideas from differential topology.

4. Manifolds and Transversality

A subset M of a finite- or infinite-dimensional linear space X is called a *manifold* in X if at each point x_0 in M the variable x can be split into (y, z) in such a way that locally (i.e. sufficiently near x_0) M is the graph of z as a differentiable function g of y (Fig. 4.1). The dimension of the space Y to which y belongs is the *dimension* of M at x_0, and the dimension of the space Z to which z belongs is the *codimension* of M at x_0: either or both of these can be infinite. Usually we insist also that M should have the same dimension at all its points. Of course Y itself is a special case of a manifold ($g = 0$), as is Z, by interchanging the rôles of y and z. Indeed, we may think of a manifold in X as a subset which is everywhere locally a curvilinear version of a given linear subspace of X.

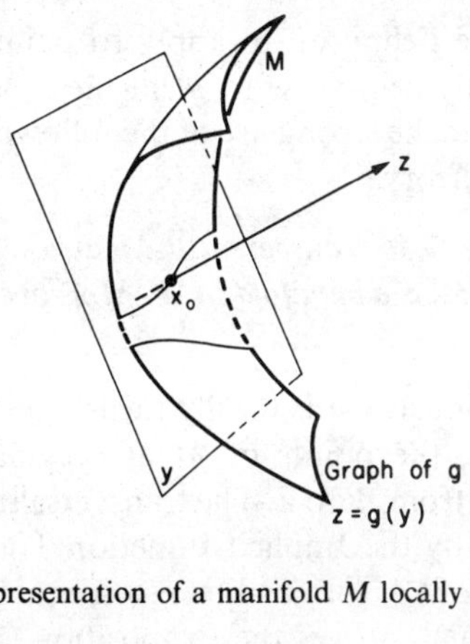

FIG. 4.1. Representation of a manifold M locally as a graph.

We have met this local picture already in Section 2 with the Implicit Function Theorem, which we now see as implying that *near a regular point the zero set* Σ *is a manifold*. Conversely, any manifold is locally the zero set for some function F from X to some Z (take $F(y, z) \equiv z - g(y)$ in local coordinates, for example) and consists of regular points.

The *tangent space* to M at x_0 is the graph of the linear approximation to g, namely the graph of the derivative of g at y_0. It is denoted by M_{x_0} and is an affine subspace of X (meaning a linear subspace translated away from the origin). Its codimension is the same as that of M. The tangent space M_{x_0} can also be thought of as the set of all initial velocity vectors in X of paths in M starting at x_0.

The importance to us of the notion of manifold is that the very well-behaved local structure allows us to develop a theory of "general position" or *transversality* for intersections of manifolds or, more usefully, for mappings and manifolds, modelled on the natural one for linear subspaces and linear operators.

4.1. DEFINITION. *Let X_1, X_2 be linear subspaces of the linear space X. Then X_1 and X_2 are* transverse *if together they span the whole of X, i.e. $X = X_1 + X_2$.*

Note that in this case the codimension of $X_1 \cap X_2$ (intersection of X_1 and X_2) in X_1 is the same as codim X_2 in X. If the dimensions are such that dim X_1 + dim X_2 is less than dim X, then X_1, X_2 cannot be transverse.

We extend this definition to apply to a linear operator and a linear subspace, rather than two subspaces.

4.2. DEFINITION. *Let X_1 be a linear subspace of X, and let L be a linear operator from another linear space into X. Then L is* transverse *to X_1 if X_2 = range L is transverse to X_1 in the above sense.*

In particular, if the range of L is all of X then L is transverse to every linear subspace of X.

The nonlinear version of this now goes as follows.

4.3. DEFINITION. *Let M be a manifold in X and let Φ be a (nonlinear) mapping from another linear space W into X. Suppose x_0 is a point in W with $\Phi(x_0) = y_0$ a point in M. Translate the origin in X to y_0 so that $M_{y_0} = M_0$ is a linear subspace of X. Then Φ is* transverse *to M at x_0 if the derivative L of Φ at x_0 is transverse to M_0 in the sense of Definition* 4.2.

It is common to extend the definition to apply to points x_0 with $\Phi(x_0)$ not in M, by declaring Φ to be automatically transverse to M at these points.

In Sections 5 and 6 we shall make strong use of the following result, which is central to the philosophy of transversality theory.

4.4. PROPOSITION. *Suppose Φ is transverse to M at x_0, in the terminology above. Then $\Phi^{-1}(M)$ in a neighbourhood of x_0 is a manifold in W. Moreover, the codimension of $\Phi^{-1}(M)$ in W is the same as that of M in X.*

Proof. As mentioned previously, M is locally (near x_0) the zero set of some mapping F from X to Z and, moreover, the points of M are regular points for F. Consider the composition $F\Phi$ as a mapping from W to Z. The transversality hypothesis implies that x_0 is a regular point for $F\Phi$, and so by the Implicit Function Theorem the zero set for $F\Phi$ is a manifold in a neighbourhood of x_0, with codimension equal to the dimension of Z. But the zero set for $F\Phi$ is $\Phi^{-1}(M)$ which proves the Proposition (Fig. 4.2).

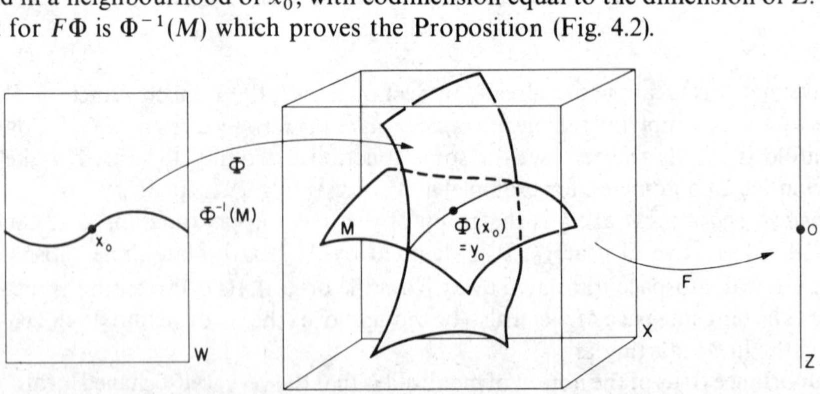

FIG. 4.2. If Φ is transverse to M at x_0, then $\Phi^{-1}(M)$ is a manifold near x_0 in W.

The transversality hypothesis for Φ at x_0 can often in practice be quite easily verified, since it uses only "linear information" about Φ at x_0 and M at y_0. Regardless of this, however, transversality also carries with it the ideas of *typicality*, *non-exceptional behaviour*, or *genericity*. In the same way that two curves in a plane "usually" meet in isolated points, two surfaces in three dimensions "usually" meet in curves, and so on, an important theorem states that given a manifold M in X, and given a region W' in the space W, *most* (*differentiable*) *mappings Φ from W' to X will be transverse to M everywhere*. This is almost the *raison d'être* for the theory of transversality. There are of course mathematical technicalities needed, concerning the global topology of M, the limitations on the degeneracy that the mapping Φ may exhibit in infinite dimensions, and the precise meaning of "most", but this is the essential content of the *Transversality Theorem* of Thom (1954), developed in an infinite-dimensional setting by Abraham and Robbin (1967) and Quinn (1970). It plays a very important rôle in singularity theory, and influences almost everything that follows in this article.

5. Application to Families of Linear Operators

Returning to Section 3 we will exploit transversality to yield some facts about R_q. Recall our decomposition of the problem into two parts, the first being to find S_q in $\mathscr{L}(X,Y)$.

A linear operator is a *Fredholm* operator if its range has finite codimension and its null-space has finite dimension. We suppose from now on that we are dealing with Fredholm operators in $\mathscr{L}(X, Y)$.

5.1. PROPOSITION. *Let L belong to S_q, and let the dimension of the null-space of L be n_L. Then near L the set S_q is a manifold in $\mathscr{L}(X,Y)$ of codimension qn_L.*

Proof. There is a splitting of x as (v, z) in X and of y as (u, w) in Y with respect to which L has a representation as a 2×2 matrix of linear operators of the form

$$L = \begin{matrix} u \\ w \end{matrix} \begin{bmatrix} 0 & 0 \\ 0 & L_0 \end{bmatrix} \qquad (5.1)$$
$$\qquad\qquad v \quad z$$

where L_0 is invertible, u-space U has dimension q, and v-space V is the null-space of L. Any other K in $\mathscr{L}(X,Y)$ has the form

$$K = \begin{bmatrix} A & P \\ Q & K_0 \end{bmatrix} \qquad (5.2)$$

and if K is sufficiently close to L then K_0 is also invertible. Composing this with the *invertible* linear operator

$$T = \begin{bmatrix} I & -PK_0^{-1} \\ 0 & I \end{bmatrix} \qquad (5.3)$$

in $\mathscr{L}(Y,Y)$ we obtain

$$TK = \begin{bmatrix} A - PK_0^{-1}Q & 0 \\ Q & K_0 \end{bmatrix} \qquad (5.4)$$

in $\mathscr{L}(X,Y)$, which lies in S_q if and only if K does. But TK is in S_q precisely when

$$A - PK_0^{-1}Q = 0 \qquad (5.5)$$

in $\mathscr{L}(U, V)$, so S_q is the zero set for Ω, where Ω is the nonlinear operator from $X' = \mathscr{L}(X, Y)$ to $Y' = \mathscr{L}(U, V)$ taking K to $A - PK_0^{-1}Q$. Now $(\partial\Omega/\partial A)(L)$ is simply the identity linear operator from $\mathscr{L}(U, V)$ to $\mathscr{L}(U, V)$, so the hypothesis of Theorem 2.1 holds for Ω at L. The theorem implies that near L the zero set S_q is a manifold of codimension = dim $\mathscr{L}(U, V) = qn_L$.

5.2. *Remarks.* 1. The Proposition shows the codimension of S_q to be locally constant, so the part of S_q corresponding to Fredholm operators consists of a collection of disjoint pieces on each of which the codimension is constant. In finite dimensions q determines n_L and so S_q has codimension qn_L everywhere.

2. When $n_L = q$ (which is the case when X, Y have the *same* finite dimension) then S_q has codimension q^2 everywhere.

3. In the case of self-adjoint operators in Hilbert space (or symmetric matrices in finite dimensions), which is relevant for gradient problems, a similar proof shows S_q to have codimension $\frac{1}{2}q(q+1)$.

This knowledge of the local structure of the S_q in $\mathscr{L}(X, Y)$, together with further use of the transversality ideas of Section 4, leads to an understanding of how the dimensions of the null-space and range of a linear operator typically vary as the operator L varies with respect to a parameter b in some finite- or infinite-dimensional space B. As in Section 3 we shall apply this to the case when $b = (\mu, x)$ and $L(b) = (\partial G/\partial x)(\mu, x)$, but first we express some results for the general case. Let P_q be the set of those b in B for which $L(b)$ has corank q. From Propositions 4.4 and 5.1 we have the following result.

5.3. THEOREM. *Suppose the mapping Φ taking b to $L(b)$ is transverse to S_q in $\mathscr{L}(X, Y)$. Then P_q is a manifold in B. The codimension of the connected component of P_q containing b_0 is equal to qn where n is the dimension of the null space of $L(b_0)$.*

In particular, if B is finite-dimensional and the transversality condition holds, then P_q is empty for $q > (1/n)$ dim B.

The transversality hypothesis is something that can be checked explicitly at any b_0 *in* P_q. It is that the Implicit Function Theorem should apply at b_0 to the mapping $\Omega\Phi$ taking b to $A(b) - P(b)K_0^{-1}(b)Q(b) = \bar{L}(b)$.

The Transversality Theorem says that, all other things being equal, we expect the hypothesis to be satisfied everywhere.

Suppose $L(b_0)$ is a Fredholm operator of index i, meaning dim (null-space) − codim (range) $= i$. This is the case if X, Y are finite-dimensional with dim $X = i +$ dim Y. Then we obtain a corollary to the foregoing theorem.

5.4 COROLLARY. *Suppose $L(b_0)$ lies in S_q, and the transversality condition on $\bar{L}(b)$ is satisfied. Then P_q near b_0 is a manifold in B of codimension $q(i+q)$.*

If dim $B < q(i+q)$ transversality cannot hold, given $L(b_0)$ is in S_q. If dim $B = q(i+q)$, then b_0 is an isolated point of P_q. For dim $B > q(i+q)$ we can split b as (b_1, b_2) where b_2 is $q(i+q)$-dimensional, b_1 is in the null-space of $(d\bar{L}/db)(b_0)$, and P_q is locally the graph of b_2 as a differentiable function of b_1. In the case $q = 1$ the transversality condition is that $(dL/db)(b_0)$ *operating on the null-space of* $L(b_0)$ *has range not lying entirely in the range of* $L(b_0)$. In this way we recover Theorem 3.1 of Shearer (1978) for the case $i = 0$, dim $B = 2$, with the exception that we have not discussed variation of the actual null-space of $L(b)$ with b. This is easy to incorporate, however, using (5.2) and (5.4).

Note that even in the absence of transversality, results about P_q could still be obtained within this framework. In a neighbourhood of b_0 the set P_q is the zero set of the map $\Psi = \Omega\Phi$ taking b to $A(b) - P(b)K_0^{-1}(b)Q(b)$, and so any results from singularity theory or elsewhere that enable this zero set to be described will solve the problem of describing P_q.

For $q = 2$ transversality cannot hold for b_0 in S_2 unless $\dim B \geqslant 2(i+2)$. If $\dim B = 2(i+2)$, then transversality implies that b_0 is an isolated point of P_2. Moreover, nearby points b belong to P_1 precisely when $A(b)$, which is a $2 \times (i+2)$ matrix, has corank 1 and hence rank 1. This occurs when one row $(u_1, u_2, \ldots, u_{i+2})$ is a multiple of the other row $(v_1, v_2, \ldots, v_{i+2})$, namely

$$\begin{aligned} u_1v_2 - u_2v_1 &= 0, \\ u_2v_3 - u_3v_2 &= 0, \\ &\cdots\cdots \\ u_{i+1}v_{i+2} - u_{i+2}v_{i+1} &= 0, \end{aligned} \tag{5.6}$$

which describes a conical algebraic locus $\mathcal{Q}$ of dimension $2(i+2)-(i+1) = i+3$ in the $2(i+2)$-dimensional space of such matrices. The transversality condition and the equality of dimensions $= 2(i+2)$ implies by the Inverse Function Theorem that mapping b to $\overline{L}(b)$ is merely an invertible (nonlinear) coordinate change, so P_1 is a "curvilinear copy" of $\mathcal{Q}$ inside B, with vertex of the cone at b_0. In particular, when $i = 0$, $\dim B = 4$ we obtain P_1 as a copy of the quadric cone $u_1v_2 - u_2v_1 = 0$ in new coordinates.

If transversality fails (which it will necessarily do if $d = \dim B < 4$) we can still read off descriptions of P_1 near b_0 from the way the d-parameter family of matrices $\overline{L}(b)$ meets $\mathcal{Q}$ in four-space. By definition, $\overline{L}(b)$ lies in $\mathcal{Q}$ precisely when

$$\phi(b) \equiv u_1(b)v_2(b) - u_2(b)v_1(b) = 0 \tag{5.7}$$

so P_1 is the zero set for ϕ. If the quadratic part of ϕ is nondegenerate at b_0 (there is no linear part because $\overline{L}(b_0) = 0$), then the Morse Lemma (Milnor, 1963, but see almost any text on singularity theory or differential topology) guarantees that a suitable invertible nonlinear coordinate change in B will transform ϕ into its own quadratic part, so the zero set for ϕ is in new coordinates the zero set of a nondegenerate quadratic form: point, pair of lines, quadric cone, pair of planes, etc. In the case $\dim B = 2$ we hence obtain most of Theorem 3.7 of Shearer (1978), incidentally without losing any differentiability at b_0.

To summarize: we have used naïve geometrical arguments, supported by analytical criteria for transversality, to derive results on the way the corank of a linear operator varies as the operator varies with respect to a multi-dimensional parameter.

In the discussion we have not mentioned behaviour of the null-space of the operator nor have we distinguished (in the case of $i = 0$) between matrices of different Jordan form but the same corank: there is much additional structure to be superimposed upon P_q. A treatment of these matters via transversality is given in the very attractive article of Arnol'd (1971); see also Arnol'd (1972b) for some applications to stability theory.

That work, as well as the main insight that we shall see singularity theory gives to bifurcation theory and which is the theme of this article, rests firmly on the important observation that the hypothesis of transversality has yet further implications which are of an essentially analytic rather than geometric type. These come from the following standard fact about corank.

5.5. LEMMA. *Two Fredholm linear operators L, L' from X to Y of the same index have the same corank if and only if there is an invertible linear operator M from X to itself and an invertible linear operator N from Y to itself such that $L' = NLM$.*

In the terminology of matrix theory, L and L' are *equivalent*. Denoting NLM by $\Lambda(L;N,M)$, define $T(b,N,M)\equiv\Lambda(L(b);N,M)$ where $L(b)$ is a parameterized family of linear operators as before, with $L(b_0)$ in S_q. Suppose the family $L(b)$ is transverse to S_q at b_0. Then T satisfies the hypothesis of the Implicit Function Theorem, as an operator from $B\times\mathscr{L}(X,X)\times\mathscr{L}(Y,Y)$ to $\mathscr{L}(X,Y)$, at the point (b_0, I_X, I_Y), because the ranges of $\partial T/\partial N$ and $\partial T/\partial M$ evaluated at (b_0, I_X, I_Y) together span the tangent space to S_q at $L(b_0)$. (This follows directly from the fact that S_q is the set of $T(b_0, N, M)$ for all N, M.) Therefore T has a right inverse, so that any K in $\mathscr{L}(X, Y)$ near $L(b_0)$ can be written as

$$K = N(K)\ L(b(K))\ M(K), \tag{5.8}$$

where $N(K)$, $b(K)$, and $M(K)$ vary differentiably with K. Replacing the single K by a family of operators $K(c)$ we obtain the next result.

5.6. THEOREM. *Let $L(b)$ be a family of linear operators transverse to S_q at b_0, and let $K(c)$ be any other family of operators with $K(c_0) = L(b_0)$. Then there exist $N(c)$, $b(c)$, and $M(c)$ varying differentiably with c such that*

$$K(c) = N(c)\ L(b(c))\ M(c). \tag{5.9}$$

The transversality of $L(b)$ at b_0 implies that each member $K(c)$ of any other family passing through $L(b_0)$ is equivalent to some $L(b)$, where b and the equivalence vary (differentiably) with c. In this way, $L(b)$ has a property of *universality*, which is algebraic and expressed using (5.9), derived from its geometrical property of transversality. As we shall now describe, this fundamental idea extends to nonlinear operators and bifurcation problems.

6. Transversality and Bifurcation Problems

To apply transversality to bifurcation problems we will initially restrict attention to finite dimensions and comment on the infinite-dimensional case later. For a bifurcation problem $G(\mu, x) = 0$ the interesting points are the singular points where $G(\mu, x) = 0$ and $(\partial G/\partial x)(\mu, x)$ has corank > 0 at the same time. An idea of this set in geometric terms is gained by using the methods of Section 5 as follows. Associated to G is the mapping

$$G^1:(\mu, x)\to\left\{G(\mu, x), \frac{\partial G}{\partial x}(\mu, x)\right\} \tag{6.1}$$

from (μ,x)-space $\mathbb{R}^k\times X$ to the product space $Y\times\mathscr{L}(X,Y)$, which is itself finite-dimensional, and the set R_q of singular points where corank $(\partial G/\partial x)(\mu, x) = q$ is $\Sigma\cap G_1^{-1}(S_q)$ which is also $(G^1)^{-1}(0\times S_q)$. Now $0\times S_q$ is a manifold in $Y\times\mathscr{L}(X, Y)$ of codimension

$$\dim Y + q(i+q) = \dim X + i(q-1) + q^2 \tag{6.2}$$

because $\dim X = i + \dim Y$. Therefore we have the following theorem.

6.1. THEOREM. *Suppose G^1 is transverse to $0\times S_q$ in $Y\times\mathscr{L}(X, Y)$. Then R_q is a manifold of dimension $k - i(q-1) - q^2$ in $\mathbb{R}^k\times X$ for $q^2 + i(q-1)\leqslant k$, and is empty if $q^2 + i(q-1) > k$.*

In the variational case $i = 0$, and we replace q^2 by $\frac{1}{2}q(q+1)$. For example, $\dim R_1 = k-1$ and so consists of isolated points when $k = 1$ (bifurcation with a single parameter). Also $\dim R_2 = k-(i+4)$, or $k-3$ in the variational case. Hence a corank-2 bifurcation can only occur transversely in k-parameter problems with $k \geqslant i+4$, or $k > 3$ in the variational case. Besides this, we know as in Section 5 that the configuration near R_q of the various R_s with $s<q$ is the pull-back via G^1 of the configuration near $0 \times S_q$ of $0 \times S_s$ for $s<q$, which data can be determined once and for all (in principle) without reference to any specific problem. Moreover, for most G the map G^1 is indeed transverse to $0 \times S_q$. This unfortunately does not follow automatically from the Transversality Theorem of Section 4, as not every perturbation of G_1 is of the form H_1 for some H. Nevertheless, the appropriate more delicate version of the theorem still holds. We shall mention it in a more general setting below.

So far we have only looked at the typical behaviour of the *linear* part of a k-parameter bifurcation problem (the quadratic part of the function in variation problems). Now we shall push the same transversality/typicality ideas further to apply to the full nonlinear behaviour, near a given point (μ_0, x_0), of a k-parameter bifurcation problem.

To try to understand how $G(\mu, x)$ behaves near (μ_0, x_0) a natural procedure is to expand the Taylor series of the function G_μ of x at the point x_0 up to some degree p in x and then stop, either hoping or proving that the higher order terms will not materially affect any conclusions about bifurcations for μ near μ_0 subsequently drawn from these terms of degree up to p. In the parlance of singularity theory this truncated Taylor series is called the *p-jet* of G_μ at x_0. Just as in Section 5 we considered simultaneously the linear behaviour of G at all points (μ, x) by thinking about the mapping $(\mu, x) \to (\partial G/\partial x)(\mu, x)$ from $\mathbb{R}^k \times X$ to $\mathscr{L}(X, Y)$, so here with one bold sweep we can consider simultaneously the behaviour up to degree p at each point by contemplating the map

$$(\mu, x) \to (p\text{-jet of } G_\mu \text{ at } x). \tag{6.3}$$

The data for the right-hand side are (apart from factorials) the various mixed partial derivatives of G_μ with respect to the components x_i of x, so formally (6.3) can be considered as

$$(\mu, x) \to \left(G, \frac{\partial G}{\partial x}, \frac{\partial^2 G}{\partial x^2}, \ldots, \frac{\partial^p G}{\partial x^p}\right)\Bigg|_{(\mu,\, x)} \tag{6.4}$$

which is a mapping G^p from $\mathbb{R}^k \times X$ to

$$J^p(X, Y) = Y \times \mathscr{L}(X, Y) \times \mathscr{L}_s^2(X, Y) \times \ldots \times \mathscr{L}_s^p(X, Y), \tag{6.5}$$

where $\mathscr{L}_s^j(X, Y)$ is the space of symmetric j-multilinear mappings from j copies of X to Y, it being the space in which $(\partial^j G/\partial x^j)(\mu, x)$ naturally lies. Equivalently, $\mathscr{L}_s^j(X, Y)$ can be thought of as the space of all m-tuples ($m = \dim Y$) of homogeneous polynomials in $x_1, x_2, \ldots, x_n$ ($n = \dim X$) of degree j. Abstract as it may look, the product space $J^p(X, Y)$ is a finite-dimensional euclidean space, called the *p-jet space* for mappings from X to Y. Moreover the mapping G^p from $\mathbb{R}^k \times X$ to $J^p(X, Y)$ is differentiable (because all derivatives of G are) and so ripe for transversality theory.

To proceed further we have to stop to ask ourselves what it is that we want to know. We are interested in finding what $G^{-1}(0)$ looks like as a subset of $\mathbb{R}^k \times X$, so to give formal meaning to that we have to provide a definition of "looking the same", or *equivalence* for our purposes. To begin with we fix μ (say $\mu = 0$) and consider what we might mean by two single mappings G_0, H_0 from X to Y being equivalent. Our main concern is with local behaviour, so without loss of generality we will work near the origin $x = 0$. Let us try a tentative definition.

6.2. Definition A. *Two mappings G_0, H_0 from X to Y are* equivalent *at* 0 *in X if there is a coordinate change on some neighbourhood of* 0 *in X which converts H_0 into G_0.*

By coordinate change we mean differentiable, possibly nonlinear coordinate change with differentiable inverse, or *local diffeomorphism* as in Section 2. Thus G_0, H_0 are equivalent at 0 if there is a local diffeomorphism Φ on a neighbourhood of 0 in X such that

$$G_0(x) \equiv H_0\{\Phi(x)\} \tag{6.6}$$

or, more symbolically, $G_0 = H_0\Phi$.

It turns out that this definition is too restrictive for our purposes because it stipulates equivalence of the functions G_0, H_0 throughout a whole neighbourhood of the origin, while we are interested only in the respective zero sets Σ_{G_0}, Σ_{H_0}. We could admit still further tinkering with the coordinates which did not interfere with Σ_{G_0}, Σ_{H_0}.

To illustrate this, consider two *gradient* mappings $G \equiv \text{grad}\, g$, $H \equiv \text{grad}\, h$ (we drop the subscript 0 for convenience). If the functions g, h are equivalent we would wish to regard their associated gradients as equivalent. But in fact from the relationship

$$g(x) = h(x'),\ x' = \Phi(x) \tag{6.7}$$

we obtain by differentiation

$$\frac{dg}{dx} = \frac{dh}{dx'} \cdot \frac{dx'}{dx} \tag{6.8}$$

or

$$G(x) = L(x)\, H(x'), \tag{6.9}$$

where $L(x)$ is the adjoint of dx'/dx. In finite dimensions this is the transpose of the matrix $[\partial x_i'/\partial x_j]$. In any case it is invertible in a neighbourhood of the origin, so $G(x) = 0$ if and only if $H(x') = 0$, which means that Φ takes Σ_G to Σ_H even though G, H are *not* necessarily equivalent by Φ in the sense of Definition A. Therefore we broaden our definition.

6.3. DEFINITION B. *Two mappings G_0, H_0 from X to Y are* equivalent *at* 0 *in X if there is a coordinate change $x \to x'$ on some neighbourhood of* 0 *in X, together with an invertible linear operator (matrix) $M(x)$ acting on Y but depending on x, satisfying*

$$G_0(x) \equiv M(x)\, H_0(x'). \tag{6.10}$$

We see that (6.9) fits into this context.

It turns out that this is the definition which bears fruit for bifurcation theory. It is the definition which in algebraic geometry is used to define formal equivalence of algebraic loci. The technical term is *contact equivalence* or *$\mathscr{K}$-equivalence*: see Gibson (1979). The term for Definition A is *right equivalence* or *$\mathscr{R}$-equivalence*. Both these notions are of great importance in singularity theory (Gibson, 1979, and Golubitsky and Guillemin, 1974), but contact equivalence is particularly appropriate to studying zero sets rather than overall geometric structure of mappings.

We return now to k-parameter bifurcation problems $G(\mu, x) = 0$. If we replace the space X by $\mathbb{R}^k \times X$ in Definition B, we immediately obtain a definition of equivalence for two mappings G, H on $\mathbb{R}^k \times X$ and under this equivalence the zero sets Σ_G, Σ_H in $\mathbb{R}^k \times X$ do correspond. However, the problem with this definition is that it obliterates the distinction in the problem between the parameters $\mu = (\lambda, \alpha_1, \ldots, \alpha_r)$ and the variable x: the two functions $x^2 - \lambda$ and $\lambda^2 - x$ are equivalent under Definition A (and hence Definition B) but obviously represent entirely different bifurcation phenomena: in the first case there is a limit point while in the second case there is no bifurcation at all (the origin is a regular point). Therefore we must refine the definition by restricting Φ to be a coordinate change that respects the differing rôles of the variable x and parameter μ. We could insist that Φ transform x and μ entirely independently, but this is too strict: it seems acceptable in engineering contexts (Thompson and Hunt, 1973) to allow the new x-coordinate to depend both on x and μ while the new μ-coordinate should depend on μ alone, so that Φ has the form

$$\Phi(\mu, x) \equiv (\mu', x'), \tag{6.11}$$

where

$$x' = x'(\mu, x), \quad \mu' = \mu'(\mu). \tag{6.12}$$

Pictorially Φ takes μ = constant slices to μ' = constant slices, while allowing unrestricted coordinate change in the x-direction (Fig. 6.1). We call Φ an *admissible coordinate change*. It is easy to see from the picture, and easy to calculate by differentiating $G = H\Phi$ by the Chain Rule, that this Φ takes the zero set for G to the zero set for H, regular points to regular points, and takes bifurcation values of μ to bifurcation values of μ', so solutions to the bifurcation problems $G = 0, H = 0$ indeed resemble each other pictorially. By slight abuse of notation let us here call a k-parameter family of mappings itself a bifurcation problem. This is to emphasize the appropriateness for bifurcation theory of the above discussion, in the light of which we now replace Definition B by:

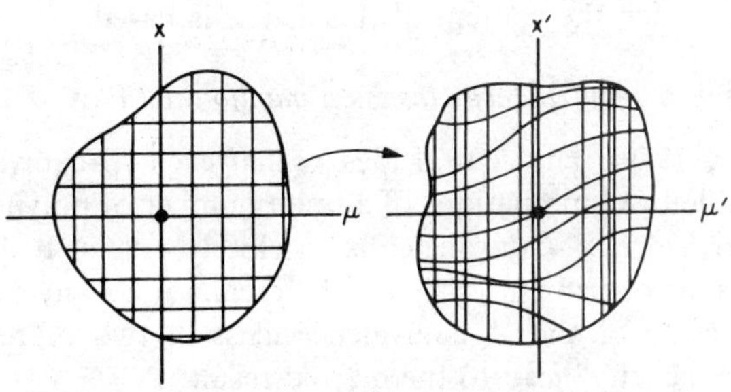

(a) Admissible coordinate change

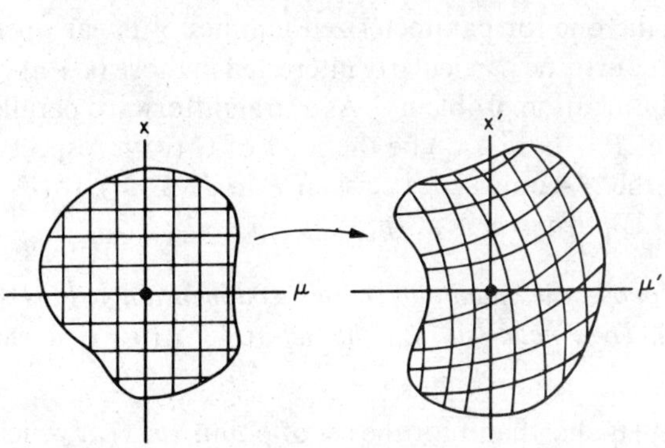

(b) Arbitrary coordinate change

FIG. 6.1. Illustration of local coordinate changes.

6.4. DEFINITION B*. *Two bifurcation problems G, H on* $\mathbb{R}^k \times X$ *with values in Y are* $\mathscr{K}$-equivalent *at* (0, 0) *if there is an admissible coordinate change* $(\mu, x) \to (\mu', x')$ *on some neighbourhood of* (0, 0) *in* $\mathbb{R}^k \times X$, *together with an invertible linear operator* (*matrix*) $M(\mu, x)$ *acting on Y but depending on* (μ, x) *satisfying*

$$G(\mu, x) \equiv M(\mu, x) H(\mu', x'). \tag{6.13}$$

Restricting $M(\mu, x)$ to be the identity operator on Y converts this into the definition of *$\mathscr{R}$-equivalence* for G, H.

Using the notation $G_\mu(x) = G(\mu, x)$, $H_{\mu'}(x') = H(\mu', x')$ we can summarize this by saying: *bifurcation problems G, H are $\mathscr{K}$-equivalent if for each μ there is a μ' such that G_μ is $\mathscr{K}$-equivalent to $H_{\mu'}$ with μ' and the equivalence depending differentiably on μ.* Similarly for $\mathscr{R}$-equivalence.

Now that we have a notion of equivalence we can define precisely what we mean by *structural stability*, or robustness against further imperfections, as promised in Section 1. A k-parameter bifurcation problem G is structurally stable if *any other k-parameter problem H sufficiently close to G has a new origin close to that for G such that G at the old origin and H at the new origin are $\mathscr{K}$-equivalent.* We shall see how to detect structural stability in Section 9.

Meanwhile, we return to p-jets.

By expanding the Taylor series of (6.10) up to degree p and comparing both sides we automatically obtain definitions of $\mathscr{K}$-equivalence for p-jets. Two points in the linear space $J^p(X, Y)$ belong to the same $\mathscr{K}$-equivalence class precisely if they are the p-jets at the origin of two mappings which are $\mathscr{K}$-equivalent. The next observation is the key to the whole of the geometric thinking on which the content of this article is based.

6.5. THEOREM. *Each $\mathscr{K}$-equivalence class is a manifold in $J^p(X, Y)$.*

Proof. See Gibson (1979). The central idea is that each $\mathscr{K}$-equivalence class is semi-algebraic, i.e. describable by combinations of a finite number of polynomial equations and inequalities. Moreover, since $\mathscr{K}$-equivalence is so closely related to looking at the same mapping in two different coordinate systems, there cannot be any important qualitative difference in the structure of the $\mathscr{K}$-equivalence class at two of its points. The semi-algebraicity and "homogeneity" lead to the desired result.

Using this we are now able to proceed with an analysis of parameterized families of p-jets directly analogous to the one for parameterized families of linear operators carried out in Section 5. We shall, of course, be particularly interested in those $(k+n)$-parameter families of the form G^p for some bifurcation problem G. As a straightforward parallel to Theorem 5.3 we have the following. Let $P = G^p(\mu_0, x_0)$ be the p-jet of G (with respect to the variable x) at (μ_0, x_0), and let $\mathscr{P}$ be the $\mathscr{K}$-equivalence class of P in $J^p(X, Y)$. If G^p is transverse to $\mathscr{P}$ at (μ_0, x_0) we say that G is *p-transverse* at (μ_0, x_0).

6.6. THEOREM. *Suppose the manifold $\mathscr{P}$ has codimension c in $J^p(X, Y)$, and G is p-transverse at (μ_0, x_0). Then near (μ_0, x_0) the set $(G^p)^{-1}(\mathscr{P})$ is a manifold in $\mathbb{R}^k \times X$ of codimension c.*

Of course $(G^p)^{-1}(\mathscr{P})$ is shorthand for the set of points (μ, x) at which the behaviour of G with respect to x is the same (in terms of $\mathscr{K}$-equivalence) as its behaviour at (μ_0, x_0), up to degree p in the respective Taylor series. Therefore we see that *if G^p happens to be p-transverse at every point*, the whole of $\mathbb{R}^k \times X$, or whatever part of it is of concern to us, is partitioned into manifolds of various (co-)dimensions, each one corresponding to local behaviour of G with respect to x of a certain $\mathcal{K}$-equivalence class up to terms of degree p. (Different $\mathcal{K}$-equivalence classes may give manifolds of the same dimension.) Moreover, if we know the local configurations of the $\mathcal{K}$-equivalence classes in $J^p(X, Y)$ we can expect to deduce a lot about the local configurations of the manifolds in $\mathbb{R}^k \times X$ of local behaviour type for G, although this is a bit more delicate than it was for linear operators in Section 5 because in general the $\mathcal{K}$-equivalence classes in $J^p(X, Y)$ do not fit together as nicely as the S_q did in $\mathcal{L}(X, Y)$.

This was all assuming transversality of G^p to the $\mathcal{K}$-equivalence classes. Is this a property that is "typically" satisfied by G? The Thom Transversality Theorem mentioned in Section 4 says that it is typical for mappings from $\mathbb{R}^k \times X$ to $J^p(X, Y)$ to be transverse to a given $\mathcal{K}$-equivalence class, but this is not quite what we want since of course not all such mappings are of the form G^p for some G. However, there is a refined version of the theorem, often called the *Jet Transversality Theorem*, which states that it *is* typical among mappings G for G^p to be transverse to a given manifold in $J^p(X, Y)$: this property holds for most G. (See Gibson, 1979, chapter II, for a clear discussion of this question for a single $F(x)$, easily extended to $G(\mu, x)$.) Transversality to all the $\mathcal{K}$-equivalence classes simultaneously is a different matter, though. This cannot usually be expected because there are too many of them. Nevertheless, by focusing attention on those of sufficiently low codimension useful results can be obtained, as will be described in Section 9.

Interesting or not as the preceding discussion may be as a general framework, it still tells us almost nothing about actual bifurcation behaviour. We need to answer a number of questions, first about each G_μ as a function of x alone:

(1) How can we find all the $\mathcal{K}$-equivalence classes in $J^p(X,Y)$?
(2) How can we identify the $\mathcal{K}$-equivalence class to which the p-jet of a given G_{μ_0} belongs at $x = x_0$?
(3) What will this tell us about the $\mathcal{K}$-equivalence class of G_{μ_0}, i.e. how justifiable is the neglect of terms of degree $> p$ in the Taylor series of G_{μ_0} at x_0?

Then will follow questions about the behaviour with respect to μ as well:

(4) How can we decide if G is p-transverse at (μ_0, x_0)?
(5) If G is p-transverse at (μ_0, x_0), what will this imply about the local behaviour of G at (μ_0, x_0)?

The next two Sections are devoted to answering these five questions.

7. Handling High Order Terms: Determinacy

A full description of the configuration of equivalence classes in $J^p(X, Y)$ is hard to envisage, since the dimensions involved, although finite, are quite high (e.g. $J^4(\mathbb{R}^3, \mathbb{R})$ has dimension 35), but just as in the case of families of matrices (Section 5) the significant thing was how the various equivalence classes abutted against each other, so it is in this case. Progress is slowly being made in obtaining this general information in $J^p(\mathbb{R}^n, \mathbb{R}^m)$ for some small values of p, n, and m with the work of Gibson and others at Liverpool. However, with the help of transversality as we shall shortly see, important conclusions can be drawn for bifurcation problems with a given number of parameters from only a limited knowledge of the equivalence classes in $J^p(X, Y)$ for small p. Therefore, what may have seemed an infinitely complicated programme becomes tractable and worthwhile. This is the response to the above Question 1.

In general Question 2 would be very hard to answer: it is a matter of skill and ingenuity with calculus and algebra. Nevertheless, in many cases, and especially those which arise from mathematical models of physical problems (for reasons see below), a "simple" representative for the $\mathscr{K}$-equivalence class of $G^p(\mu_0, x_0)$ can be found. For the sort of arguments that may be used see Gibson (1979) and Golubitsky and Schaeffer (1979a).

Question 3 is of crucial importance. It asks: *When can we rigorously reduce (in terms of $\mathscr{K}$-equivalence) a given G_{μ_0} to a polynomial mapping of degree $\leqslant p$*, i.e. to an m-vector of polynomials of degree $\leqslant p$ in $x_1, x_2, \ldots, x_n$? The answer is a slightly strange one at first sight, as we shall now see.

Let F be a differentiable mapping from X to Y, of dimensions n, m respectively, and let P be the p-jet of F at the point x_0 which for simplicity we now take to be the origin 0 in X. Write

$$F(x) = P(x) + R(x) \tag{7.1}$$

so that all derivatives of all components of R (the "remainder" of the Taylor series) at 0 vanish up to degree p.

The condition for F to be $\mathscr{K}$-equivalent to P is that there exist an invertible $m \times m$ matrix $M(x)$ and a coordinate change $x \to x'$ such that

$$P(x) + R(x) \equiv M(x)\, P(x'). \tag{7.2}$$

Suppose we make the additional assumption that $P + tR$ is $\mathscr{K}$-equivalent to P for all t with $0 \leqslant t \leqslant 1$ so that

$$P(x) + tR(x) \equiv M(t, x)\, P\,(x'(t, x)), \tag{7.3}$$

and moreover assume that M and x' can be chosen to vary smoothly with respect to t as well as x, satisfying $M(0, x) = I$ and $x'(0, x) = x$. Then differentiating (7.3) with respect to t and putting $t = 0$ gives

$$R(x) \equiv A(x)\, P(x) + DP(x)\, b(x), \tag{7.4}$$

where $A(x) = (\partial M/\partial t)\ (0, x)$ and $b(x) = (\partial x'/\partial t)\ (0, x)$. Note $b(0) = 0$ because $x'(t, 0) = 0$. Therefore, given P, our hypothesis about $\mathscr{K}$-equivalence between $F = P + R$ and P leads to the *necessary* condition that there exist an $m \times m$ matrix $A(x)$ and an n-vector $b(x)$ satisfying (7.4). This does not give a sufficient condition (it was derived only from $t = 0$) but it does provide the basis for one. The fundamental technical result is the following theorem, a sharpening of Proposition 3.11 in Golubitsky and Schaeffer (1979a). It goes back to the work of Mather (1968b).

7.1. THEOREM. *A sufficient condition for F to be $\mathscr{K}$-equivalent at 0 to P is: both $R(x)$ and $DR(x)a(x)$ for every n-vector $a(x)$ with $a(0) = 0$ should be expressible in the form*

$$A(x)P(x) + DP(x)b(x), \tag{7.5}$$

where $A(x)$ is an $m \times m$ matrix with $A(0) = 0$, and $b(x)$ is an n-vector with $b(0) = 0$ and $Db(0) = 0$.

Remember that $P(x)$ is an m-vector of polynomials in n variables so that $DP(x)$ is an $m \times n$ matrix. The expression (7.5) is an m-vector.

How would we ever know in a given case whether the conditions of Theorem 7.1 are satisfied? The most efficient answer lies in generalizing the problem.

7.2 COROLLARY. *Suppose P is given, and* every *mapping whose Taylor series at* 0 *vanishes up to degree p can be expressed as* (7.5). *Then* every *mapping whose p-jet at* 0 *is P is* $\mathscr{K}$*-equivalent at* 0 *to P*.

This follows from Theorem 7.1 as $R(x)$ and $DR(x)a(x)$ have Taylor series vanishing up to degree p. The condition here seems even harder to verify than it would be for a given R, but in fact it is easier in view of the following lemma from algebra (a case of Nakayama's lemma: see, for example, Golubitsky and Schaeffer, 1979a, Lemma 3.10, or Gibson, 1979, p. 102).

7.3. LEMMA. *If every mapping S whose Taylor series at* 0 *vanishes up to degree p can be expressed as* (7.5) when the terms of degree $p+2$ or higher are neglected, *then it can be so expressed without neglecting any terms.*

This now reduces the verification of the hypothesis of Corollary 7.2 to a problem of algebra of polynomials: *can every m-vector of homogeneous polynomials in* $x_1, x_2, \ldots, x_n$ *of degree* $p+1$ (since every S is precisely of this form, forgetting terms of degree $> p+1$) *be expressed as in* (7.5)?

7.4. *Examples.* 1. Take $n = m = 1$ and $P(x) = x^p$. Can x^{p+1} be expressed as

$$x^{p+1} = A(x)x^p + px^{p-1}b(x) \tag{7.6}$$

where $A(0) = b(0) = (db/dx)(0) = 0$? It can, and so any F with

$$F(x) = x^p + \text{higher order terms} \tag{7.7}$$

is $\mathscr{K}$-equivalent to x^p.

In fact it is a triviality to prove this directly since by Taylor's Theorem with remainder we can write

$$F(x) = x^p + x^p h(x), \tag{7.8}$$

so $F(x) = y^p$, where $y = x\{1+h(x)\}^{1/p}$ which is an invertible coordinate change near $x = 0$. However, we included the example for illustration.

2. Here $n = m = 2$; let $P(x_1, x_2) = (x_1^2 - x_2^2, -2x_1x_2)$. Now we ask whether each two-vector $v = (v_1, v_2)$ whose components are homogeneous cubics in x_1, x_2 can be written as

$$\begin{pmatrix} v_1 \\ v_2 \end{pmatrix} \equiv \begin{pmatrix} a_{11} & a_{12} \\ a_{21} & a_{22} \end{pmatrix} \begin{pmatrix} x_1^2 - x_2^2 \\ -2x_1x_2 \end{pmatrix} + \begin{pmatrix} 2x_1 & -2x_2 \\ -2x_2 & -2x_1 \end{pmatrix} \begin{pmatrix} b_1 \\ b_2 \end{pmatrix} \tag{7.9}$$

where a_{ij}, b_1, b_2 and $\partial b_i/\partial x_j$ are functions vanishing when $x_1 = x_2 = 0$. It is enough to check this for v equal to each of the eight vectors

$$\begin{pmatrix} x_1^3 \\ 0 \end{pmatrix}, \begin{pmatrix} x_1^2x_2 \\ 0 \end{pmatrix}, \begin{pmatrix} x_1x_2^2 \\ 0 \end{pmatrix}, \begin{pmatrix} x_2^3 \\ 0 \end{pmatrix}, \begin{pmatrix} 0 \\ x_1^3 \end{pmatrix}, \begin{pmatrix} 0 \\ x_1^2x_2 \end{pmatrix}, \begin{pmatrix} 0 \\ x_1x_2^2 \end{pmatrix}, \begin{pmatrix} 0 \\ x_2^3 \end{pmatrix}$$

in turn, because every other v is a linear combination of these. We leave it as an exercise to check that it can be done.

Hence any G with

$$G(x_1, x_2) = P(x_1, x_2) + \text{higher order terms} \tag{7.10}$$

is $\mathscr{K}$-equivalent to P.

A p-jet P with the property that adding arbitrary higher terms always produces mappings $\mathscr{K}$-equivalent to P is called *$\mathscr{K}$-determinate*. A mapping G whose p-jet at 0 is $\mathscr{K}$-determinate is *$\mathscr{K}$-p-determined* at 0. In order that G be $\mathscr{K}$-p-determined it is *sufficient* that the p-jet P of G at 0 should satisfy the hypothesis of Corollary 7.2, but this is not necessary. However, the sufficient condition is enough for most practical purposes, and a failure of $\mathscr{K}$-p-determinancy can often be spotted by other methods.

This completes our discussion of Question 3. Now we turn to bifurcations by involving the parameter μ again in our considerations.

8. Algebraic Implications of p-Transversality

To decide (Question 4) if G is p-transverse at (μ_0, x_0) we have to find:

(a) the tangent space $\mathscr{T}$ at $P = G^p(\mu_0, x_0)$ to the $\mathscr{K}$-equivalence class $\mathscr{P}$ of P; and
(b) the range $\mathscr{U}$ of the linear operator $DG^p(\mu_0, x_0)$ from $\mathbb{R}^k \times X$ to $J^p(X, Y)$,

and then see if their sum is the whole of $J^p(X, Y)$. To simplify matters we again take $(\mu_0, x_0) = (0, 0)$.

For (a) we seek the initial velocity vectors of all differentiable paths P_t in $\mathscr{P}$ starting at P. For such P_t we have by definition of $\mathscr{K}$-equivalence of p-jets

$$P(t, x) = P_t(x) \equiv M(t, x)\, P(x'(t, x)) \tag{8.1}$$

up to degree p, with notation as in (7.3). In this case the differentiable dependence of M and x' on t can be proved, as part of the machinery for showing $\mathscr{P}$ is a manifold (Theorem 6.5). Then again we differentiate with respect to t and put $t = 0$, obtaining

$$\frac{\partial P}{\partial t}(0, x) \equiv A(x)\, P(x) + DP(x)\, b(x) \tag{8.2}$$

up to degree p, with A, b as in (7.4). Therefore *the set of all p-jets at* 0 *of mappings from X to Y expressible as the right-hand side of* (7.4) *or* (8.2) *up to degree p* is precisely the tangent space $\mathscr{T}$ to $\mathscr{P}$ at P.

For (b) we have that $\mathscr{U}$ is just the sum of the range $\mathscr{U}_\mu$ of $(\partial G^p/\partial \mu)(0, 0)$ and the range $\mathscr{U}_x$ of $(\partial G^p/\partial x)(0, 0)$. Therefore p-transversality holds if and only if

$$J^p(X, Y) = \mathscr{T} + \mathscr{U}_\mu + \mathscr{U}_x. \tag{8.3}$$

Now a typical element w of $\mathscr{U}_\mu$ has the form

$$w = \frac{\partial G^p}{\partial \mu}(0, 0).v \tag{8.4}$$

for some constant k-vector v, and by reversing orders of differentiation we have

$$\frac{\partial G^p}{\partial \mu}(0, 0).v = (p\text{-jet at } x = 0 \text{ of } \frac{\partial G}{\partial \mu}(0, x).v). \tag{8.5}$$

Likewise a typical element of $\mathscr{U}_x$ is of the form

$$\frac{\partial G^p}{\partial x}(0,0).u = (p\text{-jet at } x = 0 \text{ of } \frac{\partial G}{\partial x}(0,x).u \equiv DG_0(x).u) \tag{8.6}$$

for a constant n-vector u. Therefore $\mathscr{T} + \mathscr{U}_x$ is the set of all p-jets of n-vectors of the form

$$A(x)G_0(x) + DG_0(x)c(x) \tag{8.7}$$

with $c(0)$ now not necessarily 0, and so we have proved the following result.

8.1. LEMMA. *The k-parameter family $G(\mu, x)$ is p-transverse at* (0, 0) *if and only if every p-jet from X to Y can be written as the p-jet at* 0 *of*

$$A(x)G_0(x) + DG_0(x)c(x) + \frac{\partial G}{\partial \mu}(0,x).v \tag{8.8}$$

where $A(x)$ is an $m \times m$ matrix, $c(x)$ is an n-vector, and v is a constant k-vector.

We shall give some examples later, after studying Question 5. This is the most important of all of them and its answer is the key to understanding the power of singularity theory as a tool in bifurcation theory. First, a definition to liberate ourselves temporarily from p-jets for any particular p.

8.2. DEFINITION. *A k-parameter family $G(\mu, x)$* is transverse *at* (0, 0) *if every differentiable map F from X to Y can be expressed as*

$$F(x) \equiv A(x)G_0(x) + DG_0(x)c(x) + \frac{\partial G}{\partial \mu}(0,x).v \tag{8.9}$$

in some neighbourhood of the origin, where A, c and v are as in Lemma 8.1.

Clearly a transverse family is p-transverse for every p by taking p-jets of (8.9), but the converse is not obvious although in fact it is true.

The condition (8.9) is not computable directly since there are too many possibilities for F, but we shall see in a moment how to overcome this problem. This transversality definition is what we might have arrived at had we been able to work sensibly with spaces of functions instead of their p-jets, but it is not easy to do that rigorously. However, the reason for making the definition is that it is the input to the following major theorem.

Recall from Section 2 that G is called a k-parameter *unfolding* of G_0, particularly when we are dealing (as we are here) with *local* behaviour near (0, 0).

8.3. THEOREM. *Suppose the k-parameter unfolding G of G_0 is a transverse unfolding. Then G is universal in the following sense: any l-parameter unfolding H of the same G_0 is $\mathscr{K}$-equivalent to a family $\overline{G}$ obtained from G by substitution of parameters.*

By this we mean H is $\mathscr{K}$-equivalent (Definition 6.4) to $\overline{G}$, where

$$\overline{G}(\nu, x) \equiv G(\mu(\nu), x) \tag{8.10}$$

for some function $\mu = \mu(\nu)$. Thus in a certain precise sense, the universal unfolding G contains all information about perturbations of G_0. From this it is straightforward to show a corollary.

8.4. COROLLARY. *Any two transverse k-parameter unfoldings of G_0 are $\mathscr{K}$-equivalent.*

Since an unfolding sufficiently close to a transverse unfolding is also transverse, but a non-transverse unfolding can be perturbed to meet $\mathscr{K}$-equivalence classes differently, we are led to the next result.

8.5. COROLLARY. A *k-parameter unfolding G is structurally stable precisely when it is transverse.*

See Gibson (1979, pp.164–167) for a treatment of these results, which stem (as does almost everything in this field) from the works of Mather (1968, *et seq.*).

Thus we have finally arrived at an analytical characterization of structurally stable families of mappings, as originally advertised in Section 1. We have yet to show that this is computable in practice. The clue to this, as with many problems of computation in singularity theory, is again Nakayama's Lemma. Suppose:

(i) G is p-transverse; and
(ii) every mapping from X to Y whose Taylor series at 0 vanishes up to degree p can be expressed as (8.7) in some neighbourhood of the origin. (8.11)

Then (i) means that (8.9) holds up to degree p, and (ii) then says that all terms of degree greater than p can be incorporated into $A(x)G_0(x) + DG_0(x)c(x)$, and so (8.9) holds. Finally, by Lemma (7.3) we see that to check (ii) it suffices to check it while neglecting terms of degree greater than $p+1$, and hence to check it for homogeneous monomials of degree $p+1$, which reduces the problem to one of finite polynomial algebra.

8.6. *Examples.* 1. $n = m = 1, k = 2$: $G(\mu_1, \mu_2, x) = x^3 + \mu_1 x + \mu_2$. Here $G_0(x) = x^3$ and so (cf. Example 7.4, 1) condition (8.11) (ii) is satisfied with $p = 3$. We have $DG_0(x) = 3x^2$, $(\partial G/\partial \mu_1)(0, x) = x$ and $(\partial G/\partial \mu_2)(0, x) = 1$. Can every differentiable function $F(x)$ be expressed as

$$F(x) \equiv A(x)x^3 + 3x^2 c(x) + x v_1 + 1 v_2 \text{ plus terms of degree} \geqslant 4 \tag{8.12}$$

in some neighbourhood of the origin? Clearly it can, so G is three-transverse and hence transverse by (8.11). By Corollary 8.5 it is structurally stable.

Observe that if μ_2 is deleted then the resulting $G(\mu_1, x) = x^3 + \mu_1 x$ is not three-transverse, so not transverse and not structurally stable—as was well known already (see Fig. 1.2).

2. $n = m = 2$, $k = 8$. Write (x, y) for (x_1, x_2). Take $G(\mu_1, \mu_2, \ldots, \mu_8, x, y) = (x^2 + y^2 - \mu_8, x^4 + \mu_1 x + \mu_2 x^2 + \mu_3 x^3 + \mu_4 y + \mu_5 xy + \mu_6 x^2 y - \mu_7)$.

$G_0(x, y) = (x^2 + y^2, x^4)$, and it is an easy exercise to check that (8.11) (ii) is satisfied with $p = 4$. For four-transversality we have to see if every $F(x, y)$ can be written as

$$F(x, y) = \begin{bmatrix} F_1(x, y) \\ F_2(x, y) \end{bmatrix} \equiv \begin{bmatrix} a_{11} & a_{12} \\ a_{21} & a_{22} \end{bmatrix} \begin{bmatrix} x^2 + y^2 \\ x^4 \end{bmatrix} + \begin{bmatrix} 2x & 2y \\ 4x^3 & 0 \end{bmatrix} \begin{bmatrix} c_1 \\ c_2 \end{bmatrix}$$
$$+ \begin{bmatrix} -\mu_8 \\ \mu_1 x + \mu_2 x^2 + \mu_3 x^3 + \mu_4 y + \mu_5 xy + \mu_6 x^2 y - \mu_7 \end{bmatrix} \text{plus terms of degree} \geqslant 5. \tag{8.13}$$

A bit of algebra shows that this can be done, so G is four-transverse, so transverse and structurally stable. The algebra also shows four-transversality and hence structural stability *fails* if any of the μ's are set equal to zero.

3. Again $n = m = 2$. Let $G_0(x, y) = (x^5, y^5)$. Then (8.11) (ii) is *not* satisfied for $p = 5$

because (among other things) there is no way in which the vector $(x^3 y^3, 0)$ can be expressed as

$$\begin{bmatrix} x^3y^3 \\ 0 \end{bmatrix} \equiv \begin{bmatrix} a_{11} & a_{12} \\ a_{21} & a_{22} \end{bmatrix} \begin{bmatrix} x^5 \\ y^5 \end{bmatrix} + \begin{bmatrix} 5x^4 & 0 \\ 0 & 5y^4 \end{bmatrix} \begin{bmatrix} c_1 \\ c_2 \end{bmatrix}. \tag{8.14}$$

However, (8.11) (ii) *is* satisfied for $p = 7$ because every monomial in x, y of degree 8 contains either x^5 or y^5 or $x^4 y^4$. It is then routine to construct a seven-transverse family, e.g. the 26-parameter family

$$G(\mu_1, \mu_2, \ldots, \mu_{26}, x, y) = \begin{bmatrix} x^5 + \mu_1 x^3 y^2 + \mu_2 x^2 y^3 + \mu_3 x y^4 + C_1 + Q_1 + L_1 + K_1 \\ y^5 + \mu_{14} x^2 y^3 + \mu_{15} x^3 y^2 + \mu_{16} x^4 y + C_2 + Q_2 + L_2 + K_2 \end{bmatrix}$$

where C_i, Q_i, L_i, K_i represent all cubic, quadratic, linear and constant terms which for two variables x, y require 4, 3, 2, 1 coefficients respectively.

4. $m = 1, n = 2$: $G_0(x, y) = x^2 y$. Here we can never hope to express an arbitrary $F(x, y)$ as

$$ax^2y + (2xy, x^2)\begin{bmatrix} c_1 \\ c_2 \end{bmatrix} + v_1 h_1(x, y) + \ldots + v_k h_k(x, y)$$

for any k and suitable choice of $h_i(x)$ because y^q for any q cannot receive any contribution from the first two terms, and will not be expressible using the last k terms for *every q*. Therefore G_0 has no transverse unfolding and no structurally stable unfolding.

9. Some General Classification Results

In the last Section we saw how to check for structural stability of k-parameter families. The rule is rather more complicated than the "rule of thumb" that suggests that the total number of parameters (including load and imperfections) needed to provide a structurally stable unfolding should be one more than the number of essential variables: cf. Thompson and Hunt (1973).

Armed with this machinery and inspired by the geometric aspect of transversality, we are now in a position to state some general theorems about bifurcations in k-parameter families. By Theorem 6.6, if G^p is transverse to every $\mathcal{K}$-equivalence class in $J^p(X, Y)$, then it will avoid all those $\mathcal{K}$-equivalence classes whose codimension is greater than $k+n$ where $n = \dim X$. We are interested in places where G vanishes, so we focus on the subspace $J_0^p(X, Y)$ of $J^p(X, Y)$ corresponding to the value 0 in the Y factor in (6.5). Clearly $J_0^p(X, Y)$ has codimension $m = \dim Y$ in $J^p(X, Y)$. The $\mathcal{K}$-equivalence classes of interest to us are those lying in $J_0^p(X, Y)$, so transversality guarantees that G^p will avoid those whose codimension in $J_0^p(X, Y)$ is greater than $k+n-m$. In particular, if $n = m$ is arbitrary, *any structurally stable (and hence p-transverse) k-parameter families can only involve $\mathcal{K}$-equivalence classes whose codimension in $J_0^p(X, Y)$ is less than or equal to k.*

A programme for classifying structurally stable k-parameter bifurcation problems is now apparent. First find the $\mathcal{K}$-equivalence classes of codimension less than or equal to $l = k+n-m$ in $J_0^1(X, Y)$: for this see Proposition 5.1, noting that $\mathcal{K}$-equivalence for 1-jets vanishing at 0 amounts to equivalence of matrices as in Lemma 5.5. Then look for $\mathcal{K}$-equivalence classes of codimension less than or equal to l in $J_0^2(X, Y)$, bearing in mind that it is only necessary to look at those whose J_0^1-parts have already been identified in the first step. Then proceed similarly in $J_0^3(X, Y)$, looking only at classes already picked out in $J_0^2(X, Y)$, and so on. These calculations involve more or less complicated algebra, and also make heavy use of p-determinacy (Section 7) to show when a class in $J^p(X, Y)$ represents also a single class in $J^{p'}(X, Y)$ for every $p' > p$. Eventually, as can be shown by an easy algebraic argument which we will not discuss (see, for example, Gibson, 1979, p. 191), at some value of p the process must terminate, in that no new $\mathscr{K}$-equivalence classes of codimension $\leqslant l$ are introduced by taking higher order terms. Taking p slightly larger if necessary it also happens automatically that the hypothesis of Corollary 7.2 is satisfied for a representative P of each $\mathscr{K}$-equivalence class, and so taking this $p = q$ we have the next theorem.

9.1. THEOREM. *A k-parameter family $G(\mu, x)$ transverse to every $\mathscr{K}$-equivalence class in $J^q(X, Y)$ will encounter only the $\mathscr{K}$-equivalence classes of codimension $\leqslant l = k+n-m$. The transversality implies the universality and hence structural stability of G at each (μ_0, x_0), and in a neighbourhood of each (μ_0, x_0) the mapping G is $\mathscr{K}$-equivalent as a bifurcation problem (Definition 6.4) to any other k-parameter problem q-transverse at (μ_0, x_0), and in particular to a polynomial problem.*

Are most mappings G transverse to every $\mathscr{K}$-equivalence class in $J^q(X, Y)$? Unfortunately not, in general, because there are too many of them: they tend to arise in continuous families of classes. However, and this is the crucial point for some classification results, *for certain low l there are only finitely many $\mathscr{K}$-equivalence classes of codimension $\leqslant l$*, and all other classes are contained in a finite collection of manifolds of codimension $> l$. *Therefore structurally stable k-parameter bifurcation problems will encounter only finitely many types of local behaviour*. We now give two instances in which this fact is usefully exploited.

The first is about variational problems. Let $G(\mu, x) = G_\mu(x) = \text{grad}\, g_\mu(x)$ where $g_\mu(x) = g(\mu, x)$ is a real-valued function on $\mathbb{R}^k \times X$, so zeros of G_μ are critical points of g_μ. We are interested in *degenerate* critical points of g_μ, i.e. singular points for the bifurcation problem $G(\mu, x) = 0$. We work with $\mathscr{R}$-equivalence of functions which, as we saw in Section 6, implies $\mathscr{K}$-equivalence of gradient mappings, although it is not strictly the same thing. The algebraic machinery of Sections 7 and 8 is much the same in this context: simply put $m = 1$, and take $M(x) = I$ and $A(x) \equiv 0$ from (7.3) onwards, allowing also the addition of arbitrary constants c_μ to g_μ which do not affect variational behaviour.

9.2. THEOREM. *For any n and for $k \leqslant 5$, any structurally stable k-parameter family of functions from $X = \mathbb{R}^n$ to $\mathbb{R}$ is at every degenerate critical point locally $\mathscr{R}$-equivalent to one of a finite list of families. Each entry on the list has the form $g_\mu(x) = Q(y) + \tilde{g}_\mu(z)$, where Q is a nondegenerate quadratic form in $(n-1)$ or $(n-2)$ variables, and z represents the remaining* 1 *or* 2 *"essential" variables. The number of $\mathscr{R}$-equivalence classes for $\tilde{g}$ is given by*

k	1	2	3	4	5
number of classes	1	3	6	10	15

If $\pm\,\tilde{g}$ are regarded as equivalent (which for applications they should not be, since the sign influences stability questions) then the numbers in the bottom row reduce to 1, 2, 5, 7, 11, respectively. For $k = 4$ these classes are the somewhat celebrated "seven elementary catastrophes" of René Thom (1975), who first propounded this theorem. It is proved in the works of Mather (1968 *et seq.*), and was also obtained by the Soviet mathematician V. I. Arnol'd (see Arnol'd, 1972a). For more accessible treatments see Wassermann (1974), Bröcker and Lander (1975), Zeeman and Trotman (1976) and Gibson (1979). See also Poston and Stewart (1978) for a full and well-illustrated discussion of the methods. We note in passing that the two $\mathscr{R}$-equivalence classes for $\tilde{g}$ when $k \leqslant 2$ are the *fold* (*limit point*) represented by $x^2 + \mu_1$ (see Example 1 in Fig. 1.1) and the ubiquitous *cusp* represented by $x^3 + \mu_1 x + \mu_2$ (cf. Example 8.6, 1). For $k \leqslant 3$ we also meet the *elliptic* and *hyperbolic umbilic*, both with two essential variables and important in coupled-mode buckling problems: see the striking pictures drawn by Thompson and Hunt (1975). The fact that for $k \leqslant 5$ there are at most two essential variables is a reflection of Remark 5.2, 3, since $q > 2$ implies codim $S_q > 5$.

Not all problems are variational. In the non-variational case there is not such a tidy theorem, but the general situation is quite well understood and has been described in some detail by Wan (1979). Here is one version of the special case $m = n$.

9.3. THEOREM. *Suppose $m = n$ and $k \leqslant 8$. Then any structurally stable k-parameter family of mappings is at every point locally $\mathscr{K}$-equivalent to one of a finite list of families.*

A full list is not explicitly available, but it could in principle be worked out in a finite amount of time. See Wan (1979) for details.

Theorems of this kind are appealing to a mathematician, since they show that what may seem to be an infinite forest of increasing complexity when approached by *ad hoc* power series methods, can be brought under control once the ingredient of transversality, with its flavours of typicality and structural stability, is introduced. Moreover, the physical translation of structural stability (robustness against small perturbations in the data) suggests that the theorems, or more probably the machinery which gives rise to them, should be of particular significance to practical bifurcation problems, such as buckling problems in engineering. Initial optimism over these prospects is soon seen to be misplaced, however, mainly because the assertions from singularity theory about what should be expected in bifurcation problems do not fit the facts of the subject. Examples such as the pitchfork in Fig. 1.2 are of fundamental importance in bifurcation theory, and yet, according to the programme we have followed in this article, they should be regarded as exceptional freaks since the structurally stable one-parameter families can involve nothing more complicated than folds (limit points). Something has gone wrong.

For a partial explanation we should return to Section 1. There we set out to deal with perturbed bifurcation theory by treating k-parameter bifurcation problems $G(\mu, x) = 0$ as the basic objects of study. This blurred the essential physical difference between a *load* λ and *imperfections* $\alpha_1, \alpha_2, \ldots, \alpha_r$. To recapture relevance to real problems we should somehow regard $G(\lambda, \alpha_1, \alpha_2, \ldots, \alpha_r, x)$ as an r-parameter unfolding of $G(\lambda, x)$, which we regard as given, rather than see the whole thing as an $(r+1)$-parameter unfolding of $G(0, x)$. Therefore we need a refined version of the algebraic techniques of Sections 7 and 8, based on a form of $\mathscr{K}$-equivalence that respects the different physical rôles of λ and the α_j.

There is another more subtle way in which transversality as we have used it is inappropriate to many bifurcation problems. In designing structures and optimizing their performance and manufacture, *symmetry* plays a very large part: it cuts costs, pleases the eye, and simplifies calculations. Although complete symmetry is unattainable in the real world, a mathematical model for a structure will usually incorporate some symmetry, which it is then left to nature to adjust. Consequently an insistence on transversality as representing *typicality* is inconsistent with this type of model, since pure symmetry is by definition atypical. What is needed here is a re-casting of the whole transversality/universality formalism *within the context of the action of a group of symmetries*, either discrete or continuous.

Happily, these two necessary advancements of the original theory of $\mathscr{K}$-equivalence have recently been successfully carried out in the two major papers of Golubitsky and Schaeffer

(1979a, b) which bring singularity theory squarely to the forefront as a tool for perturbed bifurcation theory in the possible presence of symmetries. These authors do not offer classification theorems in the style of Theorems 9.2 or 9.3, but they systematically develop the machinery from calculus and algebra which give computable criteria to:

(i) decide if a given perturbed bifurcation problem is structurally stable or not within a given load/imperfection context with or without symmetry;
(ii) decide if a finite number of additional parameters, suitably inserted into the equations, can make the problem structurally stable;
(iii) show how to do this, if it can be done;
(iv) find a polynomial problem rigorously $\mathscr{K}$-equivalent to the given problem in this case.

Remember that everything here refers to *local* behaviour near some chosen point (λ_0, x_0).

Among the results that emerge from the analysis are the following facts about structural stability of bifurcation problems with no more than two imperfections. Take $(\lambda_0, x_0) = (0, 0)$ and say that a bifurcation problem $G(\lambda, x)$ (here as in Section 6 we suppress the "= 0") is *reduced* if a Liapunov–Schmidt reduction has already been applied, so that $DG_0(0)$ is the zero matrix (G_0 has no linear terms).

9.4. THEOREM. *Any reduced bifurcation problem which is structurally stable (robust against every imperfection) must be $\mathscr{K}$-equivalent to the one-variable problem*

$$G(\lambda, x) \equiv x^2 - \lambda. \tag{9.1}$$

Of course, it is well known that limit points persist under arbitrary perturbation whereas other bifurcations do not. The formal use of $\mathscr{K}$-equivalence here adds little to a familiar scene. It contributes more in the next theorem.

9.5. THEOREM. *Any reduced bifurcation problem which needs one imperfection parameter α to make it structurally stable (robust against further imperfections) must after inclusion of such a parameter be $\mathscr{K}$-equivalent to one of the three one-variable problems:*

$$\begin{aligned} &\text{(a)}\ G(\lambda, \alpha, x) \equiv x^2 - \lambda x + \alpha, \\ &\text{(b)}\ G(\lambda, \alpha, x) \equiv x^2 + \lambda^2 + \alpha, \\ &\text{(c)}\ G(\lambda, \alpha, x) \equiv x^3 + \alpha x - \lambda. \end{aligned} \tag{9.2}$$

See Fig. 9.1 for pictorial descriptions. Note that (9.2)(c) with $\alpha = 0$ is not a bifurcation in the customary sense, but its response to varying α shows how two limit points are created, giving rise to hysteresis behaviour embryonically there in the $\alpha = 0$ picture. Also (9.2)(b) is rather unusual in representing the creation of a circle of solutions in (λ, x) as α varies from positive to negative. Such behaviour is unlikely in buckling problems but occurs in models for chemical reactions: see Golubitsky and Keyfitz (1980).

Observe that the admissible coordinate change $(x', \lambda') = (x - \frac{1}{2}\lambda, \frac{1}{2}\lambda)$ converts $x^2 - \lambda x$ into the equivalent form $x'^2 - \lambda'^2$, and so (a) and (b) are the same in a complex setting.

Again, these examples come as no surprise to a bifurcation theorist. To *prove* that any problem needing only one imperfection to realize its full behaviour must be rigorously equivalent to one of these polynomials may be a different matter, however.

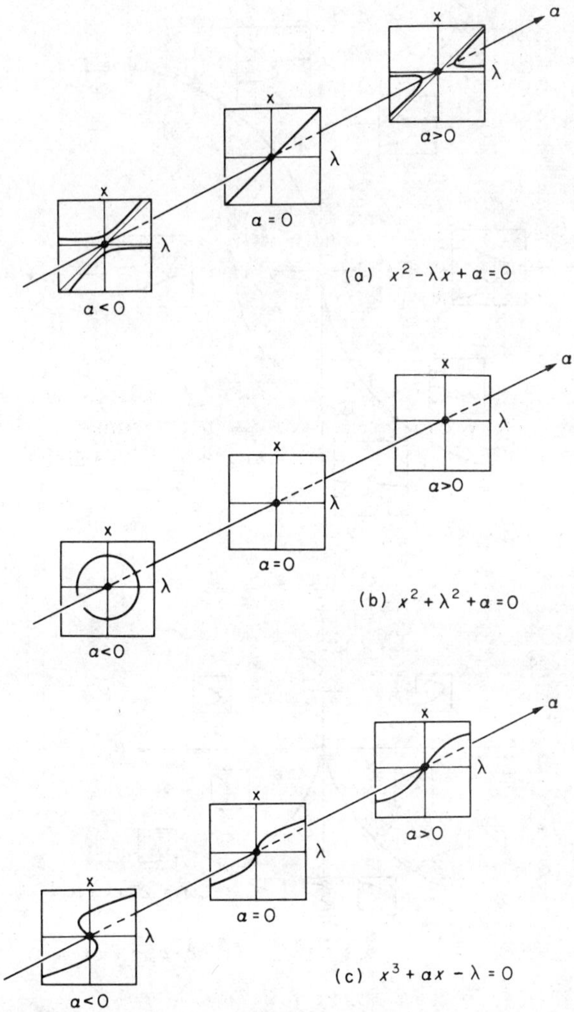

FIG. 9.1. The only three reduced bifurcation problems (up to $\mathscr{K}$-equivalence) needing one imperfection parameter α to make them structurally stable.

9.6. THEOREM. *Any reduced bifurcation problem which needs two imperfection parameters to make it structurally stable must after inclusion of such parameters be $\mathscr{K}$-equivalent to one of the three one-variable problems:*

$$\begin{aligned}
&\text{(a)}\ G(\lambda, \alpha_1, \alpha_2, x) \equiv x^3 + \alpha_2 x^2 - \lambda x + \alpha_1, \\
&\text{(b)}\ G(\lambda, \alpha_1, \alpha_2, x) \equiv x^4 + \alpha_2 x^2 + \alpha_1 x - \lambda, \\
&\text{(c)}\ G(\lambda, \alpha_1, \alpha_2, x) \equiv x^2 - \lambda^3 - \alpha_2 \lambda + \alpha_1.
\end{aligned} \tag{9.3}$$

The pictures here are more difficult to draw, needing four dimensions. A way to handle this is to sketch a (λ, x)-picture for (α_1, α_2) in various regions of the plane (Fig. 9.2).

Note that regions II, IV of Fig. 9.2(a) are only entered by perturbing $G(\lambda, x)$ along a *nonlinear* path in the (α_1, α_2)-plane, and so are relatively difficult to detect. They have been

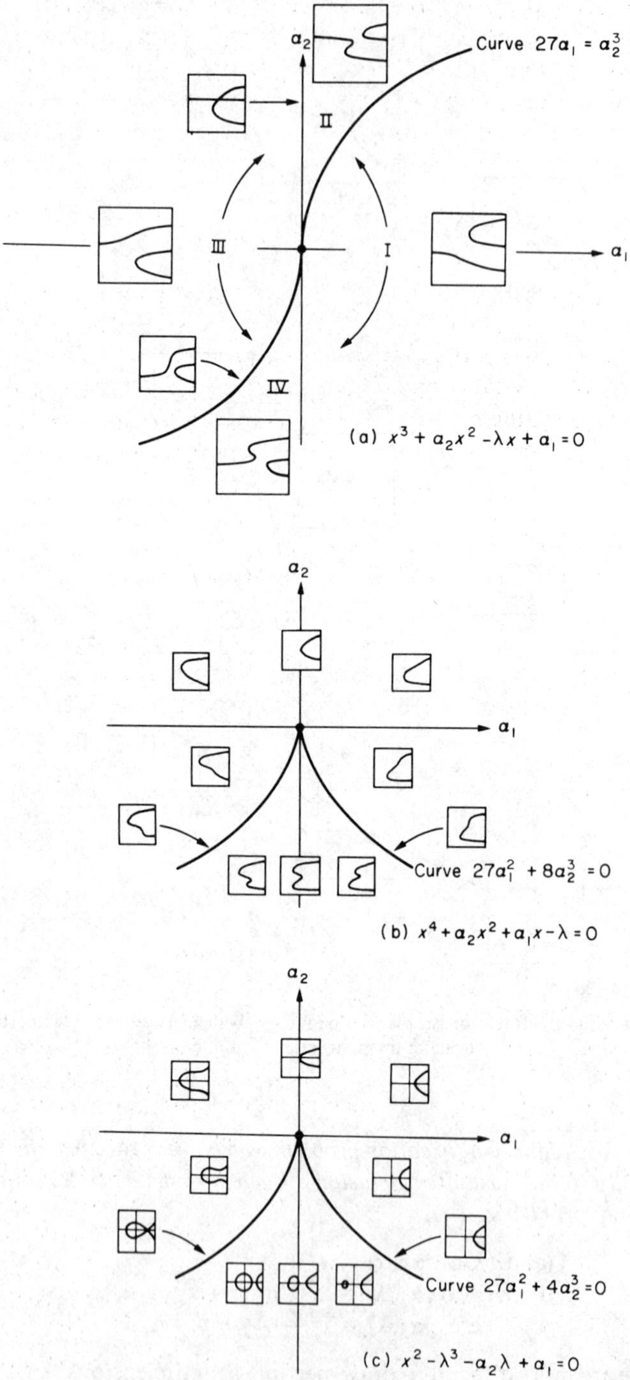

FIG. 9.2. The only three reduced bifurcation problems (up to $\mathscr{K}$-equivalence) needing two imperfection parameters α_1, α_2 to make them structurally stable.

encountered in bifurcation problems of fluid dynamics both experimentally (Benjamin, 1978) and theoretically (I. C. Walton, private communication), but without singularity theory methods the precise rôle in the universe of *all* perturbations of $x^3 - \lambda x$ may not have been clear. Also, (9.3)(b) distinguishes sharply between a quadratic fold and $x^4 - \lambda$, which "look" the same. The number of imperfection parameters needed for a full unfolding reflects the degeneracy of x^4. Case (c) arises in chemical reaction models mentioned above.

9.7. THEOREM. *Any reduced bifurcation problem* $G(\lambda, x_1, x_2)$ *in* two *variables needs at least* seven *imperfection parameters to make it structurally stable.*

This helps to explain why explicit classification theorems become less worthwhile than promotion of the tools of the theory, which can then be used in whatever ways may be appropriate to various problems in applied mathematics. The methods of Golubitsky and Schaeffer, inspired by Thom's catastrophe theory, have in a short time led to new results in bifurcation theory for buckling of plates (Schaeffer and Golubitsky, 1979), fluid dynamics (Schaeffer, 1980), and chemical reaction theory (Golubitsky and Keyfitz, 1980; Golubitsky *et al.*, 1980), and may yet lead to many more.

10. Comments on Infinite Dimensions

The general observation here is that everything goes through in direct analogy with the finite-dimensional theory, provided it is assumed that $DG_\mu(x)$ is a *Fredholm* operator for every (μ, x) in the zero set for G. This says that all degeneracies are essentially finite-dimensional in character.

The Liapunov–Schmidt reduction process converts such bifurcation problems into finite-dimensional ones, but it still remains to be asked whether p-determinacy, transversality, universality, and structural stability for the reduced problems are equivalent to these properties for the original problems.

Basic work in establishing the validity of local theory and the machinery of Sections 7 and 8 for $\mathscr{R}$-equivalence of families of scalar functions in this infinite dimensional context has been carried out by Magnus (1977, 1978, 1979) and Arkeryd (1979). (Magnus has pointed out (private communication) a mistake in the stability theorem (Theorem 5) of his paper (1979). A corrected version will be published (1981) but meanwhile see Arkeryd, 1979.) Global statements about the typicality of transversality are given by Arkeryd (1979) and Chillingworth (1980). For treatments of $\mathscr{K}$-equivalence, see Guimarães (1979) and Arkeryd (1979).

References

Abraham, R. and Robbin, J. (1967) *Transversal Mappings and Flows*, Benjamin, New York.
Arkeryd, L. (1977) Catastrophe theory in Hilbert space, Tech. Report, University of Göteborg.
Arkeryd, L. (1979a) Thom's theorem for Banach spaces, *J. London Math. Soc.* **19**, 359–370.
Arkeryd, L. (1979b) Imperfect bifurcation and Banach space singularity theory, Chalmers University of Technology and University of Göteborg preprint.
Arnol'd, V. I. (1968) Singularities of smooth mappings, *Russian Math. Surveys* **23,** 1–43.
Arnol'd, V. I. (1971) On matrices depending on parameters, *Russian Math. Surveys* **26,** 29–43.
Arnol'd, V. I. (1972a) Normal forms for functions near degenerate critical points, the Weyl Groups of A_k, D_k, E_k and Lagrangian singularities, *Funkcional. Anal. i Priložen.* **6,** 3–25; *Functional Anal. Appl.* **6,** 254–272.

Arnol'd, V. I. (1972b) Lectures on bifurcations in versal families, *Russian Math. Surveys* **27,** 54–123.
Benjamin, T. B. (1978) Bifurcation phenomena in steady flows of a viscous fluid, *Proc. R. Soc. London*, S A, **359,** 1–26, 27–43.
Bröcker, Th. and Lander, L. C. (1975) *Differentiable Germs and Catastrophes*, London Math. Soc. Lecture Notes 17, Cambridge Univ. Press.
Callahan, J. J. (1974) Singularities and plane maps, *Am. Math. Monthly* **81,** 211–240.
Cesari, L. (1976) Functional analysis, nonlinear differential equations and the alternative method, in *Nonlinear Functional Analysis and Differential Equations*, 1–198, Dekker, New York.
Chillingworth, D. R. J. (1980) A global genericity theorem for bifurcations in variational problems, *J. Func. Anal.* **35,** 251–278.
Gibson, C. G. (1979) *Singular Points of Smooth Mappings*, Research Notes in Mathematics 25, Pitman, London.
Golubitsky, M. (1978) An introduction to catastrophe theory and its applications, *SIAM Review* **20,** 352–387.
Golubitsky, M. and Guillemin, V. (1974) *Stable Mappings and their Singularities*, Graduate Texts in Mathematics. No. 14, Springer-Verlag, New York.
Golubitsky, M. and Schaeffer, D. (1979a) A theory for imperfect bifurcation via singularity theory, *Comm. Pure Appl. Math.* **32,** 21–98.
Golubitsky, M. and Schaeffer, D. (1979b) Imperfect bifurcation in the presence of symmetry, *Commun. Math. Phys.* **67,** 205–232.
Golubitsky, M. and Keyfitz, B. L. (1980) A qualitative study of the steady state solutions for a continuous flow stirred tank chemical reactor, *SIAM J. Math. Anal.* **11,** 316–339.
Golubitsky, M., Keyfitz, B. L. and Schaeffer, D. (1980) A singularity theory analysis of the thermal-chainbranching model for the explosion peninsula in the H_2-O_2 reaction, to appear.
Guimarães, L. C. (1979) *Contact Equivalence and Bifurcation Theory*, thesis, Southampton University.
Hale, J. K. (1977) Generic bifurcation with applications, in *Nonlinear Analysis and Mechanics: Heriot-Watt Symposium*, Vol. I, edited by R. J. Knops, Research Notes in Mathematics 17, Pitman, London.
Keener, J. P. (1974) Perturbed bifurcation theory at multiple eigenvalues, *Arch. Rat. Mech. Anal.* **56,** 348–366.
Keener, J. P. and Keller, H. (1973) Perturbed bifurcation theory, *Arch. Rat. Mech. Anal.* **50,** 159–175.
Koiter, W. T. (1945) *Over de Stabiliteit van het Elastich Evenwicht*, thesis, Delft, English translation Tech. Report. (1970) AFFDL-TR-70-25.
Lu, Y.-C. (1976) *Singularity Theory and an Introduction to Catastrophe Theory*, Springer-Verlag, New York.
Magnus, R. J. (1977) On universal unfoldings of certain real functions on a Banach space, *Math. Proc. Cambridge Philos. Soc.* **81,** 91–95.
Magnus, R. J. (1978) Determinacy in a class of germs on a reflexive Banach space, *Math. Proc. Cambridge Philos. Soc.* **84,** 293–302.
Magnus, R. J. (1979) Universal unfoldings in Banach spaces: reduction and stability, *Math. Proc. Cambridge Philos. Soc.* **86,** 41–45.
Magnus, R. J. (1981) Stability of universal unfoldings: a correction, *Math. Proc. Cambridge Philos. Soc.* **90,** 195–196.
Marsden, J. E. and Hughes, T. J. R. (1978) Topics in the mathematical foundations of elasticity, in *Nonlinear Analysis and Mechanics: Heriot-Watt Symposium*, Vol. II, edited by R. J. Knops, Research Notes in Mathematics 27, Pitman, London.
Mather, J., Stability of C^∞ mappings:
(1968a) The Division Theorem, *Ann. of Math.* **87,** 89–104.
(1968b) Finitely determined map germs, *Math. Publ. IHES* **35,** 127–156.
(1969a) Infinitesimal stability implies stability, *Ann. of Math.* **89,** 254–291.
(1969b) Classification of stable germs by *R*-algebras, *Math. Publ. IHES* **37,** 223–248.
(1970) Transversality, *Advances in Math.* **4,** 301–336.
(1971) The nice dimensions, *Proceedings of Liverpool Singularities Symposium* I, 207–253, edited by C. T. C. Wall, Lecture Notes in Math. 192, Springer-Verlag, Berlin.
Matkowsky, B. and Reiss, E. (1977) Singular perturbations of bifurcations, *SIAM J. Appl. Math.* **33,** 230–255.
Milnor, J. (1963) *Morse Theory*, Annals of Math. Studies 51, Princeton Univ. Press.
Poston, T. and Stewart, I. N. (1978) *Catastrophe Theory and Its Applications*, Pitman, London.
Quinn, F. (1970) Transversal approximation on Banach manifolds, *Global Analysis, Proc. AMS Symposium in Pure Math.* XV, 213–222, Am. Math. Soc., Providence, RI.
Schaeffer, D. (1980) Qualitative analysis of a model for boundary effects in the Taylor problem, *Math. Proc. Cambridge Philos. Soc.* **87,** 307–337.
Schaeffer, D. and Golubitsky, M. (1979) Boundary conditions and mode jumping in the buckling of a rectangular plate. *Commun. Math. Phys.* **69,** 209–236.
Sewell, M. J. (1966) On the connexion between stability and the shape of the equilibrium surface, *J. Mech. Phys. Solids* **14,** 203–230.
Shearer, M. (1978) On the null-spaces of linear Fredholm operators depending on several parameters, *Math. Proc. Cambridge Philos. Soc.* **84,** 131–142.
Sussmann, H. J. (1975) Catastrophe theory, *Synthèse* **31,** 229–270.

Thom, R. (1954) Quelques propriétés globales des variétés différentiables, *Comm. Math. Helv.* **28,** 17–86.
Thom, R. (1975) *Structural Stability and Morphogenesis* (trans. D. H. Fowler), Benjamin, Reading, Mass. (original French edition 1972).
Thompson, J. M. T. and Hunt, G. W. (1973) *A General Theory of Elastic Stability*, Wiley, London.
Thompson, J. M. T. and Hunt, G. W. (1975) Towards a unified bifurcation theory, *ZAMP* **26,** 581–604.
Wan, Y.-H. (1979) Generic deformations of varieties, preprint, S.U.N.Y., Buffalo, New York.
Wassermann, G. (1974) *Stability of Unfoldings*, Lecture Notes in Mathematics 393, Springer-Verlag, Berlin.
Zeeman, E. C. (1977) *Catastrophe Theory: Selected Papers 1972–1977*, Addison-Wesley, Reading, Mass.
Zeeman, E. C. and Trotman, D. J. A. (1976) The classification of elementary catastrophes of codimension $\leqslant 5$, in *Structural Stability, the Theory of Catastrophes, and Applications in the Sciences*, 263–327, edited by P. J. Hilton, Lecture Notes in Mathematics 525, Springer-Verlag, Berlin. (Also in Zeeman, 1977.)

Boundary Value Problems in Plane Strain Plasticity

by I. F. Collins

Department of Theoretical and Applied Mechanics,
University of Auckland

Summary

This article is a state of the art survey of slipline field theory techniques for solving boundary value problems for rigid/plastic solids undergoing plane strain deformations. Attention is concentrated on developments in the last 15 years, such as the use of computers to construct complex solutions to problems of "indirect type" which lead to integral equations, and of visioplastic techniques for constructing experimental slipline fields. Applications to several important forming operations such as strip rolling, extrusion, and machining are described. The problems involved in trying to include strain hardening in the analyses are discussed.

1. Introduction

The theory of quasi-static plane strain deformations of rigid/perfectly plastic materials is the best-developed branch of the mathematical theory of plasticity. It has found a large number of applications, principally to the analysis of metal-forming processes, but also to the mechanics of structures, the theory of material testing, theories of friction and wear, crack propagation in ductile materials, and various geomechanical applications such as soil and glacier mechanics. The reason for this privileged position of plane strain problems is not hard to find. In three-dimensional problems and in other two-dimensional situations, such as plane stress, axial symmetry, and plate-bending theory, several distinct plastic regimes can be distinguished in which the governing equations are variously hyperbolic, parabolic, or elliptic. However, there is only a single possible plastic regime for plane strain deformations for which the equations form a hyperbolic quasilinear system in two independent variables. Whilst the theory of characteristics for such systems is well developed, the problems met with in plane strain plasticity theory differ from those encountered in many other areas of application, such as nonlinear wave propagation, in that the solution has to be found in a *finite* region in two-dimensional physical space rather than in a semi-infinite region in two-dimensional space–time. Moreover the process of finding a solution inevitably involves a strong element of trial and error since (a) the position of the boundary between the plastically deforming and rigid regions is initially unknown, and (b) it is only rarely that one has sufficient data on a boundary in order to formulate a Cauchy problem and construct the solution by a "marching in" procedure. It is hence necessary to develop a physical intuition for these problems based on studies of existing solutions and, if possible, on experiment.

The foundations of the theory of plane strain deformations were given by Prandtl (1921), Hencky (1923), and Carathéodory and Schmidt (1923), who investigated the geometric properties of the characteristic networks (sliplines) and also discussed solutions to a number of boundary value problems, such as the loading of a half-space by a rigid punch or uniform pressure ("Prandtl's punch problem"). These early "solutions", however, were based upon analyses of the stress equations alone and did not contain considerations of whether or not compatible velocity fields (deformation modes) could be found. The basic properties of such velocity fields were found by Geiringer (1937), but it was not until R. Hill and E. H. Lee who, in a series of papers in the 1940s and culminating in the book by Hill (1950), produced solutions to a variety of problems such as indentation, forging, drawing, and notched bars under tension, which satisfied all the kinematic, as well as statical, conditions of the problems, that the proper nature of solutions to boundary value problems in plasticity was fully appreciated.

The publication of *The Mathematical Theory of Plasticity* by R. Hill in 1950 was an important landmark in the history of the development of the subject as it conclusively demonstrated to engineers the practical value of the mathematical theory of plasticity and of slipline fields in particular. The following decade saw many further fundamental developments in the theory. Hill (1951, 1956, 1957) established the basic uniqueness theorems for boundary value problems for rigid/plastic solids. These theorems are more complex than comparable theorems in other branches of continuum mechanics, such as elasticity or fluid mechanics, due to the fact that the constitutive relation between stress and strain-rate is not one-to-one. Bishop (1953) demonstrated a technique by which certain proposed solutions could be "completed" by demonstrating the existence of a self-equilibrated stress field in the proposed rigid region which does not violate the yield condition (a statically admissible stress field). This is still a difficult process today and many solutions found in the literature remain incomplete. The papers by Salençon (1967) and Grimm (1975) contain some recent examples of completed solutions. However, Hill (1954a) established some very useful criteria, which are usually sufficient for practical purposes, by which proposed rigid vertices could be tested for "overstressing".

This decade also saw the publication of a number of new solutions by A. P. Green for problems such as forging, extrusion, and notched bars under tension and bending, many of which had a more complex mathematical structure than those previously known. An important new concept which aided him in developing these new types of solution was that of the hodograph diagram for representing the velocity field, which was discovered independently by Green (1954) and by Prager (1953, 1959), who also introduced a graphical procedure for constructing the stress solution in Mohr stress space. The techniques of slipline theory were also taken up by engineers at this time and many of the existing solutions were modified and extended to meet different working conditions. For example, many forms of extrusion processes were investigated and summarized in the monograph by Johnson and Kudo (1962), whilst Alexander (1955) produced the first slipline solution for strip rolling using Prager's graphical procedure. There followed a number of textbooks aimed at introducing the subject to undergraduate engineers: particularly successful examples in the English language are those by Johnson and Mellor (1962, 1973), Ford and Alexander (1963),

Thomsen *et al.* (1965), and Rowe (1965), whilst more recent accounts have been given by Rowe (1977), Slater (1977), and Johnson and Mamalis (1977).

In the last 15 years there has been much interest in using digital computers to calculate slipline field solutions. This is not primarily to attain great accuracy, which is not justified because of the approximations already inherent in the rigid/perfectly plastic model, but because many solutions are so complex that they cannot be effectively calculated in any other way. The power series method of Ewing (1967) has proved particularly powerful in this regard and has largely replaced the finite-difference procedures previously used for desk machine calculations. Many of the more complicated types of solution lead to linear Fredholm integral equations which must first be solved before the slipline network can be calculated. Collins (1968a) has developed a systematic operational scheme for solving such problems making use of Ewing's series method and the superposition principle for slipline fields, whose practical value was first recognized by Hill (1967b). This procedure has been developed into a systematic computational procedure based on the use of a small package of standard subroutines by Dewhurst and Collins (1973) and applied to a growing number of practically important problems such as strip rolling by Dewhurst *et al.* (1973), Collins and Dewhurst (1975), and machining by Dewhurst (1978a).

Whilst the neglect of elastic strains is of little practical consequence in the analysis of many forming processes, the theory would be of much greater practical value if it could take account of variations in flow stress due to strain hardening. Whilst there have been a few theoretical investigations, e.g. by Hopkins (1972) and Collins (1978c), into the structure of the governing equations, which are still essentially hyperbolic, when strain hardening is included, there is at present no viable solution procedure comparable to classical slipline field theory and it is necessary to resort to time-consuming and expensive finite element procedures. There has, however, been a number of empirical studies, notably by Oxley (1978) and co-workers, in which the general form of the slipline fields for processes such as extrusion or steady-state machining is deduced from the observed deformation of grids.

This article is a review of the developments summarized above. It concentrates on work which has been published in the last 15 years although for completeness the classical basic theory is summarized in the next Section. An extensive account of this theory was given in the book by Johnson *et al.* (1970) together with an effectively complete list of references prior to that date. The approach adopted in this article is that of an applied mathematician interested in developing procedures for solving boundary value problems, so that the illustrative examples, though certainly of practical relevance, have been chosen primarily to illustrate theoretical points. The article by W. Johnson in this Volume discusses some of the practical applications of the recent developments in slipline field theory, whilst that by A. J. M. Spencer is concerned with the extension of classical slipline theory to granular media.

2. Fundamentals of Slipline Fields

2.1. *The geometry of Hencky–Prandtl networks*

The two families of characteristics which arise in the theory of plane strain deformations of rigid/plastic solids form an orthogonal net of curves known as Hencky–Prandtl networks. These networks had, in fact, appeared in the literature, prior to their application in plasticity theory by Prandtl (1921) and Hencky (1923), in a paper by Michell (1904) on the design of structures of minimum weight. The networks also occur in plastic plate-bending theory and various useful analogies between the three areas of application have been exploited by, for example, Hemp (1973), Johnson (1969), and Collins (1971).

A two-parameter family of curves can be represented by the equations

$$x = x(\alpha, \beta), \quad y = y(\alpha, \beta), \tag{2.1}$$

where (x, y) are cartesian coordinates. The elements of arc length along the α- and β-curves are

$$\left.\begin{aligned} ds_\alpha &= (x_{,\alpha}^2 + y_{,\alpha}^2)^{1/2}\, d\alpha = h_\alpha\, d\alpha, \\ ds_\beta &= (x_{,\beta}^2 + y_{,\beta}^2)^{1/2}\, d\beta = h_\beta\, d_\beta. \end{aligned}\right\} \tag{2.2}$$

The (α, β)-parameterization of a given network is not unique and any one-to-one mapping $\alpha \to \alpha^1 = f(\alpha)$, $\beta \to \beta^1 = g(\beta)$ defines a new parameterization. The Lame parameters (h_α, h_β) defined in (2.2) hence depend on the particular parameterization chosen.

If ϕ and $\phi + \pi/2$ are the inclinations of the α- and β-curves to the x-axis (Fig. 2.1) it follows that

$$\cos\phi = x_{,\alpha}/h_\alpha = y_{,\beta}/h_\beta \quad \text{and} \quad \sin\phi = y_{,\alpha}/h_\alpha = -x_{,\beta}/h_\beta. \tag{2.3}$$

The "orthogonality conditions" on the Lamé parameters,

$$h_{\alpha,\beta} + \phi_{,\alpha} h_\beta = 0 \quad \text{and} \quad h_{\beta,\alpha} - \phi_{,\beta} h_\alpha = 0. \tag{2.4}$$

then follow by equating the mixed second-order derivatives of x and y.

The radii of curvature ρ_α and ρ_β of the α- and β-curves are given by

$$\frac{1}{\rho_\alpha} = -\frac{\phi_{,\alpha}}{h_\alpha} = \frac{h_{\alpha,\beta}}{h_\alpha h_\beta} \quad \text{and} \quad \frac{1}{\rho_\beta} = \frac{\phi_{,\beta}}{h_\beta} = \frac{h_{\beta,\alpha}}{h_\alpha h_\beta}. \tag{2.5}$$

(N.B.—The sign convention for radii of curvature adopted in the literature is not universal and here we are following Hill (1950).) Instead of characterizing a curve by its radius of curvature it is sometimes convenient to use "moving cartesian coordinates" or "Mikhlin coordinates" $(\bar{x}, \bar{y})$ (Fig. 2.1) defined by

$$\bar{x} = x\cos\phi + y\sin\phi, \quad \bar{y} = -x\sin\phi + y\cos\phi, \tag{2.6}$$

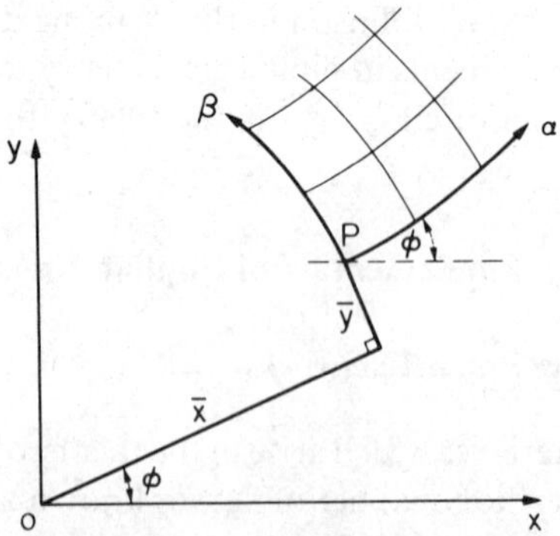

FIG. 2.1. Mikhlin coordinates $(\bar{x}, \bar{y})$.

and which satisfy the differential relations

$$\bar{x}_{,\beta} - \bar{y}\phi_{,\beta} = 0, \quad \bar{y}_{,\alpha} + \bar{x}\phi_{,\alpha} = 0, \tag{2.7}$$

$$\bar{x}_{,\alpha} - \bar{y}\phi_{,\alpha} = h_\alpha, \quad \bar{y}_{,\beta} + \bar{x}\phi_{,\beta} = h_\beta. \tag{2.8}$$

A Hencky–Prandtl network is an orthogonal network with the additional property that

$$\phi_{,\alpha\beta} = 0. \tag{2.9}$$

This equation has the general solution $\phi = f(\alpha) + g(\beta)$, but in any region where ϕ varies monotonically the parameterization can be chosen so that

$$\phi = \alpha + \beta. \tag{2.10}$$

The geometrical interpretation of this result is embodied in *Hencky's First Theorem* which states that

$$\phi_R - \phi_S = \phi_Q - \phi_P = \Delta\alpha, \quad \phi_R - \phi_Q = \phi_S - \phi_P = \Delta\beta, \tag{2.11}$$

where $PQRS$ is any curvilinear quadrilateral in the network (Fig. 2.2). Thus the angle between

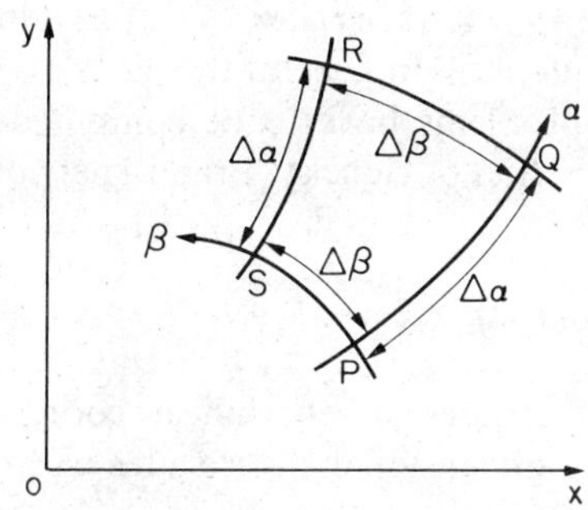

FIG. 2.2. Hencky's first theorem.

two curves of one family, at points where they are cut by a curve of the other family, is constant along their lengths. Two simple examples are "constant state regions" in which both families consist of parallel straight lines and "centred fans" consisting of radial straight lines and concentric circular arcs. It follows from (2.11) that if one curve in a network is a straight line then all the other curves belonging to the same family will be straight also. When the curves are not straight it follows from (2.10) that $\phi_{,\alpha} = \phi_{,\beta} = 1$ and hence from (2.5) that the radii of curvature can, with due consideration to sign convention, be used as the Lamé parameters of the curves. Furthermore, from the orthogonality conditions (2.4) and (2.7) it follows that $(\rho_\alpha, \rho_\beta)$ and $(\bar{x}, \bar{y})$ satisfy the systems of equations

$$\rho_{\alpha,\beta} - \rho_\beta = 0, \quad \rho_{\beta,\alpha} + \rho_\alpha = 0, \tag{2.12}$$

$$\bar{x}_{,\beta} - \bar{y} = 0, \quad \bar{y}_{,\alpha} + \bar{x} = 0. \tag{2.13}$$

Elimination between these equations shows that each of these four variables satisfies the telegraph or Klein–Gordon equation

$$f_{,\alpha\beta} + f = 0. \tag{2.14}$$

The geometrical interpretation of (2.12) is *Hencky's Second Theorem*, which states that the length of arc of one curve (PQ say) is equal to the difference in the values of the radius of curvature of the curves of the other family at P and Q. For example, if PQ is an α-line,

$$\int_P^Q ds_\alpha = -\int_P^Q \rho_\alpha \, d\alpha = \int_P^Q \rho_{\beta,\alpha} \, d\alpha = \Big[\rho_\beta\Big]_P^Q.$$

Three simple analytic solutions of (2.12) are

$$\text{(a)}\ \rho_\alpha = -R\exp(\alpha-\beta), \quad \rho_\beta = R\exp(\alpha-\beta), \tag{2.15}$$

so that both families of curves consist of equiangular spirals, and

$$\text{(b)}\ \rho_\alpha = -A\cos(\alpha+\beta), \quad \rho_\beta = A\sin(\alpha+\beta), \tag{2.16}$$

where both sets of curves are cycloids, and

$$\text{(c)}\ \rho_\alpha = (\alpha/\beta)^{n/2} I_n(2\sqrt{\alpha\beta}), \quad \rho_\beta = (\beta/\alpha)^{n/2} I_n(2\sqrt{\alpha\beta}), \tag{2.17}$$

where $I_n(\theta)$ is the modified Bessel function of the first kind.

These solutions were first discussed by Caratheodory and Schmidt (1923) who also noted that if $f(\alpha, \beta)$ is a solution of (2.14), then so is $f(\lambda\alpha, \beta/\lambda)$ (this is in fact the well-known Lorentz invariance of the telegraph equation). In general the particular networks which arise in the solution to boundary value problems have to be computed numerically as described in Section 2.4. Further basic properties of Hencky–Prandtl networks are discussed in Section 4.

2.2. *The stress equations—slipline fields*

When referred to general orthogonal curvilinear coordinates (α, β), the quasi-static equilibrium equations for two-dimensional stress states with no body forces are

$$\begin{aligned} \frac{\partial\sigma_{\alpha\alpha}}{\partial s_\alpha} + \frac{\partial\sigma_{\alpha\beta}}{\partial s_\beta} + (\sigma_{\alpha\alpha}-\sigma_{\beta\beta})\frac{\partial\phi}{\partial s_\beta} - 2\sigma_{\alpha\beta}\frac{\partial\phi}{\partial s_\alpha} &= 0, \\ \frac{\partial\sigma_{\beta\beta}}{\partial s_\beta} + \frac{\partial\sigma_{\alpha\beta}}{\partial s_\alpha} + (\sigma_{\alpha\alpha}-\sigma_{\beta\beta})\frac{\partial\phi}{\partial s_\alpha} + 2\sigma_{\alpha\beta}\frac{\partial\phi}{\partial s_\beta} &= 0, \end{aligned} \tag{2.18}$$

where $\sigma_{\alpha\alpha}$, $\sigma_{\beta\beta}$ are the normal stress components and $\sigma_{\alpha\beta}$ is the shear component. Under plane-strain conditions both the Mises and Tresca yield conditions for isotropic materials reduce to the statement that the material yields when the maximum shear stress reaches a critical value, k say. From elementary stress analysis it is known that the shear stress attains its maximum (absolute) value, $(\frac{1}{4}(\sigma_{\alpha\alpha}-\sigma_{\beta\beta})^2+\sigma_{\alpha\beta}^2)^{1/2}$, on line segments which bisect the principal stress directions. The trajectories of these two mutually orthogonal maximum shear stress directions are called sliplines, and the slipline network will be used as the basic (α, β) coordinate system. Thus $|\sigma_{\alpha\beta}| = k$ and the two normal stresses $\sigma_{\alpha\alpha}$, $\sigma_{\beta\beta}$ are equal to each other and to $-p$, the in-plane mean compressive stress (hydrostatic pressure). The two principal stresses are $-p \pm k$ and by convention the sliplines are labelled so that the greater tensile principal stress lies in the first (α, β) quadrant.

Referred to the slipline network, the equilibrium equations (2.18) reduce to

$$\frac{\partial p}{\partial s_\alpha} - \frac{\partial k}{\partial s_\beta} + 2k\frac{\partial\phi}{\partial s_\alpha} = 0, \quad \frac{\partial p}{\partial s_\beta} - \frac{\partial k}{\partial s_\alpha} - 2k\frac{\partial\phi}{\partial s_\beta} = 0. \tag{2.19}$$

These equations simplify further if k does not vary with position, as at the initial instant of deformation of a homogeneous rigid/plastic solid (hardening or non-hardening), and during the continued deformation of a homogeneous rigid/perfectly plastic material, so that

$$\frac{\partial p}{\partial s_\alpha} + 2k\frac{\partial \phi}{\partial s_\alpha} = 0, \quad \frac{\partial p}{\partial s_\beta} - 2k\frac{\partial \phi}{\partial s_\beta} = 0, \tag{2.20}$$

or, using (2.2),

$$p_{,\alpha} + 2k\phi_{,\alpha} = 0, \quad p_{,\beta} - 2k\phi_{,\beta} = 0, \tag{2.21}$$

The more general situation described by (2.19) is discussed in Section 7 on hardening materials. Since the derivatives in each of $(2.20)_{1,2}$ are in the same direction, the sliplines are the characteristics of the stress equations, which integrate to give the celebrated Hencky relations:

$$p \pm 2k\phi = \text{constant on } \alpha\text{- and } \beta\text{-lines, respectively,} \tag{2.22}$$

which enable p, and hence the complete stress state, to be evaluated at all points of a given network once p is known at one point. Eliminating p from (2.21) shows that ϕ satisfies (2.9) so that the slipline network forms a Hencky–Prandtl net.

Not infrequently in applications, (2.22) is applied along isolated sliplines separating two rigid regions. This is not justified by the usual arguments as presented above, which presume the slipline is embedded in a plastically stressed region. The result is nevertheless true, since, if it were not so, it would follow from (2.19) that the gradient of k in the direction normal to the slipline would be non-zero, so that the material in one of the adjacent rigid regions would be overstressed.

2.3. *The velocity equations—the hodograph diagram*

Under normal working conditions, the plastic deformation of a metal is isochoric so that the sum of the two in-plane normal strain-rate components is zero, and the principal strain-rates are $\pm e$, say. The magnitude of the maximum shear strain-rate is also e, and, if the material is isotropic, the trajectories of maximum shear strain-rate will coincide with the sliplines. Furthermore, the elongational strain-rates in the slipline directions will be zero, a result which is embodied in Geiringer's equations

$$u_{,\alpha} - v\phi_{,\alpha} = 0, \quad v_{,\beta} + u\phi_{,\beta} = 0, \tag{2.23}$$

where (u, v) are the velocity components in the (α, β)-directions. If the sliplines are not straight so that the parameterization (2.10) can be adopted, these equations can alternatively be expressed

$$u_{,\alpha} - v = 0, \quad v_{,\beta} + u = 0, \tag{2.24}$$

which gives the telegraph equation system (2.14) once again. There is hence a strong duality between the stress and velocity fields which is made explicit by the introduction of the hodograph diagram.

As a point P traces out a slipline field in the physical plane, its image point p, with position vector $\mathbf{v}$ (the velocity at P), traces out an image network in the velocity or hodograph plane. Since the extension-rate is zero in a slipline direction, it follows that, as P traces out an α- or

β-line, p traces out an α'- or β'-line which is everywhere *orthogonal* to the original slipline at corresponding points. It follows, therefore, that the image hodograph network also forms a Hencky–Prandtl net. (Green, 1954 established this result in an alternative way by identifying (u, v) as moving coordinates $(\bar{y}', -\bar{x}')$ for the hodograph network and comparing (2.7) and (2.23).)

The mapping between the slipline and hodograph networks is not one-to-one. Discontinuities in the tangential velocity component across a slipline are very common features of solutions. Since ϕ cannot be discontinuous across a slipline, it follows from (2.23) that the magnitude of this velocity jump is constant along the length of the slipline, so that such a curve is mapped into two parallel curves in the hodograph diagram. The hodograph image of a rigid translating region is just a single point, whilst that of a rigid rotating region is a geometrically similar region scaled by a factor of ω (the angular speed of rotation) and turned through 90° in the direction of rotation.

In the more elementary solutions the hodograph diagram is useful because it provides a simple way of representing and visualizing the deformation mode, and also for demonstrating that the energy dissipation rate is nowhere negative. However, in more complex problems, particularly those involving rotating rigid regions, the construction of the hodograph diagram is a necessary part of finding the complete solution (see Section 5).

2.4. *Numerical techniques for constructing Hencky–Prandtl networks*

Complete slipline and hodograph networks consist of a patchwork of subregions in each of which the solution is analytic. These subregions are computed piecemeal employing one of three basic constructions:

(a) *Initial characteristics or Riemann problem.* If two intersecting characteristics are known, then the field between them is determined uniquely. A very common example is the field between two circular arcs.

(b) *Cauchy problem.* If two stress or velocity components are known on an arc AB of a non-characteristic curve, then the solution is determined in the influence domain between the intersecting characteristics through A and B. A common example is when AB is a stress-free surface. (N.B.—In fact due to the quadratic form of the yield condition two stress solutions are possible, called "weak" and "strong" solutions by Prager, 1959. The correct solution is determined by the positive work-rate condition.)

(c) *Mixed or Goursat problem.* If one characteristic is given and one condition on stress or velocity is specified on a non-characteristic curve, then the field between these two curves is determined uniquely. A common example is the construction of the field between a given curve and a frictionless rigid boundary. A variant on this problem is that of constructing the field between a given slipline and a curve of unknown position, but on which two stress conditions are given, the commonest example being that of a stress-free surface.

Four techniques for actually constructing these solutions are available.

(i) *Finite differences.* Prior to 1967 the basic numerical procedure was Massau's method based on the finite difference approximation to the telegraph equation system as described by Hill (1950) or Kachanov (1974). This procedure is far less efficient than the series method (iv) below, however, when the computations are performed on a digital computer.

(ii) *The graphical procedure* introduced by Prager (1953, 1959) (see also Wang and Ku, 1979) has proved popular with many engineers and has been used to construct some very complicated solutions such as Alexander's (1955) solution for hot strip rolling. The technique is, however, very time consuming and impractical when a range of solutions is required.

(iii) *The Riemann–Green function.* The solution to the Riemann or Cauchy problems for the telegraph equation can be written in closed form. For example, if $\rho_\alpha(\alpha, 0)$ and $\rho_\beta(0, \beta)$ are the radii of curvature of the base α- and β-lines, the radii of curvature of the two families of curves at the general point (a, b) are given by

$$\rho_\alpha(a, b) = \rho_\alpha(a, 0) - \int_0^a \sqrt{\frac{b}{(a-\alpha)}} J_1[2\sqrt{(a-\alpha)b}]\rho_\alpha(\alpha, 0)\, d\alpha + \int_0^b J_0[2\sqrt{a(b-\beta)}]\rho_\beta(0, \beta)\, d\beta,$$

$$\rho_\beta(a, b) = \rho_\beta(0, b) - \int_0^a J_0[2\sqrt{(a-\alpha)b}]\rho_\alpha(\alpha, 0)\, d\alpha - \int_0^b \sqrt{\frac{a}{(b-\beta)}} J_1[2\sqrt{a(b-\beta)}]\rho_\beta(0, \beta) d\beta, \tag{2.25}$$

where $J_n(x)$ is the Bessel function of the first kind.

Whilst this form of the solution has been used by a number of authors for computing solutions to particular problems, it inevitably leads to having to evaluate numerically a number of integrals or solve an integral equation. In the writer's view there is hence no advantage therefore in using (2.25) in preference to the completely numerical algorithms developed by Ewing (1967) and Collins (1968a) described in (iv) below and in Section 4. Thomasen (1978) has recently employed (2.25) to compute the field in the neighbourhood of an elliptical hole, as he found that there was an unacceptable growth of round-off errors in the corresponding finite difference calculations.

(iv) *The series method.* In regions where it is analytic, the general solution to the telegraph equation can, as shown by Geiringer (1937), be written

$$f(\alpha, \beta) = \sum_{n=0}^{\infty} (a_n f_n(\alpha, \beta) + b_n f_{n+1}(\beta, \alpha)), \tag{2.26}$$

where

$$f_n(\alpha, \beta) = (\alpha/\beta)^{n/2} I_n\{2\sqrt{\alpha\beta}\} \tag{2.27}$$

(cf. (2.17)). The modified Bessel function I_n has the power series representation

$$I_n(2\theta) = \sum_{m=0}^{\infty} \theta_m \theta_{m+n}, \tag{2.28}$$

where $\theta_m = \theta^m/m!$ is the "reduced power" of θ. (N.B.—The arguments of these reduced powers will always be angles denoted by Greek letters, so that they should not be confused with series coefficients a_n, b_n, etc.) As Hill (1950) showed, certain relatively simple slipline networks can be represented by taking just a few terms in (2.26). Chakrabarty (1979) has recently produced some others.

In general, however, it is preferable to use (2.27) and (2.28) to express (2.26) as a double power series as shown by Ewing (1967). The solution of the telegraph equation system

$$f_{,\beta} = s_1 g, \quad g_{,\alpha} = s_2 f, \tag{2.29}$$

where $s_1, s_2 = \pm 1$ according to sign convention, is

$$f(\alpha, \beta) = \sum_{m, n=0}^{\infty} (a_n \alpha_{m+n} (s_1 s_2 \beta)_m + s_1 b_n (s_1 s_2 \alpha)_m \beta_{m+n+1}),$$
$$g(\alpha, \beta) = \sum_{m, n=0}^{\infty} (s_2 a_n \alpha_{m+n+1} (s_1 s_2 \beta)_m + b_n (s_1 s_2 \alpha)_m \beta_{m+n}), \tag{2.30}$$

where α_m, β_m are reduced powers. On the base characteristics, $\beta = 0$ and $\alpha = 0$, and (2.30) reduces to the simple power series

$$f(\alpha, 0) = \sum_{n=0}^{\infty} a_n \alpha_n, \quad g(0, \beta) = \sum_{n=0}^{\infty} b_n \beta_n, \tag{2.31}$$

so that this form of solution is particularly useful for solving the initial characteristics problem, but as shown by Ewing it can be readily adapted for solving Cauchy and mixed problems as well. The series are found to be rapidly convergent and truncation after 6–10 terms, depending on the magnitude of the slipline angles, usually gives five-figure accuracy. As well as giving the geometry of the network, Ewing showed that this series procedure can readily be extended to evaluate force and moment resultants on individual sliplines, standard subroutines for which form part of the general computer package of Dewhurst and Collins (1973).

2.5. *Uniqueness theorems for rigid/plastic solids*

In order to appreciate the problems involved in constructing solutions to boundary value problems it is necessary to discuss first the basic uniqueness theorem established by Hill (1951). The theorem is established using the standard Kirchhoff argument by considering the trace tensor product $\Delta\sigma_{ij}\Delta e_{ij}$ of the differences between two possible stress and associated strain-rate fields. It follows from the convexity of the yield surface and normal flow rule that this product is everywhere non-negative (the maximum work inequality). By the virtual work identity the integral of this product, which of course, too, must be non-negative, is equal to the integral of $\Delta t_i \Delta v_i$ over the body's surface, where Δt_i and Δv_i are the differences in the surface traction and velocity vectors. If therefore the prescribed boundary conditions are such as to make this product zero everywhere on the surface (e.g. by specifying either t_i or v_i) it follows that

$$\Delta\sigma_{ij}\Delta e_{ij} = 0 \tag{2.32}$$

at all points throughout the body. It follows from (2.32) that the stress is unique (or at least lies on a specific "flat" of the yield surface if this is not strictly convex) in any region where plastic deformation is occurring ($e_{ij} \neq 0$) in *any* of the possible deformation modes. The union of all such deforming regions is called the *deformable zone*. The stress is hence unique at all points in the deformable zone irrespective of whether the material is actually deforming at that point in the particular mode under consideration. In the remaining *rigid zone* the stress field is not unique. The value of the applied load at which plastic flow occurs is unique, however.

This lack of uniqueness in deformation mode causes most difficulty in problems where the plastic deformation is unconstrained and extends to traction-free surfaces, as in notched bars in tension or bending, as discussed in Sections 3.2 and 3.3. In steady forming problems with strong kinematic constraints, such as extrusion or strip rolling, to which the above theorem applies provided the material is isotropic and perfectly plastic, it is usually only possible to find a single deformation mode. In order to determine which mode will actually occur in practice it is necessary, as proved by Hill (1956, 1957), to specify the hardening-rate and the rates of change of the boundary tractions and velocities. However, it is very difficult to make this identification in practice due to the problem of analysing the effect of hardening in diffusing the initial singularities in the solution.

A point which is often overlooked in the literature is that in practice the boundary conditions on traction and velocity are frequently such as to make $\Delta t_i \, \Delta v_i$ non-positive rather than zero. For example, consider the frictional law, widely used in metal-forming analyses, in which the tangential traction component t_t is equal to some fraction of the yield stress, mk, say, where $0 \leqslant m \leqslant 1$, wherever slip occurs between the tool and workpiece. The precise formulation of this law is in terms of inequalities, reflecting the fact that the boundary is divided into two regions, the extent of which is initially unknown:

$$\left.\begin{aligned} &\text{(a)} \quad |t_t| = mk, \quad v_t > 0 \quad \text{(slipping region)}, \\ &\text{(b)} \quad |t_t| \leqslant mk, \quad v_t = 0 \quad \text{(sticking region)}, \end{aligned}\right\} \tag{2.33}$$

where v_t is the tangential component of relative velocity between the workpiece and tool. Alternatively (2.33) can be formulated as a variational inequality:

$$|t_t| \leqslant mk, \quad v_t \geqslant 0, \quad (mk - |t_t|)v_t = 0. \tag{2.34}$$

Since v_t and t_t must always be in opposite directions it follows that $\Delta v_t \, \Delta t_t$ and hence $\Delta v_i \, \Delta t_i$ are non-positive.

However, since $\Delta \sigma_{ij} \, \Delta e_{ij} \geqslant 0$ everywhere in the body, (2.32) and the consequent uniqueness statements still follow by applying the virtual work identity, and moreover the product $\Delta v_t \, \Delta t_t$ must in fact be zero at all points of the surface, so that the extent of the sticking region, such as the "neutral zone" in strip rolling, is unique.

The basic uniqueness theorem also applies to problems with passive support boundary conditions, in which the support can exert a pressure but not a tension on the worked material. This condition also requires a formulation in terms of inequalities:

$$\left.\begin{aligned} &\text{(a)} \; v_v = 0, \quad t_v \leqslant 0 \quad \text{(support region)}, \\ &\text{(b)} \; v_v < 0, \quad t_v = 0 \quad \text{(lift-off region)}, \end{aligned}\right\} \tag{2.35}$$

where v_v and t_v are the normal components of relative velocity and traction respectively. Arguing as above it follows that the extent of the support region is unique. An application is given in Section 5.3.

Sewell (1969) and Duvaut and Lions (1976) have made general studies of uniqueness theorems and the related extremum principles for problems in continuum mechanics involving such inequality constraints. There are no such uniqueness theorems however for problems involving Coulomb friction, which have a non-self-adjoint structure, although, as shown by Collins (1969a), it is possible to modify the upper bound theorem of limit analysis to obtain overestimates of yield point loads.

3. Slipline Field Solutions of "Direct Type"

3.1. *Classification of slipline field solutions*

Most boundary value problems met with in plasticity theory are "mixed" in the sense that the boundary conditions involve both traction and velocity components. From the point of view of constructing solution procedures it is convenient to classify solutions into two types:

(a) "Direct type" solutions in which the solution can be constructed progressively from some starting characteristic (in either the slipline or hodograph diagram) the shape of which can be deduced in advance.
(b) "Indirect type" solutions in which the solution can only be found after the shape of the starting characteristic has been found by solving an integral equation.

When presented with a new problem it is not possible, in general, to say with certainty to which class the solution will belong. Indeed, for some problems, such as drawing through wedge-shaped dies or extrusion through square dies, solutions of both types are found depending on the specified reduction ratio. "Direct type" solutions can themselves be put into various classes, representative examples of which are described in the following subsections.

3.2. *Theory of notched bars*

One of the most intensively studied areas of application of slipline theory is to the analysis of bars or plates containing one or more notches and subjected to tension and/or bending. Such problems arise in standard material-testing techniques such as the Charpy or Izod test. These solutions have also been used to provide information of the stress and strain distributions in the vicinity of a crack tip in a ductile material (see, for example, McClintock, 1971, and Knott, 1973). Since the deformation is usually localized to the vicinity of the notch, the slipline field can be constructed systematically inwards from the plastically deforming free surface, where both traction components are zero, using the standard "marching in" procedure for a Cauchy problem. The well-known solutions for symmetrically notched bars under tension with V-shaped and circular roots are shown in Fig. 3.1.

Figures 3.1(a) and (b) illustrate two different possible yield point deformation zones for a bar with V-notches. As is required by the basic uniqueness theorem (Section 2.5), the two zones are identical where they overlap. Provided the ratio of the width of the bar to that of the neck (w/a) is sufficiently large, these solutions have been shown to be complete by Ewing and Hill (1967), using the procedure originally advocated by Bishop (1953). It can further be shown using the procedure of Bishop *et al.* (1956) that the deformable zone consists of the region illustrated in Fig. 3.1(a) and outside this region the material is definitely rigid in all possible modes.

There is in fact an infinity of possible modes associated with the solution in Fig. 3.1(a) since the normal velocity distribution on AK (say) can be specified arbitrarily provided only that the associated energy dissipation rate is nowhere negative. The modes associated with Fig. 3.1(b) are a subset of the above. This lack of uniqueness for a perfectly plastic material makes it impossible to predict the exact subsequent deformation. Lee (1952) obtained one analytical solution for the continued deformation of such a bar. He assumed the basic solution is as in Fig. 3.1(b) and that the plastically deforming region remains geometrically self-similar as the deformation proceeds. Richmond (1969) has obtained an alternative analytic solution in which the free surface *AH* in Fig. 3.1(a) rotates as the deformation proceeds and produces a curved (concave inwards) neck profile in contrast to Lee's solution which predicts a straight-sided neck. Richmond showed that his predicted profile is closer to those observed in practice.

In the case of notches with circular roots, Ewing (1968) has shown that the solution in Fig. 3.1(c) is unique for the w/a-ratios for which it is valid, whilst Wang (1954) studied the continued deformation of such a bar in the absence of hardening. His analysis is approximate since he assumes that the notch roots remain approximately circular. This approximation has been vindicated by the more precise calculations of large-scale deformations by Morrison and Richmond (1972).

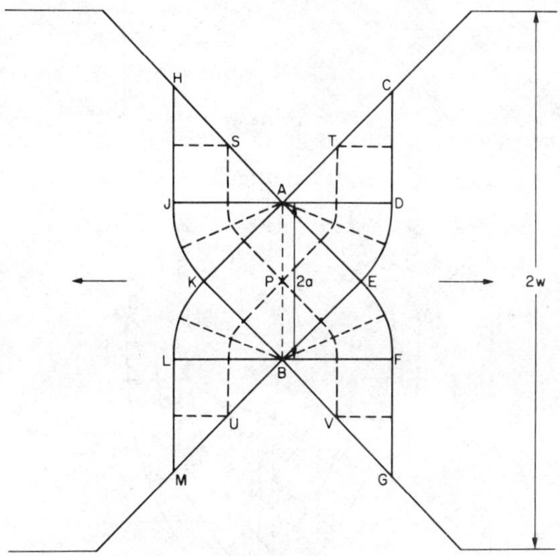

FIG. 3.1(a)

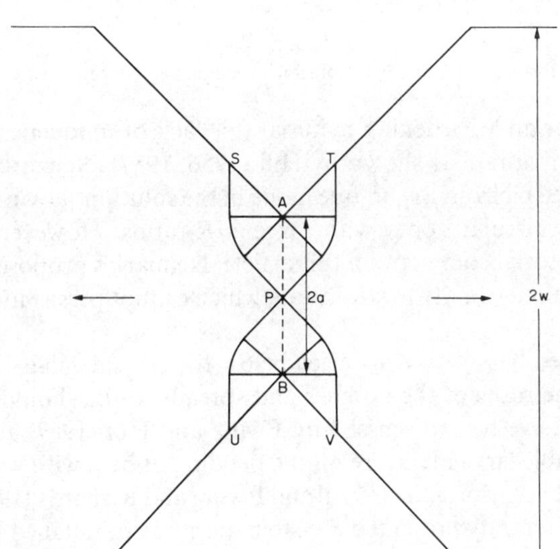

FIG. 3.1(b)

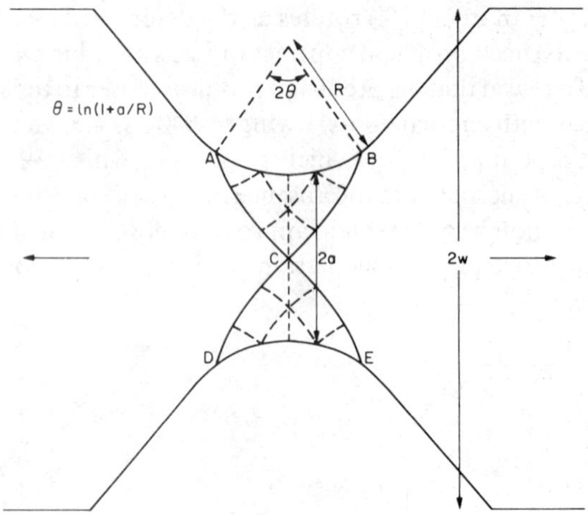

FIG. 3.1(c)

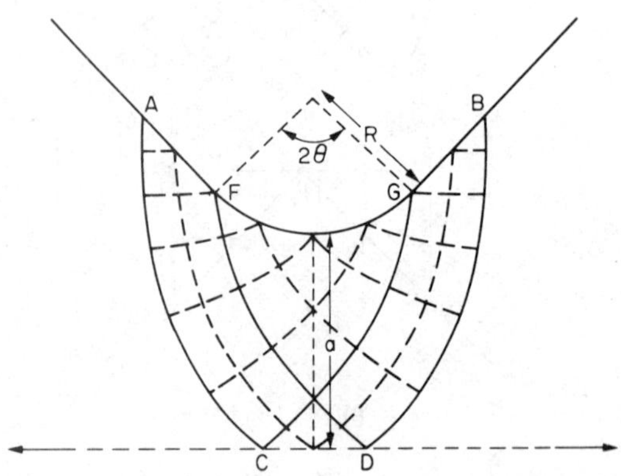

FIG. 3.1(d)

FIG. 3.1. Slipline field solutions for notched bars in tension.

As discussed earlier, in a hardening material this lack of uniqueness can be removed by specifying the hardening-rate as shown by Hill (1956, 1957). Neimark (1968) attempted to apply Hill's criterion to pick out the unique mode in the solution shown in Fig. 3.1 (d) for a bar whose notches have circular roots with large a/R-ratios. However, as pointed out by McClintock in an editorial comment on this article, Neimark's proposed solution cannot be valid as it involves a discontinuity in strain-rate which cannot persist in a hardening material (see Section 7.3).

For shallow notched bars with w/a-ratios below the critical value, the deformation is no longer localized to the roots of the notches and spreads to the shoulders of the bars. Such subcritical solutions have been discussed by Ewing and Hill (1967) and Ewing (1968).

There is a comparable large literature on the bending of bars with variously shaped single or double notches, see Johnson *et al.* (1970) and Ewing and Richards (1974), whilst Dodd and Shiratori (1978) have recently presented a systematic study of notched bars under combined tension and bending and McClintock and Clerico (1980) have found the solution for a singly notched bar subjected to transverse shear.

3.3. *Geometrically self-similar solutions*

Another class of problems in which it is usually possible to construct the slipline field by "marching in" from a free surface are the so-called "pseudo-steady" processes in which the deforming configuration grows uniformly in time remaining geometrically self-similar. These problems too lead to non-unique solutions although for rather different reasons from those for notched bars. Well-known examples of this class of solution, described in the book by Hill (1950), are the indentation of a half-space by a wedge, compression of a wedge by a flat die, and expansion of a semi-cylindrical cavity as well as Lee's notched bar solution mentioned in the previous subsection.

These problems are solved with the aid of the "unit diagram" concept. If **r** is the position vector of a point in the slipline field relative to the centre of similitude O (i.e. the point of initial contact between the tool and the workpiece (Figs. 3.2 and 3.3), and t is the scalar time variable measuring the stage of the deformation (e.g. depth of penetration of wedge), then the

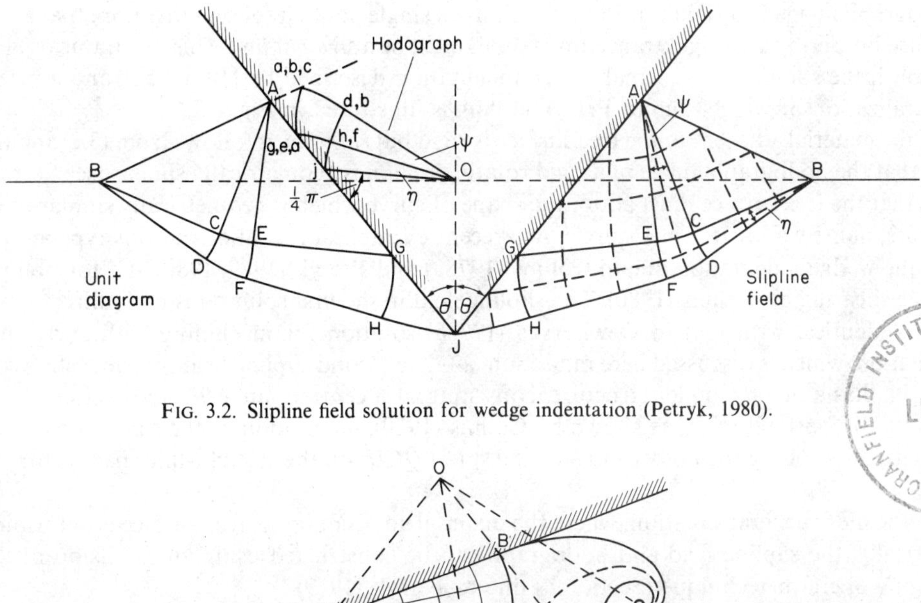

FIG. 3.2. Slipline field solution for wedge indentation (Petryk, 1980).

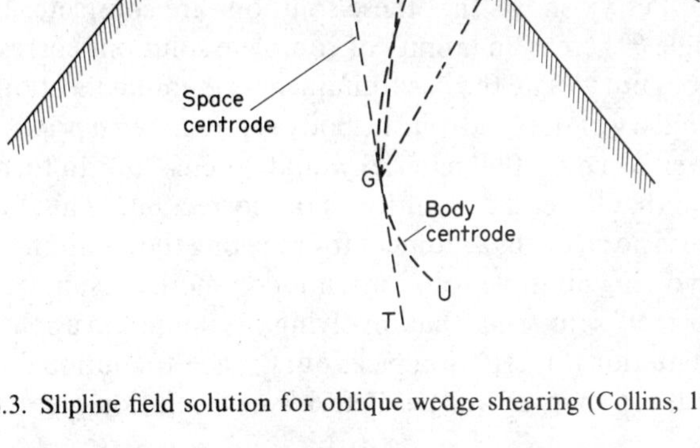

FIG. 3.3. Slipline field solution for oblique wedge shearing (Collins, 1980).

stress and velocity fields depend on $\mathbf{r}$ and t only through the scaled variable

$$\boldsymbol{\rho} = \mathbf{r}/t. \tag{3.1}$$

As $\mathbf{r}$ traces out the slipline field in physical space, $\boldsymbol{\rho}$ describes the "unit diagram" which is identical with the slipline field at "time $t = 1$". Differentiating (3.1) with respect to t yields

$$t\frac{d\boldsymbol{\rho}}{dt} = \mathbf{v} - \boldsymbol{\rho}, \tag{3.2}$$

where $\mathbf{v} = d\mathbf{r}/dt$ is the velocity of the material particle at $\mathbf{r}$. As this material particle at $\mathbf{r}$ moves in physical space, $\boldsymbol{\rho}$ traces out an image trajectory in the unit diagram whose path is described by (3.2). If the hodograph diagram is superposed on the unit diagram with origin O, (3.2) has the geometric interpretation that the tangent to the image trajectory at $\boldsymbol{\rho}$ in the unit diagram is always directed towards the corresponding image point $\mathbf{v}$ in the superposed hodograph diagram. In particular this result holds for the unit diagram images of free surfaces. Since the hodograph image of a rigid translating region is a single point, it follows therefore that a free surface bordering on a rigid translating region must be a *straight line*. This result was used to establish the solution for normal wedge indentation described by Hill (1950), and a recent generalization of which, due to Petryk (1980), is illustrated in Fig. 3.2.

If the material adjacent to the free surface is rigid but *rotating*, it follows from (3.2) and the fact that the hodograph image of a rigid rotating region is geometrically similar (see Section 2.3), that the free surface must adopt the shape of curve which is geometrically similar to its evolute, namely a logarithmic spiral. Two recent examples of solutions of this type are for oblique wedge indentation due to Collins (1978a) and Petryk (1979, 1980) and for oblique wedge shearing by Collins (1980). The slipline field in the first solution is of "indirect type" and is identical with that in Dewhurst's (1978a) solution for machining with curly chip formation, which is discussed here in Section 5.2. The second slipline field solution, shown in Fig. 3.3, has a much simpler structure, consisting of a centred fan ABC and a circular arc velocity discontinuity CD. As shown by Collins (1980), the motion of the rigid chip can be visualized by rolling the body centrode, the spiral QGU, on the straight-line space centrode OGT.

In the more general situation, where the material adjacent to the free surface is deforming plastically, the slipline field and hodograph can be constructed using an extension of the author's operator technique (Section 4) due to Petryk (1979).

Petryk's (1980) solution for normal indentation by a smooth wedge, shown in Fig. 3.2, reduces to the earlier one (Hill, 1950) when the fan angle η is zero. Provided the rigid vertex at B is not overstressed, Petryk shows that these solutions are complete, and hence that for a given semi-wedge angle θ there is an infinity of complete solutions corresponding to a range of values of η. This does not violate the basic uniqueness theorem (Section 2.5) since that only applies when the boundary of the rigid/plastic body is prescribed *a priori*. It could perhaps be argued that when given a choice, the material would "prefer" not to form a singularity on a free surface as at B, so that the earlier solution is the correct one. This, however, is an *ad hoc* assumption and there is no rigorous argument for rejecting the solutions with $\eta \neq 0$. In other similar situations, involving multiple solutions, it is commonly assumed, on general stability grounds, that the "correct" solution is that involving the minimum applied effort. In the case of normal wedge indentation this criterion picks out the $\eta = 0$ solution for semi-wedge angles greater than 47.5°, but for sharper wedges Petryk's new solution gives a lower indentation load, though by less than 2.2%. Such minimum effort criteria are, however, frequently at odds with observation. This is true, for example, in the comparable situation arising in steady-state machining and also in the case of the two notched-bar solutions of Lee and of Richmond discussed above. The former solution predicts a smaller force whereas the latter is found to correspond closer to reality. In his comprehensive paper Petryk (1980) also presents solutions for oblique indentation, with or without spiral chip formation, so generalizing the earlier work of Collins and Meguid (1977).

When the face of the wedge is rough it is found that for sufficiently large wedge angles the solution involves a rigid dead metal cap covering the tip of the indentor (see, for example, Shindo, 1972), and, as noted by Haddow (1967), it is possible for the applied load to decrease with increasing friction. This conclusion, however, may have to be modified in the light of Petryk's generalization.

An experimental study of this process was made by Roth (1978) and of the comparable problem of flattening of a wedge by Devenpeck and Weinstein (1970). Solutions to the latter problem have been used by a number of authors to study the interaction and growth of asperities on the surfaces of ductile solids in contact, with a view to developing theories of friction and wear. See, for example, Johnson (1968), Childs (1973, 1977), Wanheim (1973), Challen and Oxley (1979), and Collins (1980).

3.4. *Mixed problems with centred fans and constant state regions*

In the two previous classes of solutions it has been possible to construct the slipline field by marching in from a surface on which both traction components are known. There is, however, a large class of problems in which, although the boundary conditions are mixed, involving both traction and velocity components, it is still possible to deduce, prior to any detailed computations, that the adjacent slipline field must consist of constant state regions and centred fans. Several examples of this class of solution are discussed by Hill (1950), one of which is illustrated here in Fig. 3.4. Nearly all these solutions are for processes with *straight* tool faces and in which the rigid parts of the workpiece are at rest or are *translating* without rotation. Besides the symmetrical strip compression problem illustrated here, other examples are drawing or extrusion between symmetrical wedge-shaped or square-ended dies.

In order to demonstrate the validity of these solutions it is necessary to use a special property of the velocity fields which can be associated with a centred fan slipline field. For definiteness suppose the straight sliplines in the fan are α-lines, so that from Geiringer's equations (2.23) it follows that u is constant along each α-line and that the change in the value of v between two α-lines is constant along their length. In particular, therefore, if v is constant on one α-line it will be so on all α-lines. A rigid translational motion is hence "communicated through the fan".

Consider now the solution to the problem of symmetrical compression of a strip between two smooth dies, the upper half of which is shown in Fig. 3.4. Since the die face AB is smooth, with zero shear tractions, all the sliplines must meet AB at 45°. If the normal tractions on the die were uniform, ABD would have to be a constant state region. However, in this problem the second condition on AB is that the normal velocity, rather than traction, component is to be uniform. Nevertheless, assume for the moment that ABD is a constant state region in which the material moves down rigidly under the die so that ADE and BDF are centred fans and $DECF$ is the field between two equal circular arcs. It follows from the lemma above that the translational motion of the two rigid overhangs is consistent with the rigid motion in

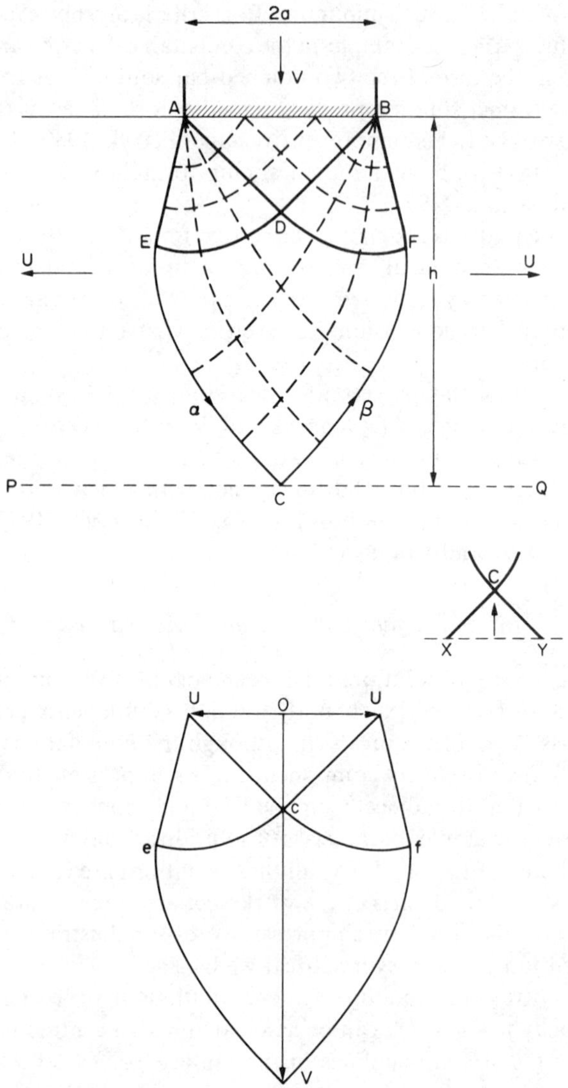

FIG. 3.4. Slipline field and hodograph for symmetrical strip compression with plane dies.

ABD irrespective of the velocity distributions on *DE* or *DF*. The velocity field in *CFDE* is determined from the prescribed components on *CE* and *CF*. There are velocity discontinuities along *AEC* and *BFC*. It has hence been shown that a velocity field satisfying all the velocity boundary conditions can be associated with the assumed slipline field. The argument depends critically on being able to link together two *translating* rigid regions by a centred fan. Problems involving *rotating* rigid regions are more complex and are discussed in Section 5.

This solution is found to be valid in the range $1 \leqslant h/a \leqslant 8.74$. The rigid vertices at *A* and *B* become overstressed at the upper limit and for thicker strips the deformation is localized under each die. The solution for $h/a < 1$ is of "indirect type" and is discussed in Section 5.1.

At first sight the solution in Fig. 3.4 may be thought to be also valid for the related problem of indentation of a strip of thickness h, resting on a frictionless foundation PCQ. However, as discussed in Section 2.5 it is necessary to allow for the possibility of the strip material lifting off from the support. Hill (1953) showed that in the range $4.75 \leqslant h/a \leqslant 8.74$ an alternative deformation mode, possessing a smaller yield point load, was possible. In this mode a triangular wedge of material CXY (Fig. 3.4, inset) lifts off from the foundation. More recently Dewhurst (1974) has shown that lower loads are obtained in the range $1 \leqslant h/a \leqslant 4.07$ if the rigid overhangs rotate upwards away from the support. Dewhurst's solutions are discussed in Section 5.3. Both these "lift off" effects have been observed in practice.

Another well-known solution belonging to this class is for drawing or extrusion through wedge-shaped dies (Fig. 3.5). The limits to the range of validity of this solution due to overstressing at the rigid vertices at A and B have been investigated by Hill and Kim (1974) and Dodd and Scivier (1975). When such overstressing occurs the solution has to be modified to include a "bow-wave" at entry or "thickness recovery" at exit. Dodd and Kudo (1980) have recently investigated the analogue of Hill's (1953) modification to the strip compression solution by allowing the drawn strip to split on the centre-line.

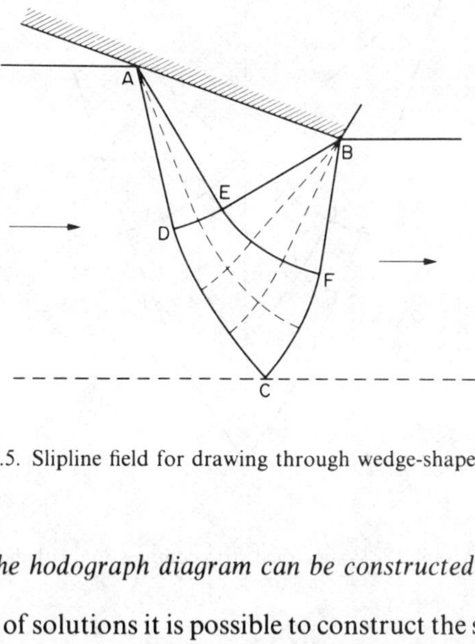

FIG. 3.5. Slipline field for drawing through wedge-shaped dies.

3.5. *Solutions in which the hodograph diagram can be constructed first*

In a further small class of solutions it is possible to construct the slipline field only after the corresponding hodograph diagram has first been found. A representative example of this class of solution, discussed by Sokolovskii (1962) and Druyanov (1972), is for the compression of a strip by a symmetrical curved smooth die shown in Fig. 3.6. The general form of the slipline field is similar to that for the flat die, but the sliplines in the singular fields emanating from A and B are no longer straight. Since the rigid overhangs are both translating rigidly, the two velocity discontinuities along $AHEC$ and $BKFC$ map into centred fans in the velocity plane, so that the hodograph image of the deforming zone is the field between two equal circular arcs just as in the straight die solution (cf. Fig. 3.4). The hodograph image $a'gb'$ of the die face AGB can be found from the conditions that all sliplines meet this face at 45° and that the normal velocity component must be consistent with the vertical motion of

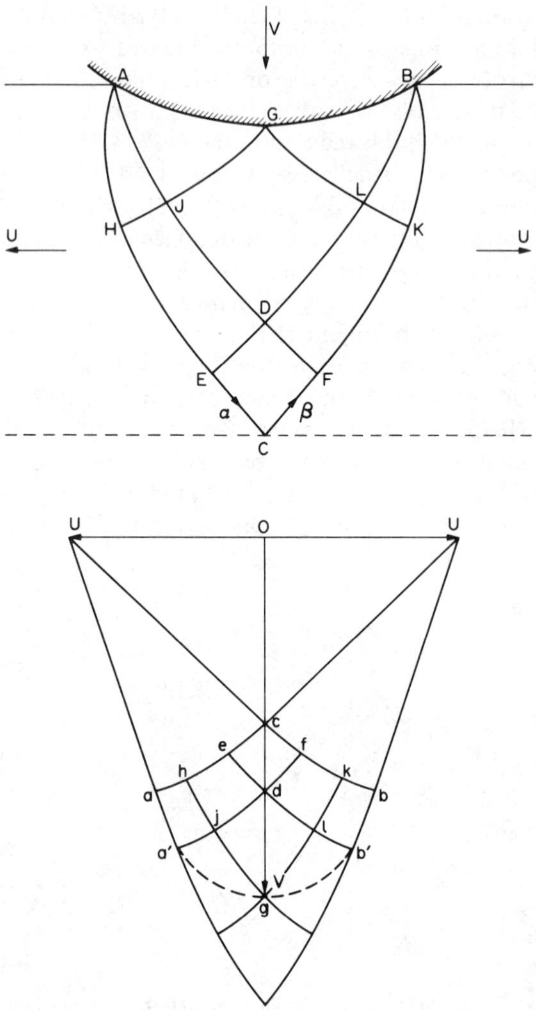

FIG. 3.6. Slipline field and hodograph for symmetrical strip compression with curved dies (Sokolovskii, 1962).

the die. Once $a'gb'$ has been found, the (α, β)-parameterization, and hence the hydrostatic pressure p, can be determined at all points on the die face. Since we now have a Cauchy problem on AGB, the slipline field can now be constructed systematically outwards from the die face starting with ABD.

Essentially the same procedure can be used to construct solutions for bar drawing or extrusion through curved dies—see Sokolovskii (1962) and Tomlenov (1971). If the shape of the die is not specified at the outset it is possible to construct solutions rather more readily using an inverse technique described by Johnson (1962), Kronsjö and Mellor (1966), and others.

3.6. *Design of ideal dies*

The vast majority of slipline field studies take the form of analyses of problems with given geometry and boundary conditions, and very few are concerned with developing design procedures. A notable exception, however, is the theory of streamlined dies for drawing and extrusion instigated by Richmond and Devenpeck (1962). In general, as a material element traverses the deformation zone in an extrusion or drawing process it undergoes shearing as well as compressive or elongational strains. Since the fundamental object of such a process is to reduce the thickness of the strip, the work done in inducing such shearing strains is "redundant". Such a process is "ideal" or 100% efficient if all the material elements are subjected to purely compressive and elongational strains with no redundant shearing, so that the work expended is the same as in uniform monotonic compression. In a drawing operation, therefore, the drawing stress is $2k \log (H/h)$, where H and h are the initial and final strip thickness.

Under plane strain conditions such "ideal" operations are realized by "diagonal streaming" deformations in which the streamlines coincide with one of the families of principal strain-rate trajectories and hence, in a rigid/perfectly plastic solid, everywhere bisect the slipline directions. If s_ξ and s_η denote the arc lengths along the principal strain-rate trajectories, the former coinciding with the streamline direction, the condition that the deformation is isochoric can be written

$$\frac{\partial}{\partial s_\xi}(\log q) + \frac{\partial \phi}{\partial s_\eta} = 0, \tag{3.3}$$

where q is the speed and ϕ is the inclination of the principal strain-rate trajectory to a fixed reference direction, whilst the condition that the shear strain-rate component relative to the s_ξ- and s_η-directions is zero is

$$\frac{\partial}{\partial s_\eta}(\log q) + \frac{\partial \phi}{\partial s_\xi} = 0. \tag{3.4}$$

Adding and subtracting these two equations gives

$$\frac{\partial}{\partial s_\alpha}(\log q) - \frac{\partial \phi}{\partial s_\alpha} = 0, \quad \frac{\partial}{\partial s_\beta}(\log q) + \frac{\partial \phi}{\partial s_\beta} = 0, \tag{3.5}$$

so that $$\log q \mp \phi = \text{constant on } \alpha\text{- and } \beta\text{-lines, respectively}, \tag{3.6}$$

whilst eliminating q yields $\phi_{,\alpha\beta} = 0$, equation (2.9), so that the (α, β)-system of maximum shear strain-rate trajectories forms a Hencky–Prandtl net. (From (3.6) it follows that q varies as $\exp(\pm \phi)$ along a slipline so that the hodograph images of the sliplines always consist of logarithmic spirals.)

It has hence been established by *purely kinematic arguments* that in "diagonal streaming" the α-, β-lines of the strain-rate field must form a Hencky–Prandtl net. Since in an isotropic, perfectly plastic material the stress equilibrium equations lead to exactly the same condition, it follows that for such materials it is possible to construct many solutions involving diagonal streaming. For other materials, however, such as rigid/hardening plastic solids, granular materials, or viscous liquids, it is only rarely that a diagonal streaming solution can be found.

The original strip drawing solution, with a sigmoidal-shaped die, given by Richmond and Devenpeck, is shown in Fig. 3.7. Both sets of sliplines in the region BCD consist of logarithmic

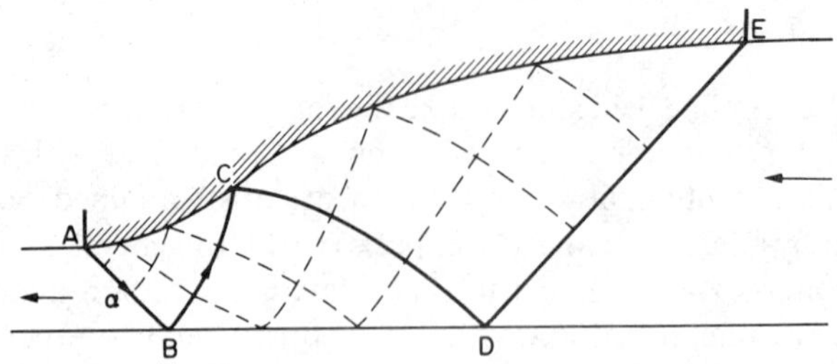

FIG. 3.7. Slipline field for drawing through streamlined dies (Richmond and Devenpeck, 1962).

spirals, whilst the streamlines are radial straight lines. The α-lines in ABC and the β-lines in CDE are straight and the die profile is obtained by constructing the streamline through C. Since the material enters and leaves the die smoothly there are no velocity discontinuities in this solution. Hill (1966) generalized this solution and showed that it is possible to construct a solution for an arbitrary choice for the shape of the curve BC. Another simple example of diagonal streaming in which the dies are in the shape of logarithmic spirals has been discussed by Richmond (1965) and is illustrated here in Fig. 7.3.

Provided one adopts the Tresca yield condition this theory can be generalized to axisymmetric situations, and streamlined dies for wire drawing have been described by Richmond and Morrison (1967), whilst the theory of completely general three-dimensional ideal forming operations has been given by Hill (1967a).

4. Algebraic Properties of Slipline Fields

The fact that when referred to the characteristic coordinates (α, β), the governing equations for Hencky–Prandtl networks, (2.12) or (2.13), are linear has, of course, been well known for many years. However, it was only fairly recently that Hill (1967b) demonstrated that the consequent superposition properties could be put to effective use in helping to solve boundary value problems. Hill formulated the superposition principle *geometrically* and showed in particular how two singular fields could be superposed, by the vectorial addition of position vectors, to give the field between two given characteristics. Collins (1968a, b) formulated these superposition properties *algebraically* using symbolic vector and linear operator notations and used them to analyse and extend some existing solutions for compression and extrusion. Both of these solutions are of "indirect type" and require the solution of linear Fredholm integral equations. If Ewing's power series (Section 2.4) method of representing the solution of the telegraph equation system is employed, Collins (1968a) showed that the construction of complete slipline field and hodograph networks could be expressed in terms of a handful of basic matrix operators, and that the determination of the "starting characteristic" in a problem of "indirect type" can be reduced to a single matrix inversion, or exceptionally to an eigenvector problem. This method has been developed into a systematic computational procedure by Dewhurst and Collins (1973) and has been successfully applied to a growing number of practical problems, five representative examples of which are described in the next Section.

Consider the initial characteristics problem of constructing the field between two sliplines OA and OB as in Fig. 4.1(a). If the radii of curvature of the two base curves are $\rho_\alpha(\alpha, 0) = f(\alpha)$

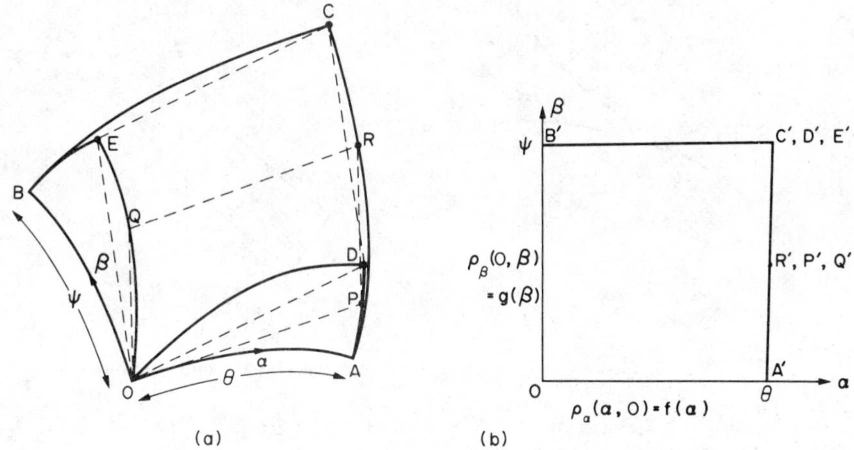

FIG. 4.1. Geometrical formulation of superposition principle (Hill, 1967b).

and $\rho_\beta(0, \beta) = g(\beta)$, then we are led to solving the equation (2.12) in the first quadrant of the characteristic plane (Fig. 4.1(b)), subject to $\rho_\alpha = f(\alpha)$ on $O\alpha$ and $\rho_\beta = g(\beta)$ on $O\beta$. It follows simply from the linearity of the equations that the solution at any point, C', say, is the sum of the solutions, at this same point, to the two initial characteristics problems: (i) $\rho_\alpha = f(\alpha)$ on $O\alpha$, $\rho_\beta = 0$ on $O\beta$, and (ii) $\rho_\alpha = 0$ on $O\alpha$, $\rho_\beta = g(\beta)$ on $O\beta$. The zero radius of curvature of a base curve signifies that this curve has degenerated into a single point, so that the corresponding fields arising in (i) and (ii) above are just the singular fields on OA and OB respectively (Fig. 4.1 (a)). In this sense, therefore, the field between OA and OB is the "sum" of the singular fields on OA and OB.

The above argument can be applied using the Mikhlin coordinates $(\bar{x}, \bar{y})$ in place of the radii of curvature, and in consequence this superposition can be accomplished by adding position vectors as shown by Hill (1967b). For example, in Fig. 4.1(a), $\overrightarrow{OC} = \overrightarrow{OD} + \overrightarrow{OE}$ and $\overrightarrow{OR} = \overrightarrow{OP} + \overrightarrow{OQ}$, etc. This superposition principle is also evident from the Riemann–Green integral form (2.25) and the Ewing double-power series form (2.30) of the solution to the initial characteristics problem, both of which take the form of the sum of two linear forms, involving the two base functions.

In the author's algebraic formulation of this principle the individual directed curves are denoted by vector symbols $\boldsymbol{\sigma}$. The theory can be developed in an abstract setting as was done in the original paper (Collins, 1968a), but for computational purposes it is convenient to regard $\boldsymbol{\sigma}$ as denoting the column vector of coefficients (a_n) in the reduced power series expansion (2.31) of the radius of curvature (or possibly a Mikhlin coordinate) of the curve. The fundamental algebraic operations are now given geometrical interpretations. The characteristic coordinate (angle turned through) is always reckoned positive, whilst the radius of curvature is taken to be positive or negative according as the slipline curves counterclockwise or clockwise. With this convention the curve $-\boldsymbol{\sigma}$ is obtained from $\boldsymbol{\sigma}$ by reflection in the tangent at the base point (Fig. 4.2(a)), whilst $\lambda\boldsymbol{\sigma}$ denotes the curve obtained from $\boldsymbol{\sigma}$ by scaling up the radius of curvature by λ at all points. The curve $\boldsymbol{\sigma}_1 + \boldsymbol{\sigma}_2$ is obtained by adding the radii of curvature of the individual curves at corresponding points. A common example of this in applications is $\boldsymbol{\sigma} + \rho\mathbf{c}$, where $\mathbf{c}$ represents an arc of the unit circle and hence

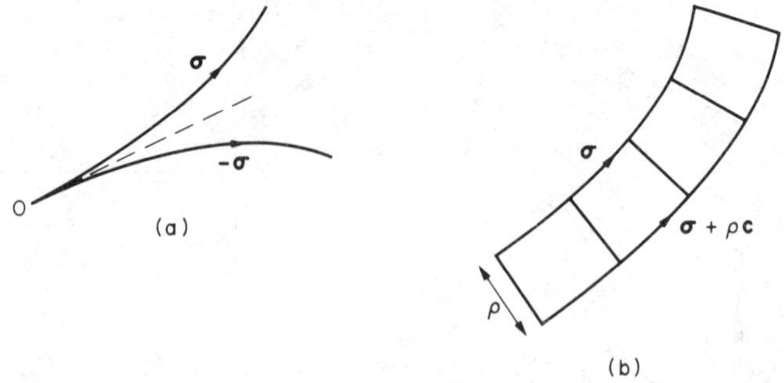

FIG. 4.2. Geometrical interpretation of basic algebraic vector operations.

has column vector representation $(1\ 0\ 0\ \ldots)^T$. Since the radii of curvature of points on $\boldsymbol{\sigma}$ have been increased by the same amount, $\boldsymbol{\sigma} + \rho\mathbf{c}$ represents a parallel curve distant ρ from $\boldsymbol{\sigma}$ (Fig. 4.2 (b)).

As has been seen, the fundamental construction is that of finding the singular field on a given slipline. If OA in Fig. 4.3(a) is denoted by $\boldsymbol{\sigma}$, the curves OD and AD in the singular field on the *convex* side of OA are denoted by $P^*_\psi\boldsymbol{\sigma}$ and $Q^*_\theta\boldsymbol{\sigma}$, where P^*_ψ and Q^*_θ are a pair of linear operators which depend respectively on the span angles ψ and θ. In the series representation

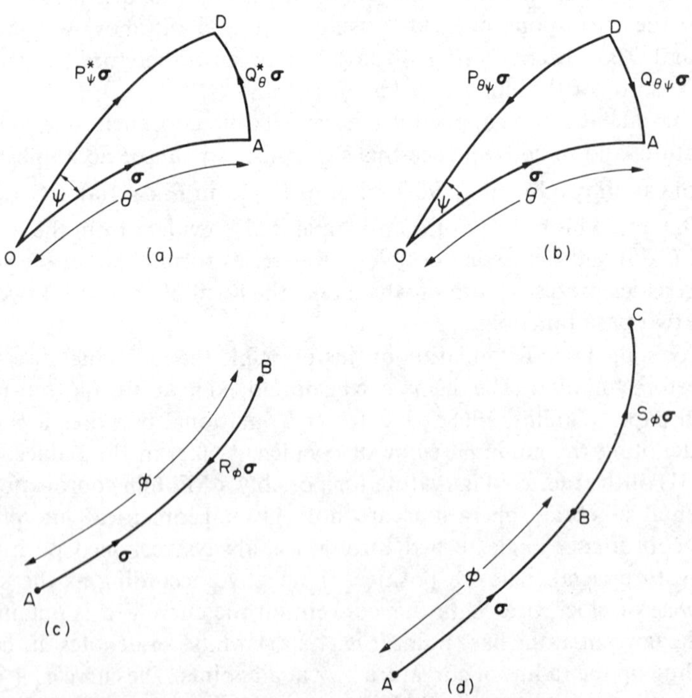

FIG. 4.3. Geometrical interpretation of basic linear operators.

in which $\boldsymbol{\sigma}$ is a column vector, these operators are represented by infinite dimensional matrices:

$$P_\phi^* = \begin{bmatrix} \phi_0 & 0 & 0 & \cdots \\ \phi_1 & \phi_0 & 0 & \cdots \\ \phi_2 & \phi_1 & \phi_0 & \cdots \\ \vdots & \vdots & \vdots & \end{bmatrix}, \quad Q_\phi^* = -\begin{bmatrix} \phi_1 & \phi_2 & \phi_3 & \cdots \\ \phi_2 & \phi_3 & \phi_4 & \cdots \\ \phi_3 & \phi_4 & \phi_5 & \cdots \\ \vdots & \vdots & \vdots & \end{bmatrix}, \tag{4.1}$$

where $\phi_n = \phi^n/n!$ are reduced powers. Alternatively, if one uses the Riemann–Green integral form of the general solution, these operators are represented by convolution integral operators:

$$\begin{aligned} P_\phi^* &: f(\eta) \to f(\eta) + \int_0^\eta \sqrt{\frac{\phi}{(\eta-\xi)}}\, I_1[2\sqrt{(\eta-\xi)\phi}]\, f(\xi)\, d\xi, \\ Q_\phi^* &: f(\eta) \to -\int_0^\phi I_0[2\sqrt{(\phi-\xi)\eta}]\, f(\xi)\, d\xi. \end{aligned} \tag{4.2}$$

The direction in which a curve is to be traced can be reversed by introducing the reversion operator $R_\phi: AB \to BA$ in Fig. 4.3 (c), whilst the position of the base point can be changed by introducing the shift operator $S_\phi: AC \to BC$ in Fig. 4.3 (d). These two operators have matrix representations:

$$R_\phi = -\begin{bmatrix} \phi_0 & \phi_1 & \phi_2 & \cdots \\ 0 & -\phi_0 & -\phi_1 & \cdots \\ 0 & 0 & \phi_0 & \cdots \\ \vdots & \vdots & \vdots & \end{bmatrix}, \quad S_\phi = \begin{bmatrix} \phi_0 & \phi_1 & \phi_2 & \cdots \\ 0 & \phi_0 & \phi_1 & \cdots \\ 0 & 0 & \phi_0 & \cdots \\ \vdots & \vdots & \vdots & \end{bmatrix}. \tag{4.3}$$

Thus the operators which map OA into DO and DA in Fig. 4.3 (b) are $P_{\theta\psi}$ and $Q_{\theta\psi}$, where

$$P_{\theta\psi} = R_\theta P_\psi^* \quad \text{and} \quad Q_{\theta\psi} = R_\psi Q_\theta^*. \tag{4.4}$$

As will be seen, these unstarred P and Q operators satisfy certain identities which are often useful in preliminary analyses. It is more efficient, however, to use the starred operators in the final computations.

The superposition principle can now be used to express the solution to the general initial characteristics problem in terms of these operators. Thus with reference to Fig. 4.4,

$$\boldsymbol{\sigma}_3 = P_{\theta\psi}\boldsymbol{\sigma}_1 + Q_{\psi\theta}\boldsymbol{\sigma}_2, \quad \boldsymbol{\sigma}_4 = P_{\psi\theta}\boldsymbol{\sigma}_2 + Q_{\theta\psi}\boldsymbol{\sigma}_1. \tag{4.5}$$

These expressions are only appropriate if the base lines $\boldsymbol{\sigma}_1$ and $\boldsymbol{\sigma}_2$ are mutually convex. If these curves are mutually concave, as when $\boldsymbol{\sigma}_3$ and $\boldsymbol{\sigma}_4$ are used as base curves, the corresponding relations are

$$\boldsymbol{\sigma}_1 = P_{\psi\theta}\boldsymbol{\sigma}_3 - Q_{\psi\theta}\boldsymbol{\sigma}_4, \quad \boldsymbol{\sigma}_2 = P_{\psi\theta}\boldsymbol{\sigma}_4 - Q_{\theta\psi}\boldsymbol{\sigma}_3. \tag{4.6}$$

Eliminating $\boldsymbol{\sigma}_3$ and $\boldsymbol{\sigma}_4$ between (4.5) and (4.6) yields the following operational identities:

$$P_{\psi\theta}Q_{\theta\psi} \equiv Q_{\theta\psi}P_{\theta\psi} \quad \text{and} \quad P_{\theta\psi}^2 - Q_{\psi\theta}Q_{\theta\psi} \equiv I. \tag{4.7}$$

The geometric interpretation of these identities is shown in Fig. 4.5, where for any base curve

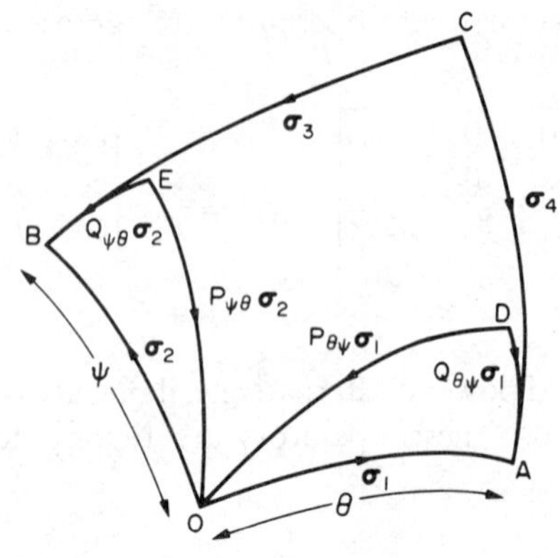

FIG. 4.4. Algebraic formulation of superposition principle (Collins, 1968a).

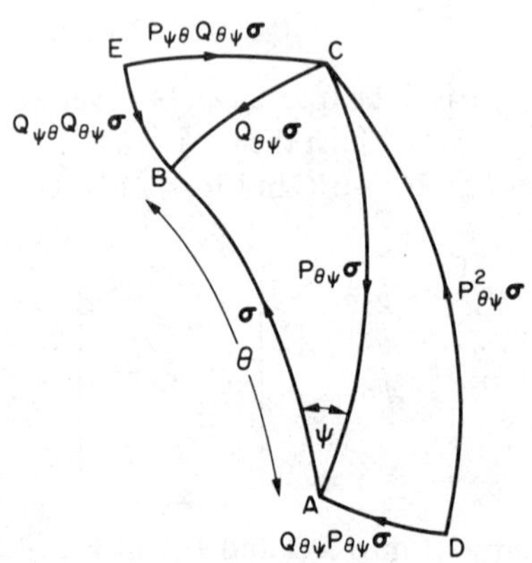

FIG. 4.5. Geometrical interpretation of basic operational identities.

AB, it follows that *EC* and *DA* are identical curves, whilst *DC* minus *EB* is equal to the original curve *AB*.

An example of the use of these operators, which is important in applications, is shown in Fig. 4.6. The problem is to construct the field between a given slipline $OA(\boldsymbol{\sigma})$ and the straight smooth boundaries *OB* and *OC*. Since all the sliplines must meet these boundaries at 45°, it follows by symmetry that these fields are in fact parts of those between $OA(\boldsymbol{\sigma})$ and its reflection *OE* or $OD(-\boldsymbol{\sigma})$ in these two boundaries. Hence from (4.5) and (4.6) the required smooth boundary operators are

$$T_\theta \equiv -P_{\theta\theta} - Q_{\theta\theta} \quad \text{and} \quad T_\theta^{-1} \equiv -P_{\theta\theta} + Q_{\theta\theta}. \tag{4.8}$$

It was shown by Dewhurst and Collins (1973) that the operator needed to construct the field between a given slipline and a straight rough boundary (on which $\tau = mk$, $0 \leqslant m \leqslant 1$) is

$$G_\lambda = Q_{\theta\theta} + P_{\theta\theta}(I\cos\lambda - J\sin\lambda)^{-1}(J\cos\lambda - I\sin\lambda), \tag{4.9}$$

where $\lambda = \frac{1}{2}\cos^{-1} m$, I is the unit operator, and J is the integration operator, which

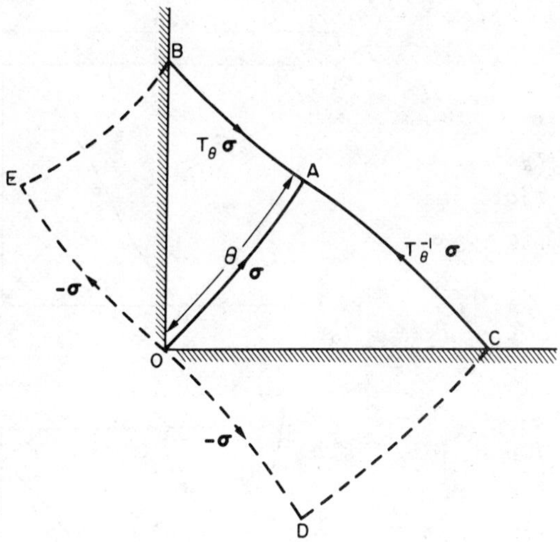

FIG. 4.6. The smooth straight boundary operator.

transforms the curve with $\rho_\alpha = f(\alpha)$ into that with $\rho_\alpha = \int_0^\alpha f(\alpha')\,d\alpha'$. In the matrix representation J is represented by the subdiagonal matrix with unit entries. If a Coulomb frictional law is assumed, the resulting operator is nonlinear and cannot be represented by a matrix. In the same paper it is shown that the operator needed to construct the field between a given slipline and a stress-free boundary (whose shape is initially unknown) is

$$F = (D+I)^{-1}(D-I), \tag{4.10}$$

where D, the differentiation operator, is the left inverse of J and whose matrix representation is the superdiagonal matrix with unit entries. Petryk (1979) has recently derived the operators needed to construct the velocity field in the neighbourhood of such a free surface, both in the case of steady and pseudo-steady deformations.

It is hence possible to express all the operations met with in solving boundary value problems in terms of a handful of basic linear operators, provided these operations, too, are linear. In practice this rules out Coulomb frictional boundary conditions and curved boundaries with the exception of stress-free surfaces. Other recent general accounts of the procedure have been given by Collins (1978a), Das and Banerjee (1978a, b), Venter and Hewitt (1978), and Venter *et al.* (1978a).

5. Slipline Field Solutions of "Indirect Type"

5.1. *Compression of strip between smooth flat dies*

The first example of a slipline field of "indirect type", involving the solution of a linear integral equation, is the rather remarkable solution for compression of a strip between smooth dies with aspect ratios $a/h \geqslant 1$ found by Green (1951). The upper left-hand quadrants of the slipline fields valid in the range $1 \leqslant a/h \leqslant 2$ are illustrated in Fig. 5.1. AB is part of the smooth die face whilst BC and CD are lines of symmetry, so that all sliplines meet

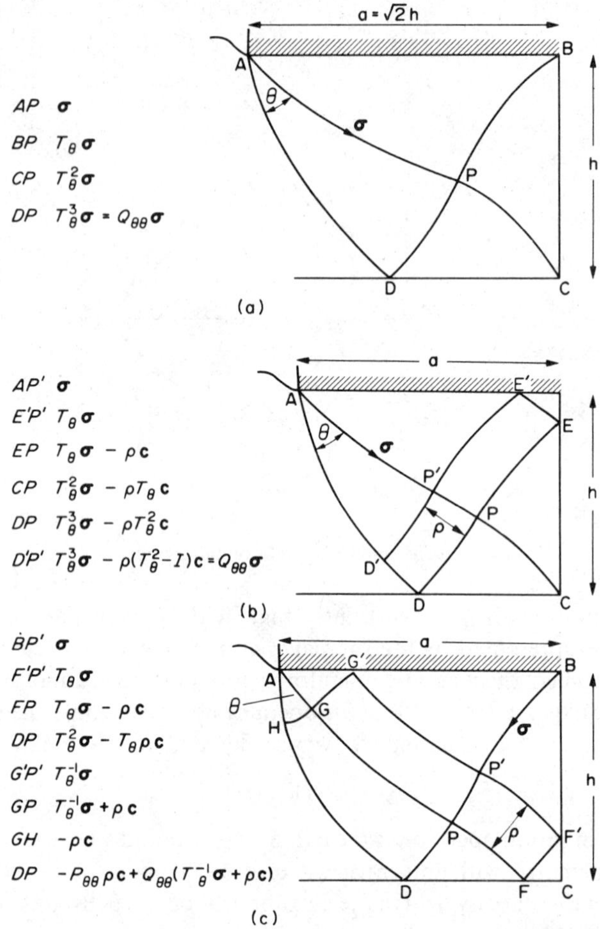

FIG. 5.1. Slipline field solution for compression of strip between smooth dies, $1 \leqslant a/h \leqslant 2$ (Green, 1951; Collins, 1968a).

these three straight lines at 45°. If the shape of AP is assumed known, the fields APB, BPC, and CPD can be successively constructed as the fields between a given slipline and a smooth straight boundary. The field between AP and PD is then determinate (initial characteristic problem), and the condition that this field should be singular at A is the "consistency condition" which determines the shape of the initial curve AP. The fields in Figs. 5.1(b) and (c) are similar except that in addition they involve regions in which one family of sliplines consists of straight lines and the other of parallel curves. The associated velocity fields involve velocity discontinuities along $AP'PC$ in Fig. 5.1(b) and $AHDPP'B$ in Fig. 5.1(c), but the velocity field in Fig. 5.1(a), which is intermediate between the other two, is continuous. As shown by Green, the associated hodograph diagrams are dual in the sense that the one representing the deformation in Fig. 5.1(b) has the form of Fig. 5.1(c) and vice versa, whilst that for Fig. 5.1(a) is identical with the slipline field.

In the original paper Green computed the shape of the initial sliplines in these solutions using an influence coefficient technique based on the finite difference approximation of the governing equations. The corresponding computations indicated a number of surprisingly simple results such that the field in Fig. 5.1(a) is valid for an aspect ratio of $\sqrt{2}$ and that the product of the aspect ratios of the other two fields, with equal singularity angles, is always equal to 2. The general theory described in the last Section was developed initially by Collins (1968a) in order to explain and amplify these results.

The operational formulation of these solutions is indicated in the tables in Fig. 5.1. Thus, for example, if AP in the first field is denoted by $\boldsymbol{\sigma}$, BP, CP, and DP are represented by $T_\theta\boldsymbol{\sigma}$, $T_\theta^2\boldsymbol{\sigma}$, and $T_\theta^3\boldsymbol{\sigma}$, respectively, where T_θ is the smooth boundary operator (4.8). However, since APD is a singular field DP is also $Q_{\theta\theta}\boldsymbol{\sigma}$, so that the equation for the base curve $\boldsymbol{\sigma}$ is

$$(T_\theta^3 - Q_{\theta\theta})\boldsymbol{\sigma} = \mathbf{0}. \tag{5.1}$$

However, using the identities (4.7) and (4.8) this equation can be reduced to

$$T_\theta\boldsymbol{\sigma} = \frac{1}{\sqrt{2}}\boldsymbol{\sigma}, \tag{5.2}$$

so that $\boldsymbol{\sigma}$ is the eigenvector of T_θ with eigenvalue $1/\sqrt{2}$. It follows therefore that the curves AP, BP, CP, and DP are geometrically similar in the ratios $1:\sqrt{2}:2:2\sqrt{2}$ and hence that the corresponding aspect ratio a/h is $\sqrt{2}$ as found by Green. The corresponding derivations of the governing equations for the base sliplines for the other two solutions are indicated in the accompanying tables. In both cases these equations can be simplified to

$$(T_\theta^2 - \tfrac{1}{2}I)\boldsymbol{\sigma} = \rho T_\theta \mathbf{c} \tag{5.3}$$

which further reduces to (5.2) in the limit as $\rho \to 0$.

Equation (5.3) is typical of the type of equation met with in these problems, namely,

$$(A - \lambda I)\boldsymbol{\sigma} = \mu B\mathbf{c}, \tag{5.4}$$

where λ, μ are parameters and A and B are composite operators consisting of various combinations of the fundamental operators defined in the last Section. When $\mu \to 0$ the equation reduces to a homogeneous eigenvector equation, the corresponding field being referred to as an "eigenfield". As in the above example, such "eigenfields" frequently occur as intermediate fields between two solutions of different types. Moreover, they always seem to involve continuous velocity fields since the parameter μ in (5.4) is a measure of the strength of the velocity discontinuity.

For computational purposes these infinite vector and matrix operators are truncated after N terms. The value of N needed to give a specified accuracy increases with the angular span of the individual sliplines, but in most cases $N = 6$ or 8 gives five-figure accuracy. An inhomogeneous equation of the type (5.4) can be solved using any standard matrix inversion routine, whilst procedures for solving the homogeneous equations for the "eigenfields" have recently been discussed by Dewhurst (1978c).

In the paper by Collins (1968a) the solutions for aspect ratios greater than 2 are also analysed and computed and it is shown that the magnitude of the velocity discontinuity reaches a peak when a/h has an integer value and vanishes at the intermediate "eigenfields" when $(a/h) = (n(n+1))^{1/2}$. This oscillatory behaviour is typical of slipline field solutions.

This solution has been generalized to take account of friction by Collins (1969b) and Das *et al.* (1979) and to the situation in which the dies rotate by Collins (1970). A very similar solution is that for drawing through long smooth dies which has been computed by Venter *et al.* (1978b).

W. A. Green (1962) has given the solution for extrusion through square dies with medium reduction, which has a mathematical structure very similar to the above compression problem. In the operational analysis of this solution, Collins (1968b) showed that the identities (4.7) are needed, not only to simplify the governing equations, but, more importantly, to establish the validity of the proposed solution in the first instance.

5.2. *Orthogonal machining*

Machining has for a long time been the *bête noire* of metal-working processes in that it has not yielded to analyses, whether these be slipline field, finite element, or simple "technological" theories, as readily as the other basic forming processes such as forging, drawing, extrusion, or strip rolling. This is basically because whereas the latter processes are governed by the strong kinematic constraints of the dies, rolls, etc., in machining the deformation occurs adjacent to an unrestrained free surface. In a steady-state operation the depth of cut will be specified initially; however, the thickness of the chip and hence the position of its free surface are not specified *a priori*, so that the basic uniqueness theorem and related extremum principles for rigid/perfectly plastic materials (Section 2.5) do not apply. The solution to the steady-state problem hence depends on the previous loading programme.

Most theories of orthogonal machining are based on the assumption that all the deformation occurs at a single velocity discontinuity (shear plane) as in Fig. 5.2. In the earliest such model due to Ernst and Merchant (1941), the inclination ϕ of the shear plane is determined by the condition that the rate of energy dissipation is a minimum which leads to the prediction that

$$\phi = \tfrac{1}{4}\pi - \tfrac{1}{2}\lambda + \tfrac{1}{2}\alpha, \tag{5.5}$$

where α is the rake angle and λ is the angle of friction on the rake face. In view of the non-applicability of the upper bound theorem of limit analysis, this minimization procedure is an *ad hoc* assumption. A rather more rational approach to determining ϕ was suggested by Lee and Shaffer (1951), who assumed that the plastic zone consists of a constant state region

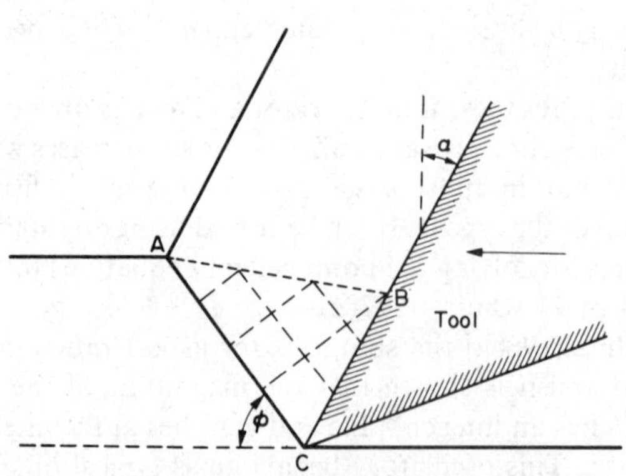

FIG. 5.2. Shear plane model of orthogonal machining.

bounded by the shear plane, the rake face and a stress-free interface with the rest of the rigid chip. With this assumption (5.5) is replaced by

$$\phi = \tfrac{1}{4}\pi - \lambda + \alpha. \qquad (5.6)$$

The fundamental lack of uniqueness of the solution to the machining problem was emphasized by Hill (1954b), who employed his overstressing criteria for proposed rigid vertices to deduce the range of allowable values of the shear plane angle ϕ for each value of $(\lambda - \alpha)$.

In all these studies it was tacitly assumed that the chip *translates* rigidly along the rake face. In practice, however, the chip is frequently found to curl away from the rake face. Tentative slipline field solutions for steady-state machining incorporating chip-curl were proposed by Kudo (1965), but he was unable to satisfy all the conditions necessary for a valid solution. Recently, however, a range of solutions involving chip-curl has been proposed by Dewhurst (1978a) and independently by Petryk (1979), and are shown here in Fig. 5.3.

A velocity discontinuity of magnitude ρ is communicated along the circular arc AB which separates the translating rigid region below this arc from the rigid material in the chip which rotates with angular speed ω. (N.B.—The reference frame chosen is that in which the tool is at rest.) DEC is a continuously deforming region in which all the sliplines meet the rake face at the constant angle ζ (or $\tfrac{1}{2}\pi - \zeta$), so that the shear traction is constant and equal to $k \cos 2\zeta$, whilst BDC is a singular field with a singularity at the tool vertex C. (There is much experimental evidence for the presence of such a singularity; see, for example, Childs, 1971.) The field has four degrees of freedom, the three angles θ, ψ, and η, and the pressure p_A at A. There are, however, only three determining conditions, stemming from the quasi-static equilibrium of the chip, so that this field provides a one-parameter family of solutions. These solutions are incomplete, although the proposed rigid vertices can be tested for overstressing. A variation on this field is to inchude a plastically deforming region AFG adjacent to the free surface in which the pressure is now determinate. Whilst this modified field only has three degrees of freedom, Dewhurst found that it always proved impossible to satisfy simultaneously all the equilibrium conditions on the chip.

In order to determine the shape of the base slipline $\boldsymbol{\sigma}$ (ED, say) it is necessary, as is invariably so in problems with rotating rigid regions, to consider the slipline and hodograph networks simultaneously. Using the rough boundary operator (4.9), CD is the curve $G_\zeta \boldsymbol{\sigma}$ and so BD is $Q_{\eta\psi} G_\zeta \boldsymbol{\sigma}$. Now since the hodograph image of a rigid rotating region is geometrically similar in the ratio $1:\omega$, the hodograph image curves ed and $b'd$ are represented by the vectors $\omega\boldsymbol{\sigma}$ and $\omega Q_{\eta\psi} G_\zeta \boldsymbol{\sigma}$ respectively. Thus $c''d$ is $\omega G_\mu^{-1} \boldsymbol{\sigma}$, where $\mu = \tfrac{1}{2}\pi - \zeta$, so that since $b'c'$ is $\rho \mathbf{c}$, application of the fundamental construction (4.5) yields the consistency equation for $\boldsymbol{\sigma}$:

$$\omega G_\mu^{-1} \boldsymbol{\sigma} = \rho P_{\eta\psi} \mathbf{c} + \omega Q_{\psi\eta} Q_{\eta\psi} G_\zeta \boldsymbol{\sigma}, \qquad (5.7)$$

which can be rearranged and put in the standard form (5.4). For each value of p_A, the angles θ, ψ, and η were found by putting to zero the two force and one moment resultants acting on the chip, by employing a standard nonlinear optimization routine as in the earlier investigations of this type by Collins and Dewhurst (1975).

The range of validity of this solution, imposed by the overstressing criterion, has been fully investigated by Dewhurst, who found that for a fixed frictional shear traction on the rake force, the extent of the range of possible solutions decreases with increasing rake angle α, and that when $\alpha + \zeta > \tfrac{1}{4}\pi$ the chip would always curl but for $\alpha + \zeta \leqslant \tfrac{1}{4}\pi$ the solution range is bounded on one side by the Lee and Shaffer solution with a straight chip. In two further

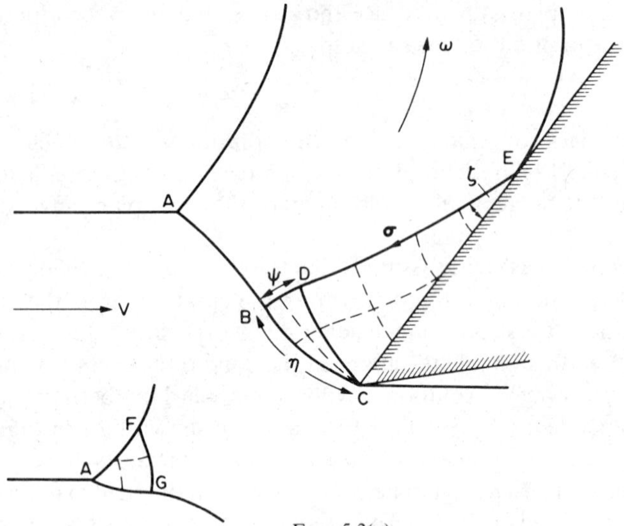

FIG. 5.3(a)

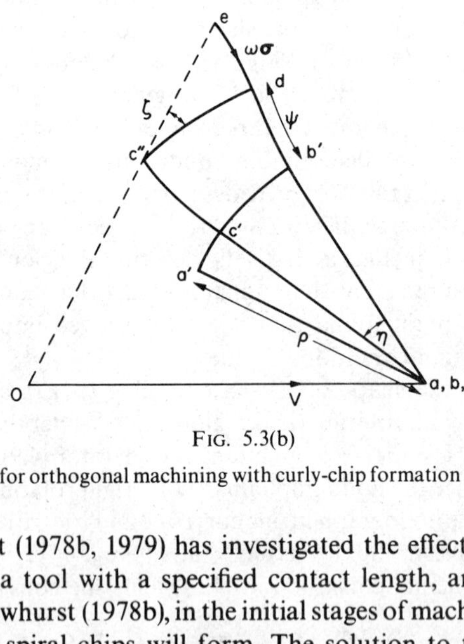

FIG. 5.3(b)

FIG. 5.3. Slipline field solution for orthogonal machining with curly-chip formation (Dewhurst, 1978a; Petryk, 1979).

recent papers Dewhurst (1978b, 1979) has investigated the effect on the range of possible solutions of (i) having a tool with a specified contact length, and (ii) introducing a chip breaker. As noted by Dewhurst (1978b), in the initial stages of machining when the tool is first pushed into the metal, spiral chips will form. The solution to this problem, which is of pseudo-steady type, has been given by Collins (1978a) and Petryk (1979, 1980) as already discussed in Section 3.3.

5.3. *Strip indentation*

As already noted in Section 3.4, when a strip resting on a smooth support is indented by a rigid die, it is frequently observed that the rigid overhangs rotate upwards away from the supports. Dewhurst (1974) proposed two solutions exhibiting this effect which are illustrated here in Fig. 5.4. They are found to be valid for h/a-ratios in the ranges 1–3.14 and 3.14–4.07

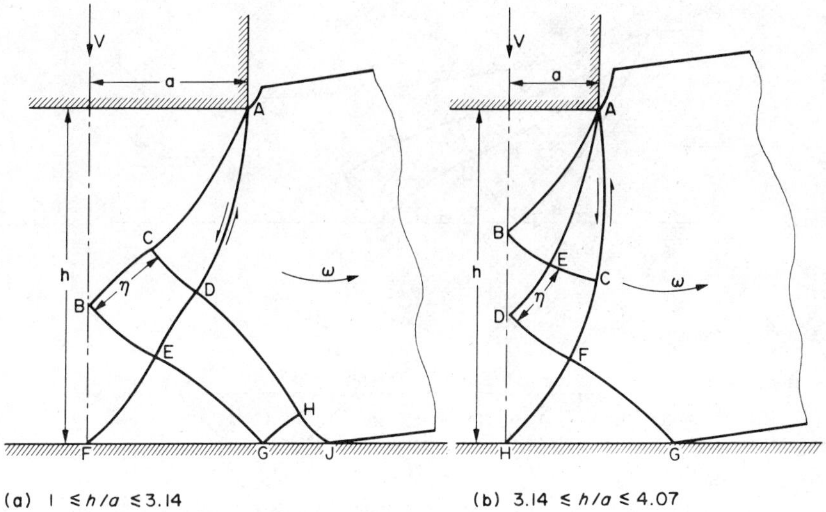

FIG. 5.4. Slipline field solution for one-sided strip indentation, $1 \leqslant h/a \leqslant 4.07$ (Dewhurst, 1974).

respectively. These solutions are structurally very similar to that discussed in the previous subsection, so that the operational analysis, which again requires the simultaneous consideration of the slipline and hodograph networks, will not be given here.

It is of interest to note that the yield point loads associated with these solutions are less than 2% smaller than those associated with the solution in Fig. 3.4. The latter, hence, has the status of an upper bound solution. The field intermediate between the two fields of Fig. 5.4, valid for $h/a = 3.14$, occurs when $\eta = 0$ in either solution, and is again an example of an "eigenfield" since the velocity discontinuities along $ADEF$ in the first solution and $ACFH$ in the second both vanish. This particular solution has been further analysed in Dewhurst (1978c).

5.4. *Strip rolling problems*

Slipline field theory is potentially very useful for the analysis of the hot strip rolling process since the two basic assumptions of a plane strain deformation and a non-hardening material are particularly appropriate. Unfortunately, however, strip rolling solutions are amongst the most complex yet encountered. The first such solution was given by Alexander (1955) who employed Prager's graphical technique together with a lengthy trial-and-error procedure to construct the slipline field for a single geometry. This procedure proved so time consuming that in subsequent papers Ford and Alexander (1964) and Crane and Alexander (1968) made a number of geometrical approximations such as approximating sliplines by circular arcs. In general these solutions involve plastically deforming zones bordering on the curved roll faces, so that the resulting integral equations are non-linear and the matrix operator procedure is not applicable. However, for small and medium size rolls the solution is of a simpler type and the strip material adjacent to the roll faces rotates rigidly with the rolls. This solution, given originally by Druyanov (1971) and illustrated here in Fig. 5.5, is linear and has been fully analysed and computed using the matrix operator procedure by Dewhurst *et al.* (1973).

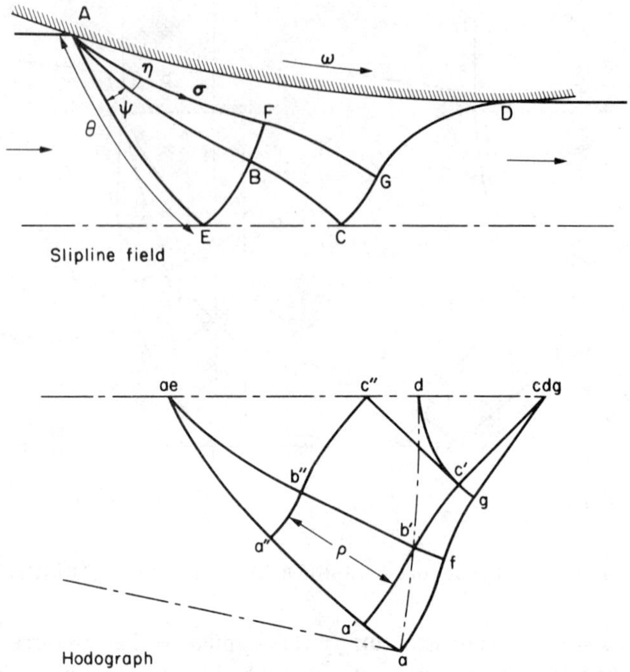

FIG. 5.5. Slipline field and hodograph for hot rolling of strip (Druyanov, 1971; Dewhurst *et al.*, 1973).

The slipline field can be constructed once the slipline AF ($\boldsymbol{\sigma}$) and the three angles θ, ψ, and η are known. As in the two previous problems, in order to construct the basic operator equation for $\boldsymbol{\sigma}$ it is necessary to consider simultaneously the slipline and hodograph diagrams. The operational analysis will not be developed here but can be found in the original paper or in Collins (1978a). The resulting equation for $\boldsymbol{\sigma}$ is of the general form (5.4) with $\lambda = 1$, $\mu = \rho/\omega$, where ρ is the magnitude of the velocity discontinuity along $ABCGD$, and

$$\left.\begin{aligned} A &= (Q_{\eta\theta}Q_{\psi\eta} + P_{\theta\eta}Q_{\psi}^{*}T_{\psi}^{-1}P_{\eta}^{*})(S_{\eta}T_{\phi}^{-1}Q_{\theta\phi}), \\ B &= [P_{\theta\eta}(I + Q_{\psi}^{*}T_{\psi}^{-1}R_{\psi}(I + Q_{\eta\psi}))] + Q_{\eta\theta}P_{\eta\psi}. \end{aligned}\right\} \quad (5.8)$$

It was found that in this problem the series expansions of the radii of curvature of the various sliplines were dominated by the first term, so that these curves do not differ significantly from circular arcs, which vindicates the earlier geometrical approximations made by Crane and Alexander. The three slipline angles and the base pressure were determined by making the front and back tensions zero and by specifying the roll radius to entry thickness and reduction ratios. The resulting predictions of roll torque were found to be in good agreement with the extant engineering theories, but those of roll separation force were found to be around 15% less.

The radius of the exit circular arc slipline GD was found to decrease with decreasing roll radius and eventually vanished simultaneously with ρ, the magnitude of the velocity discontinuity. The limiting field is hence an "eigenfield". For still smaller rolls the solution was found to involve a rigid sliver attached to the roll face and separated from the plastic deforming zone by a velocity discontinuity, which, unlike that in Fig. 5.5, does not propagate to the centre line.

This analysis has been extended in a number of directions. For example, Klarin (1979) has considered the effect of non-zero front and back tensions and produced an "atlas" of slipline field solutions for strip rolling. Dewhurst *et al.* (1974) and Collins and Dewhurst (1975) have studied the corresponding solutions when the deformation in the roll gap is asymmetrical, a situation which can arise as a result of the rolls being of different sizes or rotating with different speeds or due to asymmetrical entry conditions. (See also the article by W. Johnson in this Volume.) These analyses confirmed earlier experimental observations that the curvature of the rolled sheet is critically dependent on the overall geometry of the pass and that the sheet does not necessarily curve away from the roll with the faster peripheral speed as might seem likely intuitively. Curved sheet is also produced by extrusion through unsymmetrical square dies, and a comprehensive range of solutions, which are of indirect type and amongst the most complex yet successfully analysed, has been computed by Das *et al.* (1977).

5.5. *Rolling friction*

The solution shown in Fig. 5.6, due to Collins (1978b), is that for the problem of a rigid cylinder of radius R rolling, with angular speed ω, over a rigid/perfectly plastic half-space. This problem arises in the development of theories of rolling resistance in circumstances when the loads are sufficiently large for the elastic strains to be negligible. This solution is exact and supersedes earlier approximate analyses of this problem by Mandel (1966), Marshall (1968), and Collins (1972). In AEB the material rotates rigidly, adhering to the cylinder, whilst that in $ACDE$ deforms continuously. There is a velocity discontinuity along $CDEB$. A slipline field of this form can be constructed for any choice of the starting slipline AE; however, only one of these will correspond to a steady-state solution in which the velocity vector on CA is everywhere tangential to this free surface. As in the previous three problems, it is hence necessary to consider the corresponding hodograph diagram. The operational analysis of this problem, which makes use of the free surface operator F in (4.10), turns out to be particularly simple because, unlike all other solutions so far examined, the resultant linear integral equation is of Volterra type and can be solved analytically.

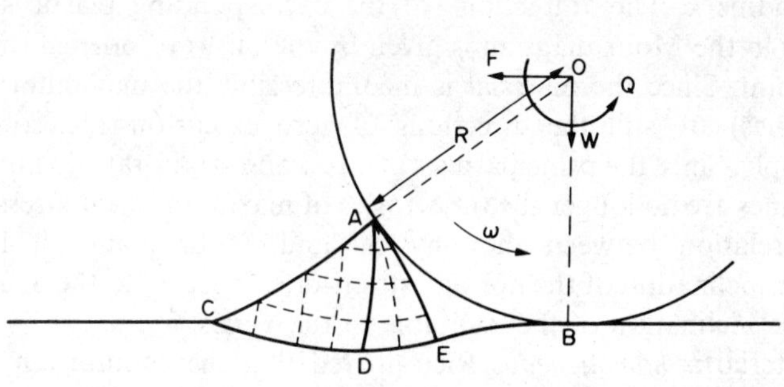

FIG. 5.6. Slipline field for cylinder rolling over a half-space (Collins, 1978b).

The computations of this solution largely confirm the rolling resistance formula suggested by Mandel (1966):

$$f+q = \tfrac{1}{2}w^2 G(f/w), \tag{5.9}$$

where $w = W/kR$ and $f = F/kR$ are the non-dimensional vertical and horizontal load coefficients and $q = Q/kR^2$ is the non-dimensional torque. The function G is monotonic increasing and has to be found numerically. Petryk (1979) has proposed a generalization of the solution of Fig. 5.6 in which a singular fan emanates from C. He does not give any computations, however.

6. Anisotropy

The most widely used yield condition for anisotropic rigid/plastic materials is Hill's generalized Mises criterion, which is appropriate for materials containing three mutually orthogonal planes of symmetry. Such orthotropic conditions occur, for example, in rolled sheet. The lines of intersection of these three planes are the principal axes of anisotropy. For plane strain deformations occurring in a plane perpendicular to one of these principal axes, this criterion can be written

$$as_{xy}^2 + b\sigma_{xy}^2 = 1, \tag{6.1}$$

where the (x, y)-axes are the in-plane principal axes of anisotropy, $s_{xy} = \frac{1}{2}(\sigma_{xx} - \sigma_{yy})$, and a and b are material constants. The theory of slipline fields appropriate for such a criterion has been given by Hill (1950) and has been used to study a number of incipient motion problems such as strip indentation by Venter *et al.* (1971) and the earing of deep drawn cups by Johnson and Sowerby (1974). The analytical details become quite complicated since the simple Hencky relations (2.22) are replaced by expressions involving elliptic integrals. Booker and Davis (1972) and Rice (1973) have independently given a generalization of Hill's theory which is applicable to any convex yield locus and in which the generalized Hencky relations are given an elegant geometrical formulation.

Rice demonstrates that under plane strain conditions the yield criterion for any incompressible rigid/plastic material, which satisfies the maximum work inequality, can be represented by a unique convex yield locus in Mohr stress space (Fig. 6.1). Hill's criterion (6.1) hence corresponds to an elliptical yield locus, whilst an isotropic material is represented by a circle of radius k. The "direction" of the corresponding plastic strain-rate vector $(e_{xx} = -e_{yy}, e_{xy})$ in the Mohr diagram is given by the outward normal to this locus at the current stress point. Since the material is incompressible the maximum shear strain-rate directions (sliplines) are still the directions of zero extension-rate, so that Geiringer's equations still apply. Since the principal axes of stress and strain-rate do not now coincide, in general, the sliplines are no longer the trajectories of maximum shear stress. From the usual "double-angle" relation between the physical and Mohr planes it follows that the counterclockwise inclination of the normal, strain-rate, "vector" to the σ_{xy}-axis is 2ϕ, where ϕ, as usual, is the inclination of the α-slipline to the x-axis.

When the material is *homogeneous* Rice proved that the equilibrium equations, when referred to the sliplines, can be integrated to give

$$p + S = \text{constant on } \alpha\text{- and } \beta\text{-lines}, \tag{6.2}$$

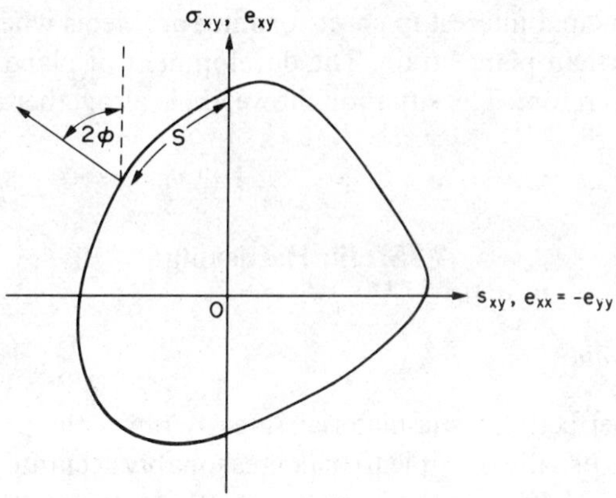

FIG. 6.1. Yield locus for general anisotropic material in Mohr space (Rice, 1973).

where S is the length of arc on the yield locus, measured positive in the counterclockwise sense. The sliplines are hence the stress characteristics. For Hill's elliptical locus the evaluation of S introduces elliptic integrals, whilst for the circular isotropic locus, $S = 2k\phi$ and (6.2) reduces to the classical Hencky relations. This result was used by Rice to produce the general solution to the Prandtl punch problem for a material with an arbitrary yield locus. The paper by Booker and Davis contains even more general results appropriate for an anisotropic granular material with pressure-dependent yielding. Chitkara and Collins (1974) have used (6.2) as a basis for generalizing Prager's graphical procedure for constructing slipline fields for isotropic materials to anisotropic materials with arbitrary convex yield loci.

More recently Hill (1980) has made a completely general study of self-equilibrated two-dimensional stress states satisfying an arbitrary scalar constraint (yield condition) $f(\sigma_{xx}, \sigma_{yy}, \sigma_{xy}) = 0$, which includes the above analyses as special cases and also covers situations of generalized plane stress. Two of the main results to come out of his analysis are that (a) a necessary and sufficient condition for the stress equilibrium equations to have real characteristics is that the principal components of the tensor $\partial f/\partial \sigma_{ij}$ (strain-rate) should have opposite signs, and (b) the stress increment tensor along one characteristic is uniaxial in the direction of the other.

All the above considerations relate to the incipient deformation of a homogeneous body. Continued deformations and steady flows of anisotropic materials are more difficult to analyse, even when hardening is neglected, because the yield locus must rotate with the material elements. As shown by Collins and Meguid (1977), the yield locus in Mohr space (Fig. 6.1) rotates at a rate of twice the local rigid body spin. In general, therefore, an initially homogeneous body will become inhomogeneous as soon as plastic deformation occurs. Since in a steady flow the orientation of the yield locus is initially unknown, the basic uniqueness theorem discussed in Section 2.5 does not apply. A problem in which it is possible to compute readily the continuing deformation is that of compression of thin strip between long, rough dies. Collins and Meguid (1977) have generalized the original Prandtl–Nádai solution to this problem to the case when the strip is made of orthotropic material, and have demonstrated how discontinuities in the orientation of the yield locus can arise as the deformation proceeds.

Anisotropy is of principal interest in sheet-forming problems where the prevailing states are plane stress rather than plane strain. The development of plane strain theory is not of great practical value therefore. The situation, however, is altogether different for hardening materials.

7. Strain Hardening

7.1. *Visioplastic techniques*

The restriction to perfectly plastic materials greatly limits the practical value of slipline field theory. Whilst it is usually possible to make reasonably accurate load estimates by using an appropriate mean yield stress in the known perfectly plastic solution to a steady forming operation, it is very much more difficult to gauge the effect of hardening on the plastic stress and deformation fields. The small amount of theoretical work that has been done in this area is summarized in the next two subsections. To date most information on the modifications to slipline fields induced by hardening has been gained by visioplastic techniques, notably by Oxley, Farmer, and coworkers. A recent summary of these investigations has been given by Oxley (1978).

The velocity, strain, and strain-rate fields are found by observing the deformation of grids inscribed in some manner on the metal surface. Such grids can now be applied most efficiently using photographic techniques as described by Farmer and Fowle (1979). The resulting deformation fields can then be evaluated using standard computer routines which include numerical smoothing techniques. Farmer and Conning (1979) describe such a procedure applied to an extrusion process, improving on earlier results obtained by Farmer and Oxley (1971). Once the streamlines have been determined, the yield stress at each point can be calculated using the experimentally determined stress–strain curve for uniaxial tension. Assuming the material is isotropic so that the maximum shear stress trajectories coincide with the known maximum shear strain-rate trajectories (sliplines), the equilibrium equations (2.19) for a non-homogeneous body can be integrated to give the pressure and hence the complete stress state at each point of the deforming region. Two examples of such experimentally determined slipline fields are shown in Figs. 7.1 and 7.2, for machining by Christopherson *et al.* (1958) and Roth and Oxley (1972), and extrusion by Farmer and Oxley (1971). The sliplines of the same family are seen to be frequently of opposite curvature in contrast to the situation for perfectly plastic materials. The main effect of hardening is seen to be the "opening up" of the velocity discontinuities of the perfectly plastic solution into regions of finite width through which the velocity varies rapidly. As will be seen, this "opening up" of velocity discontinuities is the fundamental cause of difficulty in constructing theoretical solutions to flow problems of hardening materials. Other similar experimental studies have been made by Devenpeck and Weinstein (1970) on wedge flattening and by Roth (1978) on wedge indentation.

7.2. *Steady flows*

Under plane strain conditions the governing equations for an isotropic hardening solid when referred to the α- and β-sliplines are (a) the equilibrium equations (cf. (2.19))

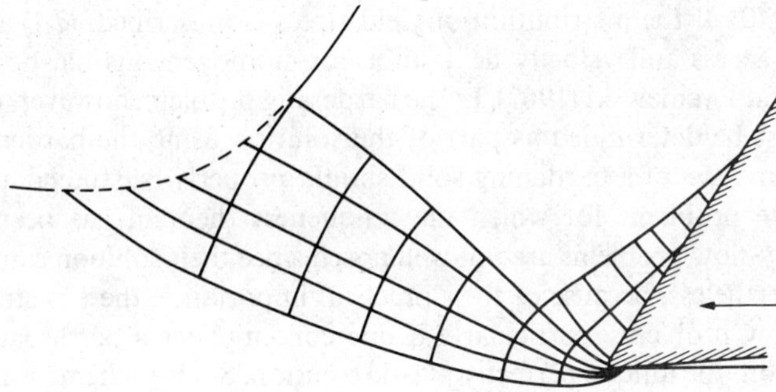

FIG. 7.1. Experimentally determined slipline field for orthogonal machining (Roth and Oxley, 1972).

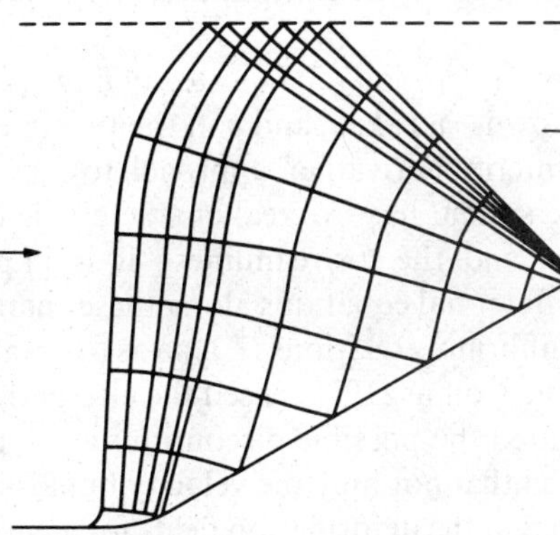

FIG. 7.2. Experimentally determined slipline field for extrusion (Farmer and Oxley, 1971).

$$\frac{\partial p}{\partial s_\alpha} + 2k\frac{\partial \phi}{\partial s_\alpha} - \frac{\partial k}{\partial s_\beta} = 0, \quad \frac{\partial p}{\partial s_\beta} - 2k\frac{\partial \phi}{\partial s_\beta} - \frac{\partial k}{\partial s_\alpha} = 0, \tag{7.1}$$

(b) Geiringer's equations (2.23), expressing the fact that the extension-rates are zero in the slipline directions,

$$\frac{\partial u}{\partial s_\alpha} - v\frac{\partial \phi}{\partial s_\alpha} = 0, \quad \frac{\partial v}{\partial s_\beta} + u\frac{\partial \phi}{\partial s_\beta} = 0, \tag{7.2}$$

and (c) the hardening rule

$$\frac{Dk}{Dt} = u\frac{\partial k}{\partial s_\alpha} + v\frac{\partial k}{\partial s_\beta} + \frac{\partial k}{\partial t} = h(2\gamma) \tag{7.3}$$

where
$$\gamma = \frac{1}{2}\left(\frac{\partial u}{\partial s_\beta} + \frac{\partial v}{\partial s_\alpha} - v\frac{\partial \phi}{\partial s_\beta} + v\frac{\partial \phi}{\partial s_\alpha}\right) \tag{7.4}$$

is the shear strain-rate and h is the hardening modulus. (N.B.—Due to the assumption of isotropy and the neglect of elastic strains it is not necessary to use the co-rotational stress-rate in (7.3).) These equations seem first to have been formulated by Richmond—see Devenpeck and Weinstein (1970). If the distribution of yield stress is prescribed, (7.1) and (7.2) can be solved to give the stress and velocity fields in a non-homogeneous plastic body—see, for example, Olszak and Rychlewski (1962). In the hardening problem, however, the distribution of yield stress must be determined as part of the solution using the hardening law.

The finite deformation of a hardening solid should properly be studied as a sequence of incremental or rate problems for which the uniqueness theorem has been given by Hill (1956, 1957). Steady-flow problems are not well posed since their solution cannot be expected to be unique. Nevertheless, because of their practical importance, there is still some merit in studying steady-flow problems, particularly if one contemplates a perturbation or iterative analysis starting from the unique perfectly plastic solution. Such a scheme was considered by Halling and Mitchell (1967) for an extrusion process. They used the strain-rate field for the known perfectly plastic solution and the hardening rule (7.3) to obtain a first-order estimate of the distribution of yield stress. However, as the authors remark, this procedure ignores the infinities in the strain-rate on the velocity discontinuities, and so fails to predict the "opening up" of such discontinuities.

Since the local time derivative is zero in (7.3) in a steady flow, (7.1)–(7.3) form a system of five equations in the five unknowns ϕ, p, k, u, and v in the two independent space variables. Hopkins (1972) adopted the "interior derivative" approach to characteristics of such systems and has shown that whilst the system has five real characteristic directions—the α- and β-sliplines, each repeated twice, and the γ-streamlines—it is impossible to decouple the equations into five ordinary differential equations along these characteristics. In particular it is impossible to rewrite the equilibrium equations (7.1) so as to obtain expressions with all the derivatives in the same direction. Collins (1978c) used the alternative "Hadamard approach" to characteristics and investigated the possible discontinuities in the derivatives of the five dependent variables. He showed that not only the velocity but also the strain-rate γ must be continuous across a slipline, so that the deformation fields for a hardening material are much smoother than those existing in the perfectly plastic limit. If $[\cdot]$ denotes the discontinuity in a quantity across a curve Γ, then the relationship between the jumps in the normal derivatives of the dependent variables across a slipline can be summarized as

$$\frac{\left[\dfrac{\partial p}{\partial n}\right]}{\pm 2k} = \frac{\left[\dfrac{\partial \phi}{\partial n}\right]}{1} = \frac{\left[\dfrac{\partial u}{\partial n}\right]}{v} = \frac{\left[\dfrac{\partial v}{\partial n}\right]}{-u} = \frac{\left[\dfrac{\partial k}{\partial n}\right]}{0}, \tag{7.5}$$

the upper and lower signs referring to α- and β-lines respectively. If θ is the counterclockwise inclination of the streamline from the α-line and q is the speed, if follows from (7.5) that

$$\left[\frac{\partial q}{\partial n}\right] = 0 \quad \text{and} \quad \left[\frac{\partial(\theta+\phi)}{\partial n}\right] = 0, \tag{7.6}$$

so that the curvature of a streamline is continuous across a slipline—a result which does not hold in the perfectly plastic limit. As a further consequence of (7.5) it follows that $\partial k/\partial s_\gamma$ is continuous across a slipline, and in particular across an unloading slipline. Moreover, since $\partial k/\partial s_\gamma$ is non-negative on the plastic side but non-positive on the rigid side, this derivative must in fact be zero on the unloading characteristic.

To date the only analytical solutions known are for essentially one-dimensional problems such as the radial flows analysed by Durban and Budiansky (1978). Because of the difficulty of dealing with velocity discontinuities, the streamline die solutions form an attractive class of perfectly plastic solutions for possible generalization to hardening materials. However, as noted in Section 3.6, kinematic requirements limit the slipline field network to a Hencky–Prandtl net, so that the equilibrium and hardening equations will only rarely be simultaneously satisfied. It would appear in fact that the only possible such solution is that with spiral dies, illustrated in Fig. 7.3, in which the sliplines consist of radial straightlines and circular arcs, and then only if the strain hardening is *linear*. This field has been investigated previously by Hill (1948) in a general study of the flow of hardening materials through

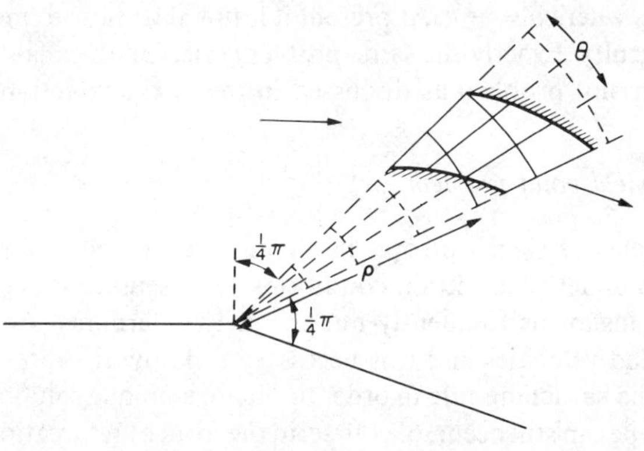

FIG. 7.3. Flow of a hardening material around a corner or through a spiral die.

centred fans around convex and concave corners, and by McClintock (1971) in a study of the propagation of a crack through a hardening material. The stress solution in the centred fan is readily shown to be given by

$$k = 2h\theta + k_0, \quad p = -2h(\log\rho + \theta^2) - 2k_0\theta + p_0, \tag{7.7}$$

where k_0 is the value of the yield stress at entry. The logarithmic infinity in pressure does not in itself invalidate the solution, although it does mean that if the vertex lies in the deforming region, such a fan can never be joined to a rigid region without overstressing occurring. Even the spiral die solution is still invalid, however, because the value of k increases linearly through the fan and hence, as proved above, equilibrium demands that it must go on increasing into the rigid region at exit. The solution cannot therefore be complete. It was found by Richmond (1979) that in experiments on spiral dies the plastic deforming region extends downstream of the exit plane in accord with the above predictions.

In order to study the "opening up" of velocity and strain-rate discontinuities in a hardening material, Collins (1978c) proposed a tentative perturbation procedure. The characteristic coordinates are used as independent variables and the slipline geometry is specified by the Lamé parameters h_α and h_β which satisfy the orthogonality conditions (2.4). These equations, together with (7.1)–(7.3), form a determinate set for the seven dependent variables h_α, h_β, ϕ, p, k, u, and v. These variables are then expanded as power series in the non-dimensional hardening modulus $\lambda = h/k_0$ (assumed $\ll 1$); the zeroth-order terms giving the base perfectly plastic solution, with constant yield stress k_0. However, unlike the regular perturbation series expansions used by Spencer (1961) and Richmond and Morrison (1968), these series will involve *fractional* powers of λ.

A velocity or strain-rate discontinuity across a slipline (α-line say) in the perfectly plastic solution is mapped into a fixed region in (α, β)-space and the condition that this region has zero thickness in the base solution is accomplished by setting the zeroth-order term of the expansion of h_β equal to zero. Comparing terms on substituting these expansions in the governing equations gives the indices of the fractional powers of λ required in each of the expansions. For example, for a velocity discontinuity it is found that p, ϕ, h_α, h_β, u, and v must all be expanded in powers of $\lambda^{1/2}$, but the first-order term in the expansion of k is $O(\lambda)$. The difference in these powers is unfortunate because it means that it is impossible to equate the first-order terms in the expansions of p and k at the point where the velocity discontinuity meets the free surface, where $p = \pm k$. At present it is not clear how to modify the procedure to overcome this difficulty. Exactly the same problem arises in the analogous procedure for the post-initial yield point problem as discussed in the next subsection.

7.3. *The post-initial yield point problem*

Consider a homogeneous rigid/isotropic hardening plastic body subjected to monotonically increasing loads under plane strain conditions. As discussed earlier the "deformation mode" at this initial instant is frequently not uniquely determined by specifying just the boundary tractions and velocities and it is necessary to know the rates-of-change of these quantities as well as the hardening-rate in order to obtain a unique solution. (Non-uniqueness of a more restricted type can still occur, of course, in the form of bifurcations.) The problem of actually identifying the unique mode is still very difficult, however, due principally to the presence of discontinuities in velocity in the initial instant. Such discontinuities can exist only instantaneously and must then diffuse out into shear bands of finite width. There is to date no theory describing this diffusion process. With the exception of the torsion problem described later, the only available analytical solutions for hardening materials undergoing plane strain deformation are those with strong symmetries in which such discontinuities do not arise. Examples are the uniform expansion of cylindrical tubes or compression of strip between very long dies as discussed in the papers by Onat (1959) and Collins and Meguid (1977). In bifurcation studies of rigid/plastic materials, either attention is restricted to continuous deformation modes as in Cowper and Onat (1962), or if discontinuous modes are envisaged their validity is restricted to the initial instant—see Hill and Hutchinson (1975).

For unsteady deformations the governing equations (7.1)–(7.3) form a quasi-linear system in three independent variables, so that the characteristics are now families of planes in (x, y, t) space. Application of the standard procedure for finding the characteristic structure of such systems shows that there are five families of characteristic planes through any point: (a) a pencil of planes through the local s_α-direction (repeated twice), (b) a pencil of planes through the local s_β-direction (repeated twice), and (c) a pencil of planes through the particle path direction in (s_α, s_β, t) space, i.e. containing the vector $(u, v, 1)$. Furthermore the jumps in the normal derivatives of the dependent variables across the planes of type (a) or (b) are found to satisfy (7.5) provided this plane is not one on which t is constant, in which case $[k_{,t}] = 0$, but the jumps in all the other quantities are unrestricted. The corollary (7.6) applies to unsteady flows as well. The discontinuities across a plane of family (c) are found to be

$$\frac{[p_{,n}]}{2k \sec 2\psi} = \frac{[\phi_{,n}]}{1} = \frac{[u_{,n}]}{v} = \frac{[v_{,n}]}{-u} = \frac{[k_{,n}]}{-2k \sec 2\psi}, \tag{7.8}$$

where ψ is the inclination of the line of intersection, Γ say, of the characteristic plane with the plane t = constant, to the local s_α-direction. By transforming these stresses it can be shown that the only possible discontinuous stress gradient across Γ is the normal derivative of the normal stress component in the direction parallel to Γ. This can be seen more directly, however, by generalizing the argument given by Hill (1950, p. 133). If Ox, Oy are cartesian axes, chosen so that Ox is locally tangential to Γ, it follows from the equilibrium equations and the yield condition that

$$(\sigma_{xx}-\sigma_{yy})\left[\frac{\partial\sigma_{xx}}{\partial y}\right]=4k\left[\frac{\partial k}{\partial y}\right]. \qquad (7.9)$$

In a homogeneous, non-hardening solid, the right-hand side is zero, so that characteristics are only possible when $\sigma_{xx}=\sigma_{yy}$, which is the condition that Γ is a slipline. For a hardening material, however, $[\partial k/\partial y]$ need not be zero and it is possible to have discontinuities in derivatives across curves which are not sliplines, although these are always accompanied by discontinuities in the derivative of the yield stress. Since such a discontinuity occurs across a plane which always contains the particle path direction, it follows that such discontinuities are material surfaces and remain embedded in the material.

The above are examples of "weak" discontinuities. As is well known, "strong" discontinuities, where some of the stress components themselves are discontinuous, are possible in classical perfectly plastic solids, although they do not feature often in solutions to specific boundary value problems as they entail a strong kinematic restriction. As proved by Hill (1950) and by Geiringer (1973), all strain-rate components must vanish on such a discontinuity. Rychlewski (1966) showed that a second form of stress discontinuity can occur in a non-homogeneous material.

With the same notation as above equilibrium demands that σ_{xy} and σ_{yy} must both be continuous across Γ but σ_{xx} need not be. If the material is plastic on both sides of Γ, the yield condition must be satisfied on each side so that

$$\sigma_{xx}^{+}=\sigma_{yy}\pm 2[(k^{+})^2-\sigma_{xy}^2]^{\frac{1}{2}} \quad\text{and}\quad \sigma_{xx}^{-}=\sigma_{yy}\pm 2[(k^{-})^2-\sigma_{xy}^2]^{\frac{1}{2}}. \qquad (7.10)$$

In a perfectly plastic material $k^{+}=k^{-}$ and the only possible discontinuity is one in which opposite signs are chosen in (7.9). However, if $k^{+}\neq k^{-}$, a second type of discontinuity is possible by taking the same signs in (7.9) so that

$$[\sigma_{xx}]=\pm 2\{[(k^{+})^2-\sigma_{xy}^2]^{\frac{1}{2}}-[(k^{-})^2-\sigma_{xy}^2]^{\frac{1}{2}}\}. \qquad (7.11)$$

The strength of this discontinuity hence vanishes in the perfectly plastic limit. Geiringer (1973) shows, using the Lévy–Mises equations, that since the signs of $(\sigma_{xx}-\sigma_{yy})$ are different on the two sides of a discontinuity of the first type, all the strain-rate components must vanish. This argument does not apply to the second form of discontinuity however, since the signs of $(\sigma_{xx}-\sigma_{yy})$ are the same on the two sides.

Since the value of k is different on the two sides of Γ such a discontinuity must necessarily be a material surface. If the jump in k is made infinitesimally small it can be readily verified that (7.11) reduces to (7.9), so that this second type of stress discontinuity is the strong form of the new characteristic discontinuity discussed above. The question as to whether such material discontinuities can arise out of strain-rate or velocity discontinuities in the initial deformation mode is still open.

An alternative, but essentially equivalent, approach is to regard the continued deformation as a sequence of rate problems. Onat (1959) considered the initial stress-rate problem for plane strain deformations and showed that the sliplines are the characteristics of the stress-rates as well as for the stresses. He did not consider velocity or strain-rate discontinuities, however.

In unpublished work the author has attempted to apply the same kind of fractional power series expansions, as described in the previous subsection, to the problem of calculating the stress and deformation fields just after the initial instant of yield. Here the small perturbation parameter is a suitably non-dimensional time variable. It is, of course, necessary to allow for the movement of the characteristics network relative to space and the body. This can be achieved by introducing $(\bar{u}, \bar{v})$, the velocity components in the slipline directions of the point (α, β) relative to space. Analytically they are defined by

$$\bar{u} = \bar{x}_{,t} - \bar{y}\phi_{,t}, \quad \bar{v} = \bar{y}_{,t} + \bar{x}\phi_{,t}, \tag{7.12}$$

where $(\bar{x}, \bar{y})$ are Mikhlin coordinates and $(\,.\,)_{,t}$ denotes the rate of change of a quantity at a point fixed in the characteristics network (i.e. at fixed (α, β)). $\bar{u}$ and $\bar{v}$ satisfy the compatibility equations

$$\left.\begin{aligned} \bar{u}_{,\alpha} - \bar{v}\phi_{,\alpha} = h_{\alpha,t}, \quad \bar{v}_{,\alpha} + \bar{u}\phi_{,\alpha} = h_{\alpha}\phi_{,t}, \\ \bar{v}_{,\beta} + \bar{u}\phi_{,\beta} = h_{\beta,t}, \quad \bar{u}_{,\beta} - \bar{v}\phi_{,\beta} = h_{\beta}\phi_{,t}. \end{aligned}\right\} \tag{7.13}$$

The resulting perturbation analysis predicts that in an expanding velocity discontinuity, k varies as $t^{2/3}$, $\bar{u}$ and $\bar{v}$ as $t^{-2/3}$ and all other variables as $t^{1/3}$. The discontinuity hence spreads out at an infinite rate, as could be expected, but since the powers in the expansions of p and k are different, one has the same problem at a free surface as in the steady flow analysis.

It is difficult to find a simple model problem on which to test any theory. However, Richmond (1979) has suggested one such problem which, although untypical in many ways, does have a simple exact solution which does lend support to the use of fractional power series as above.

Consider a hollow circular cylinder, $a \leqslant r \leqslant b$, made of rigid/hardening material subjected to monotonically increasing torsional shear traction on the inner face. With $\sigma_{r\theta} = k$, the only non-trivial equilibrium equation is

$$\frac{\partial k}{\partial r} + \frac{2k}{r} = 0, \tag{7.14}$$

which integrates to give

$$k = k_0 (c/r)^2, \tag{7.15}$$

where $c(< b)$ is the radius of the rigid/plastic boundary and k_0 is the yield stress of the undeformed material. The hardening equation is

$$\frac{\partial k}{\partial t} = -hr\frac{\partial}{\partial r}\left(\frac{1}{r}u_\theta\right), \tag{7.16}$$

where u_θ is the circumferential velocity component, which, assuming h is constant, integrates to give

$$hu_\theta = k_0\frac{dc}{dt}\left(\frac{c}{r} - \frac{r}{c}\right) \tag{7.17}$$

with $u_\theta = 0$ when $r = c$. If $u_0 = \omega a$ on the inner surface, where the angular speed ω is kept constant, then (7.17), at $r = a$, can be integrated with respect to time to give

$$\frac{ha\omega t}{k_0} = -a \log\left(\frac{c}{a}\right) + \frac{1}{2}c\left(\frac{c}{a} - \frac{a}{c}\right). \tag{7.18}$$

Putting $c/a = 1+\eta$, then for small η, (7.18) predicts that

$$\eta \approx (h\omega/k_0)^{1/2} t^{1/2}, \tag{7.19}$$

so that the shear layer grows initially as $t^{1/2}$.

The reason that whereas in this problem we have obtained a $t^{1/2}$-dependency, whilst the general theory predicts a $t^{1/3}$-law, can be attributed to the fact that in the limit as $t \to 0$ the shear layer reduces to a velocity discontinuity at the inner face of the cylinder, but that Hencky's relation does not hold on this limiting "slipline". Indeed, the pressure is constant on this line. Since this "slipline" only has material on one side, the proof of Hencky's relation for an isolated slipline given in Section 2.2 is not applicable.

8. Conclusions

In this article I have attempted to survey the current state of the art of solving boundary value problems in plane strain plasticity using analytical or semi-analytical techniques. No attempt has been made to discuss fully numerical procedures such as finite element methods. The development of the series and matrix operator procedures has appreciably widened the class of problems that can be solved and computed. For most problems which lead to linear integral equations it seems unlikely that the procedures described in Sections 4 and 5 could be improved upon significantly. However, in cases where the governing integral equation is particularly simple, as in the compression problem described in Section 5.1, it may be that the equation can be solved analytically. Certainly an investigation into the spectral properties of the fundamental operators would be valuable.

The main outstanding area of difficulty in classical slipline theory is where the governing integral equation is nonlinear as in Alexander's (1955) strip-rolling solution, or when Coulomb's frictional law is appropriate. Such problems could be solved using nonlinear optimization techniques, but a universally applicable, efficient computational scheme has yet to be devised. This paper has been restricted to plane-strain deformations. There is, however, a comparable literature on axisymmetric problems, where the governing equations are hyperbolic, provided the plastic stress state corresponds to one of the vertices of the Tresca yield locus (Haar–von Kármán hypothesis). These problems, too, frequently lead to integral equations which are invariably nonlinear due to the r^{-1}-terms in the governing equations.

The advent of finite element methods has made it possible to improve the constitutive models employed in the analysis of forming processes by including elastic, hardening, strain-rate, and thermal effects for example. Such methods, however, have not yet reached the stage of producing wholly reliable results as was evident from the round-table discussions at the recent *IUTAM Symposium on Metal Forming Plasticity* and reported by Kudo and Matsubara (1978). Since, too, these computations are extremely expensive on computer time, it is essential that they should be supported by as much analytical work as possible. Slipline solutions provide such support in giving "first-order" estimates of the inhomogeneity present in nearly all plastic deformations and hence pinpoint the areas in which grids of finer sizes are required. In addition the singularities present in the initial deformation modes of rigid/plastic materials can lead to serious numerical problems. The singular perturbation procedures outlined in Section 7 are a tentative beginning of an analysis of these singularities which would repay further study.

Acknowledgements

I am very grateful to Dr. Owen Richmond of the United States Steel Corporation, Research Laboratory, for many helpful discussions on the problems described in this paper and on the "hardening problem" in particular. I also wish to thank Mrs. M. L. Hillock for her efficient typing of the manuscript. This paper was written whilst the author was a member of the Department of Mathematics, UMIST, Manchester, England.

References

Alexander, J. M. (1955) A slipline field for the hot rolling process, *Proc. Inst. Mech. Engrs* **169,** 1021–1030.

Bishop, J. F. W. (1953) On the complete solution to problems of deformation of the plastic–rigid material, *J. Mech. Phys. Solids* **2,** 43–53.

Bishop, J. F. W., Green, A. P., and Hill, R. (1956) A note on the deformable region of a rigid–plastic body, *J. Mech. Phys. Solids* **4,** 256–258.

Booker, J. R. and Davis, E. H. (1972) A general treatment of plastic anisotropy under conditions of plane strain, *J. Mech. Phys. Solids* **20,** 239–250.

Carathéodory, C. and Schmidt, E. (1923) Über die Hencky-Prandtlschen Kurven, *Z. angew. Math. Mech.* **3,** 468–475.

Chakrabarty, J. (1979) Exact solutions for certain slipline fields, *Int. J. Mech. Sci.* **21,** 477–488.

Challen, J. M. and Oxley, P. L. B. (1979) An explanation of the different regimes of friction and wear using asperity deformation models, *Wear* **53,** 229–243.

Childs, T. H. C. (1971) A new visio-plasticity technique and a study of curly chip formation, *Int. J. Mech. Sci.* **13,** 373–387.

Childs, T. H. C. (1973) The persistence of asperities in indentation experiments, *Wear* **25,** 3–16.

Childs, T. H. C. (1977) The persistence of roughness between surfaces in static contact, *Proc. R. Soc. Lond.* A **353,** 35–53.

Chitkara, N. R. and Collins, I. F. (1974) A graphical technique for constructing anisotropic slipline fields, *Int. J. Mech. Sci.* **16,** 241–248.

Christopherson, D. G., Oxley, P. L. B., and Palmer, W. B. (1958) Orthogonal cutting of a work-hardening material, *Engineering* **186,** 113–116.

Collins, I. F. (1968a) The algebraic geometry of slipline fields with applications to boundary value problems, *Proc. R. Soc. Lond. A* **303,** 317–338.

Collins, I. F. (1968b) Geometric properties of some slip-line fields for compression and extrusion, *J. Mech. Phys. Solids* **16,** 137–152.

Collins, I. F. (1969a) The upper bound theorem for rigid/plastic solids generalised to include Coulomb friction, *J. Mech. Phys. Solids* **17,** 323–338.

Collins, I. F. (1969b) Slipline field solutions for compression and rolling with slipping friction, *Int. J. Mech. Sci.* **11,** 971–978.

Collins, I. F. (1970) Compression of a rigid–perfectly plastic strip between parallel rotating smooth dies, *Q. Jl Mech. Appl. Math.* **23,** 329–348.

Collins, I. F. (1971) On an analogy between plane strain and plate bending solutions in rigid/perfect plasticity theory, *Int. J. Solids Structures* **7,** 1057–1073.

Collins, I. F. (1972) A simplified analysis of the rolling of a cylinder on a rigid/perfectly plastic half-space, *Int. J. Mech. Sci.* **14,** 1–14.

Collins, I. F. (1978a) Integral equation formulation of slipline field problems, *Applications of Numerical Methods to Forming Processes, ASME AMD*–**28,** 129–141.

Collins, I. F. (1978b) On the rolling of a rigid cylinder on a rigid/perfectly plastic half-space, *J. Mécanique Appliquée* **2,** 431–448.

Collins, I. F. (1978c) The application of singular perturbation techniques to the analysis of forming processes for strain-hardening materials, *Proceedings IUTAM Symposium on Metal Forming Plasticity, Tutzing, Germany,* ed. by H. Lippmann, 227–243, Springer-Verlag, Berlin.

Collins, I. F. (1980) Geometrically self-similar deformations of a plastic wedge under combined shear and compression loading by a rough flat die, *Int. J. Mech. Sci.* **22,** 735–742.

Collins, I. F. and Dewhurst, P. (1975) A slipline field analysis of asymmetrical hot rolling, *Int. J. Mech. Sci.* **17,** 643–651.
Collins, I. F. and Meguid, S. A. (1977) On the influence of hardening and anisotropy on the plane-strain compression of thin metal strip, *J. Appl. Mech.* **44,** 274–278.
Cowper, G. R. and Onat, E. T. (1962) The initiation of necking and buckling in plane plastic flow, *Proceedings of the 4th U.S. National Congress of Applied Mechanics,* **2,** 1023–1029, ASME, New York.
Crane, F. A. A. and Alexander, J. M. (1968) Slipline fields and deformation in hot rolling of strip, *J. Inst. Metals* **96,** 289–300.
Das, N. S. and Banerjee, J. (1978a) Superposition method of construction of plane strain plastic slipline fields, *J. Inst. Engrs (India)* **58,** 191–197.
Das, N. S. and Banerjee, J. (1978b) Solution to some plastic slipline fields by matrix method, *J. Inst. Engrs (India)* **58,** 272–278.
Das, N. S., Chitkara, N. R., and Collins, I. F. (1977) The computation of some slipline field solutions for asymmetric extrusion, *Int. J. Num. Meth. Engng* **11,** 1379–1389.
Das, N. S., Banerjee, J., and Collins, I. F. (1979) Plane strain compression of rigid–perfectly plastic strip between parallel dies with slipping friction, *J. Appl. Mech.* **46,** 317–321.
Devenpeck, M. L. and Weinstein, A. S. (1970) Experimental investigation of work-hardening effects in wedge flattening with relation to nonhardening theory, *J. Mech. Phys. Solids* **18,** 213–232.
Dewhurst, P. (1974) Plane-strain indentation on a smooth foundation: a range of solutions for rigid perfectly plastic strip, *Int. J. Mech. Sci.* **16,** 923–930.
Dewhurst, P. (1978a) On the non-uniqueness of the machining process, *Proc. R. Soc. Lond.* **A 360,** 587–610.
Dewhurst, P. (1978b) Is the machining process uniquely defined?, *Ann. CIRP* **27,** 1–4.
Dewhurst, P. (1978c) Eigenvector slip-line field solutions, *Applications of Numerical Methods to Forming Processes, ASME AMD*–**28,** 121–127.
Dewhurst, P. (1979) The effect of chip breaker constraints on the mechanics of the machining process, *Ann. CIRP* **28,** 1–5.
Dewhurst, P. and Collins, I. F. (1973) A matrix technique for constructing slip-line field solutions to a class of plane strain plasticity problems, *Int. J. Num. Meth. Engng* **7,** 357–378.
Dewhurst, P., Collins, I. F., and Johnson, W. (1973) A class of slipline field solutions for the hot rolling of strip, *J. Mech. Engng Sci.* **15,** 439–447.
Dewhurst, P., Collins, I. F., and Johnson, W. (1974) A theoretical and experimental investigation into asymmetrical hot rolling, *Int. J. Mech. Sci.* **16,** 389–397.
Dodd, B. and Kudo, H. (1980) A slip-line field for mid-plane cracking or splitting in sheet drawing, *Int. J. Mech. Sci.* **22,** 67–72.
Dodd, B. and Scivier, D. A. (1975) On the static inadmissibility of some slipline fields for sheet drawing, *Int. J. Mech. Sci.* **17,** 663–667.
Dodd, B. and Shiratori, M. (1978) The plane-strain general yielding of notched members due to combined axial force and bending, *Int. J. Mech. Sci.* **20,** I 451–464; II 465–474.
Druyanov, B. A. (1971) Sheet rolling under maximum friction conditions, *Plastic Flow of Metals* **1,** 80–89, Consultants Bureau, New York (transl. from Russian).
Druyanov, B. A. (1972) Initial flow of a band during impression by a smooth curvilinear punch, *Plastic Flow of Metals* **2,** 75–81, Consultants Bureau, New York (transl. from Russian).
Durban, D. and Budiansky, B. (1978) Plane strain radial flow of plastic materials, *J. Mech. Phys. Solids* **26,** 303–324.
Duvaut, G. and Lions, J. L. (1976) *Inequalities in Mechanics and Physics,* Springer-Verlag, Berlin.
Ernst, H. and Merchant, M. E. (1941) Chip formation, friction and high-quality machined surfaces, *Trans. Am. Soc. Metals* **29,** 299–378.
Ewing, D. J. F. (1967) A series-method for constructing plastic slipline fields, *J. Mech. Phys. Solids* **15,** 105–114.
Ewing, D. J. F. (1968) The plastic yielding of V-notched tension bars with circular roots, *J. Mech. Phys. Solids* **16,** 81–98.
Ewing, D. J. F. and Hill, R. (1967) The plastic constraint of V-notched tension bars, *J. Mech. Phys. Solids* **15,** 115–124.
Ewing, D. J. F. and Richards, C.E. (1974) The yield-point loads of singly-notched pin-loaded tensile strips, *J. Mech. Phys. Solids* **22,** 27–36.
Farmer, L. E. and Conning, S. W. (1979) Numerical smoothing of flow patterns, *Int. J. Mech. Sci.* **21,** 577–597.
Farmer, L. E. and Fowle, R. F. (1979) An experimental procedure for studying the flow in plane strain extrusion, *Int. J. Mech. Sci.* **21,** 599–608.
Farmer, L. E. and Oxley, P. L. B. (1971) A slipline field for plane-strain extrusion of a strain-hardening material, *J. Mech. Phys. Solids* **19,** 369–388.
Farmer, L. E. and Oxley, P. L. B. (1976) A computer-aided method for calculating the distributions of strain-rate and strain from an experimental flow field, *J. Strain Analysis* **11,** 26–31.
Ford, H. and Alexander, J. M. (1963) *Advanced Mechanics of Materials,* Longmans, London.
Ford, H. and Alexander, J. M. (1964) Simplified hot rolling calculations, *J. Inst. Metals* **92,** 397–404.

Geiringer, H. (1937) Fondéments mathématiques de la théorie des corps plastiques isotropes, *Mémorial des Sciences Mathématiques*, **86,** Gauthier-Villars, Paris.
Geiringer, H. (1973) *Ideal Plasticity, Encyclopedia of Physics, Mechanics of Solids*, ed. by S. Flügge, **3,** 403–533, Springer-Verlag, Berlin.
Green, A. P. (1951) A theoretical investigation of the compression of a ductile material between smooth flat dies, *Phil. Mag.* **42,** 900–918.
Green, A. P. (1954) On the use of hodographs in problems of plane plastic strain, *J. Mech. Phys. Solids* **2,** 73–80.
Green, W. A. (1962) Extrusion through smooth square dies of medium reduction, *J. Mech. Phys. Solids* **10,** 225–233.
Grimm, J. (1975) Die Spanbildung bei Extrusion mit kleinen Reduktionen, *Ingen.-Archiv.* **44,** 209–230.
Haddow, J. B. (1967) On the wedge indentation paradox, *Int. J. Mech. Sci.* **9,** 159–161.
Halling, J. and Mitchell, L. A. (1967) The effect of variation in yield stress on slipline field solutions, *Int. J. Prod. Res.* **5,** 249–262.
Hemp, W. S. (1973) *Optimum Structures*, Clarendon Press, Oxford.
Hencky, H. (1923) Über einige statisch bestimmte Fälle des Gleichgewichts in plastischen Körpern, *Z. angew. Math. Mech.* **3,** 241–251.
Hill, R. (1948) Flow round a corner for a hardening material (unpublished note).
Hill, R. (1950) *The Mathematical Theory of Plasticity*, Clarendon Press, Oxford.
Hill, R. (1951) On the state of stress in a plastic–rigid body at the yield point, *Phil. Mag.* **42,** 868–875.
Hill, R. (1953) On the mechanics of cutting metal strips with knife-edged tools, *J. Mech. Phys. Solids* **1,** 265–270.
Hill, R. (1954a) On the limits set by plastic yielding to the intensity of singularities of stress, *J. Mech. Phys. Solids* **2,** 278–285.
Hill, R. (1954b) The mechanics of machining: a new approach, *J. Mech. Phys. Solids* **3,** 47–53.
Hill, R. (1956, 1957) On the problem of uniqueness in the theory of the rigid–plastic solid, I–IV, *J. Mech. Phys. Solids* **4,** 47–255; **5,** 1–8; **5,** 153–161; **5,** 302–307.
Hill, R. (1961) Discontinuity relations in mechanics of solids, *Progress in Solid Mechanics*, ed. by I. N. Sneddon and R. Hill, **2,** 247–276, North-Holland, Amsterdam.
Hill, R. (1962) Constitutive laws and waves in rigid/plastic solids, *J. Mech. Phys. Solids* **10,** 89–98.
Hill, R. (1966) A remark on diagonal streaming in plane plastic flow, *J. Mech. Phys. Solids* **14,** 245–248.
Hill, R. (1967a) Ideal forming operations for perfectly plastic solids, *J. Mech. Phys. Solids* **15,** 223–227.
Hill, R. (1967b) On the vectorial superposition of Hencky–Prandtl nets, *J. Mech. Phys. Solids* **15,** 255–262.
Hill, R. (1980) Basic stress analysis of hyperbolic regimes in plastic media, *Math. Proc. Camb. Phil. Soc.* **88,** 359–369.
Hill, R. and Hutchinson, J. W. (1975) Bifurcation phenomena in the plane tension test, *J. Mech. Phys. Solids* **23,** 239–264.
Hill, R. and Kim, D. W. (1974) Some theoretical aspects of hydrostatic extrusion and allied processes, *J. Mech. Phys. Solids* **22,** 73–84.
Hopkins, H. G. (1972) Mathematical methods in plasticity theory, *Problems in Plasticity, Warsaw*, ed. by A. Sawczuk, 235–260, Noordhoff, Leyden.
Johnson, K. L. (1968) Deformation of a plastic wedge by a rigid flat die under the action of a tangential force, *J. Mech. Phys. Solids* **16,** 395–402.
Johnson, W. (1962) Some slipline fields for swaging or expanding, indenting, extruding and machining for tools with curved dies, *Int. J. Mech. Sci.* **4,** 323–347.
Johnson, W. (1969) Upper bounds to the load for the transverse bending of flat rigid/perfectly plastic plates, *Int. J. Mech. Sci.* **11,** 913–938.
Johnson, W. and Kudo, H. (1962) *The Mechanics of Metal Extrusion*, University Press, Manchester.
Johnson, W. and Mamalis, A. G. (1977) *Bounds and Slip-line Fields for Plane Strain Metal Forming Engineering Plasticity, Theory of Metal Forming Processes*, ed. by H. Lippmann, CISM No. 139, 74–283, Springer-Verlag, Berlin.
Johnson, W. and Mellor, P. B. (1962) *Plasticity for Mechanical Engineers*, van Nostrand, London.
Johnson, W. and Mellor, P. B. (1973) *Engineering Plasticity*, van Nostrand Reinhold, London.
Johnson, W. and Sowerby, R. (1974) Prediction of earing in cups drawn from anisotropic sheet using slip-line field theory, *J. Strain Analysis* **9,** 102–108.
Johnson, W., Sowerby, R., and Haddow, J. B. (1970) *Plane-strain Slip-line Fields: Theory and Bibliography*, Edward Arnold, London.
Kachanov, L. M. (1974) *Fundamentals of the Theory of Plasticity* (translation of 1969 Russian Edition), MIR Pub. Co., Moscow.
Klarin, K. (1979) Slipline field atlas for hot rolling with front and back tensions (in Swedish), Beaver Report No. 4, Royal Institute of Technology, Working of Metals, Stockholm.
Knott, J. F. (1973) *Fundamentals of Fracture Mechanics*, Butterworths, London.
Kronsjö, L. and Mellor, P. B. (1966) Plane-strain extrusion through concave dies, *Int. J. Mech. Sci.* **8,** 515–524.
Kudo, H. (1965) Some new slipline solutions for two-dimensional steady state machining, *Int. J. Mech. Sci.* **7,** 43–55.

Kudo, H. and Matsubara, S. (1978) Joint examination project of validity of various numerical methods for the analysis of metal forming processes, *Proceedings IUTAM Symposium on Metal Forming Plasticity, Tutzing, Germany*, ed. by H. Lippmann, 378–403, Springer-Verlag, Berlin.
Lee, E. H. (1952) Plastic flow of a V-notched bar pulled in tension, *J. Appl. Mech.* **19**, 331–336.
Lee, E. H. and Shaffer, B. W. (1951) Theory of plasticity applied to the problem of machining, *J. Appl. Mech.* **18,** 405–413.
Mandel, J. (1966) Résistance au roulement d'un cylindre indéformable sur un massif parfaitement plastique, *Le Frottement et Usure*, 25–33.
Marshall, E. A. (1968) Rolling contact with plastic deformation, *J. Mech. Phys. Solids*, **16**, 243–254.
McClintock, F. A. (1971) Plasticity aspects of fracture, *Fracture*, ed. by H. Liebowitz, **3**, 24–225, Academic Press, New York.
McClintock, F. A. and Clerico, M. (1980) The transverse shearing of singly-grooved specimens, *J. Mech. Phys. Solids* **28,** 1–16.
Michell, A. G. M. (1904) The limit of economy of material in frame structures, *Phil. Mag.* **8,** 589–600.
Morrison, H. L. and Richmond, O. (1972) Large deformation of notched perfectly plastic tensile bars, *J. Appl. Mech.* **39,** 971–977.
Neimark, J. E. (1968) The fully plastic, plane-strain tension of a notched bar, *J. Appl. Mech.* **35,** 111–116.
Olszak, W. and Rychlewski, J. (1962) Geometrical properties of stress fields in plastically non-homogeneous bodies under conditions of plane strain, *Second Order Effects in Elasticity, Plasticity and Fluid Dynamics*, 269–281, Pergamon Press, London, Jerusalem Academic Press, Haifa.
Onat, E. T. (1959) The effects of non-homogeneity caused by strain-hardening on the small deformations of a rigid–plastic solid, *Non-Homogeneity in Elasticity and Plasticity, IUTAM Symposium, Warsaw, 1958*, ed. by W. Olszak, 171–180, Pergamon Press, London.
Oxley, P. L. B. (1978) Allowing for a variable flow stress in the analysis of metal working processes, *Proceedings IUTAM Symposium on Metal Forming Plasticity, Tutzing, Germany*, ed. by H. Lippmann, 244–259, Springer-Verlag, Berlin.
Petryk, H. (1979) On slip-line field solutions for steady-state and self-similar problems with stress free boundaries, *Arch. Mech. Stos.* **31,** 861–874.
Petryk, H. (1980) Non-unique slip-line field solutions for the wedge indentation problem, *J. Mécanique Appliquée* **4,** 255–282.
Prager, W. (1953) A geometrical discussion of the slipline field in plane plastic flow, *Trans. R. Inst. Technol. Stockholm* **65**.
Prager, W. (1959) *An Introduction to Plasticity*, Addison-Wesley, Reading, Mass.
Prandtl, L. (1921) Über die Eindringungsfestigkeit plastischer Baustoffe und die Festigkeit von Schelden, *Z. angew. Math. Mech.* **1,** 15–20.
Rice, J. R. (1973) Plane strain slipline theory for anisotropic rigid/plastic materials, *J. Mech. Phys. Solids* **21,** 63–74.
Richmond, O. (1965) Theory of streamlined dies for drawing and extrusion, *Mechanics of the Solid State*, pp. 154–167, University of Toronto Press.
Richmond, O. (1969) Plane strain necking of V-notched and un-notched tensile bars, *J. Mech. Phys. Solids* **17,** 83–90.
Richmond, O. (1979) Private communication.
Richmond, O. and Devenpeck, M. L. (1962) A die profile for maximum efficiency in strip drawing, *Proceedings of the 4th U.S. National Congress of Applied Mechanics* **2,** 1053–7, ASME, New York.
Richmond, O. and Morrison, H. L. (1967) Streamlined wire drawing dies of minimum length, *J. Mech. Phys. Solids* **15,** 195–203.
Richmond, O. and Morrison, H. L. (1968) Application of a perturbation technique based on the method of characteristics to axisymmetric plasticity, *J. Appl. Mech.* **35,** 117–122.
Roth, R. N. (1978) The experimental determination of trajectories in geometrically similar plastic flows, *J. Strain Analysis* **13,** 29–34.
Roth, R. N. and Oxley, P. L. B. (1972) Slip-line field analysis for orthogonal machining based upon experimental flow fields, *J. Mech. Engng. Sci.* **14,** 85–97.
Rowe, G. W. (1965) *An Introduction to the Principles of Metalworking*, Edward Arnold, London.
Rowe, G. W. (1977) *Principles of Industrial Metalworking Processes*, Edward Arnold, London.
Rychlewski, J. (1966) Plane plastic strain for jump non-homogeneity, *Int. J. Nonlinear Mech.* **1,** 57–78.
Salençon, J. (1967) Theory of limit loads: On the solution of the problem of plane strain indentation of a block by a rigid rectangular punch, *C. r. Acad. Sci. Paris* A **264,** 613–616.
Sewell, M. J. (1969) On dual approximation principles and optimization in continuum mechanics, *Phil. Trans. R. Soc. Lond.* A **265,** 319–351.
Shindo, A. (1972) Theoretical analysis of indentation hardness, *Mem. Fac. Eng. Kobe University, Japan* **18,** 65–88.
Slater, R. A. C. (1977) *Engineering Plasticity*, Macmillan, London.
Sokolovskii, V. V. (1962) Complete plane problems of plastic flow, *J. Mech. Phys. Solids* **10,** 353–364.
Spencer, A. J. M. (1961) Perturbation methods in plasticity. I: Plane strain of non-homogeneous plastic solids, *J. Mech. Phys. Solids* **9,** 279–288.

Thomasen, P. F. (1978) Riemann-integral solutions for the plastic slip-line fields around elliptical holes, *J. Appl. Mech.* **45,** 678–679.
Thomsen, E. G., Yang, C. T., and Kobayashi, S. (1965) *Mechanics of Plastic Deformation in Metal Processing,* Macmillan, New York.
Tomlenov, A. D. (1971) Boundary conditions in problems of plane plastic flow, *Plastic Flow of Metals* **1,** 1–9, Consultants Bureau, New York (transl. from Russian).
Venter, R. D. and Hewitt, R. L. (1978) An introduction to the matrix inversion technique for solving plane strain metal forming problems, *Proceedings of the Fourth Symposium on Engineering Applications of Solid Mechanics, Sept. 1978,* Ontario Research Foundation, Mississanga, Ontario, Canada.
Venter, R. D., Johnson, W., and de Malherbe, M. C. (1971) The plane strain indentation of anisotropic aluminium using a frictionless flat rectangular punch, *J. Mech. Engng Sci.* **13,** 416–428.
Venter, R. D., Hewitt, R. L., and Johnson, W. (1978a) An engineering approach to the matrix operator technique for slip line field construction, *Proceedings of the Sixth North American Metalworking Research Conference, April 1978,* 111–118, University of Florida, Gainsville, USA.
Venter, R. D., Hewitt, R. L., and Johnson, W. (1978b) Application of the matrix inversion technique to extrusion and drawing problems, *Applications of Numerical Methods to Forming Processes, ASME AMD-***28,** 143–154.
Wang, A. J. (1954) Plastic flow in a deeply notched bar with semi-circular root, *Q. Appl. Math.* **11,** 427–438.
Wang, Z. and Ku, C. (1979) Proof of rule of normal stress changes along slipline by means of Mohr circle of plastic stress (in Chinese), Harbin Institute of Technology, Paper No. 45.
Wanheim, T. (1973) Friction at high normal pressures, *Wear* **25,** 225–244.

Aspects of the Theory of Dislocations

by J. D. ESHELBY

Department of the Theory of Materials, University of Sheffield

Summary

Various topics in the theory of dislocations in linear elastic continua are reviewed using, wherever possible, simple physical arguments to derive the mathematical results. The elastic field of a general Somigliana dislocation is derived and specialized to the physically important case where its discontinuity vector is constant. A general account of the effect of free surfaces is illustrated by some specific examples. In discussing dislocation energetics it is emphasized that once the Peach–Koehler expression describing the interaction with an externally applied stress field has been established, its extension to the case of internal stresses and image stresses follows without further mathematics. In the final Section it is demonstrated that certain solutions for uniformly and non-uniformly moving dislocations can be derived rather simply from the corresponding static solutions.

1. Introduction

The modern mathematical theory of dislocations really begins with G. I. Taylor's (1934) recognition of the relationship between a particular one of the various kinds of elastic dislocation treated by Volterra (1907) and the line defect postulated by himself, Orowan (1934), and Polanyi (1934) to explain the plasticity of crystals. Volterra's dislocations are a special class of what have come to be called Somigliana (1914, 1915) dislocations, though they seem to have been first introduced into the theory of elasticity by Gebbia (1891) under the rather uninformative name of deformations of the third kind. To create a Somigliana dislocation we start (see Fig. 2.1) by making a cut over a surface S, bounded by a curve C, or possibly closed, in an elastic solid and giving the two faces of the cut a small relative displacement $\mathbf{b}$ which varies from point to point of the surface. A component of displacement normal to the cut will give rise to gaps and interference, and, when these have been made good by filling and scraping, the faces of the cut are welded together again, leaving the whole specimen in a state of internal stress.

Macroscopic examples of Somigliana dislocations are provided by a worked-out coal-seam whose roof and floor have been forced into contact by overburden pressure or a thin igneous intrusion between bedding planes. Or again, if, having made a cut, we apply suitable tractions to the outer surface of the specimen the faces of the cut, or crack, will separate, and we have generated a Somigliana dislocation. So long as we maintain the external stress it makes no difference whether or not we fill the gap according to prescription, but if we do so the dislocation will survive when the applied stress is removed. If the cut is plane a suitably chosen applied shear stress will make the faces of the crack slide over one another so as to make the traction on the crack zero, thus adding to the applied field the elastic field of a Somigliana dislocation which happens to have its discontinuity vector everywhere in the plane of the cut. If the faces can slip freely the dislocation will disappear when the applied stress is removed; we may say that it is merely "induced" by the applied stress. However, if we have previously smeared the crack with glue we can remove the applied stress when the glue has set, and be left with a permanent fossilized version of the dislocation and its stress field. More realistically, if slip has occurred over a finite area of a fault plane, friction and normal pressure may wholly or partially inhibit slipping back when the applied shear stress no longer acts.

To apply the theory of elasticity to the study of deformed crystal lattices we imagine that the undeformed continuum is a transparent jelly in which is marked a regular array of points to represent the centres of the atoms, and that the points move with the medium when it deforms. Suppose that in this medium we make a cut bounded by a curve C and create a Somigliana dislocation whose discontinuity vector, instead of being variable, is constant and equal to a lattice vector. Then an observer who has not watched us construct the dislocation—and who cannot distinguish one atom from another, so that for him the position of an atom is only defined modulo a lattice vector—will not notice any discontinuity at the cut and it is only the large distortion of the lattice near the bounding curve C which will tell him that something has happened.

Suppose that in the undeformed continuum we have drawn a closed circuit going from point to point of the lattice of dots and embracing the curve C. Then when we have made the dislocation the circuit will be disrupted by a gap. The vector required to close the gap is, with a suitable sign convention, the (local) Burgers vector of the dislocation. The observer who has not watched us make the dislocation can also be led to the conclusion that there is a dislocation, and can find its Burgers vector, without reference to the undeformed lattice, for if he tries to form a closed circuit, not too small, about C by going, say, ten lattice spacings north, ten west, ten south, and then ten east, he will end up one lattice spacing from his starting point. We can also look at things more physically. Suppose that, as above, we create a Somigliana dislocation in a continuum dotted with points to represent the atoms, making sure that the discontinuity exceeds a few lattice spacings over a fair fraction of the surface. Now replace the dots by atoms, at the same time abolishing the supporting elastic continuum. We shall find that over most of the former cut surface the interatomic forces will pull the atoms into perfect registry, squeezing the misfit into the neighbourhood of a set of closed curves, with each of which is associated a Burgers vector which is a lattice vector. Subject to certain constraints these lines, or, rather, narrow regions of misfit, are able to move through the lattice, and the equilibrium form, if any, which they will take up as a result of their mutual interaction and the effect of applied and internal stresses can, in simple cases, be worked out by continuum theory. The assertion that the regions of misfit, or dislocation cores, are narrow, and that they are mobile, is qualitatively confirmed by the model of Peierls and Nabarro, which uses a mixture of continuum and atomic theory.

The prime reference for the theory of dislocations is still the book by Nabarro (1967) now supplemented by a five-volume collective work under his editorship (Nabarro, 1980). The books by Hirth and Lothe (1968) and Lardner (1974) contain more mathematical details; the latter also deals with the connection between dislocations and the theory of cracks. Seeger's (1955) *Handbuch* article is still well worth reading. A recent review (Bacon *et al.*, 1979) is devoted to the application of anisotropic linear elasticity to dislocations and other lattice defects. An article by Kosevich (1979) gives references to Soviet work.

So long as the dislocation was purely a theoretical construct a fairly elementary treatment of its properties was sufficient, and was mostly supplied by physicists and metallurgists. Later they were joined by applied mathematicians, who found in dislocation theory, and also in the allied subject of fracture mechanics, a welcome addition to the rather limited stock of worthwhile topics to which the theory of elasticity could be applied. Later, the actual observation of dislocations in the electron microscope called for, and the availability of computers made possible, more elaborate calculations. The present article is pitched at the earlier level, and is confined to the linear theory. We shall concentrate on some of the concepts which are more or less foreign to the traditional theory of elasticity and have to be added to it to build up the theory of dislocations. We start (Section 2) with the calculation of the elastic field induced by establishing a displacement discontinuity, in particular a constant one, across a surface in a solid, as explained above, beginning with a medium of infinite extent. In a finite body the presence of free surfaces modifies the field of a dislocation ("image" effects, Section 3) and, in addition by deforming the surface of the body the dislocation, signals its presence to any external loading mechanism. If the dislocation deforms, the energy of the system varies, in a way which depends on the presence of other sources of internal stress, the form of the surface and the coupling to external loads. The energy changes can be picturesquely described (Section 4) in terms of a configurational or Peach–Koehler force acting on the dislocation. In this context the dislocation is supposed to deform by passing through a succession of neighbouring static configurations. However, if its speed is comparable with the velocity of elastic waves its elastic field has to satisfy the dynamical elastic equations, and in Section 5 we present a few dynamical solutions which can be deduced from static solutions without too much trouble.

2. The Elastic Field of a Dislocation

2.1. *Preliminaries*

Figure 2.1 illustrates the construction of a Somigliana dislocation in the way outlined in the Introduction. In (a) we have made a cut in the unstressed material, bounded by the curve C, and have given the labels S, S' to the two faces of the cut, at present coincident. In (b) we have given S' a small displacement $\mathbf{b}(\mathbf{r}')$ which, generally speaking, varies as $\mathbf{r}'$ moves about on S. Where there is a gap we fill it up, and where there would be interpenetration we scrape material away to avoid it. Finally, we weld the material together again. We are left with the situation of Fig. 2.1(c), a surface S with a discontinuity vector $\mathbf{b}(\mathbf{r}')$ associated with each point $\mathbf{r}'$ on it. We evidently need some sort of sign convention. Draw an arrow to represent the normal $\mathbf{n}$ at each point, and label with + and − the sides of S on, respectively, the head and tail sides of the arrow. Then we define $\mathbf{b}$ by

$$\mathbf{b}(\mathbf{r}') = \mathbf{u}^+(\mathbf{r}') - \mathbf{u}^-(\mathbf{r}'), \tag{2.1}$$

where $\mathbf{u}^+(\mathbf{r}')$ and $\mathbf{u}^-(\mathbf{r}')$ are the displacements, on the sides indicated, of two points originally coincident at $\mathbf{r}'$. The opposite convention is also used.

For brevity we shall often talk as if there was a gap everywhere in Fig. 2.1(b), leaving the reader to make the adjustments necessary to cover the case of interpenetration.

In the next Section we shall discuss the determination of the elastic field produced by a Somigliana dislocation. Here we discuss some results which depend on geometry assisted by only a minimum of the theory of elasticity—to connect (2.2), (2.3), and (2.4) below we need Hooke's law, and in connection with Fig. 2.2 we need the fact that an elastic body resumes its unstressed form when the load is removed and that a rigid body motion induces no stress in it.

Before dislocations had become associated with the physics of crystals, mathematicians had set themselves the task, one could almost say had invented the game, of determining what conditions have to be imposed on the discontinuity **b** to make the site of the cut become in some sense undetectable. In modern dislocation theory the limitation imposed on **b** comes from crystallography rather than from the investigations just mentioned, and so we shall only present a simplified intuitive account of them. (On the topic of discontinuities in general, see Hill, 1961.)

In a featureless linear isotropic continuum it is natural to start by requiring the six strain components e_{ij} to be continuous across the cut. In Fig. 2.1 let us take at each point of the cut

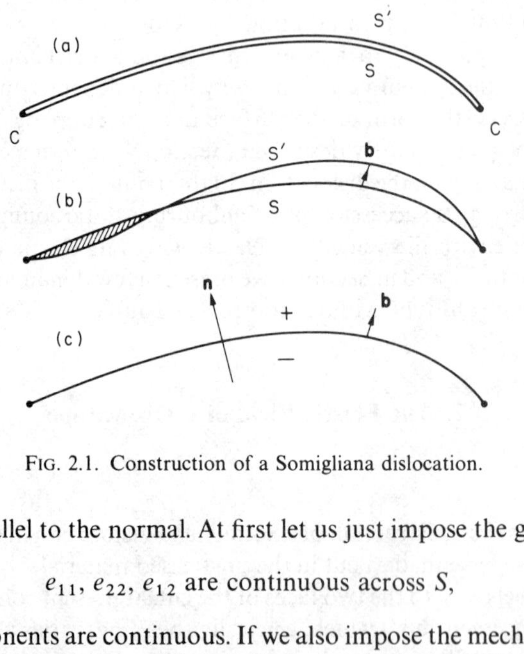

FIG. 2.1. Construction of a Somigliana dislocation.

local axes with x_3 parallel to the normal. At first let us just impose the geometrical condition

$$e_{11}, e_{22}, e_{12} \text{ are continuous across } S, \tag{2.2}$$

i.e. the in-plane components are continuous. If we also impose the mechanical condition that the traction is continuous (no layer of body force at the interface), so that

$$\sigma_{13}, \sigma_{23}, \sigma_{33} \text{ are continuous across } S, \tag{2.3}$$

(2.2) and (2.3) combined with Hooke's law then show that in addition the remaining strain components

$$e_{13}, e_{23}, e_{33} \text{ are continuous across } S, \tag{2.4}$$

as are also $\sigma_{11}, \sigma_{22}, \sigma_{12}$. If at this point we were to allow the material to be anisotropic, (2.2) and (2.3) would not imply (2.4); as there is in general a relative rotation of the material on the two sides of S the components of the modulus tensor C_{ijkl}, when referred to common axes, suffer a jump across S which must be compensated for by discontinuities of non-negligible order in e_{13}, e_{23}, e_{33}.

The requirement (2.2) imposes a simple geometrical constraint on the discontinuity **b** (Weingarten, 1901). We can imagine constructing the Somigliana dislocation in two stages. Stage 1 is essentially geometrical. Starting from the two surfaces of the crack S and S' in contact in the unstrained material, we keep S fixed with a set of infinitesimal clamps and enforce the displacement $\mathbf{b}(\mathbf{r}')$ on the points $\mathbf{r}'$ of S'. With the local axes used above we have, with an obvious notation, $e_{\alpha\beta}(S) = 0$ (where here and below α, β only take the values 1, 2) and, say, $e_{\alpha\beta}(S') = e^0_{\alpha\beta}$. After filling and scraping we weld the body together again. In stage 2 we remove the constraints and the body relaxes into a state of equilibrium. During this relaxation S and S', now joined together, each acquire the same additional strain, $e'_{\alpha\beta}$ say, so that now we have $e_{\alpha\beta}(S) = e'_{\alpha\beta}$, $e_{\alpha\beta}(S') = e^0_{\alpha\beta} + e'_{\alpha\beta}$. But according to (2.2) we must now have $e_{\alpha\beta}(S') = e_{\alpha\beta}(S)$, i.e. $e^0_{\alpha\beta}(S') = 0$. Hence the deformation which **b** induces in S' during stage 1 must produce no strain in its surface. Hence, in the language of the theory of surfaces, S' is applicable to S. This is Weingarten's result.

According to Somigliana, Weingarten's result is incomplete, and the e_{ij} would also be continuous if **b** were everywhere normal to S. His claim is refuted by the simple example of a sphere inserted into a hole slightly too small for it. We may suppose that the sphere was originally a perfect fit but was then fitted with a thin spherical jacket, so that we do in fact have a Somigliana dislocation. Also, as a sphere is not applicable to one of different radius we have the required violation of Weingarten's condition. Since the sphere is compressed and the hole is expanded, meridian circles of the two respectively decrease and increase in length so that the hoop strain, $e_{\theta\theta} = \partial u_r / r\partial\theta + u_r/r$ in the usual notation, is discontinuous. What Somigliana's analysis actually shows is that $\partial u_r / r\partial\theta$ is continuous, and similarly for more general curved surfaces. When S is flat what he says is true, but this case is already covered by Weingarten's result since, to the accuracy of the linear theory, for a plane a small variable normal displacement induces no stretching or contraction in the surface.

The simplest displacement **b** which meets the condition of applicability is the rigid-body displacement

$$\mathbf{b} = \mathbf{a} + \boldsymbol{\omega} \times \mathbf{r} \tag{2.5}$$

with constant small **a** and $\boldsymbol{\omega}$, the case studied at length by Volterra (1907); for a modern summary see Volterra and Volterra (1960). Figure 2.2 shows two surfaces S_1, S_2 bounded by the same curve C. The material in a toroidal hole containing C is removed, so rendering the body multiply-connected.

Make a cut over S_1 and remove material so as to leave a gap specified by (2.5). (The reader

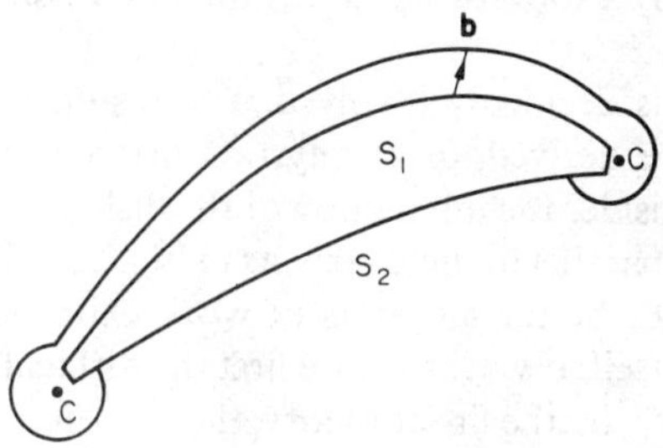

FIG. 2.2. A Volterra dislocation.

will easily make the adjustments required to cover the case where the "gap" is wholly or partly an interpenetration.) If we make a cut over S_2 and close the gap at S_1 the material between S_1 and S_2 will undergo a rigid-body motion specified by the right-hand side of (2.5) and a gap will open up over S_2, again given by (2.5). Having noted this we start again, with S_2 uncut and a gap (2.5) at S_1. If we now close this gap and weld the faces of S_1 together, the material is put into a state of stress, both between S_1 and S_2 and also outside them. If we next make a cut over S_2 this stress disappears and a gap (2.5) opens up at S_2; all we have done, in a roundabout way, is to give the piece of material between S_1 and S_2 a rigid-body displacement. If we now close the gap at S_2 the state of stress just relaxed is, naturally, restored. This shows, if it is not already obvious, that creating and closing the gap (2.5) over S_2 produces the same state of stress as creating and closing the gap (2.5) over S_1 does; the strains in the two cases are also, of course, the same, though the displacements differ by the right-hand side of (2.5) between S_1 and S_2 and are identical outside them.

It is also clear that closing a gap at S_1 will induce no discontinuity in the e_{ij}, or in their derivatives of all orders in the neighbourhood of S_2 and conversely, so that the site of the cut is undetectable by investigating the strain or its derivatives of various orders, in contrast to the situation governed by (2.2) and (2.3), where the e_{ij} are continuous, but their first derivatives are usually not.

As explained in the Introduction, when we use the continuum to simulate a crystal lattice we are limited to the case where **b** is constant and equal to a lattice vector, i.e. to (2.5) with $\boldsymbol{\omega} = \mathbf{0}$ and a suitably chosen **a**.

2.2. *The elastic field of a Somigliana dislocation*

The first person to have published an expression for the displacement field of a Somigliana dislocation seems to have been Gebbia (1902). In vector notation, and with our sign convention, it takes the form

$$\mathbf{u}(\mathbf{r}) = -\frac{1}{4\pi}\int_S \mathbf{b}\frac{\mathbf{R}\cdot d\mathbf{S}}{R^3} + \frac{1}{4\pi}\int_S \mathbf{b}\times\left(d\mathbf{S}\times\nabla'\frac{1}{R}\right)$$

$$+\frac{1}{4\pi}\frac{1}{2(1-\nu)}\nabla\int_S\left[\mathbf{b}\left(\nabla'\cdot\frac{\mathbf{R}}{R}\right)-(\mathbf{b}\cdot\nabla')\frac{\mathbf{R}}{R}\right]\cdot d\mathbf{S}, \tag{2.6}$$

where $\mathbf{R} = \mathbf{r}-\mathbf{r}'$, $R = |\mathbf{r}-\mathbf{r}'|$, ∇' denotes the gradient with respect to $\mathbf{r}'$, and $\mathbf{b}$, $d\mathbf{S}$ stand for $\mathbf{b}(\mathbf{r}')$, $dS(\mathbf{r}')$, and the material is supposed to be isotropic with Poisson's ratio ν.

No notice seems to have been taken of (2.6) in the dislocation literature, probably because Gebbia regarded it as simply a tool for use in his method of solving elastic boundary value problems.

Gebbia's proof of (2.6) is decidedly involved. His result, and also its analogue for an anisotropic medium, can be derived, or at any rate arrived at, by the following intuitive argument (Read, 1953). Consider the interaction of the dislocation with a point force applied at **r**, of unit magnitude, and parallel to the x_i-axis, as in Fig. 2.3. (The surface S' is to be ignored for the moment.). Let E_D, E_F be the amounts of work expended in introducing either the dislocation or the force by itself. If we introduce first the dislocation, then the force, the work done is simply $E_D + E_F$, since, on the linear theory, the response of the material to the force is unaffected by the prestress due to the dislocation. If, alternatively, we introduce the force first and then the dislocation, the work required will be $E_D + E_F$ plus two new terms. One is the

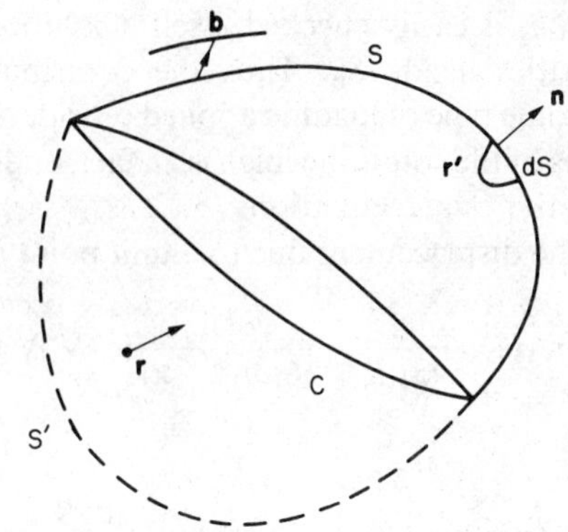

FIG. 2.3. Calculation of the elastic field of a Somigliana dislocation.

work done by the force as its point of application is shifted through the displacement induced by the dislocation. The other is the work done on the two faces of the cut by the tractions due to the point force as the faces suffer the relative displacement **b**. But these extra terms must cancel; otherwise one method of loading would require less work than the other, and by loading in the easier way and then unloading in the reverse of the harder way we should be able to create energy. This gives at once

$$u_i(\mathbf{r}) = \int_S b_j(\mathbf{r}')\sigma_{jk}^{(i)}(\mathbf{r}')n_k\, dS, \qquad (2.7)$$

where $\sigma_{jk}^{(i)}$ is the stress produced by the point force. The left-hand side is the work done by the force and the right-hand side is minus the work done by the tractions at the cut; the minus sign comes from the convention (2.5) combined with the fact that the outward normal to the material is $-\mathbf{n}$ on the positive side of the discontinuity surface.

This heuristic argument can be made respectable by using the Rayleigh–Betti reciprocal theorem which says that the integral

$$\int_S (u_i\sigma'_{ij} - u'_i\sigma_{ij})dS_j \qquad (2.8)$$

taken over the closed surface S vanishes for any two elastic fields u_i, σ_{ij} and u'_i, σ'_{ij} provided S does not embrace any body forces or singularities. The proof of (2.8) simply consists in using Gauss' theorem to convert to a volume integral over the interior of S, using the equilibrium conditions $\sigma_{ij,j} = 0, \sigma'_{ij,j} = 0$, and noting that we have $u_{i,j}\sigma'_{ij} = u_{i,j}C_{ijkl}u'_{k,l} = u'_{i,j}\sigma_{ij}$ in virtue of the symmetry relation $C_{ijkl} = C_{klij}$ satisfied by the elastic moduli. Let u_i be the dislocation displacement and u'_i the displacement due to a point force, as above. For S we may take a large surface S_0 enclosing the dislocation and force together with a small sphere excluding the force and a closely fitting jacket embracing the discontinuity surface of the dislocation. These latter surfaces reproduce the terms in (2.7), but on the right we have to add minus the expression (2.7) taken over S_0. In some presentations with pretensions to rigour, this term is simply ignored. In others it is stated to go to zero as S_0 becomes indefinitely large, but this requires a knowledge of the behaviour of u_i which we do not have until the problem is solved. There is, however, a simple way out of the difficulty. The extra term, or indeed just the contribution of a single element dS to it, is, as is easily checked, itself a solution of the elastic equilibrium equations, free of singularities inside S_0. Thus the contribution from S_0 represents a displacement produced by some type of loading applied outside S_0, and its omission does not affect the essential properties of the solution which is, in fact, undefined to within an arbitrary singularity-free solution of the elastic equations.

In an isotropic medium the displacement due to a unit point force $\mathbf{e}_i$ in the u_i-direction is

$$\mathbf{u}^{(i)}(\mathbf{r}) = \frac{\mathbf{e}_i}{4\pi\mu R} - \frac{1}{16\pi\mu(1-\nu)}(\mathbf{e}_i\cdot\nabla')\nabla' R \tag{2.9}$$

or in suffix notation

$$u_j^{(i)} = G_{ij}(\mathbf{R}) = \frac{\delta_{ij}}{4\pi\mu R} - \frac{1}{16\pi\mu(1-\nu)}\frac{\partial^2 R}{\partial x_i' \partial x_j'}, \tag{2.10}$$

where G_{ij} is the so-called elastic Green's tensor. If we calculate the stress $\sigma_{jk}^{(i)}$ from (2.10) and insert it in (2.7), the result can be manipulated to agree with (2.6).

2.3. *The field of a crystal dislocation*

For a crystal dislocation with a constant discontinuity $\mathbf{b}$ we may add to the expression in the square brackets in the last integral of (2.6) the vanishing quantity $(\mathbf{R}/R)\cdot\nabla'\mathbf{b} - (\mathbf{R}/R)(\nabla'\cdot\mathbf{b})$ which turns it into what can be recognised as the expansion of $\nabla' \times (\mathbf{b} \times \mathbf{R}/R)$, and then with the help of Stokes' theorem this term may be converted into a line integral round the curve C bounding S. The second term can be converted to a line integral directly by taking the $\mathbf{b}\times$ outside the integral and using

$$\int_S d\mathbf{S} \times \nabla'(\ldots) = \int_C d\mathbf{x}'(\ldots),$$

an alternative form of Stokes' theorem. In this way we get

$$\mathbf{u}(\mathbf{r}) = -\frac{\mathbf{b}}{4\pi}\Omega - \frac{1}{4\pi}\int_C \frac{\mathbf{b}\times d\mathbf{x}'}{R} + \frac{1}{4\pi}\frac{1}{2(1-\nu)}\nabla\int_C \frac{\mathbf{b}\times d\mathbf{x}'\cdot\mathbf{R}}{R}, \tag{2.11}$$

where

$$\Omega(\mathbf{r}) = \int_S \frac{\mathbf{R}\cdot d\mathbf{S}}{R^3} \tag{2.12}$$

is the solid angle subtended by S at $\mathbf{r}$.

This important result, due to Burgers (1939), exhibits the displacement as the sum of single-valued terms and one, the first, which produces the required cyclic behaviour. Peach and Koehler (1950) were, rather surprisingly, able to turn even the solid angle (2.12) into a line integral. It contains

$$\frac{\mathbf{R}}{R^3} = \nabla'\frac{1}{R}.$$

If we could replace the gradient by the curl of something we could use Stokes' theorem to convert (2.12) to a line integral. This is, in fact, possible (Madelung, 1950). Equation (2.12) gives the field of, say, a magnetic monopole as the gradient of a scalar potential. But the magnetic field is the same as that of a very fine semi-infinite solenoid if we choose to overlook the concentrated return field inside the solenoid. Hence we ought to be able to express the field as the curl of the vector potential of the solenoid. This leads to

$$\frac{\mathbf{R}}{R^3} = -\nabla' \times \mathbf{A} \tag{2.13}$$

with

$$\mathbf{A}(\mathbf{R}) = \frac{\mathbf{m} \times \mathbf{R}}{R(R + \mathbf{m} \cdot \mathbf{R})}, \tag{2.14}$$

where $\mathbf{m}$ is any unit vector. Equation (2.13) is easily checked for, say, $\mathbf{m} = (0, 0, 1)$. Stokes' theorem then gives

$$\Omega(\mathbf{r}) = -\int_C \mathbf{A}(\mathbf{r} - \mathbf{r}') \cdot d\mathbf{r}'. \tag{2.15}$$

This integral has the necessary discontinuity of 4π across the cylindrical surface $S_{\mathbf{m}}$ generated by drawing a semi-infinite straight line from each point of C in the direction $-\mathbf{m}$; in other words $S_{\mathbf{m}}$ is the shadow surface thrown (in a dusty atmosphere) by the loop C when it is illuminated by a lamp at a remote point in the direction of $\mathbf{m}$. It is not hard to show that the contribution of a single element $d\mathbf{r}'$ to (2.15) is the solid angle subtended at $\mathbf{r}$ by the hairpin made up of $d\mathbf{r}'$ and a pair of lines drawn from its extremities to infinity in the direction of $-\mathbf{m}$. The representation of Ω as a line integral is, therefore, perhaps a little fraudulent; it is simply (2.12) with the arbitrary cap closing the loop drawn out into the cylindrical shadow surface $S_{\mathbf{m}}$. Peach and Koehler's expression actually contains the average of (2.14) over the six vectors $\mathbf{m} = (\pm 1, 0, 0), (0, \pm 1, 0), (0, 0, \pm 1)$. Some readers may find illuminating Amar's (1980) derivation of (2.15) from (2.12) with the help of the theory of exterior forms.

For anisotropy $\sigma_{jk}^{(i)}$ is the stress derived from $u_j^{(i)} = G_{ij}$, where now $G_{ij}(\mathbf{r} - \mathbf{r}')$ is the elastic Green's function for an anisotropic medium, giving, as before, the x_j-component of the displacement at $\mathbf{r}$ produced by a unit point force at $\mathbf{r}'$ parallel to the x_i-axis. The property

$$G_{ij}(\mathbf{r} - \mathbf{r}') = G_{ji}(\mathbf{r}' - \mathbf{r}) \tag{2.16}$$

expresses the fact that if we apply forces F_i at $\mathbf{r}$ and F_i' at $\mathbf{r}'$ the total work done must be the same whether we apply F_i first, then F_i', or, conversely, so that the cross-terms in the energy, $F_i G_{ij}(\mathbf{r} - \mathbf{r}') F_j'$ and $F_i' G_{ij}(\mathbf{r}' - \mathbf{r}) F_j$, must be equal. The relation $G_{ij}(\mathbf{r}) = G_{ji}(-\mathbf{r})$ which with (2.16) also implies

$$G_{ij}(\mathbf{r}) = G_{ji}(\mathbf{r}) \tag{2.17}$$

or $u_j^{(i)} = u_i^{(j)}$ is equivalent to the statement that equal and opposite point forces produce zero total displacement at a point halfway between them. To establish it we have to use the fact that the anisotropic elastic equations are invariant under the inversion $\mathbf{r} \to -\mathbf{r}$.

It is easy enough to find a formal solution of the equation

$$C_{rsjk} G_{ij,ks}(\mathbf{r}) + \delta_{ir} \delta(\mathbf{r}) = 0$$

governing Green's function by taking its Fourier transform and writing the solution in the form of a triple integral over the transform variables (k_1, k_2, k_3); see, for example, Eshelby

(1951). To evaluate the integral explicitly one would have to be able to write down explicitly the solution of a certain sextic equation whose coefficients involve the k_i. Gebbia (1904) discussed some cases in which this is possible. (This remarkable paper contains, among other things, an anticipation of Feynman's (1949) method for dealing with integrals involving products.) Head (1979) has also discussed this point. For a hexagonal (transversely isotropic) material the factorization can be carried out and so G_{ij} can be given explicitly (Kröner, 1953). For the other case of most interest in metal physics, that of cubic symmetry, factorization is, unfortunately, not usually possible. An exception is the case where, with the usual abbreviated notation, $c_{12}+c_{44}=0$. The equations for u_1, u_2, u_3 are then uncoupled, and, after a suitable stretching of one of the coordinate axes, each equation becomes Laplace's equation. This is the continuum limit of the Rosenstock–Newell crystal, in which the three displacement equations are similarly decoupled by a suitable choice of the interatomic force law. Though the condition $c_{12}=-c_{44}<0$ is not actually ruled out by stability considerations, no known crystal satisfies it even approximately, so that this attractively simple case is not really of much value. However, Dederichs and Leibfried (1969) have examined the possibility of a perturbation treatment for the cubic case, starting from the isotropic or Rosenstock–Newell case.

When the stress $\sigma_{jk}^{(i)}$ in (2.7) with constant b_j has to be derived from the general anisotropic Green's function, we cannot isolate a term like (2.12) from the dislocation displacement, but we can show that (2.7) has the right discontinuity across S by a pretty argument due to Burgers (1939). In Fig. 2.3 let S' be a second surface bounded by C, so that $S+S'$ forms a closed surface. With the factor b_j suppressed the integral (2.7) taken not over S but over $S+S'$ is the x_j-component of the total traction due to the point force, namely zero if $\mathbf{r}$ is outside $S+S'$ and $-\delta_{ij}$ if it is inside, minus because the traction balances the force. Hence the integral over $S+S'$ increases by δ_{ij} as $\mathbf{r}$ breaches S from inside $S+S'$. But obviously there is no sudden change in the contribution from S', so that the integral over S alone suffers the jump. Hence, restoring the factor b_j, u_i jumps by b_i as $\mathbf{r}$ passes through S in the direction of the positive normal, as required. Burgers extended this argument to give a proof of (2.7). It is first necessary to verify that (2.7) actually satisfies the equilibrium equations written in terms of displacements. In fact $\sigma_{jk}^{(i)}$ by itself, when regarded as a displacement $u_i^{(jk)}$, does so, as is easily checked if we recall the symmetry relation (2.17). Burgers also felt obliged to check that (2.7) contained no unwanted singularities along C. In the energy argument we first used to derive (2.7) there is no need for this. If there had been, say, a distribution of body force along C and a line of dilatation coinciding with it, the cross-terms between these singularities and the field of the point force would have appeared in (2.7) as a line integral along C involving the displacement and displacement gradient of the field of the point force. By not allowing for such terms we unconsciously ensured that the corresponding singularities were absent.

Suppose that in the above argument we keep the point of observation $\mathbf{r}$ fixed and deform the surface S. Then except for a jump of $\pm b_i$ when S sweeps over $\mathbf{r}$, the integral over $S+S'$ is unaffected by the deformation, and so, since the integral over S', as before, does not vary, it follows that, provided it spans C, the exact form of S is, the jump apart, unobservable and irrelevant. This is Volterra's fundamental result, though here derived by a mechanical rather than a geometrical argument.

Though the displacement cannot, the displacement gradient, and hence also the strain and stress, can be written as a line integral in the anisotropic case. We shall find it convenient to write $x_{k'}$ for x'_k, $dS_{k'}$ for $dS_k(\mathbf{r}')$, $(\ldots)_{,k'}=\partial(\ldots)/\partial x_{k'}$, $(\ldots)_{,k}=\partial(\ldots)/\partial x_k$.

Differentiate (2.7) with respect to x_l and use the fact that $\sigma_{jk}^{(i)}$ depends only on $x_m-x_{m'}$ to get

$$u_{i,l} = b_j \int_S \overset{(i)}{\sigma}_{jk,l}\, dS_{k'} = -b_j \int_S \overset{(i)}{\sigma}_{jk,l'}\, dS_{k'}. \tag{2.18}$$

Apply Stokes' theorem in the form

$$\int_S \left(dS_{k'} \frac{\partial}{\partial x_{l'}} - dS_{l'} \frac{\partial}{\partial x_{k'}} \right)(\ldots) = \int_C \varepsilon_{klq}\, dx_{q'} (\ldots) \tag{2.19}$$

to get

$$u_{i,l} = -b_j \varepsilon_{klq} \int_C \overset{(i)}{\sigma}_{jk}\, dx_{q'} \tag{2.20}$$

plus the surface integral (2.18) with l', k' interchanged. This, however, is zero because we have

$$\overset{(i)}{\sigma}_{jk,k'} = -\overset{(i)}{\sigma}_{jk,k} = 0 \tag{2.21}$$

in equilibrium, and so the integral (2.20), taken round the loop, is the complete expression. Alternatively, in the first form of (2.18) write $dS_{k'}$ as $-(x_r - x_{r'})_{,k'} dS_{r'}$. By (2.21) this can be written as

$$u_{i,l} = -b_j \int_S [\overset{(i)}{\sigma}_{jk,l} (x_r - x_{r'})]_{,k'}\, dS_{r'}. \tag{2.22}$$

Stokes' theorem (2.19) converts this into the line integral

$$u_{i,l} = -b_j \varepsilon_{qkr} \int_C \overset{(i)}{\sigma}_{jk,l} (x_r - x_{r'})\, dx_{q'} \tag{2.23}$$

together with the integral (2.22) with r', k' interchanged. The latter vanishes since it contains

$$\left[3 + (x_r - x_{r'}) \frac{\partial}{\partial x_r} \right] \overset{(i)}{\sigma}_{jk,l}$$

and $\overset{(i)}{\sigma}_{jk,l}$ is homogeneous of degree -3 in the $x_r - x_{r'}$ (Indenbom and Orlov, 1967). As we shall see, the special virtue of (2.23) is that $\varepsilon_{qkr}(x_r - x_{r'})\, dx_{q'}$ is the x_k-component of $\mathbf{R} \times d\mathbf{x}'$.

2.4. *Straight dislocations*

The elastic fields of infinite straight dislocations may be found from the three-dimensional formulae, but it seems worthwhile to treat them directly.

We begin with isotropy, and go over to the elementary notation in which (x_1, x_2, x_3) and (u_1, u_2, u_3) become (x, y, z) and (u, v, w). The displacement of a screw dislocation with Burgers vector $(0, 0, b)$ has the simple form

$$w = \frac{b}{2\pi}\theta = \frac{b}{2\pi} \tan^{-1} \frac{y}{x}, \quad u = v = 0, \tag{2.24}$$

and since the dilatation div $\mathbf{u}$ is zero the non-vanishing stress components are

$$\sigma_{zx} = \mu \frac{\partial w}{\partial x} = -\frac{\mu b}{2\pi} \frac{\sin\theta}{r}, \quad \sigma_{zy} = \mu \frac{\partial w}{\partial y} = \frac{\mu b}{2\pi} \frac{\cos\theta}{r}. \tag{2.25}$$

Equation (2.24) hardly needs proving. With div $\mathbf{u} = 0$ the equilibrium equation

$$(1-2\nu)\nabla^2\mathbf{u} + \text{grad div } \mathbf{u} = 0 \tag{2.26}$$

reduces to $\nabla^2 w = 0$, so that w must be a harmonic function which changes by b when the dislocation is encircled once, and this leads directly to (2.24).

The expressions for the displacement $(u, v, 0)$ around an edge dislocation with Burgers vector $(b, 0, 0)$ are considerably more complicated:

$$\begin{aligned} u &= \frac{b}{2\pi}\theta + \frac{1}{2(1-\nu)}\frac{b}{2\pi}\frac{xy}{r^2}, \\ v &= \frac{1-2\nu}{2(1-\nu)}\frac{b}{2\pi}\ln\frac{1}{r} + \frac{1}{2(1-\nu)}\frac{b}{2\pi}\frac{y^2}{r^2}. \end{aligned} \tag{2.27}$$

The stress components σ_{zx}, σ_{zy} are zero and the rest can be derived from the simple Airy stress function

$$\chi = -\frac{\mu b}{2\pi(1-\nu)} y \ln r \tag{2.28}$$

according to the normal rule

$$\begin{aligned} \sigma_{xx} &= \partial^2\chi/\partial y^2, \quad \sigma_{yy} = \partial^2\chi/\partial x^2, \\ \sigma_{xy} &= -\partial^2\chi/\partial x\,\partial y, \quad \sigma_{zz} = \nu\nabla^2\chi. \end{aligned} \tag{2.29}$$

Newcomers to dislocation theory are sometimes disconcerted to find that in some references the y^2/r^2 in the second term of v is replaced by $-x^2/r^2$. However, the difference is a multiple of $(y^2+x^2)/r^2 = 1$, and so amounts only to an unimportant translation.

The displacements (2.27) and the associated stresses play an important part in even elementary treatments of dislocation theory, but are a good deal more complicated-looking than other formulae encountered at that level. The writer has therefore tried to provide simple proofs involving a minimum of the theory of elasticity, and in which each step of the argument appears to be reasonably inevitable. One way is to begin by first solving the simpler problem of finding the cylindrically symmetrical stress field which is set up when a narrow wedge-shaped gap is cut out and then closed up. The field of an edge dislocation can then be found from it by an essentially geometrical argument (Eshelby, 1966). Or one can start from a relationship which connects certain states of plane strain with their anti-plane analogues, and derive (2.27) from (2.24) (Bilby and Eshelby, 1969). Here we give a third method, originally devised to give a simple derivation of a standard result in fracture mechanics (Eshelby, 1971).

We start from the fact that as the dilatation

$$e = \frac{\partial u}{\partial x} + \frac{\partial v}{\partial y} \tag{2.30}$$

is harmonic we have $\nabla^2(xe) = 2\partial e/\partial x$, $\nabla^2(ye) = 2\partial e/\partial y$ so that the equilibrium equation (2.26), which becomes

$$\nabla^2 u + \frac{1}{1-2\nu}\frac{\partial e}{\partial x} = 0, \quad \nabla^2 v + \frac{1}{1-2\nu}\frac{\partial e}{\partial y} = 0,$$

may be re-written as

$$\nabla^2 u' = 0, \quad \nabla^2 v' = 0$$

with

$$u' = u + \alpha xe, \quad v' = v + \alpha ye, \quad \alpha = 1/2(1-2\nu).$$

For an edge dislocation with Burgers vector $(b,\ 0)$ the strains, being dimensionless and proportional to b, must be proportional to r^{-1}. The dilatation must therefore be

$$e = \frac{b}{2\pi}A\frac{y}{r^2} = -\frac{b}{2\pi}A\frac{\partial\theta}{\partial x}, \tag{2.31}$$

the only harmonic function proportional to r^{-1} and with the right symmetry. Likewise u', v' must each be a harmonic function whose gradient is proportional to r^{-1}, i.e. a linear combination of θ and $\ln r$. The geometry shows that u' must be $b\theta/2\pi$, while v' may be a multiple of $\ln r$. Hence

$$\frac{2\pi}{b}u = \theta - A\alpha\frac{xy}{r^2}, \quad \frac{2\pi}{b}v = B\ln r - A\alpha\frac{y^2}{r^2}. \tag{2.32}$$

The relation

$$B = 1 + (1+\alpha)A \tag{2.33}$$

between the constants A, B follows from the requirement that the e calculated from (2.30) and the displacements (2.32) agree with the e of (2.31) already incorporated in them. An easy calculation gives

$$\sigma_{yy} = \frac{\mu b}{\pi}\frac{\partial}{\partial x}\left[\left(\frac{1}{2}A - B\right)\theta + \frac{xy}{r^2}\right]. \tag{2.34}$$

Hence, integrating with respect to x between $\pm\infty$, there is a total traction $-\mu b(\frac{1}{2}A - B)$ on both the faces of any slab bounded by planes $y =$ constant above and below the dislocation, so that there must be an unwanted concentrated vertical force

$$F = 2\mu b(\tfrac{1}{2}A - B)$$

lurking at the singular point (0, 0). To get rid of it we put $\frac{1}{2}A = B$, which with (2.33) turns (2.32) into (2.27).

Two other cases are worth looking at. With $A = 0$, (2.32) gives Taylor's (1934) original solution

$$u = \frac{b}{2\pi}\theta, \quad v = \frac{b}{2\pi}\ln r \tag{2.35}$$

which actually represents the required dislocation with Burgers vector $(b, 0)$ plus an unwanted concentrated force $(0,\ -2\mu b)$. Because the dilatation (2.31) is zero, (2.35) fails to reproduce some of the properties of an edge dislocation (e.g. its interaction with point defects) even qualitatively. However, as not only the dilatation but also the rotation is zero (2.35) is a useful starting point for constructing the field of a moving dislocation (Section 5).

The case $A = -2$, $F = 2\mu b/(1-2\nu)$ is also interesting. Both σ_{yy} and σ_{xy} vanish on the x-axis and so nothing happens if we cut the material along, say, the negative x-axis. This suggests the following simple experiment. Place a block of foam rubber on a table, make a horizontal slit from the centre to the edge and cover the cut faces with a suitable glue. Pass a rod through the block at the head of the cut and press down on the ends. The two faces will slide over one another with a relative displacement which is fairly convincingly constant along the length of the cut. When the glue has set the rod can be removed to leave a permanent edge dislocation. This solution is in fact just what one gets by taking the well-known expression for a normal point load, actually $\mu b/(1-2\nu)$, acting on the edge of the half space $y>0$ and then allowing x, y to range over the whole xy-plane provided with the (mathematical) cut needed to make u single-valued.

The solution of plane elastic problems in an anisotropic medium is in a sense simpler than it is for an isotropic one. For the latter the field quantities are biharmonic functions of x_1, x_2, whereas the former involve harmonic functions of suitably chosen linear continuations of x_1 and x_2 (Lekhnitskii, 1963). To see this assume tentatively, following Eshelby *et al.* (1953),

$$u_k = A_k f(p_1 x_1 + p_2 x_2) \tag{2.36}$$

and substitute into the equilibrium equation

$$C_{ijkl} u_{k,lj} = 0 \tag{2.37}$$

to get, after cancelling out f'', the three equations

$$C_{i\beta k\alpha} p_\alpha p_\beta A_k = 0, \quad i = 1, 2, 3, \tag{2.38}$$

with α, β summed only over 1, 2. These equations can be solved for A_1, A_2, A_3 only if the determinant of the coefficients is zero.

The argument already used in the isotropic case shows that the strains behave like r^{-1}, and so we must take the function f to be $\ln(p_1 x_1 + p_2 x_2)$ or, rejecting a constant displacement proportional to $\ln p_1$, $\ln(x_1 + p x_2)$ where we have written p for p_2/p_1. Hence in the present case nothing is lost by putting $p_1 = 1$, $p_2 = p$ simply, and the equation $\det(C_{i\beta k\alpha} p_\alpha p_\beta) = 0$ is a sextic in p. Let its roots be $p_{(l)}$, $l = 1, 2, \ldots, 6$, and $A_{(l)i}$ the corresponding sets of coefficients furnished by (2.38). The remainder of the analysis depends on the $p_{(l)}$ all being complex, and never purely real. To see that this is so we compare the energy density

$$\tfrac{1}{2} C_{i\beta k\alpha} u_{i,\beta} u_{k,\alpha}$$

with the result of multiplying (2.38) by $\tfrac{1}{2}A_i$:

$$\tfrac{1}{2} C_{i\beta k\alpha} A_i p_\beta A_k p_\alpha = 0, \quad p_1 = 1,\ p_2 = p.$$

If p were real the choice $u_{k,1} = A_k p_1 = A_k$, $u_{k,2} = A_k p_2 = A_k p$ would make the energy density zero for a non-zero deformation, which is impossible in a stable elastic solid. Hence p is complex, as required.

Each of the complex displacements

$$u_i = A_{(l)i} \ln(x_1 + p_{(l)} x_2)$$

satisfies (2.36) and so does the real linear combination of them

$$u_i = \sum_{l=1,2,3} \frac{D_{(l)}}{2\pi} A_{(l)i} \ln(x_1 + p_{(l)} x_2) + \text{c.c.} \tag{2.39}$$

with constant complex coefficients $D_{(l)}$, and where c.c. denotes the complex conjugate of what precedes it, provided $p_{(1)}, p_{(2)}, p_{(3)}$ are, say, the three roots with positive real parts, so that no two of them are complex conjugates. Like the Taylor solution (2.35), the displacement (2.39) actually represents a mixture of a dislocation with Burgers vector (b_1, b_2, b_3) and the only other line singularity with stresses behaving like r^{-1}, namely a line force of say, intensity (F_1, F_2, F_3) per unit length of the x_3-axis. The six real constants contained in the three complex $D_{(l)}$ can be adjusted to make the b_i and F_i whatever we choose; in our application the b_i are dictated by the crystallography and the F_i must be zero.

The editor of Lekhnitskii's (1963) book implies that the above type of analysis uses the theory of analytic functions of several complex variables in the same way as does quantum field theory, but the comparison is quite misleading. Here we simply use a sum of functions each of which is an analytic function of a different complex variable, but quantum field theory uses something much more sophisticated, a single function of several complex variables, e.g. an analytic function of $x+iy$ which contains parameters which are themselves analytic functions of $x'+iy'$.

It is easy to see from (2.39), without working out the details, that u_i takes the form

$$u_i(r, \theta) = f_i(\theta) + C_i \ln r \tag{2.40}$$

with constant C_i, and that the associated stresses take the form

$$\sigma_{ij}(r, \theta) = \Sigma_{ij}(\theta)/r. \tag{2.41}$$

We can use this to find a quantity we shall need later, the energy E contained between two circles (strictly, cylinders of unit length) of radii $R_1, R_2 > R_1$ centred on the dislocation line. It is equal to the integral

$$E = \frac{1}{2}\int u_i \sigma_{ij}\, dS_j$$

taken along the circles and the two sides of the cut needed to make u_i single-valued. The contributions involving $f_1(\theta)$ in (2.40) are equal and opposite for the two circles, and the contributions from the logarithmic term are separately zero for each circle because there is no net traction on either of them. This leaves only the contribution from the cut. If we take it to be the radial plane θ = constant we get

$$E(\mathbf{t}) = E_0(\mathbf{t}) \ln (R_2/R_1) \tag{2.42}$$

with the so-called pre-logarithmic factor

$$E_0(\mathbf{t}) = \frac{1}{2} b_i \Sigma_{ij}(\theta) n_i(\theta), \tag{2.43}$$

where $n_i = (-\sin\theta, \cos\theta, 0)$ is the normal to the cut with, naturally, a sign convention which makes the result positive. As the position of the cut is irrelevant E_0 must, in fact, be independent of θ; that is, although the individual $\sigma_{ij}(r, \theta)$ are fairly complicated functions of θ, the traction on any radial plane has a component parallel to the Burgers vector which is independent of the inclination of the plane. (For later use we have written $E(\mathbf{t})$, $E_0(\mathbf{t})$ to emphasize that the energy (2.42) is a function of the direction $\mathbf{t}$ of the dislocation line relative to the crystal axes.)

A number of authors, beginning with Stroh (1958), have improved and extended the theory of straight dislocations in anisotropic media. We refer the reader to the recent article by Bacon *et al.* (1979) and turn to an interesting connection between the field of an infinite straight dislocation and the field of a dislocation loop, associated with the names of Lothe (1967), Brown (1967), Indenbom and Orlov (1967), and others.

There is nothing to stop us evaluating the integral (2.20) or (2.23) over an unclosed length of dislocation, with an apparent violation of Burgers vector conservation at the ends. Actually the violation is only apparent (Eshelby and Laub, 1967); the segment is surrounded by a continuous distribution of infinitesimal dislocations with their total Burgers vector equal to the Burgers vector of the segment, which fan out at one cut end, curl round and enter the other end, so ensuring a formal conservation of the Burgers vector. When several such segments, with a common Burgers vector, are joined end to end, the inflows and outflows at the junctions cancel. Although picturesque, this result is not very helpful, though it emphasizes the point that one need have no compunction in considering terminated dislocation segments so long as this is merely a preparation for fitting them together to form a closed loop, or a line going from infinity to infinity.

Figure 2.4 shows a loop in the plane $x_3 = 0$. We shall study the contribution, $du_{i,j}$ say, of a single element $dx_{q'}$ of the integral (2.23) to the displacement gradient $u_{i,j}(i, j = 1, 2, 3)$ at a point $(x_1, x_2, 0)$ in the plane of the loop. In (2.23) the factor $\varepsilon_{qkr}(x_r - x_{r'})dx_{q'}$ is the vector product of $x_r - x_{r'}$ and $dx_{q'}$ and in the present case only has an x_3-component, equal to $r \sin(\theta - \alpha) dS$ in the notation of the figure. The remaining factor $b_j \sigma^{(i)}_{jk,l}$ is the product of r^{-3} and a function of direction in the $x_1 x_2$-plane. Hence we can write

$$u_{i,j}(x_1, x_2) = \int du_{i,j} \tag{2.44}$$

with

$$du_{i,j} = (ds/r^2)\sin(\theta - \alpha)\Theta_{ij}(\theta) \tag{2.45}$$
$$= (d\theta/r)\Theta_{ij}(\theta) \tag{2.46}$$
$$= (d\theta/p)\sin(\theta - \alpha)\Theta_{ij}(\theta); \tag{2.47}$$

for the second and third forms we use the fact that $dS \sin(\theta - \alpha) = r\,d\theta$ and that p, the perpendicular from (x_1, x_2) onto the tangent AB to the loop at (x_1, x_2), is $r \sin(\theta - \alpha)$. Suppose that instead of merely being the tangent to a curved dislocation the line AB actually

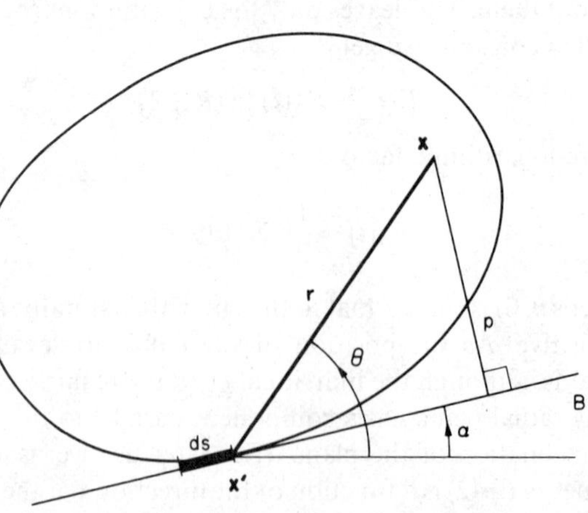

FIG. 2.4. To illustrate the Lothe–Brown formula.

represents an infinite straight dislocation. As p is constant along it we have

$$u_{i,j} = \int du_{i,j} = \frac{1}{p}\int_{\alpha}^{\alpha+\pi} \sin(\theta-\alpha)\Theta_{ij}(\theta)\,d\theta.$$

But from the two-dimensional theory we know that at a distance p from the straight dislocation its displacement gradient has the form $U_{ij}(\alpha)/p$, where $U_{ij}(\alpha)$ depends only on its orientation α, and so we have

$$\int_{\alpha}^{\alpha+\pi} \sin(\theta-\alpha)\Theta_{ij}(\theta)\,d\theta = U_{ij}(\alpha).$$

Differentiate twice with respect to α:

$$\Theta_{ij}(\alpha)+\Theta_{ij}(\alpha+\pi)-\int_{\alpha}^{\alpha+\pi} \sin(\theta-\alpha)\Theta_{ij}(\theta)\,d\theta = U_{ij}''(\theta).$$

Since G_{ij} is an even function of the vector $\mathbf{x}-\mathbf{x}'$ so is $\sigma_{jk,l}^{(i)}$ in (2.23), which is derived from it by two differentiations, and consequently $\Theta_{ij}(\alpha+\pi)$ is equal to $\Theta_{ij}(\alpha)$ so that we have

$$\Theta_{ij}(\theta) = \tfrac{1}{2}[U_{ij}(\theta)+U_{ij}''(\theta)]. \tag{2.48}$$

If we insert (2.46) into (2.44), (2.44) into (2.42), and multiply both sides by C_{klij} to obtain the stress from the displacement gradient, we get the remarkable result that the stress due to the loop, in the plane of the loop, is given by

$$\sigma_{kl}(x_1,x_2) = \frac{1}{2}\int \frac{\Sigma_{kl}(\theta)+\Sigma_{kl}''(\theta)}{|\mathbf{x}-\mathbf{x}'|}\,d\theta, \tag{2.49}$$

where $\Sigma_{kl}(\alpha)/p$ is the stress at a distance p from an infinite straight dislocation whose orientation is specified by α and which has the same Burgers vector as the loop we are interested in. (The notation is that of (2.41), but the angular argument of Σ_{kl} here has a different meaning.) Note that the associated straight dislocation is directed not, as one might perhaps have guessed, along the tangent at $x_{i'}$, but along the radius vector from ds to the point of observation. For some developments of this result see the article by Bacon *et al.* (1979).

2.5. *Treatment of the singular line*

According to the foregoing calculations the stresses at a distance r from a dislocation line behave like r^{-1}. In certain situations this physically unrealistic singularity is tolerable and in some others it can be dealt with by introducing an inner cut-off radius r_0 to the cxact choice of which the results are insensitive. In particular it may enter through a factor $\ln(R/r_0)$ with $R \gg r_0$, as in (2.42). The singularity can be removed by using nonlinear elasticity (Gairola, 1979). However, the region where nonlinearity becomes important is also more or less the same as the region where one can no longer neglect the atomic structure of the material. There have been several calculations aimed at using the theory of generalized continua to tone down the singularity without straying out of the realm of linear continuum mechanics. Cohen (1966) and Knésel and Semela (1972) treat the solid as a grade 2 material (Toupin, 1964) in which the energy density depends on strain and strain gradient. This work is an elastic counterpart of Podolsky's (1942) use of "grade 2" electrostatics to moderate the field near a point charge. Eringen (1977) used his non-local elasticity. In it the stress at a point depends linearly on the strain everywhere through an integral with a suitable kernel. Following a well-known method in theoretical physics we may say, equivalently, that the Fourier components $\sigma_{ij}(\mathbf{k}, \omega)$, $e_{kl}(\mathbf{k}, \omega)$ of stress and strain are connected by elastic constants $C_{ijkl}(\mathbf{k}, \omega)$ which themselves depend on wave number $\mathbf{k}$ and frequency $\omega/2\pi$. The application to a dislocation has a close analogue in McManus's (1949) treatment of an electron by non-local electrodynamics. Though non-local the treatment is non-localized in the sense that it cannot give an account of properties which depend on the periodicity associated with the underlying lattice. The same is, of course, the case for any true continuum theory.

The model of Peierls (1940) and Nabarro (1947) attempts to give some account of the nonlinear behaviour close to the centre of the dislocation, and of lattice periodicity. We shall give a rather oversimplified sketch of the method, treating a screw rather than, as they did, an edge dislocation. The analysis is simpler, and in addition the fact that for the screw but not the edge the atoms move at right angles to the xy-plane, so that, as always in states of anti-plane strain, x and y are indifferently Eulerian or Lagrangian coordinates, allows us to evade certain difficulties.

The model treats the crystal as two elastic blocks separated by a gap $y = \pm\frac{1}{2}a$ across which interatomic forces act, the relation between a and b depending on the crystallography. The tractions which the interatomic forces exert on the blocks are some prescribed function of the relative displacement

$$\Delta w = w(x, \tfrac{1}{2}a) - w(x, -\tfrac{1}{2}a)$$

of the blocks. Rather than set up the general formalism we shall simply ask what force law is required to maintain the simple elastic solution (2.24) outside the gap. On $y = \frac{1}{2}a$ we have

$$\sigma_{zy} = \frac{\mu b}{2\pi}\frac{\cos\theta}{r}\cdot\frac{r\sin\theta}{y} = \frac{\mu b}{2\pi a}\sin 2\theta \tag{2.50}$$

and if we eliminate θ between (2.50) and

$$\Delta w = \frac{b\theta}{\pi} = \frac{b}{\pi}\tan^{-1}\frac{a}{2x}, \tag{2.51}$$

we get

$$\sigma_{zy} = \frac{\mu b}{2\pi a}\sin\frac{2\pi}{b}\Delta w, \tag{2.52}$$

which is, in fact, the Peierls–Nabarro law, fixed by the requirements that it shall be periodic, and for simplicity sinusoidal, with period b and that it shall reduce to Hooke's law for small Δw. Thus the elastic solution itself provides the solution of (2.49). The width 2ζ of the dislocation is conventionally taken to be the distance between the points at which the displacement discontinuity attains $\frac{1}{4}b$ and $\frac{3}{4}b$ as it varies from 0 to b along the slip plane. From (2.51), $2\zeta = a$ for a screw dislocation; for an edge the result is $2\zeta = a/(1-\nu)$. These small values confirm the assertion in the Introduction that the atomic misfit is concentrated into a small region, though one cannot have much confidence in their exact values.

So far we have simulated the atomic structure by a layer of material with a non-linear periodic deformation law, but, because its properties are independent of x, we have not managed to localize the atoms. That is, our results would have been the same if we had replaced (2.24) and (2.51) by

$$w = \frac{b}{2\pi}\tan^{-1}\frac{y}{x-\xi} \quad \text{and} \quad \Delta w = \frac{b}{\pi}\tan^{-1}\frac{a}{2(x-\xi)}$$

in the elastic blocks and across the slip plane. But one would expect that if there is a solution with all atoms in stable equilibrium for, say, $\xi = 0$, there would be precisely similar solutions for $\xi = b, 2b, \ldots$, representing the same dislocation displaced by one, two, ... atomic distances, and between them solutions with $\xi = \frac{1}{2}b, \frac{3}{2}b, \ldots$, in which the atoms are in unstable equilibrium. For other values of ξ one would expect no solutions, except, perhaps, in the presence of an applied stress. In particular we should expect the energy associated with the interatomic forces to depend on ξ. According to the argument so far the energy in the gap at x is

$$\int_0^{\Delta w(x)} \sigma_{zy}\, d(\Delta w) = \frac{\mu b^2}{4\pi^2}\frac{\frac{1}{2}a}{(x-\xi)^2+(\frac{1}{2}a)^2}$$

per unit area, and so the total energy in the gap is

$$\frac{\mu b^2}{4\pi^2}\int_{-\infty}^{\infty}\frac{\frac{1}{2}a}{(x-\xi)^2+(\frac{1}{2}a)^2}\,dx, \tag{2.53}$$

which is obviously independent of ξ and is in fact equal to $\mu b^2/4\pi$. To get over this we suppose that the atom at x is acted on by a force $b\sigma_{zy}(x)$ and that the elastic field in the blocks is translated rigidly with ξ, so that the elastic energy in them, actually also $\mu b^2/4\pi$, is independent of ξ, and can be ignored in what follows. Suppose first that pairs of atoms face each other across the slip plane. Then we must modify the integral (2.52) by replacing $x-\xi$ by $nb-\xi$ and $\int \ldots dx$ by $\Sigma_n \ldots b$:

$$E_{\text{PN}} = \frac{\mu b^3}{4\pi^2}\sum_{n=-\infty}^{\infty}\frac{\frac{1}{2}a}{(nb-\xi)^2+(\frac{1}{2}a)^2}. \tag{2.54}$$

The summation is the vertical electric field due to a grid of charged wires at a point whose coordinates are $x = \xi$, $y = \frac{1}{2}a$ relative to the nearest wire, and so is made up of the constant field of a smeared-out sheet with the same overall charge plus a fluctuating term which falls off more or less exponentially with y. Its actual value (see Bromwich, 1955, or books on electrostatics and hydrodynamics) is

$$\frac{\pi \sinh(\pi a/b)}{b\cosh(\pi a/b) - b\cos(2\pi\xi/b)}.$$

If a is of the order of b the hyperbolic sine and cosine are large and nearly equal to one another and to $\frac{1}{2}\exp(\pi a/b)$ and we have

$$E_{\text{PN}} = \frac{\mu b^2}{4\pi}[1 + 2\exp(-\pi a/b)\cos(2\pi\xi/b)] \tag{2.55}$$

very nearly. Suppose that there is a uniform applied stress $\sigma_{zy} = \sigma$ acting on the material. The work which it does on the gap when ξ changes by $\delta\xi$ is the integral of $\delta\xi\sigma\partial\Delta w(\xi)/\partial\xi$ along the x-axis, i.e. $\delta\xi b\sigma$, which is equivalent to adding $E' = \xi b\sigma$ to (2.55). For the dislocation to be able to move freely through the lattice, σ must be large enough to make the curve $E_{\text{PN}}(\xi) + \xi b\sigma$

monotonic decreasing everywhere, which gives $\sigma = \mu \exp(-2\pi\zeta/b)$ with $\zeta = \frac{1}{2}a$. For $a \sim b$ this is small compared with the shear modulus, though not so small as we might like. If we suppose that the atoms instead of being in a square array in the xy-plane are piled triangularly we ought to sample the integrand of (2.53) at $\xi + nb$ in the upper plane and $\xi + (n + \frac{1}{2})b$ in the lower. To find E_{PN} we have only to add to (2.54) the same expression with ξ replaced by $\xi + \frac{1}{2}b$ and divide by 2. This doubles the argument of both exponential and cosine in (2.55), giving $\sigma = 2\mu \exp(-4\pi\zeta/b)$, which is satisfyingly smaller, but with the embarrassment that E_{PN} has period $\frac{1}{2}b$ rather than, or in addition to, the expected b. For an edge dislocation one can obtain, for example, $\sigma = [\mu/(1-\nu)] \exp(-2\pi\zeta/b)$, or $\sigma = [\mu/(1-\nu)] \exp(-4\pi\zeta/b)$, $\zeta = a/2(1-\nu)$, according to the details of the model (Huntington, 1955).

3. The Effect of Free Surfaces

3.1. *General*

The formulae we have discussed so far refer to a dislocation in an infinite medium. To emphasize this we may write u_i^∞, σ_{ij}^∞ for the elastic field. Mark out a surface S in the infinite medium and make a cut over S. We are left with a dislocation in a finite body with a stress-free surface S. The elastic field will now be, say,

$$u_i^D = u_i^\infty + u_i^{\text{im}}, \quad \sigma_{ij}^D = \sigma_{ij}^\infty + \sigma_{ij}^{\text{im}}, \tag{3.1}$$

where the additional field u_i^{im}, σ_{ij}^{im} may conveniently be called the "image" field by analogy with the terminology of electrostatics. Since the total traction $(\sigma_{ij}^\infty + \sigma_{ij}^{\text{im}})n_j$ must be zero on S, the image field is simply the field induced in a body bounded by S when surface tractions $-\sigma_{ij}^\infty n_j$ are applied to its surface. The determination of the image field thus reduces to a standard problem of elastostatics, to find the field inside a body from given surface tractions.

Quite apart from its application to the dislocation image problem the problem of finding elastic fields from surface data is itself connected, in a way which the writer finds illuminating, with the theory of Somigliana dislocations through what we shall call Gebbia's (1891, 1902) theorem. According to the theorem the elastic field of a body whose outer boundary S is subjected to a traction $\mathbf{T}$ and a displacement $\mathbf{U}$ is the same as it would be if the material inside S formed part of an infinite medium provided S, now simply a surface marked out in the infinite medium, is covered with a layer of body force of surface density $\mathbf{T}$ and is the site of a Somigliana dislocation whose variable discontinuity vector is $\mathbf{U}$. (For the sign to agree with (2.1) the normal must be directed into the body.) The part of the infinite medium outside S is actually stress-free. To see this start with the undeformed body inserted, with perfect fit, in a hole in an infinite matrix. Impose $\mathbf{T}$ and $\mathbf{U}$ and suppose, for the moment, that the surface of the solid is everywhere pulled away from the hole in the matrix, leaving a gap. Fill the gap with a thin layer of material and cement it to the body and matrix. We now have the situation envisaged in the theorem: the surface traction $\mathbf{T}$ has become built in as a layer of body force and the inserted layer constitutes the Somigliana dislocation. Also, as no traction has been applied to the surface of the hole the surrounding matrix is stress-free, as asserted. It only remains to make the usual adjustment to cover interpenetration. According to this argument the displacement at a point $\mathbf{r}$ inside S is

$$u_i(\mathbf{r}) = \int_S [u_j^{(i)}(\mathbf{r}-\mathbf{r}')T_j(\mathbf{r}') - U_j(\mathbf{r}')\sigma_{jk}^{(i)}(\mathbf{r}-\mathbf{r}')n_k(\mathbf{r}')]\, dS(\mathbf{r}'). \tag{3.2}$$

The first term gives the contribution of the layer of body force and the second the contribution of the Somigliana dislocation in (2.7) with $b_j = -U_j$; the minus is necessary as the normal is now supposed to be directed outwards from S. Equation (3.2) is just Somigliana's formula as given, e.g. by Love (1927); we have supposed there are no body forces. For more details see Jaswon and Symm's (1977) book, which will also provide formal justification for the arguments which follow. If $\mathbf{r}$ is outside S, the left-hand side of (3.2) is zero, as we have seen. If $\mathbf{r}$ is actually on S it is $\frac{1}{2}u_i(\mathbf{r})$.

Suppose that instead of filling up the gap between solid and hole we apply tractions $\mathbf{T}'\,dS$ to the surface of the hole, chosen so that each point of the hole is drawn into contact with the point on the surface of the solid with which it coincided in the undeformed uncut infinite medium. If we cement the cut together we have the same elastic field inside S as before (though the field in the matrix is no longer zero), no dislocation and a layer of body force of density $\mathbf{T}+\mathbf{T}'$. That is, we can drop the second term in (3.2) at the expense of replacing the known T_j by the unknown $T_j + T_j'$ in the first. Conversely, we could apply a force $-\mathbf{T}\,dS$ to each element of the surface of the hole, thus imposing a displacement, $\mathbf{U}'$ say, on it and then fill up the gap with the modified discontinuity $\mathbf{U}-\mathbf{U}'$. We also have a layer of force doublets formed by the built-in tractions $\pm\mathbf{T}\,dS$, but they are separated by a vector distance $\mathbf{U}-\mathbf{U}'$ which is itself of order $\mathbf{T}$ so that the effect of the layer of doublets is of second order in the loads and so has to be ignored in the linear theory. The upshot is that the first term in (3.2) may be dropped at the expense of replacing U_j by $U_j - U_j'$ in the second.

Things are particularly simple when the body is, so to speak, deformed from inside by a dislocation loop, so that there is a $\mathbf{U}$ but automatically no $\mathbf{T}$ in Gebbia's construction. In fact if we start with the loop in an infinite medium and make a cut over S a gap will open up leaving the body bounded externally by S with a traction-free surface, and with the material outside S stress-free throughout. This state of stress can be regarded as being the sum of the infinite-medium field of the loop plus that of the Somigliana dislocation with the unknown discontinuity vector $\mathbf{U}(\mathbf{r}')$. One can in principle set up an integral equation for $\mathbf{U}$ by requiring it to produce a traction which cancels the traction $\sigma_{ij}^{\infty} n_j$ at each point of S. This is in fact practicable in two dimensions, where the Somigliana dislocation can be simulated by a distribution of crystal dislocations. When we have done this we are left with an infinite block in which the material outside S is stress-free, and its removal to give a dislocation in a finite solid with a traction-free surface is then a mere formality. This rather attractive way of looking at the image problem has been strongly advocated by Marcinkowski (1979). On the other hand, the old-fashioned method of images, in the limited range of cases where it works, also has its appeal, and in Section 3.2 we consider a few simple examples.

3.2. *Some examples*

As usual it is simplest to start with a screw dislocation in an isotropic medium. When dealing with a state of anti-plane strain it is often convenient to associate with w the harmonic function ϕ conjugate to it so that $w+i\phi$ is an analytic function of the complex variable $z = x+iy$, not to be confused with the z which features as a suffix in (2.25). As, according to the Cauchy–Riemann relations, the gradients of w and ϕ are orthogonal, the traction $(0, 0, T_z)$ across a curve at a point where its normal is $(n_x, n_y, 0)$ is given by

$$T_z = \mu\frac{\partial w}{\partial x}n_x + \mu\frac{\partial w}{\partial y}n_y = \mu\frac{\partial w}{\partial n} = \mu\frac{\partial \phi}{\partial s},$$

where $\partial/\partial s$ denotes differentiation along the curve. In particular, ϕ is constant along a stress-free contour.

With (2.24) is associated

$$\phi = -\frac{b}{2\pi}\ln r, \tag{3.3}$$

and if we wish to find the field for a dislocation in an infinite rod bounded by a curve C we must add to (3.3) a harmonic function chosen so that the resulting ϕ is constant on C. The expression (3.3) can also be regarded as the scalar potential of an electrostatic line charge, the stream function of a vortex, or the magnetic vector potential of a current filament, so that we can obtain solutions by plundering other areas of mathematical physics.

For example, if we have a screw dislocation situated eccentrically at $(\xi_1, 0)$ in a circular cylinder with its circumference $x^2 + y^2 = a^2$ stress-free, plane electrostatics suggests that we should introduce a negative image at the inverse point

$$(\xi_2, 0), \quad \xi_2 = a^2/\xi_1.$$

For a point (x, y) on the circumference the quantity

$$t = r_1\xi_2^{1/2}/r_2\xi_1^{1/2} = r_1 a/r_2\xi_1 \tag{3.4}$$

(where r_1, r_2 are radii drawn from $(\xi_1, 0)$ and $(\xi_2, 0)$) has the value unity. (Consider the two similar triangles formed by connecting the ends $(0, 0)$, (x, y) of the radius vector to either the object or image.) Hence

$$\phi = -\frac{b}{2\pi}\ln t \tag{3.5}$$

has the right singularity at the object point, has a constant value (zero) on the circumference, and so solves the problem. The corresponding displacement is, of course,

$$w = \frac{b}{2\pi}\tan^{-1}\frac{y}{x-\xi_1} - \frac{b}{2\pi}\tan^{-1}\frac{y}{x-a^2/\xi_1}. \tag{3.6}$$

Actually t is constant, though not unity, on any circle with respect to which ξ_1 and ξ_2 are inverse points, so that (3.5) and (3.6) also apply to the case where the core of the dislocation is excluded by a circular hole of radius a_0, say. We might perhaps ask, with certain authors, whether the centre of the dislocation is at $(\xi_1, 0)$ or at the centre $(\xi_0, 0)$ of the hole, with which it does not exactly coincide or, perhaps, elsewhere. However, it is meaningless to ask where a dislocation is situated in a hole, circular or otherwise. The difference of the solutions for two positions would give zero stress at the hole, and there would be no Burgers vector associated with it and the difference field would be zero.

Suppose, more generally, that the dislocation is at (x', y') inside the arbitrary closed contour C and that the conformal transformation

$$\zeta = f(z)$$

maps C onto the unit circle $|\zeta| = 1$ so that

$$\overline{f(z)} = 1/f(z) \quad \text{on} \quad C. \tag{3.7}$$

Then w and ϕ are given by

$$\phi - iw = \frac{b}{2\pi} \ln F(z, z')$$

with

$$F(z, z') = \frac{1 - f(z')f(z)}{f(z) - f(z')}.$$

To see this it is only necessary to check that $\phi = (b/2\pi) \ln |F|$ behaves like $-(b/2\pi) \ln |z - z'|$ near z' and to note that $|F| = 1$ on C, in view of (3.7).

We return to the case of a screw dislocation off-centre in a circular cylinder. The energy per unit length of the cylinder can be found most easily by slitting the x-axis from the hole at the core to the outer circumference and calculating the work required to establish the displacement difference b across the cut as the stress σ_{zy} builds up to its final value:

$$E = \frac{1}{2} b \int_{\xi_0 + a_0}^{a} \sigma_{zy} \, dx = \frac{1}{2} b \int_{\xi_0 + a_0}^{a} \frac{\partial \phi}{\partial x} dx,$$

that is, $-\mu b^2/2\pi$ times the value of $\ln t$ at the lower limit. (At the upper limit it is zero.) If a_0 is comparable with a, this needs a little geometry, but for $a_0 \ll a$ we have $r_1 \sim a_0, r_2 \sim \xi_2 - \xi_1$ and

$$E = \frac{\mu b^2}{4\pi} \ln \frac{a^2 - \xi_1^2}{a_0 a}. \tag{3.8}$$

To find the plane strain field of an edge dislocation in a cylinder it is simplest to work in terms of the Airy stress function in (2.29).

If the curve C bounding the cross-section of the cylinder is stress-free the x- and y-components of traction, $-d(\partial\chi/\partial y)/ds, d(\partial\chi/\partial x)/ds$, are zero, which means that the gradient of χ is zero along C. If we add $q + lx + my$ to any χ the stresses are unaltered. So by choosing q, l, m suitably we can arrange for χ and its gradient to be zero on C, and so take

$$\chi = 0, \quad \partial\chi/\partial n = 0 \tag{3.9}$$

for the boundary conditions. (The first, of course, implies $\partial\chi/\partial s = 0$.)

For the edge dislocation $(b, 0)$ at $(\xi_1, 0)$ in the circular cylinder $x^2 + y^2 = a^2$ we may start with the function (3.5) we used for a screw dislocation in the same situation. Multiplied by y and $1/(1 - \nu)$ it gives the trial function

$$\chi = -\frac{b}{2\pi(1 - \nu)} y \ln t \tag{3.10}$$

which is biharmonic, has the right singularity near the dislocation (see (2.28)), and satisfies the first of (3.9) but not the second. To remedy this we note that since

$$\ln t = \tfrac{1}{2} \ln t^2 = \tfrac{1}{2} \ln [1 + (t^2 - 1)] = \tfrac{1}{2}(t^2 - 1) + \ldots$$

the function

$$L(t) = \ln t - \tfrac{1}{2}(t^2 - 1) \tag{3.11}$$

and its gradient with respect to t itself, or the x, y implicit in t, vanish on the circle $t = 1$, so that

$$\chi = -\frac{b}{2\pi(1-\nu)} y L(t) \tag{3.12}$$

satisfies both of (3.9), and is the required solution provided it is biharmonic. Dropping a term constant times y which does not affect the stresses, we can write

$$\chi = -\frac{b}{2\pi(1-\nu)} y \left\{ \ln \frac{r_1}{r_2} + \frac{1}{2} \frac{r_1^2 \xi_1}{r_2^2 \xi_2} \right\}. \tag{3.13}$$

The extra term is indeed of a standard biharmonic form, being the product of a harmonic function, here a multiple of y/r_2^2, and a squared radius vector, here r_1^2. It represents a collection of doublets of force doublets located at the image point which can be resolved, perhaps not very informatively, into a doublet of the simple image term $y \ln r_2$ plus a doublet of centres of dilatation. At the actual site $(\xi_1, 0)$ of the dislocation, though not elsewhere, the extra terms contribute nothing to the stresses, so that, so to speak, the dislocation thinks that, as in the screw case, its image field is just that of a dislocation of opposite sign at the image point.

The solutions (3.5) and (3.12) relate to a screw or edge dislocation in an infinite cylinder in a state of anti-plane or plane strain. If we cut out a finite rod from the cylinder the elastic state described by (3.5) or (3.12) will be maintained so long as appropriate tractions are applied to the cut ends, but if we remove them to give a rod with stress-free ends the solutions will be modified. For a long rod the new fields can be found by elementary means everywhere except in regions of the order of a extending in from the ends. On any cross-section of the circular cylinder containing the screw dislocation there is a couple about the z-axis of magnitude

$$M = \int (x\sigma_{zy} - y\sigma_{zx})\, dx\, dy = \mu \int \left[\frac{\partial}{\partial y}(xw) - \frac{\partial}{\partial x}(yw) \right] dx\, dy$$

or

$$M = \mu \int_s w(xn_y - yn_x)\, ds$$

by Gauss' theorem, where the contour s must embrace the whole area of the cross-section and respect the discontinuity in w, so that it can be taken to be made up of the circumference of the cylinder and of the hole at the core, joined by lines along the two sides of the cut used in calculating the energy (3.7). The outer circumference obviously contributes nothing and so, too, though less obviously, does the circumference of the hole. The contribution from the cut gives

$$M = \mu b \int_{\xi_1}^{a} x\, dx = \frac{1}{2} \mu b (a^2 - \xi_1^2) \tag{3.14}$$

if we neglect the radius of the hole in comparison with a.

To maintain a state of anti-plane strain in a finite rod couples $\pm M$ must be applied to the ends. If they are not the rod will develop a twist α (Mann, 1949) given, in radians per unit length, by (3.14) divided by the torsional rigidity, $\frac{1}{2}\pi\mu a^4$, of the rod:

$$\alpha = \frac{b}{\pi a^2}\left(1 - \frac{\xi_1^2}{a^2}\right). \tag{3.15}$$

Its sign is given by the following rule. When the dislocation is introduced, circles with spacing b on the surface of the rod join up to make a helix with a certain sense; α twists a generator of the cylinder into a helix of opposite hand (and much coarser pitch). Such a twist has been observed in crystal whiskers containing dislocations.

When the couples maintaining anti-plane strain are removed an amount of work $\frac{1}{2}M\alpha$ per unit length is extracted from the rod so that (3.7) must be replaced by

$$E = \frac{\mu b^2}{4\pi}\left[\ln\frac{a^2 - \xi_1^2}{a_0 a} - \frac{(a^2 - \xi_1^2)^2}{a^4}\right].$$

As a function of ξ_1, E has a minimum at $\xi_1 = 0$, a maximum at $\xi_1 = (1 - 2^{-\frac{1}{2}})^{\frac{1}{2}}\, a$ at a height $(\mu b^2/8\pi)(1 + \ln\frac{1}{2})$ above the minimum and falls to minus infinity as ξ_1 approaches a. Hence, rather surprisingly, the dislocation is in stable (strictly, metastable) equilibrium at the centre of the rod.

For a long rod the elastic field will deviate from the sum of the anti-plane solution (3.5) and the simple torsion specified by (3.15) only in the neighbourhood of the ends. However, if the rod is short enough to be called a disc the elastic field will be, so to speak, all end effect. Eshelby and Stroh (1951) have given the solution for a screw dislocation at the centre of a circular disc, with or without a cylindrical hole at the core. It appears to be the only solution for a dislocation in a finite body with the tractions exactly liquidated on *all* surfaces. We shall only quote their expression for a dislocation in an infinite plate of thickness $2d$ without a hole at the centre. In cylindrical polars (r, θ, z) the non-vanishing displacement and stress components are

$$w = u_z = \frac{b}{2\pi}\theta, \quad u_\theta = -\frac{b}{2\pi}\frac{z}{r} + \sum_n \frac{2b \sin\frac{1}{2}n\pi}{\pi^2 n} K_1(nR)\sin nZ,$$

$$\sigma_{r\theta} = \frac{\mu b_z}{\pi r^2} - \sum_n \frac{\mu b \sin\frac{1}{2}n\pi}{\pi d} K_2(nR)\cos nZ, \tag{3.16}$$

$$\sigma_{\theta z} = \sum_n \frac{\mu b \sin\frac{1}{2}n\pi}{\pi d} K_1(nR)\sin nZ, \quad R = \frac{\pi r}{2d}, \quad Z = \frac{\pi z}{2d}.$$

The summations are over positive integral n, though the terms for odd n are zero. If the terms involving the modified Bessel functions K_m are omitted, the remaining terms give zero stress on the surfaces $z = \pm d$ and the correct behaviour of w, but there is an unwanted singularity at $r = 0$, traceable to a distribution of couples along the dislocation line, which it is the business of the Bessel functions to cancel.

The dislocation interacts repulsively with a similar one in the plate through the stress $\sigma_{\theta z}$ (see Section 4). As $K_m(nR)$ falls rapidly to zero for $nR > 1$ the repulsion is negligible until the distance between them falls to about d and then rises rapidly. A set of like screw dislocations subject to some influence (e.g. local thinning of the plate) which drives them together should thus form a more or less regular lattice with a spacing of the order of the plate thickness. Something of the sort may sometimes be seen in thin film electron micrographs of metallic foils.

In contrast to the case of a screw dislocation there is no net couple (or force) on the cross-section of an infinite cylinder, circular or otherwise, containing an edge dislocation (Nabarro, 1967, p. 274) and so, apart from end effects, the field is unaffected when a finite rod is cut out. The image field of an edge dislocation $(b, 0)$ in a plate with faces parallel to the xy-plane is actually the same as that of a line force $(0, 2\mu b)$ distributed along the dislocation line. To see this note that for the solution (2.35) the out-of-plane stress components $\sigma_{zz}, \sigma_{zx}, \sigma_{zy}$ are zero. Consequently the solution represents a state of both plane strain and plane stress, and there is no relaxation when a plate is cut out. Hence the image fields of the dislocation and the line force $(0, -2\mu b)$ which the solution represents must cancel. Green and Willmore's (1948) investigation of the line force problem shows that there is no possibility of obtaining an explicit solution analogous to (3.15).

4. Energetics of Dislocations

4.1. *The force on a dislocation*

A dislocation in a crystal can, subject to certain constraints, migrate through the lattice, and we must allow the same freedom to the elastic dislocation which we use to simulate it, even though this property does not flow naturally from its definition in terms of the cutting and welding of a continuum. The dislocation will tend to adopt a configuration which minimizes the energy of the system, made up of the elastic energy of the body containing the dislocation and the potential energy of any external loading mechanism attached to the body.

The elastic field in the body can be regarded as being made up of the field of the dislocation u_i^D, σ_{ij}^D, which can be split into an infinite-medium field $u_i^\infty, \sigma_{ij}^\infty$ and an image field $u_i^{\text{im}}, \sigma_{ij}^{\text{im}}$, the externally applied field u_i^A, σ_{ij}^A, and possibly also a field, u_i^Q, σ_{ij}^Q, say, due to other sources of internal stress. In the energy we can distinguish terms quadratic in these fields individually and also cross-terms between them. In this Section we concentrate on the interaction between the dislocation field and the externally applied field. We therefore omit u_i^Q, σ_{ij}^Q for the moment. We cannot also drop the image field because with an infinite medium there would be nowhere to apply the external load. The total energy is made up of the following terms:

$$E = E_s + E_{\text{im}} + \frac{1}{2}\int_S u_i^A \sigma_{ij}^A \, dS_j - \int_S (u_i^A + u_i^D)\sigma_{ij}^D \, dS_j. \tag{4.1}$$

The first is the elastic self-energy of the dislocation loop by itself in an infinite medium. The second is the change in this energy when a closed surface S is marked out in the infinite medium, a cut is made over S, and the material outside it is removed, allowing the traction on S to relax to zero. As already explained, the elastic displacement and stress then change from $u_i^\infty, \sigma_{ij}^\infty$ to u_i^D, σ_{ij}^D as defined in (3.1).

The first integral in (4.1) is the additional energy which enters the body when it is loaded by the applied surface tractions $\sigma_{ij}^A n_j$. Here we make explicit use of the fact that, according to the linear theory, the response of a body to external forces is unaffected by the presence of internal stresses in it (Southwell, 1936). In the elastic energy of the body there is therefore no coupling cross-term between internally and externally generated stresses. The necessary coupling is provided by the second integral. It represents the potential energy of the loading mechanism which applies the traction $\sigma_{ij}^A n_j$. This integral is minus the work done by the loading mechanism as each surface element moves to its final position $x_i + u_i^A + u_i^D$ from some standard reference configuration which we take to be the unloaded and undislocated specimen. For the moment we suppose that, in engineering language, we have dead loading, i.e. the traction which the loading mechanism exerts on a particular element dS does not change when dS is displaced.

Though the point does not directly affect the present argument, one might perhaps wonder why, in the absence of the dislocation, $u_i^D = 0$, the work done by the load is not equal to the work done on the body, in which case the two integrals in (4.1) would cancel. Consider the operation of attaching a weight to a spring. When we hook the weight on to the unstretched spring we have to support the whole weight with our hand. When the weight has reached its final position it has developed a tension which is just able to support the weight and the hand can be removed; halfway between it supports half the weight, the hand the rest. Hence, of the potential energy loss of the weight, only part, actually a half, is stored in the stretched spring. The same argument can be applied to the action of the loading mechanism on each surface element dS.

The only term in (4.1) which contains the applied field and the field of the dislocation is

$$E(A, D) = -\int_S u_i^D \sigma_{ij}^A \, dS_j \tag{4.2}$$

and we wish to know how it changes when the form of the loop varies. The integral can be related more closely to the shape of the loop by a trick. Re-write (4.2) as

$$E(A, D) = -\int (u_i^D \sigma_{ij}^A - u_i^A \sigma_{ij}^D) \, dS_j. \tag{4.3}$$

The value is unchanged because the dislocation produces no traction on S, and so the added term is zero. But the integrand has now acquired the form (2.8) and so has zero divergence. Hence S can be replaced by any other surface enclosing the loop, in particular the surface shown in Fig. 4.1 made up of a tube with the loop as axis and the top and bottom surfaces of the (arbitrary) discontinuity surface.

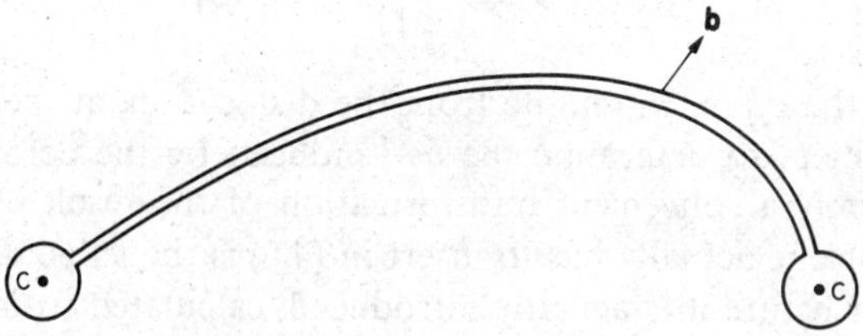

FIG. 4.1. Surface used in finding the Peach–Koehler force.

We now have to try to show that the integral over the tube is zero. In the literature this is sometimes done by supposing that the tube is actually hollow with a stress-free surface, and then shrinking it to zero, or by assuming that its cross-section is of atomic dimensions and allowing it to move with the loop when it deforms. We prefer, in the spirit of theoretical physics, to try to outface the singularity. Suppose the cross-section of the tube is everywhere a circle of radius a, though this is not really necessary. Divide the tube into a large number of short cylindrical beads threaded by the dislocation. In carrying out the integration of the second term in (4.3) over one bead we may take the slowly varying u_i^A outside the integral. As σ_{ij}^D is proportional to r^{-1}, the distance from the axis, the total traction on the curved surface of the cylinder appears to be of order $a.a^{-1}$, which does not go to zero with a. However, since the total traction on the circular ends is of order $a^2.a^{-1} = a$ and each bead is in static equilibrium, the total traction on the cylindrical surface must also be of order a, and so the second term in (4.3) goes to zero with a, as required. There is no trouble with the first term since u_i^P is only of order $\ln r$. We are left with the integral over the two sides of the cut. The second term in (4.3) is continuous across the cut and in the first u_i^P has a discontinuity b_i across it, so that with our previous sign convention we have

$$E(A, D) = -b_i \int \sigma_{ij}^A dS_j \tag{4.4}$$

taken over the positive side of the cut. Note that the image field does not affect (4.4).

If the dislocation suffers a small change of shape the change in $E(A, D)$ is just the integral (4.4) taken over the freshly formed portion of the cut. Suppose that the change of shape is carried out by giving each element dx_k of the loop a small displacement $\delta\xi_l$. Each element then sweeps out an elementary area $dS_j = \varepsilon_{jkl} dx_k \delta\xi_l$ and so we have

$$\delta E(A, D) = -\int \delta\xi_l F_l ds \tag{4.5}$$

with

$$F_l = \varepsilon_{jkl} b_i \sigma_{ij} t_k, \tag{4.6}$$

where $t_k = dx_k/ds$ is the local tangent vector to the loop and σ_{ij} is equal to σ_{ij}^A. (We do not actually write σ_{ij}^A in (4.6) because, as we shall see, it is of more general application.)

Consequently we shall get the correct energy change if we suppose that a force (4.6), the Peach–Koehler force (Peach and Koehler, 1950), acts on the dislocation. We might, perhaps, feel uncomfortable because (4.6) seems to involve the value of the applied stress at the dislocation where, in our treatment, there is a singularity due to the dislocation, or, from a more physical point of view, where there is a strong departure from linearity. However, in principle we could have calculated the energy change (4.5) directly from (4.2) as

$$\delta E(A, D) = -\int_S \delta u_i^P \sigma_{ij}^A dS_j,$$

which only involves the σ_{ij}^A on S, remote from the dislocation, at the expense of solving a boundary value problem to determine the δu_i^P induced by the deformation of the loop. Equation (4.6) is merely a convenient transformation of the result of such a hypothetical calculation, and what one actually has to insert in (4.6) is the value of σ_{ij}^A at the site of the proposed dislocation before it is actually introduced, calculated from the surface traction $\sigma_{ij}^A n_j$ on the assumption of linearity.

The component of (4.6), $F_l t_l$, parallel to the dislocation line, which would be more or less meaningless, is in fact zero. Let us take local coordinates with x_3 parallel to the dislocation line and x_2 perpendicular to the plane, the slip plane, defined by b_1 and t_k. Then if the dislocation is of pure-edge type with Burgers vector $(b_1, 0, 0)$ (4.6) gives

$$F_1 = b_1\sigma_{12}, \quad F_2 = -b_1\sigma_{11}. \tag{4.7}$$

Motion in the direction of F_1, i.e. parallel to the Burgers vector, is referred to as "glide" and in the direction of F_2 as "climb". Evidently when climb occurs the construction of the necessary extension to the discontinuity will involve filling a gap or scraping to avoid interpenetration, and our analysis assumes that the necessary material is supplied or removed. In a real crystal, as opposed to its continuum representation, this can occur by the emission or absorption of lattice vacancies. Consequently a discussion of the energetics of climb involves, for example, consideration of the work done in injecting the emitted vacancies against the osmotic pressure of the vacancies already present in the lattice. As such ideas are outside the competence of continuum mechanics we shall not pursue the matter. For a screw dislocation with Burgers vector $(0, 0, b_3)$ we have

$$F_1 = b_3\sigma_{32}, \quad F_2 = -b_3\sigma_{31}. \tag{4.8}$$

As b_i and t_i are now parallel, every plane is a potential glide plane as far as continuum theory is concerned.

Returning to the general geometry and Burgers vector of (4.6), the glide force takes the form

$$F_l^{\text{glide}} = Fm_l, \quad \text{with} \quad F = b_i\sigma_{ij}n_j, \quad m_l = \varepsilon_{lrs}n_r t_s. \tag{4.9}$$

F is the scalar product of the Burgers vector and the traction exerted by the applied stress on the glide plane (with normal n_j) and m_l is a unit vector in the glide plane and normal to the dislocation line. If we subtract (4.9) from (4.6) we are left with the climb force.

4.2. *Image force and force due to internal stresses*

It is not essential to the validity of (4.6) that the load applied to each surface element of the body should be independent of the displacement of the element. In engineering language the part of the external mechanism for loading a particular element need not be perfectly soft, but may have an arbitrary degree of stiffness. In fact, with arbitrary stiffness, when u_i^D in (4.1) suffers a change of order $\delta\xi_l$ as a result of the deformation of the loop specified by $\delta\xi_l$ the force on an element will be $\sigma_{ij}^A dS_j$ at the beginning of the change and $\sigma_{ij}^A dS_j + O(\delta\xi_l)$ during the course of it, so that the work done by the loading mechanism will only differ from its value for dead loading by a quantity of order $(\delta\xi_l)^2$, and so (4.5) is not affected.

We shall now try to show that the irrelevance of the stiffness of the loading mechanism enables us to extend the Peach–Koehler formula (4.6) to cover the image field and the internal stress field due to other defects without the use of additional mathematics. We begin with the image force. Let us start with the loop in an infinite medium. Call this state (i). Then the only non-zero term in (4.1) is E_S. Make a small change in the form of the loop, note the change $\delta E_S^{(i)}$ in E_S and restore the loop to its original form. Make a cut over S, discard the material outside it, and apply surface tractions $\sigma_{ij}^\infty n_j$ to S. The field inside S is unchanged. Call this state (ii). In state (i) we can regard the material outside S as a loading mechanism of unknown stiffness acting on S. In state (ii) we have replaced it by externally applied forces which produce the same traction on S. According to our argument this makes no difference to the change of total energy when the loop suffers a small change of form. (It is perhaps easier to see this for the reverse change from (ii) to (i).) Hence if we make the same change in the form of the loop as we did in stage (i), we shall get the same energy change $\delta E_S^{(i)}$ as before. But, in fact in stage (ii) the dislocation is acted on by two elastic fields, the image field and the field produced by applied tractions $\sigma_{ij}^A n_j = \sigma_{ij}^\infty n_j$. Yet despite this the energy change is what it would be in an infinite medium. We must conclude that the effects of these two fields cancel. In other words the image field produces the same effect as applied tractions $-\sigma_{ij}^\infty n_j$. But, by definition, it is precisely these tractions which generate the image stress σ_{ij}^{im} in the body. Hence the image stress produces a Peach–Koehler force given by (4.6) with $\sigma_{ij} = \sigma_{ij}^{\text{im}}$. (For a proof of this by an entirely different method see a paper by Gavazza and Barnett, 1975.)

We can handle in the same way a case not covered by (4.1), namely where there are other sources of internal stress in the body in addition to the dislocation loop we are interested in, for example, other dislocations, point defects, or misfitting inclusions representing various metallurgical irregularities. Suppose that they produce a stress σ^o_{ij} in the body. Draw a surface S_0 embracing the loop we are discussing and isolating it from the other sources of stress. We may now regard the interior of S_0 as our body and the self-stressed material between S_0 and S as a loading mechanism, of unknown but irrelevant stiffness, acting on S_0. It follows that the force on the dislocation, so far as it depends on the internal stress σ^o_{ij}, is given by (4.6) with $\sigma_{ij} = \sigma^o_{ij}$. In particular, we can use (4.6) to study the interaction of dislocations with one another.

An analogy is often drawn between the interactions of dislocations and current-carrying wires. The Peach–Koehler formula (4.6) is evidently roughly analogous to the expression for the force on a current element, the stress taking the role of the applied magnetic field. If we are interested in the interaction of two dislocation loops the appropriate stress is given by the integral (2.20) turned into a stress with the help of Hooke's law, and this integral is roughly similar to the Biot–Savart integral for the magnetic field. In some cases the analogy is numerically exact, e.g. if the two loops are coplanar with their Burgers vectors parallel (or anti-parallel) and in the plane of the loops, and if the material is isotropic with zero Poisson's ratio (Hart, 1953). However, since two like dislocations repel whereas two like currents attract, two loops with the same Burgers vector must be simulated by loops carrying opposite currents, and this prompts the Zen-like question, What are three like dislocation loops simulated by? The answer, based on an unpublished comment of Nabarro's is: three like loops each carrying a current $i = \sqrt{(-1)}$. For this apparently flippant idea (which can be a useful mnemonic) there is a respectable precedent in the now outdated regularization procedure of quantum electrodynamics (Jauch and Rohrlich, 1955), in which some of the coupling constants are made purely imaginary so as to change the sign of part of the interaction.

It is clear that the Peach–Koehler force (4.6), now extended to cover internal and image stresses as well as externally applied stresses, is not a straightforward force of the kind that can be balanced by a weight, but, that it simply provides a picturesque description of the energy changes associated with deformation of the dislocation. In particular, as is emphasized in the literature, it is not to be confused with a distribution of body force acting along the singular line. In a strict elastic continuum, where a dislocation is made by cutting and re-welding, and so is immobile, there is nothing objectionable about there being a non-zero Peach–Koehler force acting on it. When we use the dislocated continuum to simulate a dislocation in a crystal lattice and, so to speak, switch on crystallinity, the dislocation will try to reduce the energy by moving in the direction indicated by the Peach–Koehler force, which thus becomes a useful tool for determining equilibrium configurations.

The name configurational force is sometimes given to fictitious forces of the kind typified by the Peach–Koehler force. Recently Kléman (1977) has applied the same term to an apparently similar force acting on disclinations in nematic liquid crystals. In certain circumstances it is given by an equation like (4.6) with b_i replaced by the rotation associated with the disclination, and σ_{ij} by the torque per unit area. However, in addition to a torque there is a traction on an element of the medium given by the Ericksen stress tensor, and one can show (Eshelby, 1980a, b) that the supposedly configurational force on an element of the disclination is equal to the traction which the Ericksen stress exerts on a surface embracing the element, and the element can only be in equilibrium if an equal and opposite real force is applied to it from outside. Hence to call this force a configurational or Peach–Koehler force (Kléman, 1977) seems inappropriate. It is true that it, like the force on a disclination in an elastic continuum, is a measure of the variation of the energy of the system made up of the medium and any loading mechanism attached to its outer boundary, but this is just because the work done by or on the external balancing force is added to or extracted from the system, a state of affairs with no parallel for the dislocation. We might equally well argue that the force exerted on a charge in an electric field was configurational on the grounds that the mechanical work done in pushing it against the field is stored up in the electrostatic energy of the field. It seems to the writer that to be worthy of the name configurational a force must not be balanceable by an ordinary force and, for preference, there should also exist another, "real", force for it not to be confused with.

4.3. *The self-energy of a dislocation*

Returning to (4.1) it still remains to deal with E_S, the self-energy of the loop. Ideally we should calculate E_S and then find its variation for a small change of form. If the result could be thrown into the form (4.5), then (4.6) could be interpreted as the Peach–Koehler force which the loop exerts on itself. The problem is a difficult one, and attempts to improve existing treatments (Bacon *et al.*, 1979) are still under way. Here we shall only present a highly oversimplified treatment, based on an idea of Schoeck and Kirchner (1978).

Let $E_S(L)$ be the self-energy of a loop of a certain size, specified by some characteristic length L, say its maximum diameter. We suppose that E_S is prevented from diverging by omitting from the integration of the energy density the interior of a tube of radius r_0 embracing the singular line. Next make a magnified version of the loop in which L has become L' and associate the same strain with corresponding points $\mathbf{r}$ in the original and $\mathbf{r}' = L'\mathbf{r}/L$ in the enlarged version, so that the stresses in the enlarged version will be in static equilibrium if they were in the original. Because corresponding volume elements are in the ratio $L'^3{:}L^3$ the enlarged configuration will have the energy $E_S(L)L'^3/L^3$. We have enlarged the Burgers vector $\mathbf{b}L'/L$, but in fact we wish it to remain the same, so we must divide the above value by $(L'/L)^2$, since the energy is proportional to the square of the Burgers vector. The resulting energy $E_S(L)L'/L$ is still not quite what we want, since we have also increased the cut-off radius to r_0L'/L. To remedy this we must add the elastic energy E_{tube} within a curved tube of outer and inner radii $L'r_0/L$ and r_0 embracing the singular line. If r_0 is small compared with any characteristic lengths which enter the problem, it is highly appealing to associate with each element ds of the tube an energy $E(\mathbf{t})ds$ where $E(\mathbf{t})$ is the expression (2.42) for an infinite straight dislocation of the appropriate direction $\mathbf{t}$ and with the argument of the logarithm chosen to be $(r_0L'/L)/r_0 = L'/L$. If we admit this the elastic energy in the tube will be

$$E_{\text{tube}} = \ln(L'/L)\int E_0(\mathbf{t})\,ds = CL\ln(L'/L), \tag{4.10}$$

where

$$C = \int E_0(\mathbf{t})\,ds/L$$

is independent of L, being in fact the value of the integral in (4.10) when L happens to be unity. If we now let $L' = L + dL$ differ only slightly from L we have

$$dE_S = E_S(L') - E_S(L) = E_S(L)\,dL/L + C\ln(1 + dL/L)$$

or

$$\frac{d}{dL}\left(\frac{E_S}{L}\right) = \frac{C}{L}$$

which integrates to give

$$E_S = LC \ln (L/L_0)$$

with an unknown constant of integration L_0. We may equally well write the logarithm as $\ln (R/r_0)$ with prescribed r_0 and unknown R, and then we have

$$E_S = \int E(\mathbf{t})ds \tag{4.11}$$

with $E(\mathbf{t})$ as in (2.42) and with the logarithmic factor $\ln (R_2/R_1) = \ln (R/r_0)$ undetermined but independent of $\mathbf{t}$ and of position along the loop. For isotropy, (2.42) gives

$$E(\mathbf{t}) = \frac{\mu b^2}{4\pi}\left(\cos^2\theta + \frac{1}{1-\nu}\sin^2\theta\right)\ln\frac{R}{r_0}, \tag{4.12}$$

where θ is the angle between the Burgers vector and the dislocation line.

The shortcoming in the above argument is, of course, the identification of each element of the integral (4.10) with the corresponding element of that of a tangential infinite straight tube. If a straight dislocation is bent into a curve the elastic field is, so to speak, compressed on the inside of the curve and expanded on the outside of the curve. The resulting increase and decrease in energy density do not precisely cancel, and as a result $E(\mathbf{t})$ depends on the local curvature of the dislocation line. Indeed, strictly speaking, it is a functional which depends on the whole shape of the loop and on the point on the loop to which it applies (Stroh, 1954). The above model is, however, widely used. Exact calculations for specific cases suggest that R in the logarithmic factor is of the order of the dimensions of the loop. If we admit the validity of (4.11) we can go on to discuss the equilibrium form of a dislocation in an applied stress field.

Consider a dislocation segment confined to the xy-plane, with its ends locked at $(0, 0)$ and $(a, 0)$ and between them following the curve $y = y(x)$. If it is acted upon by an applied stress $\sigma_{ij}(x, y)$ the energy (4.1) of the system is

$$\int_0^a E(\tan^{-1} y')(1+y'^2)^{1/2}\,dx - \int_0^a dx \int_0^{y(x)} b\sigma(x, y)\,dy = \int_0^a \mathscr{E}(y, y')\,dx \tag{4.13}$$

say. The first integral in (4.13) is the self-energy and the second is the integral (4.4) extended over the area bounded by the dislocation above and closed by some fixed arbitrary curve below; we have taken it to be the x-axis. In the second integral $b\sigma$ is an abbreviation for $b_i\sigma_{ij}n_j$. The equilibrium shape which minimizes the total energy must satisfy the Euler equation

$$\frac{\partial\mathscr{E}}{\partial y} - \frac{d}{dx}\frac{\partial\mathscr{E}}{\partial y'} = 0$$

prescribed by the calculus of variations. It gives

$$\{E(\theta) + E''(\theta)\}\, y''/(1+y'^2)^{3/2} + b\sigma(y) = 0,$$

which may be written in the form

$$T(\theta)/\rho + b\sigma = 0, \tag{4.14}$$

where ρ is the radius of curvature of the dislocation and

$$T(\theta) = E(\theta) + E''(\theta). \tag{4.15}$$

The second term in (4.14) is the Peach–Koehler force due to the applied stress, and so the first term can evidently be interpreted as the Peach–Koehler force which the dislocation exerts on itself at each point. It takes the form of the product of a line tension T (de Wit and Koehler, 1959) and the curvature $1/\rho$ and is directed towards the centre of curvature so as to tend to straighten out the dislocation, supposing that T is positive. However, in an anisotropic medium it is possible for $E''(\theta)$ to be negative and exceed $E(\theta)$ in magnitude for some orientations. In that case a dislocation with fixed end points and, say, $b\sigma = 0$ may prefer to take a zigzag rather than a straight course between them. For isotropy, (4.15) applied to (4.12) gives

$$T(\theta) = \frac{\mu b^2}{4\pi}\left(\frac{1+v}{1-v}\cos^2\theta + \frac{1-2v}{1-v}\sin^2\theta\right)\ln\frac{R}{r_0}. \tag{4.16}$$

However, because the assumption that E depends only on θ but not on the form of the dislocation is not a very good one, it does not follow that the best choice of the logarithmic factor will be the same for (4.12) and (4.16). For example, Laub and Eshelby (1966) found that for a long straight dislocation deformed into a shallow sine curve the R in (4.16) was equal to the wavelength divided by π, which is comparable with the distance at which the fields of the straight and deformed dislocations become indistinguishable.

A related problem is the determination of the form of a dislocation node. The crystallographic theory of dislocations shows that several dislocations may meet at a point (node) provided that the Burgers vector observed on entering the junction along one of them is equal to the sum of the Burgers vectors observed on exiting along each of the others. One may expect that there will be some condition controlling them similar to that governing joined strings under tension in elementary statics. It may be established by a variational calculation, or by comparison with Herring's (1951, 1953) treatment of the analogous problem for grain boundaries. (We suppose that the dislocations are constrained to lie in one plane.) The result is that on the imaginary cut end of a dislocation there acts a "vector line tension" $\mathbf{T}$ given by

$$\mathbf{T} = \mathbf{t}E(\theta) + \mathbf{n}E'(\theta), \tag{4.17}$$

where, as before, E is the line energy, $\mathbf{t}$ is the tangent, and $\mathbf{n}$ is the normal to the dislocation. In equilibrium the sum of the $\mathbf{T}$'s of the dislocations terminating at a node must be zero. Equation (4.17) may also be applied to a curved dislocation. On an element ds the sum of the forces (4.17) on the ends will be $ds\, d\mathbf{T}/ds$, and from the relations

$$ds/\rho = d\theta = \mathbf{n}\cdot d\mathbf{t} = -\mathbf{t}\cdot d\mathbf{n}$$

which relate arc length s and inclination θ to $\mathbf{n}$ and $\mathbf{t}$ we easily find that its normal component is $T(\theta)ds/\rho$, which takes us back to (4.13) (Chou and Eshelby, 1962).

5. Dislocations in Motion

5.1. *Uniform motion*

So far, when we have spoken of a dislocation moving we have merely meant that it passed through a succession of neighbouring static configurations. However, if while deforming a dislocation line, or part of it, moves with a velocity comparable with the speed of elastic waves, its elastic field must be a solution of the dynamic equations of elasticity. Such solutions will not usually be very relevant to the actual behaviour of moving dislocations because they take no account of the strong retarding effect of interactions with, for example, point defects, electrons and phonons, and, indeed, with the lattice itself, regarded as a perturbation of the continuum. Some of these effects can be incorporated into the continuum description by ascribing suitable viscoelastic properties to the medium (Kosevich and Natsik, 1966; Eshelby, 1961). Here we shall simply show how certain results for moving infinite straight dislocations can be derived rather easily, starting from simple static solutions.

The substitution

$$x \to x'/\beta_n = (x - Vt)/\beta_n, \quad y \to y, \quad \beta_n = (1 - V^2/c_n^2)^{1/2}$$

converts a solution of

$$\nabla^2 f(x, y) = 0$$

into a uniformly moving solution of the wave equation

$$\{c_n^2 \nabla^2 - \partial^2/\partial t^2\} f(x'/\beta_n, y) = 0.$$

Consequently for a uniformly moving screw dislocation we have, from (2.24),

$$w = \frac{b}{2\pi} \tan^{-1} \frac{\beta_2 y}{x - Vt}, \tag{5.1}$$

where $c_2 = (\mu/\rho)^{1/2}$ is the velocity of shear waves in an isotropic medium of density ρ and shear modulus μ. One can check that (5.1), like its static counterpart, satisfies the Peierls relation (2.52).

To solve the same problem for an edge dislocation we may start from the dilatationless and irrotational solution (2.35) for a dislocation plus a vertical force and the fact that the elastic equations of motion reduce to the wave equation with c_n equal to c_2 or c_1 if, respectively, the dilatation or the rotation happens to be zero. The two quantities

$$u^{(2)}(x, y) = u_0(x'/\beta_2, y), \quad v^{(2)}(x, y) = \beta_2^{-1} v_0(x'/\beta_2, y) \tag{5.2}$$

satisfy the wave equation with $c_n = c_2$ and, thanks to the factor β_2^{-1} multiplying v_0, the vector $(u^{(2)}, v^{(2)})$, like the static (u_0, v_0), has zero dilatation, and so is a solution of the equations of motion. So too is the displacement

$$u^{(1)}(x, y) = u_0(x'/\beta_1, y), \quad v^{(1)}(x, y) = \beta_1 v_0(x'/\beta_1, y), \tag{5.3}$$

for it satisfies the wave equation with $c_n = c_1 = [(2-2\nu)\mu/(1-2\nu)\rho]^{\frac{1}{2}}$, the velocity of irrotational waves and, like (u_0, v_0), has zero rotation because of the factor β_1 before v_0.

The unwanted vertical force $-2\mu b$ of the static solution (2.35) remains unchanged for the moving version (5.3) but becomes $-2\mu b\alpha^2$, with

$$\alpha^2 = \tfrac{1}{2}(1 + \beta_2^2) = 1 - V^2/2c_2^2$$

for (5.1). So if we multiply (5.3) by $-\alpha^2$, add it to (5.2), and divide by $1 + (-\alpha^2) = V^2/2c_2^2$, we get a moving solution with the right Burgers vector $(b, 0)$ and no force, namely

$$u = \frac{b}{2\pi} \frac{2c_2^2}{V^2} \left[\tan^{-1} \frac{\beta_1 y}{x - Vt} - \alpha^2 \tan^{-1} \frac{\beta_2 y}{x - Vt} \right],$$

$$v = \frac{b}{2\pi}\frac{2c_2^2}{V^2}\left[\beta_1 \ln\{(x-Vt)^2+(\beta_1 y)^2\}^{1/2} - \frac{\alpha^2}{\beta^2}\ln\{(x-Vt)^2+(\beta_2 y)^2\}^{1/2}\right] \quad (5.4)$$

which reduces to (2.27) in the limit $V = 0$, as a rather tedious calculation will show.

The elementary solutions (5.2) and (5.3) may also be used to construct an image field which when added to (5.4) turns it into the solution for a dislocation moving parallel to the stress-free plane $y = Y$. Introduction of the geometrical image obtained by displacing (5.4) upwards so that its singular point is at a height Y above $y = Y$ is not adequate since it cancels σ_{yy} but doubles σ_{xy} on that plane, or the reverse if we change its sign. The relevant stresses corresponding to (5.2) and (5.3) are

$$\sigma_{yy}^{(2)} = \frac{b}{2\pi}\frac{2\mu\beta_2}{x'^2+(\beta_2 y)^2}, \quad \sigma_{xy}^{(2)} = \frac{b}{2\pi}\frac{2\mu\alpha^2 x'/\beta_2}{x'^2+(\beta_2 y)^2}, \quad (5.5)$$

$$\sigma_{yy}^{(1)} = \frac{b}{2\pi}\frac{2\mu\beta_1\alpha^2 y}{x'^2+(\beta_1 y)^2}, \quad \sigma_{xy}^{(1)} = \frac{b}{2\pi}\frac{2\mu\beta_1 x'}{x'^2+(\beta_1 y)^2}, \quad (5.6)$$

in terms of which the stresses associated with (5.4) are

$$\sigma_{ij} = (2c_2^2/V^2)(\sigma_{ij}^{(i)} - \alpha^2\sigma_{ij}^{(2)}). \quad (5.7)$$

If we displace, say, the (2)-solution so that its singular point is at a height $\beta_1 Y/\beta_2$ above $y = Y$ then on that plane the denominators in (5.5) will have the same form as the denominators of the undisplaced field (5.6), which occur both in the original field (5.7) and its geometrical image. Similarly, we may displace the (1)-solution so that the denominators of (5.6) agree with those of (5.5). In this way, we can arrange that not only another dilatationless field but also an irrotational one is available to help cancel the stresses due to the dilatationless part of the original field (5.4), (5.7), and similarly for its irrotational part, so adding to the richness of the solution and the chance of satisfying the boundary conditions. This suggests that we try the following form for the image field:

$$\mathbf{u}^{\text{im}} = A\mathbf{u}^{(2)}(x, y-2Y) + B\mathbf{u}^{(1)}(x, y-2Y) + C\mathbf{u}^{(2)}(x, y-Y-\beta_1 Y/\beta_2) + D\mathbf{u}^{(1)}(x, y-Y-\beta_2 Y/\beta_1) \quad (5.8)$$

with $\mathbf{u}^{(n)} = (u^{(n)}, v^{(n)})$ in the notation of (5.2), (5.3). The first and second terms are ordinary images of (5.2) and (5.3) while the last two are "extraordinary" images with their singularities shifted to points at distances $\beta_1 Y/\beta_2$ and $\beta_2 Y/\beta_1$ above $y = Y$. The four constants are fixed by the requirement that for the sum of (5.7) and the above image field, evaluated on $y = Y$, the individual parts of σ_{yy} which contain the denominators $x'^2 + (\beta_2 Y)^2$ and $x'^2 + (\beta_1 Y)^2$ (there are no others) shall vanish separately, and similarly for σ_{xy}. This gives

$$\frac{A}{\alpha^2} = B = \frac{2c_2^2}{V^2}\frac{\alpha^4+\beta_1\beta_2}{\alpha^4-\beta_1\beta_2},$$

$$\frac{C}{\beta_1\beta_2} = \frac{D}{\alpha^2} = -\frac{2c_2^2}{V^2}\frac{2\alpha^2}{\alpha^4-\beta_1\beta_2},$$

which completes the solution. The image terms become very large near the Rayleigh velocity where $\alpha^4 - \beta_1\beta_2$ vanishes.

The same device can be used to deal with a screw dislocation moving along the x-axis parallel to the interface $y = Y$ between media with parameters μ', c_2' above it and μ, c_2 below. For the undashed medium we take the field to be the sum of (5.1) and an ordinary image, of unknown strength, at the geometrical image point and for the dashed medium an image of unknown strength at a depth below the interface which makes the denominator in its stress components agree on $y = Y$ with that of the real dislocation, i.e. we now choose its position so as to compensate for the difference between the coefficients β_2 and β_2' in the two different media, instead of, as in the previous example, the difference between β_2 and β_1 in a single medium. The displacement is thus given by

$$2\pi w/b = \tan^{-1}[\beta_2 y/x'] + A\tan^{-1}[\beta_2(y-2Y)/x']$$

in the lower medium and

$$2\pi w/b = B\tan^{-1}[\beta_2'(y - Y + \beta_2 Y/\beta_2')/x']$$

in the upper. On matching stress and displacement at the interface we get

$$A = \frac{\mu'\beta_2' - \mu\beta_2}{\mu'\beta_2' + \mu\beta_2}, \quad B = \frac{2\mu\beta_2}{\mu'\beta_2' + \mu\beta_2}.$$

The analysis fails if the dislocation is supersonic with respect to either medium, so to go over to the case, $\mu' = 0$, of a stress-free surface (Hirth and Lothe, 1968) we must let ρ' tend to zero along with μ', in such a way that c_2' stays greater than V. The result is $A = -1, B = 2$, so that, as one might have guessed, there is a simple negative image. The term in B is without effect so long as we regard the upper medium as empty space, but if we think of it as a material of negligible elastic constants and density but with a definite value for c_2', the B-term gives its deformation.

The same method will handle an edge dislocation gliding parallel to an interface. Starting with the moving solution (5.4), (5.7) we use the image field (5.8) again for the lower medium, but with different A, B, C, D and for the upper medium a similar image expression involving the dashed parameters μ', c_2', c_1', and with the depths of the four image dislocations adjusted so that for each of them the denominator in its stress components agrees on $y = Y$ with one or other of the two denominators which appear in (5.5) and (5.6). The eight constants so introduced are enough to satisfy the four boundary conditions for each denominator separately, and so, fortunately, the eight simultaneous equations fall into two groups of four. It hardly seems worthwhile to write out the results in detail.

5.2. *Non-uniform motion*

More ambitiously, returning to an infinite isotropic solid, we may try to construct the elastic field of a screw dislocation moving non-uniformly along the x-axis by starting from the special case where it executes small oscillations

$$\xi = \xi_0 \exp(i\omega t) \tag{5.9}$$

about the origin. Much closer than the wavelength $2\pi c_2/\omega$ of the disturbance emitted the dislocation simply carries its static field rigidly about with it so that we have

$$w = \frac{b}{2\pi}\tan^{-1}\frac{y}{x-\xi(t)} = \frac{b}{2\pi}\theta + \delta w$$

with

$$\delta w = \frac{b}{2\pi}\frac{\sin\theta}{r}\,\xi_0 \exp(i\omega t) \tag{5.10}$$

provided we are several multiples of ξ_0 away from the origin. The solution of the two-dimensional wave equation of which (5.10) is the limiting form for small r must likewise be proportional to $\sin\theta$ and to $\xi_0 \exp(i\omega t)$ and must represent outgoing waves. The solution must therefore be (5.10) with $1/r$ replaced by some multiple of the Bessel function of the third kind (Hankel function) $H_1^{(2)}(kr)$, $k = \omega/c_2$, which represents an outgoing wave when multiplied by $\exp(i\omega t)$. The appropriate multiple can be read off from the standard expansion

$$-\frac{1}{2} i\pi k H_1^{(2)}(kr) = \frac{1}{r} - \frac{1}{2}\left[\ln(0.54kr) + \frac{1}{2}i\pi\right]k^2 r \tag{5.11}$$

for small kr (the coefficient 0.54 actually stands for $\frac{1}{2}\exp(\gamma - \frac{1}{2}) = 0.5401\ldots$, where γ is Euler's constant $0.5772\ldots$), giving

$$\delta w = -\frac{1}{4} i b k \xi_0 \sin\theta\, H_1^{(2)}(kr) \exp(i\omega t). \tag{5.12}$$

For later use we note that for (5.12) to apply within a range of a few multiples of b from the centre of the dislocation, where our interest will lie, the velocity of the dislocation must be much less than c_2. In fact we must, on the one hand, have $\xi \ll b$ for (5.10) to apply, and, on the other, the wavelength of the disturbance emitted, $c_2/2\pi\omega$, must be large compared with the lattice spacing b, and these two conditions taken together imply $\dot{\xi} \ll c_2$.

If we apply to (5.9) the operator

$$\frac{1}{2\pi i}\int_C (\ldots)\frac{d\omega}{\omega},$$

where C is the real ω-axis indented below the origin, it changes the oscillation (5.9) into the step-function

$$\xi(t) = \xi_0 H(t), \quad H(t) = 1, \quad t > 0, \quad H(t) = 0, \quad t < 0, \tag{5.13}$$

representing a sudden small shift of the dislocation's position by ξ_0. The same operator applied to (5.12) will give the corresponding change of displacement. The result,

$$\delta w = \frac{b\xi_0}{2\pi r}\sin\theta \int_1^\infty \frac{\delta(v - c_2 t/r)}{(v^2-1)^{1/2}}\, v\, dv = \frac{b\xi_0}{2\pi}\frac{y}{r^2}\frac{c_2 t}{(c_2^2 t^2 - r^2)^{1/2}} H(c_2 t - r), \tag{5.14}$$

follows easily if precariously when we use the integral representations

$$H_1^{(2)}(kr) = -\frac{2}{\pi}\int_1^\infty \frac{\exp(-ikrv)}{(v^2-1)^{1/2}}\, v\, dv$$

for the Hankel function (Watson, 1945) and

$$\delta(z) = \frac{1}{2\pi}\int_C \exp(i\omega z)\, d\omega$$

for Dirac's delta function and interchange the order of integration.

An arbitrary motion

$$\xi = \xi(t) \tag{5.15}$$

along the x-axis can be considered to be made up of jumps of amount $\dot{\xi}dt$ in successive time intervals dt. Hence replacing t, $r^2 = x^2+y^2$, ξ_0 in (5.11) by τ, $[x-\xi(\tau)]^2+y^2$, $\dot{\xi}(\tau)d\tau$ and integrating over all τ up to t, we get

$$w = \frac{b}{2\pi}\int_{-\infty}^{\tau_0} \frac{y}{[x-\xi(\tau)]^2+y^2}\,\frac{c_2(t-\tau)}{s}\,\dot{\xi}(\tau)\,d\tau \tag{5.16}$$

with

$$s^2 = c_2^2(t-\tau)^2 - [x-\xi(\tau)]^2 - y^2$$

and where τ_0 is the root of $s^2(\tau) = 0$ which is less than t. For another derivation see Nabarro (1951a).

According to (5.14) the field near the dislocation depends on the entire previous history of its motion. This is a result of the fact that, for example, even after the jump (5.13) has been completed, disturbances from remote parts of the z-axis continue to arrive at the origin. As a result the dynamic behaviour of the dislocation is somewhat peculiar.

To conclude, we return to (5.12) and try to use it to find how the dislocation responds when acted on by a small alternating applied stress, using a crude version of Nabarro's (1951b) argument. Suppose that at first we only retain the term $1/r$ in (5.11). Then the displacement is the sum of (5.10) and $b\theta/2\pi$, the field of a static dislocation rigidly transported to and fro according to the law (5.9). Because, as we have seen, we are limited to small velocities (see the remarks below (5.12)) and hence to β_2 nearly unity in (5.1), we could just as well say that the field of the oscillating dislocation agrees at each instant with that of a dislocation with the same instantaneous position and velocity, but which has been moving uniformly for all time. This means that the displacement and stress obey the Peierls relation (2.52) near the centre of the oscillating dislocation. All this is on the assumption that the second term in (5.11) has been suppressed. When it is restored there is an extra term

$$\delta w' = -\frac{b}{4\pi}k^2 y\left[\ln(0.54\,kr) + \frac{1}{2}i\pi\right]\xi_0 \exp(i\omega t) \tag{5.17}$$

which upsets the agreement with the Peierls law. We can put this right by imposing a displacement equal and opposite to (5.17) from outside. The corresponding applied stress should be

$$\sigma_{zy}^{\text{app}} = -\mu\frac{\partial\delta w'}{\partial y} = \frac{\mu b^2}{4\pi}k^2\left[\frac{y^2}{r^2} + \ln(1.47kr) + \frac{1}{2}i\pi\right]\xi_0 \exp(i\omega t). \tag{5.18}$$

Of course we cannot actually produce a non-uniform stress like (5.18) with the help of remotely applied tractions. However, because of the condition $kb \ll 1$ the logarithm is large (and negative) for $r \sim b$ and the term y^2/r^2 is unimportant in comparison with it (for neatness we shall replace it by $\frac{1}{2}$, its average value, absorb it into the logarithm and round off the coefficient) and so we may simplify (5.18) to

$$\sigma_{zy}^{\text{app}} = \frac{\mu b^2}{4\pi}\frac{\omega^2}{c_2^2}\left[-\ln\left(\frac{c_2}{\omega r}\right) + \frac{1}{2}i\pi\right]\xi_0 \exp(i\omega t). \tag{5.19}$$

Then, as the logarithm varies only slowly with r this stress can be adequately simulated over the region which matters by a uniform applied stress equal to (5.19) with r replaced by some constant value $r_0 \sim b$. The Peierls–Nabarro model suggests $r_0 = \frac{1}{2}b$, but the choice is not critical. If we multiply the stress by b to convert it to a force according to (4.8) we get the equation of motion

$$F_x = [-m(\omega) + iR]\,\omega^2\,\xi$$

which connects applied force F_x and dislocation position ξ when both vary like $\exp(i\omega t)$. The coefficient

$$m(\omega) = \rho \frac{b^2}{4\pi} \ln \frac{\omega_0}{\omega}, \quad \omega_0 = \frac{c_2}{r_0},$$

can be interpreted as a mass per unit length which depends weakly on frequency. The resistive term involving

$$R = \frac{1}{8}\rho b^2,$$

however, since it involves an odd power of i and an even power of ω, is not even approximately proportional to some time derivative of ξ. The best we can do is to say, with Nabarro (1951b), that the mass has a small imaginary component.

Our treatment of the oscillating dislocation is an example of the kind of argument, which may appear somewhat uncouth to an applied mathematician, that one often has to use to extract physical results from the continuum theory. It also illustrates a fortunate fact, which we also found in discussing line energy and tension, and which is true elsewhere in the theory of dislocations, that, even when the elastic solution has to be modified near the core, the shortcomings of the continuum model are often muffled by a merely logarithmic dependence on the details.

References

Amar, H. (1980) Elementary application of differential forms to electromagnetism, *Am. J. Phys.* **48,** 252–253.

Bacon, D. J., Barnett, D. M., and Scattergood, R. O. (1979) Anisotropic continuum theory of lattice defects, *Progress in Materials Science* **23,** 51–262.

Bilby, B. A. and Eshelby, J. D. (1969) Dislocations and fracture, *Fracture* **1,** ed. by H. Liebowitz, 99–182, Academic Press, New York.

Bromwich, T. J. I'A. (1955) *An Introduction to the Theory of Infinite Series*, 2nd edn, 314, Macmillan, London.

Brown, L. M. (1967) A proof of Lothe's theorem, *Phil. Mag.* **15,** 363–370.

Burgers, J. M. (1939) Some considerations of the field of stress connected with dislocations in a regular crystal lattice, *Proc. Konikl. Nederl. Acad. Wetenschap.* **42,** 293–325, 378–399.

Chou, Y. T. and Eshelby, J. D. (1962) The energy and line tension of a dislocation in a hexagonal crystal, *J. Mech. Phys. Solids* **10,** 27–34.

Cohen, H. (1966) Dislocations in couple-stress elasticity, *J. Math. Phys.* **45,** 35–44.

Dederichs, P. H. and Leibfried, G. (1969) Elastic Green's function for anisotropic cubic crystals, *Phys. Rev.* **188,** 1175–1183.

Eringen, C. (1977) Edge dislocation in nonlocal elasticity, *Int. J. Engng Sci.* **15,** 177–183.

Eshelby, J. D. (1951) The force on an elastic singularity, *Phil. Trans. Roy. Soc. Lond.* A **244,** 87–112.

Eshelby, J. D. (1961) Dislocations in visco-elastic materials, *Phil. Mag.* **6,** 953–963.

Eshelby, J. D. (1966) A simple derivation of the elastic field of an edge dislocation, *Br. J. Appl. Phys.* **17,** 1131–1135.

Eshelby, J. D. (1971) Fracture mechanics, *Sci. Prog. Oxf.* **59,** 161–179.

Eshelby, J. D. (1980a) The energy-momentum tensor of complex continua, *Proceedings of the Third International Symposium on Continuum Models of Discrete Systems, Freudenstadt*, ed. by E. Kröner and K.-H. Anthony, 651–665, University of Waterloo Press.

Eshelby, J. D. (1980b) The force on a disclination in a liquid crystal, *Phil. Mag.* A **42,** 359–367.

Eshelby, J. D. and Laub, T. (1967) The interpretation of terminating dislocations, *Can. J. Phys.* **45,** 887–892.

Eshelby, J. D., Read, W. T., and Shockley, W. (1953) Anisotropic elasticity with applications to dislocation theory, *Acta Metall.* **1,** 251–259.

Eshelby, J. D. and Stroh, A. N. (1951) Dislocations in thin plates, *Phil. Mag.* **42,** 1401–1405.

Feynman, R. P. (1949) Space–time approach to quantum electrodynamics, *Phys. Rev.* **76,** 769–789 (Appendix, 785).

Gairola, B. K. D. (1979) Nonlinear elastic problems, *Dislocations in Solids* **1,** ed. by F. R. N. Nabarro, 225–342, North-Holland, Amsterdam.

Gavazza, S. D. and Barnett, D. M. (1975) The image force on a dislocation loop in a bounded elastic medium, *Scripta Metall.* **9,** 1263–1265.

Gebbia, M. (1891) Formule fondamentali della statica dei corpi elastici, *Rend. Circ. Matem. di Palermo* **5,** 320–323.

Gebbia, M. (1902) Le deformazioni tipiche dei corpi solidi elastici, *Ann. Mat. pura appl.* (3) **7,** 141–230.

Gebbia, M. (1904) Le deformazioni tipiche dei solidi elastici, *Ann. Mat. pura appl.* (3) **10,** 157–200.

Green, A. E. and Willmore, T. J. (1948) Three-dimensional stress systems in isotropic plates, II, *Proc. R. Soc. Lond.* A **193,** 229–248.

Hart, E. W. (1953) A magnetic analog for the interaction of dislocation loops, *J. Appl. Phys.* **24,** 224–225.

Head, A. K. (1979) The Galois unsolvability of the sextic equation of anisotropic elasticity, *J. Elasticity* **9,** 9–20.

Herring, C. (1951) Surface tension as a motive for sintering, *The Physics of Powder Metallurgy*, ed. by W. E. Kingston, 143–179, McGraw-Hill, New York.

Herring, C. (1953) The use of macroscopic concepts in surface-energy problems, *Structure and Properties of Solid Surfaces*, ed. by R. Gomer and C. S. Smith, 5–72, University of Chicago Press.

Hill, R. (1961) Discontinuity relations in mechanics of solids, *Progress in Solid Mechanics* **2,** ed. by I. N. Sneddon and R. Hill, 248–276, North-Holland, Amsterdam.

Hirth, J. P. and Lothe, J. (1968) *Theory of Dislocations*, McGraw-Hill, New York.

Huntington, H. B. (1955) Modification of the Peierls–Nabarro model for edge dislocation core, *Proc. Phys. Soc. Lond.* B **68,** 1043–1048.

Indenbom, V. L., and Orlov, S. S. (1967) Dislocations in an anisotropic medium, *Sov. Phys. JETP Letters* **6,** 274–277.

Jaswon, M. A. and Symm, G. T. (1977) *Integral Equation Methods in Potential Theory and Elastostatics*, Academic Press, New York.

Jauch, J. M. and Rohrlich, F. (1955) *The Theory of Photons and Electrons*, 223, Addison-Wesley, Cambridge, Mass.

Kléman, M. (1977) *Points. Lignes. Parois dans les Fluides Anisotropes et les Solides Cristallins*, Les Éditions de Physique, Orsay, France.

Knesel, Z. and Semela, F. (1972) The influence of couple-stresses on the elastic properties of an edge dislocation, *Int. J. Engng Sci.* **10,** 83–91.

Kosevich, A. M. (1979) Crystal dislocations and the theory of elasticity, *Dislocations in Solids* **1,** ed. by F. R. N. Nabarro, 37–141, North-Holland, Amsterdam.

Kosevich, A. M. and Natsik, V. D. (1966) Dislocation damping in a medium having dispersion of the elastic moduli, *Sov. Phys. Solid State* **8,** 993–999.

Kröner, E. (1953) Das Fundamentalintegral der anisotropen elastischen Differentialgleichungen, *Z. Phys.* **136,** 402–410.

Lardner, R. W. (1974) *Mathematical Theory of Dislocations and Fracture,* University of Toronto Press.

Laub, T. and Eshelby, J. D. (1966) The velocity of a wave along a dislocation, *Phil. Mag.* **14,** 1285–1293.

Lekhnitskii, S. G. (1963) *Theory of Elasticity of an Anisotropic Elastic Body*, Holden-Day, San Francisco.

Lothe, J. (1967) Dislocation bends in anisotropic media, *Phil. Mag.* **15,** 353–362.

Love, A. E. H. (1927) *A Treatise on the Mathematical Theory of Elasticity*, 4th edn, Cambridge University Press.

McManus, H. (1949) Classical electrodynamics without singularities, *Proc. R. Soc. Lond.* A **195,** 323–336.

Madelung, E. (1950) *Die Mathematischen Hilfsmittel des Physikers*, 181, Springer-Verlag, Berlin.

Mann, E. H. (1949) An elastic theory of dislocations, *Proc. R. Soc. Lond.* A **199,** 376–394.

Marcinkowski, M. J. (1979) *Unified Theory of the Mechanical Behavior of Matter*, Wiley, New York.

Nabarro, F. R. N. (1947) Dislocations in a simple cubic lattice, *Proc. Phys. Soc. Lond.* **59,** 256–272.

Nabarro, F. R. N. (1951a) The synthesis of elastic dislocation fields, *Phil. Mag.* **42,** 1224–1231.

Nabarro, F. R. N. (1951b) The interaction of screw dislocations and sound waves, *Proc. R. Soc. Lond.* A **209,** 278–290.

Nabarro, F. R. N. (1967) *Theory of Crystal Dislocations*, Clarendon Press, Oxford.

Nabarro, F. R. N. (1980) *Dislocations in Solids* (editor, volumes 1–5), North-Holland, Amsterdam.

Orowan, E. (1934) Zur Kristallplastizität III: Über den Mechanismus des Gleitvorganges, *Z. Phys.* **89,** 634–659.

Peach, M. and Koehler, J. S. (1950) The forces exerted on dislocations and the stress fields produced by them, *Phys. Rev.* **80,** 436–439.

Peierls, R. (1940) Size of a dislocation, *Proc. Phys. Soc. Lond.* **53,** 34–37.

Podolsky, B. (1942) A generalized electrodynamics, Part I: Non-quantum, *Phys. Rev.* **62,** 68–71.

Polanyi, M. (1934) Über eine Art Gitterstörung, die einen Kristall plastisch machen könnte, *Z. Phys.* **89,** 660–664.

Read, Jr., W. T. (1953) *Dislocations in Crystals*, 123, McGraw-Hill, New York.
Schoeck, G. and Kirchner, H. O. K. (1978) The elastic energy of dislocation loops in anisotropic media, *J. Phys.* F. **8**, L43–L46.
Seeger, A. (1955) Theorie der Gitterfehlstellen, *Handbuch der Physik* VII/2, ed. by S. Flügge, 383–665, Springer-Verlag, Berlin.
Somigliana, C. (1914) Sulla teoria delle distorsioni elastiche: Nota I, *Rend. R. Accad. Lincei* $[5]_1$ **23,** 463–472.
Somigliana, C. (1915) Sulla teoria delle distorsioni elastiche: Nota II, *Rend. R. Accad. Lincei* $[5]_1$ **24,** 655–666.
Southwell, R. V. (1936) *An Introduction to the Theory of Elasticity for Engineers and Physicists,* 78, Clarendon Press, Oxford.
Stroh, A. N. (1954) Constrictions and jogs in extended dislocations, *Proc. Phys. Soc. Lond.* B **67,** 427–436.
Stroh, A. N. (1958) Dislocations and cracks in anisotropic elasticity, *Phil. Mag.* **3,** 625–647.
Taylor, G. I. (1934) The mechanism of plastic deformation of crystals, *Proc. R. Soc. Lond.* A **145,** 362–387.
Toupin, R. A. (1964) Theories of elasticity with couple-stress, *Arch. Rat. Mech. Anal.* **17,** 85–112.
Volterra, V. (1907) Sur l'équilibre des corps élastiques multiplement connexes, *Ani. Sci. Éc. Norm. Sup.* [3] **24**, 401–518.
Volterra, V. and Volterra, E. (1960) Sur les distorsions des corps élastiques (Théorie et applications), *Mémorial des Sciences Mathématiques*, 147, Gauthier-Villars, Paris.
Watson, G. N. (1945) *A Treatise on the Theory of Bessel Functions*, 180, Cambridge University Press.
Weingarten, G. (1901) Sulle superficie di discontinuità nella teoria della elasticità dei corpi solidi, *Atti Accad. naz. Lincei Rc.* [5] **10**, 57–60.
Wit, G. de and Koehler, J. S. (1959) Interaction of dislocations with an applied stress in anisotropic crystals, *Phys. Rev.* **116**, 1113–1125.

Superplasticity

by J. H. Gittus

Springfields Nuclear Power Development Laboratories,
United Kingdom Atomic Energy Authority

Summary

In this article we consider the phenomenon of superplasticity, outlining the mechanism and summarizing the properties of the majority of the superplastic metals, alloys, and ceramics.

In a superplastic material deformation can occur by two main processes: boundary sliding and dislocation creep. Over many orders of magnitude of strain rate the free energy of a specimen deforming by boundary sliding is lower than that for dislocation creep, and it is this stability that leads to the high ductility of alloys deforming by boundary sliding. A fine, stable grain size favours deformation by boundary sliding and is the main structural characteristic of superplastic alloys.

Two methods are used to produce a stable, fine grain size. In one case (the original superplastic materials) alloys having approximately equal amounts of two phases are worked to refine the grain size. In the other case (more modern alloys) a fine precipitate pins the grain boundaries.

In superplastic flow the grains do not exhibit the same strain as the polycrystal: instead they slide over one another like grains of sand in an hour-glass. Small, reversible changes in shape occur to obviate cavity formation. When, at high stresses (such as obtain in a necked-down region) the less-stable process of dislocation creep supervenes, the grains begin to deform to almost the same extent as the polycrystal, and failure then occurs.

1. Introduction

Most metals and alloys, when subjected to a simple tensile test, exhibit a rupture ductility of order 50%. There is a class of materials, or more strictly a class of *behaviour*, where ductilities an order of magnitude higher are found. This behaviour is called *superplasticity* and in this article we show that it is quite distinct from the ordinary processes of plastic deformation, having a different microscopic mechanism and plastic flow equation.

Superplastic flow is a form of creep. That is to say the deformation process is thermally activated and the flow rate is stress dependent. Superplastic deformation occurs by a process analogous to the flow of sand through the orifice of an hour-glass. The grains of sand move relative to one another during the flow process and retain their original shape and size whatever the deformation suffered by the body (of sand). Similarly during superplastic flow of a polycrystalline alloy, individual grains move relative to one another, and it is sliding at or near to the grain boundaries (and heterophase boundaries in a two-phased alloy) that permits this relative motion.

No dislocations are visible within the grains during classical superplastic flow. However, dislocations in (or near) the boundaries are involved in the boundary-sliding process, and we can analyse superplasticity in terms of their climb and glide motions. In the absence of intragranular dislocation activity, vacancy diffusion remains the only process by which the grains could change shape during superplastic flow. In the event, however, the grains only make small, reversible changes in shape, so as to prevent holes from forming at the boundaries as they slide over one another, and so after, for example, a tenfold increase in the length of a superplastic tensile specimen, the grains are still approximately equiaxed. This is very different to what happens in ordinary materials where the grains characteristically exhibit the same change in shape as the specimen in which they are situated.

2. Mechanisms of Superplasticity

It is easy to envisage, for a simple square grain model, how movement of a boundary dislocation (gliding on some facets of the grains, climbing up or down other facets) can produce sliding and strain. If the path of the boundary dislocation, instead of being a regular square wave, is a random one, then:

1. The total distance traversed in the climb mode will equal the distance traversed in the glide mode.
2. No voids will be created by movement of the dislocation.
3. The angles at which adjacent facets of the grain boundary meet will not change as a result of passage of the dislocation.
4. Translation will occur at the boundaries, along the path of the dislocation.
5. Shear strain will occur; the block of polycrystal on one side of the dislocation path will move bodily a distance equal to the Burgers vector of the dislocation relative to the other block.

To illustrate these facts we can represent the dislocation path by a step function (Fig. 2.1(a)). Note that this step function is quite analogous to a real boundary with its growth steps and ledges. An alternative, more general model in which the boundary makes a constantly changing angle to the Burgers vector leads, however, to precisely the same conclusions (Fig. 2.1(b)).

This Section, then, develops the theory of superplastic flow from a consideration of the climb and glide of boundary dislocations. The movements of grains (or particles of two phases) are viewed as a consequence of the movements of the network of interfacial dislocations. The problems involved when we try to describe grain translation, rotations, switching, and so on are not central. Explaining these latter events becomes a separate task and not a necessary preliminary to the formulation of an equation for superplastic flow.

The materials which exhibit what Johnson (1970) terms "fine grain superplasticity" may be single phase, but those exhibiting the highest structural stability during deformation, coupled

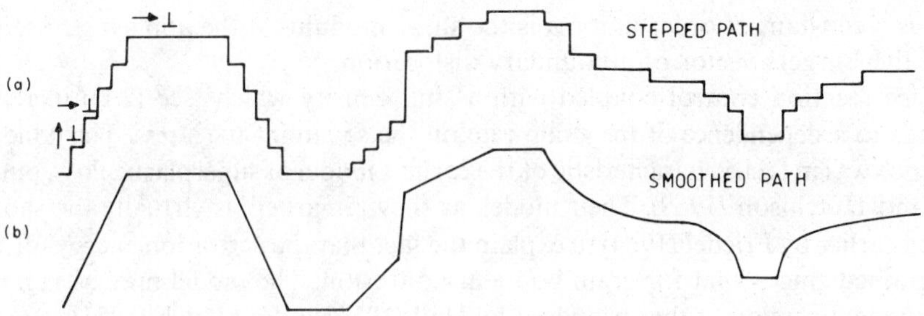

FIG. 2.1. Stepped boundary, separating grains of identical phase or dissimilar phases (a). A smoothed boundary of equivalent shape is shown at (b). A boundary dislocation moving along either boundary in a polycrystal can produce strain without causing voids to form.

with the largest extensions, are more commonly multiphased. In the latter case it is frequently observed that the two phases are present in approximately equal proportions and that they have approximately equal ductilities. Many of the superplastic materials therefore are alloys of eutectic or eutectoid composition in which the component phases have similar melting temperatures. A fine stable grain size (or particle size) is important for superplastic behaviour, and the presence of second-phase particles helps to ensure this. The effect depends on both the volume fraction of second phase and on the size of the second-phase particles. Smith (1948) shows in fact that the stable grain size equals the particle size when there are approximately equal proportions of the two phases present.

During superplastic flow, as we have noted, individual grains or particles translate large distances relative to one another (Johnson, 1970; Smith, 1948; Gittus, 1975). Simultaneously, fluctuations in particle shape occur and prevent or restrict the formation of holes at the particle boundaries. This is quite distinct from the more normal type of deformation in which each grain deforms like the polycrystal of which it comprises a part and in which each grain remains adjacent to the same neighbours throughout deformation.

Superplasticity has also been obtained in relatively dilute alloys. In the case of aluminium, for example, dilute additions of zirconium produce a precipitate of $ZrAl_3$ particles which restrict grain growth by exerting a drag on moving grain boundaries—the Zener effect (Edington, 1976). In these dilute alloys, superplastic flow does not involve sliding at heterophase boundaries. In alloys of eutectic and eutectoid composition, however, heterophase boundary sliding is observed to occur.

2.1. *Theory*

At low stresses and small grain sizes, Ashby and Verrall (1973) noted that the ability of the boundary to act as a source and sink for the diffusing point defects might become limiting. They consider the case where grain boundary dislocations are the sources and sinks of point defects and are also responsible for sliding. The creep rate then depends on the density of dislocations in the boundary. They suggested, by analogy with the spacing between dislocations in the body of a crystal, that the spacing l between boundary dislocations would be

$$l = \mu b/\beta\sigma, \tag{2.1}$$

where β is a constant of order unity, μ is the shear modulus, σ the applied stress, and b the value of the Burgers vector of a boundary dislocation.

Interface reaction control coupled with a sink density which (see (2.1)) increases with stress leads to a dependence of the strain rate on the square of the stress. This kind of stress dependence was in fact a characteristic of the earliest model of superplastic flow, put forward by Ball and Hutchison (1969). Their model, as they remarked, is virtually the same as one suggested earlier by Friedel (1964) to explain the fact that the activation energy for the creep of fine-grained zinc is that for grain boundary diffusion. The model may be regarded as a time-dependent version of that proposed by Hall (1951) and by Petch (1953) to explain the dependence of yield stress upon grain size d:

$$\sigma_e = \frac{2(\sigma - \sigma_0)^2 d}{\mu b}. \tag{2.2}$$

Here σ_e is the compressive stress at the head of a dislocation pile-up, σ_0 is the dislocation friction stress. Equation (2.2) (with σ_e a materials parameter) has been widely verified by experiments in which the value of σ at which plastic yielding occurs has been correlated with the grain size.

Following Friedel, Ball and Hutchison suggested that in superplastic flow dislocations from the head of the pile-up would climb into the boundary against which they were piled up: strain is so produced. Mohamed *et al.* (1975) have examined the Ball and Hutchison model, particularly in relation to superplasticity in the Zn–Al eutectic, and have pointed to two drawbacks: firstly, the model assumes that adjacent grains slide in groups and this is not supported by direct observations in the scanning electron microscope. Secondly, the model predicts the presence of many intergranular dislocations although transmission electron microscopy and internal marker experiments have been interpreted to indicate that there is very little within grain dislocation activity during superplastic flow.

Neither of the two theories is entirely satisfactory, even for the case of superplastic flow in a single-phase alloy. In the case of a two-phase alloy there are additional difficulties due to the nature of the sliding and climb processes at heterophase boundaries. We consider this matter in Section 2.2.

2.2. *Sliding, growth and dissolution at heterophase boundaries*

Ishida and McLean (1967) and McLean (1971) have reviewed and extended the theory of the ways in which dislocations in grain boundaries can glide and climb: the climb component leads to vacancy emission while the glide component causes grain boundary sliding. The boundaries with which these authors were concerned are those between like grains. The possibilities are, however, more limited at heterophase boundaries. It is, of course, established that particles of a second phase can precipitate and dissolve. In addition, processes such as "ripening" are observed in which material from one particle of a second phase diffuses to another through the intervening matrix. Again, the shape of a particle can change—presumably by the diffusion of matter from one region to another. These mechanisms are capable then of producing the changes in particle shape (and possibly size) which have to occur in order to preserve continuity at the contacting faces of particles and matrix during superplastic flow. Aaronson *et al.* (1970) have summarized the mechanisms of these diffusional growth processes for precipitate crystals. The orientation relationship of the precipitate lattice with respect to the heterophase boundary varies with the orientation of the boundary. Dislocation structures appear at one or two orientations and disordered structures at all of the others.

The dislocation boundaries can only migrate by the movement along their length of ledges. The narrow faces of these ledges often appear to be disordered and so there is no structural barrier to precipitation of additional material on to them. Similarly, there is no structural barrier to the growth (migration) of the disordered boundaries. In the ordered boundaries, on the other hand, addition of a new layer of atoms will generally alter the packing relationship between the two contacting lattices, introducing a measure of disorder. The energy needed to accomplish this is usually high. It is for this reason that precipitates are often in the form of long plates or rods: the long dimension is a dislocation boundary which can only thicken by the slow process of ledge growth, whereas elongation of the precipitate occurs preferentially by the addition of atoms to regions of the boundary that are already so disordered that the new atoms cannot raise their energy significantly.

Sliding at dislocation heterophase boundaries is impeded by the same type of energy barrier. Thus the passage of a glissile dislocation leaves in its wake a disordered structure of high energy. The energy increase could be so high as to prohibit the glide and with it any sliding at the heterophase boundary (HPB). That is to say, the threshold stress for HPB sliding could be very high (Ashby, 1972). Dislocation boundaries in single-phase materials can slide more easily. They do this, in most cases, by a combination of translation and migration (Ishida and McLean, 1967; McLean, 1971). Atoms are transferred across the interface during sliding to restore the boundary structure, which would otherwise be altered by the sliding process. In this way one of the grains grows at the expense of the other. If atoms are transferred across the interface between two different phases, then the grain which grows does so by acquiring the wrong kind of atoms—those which constituted the other phase. Equilibrium can be restored by the diffusional redistribution of the components, but this introduces a large viscosity term since the diffusion path is of the order of the grain size.

There is at least one possible mechanism of sliding available on dislocation HPBs. It involves the glide of a superdislocation in the boundary plane. The first of a pair of boundary dislocations creates disorder but the second dislocation, when it passes along the same path, restores order once more. Another example comprises the climb in a HPB of a faulted dislocation loop or of a superdislocation whose Burgers vector is perpendicular to the plane of the boundary. Atoms diffuse from the trailing edge to the leading edge to produce this climb motion. In the wake of the climbing dislocation order is restored to the boundary structure. Passage of such a dislocation by pure climb cannot, however, cause boundary sliding; for this to occur there must be a component of the Burgers vector oriented in the plane of the boundary. Mixed climb and glide can then occur, the glide component giving rise to sliding and the climb component involving a diffusion distance equal only to the width of the dislocation.

A specific example of a known HPB with defined dislocations occurs in the work of Nakagawa and Weatherley (1972), who have actually observed pairs of dislocations in the interface between two crystal lattices. The lattices were those of Ni_3Al and Ni_3Nb in a directionally solidified eutectic alloy. The dislocation pairs had separations of 300–400 Å. They were epitaxial dislocations with $b = a/2(110)_{Ni_3Al}$ and the orientation relationship between the phases at their common boundary was $(110)_{Ni_3Al} \| (100)_{Ni_3Nb}$, $(113)_{Ni_3Al} \| (031)_{Ni_3Nb}$. A HPB between two pure substances (α and β) may be regarded as a plane of ordered solid solution of one phase in the other. The disordered region between the leading and the trailing components of the superdislocation is therefore an antiphase boundary (APB). Gleiter (1977) in fact shows that even for grain boundaries of random orientation, the width of a boundary dislocation is theoretically always much larger than that of a dislocation of similar Burgers vector in the perfect lattice. The partial dislocations are separated by stacking fault, instead of an APB. Their movement does not disorder the structure. This may occur in superplastic alloys such as Ag–Cu where the phases are both face-centred cubic.

The equilibrium width of the disordered region can be calculated from a knowledge of the elastic constants and a fault energy Γ. Then the force with which the fault tends to contract, acting as a surface, equals Γ per unit of length, which is equivalent to a resolved stress of magnitude Γ/b pulling the partial dislocations together. On the other hand, the partials being of like orientation and on the same slip plane tend to repel one another. For reasonably large distances of separation w (the width of the disordered zone) this repulsion is, in the first approximation, due to the resolved shear stress set up by the elastic stress field of each of the two partial dislocations at the position of the other. Now outside the core, the elastic displacements, strains, stresses, and energy of the grain boundary dislocation are identical with the well-known results for a dislocation of the same Burgers vector in an isotropic elastic continuum. Providing then that the partials are sufficiently separated, that their cores do not overlap, the elastic part of their stress fields and the interaction energy between them can be calculated by treating them as elastic dislocations in a linear isotropic continuum. Of course, allowance must be made for the fact that the shear moduli of the two phases differ. We shall assume that an adequate approximation is provided by the use of an effective modulus μ defined as a function of the two moduli μ_α and μ_β:

$$\mu = f(\mu_\alpha, \mu_\beta). \tag{2.3}$$

At equilibrium a dislocation HPB will have misfit dislocations spaced out along its length. They are necessary because their presence compensates for the differences between the lattice spacings of the two phases where they contact along their boundary. If the lattice parameters were identical then the resultant coherent boundary would contain no dislocations and would have the structure of a fully ordered solid solution. That is to say the long-range order parameter S for a coherent boundary is unity. More generally, for the case where the lattice parameters differ and superdislocations occur at a separation l, we can write

$$S = 1 - 2w/l \tag{2.4}$$

so that, when the dislocations are so close together that the partials touch, $S = 0$, corresponding to the complete absence of long-range order (a disordered boundary). The width of the superdislocation is then given, to the first approximation, by an equation essentially similar to that for the width of the superdislocations in an ordered solid solution (see Haasen, 1965):

$$w = \frac{\mu b^2}{2\pi\Gamma S^2}. \tag{2.5}$$

On substituting for S from (2.4),

$$w = \left\{1 - \left[1 - \frac{8\mu b^2}{2\pi\Gamma l}\right]^{1/2}\right\}\frac{l}{4} \quad \text{for} \quad w \ll l. \tag{2.6}$$

As an example consider the case where the energy of the disordered boundary in the fault Γ is

$0.10\,J/m^2$ ($100\,erg\,cm^{-2}$) and let $b = 2\times10^{-10}\,m$ (2 Å) and $\mu = 5\times10^4$ MPa. Let $S = 0.75$. Then

$$w = 5\times10^{-9}\,\text{m}\ (50\,\text{Å}). \tag{2.7}$$

The disordered region is identical in structure with the narrow face of a growth ledge and we may in fact regard such ledges as superdislocations in which the fault and glide planes are no longer coparallel. Such "staggered" superdislocations are certainly observed in grain interiors.

2.3. *The model of superplastic flow*

In Section 2.2 we have argued that particles of differing phases can slide relative to one another by the movement of interfacial superdislocations. We have also remarked that a particle can change its shape or size by the diffusion of matter to and from disordered regions of its boundary. We are going to use these principles in a model of the superplastic flow process.

Consider the sliding process which occurs on a dislocation boundary. Suppose that superdislocations can be generated by the operation of sources. For example by the expansion of intrinsic dislocation loops under the applied stress. In addition there is the likelihood that matrix dislocations will enter the boundary. Thus the boundary dislocations have Burgers vectors that are equal to the sum or difference of those of matrix dislocations. So a matrix dislocation can enter a boundary providing that it splits into two components. One of these could then, for example, climb into a disordered region of the boundary while the other glides in a dislocation region. Once in the boundary the dislocations are effectively trapped and can only emerge into the matrix if they reassociate to form matrix dislocations of appropriate Burgers vector. The dislocation boundary is potentially a very effective pile-up of length d, therefore. From (2.2) we can calculate the stress at the head of such boundary pile-ups. μ is now defined by (2.3); b, the Burgers vector of a dislocation boundary superdislocation, equals the sum or difference of Burgers vectors for the contacting phases. The movement of dislocations in this dynamic pile-up proceeds at a rate determined by the climb of the dislocations away from the head of the pile-up. They cannot generally climb out of the boundary. Therefore they must climb in it. The most likely situation is one in which the dislocations pile up at a triple junction where a dislocation boundary, on which they are gliding, meets two disordered boundaries into which the dislocations then climb. For this climb to occur matter must diffuse to or from the dislocations (diffusion coefficient $= D_B$). Let the diffusion distance be λd. The velocity of climb is then given by the following equation due to Friedel (1964),

$$V = \frac{D_B}{\lambda d}\left(\frac{\sigma_e b^3}{kT}\right). \tag{2.8}$$

The dislocations will have to move a distance approximately equal to the diffusion distance in order to encounter others climbing away from neighbouring pile-ups. As Friedel pointed out, many of these encounters will be between boundary dislocations whose Burgers vectors add up to zero and so the dislocations will annihilate one another. The time taken for a dislocation to climb this distance λ at velocity V is $t = \lambda d/V$, i.e.

$$t = \frac{\lambda^2 d^2 kT}{D_B \sigma_e b^3}. \tag{2.9}$$

At intervals of t, therefore, two of the dislocations climbing away from each pile-up are annihilated (one as it climbs up one of the boundaries that meet at the triple junction and the other as it climbs up another). Two new dislocations will therefore need to be generated in the pile-up in order to replace those that climb out of it in pursuit of the two that have been annihilated. In this way equilibrium is maintained. The strain contributed is that which would be produced if the dislocation boundary slid a distance $2b$ at intervals of t. The strain rate is now calculated from the equation

$$\dot{\varepsilon} = \bar{u}/d. \tag{2.10}$$

Here $\bar{u}$ is the mean rate of sliding translation at the crystal boundaries.

So to produce a strain of ε, boundary sliding of εd must occur. A sliding displacement of magnitude $2b$ will therefore produce a strain of $2b/d$.

The strain rate is therefore given by

$$\dot{\varepsilon} = 4\left(\frac{d}{\lambda}\right)^2 \frac{D_B \mu b}{kT}\left(\frac{\sigma - \sigma_0}{\mu}\right)^2 \left(\frac{b}{d}\right)^2 \quad \text{for} \quad \sigma > \sigma_0. \tag{2.11}$$

2.4. *Boundary layer shear at heterophase boundaries*

Under some circumstances sliding occurs by the shear of a thin boundary layer. Recapitulating what is known about this phenomenon, Eberhardt (1979) has studied boundary sliding in two phased bicrystals of brass. One phase (alpha) is hard and non-deformable; the other phase (beta) is ductile at the temperature of his sliding experiments. Below a certain stress, sliding occurred at the alpha–beta (heterophase) boundary. Neither phase exhibited any deformation and the sliding rate was proportional to the first power of the shear stress. Sliding, he showed, is simply controlled by the rates of diffusion at these low stresses.

At higher stresses the sliding rate varied with the square of the shear stress (the class of behaviour which we examined in arriving at (2.11), in fact). Metallography showed that sliding was occurring by the movement of matrix dislocations in a thin layer of beta-phase adjacent to the alpha–beta boundary. Sliding at the boundary played a negligible role at these stresses.

The dislocations, in this instance, were piled up not in the boundary itself but in the boundary layer. Under these circumstances there is no need to invoke the presence of HPB superdislocations: sliding does not occur at the boundary and accordingly the equilibrium of the boundary structure is not perturbed by the movement of interfacial dislocations. The equation for superplastic flow does not depend on where the dislocation pile-up is situated: if it is in a thin boundary layer then this alters b and μ, but (2.11) remains valid.

2.5. *Structural effects*

The dislocation spacing on dislocation boundaries theoretically falls with increasing stress. Hence (see (2.4)) the long-range order parameter S will fall as the stress increases. Now a completely disordered boundary ($S = 0$) has no preferred relation to the lattices of the contacting phases and tends to adopt a spherical shape to minimize its area (and thus to minimize the surface energy). At the other extreme a dislocation boundary ($S \rightarrow 1$) follows preferred crystallographic directions and is essentially planar. When a material is stressed, then we expect the gradual build-up of HPB superdislocations towards the equilibrium density (and the associated fall in S) to make more and more of the boundary adopt a curved shape. This is observed (Gittus, 1975), and in the earlier literature led to the idea that the formation of "bulbous" structures was a prerequisite of superplasticity.

2.6. *Magnitude of the threshold stress*

The threshold stress due to the presence of ledges on the dislocation boundary is one source of an estimate of the value of σ_0. To traverse a ledged boundary the dislocations will have to bend. Moreover, there will be a change in fault energy of magnitude Γwh per intersection, where h is the height of a ledge, when the narrow face of a ledge is a disordered boundary. The applied stress will have to do sufficient work to overcome this energy in order to detach a dislocation from a ledge. Garmong and Rhodes (1975) saw evidence of these interactions between HPB dislocations and ledges in their transmission electron-microscope studies of Ni_3Al–Ni_3Nb composites. The interaction was manifested both as a local alteration of the line vector of the dislocation in certain circumstances and as a change in the response of the dislocation image to image contrast experiments. They remark that the presence of ledges will modify mechanical properties, such as creep, in several ways. Thus Hirth (1972) has discussed the generation of matrix dislocations at interfacial defects. There will be a simultaneous production of boundary dislocations in order to conserve Burgers vector. Garmong (1974) had observed this precise mechanism in operation in a Ni–W alloy of eutectic composition. For such a mechanism to operate the boundary dislocations produced must be glissile in the plane of the boundary, otherwise the ledge-source would not operate. Boundary ledges, by pinning the glissile boundary dislocations, could then stop the source from operating until the external stress was raised to a level at which the pinning forces could be overcome. The pinning of glissile boundary dislocations would in this manner reduce the ability of the boundary to creep, as Garmong and Rhodes pointed out. The ledges could be viewed as "keying" the boundaries, but the boundary-locking effect would arise not from physical interlocking but from the inhibition of boundary dislocation glide. They noted that in many respects the pinning of boundary dislocations by ledges, which they had observed in their work, resembles the pinning produced by boundary precipitates, illustrated by Ashby (1972, Fig. 17). Garmong and Rhodes pointed out that in cases where the narrow edges of ledges are themselves dislocation boundaries, misfit dislocations lying in the narrow face could attract misfit dislocations lying on the contacting broad face on the ledge. This is clearly a supplementary source of ledge-dislocation pinning. Aaronson *et al.* (1970) support the idea that a dislocation structure can develop at the narrow faces of ledges, and they point to the slow growth of certain ledges during recrystallization as evidence of the occasional presence of this kind of structure instead of the expected disordered boundary.

A suggested model of the ledge-dislocation network is a two-dimensional dislocation network. In proposing this we are in effect following the suggestion of Garmong and Rhodes. In order to detach itself from two ledges a boundary dislocation may be forced to adopt a curvature C as high as $2/L$, where L is the width of the broad face of a ledge (i.e. the distance between adjacent narrow faces). If the boundary defect is regarded as a flexible line whose energy per unit length is E, a potential

$$\frac{E\Omega C}{b} \tag{2.12}$$

or

$$\sim \frac{2E\Omega}{bL} \tag{2.13}$$

is required to do this (Ashby, 1972; Bardeen and Herring, 1952; Thomson and Balluffi,1962). Here Ω is the atomic volume. This then gives as an estimate of the threshold stress due to the interaction of glissile HPB superdislocations and ledges

$$\sigma_0 \leqslant \frac{2E}{bL}. \tag{2.14}$$

Precipitates lying in the boundary are another source of a threshold stress. A similar equation to (2.14) then applies but with L equal to the spacing between the precipitate particles. Solute drag may also tend to restrict the movement of the boundary superdislocations. However, this will reduce the rate of glide at all stresses rather than introduce a threshold stress.

2.7. *Comparison with the results of experiment*

Ball and Hutchison (1969) found that an equation algebraically identical with (2.11) fitted their data for the superplastic flow in an Al–Zn alloy of eutectoid composition. The factor which in our equation is $4(d/\lambda)^2$ had a value of approximately 100 for their alloy, and so $\lambda = 0.2d$, which compares with the value of $0.3d$ calculated by Friedel. They did not find any evidence of a threshold stress.

Quite generally our theoretical equation does appear to provide a reasonable fit to data with λ/d in the right order of magnitude (Gittus, 1977a, b, 1978). Geckinli and Barrett (1976) (for example) in their experimental studies of the Pb–Sn eutectic concluded that at room temperature the threshold stress was 1.3 MPa. From (2.14) the value of L is then of the order of 4×10^{-7} m (4000 Å). This is of the order of magnitude observed on dislocation boundaries. That the ledges do impede the movement of boundary dislocations is established by the electron microscopy reported by Garmong and Rhodes (1975). These workers calculated that in their Ni_3Al–Ni_3Nb composites of eutectic composition the pinning stress acting to pin the HPB dislocations to the ledges would, theoretically, be $\mu/70$. The threshold stress for dislocation movement is expected to be about h/L times this value, for the applied stress acts on the entire unpinned length of dislocation (minimum length L) while the pinning stress only acts on the length of dislocation near to the ledge (height h). So

$$\sigma_0 \sim \frac{\mu h}{70L}. \tag{2.15}$$

If this equation, derived for a different material, nevertheless holds good for a Pb–Sn alloy of eutectic composition then the observed value of the threshold stress in that material would imply that the ledges on the dislocation boundaries are about 100 times as long as they are high. This again is in reasonable accord with experiment.

For the case where the applied stress is much above the threshold stress, (2.11) predicts that the strain rate will depend on the square of the stress. In the literature on superplastic flow the reciprocal of the exponent of stress (m) is often quoted. In the case of a material conforming to (2.11), then $m = 0.5$. In the event values in the range 0.3–0.7, with a mean near to 0.5, are commonly reported as we shall see later. It is particularly to be noted that a value of $m = 1.0$, predicted and observed in the Herring–Nabarro and Coble types of diffusion creep, is rarely if ever found in superplastic materials. The characteristic of these models of diffusion creep, of course, is the assumption that the boundaries are perfect sources and sinks for point defects.

2.8. *The diffusion coefficient*

Baudelet with co-workers (Herriot *et al.*, 1976; Suery and Baudelet, 1975) found, in their studies of the two-phase Cu–P alloys, that the activation energy for superplastic flow, calculated with no endeavour to correct for change in structure, was close to that for grain boundary diffusion in copper. When, however, allowance was made for changes in structure, the activation energy for flow rose to a value quite close to that for *volume* self-diffusion in copper. On the face of it this would seem, then, to run contrary to our theory. However, Perinet (1975) and Bondy *et al.* (1971) have found that a high value of the activation energy is characteristic of self-diffusion at heterophase (as opposed to grain) boundaries. Thus Perinet found that the activation energy for diffusion in the boundary between copper and silver was 272.1 kJ/mol (65 kcal/mol) and this is more than twice the grain boundary activation energy for either of the pure elements by itself. Again, a similar finding was reported for the diffusion of iron and silver in the interface between these two metals. These findings clearly accord, on the present theory, with the high value of activation energy for superplastic flow at constant structure found by Baudelet and his collaborators. Indeed, this may be a general feature of superplastic flow in two-phase materials. It may have passed unnoticed in some investigations because care has not always been taken to compensate for the effect of changing structure when calculating the activation energy for superplastic flow. When Baudelet omitted this precaution he obtained a spuriously low activation energy which is, coincidentally it now appears, similar to that for grain boundary diffusion (as opposed to phase boundary diffusion) in one of the pure components of which his Cu–P alloy was composed.

The theory predicts that the strain rate will depend on the square of the particle size. We shall see that this is borne out in the case of the data for the superplastic flow of a number of materials.

3. Dislocation Creep in Superplastic Alloys

As the stress acting on a superplastic material is increased, dislocations, often arranged in cell boundaries, begin to appear in the grains or phase particles. Creep now occurs by the climb and glide of these dislocations as well as by grain boundary processes. Now the ductility is characteristically much lower and this is usually attributed to the associated fall in m. We shall see, however, that the fundamental reason for high ductility in the superplastic regime is a thermodynamic one.

3.1. *Model of the dislocation creep process*

Experiments show that characteristic networks of dislocations tend to form during creep. The networks outline cells within which there are relatively few free dislocations.

As the networks do not exert significant long-range stresses, most of the free dislocations within the cells are unaffected by the stresses produced by dislocations in the cell walls

(Friedel, 1964). They are still, however, subject to the internal stress produced by line tension. Thus the stress acting to cause glide of the mobile (within cell) dislocations is the difference between the applied stress σ and the internal stress $C_1 \mu b/x'$, where

$$1 \geqslant C_1 \geqslant 1/(2\pi). \tag{3.1}$$

Here x' is the mean distance between adjacent dislocations and is therefore a sub-multiple of the cell diameter L.

Holt (1970) has reviewed data on the cell structure and has advanced a theory relating the cell diameter to the mean distance between dislocations. Both the data and his theory indicate that L is of the order of the cell diameter and that the latter is proportional to the mean spacing ($\rho^{-1/2}$) of the dislocations, imagining them to be uniformly spread out throughout the volume of the material, i.e.

$$L = K\rho^{-1/2}. \tag{3.2}$$

Here ρ is the dislocation density and K a constant. For copper, deformed at temperatures between 25° and 700°C under shear stresses of from 10.69 MN/m^2 (at the highest temperature) to 78.97 MN/m^2 (at the lowest temperature), Staker and Holt (1972) later found that K had a mean value of 16. Other workers have reported values of this order of magnitude for other materials, including, for example, iron and aluminium (Gittus, 1975). We shall therefore make $x' = L$, justifying this on the grounds that this is the tendency indicated by metallography. The value of C_1 depends on the network geometry. In the one extreme case where the spacing of dislocations in the cell walls equals the cell diameter (i.e. the Frank network), the internal stress is predominantly that exerted on each dislocation by its neighbours. C_1 is then theoretically close to the lower bound set by (3.1) above. At the other extreme, where each cell contains a single dislocation which does not penetrate a cell wall but leaves lines of dislocations on the side walls as it advances, the line tension effect is the more important and in that case C_1 is near to the upper bound, i.e. close to unity. In this work, then, we take C_1 equal to unity.

The balance between the internal and external stresses can be written

$$F = \sigma - C_1 \mu b/x' \tag{3.3}$$

which for the case of cells each of which contains one mobile dislocation, and for which separation r_b of dislocations in the cell wall is much smaller than the cell diameter, becomes by the above arguments.

$$F = \sigma - \mu b/L. \tag{3.4}$$

Here F is the net stress causing the mobile dislocation to glide.

For the value of the friction stress F we consider the non-conservative motion of vacancy-adsorbing and vacancy-emitting jogs in the screw components of glide dislocations which exert a retarding effect on glide. We reviewed the models of steady-state creep which invoked this mechanism in Section 2.8 and in the case where the local vacancy concentrations are predominantly controlled by the rates at which vacancies can diffuse to or from the jogs, we then obtain for the jog velocity

$$v = \frac{b \sinh(\alpha F)}{\alpha A}, \tag{3.5}$$

which at low stresses reduces to

$$\alpha F \to 0: \quad V = \frac{Fb}{A}. \tag{3.6}$$

In these equations, following Barrett and Nix (1965),

$$\alpha = \frac{b^2 l}{kT}, \tag{3.7}$$

$$A = \frac{b^2}{20\,\alpha D_v}, \tag{3.8}$$

where, as before, b is Burgers vector, k is Boltzmann's constant, T is the absolute temperature, D_v is the volume self-diffusion coefficient, μ is the elastic modulus, and c_j is the thermal jog concentration.

The relationship between the smeared dislocation density ρ and the spacing of the cell boundary dislocations (which make up the majority of those present, if there is only a single mobile dislocation per cell) is given, to a reasonable approximation, by the expression

$$Lr_b = 1/\rho \tag{3.9}$$

whilst the expression suggested by Holt (1970), and shown by him and later workers to give a good representation of data, is

$$L = K/\sqrt{\rho}. \tag{3.10}$$

Eliminating ρ between (3.9) and (3.10) gives

$$L = K^2 r_b, \tag{3.11}$$

and substituting (3.11) in (3.4) gives

$$F = \sigma - \mu b/(K^2 r_b). \tag{3.12}$$

Our scheme is to calculate the mobile dislocation density (i.e. that of the individual mobile dislocations, one of which we assumed above to be located inside each of the cells) and the velocity v of these mobile dislocations. The mobile dislocations are on average one cell diameter apart. So the mobile dislocation density ρ_m is of order

$$\rho_m = 1/L^2. \tag{3.13}$$

The steady-state creep rate $\dot{\varepsilon}$ is then given by

$$\dot{\varepsilon} = bv\rho_m = bv/L^2. \tag{3.14}$$

Not only is glide resisted by the internal stress, it is also resisted by the friction of the screw jogs. The velocity of a dislocation subjected to a net stress F is given by (3.5). Using (3.14) and (3.5) to define a value for L and substituting this into (3.4), we find

$$\sigma - F = \mu \sqrt{\left(\frac{\alpha A \dot{\varepsilon}}{\sinh(\alpha F)}\right)}. \tag{3.15}$$

Consider now the recovery process which occurs in the cell boundary dislocation network. In many cases this network is three-dimensional and the cell boundaries have a small but finite thickness. We confine attention to such cases. As in the earlier Sections and following Friedel (1964) we consider the case in which the dislocations climb to shorten their total length and in this way reduce the energy of the network. The energy E for a dislocation in a cell boundary is of order $\mu b^2/2$ and the force causing climb is then $\mu b^2/2r_b$. The velocity of climb under this force is then given by the equation

$$\left(\frac{dr_b}{dt}\right)_{\text{recovery}} = \frac{K_1}{r_b} \tag{3.16}$$

with

$$K_1 = \frac{D_v \mu b^3 c_j}{2kT}. \tag{3.17}$$

Let us write the creep equation in the form

$$\dot{\varepsilon} = \dot{\rho} bL. \tag{3.18}$$

In this equation mobile dislocations are entering the boundary network at a frequency $\dot{\rho}$ after gliding across the cell diameter L. So

$$d\varepsilon = d\rho . bL. \tag{3.19}$$

Eliminating L between (3.9) and (3.10) we obtain

$$\rho = \frac{1}{K^2 r_b^2} \tag{3.20}$$

so that

$$d\rho = -\frac{2r_b^{-3}}{K^2} dr_b. \tag{3.21}$$

Using (3.11) to substitute for L in (3.19), and (3.21) to substitute for $d\rho$, we obtain the following equation:

$$\left(\frac{d\varepsilon}{dr_b}\right)_{\text{straining}} = -\frac{2b}{r_b^2}. \tag{3.22}$$

At an equilibrium condition, with which we are concerned, straining must generate a length of dislocation exactly equal to the length which has been eliminated, during the straining process, by the simultaneous climb controlled coarsening of the boundary network:

$$dr_b = \left(\frac{dr_b}{dt}\right)_{\text{recovery}} dt + \left(\frac{dr_b}{d\varepsilon}\right)_{\text{straining}} d\varepsilon, \tag{3.23}$$

but $dr_b/dt = 0$ when $\dot{\varepsilon} =$ constant, so

$$\dot{\varepsilon} = -\frac{(dr_b/dt)_{\text{recovery}}}{(dr_b/d\varepsilon)_{\text{straining}}}. \tag{3.24}$$

Using (3.16) and (3.22) in (3.24) we obtain

$$\dot{\varepsilon} = \frac{2K_1 b}{r_b^3}. \tag{3.25}$$

Using (3.12) to define r_b in (3.25), we find

$$\dot{\varepsilon} = 2K_1 b \left[\frac{K^2(\sigma - F)}{\mu b}\right]^3. \tag{3.26}$$

Rearranging (3.26) and using (3.15) to substitute for $(\sigma - F)$,

$$\left[\frac{\dot{\varepsilon}}{2K_1 b}\right]^{1/3} \frac{b}{K^2} = \left[\frac{\alpha A \dot{\varepsilon}}{\sinh(\alpha F)}\right]^{1/2}, \tag{3.27}$$

i.e.

$$F = \frac{1}{\alpha} \sinh^{-1}\left[\alpha A \dot{\varepsilon}^{1/3} (2K_1 b)^{2/3} K^4/b^2\right]. \tag{3.28}$$

Let

$$B = c_j \frac{D_v b}{\mu^2 kT}, \tag{3.29}$$

and then from (3.17)

$$2K_1 = \mu^3 b^2 B. \tag{3.30}$$

From (3.28) and (3.30)

$$F = \frac{1}{\alpha} \sinh^{-1}\left[\left(\frac{\dot{\varepsilon}}{B}\right)^{1/3} \mu^2 \alpha A B K^4\right], \tag{3.31}$$

From (3.26) and (3.30)

$$\sigma - F = \left(\frac{\dot{\varepsilon}}{B}\right)^{1/3} \frac{1}{K^2}. \tag{3.32}$$

From (3.31) and (3.32), finally,

$$\sigma = \frac{1}{K^2}\left(\frac{\dot{\varepsilon}}{B}\right)^{1/3} + \frac{1}{\alpha}\sinh^{-1}\left[\left(\frac{\dot{\varepsilon}}{B}\right)^{1/3} \mu^2 \alpha A B K^4\right], \tag{3.33}$$

which is the required result: the equation for dislocation creep.

3.2. *Thermodynamic calculation of K*

The approach to the calculation of K is a thermodynamic one: what is done is to calculate the value of K that minimizes the free energy of the creeping specimen. The free energy of the specimen comprises the elastic energy stored by Hookean deformation plus the energy of the dislocations. I have given the details of this calculation elsewhere (Gittus, 1978) showing that to a good approximation

$$K = 1.58\, c_j^{-0.4}. \tag{3.34}$$

4. Origin of the High Ductility in the Superplastic State

Above we showed how, during dislocation creep, the dislocations frequently arrange themselves in cells. The cell size rises if the stress causing creep is reduced and eventually the cell size becomes equal to the grain size. Significantly, when this structure is attained it appears that superplastic flow then often becomes the main deformation mechanism, and in the present Section we offer an explanation of the associated high ductility. The free energy of a specimen undergoing dislocation creep has been calculated and proves to be generally much higher than that of a specimen deforming superplastically at the same rate. Only at high strain rates does the situation reverse. Accordingly the superplastic condition, once established, persists, and neck propagation is strongly resisted until a late stage, by which time large elongations will have occurred.

Figure 4.1(a) and (b) have been prepared to illustrate the comparison that we are going to undertake. Consider Fig. 4.1(b): at low strain rates the material is superplastic and the stress

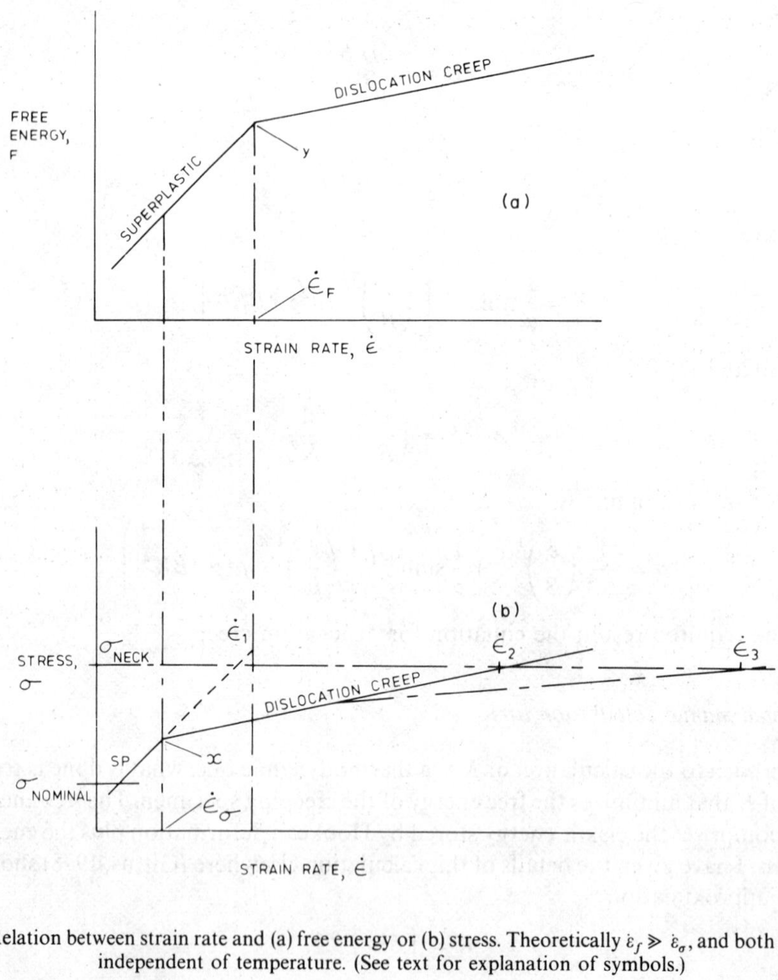

FIG. 4.1. Relation between strain rate and (a) free energy or (b) stress. Theoretically $\dot{\varepsilon}_f \gg \dot{\varepsilon}_\sigma$, and both are virtually independent of temperature. (See text for explanation of symbols.)

needed to deform it rises with increasing strain rate. At high strain rates it deforms predominantly by dislocation creep, and the strain rate dependence of applied stress is now characteristic of a material containing a dislocation cell structure. At point x the stress needed to deform the material is the same for both deformation mechanisms. The strain rate at which this equality exists is here termed $\dot{\varepsilon}_\sigma$: in effect point x is a point on a deformation map where it would mark the boundary between the stress field where dislocation creep is held to dominate and the region of lower stresses where superplasticity is said to cause most of the strain. Figure 4.1 (a) is similar except that here we plot free energy instead of stress. Then we can show that

$$\frac{\dot{\varepsilon}_F}{\dot{\varepsilon}_\sigma} = \left[1 + \left(\frac{K}{K'}\right)^2\right]^3, \tag{4.1}$$

where

$$K' = 1 + K/2. \tag{4.2}$$

We can more easily discuss the implications of (4.1) if we substitute values for K and K'. For Zircaloy at 900 K (for example) theoretical values are

$$K = 20.25 \tag{4.3}$$

and

$$K' = 11.12. \tag{4.4}$$

Using these values in (4.1), then

$$\frac{\dot{\varepsilon}_F}{\dot{\varepsilon}_\sigma} = 80.24. \tag{4.5}$$

More generally, for large K, (4.1) and (4.2) give

$$\lim_{K \to \infty} \frac{\dot{\varepsilon}_F}{\dot{\varepsilon}_\sigma} = 125. \tag{4.6}$$

What this means is that the transition to dislocation creep occurs at much higher strain rates than we would have expected. Accordingly a neck (thin region of the specimen), which locally increases the stress, will not necessarily cause dislocation creep with its (relatively) higher creep rate. Propagation of the neck is in this way resisted.

To clarify this point, consider again Fig. 4.1. Here we have marked two stresses: σ_{nominal} is the nominal stress (force divided by uniform cross-section area) causing superplastic flow in a specimen. The specimen contains a neck which locally raises the stress to σ_{neck} and should, according to the bold lines relating σ with $\dot{\varepsilon}$, produce dislocation creep at the high rate $\dot{\varepsilon}_2$. Rapid straining of the necked region would then cause the neck to become more pronounced and eventually lead to fracture. However, σ_{neck}, although a high stress, is not high enough to cause a transition to dislocation creep when we consider the free energies involved (Fig. 4.1(a)) and so the creep rate at the neck is only $\dot{\varepsilon}(\ll \dot{\varepsilon}_2)$ in Fig. 4.1(b). The neck propagates much more slowly and the ductility remains high as a consequence.

5. Superplastic Materials

Having given theories for the mechanisms of deformation and the origin of ductility in superplastic alloys we now summarize the data that have been accumulated during experimental studies of superplasticity.

5.1. *Ag–Cu: silver–copper*

Lee (1973) has examined the strength of the Ag–Cu eutectic for equiaxed and lamellar structures over a range of grain sizes, lamellar spacings, and temperatures. In the equiaxed structure at elevated temperatures the finer the grain size the weaker the material. In very-fine-grained material superplasticity becomes evident at elevated temperatures.

5.2. *Al–Cu: aluminium–copper*

One of the most careful studies of the earlier superplastic aluminium alloys is the Al–33 %Cu eutectic which can be made to yield tensile elongations of up to 2000 % at temperatures of about 500° C (Grimes *et al.*, 1976). The very-fine-grained structures are stabilized by the fine-scale, uniform distribution of $CuAl_2$ particles that results from eutectic solidification followed by hot working.

A detailed study of the superplastic properties of the Al–Cu eutectic alloy was made by Holt and Backofen (1966) and Johnson (1970). Hot working at temperatures near the melting point produces the necessary fine structure: an equiaxed aggregate of aluminium solid solution and $CuAl_2$ which is superplastic (400 % strain at fracture) for strain rates of 0.01 s^{-1} at 520° C (0.9 Tm). As with many other superplastic materials, the relationship between stress and strain rate for this alloy when plotted logarithmically appears to be sigmoidal, with the region of high strain rate sensitivity (SRS) bounded by regions of low SRS. We have already suggested that the return to low SRS at small stresses is accounted for by a back stress which has its origin in structural obstacles to grain boundary sliding, such as ledges. An approximate value of this threshold stress is indicated by the grain size dependence of the flow stress which has an intercept of 0.414 MPa at zero grain size and low strain rates. Holt and Backofen found that the strain rate was inversely proportional to the square of the grain size.

5.3. *Al–Cu–Zr: aluminium–copper–zirconium*

Grimes *et al.* (1976) have described the properties of an Al–Cu–Zr alloy. It was known that zirconium can act as a grain refiner during the casting of aluminium and also that it can have a marked effect in inhibiting the recrystallization of aluminium alloys. Zirconium-containing particles precipitated from solid solution in aluminium only coarsen very slowly at 500°C. and so it was thought likely that it would be possible to achieve reasonable grain stability and particle stability at the superplastic forming temperature. Copper was considered an essential alloying addition to give adequate service strength. It also appears that copper plays an important role in producing superplasticity, lowering the stacking fault energy of the alloy sufficiently for dynamic recrystallization to occur.

Typically elongations of 1000 % can be consistently obtained in the Al–Cu–Zr alloy, and under some conditions 2000 % has been produced. The *m*-value of this alloy is about 0.4–0.5, and at 480° C and a strain rate of $3 \times 10^{-3} s^{-1}$ the flow stress is about 9 MPa.

5.4. *Al–Zn–Zr and Al–Zn–Mg–Zr*

Zinc is one of the few alloying additions soluble in aluminium in significant amounts at or near to ambient temperature. It was therefore selected by Grimes *et al.* (1976) as a substitute for copper.

The Al–Zn–Zr alloys had good superplastic properties which increased with increasing zinc content. At 3 %, 6 %, and 10 % zinc (each with 0.5 % zirconium) elongations at fracture of 360 %, 580 % and 620 %, respectively, were obtained. The Al–Zn–Zr alloys are relatively weak and it was for this reason that magnesium was added. Al–Zn–Mg alloys are known to be relatively strong. Good superplastic elongations were obtained, an Al–5 %Zn–1 %Mg–0.5 %Zr alloy having an optimum elongation of 750 % and an Al–8 %Zn–1 %Mg–0.5 %Zr alloy having more than 1100 % elongation. In the equation $\sigma = K\dot{\varepsilon}^m$, where σ is the stress, $\dot{\varepsilon}$ is the strain rate, and K and m are constants, the Al–8 %Zn–1 %Mg–0.5 %Zr alloy had an m-value similar to that of the Al–6 %Cu–0.5 %Zr alloy (0.5–0.5 at 535°C). At a strain rate of $3 \times 10^{-3}\,s^{-1}$ the flow stress was about 7 MPa.

5.5. *Co–Al: cobalt–aluminium*

Cline (1967) studied the Co–Al eutectic. Fine-grained hot-worked material became superplastic at high temperatures with tensile elongations greater than 850 %.

5.6. *Cu–Al–Fe: aluminium-bronze*

Dunlop and Taplin (1972a, b) have studied the superplastic properties of the alloy CDA 619 (Cu–9.5 %Al–4 %Fe). There was a marked effect of both iron and aluminium content upon the strain rate at which m is a maximum (Fig. 5.1). The three-stage dependence of flow stress upon strain rate is shown in Fig. 5.2.

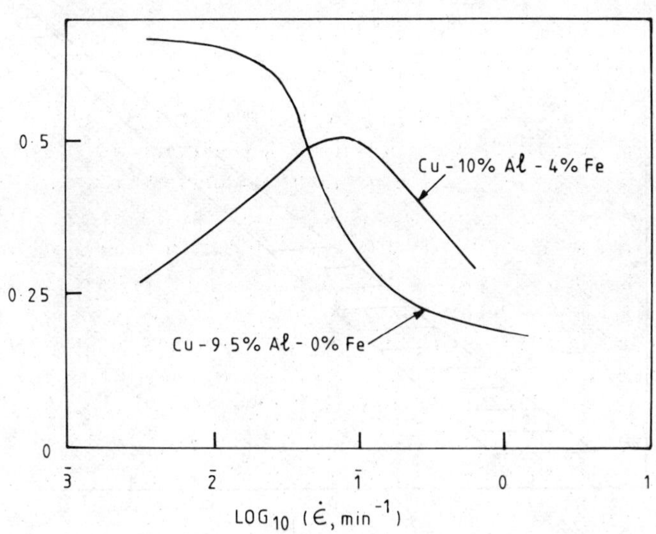

FIG. 5.1. Effects of aluminium and iron contents upon the relationship between strain rate $\dot{\varepsilon}$ and the strain rate sensitivity m. (After Dunlop and Taplin, 1972a, b.)

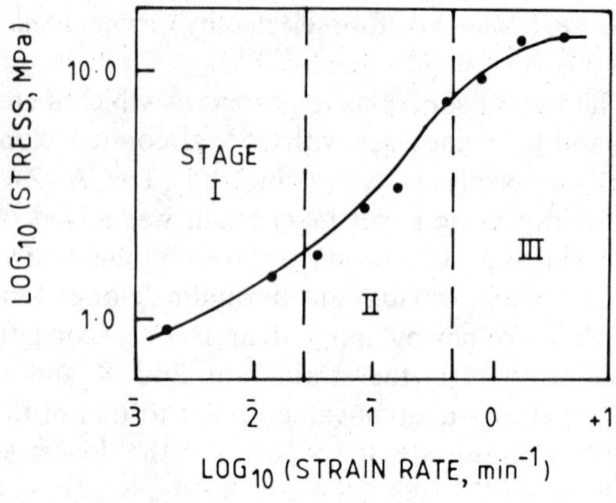

FIG. 5.2. Illustrating the three stages for deformation in aluminium–bronze containing 86.5 % Cu, 9.5 % Al, 4.0 % Fe. (After Dunlop and Taplin, 1972a, b.)

5.7. *Cu–P: copper–phosphorus*

The two different phases present in many superplastic alloys can play quite different roles. Thus texture studies on the alloys PbSn (Cutler and Edington, 1971), Al–Zn (Melton *et al.*, 1974), Cu–Al (Cutler *et al.*, 1974), and 60/40 brass (Suery *et al.*, 1974) have shown more rapid texture modification in one of the two phases, during superplastic deformation, than in the other. Herriot *et al.* (1976) made studies of the behaviour of the two phases in a Cu–P alloy. Figure 5.3 shows the correlation obtained by plotting the strain rate $\dot{\varepsilon}$ against the square of

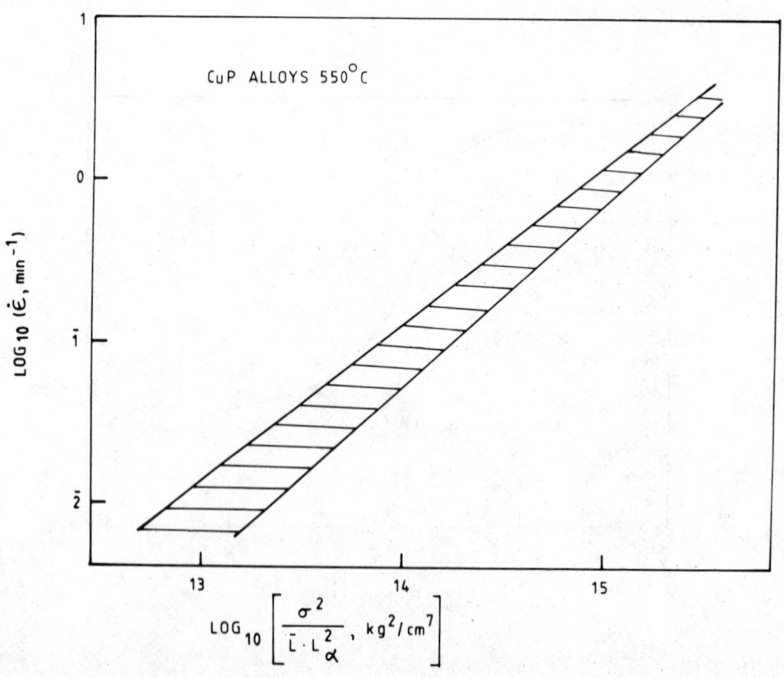

FIG. 5.3. Relationship between strain rate $\dot{\varepsilon}$ and "normalized stress", $\sigma^2/(\overline{L}.L_\alpha^2)$. (After Herriot *et al.*, 1976.)

the stress σ divided by the product of the mean intercept particle size L and the square of the alpha-particle size L_α. Irrespective of alloy composition on this plot, the data points conform to a straight line (logarithmic axes). An expression was derived for deformation by two thermally activated mechanisms operating in series: the results are consistent with a model in which the deformation of the softer (alpha) phase is rate controlling (Table 5.1).

Important studies on brass have been done by Suery (1979). In order to clarify the likely effect of the hard alpha-phase particles, he did model experiments. The ductile phase, beta, was simulated by a mixture of vaseline with a small amount of paraffin. Glass balls about 200–600 μm in diameter were used to simulate the hard alpha-phase. From the behaviour of the model material he arrived at an equation of the following type:

$$\dot{\varepsilon}_g(P) = f(P)\dot{\varepsilon}_m. \tag{5.1}$$

Here $\dot{\varepsilon}_m$ is the law of behaviour for a pure material and $f(P)$ is a function of the volumetric fraction of particles present. Figure 5.4 shows the variations on a semi-logarithmic scale of the function $f(P)$ deduced from these studies for extrusion and indentation of the mixture of vaseline, paraffin, and glass balls.

It was found for alloys A, B, and C (Table 5.2) that the coefficient n equals 2 ± 0.1, i.e. $m = 0.5$. In alloy D, which had the greatest proportion of the alpha-phase, n was approximately 2.5. The exponent of particle size can be determined for each of the alloys A, B, and C from the slope of the curve on logarithmic axes relating the quantity $\sigma^2/\dot{\varepsilon}_g$ to the quantity L_β. Inspection of these curves demonstrates that their slope is equal (and equal to 2)

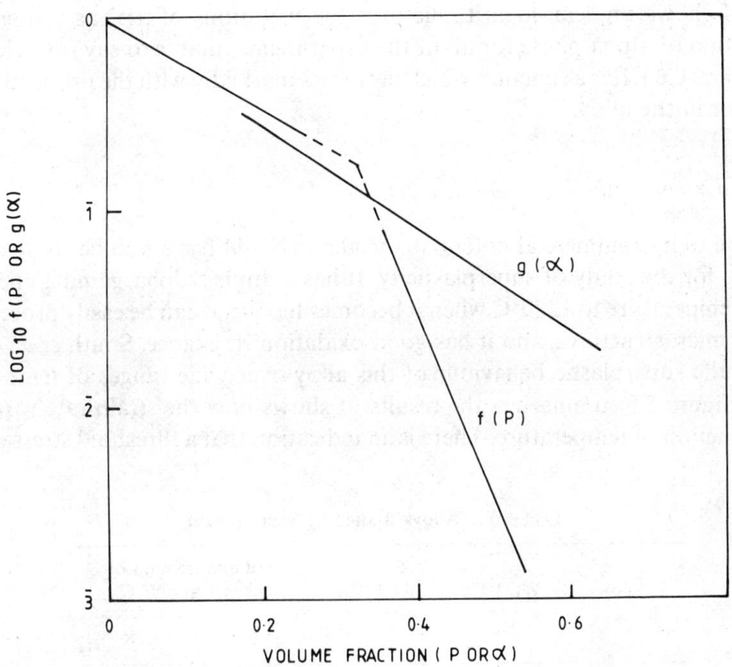

FIG. 5.4. Influence of the volume fraction P of glass balls upon the function $f(P)$ for the vaseline–paraffin mixture (broken line): influence of the volume fraction of alpha-phase (α) upon the function $g(\alpha)$ for brasses (unbroken line). The activation energy was taken to be 115 kJ mol^{-1}. (After Suery, 1979.)

TABLE 5.1. Structural conditions and test parameters studied by Herriot *et al.* (1976)

Composition (%P)	Temperature (°C)	Strain rate (min^{-1})	Strain (%)	Stress (kg/cm^2)	L_α (μm)	L_β (μm)
5	550	6.5×10^{-2} to 1.4	0 to 70	60 to 177	1.6 to 8.1	1.3 to 4.8
7	500 550 600	6.5×10^{-3} to 1.7	0 to 70	20 to 267	1.9 to 7.9	1.3 to 8.7
10	500 550 600	6.5×10^{-3} to 3.3	0 to 70	24.5 to 298	2.1 to 5.0	4.5 to 9.4

for all the alloys. The value of Q equals 115 ± 10 kJ $mol^{-1} K^{-1}$. Taking account of all the results the law for behaviour in global form is then written

$$\dot{\varepsilon}_g = g(\alpha)\left[\frac{\sigma}{L_\beta}\right]^2 \exp\left(-\frac{Q_\beta}{RT}\right), \tag{5.2}$$

where

$$Q_\beta = 115 \pm 10 \text{ kJ mol}^{-1}\text{ K}^{-1}. \tag{5.3}$$

Figure 5.4 shows on semi-logarithmic axes the variations of $g(\alpha)$ as a function of the volume fraction of alpha-phase for all of the experiments, that is to say for values of alpha between 0.2 and 0.63. It is a function which decreases markedly with the proportion of alpha-phase present in the alloy.

5.8. *Fe–Ni–Cr: microduplex stainless steel, IN 744*

In addition to its commercial potential, the alloy IN 744 has a number of features which make it ideal for the study of superplasticity. It has a duplex alpha/gamma microstructure from room temperature to 1280°C when it becomes ferritic, it can be easily processed to give fine-grained microstructures, and it has good oxidation resistance. Smith *et al.* (1976) have investigated the superplastic behaviour of this alloy over wide ranges of temperature and strain rate. Figure 5.5 summarizes the results. It shows how the strain rate varies with the stress as a function of temperature. There is an indication that a threshold stress exists below

TABLE 5.2. Alloys studied by Suery (1979)

Alloy	Wt %Cu	Wt %Zn	Volume fraction of phases at 600°C	
			α	β
A	58	42	0.24	0.76
B	59.4	40.6	0.455	0.545
C	60.6	39.4	0.63	0.37
D	60.85	39.15	0.71	0.29

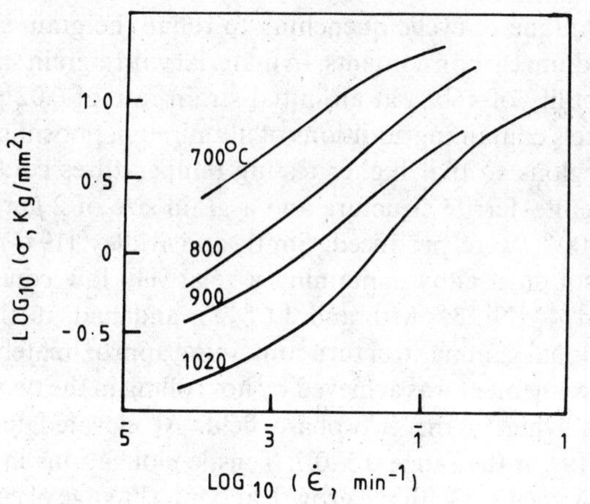

FIG. 5.5. Relationship between stress and strain rate for various temperatures. Alloy IN 744. Initial grain size is 5 μm.

which the creep rate becomes very low. In constant velocity straining tests the tensile elongation was substantial at all temperatures over appropriate ranges of strain rate. The maximum deformation at fracture exceeded 1000 % at 960°C. Figure 5.6 shows the effects of strain on strain rate sensitivity.

5.9. *Fe–Ni–Mo–Ti: microduplex steel*

Marder (1969) and Yoder and Weiss (1972) studied ferrite–cementite aggregates at 700°C. The structures which were produced by spheroidizing plain-carbon steels gave *m*-values of 0.4 or less. The elongations were rarely greater than 100 %. Superplastic behaviour in ferrite–austenite aggregates has been studied by Schadler (1968) and Morrison (1968).

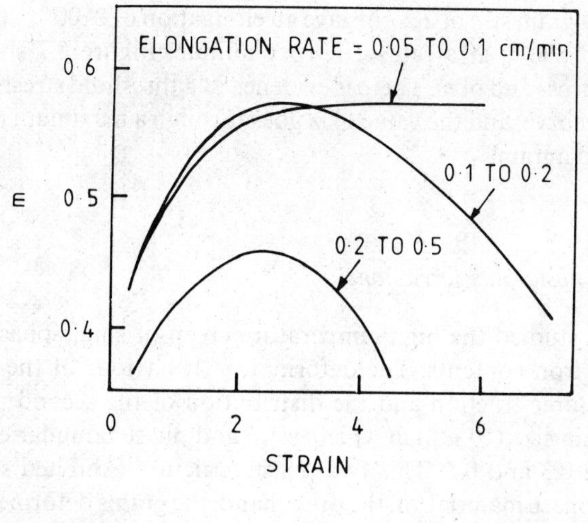

FIG. 5.6. Effect of strain and elongation rate upon the strain rate sensitivity *m* for alloy IN 744.

Schadler used the technique of cyclic quenching to refine the grain size of 2% manganese steels with low and medium carbon contents. In material with a grain size of 1.5 μm, m was 0.5 at 727°C. It has a ductility of 460% at an initial strain rate of 0.02 per minute. Morrison studied low carbon steels containing additions of aluminium, phosphorus, or silicon to raise the alpha + gamma regions so that higher testing temperatures could be used. At 900°C, material with an austenite–ferrite structure and a grain size of 2 μm had $m = 0.55$. Again, elongations of over 400% were produced. Smith and Ridley (1974) developed from first principles a superplastic iron alloy containing a relatively low concentration of alloying elements. It contained 4%Ni, 3%Mo, and 1.6%Ti, and had, in the temperature range 900–960°C, a duplex alpha–gamma structure containing approximately equal proportions of the two phases. Grain refinement was achieved by hot rolling in the two-phase field or by cold working and recrystallizing in the two-phase field. At elevated temperatures values of strain rate sensitivity lay in the range 0.5–0.7. Tensile elongations in excess of 800% were obtained. Failure was associated with the growth and interlinkage of cavities. A trace amount of boron added to this alloy led to a marked reduction in the extent of cavitation during deformation.

5.10. *Mg alloys*

Chapman and Wilson (1962–3) studied the influence of grain size upon the room temperature elongation of pure magnesium. When the grain size was less than 10 μm there was a sharp rise in tensile elongation.

An alloy of magnesium containing 6%Zn and 0.6%Zr is superplastic after double extrusion of fine-grained pellets to give a grain size of about 0.5 μm (Backofen *et al.*, 1968). Elongations exceeding 1000% between 270 and 310°C are possible with this material. At 310°C an elongation of 1700% was obtained. The apparent activation energy was 35 kcal/mol, slightly in excess of that for magnesium self-diffusion.

Lee (1969) has studied the nature of superplastic deformation in the Mg–Al eutectic. The material was extruded at 300°C; its composition was 67% Mg–balance Al. In tensile tests at 400°C material with a grain size of 2.2 μm gave an elongation of 2100% at an initial strain rate of 2 per minute and 830% at a rate of 20 per minute. Figure 5.7 shows the strain rate dependence of flow stress and of m. There is evidence of a threshold stress (sigmoidal shape of the stress–strain rate curve) and the value of m goes through a maximum in the stage II region of the deformation diagram.

5.11. *MgO–Fe: magnesia plus ferric ions*

Knoch (1975) has studied the high temperature creep of single-phase and of two-phase magnesia with high iron contents. The deformation behaviour of the two-phase material depended on the volume fraction and the distribution of the second phase. In two-phase material of small grain size (10 μm) in which grain and phase boundaries were identical, the m-value lay between 0.3 and 0.5. This two-phase structure exhibited superplastic flow. In coarse-grained two-phase material, on the other hand, the grains deformed in the same way as in single-phase material and superplasticity did not then occur.

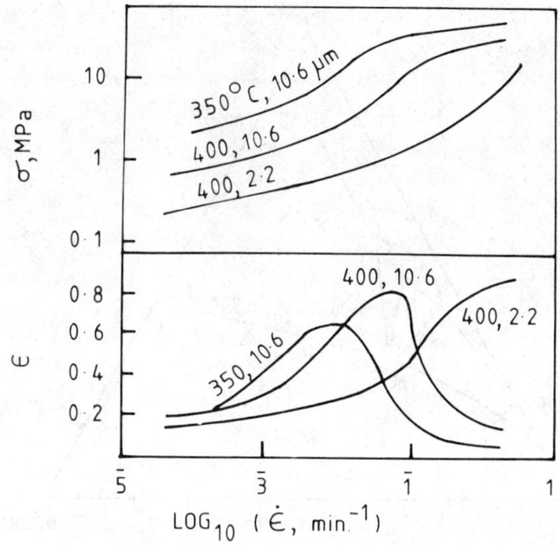

FIG. 5.7. Effect of strain rate upon flow stress σ and upon strain rate sensitivity *m* for Mg–Al eutectic alloy at 400°C and 350°C and for two grain sizes.

5.12. *Ni: pure nickel*

Floreen (1967) conducted a brief study to see if, given the proper test conditions, superplasticity could be achieved in nickel. The specimens were tested in a high-speed plastometer (GLEEBLE), in which the specimens are resistance heated at 300°C s^{-1}. After reaching the test temperature the specimens were strained at various speeds ranging from 0.1 to 300 in. per minute. There were critical values of test temperature and cross-head speed which led to maximum ductility. These presumably correspond to the conditions that lead to a fine stable recrystallized grain size.

5.13. *Ni–Fe–Cr: nickel–iron–chromium alloys*

Hayden *et al.* (1967) have studied superplasticity in the Ni–Fe–Cr system. Compositions which were studied have been plotted on ternary coordinates in Fig. 5.8. Each of the alloys also contained at least 0.6 % Ti. At each composition is written the elongation at a cross-head rate of 0.2 in. per minute. Figure 5.9 shows the variation of elongation and ultimate strength with the grain size of the gamma-phase in a 39 % Cr–8 % Fe–2 % Ti–1 % Al–balance Ni alloy. Over four orders of magnitude of strain rate the *m*-value was 0.5. No transition to an *m*-value of unity was detected even at the lowest strain rate studied and this is clear evidence that the normal Herring–Nabarro diffusion creep processes cannot account for superplastic behaviour. The apparent activation energy in the region of $m = 0.5$ was 60,000 cal/mol. This is comparable to the activation energy for volume diffusion in nickel and nickel alloys at similar temperatures. It was only when the strain was high enough to produce low *m*-values (dislocation creep) that dislocations were visible in the specimens.

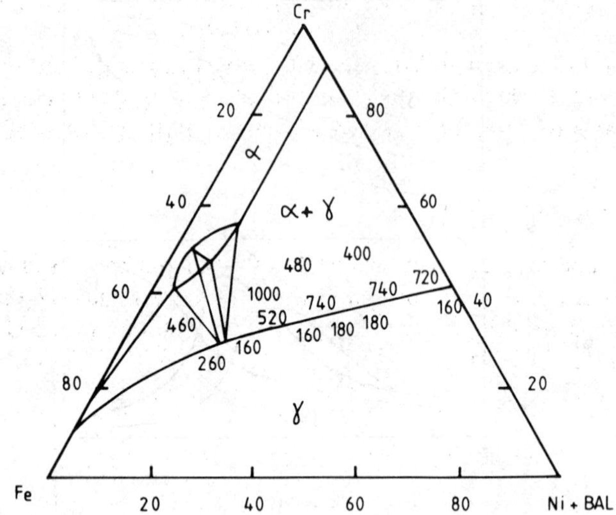

FIG. 5.8. Section through Fe–Cr–Ni phase diagram at 1000°C. Numbers are the elongations of alloys (containing in addition 0.6 % Ti). The tests were done at 0.1 mm/s elongation rate.

5.14. *Pb: pure lead*

Evidence for Newtonian flow in polycrystalline lead at room temperature and stresses below 1.7 MPa has existed for some time (Moore *et al.*, 1938; Greenwood and Cole, 1948). The flow rates are very small and this deformation process is evidently diffusion creep rather than superplastic flow. Gifkins (1958–9) found that direct application of the Nabarro–Herring and Coble models for diffusion creep produced a discrepancy of about four orders of magnitude between the experimental measurements and the calculated strain rates.

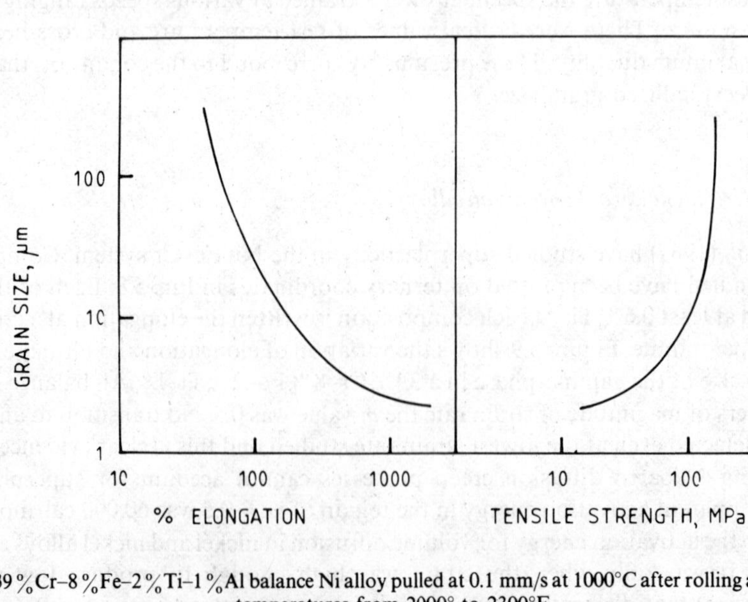

FIG. 5.9. 39 % Cr–8 % Fe–2 % Ti–1 % Al balance Ni alloy pulled at 0.1 mm/s at 1000°C after rolling and annealing at temperatures from 2000° to 2300°F.

5.15. *Pb–Cd: lead–cadmium alloys*

Sagat and Taplin (1976) have studied superplastic flow in a non-cavitating wrought Pb–Cd eutectic alloy. Figure 5.10 shows the variation of log (strain rate) with log (stress) at various strain levels. The average *m*-value was 0.34 with a standard deviation of 0.067.

5.16. *Pb–Tl: lead–tellurium alloys*

Gifkins (1977) made measurements at 290 K of grain boundary sliding in lead and a Pb–2.5 %Tl alloy over a wide range of grain sizes and of stress, including conditions which gave superplastic behaviour. Figure 5.11 is a logarithmic plot of average boundary-sliding rate, s/t *vs.* stress σ for two grain sizes.

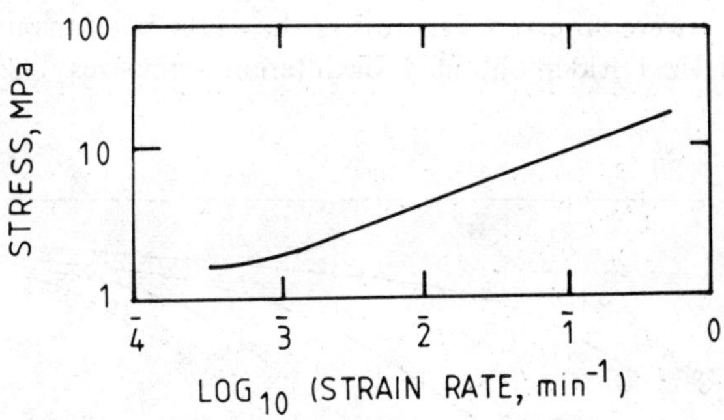

FIG. 5.10. Relationship between stress and strain rate for a lead–cadmium alloy, 100°C. Strains of up to 1.0.

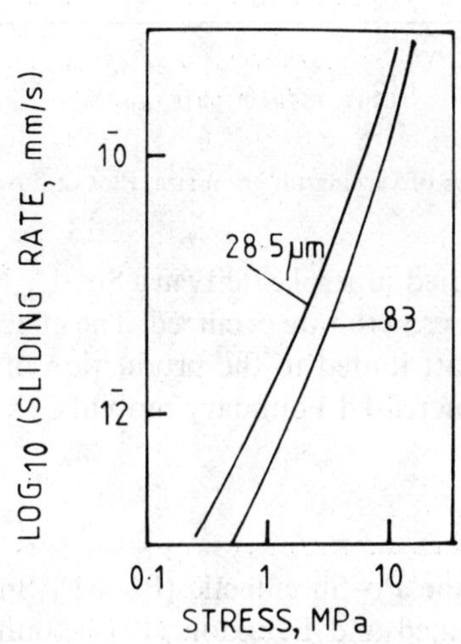

FIG. 5.11. Stress dependence of sliding rate for lead of two grain sizes.

5.17. *Sn: pure tin*

High strain rate sensitivity at room temperature ($m = 0.5$ at a strain rate of $2 \times 10^{-5}\,s^{-1}$) has been reported (Alden, 1967; Cline, 1967) for pure tin with a grain size of 7 μm. Grain growth during prolonged deformation precludes superplastic behaviour at this temperature. At coarser, more stable grain sizes, the strain rate sensitivity of tin is quite low and normal creep behaviour at these strain rates is exhibited (Cline, 1967; Breen and Weertman, 1955).

5.18. *Sn–Bi: tin–bismuth alloys*

Alden (1967) has studied superplasticity in the Sn–5%Bi alloy. He found that the flow stress and strain rate sensitivity of the flow stress are strong functions of strain rate and of the grain size of the tin. The log (stress)–log (strain rate) curve was S-shaped and the strain rate sensitivity in region II increased with decreasing grain size. In region II superplastic elongations of 1000% were observed. Figure 5.12 shows the relationship between log (strain rate) and log (stress) which Alden obtained for different grain sizes. These tests were done at room temperature.

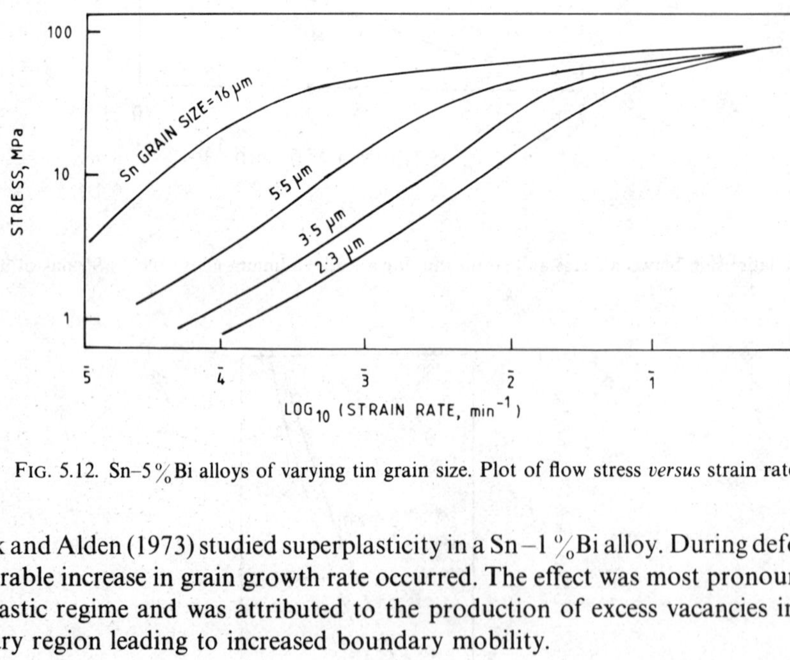

FIG. 5.12. Sn–5%Bi alloys of varying tin grain size. Plot of flow stress *versus* strain rate.

Clark and Alden (1973) studied superplasticity in a Sn–1%Bi alloy. During deformation a considerable increase in grain growth rate occurred. The effect was most pronounced in the superplastic regime and was attributed to the production of excess vacancies in the grain boundary region leading to increased boundary mobility.

5.19. *Sn–Pb: tin–lead eutectic*

Superplastic behaviour in the Pb–Sn eutectic (Pb–62%Sn) has been reported by many workers (Johnson, 1970). Ahmed and Langdon (1977) found an exceptional ductility of 4850%. The test had to be discontinued at this strain and the specimen had not failed. The material had a grain size of 6.9 μm and was pulled in tension at 413 K with an initial strain rate of $1.33 \times 10^{-4}\,s^{-1}$.

Vastava and Langdon (1979) investigated intercrystalline and interphase boundary sliding in the Pb–Sn eutectic. They found that the contribution of sliding to strain is 50 % or 60 % in region II. Sliding contributed about 20 % of the strain in regions I and III. The largest sliding offsets occurred at the Sn–Sn intercrystalline boundaries: smaller offsets were recorded at the tin HPBs and there was little or no sliding at the Pb–Pb intercrystalline boundaries.

5.20. *Ti: titanium*

In titanium (and some of its alloys) superplastic flow can be produced by stressing and by simultaneously cycling the temperature so that the material continuously changes from one metallurgical phase to another.

The strain per cycle increases linearly with applied stress independent of whether the stress is tensile (Greenwood and Johnson, 1965), torsion (Kot and Weiss, 1967), or compression (Oshida and Weiss, 1967).

5.21. *Ti–Al: titanium–aluminium alloys*

Superplasticity is exhibited by fine-structured two-phase Ti–Al alloys. One of these is an alloy of Ti: 5 %Al–2.5 %Sn. Lee and Backofen (1967) have studied this material. When heated rapidly into the two-phase region just below the beta-phase boundary (about 0.65Tm) the interparticle spacing varies between 6 and 24 μm depending on composition. *m*-values of about 0.8 are obtained associated with ductilities of more than 1000 %. At strain rates above 0.01 s^{-1}, *m* falls below 0.3 and there is no superplastic elongation. Both the flow stress and the strain rate sensitivity depend markedly upon temperature because temperature governs both the size and proportion of second phase in the two-phase structure. The *m*-value and the elongation have been correlated for temperatures between 25° and 1000°C (Lee, 1969), and in all the alloys the maximum strain rate sensitivity occurs at temperatures just below the phase boundary of the beta solid solution. The activation energy for the Ti: 5 %Al–2.5 %Sn alloy is about 50–65 kcal/mol (209–272 kJ/mol).

5.22. *U: uranium*

Thermal cycling can cause alpha-uranium to flow superplastically, producing elongations of more than 500 %. This behaviour has been achieved at 9.8 MPa with the temperature cycling between 400° and 600°C. An *m*-value of unity was obtained (Johnson and Sykes, 1966). If the stress exceeds 25 MPa, then the linear relationship between strain rate and applied stress no longer is observed. The stress largely responsible for the enhanced deformation rate is generated internally. Thus the thermal expansion of single crystals of alpha-uranium is anisotropic, and therefore temperature cycling produces changes in the shape of the individual alpha-uranium grains: changes which are resisted by surrounding grains with the consequent introduction of high internal stresses similar to those produced in other materials by temperature cycling through a phase change.

5.23. *W: tungsten alloys*

The tensile elongation of W–Re alloys increases with rhenium content above 15 wt %Re

(Garfinkle *et al.*, 1967). A W–30%Re alloy has exhibited 200% elongation. At 2000°C for alloys containing 33–35%Re an *m*-value of about 0.46 is observed. The composition of these superplastic alloys is close to the solubility limit of rhenium in tungsten indicating that small amounts of second phase are stabilizing a fine-grain size.

5.24. *Zn: pure zinc*

Superplastic effects are not well established in unalloyed polycrystalline zinc although elongations in fine-grained commercial electrolytic zinc as high as 400% are possible at room temperature (Chadwick, 1934). The requisite fine-grain structure can be produced under cold rolling conditions that ensure no temperature rise during deformation. In the case of zinc, which has been subjected to controlled rolling conditions, doubling the applied stress to 68.9 MPa increases the strain rate by a factor of 6 so that the *m*-value is 0.4.

5.25. *Zn–Al: zinc–aluminium alloys*

Ball and Hutchison (1969) studied superplasticity in the Al–Zn eutectoid alloy. On the basis of their results they proposed a theory of superplastic flow which contains many of the elements of modern theories. Figure 5.13 shows the dependence of stress upon strain rate for specimens of 2.5 μm grain size deformed at 150° and 250°C. An approximately linear relationship between log (stress) and the reciprocal of temperature was found, the slope being independent of grain size. The relationship between stress and grain size was linear at fixed temperature and strain rate. The stress depended upon the square root of the strain rate up to a limiting value of strain rate.

Mohamed *et al.* (1977) have studied factors that influence ductility in the superplastic Zn–Al eutectoid in some detail. Figure 5.14 summarizes some of their data relating

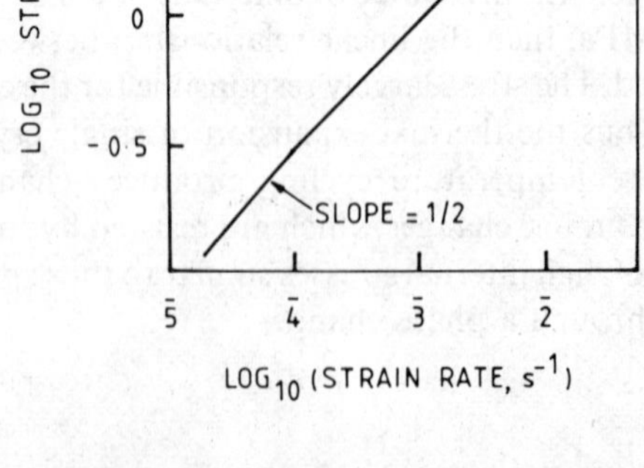

FIG. 5.13. Specimens of Al–Zn eutectoid of grain size 2.5 μm exhibit a quadratic dependence of strain rate upon stress.

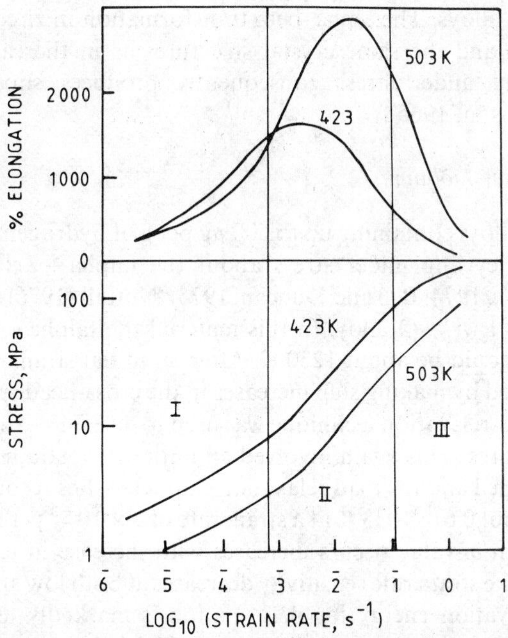

FIG. 5.14. Elongation, stress and strain rate for Zn–Al eutectoid of 2.5 μm grain size.

elongation and stress with strain rate: here the marked peak in ductility in region II is shown very clearly and they have obtained elongations of up to nearly 3000%. Metallographic examination of fracture specimens shows extensive cavitation after deformation in the low stress region I and significant cavitation after deformation to 2600% in the superplastic region II.

Naziri and Pearce (1974) have rendered the Zn–0.4%Al alloy superplastic. A fine grain size of 0.6 μm was produced by rolling a cast slab to 90% reduction at room temperature. At this temperature m was about 0.45 at an initial strain rate of 6.67×10^{-4} per second and elongations exceeding 500% were obtained.

Samuelsson *et al.* (1976) have studied the superplastic Zn–40%Al alloy and the dislocation structures in it. The material was hot extruded into bars and solution treated. It exhibits superplasticity in the temperature range 520–550 K with a maximum strain rate sensitivity of 0.48. Information on the role of grain boundary sliding derives from studies made by Melton *et al.* (1974), who studied textures in the superplastic Zn–40%Al alloy. They showed that slip occurs in both phases during superplastic flow even at maximum strain rate sensitivity. At all the strain rates that they investigated the zinc-rich phase deformed readily by slip, but large amounts of slip in the aluminium-rich phase only occurred at high non-superplastic strain rates. Grain boundary sliding reduced the overall level of texture after deformation at all strain rates. They concluded that superplastic flow in this alloy involves grain boundary sliding together with slip in both phases.

5.26. *Zr: zirconium*

The superplastic behaviour of zirconium (and zirconium alloys) is very similar to that of titanium and titanium alloys. The alpha–beta transformation in zirconium involves phases with similar strengths and the same crystal structures as in the case of titanium. Phase transformation cycling under stress consequently produces superplastic deformation (Greenwood and Johnson, 1965).

5.27. *Zr–Nb: zirconium–niobium*

In Zr–2.5 wt %Nb alloys containing up to 500 μg per g of hydrogen, superplasticity can be produced by thermal cycling under stress about the (alpha + ZrH_2) → alpha boundary (Nuttall and McCooeye, 1974; Bell and Duncan, 1975). Nuttall (1976) studied superplasticity in a Zr–2.6 wt %Nb–0.1 wt % O alloy. In this material the (alpha + beta) → beta transformation temperature should be about 1230 K. After an initial strain of 10 % the strain rate sensitivity was measured by making step increases in the cross-head speed by a factor of 2 or 2.5. In addition, a stress-relaxation technique was used to determine the strain rate sensitivity at some test temperatures. This again involved an initial 10 % strain after which the cross-head motion was stopped and the load relaxation characteristics recorded. The parameter m had a maximum of about 0.6 at 1033 K at a strain rate of 7×10^{-3} per minute. The strain rate at which the maximum m-value occurs increases with increase in test temperature. At all temperatures studied the strain rate sensitivity decreases at both low and high strain rates to a value of 0.2. The activation energy for deformation is markedly temperature dependent varying from 220 kJ/mol at lower temperatures to 92 kJ/mol at the higher temperatures studied. This is probably related to the variation in composition and volume fraction of the alpha- and beta-phases with temperature. The values of activation energy at low and high temperatures are not inconsistent with those found respectively for self-diffusion in alpha-zirconium, 188–270 kJ/mol, and in beta-zirconium at low temperatures, 75–100 kJ/mol in the temperature range 1130–1300 K. The interphase spacing after test was generally about 5 μm and there were approximately equal volume fractions of each phase dispersed fairly randomly.

5.28. *Zr–Sn: Zircaloy-4*

A typical composition of Zircaloy-4 is given in Table 5.3. The main alloying element is tin (about 1.5 %) with smaller amounts of iron and chromium. Bocek *et al.* (1976) have studied the superplastic behaviour of this material. Most of the investigations were performed in air between 750° and 1050°C at cross-head speeds ranging over three orders of magnitude. Superplastic behaviour was observed between 860° and 1000°C for a strain rate of the order

TABLE 5.3. Composition of Zircaloy-4

Element	Sn	Fe	Cr	Fe + Cr	Zr
Wt. %	1.43	0.21	0.11	0.32	Balance

Main impurities:

Element	C	HF	N	O	Pb	Si
ppm	170	46	30	980	< 50	56

of 10^{-4} per second. The strain rate sensitivity m depended upon strain. No effect of anisotropy upon the yield stress was observed above 550° C, but in the superplastic region m revealed strong dependence upon sample orientation. At higher temperatures the total elongation increased with temperature up to 850°C. The apparent activation energy for superplastic flow agreed well with the energy of grain boundary diffusion in alpha-zirconium. In Fig. 5.15 the stress dependence of strain rate is shown for Zircaloy-4 at various temperatures. In the high-temperature tests the usual sigmoidal dependence emerges. Bocek *et al.* also found convincing evidence that in the (alpha + beta) phase region, alpha-grain switching is the dominating deformation mode during superplastic flow at lower strains. Thus scanning micrographs were taken of the outer surface of internally pressurized Zircaloy-4 cladding tubes. The tubes were internally pressurized and distended by superplastic creep. To reveal the behaviour of individual grains during this ballooning or distension process, particular surface areas were provided with scratch markings. Besides simple grain translation, the grains rotated and grains from the interior appeared on the surface.

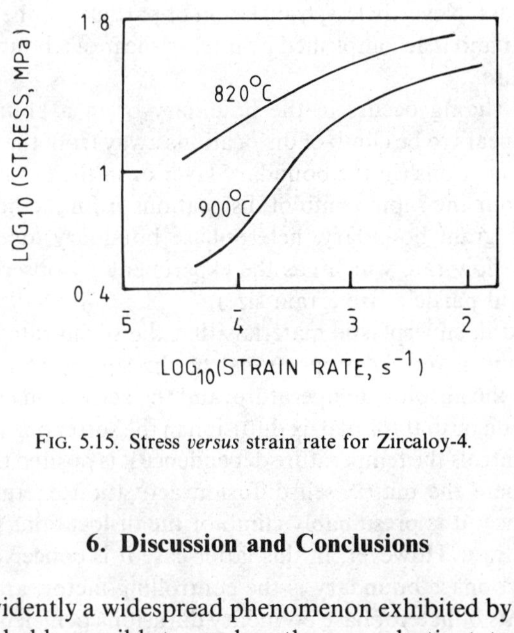

FIG. 5.15. Stress *versus* strain rate for Zircaloy-4.

6. Discussion and Conclusions

Superplasticity is evidently a widespread phenomenon exhibited by very many metals and alloys. Indeed, it is probably possible to produce the superplastic state in most common alloy systems. The prime requirement is a fine stable grain size. It can be produced in two main ways:

(1) by pinning the grain boundaries, for example, with a fine precipitate; and
(2) by having about 50 % of a second phase present in finely distributed form: the grain size at equilibrium is then approximately equal to the size of the second-phase particles.

Deformation in superplastic materials occurs by a mechanism which is quite distinct from the homogeneous straining process that characterizes normal flow. In the latter case, every grain exhibits the same strain as the polycrystal in which it is situated. Grains do not change partners during deformation by homogeneous flow. By contrast in the superplastic process the grains or particles of the alloy flow like grains of sand in an hour-glass. Individual grains make fluctuating changes in shape so as to prevent the formation of voids at the grain boundaries. But even after very large deformations the grains do not elongate and remain more or less in the same equiaxed state as at the beginning of deformation. Instead of deforming, the grains slide over one another, and it is this process of heterogeneous straining which causes the material to be in the same state after a large deformation as it was initially.

In many fine-grained single-phase superplastic alloys the deformation process is very simple. It seems to involve sliding at the grain boundaries by the movement of grain boundary dislocations. The formation of grain boundary voids is prevented by the climb of boundary dislocations on certain facets of the grains. Deposition of matter on to the climbing dislocations permits the small changes in shape that are needed to accommodate mismatch at the boundaries of grains that are translating relative to one another.

In some two-phase superplastic materials, particularly those in which one phase is very much harder than the other (e.g. brass), sliding at certain interfaces is observed to be difficult. Thus it does not occur significantly at the alpha–beta heterophase boundaries during the superplastic flow of brass. Nevertheless, translation of particles of the two phases relative to one another does occur and is accomplished by intense shear of a boundary layer about 1 μm thick in the softer phase.

Whether boundary sliding occurs in the boundary or in a boundary layer, the rate-controlling process appears to be climb of dislocations away from the heads of pile-ups. The pile-ups of gliding dislocations (in the boundary layer or in the boundary) concentrate the applied stress and favour the rapid climb of dislocations from the head of the pile-up into adjacent facets of the grain boundary, heterophase boundary, or boundary layer. This process gives rise to a flow rate which has the experimentally observed dependence upon stress, temperature, and particle size (grain size).

It is generally found in superplastic materials that the strain rate is proportional to the square of the stress and inversely proportional to the square of the grain size. It is an Arrhenius function of the absolute temperature, and the activation energy is either that for grain boundary diffusion or that for matrix diffusion in the softer phase. In cases where grain boundary diffusion controls the temperature dependence it is posited that climb occurs in the grain boundaries. Where the matrix self-diffusion activation energy seems to determine temperature dependence, it is presumably climb of the dislocations in the boundary layer that controls the flow rate. However, in this latter case it is conceivable that climb of the dislocations in a heterophase boundary is the controlling factor, and it is significant that those few measurements of heterophase boundary diffusion coefficient that have been made give values close to that for self-diffusion in the matrix of one of the contacting phases. This point awaits resolution.

Superplastic flow occurs at a lower stress than dislocation creep. For example, if a material is rendered superplastic by holding it in a two-phase field and stretched at a fixed rate, then the value of stress is lower than it would be if the temperature was raised slightly to render the material single phase. This feature has been observed in at least two alloy systems and it is predicted from theory.

Fracture, when it does finally occur in a superplastic material, does so either by necking or by the coalescence of grain boundary cavities. Necking and neck propagation are strongly resisted because of the low sensitivity of strain rate to applied stress. So initially present irregularities develop very slowly and it is only after large extensions that they become sufficiently prominent to cause terminal necking and failure. Grain boundary void formation, at least in materials having phases of similar hardness, is resisted by the very nature of the superplastic flow process which involves climb of boundary dislocations and consequently the deposition of matter on just those grain boundary regions where holes might otherwise be expected to form. However, in superplastic alloys containing a hard phase the difference in deformability between the two phases means that perfect accommodation of sliding-induced strains is not everywhere possible at all times, and so grain boundary voids do form. Many of these no doubt heal up again during the deformation process, but eventually a sufficient number coalesce to produce a critical flaw whose propagation initiates fracture.

Superplastic flow is not exhibited over the entire range of stresses and temperatures to which materials may be subjected. In almost every case if the stress is sufficiently low, flow virtually ceases: a threshold stress, perhaps. Again at high stresses dislocation creep begins to occur. Dislocations now appear in the grains or phase-particles. The grains deform homogeneously and most of the strain exhibited by the specimen is now attributable not to boundary sliding but to the uniform deformation, by dislocation creep of the individual grains. The strain rate now varies with the fourth or higher power of the applied stress. Accordingly, ductility drops precipitously and the material is no longer superplastic at these stress levels. Broadly speaking the coarser the grain size the lower the stress at which this transition to non-superplastic behaviour occurs.

References

Aaronson, H. I., Laird, C., and Kinsman, K. R. (1970) *Phase Transformations*, 313–396, American Society for Metals, Metals Park, Ohio.

Ahmed, M. M. I. and Langdon, T. G. (1977) Exceptional ductility in the superplastic Pb–62 % Sn eutectic, *Metall. Trans.* A, **8A**, 1832–1834.

Alden, T. H. (1967) The origin of superplasticity in the tin–5 % bismuth alloy, *Acta Metall.* **15**, 469–480.

Ashby, M. F. (1972) Boundary defects and atomistic aspects of boundary sliding and diffusional creep, *Surf. Sci.* **31**, 498–537.

Ashby, M. F. and Verrall, R. A. (1973) Diffusion accommodated flow and superplasticity, *Acta Metall.* **21**, 149–163.

Backofen, W. A., Azzarto, F. J., Murty, G. S., and Zehr, S. W. (1968) Superplasticity in ductility, paper presented at a seminar, Metals Park, Ohio, October 1967, 279–310, American Society for Metals, Metals Park, Ohio.

Ball, A. and Hutchison, M. M. (1969) Superplasticity in the aluminium–zinc eutectic, *Metal Sci. J.* **3**, 1–7.

Bardeen, J. and Herring, C. (1952) Diffusion in alloys and the Kirkendall effect, *Imperfections in Nearly Perfect Crystals, Symposium held at Pocono Manor, 12–14 October 1950*, edited by W. Shockley, 261–288, John Wiley, New York.

Barrett, C. R. and Nix, W. D. (1965) A model for steady state creep based on the motion of jogged screw dislocations, *Acta Metall.* **13**, 1247–1258.

Bell, L. G. and Duncan, R. G. (1975) *Hydride Orientation in Zr–2.5 % Nb*, AECL 5110, Atomic Energy of Canada Ltd.

Bocek, M., Faisst, G., and Peterson, C. (1976) Examination of the plastic properties of Zircaloy–4 at elevated temperatures in air atmosphere, *J. Nucl. Mater.* **62**, 26–36.

Bondy, A., Regnier, P., and Levy, V. (1971) Diffusion at interfaces, *Scr. Metall.* **5**, 345–350.

Breen, J. E. and Weertman, J. (1955) Creep of polycrystalline tin, *Trans. Am. Inst. Min. Metall. Engrs* **203**, 1230–1234.

Chadwick, R. (1934) *J. Inst. Metals* **54**, 131.

Chapman, J. A. and Wilson, D. V. (1962–3) The room temperature ductility of fine grain magnesium, *J. Inst. Metals* **91**, 39–40.

Clark, M. A. and Alden, T. H. (1973) Deformation enhanced grain growth in a superplastic Sn–1 %Bi alloy, *Acta Metall.* **21**, 1195–1206.

Cline, H. E. (1967) The mechanical properties of the CoAl–Co eutectic, *Trans. Metall. Soc. AIME* **239**, 710–714.

Cutler, C. P. and Edington, J. W. (1971) Textures in the superplastically deformed tin–lead eutectic alloy, *Metal Sci. J.* **5**, 201–205.
Cutler, C. P., Edington, J. W., Kallend, J. S., and Melton, K. N. (1974) Quantitative texture studies of the superplastically deformed aluminium–copper eutectic alloy, *Acta Met.* **22**, 665–671.
Dunlop, G. L. and Taplin, D. M. R. (1972a) The tensile properties of superplastic aluminium–bronze, *J. Mater. Sci.* **7**, 84–92.
Dunlop, G. L. and Taplin, D. M. R. (1972b) A metallographic study of superplasticity in a micrograin aluminium–bronze, *J. Mater. Sci.* **7**, 316–324.
Eberhardt, A. (1979) Ph.D. thesis, University of Metz, France.
Edington, J. W.(1976) Physical metallurgy of superplasticity, *Metall. Technol.* **3**, 138–151.
Floreeen, S. (1967) Superplasticity in pure nickel, *Scr. Metall.* **1**, 19–23.
Friedel, J. (1964) *Dislocations*, Pergamon Press, Oxford.
Garfinkle, M., Witzke, W. R. and Klopp, W. D. (1967) Metall. Soc. AIME, *Abstract Bulletin* **3**, 77.
Garmong, G. (1974) Science Centre, Rockwell International Corporation (unpublished work).
Garmong, G. and Rhodes, C. G. (1975) Interfacial ledge structures in Ni_3Al/Ni_3Nb eutectic composites, *Metall. Trans.* A, **6A**, 2209–2216.
Geckinli, A. E. and Barrett, C. R. (1976) Superplastic deformation of the lead–tin eutectic, *J. Mater. Sci.* **11**, 510–521.
Gifkins, R. C. (1958–9) Recrystallisation of lead during creep, *J. Inst. Metals* **87**, 255–261.
Gifkins, R. C. (1977) The effect of grain size and stress upon grain boundary sliding, *Metall. Trans.* A, **8A**, 1507–1516.
Gittus, J. H. (1975) *Creep, Viscoelasticity and Creep Fracture in Solids*, Applied Science Publishers, London.
Gittus, J. H. (1977a) Theory of superplastic flow in two-phase materials: roles of interphase boundary dislocation ledges and diffusion, *J. Engng Mater. Technol., Trans. ASME* **99**, 244–251.
Gittus, J. H. (1977b) Theoretical value of the ratio *K* of cell diameter to dislocation spacing for a material (gold) undergoing dislocation creep, *Phil. Mag.* **35**, 293–300.
Gittus, J. H. (1978) High temperature deformation of two phase structures, *Phil. Trans. R. Soc.* A, **288**, 121–146.
Gleiter, H. (1977) Private communication.
Greenwood, G. W. and Cole, J. (1948) Influence of various factors on the creep of lead, *Metallurgia* **37**, 285–289.
Greenwood, G. W. and Johnson, R. H. (1965) The deformation of metals under small stresses during phase transformations, *Proc. R. Soc. Lond.* A **283**, 403–422.
Grimes, R., Stowell, M. J., and Watts, B. M. (1976) Superplastic aluminium based alloys, *Metall. Technol.* **3**, 154–160.
Haasen, P. (1965) *Physical Metallurgy*, 867, North-Holland, Amsterdam.
Hall, E. O. (1951) The deformation and ageing of mild steel, II: Characteristics of the Lüders deformation; III: Discussion of results, *Proc. Phys. Soc.* B **64B**, 742–747, 747–753.
Hayden, H. W., Gibson, R. C., Merrick, H. F., and Brophy, J. H. (1967) Superplasticity in the Ni–Fe–Cr system, *Trans. Am. Soc. Metals* **60**, 3–14.
Herriot, G., Baudelet, B., and Jonas, J. J. (1976) Superplastic behaviour of two-phase copper–phosphorus alloys, *Acta Metall.* **24**, 687–694.
Hirth, J. P. (1972) The influence of grain boundaries on mechanical properties, *Metall. Trans.* **3**, 3047–3067.
Holt, D. L. (1970) Dislocation cell formation in metals. *J. Appl. Phys.* **41**, 3197–3201.
Holt, D. L. and Backofen, W. A. (1966) Superplasticity in the aluminium–copper eutectic alloy, *Trans. Am. Soc. Metals* **59**, 755–768.
Ishida, Y. and McLean, D. (1967) (Solid solution strengthening) Effect of nitrogen and manganese on recovery rate and friction stress during creep in iron, *J. Iron Steel Inst.* **205**, 88–93.
Johnson, R. H. (1970) Superplasticity, *Metall. Rev.* **15**, 115–134.
Johnson, R. H. and Sykes, E. C. (1966) The enhancement of ductility in alpha-uranium, *Nature* **209**, 192–193.
Knoch, H. (1975) *High Temperature Creep of Single Phase and of Two Phase MgO with High Iron Contents, ERAMS-I/TR26*, University of Erlangen-Nürnberg, West Germany.
Kot, R. and Weiss, V. (1967) Transformation plasticity of steels and titanium in torsion, *Trans. Am. Soc. Metals* **60**, 566.
Lee, D. (1969) The nature of superplastic deformation in the Mg–Al eutectic, *Acta Metall.* **17**, 1057–1069.
Lee, D. (1973) Strengthening and superplasticity in the Ag–Cu and Mg–Al eutectics, *J. Soc. Mater. Sci. Japan* **22**, 209–212.
Lee, D. and Backofen, W. A. (1967) Superplasticity in some titanium and zirconium alloys, *Trans. Metall. Soc. AIME* **239**, 1034–1040.
McLean, D. (1971) Grain boundary dislocations: their effect on vacancies and sliding, *Phil. Mag.* **23**, 467–472.
Marder, A. R. (1969) Effect of carbon content, test temperature and strain rate on the strain rate sensitivity of iron–carbon alloy, *Trans. Metall. Soc. AIME* **245**, 1337–1345.
Melton, K. N., Edington, J. W., Kallend, J. S., and Cutler, C. P. (1974) Textures in superplastic Zn–40 wt %Al, *Acta Metall.* **22**, 165–170.

Mohamed, F. A., Shei, S. A., and Langdon, T. G. (1975) The activation energies associated with superplastic flow, *Acta Metall.* **23,** 1443–1450.

Mohamed, F. A., Ahmed, M. M. I., and Langdon, T. G. (1977) Factors influencing ductility in the superplastic Zn–22%Al eutectoid, *Metall. Trans.* A, **8A,** 933–938.

Moore, H. F., Betty, B. B., and Dollins, C. W. (1938) *Univ. Ill. Bull.* **35,** 102.

Morrison, W. B. (1968) Superplasticity of low alloy steels, *Trans. Am. Soc. Metals* **61,** 423–434.

Nakagawa, Y. G. and Weatherley, G. C. (1972) The nature of interface between two ordered lattices: the Ni_3Al–Ni_3Nb eutectic, *Mater. Sci. Engng* **10,** 223–228.

Naziri, H. and Pearce, R. (1974) Superplasticity in a Zn–0.4%Al alloy, *Acta Metall.* **22,** 1321–1330.

Nuttall, K. (1976) Superplasticity in the Zr–2.5%Nb alloy, *Scr. Metall.* **10,** 835–840.

Nuttall, K. and McCooeye, D. P. (1974) Phase transformation plasticity in zirconium–hydrogen alloy, *Proc. 2nd Symp. Mech. Behav. Mater.* **2,** 129–140, Society Material Science Japan, Kyoto, Japan.

Oshida, Y. and Weiss, W. (1967) Research Inst. Rep. MET–E, 1504–1567 FR, Syracuse University, USA.

Perinet, F. (1975) Contribution to the study of the interfacial diffusion, CEA–R–4657, Saclay, Gif-sur-Yvette, France.

Petch, N. J. (1953) The cleavage strength of polycrystals, *J. Iron Steel Inst.* **174,** 23–28.

Sagat, S. and Taplin, D.M.R. (1976) The stability of plastic flow in strain-rate-sensitive materials, *Metal Sci.* **10,** 94–100.

Samuelsson, L. C. A., Melton, K. N., and Edington, J. W. (1976) Dislocation structures in a superplastic Zn–40 wt % Al alloy, *Acta Metall.* **24,** 1017–1026.

Schadler, H. W. (1968) The stress–strain rate behaviour of a manganese steel in the temperature range of the ferrite–austenite transformation, *Trans. Metall. Soc. AIME* **242,** 1281–1287.

Smith, C. I. and Ridley, N. (1974) Design of a superplastic alloy steel, *Metals Technol.* **1,** 191–198.

Smith, C. I., Norgate, B., and Ridley, N. (1976) Superplastic deformation and cavitation in a microduplex stainless steel, *Metal Sci.* **10,** 182–188.

Smith, C. S. (1948) Grains, phases and interfaces: an interpretation of microstructure, *Trans. Am. Inst. Min. Engrs* **175,** 15–51.

Staker, M. R. and Holt, D. L. (1972) The dislocation cell size and dislocation density in copper deformed at temperatures between 25 and 700°C, *Acta Metall.* **20,** 569–579.

Suery, M. (1979) Métallurgie et rhéologie des alliages superplastiques biphases, Ph.D. thesis, University of Metz, France.

Suery, M. and Baudelet, B. (1975) Theoretical and experimental constitutive equations of superplastic behaviour: discussion, *J. Mater. Sci.* **10,** 1022–1028.

Suery, M., Baudelet, B., Labulle, B., and Petipas, C. (1974) X-ray diffraction study of behaviour of 60/40 brass phases during superplastic deformation, *Scr. Metall.* **8,** 703–706.

Thompson, R. M. and Balluffi, R. W. (1962) Kinetic theory of dislocation climb, *J. Appl. Phys.* **33,** 803–825.

Vastava, R. B. and Langdon, T. G. (1979) An investigation of intercrystalline and interphase boundary sliding in the superplastic Pb–62% Sn eutectic, *Acta Metall.* **27,** 251–257.

Yoder, G. R. and Weiss, V. (1972) Superplasticity in eutectoid steel, *Metall. Trans.* **3,** 675–681.

The Theory of Finite Plastic Deformation of Crystalline Solids

by K. S. HAVNER

Department of Civil Engineering, North Carolina State University

Summary

The evolution of the continuum theory of crystal plasticity at finite strain is traced with particular reference to advances during the past decade. Included within the exposition are: the general invariance principles of Hill; analytical connections between single crystal and polycrystal deformation; a physical justification for the classical normality postulate; an overall theoretical framework for crystal plasticity, and the recently proposed simple theory of latent hardening which is embedded within that framework. Experimental foundations of the subject are taken into consideration, and the finite deformation experiments on cubic crystals of the 1960s are given prominence. Previous demonstrations of the simple theory's ability to account for significant qualitative and quantitative features of experiment are reviewed.

1. Introduction and Scope of Study

Shortly after completing his now classic monograph on the (macroscopic) theory of plasticity (Hill, 1950), which deals primarily with the plastically deformed solid as a homogeneous continuum, Rodney Hill collaborated with J. F. W. Bishop on the first rational study of the yielding of a metal as a *polycrystalline aggregate* (Bishop and Hill, 1951a, b). This work gave mathematical perspective to an insightful physical postulate of G. I. Taylor (1938a, b), known as the "principle of minimum shears," and to Taylor's theory of a polycrystalline aggregate under uniaxial loading, and substantially extended and generalized Taylor's polycrystalline model.

Bishop and Hill's papers (which continue to have widespread influence, particularly within the metallurgical literature) initiated a long line of theoretical and numerical analyses of polycrystalline solids at infinitesimal strain. Extensive bibliographies and informed commentaries on much of that work can be found in the review articles by Lin (1971) and Chin (1973), and I shall not recount any of those analyses here. Rather, to limit sufficiently the scope of this article as to afford depth in both exposition and discussion, the focus is upon an even more challenging aspect of crystal plasticity to which Hill has made seminal contributions: *finite deformation*.

The article is divided into four major Sections, each containing an introduction and three or more subsections. These major divisions are: "2. Hill's General Theory of Conjugate Stress and Strain Measures and Work Invariants;" "3. Connections Between Crystal and Aggregate Behavior;" "4. A General Theory of Finite Deformation of Elastic–Plastic Crystals;" and "5. The Simple Theory of Latent Hardening with Application to Single Crystal Experiments." Due regard is given to the experimental foundations of the mechanics of finite distortion in crystals, and significant experimental phenomena related to crystal hardening at finite strain are addressed in Section 5.

The various fundamental ideas and analyses of Hill's which have contributed to the subject, as well as influenced and guided my own work, span more than 20 years. I have endeavored to present the evolution of the theory here, however, largely in the sequence of pertinent contributions by Hill, myself, and a few others published during 1968–79. It has been my intent to integrate these contributions into a coherent whole, and the careful reader will judge for himself the degree to which this objective has been achieved.

2. Hill's General Theory of Conjugate Stress and Strain Measures and Work Invariants

Fundamental to the theoretical framework within which we undertake the rational study of finite, elastic–plastic deformation of crystalline solids is Hill's *invariant bilinear form* (Hill, 1968, 1972). This differential work invariant is expressed in terms of arbitrary, *work-conjugate* stress and strain measures **T**, **E** referred to an arbitrary reference state of density ρ_0. Specifically, Hill's result is that

$$(d_1\mathbf{T}.d_2\mathbf{E} - d_2\mathbf{T}.d_1\mathbf{E})/\rho_0 \quad \text{is invariant in value} \tag{2.1}$$

under changes of strain measure and reference state (Hill, 1972, equation (2.8)), where the respective pairs of differential variations from the current configuration of a stressed solid are arbitrary and unrelated.

Implicit within the invariance property of the bilinear form is the elegant *theory of conjugate stress and strain measures*, which was first presented in Hill (1970). These various fundamental ideas have recently been given a fresh treatment by Hill (1978), and my account of them here has been influenced by that work as well as by his earlier writings (Hill, 1968, 1970, 1972). The presentation is intended to be expository, albeit concise. From the need for brevity, however, it is necessarily less comprehensive than the original papers and the systematic development in Hill (1978), to which the reader is referred.

2.1. *A general class of conjugate stress and strain measures*

Let λ_α ($\alpha = 1, 2, 3$) denote the principal stretches at a material point X with respect to an arbitrary reference configuration, and $\boldsymbol{\mu}_\alpha^{\mathrm{L}}$, $\boldsymbol{\mu}_\alpha^{\mathrm{E}}$ denote unit vectors in the principal directions of stretch in the referential and current configurations, respectively, at X. Then $\boldsymbol{\mu}_\alpha^{\mathrm{L}}$, $\boldsymbol{\mu}_\alpha^{\mathrm{E}}$ are coincident with the axes of the *Lagrangian* (referential) and *Eulerian* (spatial) strain ellipsoids. There follows from the polar decomposition theorem (cf. Chadwick, 1976, Section 1.8)

$$\mathbf{F} = \mathbf{R}\boldsymbol{\Lambda} = \sum_\alpha \lambda_\alpha \boldsymbol{\mu}_\alpha^{\mathrm{E}} \otimes \boldsymbol{\mu}_\alpha^{\mathrm{L}}, \tag{2.2}$$

where $\mathbf{F} = \partial \mathbf{x}/\partial \mathbf{x}_0$ is the deformation gradient ($\mathbf{x}$ and $\mathbf{x}_0$ denote the current and reference positions of X) and

$$\mathbf{R} = \sum_\alpha \boldsymbol{\mu}_\alpha^{\mathrm{E}} \otimes \boldsymbol{\mu}_\alpha^{\mathrm{L}}, \quad \boldsymbol{\Lambda} = \sum_\alpha \lambda_\alpha \boldsymbol{\mu}_\alpha^{\mathrm{L}} \otimes \boldsymbol{\mu}_\alpha^{\mathrm{L}} \tag{2.3}$$

are respectively the orthogonal rotation and symmetric "right stretch" tensors (in common terminology), with $\boldsymbol{\mu}_\alpha^{\mathrm{E}} = \mathbf{R}\,\boldsymbol{\mu}_\alpha^{\mathrm{L}}$.

The general family of strain measures $\mathbf{E}(f(\lambda))$ introduced by Hill (1968) is defined, in spectral representation on the Lagrangian axes, as

$$\mathbf{E} = \sum_\alpha e_\alpha \boldsymbol{\mu}_\alpha^{\mathrm{L}} \otimes \boldsymbol{\mu}_\alpha^{\mathrm{L}}, \quad e_\alpha = f(\lambda_\alpha), \tag{2.4}$$

where $f(\lambda)$ is an arbitrary, smooth monotone function satisfying $f(1) = 0, f'(1) = 1$. A broad subclass of such measures is given by (see Seth, 1964, or Hill, 1968)

$$e(\lambda, m) = (\lambda^{2m} - 1)/(2m), \tag{2.5}$$

with the limit $\ln \lambda$ as $m \to 0$ (from L'Hôpital's rule). The traditional measures of strain which belong to this subclass (or its limit) are:

$$\left.\begin{aligned} e_{\mathrm{G}} &= \tfrac{1}{2}(\lambda^2 - 1), \quad \text{Green strain } (m = 1), \\ e_{\mathrm{N}} &= \lambda - 1, \quad \text{nominal ("engineering") strain } (m = \tfrac{1}{2}), \\ e_{\mathrm{L}} &= \ln \lambda, \quad \text{logarithmic ("true") strain } (m = 0), \\ e_{\mathrm{A}} &= \tfrac{1}{2}(1 - \lambda^{-2}), \quad \text{Almansi strain } (m = -1). \end{aligned}\right\} \tag{2.6}$$

It is our objective in the sequel to construct a general family of conjugate stress measures $\mathbf{T}$, for arbitrary measures $\mathbf{E}(f(\lambda))$, such that $\mathbf{T}.d\mathbf{E}/\rho_0$ equals differential work per unit mass, a fundamental scalar invariant.

Let $\boldsymbol{\tau} = (\rho_0/\rho)\boldsymbol{\sigma}$ denote Kirchhoff stress, where $\boldsymbol{\sigma}$ is Cauchy stress and ρ is current mass density at X. Then $\boldsymbol{\tau}$ has the representation, on the axes of the Eulerian strain ellipsoid,

$$\boldsymbol{\tau} = \sum_\alpha \sum_\beta \tau_{\alpha\beta} \boldsymbol{\mu}_\alpha^{\mathrm{E}} \otimes \boldsymbol{\mu}_\beta^{\mathrm{E}}. \tag{2.7}$$

[As our application of Hill's theory of conjugate stress and strain measures will be to metal crystals (and polycrystals), which ordinarily are both elastically and plastically *anisotropic* in a deformed state, principal axes of stress and strain cannot be expected to coincide.] For a particular conjugate ($\mathbf{T}$, $\mathbf{E}$) pair, it is necessary that, with the reference state fixed,

$$\mathbf{T}.\dot{\mathbf{E}} = \boldsymbol{\tau}.\mathbf{D}, \tag{2.8}$$

in which $\mathbf{D}$ is the rate of deformation (Eulerian strain-rate) tensor

$$\mathbf{D} = \sum_\alpha \sum_\beta d_{\alpha\beta} \boldsymbol{\mu}_\alpha^{\mathrm{E}} \otimes \boldsymbol{\mu}_\beta^{\mathrm{E}}, \quad d_{\alpha\beta} = \tfrac{1}{2}(v_{\alpha,\beta} + v_{\beta,\alpha}). \tag{2.9}$$

The $v_{\alpha,\beta}$ are components of the spatial velocity gradient $\mathbf{L} = \partial \mathbf{v}/\partial \mathbf{x}$ on the Eulerian axes. They are of course not objective (that is, observer frame-indifferent) although the $d_{\alpha\beta}$ are.

Well-known conjugate stress and deformation measures which satisfy (2.8) are (i) contravariant Kirchhoff stress $\boldsymbol{\tau}^c$ (this stress measure is also frequently named second Piola–Kirchhoff stress) and (covariant) Green strain $\mathbf{E}_{\mathrm{G}}$, (ii) covariant Kirchhoff stress $\boldsymbol{\tau}_c$ and (contravariant) Almansi strain $\mathbf{E}_{\mathrm{A}}$, and (iii) nominal stress $\mathbf{s}$ and the deformation gradient itself, although $\mathbf{F} - \mathbf{I}$ does not belong to the class (2.4) (unless $\mathbf{R} \equiv \mathbf{I}$ throughout a deformation history, in which case $\mathbf{F}-\mathbf{I}=\boldsymbol{\Lambda}-\mathbf{I}=\mathbf{E}_N$, the nominal strain). Thus (cf. Hill, 1968, Part I, Appendix),

$$\boldsymbol{\tau}^c.\dot{\mathbf{E}}_G=\boldsymbol{\tau}_c.\dot{\mathbf{E}}_A=\mathbf{s}.\dot{\mathbf{F}}=\boldsymbol{\tau}.\mathbf{D}, \tag{2.10}$$

where

$$\mathbf{E}_G=\tfrac{1}{2}(\mathbf{C}-\mathbf{I}),\quad \mathbf{E}_A=\tfrac{1}{2}(\mathbf{I}-\mathbf{C}^{-1}),\quad \mathbf{C}=\mathbf{F}^T\mathbf{F}, \tag{2.11}$$

and

$$\boldsymbol{\tau}^c=\mathbf{F}^{-1}\boldsymbol{\tau}(\mathbf{F}^{-1})^T,\quad \boldsymbol{\tau}_c=\mathbf{F}^T\boldsymbol{\tau}\mathbf{F},\quad \mathbf{s}=\mathbf{F}^{-1}\boldsymbol{\tau}. \tag{2.12}$$

In terms of the components of $\boldsymbol{\tau}$ on the Eulerian ellipsoid axes (2.7),

$$\left.\begin{aligned}\boldsymbol{\tau}^c&=\sum_\alpha\sum_\beta\lambda_\alpha^{-1}\lambda_\beta^{-1}\tau_{\alpha\beta}\boldsymbol{\mu}_\alpha^L\otimes\boldsymbol{\mu}_\beta^L,\\ \boldsymbol{\tau}_c&=\sum_\alpha\sum_\beta\lambda_\alpha\lambda_\beta\tau_{\alpha\beta}\boldsymbol{\mu}_\alpha^L\otimes\boldsymbol{\mu}_\beta^L,\\ \mathbf{s}&=\sum_\alpha\sum_\beta\lambda_\alpha^{-1}\tau_{\alpha\beta}\boldsymbol{\mu}_\alpha^L\otimes\boldsymbol{\mu}_\beta^E.\end{aligned}\right\} \tag{2.13}$$

(The spectral representations of $\mathbf{E}_G$ and $\mathbf{E}_A$ on the Lagrangian axes are given by (2.4), after $(2.6)_1$ and $(2.6)_4$ respectively are substituted for each $f(\lambda)$.)

Going far beyond these special cases, Hill (1970) has developed a completely general set of equations for $\mathbf{T}$, corresponding to arbitrary $\mathbf{E}(f(\lambda))$, by first establishing an elegant set of kinematic relations which give the spin (rate of rotation) $\boldsymbol{\Omega}^L$ of the Lagrangian strain ellipsoid about its own axes in terms of Eulerian strain-rate $\mathbf{D}$. The original arguments (Hill, 1970) which led to his equations for $\boldsymbol{\Omega}^L$ and $\mathbf{T}$ are largely geometric in nature. The derivation of Hill's equations here is in the algebraic spirit of his more recent work (Hill, 1978).

We begin with a standard result obtained by first differentiating the polar decomposition (2.2) of $\mathbf{F}$, then substituting into $\mathbf{L}=\dot{\mathbf{F}}\mathbf{F}^{-1}$ and taking the symmetric part $\mathbf{D}$ (cf. Chadwick, 1976, Section 2.5). This gives

$$\mathbf{R}^T\mathbf{D}\mathbf{R}=\tfrac{1}{2}(\dot{\boldsymbol{\Lambda}}\boldsymbol{\Lambda}^{-1}+\boldsymbol{\Lambda}^{-1}\dot{\boldsymbol{\Lambda}}), \tag{2.14}$$

in which $\boldsymbol{\Lambda}^{-1}$ has the spectral representation (from $(2.3)_2$)

$$\boldsymbol{\Lambda}^{-1}=\sum_\alpha\lambda_\alpha^{-1}\boldsymbol{\mu}_\alpha^L\otimes\boldsymbol{\mu}_\alpha^L. \tag{2.15}$$

To obtain the component representation of $\dot{\boldsymbol{\Lambda}}$ note that, from the definition of $\boldsymbol{\Omega}^L$, the triad of unit vectors $\boldsymbol{\mu}_\alpha^L$ rotates with respect to embedded material axes momentarily coincident with it according to

$$\dot{\boldsymbol{\mu}}_\alpha^L=\boldsymbol{\Omega}^L\boldsymbol{\mu}_\alpha^L=\sum_\beta\Omega_{\beta\alpha}^L\boldsymbol{\mu}_\beta^L, \tag{2.16}$$

where $\Omega_{\alpha\beta}^L$ are the skew-symmetric components of $\boldsymbol{\Omega}^L$ on the Lagrangian axes. Thus, upon differentiating $(2.3)_2$ and substituting (2.16), we find

$$\dot{\boldsymbol{\Lambda}}=\sum_\alpha\sum_\beta\{\dot{\lambda}_\alpha\delta_{\alpha\beta}+(\lambda_\alpha-\lambda_\beta)\Omega_{\beta\alpha}^L\}\boldsymbol{\mu}_\alpha^L\otimes\boldsymbol{\mu}_\beta^L, \tag{2.17}$$

which is equivalent to equations (1.46) in Hill (1978). In addition, from $(2.3)_1$ and $(2.9)_1$,

$$\mathbf{R}^T\mathbf{D}\mathbf{R}=\sum_\alpha\sum_\beta d_{\alpha\beta}\boldsymbol{\mu}_\alpha^L\otimes\boldsymbol{\mu}_\beta^L. \tag{2.18}$$

(It is to be remembered that the $d_{\alpha\beta}$ are components of **D** on the *Eulerian* ellipsoid axes.) There follows immediately, upon our substitution of (2.15), (2.17), and (2.18) into (2.14),

$$d_{\alpha\beta} = \tfrac{1}{2}(\lambda_\alpha^{-1} + \lambda_\beta^{-1})\dot{\Lambda}_{\alpha\beta} = \frac{\lambda_\alpha + \lambda_\beta}{2\lambda_\alpha\lambda_\beta}\dot{\lambda}_\alpha\delta_{\alpha\beta} + \frac{\lambda_\alpha^2 - \lambda_\beta^2}{2\lambda_\alpha\lambda_\beta}\Omega^{\mathrm{L}}_{\beta\alpha}, \tag{2.19}$$

where $\dot{\Lambda}_{\alpha\beta}$ are the Lagrangian components of $\dot{\mathbf{\Lambda}}$ in (2.17). Thus, for $\alpha = \beta$,

$$d_{\alpha\alpha} = \dot{\lambda}_\alpha/\lambda_\alpha = (\ln \lambda_\alpha)^{\dot{}}\,, \tag{2.20}$$

and, for $\alpha \neq \beta$, $\lambda_\alpha \neq \lambda_\beta$,

$$\Omega^{\mathrm{L}}_{\beta\alpha} = \frac{2\lambda_\alpha\lambda_\beta}{\lambda_\alpha^2 - \lambda_\beta^2} d_{\alpha\beta}. \tag{2.21}$$

Equation (2.20) is a standard result that holds for any material direction (not necessarily principal). Equation (2.21), however, which holds only with respect to the axes of the two strain ellipsoids, is an original discovery by Hill (1970, equation (15)) (also see Hill, 1978, equation (1.50)), and is the key to constructing the final equations for **T**.

We proceed as follows. First, by analogy with (2.17) (equivalently, differentiating $(2.4)_1$, with the aid of (2.16)), we have

$$\dot{\mathbf{E}} = \sum_\alpha\sum_\beta \{\dot{e}_\alpha\delta_{\alpha\beta} + (e_\alpha - e_\beta)\Omega^{\mathrm{L}}_{\beta\alpha}\}\,\boldsymbol{\mu}^{\mathrm{L}}_\alpha \otimes \boldsymbol{\mu}^{\mathrm{L}}_\beta. \tag{2.22}$$

Let $t_{\alpha\beta}$ denote the components of **T** on the Lagrangian ellipsoid axes:

$$\mathbf{T} = \sum_\alpha\sum_\beta t_{\alpha\beta}\,\boldsymbol{\mu}^{\mathrm{L}}_\alpha \otimes \boldsymbol{\mu}^{\mathrm{L}}_\beta. \tag{2.23}$$

Upon substituting (2.22), (2.23), and $\dot{e}_\alpha = f'(\lambda_\alpha)\dot{\lambda}_\alpha$ (from $(2.4)_2$), together with (2.7) and (2.9), into (2.8) we find

$$\sum_\alpha\sum_\beta t_{\alpha\beta}\{f'(\lambda_\alpha)\dot{\lambda}_\alpha\delta_{\alpha\beta} + (e_\alpha - e_\beta)\Omega^{\mathrm{L}}_{\beta\alpha}\} = \sum_\alpha\sum_\beta \tau_{\alpha\beta}d_{\alpha\beta}. \tag{2.24}$$

Then, either substituting (2.20) and (2.21) on the left-hand side, or (2.19) on the right, and comparing term by term, we obtain the following remarkable set of equations for the conjugate stresses $t_{\alpha\beta}$ (Hill, 1970, equations (24)):

$$t_{\alpha\alpha} = \frac{\tau_{\alpha\alpha}}{\lambda_\alpha f'(\lambda_\alpha)}, \quad \alpha = 1, 2, 3, \tag{2.25}$$

$$t_{\alpha\beta} = \begin{cases} \dfrac{\lambda_\alpha^2 - \lambda_\beta^2}{e_\alpha - e_\beta}\dfrac{\tau_{\alpha\beta}}{2\lambda_\alpha\lambda_\beta}, & \alpha \neq \beta, \quad \lambda_\beta \neq \lambda_\alpha, \\[2ex] \dfrac{\tau_{\alpha\beta}}{\lambda_\alpha f'(\lambda_\alpha)}, & \alpha \neq \beta, \quad \lambda_\beta = \lambda_\alpha. \end{cases} \tag{2.26}$$

It is worthwhile re-emphasizing that the $t_{\alpha\beta}$ are components of **T** on the *Lagrangian* axes whereas the $\tau_{\alpha\beta}$ are components of Kirchhoff stress $\boldsymbol{\tau}$ on the *Eulerian* axes (which are rotated by **R** from the Lagrangian triad).

Equations (2.25) and (2.26) complete the development of Hill's general class of conjugate stress and strain measures. The equations of course include the contravariant and covariant Kirchhoff stress components $(2.13)_{1,\,2}$ as members of the family T, which fact is obvious if one substitutes $(2.6)_1$ and $(2.6)_4$ respectively for the strain measure. Also note that the stress $\mathbf{T}_N$ conjugate to nominal strain $\mathbf{E}_N = \mathbf{\Lambda} - \mathbf{I}$ can be immediately stated, viz. $t^N_{\alpha\beta} = \frac{1}{2}(\lambda_\alpha^{-1} + \lambda_\beta^{-1})\tau_{\alpha\beta}$. Hill (1978, p. 20) has shown that $\mathbf{T}_N$ can be written

$$\mathbf{T}_N = \tfrac{1}{2}(\boldsymbol{\tau}^c\mathbf{\Lambda} + \mathbf{\Lambda}\boldsymbol{\tau}^c), \tag{2.27}$$

which in turn may be expressed (from (2.2), (2.3), and $(2.12)_1$) as the symmetric part of $\mathbf{sR}$, whence $\mathbf{s} = \mathbf{T}_N$ when $\mathbf{R} \equiv \mathbf{I}$ throughout the deformation history. (Recall the analogous relation between $\mathbf{F} - \mathbf{I}$ and $\mathbf{E}_N$.) Hill (1968, 1978) attributes the stress measure $\mathbf{T}_N$ to Biot (1965).

2.2. *The invariant bilinear form*

To establish the invariance of the bilinear differential form (2.1) under changes of strain measure and reference state, we proceed as in Hill (1968, Part II). Let e_{ij}, e^*_{ij} denote the components, on an arbitrary Cartesian reference frame, of any two strain measures $\mathbf{E}$, $\mathbf{E}^*$ belonging to the general family (2.4). Then, formally,

$$de_{ij} = \frac{\partial e_{ij}}{\partial e^*_{kl}} de^*_{kl} \tag{2.28}$$

(with the usual summation convention on repeated Latin indices). Whence, from the invariance of $\mathbf{T}.d\mathbf{E}/\rho_0$ (with $\mathbf{T}$ determined from (2.25) and (2.26)),

$$t^*_{kl}/\rho^*_0 = (t_{ij}/\rho_0)\frac{\partial e_{ij}}{\partial e^*_{kl}}. \tag{2.29}$$

Upon taking the differential of (2.29) (holding the reference states fixed) and forming the scalar product with an unrelated differential variation in strain, we obtain (Hill, 1968, p. 320, or 1972, p. 135)

$$d_1 t^*_{kl} d_2 e^*_{kl}/\rho^*_0 = d_1 t_{ij} d_2 e_{ij}/\rho_0 + (t_{ij}/\rho_0)\frac{\partial^2 e_{ij}}{\partial e^*_{kl}\partial e^*_{mn}} d_1 e^*_{kl} d_2 e^*_{mn}. \tag{2.30}$$

The invariance of the bilinear form (2.1) follows immediately by interchanging the d_1, d_2 differentials and subtracting the resulting expression from (2.30). Thus (Hill, 1972),

$$(d_1\mathbf{T}^*.d_2\mathbf{E}^* - d_2\mathbf{T}^*.d_1 E^*)/\rho^*_0 = (d_1\mathbf{T}.d_2\mathbf{E} - d_2\mathbf{T}.d_1\mathbf{E})/\rho_0 \tag{2.31}$$

under arbitrary changes in strain measure and reference state.

For the application of (2.1) (or (2.31)) to crystalline aggregates in Section 3, we require the additional equality (with fixed reference state)

$$d_1\mathbf{s}.d_2\mathbf{F} - d_2\mathbf{s}.d_1\mathbf{F} = d_1\mathbf{T}.d_2\mathbf{E} - d_2\mathbf{T}.d_1\mathbf{E}. \tag{2.32}$$

This result cannot be considered a direct consequence of (2.31) because $\mathbf{s}$ and $\mathbf{F}$ do not belong to the general class of work-conjugate measures $\mathbf{T}$, $\mathbf{E}$ defined by (2.4) and (2.25) and (2.26) (although $\mathbf{s}.d\mathbf{F}$ does equal $\mathbf{T}.d\mathbf{E}$ from (2.10)). To prove (2.32), we note that the right-hand side is invariant with respect to change in strain measure from (2.31), hence it is only necessary to demonstrate equality for a specific measure. For convenience we choose Green strain $\mathbf{E}_G$ and its conjugate stress $\boldsymbol{\tau}^c$. Then, from the standard relation (see (2.12))

$$\mathbf{s} = \boldsymbol{\tau}^c\mathbf{F}^T, \tag{2.33}$$

we find, using $\dot{\mathbf{F}} = \mathbf{L}\mathbf{F}$,

$$\dot{\mathbf{s}} = \dot{\boldsymbol{\tau}}^c \mathbf{F}^T + \mathbf{s}\mathbf{L}^T. \tag{2.34}$$

Further, with the aid of the well-known equation

$$\dot{\mathbf{E}}_G = \mathbf{F}^T \mathbf{D} \mathbf{F} \tag{2.35}$$

(which follows from (2.2), (2.9), and (2.11)), $(2.12)_3$, and the symmetry of $\boldsymbol{\tau}^c$, we obtain

$$d_1 \mathbf{s} . d_2 \mathbf{F} = d_1 \boldsymbol{\tau}^c . d_2 \mathbf{E}_G + \boldsymbol{\tau} . (\mathbf{L}_1^T \mathbf{L}_2)\, dt\, dt. \tag{2.36}$$

Finally, we write the corresponding equation for $d_2 \mathbf{s} . d_1 \mathbf{F}$ and subtract from (2.36). The second terms on the respective right-hand sides cancel from the symmetry of $\boldsymbol{\tau}$, and the equality (2.32) is established. (For other derivations of this result see Hill (1972, Section 3) and Havner (1974, Section 4.3).)

2.3. *A family of objective stress-rates in the reference state*

Because the stress $\mathbf{T}$ conjugate to an arbitrary strain measure $\mathbf{E}(f(\lambda))$ is defined *intrinsically* with respect to the deformed solid (i.e. components $t_{\alpha\beta}$ are defined explicitly by (2.25) and (2.26) on axes which have significance for the strained material, viz. the referential directions of principal stretch), the rates of change of the components of $\mathbf{T}$ on any Cartesian reference frame transform as the components themselves. Whence, $\dot{\mathbf{T}}$ (or $d\mathbf{T}$) is an objective (frame-indifferent) tensor and so is admissible, in principle, within constitutive equations of rate-type, as is $\dot{\mathbf{E}}$ (or $d\mathbf{E}$). Consequently, $\dot{\mathbf{T}}.\dot{\mathbf{E}}/\rho_0$ is a scalar invariant under change of observer frame. However, it is *not* invariant under changes in strain measure and reference state (recall (2.30)). Therefore, as first shown by Hill (1968), for the analysis of possible constitutive inequalities of type $\dot{\mathbf{T}}.\dot{\mathbf{E}} > 0$ it is necessary to have an equation for the precise way in which $\dot{\mathbf{T}}$ depends upon strain measure, *when evaluated with the current state as reference* (wherein all $\dot{\mathbf{E}}$ momentarily equal $\mathbf{D}$). This equation is (Hill, 1968, p. 231, or 1970, equation (1))

$$\dot{\mathbf{T}}^{(0)} = \frac{\mathscr{D}\boldsymbol{\tau}}{\mathscr{D}t} - m(\boldsymbol{\sigma}\mathbf{D} + \mathbf{D}\boldsymbol{\sigma}), \tag{2.37}$$

where the superscript (0) indicates the reference state and

$$\frac{\mathscr{D}\boldsymbol{\tau}}{\mathscr{D}t} \equiv \dot{\boldsymbol{\tau}} + \boldsymbol{\tau}\mathbf{W} - \mathbf{W}\boldsymbol{\tau} = \dot{\boldsymbol{\sigma}} + \boldsymbol{\sigma}\,\mathrm{tr}\,\mathbf{D} + \boldsymbol{\sigma}\mathbf{W} - \mathbf{W}\boldsymbol{\sigma} \tag{2.38}$$

is the Jaumann–Zaremba, or material co-rotational, flux of Kirchhoff stress $\boldsymbol{\tau}$ $(=\boldsymbol{\sigma}$ momentarily). $\mathbf{W}$ denotes the material spin (rate of rotation) tensor on the arbitrary reference frame, and (consistent with both the general strain measure $f(\lambda)$ and the subclass (2.5)) m is defined as (Hill, 1972, equation (2.1))

$$m = -\tfrac{1}{2}\{1 + f''(1)\}. \tag{2.39}$$

The following proof of (2.37) is patterned after the original derivation by Hill (1968, Part I).

Consider a power series expansion of $f(\lambda)$ in terms of the logarithmic measure $\ln \lambda$ (holding the arbitrary reference state fixed):

$$f(\lambda) = \ln \lambda + \tfrac{1}{2}\{1 + f''(1)\}(\ln \lambda)^2 + \tfrac{1}{6}\{1 + 3f''(1) + f'''(1)\}(\ln \lambda)^3 + \dots . \tag{2.40}$$

This identically satisfies the conditions $f(1) = 0, f'(1) = 1$ and converges in some neighborhood of $\lambda = 1$ (everywhere for some measures, as remarked by Hill (1972, p. 133)). [The first three terms of the expansion (2.40) give an accurate representation of each of the common strain measures through relatively large deformations. At 50% extension ($\lambda = 1.5$) the exact values range from 0.278 (Almansi) to 0.625 (Green), with a maximum error in the three-term approximation of only 2.8% (for the Almansi measure). Similarly, for a one-third reduction in length ($\lambda = 2/3$), the exact range is -0.625 (Almansi) to -0.278 (Green), with the maximum (three-term) error again 2.8% (for the Green measure).] Equivalently, from $(2.4)_2$ and the definition (2.39), $e_\alpha = \ln \lambda_\alpha + m(\ln \lambda_\alpha)^2 + \ldots$. Hence, as all measures of the family (2.4) are coaxial,

$$\mathbf{E} = \mathbf{E}_\mathrm{L} + m\mathbf{E}_\mathrm{L}^2 + \ldots, \tag{2.41}$$

where $\mathbf{E}_\mathrm{L}$ denotes the "logarithmic strain" whose components are $\ln \lambda_\alpha$ in spectral representation on the Lagrangian axes.

Let e_{ij}^L denote the components of $\mathbf{E}_\mathrm{L}$ on the Cartesian reference frame (O, X_i):

$$e_{ij}^\mathrm{L} = \sum_\alpha Q_{i\alpha} Q_{j\alpha} \ln \lambda_\alpha, \quad Q_{i\alpha} = \cos(X_i, \boldsymbol{\mu}_\alpha). \tag{2.42}$$

Then, from (2.41),

$$e_{ij} = e_{ij}^\mathrm{L} + m e_{in}^\mathrm{L} e_{nj}^\mathrm{L} + O(e_\mathrm{L}{}^3), \tag{2.43}$$

so that $\partial e_{ij}/\partial e_{kl}^\mathrm{L} = \delta_{ik}\delta_{jl} + m(\delta_{ik} e_{jl}^\mathrm{L} + e_{ik}^\mathrm{L}\delta_{jl}) + O(e_\mathrm{L}{}^2)$. Upon substituting this result into (2.29) (with the starred quantities there replaced by the logarithmic measures and the reference state densities set equal) we obtain

$$\mathbf{T}_\mathrm{L} = \mathbf{T} + m(\mathbf{T}\mathbf{E}_\mathrm{L} + \mathbf{E}_\mathrm{L}\mathbf{T}) + O(\mathbf{E}_\mathrm{L}{}^2), \tag{2.44}$$

where $\mathbf{T}_\mathrm{L}$ denotes the stress conjugate to $\mathbf{E}_\mathrm{L}$. After differentiation, there follows

$$\dot{\mathbf{T}}_\mathrm{L} = \dot{\mathbf{T}} + m(\mathbf{T}\dot{\mathbf{E}}_\mathrm{L} + \dot{\mathbf{E}}_\mathrm{L}\mathbf{T}) + O(\mathbf{E}_\mathrm{L}). \tag{2.45}$$

Now choose the current state as reference; whence, momentarily, all strain measures vanish, all strain-rates equal the rate of deformation (Eulerian strain-rate) **D**, and all stress measures equal Cauchy stress $\boldsymbol{\sigma}$. Thus, *exactly* (Hill, 1968, Part I, equation (5))

$$\dot{\mathbf{T}}_\mathrm{L}^{(0)} = \dot{\mathbf{T}}^{(0)} + m(\boldsymbol{\sigma}\mathbf{D} + \mathbf{D}\boldsymbol{\sigma}). \tag{2.46}$$

To complete the proof of (2.37), we determine $\dot{\mathbf{T}}_\mathrm{L}^{(0)}$ from (2.46) and either of the standard equations for the fluxes of contravariant and covariant Kirchhoff stress $\boldsymbol{\tau}^c$, $\boldsymbol{\tau}_c$ (which measures of course belong to the general family **T**). These fluxes are

$$\dot{\boldsymbol{\tau}}^c = \mathbf{F}^{-1}\frac{\delta^c\boldsymbol{\tau}}{\delta t}(\mathbf{F}^{-1})^\mathrm{T}, \quad \dot{\boldsymbol{\tau}}_c = \mathbf{F}^\mathrm{T}\frac{\delta_c\boldsymbol{\tau}}{\delta t}\mathbf{F}, \tag{2.47}$$

where $\delta^c/\delta t$, $\delta_c/\delta t$ respectively denote contravariant and covariant (convected) derivatives on embedded material axes momentarily coincident with the reference frame (O, X_i) (cf. Hill, 1972, equation (2.7)):

$$\frac{\delta^c\boldsymbol{\tau}}{\delta t} = \frac{\mathscr{D}\boldsymbol{\tau}}{\mathscr{D}t} - (\boldsymbol{\tau}\mathbf{D} + \mathbf{D}\boldsymbol{\tau}), \quad \frac{\delta_c\boldsymbol{\tau}}{\delta t} = \frac{\mathscr{D}\boldsymbol{\tau}}{\mathscr{D}t} + (\boldsymbol{\tau}\mathbf{D} + \mathbf{D}\boldsymbol{\tau}). \tag{2.48}$$

(Equations (2.47) are easily derived, with the aid of definitions (2.38) and (2.48), by differentiating $(2.12)_{1,2}$.) Thus, recalling that $\boldsymbol{\tau}^c$ is conjugate to the Green measure ($m = 1$) while $\boldsymbol{\tau}_c$ is conjugate to the Almansi measure ($m = -1$), we substitute either of $(2.47)_{1,2}$ *evaluated in the reference state* ($\mathbf{F} = \mathbf{I}$), and the corresponding value of m into (2.46) to obtain (using (2.48))

$$\dot{\mathbf{T}}_{\mathrm{L}}^{(0)} = \frac{\mathscr{D}\boldsymbol{\tau}}{\mathscr{D}t} = \frac{\mathscr{D}\boldsymbol{\sigma}}{\mathscr{D}t} + \boldsymbol{\sigma}\,\mathrm{tr}\,\mathbf{D}. \tag{2.49}$$

Hill's result (2.37) is immediately evident from (2.46) and (2.49).

By (2.37) and the equality $\dot{\mathbf{E}}^{(0)} = \mathbf{D}$, we have the following equation (Hill, 1978, Part I, equation (14)) for the dependence of $\dot{\mathbf{T}}.\dot{\mathbf{E}}$ *in the reference state* on the choice of strain measure $f(\lambda)$ (through the definition (2.39)):

$$(\dot{\mathbf{T}}.\dot{\mathbf{E}})^{(0)} = \frac{\mathscr{D}\boldsymbol{\tau}}{\mathscr{D}t}.\mathbf{D} - 2m\boldsymbol{\sigma}.\mathbf{D}^2. \tag{2.50}$$

Hill (1968, Part II, Section 3) has given convincing arguments in support of the *logarithmic measure as the norm* for finite plastic straining of metals. Hill's arguments lead to the conclusion that, if *any* inequality of the type $\dot{\mathbf{T}}.\dot{\mathbf{E}} > 0$ generally applies, it is

$$(\dot{\mathbf{T}}_{\mathrm{L}}.\dot{\mathbf{E}}_{\mathrm{L}})^{(0)} = \frac{\mathscr{D}\boldsymbol{\tau}}{\mathscr{D}t}.\mathbf{D} > 0. \tag{2.51}$$

This inequality will be considered further in the context of elastic–plastic deformation of crystals in Section 4.3.

3. Connections Between Crystal and Aggregate Behavior

As shown by Hill (1972), a number of important connections between macroscopic measures of stress and strain in polycrystalline metals (or other heterogeneous media) and micro-fields within a representative volume element can be established without regard to specific constitutive properties. Prominent in the working-out of these connections is the "averaging theorem," which was introduced and first applied in a classic paper by Bishop and Hill (1951a). The theorem also has been argued by Kocks (1958), Hill (1963, 1967), and Havner (1971), all at infinitesimal strain. A proof of the theorem at *finite* deformation was first presented by Hill (1972). An alternative proof (at finite strain) has been given by Havner (1974), and both these arguments are included here.

In regard to *plastic* deformation of metals, the invariant bilinear form (2.1) plays an equally prominent role in the analysis of the transition from crystal to aggregate behavior. The bilinear form and averaging theorem together afford a physically significant connection between crystalline slip-system unloading (as the criterion for local *elastic* changes) and the classical "normality postulate" of macroscopic plasticity theory. Derivations of the averaging theorem and of this connection, preceded by a needed perspective on the physical scale of continuum modeling of crystal plasticity, are presented in the following Sections.

3.1. *Physical considerations and kinematics of crystals*

In adopting a continuum mechanics description of finite deformation processes in metal crystals, one is limited as to the minimum level at which physical features of inelastic behavior can be represented. Taking *gross crystallographic slip* to be the dominant mechanism of this behavior (at other than very high temperatures), the following discussion (from Havner, 1973a, b) is pertinent to an assessment of the minimum scale.

To an observer who can resolve distances to 10^{-3} mm, the deformation of crystal grains (of linear dimensions of order 10^{-1} mm, say) within permanently strained metal polycrystals is relatively smooth. At this level of observation (called "microscopic" here) one can just distinguish between slip lines on crystal surfaces after extensive permanent distortion of an aggregate specimen. In contrast, a *sub*microscopic observer, resolving distances to 10^{-5} mm (the order of 100 atomic spacings), is aware of highly inhomogeneous and discontinuous deformation within each grain. The microscopic observer's slip "lines" appear to the submicroscopic observer as slip *bands* (or packets) of order 10^{-4} mm width at finite distortions, containing numerous glide lamellae between which amounts of slip as great as 10^3 lattice spacings have occurred (cf. Heidenreich, 1949). Thus, if we wish to admit the concept of a *uniformly strained* crystal within our mathematical framework, the physical dimension of a continuum "material point" in an individual grain cannot be less than 10^{-3} mm (i.e. $> 10^3$ lattice spacings). This minimum "point size" is consistent with the conclusion of Hill (1956, p. 8) that "the linear dimension of the smallest crystal whose behaviour can legitimately be considered from the standpoint of the theory of plasticity is probably of order 10^{-3} cm" (that is, an order of magnitude greater than the physical scale of a crystal material point as defined here). Crystal grain sizes *in situ* typically are in the range 10^{-3}–10^{-2} cm for metals, hence the application of a continuum mechanics perspective at this microscale is physically appropriate. With this limitation in mind, we turn attention to the representation of the *kinematics* of crystal deformation.

The single most important phenomenological aspect of gross inelastic behavior in metal crystals (indeed, distinguishing them by kinematics alone from polymers and soils) is *the movement of the macroscopic material relative to the underlying crystalline structure.* Roughly speaking, material "lines" and "planes" translate and rotate with respect to the averaged lattice orientation over volumes of typical dimensions of order 10^{-2} mm. The movement of dislocations at any instant takes place *through* the current deformed lattice, but the averaged lattice strain over a "material point" (of dimension $\approx 10^{-3}$ mm) is only slightly affected by the dislocations present. In consequence, the appropriate gross kinematic representation of incremental motion from the current deformed state is taken to be a *linear superposition of additional lattice straining with incremental slips on distinct crystallographic glide systems.* Each of these latter deformation mechanisms translates material "lines" of points (i.e. glide packets) relative to one another, resulting in gross incremental shearing, but leaves the averaged crystal structure unchanged. Thus, the spatial gradient of incremental displacement at a crystal material point X can be written (cf. Havner, 1972)

$$\mathbf{L}\,dt = (\boldsymbol{\varepsilon} + \boldsymbol{\omega})dt + \sum_k (\mathbf{b}\otimes\mathbf{n})_k\, d\gamma_k \quad (k = 1, \ldots, n). \tag{3.1}$$

Here $\boldsymbol{\varepsilon}$ is the (Eulerian) lattice strain-rate; $\boldsymbol{\omega}$ is the spin tensor of the lattice relative to an arbitrary reference frame (O, X_i); $\mathbf{n}_k$, $\mathbf{b}_k$ denote unit vectors in the current normal and glide directions respectively of the kth slip system; and $d\gamma_k$ is the incremental slip in that system, measured relative to the (current) averaged lattice configuration at the material point. Opposite senses of slip in the same crystallographic glide system are represented by distinct k's, so that $d\gamma_k$ is always non-negative; and the number of locally *critical* (i.e. potentially active) slip systems is designated by n.

Decomposing $\mathbf{L}$ into the symmetric and skew-symmetric tensors $\mathbf{D}$ and $\mathbf{W}$, we have

$$(\mathbf{D}-\boldsymbol{\varepsilon})\,dt = \sum_k \mathbf{N}_k\,d\gamma_k, \quad (\mathbf{W}-\boldsymbol{\omega})\,dt = \sum_k \boldsymbol{\Omega}_k\,d\gamma_k, \tag{3.2}$$

where $\mathbf{N}_k$, $\boldsymbol{\Omega}_k$ are respectively the symmetric and skew-symmetric parts of $(\mathbf{b}\otimes\mathbf{n})_k$ and the separate sides of the equations are objective under observer transformations. Taking the trace of $(3.2)_1$, we find that $\mathrm{tr}\,\mathbf{D} = \mathrm{tr}\,\boldsymbol{\varepsilon}$ from the orthogonality of each $(\mathbf{b}_k, \mathbf{n}_k)$ pair. Hence, the volumetric "natural" strain-rate of the crystal in the kinematic model (3.1) is attributable solely to the lattice.

3.2. *Polycrystalline metals: the averaging theorem*

Before we proceed to the derivation of an analytical connection between the local deformation of individual crystals *in situ* and the gross behavior of metals at the (macroscopic) level of ordinary stress and strain analysis, consider the following with regard to the physical scale of a representative "macro-element." The wall thickness of thin-walled tubes used for standard combined-stress tests (say, axial load and torsion) is often in the range 1–2 mm, with from 10 to 30 crystal grains through the thickness (cf. Mair and Pugh (1964) and Ronay (1968)). Therefore, it is reasonable to identify a representative material volume, at the level of overall aggregate behavior, to be a "unit cube" in the unstressed reference state (of linear dimension 1 mm, say, for moderately fine-grained metals), containing a minimum of 1000 crystal grains. (The concept of the "unit cube" in this context was introduced by Bishop and Hill, 1951a.) Precise boundary data for the unit cube in our mathematical idealization of the crystal-to-aggregate transition problem will be introduced in the sequel.

To advance analytically, we restrict consideration to *quasi-static* deformation processes and suppose body forces negligible. Therefore, the nominal stress field $\mathbf{s}(X)$ (across interior faces between grains, $\mathbf{s}$ need only satisfy the appropriate traction conditions) within the macro-element must satisfy

$$\mathrm{div}\,\mathbf{s} = \mathbf{0}, \quad \mathrm{div}\,\dot{\mathbf{s}} = \mathbf{0} \quad \text{in} \quad V_0, \tag{3.3}$$

where V_0 denotes the reference volume. (The divergence is of course taken with respect to the reference position $\mathbf{x}_0$ of material point X, which operation commutes with time differentiation.) Let $\mathbf{t}_N$ denote the nominal traction vector across the surface of the deformed unit cube, so that

$$\mathbf{n}_0.\mathbf{s} = \mathbf{t}_N, \quad \mathbf{n}_0.\dot{\mathbf{s}} = \dot{\mathbf{t}}_N \quad \text{on} \quad \partial V_0, \tag{3.4}$$

where $\mathbf{n}_0$ is the unit normal to the undeformed cube at a point on its surface ∂V_0. Further, let $\langle \ldots \rangle$ denote volume average per unit reference volume, with $\int dV_0 = 1$ to some scale (of order 1 mm^3). Then, from $(3.3)_1$, $(3.4)_1$, and the Green–Gauss transformation, we obtain for $\mathbf{F}$, $\mathbf{s}$ and their tensor product $\boldsymbol{\tau}$ (recall $(2.12)_3$)

$$\langle \mathbf{F} \rangle = \int \mathbf{x}\otimes\mathbf{n}_0\,dA_0, \quad \langle \mathbf{s} \rangle = \int \mathbf{x}_0\otimes\mathbf{t}_N\,dA_0, \tag{3.5}$$

and
$$\langle \boldsymbol{\tau} \rangle = \langle \mathbf{Fs} \rangle = \int \mathbf{x} \otimes \mathbf{t}_{\mathrm{N}} \, dA_0, \tag{3.6}$$

in which the integrals are taken over the reference surface ∂V_0 of the unit cube (of area A_0). The *averaging theorem at finite deformation* can be stated

$$\langle \boldsymbol{\tau} \rangle \equiv \langle \mathbf{Fs} \rangle = \langle \mathbf{F} \rangle \langle \mathbf{s} \rangle, \tag{3.7}$$

which is identical to equation (3.8) in Hill (1972). Specifications of boundary conditions on the macro-element, for which the theorem holds rigorously, follow.

We first consider the boundary data of Hill (1972, p. 139), who supposes that either (i) the deformed macro-element (Hill does not specifically define his "macro-element" as a cube in the reference configuration, but merely as a unit volume to some scale) is (in effect) embedded in a uniformly strained continuum with macroscopic deformation gradient $\overline{\mathbf{F}} = \langle \mathbf{F} \rangle$ or (ii) the nominal traction $\mathbf{t}_{\mathrm{N}}(X)$ on the element surface is in equilibrium with a uniform field of macroscopic nominal stress $\bar{\mathbf{s}} = \langle \mathbf{s} \rangle$. Thus, either (Hill, 1972, equation (3.7))

$$\mathbf{x} = \langle \mathbf{F} \rangle \mathbf{x}_0 \quad \text{or} \quad \mathbf{t}_{\mathrm{N}} = \mathbf{n}_0 . \langle \mathbf{s} \rangle \quad \text{on} \quad \partial V_0. \tag{3.8}$$

Upon substituting $(3.8)_1$ or $(3.8)_2$ into the right-hand side of (3.6) and comparing $(3.5)_2$ or $(3.5)_1$ respectively, we obtain the stated decomposition (3.7) of $\langle \mathbf{Fs} \rangle$ into the tensor product of separate averages.

The alternative surface conditions for the unit cube proposed by Havner (1971), and extended to finite deformations in Havner (1974), are in part less restrictive and in part more so than (either of) the requirements (3.8). Consider as a model an ideal, macroscopically homogeneous aggregate consisting of a very large number of *identical* polycrystalline "unit cubes" in the reference state. Consistent with this model, macroscopically uniform fields of deformation and stress must satisfy (Havner, 1974, equations (4.1))

$$\mathbf{x}(A_i^+) = \mathbf{x}(A_i^-) + \overset{i}{\mathbf{c}}, \quad \mathbf{t}_{\mathrm{N}}(A_i^+) = -\mathbf{t}_{\mathrm{N}}(A_i^-) \quad (i = 1, 2, 3), \tag{3.9}$$

where A_i^+, A_i^- are the pair of "cube" material faces (initially of unit area) which were normal to the Cartesian reference axis X_i before deformation, and $\overset{i}{\mathbf{c}}$ is the position of face A_i^+ relative to face A_i^- of the deformed cube ($\overset{i}{\mathbf{c}}$ is independent of position over A_i because the cubes are equally deformed). The notation $\mathbf{x}(A_i^+)$, $\mathbf{t}_{\mathrm{N}}(A_i^+)$, etc. signifies a possible point-dependence over the indicated face.

From (3.5) and (3.9) we obtain (Havner, 1974, equations (4.3))

$$\left.\begin{aligned} \langle F_{im} \rangle &= \int x_i n_m^0 \, dA_0 = \int (x_i^+ - x_i^-) dA_m^+ = \overset{m}{c}_i, \\ \langle s_{mj} \rangle &= \int x_m^0 t_j^{\mathrm{N}} \, dA_0 = \int t_j^{\mathrm{N}} \, dA_m^+, \end{aligned}\right\} \tag{3.10}$$

in which the integrals are taken over the (positive) unit reference face A_m^0 of the cube. Moreover, from (3.6) and (3.9),

$$\langle \tau_{ij} \rangle = \langle F_{im} s_{mj} \rangle = \int x_i t_j^{\mathrm{N}} \, dA_0 = \overset{m}{c}_i \int t_j^{\mathrm{N}} \, dA_m^+ \tag{3.11}$$

(with summation on the repeated index and the integration again taken over the positive reference face A_m^0 for each m). The averaging theorem (3.7) follows immediately from (3.10) and (3.11).

Each of the three proposals regarding boundary conditions on a macro-element (viz. uniform constraint $(3.8)_1$, uniform loading $(3.8)_2$, and equal deformation of identical cubes (3.9)) is of course a *mathematical* conception, and as remarked by Hill (private correspondence, May 4, 1971), "none of these exactly reproduce the actual situation in materials testing." However, they remain (equally) useful models, and Hill (1967, p. 81) has given convincing physical arguments, relating to an assessment of the expected "wavelike" fluctuations of stress and strain about a mean in a macroscopically homogeneous system, which support the contention (Hill, 1972, p. 139) "that derived constitutive properties are in the limit independent of the data, so long as this is merely 'macroscopically uniform'." In this sense (3.7) *alone* can be taken as a precise definition of macroscopically homogeneous fields in polycrystalline metals.

Consider now changes with respect to the current deformed state of the macro-element. From $(3.3)_2$, $(3.4)_2$, and the Green–Gauss transformation we can write, analogous to (3.5) and (3.6) (with $\mathbf{v} \equiv \dot{\mathbf{x}}$),

$$\langle \dot{\mathbf{F}} \rangle = \int \mathbf{v} \otimes \mathbf{n}_0 \, dA_0, \quad \langle \dot{\mathbf{s}} \rangle = \int \mathbf{x}_0 \otimes \dot{\mathbf{t}}_N \, dA_0, \tag{3.12}$$

and

$$\langle \dot{\mathbf{F}} \dot{\mathbf{s}} \rangle = \int \mathbf{v} \otimes \dot{\mathbf{t}}_N \, dA_0. \tag{3.13}$$

Then, from any one of $(3.8)_{1,2}$ or (3.9), assumed to hold throughout the deformation, there follows an equivalent decomposition to (3.7) in terms of the rates of change of $\mathbf{F}$ and $\mathbf{s}$ (cf. Hill, 1972, equation (3.10), and Havner, 1974, equation (4.6)),

$$\langle \dot{\mathbf{F}} \dot{\mathbf{s}} \rangle = \langle \dot{\mathbf{F}} \rangle \langle \dot{\mathbf{s}} \rangle, \tag{3.14}$$

with the further proviso that these may be *unrelated* changes (say $d_1 \mathbf{s}, d_2 \mathbf{F}$) from the current state. (This is readily proved by a parallel of the previous arguments.) We also note, again adopting any one of the three proposals for boundary conditions, that the work-rate per unit reference volume, integrated over the unit cube, can be written (cf. Hill, 1972, equation (3.11))

$$\langle \mathbf{T} . \dot{\mathbf{E}} \rangle = \langle \dot{\mathbf{s}} . \dot{\mathbf{F}} \rangle = \int \mathbf{t}_N . \mathbf{v} \, dA_0 = \langle \mathbf{s} \rangle . \langle \dot{\mathbf{F}} \rangle. \tag{3.15}$$

Henceforth we take the averaging theorem for a polycrystalline macro-element to hold in $\mathbf{F}$, $\mathbf{s}$ and their rates without further regard to the precise specification of boundary data.

3.3. *Macroscopic variables and work invariants*

As the volume averages of $\mathbf{F}$, $\mathbf{s}$ and $\dot{\mathbf{F}}$, $\dot{\mathbf{s}}$ are uniquely determined by surface data alone (from (3.5) and (3.12)), and as the work-rate over the macro-element has the decomposition (3.15) (for any of the boundary conditions leading to the averaging theorem), $\langle \mathbf{F} \rangle$ *and* $\langle \mathbf{s} \rangle$ *are suitable as macroscopic measures of stress and deformation.* Henceforth all macroscopic measures will be distinguished by a bar. Thus (cf. Hill, 1972, equation (4.1)),

$$\bar{\mathbf{F}} = \langle \mathbf{F} \rangle, \quad \bar{\mathbf{s}} = \langle \mathbf{s} \rangle. \tag{3.16}$$

The interpretation of $\langle F_{im} \rangle$ and $\langle s_{mj} \rangle$ as macroscopic measures is particularly clear from (3.10), in which they are seen to equal the natural, *operational* definitions of overall deformation gradient and nominal ("engineering") stress for the unit cube (that is, projected length per unit reference length and force component per unit reference area respectively).

Other macroscopic variables can be defined in their usual way from the operationally determined values $\bar{\mathbf{s}}, \bar{\mathbf{F}}$. In particular, macroscopic Kirchhoff stress $\bar{\boldsymbol{\tau}}$ is given by the averaging theorem (3.7) (cf. Hill, 1972, equation (4.8), and Havner, 1974, equation $(4.7)_1$),

$$\bar{\boldsymbol{\tau}} = \langle \boldsymbol{\tau} \rangle = \bar{\mathbf{F}}\bar{\mathbf{s}}, \tag{3.17}$$

and the macroscopic rate of deformation and material spin tensors $\bar{\mathbf{D}}, \bar{\mathbf{W}}$ are determined as the symmetric and skew-symmetric parts of $\bar{\mathbf{L}}$, calculated from (Havner, 1974, equation $(4.7)_2$)

$$\bar{\mathbf{L}} = \frac{d\bar{\mathbf{F}}}{dt}(\bar{\mathbf{F}})^{-1}. \tag{3.18}$$

(Note that $(\bar{\mathbf{F}})^{-1}$ is uniquely defined because $\det \bar{\mathbf{F}} = \langle \rho_0/\rho \rangle = V$, the non-zero volume of the deformed macro-element.) Further, macroscopic rotation and right-stretch tensors $\bar{\mathbf{R}}, \bar{\boldsymbol{\Lambda}}$ can be calculated from the polar decomposition theorem (recall (2.2)):

$$\bar{\boldsymbol{\Lambda}}^2 = \bar{\mathbf{F}}^{\mathrm{T}}\,\bar{\mathbf{F}}, \quad \bar{\mathbf{R}} = \bar{\mathbf{F}}(\bar{\boldsymbol{\Lambda}})^{-1}. \tag{3.19}$$

Thus, the components of *any* pair of macro-strain and conjugate macro-stress measures $\bar{\mathbf{E}}(f(\bar{\lambda}))$ and $\bar{\mathbf{T}}$ are determinable from $\bar{\mathbf{F}}, \bar{\mathbf{s}}$ using (3.17)–(3.19), (2.4), (2.25), and (2.26). Moreover, from (3.15) these macroscopic measures will satisfy (cf. Hill, 1972, equation (4.4))

$$\bar{\mathbf{T}}.d\bar{\mathbf{E}} = \langle \mathbf{T}.d\mathbf{E} \rangle, \tag{3.20}$$

a scalar invariant under change in strain measure. (Equation (3.20) does *not* imply $\bar{\mathbf{T}} = \langle \mathbf{T} \rangle$ and $\bar{\mathbf{E}} = \langle \mathbf{E} \rangle$ for the same choice of measure at the micro- and macro-levels, however, as pointed out by Hill (1972, pp. 140–141).)

Consider now the invariant bilinear form (2.1) and the equality (2.32), written in terms of *macroscopic* variables $\bar{\mathbf{s}}$, $\bar{\mathbf{F}}$, and $\bar{\mathbf{T}}$, $\bar{\mathbf{E}}$:

$$d_1\bar{\mathbf{s}}.d_2\bar{\mathbf{F}} - d_2\bar{\mathbf{s}}.d_1\bar{\mathbf{F}} = d_1\bar{\mathbf{T}}.d_2\bar{\mathbf{E}} - d_2\bar{\mathbf{T}}.d_1\bar{\mathbf{E}}. \tag{3.21}$$

From (3.14) and (3.16), the left side can be replaced by the volume average (over the macro-element) of the *local* bilinear form (i.e. $\langle d_1\mathbf{s}.d_2\mathbf{F} - d_2\mathbf{s}.d_1\mathbf{F} \rangle$). Thus, there follows (Hill, 1972, p. 142)

$$d_1\bar{\mathbf{T}}.d_2\bar{\mathbf{E}} - d_2\bar{\mathbf{T}}.d_1\bar{\mathbf{E}} = \langle d_1\mathbf{T}.d_2\mathbf{E} - d_2\mathbf{T}.d_1\mathbf{E} \rangle, \tag{3.22}$$

"and so the bilinear invariant has the notable property that its macroscopic value is an unweighted average over the representative volume." Equation (3.22) is the central result which enables us to investigate the connection, mentioned in the introductory paragraphs of Section 3, between local crystalline behavior and the normality postulate of macroscopic plasticity theory.

3.4. *Elastic unloading and the normality postulate*

We presume that continuum stresses at a crystal material "point" can be sustained only through interatomic forces averaged over the underlying lattice volume. Consequently, the stress state is taken to be directly connected to the gross *lattice* strain, which we suppose is governed by *nonlinear thermoelasticity theory*, as argued in Havner (1973b, 1974). Then, in an isothermal deformation process (wherein all changes are measured after heat flow and energy dissipation have taken place and local temperature has returned to its equilibrium value), the lattice is Green-elastic and any fourth-rank tensor $\mathscr{L}$ of instantaneous elastic moduli has the symmetries $\mathscr{L}_{ijkl} = \mathscr{L}_{jikl} = \mathscr{L}_{klij}$.

Let $\hat{d}\boldsymbol{\tau}$ denote the *lattice* co-rotational increment in Kirchhoff stress. From (2.38) and $(3.2)_2$ (cf. Havner, 1972, equation (11), or Hill and Rice, 1972, equation (3.9)),

$$\frac{\hat{d}\boldsymbol{\tau}}{dt} \equiv \hat{\boldsymbol{\tau}} = \dot{\boldsymbol{\tau}} + \boldsymbol{\tau}\boldsymbol{\omega} - \boldsymbol{\omega}\boldsymbol{\tau} = \frac{\mathscr{D}\boldsymbol{\tau}}{\mathscr{D}t} - \sum_k \boldsymbol{\alpha}_k \dot{\gamma}_k, \tag{3.23}$$

where the symmetric tensor $\boldsymbol{\alpha}_k$ is given by

$$\boldsymbol{\alpha}_k = \boldsymbol{\tau}\boldsymbol{\Omega}_k - \boldsymbol{\Omega}_k\boldsymbol{\tau}. \tag{3.24}$$

We denote by $\mathscr{L}_0$ the tensor of instantaneous elastic moduli connecting $\hat{\boldsymbol{\tau}}$ and the lattice strain-rate $\boldsymbol{\varepsilon}$ during a purely *elastic* change, with $\mathscr{L}_0 = \mathscr{L}_0^{\mathrm{T}}$ (as previously noted for any $\mathscr{L}$). We also assume that $\hat{\boldsymbol{\tau}} = \mathscr{L}_0 . \boldsymbol{\varepsilon}$ when there is simultaneous incremental slipping. (The probable error in this assumption has been assessed by Hill and Rice (1972) and Havner (1973b) and is analyzed in Section 4.1.) Thus, from $(3.2)_1$ and (3.23),

$$\frac{\mathscr{D}\boldsymbol{\tau}}{\mathscr{D}t} = \mathscr{L}_0 . \mathbf{D} - \sum_k \boldsymbol{\Lambda}_k \dot{\gamma}_k, \tag{3.25}$$

where

$$\boldsymbol{\Lambda}_k = \mathscr{L}_0 . \mathbf{N}_k - \boldsymbol{\alpha}_k. \tag{3.26}$$

(Hill and Rice, 1972, equation (3.5), have given a slightly more general expression for the $\boldsymbol{\Lambda}_k$ which takes into account a possible difference between $\hat{\boldsymbol{\tau}}$ and $\mathscr{L}_0 . \boldsymbol{\varepsilon}$.) Upon substituting (3.25) into (2.27), we obtain an equation for the materially objective increment $d\mathbf{T}^{(0)}$ in an arbitrary stress measure $\mathbf{T}$, evaluated with the current state as reference (Hill and Rice, 1972, equation (2.1_1)):

$$d\mathbf{T}^{(0)} = \mathscr{L} . d\mathbf{E}^{(0)} - \sum_k \boldsymbol{\Lambda}_k d\gamma_k, \tag{3.27}$$

in which (cf. Hill, 1968, or Hill and Rice, 1972)

$$\left.\begin{aligned} &\mathscr{L} = \mathscr{L}_0 - m\mathscr{T}, \\ &\mathscr{T}_{ijkl} = \tfrac{1}{2}(\sigma_{ik}\delta_{jl} + \sigma_{il}\delta_{jk} + \sigma_{jk}\delta_{il} + \sigma_{jl}\delta_{ik}), \end{aligned}\right\} \tag{3.28}$$

and $\mathscr{L}_0 = \mathscr{L}_L$, the tensor modulus for the logarithmic measure in the reference state. Henceforth we take $\mathscr{L}$ (for any m from -1 to $+1$, say) to be *positive-definite*, hence invertible, and so also write (3.27) in the alternative form

$$d\mathbf{E}^{(0)} = \mathscr{L}^{-1} . d\mathbf{T}^{(0)} + \sum_k \mathbf{M}_k d\gamma_k, \quad \mathbf{M}_k = \mathscr{L}^{-1} . \boldsymbol{\Lambda}_k. \tag{3.29}$$

(It is of course understood that the corresponding quadratic form in $\mathscr{L}$ is positive-definite only for *symmetric* second-rank tensors.)

As in Hill and Rice (1972), we observe from (3.27) that $\sum \mathbf{\Lambda}_k d\gamma_k$ is the *stress decrement* after an infinitesimal *strain* cycle (from the current state) whereas (from (3.29)) $\sum \mathbf{M}_k d\gamma_k$ is the *strain increment* after an infinitesimal *stress* cycle. Correspondingly, we call these residuals the "plastic stress decrement" $d_p\mathbf{T}^{(0)}$ and "plastic strain increment" $d_p\mathbf{E}^{(0)}$ respectively:

$$d_p\mathbf{T}^{(0)} = \mathscr{L}.d\mathbf{E}^{(0)} - d\mathbf{T}^{(0)}, \quad d_p\mathbf{E}^{(0)} = d\mathbf{E}^{(0)} - \mathscr{L}^{-1}.d\mathbf{T}^{(0)}. \tag{3.30}$$

By analogy, we define *macroscopic* plastic changes over a representative volume element as

$$d_p\overline{\mathbf{T}}^{(0)} = \overline{\mathscr{L}}.d\overline{\mathbf{E}}^{(0)} - d\overline{\mathbf{T}}^{(0)}, \quad d_p\overline{\mathbf{E}}^{(0)} = d\overline{\mathbf{E}}^{(0)} - \overline{\mathscr{L}}^{-1}.d\overline{\mathbf{T}}^{(0)}, \tag{3.31}$$

where $\overline{\mathscr{L}}$ is the tensor of instantaneous macroscopic elastic moduli on scale m (i.e. $\overline{\mathscr{L}} = \overline{\mathscr{L}}_0 - m\overline{\mathscr{T}}$) which we take to be determinable, in principle, from experiment. [As is well known, $\overline{\mathscr{L}}$ is *not* the unweighted volume average of $\mathscr{L}$. See, for example, Hill (1967) and Havner (1971).]

Let $d^*\overline{\mathbf{T}}, d^*\overline{\mathbf{E}}$ denote purely elastic increments from the current deformed state of the (aggregate) macro-element, so that $d^*\overline{\mathbf{T}} = \overline{\mathscr{L}}.d\overline{\mathbf{E}}$, while $d\overline{\mathbf{T}}, d\overline{\mathbf{E}}$ are arbitray as in (3.31). For this unrelated pair of differentials, the bilinear form (3.21) in macroscopic variables can be equivalently expressed, with the current state as reference,

$$d^*\overline{\mathbf{T}}^{(0)}.(d\overline{\mathbf{E}} - \overline{\mathscr{L}}^{-1}.d\overline{\mathbf{T}})^{(0)} = d^*\overline{\mathbf{E}}^{(0)}.(\overline{\mathscr{L}}.d\overline{\mathbf{E}} - d\overline{\mathbf{T}})^{(0)}. \tag{3.32}$$

Comparing (3.31), (3.32), and (3.21), we conclude that the classical normality postulate of (macroscopic) plasticity theory, in either stress or strain space, is synonymous with the proposition that the bilinear form (2.1) (in macroscopic measures) is *non-positive* under purely elastic d_1-variations. The dual statements of normality are (cf. Hill, 1967, equation (9))

$$(d^*\overline{\mathbf{T}}.d_p\overline{\mathbf{E}})^{(0)} \leqslant 0, \quad (d^*\overline{\mathbf{E}}.d_p\overline{\mathbf{T}})^{(0)} \leqslant 0, \tag{3.33}$$

and either both hold or neither does.

[Observe from the fundamental invariance of (2.1) (under changes in both strain measure and reference state) that writing (3.32) in the undeformed rather than the current state (of mean density $\bar{\rho}$) merely multiplies its value by $\bar{\rho}/\rho_0$, leaving its sign unchanged. Consequently, equations (3.33) may just as well be written with the *unstressed* state of the aggregate volume as reference. Indeed, this is necessary for the direct application of (3.22) in the sequel.]

In regard to a *yield surface* in *macro-stress space*, the domain of all materially objective, elastic stress increments $d^*\overline{\mathbf{T}}$ defines the cone (or pyramid) which is tangent to this surface at a vertex $\overline{\mathbf{T}}$ (see Hill, 1967, Section 3(ii)). At a smooth point the domain of the $d^*\overline{\mathbf{T}}$ is the half-space which contains (and is tangent to) the yield surface at $\overline{\mathbf{T}}$. Equation $(3.33)_1$ states that the plastic strain increment $d_p\overline{\mathbf{E}}$ lies within the outward cone (or pyramid) of normals to the yield surface at a vertex and is the unique outward normal at a smooth point. Similarly, in macro-*strain* space, the domain of all elastic strain increments $d^*\overline{\mathbf{E}}$ defines the tangent cone to the yield surface at a vertex $\overline{\mathbf{E}}$; and $(3.33)_2$ states that the plastic stress decrement $d_p\overline{\mathbf{T}}$ is contained within the outward cone of normals to the surface at $\overline{\mathbf{E}}$ (with comparable statements as before at a smooth point). Equivalently, $(3.33)_1$ states that the domains of elastic stress and plastic strain increments are orthogonal, whereas its dual $(3.33)_2$ expresses the orthogonality of the domains of elastic strain increment and plastic stress decrement. (Obviously, from the invariance property of the bilinear form, all the preceding statements are independent of the choice of strain measure $\overline{\mathbf{E}}(f(\bar{\lambda}))$.)

We now connect (3.33) with local crystalline behavior through the central result (3.22). We can immediately write (using (3.30) and taking into account the note in square brackets below (3.33))

$$\langle d^*\mathbf{T}.d_p\mathbf{E}\rangle^{(0)} \leqslant 0 \quad \text{or} \quad \langle d^*\mathbf{E}.d_p\mathbf{T}\rangle^{(0)} \leqslant 0 \tag{3.34}$$

as fully equivalent statements of the macroscopic normality postulate. Thus, from (3.27) and the non-negative values of the $d\gamma$'s, *the classical structure of the mathematical theory of plasticity can be physically motivated by the suitability of*

$$\mathbf{\Lambda}_k . d^*\mathbf{E}^{(0)} \leqslant 0 \quad (k = 1, \ldots, n), \tag{3.35}$$

as the elastic unloading criterion in the local critical slip systems (Hill and Rice, 1972). Suffice it to say here that (3.35) *is acceptable* (at constant pressure) for that purpose. The analysis that supports this contention, and a general theory of finite elastic–plastic deformation of crystals which encompasses it, are given in Section 4.

4. A General Theory of Finite Deformation of Elastic–Plastic Crystals

In the following Sections a broad mathematical theory of crystal plasticity is presented whose evolution can be traced through the work of Hill (1966, 1972), Rice (1971), Hill and Rice (1972), Sewell (1969, 1972, 1974), and Havner (1973b, 1974, 1977a, b). We first address two issues raised in Section 3.4: (i) the justification for using $\hat{\boldsymbol{\tau}} = \mathscr{L}_0 . \boldsymbol{\varepsilon}$ as the (incremental) constitutive equation of the lattice during continued crystallographic slip; and (ii) the physical appropriateness of $\mathbf{\Lambda}_k . d^*\mathbf{E}^{(0)} \leqslant 0$ as the local elastic unloading criterion at constant pressure. Subsequently, we present a physical motivation for the constitutive equations of Hill and Rice (1972) and introduce a symmetry postulate (Havner, 1977b; Havner and Shalaby, 1977) which enables the incorporation of these equations into the theoretical framework of "saddle-shaped generating functionals" of Sewell (1974). We then explore certain consequences of the general theory for the formulation and solution of boundary value problems, following developments in Havner (1977b).

4.1. *Crystal vs. lattice deformations: thermoelastic theory of the lattice*

We begin by expanding upon the idea, introduced in Section 3.4, that the crystal lattice deformation is Green-elastic. Let $\mathbf{e}_i$ ($i = 1, 2, 3$) denote orthonormal lattice directors ($\mathbf{e}_i . \mathbf{e}_j = \delta_{ij}$) which define the underlying lattice cell at a crystal material point X in an *unstressed* reference state, corresponding to a specified equilibrium temperature T_0. (As we have adopted crystallographic slip as the sole mechanism of gross inelastic deformation, the current temperature T should never exceed about 40% of the absolute melting temperature (see, for example, Hayden *et al.*, 1965).) The deformed lattice cell at X in the current configuration is, in general, a parallelepiped defined by three edge vectors $\mathbf{h}_i$. The *lattice deformation gradient* $\mathbf{J}$ is the transformation which maps $\mathbf{e}_i$ into $\mathbf{h}_i$ ($i = 1, 2, 3$): $\mathbf{h}_i = \mathbf{J}\mathbf{e}_i$. Further, the *lattice Green-strain* relative to the unstressed configuration is given by (compare (2.11) for total Green-strain $\mathbf{E}_G$)

$$\boldsymbol{\eta} = \tfrac{1}{2}(\mathbf{J}^T\mathbf{J} - \mathbf{I}), \tag{4.1}$$

or, equivalently, referred to the $\mathbf{e}_i$-directions,

$$\eta_{ij} = \tfrac{1}{2}(\mathbf{h}_i \cdot \mathbf{h}_j - \delta_{ij}) = \tfrac{1}{2}(\lambda_i \lambda_j \cos\theta_{ij} - \delta_{ij}), \tag{4.2}$$

where the λ's are the averaged lattice stretches (i.e. the lengths of the **h**'s) along the (initially orthogonal) crystal axis directions and θ_{ij} is the angle between $\mathbf{h}_i$, $\mathbf{h}_j$ of the deformed cell. (In the testing of single crystals, the λ's and θ's, as well as the directions of the **h**'s, can be determined as averaged values over the crystal material point by X-ray diffraction techniques, at least in principle; hence **J** is fully determined.) The rate of change of $\boldsymbol{\eta}$ is connected to the (Eulerian) lattice strain-rate $\boldsymbol{\varepsilon}$, introduced in (3.1), by

$$\dot{\eta} = \mathbf{J}^{\mathrm{T}} \boldsymbol{\varepsilon} \mathbf{J} \tag{4.3}$$

(compare (2.35) for $\dot{\mathbf{E}}_{\mathrm{G}}$).

Let **y** denote the position vector of X as an infinitesimal neighborhood $\mathcal{N}(X)$ momentarily passes through an unstressed state, at the reference equilibrium temperature T_0, during some deformation history from another unstressed state $\mathbf{x}_0$. Then $\mathbf{J} = \mathbf{Q}$ momentarily, where **Q** is a proper orthogonal transformation; but the *overall* (total) deformation gradient is **P**, which we shall call the *plastic deformation gradient*. ($\mathbf{P} = \mathbf{Q}$ if the deformation history from $\mathbf{x}_0$ to **y** is purely elastic.) Thus, $d\mathbf{y} = \mathbf{P}\,d\mathbf{x}_0$, while for the underlying crystalline lattice $\mathbf{h}_i(\mathbf{y}) = \mathbf{Q}\mathbf{e}_i(\mathbf{x}_0)$. If further slipping is not imminent as X passes through **y** [this requires that the local (crystal) elastic domain in (micro) stress space, bounded by the multi-faceted current yield surface associated with the N slip systems, contain the origin for arbitrary measure **T**], the material within $\mathcal{N}(X)$ can be *thermoelastically* deformed to some other position **x**, so that (cf. Havner, 1974, equations (2.1))

$$d\mathbf{x} = \mathbf{J}\,d\mathbf{y} = \mathbf{J}\mathbf{P}\,d\mathbf{x}_0 = \mathbf{F}\,d\mathbf{x}_0. \tag{4.4}$$

Therefore, if an unstressed state of $\mathcal{N}(X)$ is chosen for the reference configuration $\mathbf{x}_0$, and if some other unstressed state **y** at the same reference temperature can be attained thermoelastically from the current deformed state **x**, then the total deformation gradient **F** has the decomposition **JP** of (4.4). (A decomposition of this type was introduced into the theory of plasticity by Lee and Liu (1967), although not in the specific context of crystal deformation.) Moreover, from $(d\mathbf{x})^{\cdot} = \mathbf{L}\,d\mathbf{x}$, it is possible to connect **J**, **P** and their rates to the fundamental kinematic decomposition (3.1) (see Havner, 1974, Section 2.1). The corresponding equations do not appear to be particularly useful, however, and (4.4) itself (although adopted by many writers in both crystal and macroscopic plasticity) is of questionable value if the current (crystal) yield surface in stress space does not contain the origin. (There is no reason from thermostatic laws that it must, as discussed by Rice, 1971 and Havner, 1974.) Here, we shall use only (4.3) relating lattice strain-rates.

To define the thermoelastic theory, we introduce contravariant Kirchhoff stress $\boldsymbol{\tau}^*$ convected with respect to the *lattice*, given by

$$\boldsymbol{\tau}^* = \mathbf{J}^{-1}\boldsymbol{\tau}(\mathbf{J}^{-1})^{\mathrm{T}} \tag{4.5}$$

such that $\boldsymbol{\tau}^* = \boldsymbol{\tau}^{\mathrm{c}}$, from $(2.12)_1$, in any *thermoelastic* deformation from an unstressed state **y** to **x** (as in $(4.4)_1$), but not otherwise. Further, we consider that the current (thermodynamic) state of $\mathcal{N}(X)$ in configuration **x** is characterized by the underlying lattice strain $\boldsymbol{\eta}$, the equilibrium temperature T, and a set of microstructural variables ξ_m ($m = 1, \ldots, M$) related to the ongoing *inelastic* deformation at X (Havner, 1973b, 1974). We shall not attempt here to specify the ζ's, other than to remark that they perhaps can be related to dislocation densities and dislocation moment tensors and, conceivably, are roughly representable by the gross crystallographic shears. [There is some similarity between this view and the choice of total shape-change strain as a basic thermodynamic variable (see Rice (1971, 1975) and Kratochvil (1973), for example), although the formal arguments are somewhat different. However, Kocks (1975, Section 3.6) presents objections to the use of total strain as a state variable (also see Havner, 1973b, Section 4).] The ζ's are manifest in the differences between crystal yield surfaces at X in different unstressed states $\mathbf{x}_0$ and $\mathbf{y}$, at equilibrium temperature T_0, whenever there has been intervening crystallographic slip. From rather standard thermostatic arguments, the *thermoelastic theory of the crystal lattice* gives (cf. Havner, 1973b, 1974)

$$\boldsymbol{\tau}^* = \rho_0 \frac{\partial \psi}{\partial \boldsymbol{\eta}}, \quad \psi = \psi(\boldsymbol{\eta}, T, \zeta_m, m = 1, \ldots, M), \tag{4.6}$$

where the potential function ψ is (Helmholtz) free energy per unit mass.

Consider an *isothermal* change from the current state which includes incremental slip as well as incremental lattice strain:

$$d\boldsymbol{\tau}^* = \rho_0 \left(\frac{\partial^2 \psi}{\partial \boldsymbol{\eta} \otimes \partial \boldsymbol{\eta}} \right) . d\boldsymbol{\eta} + \rho_0 d \left(\frac{\partial \psi}{\partial \boldsymbol{\eta}} \right) \bigg|_{\boldsymbol{\eta}}, \tag{4.7}$$

with the second term on the right side arising from changes in the microstructural variables. From (4.3), (4.5), and (4.7) (cf. Havner, 1973b, Section 6; 1978, equation (25)),

$$\hat{d}\boldsymbol{\tau} - \mathscr{L}_0 . \boldsymbol{\varepsilon} dt = \rho_0 \mathbf{J} d \left(\frac{\partial \psi}{\partial \boldsymbol{\eta}} \right) \bigg|_{\boldsymbol{\eta}} \mathbf{J}^{\mathrm{T}}, \tag{4.8}$$

where

$$(\mathscr{L}_0)_{ijkl} = \rho_0 J_{im} J_{jn} J_{kp} J_{lq} \frac{\partial^2 \psi}{\partial \eta_{mn} \partial \eta_{pq}} + (\rho_0/\rho) \mathscr{T}_{ijkl} \tag{4.9}$$

with the evident Green-elastic symmetries previously stated (Section 3.4). The right side of (4.8) (alternatively expressed $\sum_k \beta_k d\gamma_k$ by Hill and Rice, 1972) was assumed negligible in Section 3.4. We now present a justification for that assumption.

The case of greatest practical interest involving finite lattice strain is that of additional, infinitesimal (lattice) deformations superimposed upon a state of large hydrostatic compression. The corresponding Cauchy stress can be expressed, to first order in superimposed lattice strain $\boldsymbol{\xi}$,

$$\boldsymbol{\sigma} = -p(1 - \operatorname{tr} \boldsymbol{\xi})\mathbf{I} + \mathscr{L}(p) . \boldsymbol{\xi}, \tag{4.10}$$

where p is the (large) pressure and $\mathscr{L}(p)$ represents the tensor of pressure-dependent elastic moduli associated with incremental straining of the lattice from the hydrostatic state $\boldsymbol{\sigma} = -p\mathbf{I}$ (see, for example, Havner, 1973b, equation (4.2)). The tensor $\mathscr{L}(p)$ also depends to some extent upon the previous inelastic deformation at X through the state variables ζ_m. To focus upon this dependence we consider isothermal changes *at constant pressure* from the current deformed state; whence $\varepsilon = \dot{\boldsymbol{\xi}}$, to first order in $\boldsymbol{\xi}$, and

$$(\rho/\rho_0) \hat{d}\boldsymbol{\tau} = \hat{d}\boldsymbol{\sigma} + \boldsymbol{\sigma} \operatorname{tr} d\boldsymbol{\xi} = \mathscr{L}(p) . d\boldsymbol{\xi} + d\mathscr{L}\big|_p . \boldsymbol{\xi}, \tag{4.11}$$

where $d\mathscr{L}|_p$ represents changes in the instantaneous elastic moduli (at constant pressure) caused by the incremental slips. Thus, the right side of (4.8) is equivalent to $(\rho_0/\rho)d\mathscr{L}|_p.\xi$, again to first order in $\boldsymbol{\xi}$ (Havner, 1973b, equations (6.3) and (6.4)), and $\mathscr{L}_0 = (\rho_0/\rho)\mathscr{L}(p)$.

From experimental evidence (see, for example, McLean, 1962) the changes in elastic properties of single crystals, after they have been severely deformed to overall strains of order 1, are only a few percent. Thus, we expect that $d\mathscr{L}|_p = 10^{-2}\mathscr{L}\,d\gamma$ (element by element magnitudes), where γ is a representative value of plastic shear. Further, letting ξ represent the magnitude of lattice strain with respect to the hydrostatic reference state, we typically expect $\xi = 10^{-4}$ to 10^{-3} in metals. Thus, at finite distortions ($\gamma = 10^{-1}$ to 1, say), with continued inelastic straining dominant, $d\gamma = (10^2$ to $10^3)d\xi$ and the second term on the right side of (4.11) is of *maximum* order 10^{-3} to 10^{-2} times the first term. Therefore, the assumption that $\hat{\boldsymbol{\tau}} = \mathscr{L}_0.\boldsymbol{\varepsilon}$ during continued elastic–plastic deformation is justified.

4.2. *Elastic unloading at large pressure*

We now return to the question of the suitability of (3.35), $\boldsymbol{\Lambda}_k.d^*\mathbf{E}^{(0)} \leqslant 0$, as the (local) elastic unloading criterion for the case of finite lattice strain. In investigating this matter, it is useful to have an equation for $\boldsymbol{\Lambda}_k.\boldsymbol{\varepsilon}$ in terms of the (true) resolved shear stress $\mathbf{N}_k.\boldsymbol{\sigma}$ and its rate of change. This relation is developed as follows.

Consistent with (4.1) and (4.3), the lattice co-rotational derivative of the underlying lattice deformation gradient $\mathbf{J}$ is given by $\hat{\mathbf{J}} = \boldsymbol{\varepsilon}\mathbf{J}$. Thus, with the aid of Nanson's equation of vectorial areas, written for any (lattice) vectorial reference area $\mathbf{n}\delta A$ within the lattice volume at material point X, we obtain

$$(\delta \mathrm{A})^{\cdot}/\delta \mathrm{A} = \mathrm{tr}\,\boldsymbol{\varepsilon} - \mathbf{n}.\boldsymbol{\varepsilon}\mathbf{n}, \quad \hat{\mathbf{n}} = (\mathbf{n}.\boldsymbol{\varepsilon}\mathbf{n})\mathbf{n} - \mathbf{n}.\boldsymbol{\varepsilon}, \tag{4.12}$$

where we have used $\mathrm{tr}\,\boldsymbol{\varepsilon} = (\delta V)^{\cdot}/\delta V$, the natural (logarithmic) volumetric strain-rate (recall $(3.2)_1$ and the comment at the end of Section 3.1). Also, for any lattice vectorial line segment $\mathbf{b}\lambda_b$ at X (given by $\mathbf{b}\lambda_b = \mathbf{J}\mathbf{b}_0$, where $\mathbf{b}_0$ is the direction of $\mathbf{b}$ in an unstressed reference state and λ_b is the lattice stretch),

$$\dot{\lambda}_b/\lambda_b = \mathbf{b}.\boldsymbol{\varepsilon}\mathbf{b}, \quad \hat{\mathbf{b}} = \boldsymbol{\varepsilon}\mathbf{b} - (\mathbf{b}.\boldsymbol{\varepsilon}\mathbf{b})\mathbf{b}. \tag{4.13}$$

Applying (4.12) and (4.13) to the pair of orthonormal vectors $\mathbf{b}_k$, $\mathbf{n}_k$ of the kth slip system and using $\hat{\boldsymbol{\tau}} = \mathscr{L}_0.\boldsymbol{\varepsilon}$, together with (3.23) and (3.24), and (3.26), we find

$$\boldsymbol{\Lambda}_k.\boldsymbol{\varepsilon}dt = d(\mathbf{N}_k.\boldsymbol{\sigma}) + \left(\frac{d(\delta A)}{\delta A} + \frac{d\lambda_b}{\lambda_b}\right)(\mathbf{N}_k.\boldsymbol{\sigma}), \tag{4.14}$$

which is the desired relation (Havner, 1973b, equation (5.11)).

Again consider the case of infinitesimal lattice strain $\boldsymbol{\xi}$ superimposed upon a state of hydrostatic compression. From (4.10) and the identity $\mathrm{tr}\,\mathbf{N}_k = 0$, the resolved shear stress is, to first order in $\boldsymbol{\xi}$,

$$\mathbf{N}_k.\boldsymbol{\sigma} = \mathbf{N}_k.\mathscr{L}(p).\boldsymbol{\xi}. \tag{4.15}$$

Consequently, at constant pressure, the first term on the right side of (4.14) is of order $\mathscr{L}d\xi$, whereas the second term is of order $\mathscr{L}\xi d\xi$, hence of maximum value $\approx 10^{-3}$ times the first term (equivalently, of second order in ξ). Therefore, during elastic unloading (in which

$\boldsymbol{\varepsilon} dt = d\boldsymbol{\xi} = d^*\mathbf{E}^{(0)}$), $\boldsymbol{\Lambda}_k . d^*\mathbf{E}^{(0)} = d(\mathbf{N}_k . \boldsymbol{\sigma})$ to first order in infinitesimal strain $\boldsymbol{\xi}$. Thus, for practical purposes, the unloading criterion (3.35) is equivalent *at constant pressure* to the statement that *the resolved shear stress in every critical slip system is decreasing.* As it is a critical value τ_k of $\mathbf{N}_k . \boldsymbol{\sigma}$, commonly called the "flow stress" (henceforth we shall call τ_k the *critical strength* of the kth slip system), which is generally considered to govern the motion of dislocations in a crystallographic slip system, we conclude that the criterion (3.35) is physically acceptable at constant pressure.

Before leaving this matter it is worthwhile to investigate the likelihood of macroscopic normality, based upon the local inequality (3.35), when a plastically deforming metal is subject to *large changes in pressure.* The necessary analysis for the case of *cubic* crystals is given in Havner (1973b) and outlined briefly in the following.

For a changing pressure state, to first order in $\boldsymbol{\xi}$,

$$\boldsymbol{\varepsilon} = (\dot{\lambda}/\lambda)\mathbf{I} + \dot{\boldsymbol{\xi}}, \tag{4.16}$$

where λ (< 1) is the cubic lattice "stretch" under hydrostatic stress $-p\mathbf{I}$. Upon substituting (4.16) into (4.14), we obtain for elastic unloading, again retaining only first-order terms in $\boldsymbol{\xi}$ (taking into account (4.15) as before),

$$\boldsymbol{\Lambda}_k . d^*\mathbf{E}^{(0)} = d(\mathbf{N}_k . \boldsymbol{\sigma}) + 3\frac{d\lambda}{\lambda}(\mathbf{N}_k . \boldsymbol{\sigma}), \quad \lambda = \lambda(p). \tag{4.17}$$

Hence, in a momentarily critical system (with $\mathbf{N}_k . \sigma = \tau_k$), normality requires $d\tau_k \leqslant -3(d\lambda/\lambda)\tau_k$ during elastic unloading (Havner, 1973b, equation (6.6)). If the critical strengths, as the elastic moduli, are considered to be monotone functions of large pressure, this relation must be an *equality* in order to recover the previous value of τ_k in an elastic, unloading–reloading pressure cycle. [If the pressure is cycled from a combined stress state for which slip is imminent in one or more systems, changes in $\boldsymbol{\xi}$ may be necessary to ensure elastic behavior throughout the cycle. The general requirement, from (4.15), is that $\mathbf{N}_k . \mathscr{L}(p) . \boldsymbol{\xi} \leqslant \tau_k(p)$ in each system at all pressures. Specific equations for cubic crystals are given in Havner (1973b).]

Thus, denoting $\Delta\tau_k$ as the *decrease* in the critical strength upon unloading to zero pressure along an elastic path, we find from the thermoelastic theory of the cubic lattice (refer to Havner, 1973b, for a detailed development) that

$$(\Delta\tau/\tau)_k = 3(p/c_1) - \tfrac{3}{2}(3 - C_1/c_1)(p/c_1)^2 + \ldots, \tag{4.18}$$

where

$$c_1 = (c_{11} + 2c_{12})_0, \quad C_1 = (C_{111} + 6C_{112} + 2C_{123})_0. \tag{4.19}$$

The c_{ij} and C_{ijk} are second- and third-order (isothermal) elastic moduli from a Taylor series expansion of (4.6) about an unstressed reference state, using the generalized Voigt notation introduced by Brugger (1964). [Subscript pairs in the respective fourth- and sixth-rank tensors are replaced by single indices according to the standard scheme (11) → 1, (22) → 2, (33) → 3, (23, 32) → 4, (31, 13) → 5, and (12, 21) → 6.] We shall refer to (4.18) as the "normality criterion" for the elastic dependence of critical strengths in cubic crystals on large pressure changes.

Isothermal values of c_1 and C_1 may be computed for a variety of cubic metals from the second- and third-order adiabatic and isothermal elastic moduli tabulated in Barsch and Chang (1967). Consequently, the predicted changes in the τ_k (corresponding to normality) can be determined from (4.18) to second order in p/c_1. Unfortunately, experimental information of these changes is rather scarce in the literature. The only pertinent data that I have found were obtained by Haasen and Lawson (1958), who conducted full-range tensile tests on several metal single crystals at 5×10^3 atm as well as at zero pressure. They concluded that average decreases in the critical strengths of active systems in copper and aluminum are approximately 1.9 and 6% respectively at 300 K. For these metals the isothermal values of c_1 and C_1, in units of 10^5 MN m^{-2}, are computed to be (Havner, 1973b): Cu, $c_1 = 3.99$ and $C_1 = -67.7$; Al, $c_1 = 2.195$ and $C_1 = -36.6$. Therefore, the decreases in τ_k according to (4.18), in going from 5×10^3 to 0 atm, would be 0.38% for Cu and 0.68% for Al. Thus, the "normality criterion" is *qualitatively* correct in that (i) it predicts increases in critical strengths with increasing pressure and (ii) gives the relative changes for the two metals in the correct order and roughly the correct proportion (3 to 1 experimentally vs. 2 to 1 theoretically for aluminum compared with copper). *Quantitatively*, however, the agreement is less satisfactory. It is conceivable that at very large pressures (of the order 10^5 atm) the theoretical values might be closer to the actual elastic changes, but data apparently are not available. The predicted decreases, from (4.18), corresponding to pressure reduction from 10^5 to 0 atm are 5.7% and 7.6% respectively for copper and aluminum, to second order in p/c_1.

4.3. *The constitutive equations of Hill and Rice*

To summarize our progress towards a rational theory of finite plastic deformation of crystals at constant pressure, we restate the following basic relationships:

$$\boldsymbol{\varepsilon} = \mathbf{D} - \sum_k \mathbf{N}_k \dot{\gamma}_k, \quad \mathbf{W} - \boldsymbol{\omega} = \sum_k \boldsymbol{\Omega}_k \dot{\gamma}_k, \quad \dot{\gamma}_k \geqslant 0 \quad (k = 1, \ldots, n), \tag{4.20}$$

$$\frac{\mathscr{D}\boldsymbol{\tau}}{\mathscr{D}t} = \mathscr{L}_0 . \mathbf{D} - \sum_k \boldsymbol{\Lambda}_k \dot{\gamma}_k, \quad \boldsymbol{\Lambda}_k = \mathscr{L}_0 . \mathbf{N}_k - \boldsymbol{\alpha}_k, \quad \boldsymbol{\alpha}_k = \boldsymbol{\tau}\boldsymbol{\Omega}_k - \boldsymbol{\Omega}_k\boldsymbol{\tau}, \tag{4.21}$$

and $\mathbf{N}_k . \boldsymbol{\sigma} \leqslant \tau_k$ in each of the N systems, with the *unloading* criteria for the n critical systems (corresponding to $\dot{\gamma}_k = 0$ for all k) given by

$$\boldsymbol{\Lambda}_k . \mathbf{D} \leqslant 0 \quad \text{when} \quad \mathbf{D} = \boldsymbol{\varepsilon} \quad \text{and} \quad \mathbf{N}_k . \boldsymbol{\sigma} = \tau_k. \tag{4.22}$$

We re-emphasize that these equations are compatible with both the classical normality postulate of macroscopic plasticity theory (through the averaging theorem and Hill's invariant bilinear form) and the thermoelastic theory of the lattice, as has been shown. The physical justification for (4.20)–(4.22) has been argued in Sections 3.1, 4.1 and 4.2. [Note that each side of each equation is objective under observer transformations, and note particularly that the *relative spin* $\mathbf{W} - \boldsymbol{\omega}$ of material and lattice, but neither spin separately, is admissible as a constitutive variable. This relative rotation will play an important role in the simple theory of latent hardening to be presented in Section 5.]

With the current state as reference, $\boldsymbol{\tau} = \boldsymbol{\sigma}$ and $\mathscr{L}_0 = \mathscr{L}(p)$, which we consider known for a given metal and pressure. (At zero or very small pressure, the constitutive equations (4.21) are correct to first order in infinitesimal lattice strain if we simply replace $\mathscr{L}(p)$ by the tensor of standard second-order (isothermal) elastic moduli for the crystal.) To complete the theory, (4.20)–(4.22) must be augmented by evolutionary equations for the τ_k such that their present values (for a particular strain-rate and equilibrium temperature) are known in all slip systems, given the deformation history. Toward that end we take into account the following experimental considerations.

Suppose that in a test of a uniformly strained crystal an experimentalist can impose (or measure) a particular incremental deformation gradient $d\mathbf{F} = \mathbf{L}\, dt$ (with the current state as reference), so that $\mathbf{D}$ and $\mathbf{W}$ are known (the latter with respect to some arbitrarily chosen reference frame). Further suppose that $\boldsymbol{\varepsilon}\, dt$ and $(\boldsymbol{\omega} - \mathbf{W})dt$, giving the incremental lattice strain and lattice rotation relative to the gross material, can be determined by X-ray diffraction techniques. Then, the incremental slips $d\gamma_k$ can be calculated from (3.1) or (4.20) (viz. $\sum_k (\mathbf{b}\otimes\mathbf{n})_k\, d\gamma_k = \{\mathbf{D} - \boldsymbol{\varepsilon} + (\mathbf{W} - \boldsymbol{\omega})\}dt$) if there are no more than *eight* simultaneously critical systems (noting that $\operatorname{tr}(\mathbf{b}\otimes\mathbf{n})_k \equiv 0$). Because the maximum symmetry of the *cubic* crystallographic system is eight-fold, and this only for particular loadings of crystals in their *plastically isotropic* virgin state, we consider the possibility of more than eight critical systems highly unlikely. (There are even fewer probable simultaneous slip systems in the hexagonal close-packed (h.c.p.) lattice system.) The problem comes, experimentally, in not knowing which of, say, a dozen potential systems (in a face-centered cubic (f.c.c.) crystal, for example) *are* critical because of deformation-induced anisotropy after finite deformation. However, the experimentalist can be guided by other information: in particular, traces of slip lines on polished crystal faces. This information may enable him to reduce the possible set of simultaneous slip systems to eight or less. Moreover, he is not actually concerned with *critical* systems, but only with *active* ones. Therefore, we take it to be generally possible that $\boldsymbol{\varepsilon}\, dt$, $(\mathbf{W} - \boldsymbol{\omega})dt$, and all nonzero $d\gamma$'s can be determined from experiment, given all components of an imposed $d\mathbf{E}^{(0)} = \mathbf{D}\, dt$ alone.

Consistent with the foregoing, *a rational theory of crystal plasticity should provide a unique solution for the* $\dot{\gamma}$'s *corresponding to a prescribed* $\mathbf{D}$. Then $\boldsymbol{\varepsilon}$, $\mathbf{W} - \boldsymbol{\omega}$, and $\mathscr{D}\boldsymbol{\tau}/\mathscr{D}t$ can be determined from (4.20) and (4.21), the current stress being known through the evolved history of the deformation. Hill and Rice (1972) have presented a set of *rate-independent* constitutive equations which assure this objective and which also incorporate the (normality) unloading criterion (4.22) in critical systems. With the current state as reference, these equations can be written (Hill and Rice, 1972, equations (6.1))

$$\left.\begin{aligned} \boldsymbol{\Lambda}_k \cdot \mathbf{D} &\leqslant \sum_j g_{kj}\dot{\gamma}_j \quad (\dot{\gamma}_k \geqslant 0), \\ \dot{\gamma}_k \boldsymbol{\Lambda}_k \cdot \mathbf{D} &= \dot{\gamma}_k \sum_j g_{kj}\dot{\gamma}_j \quad (k = 1, \ldots, n), \end{aligned}\right\} \tag{4.23}$$

where the moduli g_{kj} are invariant with respect to change in strain measure. Equations (4.23) constitute a "linear complementarity problem" in the incremental slips (or $\dot{\gamma}$'s) for a specified total strain increment $d\mathbf{E}^{(0)} = \mathbf{D}\, dt$. As shown by Hill and Rice (1972) in this context, and as is also known from the quadratic programming literature, a sufficient (and apparently necessary) condition for uniqueness of solution to these equations is that matrix (g_{kj}) be *positive-definite*. Sufficiency follows from the inequality (with Δ denoting the difference between assumed distinct pairs of corresponding quantities)

$$\sum_k \Delta\dot{\gamma}_k \boldsymbol{\Lambda}_k \cdot \Delta\mathbf{D} \geqslant \sum_k \sum_j g_{kj}\Delta\dot{\gamma}_k \Delta\dot{\gamma}_j, \tag{4.24}$$

proofs of which may be found in Hill and Rice (1972) and Sewell (1972) (also see Havner, 1978). The specification of positive-definite moduli g_{kj} is not at all a severe restriction on mechanical response. This assumption encompasses both strain-hardening and strain-softening behavior and merely serves to exclude so-called "locking" materials (see Havner, 1977a).

Continuing with the current state as reference, we redefine the symmetric tensor $\mathbf{M}_k$ of (3.29) specifically for the *logarithmic* measure of strain:

$$\mathbf{M}_k = \mathscr{L}_0^{-1} \cdot \mathbf{\Lambda}_k = \mathbf{N}_k - \mathscr{L}_0^{-1} \cdot \boldsymbol{\alpha}_k. \tag{4.25}$$

Then, from $(4.21)_1$, the constitutive equations (4.23) can be expressed

$$\left.\begin{aligned} \mathbf{M}_k \cdot \mathscr{D}\boldsymbol{\tau} &\leqslant \sum_j h_{kj}\, d\gamma_j \quad (d\gamma_k \geqslant 0), \\ d\gamma_k \mathbf{M}_k \cdot \mathscr{D}\boldsymbol{\tau} &= d\gamma_k \sum_j h_{kj}\, d\gamma_j \quad (k = 1, \ldots, n), \end{aligned}\right\} \tag{4.26}$$

which are equivalent to equations (6.3) in Hill and Rice (1972). The coefficients

$$h_{kj} = g_{kj} - \mathbf{\Lambda}_k \cdot \mathscr{L}_0^{-1} \cdot \mathbf{\Lambda}_j \tag{4.27}$$

will be called the *effective slip-system hardening moduli* (Havner, 1977b). If positive-definite, they assure a unique set of incremental shears and a unique $d\mathbf{E}^{(0)}$ for any material co-rotational Kirchhoff stress increment $\mathscr{D}\boldsymbol{\tau}$. In contrast, if the h_{kj} are *negative*-definite a stress *decrement* may result from either an increase or a decrease in total strain, corresponding to strain-softening behavior.

Recall (2.51), $(\mathscr{D}\boldsymbol{\tau}/\mathscr{D}t) \cdot \mathbf{D} > 0$, which is the *conjectural* inequality of Hill (1968, Part II). For elastic–plastic crystals, this becomes (from $(4.21)_1$, (4.25) and $(4.26)_2$)

$$\frac{\mathscr{D}\boldsymbol{\tau}}{\mathscr{D}t} \cdot \mathbf{D} = \frac{\mathscr{D}\boldsymbol{\tau}}{\mathscr{D}t} \cdot \mathscr{L}_0^{-1} \cdot \frac{\mathscr{D}\boldsymbol{\tau}}{\mathscr{D}t} + \sum_k \sum_j h_{kj} \dot{\gamma}_k \dot{\gamma}_j > 0. \tag{4.28}$$

From the (expected) positive-definiteness of the elastic moduli $\mathscr{L}_0$, it is evident that (4.28) is satisfied if only the matrix (h_{kj}) is *non-negative-definite*, which may be considered representative of "non-softening" behavior. As we shall see, non-negative-definiteness is a quite reasonable constraint to impose upon the effective hardening moduli; moreover, this condition generally is sufficient to ensure the positive-definiteness of the matrix (g_{kj}) from (4.27). Consequently, we consider Hill's inequality to be satisfied in elastic–plastic crystals and henceforth take the matrix (h_{kj}) to be non-negative-definite.

4.4. *The symmetry postulate: uniqueness criteria and minimum principles for boundary value problems*

Following Havner and Shalaby (1977) (also see Havner, 1977b) *we postulate that the matrix of effective hardening moduli h_{kj} is symmetric*, whence the moduli g_{kj} also are symmetric, from (4.27), corresponding to the Green-elasticity of the crystalline lattice. This postulate enables the basic constitutive equations $(4.21)_1$ and (4.23) to be expressed in terms of a single scalar potential function, in the tradition of macroscopic continuum theories of plasticity (cf. von Mises, 1928, and Hill, 1961). Moreover, the postulate has desirable consequences from the standpoint of the mathematical characteristics of quasi-static, incremental boundary value problems in crystalline solids, as will be shown. [Perhaps of greater import, the symmetry hypothesis is an essential part of a remarkably simple yet demonstrably useful theory of latent hardening conceived in late 1975 and first published in Havner and Shalaby (1977). This simple hardening theory is the subject of Section 5.]

Consider the incremental, stress-like "slack" variables

$$d\mu_k = \mathbf{\Lambda}_k \,.\, \mathbf{D}\,dt - \sum_j g_{kj}\,d\gamma_j \quad (k = 1, \ldots, n), \tag{4.29}$$

which are equivalent to those suggested by Sewell (1974). Each $d\mu_k$ is non-positive from $(4.23)_1$, and its magnitude at constant pressure equals the amount by which the increment $d\tau_k$ in shear strength of a critical slip system exceeds the increment in resolved shear stress on that system, to first order in infinitesimal lattice strain $\boldsymbol{\xi}$. (This will become apparent in Section 5.1.) Further, consider the "saddle function" introduced by Sewell (1974), defined in terms of $\mathbf{D}$ and the $\dot{\mu}$'s:

$$2W(\mathbf{D}, \dot{\mu}) = \mathbf{D} \,.\, \mathscr{L}_0 \,.\, \mathbf{D} - \sum_k \sum_j g_{kj}^{-1}\,(\mathbf{\Lambda}_k \,.\, \mathbf{D} - \dot{\mu}_k)(\mathbf{\Lambda}_j \,.\, \mathbf{D} - \dot{\mu}_j), \tag{4.30}$$

where the inverse moduli g_{kj}^{-1} also are symmetric, positive-definite. It is straightforward to show that W is a *potential function* for both material co-rotational stress-rate $\mathscr{D}\boldsymbol{\tau}/\mathscr{D}t$ and crystallographic shear-rates $\dot{\gamma}_k$, so that the constitutive equations $(4.21)_1$ and (4.23) (or (4.26)) are equivalent to the relations (in critical systems)

$$\frac{\mathscr{D}\boldsymbol{\tau}}{\mathscr{D}t} = \frac{\partial W}{\partial \mathbf{D}}, \quad \dot{\gamma}_k = \frac{\partial W}{\partial \dot{\mu}_k} \geqslant 0, \quad \dot{\mu}_k \leqslant 0, \quad \dot{\mu}_k \dot{\gamma}_k = 0. \tag{4.31}$$

Equations (4.30) and (4.31), together with definitions $(4.21)_{2,\,3}$ and kinematic relations (4.20), represent the final form of the general mathematical theory of finite, elastic–plastic deformation of crystals at constant pressure as presented in Havner (1977b). [It will be obvious to the careful reader that the theory is not yet complete, even in formal terms, because the required evolutionary equations for the critical strengths τ_k have not been presented. These equations were first given in Havner and Shalaby (1977, equation (18)) and are included in Section 5.1.] This structure of the theory, which is dependent upon the symmetry postulate, merges the constitutive analyses of Hill and Rice (1972) and Havner (1973b) with the theoretical framework of "saddle-shaped generating functionals" of Sewell (1974).

We now direct our attention to *incremental boundary value problems* in crystalline solids, restricting consideration to the class of quasi-static problems studied by Hill (1957, 1958, 1959, 1961) in a fundamental series of papers dealing with macroscopic elasticity and plasticity. In the considered problem, velocities $\mathbf{v}$ (displacement increments) are prescribed on a portion A_{D} of the current surface ∂V of a deformed crystal or crystalline aggregate, and nominal traction rates $\dot{\mathbf{t}}$ (force increments) are prescribed on the remainder A_{F}. Body force rates $\dot{\mathbf{f}}$ (per unit current volume V) are prescribed within V, and the current stress and deformation states are supposed known, together with all constitutive properties. The resultant velocity distribution $\mathbf{v}(\mathbf{x})$ is to be determined throughout V, from which $\mathbf{D}(\mathbf{x})$ and $\mathbf{W}(\mathbf{x})$ follow by definition and the distributions of the $\dot{\gamma}$'s, $\mathscr{D}\boldsymbol{\tau}/\mathscr{D}t$, $\boldsymbol{\varepsilon}$ and $(\mathbf{W} - \boldsymbol{\omega})$ follow from the constitutive and kinematic relations. In Havner (1977b), uniqueness criteria and associated minimum principles are established for this problem, based upon the general theory, (4.30) and (4.31). These results are reviewed in brief in the sequel.

We begin with the *general uniqueness theorem of Hill* (1958), taking the current configuration as reference. *If* $\int \Delta\dot{\mathbf{s}} \,.\, \Delta\mathbf{L}\,dV > 0$ *for all pairs of continuous, piecewise differentiable velocity fields whose difference* $\Delta\mathbf{v}$ *vanishes on* A_D, *where each* $\dot{\mathbf{s}}(\mathbf{x})$ *is determined from the corresponding* $\mathbf{v}(\mathbf{x})$ *through the constitutive equations, the incremental solution to the considered boundary value problem is unique.* From (2.34) and (2.47) and (2.48) (with $\mathbf{s} = \boldsymbol{\tau} = \boldsymbol{\sigma}$ and $\mathbf{F} = \mathbf{I}$ momentarily),

$$\dot{\mathbf{s}} = \frac{\mathscr{D}\boldsymbol{\tau}}{\mathscr{D}t} + \boldsymbol{\sigma}\mathbf{L}^{\mathrm{T}} - (\boldsymbol{\sigma}\mathbf{D} + \mathbf{D}\boldsymbol{\sigma}) = \frac{\mathscr{D}\boldsymbol{\tau}}{\mathscr{D}t} - (\boldsymbol{\sigma}\mathbf{W} + \mathbf{D}\boldsymbol{\sigma}). \tag{4.32}$$

Whence the uniqueness condition can be alternatively expressed (Hill, 1958)

$$\int \Delta\frac{\mathscr{D}\boldsymbol{\tau}}{\mathscr{D}t} \, . \, \Delta\mathbf{D}\, dV - b(\Delta\mathbf{v}, \Delta\mathbf{v}) > 0, \tag{4.33}$$

where the quadratic form in $\Delta\mathbf{v}$ is defined by

$$b(\Delta\mathbf{v}, \Delta\mathbf{v}) = \int \boldsymbol{\sigma} \, . \, (2\Delta\mathbf{D}\,\Delta\mathbf{D} - \Delta\mathbf{L}^{\mathrm{T}}\Delta\mathbf{L})dV = \int \boldsymbol{\sigma} \, . \, (\Delta\mathbf{W}\,\Delta\mathbf{L} + \Delta\mathbf{L}\,\Delta\mathbf{D})\,dV. \tag{4.34}$$

For the present theory, (4.30) and (4.31), with the aid of (4.24), we can establish the *continued inequality* (Havner, 1977a, b)

$$\begin{aligned}\Delta\frac{\mathscr{D}\boldsymbol{\tau}}{\mathscr{D}t} \, . \, \Delta\mathbf{D} &\geqslant \Delta\frac{\mathscr{D}\boldsymbol{\tau}}{\mathscr{D}t} \, . \, \mathscr{L}_0^{-1} \, . \, \Delta\frac{\mathscr{D}\boldsymbol{\tau}}{\mathscr{D}t} + \sum_k \sum_j h_{kj}\Delta\dot{\gamma}_k\Delta\dot{\gamma}_j \\ &\geqslant \Delta\mathbf{D} \, . \left(\mathscr{L}_0 - \sum_k \sum_j g_{kj}^{-1}\, \boldsymbol{\Lambda}_k \otimes \boldsymbol{\Lambda}_j\right) . \, \Delta\mathbf{D}. \end{aligned} \tag{4.35}$$

The left-hand inequality in (4.35) was first given explicitly in Havner (1977a). However, it is equivalent in form to an inequality established by Hill (1966, equation (9)) in terms of *lattice* co-rotational stress-rates and yet another set of slip-system moduli which will be introduced in Section 5.1. The inequality relating the extreme left-hand and right-hand terms is due to Sewell (1972). The *relative ordering* of these inequalities was first given in Havner (1977a). (Also, see Hill, 1978, equations (2.60) and (2.61).) The strict inequality signs apply in (4.35) unless the same critical slip systems are activated by the distinct pairs of stress- and strain-rates.

Upon combining (4.35) and (4.34) we obtain the obvious *sufficient condition for uniqueness* (Havner, 1977b, equation (4.5)) that

$$\int \left(\Delta\frac{\mathscr{D}\boldsymbol{\tau}}{\mathscr{D}t} \, . \, \mathscr{L}_0^{-1} \, . \, \Delta\frac{\mathscr{D}\boldsymbol{\tau}}{\mathscr{D}t} + \sum_k \sum_j h_{kj}\,\Delta\dot{\gamma}_k\Delta\dot{\gamma}_j\right) dV - b(\Delta\mathbf{v}, \Delta\mathbf{v}) > 0 \tag{4.36}$$

for all kinematically admissible velocity pairs. Alternatively, a *stricter* uniqueness criterion is (Sewell, 1972, equation (4.12))

$$\int \Delta\mathbf{D} \, . \left(\mathscr{L}_0 - \sum_k \sum_j g_{kj}^{-1}\, \boldsymbol{\Lambda}_k \otimes \boldsymbol{\Lambda}_j\right) . \, \Delta\mathbf{D}\, dV - b(\Delta\mathbf{v}, \Delta\mathbf{v}) > 0. \tag{4.37}$$

If (4.36) *holds* (but not necessarily (4.37)), the unique solution is characterized by the following *minimum principle* (Havner, 1977b). *The statically admissible true solution* (which satisfies the equations of continuing equilibrium div $\dot{\mathbf{s}} + \mathbf{f} = \mathbf{0}$ within V and the prescribed nominal traction rates $\dot{\mathbf{t}}$ on A_F) *minimizes the scalar functional*

$$2I(\mathbf{v}) = \int \frac{\mathscr{D}\boldsymbol{\tau}}{\mathscr{D}t}(\mathbf{v}) \, . \, \mathbf{D}(\mathbf{v})\,dV - b(\mathbf{v}, \mathbf{v}) - 2\int \dot{\mathbf{f}} \, . \, \mathbf{v}\,dV - 2\int \dot{\mathbf{t}} \, . \, \mathbf{v}\,dA_F \tag{4.38}$$

on the class of kinematically admissible velocity fields $\mathbf{v}(\mathbf{x})$ (for proof refer to Havner, 1977b, Section 6.1). Although the continued inequality (4.35) does not, in fact, require symmetry of the moduli h_{kj}, g_{kj}, *the minimum principle is strictly dependent upon this symmetry.*

There is an additional connection between uniqueness of solution and an extremum principle in the special circumstances where the two dominant principal stresses are everywhere tensile. Space limitations prevent the presentation of details, but it turns out that *locally* sufficient uniqueness criteria (as opposed to the *global* criteria of (4.36) and (4.37)) can be established in this case. Moreover, a related minimum principle can be proved which characterizes the solution for the *independent* variation of velocity and slip-rate fields (Havner, 1977b, Sections 5 and 6.2). Again, the minimum principle requires symmetric moduli h_{kj}.

5. The Simple Theory of Latent Hardening with Application to Single Crystal Experiments

In this final Section of our rational study of crystal and polycrystalline plasticity, we focus upon the *simplest possible* theory of finite distortional crystal hardening consistent with the foregoing general mathematical structure. This is the theory of *rotation-dependent* plastic anisotropy which I introduced in Havner and Shalaby (1977) and have called "the simple theory" of latent hardening [here, and elsewhere, is meant the rotation of gross crystalline material *relative to* the underlying lattice (i.e. the relative spin $(\mathbf{W}-\boldsymbol{\omega})$, which is an objective tensor)]. The corresponding evolutionary equations for the critical strengths τ_k at constant pressure are (Havner and Shalaby, 1977, equation (23))

$$d\tau_k = \sum_j (h - \mathbf{N}_k \,.\, \boldsymbol{\alpha}_j)\, d\gamma_j, \quad (k = 1, \ldots, N), \tag{5.1}$$

where the single material parameter h may be a functional of deformation history. The symmetric, positive-definite moduli g_{kj} are then determined as $g_{kj} = h + \boldsymbol{\Lambda}_k \,.\, \mathscr{L}_0^{-1} \,.\, \boldsymbol{\Lambda}_j$.

The consequences of (5.1) as regards latent hardening in axially loaded crystals, of both the face-centered cubic (f.c.c.) and body-centered cubic (b.c.c.) classes, have been explored in Havner and Shalaby (1977, 1978), Havner *et al.* (1979), Havner and Baker (1979), and Vause and Havner (1979). The simple theory has been shown to account for a greater variety of experimental phenomena in tension and compression tests, to large strains, than any other hardening law presently available. Here we develop the general set of evolutionary equations for critical strengths from which (5.1) emerged (Havner and Shalaby, 1977) and present a brief exposition of the cited latent hardening studies in cubic crystals. The article ends with suggestions for future investigation and applications of the simple theory to crystalline aggregates.

5.1. *Evolutionary equations for critical strengths: the general and simple theories of crystal hardening*

From (4.25), $(4.21)_{1,2}$, (4.20), and (4.14), the incremental quantity $\mathbf{M}_k \,.\, \mathscr{D}\boldsymbol{\tau}$ is precisely given in terms of the resolved shear stress $\mathbf{N}_k \,.\, \boldsymbol{\sigma}$ and the incremental slips by

$$\mathbf{M}_k \,.\, \mathscr{D}\boldsymbol{\tau} = d(\mathbf{N}_k \,.\, \boldsymbol{\sigma}) + \sum_j (\mathbf{N}_k \,.\, \boldsymbol{\alpha}_j - \boldsymbol{\alpha}_k \,\mathscr{L}_0^{-1} \,.\, \boldsymbol{\alpha}_j)\, d\gamma_j + \left(\frac{d(\delta A)}{\delta A} + \frac{d\lambda_b}{\lambda_b}\right) \mathbf{N}_k \,.\, \boldsymbol{\sigma}. \tag{5.2}$$

Consequently, to assure that the constitutive inequality $(4.26)_1$, $\mathbf{M}_k . \mathscr{D}\boldsymbol{\tau} \leqslant \sum_j h_{kj} d\gamma_j$, is satisfied whenever $\mathbf{N}_k . \boldsymbol{\sigma} = \tau_k$ (with the further requirement that the resolved shear stress never exceed the critical strength), we adopt the following set of *hardening laws* for all slip systems, whether or not critical (Havner and Shalaby, 1977):

$$d\tau_k = \sum_j (h_{kj} - \mathbf{N}_k . \boldsymbol{a}_j + \boldsymbol{a}_k \mathscr{L}_0^{-1} . \boldsymbol{a}_j) d\gamma_j - \left(\frac{d(\delta A)}{\delta A} + \frac{d\lambda_b}{\lambda_b}\right)\tau_k \quad (k = 1, \ldots, N). \tag{5.3}$$

These are the desired evolutionary equations for the τ_k. [Recall from Section 4.2 that, for cubic crystals, $d(\delta A)/\delta A + d\lambda_b/\lambda_b = 3d\lambda/\lambda$ to first order in $\boldsymbol{\xi}$. Thus, (5.3) is consistent with (4.18), which expresses the elastic dependence of critical strengths on large pressure changes (in cubic crystals) according to the normality criterion.]

To advance further, we consider the relative magnitudes of the various contributions in (5.3), following Havner and Shalaby (1977, Section 4). For this purpose we express the equation in words, distinguishing among (in their sequence on the right side) the three terms involving $d\gamma$'s and the term which includes τ_k. Thus we may write (Havner and Shalaby, 1977, equation (19)):

$$\begin{aligned} \text{incremental strength} = {} & \text{effective moduli} \times \text{incremental slips} \\ & + O\ (\text{stress} \times \text{incremental slips}) \\ & + O\ (\text{stress} \times \text{lattice strain} \times \text{incremental slips}) \\ & + O\ (\text{stress} \times \text{incremental lattice strain}). \end{aligned} \tag{5.4}$$

At constant pressure, to which the analysis henceforth is restricted, "lattice strain" means the superimposed infinitesimal strain $\boldsymbol{\xi}$, and "stress" is the corresponding superimposed stress $\mathscr{L}(p) . \boldsymbol{\xi}$ (that is, the hydrostatic state $-p\mathbf{I}$ does not appear in any of the terms of (5.4)). From the previous discussion (Section 4.1) of expected magnitudes $\gamma (= 10^{-1}$ to $1)$ and $\xi (= 10^{-4}$ to $10^{-3})$ of plastic shears and lattice strain, $\xi d\gamma < 10^{-3} d\gamma$ and $d\xi < 10^{-2} d\gamma$. Thus, the third and fourth terms in (5.3) and (5.4) are of maximum order 10^{-3} to 10^{-2} times the second term, and neglecting them is consistent with a simplification which already has been made (viz. taking $\hat{\boldsymbol{\tau}} = \mathscr{L}_0 . \boldsymbol{\varepsilon}$, which disregards the plastic-deformation dependence of crystal elastic moduli assessed in Section 4.1). Consequently, we adopt

$$d\tau_k = \sum_j (h_{kj} - \mathbf{N}_k . \boldsymbol{a}_j) d\gamma_j, \quad h_{kj} = h_{jk}, \tag{5.5}$$

as the "*general hardening law at constant pressure, incorporating the symmetry postulate and correct to first order in infinitesimal lattice strain*" (Havner and Shalaby, 1977, p. 54). [The physical definition of the "slack" variable $d\mu_k$ in Section 4.4 can now be justified. From the comparative magnitudes of terms in (5.2)–(5.4), $\mathbf{M}_k . \mathscr{D}\boldsymbol{\tau} = d(\mathbf{N}_k . \boldsymbol{\sigma}) + \sum_j \mathbf{N}_k . \boldsymbol{a}_j d\gamma_j$ to first order in $\boldsymbol{\xi}$. Thus, from (4.25), (4.27), (4.29), and (5.5), $d\mu_k = \mathbf{M}_k . \mathscr{D}\boldsymbol{\tau} - \sum_j h_{kj} d\gamma_j = -\{d\tau_k - d(\mathbf{N}_k . \boldsymbol{\sigma})\}$ as previously stated.]

As in Havner and Shalaby (1977), the set of material parameters H_{kj} which directly connect the incremental shear strengths and the incremental crystallographic slips will be called the *physical* slip system hardening moduli (cf. Hill, 1966):

$$d\tau_k = \sum_j H_{kj} d\gamma_j, \quad (k = 1, \ldots, N). \tag{5.6}$$

Within the scope of the present theory, the H_{kj} are (in general) rate-independent functionals of the deformation history. Obviously,

$$H_{kj} = h_{kj} - \mathbf{N}_k . \boldsymbol{\alpha}_j; \tag{5.7}$$

hence the H_{kj} *are not symmetric.* In particular, *Taylor's classical isotropic hardening rule,* $H_{kj} = H$ for all k, j (Taylor and Elam, 1923, 1925), "*is not admissible within the framework of the general mathematical theory*" (Havner and Shalaby, 1977, p. 55).

Adopting a theoretical framework which excludes Taylor hardening, even as a special case, is not done lightly, for Taylor's rule is appealing in its simplicity and is considered a reasonable approximation to the complex realities of crystal hardening at small strain (cf. Kocks, 1970). Moreover, it has been by far the most widely used hardening law in polycrystal calculations representing various aspects of small deformation behavior of metals (see, for example, Taylor, 1938a; Bishop and Hill, 1951b; Kocks, 1958; Hutchinson, 1964, 1970; Lin, 1971; Chin, 1973; Havner and Varadarajan, 1973; Havner *et al.*, 1974; Lin *et al.*, 1974; and Havner and Singh, 1977). On the other hand, Taylor's equal hardening rule is not generally applicable at *finite* strain because it fails to predict the following well-known experimental phenomenon from single crystal tests.

According to Taylor's law, double slip begins in the uniaxial test of a crystal specimen initially oriented for *single* slip when the loading axis reaches a crystallographic "symmetry line" (distinguished by equal resolved shear stresses on the primary and secondary slip systems). (This position is attained through rotation of the loading axis, relative to the crystal lattice axes, toward the slip direction in tension or the slip plane normal in compression.) The weight of experimental evidence, however, indicates that finite single slip continues as the dominant mode of deformation, well beyond the symmetry line, in tensile and compressive tests of both f.c.c. and b.c.c. crystals (cf. Fahrenhorst and Schmid, 1932; Piercy *et al.*, 1955; Kocks, 1964; Keh, 1965; and Bell and Green, 1967). This phenomenon, commonly called "overshooting," is related to, consistent with, and caused by the *anisotropy* of finite-distortional latent hardening found by Kocks (1964), Ramaswami *et al.* (1965), Nakada and Keh (1966), Kocks and Brown (1966), and Jackson and Basinski (1967). (Refer to Bell and Green, 1967, Havner and Shalaby, 1977, and Havner *et al.*, 1979 for detailed discussions.)

Remarkably, *the simple theory of latent hardening reconciles the use of Taylor's rule at infinitesimal strain with the phenomenon of "overshooting" at finite deformation.* (The simple theory (5.1) obviously follows from the general law (5.5) by taking the effective hardening moduli h_{ki} to be equal.)

Observe, from $(4.20)_2$ and $(4.21)_3$, that the non-symmetric terms $\sum_j \mathbf{N}_k . \boldsymbol{\alpha}_j d\gamma_j$ in (5.1) and (5.5) involve geometric rotational effects that are negligible at very small strains. Moreover, these terms are zero whenever the relative spin $\mathbf{W} - \boldsymbol{\omega} = \mathbf{0}$. Thus, *the simple theory* (5.1) *effectively reduces to Taylor's law at infinitesimal overall deformations or when there is no relative rotation of material and lattice* (the latter case includes four-, six-, and eight-fold symmetric multiple slip in cubic crystals). In *single* slip the rotational terms play a prominent role at finite strain, as will be shown in the sequel, and the simple hardening law (5.1) is a *universal theory of "overshooting"* in axially loaded cubic crystals.

5.2. *The theoretical latent hardening law in single slip*

The following derivation of the general equations for latent hardening in single slip (according to the simple theory) is condensed from the detailed development in Havner *et al.* (1979, Sections 3 and 4). The analysis is applicable to any crystal class, in either tension or compression, and applies to arbitrary latent slip systems.

Let λ and ψ denote the angles between vector $\mathbf{l}$, coincident with a chosen line segment of crystalline material, and unit vectors $\mathbf{b}$, $\mathbf{n}$ in the slip and normal directions of the active system. Also let χ, ϕ denote the angles between areal vector $\mathbf{a}$, normal to a particular material plane, and the $\mathbf{b}$, $\mathbf{n}$ directions. As regards kinematic relations alone, $\mathbf{l}$ and $\mathbf{a}$ may represent any material line and material plane within the crystal; but in application to tension and compression tests, we take $\mathbf{l}$ to coincide with a segment of specimen axis and $\mathbf{a}$ to be the areal vector of an end-plane. Then $\mathbf{a} \cdot \mathbf{l} = aD = V$, where V is the segment volume and D is the (current) distance between planes.

Henceforth we neglect infinitesimal lattice strain in the geometric relations. The deformation gradient for finite shear γ is then

$$\mathbf{F} = \mathbf{I} + \gamma\, \mathbf{b} \otimes \mathbf{n}, \tag{5.8}$$

whence $\det \mathbf{F} = 1$. Also, denoting by $\mathbf{l}_0$, $\mathbf{a}_0$ the position and areal vectors at the beginning of single slip, we have (with the aid of Nanson's equation)

$$\mathbf{l} = \mathbf{l}_0 + \gamma(\mathbf{n} \cdot \mathbf{l}_0)\mathbf{b}, \quad \mathbf{a} = \mathbf{a}_0 - \gamma(\mathbf{a}_0 \cdot \mathbf{b})\mathbf{n}. \tag{5.9}$$

It is straightforward to show, from (5.9) and the definitions of angles λ, ψ, χ, and ϕ, that

$$\frac{\cos\lambda}{\cos\psi} = \frac{\cos\lambda_0}{\cos\psi_0} + \gamma, \quad \frac{\cos\phi}{\cos\chi} = \frac{\cos\phi_0}{\cos\chi_0} - \gamma. \tag{5.10}$$

Let $\mathbf{m} = \mathbf{b} \wedge \mathbf{n}$ and denote the angles between this unit vector and $\mathbf{l}$, $\mathbf{a}$ by η, ζ respectively. Further, define $\boldsymbol{\nu}$, $\boldsymbol{\mu}$ as unit vectors in directions $\mathbf{l}$, $\mathbf{a}$. Then, from (5.9) and (5.10), after some algebra (refer to Havner *et al.*, 1979, Section 3 for details), we obtain

$$\left.\begin{aligned} \boldsymbol{\nu} &= \mathbf{b}\cos\lambda + \mathbf{p}\cos\psi, \quad \mathbf{p} = \mathbf{n} + (\cos\eta_0/\cos\psi_0)\mathbf{m}, \\ \boldsymbol{\mu} &= \mathbf{n}\cos\phi + \mathbf{q}\cos\chi, \quad \mathbf{q} = \mathbf{b} + (\cos\zeta_0/\cos\chi_0)\mathbf{m}. \end{aligned}\right\} \tag{5.11}$$

For axially loaded crystals, (Cauchy) stress $\boldsymbol{\sigma}$ can be expressed in terms of the unit vector ($\boldsymbol{\nu}$ or $\boldsymbol{\mu}$) in the loading direction as

$$\boldsymbol{\sigma} = \left\{\begin{array}{lll} \sigma\boldsymbol{\nu}\otimes\boldsymbol{\nu}, & \sigma > 0, & \text{in tension,} \\ \sigma\boldsymbol{\mu}\otimes\boldsymbol{\mu}, & \sigma < 0, & \text{in compression.} \end{array}\right\} \tag{5.12}$$

The resolved shear stress $\mathbf{N} \cdot \boldsymbol{\sigma}$ in the active system must equal the critical strength $\tau = \tau_0 + \int H d\gamma$ (> 0) during single slip, where H is the "physical" hardening modulus (i.e. the slope of the resolved shear stress vs. slip curve). Thus,

$$\sigma = \begin{cases} \dfrac{\tau_0 + \int H d\gamma}{\cos\psi\cos\lambda} > 0, & \text{tension,} \\[2ex] \dfrac{\tau_0 + \int H d\gamma}{\cos\psi\cos\phi} < 0, & \text{compression (as } \cos\chi < 0). \end{cases} \tag{5.13}$$

From (5.1), the critical strength of the kth latent system after finite shear γ becomes, according to the simple theory,

$$\tau_k = (\tau_0)_k + \int_0^\gamma H d\gamma + \int_0^\gamma (\mathbf{N} - \mathbf{N}_k) \cdot \boldsymbol{a}\, d\gamma. \tag{5.14}$$

The tensors $\mathbf{N}$, $\boldsymbol{\Omega}$ in the active system and $\mathbf{N}_k$ in the latent systems are fixed relative to any lattice co-rotational frame; but the symmetric tensor $\boldsymbol{\alpha} = \boldsymbol{\sigma}\boldsymbol{\Omega} - \boldsymbol{\Omega}\boldsymbol{\sigma}$ is a nonlinear function of γ through (5.10)–(5.13). There follows from (5.14), after a considerable amount of algebra, the *general equation for theoretical latent strength* τ_k *in single slip* (Havner *et al.*, 1979, equation (4.11)):

$$\tau_k = (\tau_0)_k + \{A_k\gamma + \tfrac{1}{2}B_k\gamma^2 + C_k \ln(1+\gamma/c_2)\}\tau_0 + \int H d\gamma + A_k \int \left(\int H d\gamma \right) d\gamma$$
$$+ B_k \int \left(\int H d\gamma \right) \gamma d\gamma + C_k \int \left(\int H d\gamma \right) \Big/ \left(c_2 + \gamma \right) d\gamma, \tag{5.15}$$

where

$$\left. \begin{aligned} \textit{Tension:}\ A_k &= c_2 B_k - \mathbf{N}_k . (\mathbf{p}\otimes\mathbf{n} - \mathbf{b}\otimes\mathbf{b}), \\ B_k &= \tfrac{1}{2} - \mathbf{N}_k . (\mathbf{b}\otimes\mathbf{n}), \quad C_k = \mathbf{N}_k . (\mathbf{p}\otimes\mathbf{b}) - \tfrac{1}{2}, \end{aligned} \right\} \tag{5.16}$$

and

$$\left. \begin{aligned} \textit{Compression:}\ A_k &= c_2 B_k - \mathbf{N}_k . (\mathbf{n}\otimes\mathbf{n} - \mathbf{q}\otimes\mathbf{b}), \\ B_k &= \tfrac{1}{2} - \mathbf{N}_k . (\mathbf{n}\otimes\mathbf{b}), \quad C_k = N_k . (\mathbf{q}\otimes\mathbf{n}) - \tfrac{1}{2}, \end{aligned} \right\} \tag{5.17}$$

with the positive constant c_2 (equivalently defined for the tensile test in Havner and Shalaby, 1977) given by (Havner *et al.*, 1979, equation (4.6))

$$c_2 = \left\{ \begin{array}{ll} \cos\lambda_0/\cos\psi_0 > 0, & \text{tension,} \\ -\cos\phi_0/\cos\chi_0 > 0, & \text{compression.} \end{array} \right\} \tag{5.18}$$

5.3. *Cubic crystals in tension and compression*

In discussing applications of the theoretical hardening law (5.15) to cubic crystals in single slip, we shall find it convenient to refer to the standard [001] stereographic projection for the *tensile* test shown in Fig. 5.1. For f.c.c. crystals, the letters a, b, c, d represent the four slip planes {111} (Miller index notation), and the indices 1, 2, 3 designate the three slip directions ⟨110⟩ in each plane (refer to Table 5.1) with a bar above a number denoting the opposite slip system. In b.c.c. crystals (Table 5.2) the letters represent the four slip directions ⟨111⟩, and the indices designate the three {110} planes which contain each slip direction. (The case of slip on other planes in b.c.c. crystals will be discussed and illustrated later in this Section.) For the *compression* test each system *exchanges position* on the stereographic projection with its opposite sense system. In a virgin crystal having a statistically isotropic distribution of dislocations (relative to crystallographically equivalent planes and directions) all initial critical strengths equal τ_0. The primary slip system is then determined by the stereographic triangle within which the loading axis lies at the initiation of slip (corresponding to the greatest resolved shear stress criterion).

The simple theory was shown to account for the perpetuation of single slip (beyond the symmetry line) in *f.c.c. crystals in tension* by Havner and Shalaby (1977, 1978). (These two papers treat linear active hardening $\tau = \tau_0 + H\gamma$ and parabolic hardening $\tau = A\gamma^{\frac{1}{2}}$ respectively.) This was done graphically via computer plots demonstrating that $\tau_k > \mathbf{N}_k . \boldsymbol{\sigma}$, in all latent systems, for each of five initial positions chosen to span the standard stereographic triangle $a1$ (arbitrarily selected as the active system). In addition, the simple theory was found

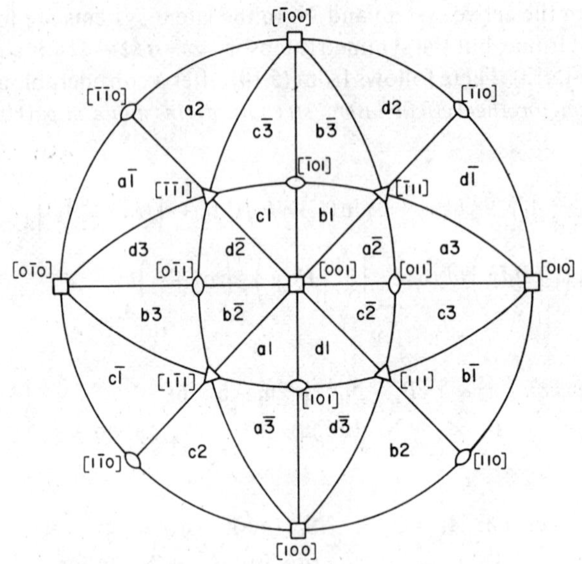

FIG. 5.1. Standard [001] stereographic projection of cubic crystals.

TABLE 5.1. Designation of slip systems in f.c.c. crystals

Plane	(111)			$(\bar{1}\bar{1}1)$			$(\bar{1}11)$			$(1\bar{1}1)$		
Direction	$[0\bar{1}1]$	$[10\bar{1}]$	$[\bar{1}10]$	[011]	$[\bar{1}0\bar{1}]$	$[\bar{1}10]$	$[0\bar{1}1]$	$[\bar{1}0\bar{1}]$	[110]	[011]	$[10\bar{1}]$	$[\bar{1}\bar{1}0]$
System	*a*1	*a*2	*a*3	*b*1	*b*2	*b*3	*c*1	*c*2	*c*3	*d*1	*d*2	*d*3

to be in qualitative agreement with experimental results of Jackson and Basinski (1967) for copper crystals. Comparisons with these experiments were explored more fully in Havner *et al.* (1979), and Fig. 5.2 is taken from that work.

The virgin specimens ("parent" crystals) of Jackson and Basinski were initially oriented on the [001] stereographic projection (Fig. 5.1) 10° from [011] on the great circle connecting [011] and [$\bar{1}01$]; hence $a\bar{2}$ was the primary slip system. Of the six "parent" specimens prestrained to shears γ between 5 and 25% (stage II hardening), Jackson and Basinski (1967, table I) obtained latent hardening information about eight different secondary slip systems, several at more than one level of prestrain. From an examination of their results (refer to Havner *et al.*, 1979, table 2 and fig. 3), the rank of the nine systems (including $a\bar{2}$) is, in descending order of hardening, $b1$, $c\bar{3}$, $c\bar{1}$, $d\bar{1}$, $d2$, $c2$, $b\bar{2}$, $a1$, and $a\bar{2}$. The rank of the *theoretical* hardening curves (Fig. 5.2) for these systems is, within the range of the data, $c\bar{3}$, $b1$, $c2$, $d\bar{1}$, $d2$, $b\bar{2}$, $a1$, $a\bar{2}$, and $c\bar{1}$. [The curve for primary system $a\bar{2}$ of Fig. 5.2 is actually *piecewise-linear*, based upon many small $\Delta\gamma$-intervals. The theoretical latent strengths (5.15) were integrated and then calculated using this piecewise-linear relation.] Thus, only two of the nine predictions are significantly out of order: $c2$ and $c\bar{1}$.

In the strain range 0.0–0.25, curve $a\bar{2}$ of Fig. 5.2 represents the (averaged) primary system curve from the actual data. In the stage III range, $\gamma > 0.2$, the curve for $a\bar{2}$, given by $\tau = 7.56(\gamma - 0.111)^{1/2}$ (kg mm^{-2}), is *extrapolated* from the data using a procedure whose accuracy for copper has been demonstrated by Bell (1964) (refer to Havner *et al.*, 1979, Section 5). The empirical mean of the hardening data for latent systems $b1$, $c\bar{1}$, $c\bar{3}$, $d\bar{1}$, and $d2$

TABLE 5.2. Designation of slip systems in b.c.c. crystals

Plane	$(0\bar{1}1)$	$(10\bar{1})$	$(\bar{1}10)$	(011)	$(\bar{1}0\bar{1})$	$(1\bar{1}0)$	$(0\bar{1}1)$	$(\bar{1}0\bar{1})$	(110)	(011)	$(10\bar{1})$	$(\bar{1}\bar{1}0)$
Direction		$[111]$			$[\bar{1}\bar{1}1]$			$[\bar{1}11]$			$[1\bar{1}1]$	
System	$a1$	$a2$	$a3$	$b1$	$b2$	$b3$	$c1$	$c2$	$c3$	$d1$	$d2$	$d3$

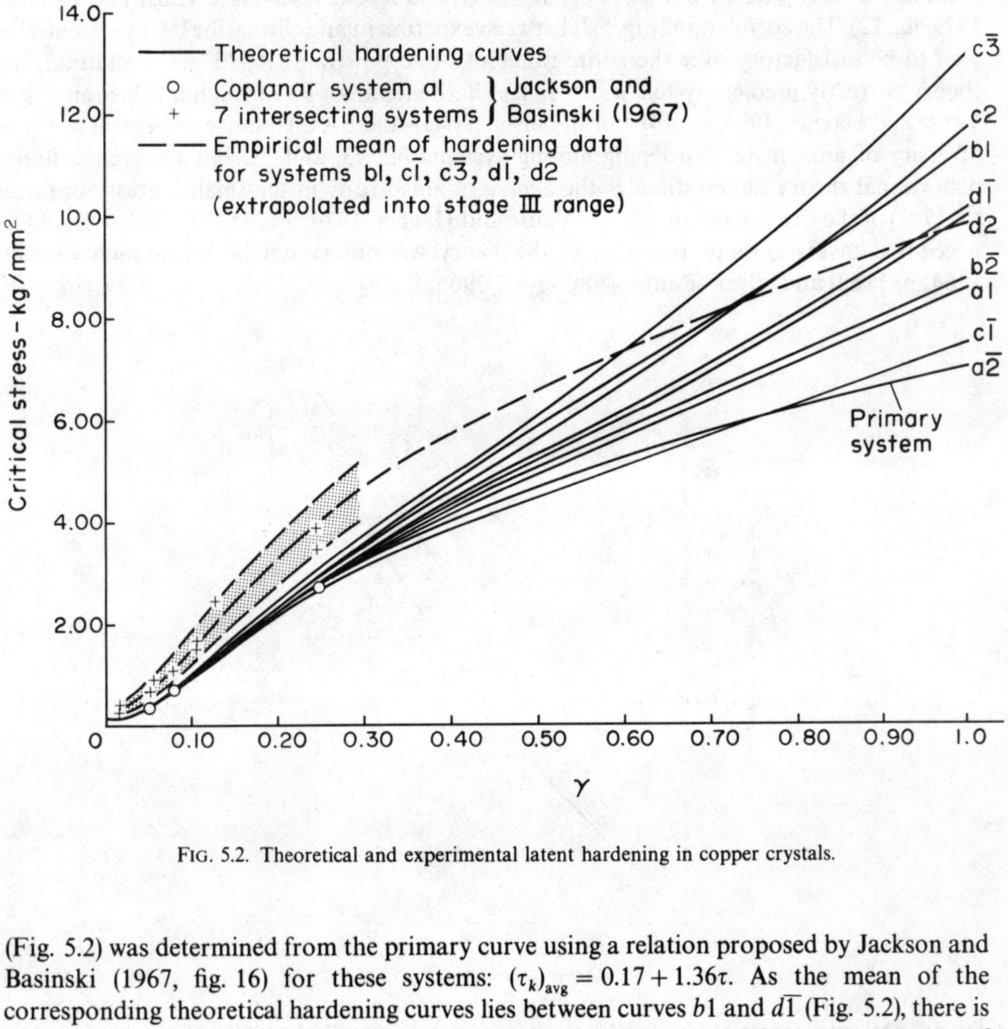

FIG. 5.2. Theoretical and experimental latent hardening in copper crystals.

(Fig. 5.2) was determined from the primary curve using a relation proposed by Jackson and Basinski (1967, fig. 16) for these systems: $(\tau_k)_{avg} = 0.17 + 1.36\tau$. As the mean of the corresponding theoretical hardening curves lies between curves $b1$ and $d\bar{1}$ (Fig. 5.2), there is moderately good agreement in stage III between the simple theory and the extrapolation of experiment.

For f.c.c. crystals in *compression*, the simple hardening law was shown to be a theory of "overshooting" by Vause and Havner (1979). They analyzed seven initial positions spanning the stereographic triangle of slip system $a2$ (which in compression occupies triangle $a\bar{2}$ of Fig. 5.1 as previously remarked). Vause and Havner also made comparisons with latent hardening data for aluminum and silver crystals from compression experiments of Kocks (1964) and Ramaswami *et al.* (1965). (The initial loading-axis orientation of the virgin crystals in these experiments corresponded to the maximum value 0.5 of the "Schmid factor" $|\cos\phi_0 \cos\chi_0|$.)

Kocks (1964) investigated latent hardening of systems $b\bar{3}$ and $d\bar{3}$ (triangles $b3$, $d3$ of Fig. 5.1) in both aluminum and silver; Ramaswami *et al.* (1965) added latent hardening data for system $d\bar{1}$ (triangle $d1$) in silver crystals.

A comparison between the theoretical curve τ_k vs. τ and Kocks' (1964) data for system $b\bar{3}$ in aluminum is shown in Fig. 5.3 (from Vause and Havner, 1979). The active-system hardening curve is $\tau = 2\gamma^{1/2}$ (kg mm^{-2}), which is a reasonable approximation to resolved shear stress vs. slip data from Kocks *et al.* (1964) and Kocks (1964) (see Vause and Havner, 1979, fig. 12). The correlation (Fig. 5.3) between experiment and theory for latent system $b\bar{3}$ is seen to be satisfactory over the entire range of finite prestrains (6–70%). In addition, the theory correctly predicts system $d\bar{3}$ to be less hardened than $b\bar{3}$ throughout this range (see Vause and Havner, 1979, fig. 14). For the silver crystals there is qualitative agreement as to the diversity of anisotropic hardening among systems $b\bar{3}$, $d\bar{3}$, and $d\bar{1}$, but the simple finite-distortional theory underestimates the *degree* of anisotropy in the smaller prestrain range (5–17%) of the silver experiments (see Vause and Havner, 1979, figs. 15–17). It also should be noted that "overshooting," predicted by the theory, was observed in both aluminum (Kocks, 1964, p. 1162) and silver (Ramaswami *et al.*, 1965, fig. 1).

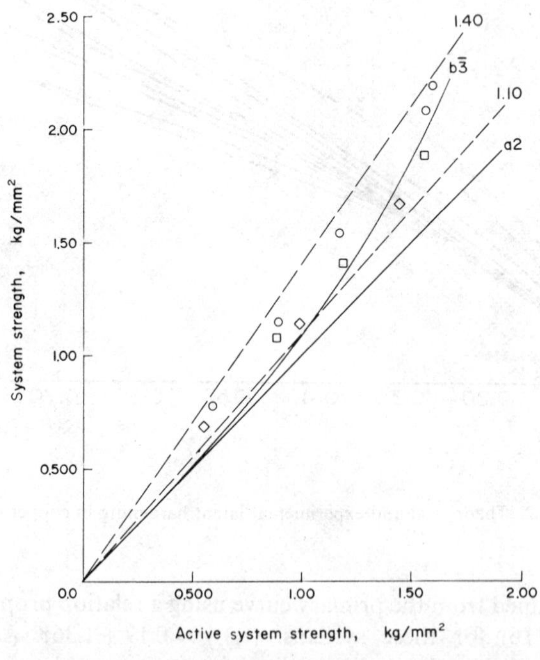

FIG. 5.3. Theoretical hardening curve (solid line) and experimental results (Kocks, 1964) for latent slip system $b\bar{3}$ in aluminum crystals.

In b.c.c. metals there are historical ambiguities regarding identification of the slip plane which are discussed at length in Havner and Baker (1979, Section 3). In that work parallel analyses are presented corresponding to two separate hypotheses: (i) slip on {110}, {112}, and {123} crystallographic planes; and (ii) (initial) slip on the arbitrary plane of maximum resolved shear stress containing a ⟨111⟩ direction. (The latter criterion is the *experimental* law for the (mean) plane of slip found by Taylor and Elam (1926) in their classic work on iron crystals.) Havner and Baker (1979) established that the simple latent hardening law (5.15) is a theory of "overshooting" in b.c.c. crystals (for both tension and compression) according to *either* specification of the slip plane. An example of the increased complexity associated with the first of these hypotheses is shown in Fig. 5.4. There the three families of planes $\{110\}$, $\{112\}$, $\{123\}$ are considered equal as potential slip planes in the tensile test. Correspondingly, the stereographic triangles $a\overline{2}$ and $c\overline{2}$ of Fig. 5.1 must be subdivided as shown. The indicated rotations from the eight initial positions (Fig. 5.4) are those predicted by the simple theory (Havner and Baker, 1979). They may be compared with similar rotations for iron crystals found experimentally by Fahrenhorst and Schmid (1932, fig. 1) and Keh (1965, fig. 6).

Also of interest with regard to theoretical hardening predictions is the following *theorem* (Havner and Baker, 1979, p. 292): *If only one of two equally stressed systems having a common slip direction is initially activated in a tensile test, it will remain the only active system as the tension axis rotates along the symmetry line.* For example, if slip is initiated in either of systems $a\overline{2}$, $a\overline{4}$ (Fig. 5.4) from an axis position on their common boundary (because of a slightly smaller shear strength in the active system), the other system will harden at a greater rate with increasing slip, thus remaining latent. This theoretical prediction has some experimental support (see Havner and Baker, 1979, Section 4.1) as well as a counterpart in *compression* (Vause and Havner, 1979).

5.4. *Closing remarks*

As discussed and illustrated in Sections 5.1–5.3, the simple hardening law (5.1) plays a useful role within the theoretical framework constructed for crystal plasticity. Moreover, the

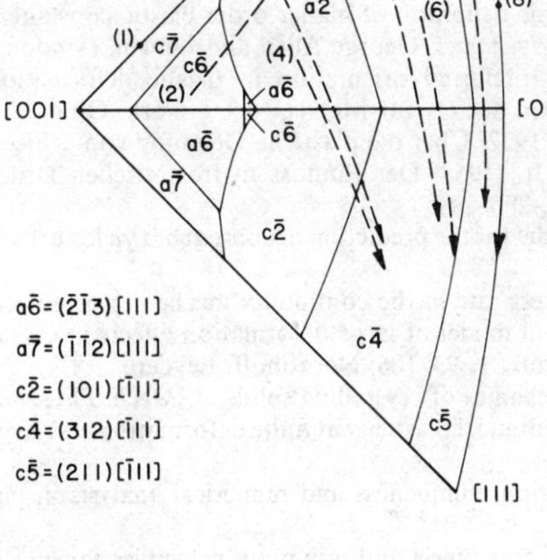

FIG. 5.4. Initial tensile axis positions and axis rotations to $\gamma = 1.5$ for single slip on $\{110\}$, $\{112\}$, and $\{123\}$ planes in b.c.c. crystals.

simple theory is closely related to (and in a sense emerged from) the fundamental contributions of Hill recounted earlier in this article. In particular, the theory is consistent with classical, experimentally founded concepts of *macroscopic* metal plasticity (theory) through the averaging theorem and Hill's invariant bilinear form. Consequently, at the time of writing (January 1980) there are reasonable prospects that the simple theory will prove valuable to the understanding and calculation of finite plastic deformation of *crystalline aggregates*. A significant undertaking along those lines would be the prediction of texture development in sheet rolling. Although clearly involving finite deformation, this problem has been studied almost exclusively within the context of infinitesimal strain (an exception being the work of Kallend and Davies, 1972). There are substantial complexities of (finite) multiple shears to contend with in such investigations, and the *double*-slip kinematic solutions of Shalaby and Havner (1978) and Havner (1979) may prove a useful starting-point.

Acknowledgement

This work was partially supported by the United States National Science Foundation, Solid Mechanics Program, through Grant ENG78–08154.

References

Barsch, G. R. and Chang, Z. P. (1967) Adiabatic, isothermal, and intermediate pressure derivatives of the elastic constants for cubic symmetry, II: Numerical results for 25 materials, *Phys. Status Solidi* **19,** 139–151.

Bell, J. F. (1964) A generalized large deformation behaviour for face-centred cubic solids—high purity copper, *Phil. Mag.*, Ser. 8, **10,** 107–126.

Bell, J. F. and Green, R. E., Jr. (1967) An experimental study of the double slip deformation hypothesis for face-centred cubic single crystals, *Phil. Mag.*, Ser. 8, **15,** 469–476.

Biot, M. A. (1965) *Mechanics of Incremental Deformations*, Wiley, New York.

Bishop, J. F. W. and Hill, R. (1951a) A theory of the plastic distortion of a polycrystalline aggregate under combined stresses, *Phil. Mag.*, Ser. 7, **42,** 414–427.

Bishop, J. F. W. and Hill, R. (1951b) A theoretical derivation of the plastic properties of a polycrystalline face-centred metal, *Phil. Mag.*, Ser. 7, **42,** 1298–1307.

Brugger, K. (1964) Thermodynamic definition of higher order elastic constants, *Phys. Rev.* A **133,** 1611–1612.

Chadwick, P. (1976) *Continuum Mechanics*, George Allen and Unwin, London; Wiley, New York.

Chin, G. Y. (1973) The role of preferred orientation in plastic deformation, *The Inhomogeneity of Plastic Deformation*, 83–112, American Society for Metals, Metals Park, Ohio.

Fahrenhorst, W. and Schmid, E. (1932) Über die plastiche Dehnung von α-Eisenkristallen, *Z. Phys.* **78,** 383–394.

Haasen, P. and Lawson, A. W., Jr. (1958) Der Einfluss hydrostatischen Druckes auf die Zugverformung von Einkristallen, *Z. Metallk.* **49,** 280–291.

Havner, K. S. (1971) A discrete model for the prediction of subsequent yield surfaces in polycrystalline plasticity, *Int. J. Solids Struct.* **7,** 719–730.

Havner, K. S. (1972) On Hill's stress rate in the continuum mechanics of polycrystals, *Acta Mech.* **14,** 183–187.

Havner, K. S. (1973a) An analytical model of large deformation effects in crystalline aggregates, *Foundations of Plasticity*, edited by A. Sawczuk, **1,** 93–106, Noordhoff, Leyden.

Havner, K. S. (1973b) On the mechanics of crystalline solids, *J. Mech. Phys. Solids* **21,** 383–394.

Havner, K. S. (1974) Aspects of theoretical plasticity at finite deformation and large pressure, *Z. angew. Math. Phys.* **25,** 765–781.

Havner, K. S. (1977a) On unification, uniqueness and numerical analysis in plasticity, *Int. J. Solids Struct.* **13,** 625–635.

Havner, K. S. (1977b) On uniqueness criteria and minimum principles for crystalline solids at finite strain, *Acta Mech.* **28,** 139–151.

Havner, K. S. (1978) On unifying concepts in plasticity theory and related matters in numerical analysis, *Nucl. Eng. Design* **46,** 187–201.

Havner, K. S. (1979) The kinematics of double slip with application to cubic crystals in the compression test, *J. Mech. Phys. Solids* **27**, 415–429.

Havner, K. S. and Baker, G. S. (1979) Theoretical latent hardening in crystals—II. BCC crystals in tension and compression, *J. Mech. Phys. Solids* **27**, 285–314.

Havner, K. S., Baker, G. S., and Vause, R. F. (1979) Theoretical latent hardening in crystals—I. General equations for tension and compression with application to f.c.c. crystals in tension, *J. Mech. Phys. Solids* **27**, 33–50.

Havner, K. S. and Shalaby, A. H. (1977) A simple mathematical theory of finite distortional latent hardening in single crystals, *Proc. R. Soc. Lond.* A **358,** 47–70.

Havner, K. S. and Shalaby, A. H. (1978) Further investigation of a new hardening law in crystal plasticity, *J. Appl. Mech.* **45,** 500–506.

Havner, K. S. and Singh, C. (1977) Application of a discrete polycrystal model to the analysis of cyclic straining in copper, *Int. J. Solids Struct.* **13,** 395–407.

Havner, K. S., Singh, C., and Varadarajan, R. (1974) Plastic deformation and latent strain energy in a polycrystalline aluminum model, *Int. J. Solids Struct.* **10,** 853–862.

Havner, K. S. and Varadarajan, R. (1973) A quantitative study of a crystalline aggregate model, *Int. J. Solids Struct.* **9,** 379–394.

Hayden, H. W., Moffatt, W. G., and Wulff, J. (1965) *The Structure and Properties of Materials*, Vol. III, *Mechanical Behavior*, Ch. 6. Wiley, New York.

Heidenreich, R. D. (1949) Structure of slip bands and cold-worked metal, *Cold Working of Metals*, 57–64, American Society for Metals, Metals Park, Ohio.

Hill, R. (1950) *The Mathematical Theory of Plasticity*, Clarendon Press, Oxford.

Hill, R. (1956) The mechanics of quasi-static plastic deformation in metals, *Surveys in Mechanics: G. I. Taylor* 70th *Anniversary Volume*, edited by G. K. Batchelor and R. M. Davies, 7–31, Cambridge University Press.

Hill, R. (1957) On uniqueness and stability in the theory of finite elastic strain, *J. Mech. Phys. Solids* **5,** 229–241.

Hill, R. (1958) A general theory of uniqueness and stability in elastic–plastic solids, *J. Mech. Phys. Solids* **6,** 236–249.

Hill, R. (1959) Some basic principles in the mechanics of solids without a natural time, *J. Mech. Phys. Solids* **7,** 209–225.

Hill, R. (1961) Bifurcation and uniqueness in non-linear mechanics of continua, *Problems of Continuum Mechanics*, edited by M. A. Lavrent'ev *et al.*, 155–164, SIAM, Philadelphia.

Hill, R. (1963) Elastic properties of reinforced solids: some theoretical principles, *J. Mech. Phys. Solids* **11,** 357–372.

Hill, R. (1966) Generalized constitutive relations for incremental deformation of metal crystals by multislip, *J. Mech. Phys. Solids* **14,** 95–102.

Hill, R. (1967) The essential structure of constitutive laws for metal composites and polycrystals, *J. Mech. Phys. Solids* **15,** 79–95.

Hill, R. (1968) On constitutive inequalities for simple materials, *J. Mech. Phys. Solids* **16,** I, 229–242, II, 315–322.

Hill, R. (1970) Constitutive inequalities for isotropic elastic solids under finite strain, *Proc. R. Soc. Lond.* A **314,** 457–472.

Hill, R. (1972) On constitutive macro-variables for heterogeneous solids at finite strain, *Proc. R. Soc. Lond.* A **326,** 131–147.

Hill, R. (1978) Aspects of invariance in solid mechanics, *Advances in Applied Mechanics*, edited by C.-S. Yih, **18,** 1–75, Academic Press, New York.

Hill, R. and Rice, J. R. (1972) Constitutive analysis of elastic–plastic crystals at arbitrary strain, *J. Mech. Phys. Solids* **20,** 401–413.

Hutchinson, J. W. (1964) Plastic stress–strain relations of f.c.c. polycrystalline metals hardening according to Taylor's rule; Plastic deformation of b.c.c. polycrystals, *J. Mech. Phys. Solids* **12,** 11–24, 25–33.

Hutchinson, J. W. (1970) Elastic–plastic behaviour of polycrystalline metals and composites, *Proc. R. Soc. Lond.* A **319,** 247–272.

Jackson, P. J. and Basinski, Z. S. (1967) Latent hardening and the flow stress in copper single crystals, *Can. J. Phys.* **45,** 707–735.

Kallend, J. S. and Davies, G. J. (1972) A simulation of texture development in f.c.c. metals, *Phil. Mag.*, Ser. 8, **25,** 471–490.

Keh, A. S. (1965) Work hardening and deformation sub-structure in iron single crystals deformed in tension at 298°K. *Phil. Mag.*, Ser. 8, **12,** 9–30.

Kocks, U. F. (1958) Polyslip in polycrystals, *Acta Metall.* **6,** 85–94.

Kocks, U. F. (1964) Latent hardening and secondary slip in aluminum and silver, *Trans. Met. Soc. AIME* **230,** 1160–1167.

Kocks, U. F. (1970) The relation between polycrystal deformation and single-crystal deformation, *Met. Trans.* **1,** 1121–1143.

Kocks, U. F. (1975) Constitutive relations for slip, *Constitutive Equations in Plasticity*, edited by A. S. Argon, 81–115, MIT Press, Cambridge, Mass.

Kocks, U. F. and Brown, T. J. (1966) Latent hardening in aluminum, *Acta Metall.* **14,** 87–98.

Kocks, U. F., Nakada, Y., and Ramaswami, B. (1964) Compression testing of fcc crystals, *Trans. Met. Soc. AIME* **230,** 1005–1009.

Kratochvil, J. (1973) Finite-strain theory of inelastic behavior of crystalline solids, *Foundations of Plasticity*, edited by A. Sawczuk, **1,** 401–415, Noordhoff, Leyden.
Lee, E. H. and Liu, D. T. (1967) Finite-strain elastic–plastic theory with application to plane-wave analysis, *J. Appl. Phys.* **38,** 19–27.
Lin, T. H. (1971) Physical theory of plasticity, *Advances in Applied Mechanics* **11,** edited by C.-S. Yih, 255–311, Academic Press, New York.
Lin, T. H., Ito, Y. M., and Yu, C. L. (1974) A new slip theory of plasticity, *J. Appl. Mech.* **41,** 587–592.
McLean, D. (1962) *Mechanical Properties of Metals*, Ch. 1, Wiley, New York.
Mair, W. M. and Pugh, H. Ll. D. (1964) Effect of pre-strain on yield surfaces in copper, *J. Mech. Engng Sci.* **6,** 150–163.
Mises, R. von (1928) Mechanik der plastischen Formanderung von Kristallen, *Z. angew. Math. Mech.* **8,** 161–185.
Nakada, Y. and Keh, A. S. (1966) Latent hardening in iron single crystals, *Acta Metall.* **14,** 961–973.
Piercy, G. R., Cahn, R. W., and Cottrell, A. H. (1955) A study of primary and conjugate slip in crystals of alpha-brass, *Acta Metall.* **3,** 331–338.
Ramaswami, B., Kocks, U. F., and Chalmers, B. (1965) Latent hardening in silver and an Ag–Au alloy, *Trans Met. Soc. AIME* **233,** 927–931.
Rice, J. R. (1971) Inelastic constitutive relations for solids: an internal-variable theory and its application to metal plasticity, *J. Mech. Phys. Solids* **19,** 433–455.
Rice, J. R. (1975) Continuum mechanics and thermodynamics of plasticity in relation to microscale deformation mechanisms, *Constitutive Equations in Plasticity*, edited by A. S. Argon, 23–79, MIT Press, Cambridge, Mass.
Ronay, M. (1968) Second-order elongation of metal tubes in cyclic torsion, *Int. J. Solids Struct.* **4,** 509–516.
Seth, B. R. (1964) Generalized strain measure with applications to physical problems, *Second-Order Effects in Elasticity, Plasticity and Fluid Dynamics*, edited by M. Reiner and D. Abir, 162–171, Macmillan, New York.
Sewell, M. J. (1969) On dual approximation principles and optimization in continuum mechanics, *Phil. Trans. R. Soc. Lond.* A **265,** 319–351.
Sewell, M. J. (1972) A survey of plastic buckling, *Stability*, edited by H. Leipholz, 85–197, University of Waterloo Press.
Sewell, M. J. (1974) On applications of saddle-shaped and convex generating functionals, *Physical Structure in Systems Theory*, edited by J. J. van Dixhoorn and F. J. Evans, 219–245, Academic Press, London.
Shalaby, A. H. and Havner, K. S. (1978) A general kinematical analysis of double slip, *J. Mech. Phys. Solids* **26,** 79–92.
Taylor, G. I. (1938a) Plastic strain in metals, *J. Inst. Metals* **62,** 307–324.
Taylor, G. I. (1938b) Analysis of plastic strain in a cubic crystal, *Stephen Timoshenko 60th Anniversary Volume*, 218–224, Macmillan, New York.
Taylor, G. I. and Elam, C. F. (1923) The distortion of an aluminium crystal during a tensile test, *Proc. R. Soc. Lond.* A **102,** 643–667.
Taylor, G. I. and Elam, C. F. (1925) The plastic extension and fracture of aluminium crystals, *Proc. R. Soc. Lond.* A **108,** 28–51.
Taylor, G. I. and Elam, C. F. (1926) The distortion of iron crystals, *Proc. R. Soc. Lond.* A **112,** 337–361.
Vause, R. F. and Havner, K. S. (1979) Theoretical latent hardening in crystals, III: F.c.c. crystals in compression, *J. Mech. Phys. Solids* **27,** 393–414.

The Mechanics of Some Industrial Pressing, Rolling, and Forging Processes

by W. Johnson

University Engineering Department, Cambridge

Summary

For the following little-researched industrial plastic metal-forming processes, brief descriptions are outlined and short accounts of the mechanics of important aspects of them are presented: the pressing of plate in circular or spherical dies, curved plate forming by peening or rolling, ring rolling, rotary forging, ball forming by forging, rolling, or the use of a shot-tower, and sheet perforation or neck extruding. These operations are chosen as being complex and to indicate some likely practical outlets for plasticity theory. For all of them the plastic deformation encountered is described.

Results for machine or tool loads and torque are referred to and some discussion of processing defects is presented. Elastic springback in pressed plates is considered especially.

1. Introduction

In 1981 the University of Cambridge will graduate for the first time engineers prepared specifically for entering manufacturing industry. Their two-year Cambridge course, the Production Engineering Tripos (post-Part IB, Engineering Tripos), was started in 1979. Part of the course will have been devoted to studying the design of machines and the production of goods, and they will have had courses in *Metal-forming Plasticity*. The year 1981 thus combines Rodney Hill's anniversary with a recognition of the importance and industrial applicability of some of the work he has produced over a lifetime, by its being fully assimilated into the corpus of high-level undergraduate engineering teaching in his own University.

In the face of intense international economic rivalries, increasing technical sophistication, and the relative decline of the importance of industrial craftsmanship, it has become imperative to ensure that, among other things, the education of engineers should include a central and substantial course on metal-processing theory. The need to do this has become the more urgent, the more traditional metal craft-training has been seen to be unequal to present-day demands. Properly to supply the professional and intellectual requirements of young mechanical and manufacturing engineers, the area of metal forming now has at hand an international body of teachers who have researched the area and thoroughly developed an acquaintance with its principles and its technical complexities. They have also concurrently developed sound scientific courses in the subject. A large fraction of their various kinds of work derives directly from Rodney Hill's *The Mathematical Theory of Plasticity* (1950) and from his subsequent papers.

The article as contributed here consists of process descriptions and, hopefully, illuminating comments in terms of the engineering mechanics of a wide variety of metal-forming operations. These have been chosen for the following reasons.

1. They illustrate the industrial outlets of such work as is found in *The Mathematical Theory of Plasticity*.
2. They are relatively little-researched subjects and besides being of interest in themselves are in need of greater technological attention.
3. They are complex processes, and this thus demonstrates some of the difficult problems a research and development engineer must face—and cannot turn away from.
4. Challenging questions of principle are embedded in the treatments of the topics discussed. These are necessarily passed over by engineers who must generally somehow arrive at a technological solution. Perhaps applied mathematicians can be attracted to devote their attention to some of them or their like.

2. Plate Pressing

2.1. *Industrial background*

Figure 2.1 shows the typical form of an elevated tank intended to store about 500 000 gallons of water. Such tanks are about 60 ft in diameter and 40 ft in depth, and are supported on a tubular steel structure of height 50 ft. It is planned to build similar tanks of two million gallons capacity. The tank shapes vary and are spheroidal (Fig. 2.1) and ellipsoidal and toroidal (Fig. 2.2). The steel plates of which such tanks are made-up are initially flat but cut to be of various non-rectangular shapes so that after pressing they fit together to give the required form (Fig. 2.3). Such shapes are usually encompassed within plate dimensions of about 15 ft by 10 ft. Plates have thicknesses which range between 9 mm and 30 mm and are obviously required to possess, on assembly, curvatures which are different in perpendicular planes, though some of them (for the case of the ellipsoidal tank) need only to be given a cylindrical form. Figure 2.4 shows a tank in process of erection.

Plates are cold- or hot-pressed to the required shape in a suitably curved die which has egg-box ribs, by a form of mating punch in a large press. Aside from the difficulties of accurately welding together over one hundred such plates, a major difficulty arises in seeking to press plastically the initially flat plates to achieve just the required magnitudes of double curvature because a significant amount of elastic springback must be allowed for. Typically, after pressing in a four-column press mild steel plate of 16 mm thickness with a *tool of radii 19 ft and 33 ft*, the plate on its centre lines was found to possess *radii of 32 ft and 50 ft*, respectively. The force required to achieve a full pressing of this plate was not large: it was scarcely more than simply the weight of the pressing tool. It is noteworthy that in some of these workshop circumstances the self-weight of flat plate causes significant elastic deflection and hence curvature even before pressing begins. The self-weight contribution to any measured springback plate radii of curvature after pressing may well be important, and this

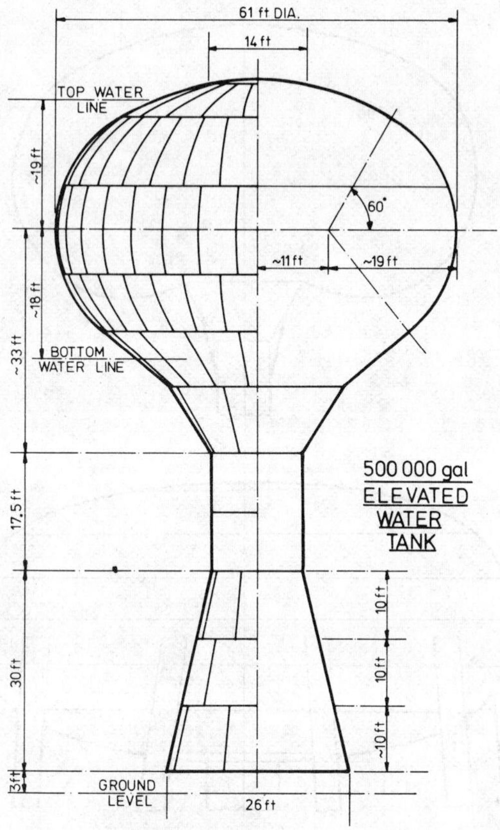

FIG. 2.1. A half-million-gallon elevated spherical watertank for erection in Arabia.

does, in fact, somewhat depend upon the plate attitude with respect to the direction in which gravity acts; this relates also to the welding-up of pressed plate in the various locations, as will be evident from Figs. 2.1 and 2.4.

The writer was unable to trace any published paper which analysed the cold (or hot) pressing of plates into a die to take up a *prescribed double curvature* and that facilitated calculations to allow for elastic springback or residual stress distributions. (Thermal strains and changes of curvature are nowhere considered in this context.) Nearly all theoretical work on the bending of plate refers to cylindrical bending and *does not contain reference to the tooling or method which brings it about*; it is nearly always assumed that the conditions which apply are those of pure bending, that there are no edge effects (i.e. any requirement for stresses at the edges to be zero is neglected), that no special problems are raised by the tooling, and that any complications due to self-weight are disregarded. Analyses also nearly always assume that the material is isotropic and that unloading is entirely linear elastic.

There is, of course, much general descriptive literature on cylindrical sheet bending in dies and over formers, with and without tensions, but as recourse to well-known monographs and conference proceedings will show, it all refers to circumstances in which the plate extends outside the curved die surface.

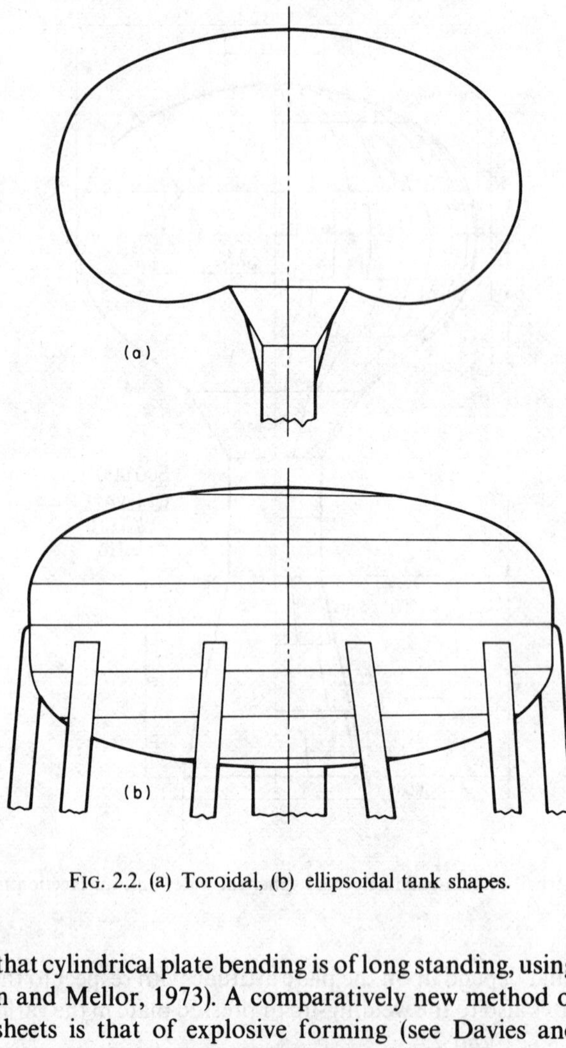

FIG. 2.2. (a) Toroidal, (b) ellipsoidal tank shapes.

It may be noted that cylindrical plate bending is of long standing, using pinch or pyramidal type rolls (Johnson and Mellor, 1973). A comparatively new method of fashioning doubly curved plates or sheets is that of explosive forming (see Davies and Austin, 1970); for axisymmetric sheet and plate shaping the historical methods of spinning and flow-forming are used; both are currently receiving much attention (see, for example, Slater and Joorabchian, 1976). The bending of plates to a spherical form in a vertical press (somewhat similar to a pyramidal machine but with the rolls vertical) with auxiliary inexpensive dies (and the subject of patents) is currently being referred to in some advertising literature.

In view of the paucity of information about the springback resulting from the pressing of sheet and plate into an enveloping die, the following results and observations are summarized (Johnson and Singh, 1981).

2.2. *The cylindrical pressing of plate or strip*

Figure 2.5(a) is a diagram of a cylindrical punch pressing a rectangular strip into a cylindrical die. When the punch has fully advanced, the strip will possess the punch curvature over its concave surface and the die curvature over its convex surface. After releasing the

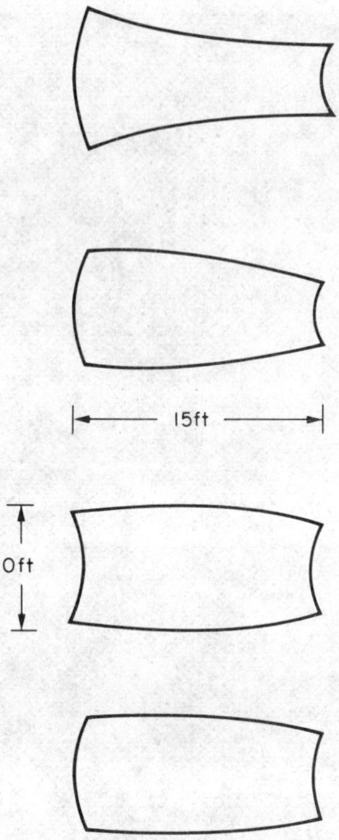

FIG. 2.3. Illustrating the shapes to which individual plates are cut before pressing and assembly.

applied load, the contour of the strip is changed because of the elastic unloading.

A phenomenon observed during the bending or punch advance process is the development of a gap between the strip and the punch over the punch crown; the magnitude of this separation increases with punch advance until the strip bottoms on the die. The steady increase in magnitude of this gap is clearly shown in Fig. 2.6. After bottoming occurs the process then enters another phase which begins as a five-point symmetrical bending process. This is evident from the punch-load vs. punch-travel diagram of Fig. 2.7. Each concentrated point-load rapidly spreads to become an area of loading, and eventually there is, nominally, complete contact between the workpiece surfaces and the tools. In summary, if we neglect any initial elastic loading, the first phase of this plastic bending operation is one of four-point loading with the punch load relatively small. Very late in the process a second phase is developed with the onset of bottoming and when a further central point-load is brought about; the five-point loads then quickly transform to distributed ones and the punch load rapidly increases.

If there is entirely frictionless bending, the four forces on the strip in the first phase are S_R and N_R (Fig. 2.5(a)). At C and C' it is reasonable to assume that the bending moment magnitude due to S_R is just such as will create a plastic hinge. The whole process may be conceived as one in which a plastic hinge, first initiated at the crown, then steadily travels outwards until the strip is fully trapped between the punch and die. It is easy to arrive at expressions relating the punch force P to its advance δ (after first contact), in terms of the

FIG. 2.4. A tank in process of erection.

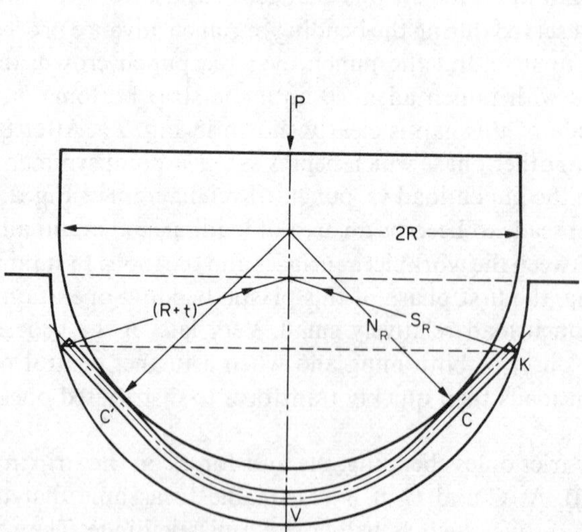

FIG. 2.5(a). A punch pressing a strip into an enveloping die.

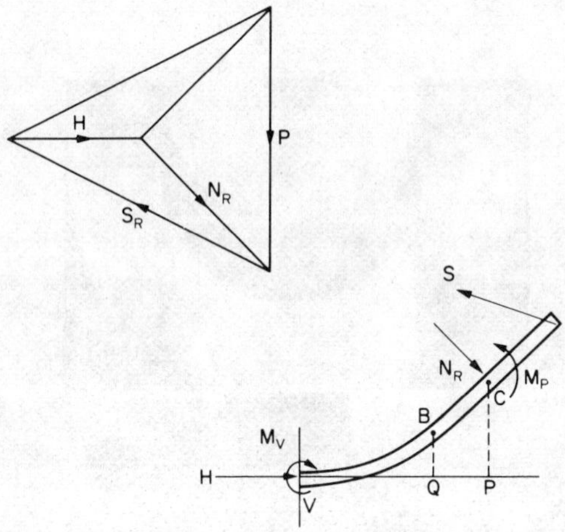

FIG. 2.5(b). Forces and moments imposed on the central portion of the strip.

radius of the punch R, the strip length L, the thickness t, and the yield stress of the (rigid–perfectly plastic) material Y. Typical non-dimensionalized punch-load vs. punch-travel curves as calculated compare well with those found experimentally except for very small values of δ (see the example in Fig. 2.8). Figure 2.9 shows how the measured springback percentage $[100(R_s - R_d)/R_d]$ (where R_s is springback radius and R_d the die radius) varies with strip-length to die-diameter ratio. Observe especially that *negative* springback was encountered; this applied to both steel and aluminium specimens. The curvature of an unloaded bent strip was also found to vary along its length (Fig. 2.10); note that the ends of the strip can only be deformed elastically. That the final shape of plate is not only a function of the tool radii but is also a function of length is made clear by the non-conforming strips shown in Fig. 2.11, where all the strips in each group were formed in the same die by the same punch.

The few results presented show that our physical appreciation of the factors in cylindrical plate bending into a die of relatively small radius is limited. Not only are geometrical factors of importance but the physical sequence of events is different from what would have been supposed. Recent experiments (Singh, 1979, unpublished) into pressing strips into dies whose radius of curvature was many times larger than those used in the above reported work have shown the phenomena just described to be much less conspicuous.

2.3. *Disc pressing*

Experiments have been performed in which flat circular blanks or discs were pressed into a hemispherical die using a hemispherical punch to form part-spherical caps; the kind of material and many other material and tool dimensions were the same as referred to in Section 2.2. It transpires that most of the results were very similar in nature to those for the cylindrical strip (Johnson and Singh, 1980). It was found, for instance, that the punch-load vs. punch-travel curve is of the same form and that there is a reduction of springback percentage with

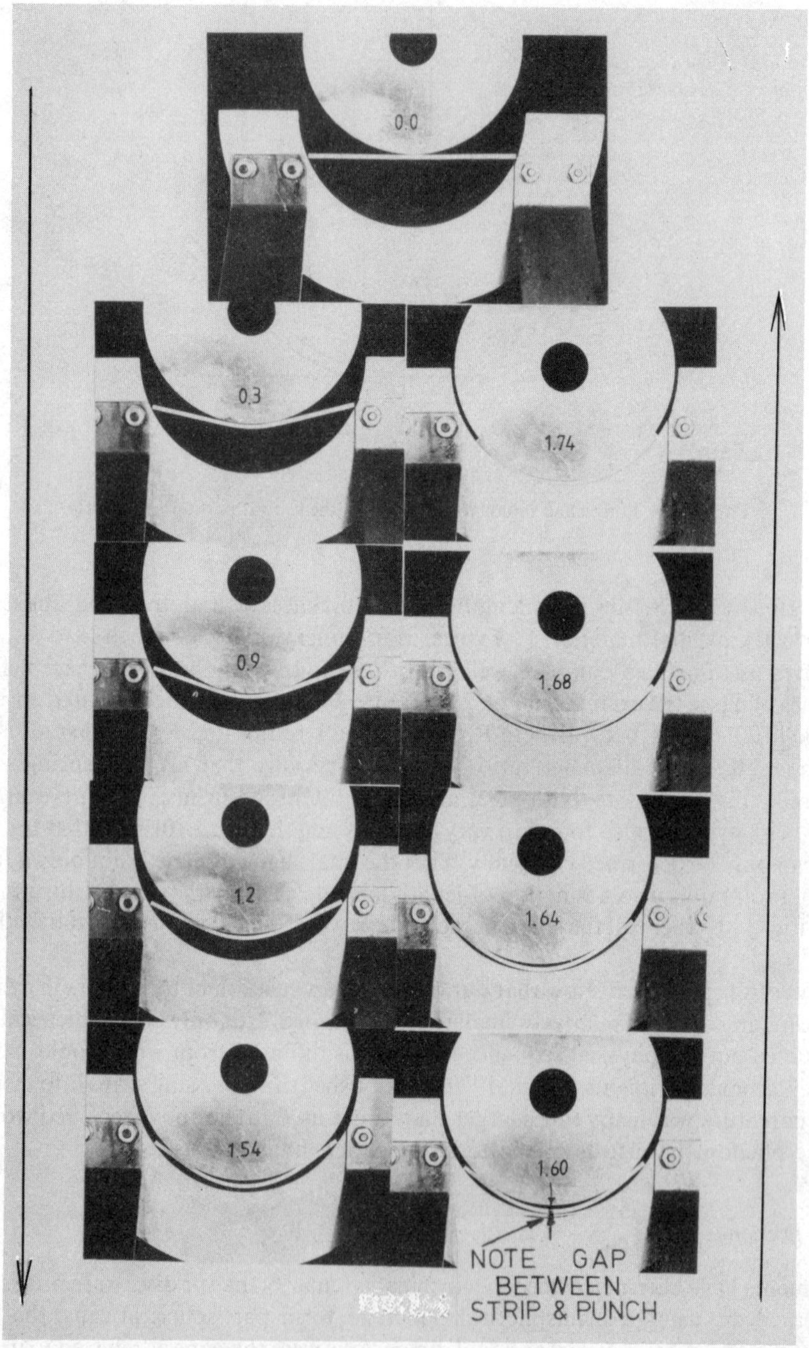

FIG. 2.6. Showing the development of the gap or clearance between the workpiece and the punch, before and after bottoming.

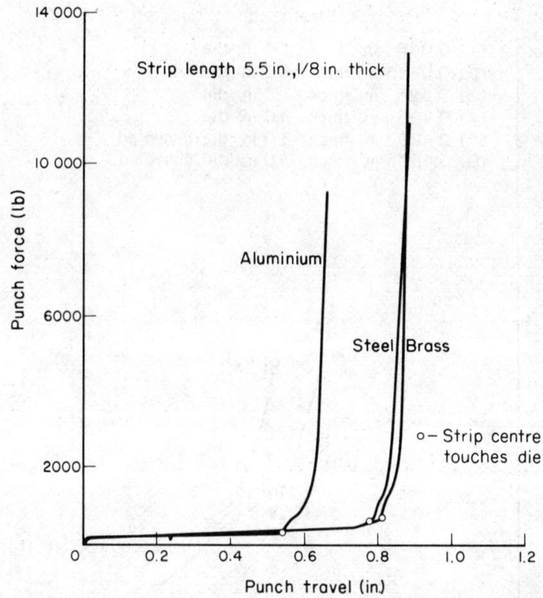

FIG. 2.7. Showing a typical punch-load vs. punch-travel diagram and identifying the rapid load increase at the bottoming point.

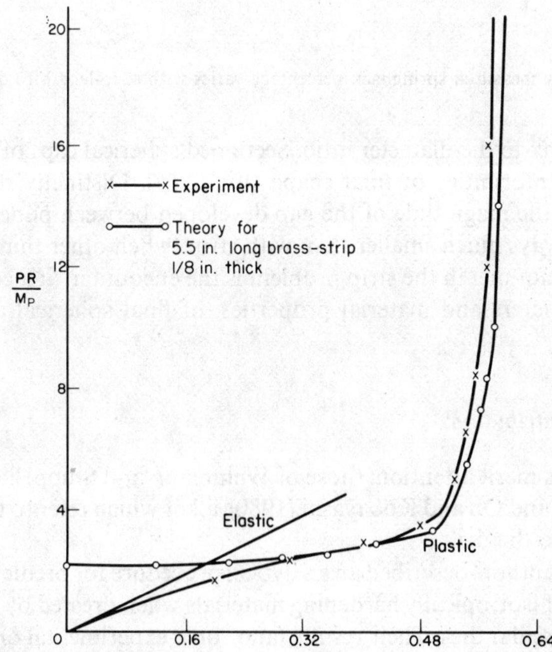

FIG. 2.8. Showing how the non-dimensionalized load PR/M_P (M_P is the plastic bending moment per unit width of strip) varies with the non-dimensionalized punch advance δ/R.

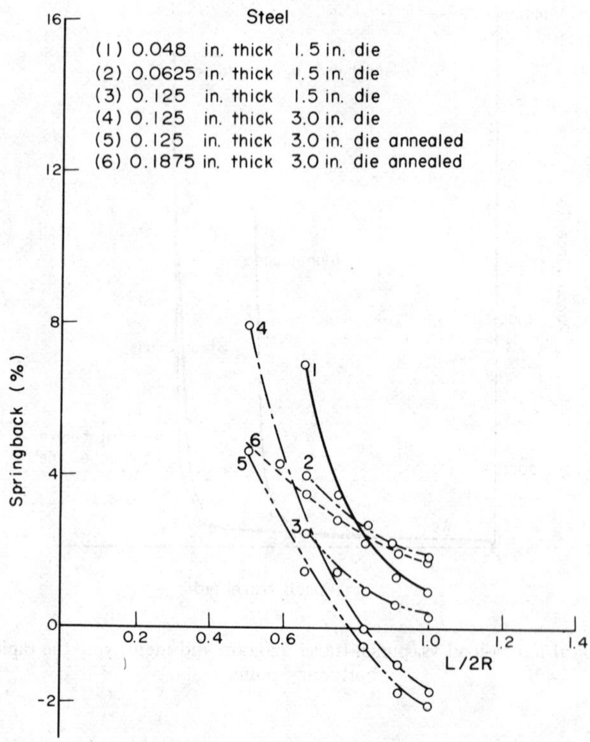

FIG. 2.9. Showing how measured springback percentage varies with strip-length to die-diameter ratio.

increase in initial–blank to die-diameter ratio. Sectioned spherical caps of different diameters again display non-conformities of final shape (Fig. 2.12). Distinctly different from strip pressing, however, is the magnitude of the gap developed between punch and cap over the punch crown; it is always much smaller than with strip (when other things are the same). A complicating feature not met in the strip problem is the encounter, at a certain combination of geometrical parameters and material properties, of final spherical caps which emerge plastically buckled.

2.4. *Three related contributions*

Three recent papers merit attention: those of Weinmann and Shippell (1978), Sidebottom and Gebhardt (1979), and Oh and Kobayashi (1980), all of which refer to the bending of strip which extends outside the die.

The second pair of authors described an analytical procedure for predicting the springback of plates and beams of isotropically hardening materials when pressed by a cylindrical punch in V-shaped and in circular dies. Their results (and some experimental ones on beams) were compared with experimental and theoretical ones provided by Queener and de Angelis (1968), and both indicated that the springback radii of curvature were less than predicted values. However, in particular, Sidebottom and Gebhardt noted that with hot-rolled *steel* plates, the plate pulled away from the punch over its centre portion as in Fig. 2.6. (Their Fig. 9

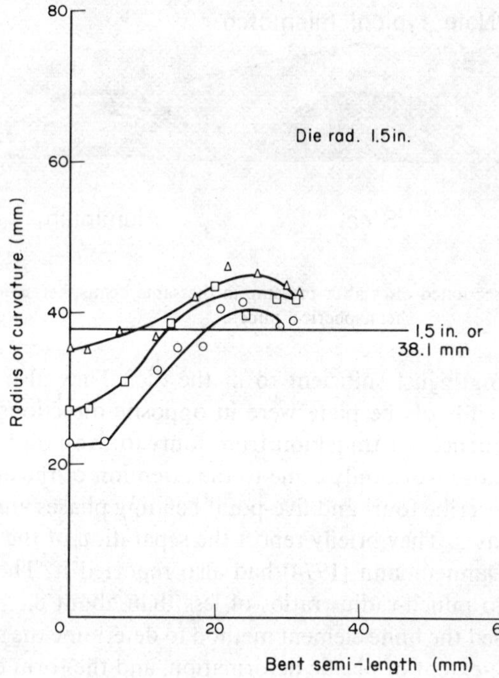

FIG. 2.10. Showing how the curvature of an unloaded pressed strip may vary along its length.

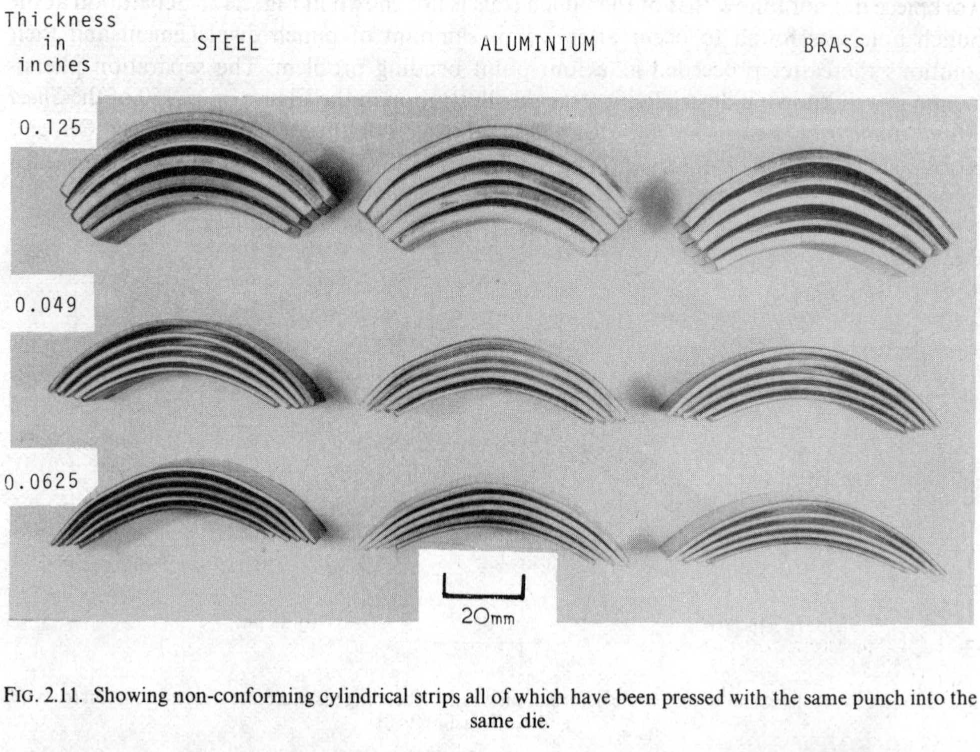

FIG. 2.11. Showing non-conforming cylindrical strips all of which have been pressed with the same punch into the same die.

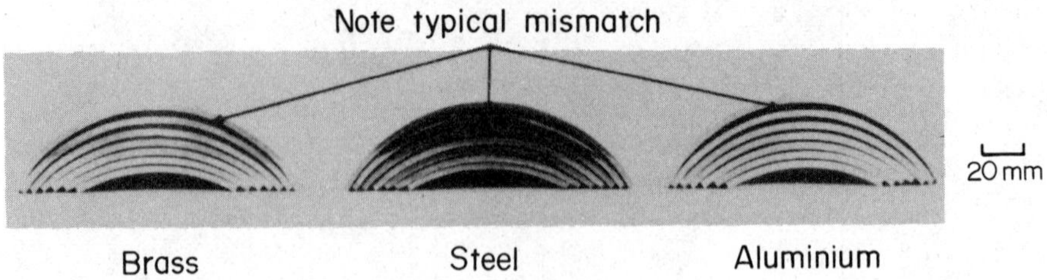

FIG. 2.12. Showing non-conforming sectioned discs after pressing in the same hemispherical die with the same hemispherical punch.

shows the plate to be of a length just sufficient to fit the die.) They also observed that springback at the centre and ends of the plate were in opposite directions. They do not mention or discuss the consequences of transition from four- to five-point loading.

The first pair of authors (whose work only came to the attention of the author after the completion of Singh's work) describe four- and five-point bending phases and refer to them as air-bending and coining phases. They briefly report the separation of the plate from the punch crown and note that Dannenmann (1974) had also reported it. They associate its occurrence with die-opening to punch-radius ratios of less than about 8.

Oh and Kobayashi (1980) used the finite element method to determine the springback, the residual stress distribution, the extent of plastic deformation, and the form of deflection in plane-strain *aluminium alloy* sheet bending for the die and tool arrangement evident in Fig. 2.13. This purely theoretical analysis very clearly showed that the bend radius of the workpiece did not follow that of the punch (this is not shown in Fig. 2.13). Separation at the punch polę was found to occur after a finite amount of punch displacement and their solutions thereafter proceeded as a four-point bending problem. The separation phenomenon is well known industrially (e.g. see the illustration in the Figure on p. 150 of the *Sheet Metal Industries Yearbook for 1972–3*), and was obtained theoretically by Oh and Kobayashi (1980) and by observation by Sidebottom and Gebhardt (1979) in the one case for

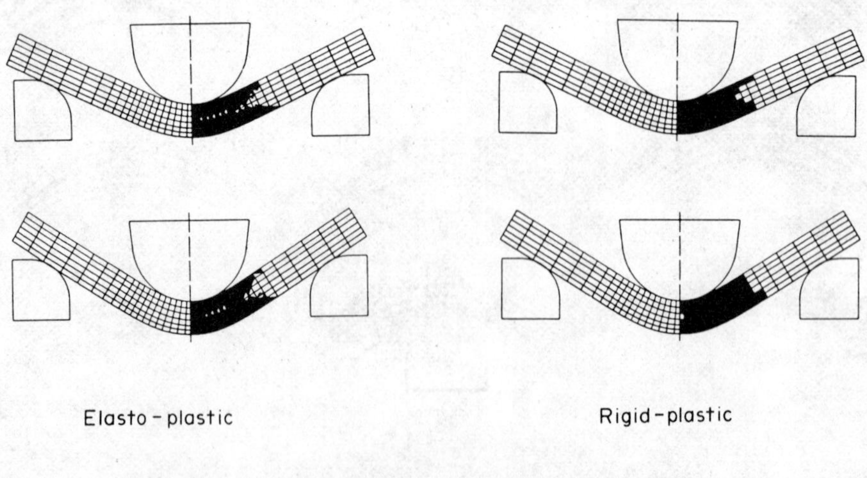

FIG. 2.13. Showing the extent of the plastic deformation when air-bending plate: a finite element solution from Oh and Kobayashi (1980).

an aluminium alloy and in the other for steel plate. It was well confirmed by the several experimental results on several different materials of Johnson and Singh (1980).

The origin of the separation can be understood through Fig. 2.5(b). If the bending moment at C just attains the full plastic moment M_P, then at any point between points C and V the bending moment must exceed M_P by $H.QB$, where H is the horizontal thrust acting through the strip at V and QB the moment arm; in particular, $M_V = M_P + H.CP$. If the strip being formed is of a strain-hardening material, then its curvature from C to V will increase as does the moment. As the curvature at C must be the same as that of the punch, the workpiece must thus separate from it. Alternatively, the thrust H will tend to cause buckling (and simple compression), and this will be the greater, the greater is H and the longer the column length CV. The magnitude of H alters with the degree of deformation imposed. If the reactions N_R and S_R are parallel, then of course the moment between C and V would be constant.

Of considerable interest are the four diagrams from Oh and Kobayashi's paper which show the form and extent of plastic deformation under the punch near to the beginning and end of the operation for elasto-plastic and rigid–plastic (work-hardening) materials (Fig. 2.13). Two observations on these results follow. (i) The form of the zone, with plastic deformation more extensive at the surfaces than the axis and projecting out beyond the region of contact with the die, corroborates what would be expected to apply by recalling knowledge about plane-strain slipline field results for the end loading of stubby cantilevers (Fig. 2.14). This was intimated in the book by Johnson and Mellor (1973). (They did not, however, expect or have need of separation phenomena.) (ii) The degree of exactness with which springback can be calculated remains to be tested. Discrepancies with experimental work were reported by Sidebottom and Gebhardt (1979), though in it much analytical care and effort was obviously devoted to obtaining rigorous results. These latter authors, among other things, carefully note the role played by real material behaviour in unloading and reversed loading (Bauschinger effect) operations; e.g. that at the start of unloading there is first a further small amount of plastic deformation and that departure from a linear unloading line can arise even before unloading is completed. It may well be that these small departures from straight-line unloading play a significant part in denying springback analysis precision when making calculations.

2.5. *Concluding remarks*

The results of two predominantly experimental investigations have been described. The cases of cylindrical and spherical forming represent the extreme radius ratios of doubly-curved sheets, i.e. zero and unity (though the individual magnitudes or degrees of bending are

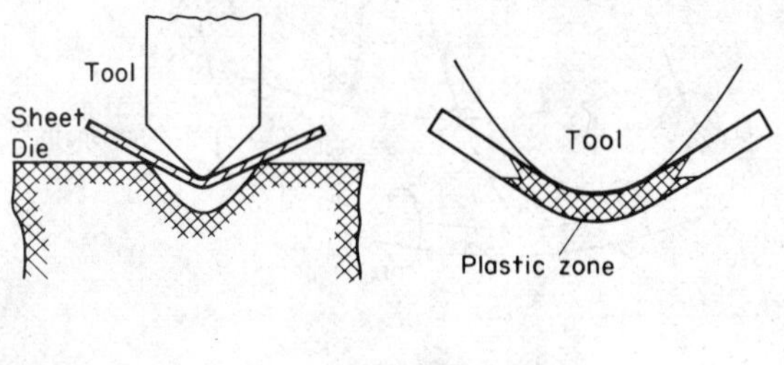

FIG. 2.14. The form of the edge of the plastic deformation suggested by slipline field theory (for comparison with Fig. 2.13).

important) and it is between these values that results about springback magnitude for unequal biaxial bending for use in structural designs will lie.

If for a plate subjected to plane stress elastic–plastic bending due to moment M_X per unit length only (Fig. 2.15(a)), the radius of curvature is R_X, then after unloading the radius to which it springs back, R_{F_X}, is given by (Johnson and Mellor, 1973)

$$\frac{R_X}{R_{F_X}} = 1 - 3X + 4X^3,$$

where $X = YR_X/Eh$; Y denotes the yield stress of the material, E is Young's modulus, and h is the plate thickness.

If now a plate is plastically bent by edge couples M_X and M_Y per unit length (Fig. 2.15(b)), which induce plasticity, with radii R_X and R_Y, then, using the Tresca yield criterion, the springback radius R_{F_X}, assuming non-proportional loading and small transverse deflections, is given by (unpublished work)

$$\frac{R_X}{R_{F_X}} = 1 - 3X + 4X^3 A_X,$$

where

$$X = YR_X(1-\nu)/Eh,$$

$$A_X = \frac{(1+\nu)^2}{1-\nu}\left[\frac{1}{(1+\nu q)^2} - \frac{\nu}{(\nu+q)^2}\right],$$

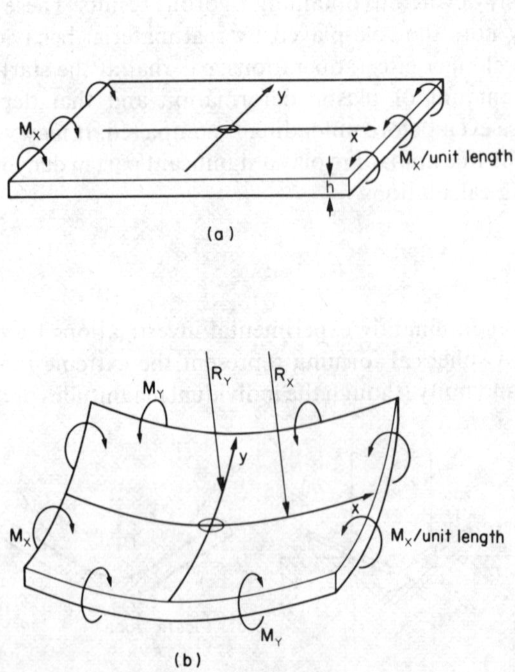

FIG. 2.15. Notation for springback in plates.

$$q = R_X/R_Y$$

and v is Poisson's ratio. A similar expression applies concerning R_Y/R_{F_Y}, except that for R_X and q in these last definitions of X and q, R_Y and $1/q$, respectively, must be substituted. Experiments and analysis to interpolate values are presently in progress in the author's laboratory.

However, it is evident that this plate-pressing process contains many features that will require to be examined and analysed in detail before analysts are in a position to be able to provide manufacturers with a finite element method programme or "package" that will supply them with all the detail they will need to have in order to design optimally and efficiently.

A clear lesson administered by the operation of plate pressing with a punch and die—and one typical of many from the forming field—is to associate closely the analysis with a detailed knowledge of the actual operation.

3. Forming by Peening and Rolling

3.1. *Peening*

3.1.1. Shot peen forming

This process is derived from shot peening which appears first to have been introduced in about 1927 when a method was developed of projecting a "cloudburst" of small steel balls or shot against a metal surface and so hardening the metal by dynamically compressing or peening it. Knowledge of the beneficial effects of cold hammering to harden metal (as with military implements such as swords) has long been common. The balls, typically of 0.01–0.1 in. diameter, are usually of a hard steel—but not so hard that they fracture on impact—though small *glass* shot is used for some purposes.

Shot is usually thrown by a centrifugal impeller (wheel-peening) or lifted, carried, and projected by means of air pressure. The speed of shot is generally a few hundred ft/sec, so that a surface after bombardment for times of the order of one minute is roughened and hardened with the metal grains heavily distorted down to a depth of perhaps 0.01 in. The surface layer, which will have been compressed and therefore stretched, ends by containing large residual compressive stresses; these are equilibrated by small tensile stresses in the lower (only elastically disturbed) material (Fig. 3.1). The surface stresses in any engineering component

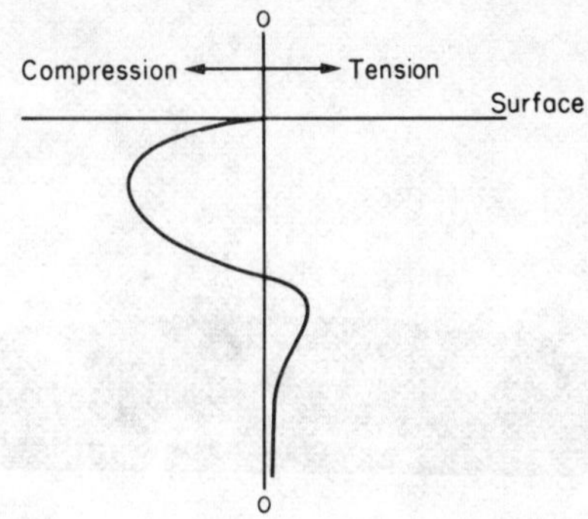

FIG. 3.1. Showing how the residual stresses parallel to the surfaces vary with depth in a peened sheet.

which has been peened (e.g. coiled springs) are thus prepared to offset the deleterious effect of any tensile stress parallel to the surface which may be brought to bear when the component is in service. Shot peening is thus especially a method of preparing engineering items for resisting cyclic stressing (i.e. which would develop tensile stresses that would quickly open up surface cracks) and hence of improving fatigue resistance. The two volumes by Plaster (1973) describe many aspects of industrial shot blasting. Japanese research in peen forming has been reviewed by Kondo *et al.* (1979).

Since shot peening just one side of a not-too-thick plate confers a mildly double curvature, it is obviously a potential forming method, and to a much less degree, a "correcting" or straightening method. A remarkable example of using the shot-peening process to good effect for several purposes is that of the 747 Boeing aluminium alloy wing panel, 105 ft × 7 ft and thickness $\frac{1}{4}-\frac{3}{4}$ in. (Fig. 3.2) (Straub, 1970). Mainly, the peening (at a rate of 500 lb/min) was intended to improve resistance to fatigue and stress corrosion cracking, but the correct amount of curvature was obtained (note, without dies) by applying a higher ball impact speed to the side which was required to be convex.

FIG. 3.2. A long wing panel after peening and forming.

3.1.2. The arc height vs. time curve

A flat strip of sheet or plate lying on a solid foundation and bombarded from one side with shot will slowly rise or curve into the on-coming stream. The rate of rise falls off with time and approaches a limiting value.

Typical arc height h and time of impact or exposure t characteristics are shown in Fig. 3.3 (Meguid *et al.*, 1976a). They have a form

$$h/h_f = 1 - e^{-k\dot{N}t},$$

where h_f is the (asymptotic) arc height reached after a relatively long time of impact, i.e. $t \to \infty$; this supposes that h_f is independent of shot delivery rate N and that the constant k has the dimensions of mass. The equation properly indicates that the greater is N, the more rapidly does $h \to h_f$. For the curves in Fig. 3.3, kN is of the order $10^{-1}\,\text{sec}^{-1}$.

Asymptotic arc height h_f is a function of $mv^2/\bar{Y}$, i.e. it is the greater, the greater is the individual shot mass m and its speed v, and is the less, the greater is the mean flow stress of the peened metal $\bar{Y}$. These parameters are not the only determinants of the limiting arc height. Among others are the angle of obliquity with which the mass of the jet of balls strikes the surface, the jet nozzle stand-off distance from a surface, and the geometrical inclination between the jet and the metal surface; this inclination changes as h increases since the angle of the jet is fixed.

Results and references to papers on the effects of the various parameters of this process on arc height and depth of deformation zone will be found in the references given in Meguid *et al.* (1976b). (I have recently learned that an equation similar in form to that given above was used by Hirose (1975).) It may be noted that shot bombardment also causes a small degree of biaxial curvature (see Johnson *et al.*, 1981b, who also report some recent German work).

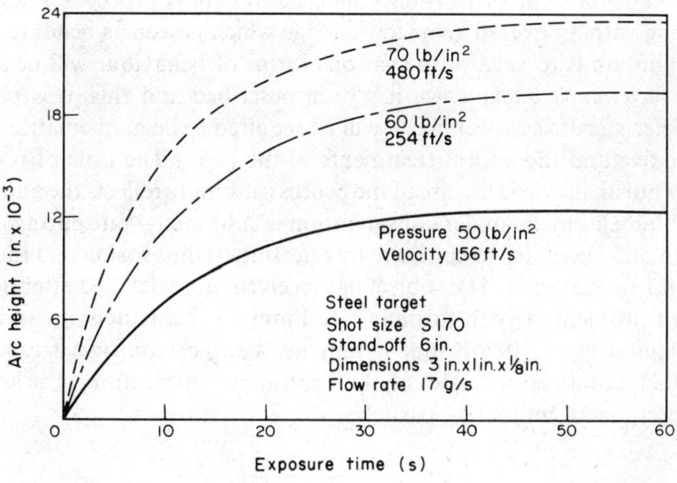

FIG. 3.3. Typical arc height h vs. time of exposure t curve for a peened sheet.

3.1.3. Deformation mechanism

From the above description of this forming process it will be appreciated that analytical treatments aimed at ultimately correlating all the process parameters will be difficult to achieve. Nearly all the investigations reported so far confine themselves to being simple descriptions and presentations of purely empirical results.

To appreciate fully this process it will not be sufficient simply to study and understand the consequences of the impact and indentation of one single spherical shot. There are many repeated loadings and successive craters or indentations, and plastic zones overlap. Even the hope of making profound, and useful, inferences by studying simultaneously created neighbouring plastic indentations from two balls in terms of ideal plastic theory is not fulfilled because the situation has been found not to yield to analysis in terms of more than simple upper bounds (Meguid *et al.*, 1977). Thus expectations of treating accurately the multiple impact situation of a progressively hardening material are hardly likely soon to be forthcoming.

When a small hard sphere moving at a few hundred ft/sec impinges elastically on a ductile metal surface, it creates a shallow crater and enforces plastic deformation around it. Some fraction of the initial kinetic energy of the sphere f_1 is dissipated in plastic deformation of the surface layer and the remainder will be retained almost entirely by the sphere in elastic rebound. A second identical impact on the same surface will enforce further plastic deformation but the fraction of kinetic energy now dissipated f_2 will be less than f_1; this arises because of the hardening enforced at the first impact. Subsequent impacts will contribute to further plastic deformation but smaller and smaller fractions of the initial kinetic energy will be absorbed. After very many impacts, the process will be one of almost complete elastic restitution for the sphere with virtually no further plastic deformation being conferred on the metal.

Applied to the peen forming of a strip by a "rain" of balls, this notion suggests that over a given interval of time, an increment of plastic deformation is conferred, but that this increment is the smaller, the later on in the process it takes place. Each successive increment of plastic work done will be accompanied by increased hardening of individual surface layers. The propagation of the zone of working to increased depths proceeds at a decreasing rate and occurs along with an overall geometry change which is seen as bending. These notions imply that certain limits to each of the various forms of behaviour will be approached. In effect, a form of *dynamic shake-down* has been described and this, it is proposed, is the essential characteristic of shot peening. It will be required to be mathematically expressed to permit quantitative and theoretical treatments of the topic. The time of bombardment in peen forming is not an intrinsic feature of the process; it simply reflects the number of "cycles" to which the volume of metal under consideration is subjected—rate effects being excluded.

It is pertinent and necessary to conclude by referring to the erosion and the blast cleaning of ductile and brittle materials. The subject has received more detailed attention than that of this interrelated problem of peen forming (see Finnie, 1972; Hutchings and Winter, 1975; Tilly, 1973; Meguid *et al.*, 1976b) but it remains a topic requiring a fuller quantitative description which embraces a model of the continuous interaction of a large number of identical indenters or of impinging particles.

3.1.4. Negative curvatures

After peen forming circular blanks of aluminium which were firmly clamped around their periphery and of up to 1.5 mm thickness, by using shot of about the same diameter as the thickness (in a wheel-peening machine) it has been found by Kondo *et al.* (1979) that instead of bulging (i.e. rising towards the shot) the bulge developed negatively. This phenomenon was associated with shot speeds which were too high and the creation of a region of plasticity which passes through the whole sheet thickness and exceeds in lateral area at the bottom surface that at the impact surface.

Experiments using the ball-forming method (see Section 3.1.5) have also produced negative curvatures in recent experiments on initially flat strips held firmly at their ends during bombardment (unpublished work, Cambridge University Engineering Department).

3.1.5. Ball forming

This process is in effect shot peening with massive balls impinging at low speed. It is a patented process that has been used for contouring wing panels for the Lockheed L-1011 aircraft; many steel balls all of the same diameter, but several times larger than that normally employed in shot peening, are dropped under gravity from only a few feet and typically confer on aluminium alloy sheet a curvature of about 100-in. radius. Using balls at small velocity avoids fracturing them and hence eliminates the need for separation equipment which removes all the particles possessing jagged edges; arrangements for containing the impinging balls are also less demanding. Aircraft structures up to 85 ft long have been treated in this way, the delivery of the balls being at a rate of more than one ton per minute with the component passing under the stream at the rate of 2–6 ft/min.

Stress peen forming, i.e. elastically pre-stressing parts and then subjecting them to the ball-forming or peening process, is reported to give rise to greater contour and surface smoothness. Microfissures, folds at the impacted surfaces, and excessive surface roughness (e.g. exceeding aerodynamic requirements) are frequently mitigated by using balls at lower speeds but possessing greater mass. Conventional shot peen forming and other sheet and plate, press- and stretch-forming methods, in the circumstances of the treatment of the structure referred to, are said to confer undesirable mechanical effects (Brandel and Klass, 1971).

Massive ball, slow-speed impact may cause the zone of plastic deformation to penetrate the whole thickness of a sheet (Johnson and Mamalis, unpublished work). Thus of direct interest in connection with slow-speed peening using large balls are the studies of the one-sided plane strain indentation by a rigid punch of sheet lying on a rigid foundation; see the early slipline field analyses of Hill (1950) and Johnson and Woo (1958), and the later studies of the bending-up phenomenon of Dodd and Osakada (1974) and Dewhurst (1974) as initiated by Dugdale (1964).

3.2. *Asymmetrical flat rolling*

There are many circumstances in which the strip or plate emerging from a roll gap is found to be uniformly curved. This may arise from differences in roll radii, roll speeds, roll surface finish, lubrication, and the initial non-symmetrical presentation of the strip to the pass. These situations describe asymmetrical rolling; they usually arise unintentionally but are implicit in some forming processes such as ring rolling and the straight rolling of non-symmetrical sections. In so far as there is a demand for curved plate, it is understandable that the thought occurs to use the asymmetrical rolling process constructively. Reflection shows, however, that the curvature would be very difficult to control, especially since friction between the plate and a roll too easily changes or becomes non-uniform; however, to propose a method based on it, one can come to the kind of system described in Section 3.3.

It is sometimes required to determine the load and torque in asymmetrical rolling, and this does not present too difficult a problem provided a high degree of accuracy is not sought. Reasonable approximations can be made to facilitate arriving at soluble elementary equilibrium equations for an element of material (which are independent of thickness) with the help of a yield criterion. After appreciating the existence of a cross-rolled zone, progress towards a final solution can be made, and, indeed, results of a few calculations of this kind have long been available. The inadequacy of the latter approach (including the von Kármán, the Orowan, and the Ford and Bland analyses), as, indeed, it is of all "slab methods", is their inability to be able to describe or predict the details of the internal deformation of the material and, for most kinds of asymmetric cases, the final velocity of the formed material. In predicting the angle of emergence or of curvature in formed strip in asymmetric rolling (and eccentric extrusion) the slipline field method is here supreme and probably still a unique instrument for accurate analysis. (Though even here it is practically limited by inability to take account of strain-hardening, self-heating of the work material, and uncertainties about the frictional conditions in the roll gap; it is, of course, also restricted to the rolling of wide sheet.) It is only by employing slipline field theory that curvature of the rolled strip can be determined. A few authors have endeavoured to do this without slipline field theory by making well-chosen hypotheses, but none of them are more than partly adequate to the situation.

The previous English-language literature on this topic is not large but it is easily available through reference to the papers by Johnson and Needham (1966) and Collins and Dewhurst (1975). The most comprehensive experimental study appears to have been published by the Russian workers, Chekmarev and Nefedov (1956), with less extensive results being given by Johnson and Needham (1966), Buxton and Browning (1972), and Dewhurst *et al.* (1974).

As pointed out by Collins and Dewhurst (1975), after reviewing the literature, it is not immediately obvious which way the strip will curve in any given asymmetric rolling situation. One common supposition has been that it will always curve away from the roll which has the greater peripheral speed, but this is not so. In fact it is not possible to make a simple statement concerning curvature for all kinds of asymmetric rolling situation.

For even wide hot-strip symmetrical rolling the solution using slipline field theory is not a simple and straightforward matter, as the first solution obtained by Alexander (1955) showed. The present-day fortunate situation of using a matrix technique for arriving at the correct form and proportion of classes of slipline fields for hot asymmetric rolling is principally due to Collins (1968), though deriving directly from earlier work by Hill and Ewing. (This subject is dealt with at length by I. F. Collins elsewhere in this Volume.) Using this technique, Collins and Dewhurst (1975) were able to give the slipline fields and thence the results for roll torques and degree of curvature for a few combinations of the parameters in the asymmetric rolling situation (see the typical slipline field in Fig. 3.4). In brief, the latter authors concluded that strip curvature is very sensitive to changes in roll geometry and that it may be upwards or downwards; variation in strip reduction and radii ratio (when other things remain constant) may result in an oscillation of the curvature of the strip (see the example in Fig. 3.5). There is much rolling mill experience and experimental evidence to corroborate their theoretical findings. Whether useful advantage can be taken of these often

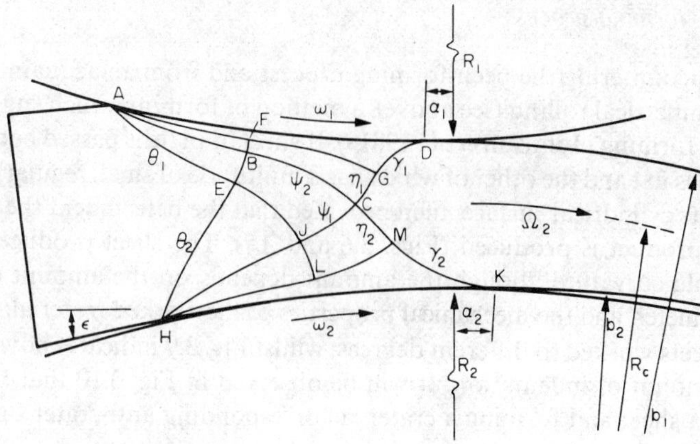

FIG. 3.4. A slipline field for asymmetrical rolling.

unwanted asymmetries in rolling by deliberately introducing them and so producing specially curved strip remains to be seen.

It has also to be noted that the slipline field solutions of Collins and Dewhurst (1975) pertain only to hot rolling where the non-deforming blocks of the workpiece are supposed at rest relative to the rolls and to extend over the whole arc of contact. For non-sticking friction, slipline field solutions have yet to be proposed. In the practical circumstances of real rolling mills the variability of frictional conditions across a workpiece surface (with random oxide entrainment, if hot) of itself would necessitate the employment of external devices for securing constancy of form in the product if specific plate curvatures were aimed for. Furthermore, in the warm- or hot-rolling case, when operating with horizontal rolls, the bulk weight of as-produced unsupported strip would introduce bending moments at the roll throat that would aggravate an already sensitive situation.

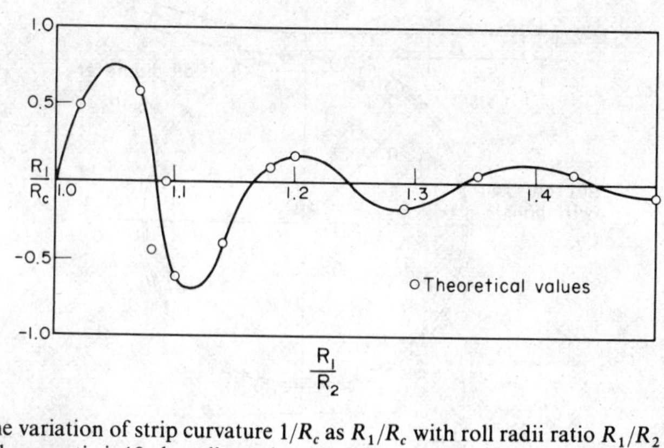

FIG. 3.5. Showing the variation of strip curvature $1/R_c$ as R_1/R_c with roll radii ratio R_1/R_2 when the upper roll to initial workpiece thickness ratio is 10, the roll angular velocities are equal, and the percentage reduction is 20%. (R_c is the mean radius of curvature of the rolled strip; refer to Fig. 3.4.) (From Collins and Dewhurst, 1975.)

3.3. *A roll "peen" forming process*

As a result of encountering the peen-forming process and from an acquaintance with the mechanics of asymmetrical rolling (see above), a method of forming which suggested itself is that of roll "peen" forming (Johnson *et al.*, 1981a). If sheet or plate is passed between a pair of rolls, one of which is flat and the other of which has a multitude of small regularly spaced balls protruding from its cylindrical surface, then provided that the balls indent the sheet to some degree, a curved product is produced (Figs. 3.6 and 3.7). The sheet produced possesses a highly reproducible curvature, though the amount depends on the amount of ball indentation, the roll diameter, and the mechanical properties of the worked material. Figure 3.8 is a photograph of sheets worked to different degrees, whilst Fig. 3.9 indicates how the curvature varies with the amount of indentation. It will be observed in Fig. 3.10 that by penetrating from one side of a sheet and forming a crater, a corresponding anti-crater or depression is formed on the other.

This method of forming is clearly capable of quickly forming plate to a reproducible curvature and suffers less from "edge" effects than does the conventional pyramidal roll arrangement. However, besides the conspicuous indentation of the top and bottom surfaces, the many deformation zones could be sites for corrosion and stress concentrations. By

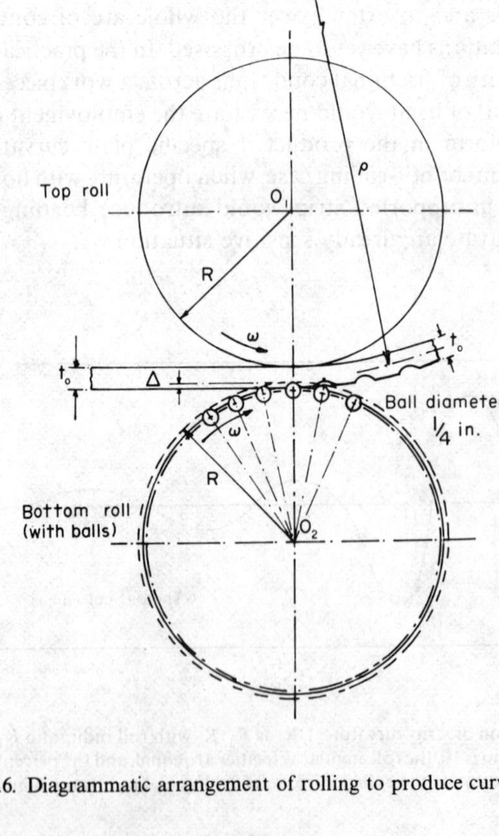

FIG. 3.6. Diagrammatic arrangement of rolling to produce curved plate.

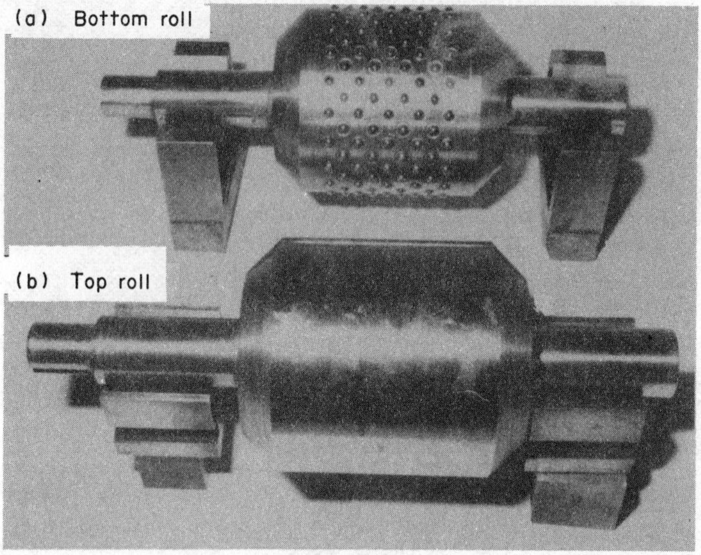

FIG. 3.7. Rolls for forming curved plate.

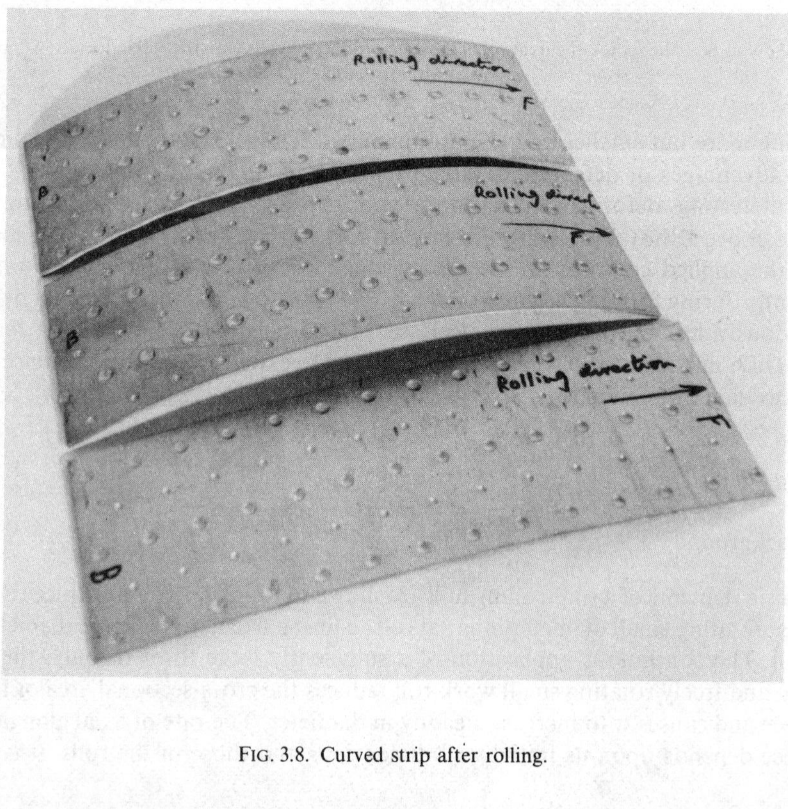

FIG. 3.8. Curved strip after rolling.

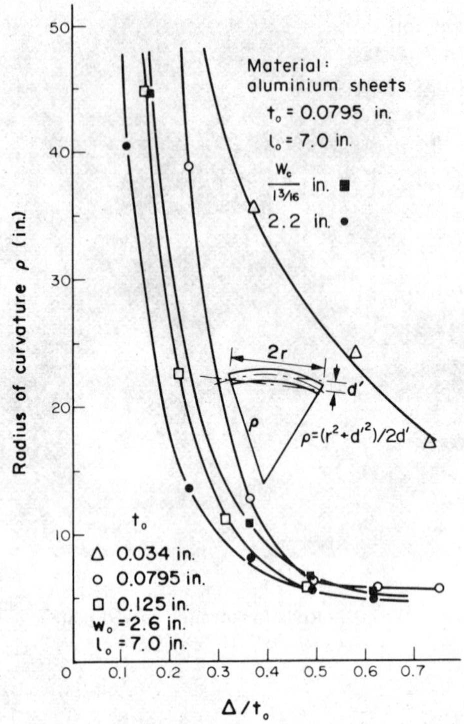

FIG. 3.9. Showing how the radius of curvature of the strip produced varies with the ball indentation depth Δ (t_0 is the initial uniform strip thickness).

implanting more but smaller and less prominent steel spheres into a cylindrical roll surface these disadvantages or defects should be capable of being greatly reduced.

The interesting deformation response common to the two ball-forming processes described above is the taking-on by the worked surface of the convex side of the shaped plate. Due to the applied compression there is a tendency to stretch the surface; this happens significantly during forming because the surface layers are kept in compression by the deeper and obviously less strained layers. Release of the plastic-surface-forming load (i.e. by cessation of bombardment or indentation) allows it to extend or spring and hence to become the convex side of a curved plate.

3.4. *Ring rolling*

3.4.1. Background

The basic elements of a ring-rolling mill are shown in Fig. 3.11. The workpiece is pressed by a moving, rotating small diameter internal roll against a fixed, driven large diameter external main roll. The continuous application of a sufficiently large force through the advancing undriven and freely rotating small work roll reduces the cross-sectional area of the rotating workpiece and causes it to increase steadily in diameter. The rate of axial elongation of the workpiece depends upon its initial wall dimensions and those of the rolls. It is sometimes

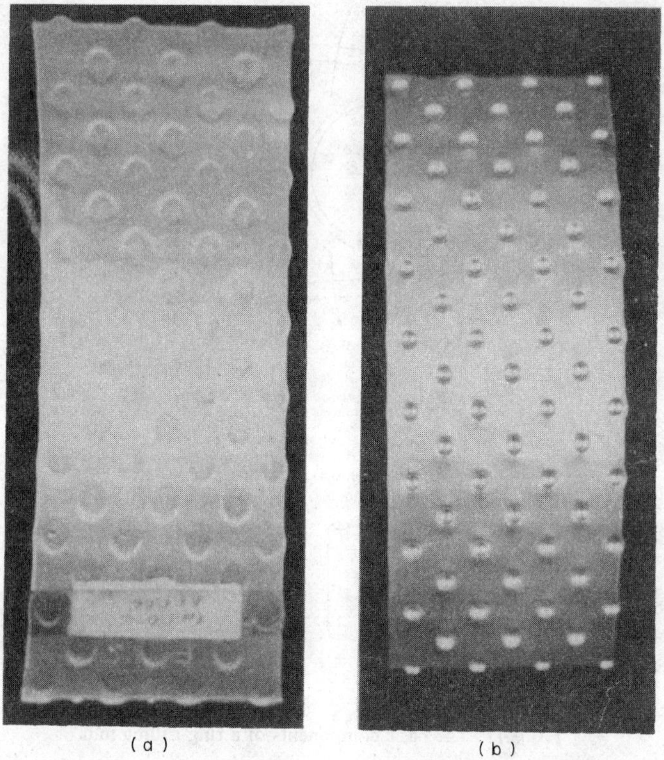

FIG. 3.10. Strip produced: (a) bottom surface, (b) top surface.

necessary to introduce a pair of conical edge rolls if this axial spread must be restricted. Many rotations of the workpiece-ring are usually required because the work-roll feed rate is generally small. The use of the two guide rolls is imperative both for maintaining the circularity and the central position of a ring; their location with respect to the roll is critical for the forming of a good product.

The first ring-rolling mill was apparently designed, manufactured, and operated in Manchester (England) in 1842. Ring-rolling mills were developed slightly later in Europe by, for example, Thyssen in Germany in 1858. The hot rolling of very heavy pieces weighing many tons is now carried out frequently. It is usual to start from a large forged blank or "cheese", formed by pushing out the central portion of an upset cylinder. Large railway tyres are typical hot-rolling products; often they are expanded in diameter from that of the initial blank by five times. At the other extreme there is the precision profile rolling of rings, e.g. as for bicycle hubs, and this is carried out cold, starting from a short length of tube, or from strip, the ends of which have been welded to form a flat ring. Renewed industrial interest in ring rolling has come especially from the need to form rings in high-strength and costly materials, e.g. titanium rings for aircraft engines, and in an endeavour to reduce the waste of expensive materials such as would occur if they were machined from a solid block. There are also several important advantages to ring rolling, as opposed to forging or machining—aspects that seldom enter into the consideration of researchers in applied mechanics—and these are that the process gives rise to a product of a uniform quality, one possessing a

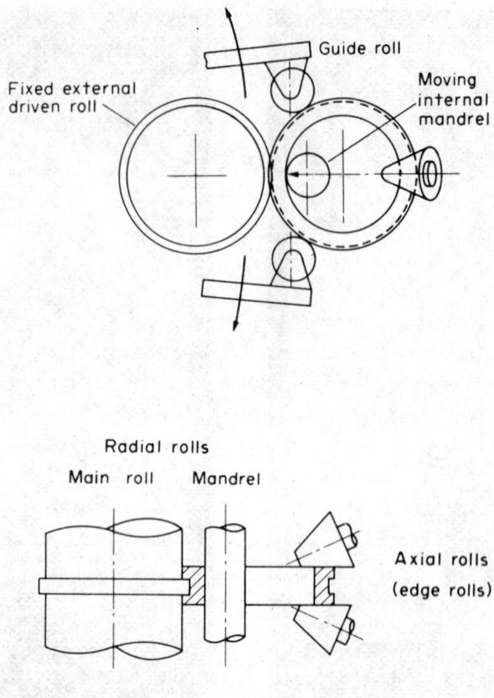

FIG. 3.11. The basic components of a ring-rolling mill.

favourable grain flow that may lead to improved material properties and a smoother surface finish.

Several different forms of ring-rolling mill exist, e.g. those having four mechanical mandrels, but these are not considered here and for further details and bibliography reference can be made to Johnson and Mamalis (1979). This reference also notices Russian literature, which contains many very useful studies of rolling, especially for instance those of Tselikov, Smirnov, and Polukhin.

3.4.2. The mechanics of ring rolling

Basically, flat ring rolling is achieved as for small reduction rolling but between rolls of very different diameters, only one of which is driven. It is also obvious that the proportions of the many passes to which a ring is subjected during a complete forming operation is a continuously changing one. Despite the formidable-seeming number of variables involved—and note in terms of the mechanics the similarities here with those of asymmetric rolling (see Section 3.2)—it has proved possible to calculate operating roll forces and torques by considering the plastic deformation zone in the roll gap to be defined by the kind of slipline field (albeit one of continuously changing proportions) well known to apply for flat-end rigid tools when forging a block whose depth is several times that of the tool width (Fig. 3.12). After using the results obtained by Hill (1950) for non-dimensionalized indentation-pressure vs. tool-geometry ratio (and making allowances for the different indenter widths l_1 and l_2), a

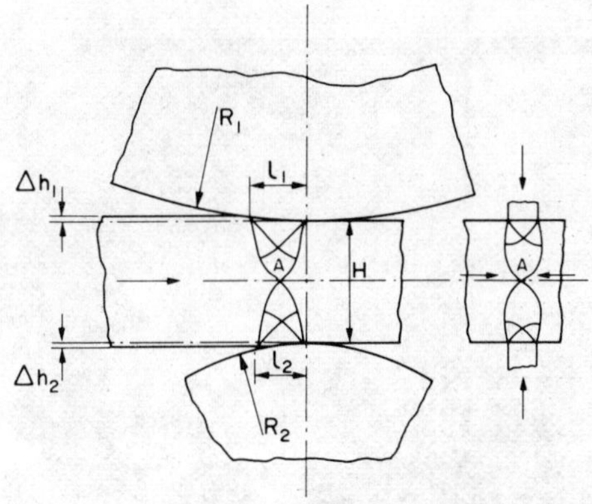

FIG. 3.12. Assumed slipline field in the roll gap.

helpful nomogram can be provided for designers for predicting forces and torques, which embraces most of the geometrical and machine parameters and is reliable (i.e. experimentally verified) to better than 15% (Hawkyard *et al.*, 1973).

It is interesting to note that for sufficiently large ratios of l/h, i.e. when plastic deformation from roll to roll does not penetrate the whole ring thickness, and the roll pressure $\bar{p}$ equals $2k(1 + \frac{1}{2}\pi)$ (k is the yield shear stress), that a limitation to forming is encountered and there is then a tendency to roll "square" rings.

The roll pressure distributions (and contact lengths) on both forming rolls are not of course identical, but by inserting strain-gauged pressure pins into both of them, Mamalis was able to determine the pressure distributions for a wide range of parameters (Mamalis *et al.*, 1976a). Though the pressure distributions on the two rolls at an instant are unalike, particularly noteworthy were the several occurrences of two (but unequal) pressure maxima when rolling tellurium lead which simulates hot-rolling circumstances; usually these were the more conspicuous, the smaller the reduction attempted and the longer the workpiece–roll contact length.

Though the spread of material on emerging from the roll gap in conventional flat symmetric rolling has become calculable during the last few years, particularly by using Hill's general method of analysis (see, for example, the work of Kobayashi, 1977, and his co-workers), experimental and calculated values frequently do not match to some degree. This is due, among other things, to an inability properly to take account of changes in material properties due to working the metal, e.g. strain-hardening, and to unavoidable (and often unknown) non-uniform frictional conditions in the roll gap which arise due to the stretching and thinning of lubricant films on spreading surfaces. Such calculations have not yet been attempted for ring rolling. However, many experimental results for spread factors have been given by Mamalis *et al.* (1976b).

Fishtailing is due to unconstrained axial flow (and to prevent it, edge rolls are used) (see Fig. 3.13 which also shows the lateral grain flow in a section of a ring). Figure 3.14 shows how initially radial lines on the edge of a ring become deformed after several passes.

Difficult to perform accurately in practice—because the material flows in unexpected directions and so does not give the cross-sectional shape which is sought—is profile ring

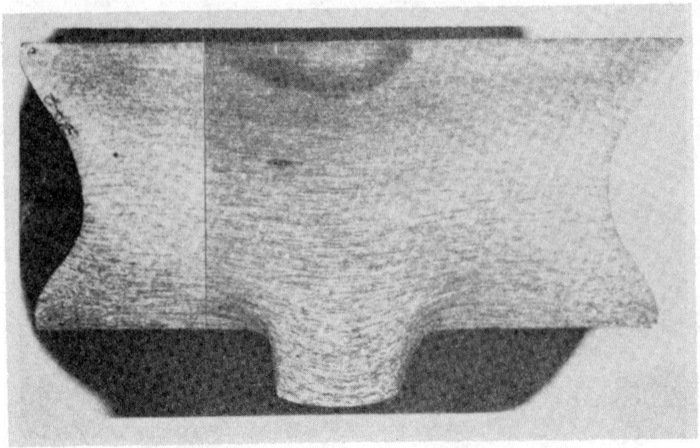

FIG. 3.13. Showing fishtail (the curved sides) and the grain flow in a rolled T-section aluminium ring.

rolling, e.g. the rolling into steel tube of profiles typical of inner and outer ball-bearing races. Flat section rolling is still a very empirically treated subject (full of art and experience rather than science)—one little amenable to technological, much less to sophisticated, theory, so that profile ring rolling is expected to be even less amenable to analysis. One frequently rolled-ring profile is that seen in Fig. 3.15. Notably, when the profile flange is rolled sufficiently thin, a defect arises on its outer surface—analogous to the well-known back-end extrusion defect. When this T-section is rolled, it is clearly not obvious how to proceed in predicting the amount of metal that will flow into the web from a reducing area of flange; there is diametral ring expansion and transverse flow following from the application of a predetermined feed rate of the work roll to compress the flange of the ring. However, Hawkyard and Ingham (1979) have carefully examined the development of the closed-pass profiles seen in Fig. 3.16. They have also shown that in rolling T-profiles the main mass, i.e. the flange A_1, can be conceived to expand hoopwise only (no lateral strain) and to pull or carry with it the less constrained minor web portion A_2 (Fig. 3.16). The latter is subjected to hoop tension and thus has created in it more or less equal transverse contractile strains. The rate of progress into the minor gap and its shape at an instant can then be successfully assessed on the basis of these flow premises. According to the geometry of the pass and other mill variables, it thus may or may not be possible to achieve the required profile by the time the sought ring diameter is arrived at. A recent paper on this topic is that of Yang *et al.* (1981).

4. Rotary Forging

In a remarkable and virtually unknown paper, read before the Manchester Association of Engineers in 1921, H. F. Massey (then Chief Engineer of B. & S. Massey Ltd., Openshaw, Manchester, a company which recently became part of the Davy–Ashmore Group) described his researches into the study of the plastic flow of metal during forging. He used layered blocks of plasticine and demonstrated modes of flow and deformation which, since about 1960, have been common-place knowledge to academics engaged in the field of metal-

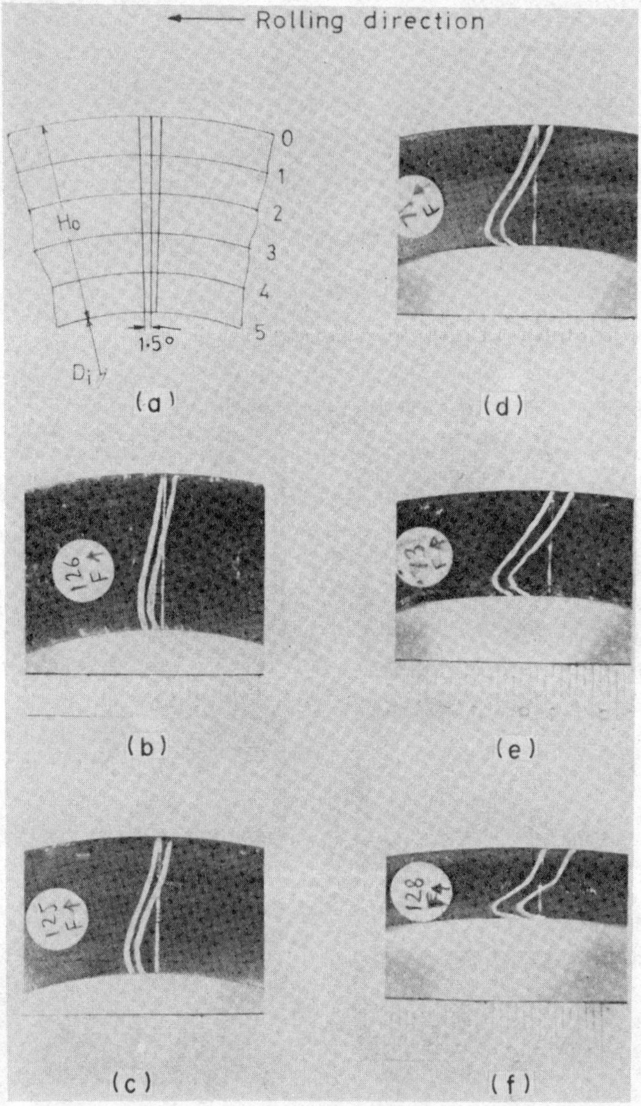

FIG. 3.14. Showing how radial lines in a tellurium lead ring are increasingly deformed as the reduction increases: (b) 20%, (c) 30%, (d) 39%, (e) 47%, (f) 60%.

forming plasticity. Massey in the same paper recorded his observations of the patterns of heat lines—40 years later shown to be associated with the existence of velocity discontinuities (see, for instance, Slater, 1966)—in metal undergoing rapid plastic deformation. (These lines can be made clearly visible on the floor of any hot forging shop when relatively low temperature side pressing is performed. For the same basic reason, short, white, spiral lines of martensite are left behind at the edge of shot holes in perforated sheet plate.)

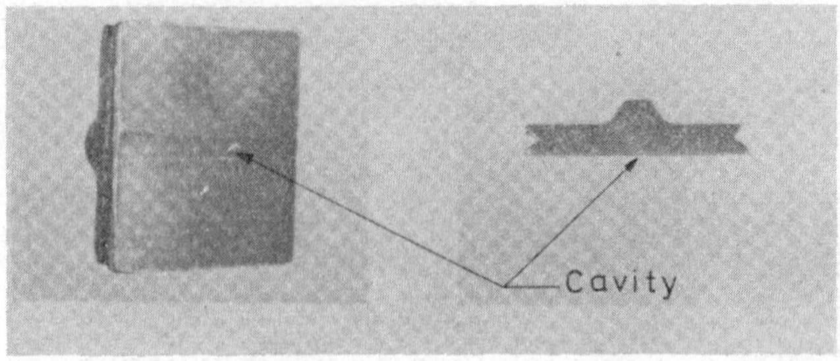

FIG. 3.15. Showing the cavity which is developed through rolling the flange of a T-section ring excessively.

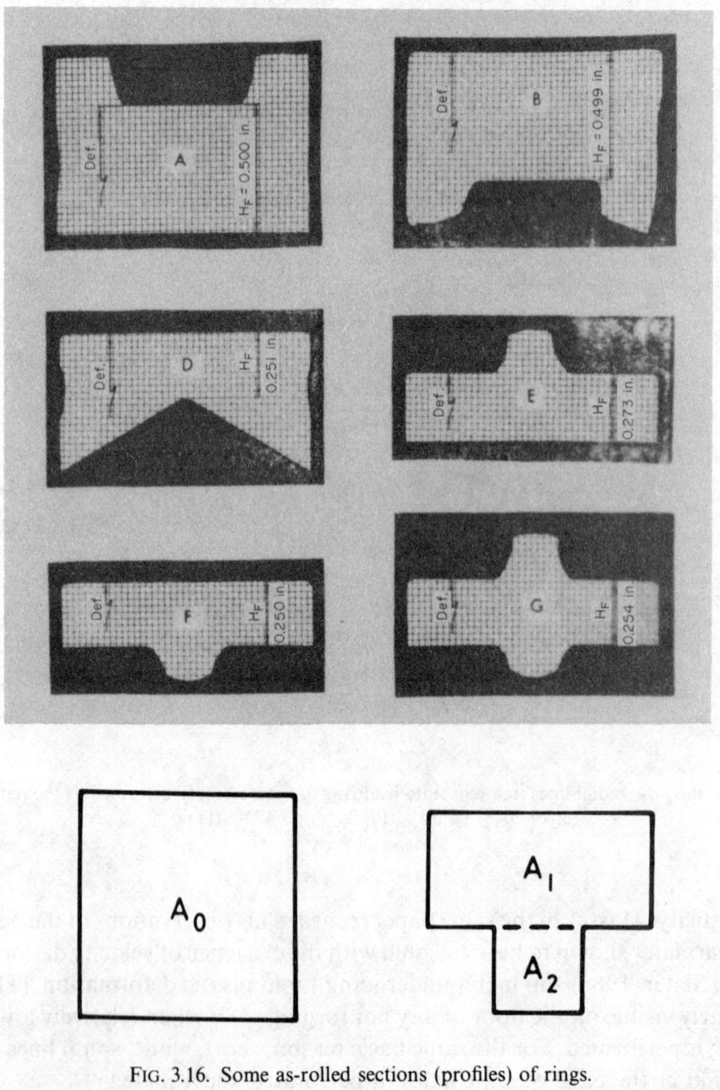

FIG. 3.16. Some as-rolled sections (profiles) of rings.

Massey also proposed and patented in 1929 a vertically operating rocking-die forging press which could be used to produce short, stubby, axially symmetric components such as bevel gears and spur wheels. The potentialities of this form of press were not fully investigated until about 1964 when, after building a model, Massey Ltd. began to develop it in 1966 as the Rotaform. Technical details of this press and an account of its progress during development for the forging industry in the 1970s can be found in the paper by Standring and Appleton (1979) and in the references to many of the papers mentioned below in this Section. During this same period of about 15 years a rocking-die forging machine, which contains various novel features, was developed in Poland by Marciniak and his co-workers. Both types of machine are now available on the international market.

The paper by Standring and Appleton also gives a considerable discussion of the early but closely related Slick wheel mill (Fig. 4.1) first designed in about 1910. In principle, it is similar to many of the current designs of rotary-forming machines, except that the axis of the working die remains fixed at a given obliquity to the vertical. It was pointed out to the author during a visit to India in December 1979 that the principle of rotary forging has been in use by Indian housewives for many centuries for chipati or bread making. (These are thin, circular, and of about 20 cm diameter.) I have also since found this same fact to have been noted in a paper by Appleton *et al.* (1976).

The basic operating principle of the rotary forging machine is shown in Fig. 4.2(a); in Fig. 4.3 it is shown incorporated into an early model mill designed by Slater *et al.* (1969). The workpiece–die contact is approximately sectorial in plan (Fig. 4.2(b)) and typically is about one-fifteenth of the area of the upper workpiece circular end. Precessional motion of the die axis enforces a progressive, rolling-die action with incremental working of the workpiece. Thus, for the forming of shallow blocks of metal, machine loads are less than about one-tenth that which would be necessary if load was applied at once to the whole end-face of the billet. Two beneficial consequences of this system among others are (i) the reduced physical size of the press required to form a workpiece, and hence the cost of a press to forge a given product can be greatly reduced, and (ii) that noise due to the slam of a ram in straight-

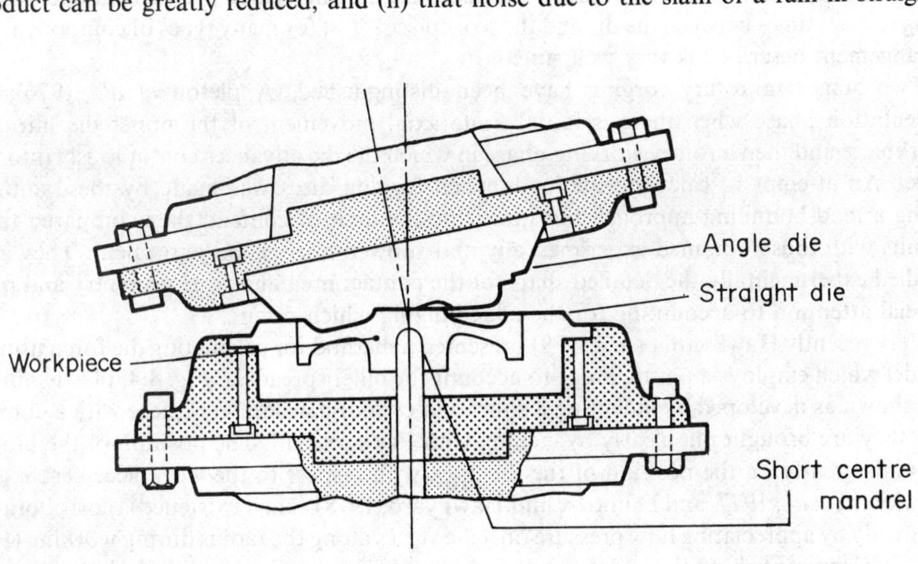

FIG. 4.1. The principle of the Slick mill.

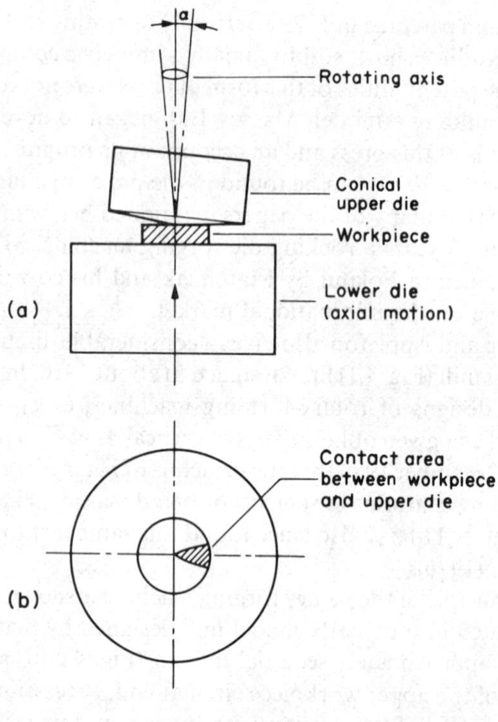

FIG. 4.2. The principle of rotary forging.

forward pressing is obviated, as is the deleterious tremor in the surrounding earth. There are, of course, disadvantages to this system—reduced die life due to high pressures and relatively long contact times between the die and the workpiece—but for many types of component the arrangement described is very well suited.

Two stages in rotary forging have been distinguished (Appleton *et al.*, 1976); an indentation phase when there is initial finite axial movement of the upper die into the workpiece and then a rotary working phase in which the die advances continuously into the billet. An attempt to calculate die loads when forging discs was made by these authors using a load-bounding approach but idealizing the area of contact; they compared their results with those obtained experimentally and showed fairly good agreement. They also studied experimentally the detailed shape of the contact area and zone geometry and paid special attention to accounting for the edge bulging which occurs.

Very recently Hawkyard *et al.* (1979) presented a method for calculating die force using a model which employs a plastic hinge to account for billet spread. In Fig. 4.4, plastic hinges are shown as developed by indenting a short, flat cylindrical steel workpiece with a conical die; they are brought out clearly by etching with Fry's reagent. The position of the hinges depends greatly on the position of the die apex with respect to the workpiece centre (see Hawkyard *et al.*, 1977, and Dimitrov and Hawkyard, 1978). Their existence is corroborated indirectly by appreciating how pressure on a die varies along the radius during working (Fig. 4.5); it is largest at about the mid-point of a radius and is almost zero at the workpiece centre. This was determined using a strain-gauged plug in the lower die. It may easily be imagined

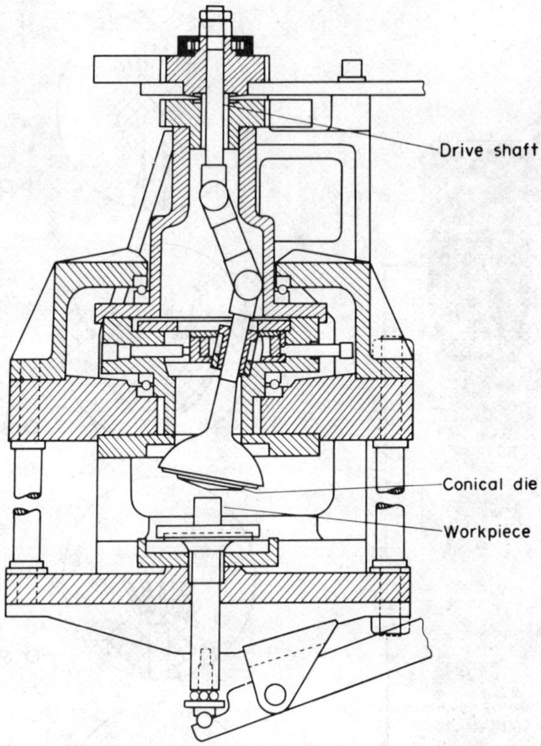

FIG. 4.3. An early model of a rotary forging machine.

that under axial thrust distributed along a radius, the predominantly sideways expulsion of metal below the conical die generates a bending situation which involves one or two plastic hinges on the opposite side of the workpiece; this mechanism also explains the very low stress at the disc centre near to the die apex and why on occasion the metal in that region can be found to thin and indeed to crack. Satisfactory estimates of the forging load for billet height to diameter ratios of less than 1 to 4 can be determined after accepting that (i) the deformation zone through the billet thickness has the shape shown in Fig. 4.6, and (ii) the geometry of the surface contact zones is such that Hill's die-pressure vs. depth-to-indenter ratio curve (as found using plane-strain slipline field theory) is reasonably applicable (Hawkyard *et al.*, 1979).

In a recent interesting Chinese investigation, Wang *et al.* (1979) described many of the above features of rotary forging including those of central tensile fracturing (see the top diagram in Fig. 4.7), plastic hinges, and the spread mechanism of deformation. Figure 4.7, taken from the work of these authors, also shows how a short billet plastically deforms during working.

The plastic forming operation described above illustrates just how complex an initially attractive industrial process can be. It indicates the amount of time needed to develop a machine from the initial idea to the point at which it can be made available as a piece of industrial equipment. A variety of specific modes of plastic deformation are seen to be

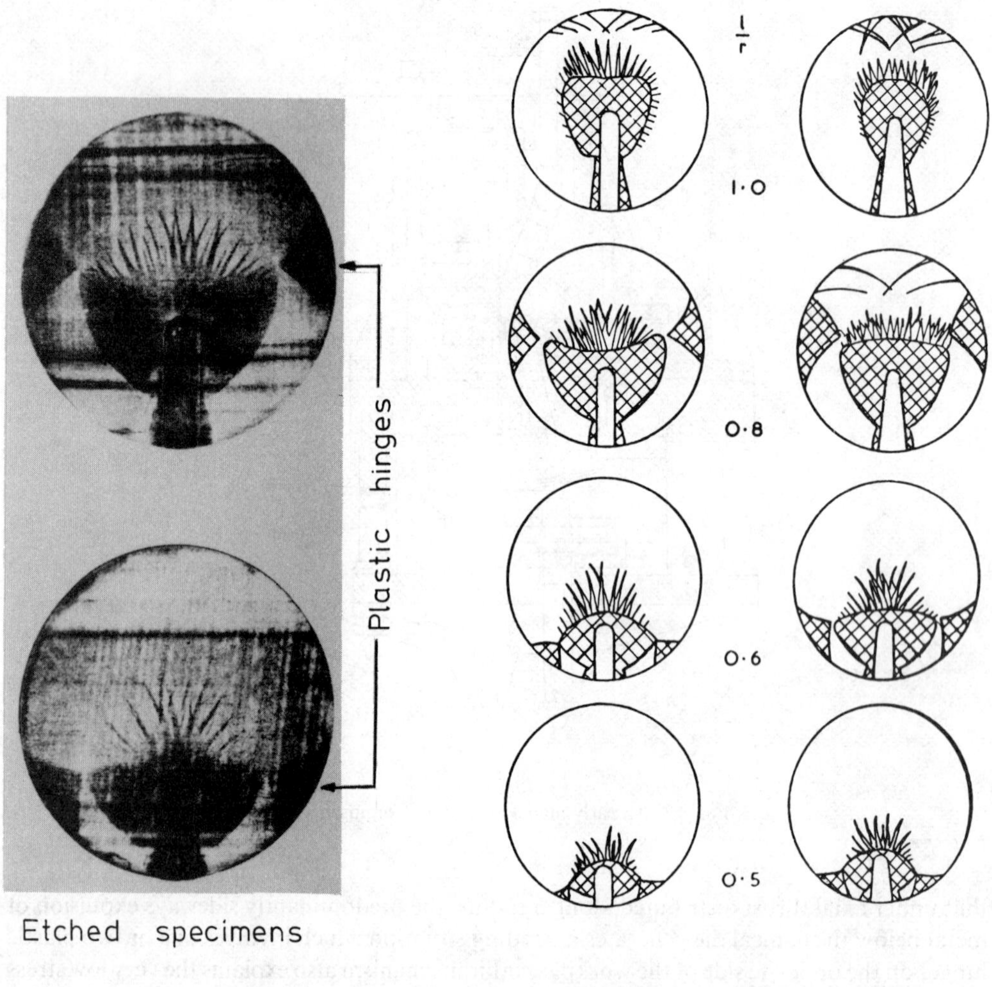

FIG. 4.4. Showing the plastic hinges which are developed in the rotary forging of a disc.

present when forming occurs but none are straightforwardly interpretable. It well illustrates how valuable is a knowledge of plasticity theory and it demonstrates that its use for engineering design will frequently require imagination, intuition, and a willingness to compromise when establishing models for calculation, from all of which adequate but not high accuracy can be expected.

[An up-to-date comprehensive list of references to papers and patents for the rotay forging process is available from Dr. P. M. Standring of Nottingham University Production Engineering Department. It includes, for example, Standring *et al.* (1980).]

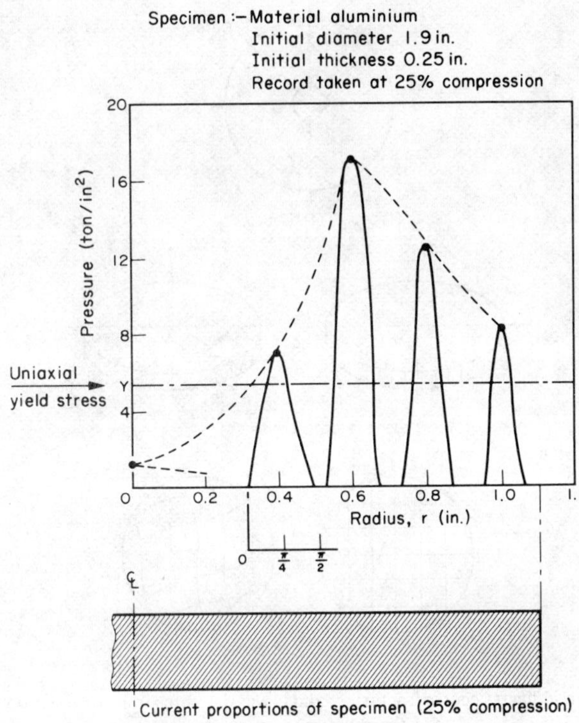

FIG. 4.5. Pressure distribution at certain radii (and along the arc of contact) in rotary forging a short billet.

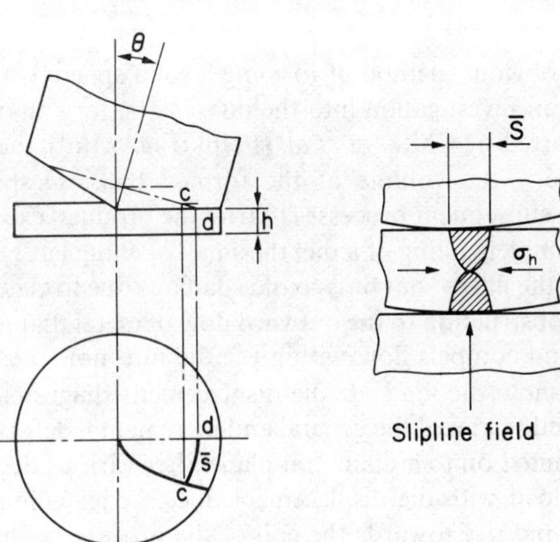

FIG. 4.6. The deformation zone—a slipline field—through the billet thickness in rotary forging.

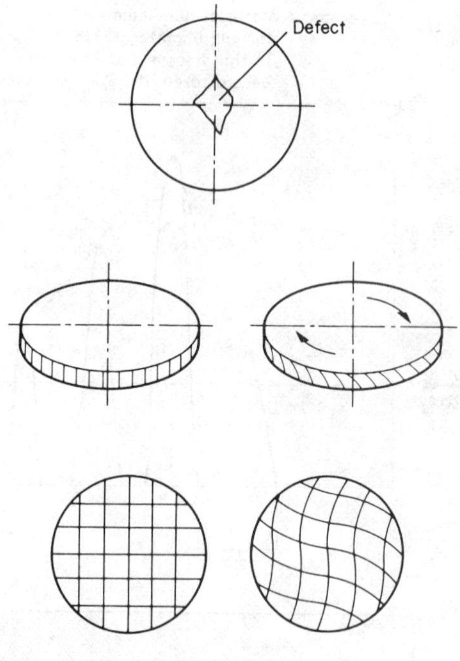

FIG. 4.7. The top diagram indicates centre fracturing, and lower diagrams indicate the mode of plastic deformation of the billet mass in rotary forging.

5. The Forming of Balls

5.1. *Ball forging*

A common and most obvious method of forming a solid sphere is to upset-forge it from cylindrical bar stock. In an investigation into the quasi-static forging of $\frac{3}{4}$-in. diameter balls from such stock, as reported by Chitkara *et al.* (1976) (Fig. 5.1(a)), the volume of the upset cylinders exceeded by 5% the volume of the formed balls. Despite wastage and the consequent need for any subsequent processes to trim the product, excess metal is necessary in order to ensure the complete filling of a die; the small opening left between the dies at the conclusion of forming—the flash—not only avoids damage due to clashing of the dies but it also ensures a degree of obstruction to the outward flow of metal that generates a restricting pressure which assists and compels flow within the die into not so easily reached regions. Figures 5.1(b) and (c) show die-load vs. die-displacement diagrams for the forming of spheres. Figures 5.1(d) and (e) reveal the general and incremental deformation undergone by a square grid of lines printed on to a diametral plane after various degrees of compression.

The slow increase in load with die displacement due to edge coining is evident in Figs. 5.1(b) and (c) as is its rapid rise towards the end of the process as final sizing occurs. The concentration of the largest strains imposed in the neighbourhood of a horizontal diameter through the flash region and down the sides of the deforming cylinder is seen in Fig. 5.1(d); this is entirely as would be expected from an elementary knowledge of plane flow.

The mode of deformation to secure a spherical form, as here, should be compared with the

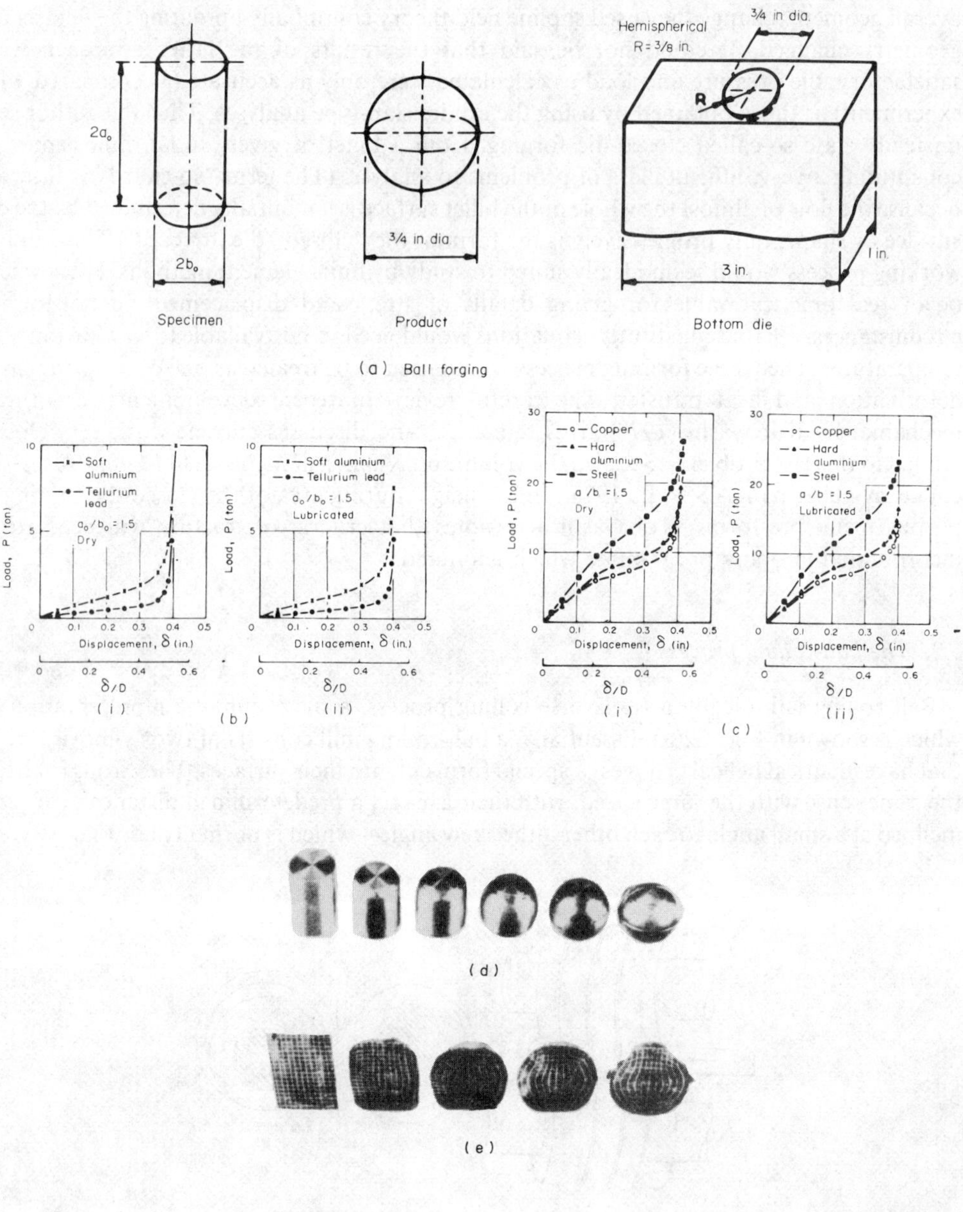

FIG. 5.1. (a) Workpiece, product, and bottom die for ball forging. (b) and (c) Die-displacement vs. die-load diagrams. (d) Progressive general deformation from a cylinder to a sphere. (e) The incremental internal deformation.

very different one which applies for the ball-rolling process. This gives some insight into expectations about the different mechanical properties of balls as manufactured by the two processes.

The forging process here is continuously one of unsteady state and ill-defined boundary conditions. In the plane strain forging of turbine blade sections from round bar, Aksenov *et al.* (1975) encountered a similar situation. To monitor the details of internal plastic flow and overall geometry change they used slipline field theory continually up-dating the field as the geometry changed. It could not be said that the results of the latter approach were satisfactory; die pressure and load as calculated were only as accurate (as compared with experiment) as those obtained by using the cruder slab-type analyses. Thus the author sees unsteady state so-called closed-die forging, from a billet of given initial dimensions, as constituting a very difficult class of problems to analyse. (The term "so-called" is inserted because the flow of almost the whole of the billet surface is *not* initially determined by the die surface or shape; only progressively is the form of the "closed" die imposed.) This kind of working process would seem ideally suited to study by finite element methods. However, to be of real practical value for giving details of stress and displacement in hot-forging circumstances, reliable constitutive equations would need to be available for a wide range of temperatures. The whole forming process would need to be treated as one of coupled plastic deformation and heat transfer. The careful review of recent developments in forming mechanics by Kobayashi (1979) gives references and discusses current work very closely related to this last problem. (See also the volume on *Numerical Methods in Metal Processing*, edited by H. Arman, A.S.M.E., 1978.) In the neighbouring area of the closed-die forging of porous metal pre-forms, Kobayashi also notes that criteria of fracture will need to be incorporated into any programme which is offered.

5.2. *The ball-rolling process*

Ball rolling is basically a transverse rolling process, a diagrammatic representation of which is shown in Fig. 5.2(a). Essentially, a ball-rolling mill consists of two cylindrical rolls that have identical helical grooves of special form cut into their surfaces. These rolls rotate in the same sense with the same speed, with their axes set a predetermined distance apart and inclined at a small angle to each other—the skew angle—which is normally less than 10° and

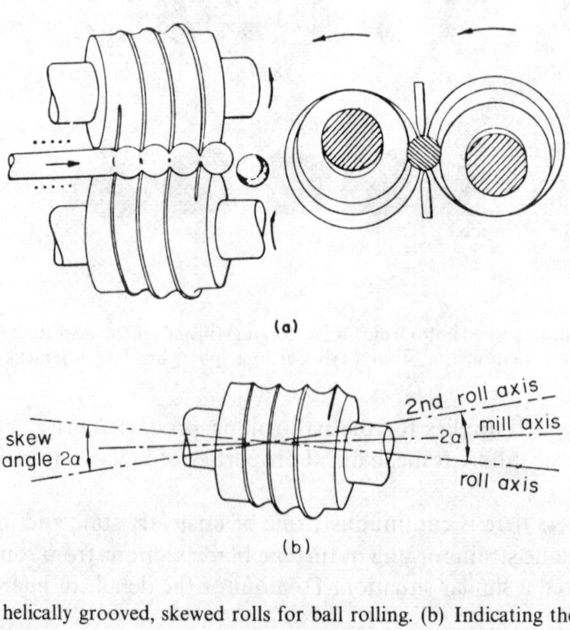

FIG. 5.2. (a) A pair of helically grooved, skewed rolls for ball rolling. (b) Indicating the skew angle of the rolls.

which depends on the ball size required (Fig. 5.2(b)). Contact with the rolls causes the bar to rotate in the opposite sense to that of the rolls whilst a ram pushes the bar forward axially until the helical thread on the rolls bites into the bar and produces an indentation. The bite causes intense local plastic deformation, the extent of which increases as the thread penetrates deeper with axial travel of the bar. The bar passes through the mill in a screwing motion and takes up the shape of the roll pass. In each revolution of the rolls a small mass of the bar is deformed into a ball-like shape until the bar becomes a string of ball-like blocks joined by "pips". The pip becomes thinner at each pitch distance along the roll, and it is finally sheared through at the last sharp-edged thread when a ball emerges.

The rolling of balls from bar stock in a specially designed mill and the first commercial installation of such a mill seems to have first been achieved in the U.S.S.R, in about 1950. Much of the background to ball rolling is surveyed in a paper by Chitkara *et al.* (1975). Rolling balls, as opposed to forging and machining them, is advantageous because it results in a considerable saving of material, labour, and secondary processing. It is obviously attractive as a fast, large production process.

In a paper entitled the "Mastery of ball-rolling machines", Kurganov, Nikolau and Masharov (1962; see Chitkara *et al.*, 1975) mentioned the possibility of doubling the rate of production of balls from round billets by using double-threaded helical rolls; balls with a diameter of 40 mm were made and later this was increased to 50 and 60 mm. About the same time, Masharov and Nikolau claimed to be able to roll balls of different sizes in four-start threaded rolls with a minimum diameter of ball of 62 mm. As well as reviewing the ball-rolling process, these authors discussed the quintessential problem of evolving a successful machine, namely that of determining the roll-pass shape. Indeed, they did not indicate clearly how groove shapes were arrived at, but stated that force and torque measurements were used to correct the design of the roll shape. Mention was made of the fact that flange width was calculated on the basis of constancy of volume; they did remark, however, that they could predict correctly the elongation which takes place in the neck of the multi-threaded grooved rolls, which helped in selecting different pitch distances between the finishing and the forming zones of the pass. Tselikov *et al.* (1970) have discussed and compared the practical characteristics of special rolling mills for rolling periodic or complex profiles, noting how economical they were and that they saved about 500,000 tons of steel per annum in the U.S.S.R.

These few remarks are sufficient to indicate the complexity and usefulness of this process and to underline the high degree of empiricism which is used in arriving at an economic and industrial plastic deformation process solution. Some results are therefore now presented in the hope that it will encourage more profound research and understanding, the better to facilitate the design of ball-rolling mills.

An approximate estimate of the roll load and torque in ball rolling has been made as follows. Consider a simple compression type analysis in which, during rolling, a peak force P occurs. Then, we assume simply that $P = AY_m$, where A is the experimentally measured projected area of contact between one of the rolls and the bar and Y_m is the mean yield stress of the material. At any one time if there are n balls in the pass, the maximum projected area of the bar in contact with the rolls (assuming no increase in ball diameter) is $A_{max} = \frac{1}{4}n\pi d^2$, and, of course, only a proportion c of this area is in contact with the roll at any one time. The roll separating force can be written

$$P = \tfrac{1}{4}n\pi d^2 c Y_m,$$

with c found from simple experiments. As a first approximation, the roll-torque per roll relationship is then

$$T = ARk = \tfrac{1}{2}ARY_m,$$

where R is the average radius of the roll and k is the yield shear stress of the (hot) material taken as $\frac{1}{2}Y_m$; it might be more appropriate to use the curved area of the ball in contact with a roll instead of the projected area.

The above simplistic approach neglects considerations of redundant work, variations of roll radius, pressure distribution, and change in pressure angle from section to section across the roll.

A small experimental ball mill for rolling plasticine and a power-driven one for rolling tellurium-lead balls, as used by the author and his co-workers, are described elsewhere. Trial-and-error procedures were adopted to select satisfactory skew angles and pass shape in order to secure the correct ball form (Fig. 5.3).

Calculated values of load and torque to roll lead balls using the projected area approach based on the above two equations were found to be surprisingly close to corresponding experimental values.

Typical plasticine balls rolled from specially prepared, layered bar stock are shown in Fig. 5.4. Large amounts of local deformation are seen to occur in regions where a bar is indented by roll threads, but it does not penetrate right through the body. The material, when indented, is squeezed out on either side of the thread. A small inner core of the material of any ball remains almost undisturbed during its passage through the pass. Surface shear and twist clearly take place, and there is a large surface shearing action in the zone where a ball is finally separated from the bar stock; during forming, a ball spins and is wrung away from the body of a bar. Shearing action also increases as the skew angle of a roll increases. When a formed ball diameter is compared with the original bar diameter, it is slightly larger by about 2.5%.

Some balls emerge having axial ruptures or internal cracks—observations also reported by others and applying when the billets are of too large a diameter for normal roll settings. This indicates the possibility of certain conditions arising at certain stages that must be avoided if balls of the correct shape and integrity are to be produced.

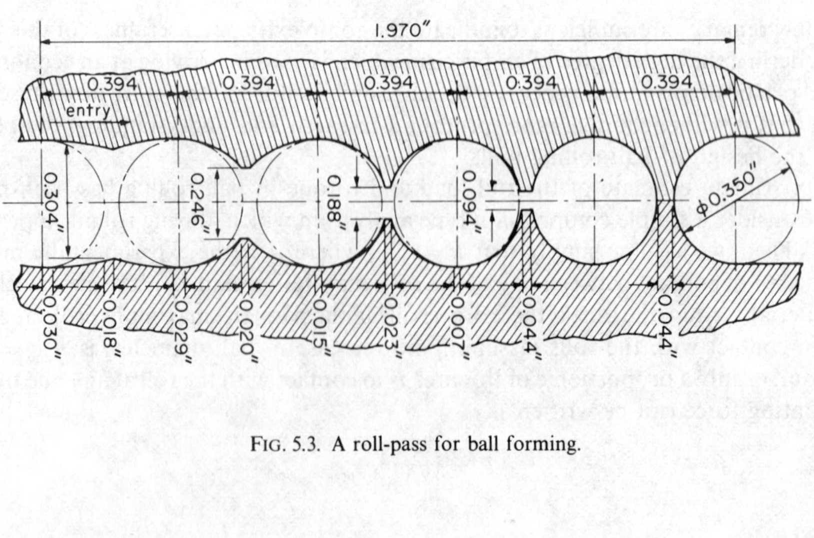

FIG. 5.3. A roll-pass for ball forming.

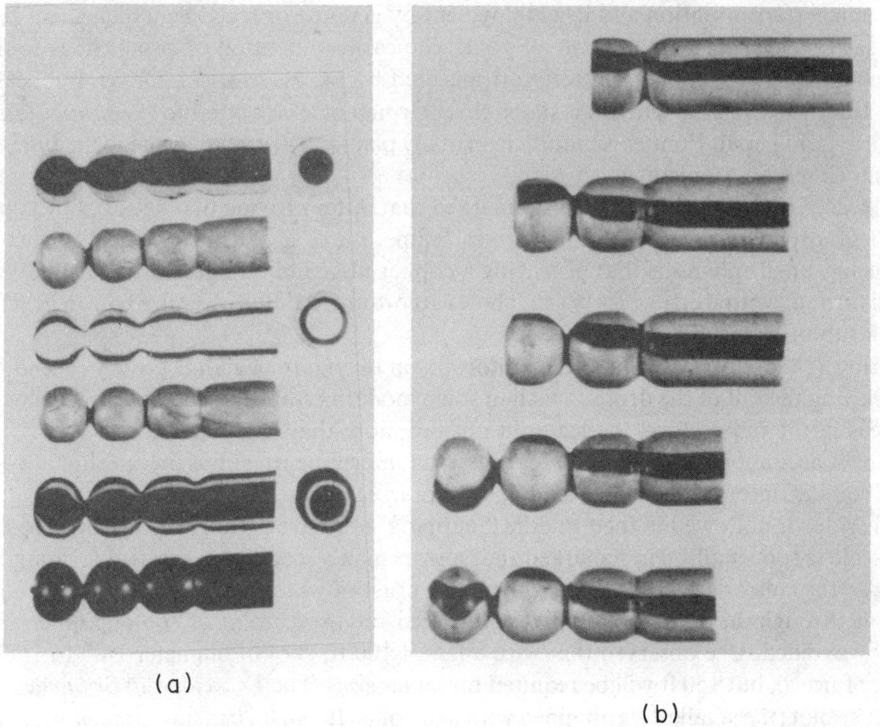

FIG. 5.4. Layered plasticine bars rolled into balls, showing progressive (a) indentation and (b) wringing action.

5.3. *Small, spherical lead-shot forming from the liquid, using the shot-tower method*

An ancient method of forming spherical lead shot was by allowing liquid lead to fall down a shot-tower. The topic is a fascinating one and the only attempt at an extensive account of this now obsolete process (to the best knowledge of the author) in terms of some applied mechanics is that by Johnson *et al.* (1976). (All further references in this subsection will be found in their paper.)

Small lead-shot forming from the liquid state is described in Robert Hooke's *Micrographia* (1665). The technique of alloying the lead with arsenic trisulphide (As_2S_3) and permitting the molten lead to fall through a sieve-like pan from a small height and into a tub of water was communicated to Hooke by Prince Rupert who is said to have invented the method in about 1650. After removing black scum from molten lead pig, powdered auripigmentum (literally gold pigment and also called orpiment) was added; the alloy was tested by flicking some single drops into water and if they were "round without tails", then the alloy was right for shot forming. Interesting is the observation that "the chief cause of this globular figure of the shot seems to be the auripigmentum. . . . When the air comes at the falling drops of the melted lead, the skin constricts them everywhere equally. . . ."

Knowledge of the useful capacity of orpiment to change the "fluidity" of lead and thus make shot forming feasible was probably owed originally to an early belief in the possibility of chemical transmutation and its employment by genuine or fraudulent alchemists to give the heavy element lead the colour of gold. The earliest mention of orpiment is found in Vitruvius's *Ten Books on Architecture* (translated by M. H. Morgan, Dover Publications, New York, 1960, p. 214), where it is stated that "Orpiment which is termed *ἀρσενικόν* (arsenic) in Greek is dug up in Pontus, Sandorach, in many places". (Vitruvius wrote in the time of the Roman Emperor Augustus.)

In *Rees' Encyclopedia* of 1819 it is remarked that, after proving that a lead alloy contains sufficient orpiment and is of the correct temperature, a satisfactory arrangement for producing small spheres is that of placing a copper plate hollowed towards the middle and bored through with thirty or forty small holes according to the size required on an iron frame over a tub of water and 4 in. above it.

Barlow (1836) observed that: "the uniformity in the figure of shot is greatly promoted by lengthening the fall of the drops, which in some modern establishments is very considerable whereas in the old method the lead did not fall more than a few inches."

A fuller account given by Ure (1875) noted that most imperfections in lead shot are due to rapid cooling, internal shrinkage leading to shot irregularities, and that if they are flattened on one side, or hollowed in their middle ("cupping") or have a "tail", then "the proportion of arsenic is too small". He remarked that a workman's secret of successful forming is the lining of the cullender with "cream", the oxide crust of which is white and spongy, so that running through the holes too rapidly is prevented and the forming of "oblong spheroids" is thereby avoided. Ure observed that with a fall of 100 ft, shot of diameter 1/80 to 1/500 in. may be formed, but 150 ft will be required for larger sizes. The *Encyclopedia Britannica* states that dust shot (the smallest) is obtained with a fall of 40 ft and 0.20 in. in diameter when there is a drop of 150 ft.

The manufacture of powders by the disintegration of liquid metals—atomization—is, of course, considered at length in books devoted to chemical engineering, but there is virtually no specific treatment of the shot-tower shot-forming process. It is typically not pursued after remarking that it concerns "heat and mass transfer in falling drops".

Price (1963), writing in the *Northern Architect*, noted that only four shot-towers then remained in England. (The famous London Shot Tower, incorporated into the South Bank Festival Hall Site for the 1951 Exhibition, had been demolished by that date.) A Newcastle tower was erected in 1797 (30 in. in diameter and 175 ft high), but the oldest tower was one in Bristol, founded in about 1783. The erection of these various towers was stimulated by the Napoleonic wars. Only the tower in Bristol, belonging to the Sheldon Bush and Patent Shot Co., seems still to be functioning in England.

In some experiments in which the author participated, with the aim of forming small spheres, a very thin brass dish or sieve of about 150 mm diameter perforated with holes of 1 mm diameter was used, the holes being circular and free from burrs; their profile is illustrated in Fig. 5.5(a). The dish was placed on an iron frame over a tub of water for receiving the lead at a selected height and was kept at an appropriate temperature by means of a burner flame. Onto it, a certain quantity of molten alloy was poured so that it made its way through the holes and fell into the water, the shot thereby being formed. The lead alloy was earlier prepared.

The influence of the amount of orpiment is well illustrated in Fig. 5.5(b). Other factors remaining the same, a shape closer to the spherical one required is formed as the orpiment percentage increases. The influence of orpiment is also made clear in Fig. 5.5(c), where the effect of less than 1.5 % orpiment is compared with no orpiment, i.e. pure lead; the height of

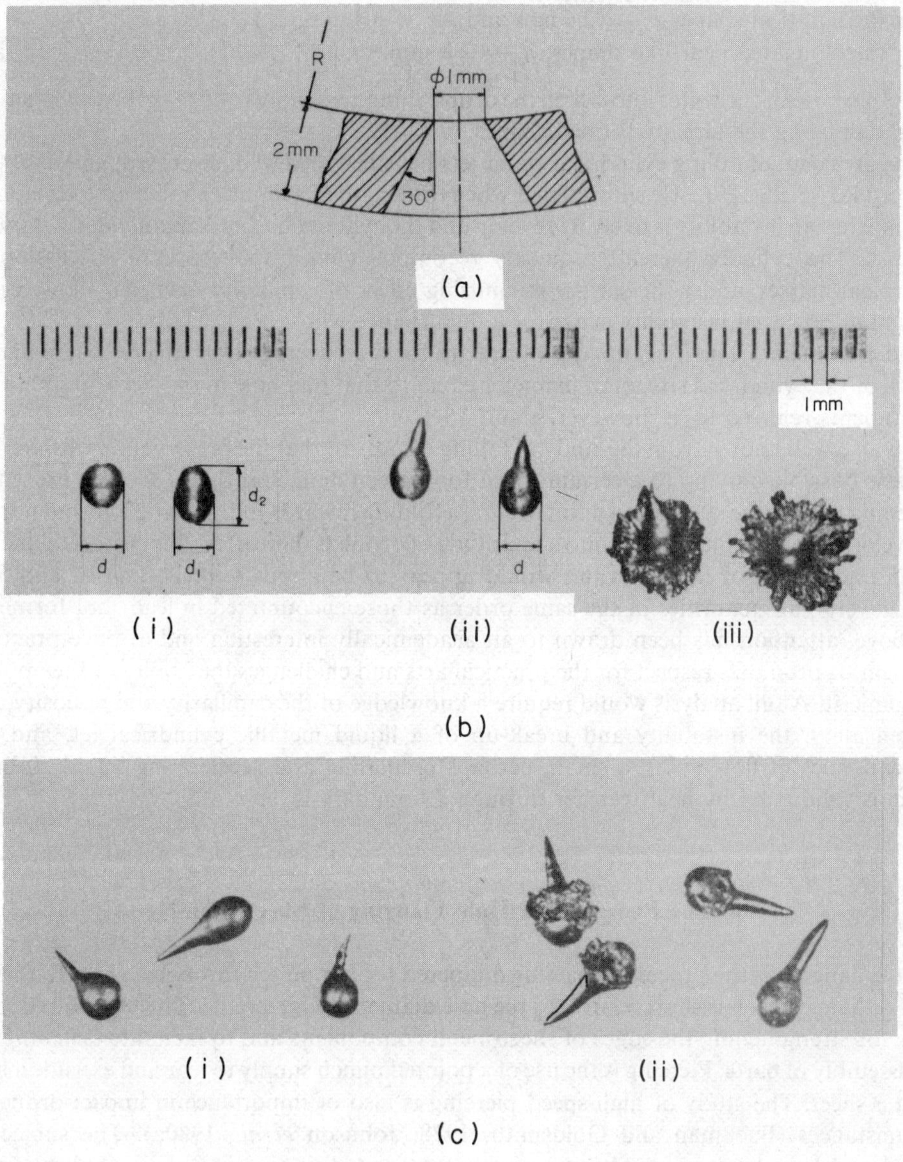

FIG. 5.5. (a) Hole profile. (b) Shapes formed after a fall of 8 m for an alloy of (i) lead with 2 % orpiment and (ii) lead with 1.5 % orpiment and for (iii) pure lead. (c) Forms obtained after a fall of 31 m with (i) lead with orpiment of less than 1.5 % and (ii) pure lead. Alloy temperature 400°C.

fall was 31 m in both cases. Clearly, the quantity of the orpiment added to the molten lead influences its fluidity by increasing its surface tension.

The final dimensions of various regular specimens were measured and averaged values of diameters are given overleaf (Fig. 5.5(b)):

for the round shape, $d_{av} = 2.8$ mm,
for the elliptical shape, $d_{1av} = 2.6$ mm and $d_{2av} = 3.4$ mm,
for the elongated pear-like shapes, $d_{av} = 2.8$ mm.

The purpose of the water into which the drops plunge is simply to "break their fall" and to avoid damaging the already formed shapes.

The break-up of falling cylindrical water jets has been reviewed, discussed, and analysed by Lord Rayleigh (1896). He showed that when the length of a liquid cylinder exceeds π times its diameter, an instability is likely to develop and the cylinder becomes unduloid, i.e. it forms a waist. The cylinder thereafter quickly recomposes itself to become a succession of spheroidal masses, under the surface-minimizing effect of capillarity. Rayleigh showed that maximum potential instability exists for a disturbance whose wavelength is $4.51d$ (d is the cylinder diameter). The l/d ratio appropriate to form an average lead sphere, where l is the length of the cylindrical stream of diameter equal to that of a hole in the sieve, found in the experiments referred to, is, however, about 15/1.

It is also well known from the study of falling raindrops that the preservation of spherical shape is possible only up to a certain speed for a given diameter. If this speed is exceeded, flattening near to the stagnation point occurs, a folding inwards follows, and a toroidal form is developed with disintegration into a multitude of droplets thereafter. The critical diameter, speed, and height of fall for water would appear to be about 4 mm, 9 m/sec, and 5 m respectively—all quantities of the same order as those encountered in lead-shot forming.

Above, attention has been drawn to an academically interesting and at once practical problem. It promotes respect for the practical arts and challenges the ability of the applied mechanicist. A full analysis would require a knowledge of the capillarity and viscosity of a molten alloy, the instability and break-up of a liquid metallic cylindrical jet, and the aerodynamics of flow about a small, perhaps oscillating and accelerating liquid globule, which is solidifying by heat transfer through a boundary layer.

6. The Piercing and Hole Flanging of Sheet Metal

Hole flanging is the process of passing a tapered tool or punch through a sheet to form a neck or flange which is short relative to the hole diameter (Fig. 6.1(a)). This is required to be done for strengthening the edges of sheet-metal components and to facilitate fastening for the assembly of parts. Piercing is the use of a pointed punch simply to tear and extrude a hole from a sheet. The study of high-speed piercing is also of importance in impact-ordnance circumstances (Backman and Goldsmith, 1978; Johnson *et al.*, 1980a). The subject is introduced here to draw attention to important and diverse areas of its application; it includes not only quasi-static and high-speed plastic deformation operations, but ductile fracture initiation and propagation, tensile instability, and the effects of any initial directional forming present in the sheet. Since articles in this Volume are addressed to some of the latter topics, it seemed worth while to draw attention to a process in which they could all simultaneously find application.

In passing a 30° semi-angle punch, at slow speed, through a sheet containing an initial hole, a neck is extruded; if the punch angle is much larger, a portion of the sheet is blanked-out and no neck is formed (Fig. 6.1(d)). A punch-load vs. punch-advance diagram is shown in Fig. 6.1(b)(i), and on it are noted salient points associated with various stages of punch

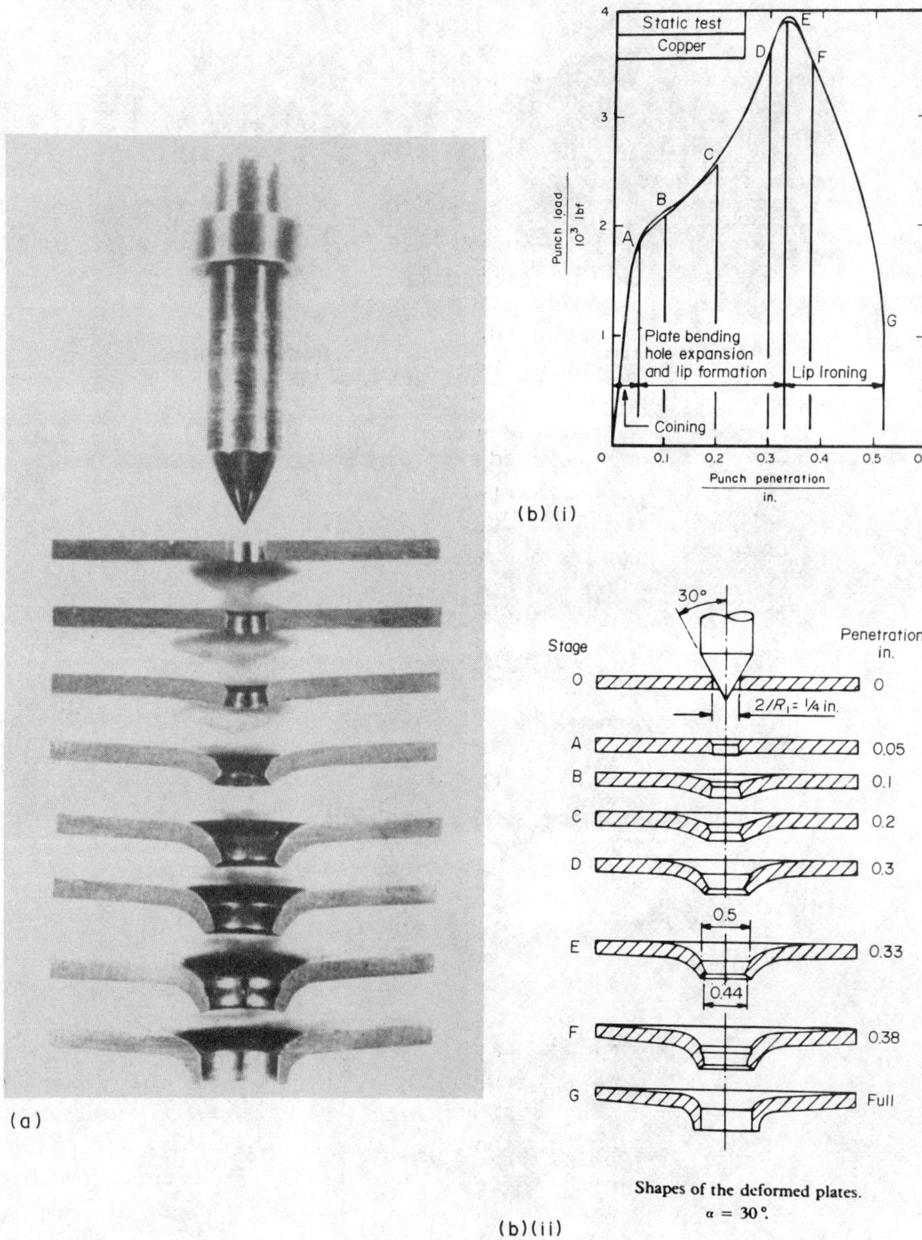

FIG. 6.1. (a) Sectional views of plates deformed during progressive penetration by a punch of semi-angle 30° showing neck formation. (b) (i) Variation of punch load with punch advance for (a); salient points on the curve are identifiable by reference to (ii), (ii) stages of plate deformation.

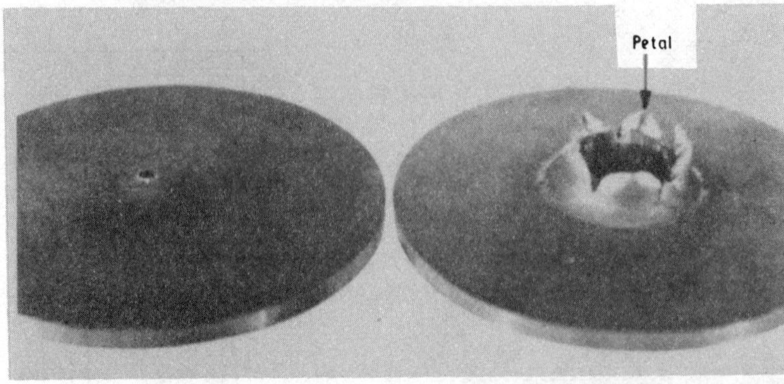

Start of fracture Fracture with petals

Brass. $\alpha = 22\frac{1}{2}°$. Hole size $= \frac{1}{16}$ in.

(c)

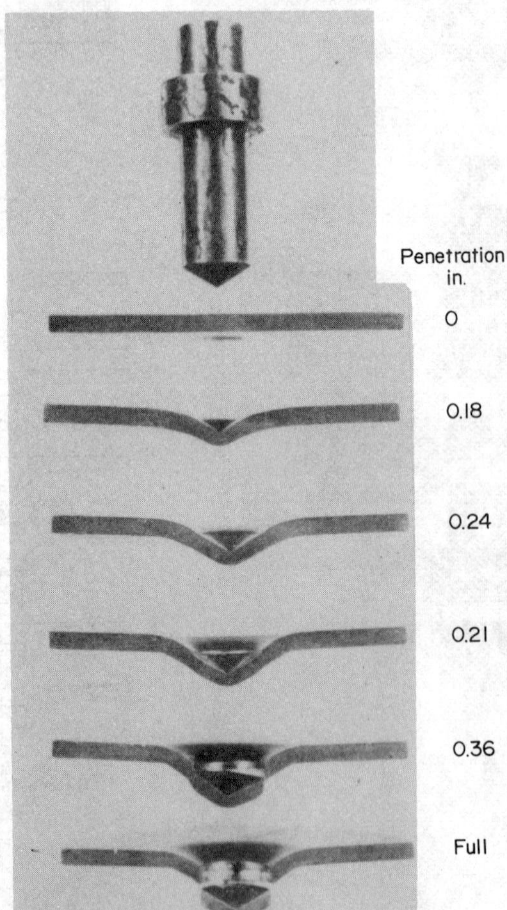

$\alpha = 60°$

(d)

FIG. 6.1. (c) Brass specimens showing start and finish of fracturing with petal formation. (d) Sectional views of deformed plates during blanking.

penetration, as in Fig. 6.1(b)(ii). If the initial hole diameter is too small (or the sheet is solid), four fractures usually form—along and perpendicular to the direction in which the sheet was rolled. The sharp-ended punch thus develops four petals; this phenomenon is well known from ordnance studies except that several petals are commonly observed with high-speed penetrators due to the remoteness of plate restraints and a very large ratio of penetrator to plate thickness (Johnson and Ghosh, 1980). The development of tensile instability necks at the tip of an extruded hole during its slow speed formation, which leads to fracture and thus petalling, was carefully studied by van Minh (Johnson *et al.*, 1977), using Hill's anisotropic yield criterion. Wang and Wenner (1974) made a most useful analysis to estimate the maximum strain that a flange could sustain during a drawing process. It remains a surprising fact that there is virtually no systematic study and rationalization of piercing phenomena available from ordnance or defence establishments.

Important parameters in this operation are sheet thickness/punch diameter ratio, die support diameter (in slow-speed work), and punch-nose angle. Figure 6.2 shows curves of maximum punch-load (divided by the ultimate strength of each metal) vs. punch-nose semi-angle for five different materials of two thicknesses; they demonstrate clearly the change-over from a petalling mode of behaviour to a plugging or blanking mode, at a nose semi-angle of between 45° and 60°, and that this angle is the higher, the thinner is the sheet.

Recently the study of the piercing of pipes and tubes has assumed importance because of

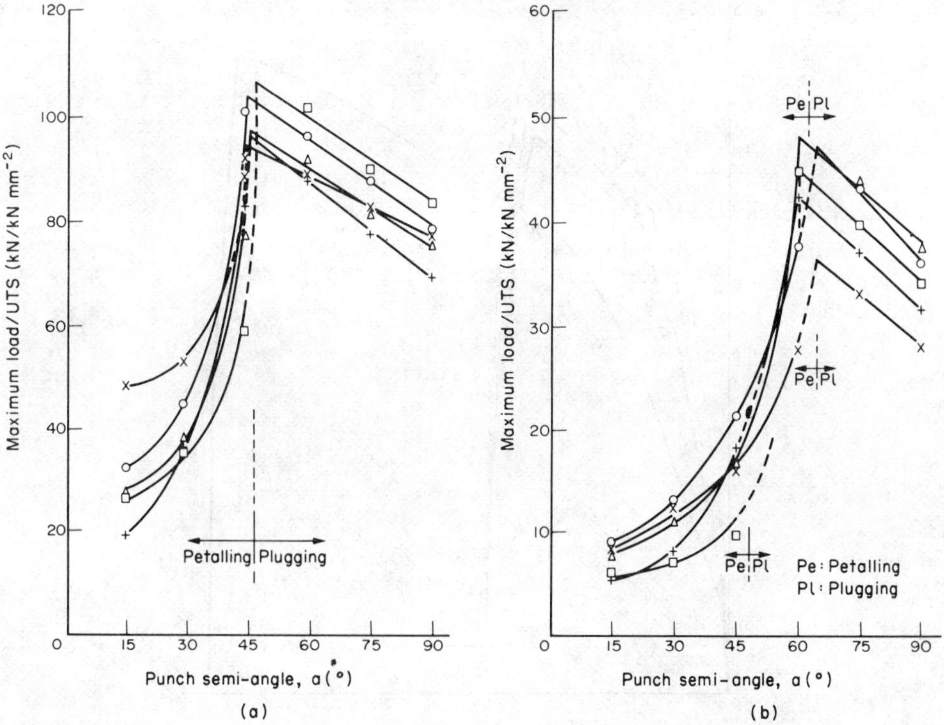

FIG. 6.2. Variation of maximum load (divided by mean ultimate tensile strength) with punch semi-angle for penetrated plates of thickness: (a) 3.2 mm and (b) 1.6 mm. (×, aluminium; o, copper; Δ, mild steel; +, brass 70/30; □, brass 60/40.)

the need for such information by the designers of protective shields, tank container heads and energy absorbing devices, especially in relation to the two considerations of pipe-whip accidents (Reid *et al.*, 1980) and the containment of the transport of hazardous materials. Figure 6.3(a) shows the normal penetration, with a square punch, of a model reinforced tank vessel. Figure 6.3(b) compares its punch-load vs. punch-penetration characteristics with those of similarly loaded and unloaded structures (Johnson *et al.*, 1980a, b).

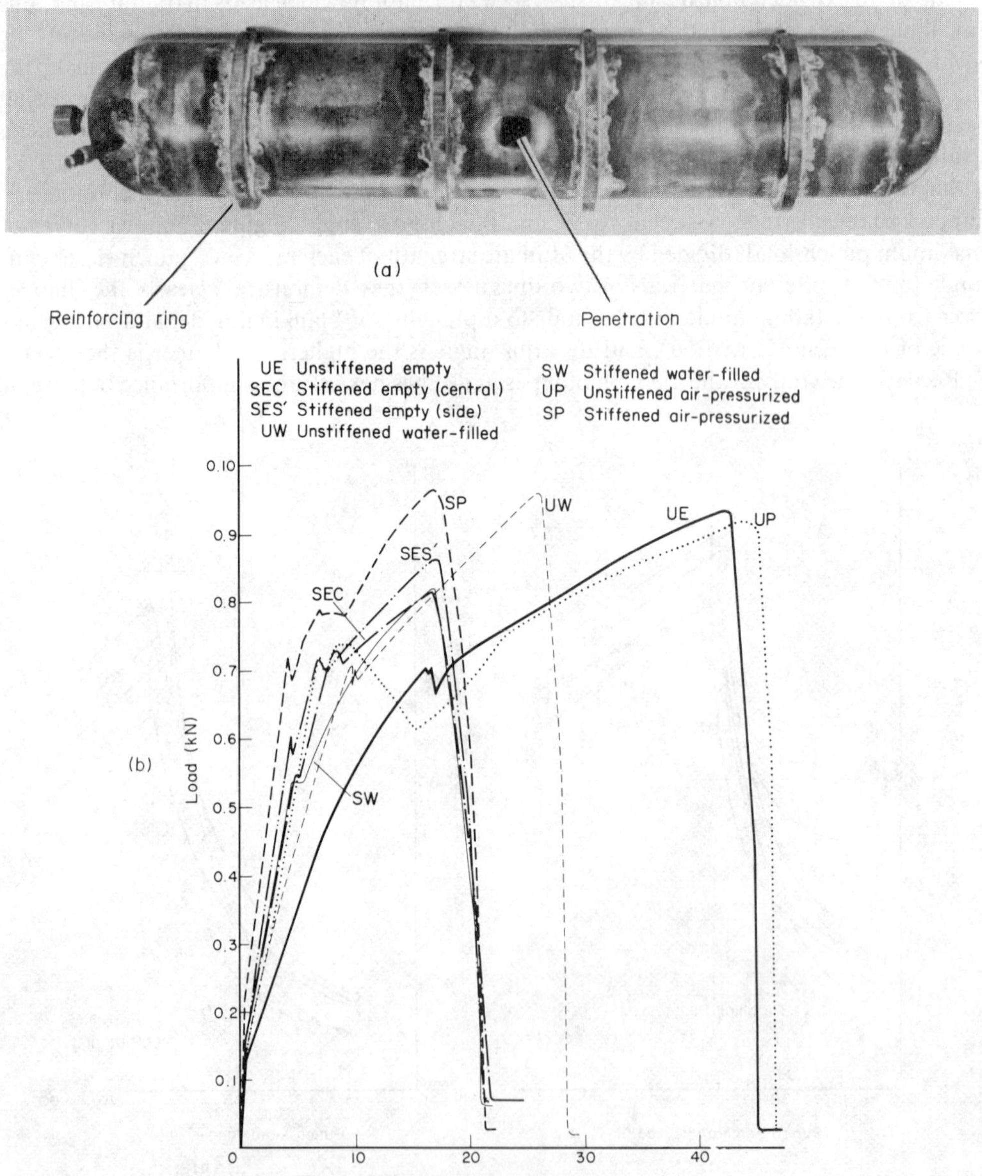

FIG. 6.3. (a) A model tank container pierced by a square punch. (b) Punch-load vs. punch-penetration curves for various kinds and conditions of container.

7. Defects

7.1. *Crop-loss in rolling ingots to slabs*

A large tonnage of ingots is rolled each year in blooming and slabbing mills and significant defects are found to occur in the resulting slabs at the front and rear ends (see the diagrams in Figs. 7.1(a) and (b)). The actual proportions of the defect shape depend on the rolling schedule which has been followed and whether or not edging rolls are used. These defective end-regions are cut off or cropped from rolled slabs and each may account for up to about

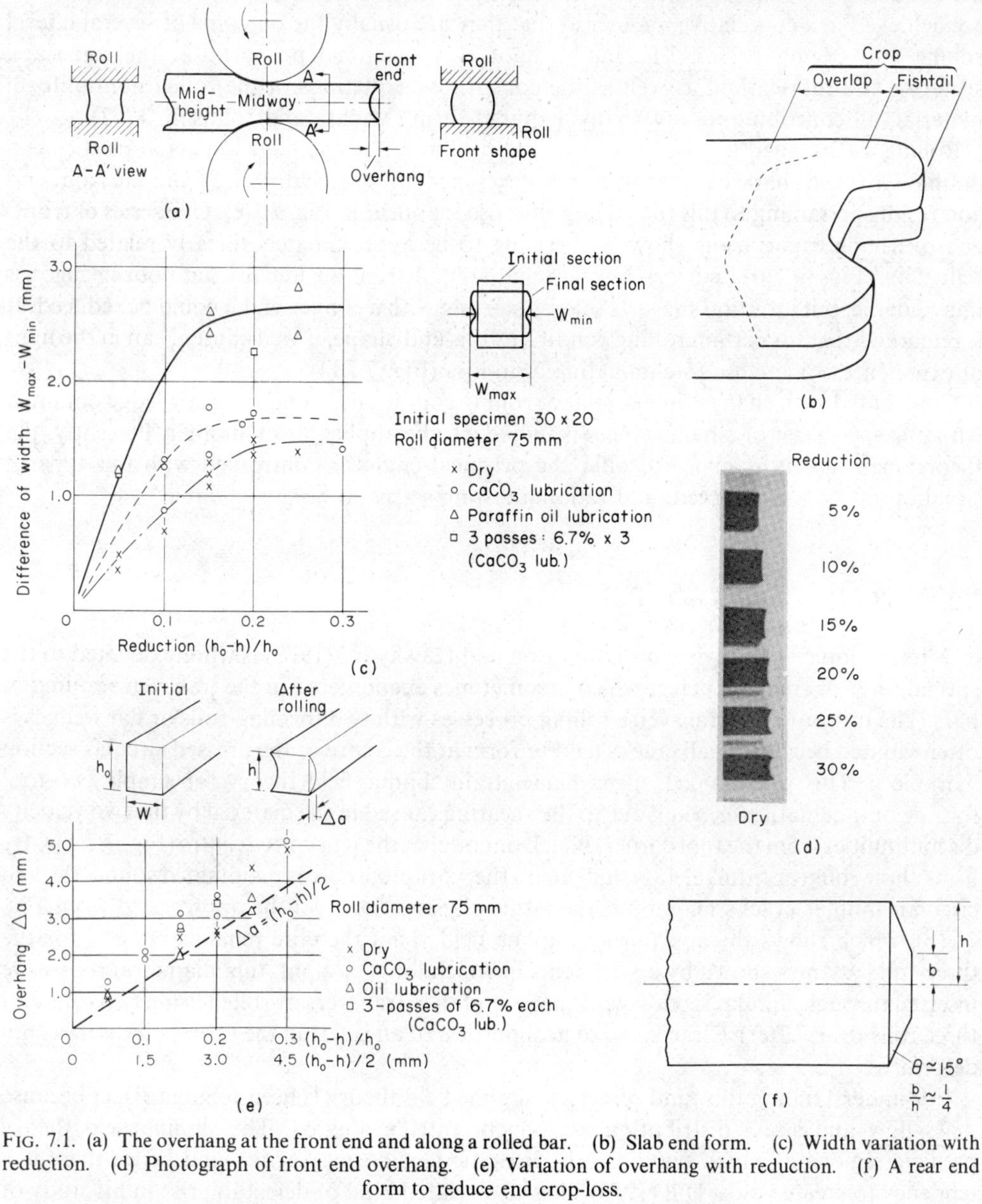

FIG. 7.1. (a) The overhang at the front end and along a rolled bar. (b) Slab end form. (c) Width variation with reduction. (d) Photograph of front end overhang. (e) Variation of overhang with reduction. (f) A rear end form to reduce end crop-loss.

4 % of the total rolled weight. The international slab yield is said to be about 90–92 %. Evidently, if the yield can be increased by only a small percentage by reducing the crop-loss, then the economic savings when there is a large volume of production would be great.

Very little academic attention has been given to studying the front- and rear-end changes in rolled slabs, in strong contrast to that devoted to the study of steady-state conditions for which theory and experiment abound for roll torque, load, pressure, and, less so, for spread. Such end-shape changes as occur will, of course, be difficult to predict by starting from fundamental considerations, since they occur in the context of unsteady-state conditions and are of a complicated three-dimensional character. Furthermore, such final shapes in practice are achieved in a cumulative manner in that they are usually the outcome of several lateral rolling and edging passes. The initial ingot cross-sectional proportions, the roll sizes, surfaces, and lubrication, as well as the constitutive equations of the rolled non-uniform material, all contribute to any finally acquired form (Mihara and Johnson, 1977).

Rolling a flat-ended ingot of height to width ratio 3/2 in the model material plasticine, a double-barrelled shape of cross-section is developed after some distance from the front end and results pertaining to this (no edging rolls used) appear in Fig. 7.1(e). One series of front-end overhang experiments showed overhang to be approximately linearly related to the reduction (Figs. 7.1(d) and (e)). Overlap and fishtail from an initially flat-bottom ingot is unavoidable, but intuition suggests that by starting with a convex end it could be reduced. It is remarked that for certain rolling conditions this end-shape, if well chosen, can in the light of experience come close to eliminating crop-loss (Fig. 7.1(f)).

Confronted with changes in material or rolling conditions, empirical knowledge obtained with one special set of circumstances is unlikely to be applicable in another. Presently, the theoretician has little advice to offer the practical engineer confronted with this type of situation on how to proceed, and rational testing seems to be unavoidable.

7.2. *Cracks in transverse rolling*

A recent interesting review by Thompson and Hawkyard (1979) is wholly devoted to the special form of cracking defect which is sometimes encountered in the transverse rolling of bars. The utilization of transverse rolling processes with two forming-rolls or flat wedges is often vitiated because small cracks tend to form at the centre of the pressed circular section workpiece. This is explained, using plane-strain slipline field theory for simple two-tool forging or indentation as being due to the shearing caused in the material by the two velocity discontinuities from the tool corners which intersect at the workpiece centre (Fig. 7.2(a)). By using three rolls (or, equivalently, indenters) the workpiece core is maintained sound, though circular annular cracks may arise; the latter physical facts will be appreciated from Fig. 7.2(b), which shows the appropriate slipline field. That the core remains integral for the three-roll system is shown by experiments in which long, straight, thin diametral rods were inserted through circular section workpieces and then transversely rolled using two rolls and three rolls (Fig. 7.2(c)). Clearly, the central portion of an insert in the three-roll case was not deformed.

For understanding this kind of defect, slipline field theory is used to great effect because mass flow and stress distribution are concurrently evaluated. The slipline field theory predicts regions in which significant hydrostatic tension may arise—and hence there is a tendency to create voids. Hill (1953) showed a related kind of defect to arise in his study of

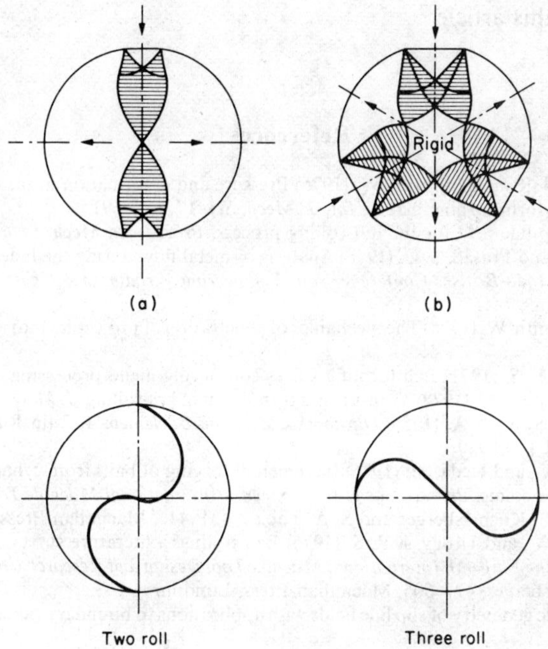

FIG. 7.2. (a) Slipline field for indentation with two rolls. (b) Slipline field for indentation with three rolls. (c) Comparing the deformation imposed during rolling by fields (a) and (b) and demonstrating *no* deformation at the bar centre in the case of (b).

the cutting of strips lying on a rigid foundation using a knife-edge tool. After pressing in the tool a portion of the strip adjacent to the foundation rises and separates from it, provided certain dimension ratios prevail. This same notion may be seen to suggest void formation when forging a block if it originally contained a small horizontal crack at its centre.

Crop-loss and cracking due to pressing are but two examples of the very many defects encountered in the field of metal forming, as reference to a previous review by Johnson and Mamalis (1977) will show. Their economic importance is very great and as scientific problems they are deserving of much greater attention than they currently receive. A comprehensive bibliography in monograph form about common kinds of metal forming defects is soon to be published (Johnson *et al.*, 1982).

Acknowledgements

The author is indebted to Mrs. H. M. Johnson for help in preparing and typing his manuscript and to Dr. S. K. Ghosh for assistance with the Figures. He is also grateful to the University of Cambridge for allowing him to take up a visiting Springer Professorship at the

University of California, Berkeley, during the winter quarter of 1980 which allowed time and opportunity to write this article.

References

Aksenov, L. B., Chitkara, N. R., and Johnson, W. (1975) Pressure and deformation in the plane strain pressing of circular section bar to form turbine blades, *Int. J. Mech. Sci.* **17,** 681–691.

Alexander, J. M. (1955) Slip line field for the hot rolling process, *Proc. Instn Mech. Engrs* **169,** 1021–1030.

Appleton, E., Sinha, K. P., and Prasad, S. C. (1976) Analysis of metal flow during the indentation phase of rotary forging, *Proceedings Indo-British Conference on Engineering Production*, C67-83, Indian Institute of Technology, New Delhi.

Backman, M. E. and Goldsmith, W. (1978) The mechanics of penetration of projectiles into targets, *Int. J. Engng Sci.* **16,** 1–99.

Brandel, W. W. and Klass, L. S. (1971) Ball forming solves contouring metal processing, *Metal Prog.* **99,** 56–57.

Buxton, S. A. E. and Browning, S. C. (1972) Turn-up and turn-down in hot rolling, *J. Mech. Engng Sci.* **14,** 245–254.

Chekmarev, A. P. and Nefedov, A. A. (1956) *Obrabotka Metallov Davleniem* **4,** 2 (in Russian), British Lending Library Translation, R.T.S. 8939.

Chitkara, N. R., Johnson, W., and Ueda, M. (1976) Incremental forging of balls from cylindrical specimens and of rollers from rectangular solids, *Proceedings of the Sixteenth International Machine Tool Design and Research Conference*, edited by F. Koenigsberger and S. A. Tobias, 431–446, Macmillan Press, London.

Chitkara, N. R., Johnson, W., and Uttley, J. R. S. (1975) Ball rolling: a literature survey and some experimental results, *Proceedings of the Fifteenth International Machine Tool Design and Research Conference*, edited by S. A. Tobias and F. Koenigsberger, 497–505, Macmillan Press, London.

Collins, I. F. (1968) Algebraic geometry of slip line fields with applications to boundary value problems, *Proc. R. Soc. Lond.* A **303,** 317–338.

Collins, I. F. and Dewhurst, P. (1975) Slip line field analysis of asymmetrical hot rolling, *Int. J. Mech. Sci.* **17,** 643–651.

Dannenmann, E. (1974) Werkzeug Geometrie und Stempelkraft beim Biegen in V-gesenk, *Werkstattstechnik* **64,** 527–531.

Davies, R. and Austin, E. R. (1970) *Developments in High Speed Metal Forming*, Machinery Publishing Co. Ltd., London.

Dewhurst, P. (1974) Plane-strain indentation on a smooth foundation, *Int. J. Mech. Sci.* **16,** 923–930.

Dewhurst, P., Collins, I. F., and Johnson, W. (1974) A theoretical and experimental investigation into asymmetrical hot rolling, *Int. J. Mech. Sci.* **16,** 389–397.

Dimitrov, D. and Hawkyard, J. B. (1978) Investigation of yield in rotary forging by the indentation of thin discs, *Proceedings of the Eighteenth International Machine Tool Design and Research Conference*, edited by J. M. Alexander, 133–136, Macmillan Press, London.

Dodd, B. and Osakada, K. (1974) Bending-up during one-sided flat punch indentation, *Memoirs of the Faculty of Engineering, Kobe University*, No. 20.

Dugdale, D. S. (1964) Indentation of strips with flat dies on a flat anvil, *Int. J. Prod. Res.* **3,** 141–151.

Finnie, I. (1972) Observations on the erosion of ductile metals, *Wear* **19,** 81–90.

Hawkyard, J. B. and Ingham, J. B. (1979) An investigation into profile ring rolling, *Proceedings of the First International Conference on Rotary Metal Working Processes*, edited by E. Appleton, 309–320, I.F.S. (Conferences) Ltd., Kempston, Bedford.

Hawkyard, J. B., Gurnani, C. K. S., and Johnson, W. (1977) Pressure distribution measurements in rotary forming, *J. Mech. Engng Sci.* **19,** 135–142.

Hawkyard, J. B., Yunus, Y. M., and Gurnani, K. S. (1979) Predicted force in the rotary forging of short circular cylinders, *Proceedings of the First International Conference on Rotary Metal Working Processes*, edited by E. Appleton, 111–124, I.F.S. (Conferences) Ltd., Kempston, Bedford.

Hawkyard, J. B., Johnson, W., Kirkland, J., and Appleton, E. (1973) Analyses for roll force and torque in ring-rolling, with some supporting experiments, *Int. J. Mech. Sci.* **15,** 873–893.

Hill, R. (1950) *The Mathematical Theory of Plasticity*, Clarendon Press, Oxford.

Hill, R. (1953) On the mechanics of cutting metal strips with knife-edged tools, *J. Mech. Phys. Solids* **1,** 265–270.

Hirose, S. (1975) *Simulation of Shot Peening Coil Springs*, Report of the Casting Laboratory, Waseda University (Japan), No. 26, 11–19.

Hutchings, I. M. and Winter, R. E. (1975) Erosion of ductile metals by spherical particles, *J. Phys.* D, *App. Phys.* **8,** 8–14.

Johnson, W. and Ghosh, S. K. (1980) The quasi-static and dynamic penetration of thin aluminium plates, *Aluminium* **56,** 142–146.

Johnson, W. and Mamalis, A. G. (1977) A survey of some physical defects arising in metal working processes,

Proceedings of the Seventeenth International Machine Tool Design and Research Conference, edited by S. A. Tobias, 607–621, Macmillan Press, London.

Johnson, W. and Mamalis, A. G. (1979) Ring rolling: a review, *Metall. Rev., Inst. Metals* **4**, 137–148.

Johnson, W. and Mellor, P. B. (1973) *Engineering Plasticity*, van Nostrand-Reinhold, London.

Johnson, W. and Needham, G. (1966) An experimental study of asymmetrical rolling, *Proc. Instn Mech. Engrs* **180**, 270–281.

Johnson, W. and Singh, A. N. (1980) Elastic springback in circular blanks formed by bending with a hemi-spherical punch and die, *Metall. Met. Form.* **47**, 275–280.

Johnson, W. and Singh, A. N. (1981) Springback after cylindrically bending metal strips, *Conference on Large Deformations*, edited by N. K. Gupta, Indian Institute of Technology, Delhi.

Johnson, W. and Woo, D. M. (1958) Pressure for indenting material resting on a rough foundation, *J. Appl. Mech.* **80**, 64–66.

Johnson, W., Chitkara, N. R., and van Minh, H. (1977) Deformation modes and lip fracture during hole flanging of circular plates of anisotropic materials, *J. Engng Ind.* **99**, 738–748.

Johnson, W., Chitkara, N. R., and Parker, K. J. (1981a) Roll peen-forming, *Proceedings of the Twenty-first International Machine Tool Design and Research Conference*, edited by J. M. Alexander, 75–84, Macmillan Press, London.

Johnson, W., Ghosh, S. K., and Reid, S. R. (1980a) Piercing and hole-flanging of sheet metals: a survey, *Mémoire Scientifique Revue Métallurgie*, 585–606.

Johnson, W., Mamalis, A. G., and Ghosh, S. K. (1981b) On the peen-forming of metals, *Proceedings of the Twenty-first International Machine Tool Design and Research Conference*, edited by J. M. Alexander, 85–93, Macmillan Press, London.

Johnson, W., Mamalis, A. G., and Ghosh, S. K. (1982) *Defects in Metal Processing*, Applied Science Publishers, Barking (to be published).

Johnson, W., Mamalis, A. G., and Hunt, H. (1976) Small spherical lead shot forming from the liquid, using a shot tower, *Metall. Met. Form.* **43**, 68–72.

Johnson, W., Reddy, T. Y., and Reid, S. R. (1980b) Model road-tank metal vessels subject to internal explosion or static penetration, *J. Strain Analysis* **15**, 225–233.

Kobayashi, S. (1977) A theory on uncontained steady-state processes, *Engineering Plasticity: Theory of Metal Forming Processes* I, edited by H. Lippmann, 65–79, Springer-Verlag, Berlin.

Kobayashi, S. (1979) The mechanics of plastic deformation in metal forming, *Proceedings of the 8th U.S. National Congress of Applied Mechanics*, edited by R. E. Kelly, 115–146, Western Periodicals Co., North Hollywood, California.

Kondo, K., Matsuzaki, S., Hiraiwa, M., and Ohga, K. (1979) Investigations of peen-forming, *Bull. JSME* **22**, 893–900.

Mamalis, A. G., Johnson, W., and Hawkyard, J. B. (1976a) Pressure distribution, roll force and torque in cold ring rolling, *J. Mech. Engng Sci.* **18**, 196–204.

Mamalis, A. G., Johnson, W., and Hawkyard, J. B. (1976b) Spread and flow patterns in ring rolling, *Int. J. Mech. Sci.* **18**, 11–16.

Meguid, S. A., Collins, I. F., and Johnson, W. (1977) Co-indentation of a layer by two flat plane or spherical-headed rigid punches, *Int. J. Mech. Sci.* **19**, 1–9.

Meguid, S. A., Johnson, W., and Al Hassani, S. T. S. (1976a) Some factors in the shot peening and peen-forming processes, *Proceedings of the Seventeenth International Machine Tool Design and Research Conference*, edited by S. A. Tobias, 653–659, Macmillan Press, London.

Meguid, S. A., Johnson, W., and Al Hassani, S. T. S. (1976b) Multiple impact erosion of ductile metals by spherical particles, *Proceedings of the Seventeenth International Machine Tool Design and Research Conference*, edited by S. A. Tobias, 661–668, Macmillan Press, London.

Mihara, Y. and Johnson, W. (1977) Crop loss: front and back end deformation during slab and bloom rolling, *Metall. Met. Form.* **44**, 332–339.

Oh, S. I. and Kobayashi, S. (1980) Finite element analysis of plane strain sheet bending, *Int. J. Mech. Sci.* **22**, 583–594.

Plaster, H. J. (1973) *Blast Cleaning and Allied Processes*, Vols. I and II, Industrial Newspapers Ltd., London.

Queener, C. A. and de Angelis, R. J. (1968) Elastic springback and residual stresses in the sheet metal formed by bending, *Trans. Am. Soc. Metals* **61**, 757–768.

Reid, S. R., Johnson, W., and Reddy, T. Y. (1980) Pipe whip, *Chart. Mech. Engr* **27**, 55–60.

Sidebottom, O. M. and Gebhardt, C. P. (1979) Elastic springback in plates and beams formed by bending, *Experimental Mechanics* **19**, 371–377.

Slater, R. A. C. (1966) *Velocity and Thermal Discontinuities in the Forging of Steels*, Manchester Association of Engineers, No. 5.

Slater, R. A. C. and Joorabchian, A. C. (1976) Spin forging of sheet metal cones, *Proceedings of the Seventeenth International Machine Tool Design and Research Conference*, edited by S. A. Tobias, 531–537, Macmillan Press. London.

Slater, R. A. C., Barooah, N. K., Appleton, E., and Johnson, W. (1969) Rotary forging concept and initial work with an experimental machine, *Proc. Instn Mech. Engrs* **184**, 577–592.

Standring, P. M. and Appleton, E. (1979) Development of rotary forging machine design, *Chart. Mech. Engr* **26,** 44–50.

Standring, P. M., Moon, J. R., and Appleton, E. (1980) Plastic deformation produced during the indentation phase of rotary forming, *Metals Technol.* **7,** 159–166.

Straub, J. (1970) Shot peening wing panels for the 747, *Mfg Engng Management* 27–29.

Thompson, G. and Hawkyard, J. B. (1979) Crack formation in transverse rolling—a review, *Proceedings of the First International Conference on Rotary Metal-working Processes*, edited by E. Appleton, 171–183, I.F.S. (Conferences) Ltd., Kempston, Bedford.

Tilly, G. P. (1973) A two stage mechanism of ductile erosion, *Wear* **23,** 87–96.

Tselikov, A. I., Milyutin, S. P., Kazankaya, I. I., and Levin, E. I. (1970) Technology and rolling mills for rolling round periodic profiles, *Metallurgy* 40–43 (in Russian).

Wang, N.-M. and Wenner, M. L. (1974) An analytical and experimental study of stretch flanging, *Int. J. Mech. Sci.* **16,** 135–143.

Wang, Z., Pei, X., Zhou, D., Chao, J., Chung, M., Chung, W., and Chang, X. (1979) Experimental study of some problems on rotary forging technology, Harbin Institute of Technology, Paper No. 54, English translation available from P. M. Standring.

Weinmann, K. J. and Shippell, R. J. (1978) Effect of tool-workpiece geometries upon bending forces and springback in 90° V-die bending of HSLA steel plate, *Sixth North American Metal Working Research Conference*, 220–237, Society of Manufacturing Engineers, Dearborn, Michigan.

Yang, D.-Y., Ryoo, J. S., Choi, J. C., and Johnson, W. (1981) Analysis of roll torque in profile ring-rolling of L-sections, *Proceedings of the Twenty-first International Machine Tool Design and Research Conference*, edited by J. M. Alexander, 69–74, Macmillan Press, London.

Instability and the Ill-posed Cauchy Problem in Elasticity

by R. J. Knops

Department of Mathematics, Heriot–Watt University

Summary

This article is concerned with the problem of instability of the null solution in the theory of hyperelasticity. The problem is equivalent to an ill-posed Cauchy problem and in the linear theory the existence, uniqueness, and evolutionary behaviour of solutions are studied. For instance, by means of the closed-graph theory, it is shown that a solution does not exist for general data. On the other hand, by means of the theory of adjoint operators, it is shown that data for which a solution does exist are dense in L^2. Certain of these properties are described also for linear thermoelasticity and for linear elasticity in an exterior domain. Finally, a brief survey of the nonlinear problem is presented, although the results obtained so far are still comparatively fragmentary.

1. Introduction

A problem is said to be well posed in the sense of Hadamard when the solution exists, is unique, and depends continuously upon the data. Hadamard conjectured that when these properties were absent the problem would not be physically relevant and hence hardly worthy of study. This view prevailed until about 30 years ago, when it became realized that ill-posed problems occurred in several situations of practical importance. Now the subject is one of intense activity.

An ill-posed problem is one that is not well posed. In general, the solution to an ill-posed problem fails to exist, does not depend continuously on the data, but is usually unique. Much of the previous effort has been devoted to establishing uniqueness and *a priori* conditions on the solution in order to recover continuous dependence. However, the question of existence appears to have received little attention.

Although in this article we study exclusively the Cauchy problem, it must be remarked that this is not the only ill-posed problem that arises in either elasticity or other theories of continuum physics. For instance, the problem of phase transitions involves the Dirichlet problem for non-(strongly) elliptic equilibrium equations. Problems in which part of the boundary is inaccessible to observation are ill posed in the sense that Cauchy data are specified for elliptic or parabolic equations. Inverse problems also fall within the category of ill-posed problems. The entire field has recently been extensively surveyed by Payne (1975) and the interested reader may consult this article not only for the different kinds of possible applications but also for the various techniques employed.

The Cauchy ill-posed problem in elastodynamics arises naturally in the study of sufficient conditions for instability of the null solution. A study of this kind helps to establish necessary conditions for stability. Now sufficient conditions for stability in the linearized theory include the positive-definiteness of the elasticities. Hence, when treating instability, the elasticities must cease to be positive-definite and might even fail to be strongly elliptic so that hyperbolicity is lost. The usual theorems governing existence, uniqueness, and regularity of solutions no longer apply, and in order to avoid a purely formal analysis others must be found. Indeed, the absence of such theorems has lent considerable justification to the adoption of the energy criterion *per se* as a test for stability without attempting its verification based upon dynamics. The energy criterion corresponds to the equilibrium solution being a minimum of the potential energy, and is assumed to be both necessary and sufficient for stability. Thus the criterion is expressed entirely in terms of the statics of the problem, and since the equilibrium equations are by hypothesis elliptic, the solutions used in the criterion exist and are regular. In this respect, at least, the criterion is meaningful. However, stability as used in the present context is a concept defined in terms of the dynamics of a problem, and therefore many feel that its analysis can be satisfactorily conducted only by means of the dynamical solutions. But then full knowledge is required of the existence, uniqueness, regularity and evolutionary properties of these solutions.

There remains, of course, the question as to which of the several possible dynamical theories is to be selected as best appropriate to a given equilibrium state. Consideration of even simple examples quickly demonstrates that results of a dynamical analysis depend crucially on the particular model chosen. A static criterion has the advantage of avoiding this particular issue altogether, but perhaps a more mathematical viewpoint should be adopted in which the behaviour is analysed of solutions to *all* possible dynamical theories and the final choice left to the practitioner.

Let us also note in passing another reason for studying instability by means of dynamics. In, for example, the problem of buckling, it is important not only to identify the bifurcated equilibrium states but also to establish how, at criticality, the system can move from one position to another. This clearly necessitates investigation of the nonlinear dynamics, although the linear theory will assist the analysis of local behaviour.

A fuller discussion of these and related points may be found in the articles by Knops and Wilkes (1973) and by Ericksen (1977), and it is not the intention to dwell further on them here. Rather, the main theme of the paper is devoted to the particular theory of linear elastodynamics in the case when the elasticities fail to be positive-definite. A preliminary study is given of existence, while a somewhat more complete treatment is presented of uniqueness and evolutionary properties. Similar topics are then discussed for linear thermoelasticity and for linear elasticity in exterior domains. A final Section provides a brief survey of the comparatively few results so far established in the nonlinear theory.

No typographical distinction is made between scalar, vector, and tensor quantities. However, in general we use lower case letters to denote scalar and vector quantities, while upper case letters are reserved for matrices and tensors. All quantities are referred to the same fixed cartesian coordinate system. Only occasionally is suffix notation employed, but then lower and upper case suffixes represent respectively positions in the current and reference configurations. We also use the summation convention and the usual comma notation for partial differentiation with respect to the spatial variables.

2. Notation and Basic Equations

In this Section the initial boundary value problem is stated for the theory of three-dimensional nonlinear elasticity under the assumption that a strain-energy function exists. Specialization and extensions to other theories are presented later as needed.

Consider a homogeneous elastic body which at time $t = 0$ occupies a region Ω of euclidean three-space which may be unbounded. The (interior) surface $\partial\Omega$ of the region is always assumed bounded and closed and of sufficient smoothness to permit applications of the divergence theorem. The particles of the body are identified with their positions x in Ω, which is taken to be the reference configuration. Furthermore, we suppose that the body in the configuration Ω is in equilibrium under the action of specified body forces, surface tractions, and surface displacements. Equilibrium is destroyed at $t = 0$ by instantaneously imparting to all points of Ω a given displacement and velocity but keeping the boundary $\partial\Omega$ fixed throughout the ensuing motion. Other kinds of boundary conditions are permissible (e.g. mixed) but are excluded for simplicity.

Let $u(x, t)$ be the displacement at time t of the point $x \in \Omega$, and let T denote the Piola–Kirchhoff stress tensor. For a homogeneous hyperelastic medium, T satisfies the constitutive relation

$$T = \frac{\partial W}{\partial \nabla u}, \tag{2.1}$$

where

$$W = W(\nabla u) \tag{2.2}$$

is the strain-energy function per unit volume of Ω and ∇u is the displacement gradient whose components are given by

$$\nabla u = (\partial u_i/\partial x_j) \quad i, j = 1, 2, 3. \tag{2.3}$$

In order to preserve orientation and prevent local interpenetration of material, all deformations must satisfy

$$J = \det \nabla y > 0, \quad y = u + x, \tag{2.4}$$

but in what follows no explicit use is made of this constraint.

The governing equations of motion are given by

$$\operatorname{Div} T = \rho \ddot{u}, \quad (x, t) \in \Omega \times [0, T), \tag{2.5}$$

where a superposed dot denotes differentiation with respect to time, $\rho(x) (> 0)$ is the density of the body in the configuration Ω, and we have assumed the extra body force to be identically zero. The divergence operator, indicated by Div, is taken with respect to the x-coordinates, and the maximal interval of existence is given by $[0, T)$. When the strain-energy function W is twice continuously differentiable, we may replace (2.5) by the quasilinear system of equations

$$A_{\alpha\beta} \frac{\partial^2 u}{\partial x_\alpha \partial x_\beta} = \rho \ddot{u}, \quad (x, t) \in \Omega \times [0, T), \tag{2.6}$$

where $A_{\alpha\beta}$ are matrices defined by

$$A_{\alpha\beta}(u) = \frac{\partial^2 W}{\partial \nabla u \partial \nabla u} (\nabla u), \tag{2.7}$$

$$(A_{\alpha\beta})_{ij} = \frac{\partial^2 W}{\partial u_{i,\alpha}\partial u_{j,\beta}} = \frac{\partial^2 W}{\partial u_{j,\beta}\partial u_{i,\alpha}} = (A_{\beta\alpha})_{ji}. \tag{2.8}$$

We see from (2.8) that the hypothesis of hyperelasticity implies, formally at least, that the coefficient matrices $A_{\alpha\beta}$ are symmetric, or that the system (2.6) is formally self-adjoint. However, the standard formulation of the basic theory provides no further information on the form of equations (2.6), nor on the form of the strain-energy, and a major open problem is the determination of conditions on W that produce physically acceptable behaviour. One approach is to explore the qualitative properties of solutions under a variety of assumptions on the strain-energy, rejecting those assumptions that lead to unacceptable behaviour. In this direction, contributions to the nonlinear theory include those by Hill (e.g. 1957, 1970), Wang and Truesdell (1973), and Rivlin (1973), with that by Ball (1977) being of especial importance for existence in the static theory. We follow the same general approach, but mainly concentrate on certain aspects of the linear theory.

To derive the equations of the linearized theory, we separate the linear part of the stress in (2.1) and (2.6) by introducing the elasticities $\mathbb{C}$ of the body in its configuration Ω. The elasticities are defined to be

$$\mathbb{C} = \left.\frac{\partial T}{\partial \nabla u}\right|_{u=0}, \tag{2.9}$$

from which it is clear that any sign-definiteness property depends upon the deformation immediately prior to time $t = 0$. Equation (2.6) now reads

$$\operatorname{Div} \mathbb{C}\ \nabla u + \operatorname{Div} \tilde{T} = \rho \ddot{u}, \quad (x, t) \in \Omega \times [0, T), \tag{2.10}$$

where $\tilde{T}$ represents the remaining part of the stress T. Equations for the linearized problem (corresponding to small deformations superposed upon large) are then obtained by setting $\tilde{T} = 0$:

$$\operatorname{Div} \mathbb{C}\ \nabla u = \rho \ddot{u}, \quad (x, t) \in \Omega \times [0, T). \tag{2.11}$$

It is assumed that the data are small in an appropriate sense,

For simplicity, we suppose that the boundary is held fixed throughout the motion, and then the boundary condition for both the linear and nonlinear problem is

$$u(x, t) = 0, \quad (x, t) \in \partial\Omega \times [0, T). \tag{2.12}$$

The associated initial conditions in both problems are

$$u(x, 0) = u_0(x), \quad \dot{u}(x, 0) = u_1(x), \quad x \in \Omega. \tag{2.13}$$

For unbounded regions Ω the asymptotic behaviour of the displacement and stress at large spatial distances must be included. We adjoin these conditions at the appropriate place in the text, but meanwhile we suppose that Ω is bounded.

Substitution of (2.1) into (2.5), followed by multiplication by u, and integration and use of (2.12), leads to the conservation of total energy, expressed formally as

$$E(t) \equiv \frac{1}{2}\int_\Omega \rho \dot{u}\dot{u}\, dx + \int_\Omega W(\nabla u(x, t))dx = E(0). \tag{2.14}$$

This last equation is valid for smooth, or classical, solutions to the initial boundary value problem, but since it is likely that shock waves and other singular surfaces will develop, which probably will serve to dissipate energy, we prefer to replace (2.14) in both the linear and nonlinear problems by the energy inequality

$$E(t) \leqslant E(0), \quad t \in [0, T). \tag{2.15}$$

Of course, strictly speaking, we should combine with (2.15) a weak formulation of the problem, but since many of the techniques we describe for uniqueness, continuous dependence, and growth of solutions are equally valid for both classical and weak solutions, we shall content ourselves with deriving results only for the smooth solution. An exception occurs in the discussion of existence in the linear problem. Here, it becomes essential to work with weak solutions, and we shall give a precise definition in the next Section. For other weak formulations of the nonlinear and linear elastodynamic problems relevant to the present approach see, for example, Knops and Payne (1971), Levine and Payne (1975a, b), and Knops *et al.* (1974).

3. Linear Elasticity

In this Section we describe some properties of the solution to the initial boundary value problem of linear elastodynamics when the elasticities are not positive-definite. We start by recalling how the method of logarithmic convexity may be applied to establish uniqueness, continuous dependence on the Cauchy initial data and growth estimates, as well as a further property not previously noted. These results are based on the assumption that a solution exists. We then use the growth estimate to prove from the closed-graph theorem that, in fact, a solution cannot exist for general data. Next, a weak solution to the problem is defined and by means of the theory of adjoint operators it is shown that provided the elasticities are negative-definite this solution always exists for a dense set of data. Expressed otherwise, a solution always exists for some data in a neighbourhood of a given set of data. This result appears to be a reasonable approximation to what happens in any actual experiment. The apparatus is insufficiently accurate to be set at the precise data specified and instead automatically selects some nearby data which can initiate motion. Hence, in the particular case of buckling, although the non-existence result at first suggests that when criticality is reached the rod or system will be unable to move away from the undeformed shape unless exactly the correct combination of data is applied, the existence result shows that such combinations form a dense set, and hence motion occurs for arbitrary data. While this conclusion is proved only for negative-definite elasticities, it will be shown elsewhere to hold in the general case of indefinite elasticities. The conclusion on existence also has significance for numerical computations of the solution.

Of course, local existence can always be proved using the Cauchy–Kowelevski theorem provided the initial manifold is non-characteristic and certain other conditions are satisfied. Furthermore, when the elasticities are positive-definite or strongly elliptic, so that the linear elastodynamic equations are hyperbolic, global existence of the solution has been established by Hughes and Marsden (1978), using symmetric hyperbolic operators, by Marsden and Hughes (1978), using semi-groups, and by Fichera (1972), using function analytic methods.

3.1. *Logarithmic convexity*

We consider the formal initial boundary value problem for linear elasticity given by

(2.11)–(2.13) and, under the assumption that a solution exists, derive some properties of the solution by means of logarithmic convexity arguments. As already noted, this technique applies equally well to strong or weak versions of the problem. For a fuller account of some of the results presented here, the interested reader may consult Knops and Payne (1971) and Payne (1975).

Thus, set

$$F(t) = \int_\Omega \rho u(x,t)u(x,t)\,dx + \alpha(t+t_0)^2, \tag{3.1}$$

where α, t_0 are positive constants to be determined. Differentiation of (3.1) immediately gives

$$\dot{F}(t) = 2\int_\Omega \rho u\dot{u}\,dx + 2\alpha(t+t_0), \tag{3.2}$$

while a further differentiation, followed by an application of the equations of motion (2.11) to replace the inertia term and the energy inequality (2.15) to replace the potential energy, leads to

$$\ddot{F}(t) \geqslant 4\left[\int_\Omega \rho\dot{u}\dot{u}\,dx + \alpha\right] - 2(\alpha + 2E(0)). \tag{3.3}$$

It then follows from (3.1), (3.2), and (3.3), together with Schwarz's inequality, that

$$F\ddot{F} - (\dot{F})^2 \geqslant -2(\alpha + 2E(0))F(t). \tag{3.4}$$

We assume that the elasticities are indefinite and that the initial data are chosen to satisfy $E(0) < 0$. We set $\alpha + 2E(0) = 0$, so that the right side of (3.4) becomes zero. An argument due to Levine (1970b) shows that unless $F(t)$ vanishes at one point it is non-zero for all t (which, of course, asserts the *uniqueness* of the solution), and hence for non-zero initial data, we may rewrite (3.4) as

$$[\ln F(t)]^{\cdot\cdot} \geqslant 0. \tag{3.5}$$

A convex function lies beneath the join of any two points on its curve, and hence from (3.5) we have

$$F(t) \leqslant [F(0)]^{1-(t/T)}[F(T)]^{t/T}, \quad t \in [0, T]. \tag{3.6}$$

This inequality shows that in the class of functions given by

$$\mathcal{M} = \left\{v\colon \int_\Omega \rho v^2\,dx \leqslant M^2, \quad M \text{ constant}\right\}, \tag{3.7}$$

the solution depends Hölder continuously upon the initial data as measured by $F(0)$.

Thus, provided a solution exists and the elasticities are indefinite or negative-definite, uniqueness and (Hölder) continuous dependence are easily proved for initial data with negative total energy. Similar results hold for initial data with zero or positive total energies and are derived by Knops and Payne (1971).

We may integrate (3.5) in another way by noting that a convex function also lies above the tangent at any point of its curve. Thus we have

$$F(t) \geqslant F(0)\exp(\dot{F}(0)t/F(0)), \quad t \geqslant 0. \tag{3.8}$$

We may choose t_0 to ensure $\dot{F}(0) > 0$ irrespective of the values of the initial data, and hence we see that solutions to our problem have the further property of increasing at least exponentially in time. An immediate deduction is that the null solution (and hence by linearity, any solution) is unstable in the sense of Liapunov. However, we wish to emphasize that so far it is unknown whether a solution exists to the problem under the supposed conditions, and so our conclusions are still to be regarded as formal. It becomes essential therefore to consider the question of existence, but before doing so we note some other properties that may be extracted from (3.6) and (3.8). These are suggested by some results due to L. Tartar for the equation $\dot{u} = Au$, where A is a self-adjoint unbounded linear operator.

(i) *$F(t)$ is a non-decreasing function of t*

Assume the contrary and that

$$F(0) > F(T).$$

Then, from (3.6), we obtain

$$F(t) < F(0), \quad t \in [0, T], \tag{3.9}$$

which on combining with (3.8) enables us to write

$$\exp \frac{\dot{F}(0)t}{F(0)} < 1, \quad t \in [0, T]. \tag{3.10}$$

Hence, we must have

$$\dot{F}(0) < 0. \tag{3.11}$$

But t_0 may always be chosen to make $\dot{F}(0) \geqslant 0$, in contradiction to (3.11) and so the assertion follows.

(ii) *Upper and lower bounds for $F(t)$*

We may clearly use (3.6) and (3.8) to give the bounds

$$F(0) \exp \frac{\dot{F}(0)t}{F(0)} \leqslant F(t) \leqslant F(0) \left\{ \frac{F(T)}{F(0)} \right\}^{t/T}, \quad t \in [0, T], \tag{3.12}$$

which, provided $F(T)$ is known, may be useful in obtaining numerical estimates of the solution.

Let us observe that for ill-posed problems in partial differential equations, it is generally the case that solutions, although unique, do not depend continuously on the data unless restricted by an *a priori* condition (cf. (3.7)) which is usually equivalent to a compactness assumption. Several examples are discussed by John (e.g. 1955, 1960), who also remarks that to recover continuous dependence may sometimes need a concept of continuity much weaker than the standard one. Thus, Hölder continuity is weaker than the usual concept, but John (1960) shows that for the linear wave equation, the solution outside a cylinder does not even depend Hölder continuously on its value inside. Instead, logarithmic continuity must be employed.

3.2. *Non-existence of the solution*

We employ the closed-graph theorem to prove that a solution cannot exist to our problem for general data. The technique has been previously used in partial differential equations by such authors as Banach (1932), Lax (1957), and Fichera (1965), who use it to relate existence and continuous dependence. By means of inequality (3.8) and a scaling procedure, we prove that at finite time the solution is unbounded and hence cannot exist.

Thus, let us assume that $\rho = 1$ and select the function w to have compact support in Ω and negative potential energy so that

$$\int_\Omega \nabla w . \mathbb{C} \nabla w \, dx < 0. \tag{3.13}$$

Without loss we may suppose the origin is located inside Ω, and then we define

$$u_0(x) = h^{3/2} w(hx), \quad h > 0. \tag{3.14}$$

An easy calculation shows that

$$\int_\Omega u_0 u_0 \, dx = \int_\Omega w w \, dx, \tag{3.15}$$

and

$$\int_\Omega \nabla u_0 . \mathbb{C} \nabla u_0 \, dx = h^2 \int_\Omega \nabla w . \mathbb{C} \nabla w \, dx. \tag{3.16}$$

As initial data to the problem, we set

$$u(x, 0) = u_0(x), \quad \dot{u}(x, 0) = u_1(x),$$

and choose $u_1(x)$ to make

$$\dot{F}(0) = \int_\Omega u_0 u_1 \, dx$$

arbitrarily large. Then h is selected to give negative initial total energy. We conclude from (3.8) that the solution is unbounded for each finite t, and so does not depend continuously upon the initial data. Thus, by the closed graph theorem, the solution cannot exist for general data.

The above scaling procedure and demonstration of unboundedness has been used by Marsden and Hughes (1978) and Wilkes (1980) to prove that strong ellipticity is a necessary condition for the equations of linear elastodynamics to generate a contractive semi-group. Wilkes (1980) has extended the argument to linear thermoelastodynamics and linear viscoelasticity, as well as to certain other dissipative equations. We discuss certain aspects of his work in Section 3.3.

We emphasize that our conclusion does not say that a solution does not exist for *all* initial data. As illustration, let us recall the observation, originally due to Hadamard (1923, p. 26), concerning the following Cauchy problem for Laplace's equation:

$$\frac{\partial^2 u}{\partial x^2} + \frac{\partial^2 u}{\partial t^2} = 0, \quad 0 \leqslant x \leqslant \pi, \quad t \geqslant 0.$$

$$u(0, t) = u(\pi, t) = 0,$$

$$u(x, 0) = 0, \quad \frac{\partial u}{\partial t}(x, 0) = g(x),$$

where $g(x)$ is given. It is well known that solutions to Laplace's equation may be continued by reflection across $t = 0$ into $t < 0$, and hence must be analytic in $(0, \pi) \times (-\infty, \infty)$. Thus a solution cannot exist unless the data $g(x)$ are analytic.

More generally, a solution may be shown to exist a provided the Cauchy data satisfy a compatibility relation. Such relations have been derived by Hadamard (1923) for the Laplace and heat conduction equations in non-cylindrical domains, while Sather and Sather (1968) have treated an elliptic parabolic operator. However, these compatibility relations are fairly restrictive and will be satisfied only by data of a special kind. Thus it appears that the number of solutions that can exist is limited, and can occur only in the exceptional circumstances when compatible data are imposed. Furthermore, it is hardly feasible that the precise relations can be realized in practice and hence it is difficult to resist the conclusion that in practical situations a solution rarely, if ever, exists. Is Hadamard's distaste for ill-posed problems after all justified? Fortunately, for our problem at least, the density results of Section 3.3 appear to indicate otherwise.

3.3. *Existence*

As already mentioned, in order to describe the existence of solutions, we must formulate the problem in a weak sense. We suppose, without loss, that the initial data are homogeneous, but that the body force is now non-zero. We denote it by f. Let $Q = \Omega \times [0, T)$ and assume $f \in L^2(Q)$. Then we define $u \in L^2(Q)$ to be a weak solution to our problem provided u satisfies the equation

$$\int_Q uCv\,dx\,d\eta = \int_Q fv\,dx\,d\eta \tag{3.17}$$

for all $v \in \mathscr{D}(C)$, the domain of the linear operator $C: \mathscr{D}(C) \to L^2(Q)$, where

$$Cv = \ddot{v} - \mathrm{Div}\,\mathbb{C}\,\nabla v, \tag{3.18}$$

and

$$\mathscr{D}(C) = \{v \in W^{2,2}(Q): v(T) = \dot{v}(T) = O,\ v|_{\partial\Omega} = 0\}. \tag{3.19}$$

Here, $W^{m,p}(Q)$ denotes the usual Sobolev space of functions which, together with their partial derivatives up to and including those of mth order, are in $L^p(Q)$.

The right side of (3.17) is a bounded linear functional and so defines the adjoint $C^*: \mathscr{D}(C^*) \to L^2(Q)$ with $u \in \mathscr{D}(C^*)$. Hence,

$$\int_Q vC^*u\,dx\,d\eta = \int_Q fv\,dx\,d\eta, \tag{3.20}$$

and thus in a distributional sense, u satisfies the equation

$$C^*u = f. \tag{3.21}$$

The conclusion on existence is contained in the following theorem.

Theorem. *Suppose the elasticities are negative-definite in the sense that*

$$-\xi.\mathbb{C}\xi \geqslant c\xi\xi, \quad c > 0, \quad \forall\ \xi,\ x \in \Omega. \tag{3.22}$$

Then the range of C^* is dense in $L^2(Q)$.

Proof. Let us assume for the moment that C is closed, and let $\mathscr{R}(C)$ and $\mathscr{N}(C)$ denote, respectively, the range space and null space of C. Now, since C is densely defined, it follows by a standard result that

$$\mathscr{R}(C^*)^{\perp} = \mathscr{N}(C^{**}), \tag{3.23}$$

where the superscript $\perp$ denotes the orthogonal space. By means of the closedness of C, we then have

$$\mathscr{N}(C^{**}) = \mathscr{N}(C). \tag{3.24}$$

The conclusion of the theorem then follows provided $\mathscr{N}(C) = 0$. This condition is equivalent to the uniqueness of the solution to (3.17) and (3.18), and hence to the uniqueness of the solution to (2.11), (2.12), (2.13). But we have seen in Section 3.1 how uniqueness may be established by means of logarithmic convexity arguments and the energy inequality (2.15). Note, however, that (2.15) has not been proved to hold for solutions of (3.17) and (3.18), and so we prefer either to postulate separately the validity of (2.15), or better still, avoid its use altogether by appealing instead to the uniqueness theorem of Levine (1977) which relies solely upon the symmetry of $\mathbb{C}$ and makes no use whatsoever of energy conservation. All three possibilities ensure uniqueness and thus we have $\mathscr{N}(C) = 0$ and hence

$$\mathscr{R}(C^*)^{\perp} = 0, \tag{3.25}$$

which proves the conclusion.

It remains to show that C is closed, and in order to do this we must prove that given the sequence $v_n \in \mathscr{D}(C)$ such that

$$v_n \xrightarrow{L^2(Q)} v, \tag{3.26}$$

$$Cv_n \xrightarrow{L^2(Q)} w, \tag{3.27}$$

then $v \in \mathscr{D}(C)$ and $Cv = w$. We suppose that negative-definiteness of the elasticities implies the coerciveness of $-\mathbb{C}$ on $W^{1,2}(\Omega)$. Hence, the operator (3.18) is elliptic, and thus by well-known *a priori* estimates, we have

$$\| v_n \|_{W^{2,2}(Q)} \leqslant c\Big(\| v_n \|_{L^2(Q)} + \| C v_n \|_{L^2(Q)} \Big). \tag{3.28}$$

But by hypothesis, v_n, $Cv_n \in L^2(Q)$, so that $v_n \in W^{2,2}(Q)$. Hence, there exists a subsequence v_{n_e} such that

$$v_{n_e} \xrightarrow{W^{2,2}(Q)} v, \tag{3.29}$$

and thus $v \in W^{2,2}(Q)$. Furthermore, by trace theorems

$$v_n(T) \to v(T), \quad \dot{v}_n(T) \to \dot{v}(T), \tag{3.30}$$

and thus $v \in \mathscr{D}(C)$.

Finally, $Cv_n \xrightarrow{L^2(Q)} Cv$, but since, by hypothesis, $Cv_n \xrightarrow{L^2(Q)} w$, we must have

$$Cv = w,$$

and the theorem is proved.

Let us now return to the initial boundary value problem with zero body force and homogeneous boundary conditions. The theorem corresponding to the one just proved states that for this problem a solution exists for initial data, u_0, u_1, dense in $W^{1,2}(\Omega) \times L^2(\Omega)$. Simple density arguments can then be used to show that for specified initial data with negative total energy, a solution exists with initial values arbitrarily close to the specified data in $W^{1,2}(\Omega) \times L^2(\Omega)$, and also with negative total energy. Hence, this solution possesses the properties established in Section 3.1 and, in particular, is arbitrarily close to its predicted values provided that *a priori* condition (3.7) is satisfied. More generally, the solution grows at least exponentially in time.

The existence theorem can be extended to indefinite elasticities and hence gives meaning to other continuous dependence and growth estimates that can be similarly obtained by logarithmic convexity methods. Density arguments may again be used to obtain growth estimates on the actual solution, and so we have the following theorem for instability.

Theorem. The null solution for the initial boundary value problem (2.11)–(2.13) *is unstable provided the elasticities are indefinite or negative-definite.*

This concludes our study of the linear elastic problem. The next two Sections describe similar properties for the solution in the exterior problem, and also for the solution to the equations of linear elastodynamics. However, we do not treat questions of existence but content ourselves with a discussion of uniqueness, continuous dependence, and growth estimates.

4. Linear Thermoelasticity in Bounded Regions

We now consider linear thermoelasticity in a bounded region with the object of deriving inequalities similar to (3.6) and (3.8) for linear elastodynamics. We shall briefly review some results obtained by Knops and Payne (1970) and then discuss estimates recently derived by Wilkes (1979).

The equations of linear thermoelasticity are given by:

$$\operatorname{Div}[\mathbb{C}\,\nabla u - F\theta] = \rho\ddot{u}, \qquad (x, t) \in \Omega \times [0, T), \tag{4.1}$$

$$\operatorname{div}(\mathbb{a}\nabla\theta) - c \operatorname{tr} F \nabla \dot{u} = \dot{\theta}, \qquad (x, t) \in \Omega \times [0, T), \tag{4.2}$$

where, in addition to the notation of Section 2, we have that $\mathbb{a}$ is the thermal conductivity, assumed symmetric,

$$a_{ij} = a_{ji}, \tag{4.3}$$

and positive-definite,

$$a_{ij}(x)\xi_i\xi_j \geqslant a_0\xi_i\xi_i, \quad a_0 > 0, \quad x \in \Omega. \tag{4.4}$$

The temperature deviation is denoted by θ, while the tensor function $F(x)$ and non-zero scalar function $c(x)$ are assumed prescribed. As in Section 3, we continue to suppose that the elasticities are symmetric, but we do not impose any definiteness condition.

The associated boundary and initial conditions are taken to be

$$u = 0, \quad \theta = 0, \quad (x, t) \in \partial\Omega \times [0, T), \tag{4.5}$$

$$\left.\begin{aligned} u(x,0) &= u_0(x), \quad \dot{u}(x,0) = u_1(x), \\ \theta(x,0) &= \theta_0(x), \end{aligned}\right\} \tag{4.6}$$

where u_0, u_1, θ_0 are prescribed functions.

Because the methods discussed here continue to extend easily to an abstract version of the problem, we consider only the classical formulation. Those interested in an abstract derivation should consult, for instance, Levine (1970a), Bloom (1975), and Marsden and Hughes (1978).

About 10 years ago Knops and Payne (1970) showed that when the given tensor function F is restricted by the inequalities

$$\operatorname{tr} FF^{\mathrm{T}} \leqslant M_1^2, \quad \operatorname{tr} \operatorname{Div} F \operatorname{Div} F \leqslant M_2^2, \tag{4.7}$$

for prescribed constants M_1, M_2, then the space–time norm of the solution given by

$$G(t) = \int_0^t \int_\Omega \rho u(x,\eta) u(x,\eta)\, dx\, d\eta + (T-t) \int_\Omega u_0(x) u_0(x)\, dx + \gamma, \tag{4.8}$$

where γ is a constant depending upon the initial data, satisfies the differential inequality

$$G\ddot{G} - (\dot{G})^2 \geqslant -k_2 G\dot{G} - 2k_1 G^2 \quad t \in [0, T). \tag{4.9}$$

Here k_1 and k_2 are computable positive constants.

It follows in a standard manner that (4.9) may be integrated to yield an inequality from which may be deduced uniqueness on the half-line. Also, for bounded solutions in the class

$$\int_0^T \int_\Omega \rho u(x,\eta) u(x,\eta)\, dx\, d\eta \leqslant M^2, \tag{4.10}$$

for prescribed constant M, (4.9) may be similarly integrated to give

$$G(t) \leqslant K_1 M^\delta \left\{ K_2 \int_\Omega \rho u_0 u_0\, dx + K_3 \int_\Omega \rho u_1 u_1\, dx + K_4 \left| \int_Q \nabla u_0 \mathbb{C} \nabla u_0\, dx \right| \right.$$

$$\left. + K_5 \int_Q \theta_0^2\, dx \right\}^{1-\delta}, \quad 0 \leqslant \delta \leqslant 1, \tag{4.11}$$

where K_i, $i = 1, 2, \ldots, 5$, are computable positive constants. Clearly, we may easily deduce from (4.11) the Hölder continuous dependence of the solution on the initial data.

Abstract versions of these two results have been obtained by Levine (1970a), while Bloom (1975) has extended them to linear thermoelastic materials with continuous distributions of dislocations and temperature-dependent elasticities. He has also shown how for the uncoupled equations ($c = 0$) and for temperature-dependent elasticities the displacement depends Hölder continuously on θ under appropriate smoothness assumptions.

Inequality (4.9) may also be integrated to provide a lower bound for $G(t)$ similar to the estimate (3.8) in linear elasticity. However, here the lower bound is decaying, and hence does not yield anything of significant interest. (We observe that for linear viscoelasticity, Bloom, 1977, has demonstrated how the corresponding lower bound may be used to establish non-existence of a certain class of solutions.)

Yet in order to employ the approach of Section 3 to prove non-existence of a wide class of solutions, and also to prove that the governing equations do not lead to a semi-group, an estimate establishing unboundedness of the solution is required. Physically, it is unclear whether the displacement can possess such singular behaviour since the temperature, for positive-definite conductivity, may provide sufficient dissipation to prevent breakdown. However, we now describe a device, due to Wilkes (1979), by which exponential growth of the $L^2(\Omega)$-integral of the displacement may be obtained.

Define the norm $F(t)$ by

$$F(t) = \int_\Omega \rho u(x,t)u(x,t)\,dx + \frac{1}{c}\int_0^t \int_\Omega \Phi(x,\eta)\mathfrak{a}\Phi(x,\eta)\,dx\,d\eta + b(t+t_0)^2, \tag{4.12}$$

where b, t_0 are positive constants to be determined, and

$$\Phi(x,t) = \int_0^t \nabla\theta(x,s)\,ds + \nabla h(x) \tag{4.13}$$

with the scalar function $h(x)$ determined from the boundary value problem

$$\left.\begin{aligned} \operatorname{div}(\mathfrak{a}\nabla h) &= \theta_0(x) + c\,\mathrm{tr}(F\nabla u_0), \quad x\in\Omega, \\ h &= 0, \quad x\in\partial\Omega. \end{aligned}\right\} \tag{4.14}$$

The theory of elliptic equations ensures the existence of $h(x)$.

On selecting initial data to satisfy

$$\int_\Omega \theta_0^2\,dx + c\int_\Omega (\rho u_1 u_1 + \nabla u_0 \mathbb{C}\nabla u_0)\,dx < 0, \tag{4.15}$$

it may then be proved that (for details, see Wilkes, 1979)

$$F(t) \geqslant \frac{F(0)\dot F(0)}{(\dot F(0)-L)}\exp\left\{\frac{\dot F(0)-L}{F(0)}\right\}t - \frac{LF(0)}{(\dot F(0)-L)}, \tag{4.16}$$

$$L = \frac{2}{c}\int_\Omega \nabla h\,.\,\mathfrak{a}\nabla h\,dx.$$

Hence, provided initial data are chosen to satisfy $\dot F(0) > L$, we obtain from (4.16) an exponentially increasing lower bound for $F(t)$. We therefore know that the sum of the integrals in (4.12) becomes unbounded in time, but it is important to determine whether the unboundedness is due to one or both integrals. In order to resolve this question, we introduce the function $P(t)$ defined by

$$P(t) = \int_\Omega \rho uu\,dx - \frac{1}{c}\int_0^t\int_\Omega \Phi\mathfrak{a}\Phi\,dx\,d\eta + \alpha t, \tag{4.17}$$

where α is another positive constant. We further suppose that the elasticities are negative-definite and then it may be shown that

$$\ddot P(t) \geqslant 0. \tag{4.18}$$

On integrating twice and choosing α sufficiently large so that $\dot P(0) > 0$, we see that $P(t)$ increases linearly in t. Hence, by addition of (4.12) and (4.17) we conclude that $\int_\Omega \rho uu\,dx$ has at least an exponentially increasing lower bound.

By a scaling procedure analogous to that described in Section 3.2, we can now show that the norm of the displacement becomes unbounded at finite time, and hence the closed-graph theorem may be used to prove non-existence of solutions for general data. On the other hand, it should be possible to prove an existence theorem similar to that of Section 3.3, but this is not attempted here.

Wilkes (1978) has applied similar computations to linear viscoelasticity and obtained corresponding time-increasing lower bounds, which he uses to demonstrate non-existence of a contractive semi-group (1980). He has also discussed continuous dependence in both linear theories without imposing restrictions of the kind (4.7), and also treats certain other results.

5. Linear Hyperelasticity in Exterior Domains

In this Section we treat the linear equations of hyperelasticity, given in (3.1), in the domain Ω^* exterior to Ω:

$$\Omega^* = \mathbb{R}^n \backslash \Omega.$$

We set down the inequalities analogous to (3.6), (3.8) and show how far the assumption of positive-definite elasticities may be relaxed. A full account of these results will be provided in a forthcoming paper by Knops and Payne. Of course, the computations of Section 3 leading to the original inequalities (3.6), (3.8) can be easily adapted to the present exterior problem by merely prescribing suitable asymptotic decay rates at large distances to ensure convergence of the appropriate integrals. However, we wish to admit weaker asymptotic behaviour which allows some growth at large distances. To be precise, we suppose that the displacement and its first-order derivatives behave as $o(e^{kr})$ at large distances, where k is some positive constant and r is the distance from some fixed origin in Ω^*.

Uniqueness of the solution to the linear exterior problem under the assumption of positive-definite elasticities has been treated by Wheeler and Sternberg (1968) and Wheeler (1970, 1971). Continuous dependence of the solution upon the initial data, again for positive semi-definite elasticities, has been considered by Knops and Straughan (1976a, b) and by Galdi and Rionero (1979), who admit asymptotic behaviour corresponding to (H3).

For simplicity, we suppose the boundary and initial conditions to our problem are given by (2.12), (2.13), and then our results are valid under the hypotheses:

(H1) $\mathbb{C} = \mathbb{C}^T$.

(H2) The displacement is piecewise smooth in $\Omega^* \times [0, T)$.

(H3) u, ∇u, $\dot{u} = o(e^{kr})$, for some positive constant k, as $r = (x_i x_i)^{1/2} \to \infty$.

(H4) There exists a positive constant R, such that

$$\xi \mathbb{C} \xi \geqslant c_0 \xi\xi, \qquad \forall\, \xi, r \geqslant R. \tag{5.1}$$

In order to state the inequalities we introduce the functions ω and ϕ defined by

$$\omega(r) = \begin{cases} 1 & r \leqslant R_1, \\ e^{-k(r-R_1)}, & r > R_1 > R, \end{cases} \tag{5.2}$$

$$\phi(t) = \int_0^t \int_{\Omega^*} e^{-\lambda\eta}\, \omega(r) uu \, dx\, d\eta + k_1 S, \tag{5.3}$$

where k_1 is a positive constant and S, a quantity depending upon the initial data, vanishes when $u_0 = u_1 = 0$.

In terms of these functions, the inequality corresponding to (3.6) becomes

$$\Phi(t) \leqslant \sigma^{\gamma_1}\left(\frac{k_1 S}{\sigma_0^{\gamma_1}}\right)^{\frac{\sigma-\sigma_1}{\sigma_0-\sigma_1}}\left(\frac{\phi(t_1)}{\sigma_1^{\gamma_1}}\right)^{\frac{\sigma_0-\sigma}{\sigma_0-\sigma_1}}, \tag{5.4}$$

where $\sigma = \left(\frac{1}{\gamma_2}\right)e^{-\gamma_2 t}$, γ_1 and γ_2 are computable positive constants and $0 \leqslant t \leqslant t_1 < T$. Uniqueness and continuous dependence of the solution on initial data are immediate consequences.

It appears that condition (5.1) is crucial, since in its absence simple examples can be constructed for which uniqueness of solution fails.

The inequality analogous to (3.8) is the content of the next proposition.

Proposition. Under Hypotheses (1)–(4), together with certain restrictions on the initial data of type

$$\int_{\Omega^*} \omega(r)\,(u_1 u_1 + \nabla u_0 . \mathbb{C}\, \nabla u_0)\, dx + \sqrt{\lambda k}\int_{\Omega^*\setminus B(R_1)} e^{-k(r-R)} u_0 u_0\, dx \leqslant 0, \tag{5.5}$$

the function $H(t)$, defined by

$$H(t) = \int_{\Omega^*} \omega(r) u(x,t) u(x,t)\, dx,$$

satisfies

$$H(t) \geqslant H(0)\exp\left\{\frac{I(0)}{H(0)} - \lambda\right\}t$$

where

$$I(0) = 2\int_{\Omega^*} \omega(r) u_0 u_1\, dx - \left(\frac{1}{2} + \frac{k\beta}{\sqrt{\lambda}}\right)\alpha(R_1)$$

with β and $\alpha(R_1)$ being constants chosen to satisfy

$$I(0) > \lambda H(0),$$

and

$$\int_0^\infty \int_{\Omega^*} e^{-\lambda\eta - k(r-R)}\{\dot{u}(x,\eta)\dot{u}(x,\eta) + \nabla u(x,\eta) . \mathbb{C}\, \nabla u(x,\eta)\}\, dx\, d\eta \leqslant \alpha(R_1).$$

Here, $B(R_1)$ is a ball of radius R_1 and λ is a positive constant satisfying (5.5).

The underlying method of proof uses integration by parts on certain weighted integrals and is similar to the method of weighted energy employed by Galdi and Rionero.

6. Nonlinear Hyperelasticity

6.1. *Introduction*

So far we have described results only for linear theories. We turn now to a consideration of nonlinear hyperelasticity and, with the exception of existence, describe properties of the solution similar to those already treated in Section 3 for linear elasticity. We conclude with some very brief remarks on Cauchy elasticity. As might be expected, there is not such an extensive set of results, even of a formal nature, nor are the conditions under which some hold so obvious.

The initial boundary value problem of nonlinear hyperelasticity is given by (2.6), (2.12), and (2.13), with the elasticities satisfying the symmetry relations (2.8). We recall that the strain-energy W does not satisfy any *a priori* condition such as convexity so that, in particular, the quasilinear system (2.6) is of no definite type, e.g. it is neither hyperbolic nor elliptic. Nevertheless, as already mentioned, the intention here is to study properties of the solution to (2.6) under selected hypotheses on W, and hence we start by recalling some of these when equations (2.6) are hyperbolic. It is reasonable to suppose that under weaker hypotheses the solution might retain some of the qualitative behaviour possessed by that of the hyperbolic system.

It is well known that classical solutions to one-dimensional hyperbolic conservation laws do not exist for all time (cf. Lax, 1964, Dafermos, 1974), and hence it can be expected that the same will hold for solutions of nonlinear elastodynamics. This has, in fact, been demonstrated by John (e.g. 1976a, b, c) for certain plane waves in the three-dimensional theory in unbounded regions. A shock wave is shown to develop after finite time from initial data that are arbitrarily smooth. The existence of these and other singular surfaces gives rise to the study of generalized solutions, and in the case of systems of nonlinear hyperbolic conservation laws in one independent space variable, Glimm (1965) has proved global existence for initial data of sufficiently small oscillation and total variation. Furthermore, the precise structure of generalized solutions has been investigated within the class of functions of bounded variation in the sense of Tonelli and Cesari. This appears to be the natural class of functions in which to undertake the study, since geometrically these functions resemble piecewise smooth functions. Detailed accounts of this work are provided in the articles by Dafermos (1977) and DiPerna (1979), where references to original contributions may be found.

Of course, under smooth initial data, singular behaviour cannot occur instantaneously, and there is always an interval of time in which a smooth (classical) solution exists. Estimates of the length of the maximal time interval of existence of a smooth solution have been obtained by various authors, and show generally that it is at least of order ε^{-1} when the initial data are εu_0, εu_1 where u_0, u_1, are given smooth functions with compact support, and the region is unbounded. For one-dimensional hyperbolic conservation laws that are genuinely nonlinear, the estimate has been proved exact (Lax, 1964), but in higher spatial dimensions it has been conjectured, notably by John (e.g. 1976a, b, c), that the increased space may allow energy adequately to dissipate and delay, or avoid altogether, the development of singularities. This result has been established for spaces of at least six dimensions (Klainerman, 1980, 1981), but recently John (1979) has proved for a special example that in three spatial dimensions, which is the situation of immediate interest for elasticity, singularities will occur after finite time from sufficiently small and smooth data with compact support. This author (1976) has also provided lower bounds for the maximal interval of existence of smooth solutions in unbounded regions.

Of course, it must not be concluded that in three dimensions smooth solutions will always have a finite time interval of existence. Carroll (1977a, b), for instance, presents some counter examples in elasticity for special choices of the strain-energy (see also Antman and Liu, 1979). On the other hand, Hughes *et al.* (1977) (see also Hughes and Marsden, 1978) have established that smooth solutions exist, are unique, and depend continuously upon initial data locally in time for unbounded three-dimensional regions under a strong ellipticity assumption on W and for small initial data that are smooth. For the corresponding initial boundary value problem for a bounded region, local existence and uniqueness has been proved by Kato (1976), but continuous dependence of the solution upon the initial data is still open. We mention also that Wheeler (1976), again under the assumption of strong ellipticity, derived uniqueness of smooth solutions in the initial displacement boundary value problem, while Gurtin and Spector (1979) in the static theory have related uniqueness to positivity of the incremental stress power (interpreted to mean stability). Existence of weak solutions in the static theory under mild assumptions on the strain-energy has been established by Ball (1977, 1980).

In view of these remarks for the dynamic theory, it is clear that generalized solutions must be admitted, even when the quasilinear system (2.6) is not hyperbolic. However, in this case some further conditions must be imposed. Generalized solutions may be non-unique and hence lead automatically to instability of, say, the null solution, unless a selection procedure, such as the viscosity method is employed to recover uniqueness. Within the purely mechanical theory such a criterion is partially supplied by the energy inequality (2.15), but what other conditions, if any, are additionally required is still unclear. However, by appealing to the second law of thermodynamics in the form of the Clausius–Duhem inequality, Dafermos (1979) has proved that for thermoelasticity, smooth adiabatic solutions for which the strain-energy (strictly, internal energy) is convex depend continuously upon the initial data, in the class of solutions that may develop singularities. Hence, in this sense, uniqueness is regained. Dafermos also shows that in certain circumstances, the convexity condition on the strain-energy may be replaced by strong ellipticity, and continuous dependence continues to hold as before. In a later paper, Dafermos (1981) shows that the second law guarantees the stability of weak shock waves in the sense that Lax's E-condition is satisfied; neither convexity nor strong ellipticity is required.

However, we wish to remain with a purely mechanical theory of elasticity as represented by equations (2.6), and we shall adopt inequality (2.15) as a postulate, without attempting here any exact derivation from thermodynamics. In view of the paucity of existence results, subsequent discussion must be regarded as formal. Indeed, we set aside completely the question of existence and return once more to a consideration of stability. Even here, we shall see that unfortunately very few results appear to be available.

6.2. *Stability*

As in the linear theory, it is fairly popular in the nonlinear theory to adopt the energy criterion as a test for the stability of an equilibrium solution despite the fact that on the basis of dynamics it has not yet been confirmed as either a sufficient or necessary condition.

As regards sufficiency, it has been shown that when the requirement of a minimum of the strain-energy is replaced by that of a potential well, the equilibrium solution is stable in the sense of Liapunov. Proofs have been provided, for instance, by Koiter (1965), Coleman (1973), Gurtin (1975), and Hughes and Marsden (1978). However, there then arises the problem of determining under what conditions the strain-energy W possesses a potential well. By means of particular examples (e.g. Knops and Payne, 1978, Ball *et al.*, 1978) it is clear that restrictions on the quadratic part of W are insufficient to guarantee a well in the spaces $W^{1,p}$, $1 \leqslant p \leqslant \infty$, and thus the problem is still open. Convexity assumptions (including convexification) or strong ellipticity combined with thermodynamics (as demonstrated by Dafermos,

1979) may be adequate but are unlikely to be universally acceptable in view, for example, of the recent applications by Ericksen (1975) and James (1979, 1980) of non-convex strain-energies to describe certain phase transitions in elasticity (see also Ericksen, 1977). Other devices include the addition of dissipative terms in the equation (see, for example, Hughes and Marsden, 1978, and Holmes and Marsden, 1978), or adopting a semilinear or finite dimensional approximation for which verification of the energy criterion is trivial (see, for example, Marsden, 1973). Yet another, but related, technique is to place an *a priori* restriction on the solution. We illustrate this remark by means of a concrete example and examine the total strain-energy

$$V(u) = \int_0^1 (u_x^2 - u_x^4)\,dx \tag{6.1}$$

for functions $u: \mathbb{R} \to \mathbb{R}$ satisfying

$$u(0) = u(1) = 0.$$

It has been shown (e.g. Ball *et al.*, 1978) that, in general, a potential well for (6.1) does not exist in the spaces $W^{1,p}(0, 1)$, $1 \leqslant p \leqslant \infty$. Now impose the constraints

$$|u_x| \leqslant \delta < \frac{1}{2}, \quad |u_{xx}| < c < \infty, \quad x \in [0, 1]. \tag{6.2}$$

We shall show that, subject to (6.2), we have

$$V(u) \geqslant \varepsilon > 0 \quad \text{when} \quad \|u\|_{W^{1,\infty}} = \delta, \tag{6.3}$$

so that a potential well has been recovered in $W^{1,\infty}(0, 1)$. By (6.2) we have

$$V(u) \geqslant \frac{3}{4}\int_0^1 u_x^2\,dx.$$

Let

$$u_x(x_0) = \frac{1}{2}, \quad x_0 \in (0, 1).$$

Then, for $\eta < \delta$,

$$|u_x(x)| > \eta, \quad |x - x_0| < \frac{\delta - \eta}{c},$$

and so

$$V(u) \geqslant \frac{3}{2}\eta^2\left(\frac{\delta - \eta}{c}\right) > 0,$$

which proves (6.3).

Whether conditions (6.2) are satisfied depends upon the properties of the solution to the dynamic problem. For instance, the addition of a linear dissipative term to the one-dimensional nonlinear wave equation may be shown by the methods of Riemann invariants (see, for example, Slemrod, 1979) to produce the desired estimates.

We comment next on the energy criterion as a necessary condition for stability. Within the context of thermoelasticity, Ericksen (1965) has shown that when the equilibrium solution is asymptotically stable, then necessarily it must be a minimum of the internal energy. However, in a purely mechanical theory of elasticity the situation is less clear. What we would like to show is that when an equilibrium solution is not a minimum of the strain-energy then it is Liapunov unstable, this being the strict equivalent of the proof for necessity. Since a result of this kind has so far not been established for the discrete case (it has been listed by Arnold, 1976, as one of the modern Hilbert problems), it is not surprising that a proof of the complete statement appears still to be lacking in the literature. However, the following result helps to bring elasticity more into line with what is known for discrete systems (cf. also Knops, 1976).

Proposition. For the initial boundary value problem (2.6), (2.12) *and* (2.13), *suppose* $W \in C^2$, *and that the total strain-energy, given by*

$$V(u) = \int_\Omega W(\nabla u(x,t))dx, \tag{6.4}$$

possesses a strict maximum at the equilibrium solution $v(x)$. *Then* $v(x)$ *is Liapunov unstable.*

Proof. We proceed by a contradiction, using an idea essentially due to Duhem. Thus, let us assume that $v(x)$ is stable in the sense that for all data sufficiently small, there holds

$$\sup_{t\in[0,\infty)} \|u(x,t) - v(x)\|_{W^{1,\infty}(\Omega)} \leqslant \varepsilon. \tag{6.5}$$

Then, the strict maximum of $V(u)$ at $v(x)$ implies that there exists a constant $a < 0$ such that

$$\int_\Omega \nabla u\, \mathbb{C}\, \nabla u\, dx \leqslant a \int_\Omega \nabla u . \nabla u\, dx, \qquad \|u\|_{W^{1,\infty}(\Omega)} < \varepsilon, \tag{6.6}$$

where $\mathbb{C}$ is given by (2.9), and without loss we may take $v(x)$ to be identically zero. Furthermore, in view of assumption (6.5), we have by Taylor's theorem that

$$\begin{aligned}\int_\Omega \nabla u \frac{\partial W}{\partial \nabla u}(\nabla u)\, dx &= \int_\Omega \int_0^1 \nabla u \frac{\partial^2 W}{\partial \nabla u\, \partial \nabla u}(s\nabla u)\nabla u\, ds\, dx \\ &\leqslant C \int_\Omega \nabla u \nabla u\, dx,\end{aligned} \tag{6.7}$$

where $C \leqslant 0$, as ∇u is uniformly bounded (from (6.5)) and $W \in C^2$.

Next, we define

$$F(t) = \int_\Omega \rho u u\, dx, \tag{6.8}$$

and by differentiation and use of (2.6) and (2.12) obtain

$$\begin{aligned}\ddot{F}(t) &= 2\int_\Omega \rho \dot{u}\dot{u}\, dx - 2\int_\Omega \nabla u \frac{\partial W}{\partial \nabla u} dx \\ &\geqslant 2\int_\Omega \rho \dot{u}\dot{u}\, dx,\end{aligned} \tag{6.9}$$

where we have taken into account (6.5) and (6.7). It then follows by means of Schwarz's inequality that

$$(F^{1/2})^{\cdot\cdot} \geqslant 0,$$

so that

$$F^{1/2}(t) \geqslant \frac{\dot{F}(0)t + 2F(0)}{2F^{1/2}(0)}. \tag{6.10}$$

Let us choose initial data u_0, u_1 such that

$$\| u_0 \|_{W^{1,2}(\Omega)} + \| u_1 \|_{L^2(\Omega)} \tag{6.11}$$

is arbitrarily small, but

$$\dot{F}(0) = 2 \int_\Omega \rho u_0 u_1 \, dx > 0. \tag{6.12}$$

Then (6.10) shows that there exists t_1 such that $F(t_1) > \varepsilon_1$. We may then use Poincaré's inequality to obtain a contradiction to (6.5), and so the assertion is proved.

The condition of a strict maximum may be relaxed in the Proposition provided some restriction is placed on the nonlinearity of the strain-energy. For instance, the Proposition is still true if instead of a strict maximum we have

$$\int_\Omega \nabla u . \frac{\partial W}{\partial \nabla u}(\nabla u) \, dx \leqslant 0, \quad \| u \|_{W^{1,\infty}} < \varepsilon, \tag{6.13}$$

where again without loss the equilibrium solution is taken as the trivial one. That some such condition is required may be illustrated by the one-dimensional example in which

$$W(\nabla u) = u_x^4.$$

Thus, the linearized problem yields instability of the null solution, while the nonlinear problem yields stability.

In Section 6.3 we discuss other conditions on the nonlinearity that produce instability. Some of the results described help illuminate another aspect of stability theory already apparent in the treatment of potential wells and the necessary conditions for stability. We refer to conditions under which stability of the linearized problem guarantees stability in the nonlinear one. We show that a result of this kind cannot be valid with restrictions placed only on the linearized equations.

6.3. *Concavity and other arguments*

Strengthening condition (6.13) leads to different kinds of behaviour for the dynamic solutions. We analyse the problem for two other conditions, the first of which is given by

$$2(1+2\alpha) \int_\Omega W(\nabla u(x)) \, dx - \int_\Omega \nabla u \frac{\partial W}{\partial \nabla u} dx \geqslant 0, \quad \alpha > 0, \tag{6.14}$$

which must hold for all u. The method of proof is a generalization of the logarithmic convexity argument contained in Section 3.1, and may be applied to an appropriate weak formulation. Details may be found in the paper by Knops *et al.* (1974). Similar methods have been used by Kalantarov and Ladyzhenskaya (1977) in a related study.

We set

$$F(t) = \int_\Omega \rho u u \, dx + \beta (t + t_0)^2, \tag{6.15}$$

where, as before, β, t_0 are positive constants to be determined. On differentiating (6.15) twice with respect to time, eliminating the inertial term by means of (2.6), integrating by parts, and using the energy inequality (2.15) together with (6.14), we obtain

$$\ddot{F}(t) \geqslant 4(1+\alpha)\int_\Omega \rho \dot{u}\dot{u}\,dx - 2(1+2\alpha)(2E(0)+\beta), \quad t \in [0, T). \tag{6.16}$$

Hence, from Schwarz's inequality it follows that

$$F\ddot{F} - (1+\alpha)(\dot{F})^2 \geqslant -2(1+2\alpha)(2E(0)+\beta)F(t). \tag{6.17}$$

Let us suppose that initial data are chosen such that $E(0) < 0$, and then set $2E(0)+\beta = 0$. We have from (6.17),

$$(F^{-\alpha})^{\cdot\cdot} \leqslant 0, \quad t \in [0, T), \tag{6.18}$$

which shows that $F^{-\alpha}$ is a concave function of t on the maximum time interval of existence. We integrate (6.18) in two different ways, corresponding to the chord and tangent properties of concave functions. Thus, we have first,

$$F^\alpha(t) \leqslant F^\alpha(0) \bigg/ \left[1 + \frac{t}{T}\,\frac{F^{(\alpha)}(0) - F^\alpha(T)}{F^\alpha(T)}\right], \tag{6.19}$$

from which may be deduced uniqueness of the null solution, and also continuous dependence of the solution upon the initial data, as measured by $F(t)$. Secondly, we have

$$F(t) \geqslant F(0) \bigg/ \left[1 - \alpha t\frac{\dot{F}(0)}{F(0)}\right]^{1/\alpha}, \quad t \in [0, T), \tag{6.20}$$

which by choosing t_0 such that

$$\dot{F}(0) = 2\int_\Omega \rho u_0 u_1\,dx + 2\alpha t_0 > 0, \tag{6.21}$$

provides a lower bound for $F(t)$, which becomes unbounded in finite time. Hence (6.20) becomes the basis of a proof that solutions to the initial displacement boundary value problem of nonlinear elasticity satisfying the conditions of this Section cannot exist globally in time. It remains open whether solutions cease to exist by blowing up in finite time. Some results of this kind have been derived by Moser for a discrete system, and also by Ball (1979) for certain semi-linear wave equations. One-dimensional problems in elasticity have been examined by Calderer (1980).

The same kind of results hold also for initial data which as well as other conditions also satisfy $E(0) \geqslant 0$. In all cases, however, the imposed conditions require the total strain-energy to have a non-positive initial value.

Let us observe that a solution satisfying (6.19) and (6.20) is always non-decreasing in time. To see this, let us suppose the contrary and without loss take

$$F(0) > F(T). \tag{6.22}$$

Then (6.19) shows that

$$F(t) < F(0), \quad t \in [0, T),$$

and hence from (6.20) we must have $\dot{F}(0) < 0$ in contradiction to (6.21). Thus (6.22) cannot hold, and so the result follows.

We may also use (6.19) and (6.20) to furnish upper and lower bounds on the solution.

Conditions (6.14) may be interpreted in a variety of ways. Ball (1979) has shown that materials for which it holds are "weak" in the sense that a cube composed from the material has a decreasing strain-energy for increasing deformation. Ball has also given an example of an elastic body under pressure in which concavity and other arguments are used to prove global non-existence of the solution.

Another interpretation of (6.14), based on an idea of N. S. Wilkes, is to regard (6.14) as stating that the functional

$$S(u) = 2(1+2\alpha)\int_\Omega W(\nabla u)\,dx - \int_\Omega \nabla u \frac{\partial W}{\partial \nabla u}\,dx \tag{6.23}$$

possesses a minimum at $u = 0$, i.e.

$$S(u) \geqslant 0$$

with equality if and only if $u = 0$. Thus it is supposed that $S(u)$ is not identically zero. It then necessarily follows that

$$\int_\Omega \nabla u \,.\, \mathbb{C}\, \nabla u\,dx \geqslant 0, \tag{6.24}$$

where $\mathbb{C}$ is given by (2.9). By well-known results (e.g. Wang and Truesdell, 1973), the last condition implies the weak form of the strong ellipticity condition. Thus when $S(u)$ is not identically zero we see that (6.14) precludes the existence of the null solution being a strict maximum of the total strain-energy.

Other interpretations and implications of (6.14) have been studied by Andreussi (1974, 1976).

Condition (6.14) is not the only one capable of implying global non-existence of solution, and we refer to the work by, for example, Levine and Payne (1975a, b), Hills and Knops (1974, 1976), and Knops and Straughan (1976c) for a discussion of related results. Again, modified versions of (6.14) also lead to decay to zero in finite time. We describe one such modification (cf. Knops and Straughan, 1981), but others are discussed by, for example, Sleeman (1976) and Hills and Knops (1976).

Theorem. For the initial boundary value problem (2.10), (2.12) *and* (2.13), let

H(1) $\xi \mathbb{C}\, \xi \geqslant 0$;

H(2) $2(2\alpha-1)\displaystyle\int_\Omega \tilde{W}(\nabla u(x))dx + \int_\Omega \nabla u \frac{\partial \tilde{W}}{\partial \nabla u}dx \geqslant -\mu\left[\int_\Omega \rho u u\,dx\right]^\gamma,$

where $\alpha > 1$, $\mu > 0$, *and* $0 < \gamma < 1$, *and* $\tilde{W}$ *is the nonlinear part of the strain-energy corresponding to* T;

H(3) $E(0) \leqslant 0$;

H(4) $\alpha\displaystyle\int_\Omega \rho u_0 u_1\,dx + \beta\left[\int_\Omega \rho u_0 u_0 dx\right]^{\frac{\gamma+1}{2}} \leqslant 0, \quad \beta = 2\alpha[\mu/(2\alpha+\gamma-1)]^{1/2}.$

Then the solution decays to zero in finite time.

Proof. We set $F(t) = \displaystyle\int_\Omega \rho u u\,dx$ and by differentiation and means of techniques similar to those used in deriving (6.17) combined with (H1)–(H3), we obtain the differential inequality

$$(F^{\alpha})^{\cdot\cdot} \leqslant 2\mu\alpha F^{\alpha+\gamma-1}. \tag{6.25}$$

But from (H4) we have $\dot{F}(0) < 0$, and hence, by continuity, there exists t_1 such that

$$\dot{F}(t) < 0, \quad t \in [0, t_1) \tag{6.26}$$

so that (6.25) may be multiplied by $(F^{\alpha})^{\cdot}$ and integrated to give

$$P(t) \equiv [(F^{\alpha})^{\cdot} - \beta F^{\delta}][(F^{\alpha})^{\cdot} + \beta F^{\delta}] \geqslant P(0) > 0, \quad t \in [0, t_1], \tag{6.27}$$

where $\delta = (2\alpha + \gamma - 1)/2$, and the final inequality holds by virtue of (H4).

Now suppose $\dot{F}(t_1) = 0$, so that $P(t_1) = -\beta^2 F^{2\delta} \leqslant 0$, in contradiction to (6.27). Hence, $\dot{F}(t) < 0$, $t \in [0, \infty)$, and so $P(t) > 0$, $t \in [0, \infty)$. Since neither factor in (6.27) can change sign, we necessarily have

$$(F^{\alpha})^{\cdot} + \beta F^{\delta} < 0, \quad t \in [0, \infty),$$

integration of which proves the theorem.

6.4. *Cauchy elasticity and non-self-adjoint problems*

The previous developments have been based upon hyperelasticity which, as already remarked, leads to a formal self-adjoint problem. However, an alternative formulation of the field equations, based entirely upon balance laws, produces a non-self-adjoint problem in which the equations of motion have the same form as (2.6) and (2.10) except that the coefficients $A_{\alpha\beta}$ no longer possess the symmetry (2.8). Furthermore, the theory, which is called Cauchy elasticity, lacks the notion of a strain-energy and an energy conservation law is absent.

While it is well known that many of the results of hyperelasticity hold also for Cauchy elasticity, it has been shown by Green and Naghdi (1971) that the latter theory gives rise to the production of infinite work in closed strain cycles. On these and similar grounds, the theory is felt to be unacceptable. However, perhaps this rejection is premature, since recent work by, for example, Carr and Mulhardeen (1980) has shown that in the corresponding linear problem the null solution is Liapunov stable for perturbations of a positive self-adjoint operator. It would be of interest to discover conditions under which the null solution is Liapunov unstable. Even the full set of conditions is unknown which guarantee uniqueness in linearized problems, although partial results are due principally to Murray (1972). In the nonlinear theory, Knops and Payne (1979) have studied conditions under which the solution fails to exist globally.

Apart from Cauchy elasticity, self-adjointness may also be lost by the application of non-dead loads. Several examples may be found in the literature, but we mention especially the study by Carr and Mulhardeen (1979) of Beck's problem in which a beam is loaded at one end by a follower force. Such non-conservative problems represent a challenging area for further study.

Acknowledgements

It is a pleasure to record helpful discussions with Professors C. M. Dafermos, J. E. Marsden, L. E. Payne, M. Slemrod, C. A. Stuart, and L. Tartar during the preparation of this paper. The method of studying existence in the linear problem was suggested by Professor Dafermos, and the author is grateful for his assistance on this and other topics. Part of the work was completed while the author held a Visiting Professorship at the Ecole Polytechnique Fédêrale, Lausanne.

References

Andreussi, F. (1974) *Su certe proprietá di evoluzione della soluzione in Elastodinamica non lineare*, Nota Interna B75–17, Istituto di Mathematica, Universitá di Pisa.

Andreussi, F. (1976) *On Certain Evolution Properties of the Solution in Nonlinear Elastodynamics*, Istituto di Mathematica, Universitá di Pisa.

Antman, S. S. and Liu, T.-P. (1979) Travelling waves in hyperelastic rods, *Q. Appl. Math.* **36,** 377–399.

Arnold, V. (1976) Mathematical developments arising from Hilbert problems, *Proceedings of Symposia in Pure Mathematics* **28,** 59, edited by F. E. Browder, American Mathematical Society, Providence, R.I.

Ball, J. M. (1977) Convexity conditions and existence theorems in nonlinear elasticity, *Arch. Ration. Mech. Analysis* **63,** 337–403.

Ball, J. M. (1979) Finite time blow-up in nonlinear problems, *Proceedings of Symposium on Nonlinear Evolution Equations, University of Wisconsin,* edited by M. G. Crandall, Academic Press, New York.

Ball, J. M. (1980) Strict convexity, strong ellipticity, and regularity in the calculus of variations, *Math. Proc. Camb. Phil. Soc.* **87,** 501–513.

Ball, J. M., Knops, R. J., and Marsden, J. E. (1978) Two examples is nonlinear elasticity, *Journée d'Analyse Nonlinéaire, Proceedings 1977*, 47–48, edited by P. Benilan and J. Robert, Lecture Notes in Mathematics **665,** Springer-Verlag, Berlin.

Banach, S. (1932) *Théorie des Opérations Linéaires*, 44–47, Warsaw (Chelsea Reprint, 1963),

Bloom, F. (1975) On continuous dependence for a class of dynamical problems in linear thermoelasticity, *Z. angew. Math. Phys.* **26,** 569–579.

Bloom, F. (1977) Growth estimates for solutions to initial boundary value problems in viscoelasticity, *J. Math. Analysis Applic.* **59,** 469–487.

Calderer, C. (1980) The dynamical behaviour of nonlinear elastic spherical shells, *J. Elasticity* (submitted).

Carroll, M. M. (1977a) Plane circular shearing of incompressible fluids and solids, *Q. Jl Mech. Appl. Math.* **30,** 223–234.

Carroll, M. M. (1977b) Plane elastic standing waves of finite amplitude, *J. Elasticity* **7,** 425–436.

Carr, J. and Mulhardeen, M. Z. M. (1979) Beck's problem, *SIAM J. Appl. Math.* **37,** 261–262.

Carr, J. and Mulhardeen, M. Z. M. (1980) Stability of nonconservative linear systems, *Functional Differential Equations and Bifurcation, Proceedings*, 45–68, edited by A. F. Ize, Lecture Notes in Mathematics **799,** Springer-Verlag, Berlin.

Coleman, B. D. (1973) The energy criterion for stability in continuum thermodynamics, *Rc. Semin. mat. fis. Milano* **43,** 85–99.

Dafermos, C. M. (1974) Quasilinear hyperbolic systems, *Nonlinear Waves*, 83–100, Cornell University Press.

Dafermos, C. M. (1977) Characteristics in hyperbolic conservation laws: a study of the structure and the asymptotic behaviour of solutions, *Nonlinear Analysis and Mechanics: Heriot-Watt Symposium* **I,** 1–56, edited by R. J. Knops. Research Notes in Mathematics **17,** Pitman, London.

Dafermos, C. M. (1979) The second law of thermodynamics and stability, *Archs Ration. Mech. Analysis* **70,** 167–180.

Dafermos, C. M. (1981) The equations of elasticity are special, *Trends in the Applications of Mathematics to Mechanics*, Vol. III, edited by R. J. Knops, Pitman, London.

DiPerna, R. J. (1979) The structure of solutions to hyperbolic conservation laws, *Nonlinear Analysis and Mechanics: Heriot- Watt Symposium* IV, 1–16, edited by R. J. Knops, Research Notes in Mathematics **39,** Pitman, London.

Ericksen, J. L. (1965) A thermo-kinetic view of elastic stability theory, *Int. J. Solids Structures* **3,** 573–580.

Ericksen, J. L. (1975) Equilibrium of bars, *J. Elasticity* **5,** 191–201.

Ericksen, J. L. (1977) Special topics in elastostatics, *Adv. Appl. Mech.* **17,** 189–244.

Fichera, G. (1965) Linear elliptic differential systems and eigenvalue problems, Lecture Notes in Mathematics **8,** Springer-Verlag, Berlin.

Fichera, G. (1972) Existence theorems in elasticity, *Handbuch der Physik*, **VIa/2,** 347–389, edited by S. Flügge and C. Truesdell, Springer-Verlag, Berlin.

Galdi, G. P. and Rionero, S. (1979) Continuous dependence theorems in linear elasticity on exterior domains, *Int. J. Engng Sci.* **17,** 521–526.

Glimm, J. (1965) Solutions in the large for nonlinear hyperbolic systems of equations, *Com—uns Pure Appl. Math.* **18,** 697–715.

Green, A. E. and Naghdi, P. M. (1971) On thermodynamics, rate of work and energy, *Arch. Ration. Mech. Analysis* **40,** 37–49.

Gurtin, M. E. (1975) Thermodynamics and stability, *Arch. Ration. Mech. Analysis* **59,** 63–96.

Gurtin, M. E. and Spector, S. J. (1979) On stability and uniqueness in finite elasticity, *Arch. Ration. Mech. Analysis* **70,** 153–166.

Hadamard, J. (1923) *Lectures on Cauchy's Problems in Linear Partial Differential Equations*, Yale University Press (Dover Reprint, 1952).

Hill, R. (1957) On uniqueness and stability in the theory of finite elastic strain, *J. Mech. Phys. Solids* **5,** 229–241.

Hill, R. (1970) Constitutive inequalities for isotropic elastic solids under finite strain, *Proc. R. Soc. Lond.* A **314,** 457–472.

Hills, R. N. and Knops, R. J. (1974) Concavity and the evolutionary properties of a class of general materials, *Proc. R. Soc. Edin.* **72A,** 239–243.

Hills, R. N. and Knops, R. J. (1976) Qualitative results for a general class of material behaviour, *SIAM J. Appl. Math.* **30,** 424–451.

Holmes, P. and Marsden, J. E. (1978) Bifurcations of dynamical systems and nonlinear oscillations in engineering systems, *Proceedings of Conference on Nonlinear Partial Differential Equations* 163–206, edited by J. Chadam, Lecture Notes in Mathematics **648,** Springer-Verlag, Berlin.

Hughes, T. J. R., Kato, T., and Marsden, J. E. (1977) Well-posed quasi-linear second-order hyperbolic systems with applications to nonlinear elastodynamics and general relativity, *Arch. Ration. Mech. Analysis* **63,** 273–294.

Hughes, T. J. R. and Marsden, J. E. (1978) Classical elastodynamics as a symmetric hyperbolic system, *J. Elasticity* **8,** 97–110.

James, R. D. (1979) Co-existent phases in the one-dimensional static theory of elastic bars, *Arch. Ration. Mech. Analysis* **72,** 99–140.

James, R. D. (1980) The propagation of phase boundaries in elastic bars, *Arch. Ration. Mech. Analysis* **73,** 125–158.

John, F. (1955) A note on "improper" problems in partial differential equations, *Communs Pure Appl. Math.* **7,** 591–594.

John, F. (1960) Continuous dependence on data for solutions of partial differential equations with a prescribed bound, *Communs Pure Appl. Math.* **13,** 551–585.

John, F. (1976a) Wave propagation, *Trends in the Applications of Pure Mathematics to Mechanics* 175–185, edited by G. Fichera, Pitman, London.

John, F. (1976b) Delayed singularity formation for solutions of nonlinear partial differential equations in higher dimensions, *Proc. Natn. Acad. Sci. U.S.A.* **73** (2) 281–282.

John, F. (1976c) Delayed singularity formation in solutions of nonlinear wave equations in higher dimensions, *Communs Pure Appl. Math.* **29,** 649–681.

John, F. (1979) Blow-up of solutions of nonlinear wave equations in three space dimensions, *Manuscripta Math.* **28,** 235–268.

Kalantarov, V. K. and Ladyzhenskaya, O. A. (1977) Blow-up theorems for quasilinear parabolic and hyperbolic equations, *Zap. Nauk. Sem. Leningrad. Otdel. Mat. Inst. Steklov.* **69,** 72–102.

Kato, T. (1976) *Linear and Quasilinear Equations of Evolution and Hyperbolic Type*, CIME, Bressanone.

Klainerman, S. (1980) Global existence for nonlinear wave equations, *Communs Pure Appl. Math.* **23,** 43–101.

Klainerman, S. (1981) Classical solutions to nonlinear wave equations and nonlinear scattering, *Trends in the Applications of Mathematics to Mechanics*, Vol. III, edited by R. J. Knops, Pitman, London.

Knops, R. J. (1976) Comments on nonlinear elasticity and stability, *Ordinary and Partial Differential Equations*, edited by W. N. Everitt and B. D. Sleeman; Lecture Notes in Mathematics **564,** Springer-Verlag, Berlin.

Knops, R. J. and Payne, L. E. (1970) On uniqueness and continuous dependence in dynamical problems of linear thermoelasticity, *Int. J. Solids Structures* **6,** 1173–1184.

Knops, R. J. and Payne, L. E. (1971) Growth estimates for solutions of evolutionary equations in Hilbert space with applications in elastodynamics, *Arch. Ration. Mech. Analysis* **41,** 363–398.

Knops, R. J. and Payne, L. E. (1978) On potential wells and stability in nonlinear elasticity, *Math. Proc. Camb. Philos. Soc.* **84,** 177–190.

Knops, R. J. and Payne, L. E. (1979) Nonexistence of global solutions in nonlinear Cauchy elastodynamics, *Arch. Ration. Mech. Analysis* **70,** 125–133.

Knops, R. J. and Straughan, B. (1976a) Convergence of solutions of equations of dynamical linear dipolar elasticity to solutions of classical elastodynamics, *Archum. Mech. Stos.* **28,** 431–441.

Knops, R. J. and Straughan, B. (1976b) Continuous dependence theorems in the theory of linear elastic materials with microstructure, *Int. J. Engng Sci.* **14,** 555–565.

Knops, R. J. and Straughan, B. (1976c) Non-existence of global solutions to nonlinear Cauchy problems arising in mechanics, *Trends in the Applications of Pure Mathematics to Mechanics* 187–206, edited by G. Fichera, Pitman, London.

Knops, R. J. and Straughan, B. (1981) Decay to zero for forced systems in continuum mechanics (to appear).

Knops, R. J. and Wilkes, E. W. (1973) Theory of elastic stability, *Handbuch der Physik* **VIa/3,** 125–302, edited by S. Flügge and C. Truesdell, Springer-Verlag, Berlin.

Knops, R. J., Levine, H. A., and Payne, L. E. (1974) Nonexistence, instability and growth theorems for solutions of a class of abstract nonlinear equations with applications to nonlinear elastodynamics. *Arch. Ration. Mech. Analysis* **55,** 52–72.

Koiter, W. T. (1965) The energy criterion of stability for continuous elastic bodies, *Proc. K. Ned. Akad. Wet.* **B68,** 178–202.

Lax, P. (1957) Asymptotic solutions of oscillatory initial value problems, *Duke Math. J.* **24,** 627–646.

Lax, P. (1964) Development of singularities of solutions of nonlinear hyperbolic partial differential equations, *J. Math. Phys.* **5,** 611–613.

Levine, H. A. (1970a) On a theorem of Knops and Payne in dynamical linear thermoelasticity, *Arch. Ration. Mech. Analysis* **38,** 290–319.

Levine, H. A. (1970b) Logarithmic convexity and the Cauchy problem for some abstract second order differential inequalities, *J. Differential Eqns* **8,** 34–55.

Levine, H. A. (1977) An equipartition of energy theorem for weak solutions of evolutionary equations in Hilbert space: The Lagrange identity method, *J. Differential Eqns* **24,** 197–210.

Levine, H. A. and Payne, L. E. (1975a) On the nonexistence of global solutions to some abstract Cauchy problems of standard and non-standard types, *Re. Mat. Stud. Roma* **8** (Ser. VI), 413–428.

Levine, H. A. and Payne, L. E. (1975b) Growth estimates and lower bounds for solutions in nonlinear elastodynamics with indefinite strain energy, *J. Elasticity* **5,** 273–286.

Marsden, J. E. (1973) On global solutions for nonlinear Hamiltonian evolution equations, *Communs Math. Phys.* **30,** 79–81.

Marsden, J. E. and Hughes, T. J. R. (1978) Topics in the mathematical foundations of elasticity, *Nonlinear Analysis and Mechanics: Heriot-Watt Symposium* II, 30–269, edited by R. J. Knops, Research Notes in Mathematics **27,** Pitman, London.

Murray, A. (1972) Uniqueness and continuous dependence for the equations of elastodynamics without strain energy function, *Arch. Ration. Mech. Analysis* **47,** 195–204.

Payne, L. E. (1975) *Improperly Posed Problems in Partial Differential Equations*, Regional Conference Series in Applied Mathematics, Society for Industrial and Applied Mathematics, Philadelphia.

Rivlin, R. S. (1973) Some restrictions on constitutive equations, *Proceedings of the International Symposium on the Foundations of Continuum Thermodynamics*, Bressanone.

Sather, D. and Sather, J. (1968) The Cauchy problem for an elliptic parabolic operator, *Annali Mat. pura appl.* **30,** 197–214.

Sleeman, B. D. (1976) Global estimates for nonlinear wave equations and linear wave equations with nonlinear boundary constraints, *Ordinary and Differential Equations*, 438–451, edited by W. N. Everitt and B. D. Sleeman, Lecture Notes in Mathematics **564,** Springer-Verlag, Berlin.

Slemrod, M. (1979) Damped conservation laws in continuum mechanics, *Nonlinear Analysis and Mechanics: Heriot-Watt Symposium* **III** 135–173, edited by R. J. Knops, Research Notes in Mathematics **39,** Pitman, London.

Wang, C. C. and Truesdell, C. (1973) *Introduction to Rational Elasticity*, Noordhoff, Leyden.

Wheeler, L. T. (1970) Some results in the linear dynamical theory of anisotropic elastic solids, *Q. Appl. Math.* **28,** 91–101.

Wheeler, L. T. (1971) On the uniqueness of solutions to the displacement problem in linear elastodynamics, *J. Elasticity* **1,** 121–124.

Wheeler, L. T. (1976) A uniqueness theorem for the displacement problem in finite elastodynamics, *Arch. Ration. Mech. Analysis* **63,** 183–189.

Wheeler, L. T. and Sternberg, E. (1968) Some theorems in classical elastodynamics, *Arch. Ration. Mech. Analysis* **31,** 51–90.

Wilkes, N. S. (1978) Continuous dependence and instability in viscoelasticity, *J. de Mecanique* **17,** 717–726.

Wilkes, N. S. (1979) Continuous dependence and instability in linear thermo-elasticity, *SIAM J. Appl. Math.* (submitted).

Wilkes, N. S. (1980) On the non-existence of semi-groups for some equations of continuum mechanics, *Proc. Roy. Soc. Edin.* A **86,** 303–306.

Experimental Studies of Plastic Anisotropy in Sheet Metal

by P. B. MELLOR

Postgraduate School of Mechanical and Manufacturing Systems Engineering, University of Bradford

Summary

When the mathematical theory of plasticity is applied to sheet-metal forming it should be based on an experimental yield locus for the material concerned. The paper points out the difficulties of obtaining expanding yield loci for sheet metals and suggests that a sound understanding of forming processes is being held back by a lack of such experimental data. A hole-expanding test is put forward as a possible indirect method of checking the constitutive relations of cubic metals when planar anisotropy is small.

Predictions of limiting drawing ratio in cup drawing are compared with experimental results, and it is suggested that lack of correlation between theory and experiment may be caused by inadequate modelling of the process rather than by inappropriate constitutive equations. In stretch forming the exact shape of the yield locus appears to be very important in controlling the limit strains.

1. Introduction

In recent years great improvements have been made in the production of sheet metals, especially in the control of cleanliness, grain size, crystallographic texture, surface finish, and gauge. Similarly, improvements have been made in tool steels, in the design and control of presses, in feeding of coiled strip, and in lubrication. These very improvements highlight the fact that only limited progress has been made towards understanding the basic formability process, and where advances have been made they have usually been diagnostic, not predictive in nature. Keeler (1978), for example, has pointed out that our basic understanding of metal-forming processes has advanced little since the subject was so ably discussed by Swift (1952a). The sheet-metal forming process, in particular, still depends almost entirely on the skills of talented craftsmen, and this at a time when the artisan system is breaking down, when there is a growing demand for increased part complexity, and when there is a move towards thinner-gauge higher-strength materials in order to conserve natural resources.

If sheet-metal forming processes are to be controlled efficiently it will be necessary to predict the evolving strain distribution, which means that the constitutive equations of the sheet metals must be based on experimentally determined expanding yield loci to large plastic strains. Surprisingly, relatively little attention has been paid to this vital aspect of sheet-metal forming. There are of course certain inherent difficulties, not least being that of adequate testing in the thickness direction. The classical experimental work on yielding, e.g. by Taylor and Quinney (1931), has normally been carried out on thin-walled tubes where an initial yield point is defined by the backward extrapolation of the work-hardening characteristics to zero plastic strain. The use of thin-walled tubes is restricted to small plastic strains (but see Holloway and Shelton, 1979) because of buckling failure and therefore even if it were practicable to wrap the sheet into a tube following the method of Babel *et al.* (1966) the results would be of very limited value to sheet-metal forming.

Lee and Backofen (1966) have suggested simple loading paths for determining the main points on an experimental yield locus of a sheet metal (Fig. 1.1). For certain loading paths this may involve glueing several plane specimens together in order to obtain a suitable thickness for testing. Where the paths are compressive it is impossible to obtain an accurate stress–strain curve at low plastic strains and therefore any backward extrapolation to zero plastic strain will lead to uncertain and therefore unacceptable values of yield stress. Several ways have been suggested for overcoming this. Lee and Backofen plotted the yield locus points at a constant largest principal strain value for each loading path, but, as they realized,

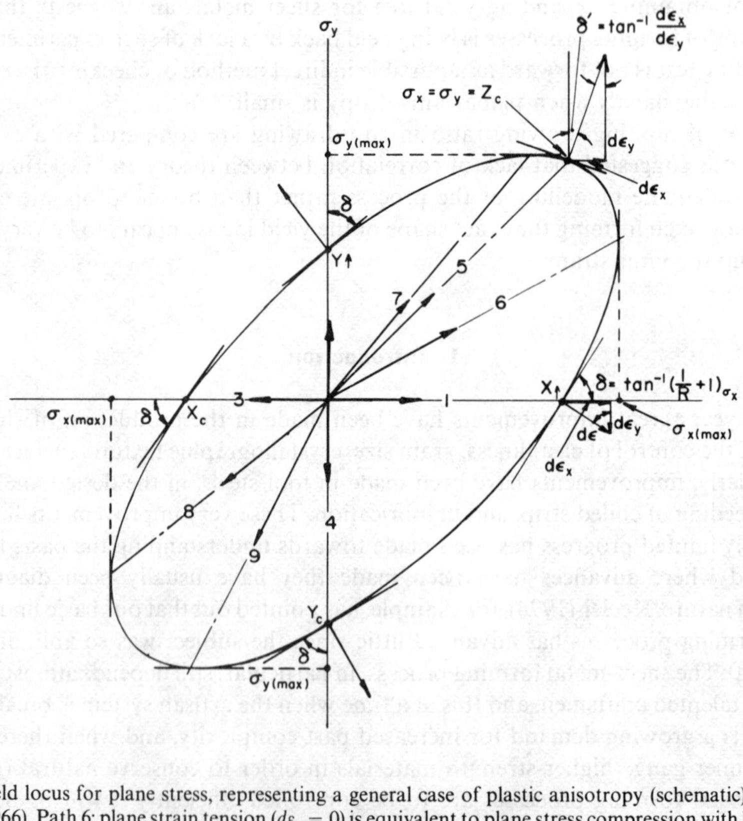

FIG. 1.1. Yield locus for plane stress, representing a general case of plastic anisotropy (schematic), after Lee and Backofen (1966). Path 6: plane strain tension ($d\varepsilon_y = 0$) is equivalent to plane stress compression with loading normal to plane sheet and $d\varepsilon_y = 0$. Path 7: similar case with $d\varepsilon_x = 0$. Path 8: plane strain compression $d\varepsilon_y = 0$, loading in x-direction. Path 9: plane strain compression $d\varepsilon_x = 0$, loading in y-direction.

this also leads to errors. Tozawa (1978) appears to have made the most detailed experimental investigations of expanding yield loci for sheet metals. However, each yield locus was plotted for an effective strain based on isotropic behaviour, and it is therefore difficult to interpret the significance of his results. Dillamore *et al.* (1971) have given a sound commentary on the experimental determination of yield loci based on their work on thin-walled tubes of anisotropic metals. They studied three metals: mild steel, representing the body-centred cubic structure; austenitic stainless steel, which is face-centred cubic; and magnesium, which is close-packed hexagonal. The cubic metals showed constant strain ratios for constant loading paths and, to a reasonable approximation, the hardening was found to be a function of the plastic work. With magnesium, however, the strain path varied for constant loading paths, and work hardening did not obey a simple rule. They concluded that for cubic metals experimental yield loci could be constructed by plotting each locus at the same level of plastic work; but that yield loci obtained for metals in which twinning is an important deformation mechanism can only be approximate.

The most generally used constitutive relations for the plastic yielding and deformation of anisotropic metals at a macroscopic level were proposed by Hill (1948). The theory was the simplest conceivable and the following conditions are implicit in it: the state of anisotropy must possess three mutually orthogonal planes of symmetry at every point; hydrostatic stress is without influence on yielding; there is no Bauschinger effect; and the yield strength is insensitive to the sign of stress. The last condition is perhaps the most serious for hexagonal close-packed materials in which deformation by twinning responds to the sign of the stress. Hill (1950) emphasized that *where the theory is applicable* measurements of the strain ratios in tension specimens cut parallel to and at right angles to an axis of anisotropy provide an indirect method for determining the ratios of the three tensile yield stresses. Unfortunately many research workers accepted the proposed method for determining the three tensile yield stresses without checking whether or not the theory was applicable to their particular material. When applied in this way the theory appeared to give the correct trend for the prediction of the limiting drawing ratio in the deep drawing of a flat-headed cup, but not for the limit strains in stretch forming. Since these two processes are essentially controlled by different parts of the yield locus, it is of interest to review the effect of the shape of the yield locus on the ensuing plastic deformation. In the following, experimental results for deep drawing and stretch forming will be discussed in relation to the macroscopic theory of plasticity.

2. The *R*-value

Hill's anisotropic yield criterion of 1948 is most widely known in the simplified form for transversely isotropic materials

$$(\sigma_1 + \sigma_2)^2 + (1 + 2R)(\sigma_1 - \sigma_2)^2 = 2(1 + R)\sigma_u^{\,2}, \tag{2.1}$$

where σ_1 and σ_2 are the principal stresses in the plane of the sheet, σ_u is the yield stress in uniaxial tension, and R is the ratio of the plastic strain increment in the transverse direction to the plastic strain increment in the thickness direction for uniaxial tension. The effect of the R-value on the shape of the yield locus is shown in Fig. 2.1. Experimental determination of R is, in fact, most commonly based on plastic strains and it is implicit in such work that the ratio remains constant with increasing plastic strain. There appears to be a paucity of information

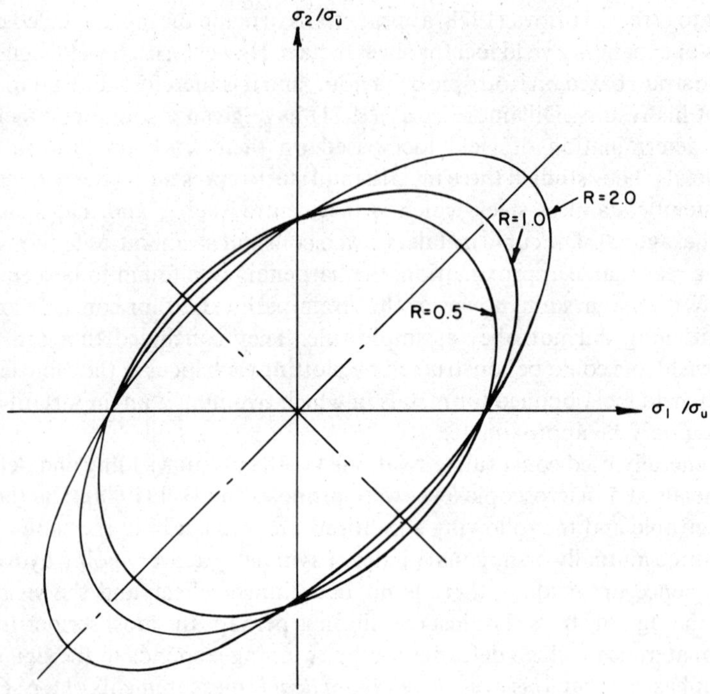

FIG. 2.1. Yield loci based on (2.1).

on the strain increment ratio, especially at initial yielding. Available results are for hexagonal close-packed materials which are of interest mainly because of their increased strength under biaxial tension. Babel *et al.* (1966) pressurized thin-walled cylinders of titanium alloys to burst and by comparing their stress results with Hill's theory concluded that the *R*-value of the material increased with strain. This must of course be treated with caution since the anisotropic yield criterion is not appropriate to hexagonal metals. Avery *et al.* (1965) measured *R*-values in magnesium alloy sheets and found that in these hexagonal materials the *R*-values changed with increasing plastic strain and that this change was still rapid even to total axial strains of 0.15. From measurements of width and thickness after unloading a tensile specimen they obtained values of $R_s = \varepsilon_w^p/\varepsilon_t^p$. Since these values varied with strain they also derived the incremental *R*-values from the following equation:

$$R = \frac{d\varepsilon_w^p}{d\varepsilon_t^p} = R_s + \varepsilon_t^p \frac{dR_s}{d\varepsilon_t^p}, \tag{2.2}$$

where ε_w^p and ε_t^p are the plastic width strain and plastic thickness strain respectively. $R > R_s$ and it was found that *R* increased more rapidly with strain so that in a plot of the two parameters against longitudinal strain the two curves diverged. These results indicate a marked change in the gradient of the yield locus at uniaxial tension with increasing plastic strain. Taghvaipour *et al.* (1972) have reported measurements of width and thickness for tensile specimens of titanium 115. However, since the results were reported in terms of total strain they need further manipulation to be of value to the present discussion. This is possible by re-interpreting the results of Taghvaipour (1970). Figure 2.2 gives a plot of total width strain against total thickness strain for uniaxial straining of titanium 115. The strains were measured under load by high elongation electric resistance strain gauges on a tensile specimen of initial width 0.5 in., with a longitudinal parallel length of 2.5 in. The initial thickness of the specimen was chosen to be 0.064 in., which was the minimum thickness that could conveniently be used with the currently available strain gauges. The gauge length was 0.25 in. for longitudinal and width measurements and 0.031 in. for thickness measurements. The tensile specimen was prepared extremely carefully to ensure a parallel gauge length and sides that were orthogonal to the plane of the sheet. Longitudinal and width gauges were bonded to both surfaces of the specimen to check that bending did not occur. It is seen from Fig. 2.2 that, as expected, the ratio of width to thickness strain is constant in the elastic region but that at yielding there is a rapid change until, after a certain strain, the ratio again becomes constant. Larson (1964) gives a schematic diagram for variation of width strain with longitudinal strain for a different titanium alloy, but he indicates a much sharper "knee" between the two linear portions of the graph. The ratios of the total width strain to total thickness strain are shown in Fig. 2.3 for two strain ranges.

The ratios of the plastic strain were determined by subtracting the elastic components of strain using the measured elastic constants of Table 2.1 and an experimental stress–strain

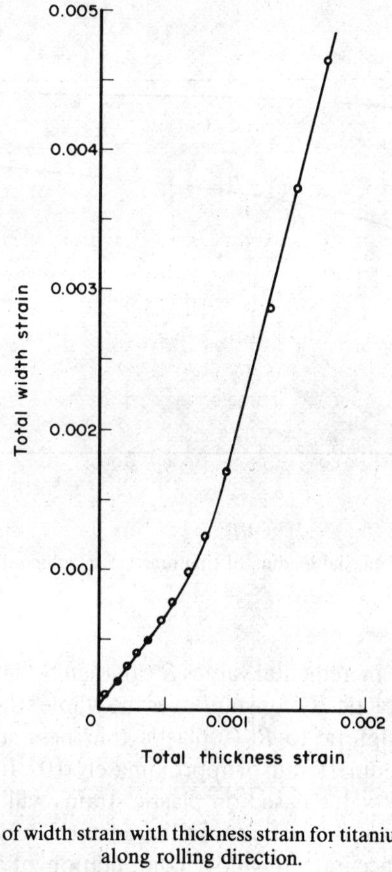

FIG. 2.2. Experimental variation of width strain with thickness strain for titanium 115. Loading in uniaxial tension along rolling direction.

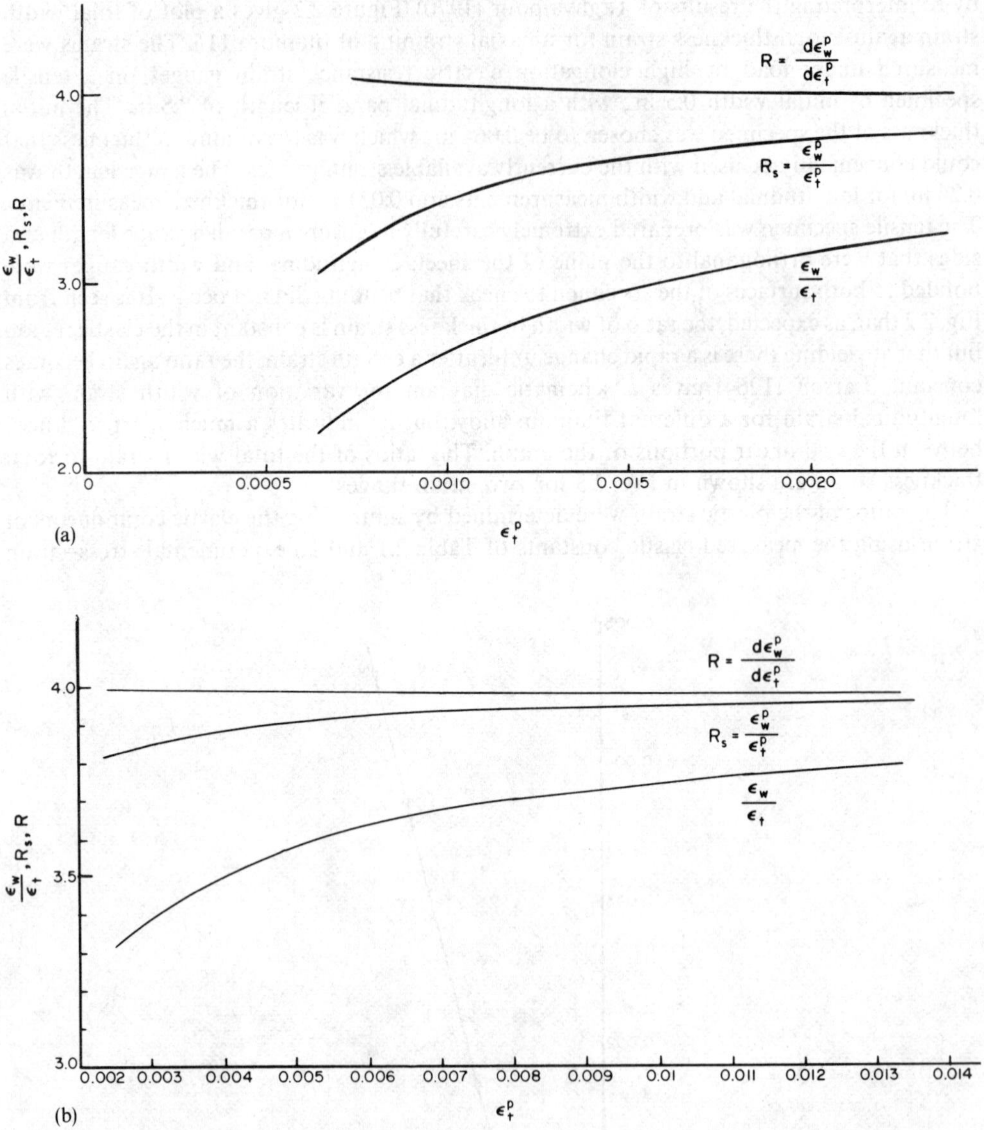

FIG. 2.3. Uniaxial loading of titanium 115 in rolling direction.

curve for simple tension. The incremental values R can then be determined numerically from (2.2). For this material R is practically constant over the whole strain range while R_s increases with strain to become asymptotic to R. A plastic thickness strain of 0.013 (Fig. 2.3(b)) corresponds to a total longitudinal strain of approximately 0.07. It seems clear, therefore, that above this latter strain the R-value based on plastic strains will be little different from the more fundamental value based on plastic strain increment. Unfortunately the experimental results are not sufficiently accurate to allow computation of the ratio of plastic strain

TABLE 2.1. Elastic constants for titanium 115

Inclination to rolling direction	Young's modulus (lbf/in^2)	Poisson's ratio	
0°	$E_1 = 12.8 \times 10^6$	$\nu_{12} = 0.24$	$\nu_{13} = 0.21$
90°	$E_2 = 16.0 \times 10^6$	$\nu_{21} = 0.39$	$\nu_{23} = 0.27$

increment closer to initial yielding. The curve for R (Fig. 2.3(a)) rises slightly with decreasing strain, but it would be unwise to extrapolate this curve to zero strain.

When measurements have been made of the variation of R_s with strain they have usually been restricted to longitudinal strains greater than 0.05. At this level the values of R_s are constant for the more common materials such as steel, aluminium, and 70/30 brass, as well as for titanium 115 and zinc (Bramley and Mellor, 1968), although it should be noted that Naziri and Pearce (1968) measured the R_s values for titanium 115 at three different strain levels and observed large variations with strain. In the following discussion the values of R and R_s will be assumed identical. This is reasonable for the range of plastic strain and materials that will be studied.

3. Work-hardening Characteristics for Sheet Metals

Because of the difficulty of obtaining yield loci for sheet metals, indirect methods have been adopted for checking hypotheses of yielding and the equations of plastic flow. Following the early work of Jackson *et al.* (1948) it has become common to obtain work-hardening characteristics over simple strain paths and to attempt correlation between them by means of anisotropic theory using an effective strain based on the equivalence of plastic work. As discussed earlier, this latter assumption will be more correct for cubic metals than for hexagonal metals. The three tests usually adopted are uniaxial tension, balanced biaxial tension, and plane strain compression.

Of the three tests balanced biaxial tension using the circular diaphragm method is the most indirect, the most difficult to control at constant strain-rate, and usually an average value is adopted for the properties in the plane of the sheet. With this latter approximation the relations between through thickness strength and uniaxial properties, based on Hill's yield criterion of 1948, are given by

$$\frac{\sigma_b}{\sigma_u} = \left(\frac{1+R}{2}\right)^{1/2}, \quad \frac{\varepsilon_t}{\varepsilon_u} = \left(\frac{2}{1+R}\right)^{1/2}, \tag{3.1}$$

where σ_b is the stress at the pole of the diaphragm and ε_t the corresponding thickness strain. σ_u is the average stress in uniaxial tension, ε_u the corresponding longitudinal strain, and R the average of the R-values determined at 0°, 45°, and 90° to the direction of rolling.

Friction is the most likely source of error in the plane strain compression test, and the most reliable results are obtained by using incremental loading and by placing new PTFE film between the dies and the sheet metal on each loading. Provided loading is along the anisotropic axes, planar anisotropy presents no difficulties when relating plane strain compression to uniaxial tension. Adopting Hill's yield criterion of 1948 and the equivalence of plastic work, we have

$$\left.\begin{aligned}\frac{p_0}{\sigma_0} &= \left[\frac{(1+R_0)\left(R_{90}+\dfrac{R_{90}}{R_0}\right)}{1+R_0+R_{90}}\right]^{1/2}, \\ \frac{\varepsilon_{z_0}}{\varepsilon_0} &= \left[\frac{1+R_0+R_{90}}{(1+R_0)\left(R_{90}+\dfrac{R_{90}}{R_0}\right)}\right]^{1/2},\end{aligned}\right\} \tag{3.2}$$

where suffix 0 indicates the rolling direction and suffix 90 indicates the transverse direction in the plane of the sheet. The plane strain dies are here aligned along the rolling direction of the sheet. p_0 is the normal stress, ε_{z_0} the thickness strain in plane strain, and σ_0 and ε_0 are the longitudinal stress and strain in simple tension.

Most of the early experimental investigations were concerned with low carbon steel, which is still the most important material today in terms of tonnage used. The work by Jackson *et al.* (1948) on circular diaphragms was later repeated for four aluminium killed steels by Bramley and Mellor (1966) and the results for one of the steels is shown in Fig. 3.1(a). Taghvaipour (1970) tested the same batch of steel sheet in plane strain compression and his results are shown in Fig. 3.1(b). In these diagrams the predicted results for balanced biaxial tension and plane strain compression are based on (3.1) and (3.2), respectively. The two predicted curves

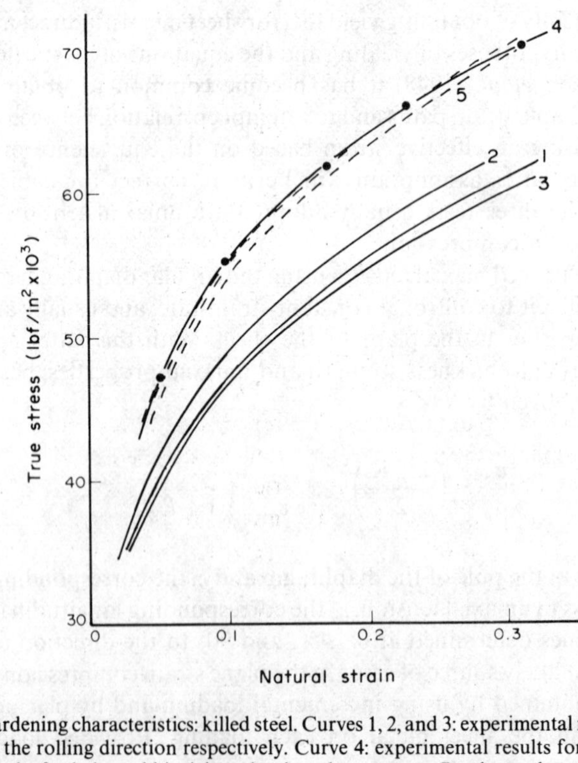

FIG. 3.1(a). Work-hardening characteristics: killed steel. Curves 1, 2, and 3: experimental results for uniaxial tension at 0°, 45°, and 90° to the rolling direction respectively. Curve 4: experimental results for balanced biaxial tension. Curve 5: predicted results for balanced biaxial tension based on average R-value and corresponding uniaxial curve.

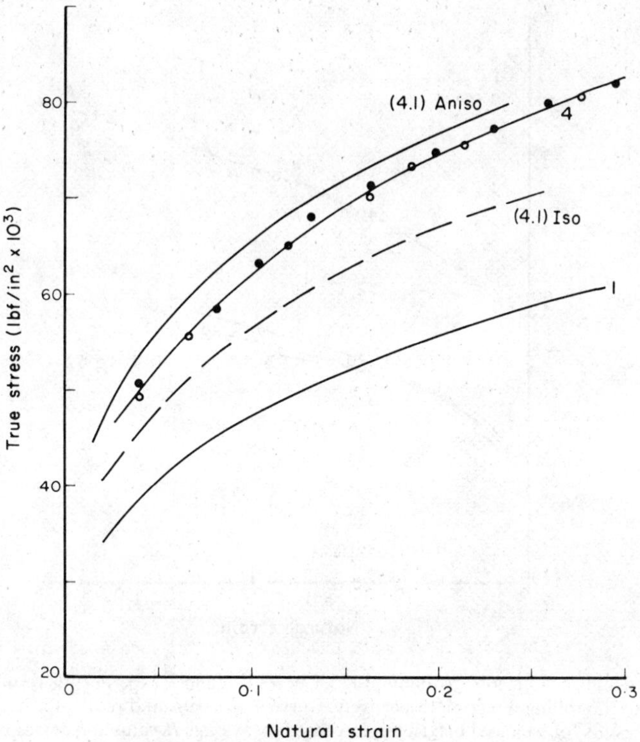

FIG. 3.1(b). Work-hardening characteristics: killed steel. Curve 1: experimental results for uniaxial tension in rolling direction. Curve 4: experimental results for plane strain compression (zero strain in rolling direction). Curve (4.1) anisotropic: plane strain compression, predicted by anisotropic theory. Curve (4.1) isotropic: plane strain compression, predicted on basis of isotropic theory.

marked 5 in Fig. 3.1(a) arise because the average *R*-value can be identified with two orientations, each having a slightly different stress level. The correlations between theory and experiment are considered reasonable for this cubic material, and it should be noted that the *R*-values remain constant with increasing plastic strain and that the work-hardening index *n* does not vary greatly with orientation. In this latter respect it is interesting to make a comparison with the behaviour of a close-packed hexagonal metal, titanium 115. The results in Fig. 3.2 for the diaphragm test were obtained by Bramley and Mellor (1968) and show good correlation with theory despite the large difference in the slopes of the uniaxial curves. However, if we consider results for plane strain compression (Fig. 3.3), where planar anisotropy can be accommodated within the restrictions of the theory, the individual uniaxial curves assume a greater importance. The experimental results for plane strain compression are similar for the two orientations, but the chosen methods of predicting the plane strain compression curves lead to a large discrepancy between theory and experiment in Fig. 3.3(b) (Taghvaipour and Mellor, 1971). In terms of fundamental research the correlation between theory and experiment must be regarded as poor, but in practical engineering terms such predictions can be regarded sufficiently accurate and extremely useful for predicting loads required in deformation processing. Rogers and Roberts (1968) have interpreted the anisotropic behaviour of this particular titanium sheet in crystallographic terms.

The above method of predicting through thickness strength in balanced biaxial tension

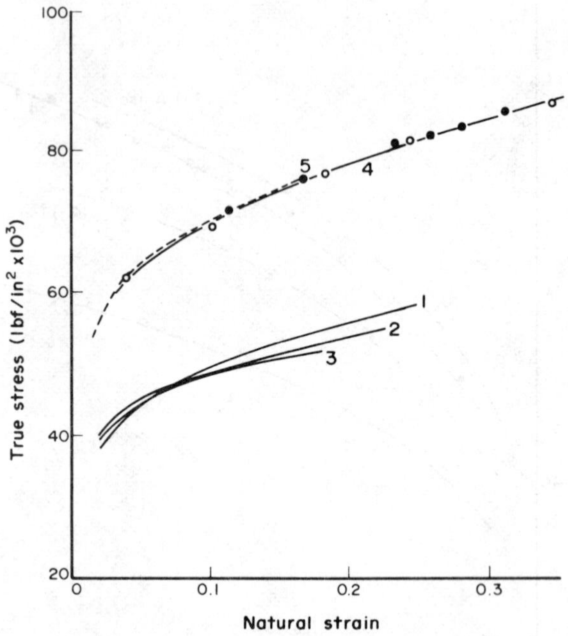

FIG. 3.2. Work-hardening characteristics: titanium 115. Curves 1, 2, and 3: experimental results for uniaxial tension at 0°, 45°, and 90° to the rolling direction respectively. Curve 4: experimental results for balanced biaxial tension. Curve 5: predicted results for balanced biaxial tension based on average R-value and corresponding uniaxial curve.

was adopted by Pearce (1968), who found from his experimental investigations that the theory was not appropriate for materials having $R < 1$. The case of commercially pure annealed aluminium is interesting since other workers including Mellor (1956), and Venter *et al.* (1971), have found that the work-hardening characteristics for balanced biaxial tension and unaxial tension are almost coincident. More recent data for aluminium, where the R-value was 0.635 (Ilahi, 1977), is shown in Fig. 3.4(a) and confirms this view. Curve 1 is the average of the uniaxial characteristics and Curve 2 is the balanced biaxial characteristic. It is seen that application of (3.1) predicts a much lower balanced biaxial characteristic, Curve 4. However, results obtained for plane strain compression of aluminium by Taghvaipour and Mellor (1971) (Fig. 3.4(b)) show reasonable correlation with theory based on Hill's yield criterion (2.1). Again, Wagoner and Wang (1979) found that Hill's theory gave good correlation with experimental results for the case of in-plane deformation of 2036–T4 aluminium alloy between the strain states of plane strain tension and uniaxial tension. The average R-value for this material was 0.707. Isotropic theory and anisotropic theory based on (2.1) both assume that the yield locus can be specified from a knowledge of material behaviour under uniaxial stress. It is clear that a theoretical yield locus which is based on measured yield stress at uniaxial tension and at balanced biaxial tension, and also takes into account the correct slope of the yield locus at uniaxial stress, will be a distinct improvement. Hill (1979), and prior to this date in private communications, put forward the following yield function:

$$|\sigma_1 + \sigma_2|^m + (1+2R)|\sigma_1 - \sigma_2|^m = 2(1+R)\sigma_u{}^m, \tag{3.3}$$

where m is a new index which must be determined experimentally. The yield locus described by (3.3) is convex for $m \geqslant 1$, for values of $m < 2$ the locus is elongated in the direction of balanced biaxial tension and when $m = 2$ the equation reduces to Hill's original yield function for anisotropic behaviour (2.1). Figure 3.5 shows the variation of the ratio of balanced biaxial stress to uniaxial stress σ_b/σ_u and the ratio of the stress in plane strain to uniaxial stress σ_{ps}/σ_u. Clearly the stress in plane strain is always greater than the uniaxial stress, but theoretically it is possible for the balanced biaxial stress to be less than the uniaxial stress. Whether or not the latter occurs in practice is a matter for experiment.

From the yield function of (3.3), its associated flow rule and an assumption of the equivalence of plastic work, the relations between stress and strain in balanced biaxial and uniaxial tension can be written as

$$\frac{\sigma_b}{\sigma_u} = \frac{[2(1+R)]^{\frac{1}{m}}}{2} \quad \text{and} \quad \frac{\varepsilon_t}{\varepsilon_u} = \frac{2}{[2(1+R)]^{\frac{1}{m}}}. \tag{3.4}$$

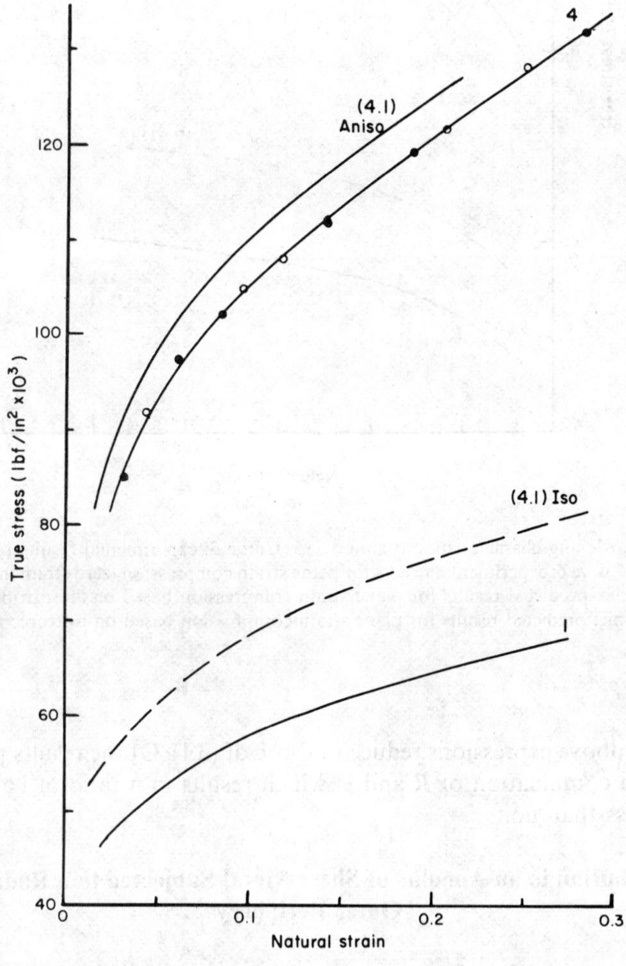

FIG. 3.3(a). Work-hardening characteristics: titanium 115. Curve 1: experimental results for uniaxial tension in rolling direction. Curve 4: experimental results for plane strain compression (zero strain in rolling direction). Curve (4.1) anisotropic: predicted results for plane strain compression based on anisotropic theory. Curve (4.1) isotropic: predicted results for plane strain compression based on isotropic theory.

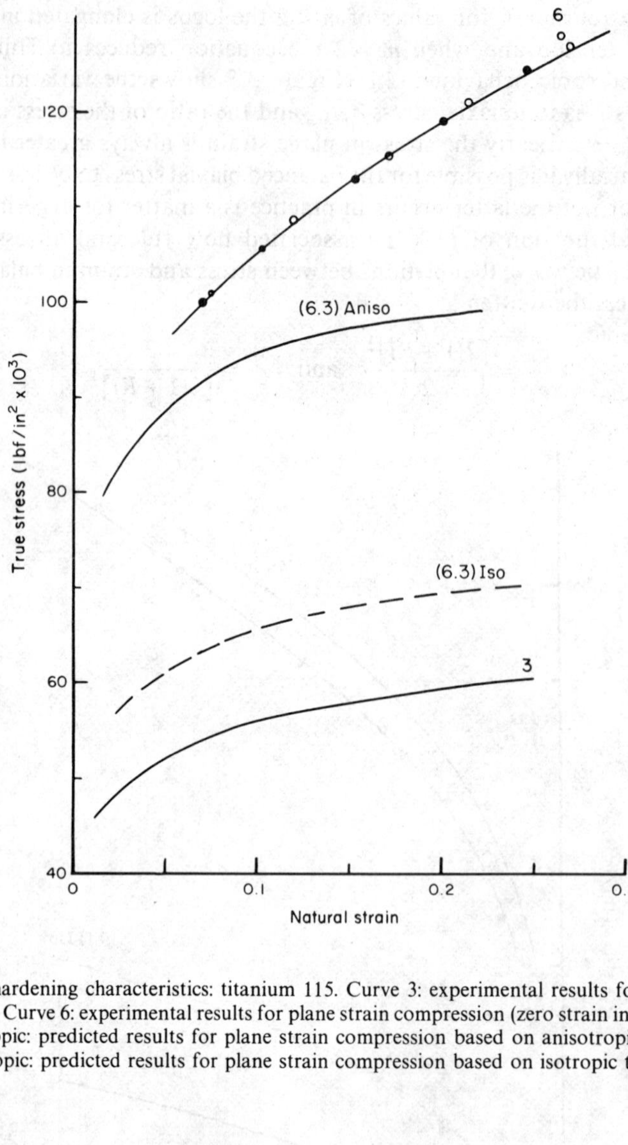

FIG. 3.3(b). Work-hardening characteristics: titanium 115. Curve 3: experimental results for uniaxial tension in transverse direction. Curve 6: experimental results for plane strain compression (zero strain in transverse direction). Curve (6.3) anisotropic: predicted results for plane strain compression based on anisotropic theory. Curve (6.3) isotropic: predicted results for plane strain compression based on isotropic theory.

When $m = 2$ the above expressions reduce to those of (3.1). Of the results published to date none has given a combination of R and m which results in a ratio of balanced biaxial to uniaxial stress less than unity.

4. Strain Distribution in an Annulus of Sheet Metal Subjected to a Radial Tension at Its Outer Periphery

If a yield function is to be of practical importance in sheet-metal forming the flow rule associated with it by the standard normality hypothesis should be capable of being readily manipulated. The yield function of (3.3) meets this requirement and has been used by Parmar and Mellor (1978a) to predict the strain distributions around a small hole in a sheet metal subjected to uniform radial tension at a great distance away from the hole. The stress system ranges from simple circumferential tension at the hole, through plane strain tension, to balanced biaxial tension. The theory was checked over the strain range in a series of experiments using the well-known stretch-forming technique of Marciniak and Kuczyński (1967). Typical experimentally determined strain distributions are shown in Fig. 4.1 for killed steel and annealed commercial purity aluminium. The ratio of current bore diameter a to initial bore diameter a_0 is 1.2 in each case. The circumferential and thickness strains were measured and the radial strains derived from an assumption of volume constancy. Strain measurements were made at 0°, 45°, and 90° to the direction of rolling, and it is important to note any variation of strain with orientation. For the sake of clarity the strain distributions presented are based on measurements taken in the rolling direction of the sheet, and values of strain at 45° and 90° are only included when they are significantly different from the above. It was found that the bore remained circular, which means that the circumferential strains are little affected by any planar variation of material properties. However, planar anisotropy is manifest in the measured thickness strains in the region of the bore and reflected in the derived radial strains. There is very little variation in any of the strain components with orientation as we move away from the bore. The significant material properties are given in Table 4.1.

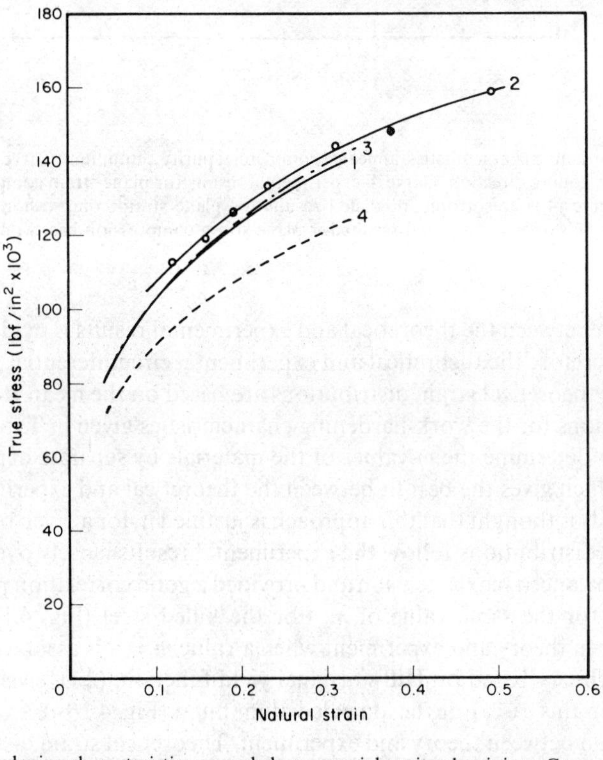

FIG. 3.4(a). Work-hardening characteristics: annealed commercial purity aluminium. Curve 1: experimental results for uniaxial tension, average results in plane of sheet. Curve 2: experimental results for balanced biaxial tension. Curve 3: predicted curve for balanced biaxial tension, $m = 1.7$ in (3.4). Curve 4: predicted curve for balanced biaxial tension, $m = 2$ in (3.4), which then reduces to (3.1).

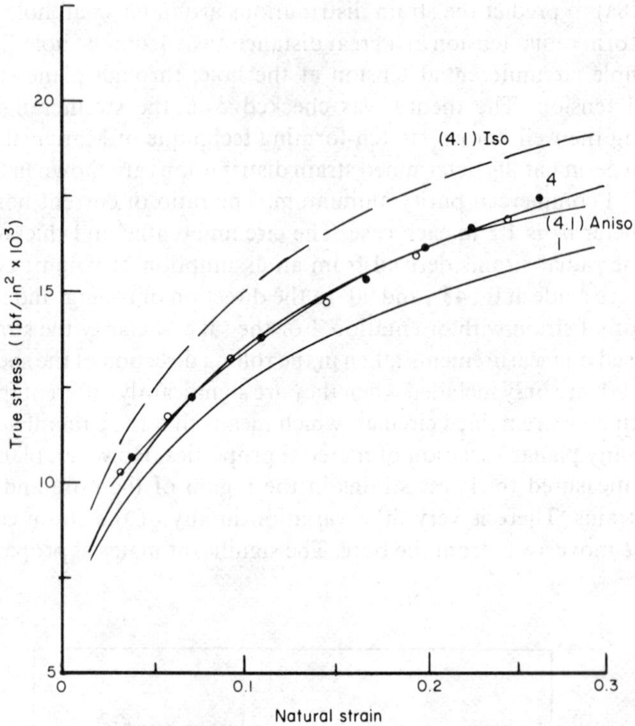

FIG. 3.4(b). Work-hardening characteristics: annealed commercial purity aluminium. Curve 1: experimental results for uniaxial tension in rolling direction. Curve 4: experimental results for plane strain compression (zero strain in rolling direction). Curve (4.1) anisotropic: predicted results for plane strain compression based on anisotropic theory. Curve (4.1) isotropic: predicted results for plane strain compression based on isotropic theory.

The correlation between the theoretical and experimental results is made at the same bore expansion and therefore the theoretical and experimental circumferential strains at the bore are identical. The theoretical strain distributions are based on the mean R-values and on the empirical expressions for the work-hardening characteristics given in Table 4.1. No attempt has been made to determine the m-values of the materials by separate experiments. Instead the value of m, which gives the best fit between the theoretical and experimental results, has been determined. It is thought that this approach is justified if, for a given bore expansion, the theoretical strain distributions follow the experimental results closely over the strain range from uniaxial to balanced biaxial tension and provided a good correlation persists at different bore expansions for the same value of m. For the killed steel (Fig. 4.1(a)) there is good correlation between theory and experiment when a value $m = 2$ is assumed in the theory. In other words the theory based on Hill's original yield function, (2.1), gives good correlation with experiment in this case. For the annealed aluminium (Fig. 4.1(b)), a value $m = 1.7$ gives the best correlation between theory and experiment. Theoretical strain distributions are also given assuming a value $m = 2$. This demonstrates very clearly that the original anisotropic theory is inappropriate for this aluminium sheet especially in the strain range between plane strain tension and balanced biaxial tension. Evidently the effect of the R-value, which defines

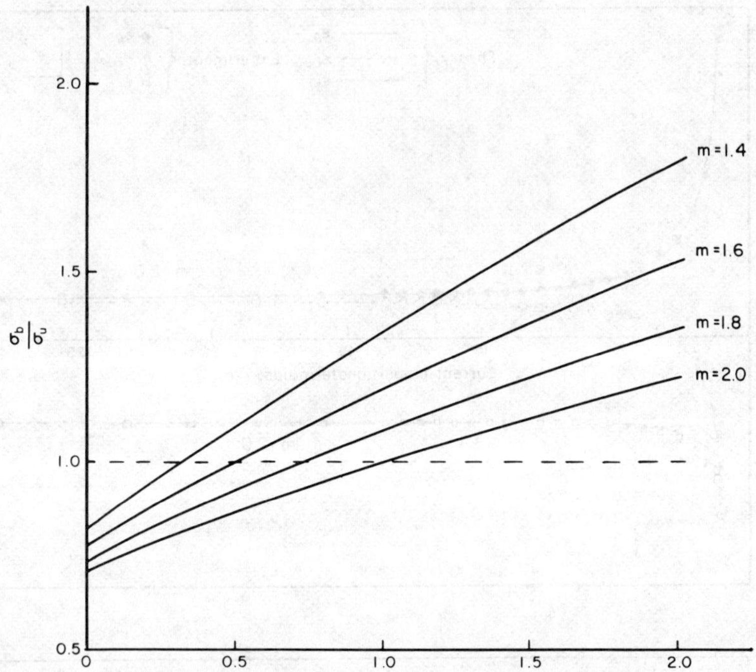

(a)

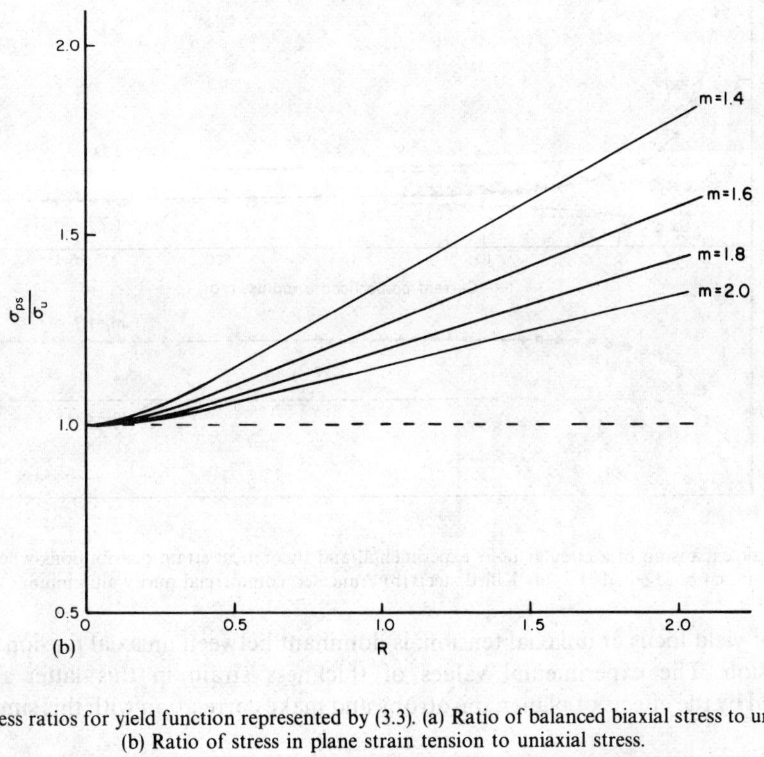

(b)

FIG. 3.5. Stress ratios for yield function represented by (3.3). (a) Ratio of balanced biaxial stress to uniaxial stress. (b) Ratio of stress in plane strain tension to uniaxial stress.

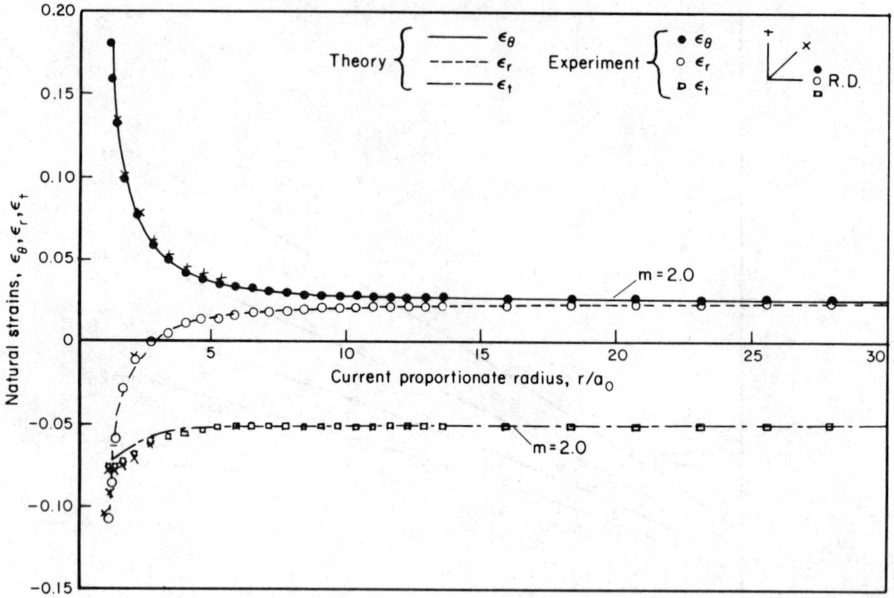

(a)

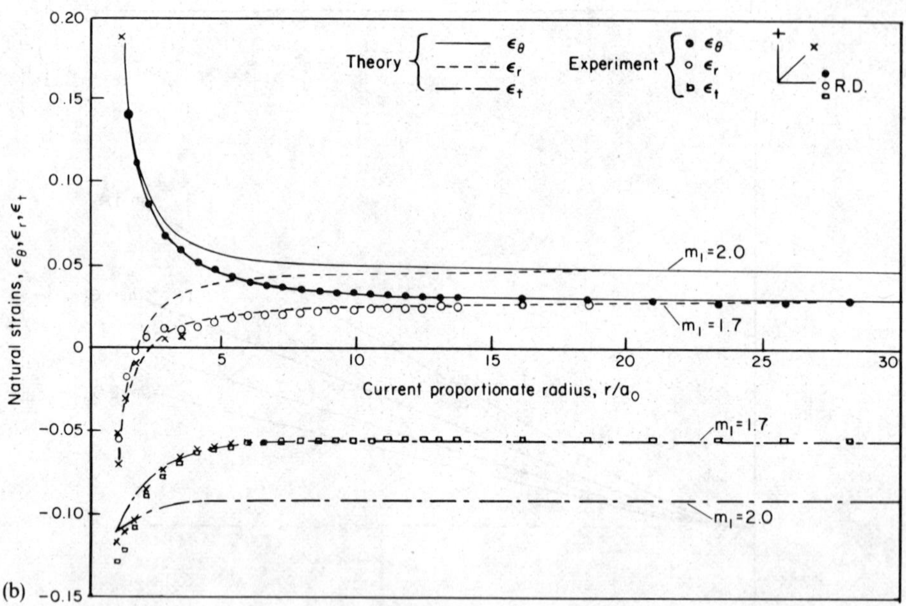

(b)

FIG. 4.1. Plastic expansion of a circular hole: experimental and theoretical strain distributions when current to initial bore equals 1.2. (a) Killed steel. (b) Annealed commercial purity aluminium.

the slope of yield locus at uniaxial tension, is dominant between uniaxial tension and plane strain tension. The experimental values of thickness strain in this latter region are complicated by the effects of planar anisotropy and make correlation with this simple theory difficult.

TABLE 4.1. Material properties in hole expansion tests

	R-values	
	A–K steel	Annealed aluminium
R_0	1.640	0.620
R_{45}	1.235	0.581
R_{90}	1.885	0.956
Average *R*	1.500	0.635
Mean work-hardening characteristics:		
A–K steel	$\sigma = 71\,768\,\varepsilon^{0.222}$ lbf/in^2	
Annealed aluminium	$\sigma = 19\,166\,\varepsilon^{0.269}$ lbf/in^2	

So far as is known, Lee and Kobayashi (1975) are the only workers who have made a theoretical study of bore expanding and flange drawing taking the effects of planar anisotropy into account. Their analysis was based on Hill's yield function of 1948, and using a matrix method of solution they showed that planar anisotropy had little effect on the shape of the bore but greatly affected the circumferential distribution of the thickness strains.

The value of $m = 1.7$ obtained from Fig. 4.1(b) was used to predict the work-hardening characteristic for the same aluminium sheet of Fig. 3.4(a). The results of the plane strain compression tests given in Fig. 3.4(b) are for a different batch of aluminium where the *m*-value is unknown, and therefore the results in Fig. 3.4(a) and (b) cannot be compared. There is obviously a need for much more experimental work to be directed towards obtaining yielding data. However, it is felt that the hole-expanding test is a sensitive method of checking constitutive equations, including the effects of planar anisotropy.

5. Stretch-forming

Forming limit diagrams are empirically constructed to describe the combinations of major and minor principal surface strains at which a highly localized zone of thinning or necking becomes visible in the surface of the sheet metals. The diagrams arose from the work of Keeler and Backofen (1963) and were given the form in which they are presently used by Keeler (1965). Knowledge of the peak strain limit described by the forming limit diagram is insufficient to completely characterize the formability of a sheet metal. Equally important is the ability of a sheet metal to distribute the strain as uniformly as possible, and this depends on the bulk material characteristics, the characteristics at the sheet–tooling interface, the tooling design, and press parameters. At the present time there is no generally accepted explanation of the deformation leading up to fracture and there is no theoretical analysis which can be used with such confidence that experimental determination of the forming limits can be dispensed with.

The most discussed theoretical analysis for the prediction of limit strain is the formulation of Marciniak and Kuczyński (1967) for in-plane stretching which applies when both surface strains are positive. They assume an initial inhomogeneity in the original sheet which develops with increasing plastic strain into a localized neck running in a direction perpendicular to the larger principal stress. Sowerby and Duncan (1971) have given a particularly clear and detailed explanation of the mechanics of deformation in this analysis and the results, for a material with properties based on (2.1), which are shown in Fig. 5.1(a), are quoted from their work. Since the initial inhomogeneity, caused by geometrical or metallurgical effects, is assumed to grow continuously with deformation, the limit strain under plane strain tension is less than the instability strain at this point predicted by the well-known analysis of Hill (1952). In balanced biaxial tension the straining is greatly prolonged by the need for the deformation to change to plane strain tension before a localized neck can form. According to the analysis the limit strains under balanced biaxial tension will be greatly influenced by the shape of the yield locus, and in the case of Fig. 5.1(a) this means that the limit strains will be very dependent on the *R*-values of the material. In particular it is seen from Fig. 5.1(a) that the predicted limit strain at balanced biaxial tension increases rapidly as the *R*-value decreases. Such a high dependence on *R*-value has not been observed experimentally. Woodthorpe and Pearce (1969), for example, have established for two different steel sheets that the effect of *R*-value on the limit strains is very small. This, however, does not necessarily mean that the Marciniak–Kuczyński analysis is incorrect. It is much more likely that the low *R*-value material tested had a yield locus whose shape could not be described solely in terms of *R*. As already discussed, several materials with $R < 1$ have yield loci that can be better described by Hill's yield function in (3.3). Parmar and Mellor (1978b) have applied this yield function in the analysis of a material with $R = 0.5$. The predicted limit strains are shown in Fig. 5.1(b). The value $m = 2$ gives the original anisotropic yield locus. The effect of reducing m below this value is to dramatically reduce the predicted limit strain in the region near to balanced biaxial tension.

The choice of an initial inhomogeneity factor $f_0 = 0.98$ in Fig. 5.1 is arbitrary and it leads to an underestimate of the limit strain in plane strain tension. However, if $f_0 > 0.98$ the limit strain at balanced biaxial tension will be greater than most experimentally observed values. Azrin and Backofen (1970) carried out a series of experiments using in-plane testing to determine

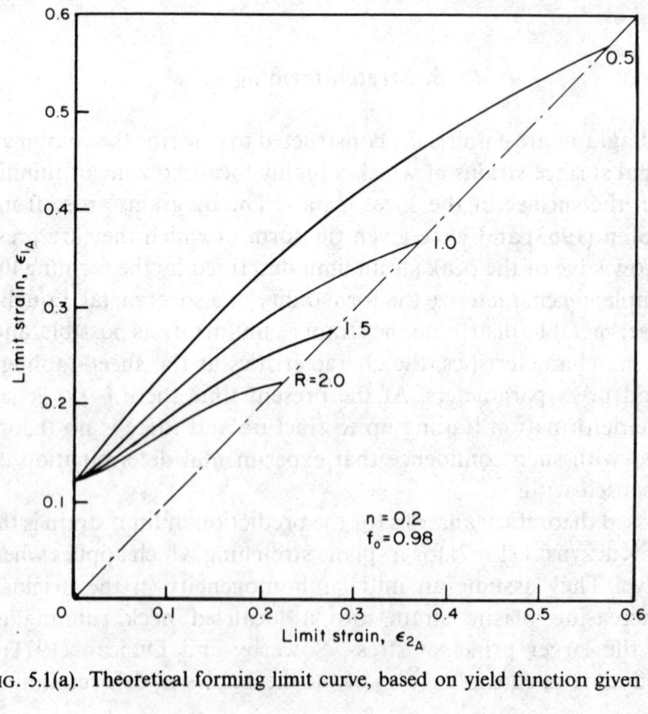

FIG. 5.1(a). Theoretical forming limit curve, based on yield function given by (2.1).

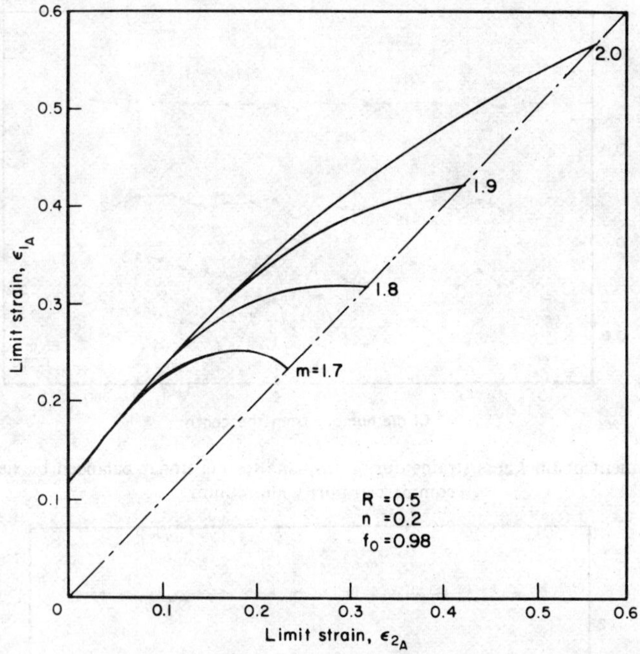

FIG. 5.1(b). Theoretical forming limit curve, based on yield function given by (3.3).

limit strains and they concluded that f_0 was a function of the accumulated strain and the ratio of the surface strains. Tadros and Mellor (1978), using the stretching technique of Marciniak and Kuczyński (1967), measured the development of strains up to fracture for killed steel, aluminium, and 70/30 brass. It is clear from the thickness strains shown in Fig. 5.2 that as deformation proceeds the thickness of the sheet becomes more and more non-uniform, and several "troughs" occur across the specimen. Such observations had led Tadros and Mellor

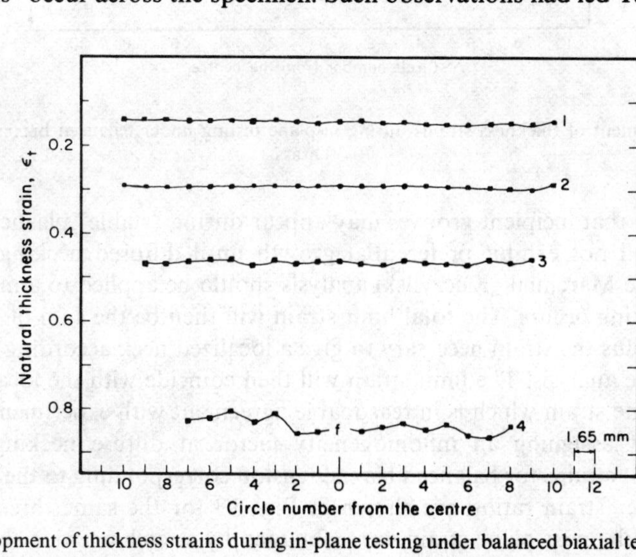

FIG. 5.2(a). Development of thickness strains during in-plane testing under balanced biaxial tension. Rimming steel.

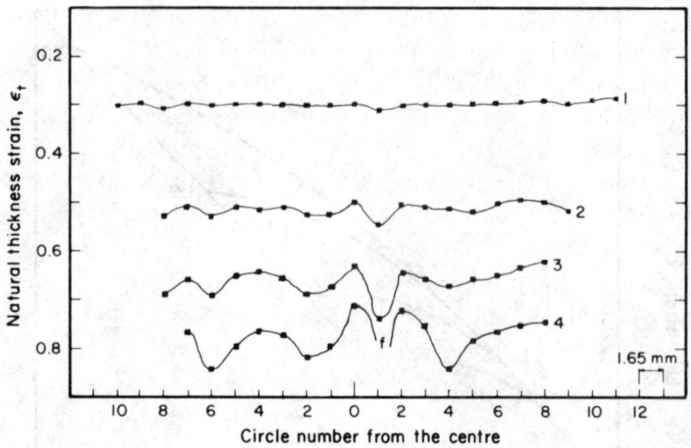

FIG. 5.2(b). Development of thickness strains during in-plane testing under balanced biaxial tension. Annealed commercial purity aluminium.

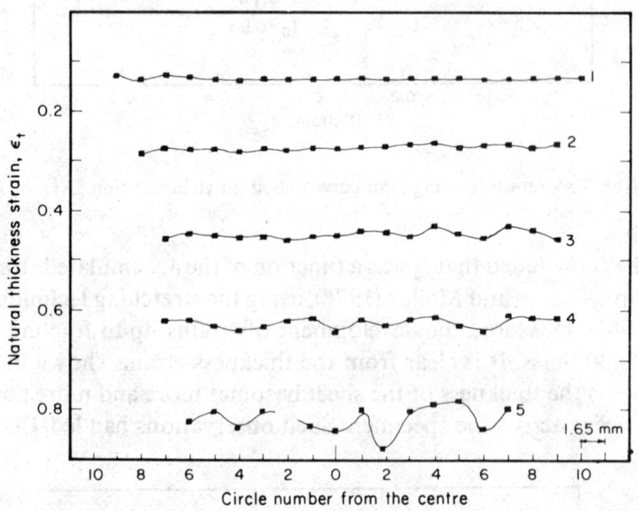

FIG. 5.2(c). Development of thickness strains during in-plane testing under balanced biaxial tension. Annealed 70/30 brass.

(1975) to suggest that incipient grooves may appear during "stable" plastic flow but that a single groove will not exhibit preferential growth until diffused necking occurs. It was proposed that the Marciniak–Kuczyński analysis should be applied to a material from the time diffuse necking occurs. The total limit strain will then be the sum of the strain up to diffuse necking plus the strain necessary to give a localized neck according to the modified application of the analysis. The limit strain will then coincide with the localized instability prediction at plane strain which is in reasonable agreement with experimental observation. The difficulty of assigning an inhomogeneity factor at diffuse necking remains. The experimental limit strains for balanced biaxial tension corresponding to the tests of Fig. 5.2, and for two other strain ratios, are shown in Fig. 5.3 for the same three materials. The experimental results near plane strain were obtained by stretching wide sheet over curved tooling. Despite the wide differences in the mechanical properties (Table 5.1), the values for the limit strain are remarkably similar. The general instability strains (Swift, 1952b) are also shown on each diagram. For steel and aluminium the experimental limit strains are greater than the predicted instability strains, but for 70/30 brass the position is reversed. The behaviour of the three materials is so different as to warrant separate discussion.

The rimming steel (Fig. 5.3(a)) presents the more normal situation with a rising characteristic as the strain ratio increases. The magnitude of the thickness strain at the general instability strain for this material would be 0.42, and it is seen from Fig. 5.2(a) that at a thickness strain of approximately 0.47 the strain is much more uniform than for the aluminium or brass. Near fracture the steel surface had a crazed appearance and the thickness

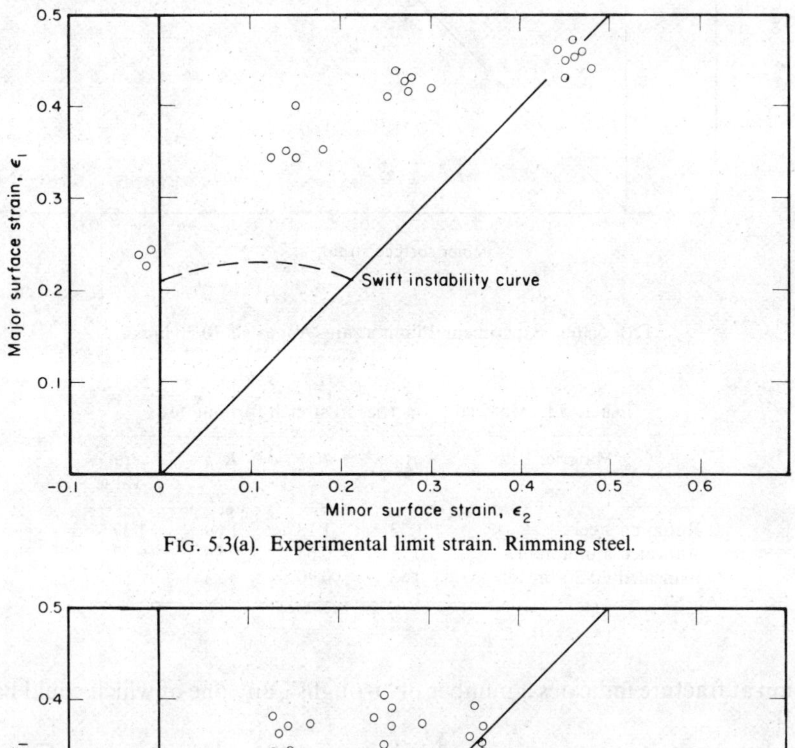

FIG. 5.3(a). Experimental limit strain. Rimming steel.

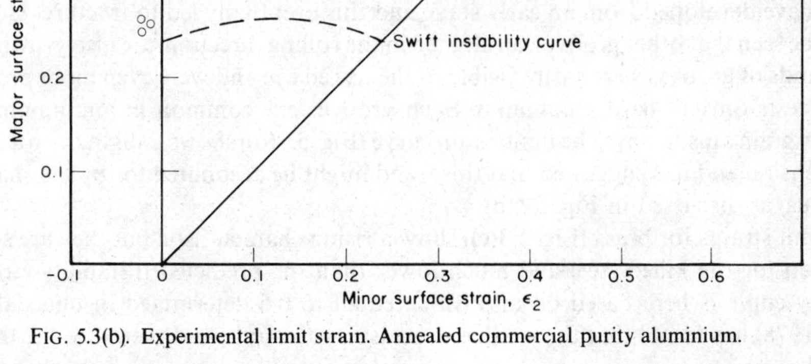

FIG. 5.3(b). Experimental limit strain. Annealed commercial purity aluminium.

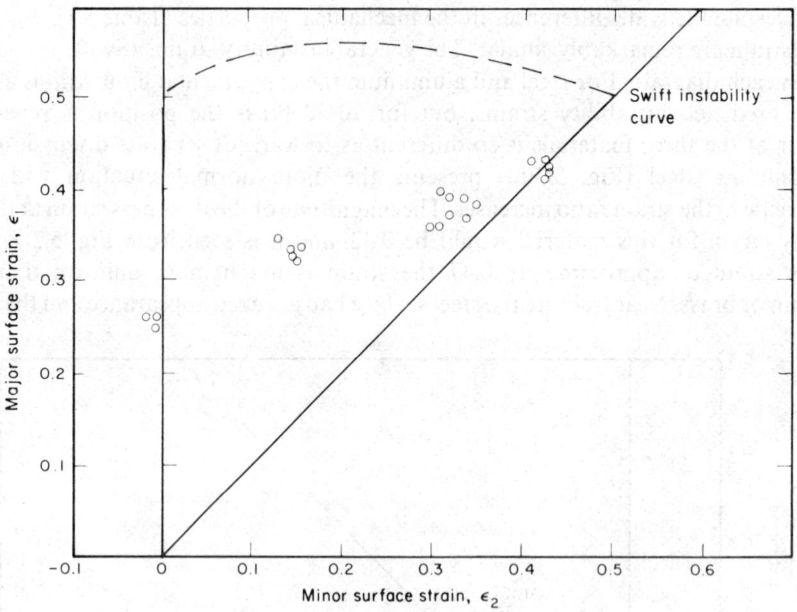

FIG. 5.3(c). Experimental limit strain. Annealed 70/30 brass.

TABLE 5.1. Material properties in stretch-forming tests

Material	R_0	R_{45}	R_{90}	Average R-value
Rimming steel	1.53	1.13	1.69	1.37
Annealed aluminium	0.75	0.64	0.86	0.72
Annealed 70/30 brass	0.82	0.90	0.83	0.86

distribution at fracture indicates a number of "troughs", any one of which could have led to fracture.

The predicted instability strain for the aluminium is 0.48, and it is clear from Fig. 5.2(b) that definite grooves have already developed at this stage. Indeed, in this case a single groove does seem to have developed from an early stage and this eventually led to fracture. However, it can also be seen that other grooves, all aligned in the rolling direction, are also growing in size. These bands of grooves were easily visible to the naked eye and were even more pronounced in some tests on half-hard aluminium. Such grooves are common in aluminium and are caused by grain anisotropy. The limit strain curve (Fig. 5.3(b)) shows a slight drop at balanced biaxial tension and it is suggested that this trend might be accounted for by the shape of the yield locus, as discussed in Fig. 5.1(b).

The limit strains for brass (Fig. 5.3(c)) show a rising characteristic but they are somewhat lower than for the killed steel and much lower than the predicted instability strains. The instability curve is here based on an n-value equal to 0.5 determined in uniaxial tension. Ghosh (1978) has found that the n-value decreases with strain and therefore the theoretical instability curve presented in Fig. 5.3(c) may somewhat overestimate the theoretically attainable strain. Azrin and Backofen (1970) in their experiments found that the limit strain for 70/30 brass was independent of the strain ratio. Ghosh and Hecker (1974) tested annealed 70/30 brass in both in-plane stretching and punch stretching and found that punch stretching gave a rising characteristic while the in-plane test gave results similar to those of Azrin and Backofen. However, it must be noted that Kleemola and Pelkkikangas (1977) in punch-stretching tests on annealed 70/30 brass obtained a result similar to that of Azrin and Backofen and they demonstrated that prior cold rolling of the brass led to a rising characteristic, a result which had also been noted by Ghosh and Backofen (1973). Annealed 70/30 brass is subject to jerky flow and the non-uniform extension in a simple tensile test is low compared with killed steel. Ghosh (1977) has explained such differences in terms of the strain-rate sensitivity and anisotropy of the materials. Although the results of Fig. 5.3(c) appear detrimental to annealed 70/30 brass, it must be remembered that brass has a high rate of work hardening which leads to a uniform strain distribution and therefore the possibility of a deep pressing, and it is also easier to obtain a low coefficient of friction between sheet and tooling which again reduces unwanted strain gradients. Parmar *et al.* (1977) have attempted to explain the discrepancy between the predicted instability strains and experimental limit strains by taking into account possible effects of surface roughness, and useful trends are indicated. However, it must be concluded that the necessary input data for the analysis would involve much more experimental work than the direct experimental determination of the forming limit diagram of the same material.

Acselrad and Wilson (1979) have attempted to explain the failure of sheet metal solely in terms of the microstructure. They propose that the deformation consists of two stages. In the first stage, which they term the microstructural stage, the individual regions of strain concentration are small compared with the sheet thickness. This stage is terminated when growth and coalescence of the individual regions of strain concentration have developed far enough to form a coherent region of strain concentration which spans the sheet thickness. This may then either lead directly to a groove type failure or such weak regions may link up after further deformation to form extensive surface grooves. While it has long been recognized that the expansion and coalescence of voids takes place in a necked region, there appears to be no direct evidence that the expansion and coalescence of voids is responsible for the initiation of a localized neck. Acselrad and Wilson have investigated the effect of grain size on the behaviour of low carbon steel. They found that for grain sizes in the range 20–45 mm, which are typical of good quality commercial sheets, the difference in inclusion content is of principal importance but, with appreciably coarser grain sizes, grain size itself appears to exert the dominating influence. However, tests on 70/30 brass raised the possibility that for thin materials grain anisotropy is the dominant source of inhomogeneity down to quite fine grain sizes.

Marciniak *et al.* (1973) introduced the effect of strain-rate sensitivity into their analysis for limit strain. While this indicates that increasing strain-rate sensitivity increases the value of the limit strain the essential shape of the predicted limit curve is unaltered. Hutchinson and Neale (1978) have also considered the effects of strain-rate sensitivity and have based constitutive equations on both a flow theory and a deformation theory. The latter theory was seen to be more acceptable since it was in better agreement with some experimental results. In particular it could predict a falling curve for the higher n-values ($\geqslant 0.5$), which had been observed experimentally. However, there appears to be no basic experimental data which would justify the use of a deformation theory. Bassani *et al.* (1979) studied the effects of various constitutive laws on the predicted limit strains and concluded that the forming limit curves are strongly dependent on the shape of the anisotropic yield surface.

6. Deep-drawing of a Circular Blank

The suitability of sheet materials for presswork has been assessed for many years by deep drawing a circular blank through a cylindrical die with a flat-headed cylindrical punch, and the popularity of this test increased sharply when Whiteley (1960) found experimentally that the correlation of limiting drawing ratio with *R*-value was highly significant. The value of the limiting drawing ratio is also a function of the tooling geometry, absolute thickness of sheet metal blank, and of the frictional forces between the deforming blank and tooling. In practice the limiting drawing ratio is particularly sensitive to lubrication and to blank-holding pressure.

The theoretical analyses which have been proposed for predicting the limiting drawing ratio are extremely simple in nature, and great caution is needed when relating them to practice. Such analyses ignore the effects of punch and die radii and simply compare the strength of the material in the flange with the strength of the material in the partially drawn cup-wall. Budiansky and Wang (1966) based their analysis on Hill's original anisotropic yield criterion (2.1), and an assumption that failure occurred under plane strain tension in material in contact with the punch stem. The effect of *R*-value and *n*-value on the predicted limiting drawing ratios is shown in Fig. 6.1, where it can be seen that this type of failure is

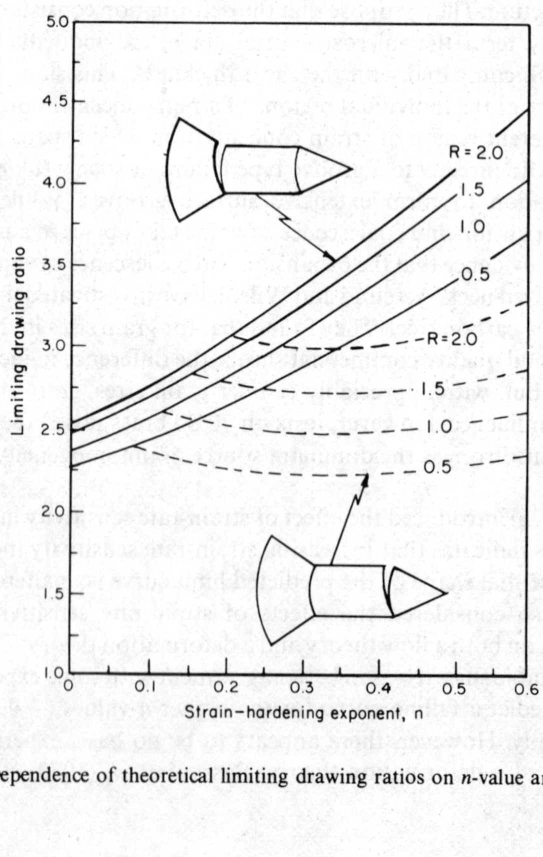

FIG. 6.1. Dependence of theoretical limiting drawing ratios on *n*-value and *R*-value.

largely independent of n-value over the range of applicability. El-Sebaie and Mellor (1972, 1973) using a similar analysis showed that under certain circumstances the mode of failure would change to one of necking under simple tension in material emerging from the flange (Fig. 6.1), and this mode of failure is very dependent on the n-value and less dependent on the R-value. According to these simple analyses the applicable mode is the one that predicts the lower limiting drawing ratio. The theoretically predicted limiting drawing ratio over-estimates what can be achieved in practice because the analyses neglect the bending and unbending over the tooling profiles and do not take explicit account of frictional forces. However, it is implicit in the plane strain mode of failure in the cup-wall that there is sufficient friction between cup and punch to restrict thinning over the punch profile radius. El-Sebaie and Mellor (1972) in a detailed experimental investigation showed that the location of necking failure is very dependent on the n-value and on lubrication conditions. A high n-value and poor lubrication between punch-head and blank moves the neck towards the junction of punch stem and punch profile radius while a low n-value and good lubrication moves the neck towards the flat-head of the punch. For example, annealed 70/30 brass ($n \simeq 0.5$) with a roughened punch gives a neck location very close to the junction of punch stem and punch profile radius, while half-hard aluminium ($n \simeq 0.04$) with good lubrication gives a neck location very close to the flat-head of the punch. Friction in the flange area, by increasing the radial load required to draw the material, will lower the limiting drawing ratio below the values given in Fig. 6.1.

Ng *et al.* (1976) have shown how different lubrication conditions can affect the value of the limiting drawing ratio and sometimes also the mode of failure. They used two types of punches: one knurled all over to give an extremely rough finish, the other with a good ground and polished finish. The following lubrication conditions were used.

Type A lubrication: both sides of the blank were lubricated with "Droyt-Sol 4M". This is a water-soluble emulsion of mineral oil which was mixed two parts of water to one of oil. This was used with the knurled punch.

Type B lubrication: the top face of the blank, which was in contact with the die, was covered with a full disc of PTFE film. The bottom face of the blank was covered with an annular PTFE disc, which prevented metal-to-metal contact in the flange area while ensuring such contact between the head of the punch and the blank. This restricted plastic flow over the punch-head region while maintaining good lubrication in the flange. This arrangement was used only with the knurled punch.

Type C lubrication: both sides of the blank were covered with full discs of PTFE film. This was used only with the "smooth" punch.

The limiting drawing ratios for four different sheet materials and for different methods of lubrication are given in Table 6.1. It is seen that in each case Type B lubrication yields a higher limiting drawing ratio than Type A lubrication. This results mainly from the lower radial drawing load brought about by improved lubrication in the flange and over the die profile radius when using Type B lubrication. Thus for steel and a drawing ratio of 2.2 the maximum drawing load was reduced by 20% when the lubrication was changed from Type A to Type B. Note that for a particular sheet material the mode of failure was the same for Type A and Type B lubrication. For soft aluminium, 70/30 brass, and killed steel the position of the neck was near the junction of the punch profile radius and punch stem and the failure was judged to have occurred under plane strain tension. With half-hard aluminium the neck was near to the flat-head of the cup and it was judged that in the limiting case failure occurred, under balanced biaxial tension, at an early stage in the draw. The limiting drawing ratio for half-

TABLE 6.1. Effect of lubrication on limiting drawing ratio

Material	Soft Al	$\frac{1}{2}$H Al	70/30 brass	Killed steel
LDR, Type A lubrication	1.90	1.88	2.08	2.32
Approximate mode of failure	PST	BBT	PST	PST
LDR, Type B lubrication	2.25	2.22	2.42	2.67
Approximate mode of failure	PST	BBT	PST	PST
LDR, Type C lubrication	2.13	2.07	—	—
Approximate mode of failure	BBT	BBT	—	—

hard aluminium was 2.22 when using Type B compared with 1.88 when using Type A lubrication. This increase in drawing ratio was caused by improved lubrication in the flange and over the die profile radius. The use of Type C lubrication (full discs of PTFE film on both sides of the blank) should give the same frictional conditions over the flange and die profile radius as for Type B lubrication but much improved lubrication over the punch head. It would therefore be expected that failure in material over the punch head would occur at a smaller drawing load and therefore at a reduced limiting drawing ratio. This proved to be the case for half-hard aluminium where the limiting drawing ratio using Type C lubrication was 2.07. (Table 6.1.) A similar reduction in limiting drawing ratio occurred where Type C lubrication was used with soft aluminium and the location of the neck was now close to the flat-head of the punch. There was also increased stretching over the flat-head of the punch.

Ng *et al.* (1976) also predicted the limiting drawing ratios theoretically when the mode of failure was balanced biaxial tension. The critical load for the failure in the blank at a radius equal to the punch-stem radius is equated to the maximum radial drawing load. Basing the calculation on Hill's yield criterion, (2.1), the theory predicts that when the R-value is less than unity, and there is lubrication over the punch head, the limiting drawing ratios are slightly less than the values for the other two modes given in Fig. 6.1. However, it is now clear that (2.1) is not appropriate for aluminium especially in the region between plane strain tension and balanced biaxial tension.

Hill's original anisotropic yield criterion, (2.1), together with its associated flow rule and the added assumption of equivalence of plastic work, describes the plastic deformation of low-carbon steels, $1 < R < 2$, reasonably well. However, the theoretical results for the variation of limiting drawing ratios with R-value when the mode of failure is plane strain tension (Fig. 6.1) indicates a far greater dependence on R-value than actually occurs in practice. This is clear from Fig. 6.2(a) which compares the variation of theoretical limiting drawing ratio with R-value, for $n = 2$, with the general relationship proposed by Atkinson (1972) from an analysis of experimental results over a wide range of materials. A difference in the absolute values would be expected because of frictional forces in the flange. The main point of interest lies in the increasing difference between theory and experiment with increasing R-value. This point has been stressed by Hosford and Kim (1976), but it seems clear that in the case of killed steel the lack of correlation between the theoretical analysis and experiment is caused by an inadequacy in the modelling of the deep-drawing process rather than by any inadequacy in the constitutive equations of the material.

The second mode of failure, necking under uniaxial tension in material emerging from the flange, gives better correlation with practice. The theory can be applied only to materials that

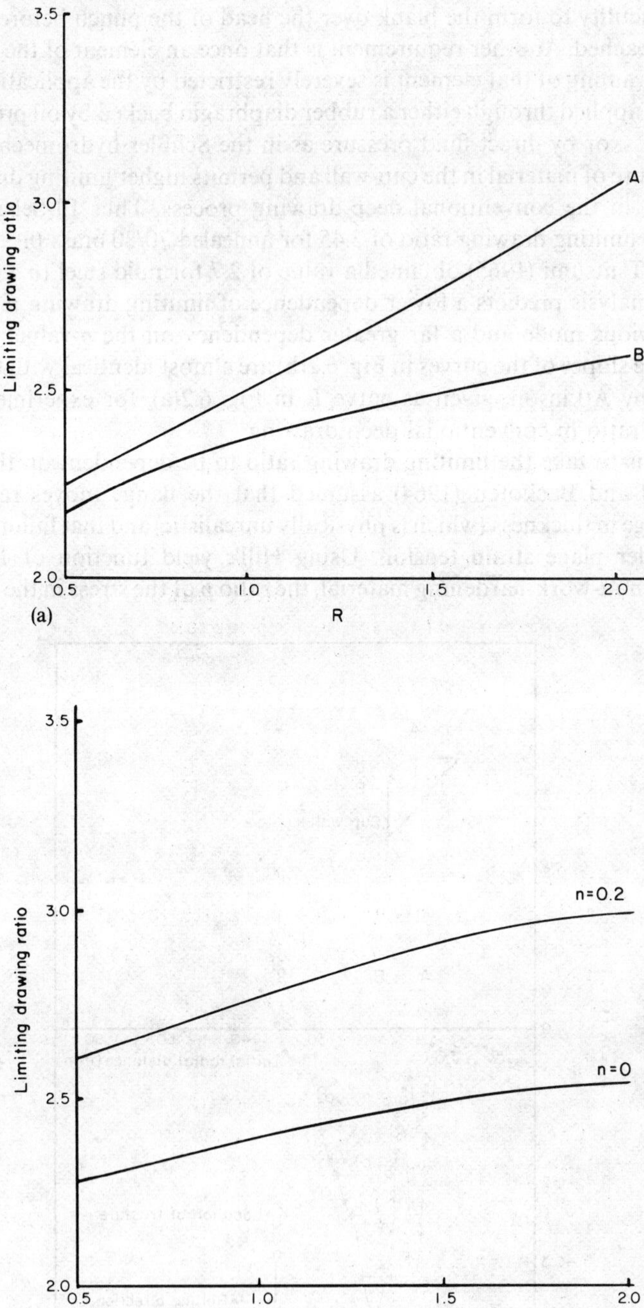

FIG. 6.2. Limiting drawing ratios. (a) Curve A: theoretical limiting drawing ratio for $n = 0.2$ when failure under plane strain tension near punch head profile radius. Curve B: trend of experimental results. (b) Variation of theoretical limiting drawing ratio with R-value when failure under uniaxial tension in material emerging from the flange.

have sufficient ductility to form the blank over the head of the punch before the maximum drawing load is reached. Another requirement is that once an element of the blank contacts the punch then straining of that element is severely restricted by the application of a normal force. This can be applied through either a rubber diaphragm backed by oil pressure, as in the hydroform process, or by direct fluid pressure as in the Schuler hydromechanical process. This prevents failure of material in the cup-wall and permits higher limiting drawing ratios to be achieved than in the conventional deep-drawing process. Thus El-Sebaie and Mellor (1973) obtained a limiting drawing ratio of 3.45 for annealed 70/30 brass ($n \simeq 0.49$, $R \simeq 0.83$) and Kasuga and Tsutsumi (1965) obtained a value of 2.7 for mild steel ($n \simeq 0.22$, $R \simeq 1.4$). The theoretical analysis predicts a lower dependence of limiting drawing ratio on R-value than for the previous mode and a far greater dependency on the n-value. It would seem fortuitous that the slopes of the curves in Fig. 6.2(b) are almost identical with the slope of the curve proposed by Atkinson, given as curve B in Fig. 6.2(a), for experimental results of limiting drawing ratio in conventional deep drawing.

A simple view is to take the limiting drawing ratio to be dependent on the ratio of two stresses. Hosford and Backofen (1964) assumed that the flange moves radially inwards without any change in thickness (which is physically unrealistic) and that failure takes place in the cup-wall under plane strain tension. Using Hill's yield function of 1948 they then determined, for a non-work-hardening material, the ratio β of the stress in the cup-wall to the

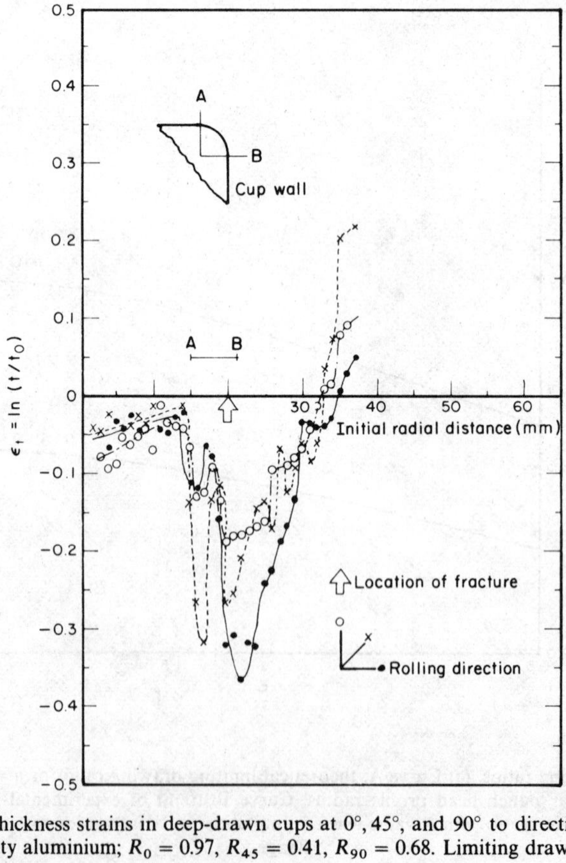

FIG. 6.3(a). Measured thickness strains in deep-drawn cups at 0°, 45°, and 90° to direction of rolling. Annealed commercial purity aluminium; $R_0 = 0.97$, $R_{45} = 0.41$, $R_{90} = 0.68$. Limiting drawing ratio = 1.96.

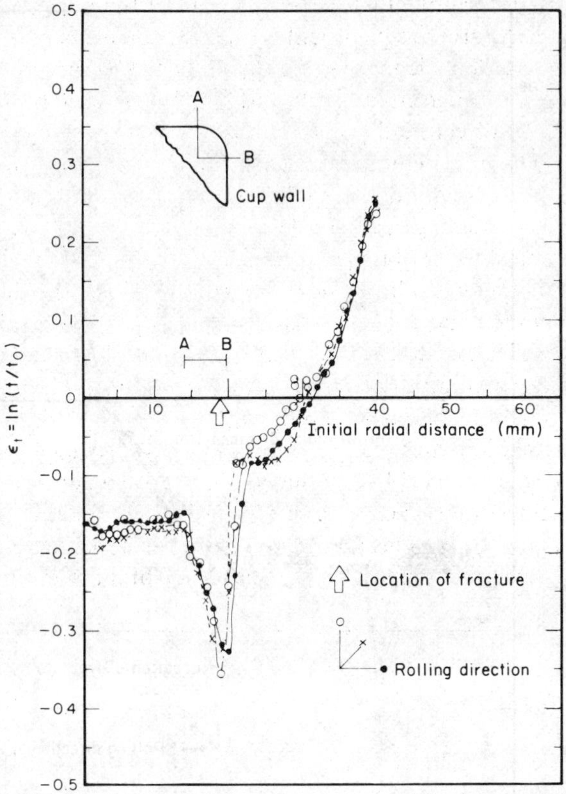

FIG. 6.3(b). Measured thickness strains in deep-drawn cups at 0°, 45°, and 90° to direction of rolling. Annealed 70/30 brass; $R_0 = 0.86$, $R_{45} = 0.85$, $R_{90} = 0.78$. Limiting drawing ratio = 2.12.

stress at the periphery of the flange as

$$\beta = \left(\frac{1+R}{2}\right)^{1/2} \tag{6.1}$$

They also determined from crystallographic analysis of cubic materials that the highest possible β-value is about 1.24 and from (6.1) this should correspond to an R-value of approximately 2.0. As pointed out by Sowerby and Johnson (1975), the implication here is that a further increase in R will yield no significant benefits as regards the limiting drawing ratio. Dillamore (1974) has put forward a similar argument on the basis of the ratio of balanced biaxial stress to uniaxial stress, which in fact leads to the same expression as (6.1); see also (3.1). From his crystallographic analysis Dillamore obtained a maximum ratio of stresses of approximately 1.18 and he concluded that the pursuit of R-values above a value of about 3.0 would be unrewarding. It must be questioned whether such attempted correlations between crystallographic and macroscopic theory are of significance. A mathematical theory of plasticity has no standing for a real material unless it can be shown to fit the experimentally determined expanding yielding loci of that material. It would, therefore, be more meaningful to compare any prediction of the relative strengths and R-values from crystallographic analysis with experimentally determined values of the same quantities. The inadequacy of a model of deep drawing based on failure in the cup-wall has already been discussed.

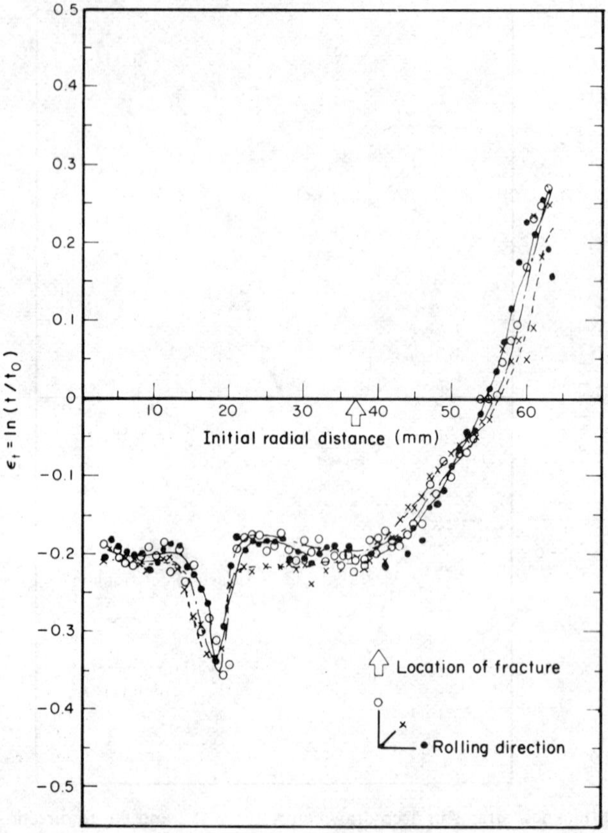

FIG. 6.3(c). Measured thickness strains in deep-drawn cups at 0°, 45°, and 90° to direction of rolling. Annealed 70/30 brass, drawing into high fluid pressure, drawing ratio = 3.34.

The above discussion has been based on average R-values. In practice planar anisotropy affects the developing strain distribution in deep-drawing. This is highlighted in Fig. 6.3(a) for annealed commercial purity aluminium where there is appreciable planar anisotropy. This should be compared with the good uniformity of straining in annealed 70/30 brass (Fig. 6.3(b)). For the brass this uniformity is still evident when drawing into a high-pressure medium to a much higher limiting drawing ratio (Fig. 6.3(c)).

In cup drawing planar anisotropy is also manifest in the formation of ears. The relation between earing and anisotropy was studied by Bourne and Hill (1950) in tension and cupping tests. The theory of plastic anisotropy (Hill, 1948), was found to be in good agreement with the experimental data for materials producing four ears. The theory was extended to describe more complex states of anisotropy, e.g. a brass giving six ears. The variation of R-value with orientation to the rolling direction could then be described by assuming a homogeneous cubic polynomial for the plastic potential. Gotoh (1977) and Gotoh and Ishisé (1978) have introduced a yield function of fourth order and have studied the deformation of a flange in deep-drawing using a finite element analysis.

7. Concluding Remarks

The plastic deformation of metals is extremely complex and it is perhaps not reasonable to seek a macroscopic theory which has such generality that it can be applied to any material or to any process. Some limitations of present theories have been touched on here where the discussion has been limited to simple loading paths. When the deformation follows more complex loading paths more experimental data will be needed before a macroscopic theory can be applied successfully. Even a simple change in strain path can lead to a significant change in anisotropy which can only be measured by experiment. For example, Ghosh and Laukonis (1976) demonstrated that when steel sheets are prestrained in balanced biaxial tension and then tested in simple tension, the measured R-values decrease with the amount of prestrain. Each material and each loading path must therefore be considered separately. This approach requires a great deal of effort and one must learn and accept the restrictions that this implies.

References

Acselrad, O. and Wilson, D. V. (1979) Effects of microstructure on forming limits in biaxial stretching, *Sh. Metal Inds* **56**, 408–412.

Atkinson, M. (1972) Assessing normal anisotropic plasticity of sheet metal, *Sh. Metal Inds* **44**, 167–178.

Avery, D. H., Hosford, W. F., and Backofen, W. A. (1965) Plastic anisotropy in magnesium alloy sheets, *Trans. Metall. Soc. AIME* **233**, 71–78.

Azrin, M. and Backofen, W. A. (1970) The deformation and failure of a biaxially stretched sheet, *Metall. Trans.* **1**, 2857–2865.

Babel, H. W., Eitman, D. A., and McIver, R. W. (1966) The biaxial strengthening of textured titanium, *Trans. ASME*, Paper 66, Met 6, 1–9.

Bassani, J. L., Hutchinson, J. W., and Neale, K. W. (1979) On the prediction of necking in anisotropic sheets, *Metal Forming Plasticity*, 1–13, edited by H. Lippmann, Springer-Verlag, Berlin–Heidelberg.

Bourne, L. and Hill, R. (1950) On the correlation of the directional properties of rolled sheet in tension and cupping tests, *Phil. Mag.* **41**, 671–681.

Bramley, A. N. and Mellor, P. B. (1966) Plastic flow in stabilized sheet steel, *Int. J. Mech. Sci.* **8**, 101–114.

Bramley, A. N. and Mellor, P. B. (1968) Plastic anisotropy of titanium and of zinc sheet: macroscopic approach, *Int. J. Mech. Sci.* **10**, 211–219.

Budiansky, B. and Wang, N. M. (1966) On the Swift cup test, *J. Mech. Phys. Solids* **14**, 357–374.

Dillamore, I. L. (1974) The relevance of tensile properties to sheet formability, *J. Phys.* D: *Appl. Phys.* **7**, 979–991.

Dillamore, I. L., Hazel, R. J., Watson, T. W., and Hadden, P. (1971) An experimental study of the mechanical anisotropy of some common metals, *Int. J. Mech. Sci.* **13**, 1049–1061.

El-Sebaie, M. G. and Mellor, P. B. (1972) Plastic instability conditions in the deep-drawing of a circular blank of sheet metal, *Int. J. Mech. Sci.* **14**, 535–556.

El-Sebaie, M. G. and Mellor, P. B. (1973) Plastic instability conditions when deep-drawing into a high pressure medium, *Int. J. Mech. Sci.* **15**, 485–501.

Ghosh, A. K. (1977) The influence of strain hardening and strain rate sensitivity on sheet metal forming, *J. Engng Mat. Tech. (Trans. ASME)* **99**, 264–274.

Ghosh, A. K. (1978) Plastic flow properties in relation to localized necking in sheets, *Mechanics of Sheet Metal Forming: Material Behavior and Deformation Analysis*, 289–312, edited by D. P. Koistinen and N. M. Wang, Plenum Press, New York.

Ghosh, A. K. and Backofen, W. A. (1973) Strain hardening and instability in biaxially stretched sheet, *Metall. Trans* **4**, 1113–1123.

Ghosh, A. K. and Hecker, S. S. (1974) Stretching limits in sheet metals: in-plane versus out-of-plane deformation, *Metall. Trans.* **5**, 2161–2164.

Ghosh, A. K. and Laukonis, J. V. (1976) The influence of strain-path changes on the formability of sheet steel, *9th Biennial Congress of the International Deep-Drawing Research Group*, ASM, 167–178.

Gotoh, M. (1977) A theory of plastic anisotropy based on a yield function of fourth order (plane stress state), *Int. J. Mech. Sci.* **19**, 505–520.

Gotoh, M. and Ishisé, F. (1978) A finite element analysis of rigid-plastic deformation of the flange in a deep-drawing process based on a fourth-degree yield function, *Int. J. Mech. Sci.* **20**, 423–435.

Hill, R. (1948) A theory of the yielding and plastic flow of anisotropic metals, *Proc. R. Soc. London*, A **193**, 281–297.

Hill, R. (1950) *The Mathematical Theory of Plasticity*, Clarendon Press, Oxford.

Hill, R. (1952) On discontinuous plastic states, with particular reference to localized necking in thin sheets, *J. Mech. Phys. Solids* **1,** 19–30.
Hill, R. (1979) Theoretical plasticity of textured aggregates, *Math. Proc. Camb. Phil. Soc.* **85,** 179–191.
Holloway, G. and Shelton, A. (1979) The effect of cold-drawing on the yield and flow behaviour of O. F. H. C. copper under combined stress, *J. Mech. Engng Sci*, **21,** 235–245.
Hosford, W. F. and Backofen, W. A. (1964) Strength and plasticity of textured materials, *Fundamentals of Deformation Processing*, 259–292, edited by W. A. Backofen, J. J. Burke, L. F. Coffin, N. L. Reed, and V. Weiss, Syracuse University Press.
Hosford, W. H. and Kim, C. (1976) The dependence of deep-drawability on normal anisotropy: crystallographic analysis, *Metall. Trans.* **7A,** 468–472.
Hutchinson, J. W. and Neale, K. W. (1978) Sheet necking–time independent behaviour, strain-rate effects, *Mechanics of Sheet Metal Forming: Material Behaviour and Deformation Analysis*, 127–153 and 269–285, edited by D. P. Koistinen and N. M. Wang, Plenum Press, New York.
Ilahi, M. F. (1977) Plastic deformation of circular metal diaphragms, Ph.D. thesis, University of Bradford.
Jackson, L. R., Smith, K. F. and Lankford, W. T. (1948) *Plastic Flow in Anisotropic Steel Sheet*, Metals Technology Tech. Pub. No. 2440.
Kasuga, Y. and Tsutsumi, S. (1965) Pressure lubricated deep-drawing, *Bull. JSME* **8,** 120–126.
Keeler, S. P. (1965) Determination of forming limits in automotive stampings, SAE Paper No. 650535, 1–9.
Keeler, S. P. (1978) Sheet metal stamping technology—need for fundamental understanding, *Mechanics of Sheet Metal Forming: Material Behaviour and Deformation Analysis*, 3–18, edited by D. P. Koistinen and N. M. Wang, Plenum Press, New York.
Keeler, S. P. and Backofen, W. A. (1963) Plastic instability and fracture in sheets stretched over rigid punches, *Trans. ASM* **56,** 25–48.
Kleemola, H. J. and Pelkkikangas, M. T. (1977) Effect of predeformation and strain path on the forming limits of steel, copper and brass, *Sh. Metal Inds* **54,** 591–599.
Larson, F. R. (1964) Anisotropy of titanium sheet in uniaxial tension, *Trans. ASM* **57,** 620–631.
Lee, D. and Backofen, W. A. (1966) An experimental determination of the yield locus of titanium and titanium-alloy sheet, *Trans. Metall. Soc. AIME* **236,** 1077–1084.
Lee, S. H. and Kobayashi, S. (1975) Rigid plastic analysis of bore expanding and flange drawing with anisotropic sheet metals by the matrix method, *Proceedings of the 15th International Machine Tool Design Research Conference*, 561–569, edited by S. A. Tobias and F. Koenigsberger, Macmillan Press, London.
Marciniak, Z. and Kuczyński, K. (1967) Limit strains in the processes of stretch-forming sheet metal, *Int. J. Mech. Sci.* **9,** 609–620.
Marciniak, Z., Kuczyński, K. and Pokora, T. (1973) Influence of the plastic properties of a material on the forming limit diagram for sheet metal in tension, *Int. J. Mech. Sci.* **15,** 789–805.
Mellor, P. B. (1956) Stretch forming under fluid pressure, *J. Mech. Phys. Solids* **5,** 41–56.
Naziri, H. and Pearce, R. (1968) The effect of plastic anisotropy on flange-wrinkling behaviour during sheet metal forming, *Int. J. Mech. Sci.* **10,** 681–694.
Ng, P., Chakrabarty, J. and Mellor, P. B. (1976) Instability modes in deep-drawing, *Proceedings of the 15th International Machine Tool Design and Research Conference*, 579–586, edited by S. A. Tobias and F. Koenigsberger, Macmillan Press, London.
Parmar, A. and Mellor, P. B. (1978a) Plastic expansion of a circular hole in sheet metal subjected to biaxial tensile stress, *Int. J. Mech. Sci.* **20,** 707–720.
Parmar, A. and Mellor, P. B. (1978b) Predictions of limit strains in sheet metal using a more general yield criterion, *Int. J. Mech. Sci.* **20,** 385–391.
Parmar, A., Mellor, P. B. and Chakrabarty, J. (1977) A new model for the prediction of instability and limit strains in thin sheet metals, *Int. J. Mech. Sci.* **19,** 389–398.
Pearce, R. (1968) Some aspects of anisotropic plasticity in sheet metals, *Int. J. Mech. Sci.* **10,** 995–1005.
Rogers, D. H. and Roberts, W. T. (1968) Plastic anisotropy of titanium and zinc sheet: crystallographic approach, *Int. J. Mech. Sci.* **10,** 221–229.
Sowerby, R. and Duncan, J. L. (1971) Failure in sheet metal in biaxial tension, *Int. J. Mech. Sci.* **13,** 217–230.
Sowerby, R. and Johnson, W. (1975) A review of texture and anisotropy in relation to metal forming, *Mater. Sci. Engng* **20,** 101–111.
Swift, H. W. (1952a) On the foot-hills of the plastic range, *J. Inst. Metals* **81,** 109–120.
Swift, H. W. (1952b) Plastic instability under plane stress, *J. Mech. Phys. Solids* **1,** 1–18.
Tadros, A. K. and Mellor, P. B. (1975) Some comments on the limit strains in sheet metal stretching, *Int. J. Mech. Sci.* **17,** 203–210.
Tadros, A. K. and Mellor, P. B. (1978) An experimental study of the in-plane stretching of sheet metal, *Int. J. Mech. Sci.* **20,** 121–134.
Taghvaipour, M. (1970) Normal anisotropy in sheet metals, Ph.D. Thesis, University of Birmingham.
Taghvaipour, M. and Mellor, P. B. (1971) Plane strain compression of anisotropic sheet metal, *Proc. Instn Mech. Engrs* **185,** 593–606.

Taghvaipour, M., Chakrabarty, J., and Mellor, P. B. (1972) The variation of the *R*-value in titanium with increasing strain, *Int. J. Mech. Sci.* **14,** 117–124.

Taylor, G. I. and Quinney, H. (1931) The plastic distortion of metals, *Phil. Trans. R. Soc. London* A **230,** 323–362.

Tozawa, Y. (1978) Plastic deformation behavior under conditions of combined stress, *Mechanics of Sheet Metal Forming: Material Behavior of Deformation Analysis*, 81–110, edited by D. P. Koistinen and N. M. Wang, Plenum Press, New York.

Venter, R., Johnson, W., and de Malherbe, M. C. (1971) The limit strains of inhomogeneous sheet metal in biaxial tension, *Int. J. Mech. Sci.* **13,** 299–308.

Wagoner, R. H. and Wang, N. M. (1979) An experimental and analytical investigation of in-plane deformation of 2036-T4 aluminium sheet, *Int. J. Mech. Sci.* **21,** 255–264.

Whiteley, R. L. (1960) The importance of directionality in drawing quality sheet steel, *Trans. ASM* **52,** 154–169.

Woodthorpe, J. and Pearce, R. (1969) The effect of *R* and *n* upon the forming limit diagrams of sheet metal, *Sh. Metal Inds* **46,** 1061–1067.

Crystal Elasticity

by F. Milstein

Department of Mechanical and Environmental Engineering,
University of California, Santa Barbara

Summary

Research in crystal elasticity is reviewed, with emphasis upon theoretical mechanical behavior at finite strains. Characteristic theoretical responses of simple crystals to a variety of different modes of loading are examined. Particular attention is given to the topic of bifurcations and branchings from primary paths under prescribed programs of loading. Specifically included are lattice models and computation techniques, elasticity and stability under hydrostatic pressure, the definition and computation of elastic moduli under load, and a detailed exposition of branching under [100] loading. Numerous computational results are presented. Examples are given of how finite strain considerations provide insight into aspects of crystal elasticity at infinitesimal strain.

1. Introduction

The present article reviews current research in crystal elasticity; the emphasis is upon work aimed at exploring and characterizing the theoretical mechanical behavior of crystals at finite strains, under various modes of homogeneous deformation. It is recognized that elastic deformation may be large and nonlinear in a variety of circumstances, particularly if (i) it occurs very rapidly (e.g. in shock loading, martensitic phase transformations, or mechanical twinning), (ii) it is highly localized (i.e. near cracks or other stress-raisers), or (iii) the deformed specimen itself is very small (e.g. a whisker or integrated circuit structure). Hill (1975) observed that "Single crystals free from lattice imperfections are used increasingly as micro-structural components. Perfect crystals are capable of elastic strains well beyond what can properly be treated as infinitesimal. Their response to general loading is virtually unknown and is doubtless complex, so experimentation will have to be conducted within some plausible theoretical framework."

Thus the study of the theoretical elastic behavior of crystals at finite strain is enhanced with a special significance (particularly in view of experimental difficulties associated with attempting to characterize deformations occurring over very short times or in very small specimens). General topics of interest include (i) the definition of elastic moduli of crystals under load, (ii) the development of principles and methodology of stability analysis for loaded crystals, and (iii) the computation of loads, lattice strains, elastic moduli, and stability domains within the framework of particular lattice models representing crystal binding. The significance of such computations often extends beyond that of the immediate results for the *particular* model since many *phenomena* (revealed by the computations) are dependent largely upon crystal symmetry (and are therefore, by and large, model independent). Recently, special attention has been given to the topics of (a) crystal elasticity and stability under hydrostatic loading, (b) the detailed nature of bifurcations from a primary to a secondary path (under a prescribed mode of loading), and (c) the nature of the secondary paths themselves. It might be noted that these topics are fundamental to areas of metallurgy, solid state physics, solid mechanics, and geophysics concerned with crystal stability and its relation to structure.

2. Crystal Elasticity at Infinitesimal Strain as Revealed by Finite Strain Considerations

We begin by mentioning that, as an "additional bonus," studies of lattice behavior at finite strain have provided deeper insight into the bases for certain aspects of crystal elasticity at infinitesimal strain. Three examples are presented in Sections 2.1–2.3.

2.1. *Existence of unstressed tetragonal crystals*

A number of elements (including boron, indium, mercury, tin, and uranium) can exist in a tetragonal crystal structure (lattice parameters $a_1 \neq a_2 = a_3$, with their included angles equal to 90°) at atmospheric pressure (Barrett and Massalski, 1966). A basic question in the subject of crystal physics is whether or not a collection of atoms with a particular form of interatomic interactions admits an *unstressed*, equilibrium, tetragonal crystal structure. By considering the general behavior of face-centered cubic (fcc) or body-centered cubic (bcc) crystals loaded uniaxially along a cube edge (i.e. along a [100] direction in Miller indices notation), a simple general proof can be given (Milstein, 1980) that any monatomic collection of atoms does admit the existence of a stress-free lattice possessing a body-centered (or equivalently, face-centered) tetragonal configuration (Hermann–Mauguin lattice symbol, tetragonal I). The collection is assumed to be monatomic, with the only physical restriction being that as the distance between atoms becomes very large or very small there is mutual attraction or repulsion, respectively. Let the atoms be arranged initially in a fcc lattice configuration (i.e. atoms are at the corners and in the centers of the faces of a cubic unit cell). From lattice symmetry, it follows that the state of loading is hydrostatic. From the nature of crystal binding, at small values of lattice parameter the crystal is under compression; at large values it is in a state of hydrostatic tension. Thus the lattice parameter can be adjusted to a state of zero loading (pressure equals zero). From this state, compress the crystal uniaxially along a cube edge (or [100] direction). By definition, along this path of loading, the transverse lattice parameters vary in such a manner that the transverse loads are zero at any stage. For convenience, define the stretch λ_i of any embedded fibre as its final length divided by its initial length; in the unstressed fcc state, $\lambda_1 = \lambda_2 = \lambda_3 = 1$; along the primary path of [100] loading, $\lambda_1 \neq \lambda_2 = \lambda_3$, in general; the [100] axial stretch is λ_1 and the axisymmetric transverse stretches are λ_2 and λ_3. With full generality, we can let the lattice be compressed on a primary [100] loading path to a state where $\lambda_1 = \lambda_2/\sqrt{2}$. Figure 2.1 illustrates the lattice structure at any stage on the primary path; it is readily seen that either a body-centered tetragonal (bct) or face-centered tetragonal (fct) fundamental cell can be defined. In the state where $\lambda_1 = \lambda_2/\sqrt{2}$, the bct cell is a cube (i.e. the lattice is bcc). Since the transverse loads are zero (at any stage on the primary path), crystal symmetry implies that the axial load is also zero at $\lambda_1 = \lambda_2/\sqrt{2}$; i.e. the crystal is *unstressed* bcc. According to the nature of atomic binding, the axial load must be tensile as $\lambda_1 \to \infty$ and compressive as $\lambda_1 \to 0$; it follows that the existence of two zeros (i.e. at fcc and bcc) in the primary path of [100] loading also implies a third. Thus a third unstressed equilibrium crystal structure not possessing any higher symmetry than tetragonal must exist in general.

The predicted form of the [100] loading path is thus as shown in Fig. 2.2. [For completeness, it is noted that only two load zeros would occur for a "special case" of interatomic interactions for which the load would be tangent to the abscissa at a cubic (unstable) state.] The zeros on the loading path correspond to extrema of potential energy (with the central zero corresponding to an unstable, local energy maximum). Actual computations of such paths (i.e. exhibiting multiple loading zeros) were first carried out by Milstein (1970a, 1971, 1973b). His initial calculations were for lattice models of iron (Fe) and

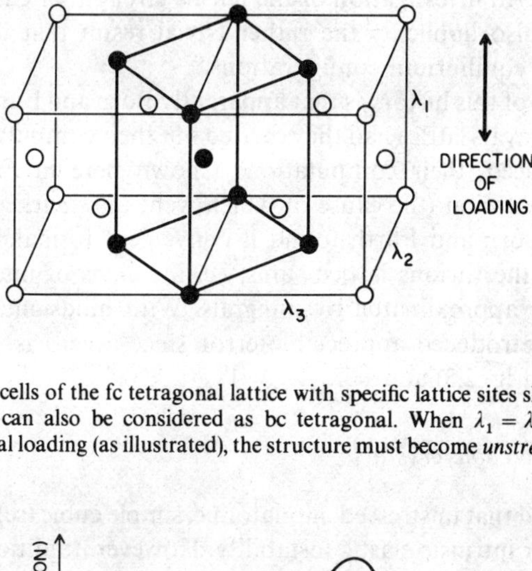

FIG. 2.1. Two fundamental cells of the fc tetragonal lattice with specific lattice sites shown with "solid" centers to indicate that the structure can also be considered as bc tetragonal. When $\lambda_1 = \lambda_2 = \lambda_3 = 1$, the structure is unstressed fcc. Under uniaxial loading (as illustrated), the structure must become *unstressed* bcc at $\lambda_2 = \lambda_3 = \lambda_1\sqrt{2}$.

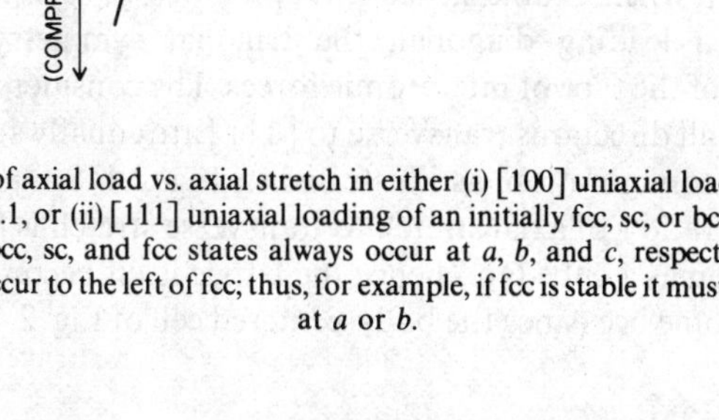

FIG. 2.2. General behavior of axial load vs. axial stretch in either (i) [100] uniaxial loading of an initially fcc or bcc lattice, as illustrated in Fig. 2.1, or (ii) [111] uniaxial loading of an initially fcc, sc, or bcc lattice, as illustrated in Fig. 2.4. In the [111] case, the bcc, sc, and fcc states always occur at *a*, *b*, and *c*, respectively. In the [100] case, the restriction is that bcc must occur to the left of fcc; thus, for example, if fcc is stable it must occur at *c*; then bcc could be at *a* or *b*.

nickel (Ni); in both cases, the fcc state was found at the right-hand zero. For Fe, the bcc state was at the left-hand zero, while for Ni it was in the central (and hence unstable) unloaded position. More recently, such paths have been computed for the complete family of Morse function lattices, including the "special case" where the load is tangent to the abscissa at the bcc state (cf. Section 5).

For historical purposes, it is interesting to mention that Born and co-workers concluded that the central force model of a crystal did *not* exhibit a stress free, tetragonal equilibrium state. This conclusion apparently was based upon Born and Fürth's (1940) study of [100] uniaxial loading of a fcc crystal and a study of rhombohedral lattices by Peng and Power (1942) as communicated by Born. Indeed, Peng and Power (1942) state: "There [i.e. in Born and Fürth, 1940] it was shown, or at least made plausible by numerical calculation, that the one-parametric set of configurations with a rectangular prismatic cell, which we obtain by extending a cubic cell parallel to one side, contains no other equilibrium positions." Also, they conclude: ". . . with the help of a certain characteristic function, it is shown by numerical calculation that the fcc lattice corresponds to the absolute minimum of potential energy, and that no extrema, other than fcc, sc, and bcc, exist." In listing the significant results of his group's pioneering investigations of the behavior of central force lattices, Born (1943) includes the comment: "An investigation of the tensile strength of cubic crystals (Born and Fürth, 1940) contains also implicitly the rather trivial result that tetragonal (non-cubic) Bravais lattices are not equilibrium configurations."

A final curious aspect of this history is that apparently Born and Fürth *would have* seen the stress-free tetragonal Bravais lattice had they carried out their computations to greater [100] lattice contraction. Indeed, their computations (shown here in Fig. 2.3) did exhibit a minimum load in compression (the cause of which went unremarked). Without the aid of electronic computers, Born and Fürth found it convenient to make simplifying approximations in calculating the various lattice sums; i.e. the sums of interactions beyond first nearest neighbors were approximated by integrals. With hindsight, we can see that this technique apparently introduced appreciable error since a zero is required in Born and Fürth's loading curve (Fig. 2.3) at $\alpha = \lambda_1/\lambda_2 = 1/\sqrt{2} \approx 0.71$.

2.2. *Instability of simple cubic crystals*

It is generally assumed that unstressed, monatomic, simple cubic (sc) crystals are not found in nature owing to their intrinsic elastic instability. However, traditional theoretical proofs (that sc lattices are unstable) rely upon central force models of bonding and speculations about the nature of the interatomic potential (Born and Huang, 1954). A simpler, more general proof (free from assumptions about the details of the specific atomic binding) was recently given by Milstein *et al.* (1980). Their proof emerged as a consequence of their study of cubic crystals in uniaxial loading along a leading diagonal (i.e. in unconstrained [111] uniaxial loading). Briefly, when a cubic lattice (either fcc, bcc, or sc) is loaded homogeneously and uniaxially along a leading diagonal, the trigonal symmetry can be maintained throughout, regardless of the type of interatomic forces. The considered path of deformation is axisymmetric, in that all directions transverse to [111] are equally stretched. Let the lattice reside initially in a fcc configuration, as illustrated in Fig. 2.4. Subject the lattice to [111] compression; when the ratio r of axial stretch to transverse stretch is halved, the Bravais cell (shown in Fig. 2.4) becomes a cube (and hence the lattice itself becomes sc). Halving r again causes the lattice to become bcc (since the body-centered cell of Fig. 2.4 then becomes a cube).

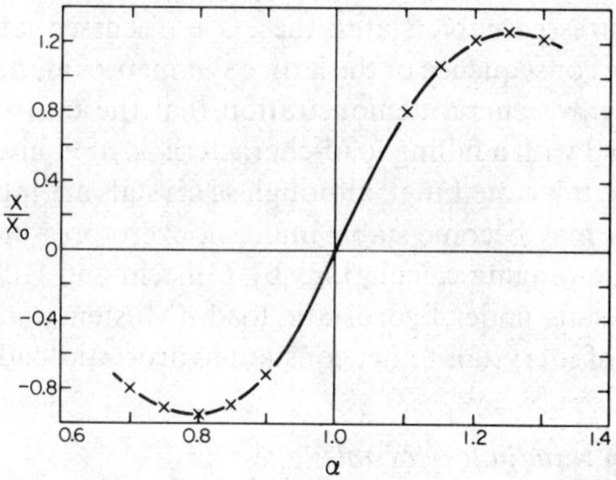

FIG. 2.3. [100] loading curve of Born and Fürth (1940) for a fcc Lennard-Jones-model crystal. $\alpha = \lambda_1/\lambda_2$ (as defined by their equation 2). There is some confusion over the ordinate variable, as previously noted by Hill and Milstein (1977), viz. Born and Fürth "define their conjugate variable as 'the external force X (per unit cross section)' or 'the stress (or the pressure) X/X_0.' However, examination of the formulae that define these quantities (their equations 12, 22, and 29) reveals that both X and X/X_0 are actually the Green's conjugate stress l_1/λ_1, and not the force l_1 or the 'pressure' $l_1/\lambda_2\lambda_3$."

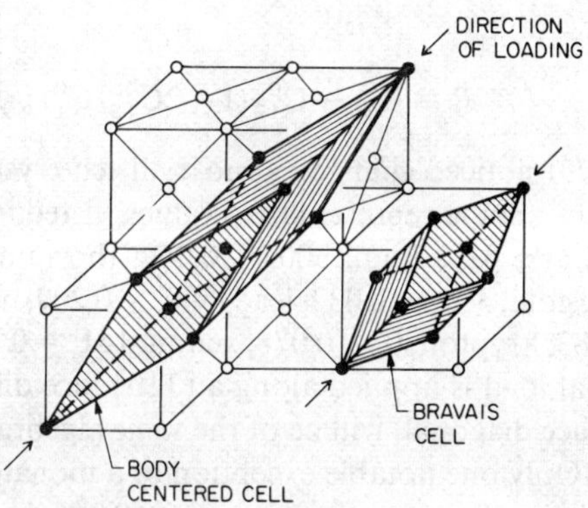

FIG. 2.4. Illustration of cells that become cubic in [111] loading of an initially fcc crystal.

The axisymmetric path of deformation under [111] loading consequently passes through all three cubic configurations, and these occur in the sequence fcc, sc, bcc as r decreases between ∞ and 0. Moreover, since there is no lateral load, the cubic symmetry in each configuration means that the axial load vanishes also. Since the load is necessarily compressive as $r \to 0$ and tensile as $r \to \infty$, the qualitative form of the response must also be as depicted in Fig. 2.2, given that the load vanishes thrice only (five zeros are of course geometrically possible, but scarcely conceivable within the constraint of known crystal binding). Milstein *et al.* (1980) carried out actual computations of a [111] loading path spanning the three unstressed cubic states; these are discussed later. Since the qualitative response is an inevitable consequence of the lattice symmetries alone, it therefore amounts to a rigorous and thoroughly general demonstration that the unstressed sc lattice is always unstable, since associated with a falling load-characteristic or equivalently a local maximum in strain energy. Finally, it is noted that, although sc crystals are intrinsically unstable in the unstressed sc state, they may become stable under nonzero pressure. These conclusions are consonant with the wide-ranging calculations by Milstein and Hill (1977, 1978, 1979a) for Morse-model cubic crystals under hydrostatic load. (Milstein and Hill, 1979a, did indeed compute a stable range of sc crystals under constant hydrostatic loading, as is discussed later.)

2.3. *A negative Poisson ratio in fcc crystals*

For a cubic crystal in [110] loading (where a uniaxial load acts parallel to the [110] face diagonal), the initial Poisson ratios υ_{110} and υ_{001} can be defined along two natural principal axes perpendicular to [110], viz. the face diagonal [110] and the cube edge [001], respectively. The Poisson ratios are readily computed from experimental values of elastic moduli C^0_{ij} of unstressed cubic crystals according to

$$\upsilon_{1\bar{1}0} = (R - 2C^0_{44})/(R + 2C^0_{44}),$$

and

$$\upsilon_{001} = 4C^0_{12}C^0_{44}/C^0_{11}(R + 2C^0_{44}), \tag{2.1}$$

where

$$R \equiv C^0_{11} + C^0_{12}(1 - 2C^0_{12}/C^0_{11}).$$

Milstein and Huang (1979) noticed that for almost all fcc crystals, over a wide range of temperature $\upsilon_{1\bar{1}0}$ is negative. Some representative values (listed in the sequence "substance, temperature in K, $\upsilon_{1\bar{1}0}$") are: palladium, 300, −0.049; thorium, 300, −0.223; copper, 0, −0.138; silver, 0, −0.093; gold, 300, −0.032; lead, 300, −0.209; nickel, 0, −0.051; neon, 4.7, −0.13; argon, 82.3, −0.083; krypton, 0, −0.078; xenon, 151, −0.10. The implication is that, in general, when a uniaxial load is applied along a [110] face diagonal, the strain $\varepsilon_{1\bar{1}0}$ (i.e. along the perpendicular face diagonal) will be of the *same* algebraic sign as the strain ε_{110} in the direction of loading. (Only one notable exception of a monatomic fcc crystal with large, positive $\upsilon_{1\bar{1}0}$ was found; i.e. for aluminum $\upsilon_{1\bar{1}0}$ is 0.266 and 0.272 at 0 and 300 K, respectively.)

A theoretical basis for the above phenomenon is discussed in detail in a later Section where we examine the nature of branching, under a strict uniaxial load, from a tetragonal path of [100] loading (of an initially bcc or fcc crystal) to an orthorhombic path. Loading on the tetragonal path is discussed in Section 2.1 and represented in Fig. 2.2. On the orthorhombic path, a fundamental cell with mutually orthogonal, but generally unequal, edges exists at any stage; the uniaxial load is parallel to one of the edges. Here it is mentioned briefly that the negative values of υ_{110} can be understood as a general consequence of such branching, as demonstrated by Milstein and Huang (1979). In particular, the lattice model computations of Milstein and Huang (1978) showed that a path of [110] uniaxial loading of an initially fcc crystal corresponds with the orthorhombic path (reckoned relative to a body-centered fundamental cell) that branches (under [100] uniaxial load) from the tetragonal path. After branching, the theoretical Poisson ratios along the two principal transverse directions are of opposite algebraic sign (at any stage); furthermore, their magnitude becomes infinite at the branch point. The generality (i.e. independence of crystal model) of the behavior (revealed by the theoretical computations) is supported by examination of the relevant Poisson ratios of actual crystals.

3. Lattice Models

In order to perform computations of load, strain, and elastic moduli, a crystal model is required for computing the internal energy W per atom as a function of geometric variables q_r that define the state of homogeneous strain of the crystal:

$$W = W(q_1, q_2, \ldots, q_6). \tag{3.1}$$

Generally W is taken as a pairwise sum of interactions ϕ over a large enough region of space to obtain convergence to the desired number of significant figures. The loads, stresses, and elastic moduli are then evaluated as indexed lattice summations of first and second derivatives of ϕ; additional computational details are given in later Sections.

In the pioneering work of Born and co-workers (Born, 1940; Misra, 1940; Born and Furth, 1940; Born and Misra, 1940; Furth, 1941a, b; Peng and Power, 1942), the potential energy $\phi(r)$ between two atoms (in a crystal) separated by a distance r was taken to be of the "inverse power" or "Mie" form, viz.

$$\phi(r) = D\frac{nm}{n-m}\left[\frac{1}{n}\left(\frac{r_0}{r}\right)^n - \frac{1}{m}\left(\frac{r_0}{r}\right)^m\right], \tag{3.2}$$

where $n > m$. $d\phi/dr$ vanishes for $r = r_0$; $\phi(r_0) = -D$, the dissociation energy for a given pairwise "bond"; and D, n, m, and r_0 are empirical potential parameters determined from the experimental crystal properties (e.g. lattice parameter, compressibility, cohesive energy, etc. of the actual crystal). The particular case of $m = 6$, $n = 12$ is the "well-known" Lennard-Jones function. In the absence of electronic computers, the numerical techniques for determining the required lattice summations were complicated and tedious (or, as mentioned earlier in connection with Fig. 2.3, inaccurate). This might explain why Born and Furth (1940) limited their non-hydrostatic loading studies to the one example of [100] loading of an initially fcc crystal with Lennard-Jones interatomic interactions. Nevertheless, the inverse power function (3.2) was the most mathematically tractable (among various types of two-body potential functions that could have been selected); although other functions, including the Morse function

$$\phi(r) = D\{\exp[-2\alpha(r-r_0)] - 2\exp[-\alpha(r-r_0)]\}, \tag{3.3}$$

have certain inherent theoretical advantages (see, for example, Girifalco and Weizer, 1959a, b). Lattice summations of exponentials are of course readily evaluated on electronic computers. Such considerations prompted Milstein (1973a) to suggest a generalization of (3.3), viz.

$$\phi(r) = \phi_m(r) = \frac{D}{m-1}\{\exp[-m\alpha(r-r_0)] - m\exp[-\alpha(r-r_0)]\}. \tag{3.4}$$

D, α, r_0, and m are empirical parameters; $(d\phi/dr) = 0$ at $r = r_0$; and $\phi(r_0) = -D$, the dissociation energy of two atoms. (The generalized Morse function reduces to the usual form for the particular case of $m = 2$.)

A wide variety of computations of crystal behavior at finite strain have employed (3.4).

Milstein (1973a) determined sets of explicit functions of the form (3.4) for a number of cubic crystals from experimental values of the stress-free lattice parameter and elastic constants C_{11}^0 and C_{12}^0 (Tables 3.1 and 3.2). The particular set of functions ϕ that was determined for the element Ni is of special interest in that (i) among the fcc metals that were examined, Ni comes the closest to obeying the Cauchy condition $C_{12}^0 = C_{44}^0$ and (ii) reasonably good agreement was obtained between theoretical and experimental pressure–volume relations in the region of anharmonic (i.e. nonlinear) behavior. The theoretical crystal model of course exhibits a reasonably accurate stress–strain curve in the linear region (since experimental values of elastic moduli were used to determine the atomic parameters defining ϕ). Included in this set of functions ϕ are both short-range, steep functions and long-range, shallow functions. For example, among the functions appropriate to Ni, with the "steepest" potential ($\phi = \phi_{1.25}$) about 92% of the cohesive energy per atom in the crystal comes from nearest-neighbor interactions and less than 0.1% from, say, all of the sixth or seventh near-neighbor interactions, whereas with the "shallowest" potential ($\phi = \phi_6$) only about 51% of this cohesive energy can be attributed to nearest-neighbor interactions while as much as 3.1% comes from seventh near-neighbors. Nevertheless, the model was found to be self-consistent in the sense that the response of the crystal to various modes of loading, even after very large strains, was surprisingly insensitive to the details of the function ϕ (assuming, of course, that the same empirical data were used to specify each function ϕ in the set).

Examples of computational results are shown in Figs. 3.1–3.3. The computations employ the generalized Morse function interactions over "all possible pairs." (In practice, the lattice summations are performed over a sufficient number of interactions to obtain convergence to at least five significant figures.) Figures 3.1 and 3.2 show computations for [100] loading (the general behavior of which is discussed in Section 2.1). Figure 3.3 is for a shear-like mode of deformation in which the load is constrained to remain parallel to the (100) planes and in the [010] direction so that no other load acts on the (100) planes (i.e. the load is maintained parallel to a cell face and to a cell edge of the (initially cubic) unit cell). The crystal loading is indicated schematically in Fig. 3.4. This mode exhibited a variety of interesting features, which are discussed later, including the unexpected result that the shear angle θ_F (between the [100] and [010] directions), as well as all other lattice parameters, could be varied continuously up to a maximum critical variation occurring at $\Delta\theta_F \approx 7.6°$; further continuous

TABLE 3.1. Experimental data used in determining the generalized Morse parameters listed in Table 3.2

Element	Lattice parameter a_0 (Å)	Elastic constants C_{11}^0 (10^{12} dyn cm^{-2})	C_{12}^0
Aluminum	4.0490	1.143	0.619
Nickel	3.5238	2.612	1.508
Copper	3.6153	1.762	1.249
Silver	4.0856	1.315	0.973
Palladium	3.8902	2.341	1.761
Lead	4.9495	0.555	0.454
Gold	4.0783	2.016	1.697
Potassium	5.344	0.0416	0.0341

TABLE 3.2. Potential parameters for the families of functions $\phi_m(r) = [D/(m-1)]\{\exp[-m\alpha(r-r_0)] - m\exp[-\alpha(r-r_0)]\}$ determined for several elements from experimental values of the elastic moduli C^0_{11} and C^0_{12} and the condition of stress-free equilibrium at the value of lattice parameter equal to a_0 (from Milstein, 1973a)

Element	m	D (10^{-12} erg)	αa_0	r_0/a_0
Aluminum	6.00	0.13911	5.7635	0.71859
	4.00	0.14749	7.4856	0.71410
	3.00	0.15099	8.8354	0.71225
	2.00	0.15437	10.9792	0.71069
	1.50	0.15591	12.7273	0.71010
	1.25	0.15644	13.9588	0.70989
Nickel	6.00	0.28925	3.9802	0.74555
	4.00	0.33460	5.5272	0.72925
	3.00	0.34676	6.7847	0.72255
	2.00	0.35059	8.7660	0.71727
	1.50	0.35018	10.3296	0.71539
	1.25	0.35037	11.3840	0.71479
Copper	6.00	0.22147	1.7948	0.98029
	4.00	0.37293	2.6586	0.89154
	3.00	0.47499	3.4637	0.84309
	2.00	0.56774	4.8418	0.79800
	1.50	0.59623	5.9297	0.78074
	1.25	0.60329	6.6431	0.77466
Silver	6.00	0.22790	1.5179	1.08791
	4.00	0.40571	2.2572	0.97494
	3.00	0.54151	2.9571	0.91032
	2.00	0.68561	4.1332	0.84658
	1.50	0.73828	5.1629	0.82102
	1.25	0.75304	5.8051	0.81174
Palladium	6.00	0.34358	1.4142	1.14235
	4.00	0.62347	2.1067	1.01775
	3.00	0.84651	2.7654	0.94551
	2.00	1.09678	3.9289	0.87269
	1.50	1.19384	4.8649	0.84282
	1.25	1.22044	5.4811	0.83110
Lead	6.00	0.15617	0.8781	1.68561
	4.00	0.30446	1.3236	1.45605
	3.00	0.44469	1.7512	1.31973
	2.00	0.64609	2.5336	1.16881
	1.50	0.74816	3.1993	1.09546
	1.25	0.79350	3.5815	1.07928
Gold	6.00	0.33451	0.6539	2.21933
	4.00	0.64861	1.0059	1.86974
	3.00	0.96564	1.3300	1.68710
	2.00	1.44487	1.9439	1.46740
	1.50	1.70619	2.4783	1.35491
	1.25	1.87218	2.7277	1.35346
Potassium	6.00	0.03098	0.9186	1.67614
	4.00	0.05881	1.3756	1.47097
	3.00	0.08297	1.8146	1.34898
	2.00	0.11375	2.6061	1.21963
	1.50	0.12736	3.2608	1.16206
	1.25	0.13331	3.6414	1.14917

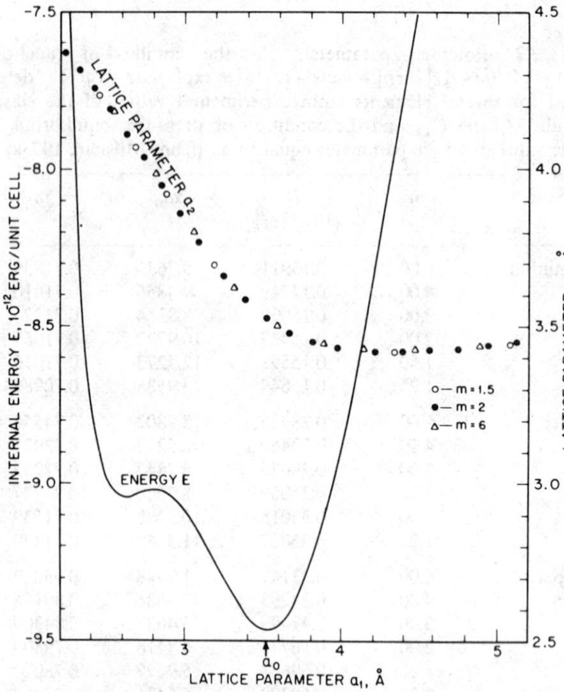

FIG. 3.1. Computations of transverse lattice parameter a_2 (where $a_2 = a_3$) as a function of axial lattice parameter a_1 in [100] uniaxial loading of crystal models of fcc Ni; as a_1 is varied, the value of a_2 is also varied in such a manner as to keep the transverse stresses σ_2, σ_3 zero. (In practice, this is accomplished by an iterative technique.) The variation of the energy per unit cell E (for the case $\phi_m = \phi_2$) is also shown. (From Milstein, 1973b.)

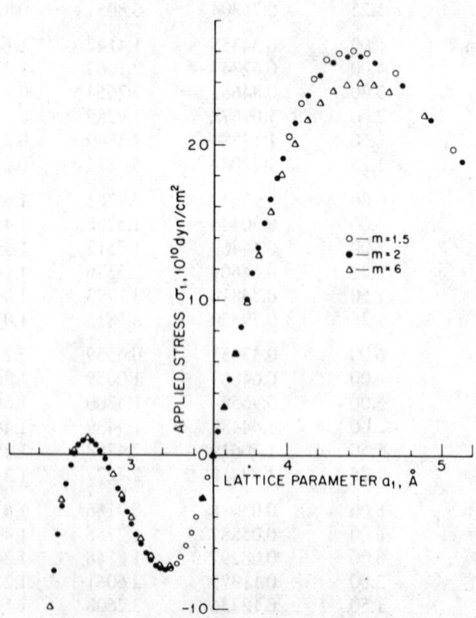

FIG. 3.2. Computations of true axial stress σ_1 as a function of axial lattice parameter a_1 in [100] uniaxial loading of crystal models of fcc Ni. (From Milstein, 1973b.)

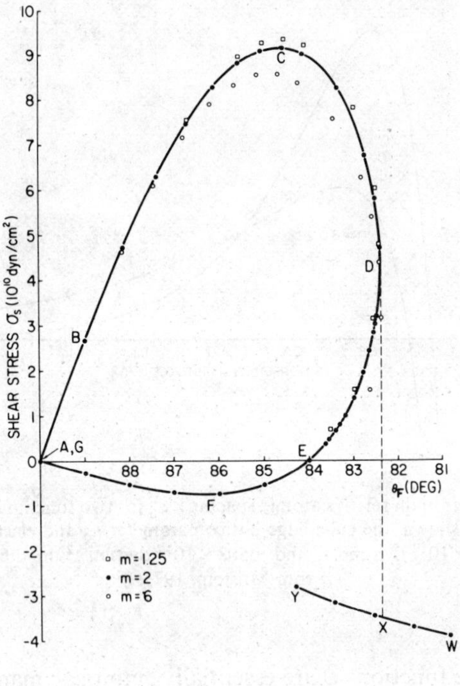

FIG. 3.3. Theoretical values of shear stress σ_s vs. shearing angle θ_F. (From Huang *et al.*, 1974.) The state of loading is shown schematically in Fig. 3.4, and is explained in greater detail in Section 7.

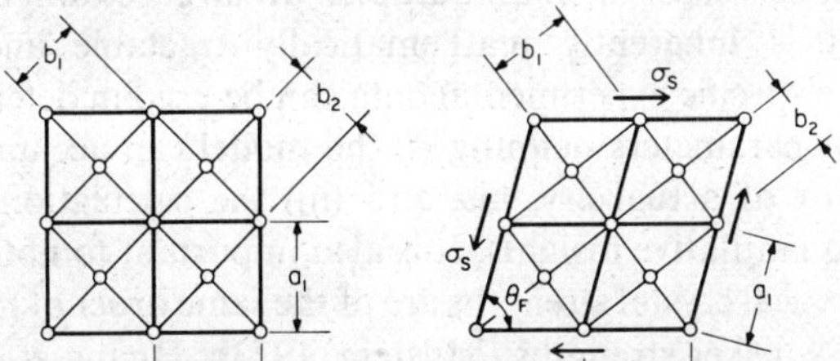

FIG. 3.4. Schematic representation of the shear-like loading for which the computations of σ_s vs. θ_F are given in Fig. 3.3.

variation of θ_F required a first-order transition (i.e. a discontinuity in the other lattice parameters) since a continuous path taking θ_F beyond $\approx 82.4°$ did not exist. (There did exist a continuous path taking θ_F back to 90°, transforming the initial fcc structure into a bcc structure, and passing through an unstressed tetragonal configuration on the way.) What concerns us here, however, is the remarkable self-consistency of the computations; in particular, the maximum theoretical stresses (Figs. 3.2 and 3.3) vary by only about 10% for the different interactions ϕ_m appropriate to fcc Ni, even though the maximum possible attractive interatomic forces vary by about 100% among the ϕ_m (Fig. 3.5). Similar self-consistency was also obtained in the computations of the elastic moduli.

There are, of course, a number of limitations inherent in the central potential model of a

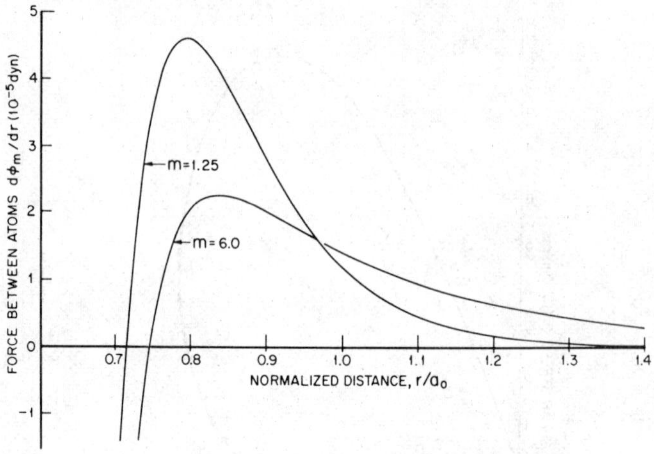

FIG. 3.5. Dependence of $d\phi_m/dr$ upon relative atomic spacing r/a_0 for two functions ϕ_m appropriate to fcc Ni. For crystals with the interactions shown, the cube edge lattice parameter a_0 and elastic constants C_{11}^0 and C_{12}^0 are, respectively, 3.5238 Å, 2.612×10^{12} dyn cm^{-2}, and 1.508×10^{12} dyn cm^{-2} in the unstressed fcc configuration. (From Milstein, 1973a.)

crystal. In particular, the functions ϕ are essentially empirical; many-body interactions may play an important role; and the elastic moduli of the theoretical crystals possess Cauchy symmetry while real crystals often do not. On the other hand, the model has found widespread use in studies of the solid state and, with but few exceptions, has been the only reasonable atomic model used in computations of large strain crystal mechanics. Its advantages are (i) it is inherently mathematically tractable and facilitates detailed computations; (ii) since specific experimental data can be used in determining the numerical values of the potential parameters defining ϕ, the model can accurately reflect important aspects of the behavior of actual crystals; and (iii) the calculations can lead to greater qualitative and often quantitative insights. It is also important to note that Morse function computations of theoretical crystal strengths are of the same order of magnitude as the upper limits of experimental whisker strengths (Milstein, 1973b; Huang *et al.*, 1974; Milstein and Huang, 1978; Milstein and Farber, 1980b; Milstein *et al.*, 1980).

Little has been done in the area of "first principles" computations of the mechanics of strained crystals, owing mainly to the complexity of quantum-mechanically based crystal models. In this regard, recent work by Esposito *et al.* (1980) is noteworthy. They used a variety of techniques to compute curves of axial stress versus elongation for copper (Cu) in [100] tensile *deformation* (as distinct from [100] *loading*; see below). Their computations employed two distinct band-theoretic approaches (viz. non self-consistent KKR and self-consistent ASW) and a new non-empirical pair potential derived "from a first-principles calculation of the cohesive energy as a function of volume without recourse to experimental information." Their crystal was deformed according to $\lambda_1 \geqslant 1; \lambda_2 = \lambda_3 = 1$, where λ_1, λ_2, and λ_3 are the stretches in the directions [100], [010], and [001], respectively. Thus, their path of loading is triaxial, with the stresses in [100], [010], and [001], respectively, varying according to $\sigma_1 \neq \sigma_2 = \sigma_3 \neq 0$ (in general). Their computations were of σ_1 vs. λ_1. For purposes of comparison, Milstein and Farber (1980b) computed the mechanical response of a Morse model of Cu using the empirical parameters determined earlier by Milstein (1973a). Their computations were of σ_1 and σ_2 on a path of [100] *deformation* (i.e. $\lambda_2 \equiv \lambda_3 \equiv 1$ at any stage) and of σ_1 and λ_2 ($= \lambda_3$) in [100] *loading*. Computations in the latter mode are more complicated since the transverse stretches λ_2 and λ_3 must be "adjusted" to obtain the state $\sigma_2 = \sigma_3 = 0$ at each stage of loading; i.e. for each $\lambda_1 \neq 1$, values of $\lambda_2 = \lambda_3$ are computed by an iterative technique to provide that $\sigma_2 = \sigma_3 = 0$; the values of λ_i so determined are then employed in the computations of σ_1 and the elastic moduli.

In the Morse-model computations of Milstein and Farber (1980b), the elastic moduli were computed in order to locate the instability associated with the invariant eigenstate $C_{22} = C_{23}$ (Hill and Milstein, 1977); the eigenstate was found to occur in tensile [100] *loading* at about 20% of the maximum theoretical stress, thus lowering the estimate of the "theoretical tensile strength" of Cu to within the range of experimentally based estimates. It is not particularly meaningful to speak of an equivalent state of instability in [100] *deformation*; indeed, the state $C_{22} = C_{23}$ is not even coordinate independent in [100] *deformation* in the sense that its location on the primary path depends upon the choice of geometric variables q_r employed in the computations of the C_{rs}.

As noted earlier, a primary path of [100] *loading* passes through three zeros; Milstein and Farber (1980b) demonstrated that three corresponding states of hydrostatic loading (one of which is unstressed) are reached on a path of [100] *deformation.*

Figure 3.6 compares the various crystal model computations of σ_1 vs. λ_1 for Cu. The results of Esposito *et al.* (1980) are in excellent agreement with the [100] tensile *deformation* computations of Milstein and Farber (1980b). This agreement is consonant with the findings of Milstein (1973a) and Huang *et al.* (1974), mentioned earlier, that the mechanical response computations were remarkably insensitive to the quantitative details of the potential function (provided the fcc crystals had the same elastic constants and lattice parameter in the unstressed state). The self-consistency of these combined results thus lends a large degree of credibility to the quantitative as well as qualitative phenomena revealed by such computations. [Some differences are observed in the initial slopes of the theoretical *deformation* curves in

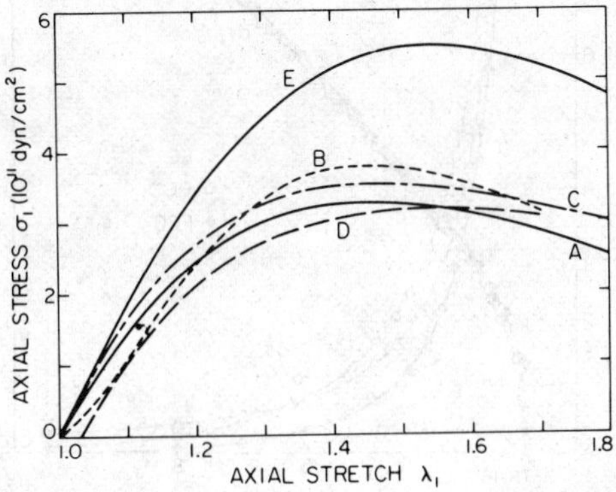

FIG. 3.6. Comparison of various computations of tensile axial stress σ_1 vs. stretch λ_1 for fcc Cu. *A* and *B* are for the Morse model in [100] *deformation* and [100] *loading*, respectively. *C*, *D*, and *E* represent the [100] *deformation* computations of Esposito *et al.* (1980) for the crystal modeled according to their "*ab initio* potential" (*C*) and augmented spherical wave (ASW) with Ewald correction (*D*) and non-self consistent KKR (*E*) band-theoretic computations. (From Milstein and Farber, 1980b.)

Fig. 3.6 owing to the fact that the theoretical values of the initial (i.e. unstressed) elastic constants differ slightly from the experimental values in the Esposito *et al.* computations, whereas the initial slopes in the computations of Milstein and Farber are in agreement with experiment (since experimental values of the elastic constants C^0_{11} and C^0_{12} are included in determining the empirical parameters of the lattice models). The initial slopes are, of course, given by C^0_{11} in [100] *deformation* and $C^0_{11}-2(C^0_{12})^2/(C^0_{11}+C^0_{12})$ in [100] *loading.*]

Another useful and interesting characteristic of the Morse function is that it conveniently permits the modeling of interatomic interactions that range from the "nearest neighbor" limit (for the case $\beta \equiv \exp(\alpha r_0) \to \infty$) to the "continuum" limit (for $\beta \to 16$) (Milstein and Hill, 1977). This is readily seen in Fig. 3.7, where $r_0/r_1 \to 1$ as $\beta \to \infty$ and $r_0/r_1 \to \infty$ as $\beta \to 16$; r_1 is the nearest neighbor distance in the unloaded crystal; r_0 is the value of r at the minimum of ϕ. Thus the parameter $\log \beta$ serves as an effective potential range indicator; larger values of $\log \beta$ mean shorter range and steeper functions $\phi(r)$. That is, as β becomes large, the nearest neighbor distance in the unstressed crystal approaches the value that would exist in the limiting approximation of nearest neighbor interactions only. [This limit is approached most rapidly for the fcc crystal (which has the greatest number of nearest neighbors) and most slowly for the bcc crystal (in which the second nearest neighbors play a more important role; i.e. relative to the nearest neighbor distance, the second nearest neighbor distance is smallest for the bcc crystal).] As β decreases, the ratio r_0/r_1 increases; thus, the distance from the origin to the potential minimum becomes larger, relative to r_1. In other words, the potential

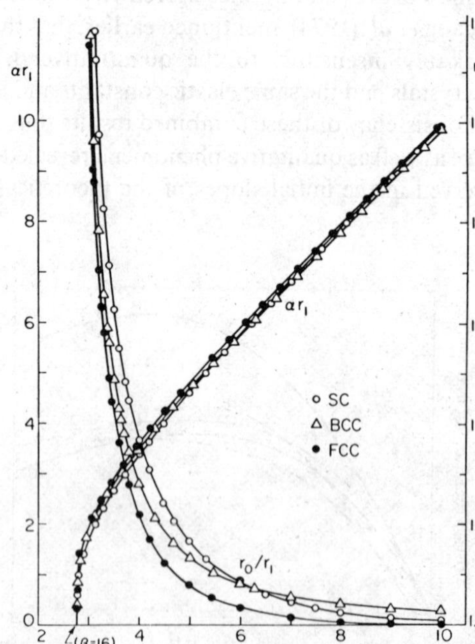

FIG. 3.7. Dimensionless quantities αr_1 and r_0/r_1 vs. $\log \beta$ for the three cubic Bravais lattices in the Morse model. The nearest neighbor distance at zero pressure is r_1 and the distance from the origin to the potential minimum is r_0. Thus, the limits $\beta \to 16$ and $\beta \to \infty$ correspond, respectively, to the limits of arbitrarily long range interactions (i.e. the "continuum" limit) and nearest neighbor interactions only (or the "hard sphere" limit). (From Milstein and Hill, 1977.)

becomes "longer range" and "shallower." The intermediate values of $\log \beta$, say from about 3 to 8, provide the most realistic models of actual crystals (Girifalco and Weizer, 1959a, b; Milstein, 1970b, 1973a). The computation of Fig. 3.7 is discussed briefly in Section 4.

4. Hydrostatic Loading

Milstein and Hill (1977, 1978, 1979a) recently completed a three-part series of papers on the elastic properties of simple, monatomic, cubic crystals at arbitrary pressure. A brief review of this work is presented here.

4.1. *Bulk and shear moduli*

In the first two papers, Milstein and Hill (1977, 1978) derive expressions for calculating (as functions of all-round stretch λ) the pressure $P(\lambda)$, the bulk modulus $\kappa(\lambda)$, and the shear moduli $\mu(\lambda)$ and $\mu'(\lambda)$ for cubic crystals in which the atomic bonds are modeled by pairwise interaction energies ϕ depending only on current separation between the atoms, viz.

$$P(\lambda) = -\frac{1}{3V\lambda}\sum \delta^2\phi'(r^2), \tag{4.1}$$

$$3\kappa(\lambda) = \frac{1}{3V\lambda}\left[2\lambda^2\sum\delta^4\phi''(r^2) - \sum\delta^2\phi'(r^2)\right], \tag{4.2}$$

$$\mu(\lambda) = \frac{\lambda}{V}\sum[(\delta_1^4 - \delta_1^2\delta_2^2)\phi''(r^2)] - P(\lambda), \tag{4.3}$$

$$\mu'(\lambda) = \frac{2\lambda}{V}\sum[\delta_1^2\delta_2^2\phi''(r^2)] - P(\lambda), \tag{4.4}$$

where $\phi'(r^2)$ and $\phi''(r^2)$ are, respectively, the first and second derivatives of ϕ with respect to r^2, and r^2 is the square of the magnitude of the lattice vector $\mathbf{r}$ connecting two lattice sites in the crystal. [Equation (4.2) is equivalent to

$$3\kappa(\lambda) = \frac{2\lambda}{V}\sum[(\delta_1^4 + 2\delta_1^2\delta_2^2)\phi''(r^2)] + P(\lambda);$$

thus, in the central potential model,

$$\kappa - \tfrac{2}{3}\mu = \mu' + 2P,$$

as given in Milstein and Hill (1978, equation (2.6)).]

The undeformed cubic configuration at pressure $P = 0$ is taken to be the reference state ($\lambda = 1$); in this state, the volume per atom is V and $\mathbf{r} \equiv \boldsymbol{\delta}$. The components of $\boldsymbol{\delta}$ can be expressed as $\delta_i = cm_i$ ($i = 1, 2, 3$), where $2c$ is the length of the edge of the fundamental cubic cell in the reference state, and the m_i are integer indices that identify the particular site. Useful algorithms for performing the lattice summations in (4.1)–(4.4) are: $m_1 + m_2 + m_3$ must be even for the fcc lattice; the m_i are either all even or all odd for a bcc lattice; and the m_i are all even for the sc lattice.

Born and his associates approached the subject of crystal elasticity from the viewpoint of the Green moduli. This approach seems to have set the stage for many subsequent investigations of the elasticity of crystals (primarily cubic) at finite deformation. Milstein and Hill (1978), however, adopted the view that the shear moduli $\mu(\lambda)$ and $\mu'(\lambda)$ and bulk modulus $\kappa(\lambda)$ provide a more direct, physically meaningful measure of the elastic response of cubic crystals at nonzero pressure, including elastic stability (Hill and Milstein, 1977; Milstein and Hill, 1979a) under constant hydrostatic loading. In (4.2)–(4.4), κ is the ordinary bulk modulus and μ and μ' are the usual shear moduli in the relation between the cubic-axes components of the Cauchy stress increment $\delta\sigma_{ij}$ and the rotationless strain increment $\delta\varepsilon_{ij}$ (reckoned relative to the current configuration under P). Apparently no such simple, direct mechanical interpretation applies to the Green moduli (or linear combinations thereof) in a current state of nonzero pressure, as is evident from the relations (Hill and Milstein, 1977)

$$\left.\begin{aligned} C_{11} &= \left(\kappa + \frac{4}{3}\mu + P\right)\Big/\lambda, \\ C_{12} &= \left(\kappa - \frac{2}{3}\mu - P\right)\Big/\lambda, \\ C_{44} &= (\mu' + P)/\lambda. \end{aligned}\right\} \tag{4.5}$$

Thus (4.2)–(4.4) are not simply generalizations of formulae derived by Born (1940) for the Green moduli at zero pressure ($\lambda = 1$) since the physical implications are more significant than just "a generalization." [Elastic moduli in general can be defined according to $C_{rs} \equiv \frac{1}{V}(\partial^2 W/\partial q_r \partial q_s)$, where W is the energy per atom and V is the reference volume per atom. In (4.5), the q_r are specifically the Green variables. (See Hill and Milstein, 1977, for further detail.)]

Milstein and Hill (1977, 1978, 1979a) employed (4.1)–(4.4) to compute the pressure and the bulk and shear moduli of the entire Morse-model family of fcc, bcc, and sc monatomic crystals under arbitrary volumetric deformation. These apparently are the first theoretical computations of the shear moduli of cubic crystals under nonzero hydrostatic loading (based on any reasonable atomic model) to have appeared in the literature. Milstein and Hill characterized the essential elastic behavior in terms of the roles of (i) the "potential range indicator," β (or $\log\beta$), (ii) the crystal structure, and (iii) the all-round stretch λ.

For the Morse model, the condition that $P(1) = 0$ can be expressed (Milstein and Hill, 1977) as $\beta = \sum me^{-mA}/\sum me^{-2mA}$, where $m = (m_1^2 + m_2^2 + m_3^2)^{1/2}$ and $A \equiv \alpha c$. ($\alpha r_1 = A\sqrt{2}$ for the fcc lattice, $A\sqrt{3}$ for the bcc lattice, and $2A$ for the sc lattice.) This enables αr_1 (and $r_0/r_1 = (\log\beta)/\alpha r_1$) to be computed versus $\log\beta$ (Fig. 3.7); the computations, in turn, provide the interpretation of $\log\beta$ as the potential range indicator.

4.2. *Stability*

In the third paper in the series, Milstein and Hill (1979a) present a complete account of the application of the principles of bifurcation analysis for general materials to the particular case of cubic crystals subjected to hydrostatic loading. The analyses are carried out in a manner equivalent to Hill (1978, Chapter III, Section C2) but without recourse to the general mathematical apparatus for handling follower-loadings. Their treatment of crystal stability is classical in that (i) the loading environment is fully specified, to sufficient order and in both its active and passive modes, and (ii) the potential energy of the system as a whole is examined in all the nearby, possibly inhomogeneous, configurations allowed by the kinematic constraints, if any. Under a hydrostatic pressure that does vary during any departure from a considered configuration of equilibrium, classical elastic stability is guaranteed if (Milstein and Hill, 1979a, equation (2.16); see also Hill and Milstein, 1977, equation (42))

$$\kappa(\varepsilon_{11}+\varepsilon_{22}+\varepsilon_{33})^2+\tfrac{2}{3}\mu[(\varepsilon_{11}-\varepsilon_{22})^2+.+.]+4\mu'(\varepsilon_{12}^2+.+.)>0, \qquad (4.6)$$

where ε_{ij} denotes the Eulerian strain-rate in components on the cubic axes. Since the three terms are independently variable, necessary and sufficient conditions for the inequality are that

$$\kappa(P)>0, \quad \mu(P)>0, \quad \mu'(P)>0, \qquad (4.7)$$

simultaneously. Milstein and Hill (1979a) identified the primary eigenstates and corresponding eigensolutions η_{ij} associated with loss of stability on a fundamental path at a pressure $P = Q$ as follows. (i) $\kappa(Q) = 0$, $\mu(Q) > 0$, $\mu'(Q) > 0$ with eigensolution $\eta_{11} = \eta_{22} = \eta_{33} \neq 0$; $\eta_{12} = \eta_{23} = \eta_{31} = 0$ (the eigenmode is necessarily homogeneous and purely volumetric, coincident with $dP/d\lambda = 0$). (ii) $\mu(Q) = 0$, $\kappa(Q) > 0$, $\mu'(Q) > 0$ with solutions such that $\eta_{11}+\eta_{22}+\eta_{33} = 0$; $\eta_{12} = \eta_{23} = \eta_{31} = 0$; (iii) $\mu'(Q) = 0$, $\kappa(Q) > 0$, $\mu(Q) > 0$ with solutions such that $\eta_{11} = \eta_{22} = \eta_{33} = 0$; any ratios $\eta_{12}:\eta_{23}:\eta_{31}$.

Milstein and Hill (1979a) used their bulk and shear moduli computations to determine the stable ranges (and potential bifurcations at the range limits) of the entire Morse-model family of fcc, bcc, and sc monatomic crystals under constant hydrostatic compression and tension. They showed, in particular, that the positive-valued range of each modulus (i.e. the range of λ-values over which each modulus is positive) depends only on the value of the potential range indicator $\log\beta$ and the particular lattice structure; it thus follows that the stable range of each lattice depends only on $\log\beta$ and crystal structure. Table 4.1 indicates the domains of positive moduli for the three cubic crystals, and Figs. 4.1 and 4.2 show the dependence of these domains upon $\log\beta$. The application of (4.7) to the data in Figs. 4.1 and 4.2 and Table 4.1 leads to the ranges of stability of the three cubic crystal lattices, as are summarized in Fig. 4.3. (This work is apparently the first where a theoretical range of elastic stability has been revealed for sc crystals.)

TABLE 4.1. Qualitative summary of the domains of all-round stretch over which the individual elastic moduli are positive. The subscripts F, B and S refer to the fcc, bcc, and sc lattices, respectively (the domains of positive κ were virtually the same for all three lattices) (from Milstein and Hill, 1979a)

Modulus	Domain of positive modulus
κ	$\lambda < \Lambda_\kappa$, $\Lambda_\kappa > 1$
μ_F	$\lambda < \Lambda_{\mu F}$, $\Lambda_{\mu F} > 1$
μ'_F	$\lambda < \Lambda_{\mu' F}$, $\Lambda_{\mu' F} > 1$
μ_B	$\lambda < \Lambda_{\mu B(L)}$ or $\lambda > \Lambda_{\mu B(R)} > 1$
μ'_B	$\lambda < \Lambda_{\mu' B}$, $\Lambda_{\mu' B} > 1$
μ_S	$\lambda < \Lambda_{\mu S}$, $\Lambda_{\mu S} > 1$
μ'_S	$\lambda > \Lambda_{\mu' S} > 1$

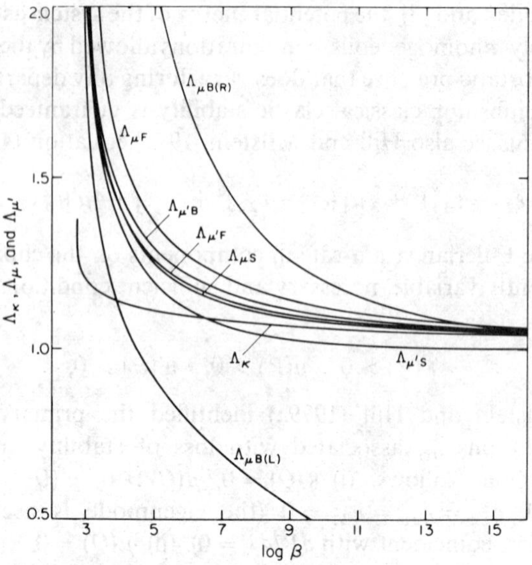

FIG. 4.1. Values of stretch at which the bulk and shear moduli of Morse-function cubic crystals undergo changes in algebraic sign as functions of the potential range indicator log β. (From Milstein and Hill, 1979a.)

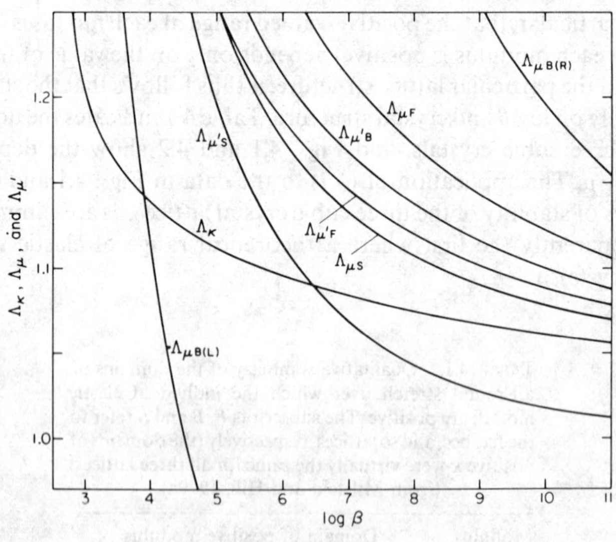

FIG. 4.2. Expanded view of a selected region of Fig. 4.1. (From Milstein and Hill, 1979a.)

4.2.1. Notional criteria of stability

Milstein and Hill (1979a, b) distinguished their classical treatment of stability from "a simplistic notion of material stability introduced by Born, and variously implemented by later writers on lattice elasticity." In the Born criterion, positive definiteness of the matrix of

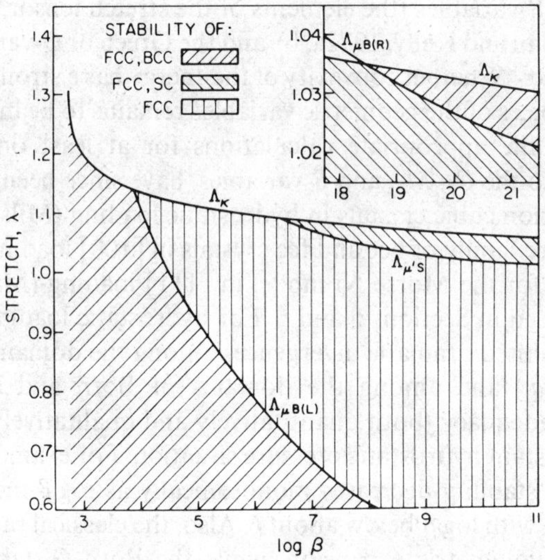

FIG. 4.3. Ranges of stability of the three cubic lattices for the Morse-model family of fcc, bcc, and sc monatomic crystals under constant hydrostatic loading. (The region in the upper "inset" indicating stability of fcc and bcc below Λ_K and above $\Lambda_{\mu B(R)}$ is also *SC*-stable.) (From Milstein and Hill, 1979a.)

elastic moduli, $C_{rs} = \frac{1}{V}(\partial^2 W/\partial q_r \partial q_s)$, evaluated in the current state of loading, is taken to be synonymous with stability. However, as first noted by Hill (1975), the predicted onset of instability in the Born criterion depends upon the choice of geometric variables q_r used to define the state of homogeneous strain of the crystal. Equivalently, in a crystal under load, convexity of the internal-energy function is not coordinate invariant. Consistency of the Born criterion with the classical criterion requires a special environment in which the loads "follow" the material during any disturbance so as to keep fixed the current values of the conjugate forces $\partial W/\partial q_r$; they may consequently become different in kind from those in the state of equilibrium whose stability is under test. In practice the Born school automatically took q_r to be the components of Green strain, and there was no attempt to match the analysis to a practicable physical situation; the same can be said of other popular choices in the literature. (See Hill, 1975; Hill and Milstein, 1977; Milstein and Hill, 1979a, b.) For cubic crystals under hydrostatic loading, the Born criterion yields the well-known conditions

$$C_{11} + 2C_{12} > 0, \quad C_{11} - C_{12} > 0, \quad C_{44} > 0. \tag{4.8}$$

Although the conditions (4.8) are naturally equivalent to the classical conditions (4.7) at zero load, divergences between the two are clearly present under non-zero loading (as is immediately evident from (4.5) for the particular case of the Green variables). This topic is explored and explained in detail in a number of recent papers, particularly Hill (1975), Hill and Milstein (1977), Parry (1978), and Milstein and Hill (1979a, b). In their 1977 paper, Hill and Milstein illustrated several qualitative divergences, in estimates of the domains of Born stability, with the aid of some general comparison theorems that they developed. This was done for three choices of strain variables q_r that have been popular in the literature, viz. the M-variables (consisting of the lengths of the edges of a fundamental cell and the included angles), the stretch or S-variables (the elements of the stretch tensor, as introduced in lattice mechanics by Macmillan and Kelly, 1972a, b), and the Green or G-variables. At that time Hill and Milstein also noted: "Whether convexity of the energy has a strong or a weak dependence on any reasonable choices of the geometric variables remains to be investigated; we intend in due course to make the appropriate calculations for at least one crystal model." The calculations, relative to the G-, M-, and S-variables, have since been made for the complete family of Morse function cubic crystals in hydrostatic loading (Milstein and Hill, 1979a, b), for the complete Morse family of bcc and fcc crystals in [100] loading (Milstein and Farber, to be published), and for the Morse Ni model in [111] loading (Milstein *et al.*, 1980). The latter cases are discussed in Sections 5 and 6. For hydrostatic loading, the computations of Milstein and Hill exhibited dramatic divergences among the domains of Born stability (for different choices of q_r) and among the domains of Born and classical stability, thus demonstrating the inadequacy (both quantitatively and qualitatively) of the Born criterion when applied to a constant hydrostatic pressure situation. For example, for the fcc structure, the range of classical stability decreases monotonically as $\log \beta$ increases, whereas the G-stable range increased with $\log \beta$ below about 6. Also, the classical range of stability of the sc lattice exists only in volumetric expansion, while the Born G-stable range was found in compression.

5. [100] Loading

General behavior on a primary path of [100] loading of an initially fcc or bcc crystal is discussed in Section 2.1. Specific computations (including elastic moduli) were carried out over a wide range of compression and tension for various Morse and generalized Morse models of crystals by Milstein (1970a, 1971, 1973b) and Milstein and Farber (1980b). Some computational results are illustrated in Figs. 3.1, 3.2, and 3.6. There are, of course, important qualitative differences between behavior in [100] loading and [100] deformation, as already mentioned in Section 3. Results of computations in both [100] loading and deformation, in tension and compression, are presented in Milstein and Farber (1980b) for a Morse-model Cu crystal. In addition to the work of Born and Fürth (Fig. 2.3), Lennard-Jones-model computations of [100] tensile loading of a fcc crystal were also made by Tyson (as reviewed by Macmillan, 1972) and by Macmillan and Kelly (1972a, b); apparently neither Tyson nor Macmillan and Kelly examined the compressive range (i.e. the region to the left of c in Fig. 2.2, where the additional load-zeros are found).

Milstein and Farber (to be published) have recently completed a study of the mechanical behavior of the entire Morse family of bcc and fcc crystals in [100] loading. The theoretical stress–strain curves in tension are shown in Fig. 5.1. The reference state ($\lambda_1 = \lambda_2 = \lambda_3 = 1$) is unstressed fcc; the plots of true stress σ_1 (when normalized by the "initial Young's modulus" $C_{11}^0 - 2(C_{12}^0)^2/(C_{11}^0 + C_{12}^0)$) depend uniquely upon the potential range indicator, $\log \beta$. The values of axial stretch λ_1 at the maxima of tensile stress σ_1 vary from about 1.7 to 1.15 as $\log \beta$ decreases from 3 to 10; however, an instability under uniaxial "dead load" (Hill, 1975) occurs well before the maximum stress is reached (as is discussed in the following subsection). An interesting property of the complete family of theoretical [100] loading curves is their upward concavity at the origin (specifically both the true stress and the load versus λ_1 are concave upward at $\lambda_1 = 1$). This property is closely linked with the necessary passage through unstressed bcc on the primary path of [100] compression. Born and Fürth (1940)

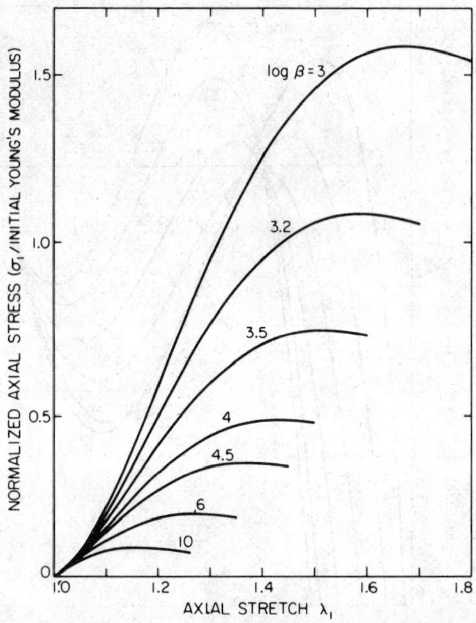

FIG. 5.1. Normalized tensile stress σ_1 vs. axial stretch λ_1 in [100] loading of the Morse family of fcc crystals with the potential range indicator log β as the sole empirical parameter. The crystals are unstressed fcc at $\lambda_1 = \lambda_2 = \lambda_3 = 1$; at any other stage, $\lambda_1 > 1$; $\lambda_2 = \lambda_3 < 1$; $\sigma_2 = \sigma_3 = 0$.

described their [100] loading curve as follows: "Up to deformations of about 10% the curve is practically a straight line, i.e. Hooke's law is satisfied." However, according to their data, they apparently *began* their computations at a 10% "deformation," and at large strain, they evidently lacked the numerical accuracy necessary to establish linearity over the "10% range" (see Section 2.1).

The theoretical stress–strain curves for the Morse family of fcc and bcc crystals are shown in Fig. 5.2 in the region $\lambda_1 \leqslant 1$. Here again the crystals are unstressed fcc at $\lambda_1 = \lambda_2 = \lambda_3 = 1$. The bcc states are found in a fairly narrow range of values of λ_1, i.e. about 0.79–0.81; the unstressed tetragonal states (indicated by the symbol T) span a much wider range of values of λ_2. We can make the connection between the condition for bcc stability at allround stretch $\lambda = 1$ in Fig. 4.3 and the locations of the bcc states in Fig. 5.2. Specifically, if log $\beta \gtrsim 4.517$, bcc is unstable at $\lambda = 1$ (Fig. 4.3) and the bcc states in [100] loading are found on the central zeros (Fig. 5.2). For the particular case of log $\beta \approx 4.517$, the shear modulus μ of the bcc crystal is zero at $\lambda = 1$ (Fig. 4.3); for the same value of log β, the [100] loading curve is tangent to the abscissa (Fig. 5.2) at a point where, in effect, the unstressed bcc and tetragonal states coincide. The behavior of this particular curve is readily understood; since $\mu = 0$ is equivalent to $C_{11} = C_{12}$ (in an unstressed cubic configuration), the slope of the [100] loading curve at the bcc state (given by $C_{11} - 2(C_{12})^2/(C_{11} + C_{12})$, relative to the bc cubic axes) is likewise zero.

5.1. *Branching from tetragonal to orthorhombic under uniaxial load;* [100] *and* [110] *loading*

We now come to the topic of branching from a primary path of [100] loading. As noted by Hill and Milstein (1977): "Branching of a primary path of deformation under a prescribed

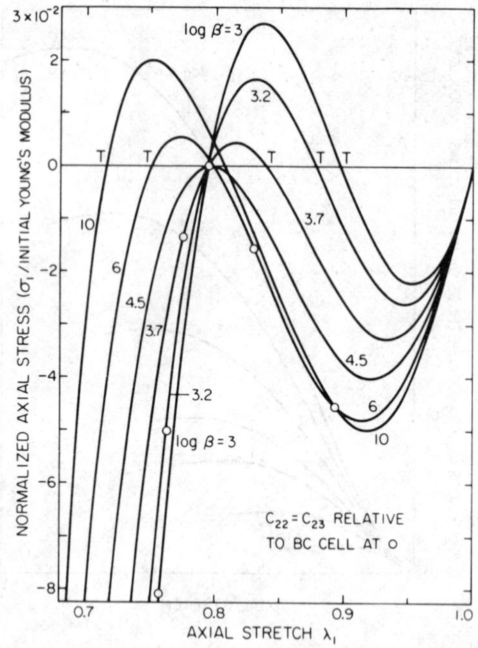

FIG. 5.2. Normalized axial stress σ_1 vs. axial stretch λ_1 in [100] loading of the Morse family of fcc crystals in the range $\lambda_1 \leqslant 1$. The crystals are unstressed fcc at $\lambda_1 = \lambda_2 = \lambda_3 = 1$; they pass through the unstressed tetragonal state at T; and they are unstressed bcc at the remaining zeros. The curves are therefore equivalent to plots of σ_1 vs. λ_1 for the bcc crystals (with but a small adjustment of the λ_1-scale).

loading program is well known to be closely associated with loss or exchange of stability . . . branching is . . . not coordinate invariant in general, when the criterion for its inception is stationarity of the conjugate forces during some virtual increment of deformation. On some paths, however, there are exceptional bifurcations that *are* substantially coordinate invariant on this criterion." Hill and Milstein (1977) identified one such coordinate invariant bifurcation on the [100] primary loading path at the point at which $C_{22} = C_{23}$. The bifurcation is coordinate invariant in the sense that it occurs at the same point on the primary path, irrespective of the choice of geometric variables or coordinates q_r used in specifying the strain and elastic moduli (assuming a "reasonable" choice with regard to lattice symmetry). The elastic moduli naturally vary with lattice deformation and thus, in general the $C_{22} = C_{23}$ eigenstate exists at some definite stage on the *primary* loading path. In fact, two such states exist in general on a [100] primary loading path; for one, the q_r (and λ_i) and hence the C_{rs} are reckoned relative to the face-centered (fc) cell (as illustrated in Fig. 2.1); for the other, these quantities are reckoned relative to the body-centered (bc) cell. In actual computations, the invariant eigenstate $C_{22} = C_{23}$ relative to the fc cell (or simply $C^F_{22} = C^F_{23}$) occurred in tension (i.e. to the right of c in Fig. 2.2) for lattice models of argon (Macmillan and Kelly, 1972a, b), iron (Milstein, 1970a, 1971), nickel (Milstein, 1973b), and copper (Milstein and Farber, 1980b). More generally, in the complete Morse function [100] loading computations of Milstein and Farber (to be published), the $C^F_{22} = C^F_{23}$ states were found at values of λ_1 that decreased from about 1.12 to about 1.04 as $\log \beta$ decreased from 10 to 3. It is thus interesting to note that the values of axial stretch λ_1 at the stress maxima vary inversely with those at $C^F_{22} = C^F_{23}$.

The state $C_{22} = C_{23}$ relative to the bc cell (or simply $C^B_{22} = C^B_{23}$) occurred in the region between b and the local load minimum (Fig. 2.2) for the lattice model of Ni (and hence was embedded in an unstable region) in the computations of Milstein (1973b), whereas it was to the left of a, in compression, in the lattice model of Fe (Milstein, 1970a, 1971). The variation of the position of the $C^B_{22} = C^B_{23}$ state for the complete Morse family is shown in Fig. 5.2; for $\log \beta \gtrsim 4.517$, $C^B_{22} = C^B_{23}$ is found to occur in the region of "falling load" (i.e. between the local maximum and minimum of load); otherwise it terminates a stable range of G-, M-, and S-strength in compression (in the sense of Hill and Milstein, 1977). Likewise, in tension, the states $C^F_{22} = C^F_{23}$ always terminated these "stable" ranges. (The state $C_{22} = C_{23}$, relative to the fc cell, occurs where $C_{44} = 0$, relative to the bc cell, and vice versa.)

We consider first the branching from the states at $C^F_{22} = C^F_{23}$. Start with the crystal in state c, which is a stable, unstressed fcc configuration ($\lambda_1 = \lambda_2 = \lambda_3 = 1$); under an *initial* uniaxial [100] tensile load, the crystal follows a *primary* path, viz. $\lambda_2 = \lambda_3 < 1$. In general, a departure from $\lambda_2 = \lambda_3$ (i.e. from tetragonality) implies the presence of transverse loading; however, the special bifurcation existing at the "invariant eigenstate" $C^F_{22} = C^F_{23}$ enables the crystal to depart upon a *secondary* path (where $\lambda_2 \neq \lambda_3$) under a program of strict uniaxial load (Hill and Milstein, 1977). The nature of the bifurcation itself is as follows: all loads remain stationary (no loads act on the (010) or (001) faces of the fc cell and the uniaxial [100] load remains *dead* (Hill, 1975)); λ_1 remains stationary ($\delta\lambda_1 = 0$); λ_2 and λ_3 vary according to $\delta\lambda_2 = -\delta\lambda_3$ (the δ symbol indicates an incremental change); the fc cell edges remain orthogonal, thus taking the fct configuration into fc orthorhombic. Once the bifurcation from the tetragonal to the orthorhombic path occurs, the lattice can follow an equilibrium path of [100] uniaxial loading wherein orthorhombic symmetry is maintained (fc cell edges remain orthogonal and $\lambda_2 \neq \lambda_3$).

Milstein and Farber (1980a) recently showed by theoretical arguments and lattice model computations (for Ni and Fe) that the load drops rapidly on the secondary path branching from the invariant eigenstate $C^F_{22} = C^F_{23}$ (and thus the secondary branch has a lower energy than the primary); they also showed that the secondary branch passes through the stress-free bcc configuration (as well as the unstressed tetragonal state). The bc cell that becomes unstressed bcc (and also tetragonal) on this path is shown in Fig. 5.3 (here the bifurcation from the primary to the secondary branch is arbitrarily assumed to occur with $\delta\lambda_2 > 0$). The fcc → bcc transition under [100] uniaxial tension is understood as follows. Immediately after the bifurcation, λ_2 increases much faster than λ_1 (since the crystal bifurcates with $\delta\lambda_1 = 0$).

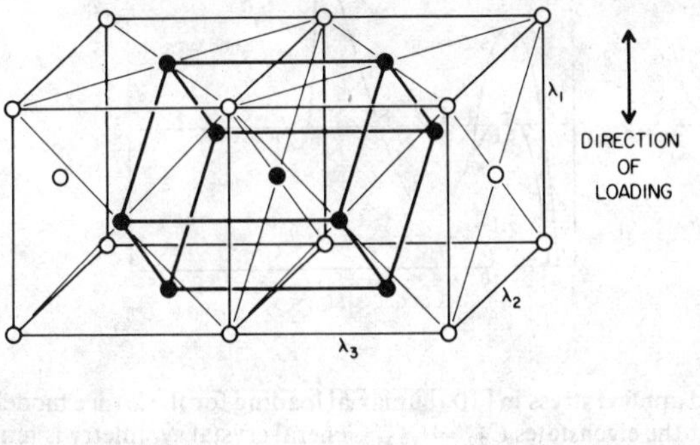

FIG. 5.3. Illustration of a portion of the lattice in [100] loading, showing (in heavy lines) the bc cell that becomes unstressed bcc on the secondary path branching from $C^F_{22} = C^F_{23}$.

Thus it is reasonable that the state $\lambda_2 = \lambda_1$ is reached on the secondary path (two such states were indeed reached in the lattice-model computations). However, if $\lambda_1 = \lambda_2$, the bc cell in Fig. 5.3 becomes tetragonal; furthermore, since no transverse loads act on the crystal at any stage of the secondary (as well as primary) equilibrium path, crystal symmetry at $\lambda_1 = \lambda_2$ implies no axial load either. Thus, the load is predicted to drop to zero along the secondary branch. If one state $\lambda_1 = \lambda_2$ is reached on the secondary path, as predicted, a second such state will also be reached in general (since, as $\lambda_1 \to \infty$, λ_2 approaches a constant, finite value). Thus, the existence of one unstressed state on the secondary path generally implies a second. Since λ_2 initially increases on the secondary path (by definition), λ_3 must decrease (according to the nature of the bifurcation); if $\lambda_3 = \lambda_2/\sqrt{2}$ coincident with $\lambda_1 = \lambda_2$ (i.e. at one of the two unstressed states), the crystal will be unstressed bcc. Given that an *unstressed* bc lattice (apparently) must exist in one of *three* configurations (one of which is bcc), the latter presumption is more than reasonable.

Examples of computational results are shown in Figs. 5.4–5.8 for the lattice models of Ni and Fe. As discussed above, the secondary path from $C_{22}^F = C_{23}^F$ is seen to branch to the right, with the uniaxial load falling rapidly after branching. In this sense, the location of $C_{22}^F = C_{23}^F$ on the primary path represents an unambiguous upper limit to the theoretical strength in [100] loading. In the Fe model, since the bcc state is stable ($\log \beta < 4.517$), it naturally occurs after the tetragonal state on the secondary branch (as λ_1 increases); the opposite is true for the Ni model. Finally, from the cell geometry shown in Fig. 5.3, we can also see that the secondary path branching from $C_{22}^F = C_{23}^F$ under [100] loading (i.e. the path on which the fc cell remains orthorhombic) can itself be considered as a primary path under [110] loading of an initially bcc crystal, since the path passes through unstressed bcc.

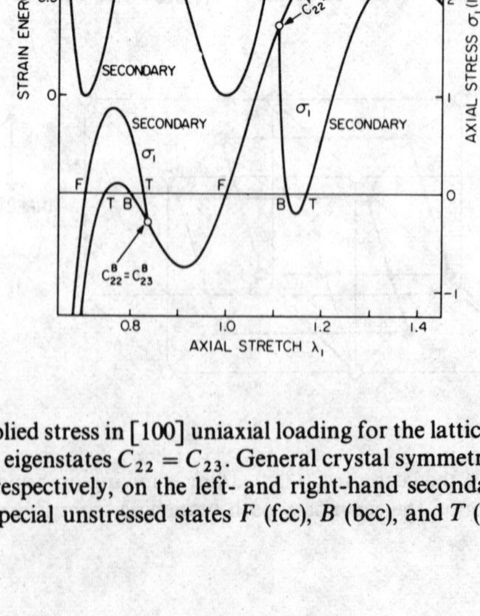

FIG. 5.4. Strain energy and applied stress in [100] uniaxial loading for the lattice model of Ni. The secondary paths branch from the primary at the eigenstates $C_{22} = C_{23}$. General crystal symmetry is tetragonal on the primary path and bc and fc orthorhombic, respectively, on the left- and right-hand secondary branches (exceptions being the special unstressed states F (fcc), B (bcc), and T (bct)).

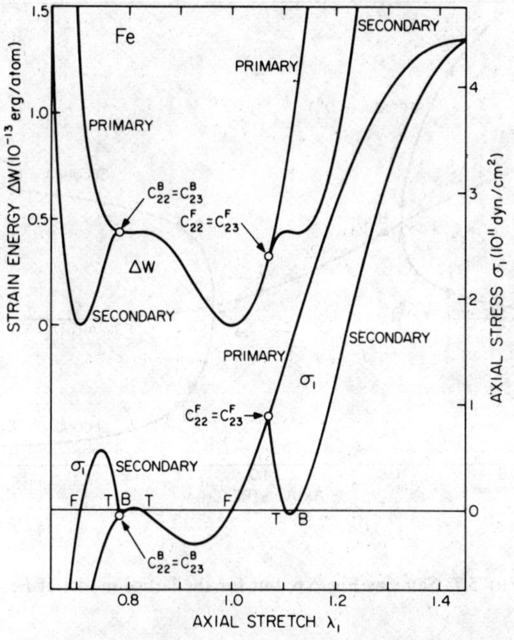

FIG. 5.5. Same as Fig. 5.4, but for the lattice model of Fe.

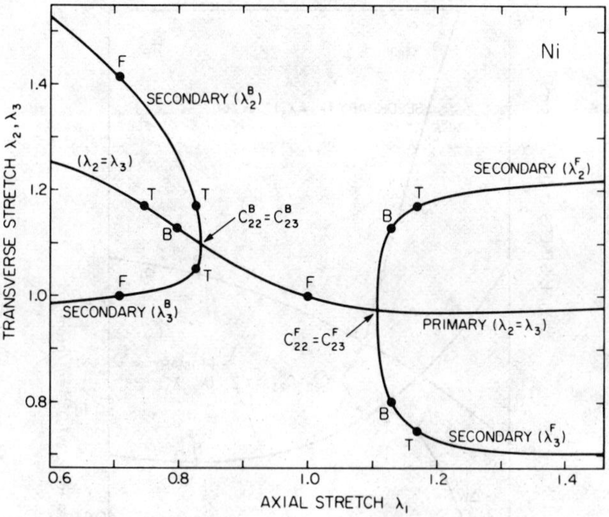

FIG. 5.6. Transverse stretches λ_2 and λ_3 vs. axial stretch λ_1 along the primary and secondary paths of [100] loading for the lattice model of Ni. The superscripts B and F indicate that the stretch is reckoned relative to the bc and fc cell axes, respectively. The unstressed states are F (fcc), B (bcc), and T (tetragonal).

Previously, the path branching from the $C_{22}^B = C_{23}^B$ eigenstate under uniaxial loading was studied in detail by Milstein and Huang (1978) for the lattice model of Ni; they showed this

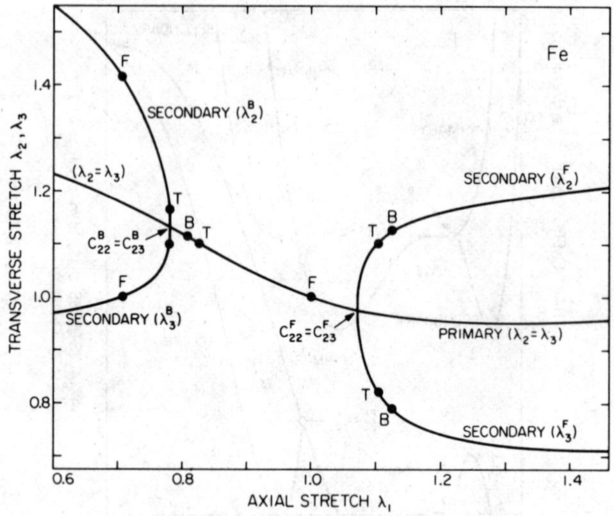

FIG. 5.7. Same as Fig. 5.6, but for the lattice model of Fe.

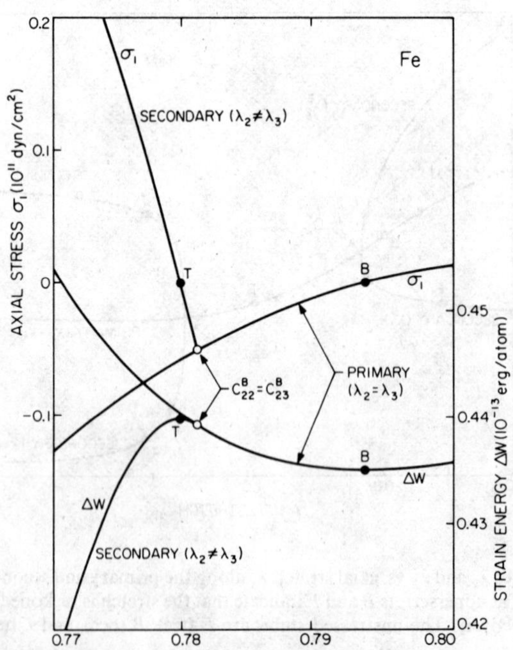

FIG. 5.8. Expanded view of the "left-hand" bifurcation shown in Fig. 5.5.

path to be identical to the path of [110] (i.e. face diagonal) uniaxial loading of the fcc crystal. More recently Milstein and Farber computed the analogous path for the Fe lattice model; as in the case of the Ni model, the secondary path branches to the left and passes through the stable, unstressed fcc state. On the secondary path branching from $C_{22}^B = C_{23}^B$, the edges of the bc cell remain orthogonal; the crystal is thus bc orthorhombic on that path. Figure 5.9 shows four bc orthorhombic cells and (in heavy lines) the fc cell that becomes unstressed fcc on the bc orthorhombic path. (The lateral stretches on that path are defined as $\lambda_2^B = b_2/b_2^0$ and $\lambda_3^B = b_3/b_3^0$, where b_2 and b_3 are as indicated in Fig. 5.9 and $b_2^0 = b_3^0$ equals the length of the 2- and 3-edges of the bc cell when the crystal is in the reference fcc state, i.e. $\lambda_1 = \lambda_2 = \lambda_3 = 1$. (The orientation of the bc cell with respect to the fcc cell at $\lambda_1 = \lambda_2 = \lambda_3 = 1$ is as shown in Fig. 2.1.) If a_0 is the length of the reference fcc cube edge, $b_2^0 = a_0/\sqrt{2}$. On the primary path, $\lambda_2 = \lambda_3 = a_2/a_0 = b_2/b_2^0$.) At the point of bifurcation from the tetragonal to the orthorhombic path it is, of course, arbitrary whether $\delta\lambda_2 > 0$ or $\delta\lambda_2 < 0$. Thus, for example, if $\delta\lambda_2^B > 0$, as in Figs. 5.6 and 5.7, the fcc state occurring at F on the secondary path is oriented as in Fig. 5.9. If, however, $\delta\lambda_2^B < 0$ (i.e. $\delta\lambda_3^B > 0$), the fcc state, of course, still occurs at the same value of λ_1, but the orientation of the fcc cell with respect to the bc cell axes is as shown in Fig. 5.10.

In summary, the complete paths of [100] loading, including the primary and two secondary branches, for the lattice models of Fe and Ni are as indicated in Figs. 5.4–5.8. For Fe, we thus see the remarkable result that the bcc crystal, when subjected to continual

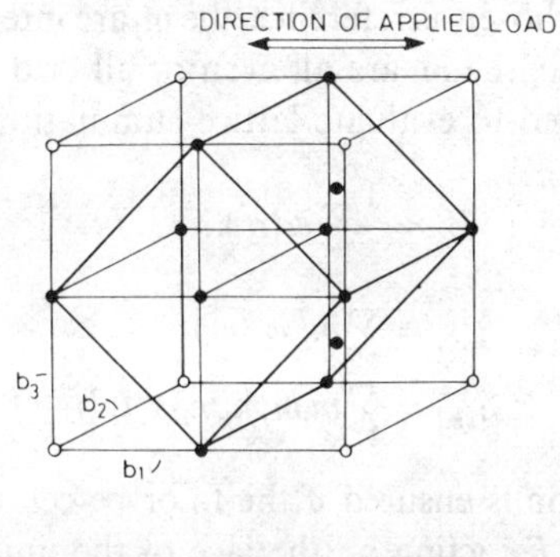

FIG. 5.9. Illustration of a portion of the lattice in [100] loading, showing (in heavy lines) the fc cell that becomes unstressed fcc on the secondary path branching from $C_{22}^B = C_{23}^B$ (assuming the bifurcation occurs with $\delta b_2 > 0$).

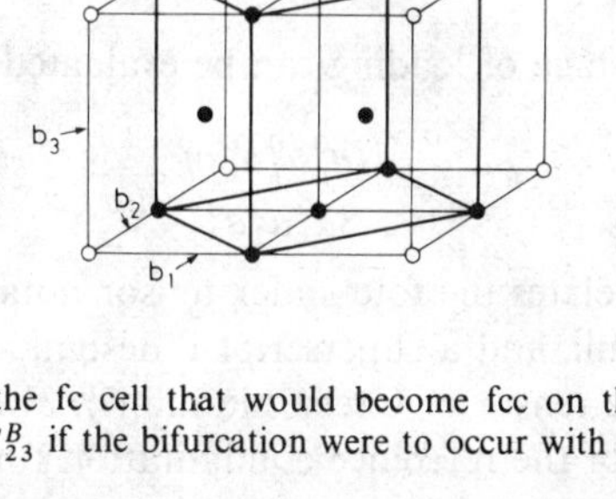

FIG. 5.10. Relative orientation of the fc cell that would become fcc on the secondary path branching from $C_{22}^B = C_{23}^B$ if the bifurcation were to occur with $\delta b_2 < 0$.

contraction (λ_1 decreasing) *under uniaxial load*, passes through seven zeros (i.e. seven unstressed states) on an equilibrium loading path (The path taking Fe through the seven stress-free states is the one of minimum energy under the uniaxial load.) For Fe, there are four stable states: two bcc (the same lattice parameter, but differently oriented) and two fcc (same but differently oriented); the three unstressed, unstable states are tetragonal, without higher symmetry (again, the same, but differently oriented). For both Fe and Ni, the secondary paths branching from the $C^B_{22} = C^B_{23}$ states on the primary [100] loading paths are also equivalent to primary paths of [$1\bar{1}0$] uniaxial loading of fcc crystals while the secondary paths branching from the $C^F_{22} = C^F_{23}$ states are equivalent to primary paths of [110] uniaxial loading of bcc crystals.

5.2. *Computation procedure*

On either a primary or secondary path of [100] loading, a lattice vector **r** connecting the origin atom with any other site in the crystal can be written as

$$\mathbf{r} = \frac{1}{2}(m_1\mathbf{a}_1 + m_2\mathbf{a}_2 + m_3\mathbf{a}_3), \tag{5.1}$$

where it is convenient here to let the $\mathbf{a}_i$ represent vectors coincident with the edges of either the fc or the bc cell; if the fc cell is chosen as reference, the m_i are integers such that $m_1 + m_2 + m_3$ is even; if the reference cell is bc, the m_i are all even or all odd integers at any lattice site. A computer program can be used to evaluate lattice summations of the form

$$W = \frac{1}{2}\sum \phi(r^2), \tag{5.2}$$

$$s_i = \sum m_i^2 \phi'(r^2), \tag{5.3}$$

$$S_{ijkl} = \sum m_i m_j m_k m_l \phi''(r^2). \tag{5.4}$$

Uniaxial load in the 1-direction is ensured if the fc or bc cell edges remain orthogonal and $s_2 = s_3 = 0$; the uniaxial load F_1 acting on the face of the unit cell is

$$F_1 = N\frac{\partial W}{\partial a_1} = \frac{N}{4}a_1 s_1, \tag{5.5}$$

where N is the number of atoms per unit cell (4 for fc and 2 for bc). The true or Cauchy stress is then

$$\sigma_1 = F_1/(a_2 a_3). \tag{5.6}$$

The Green moduli at any stage of loading can be evaluated (Hill, 1975) from

$$C^G_{ijkl} = \frac{N a_i^0 a_j^0 a_k^0 a_l^0}{8 a_1^0 a_2^0 a_3^0} S_{ijkl} \tag{5.7}$$

where the usual convention relates the four-index tensor notation to the two-index matrix notation for the Green moduli, and a superscript 0 designates the reference state. It was shown by Hill (1975) that the tensor of Green moduli C^G_{ij} always has "Cauchy symmetry" regardless of the loading or of the reference configuration for the strain, in the sense that

$$C_{23}^G = C_{44}^G, \quad C_{13}^G = C_{55}^G, \quad C_{12}^G = C_{66}^G,$$
$$C_{14}^G = C_{56}^G, \quad C_{25}^G = C_{46}^G, \quad C_{36}^G = C_{45}^G,$$

for any choice of coordinate axes, under the assumption of pairwise interactions. On any orthorhombic path on which a bc or fc cell can be defined, the elastic moduli relative to the M- and S-variables are readily evaluated from transformation formulae found in Hill and Milstein (1977). For the M-moduli,

$$C_{iijj}^M = \lambda_i \lambda_j C_{iijj}^G + \delta_{ij} l_i / \lambda_i, \tag{5.8}$$
$$C_{ijkl}^M = \lambda_i \lambda_j \lambda_k \lambda_l C_{ijkl}^G \quad (i \neq j;\ k \neq l), \tag{5.9}$$

where δ_{ij} is the Kronecker delta. The S-moduli are then given by

$$C_{iijj}^S = C_{iijj}^M, \tag{5.10}$$
$$C_{ijik}^S = \tfrac{1}{4}(\lambda_i + \lambda_j)(\lambda_i + \lambda_k) C_{ijik}^G + \tfrac{1}{4}\delta_{kj}[l_i/\lambda_i + l_j/\lambda_j] \quad (i \neq j;\ i \neq k). \tag{5.11}$$

(No summation convention is intended in (5.7)–(5.11).) Equivalently, the M-moduli can be evaluated directly from formulae found in Milstein (1970a, 1973a) and Huang *et al.* (1974); relations (5.8)–(5.11) could then be used to compute the Green and stretch moduli. Non-zero moduli of the type $C_{iijk} (i \neq j \neq k)$ were also evaluated in [111] loading by Milstein *et al.* (1980) (specifically $C_{14} = C_{1123}$); the transformation formulae are then

$$C_{iijk}^M = -\lambda_i \lambda_j \lambda_k C_{iijk}^G \quad \text{and} \quad C_{iijk}^S = \tfrac{1}{2}\lambda_i(\lambda_j + \lambda_k) C_{iijk}^G.$$

Computations on a path of uniaxial loading are performed by incrementing the lattice parameter a_1 by some fixed amount and then employing an iterative technique to determine the values of a_2 and a_3 causing s_2 and s_3 to vanish. A two-variable iteration is of course required on the secondary or orthorhombic path where $a_2 \neq' a_3$. After so locating the state of uniaxial loading, the energy, load, stress, and various elastic moduli are computed via (5.2)–(5.11).

6. Uniaxial Loading Normal to Close-packed Planes

Close-packed crystallographic planes (in which a given atom has six equidistant, coplanar, nearest neighbors) occur in two types of monatomic crystals, viz. fcc and hexagonal close-packed (hcp). The well-known "stacking sequences" of close-packed planes in fcc and hcp crystals are illustrated in Fig. 6.1. Here we consider uniaxial loading perpendicular to the planes of Fig. 6.1 for both the *ABABAB* . . . and the *ABCABC* . . . sequences; the latter of course is [111] loading of an initially fcc crystal; we refer to the former as "*c*-axis loading." The similarity between the crystal structures (when viewed from the perspective of Fig. 6.1) suggests the likelihood of a corresponding similarity in mechanical behavior on the two distinct loading paths. Since the [111] loading path must pass successively through the unstressed fcc, sc, and bcc states (Milstein *et al.*, 1980), as discussed in Section 2.2, a question of particular interest is whether or not "corresponding" unstressed states exist on the path of *c*-axis loading of the hexagonal crystal.

The computations of energy, load, stress, and lateral strain of Milstein *et al.* (1980) for the lattice model of Ni in [111] loading are shown in Fig. 6.2. (Also, in their study, general elastic moduli (taken as any fourth-rank tensor of moduli) were characterized according to symmetry along the path of [111] loading. Three sets of moduli (two of which transform tensorially) were discussed, both for general crystals and for lattice models employing pairwise interactions, and formulae were presented for transforming from one set to another or from one reference state to another along the path. Actual computations of the three sets of moduli were made for the particular lattice model. The Born criterion was particularized to the specific loading path; in the region of the stable fcc crystal, only small differences were observed among the predictions of the various notional or Born stability criteria.)

More recently Milstein, Huang, and Hsu (to be published) studied the theoretical behavior of the *ABABAB* ... sequence in *c*-axis loading, modeled under the same function ϕ. Their computations of energy, load, stress, and lateral strain are shown in Fig. 6.3. The anticipated similarity to the [111] loading behavior is evident (including the existence of the three unstressed states). These computations thus demonstrate that planes of atoms possessing hexagonal symmetry (within each plane) arranged in a stacking sequence *ABABAB* ... *can* exhibit three distinct unstressed equilibrium structures, one of which is necessarily unstable. We intend to study the influence of the potential range indicator $\log\beta$ upon the phenomena shown in Figs. 6.2 and 6.3; in particular, since the bcc state in Fig. 6.2 must be stable for $\log\beta < 4.517$, it will be of interest to determine whether the "left-hand" unstressed state in Fig. 6.3 likewise becomes stable for small $\log\beta$.

7. Other Modes of Loading

We conclude by considering two examples which provide an illustration of (i) the rich variety of phenomena to be found in studies of large strain elastic deformation of crystals, and (ii) how two seemingly similar modes of loading result in markedly different crystal responses for the same crystal model. We consider first the case mentioned earlier in connection with Figs. 3.3 and 3.4, as studied by Huang *et al.* (1974). With reference to Fig. 7.1, the mode of loading is such that the 1–2 face of the fc (or bc) cell is stress free; the only force acting on the 1–3 face of the fc cell acts parallel to this face and in the direction of $\mathbf{a}_1$; likewise, on the 2–3 face of the fc cell, the only force is parallel to the face and to $\mathbf{a}_2$. By symmetry of the loading, $a_1 = a_2$ at any stage; so the bc cell is orthorhombic; if F_s is the force acting on the 1–3 face of the fc cell, $\sigma_s = F_s/a_1a_3$. Figure 3.4 thus represents a top view of the cells shown in Fig. 7.1 in (a) the unstressed, and (b) the stressed states. The stated mode of loading is ensured if the loads

$$F_3 \equiv 2(\partial W/\partial b_3) \equiv 0 \text{ and } F_1/F_2 = -\cot(\tfrac{1}{2}\theta_F), \text{ where } F_1 = 2(\partial W/\partial b_1) \text{ and } F_2 = 2(\partial W/\partial b_2).$$

Under these conditions, $F_s = F_1 \cos(\tfrac{1}{2}\theta_F) - F_2 \sin(\tfrac{1}{2}\theta_F)$. The method used to determine the values of the lattice parameters was first to vary a particular one of them by a small fixed amount. Then, while holding that value constant, the remaining lattice parameters were simultaneously "relaxed," using a generalized Newton–Raphson iteration method, in order to find the set b_1, b_2, and b_3 for which the above conditions on the F_i were satisfied simultaneously. (The forces F_i were evaluated by lattice summations, as discussed in Section 5.2.) The remarkable fact that θ_F could not be varied continuously beyond about 82.4° without sudden discontinuities in the loads and lattice parameters was mentioned in Section

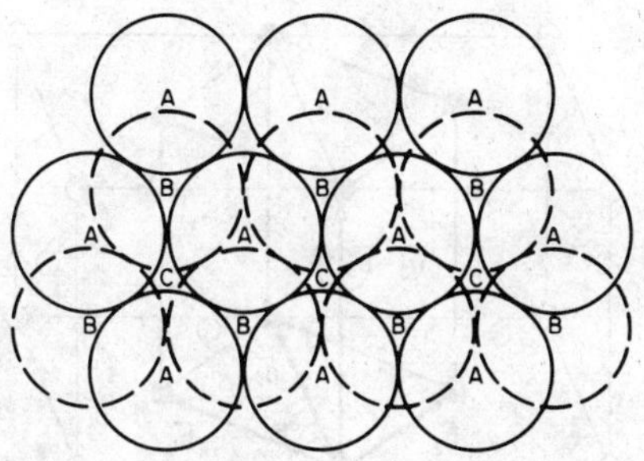

FIG. 6.1. Illustration of the differences between the stacking of close-packed planes in fcc and hcp structures. The points marked A represent the centers of a close-packed layer of spheres (shown as solid circles); a second, identical layer of spheres with their centers at points marked B is placed over the initial layer. After the first and second layers are formed, there are two non-equivalent choices for the third layer; it can go in over points marked either A or C. If it goes in over C and if the sequence is then repeated as $ABCABC$ . . . , the structure is fcc and the close-packed family of planes so generated is (111). If the third layer goes in over B and the sequence is $ABABAB$ . . . , the structure is hcp.

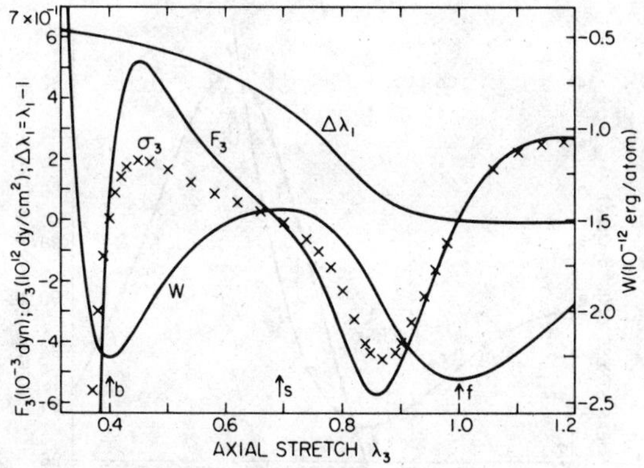

FIG. 6.2. Internal energy W, load F_3, stress σ_3, and incremental lateral stretch ($\Delta\lambda_1$) vs. axial stretch (λ_3) in [111] loading of cubic crystals. The locations of the fcc, sc, and bcc states are indicated as f, s, and b, respectively. (From Milstein *et al.*, 1980.)

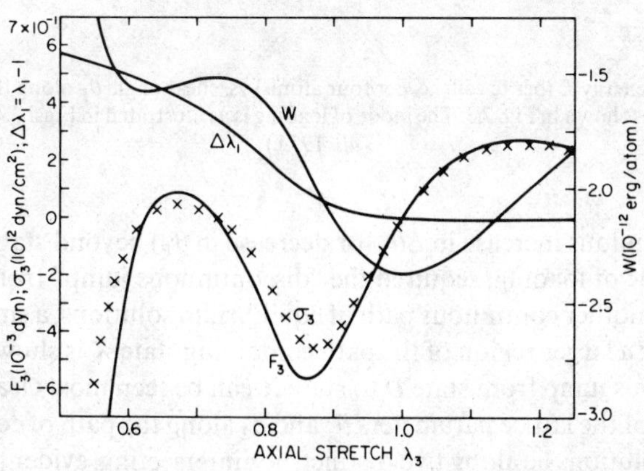

FIG. 6.3. Internal energy W, load F_3, stress σ_3, and incremental lateral stretch $\Delta\lambda_1$ in uniaxial loading perpendicular to "close-packed" planes in the stacking order $ABABAB$ . . . (see Fig. 6.1).

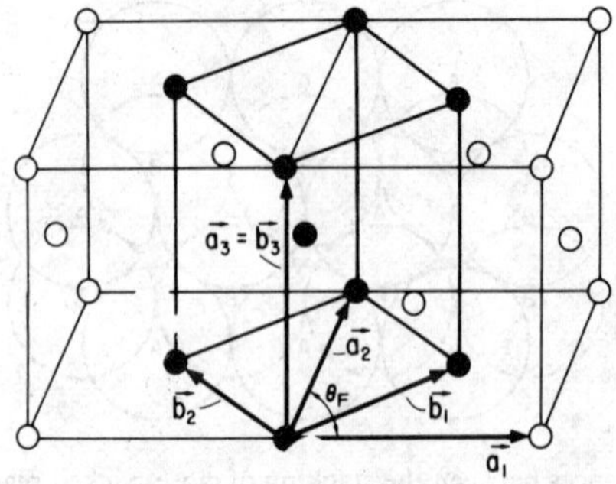

FIG. 7.1. Illustration of the bc and fc cells and definitions of the corresponding cell axes employed in the computations discussed in Section 7.

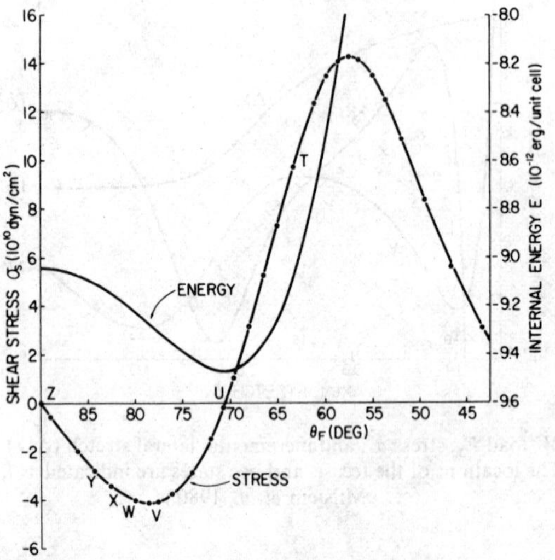

FIG. 7.2. Stress σ_s and energy E (per fc cell, i.e. per four atoms) vs. shear angle θ_F along the hyperbolic-like branch of the equilibrium curves shown in Fig. 7.3. The mode of loading is as illustrated in Figs. 3.4 and 7.4. (From Huang *et al.*, 1974.)

3. That is, the continuous increase in $\Delta\theta_F$ (or decrease in θ_F) beyond state D in Fig. 3.3, while preserving the mode of loading, required the "discontinuous jump" from state D to state X. State X lies along another continuous path of equilibrium solutions, a small portion of which is shown in Fig. 3.3; a larger region of the path containing state X is shown in Fig. 7.2. A basis for the discontinuous jump from state D to state X can be seen most clearly in Fig. 7.3 which gives the variation of the lattice parameters b_1 and b_2 along the path of equilibrium solutions. It is seen that the solutions lie along two distinct, nonintersecting, evidently continuous paths. The initial path, which includes the region $ABCDEFG$, is a closed curve; the other path, which includes the region $TUVWXYZ$, is hyperbolic-like in behavior. Since the initial curve is

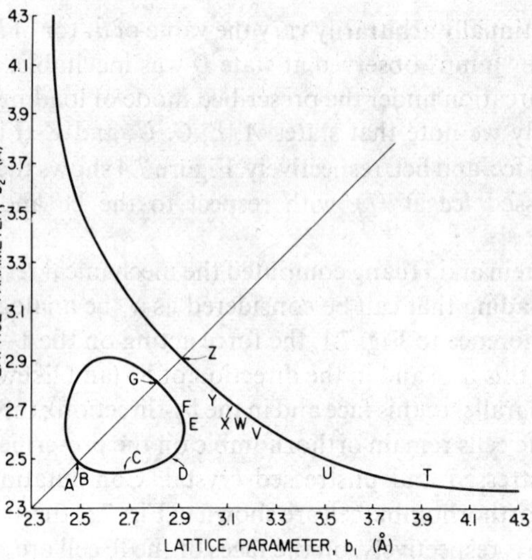

FIG. 7.3. Variation of the bc lattice parameters b_1 and b_2 on the two nonintersecting equilibrium paths for the mode of loading shown in Figs. 3.4 and 7.4. (From Huang *et al.*, 1974.)

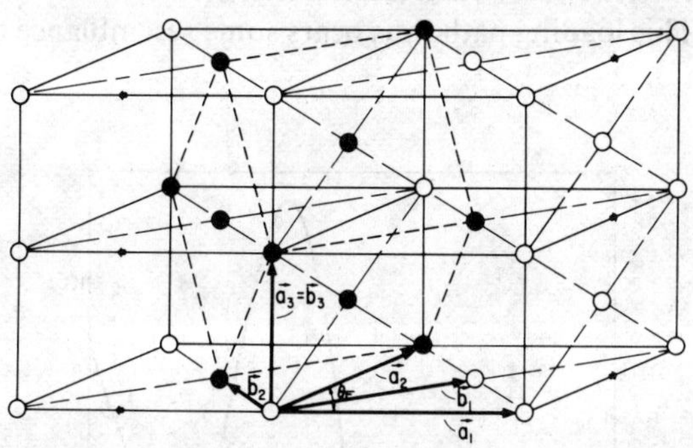

FIG. 7.4. Crystallographic planes and directions of applied shear stress, with specific atoms shown in "solid" centers to indicate a fc cell that becomes cubic and stress free when the lattice is in the state indicated by the letter *U* in Figs. 7.2 and 7.3. The directions of applied shear stress are indicated by "short arrows" acting in the "shear planes." For clarity, not all atoms are shown.

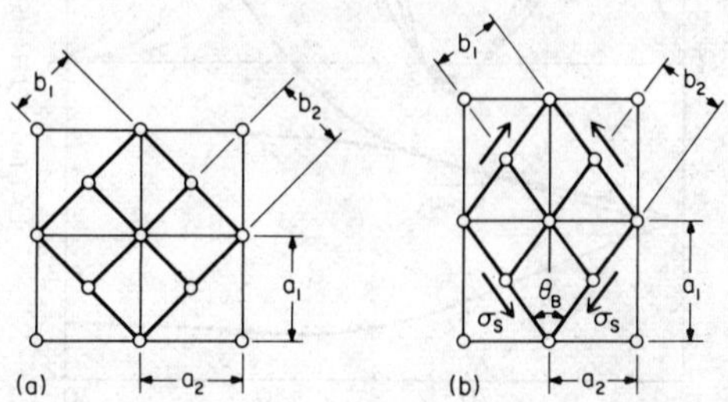

FIG. 7.5. Top views of the cells shown in Fig. 7.1: (a) cells are unstressed; (b) stresses act parallel to the faces of the originally bct cell, as illustrated.

closed, one cannot continually arbitrarily vary the value of θ_F (or of b_1 or b_2) and still remain on the curve; hence the "jump" observed at state D was inevitable. If the two paths were to have intersected, a bifurcation under the prescribed mode of loading would have occurred at the intersection. Finally we note that states A, E, G, U, and Z (Figs. 3.3, 7.2, and 7.3) are unstressed fcc, bct, bcc, fcc, and bct, respectively. Figure 7.4 shows the orientation of the fc cell (that becomes unstressed fcc at U), with respect to the bc and fc cell axes $\mathbf{b}_i$ and $\mathbf{a}_i$, respectively.

More recently, Milstein and Huang computed the mechanical response of the same crystal model in a mode of loading that can be considered as a "bc analog" to the mode discussed above. That is, with reference to Fig. 7.1, the force acting on the 1—3 face of the bc cell was maintained parallel to this face and in the direction of $\mathbf{b}_1$ (and likewise the force on the 2—3 face of the bc cell was parallel to this face and in the $\mathbf{b}_2$-direction); the 1—2 face was stress free. Thus $b_1 = b_2$ and the fc cells remain orthorhombic on the prescribed path. Figure 7.5 shows the top view of the stressed and unstressed crystal. Computations of σ_s and the lattice parameters a_i of the fc orthorhombic cell are shown in Fig. 7.6; the axial stresses σ_1 and σ_2 that act parallel to a_1 and a_2, respectively, on the faces of the fc cell are also shown. We first note that the lattice does not undergo a discontinuous jump (as in Fig. 3.3) as the shear angle θ_B is varied. However, the lattice does pass through two states where $a_1 = a_3$; since $\sigma_3 = 0$ at any stage, it follows from symmetry that $\sigma_1 = 0$ when $a_1 = a_3$; then static equilibrium implies $\sigma_s = \sigma_2 = 0$ also. In the first of these unstressed states, $a_1 = a_3 = a_2\sqrt{2}$, so the crystal is bcc; in the second it is bct. This loading path thus bears some resemblance to the primary path of [100] loading.

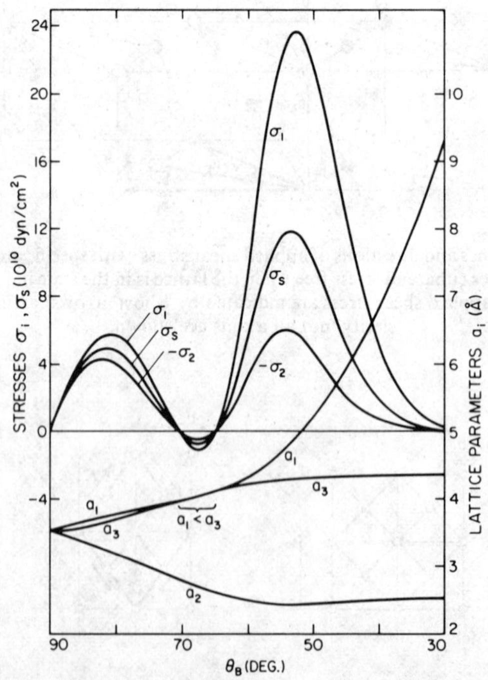

FIG. 7.6. Stresses and fc lattice parameters a_i in the mode of loading shown in Fig. 7.5.

Acknowledgement

This work was supported by the National Science Foundation under Grant No. DMR78—06865.

References

Barrett, C. S. and Massalski, T. B. (1966) *Structure of Metals*, McGraw-Hill, New York.

Born, M. (1940) On the stability of crystal lattices, I, *Proc. Camb. Philos. Soc.* **36,** 160–172.

Born, M. (1943) The thermodynamics of crystal lattices, I: Discussion of the methods of calculation, *Proc. Camb. Philos. Soc.* **39,** 100–103.

Born, M. and Fürth, R. (1940) The stability of crystal lattices, III: An attempt to calculate the tensile strength of a cubic lattice by purely static considerations, *Proc. Camb. Philos. Soc.* **36,** 454–465.

Born, M. and Huang, K. (1954) *Dynamical Theory of Crystal Lattices*, Clarendon Press, Oxford.

Born, M. and Misra, R. D. (1940) On the stability of crystal lattices, IV, *Proc. Camb. Philos. Soc.* **36,** 466–478.

Esposito, E., Carlsson, A. E., Ling, D. D., Ehrenreich, H., and Gelatt, C. D., Jr. (1980) First principles calculations of the theoretical tensile strength of copper, *Phil. Mag.* A **41,** 251–257.

Fürth, R. (1941a) On the stability of crystal lattices, V: Experimental evidence on recent theories of the equation of state and the melting of solids, *Proc. Camb. Philos. Soc.* **37,** 34–54.

Fürth, R. (1941b) On the stability of crystal lattices, VI: The properties of matter under high pressure and the lattice theory of crystals, *Proc. Camb. Philos. Soc.* **37,** 177–185.

Girifalco, L. A. and Weizer, V. G. (1959a) Application of Morse potential function to cubic metals, National Aeronautics and Space Administration Technical Report R–5.

Girifalco, L. A. and Weizer, V. G. (1959b) Application of the Morse potential function to cubic metals, *Phys. Rev.* **114,** 687–690.

Hill, R. (1975) On the elasticity and stability of perfect crystals at finite strain, *Math. Proc. Camb. Philos. Soc.* **77,** 225–240.

Hill, R. (1978) Aspects of invariance in solid mechanics, *Adv. Appl. Mech.* **18,** 1–75.

Hill, R. and Milstein, F. (1977) Principles of stability analysis of ideal crystals, *Phys. Rev.* B **15,** 3087–3096.

Huang, K., Milstein, F., and Baldwin, J. A., Jr. (1974) Theoretical strength of a perfect crystal in a state of simple shear, *Phys. Rev.* B, **10,** 3635–3646.

Macmillan, N. H. (1972) Review: The theoretical strength of solids, *J. Mater. Sci.* **7,** 239–254.

Macmillan, N. H. and Kelly, A. (1972a) The mechanical properties of perfect crystals, I: The ideal strength, *Proc. R. Soc. London*, A, **330,** 291–308.

Macmillan, N. H. and Kelly, A. (1972b) The mechanical properties of perfect crystals, II: The stability and mode of fracture of highly stressed ideal crystals, *Proc. R. Soc. London*, A, **330,** 309–317.

Milstein, F. (1970a) Theoretical strength of perfect crystalline materials, United States Air Force Project RAND, RM–6379–PR.

Milstein, F. (1970b) Mechanical stability of crystal lattices with two-body interactions, *Phys. Rev.* B **2,** 512–518.

Milstein, F. (1971) Theoretical strength of a perfect crystal, *Phys. Rev.* B **3,** 1130–1141.

Milstein, F. (1973a) Applicability of exponentially attractive and repulsive interatomic potential functions in the description of cubic crystals, *J. Appl. Phys.* **44,** 3825–3832.

Milstein, F. (1973b) Theoretical strength of a perfect crystal with exponentially attractive and repulsive interatomic interactions, *J. Appl. Phys.* **44,** 3833–3840.

Milstein, F. (1980) A simple proof for the existence of unstressed tetragonal crystals for arbitrary interatomic interactions, *Solid State Communs* **34,** 653–654.

Milstein, F. and Farber, B. (1980a) A theoretical fcc → bcc transition under [100] tensile loading, *Phys. Rev. Lett.* **44,** 277–280.

Milstein, F. and Farber, B. (1980b) On the theoretical strength of copper, *Phil. Mag.* A **42,** 19–29.

Milstein, F. and Hill, R. (1977) Theoretical properties of cubic crystals at arbitrary pressure, I: Density and bulk modulus, *J. Mech. Phys. Solids* **25,** 457–477.

Milstein, F. and Hill, R. (1978) Theoretical properties of cubic crystals at arbitrary pressure, II: Shear moduli, *J. Mech. Phys. Solids* **26,** 213–239.

Milstein, F. and Hill, R. (1979a) Theoretical properties of cubic crystals at arbitrary pressure, III: Stability, *J. Mech. Phys. Solids* **27,** 255–279.

Milstein, F. and Hill, R. (1979b) Divergences among the Born and classical stability criteria for cubic crystals under hydrostatic loading, *Phys. Rev. Lett.* **43,** 1411–1413.

Milstein, F. and Huang, K. (1978) Theory of the response of an fcc crystal to [110] uniaxial loading, *Phys. Rev.* B **18,** 2529–2541.

Milstein, F. and Huang, K. (1979) Existence of a negative Poisson ratio in fcc crystals, *Phys. Rev.* **B 19,** 2030–2033.
Milstein, F., Hill, R., and Huang, K. (1980) Theory of the response of cubic crystals to [111] loading, *Phys. Rev.* **B 21,** 4282–4291.
Misra, R. D. (1940) On the stability of crystal lattices, II, *Proc. Camb. Philos. Soc.* **36,** 173–182.
Parry, C. P. (1978) On the relative strengths of intrinsic stability criteria, *Q. Jl Mech. Appl. Math.* **31,** 1–7.
Peng, H. W. and Power, S. C. (1942) On the stability of crystal lattices, VIII: Stability of rhombohedral Bravais lattices, *Proc. Camb. Philos. Soc.* **38,** 67–81.

Aspects of Plastic Postbuckling Behavior

by A. NEEDLEMAN

Division of Engineering, Brown University

and V. TVERGAARD

Department of Solid Mechanics, Technical University of Denmark

Summary

The general theory of bifurcation and uniqueness for elastic–plastic solids and the asymptotic theory of initial postbifurcation behavior in the plastic range are briefly reviewed. A column with an asymmetric cross-section is used to illustrate some of the interactions between geometrical and material nonlinearities that occur in plastic buckling problems. Then, for the cruciform column, the influence of a corner theory of plasticity on postbifurcation behavior and imperfection sensitivity is analyzed in detail.

An asymptotic analysis of imperfection sensitivity in the plastic range, which neglects elastic unloading, is presented. Some of the results obtained by this theory are discussed and compared with results of full numerical solutions accounting for the actual elastic–plastic material behavior. Finally, the localization of initially periodic buckling patterns is considered, and numerical computations of the actual localization process are presented for several elastic–plastic structures.

1. Introduction

Although the roots of plastic buckling theory extend back to the nineteenth century, it was not until the late 1940s that the fundamental distinction, in the plastic range, between loss of uniqueness and loss of stability was elucidated by Shanley (1947). Subsequently, Hill (1958, 1961) developed a general theory of bifurcation and uniqueness for elastic–plastic solids, subject to conservative loading, which not only provides a mathematical foundation for plastic buckling theory, but also underlies much recent work on tensile plastic instabilities. Based largely on Hill's work a rather complete understanding of bifurcation behavior in the plastic range has been achieved. However, the appropriate constitutive law to employ in plastic bifurcation analyses has remained a controversial issue for over 30 years.

A similarly comprehensive framework for the analysis of postbifurcation behavior and imperfection sensitivity in the plastic range does not yet exist. Nevertheless, a coherent, if not yet complete, picture of plastic postbifurcation behavior has emerged. Fairly general results characterizing the initial postbifurcation behavior in the plastic range are available (Hutchinson, 1973, 1974). On the other hand, no rigorous relationship has been established between the initial postbifurcation behavior of a perfect structure in the plastic range and the behavior of a corresponding imperfect structure. For elastic structures such a relationship is given by Koiter's (1945) general theory of elastic stability. The major impediment to an analogous development in plastic buckling theory is the complex (and highly nonlinear) interaction between material and geometric nonlinearities that occurs in plastic buckling problems.

Here, a combination of analytical and numerical studies will be drawn on to illustrate the sorts of nonlinear interactions that occur between material and geometric nonlinearities in representative situations. To set the stage we shall briefly review Hill's (1958, 1961) theory of bifurcation and uniqueness and Hutchinson's (1973, 1974) analysis of initial postbifurcation behavior, within the context of the widely employed Donnell–Mushtari–Vlasov (DMV) theory of plates and shells. A more comprehensive review of Hill's theory—in addition to Hill's original papers—is given by Hill (1978) and by Sewell (1972), who also provides an organized bibliography of the extensive literature on plastic buckling. Also, Hutchinson (1974) presents the specialization of Hill's analysis within the DMV framework in more detail than given here as well as presenting the full analysis of initial postbifurcation behavior.

We have selected two structures to detail the nonlinear effects alluded to above. One structure is an asymmetric column analyzed by Tvergaard and Needleman (1975a). The other example is the celebrated cruciform column, which played a central role in discussions of the so-called "plastic buckling paradox." In the late 1940s and early 1950s it was noted that bifurcation calculations based on the simplest deformation theory of plasticity gave much better agreement with experimentally obtained buckling loads than did similar calculations based on the simplest flow theory. Very early, Batdorf (1949) realized that the bifurcation predictions of deformation theory could be rigorously justified by appealing to a more sophisticated flow theory with a vertex on the yield surface. Physical models of polycrystalline metal plasticity, based on the concept of single crystal slip (Hill, 1966; Hutchinson, 1970; Lin, 1971), unanimously imply the existence of such vertices, but experimental evidence is inconclusive (Hecker, 1976). Another line of attack on this paradox attributed the discrepancy to unavoidably small imperfections (Onat and Drucker, 1953). A comprehensive discussion of this controversy is given by Hutchinson (1974). Here, we employ the recently developed phenomenological vertex theory of Christoffersen and Hutchinson (1979) to investigate the postbifurcation behavior and imperfection sensitivity of the cruciform column for a range of vertex models.

The controversy generated by the plastic buckling paradox emphasized the presence of imperfection sensitivity in the plastic range when a discrepancy exists between the bifurcation predictions of simplest flow theory and simplest deformation theory. But, in fact, significant imperfection sensitivity can occur even when the bifurcation predictions of these two theories agree; the simplest example being a column subject to axial compression. The column has a fully stable elastic postbuckling behavior, but exhibits imperfection sensitivity when bifurcation occurs in the plastic range although the bifurcation predictions of flow and deformation theory coincide trivially. Likewise some plate and shell structures, which exhibit a stable postbifurcation behavior in the elastic range, may exhibit imperfection sensitive behavior in the plastic range. An asymptotic analysis which has proven capable, at least in certain cases, of distinguishing between imperfection sensitive and insensitive situations is presented here in some detail. This asymptotic analysis is carried out for a hypoelastic material, so that the effect of elastic unloading is neglected. Some predictions of maximum support loads obtained from this analysis are compared with the maximum support loads obtained from full numerical solutions which include the effects of elastic unloading.

We close with an analysis of the localization of buckling patterns which is frequently observed in plastic buckling experiments. Quite often in such experiments, e.g. Moxham (1971a, b), the final buckled configuration involves a localized deformation pattern in contrast to the periodic pattern associated with the critical buckling mode. We follow the development in Tvergaard and Needleman (1980) to show that the basic mechanism involved in such a localization is a bifurcation associated with the maximum load point, analogous to the phenomenon of necking in tensile bars. Additionally, we present numerical computations which illustrate the development of the localized deformation pattern in realistic elastic–plastic structural models.

2. Bifurcation Analysis

The bifurcation criterion to be discussed in this Section is a specialization of Hill's (1958, 1961) general theory of uniqueness and bifurcation in elastic–plastic solids within the context of the DMV theory of plates and shells. The relatively simple DMV equations, which presume small strains and moderately small rotations, are sufficient for the vast majority of buckling problems. However, it should be noted that there are problems in which the use of more accurate shell equations (Koiter, 1966) is essential, either because the wavelength of the characteristic deformation pattern is not small compared with the radii of curvature of the middle surface, or because nonlinearities in terms of the in-plane displacements play a significant role.

2.1. *Basic equations of DMV theory*

A point on the shell middle surface is identified by the coordinates x^1 and x^2, while x^3 is the coordinate along the normal that measures the distance from the middle surface. The displacements of the shell middle surface are u_α in the directions of the surface base vectors and w in the direction of the surface normal. Then, within the context of DMV shell theory the increment of the membrane strain tensor $\dot{\varepsilon}_{\alpha\beta}$ and the increment of the bending strain tensor $\dot{\kappa}_{\alpha\beta}$ are

$$\left.\begin{aligned} \dot{\varepsilon}_{\alpha\beta} &= \tfrac{1}{2}(\dot{u}_{\alpha,\beta} + \dot{u}_{\beta,\alpha}) + b_{\alpha\beta}\dot{w} + \tfrac{1}{2}(w_{,\alpha}\dot{w}_{,\beta} + \dot{w}_{,\alpha}w_{,\beta}), \\ \dot{\kappa}_{\alpha\beta} &= -\dot{w}_{,\alpha\beta}. \end{aligned}\right\} \tag{2.1}$$

Here increments are denoted by $(\dot{\ })$, $b_{\alpha\beta}$ are the covariant components of the curvature tensor for the undeformed middle surface, and $(\ )_{,\alpha}$ denotes covariant differentiation. The in-plane Lagrangian strain-rate components are approximated by

$$\dot{\eta}_{\alpha\beta} = \dot{\varepsilon}_{\alpha\beta} + x^3\dot{\kappa}_{\alpha\beta}. \tag{2.2}$$

The three-dimensional stress rates $\dot{\sigma}^{ij}$ and strain rates $\dot{\eta}_{kl}$ in the shell material are assumed to be related by the equation

$$\dot{\sigma}^{ij} = L^{ijkl}\dot{\eta}_{kl}, \tag{2.3}$$

where $L^{ijkl} = L^{jikl} = L^{klij}$. Here, Latin indices range from 1 to 3, while Greek indices range from 1 to 2, and the summation convention is used for repeated indices. For an elastic–plastic solid the instantaneous moduli, although homogeneous of degree zero in the stress rate, do depend on the direction of the stress rate.

The stress state in the shell is approximately plane. Thus, only the in-plane stresses enter into the stress–strain relations, and we can write

$$\dot{\sigma}^{\alpha\beta} = \hat{L}^{\alpha\beta\gamma\delta}\dot{\eta}_{\gamma\delta}, \tag{2.4}$$

where the in-plane moduli are given by

$$\hat{L}^{\alpha\beta\gamma\delta} = L^{\alpha\beta\gamma\delta} - \frac{L^{\alpha\beta 33}L^{\gamma\delta 33}}{L^{3333}}. \tag{2.5}$$

The membrane stress tensor $N^{\alpha\beta}$ and the moment tensor $M^{\alpha\beta}$ are taken to be

$$N^{\alpha\beta} = \int_{-h/2}^{h/2} \sigma^{\alpha\beta}\, dx^3, \quad M^{\alpha\beta} = \int_{-h/2}^{h/2} \sigma^{\alpha\beta} x^3\, dx^3, \tag{2.6}$$

where h is the shell thickness. Thus, using (2.3) and (2.5) in (2.6) the following incremental constitutive relations are found for the shell theory

$$\dot{N}^{\alpha\beta} = H_{(1)}^{\alpha\beta\gamma\delta}\dot{\varepsilon}_{\gamma\delta} + H_{(2)}^{\alpha\beta\gamma\delta}\dot{\kappa}_{\gamma\delta}, \quad \dot{M}^{\alpha\beta} = H_{(2)}^{\alpha\beta\gamma\delta}\dot{\varepsilon}_{\gamma\delta} + H_{(3)}^{\alpha\beta\gamma\delta}\dot{\kappa}_{\gamma\delta}, \tag{2.7}$$

where

$$H_{(i)}^{\alpha\beta\gamma\delta} = \int_{-h/2}^{h/2} \hat{L}^{\alpha\beta\gamma\delta}(x^3)^{i-1}\, dx^3. \tag{2.8}$$

Within the context of DMV shell theory the incremental principle of virtual work takes the form

$$\int_A [\dot{N}^{\alpha\beta}\delta\dot{\varepsilon}_{\alpha\beta} + \dot{M}^{\alpha\beta}\delta\dot{\kappa}_{\alpha\beta} + N^{\alpha\beta}\dot{w}_{,\alpha}\delta\dot{w}_{,\beta}]\, dA$$
$$= \int_A [\dot{p}^{\alpha}\delta\dot{u}_{\alpha} + \dot{p}\delta\dot{w}]\, dA + \int_S [\dot{T}^{\alpha}\delta\dot{u}_{\alpha} + \dot{T}\delta\dot{w} + \dot{M}\delta\dot{w}_{,\alpha}n^{\alpha}]\, dS. \tag{2.9}$$

Here, A is the area of the undeformed middle surface and S is the edge of the shell. The prescribed load increments $\dot{p}^{\alpha}$, $\dot{p}$ per unit initial area and $\dot{T}^{\alpha}$, $\dot{T}$ per unit initial length of the edge are referred to the initial base vectors, and $\dot{M}$ is the prescribed increment of the bending moment per unit initial length of an edge with normal n^{α}.

2.2. *Solids with smooth yield surfaces*

Bifurcation will first be discussed for solids governed by the general smooth yield surface constitutive relation considered by Hill (1958), as was also done in the presentation of Hutchinson (1974). For linearly elastic response the instantaneous moduli in (2.3) are denoted by $\mathscr{L}^{ijkl}$. If the current stress state is on the yield surface at a point in which the outward unit normal to the yield surface in strain-rate space is m^{ij}, the instantaneous moduli are

$$L^{ijkl} = \mathscr{L}^{ijkl} - \alpha g^{-1} m^{ij} m^{kl}. \tag{2.10}$$

Here g depends on the deformation history and on the current stress state, and $\alpha = 1$ for $m^{ij}\dot{\eta}_{ij} \geqslant 0$, while $\alpha = 0$ for $m^{ij}\dot{\eta}_{ij} < 0$.

In particular for the simplest flow theory of plasticity, J_2 flow theory, these quantities are given by

$$m^{ij} = \sqrt{\frac{3}{2}}\frac{s^{ij}}{\sigma_e}, \quad g^{-1} = \left(\frac{E}{1+\nu}\right)\frac{E/E_t - 1}{E/E_t - (1-2\nu)/3}. \tag{2.11}$$

Here $s^{ij} = \sigma^{ij} - g^{ij}\sigma^k_k/3$ is the stress deviator (g^{ij} being the metric tensor), $\sigma_e = (\frac{3}{2}s^{ij}s_{ij})^{1/2}$ is the Mises effective stress, E is Young's modulus, ν is Poisson's ratio, and the tangent modulus E_t is the slope of the uniaxial stress–strain curve at stress level σ_e.

Based on (2.10) the plane stress moduli can be written in the form (Hutchinson, 1974)

$$\hat{L}^{\alpha\beta\gamma\delta} = \hat{\mathscr{L}}^{\alpha\beta\gamma\delta} - \alpha\hat{g}^{-1}\hat{m}^{\alpha\beta}\hat{m}^{\gamma\delta}, \tag{2.12}$$

where $\hat{\mathscr{L}}^{\alpha\beta\gamma\delta}$ are the in-plane elastic moduli and

$$\hat{m}^{\alpha\beta} = m^{\alpha\beta} - m^{33}\frac{\mathscr{L}^{\alpha\beta 33}}{\mathscr{L}^{3333}}, \quad \hat{g}^{-1} = g^{-1}\frac{\mathscr{L}^{3333}}{L^{3333}}. \tag{2.13}$$

In (2.12) $\alpha = 1$ for $\hat{m}^{\alpha\beta}\dot{\eta}_{\alpha\beta} \geqslant 0$ and $\alpha = 0$ for $\hat{m}^{\alpha\beta}\dot{\eta}_{\alpha\beta} < 0$.

Now, consider the question of uniqueness. The prescribed dead loads or displacements are taken to be proportional to a single parameter λ. If two different solution increments that each correspond to a given load increment $\dot{\lambda}$ are denoted by $\dot{u}^a_\alpha$, $\dot{w}^a$, etc. and $\dot{u}^b_\alpha$, $\dot{w}^b$, etc., and their differences are denoted by $\tilde{u}_\alpha = \dot{u}^b_\alpha - \dot{u}^a_\alpha$, $\tilde{w} = \dot{w}^b - \dot{w}^a$, etc., then according to (2.1) the differences between the increments of the DMV strain measures are

$$\begin{aligned} \tilde{\varepsilon}_{\alpha\beta} &= \tfrac{1}{2}(\tilde{u}_{\alpha,\beta} + \tilde{u}_{\beta,\alpha}) + b_{\alpha\beta}\tilde{w} + \tfrac{1}{2}(w^0_{,\alpha}\tilde{w}_{,\beta} + \tilde{w}_{,\alpha}w^0_{,\beta}) \\ \tilde{\kappa}_{\alpha\beta} &= -\tilde{w}_{,\alpha\beta}, \end{aligned} \tag{2.14}$$

where $(\ )^0$ denotes the fundamental state of deformation. By the definition of λ the load differences $\tilde{p}^\alpha$, $\tilde{p}$, etc. vanish where loads are prescribed, and $\tilde{u}_\alpha$, $\tilde{w}$ vanish where displacements are prescribed. Then according to (2.9) the following equation must be satisfied:

$$H \equiv \int_A [\tilde{N}^{\alpha\beta}\tilde{\varepsilon}_{\alpha\beta} + \tilde{M}^{\alpha\beta}\tilde{\kappa}_{\alpha\beta} + N_0^{\alpha\beta}\tilde{w}_{,\alpha}\tilde{w}_{,\beta}]\,dA = 0. \tag{2.15}$$

Following Hill (1958, 1961), define at the current load level a *linear comparison solid* for which the moduli L_c^{ijkl} equal the plastic moduli wherever the stress state is on the yield surface (independent of the sign of $m^{ij}\dot{\eta}_{ij}$) and the elastic moduli elsewhere. Thus, the linear comparison moduli are independent of the direction of the stress rate. The linear comparison moduli are used in (2.5) and (2.8) to obtain the DMV moduli $H^{\alpha\beta\gamma\delta}_{c(i)}$ for the comparison solid and the following quadratic functional is introduced:

$$F = \int_A [H^{\alpha\beta\gamma\delta}_{c(1)}\tilde{\varepsilon}_{\alpha\beta}\tilde{\varepsilon}_{\gamma\delta} + 2H^{\alpha\beta\gamma\delta}_{c(2)}\tilde{\varepsilon}_{\alpha\beta}\tilde{\kappa}_{\gamma\delta} + H^{\alpha\beta\gamma\delta}_{c(3)}\tilde{\kappa}_{\alpha\beta}\tilde{\kappa}_{\gamma\delta} + N_0^{\alpha\beta}\tilde{w}_{,\alpha}\tilde{w}_{,\beta}]\,dA. \tag{2.16}$$

For positive g in (2.10) it can be proven that the integrand of H is nowhere smaller than that of F (Hill, 1958, 1961; Hutchinson, 1974), and thus

$$H \geqslant F. \tag{2.17}$$

Consequently $F > 0$ for any admissible nonvanishing displacement fields is a sufficient condition for uniqueness, as shown by Hill (1958).

Bifurcation at the lowest eigenvalue λ_c of F occurs only if $H = 0$ simultaneously with $F = 0$, and this requires that both $m^{ij}\dot{\eta}^a_{ij} \geqslant 0$ and $m^{ij}\dot{\eta}^b_{ij} \geqslant 0$ wherever the current stress state is on the yield surface.

Denote the eigenmode obtained from (2.16) by $\overset{(1)}{u}_\alpha$ and $\overset{(1)}{w}$ and let $\xi\overset{(1)}{u}_\alpha$, etc. be the contribution of this eigenmode to the bifurcation mode quantities, u^b_a and w^b, where $\overset{(1)}{u}_\alpha$ and $\overset{(1)}{w}$ are normalized in some convenient fashion, but such that the amplitude ξ is positive. Then we write

$$\lambda = \lambda_c + \lambda_1 \xi + \ldots, \quad \frac{du^b_\alpha}{d\lambda} = \frac{du^0_\alpha}{d\lambda} + \frac{d\xi}{d\lambda}\overset{(1)}{u}_\alpha + \ldots. \tag{2.18}$$

By identifying $\dot{u}^a_\alpha$ with $\dot{u}^0_\alpha$ and $\dot{w}^a$ with $\dot{w}^0$, etc., the condition $m^{ij}\dot{\eta}^a_{ij} \geqslant 0$ is satisfied, since it is presumed that the fundamental solution involves plastic loading everywhere in the current plastic zone. Then λ_1 has to be chosen sufficiently large so that $m^{ij}\dot{\eta}^b_{ij} \geqslant 0$ everywhere in the current plastic region, presuming $m^{ij}\overset{(1)}{\eta}_{ij} < 0$ in part of the current plastic zone. In fact, as will be shown in Section 3, the plastic loading condition $m^{ij}\dot{\eta}^b_{ij} \geqslant 0$ generally determines a unique value of λ_1. Furthermore, $\lambda_1 > 0$ implies that bifurcation takes place under increasing load in agreement with the Shanley condition for column buckling (Shanley, 1947). We also note that to obtain the behavior corresponding to the opposite signed contribution of the eigenmode, the sign of the mode $\overset{(1)}{u}_\alpha, \overset{(1)}{w}$ is changed, not the sign of ξ.

2.3. *Solids that develop a vertex on subsequent yield surfaces*

Already in the late 1940s and early 1950s it was noted that buckling predictions based on the simplest deformation theory of plasticity gave much better agreement with experimental results than did similar predictions based on the simplest flow theory (Bijlaard, 1949; Stowell, 1948; Gerard and Becker, 1957). However, the fact that deformation theory is a purely elastic constitutive relation, which does not in general represent a physically acceptable theory of plasticity, has been a major objection to bifurcation predictions based on this theory.

The justification of the bifurcation predictions of deformation theory is related to the possibility that a vertex may develop on subsequent yield surfaces at the point of loading, as illustrated in Fig. 2.1. In incremental theories of plasticity based on a system of linear loading functions, a pyramidal corner develops on the yield surface and Sanders (1954) has shown that an incremental theory with infinitely many loading functions can be constructed which coincides exactly with J_2 deformation theory for total loading (i.e. when no activated loading function has unloaded). More recently, Sewell (1974) has constructed an incremental theory with a pyramidal vertex having three facets for which the total loading moduli can also be fitted to those of J_2 deformation theory. For these incremental theories the J_2 deformation theory predictions of bifurcations are valid as long as the total loading condition is met.

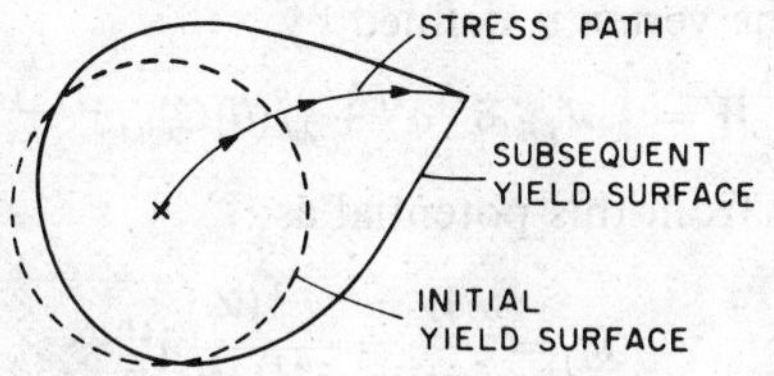

FIG. 2.1. Development of a vertex on a subsequent yield surface.

The occurrence of vertices on yield surfaces is supported by models of the elastic–plastic behavior of polycrystalline metals based on single crystal slip. For a self-consistent model of a polycrystal, Hutchinson (1970) has found that the shear moduli at the vertex, subsequent to uniaxial tension, are rather close to the predictions of deformation theory. On the other hand, in experimental investigations of material behavior the evidence of vertex formation on yield surfaces is conflicting (Hecker, 1976).

A phenomenological corner theory of plasticity, called J_2 corner theory, has recently been proposed by Christoffersen and Hutchinson (1979). In this theory the instantaneous moduli for nearly proportional loading are chosen equal to the J_2 deformation theory moduli and for increasing deviation from proportional loading the moduli increase smoothly until they coincide with the elastic moduli for stress increments directed along or within the corner of the yield surface.

In the following M^0_{ijkl} is taken to denote the J_2 deformation theory compliances, so that $\dot{\varepsilon}_{ij} = M^0_{ijkl}\dot{\sigma}^{kl}$. With $\mathscr{M}_{ijkl}$ denoting the elastic compliances, the plastic compliances for J_2 deformation theory are $C_{ijkl} = M^0_{ijkl} - \mathscr{M}_{ijkl}$.

In J_2 corner theory the yield surface in the neighborhood of the loading point is a cone in stress deviator space (Fig. 2.2) with axis in the direction

$$\lambda^{ij} = s^{ij}(C_{mnpq}s^{mn}s^{pq})^{-1/2}, \tag{2.19}$$

where s^{ij} is the stress deviator. The angular measure θ of the stress-rate direction relative to the cone axis is defined by

$$\cos\theta = C_{ijkl}\lambda^{ij}\dot{s}^{kl}(C_{mnpq}\dot{s}^{mn}\dot{s}^{pq})^{-1/2}. \tag{2.20}$$

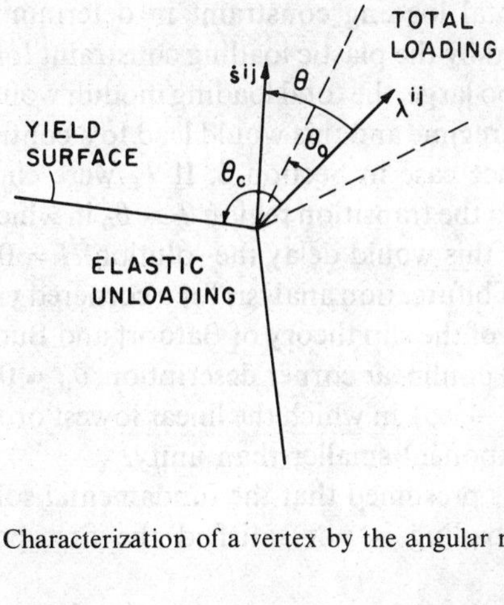

FIG. 2.2. Characterization of a vertex by the angular measure θ.

A stress-rate potential at the vertex is defined by

$$W = \tfrac{1}{2}\mathcal{M}_{ijkl}\dot{\sigma}^{ij}\dot{\sigma}^{kl} + \tfrac{1}{2}f(\theta)C_{ijkl}\dot{\sigma}^{ij}\dot{\sigma}^{kl}. \tag{2.21}$$

The strain-rate is obtained from this potential as

$$\dot{\varepsilon}_{ij} = \frac{\partial W}{\partial \dot{\sigma}^{ij}} = \frac{\partial^2 W}{\partial \dot{\sigma}^{ij}\dot{\sigma}^{kl}}\dot{\sigma}^{kl} \tag{2.22}$$

which defines the corner theory compliances $M_{ijkl}(\theta)$. Inverting (2.22) gives the corner theory moduli $L^{ijkl}(\theta)$ to be used in (2.3).

The angle of the yield surface cone is denoted by θ_c, as shown in Fig. 2.2, and thus the plastic strain-rate vector falls within the range $0 \leqslant \theta \leqslant \theta_n$, where $\theta_n = \theta_c - \frac{1}{2}\pi$. The angle θ_0 limiting the total loading range is taken to be some fraction of θ_n. Now, in the total loading range, $0 \leqslant \theta \leqslant \theta_0$, where $f(\theta) = 1$, the instantaneous moduli $L^{ijkl}(\theta)$ coincide with the deformation theory moduli. In the transition region, $\theta_0 \leqslant \theta \leqslant \theta_c$, $f(\theta)$ is taken to decrease monotonically and smoothly from unity to zero in a way which ensures convexity of the incremental relation. Thus, in the transition region the moduli increase monotonically from the deformation theory moduli at $\theta = \theta_0$ to the elastic moduli at $\theta = \theta_c$. For $\theta_c \leqslant \theta \leqslant \pi$ elastic unloading occurs with $f(\theta) = 0$.

The question of uniqueness is again analyzed by considering the possibility that a given load increment can result in two different solution increments. The difference between these solution increments is denoted by ($\tilde{\ }$) with corresponding strain measures $\tilde{\varepsilon}_{\alpha\beta}$ and $\tilde{\kappa}_{\alpha\beta}$ given by (2.14), and bifurcation is possible if a nonzero solution exists to (2.15). For J_2 corner theory, provided that the fundamental solution satisfies nearly proportional loading everywhere in the current plastic zone with $\theta \leqslant \theta_0$, the total loading (i.e. deformation theory) moduli can be used as the moduli of the comparison solid. For deformation theory the lowest eigenvalue of $F = 0$ (F given by (2.16)) governs bifurcation. We must then ensure that the bifurcation mode, the linear combination of the fundamental solution and the eigenmode obtained from (2.16), satisfies total loading. For λ_1 sufficiently large in (2.18) the bifurcation mode will satisfy total loading, where the condition to be satisfied, in terms of (2.20), is

$$\cos\theta \geqslant \cos\theta_0 \quad \text{for} \quad \frac{ds^{ij}}{d\lambda} = \frac{ds_0^{ij}}{d\lambda} + \frac{d\xi}{d\lambda}\overset{(1)}{s^{ij}}. \tag{2.23}$$

The role played by the total loading constraint in determining a unique value of λ_1 is analogous to the role played by the plastic loading constraint for a solid with a smooth yield surface. If λ_1 were chosen too large, the total loading moduli would govern the response in the immediate postbifurcation regime and this would lead to a contradiction as will be discussed for the smooth yield surface case in Section 3. If λ_1 were chosen too small, some stress increments would point into the transition region $\theta > \theta_0$ in which the moduli are larger than those for total loading and this would delay the solution $H = 0$ to a point after $F = 0$. The total loading constraint in a bifurcation analysis for a cornered yield surface was discussed by Batdorf (1949) on the basis of the slip theory of Batdorf and Budiansky (1949). We note that in the limit of a thoroughly nonlinear corner description, $\theta_0 = 0$, the condition (2.23) results in a smooth bifurcation ($\lambda_1 \to \infty$), in which the linear lowest order ξ contribution in (2.18) is replaced by a term with exponent smaller than unity.

The above discussion has presumed that the fundamental solution satisfies total loading everywhere. When total loading is not satisfied the first possible bifurcation may be considerably delayed, since then bifurcation will usually be governed by the stiffer moduli of the transition range.

3. Initial Postbifurcation Behavior

An asymptotic theory of the postbifurcation behavior of structures in the plastic range has been developed by Hutchinson (1973, 1974). This theory is an extension into the initial postbifurcation range of the bifurcation theory of Hill (1958, 1961). As in the elastic postbuckling theory of Koiter (1945) the equilibrium load in the vicinity of the bifurcation point is expanded in terms of the bifurcation mode amplitude. A major difference from the elastic analysis is, however, the necessity to account for elastic unloading, which plays a crucial role after bifurcation.

In the present article the postbifurcation analysis will be summarized rather briefly in the context of DMV shell theory (Hutchinson, 1974). As in Hutchinson's analysis, the treatment is limited to solids characterized by a smooth yield surface, with instantaneous moduli of the form (2.10). The fundamental solution is taken to be associated with a monotonically increasing load parameter λ in the range of interest, and for simplicity consideration is restricted to cases in which the body has become fully plastic prior to bifurcation.

In the vicinity of the critical bifurcation load λ_c the asymptotically exact expression for the load parameter λ in terms of the bifurcation mode amplitude ξ is obtained of the form

$$\lambda = \lambda_c + \lambda_1 \xi + \lambda_2 \xi^{1+\beta} + \dots \tag{3.1}$$

for $\xi \geqslant 0$. For mode displacements in the opposite direction, the sign of ξ is changed, and then (3.1) still applies with different constants λ_1, λ_2, and β. In (3.1) λ_1 is positive, λ_2 is negative, and $0 < \beta < 1$.

The expansion corresponding to (3.1) of the increments of the field quantities takes the form

$$\left\{\begin{matrix} \dot{u}_{\gamma,\delta} \\ \dot{\eta}_{\gamma\delta} \\ \dot{\sigma}^{\gamma\delta} \\ \dot{N}^{\gamma\delta} \\ \vdots \end{matrix}\right\} = \left\{\begin{matrix} \dot{u}^0_{\gamma,\delta} \\ \dot{\eta}^0_{\gamma\delta} \\ \dot{\sigma}^{\gamma\delta}_0 \\ \dot{N}^{\gamma\delta}_0 \\ \vdots \end{matrix}\right\} + \left\{\begin{matrix} \overset{(1)}{u}_{\gamma,\delta} \\ \overset{(1)}{\eta}_{\gamma\delta} \\ \overset{(1)}{\sigma}{}^{\gamma\delta} \\ \overset{(1)}{N}{}^{\gamma\delta} \\ \vdots \end{matrix}\right\} + \xi^\beta \left\{\begin{matrix} \overset{*}{u}_{\gamma,\delta} \\ \overset{*}{\eta}_{\gamma\delta} \\ \overset{*}{\sigma}{}^{\gamma\delta} \\ \overset{*}{N}{}^{\gamma\delta} \\ \vdots \end{matrix}\right\} + \left\{\begin{matrix} \overset{d}{u}_{\gamma,\delta}(\xi, x^\alpha) \\ \overset{d}{\eta}_{\gamma\delta}(\xi, x^i) \\ \overset{d}{\sigma}{}^{\gamma\delta}(\xi, x^i) \\ \overset{d}{N}{}^{\gamma\delta}(\xi, x^\alpha) \\ \vdots \end{matrix}\right\} \tag{3.2}$$

where, here, $(\dot{\ }) \equiv d(\ \ )/d\xi$ and, since the fundamental solution is a function of λ, $(\dot{\ })^0 \equiv (d(\ \ )^0/d\lambda)(d\lambda/d\xi)$. The quantities denoted by $^{(1)}$ in (3.2) are the eigenmodal quantities associated with the lowest eigenvalue λ_c of $F = 0$. This mode also satisfies the variational equation

$$\int_A \left\{ \overset{(1)}{N}{}^{\alpha\beta} \delta^c \varepsilon_{\alpha\beta} + \overset{(1)}{M}{}^{\alpha\beta} \delta\kappa_{\alpha\beta} + N^{\alpha\beta}_{0c} \overset{(1)}{w}_{,\alpha} \delta w_{,\beta} \right\} dA = 0, \tag{3.3}$$

where

$$\delta^c \varepsilon_{\alpha\beta} \equiv \tfrac{1}{2}(\delta u_{\alpha,\beta} + \delta u_{\beta,\alpha}) + b_{\alpha\beta}\delta w + \tfrac{1}{2}(w,_\alpha^{0c}\,\delta w,_\beta + w,_\beta^{0c}\,\delta w,_\alpha). \tag{3.4}$$

The quantities denoted by (*) in (3.2) result from boundary layer terms due to elastic unloading, and (^{d}) denotes higher order terms that satisfy $\xi^{-\beta}(^d) \to 0$ for $\xi \to 0$.

As mentioned in connection with (2.18), the initial slope λ_1 must be chosen large enough so that no elastic unloading occurs at bifurcation at any point of the body:

$$\hat{m}_c^{\alpha\beta}\left(\lambda_1 \frac{d\eta^0_{\alpha\beta}}{d\lambda}\bigg|_c + \overset{(1)}{\eta}_{\alpha\beta}\right) \geq 0. \tag{3.5}$$

To determine the actual slope we first note that for the linear comparison problem, which results from neglecting the possibility of elastic unloading (see Section 5.1) the expansion (3.1) is replaced by $\lambda = \lambda_c + \lambda_1^{he}\xi + O(\xi^2)$. If the minimum value of λ_1 that is needed to satisfy the equality in (3.5) is less than λ_1^{he}, then in the immediate vicinity of the bifurcation point no elastic unloading will occur and, in this vicinity, the elastic–plastic solid behaves as the linear comparison solid. The more usual case is that the minimum value of λ_1 that satisfies the equality in (3.5) is greater than λ_1^{he}. Clearly, choosing a smaller value of λ_1 than this minimum violates (3.5). Now suppose a value of λ_1 greater than this minimum were chosen. Then, no elastic unloading would occur in some vicinity of the bifurcation point. Hence, in this vicinity the elastic–plastic solid behaves as the linear comparison solid for which $\lambda_1 = \lambda_1^{he}$, but this choice of λ_1, by supposition, violates (3.5), leading to a contradiction. Therefore, λ_1 must be taken as the minimum value that satisfies (3.5). In the following discussion attention is confined to the usual case where $\lambda_1 > \lambda_1^{he}$.

Neutral loading at bifurcation and subsequent elastic unloading may start at one isolated point, at several points simultaneously, or along a curve. In a case where unloading starts at one isolated point x_c^i on a smooth shell surface (Fig. 3.1(a)) we shall introduce a set of local Cartesian coordinates z_i centered at x_c^i such that z_3 is directed along the outward normal to the surface. In the analysis we then need a Taylor expansion of the function on the left-hand side in (3.5),

$$\hat{m}_c^{\alpha\beta}\left(\lambda_1 \frac{d\eta^0_{\alpha\beta}}{d\lambda}\bigg|_c + \overset{(1)}{\eta}_{\alpha\beta}\right) = C_3 z_3 + C_{\alpha\beta} z_\alpha z_\beta + \ldots, \tag{3.6}$$

and of the shell surface

$$D_3 z_3 + D_{\alpha\beta} z_\alpha z_\beta + \ldots = 0. \tag{3.7}$$

In the following $C_3 < 0$ *is* assumed, and we note that there are cases in which some of the linear terms with coefficients C_1, C_2, D_1, and D_2 in (3.6) and (3.7) do not vanish.

Now, stretched boundary layer coordinates $\overset{*}{z}_i$ are chosen such that the surfaces enclosing the elastically unloaded region are independent of ξ to lowest order when written in terms of $\overset{*}{z}_i$. Anticipating $\lambda_2 < 0$, we choose

$$\overset{*}{z}_i = z_i \left[\frac{\xi^{-\beta}}{-\lambda_2(1+\beta)}\right]^{a_i}, \tag{3.8}$$

where $a_i = 1$ if lowest order z_i-term in (3.6) or (3.7) is linear, $a_i = \tfrac{1}{2}$ if the lowest order z_i-term is quadratic, and $a_i = 0$ if the two surfaces (3.6) and (3.7) are parallel in the z_i-direction.

Expanding the expression $\hat{m}^{\gamma\delta}\dot{\eta}_{\gamma\delta} = 0$ for the instantaneous neutral-loading surface and introducing the stretched coordinates (Hutchinson, 1974) the following lowest order expression appears:

$$0 = \hat{m}^{\gamma\delta}\dot{\eta}_{\gamma\delta} = \xi^{\beta}[(1+\beta)\lambda_2 f(\overset{*}{z}_i) + \hat{m}_c^{\gamma\delta}\overset{*}{\eta}_{\gamma\delta}] + \ldots, \tag{3.9}$$

where the terms in square brackets have values, for fixed $\overset{*}{z}_i$, independent of ξ. In particular, for (3.6)

$$f(\overset{*}{z}_i) = \left[\hat{m}_c^{\gamma\delta} \left.\frac{d\eta_{\gamma\delta}^0}{d\lambda}\right|_c\right]_{x_c^i} - C_3 \overset{*}{z}_3 - C_{\alpha\beta}\overset{*}{z}_\alpha \overset{*}{z}_\beta. \tag{3.10}$$

The strains are constrained by (2.2) to vary linearly through the thickness, which means that a nonzero $\overset{*}{\eta}_{\gamma\delta}$, resulting from nonzero $\overset{*}{\varepsilon}_{\gamma\delta}$ and $\overset{*}{\kappa}_{\gamma\delta}$, would vary linearly through the entire thickness, and also outside the region of elastic unloading. Therefore, we must require $\overset{*}{\eta}_{\gamma\delta} = \overset{*}{\varepsilon}_{\gamma\delta} = \overset{*}{\kappa}_{\gamma\delta} = 0$, and consequently the lowest order expression (3.9) for the neutral loading surface reduces to $f(\overset{*}{z}_i) = 0$ in stretched coordinates.

Hutchinson's analysis of the lowest order changes of the stress rate due to elastic unloading leads to the expression

$$\overset{*}{\sigma}^{\gamma\delta} = (1+\beta)\lambda_2[\hat{g}_c^{-1}\hat{m}_c^{\gamma\delta}]_{x_c^i} f(\overset{*}{z}_i) \tag{3.11}$$

inside the unloading region and $\overset{*}{\sigma}^{\gamma\delta} = 0$ outside. The contribution of these stresses to the increment of the membrane stress tensor is

$$\overset{*}{N}^{\gamma\delta} = \int_{-h/2}^{h/2} \overset{*}{\sigma}^{\gamma\delta}\, dx^3 = -(1+\beta)\lambda_2\xi^{\beta}\int \overset{*}{\sigma}^{\gamma\delta}\, d\overset{*}{z}_3 + \ldots$$

$$= -(1+\beta)^2\lambda_2^2\xi^{\beta}[\hat{g}_c^{-1}\hat{m}_c^{\gamma\delta}]_{x_c^i}\int f(\overset{*}{z}_i)\, d\overset{*}{z}_3 + \ldots \tag{3.12}$$

and similarly for $\overset{*}{M}^{\gamma\delta}$, where the last integrals extend through the unloading region.

The incremental principle of virtual work (2.9) is satisfied by the fundamental solution as well as by the bifurcated solution. When the fundamental solution is used to eliminate the external virtual work from (2.9), and the expansion (3.2) is substituted into the resulting equation, using (3.1), (3.3), and (3.4), we obtain

$$\int_A \left\{\overset{d}{N}^{\gamma\delta}\delta^c\varepsilon_{\gamma\delta} + \overset{d}{M}^{\gamma\delta}\delta\kappa_{\gamma\delta} + N_{0c}^{\gamma\delta}\overset{d}{w}_{,\gamma}\delta w_{,\delta}\, dA\right\} + \xi^{\beta}\int_A \left\{\overset{*}{N}^{\gamma\delta}\delta^c\varepsilon_{\gamma\delta} + \overset{*}{M}^{\gamma\delta}\delta\kappa_{\gamma\delta}\right\} dA$$

$$+\,\xi\int_A \left\{2\lambda_1 \overset{(1)}{N}^{\gamma\delta}\left.\frac{dw_{,\gamma}^0}{d\lambda}\right|_c \delta w_{,\delta} + 2\left(\lambda_1 \left.\frac{dN_0^{\gamma\delta}}{d\lambda}\right|_c + \overset{(1)}{N}^{\gamma\delta}\right)\overset{(1)}{w}_{,\gamma}\delta w_{,\delta}\right\} dA + \ldots = 0. \tag{3.13}$$

This variational equation for the $(^d)$-quantities is used to identify the value of β and to determine λ_2.

The second term in (3.13) can be rewritten using (3.8) and (3.12) as

$$\xi^{\beta}\int_A \left\{\overset{*}{N}^{\gamma\delta}\delta^c\varepsilon_{\gamma\delta} + \overset{*}{M}^{\gamma\delta}\delta\kappa_{\gamma\delta}\right\} dA =$$

$$-[\xi^{\beta}(1+\beta)(-\lambda_2)]^{(2+a_1+a_2)}[\hat{g}_c^{-1}\hat{m}_c^{\gamma\delta}\delta\eta_{\gamma\delta}]_{x_c^i}\int_{V^*} f(\overset{*}{z}_i)\, d\overset{*}{V} + \ldots. \tag{3.14}$$

Here, $d\overset{*}{V} = d\overset{*}{z}_1 d\overset{*}{z}_2 d\overset{*}{z}_3$, and $\overset{*}{V}$ is the elastic unloading region enclosed by (3.10) and the stretched coordinate version of (3.7). Now substituting (3.14) into (3.13) we can argue that all three terms must be of order ξ, with $\beta = (2 + a_1 + a_2)^{-1}$. Suppose $\beta > (2 + a_1 + a_2)^{-1}$, so that (3.14) is of order higher than ξ. Then the first term of (3.13) must be of order ξ and balance the last term; but this is only possible if $\lambda_1 = \lambda_1^{he}$, i.e. without elastic unloading. If $\beta < (2 + a_1 + a_2)^{-1}$, the first two terms in (3.13) must balance each other and the $(^d)$ - terms be of order $\xi^{\beta(2 + a_1 + a_2)}$; but then the bifurcation mode would be the only solution for $(^d)$ and $\lambda_2 = 0$ would be required (implying that the boundary terms vanish).

Thus, $\beta = (2 + a_1 + a_2)^{-1}$ must be chosen, and (3.13) leads to the following equation from which λ_2 is determined:

$$\{-\lambda_2(1+\beta)\}^{1/\beta}[\hat{g}_c^{-1}\hat{m}_c^{\gamma\delta}\overset{(1)}{\eta}_{\gamma\delta}]_{x_c^i}\int_{V^*} f(\overset{*}{z}_i)\,d\overset{*}{V} = \mathscr{A} + \lambda_1\,\mathscr{B}. \tag{3.15}$$

Here, $\mathscr{A}$ and $\mathscr{B}$, given by (5.21) and (5.22), are determined from a hypoelastic analysis.

In the most common case, in which unloading starts at an isolated point on a smooth surface, as anticipated in (3.6), (3.7), and Fig. 3.1(a), $a_1 = a_2 = \frac{1}{2}$, which leads to $\beta = \frac{1}{3}$. If unloading starts simultaneously along every point of the z_2-axis, such as on a flat surface of a column (Fig. 3.1(c)), or in axisymmetric bifurcation of a circular cylindrical shell, $a_1 = \frac{1}{2}$, $a_2 = 0$ results in $\beta = \frac{2}{5}$ (Hutchinson, 1974). A wedge-shaped unloading region (Fig. 3.1(b)), due to linear z_2-terms in the expression corresponding to either (3.6) or (3.7), requires $a_1 = \frac{1}{2}$, $a_2 = 1$, and thus $\beta = \frac{2}{7}$ (Tvergaard and Needleman, 1975a, b).

Since $\lambda_2 < 0$, the first three terms of (3.1) can be employed to obtain the following estimates of the maximum support load $\lambda_{\max}$ and the corresponding bifurcation mode amplitude $\xi_{\max}$:

$$\xi_{\max} = \left(\frac{\lambda_1}{-\lambda_2(1+\beta)}\right)^{1/\beta}, \quad \lambda_{\max} = \lambda(\xi_{\max}). \tag{3.16}$$

It must be emphasized, however, that although (3.1) is an asymptotically exact expression, the values given by (3.16) are not asymptotic in any sense, since the maximum occurs at a finite (perhaps small) value of ξ. In most cases numerical computations of postbifurcation behavior show that both $\xi_{\max}$ and the elevation of $\lambda_{\max}$ over λ_c are somewhat larger than predicted by (3.16).

One of the most important outcomes of the asymptotic postbifurcation analysis is the detailed description it gives of the influence of elastic unloading just after bifurcation. This forms the main basis for understanding the difference between elastic and elastic–plastic postbifurcation behavior.

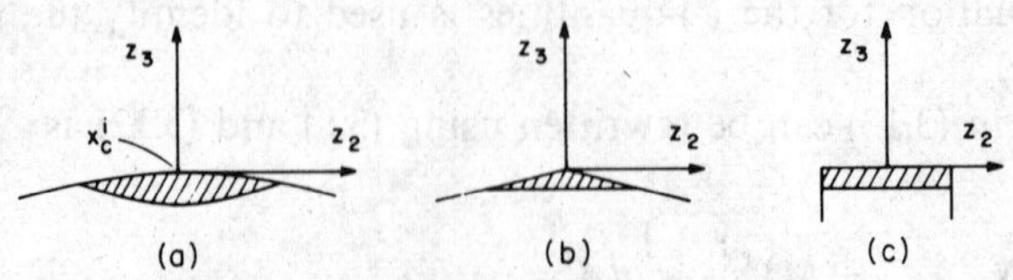

FIG. 3.1. Cross-sections of elastic unloading regions (hatched areas). (a) Unloading starts at a single point of smooth surface. (b) Unloading starts at the top of a sharp edge on the surface. (c) Unloading starts simultaneously along the z_2-axis.

4. Examples of Postbuckling Behavior in the Plastic Range

We employ two structures to highlight certain characteristic features of plastic postbuckling behavior. An asymmetric column studied by Tvergaard and Needleman (1975a) illustrates the interaction of material nonlinearity and structural geometry. For this simple structure, the effects of the two relevant sorts of material nonlinearity—the decreasing stiffness with increasing deformation and the onset of elastic unloading at bifurcation—are isolated and illustrated explicitly.

The other structure discussed is the cruciform column which has played such a prominent role in the history of plastic buckling theory. The details of this history will not be recounted here except to note that, in the plastic range, the bifurcation stresses predicted by the simplest flow theory can be more than a factor of two greater than the corresponding bifurcation stresses obtained from the simplest deformation theory and that experimental buckling loads unequivocally correspond with the deformation theory predictions; see, for example, Gerard and Becker (1957) and Hutchinson (1974). Here, we employ the phenomenological J_2 vertex model of Christoffersen and Hutchinson (1979) in an analysis of an imperfect cruciform column to illustrate the interaction between material path dependence and imperfection sensitivity.

4.1. *Asymmetric column*

The plastic buckling behavior of columns subject to axial compression has been extensively investigated and has been reviewed by Sewell (1972). An investigation of the postbuckling behavior of asymmetric columns by Tvergaard and Needleman (1975a) provides an example where the postbuckling behavior in the plastic range is entirely different from that in the elastic range. To be specific attention will be focused on the U-shaped column depicted in Fig. 4.1, although many of the subsequent formulae remain valid for more general cross-sections. The one-dimensional analog of the DMV equations gives the stretching strain rate of the center line $\dot{\varepsilon}$ and the rate of curvature $\dot{\kappa}$ as

$$\dot{\varepsilon} = \dot{u}_{,1} + w_{,1}\dot{w}_{,1}, \quad \dot{\kappa} = -\dot{w}_{,11}, \tag{4.1}$$

where u and w are displacements of the center line in the x_1 and x_2 directions respectively.

The column is subject to axial compression so that for a perfect column, the prebifurcation stress state can be written as $\sigma = \lambda\sigma_0$, where λ is a positive load parameter and σ_0 is a unit compressive stress. Bifurcation occurs with the critical value of the load parameter λ_c and the eigenmode w_c given by

$$\lambda_c = \left(\frac{\pi r}{l}\right)^2 \frac{E_t}{(-\sigma_0)}, \quad w_c = r \sin\frac{\pi x_1}{l}, \tag{4.2}$$

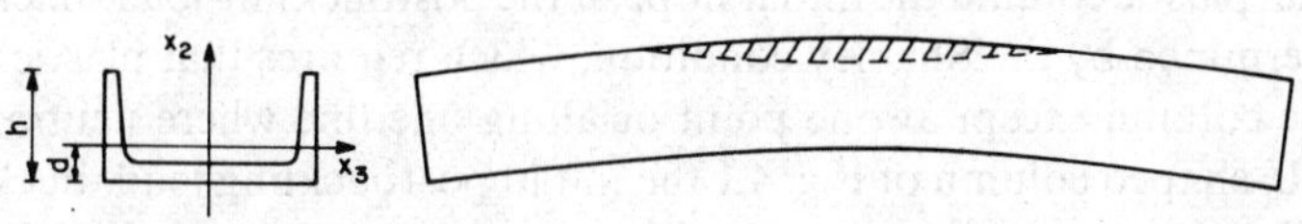

FIG. 4.1. A column with a U-shaped cross-section. The elastic unloading region at bifurcation is indicated by the hatched area.

where E_t is the critical tangent modulus, i.e. the slope of the stress–strain curve at bifurcation, l is the length of the column, and r is the radius of gyration, defined by $r = \sqrt{I/A}$, with A being the cross-sectional area and I the area moment of inertia.

A linear elastic column ($E_t = E$, Young's modulus) has a symmetric postbuckling behavior even though the cross-section is asymmetric. In fact, as is well known, the stable symmetric postbuckling behavior of such a column is revealed only by a more sophisticated description of the change in curvature than that embodied in (4.1). On the other hand, (4.1) is adequate to exhibit the initial postbifurcation behavior of an asymmetric nonlinearly elastic column.

For such a nonlinear elastic column, application of Koiter's (1945) general theory of elastic stability gives

$$\lambda = \lambda_c + \lambda_1^e \xi + O(\xi^2), \tag{4.3}$$

where λ_c is given by (4.2) and the initial slope of the postbifurcation curve λ_1^e is

$$\frac{\lambda_1^e}{\lambda_c} = \frac{4\rho}{3\pi}\left(\frac{q}{1+q}\right) \tag{4.4}$$

Here

$$\rho = \frac{1}{rI}\int_A x_2^3 dA, \quad q = \left|\left(\frac{\sigma_0 \lambda_c}{E_t^c}\frac{dE_t}{d\sigma}\right)_c\right|. \tag{4.5}$$

The non-dimensional parameter ρ is a measure of the asymmetry of the cross-section which vanishes for symmetric cross-sections while the parameter q is a measure of material non-linearity. For a piecewise power law of the form

$$\varepsilon = \begin{cases} \dfrac{\sigma}{E}, & \sigma \leqslant \sigma_y, \\ \dfrac{\sigma_y}{E}\left[\dfrac{1}{n}\left(\dfrac{\sigma}{\sigma_y}\right)^n - \dfrac{1}{n} + 1\right], & \sigma > \sigma_y, \end{cases} \tag{4.6}$$

with E being Young's modulus and σ_y the uniaxial yield stress, q is simply $(n-1)$ for $|\lambda_c \sigma_0| > \sigma_y$.

If the cross-section is symmetric ($\rho = 0$) or if the material is linear ($n = 1$), λ_1^e vanishes and a further analysis is required to determine the initial postbuckling behavior. On the other hand, for the U-shaped cross-section depicted in Fig. 4.1 λ_1^e is positive so that the load rises for buckling toward the open end of the U ($w > 0$) and the load falls for buckling toward the closed end of the U ($w < 0$).

For an elastic–plastic column the initial slope of the postbuckling load–buckling deflection curve λ_1 is determined by the Shanley condition, which requires that plastic loading occurs throughout the column except at one point or along one line where neutral loading takes place. For the U-shaped column of Fig. 4.1 the initial postbuckling load–buckling deflection curve is given by

$$\frac{\lambda}{\lambda_c} = 1 + \frac{\lambda_1}{\lambda_c}\xi + \cdots, \tag{4.7}$$

where

$$\frac{\lambda_1}{\lambda_c} = \begin{cases} \dfrac{h-d}{r}, & \text{for } w > 0, \\[2ex] \dfrac{d}{r}, & \text{for } w < 0, \end{cases} \tag{4.8}$$

with h and d defined in Fig. 4.1. The initial slope λ_1 for the elastic–plastic column is positive regardless of the sign of w. Hence, buckling takes place under increasing load in accord with Shanley's (1947) concept.

The expansion (4.7) applies so long as $\lambda_1 > \lambda_1^e$; if $\lambda_1 < \lambda_1^e$ for some case then (4.7) is replaced by (4.3) and no elastic unloading takes place in the vicinity of the bifurcation point for buckling in that direction.

Provided that $\lambda_1 > \lambda_1^e$, the next term in the postbifurcation expansion is determined by application of Hutchinson's (1973, 1974) theory. For the U-shaped column unloading begins along a line of points so that β in (3.15) is 2/5 and the first three terms of the postbifurcation expansion are given by

$$\frac{\lambda}{\lambda_c} = 1 + \frac{\lambda_1}{\lambda_c}\xi + \frac{\lambda_2}{\lambda_c}\xi^{7/5}. \tag{4.9}$$

An explicit analytical expression for λ_2 is given by Tvergaard and Needleman (1975a). The value of λ_2 is always negative and from (4.9) an approximate value (3.16) of the maximum support load of the column can be calculated. Numerical values of this maximum support load λ_{max} for a specific case considered by Tvergaard and Needleman (1975a) are $\lambda_{max}/\lambda_c = 1.0096$ for $w < 0$ and $\lambda_{max}/\lambda_c = 1.0220$ for $w > 0$, exhibiting the asymmetric postbifurcation behavior with the greatest load carrying capacity for buckling toward the open end of the U, as also exhibited by the nonlinear elastic column.

Numerical solutions to the column equations were also obtained by Tvergaard and Needleman (1975a). Figure 4.2 depicts the numerical results for a perfect U-shaped column

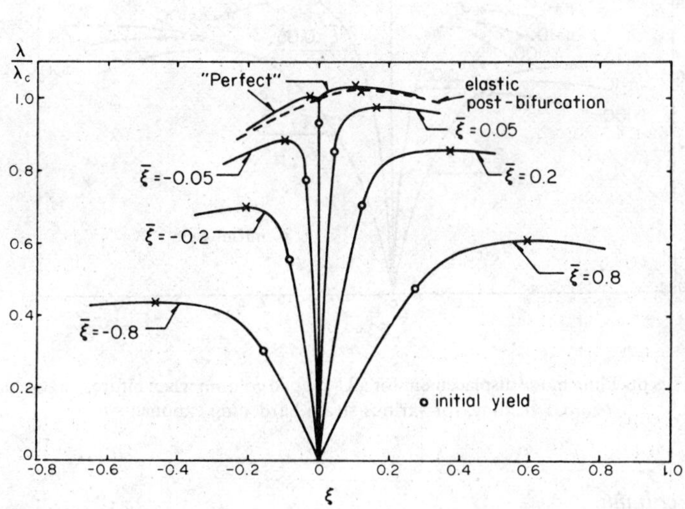

FIG. 4.2. Load *versus* buckling mode displacement for a U-shaped column. Bifurcation takes place in the plastic region and the dashed curve shows the postbifurcation behavior when unloading is suppressed.

that possesses the initial postbifurcation behavior quoted above and for imperfect columns. The imperfection is taken to be in the shape of the critical mode (4.2) and the imperfection magnitude $\bar{\xi}$ is normalized by the radius of gyration. The relation between the additional growth of the buckling mode amplitude ξ and the load parameter λ is shown in Fig. 4.2. Although there is general good agreement between the predictions of the asymptotic analysis for the postbifurcation behavior of a perfect column and the corresponding numerically computed curve, the numerical results exhibit significantly more asymmetry than is indicated by the asymptotic analysis.

The computations for imperfect columns show considerable sensitivity to negative imperfections, which reflects the asymmetry in the initial postbifurcation behavior.

The dotted curve in Fig. 4.2 shows the results of a numerical computation for the behavior of a perfect nonlinear elastic column. Relative to this curve it can be seen that the principal effect of elastic unloading on the initial postbifurcation behavior is to provide a slight stabilizing influence. For imperfect columns, the results of nonlinear elastic computations carried out by Tvergaard and Needleman (1975a) were indistinguishable from the elastic–plastic results in Fig. 4.2.

Figure 4.3 shows the behavior of a U-shaped column designed so that bifurcation of the perfect column takes place slightly before plastic yielding. The initial postbifurcation behavior is governed by the elastic response and is symmetric. However, the imperfection sensitivity is governed by the plastic response and is asymmetric, as shown in Fig. 4.3 for strain hardening exponents n of 4, 10, and 100. This asymmetry would also be revealed by a Perry first-yield formula, as can be seen from the onset of initial yielding marked in Fig. 4.3.

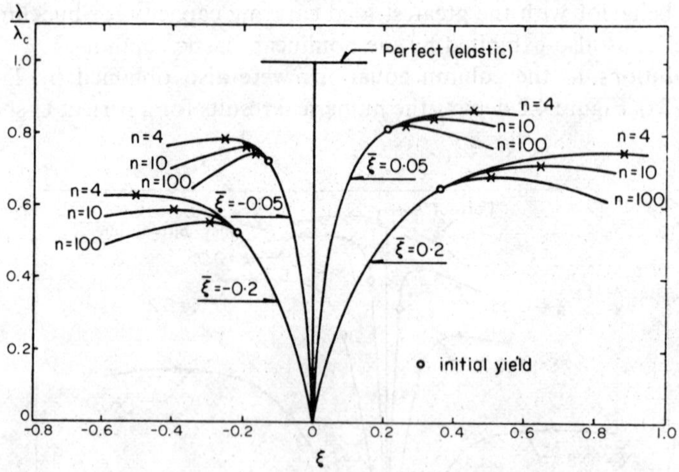

FIG. 4.3. Load *versus* buckling mode displacement for a U-shaped column when bifurcation occurs slightly prior to initial yielding, for various strain hardening exponents n.

4.2. *Cruciform column*

The cruciform column (Fig. 4.4) is perhaps the most extreme example of the discrepancy

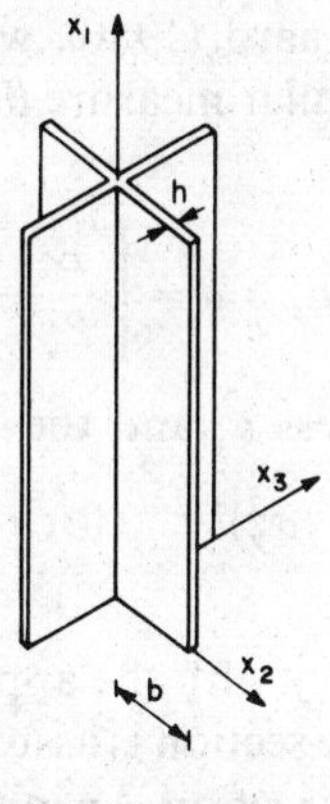

FIG. 4.4. The cruciform column.

between flow theory and deformation theory bifurcation predictions. In the elastic range the critical stress for torsional buckling is given by

$$\sigma_E = G\left[\frac{h}{b}\right]^2, \tag{4.10}$$

where G is the elastic shear modulus, h is the plate thickness, and b is the plate width. Even in the plastic range the expression (4.10) for the critical stress applies according to any isotropic flow theory with a smooth yield surface, whereas according to any deformation theory for an initially isotropic solid the elastic shear modulus in (4.10) has to be replaced by a modulus G_s given by

$$G_s = G/(1+3G(1/E_s - 1/E)), \tag{4.11}$$

where E_s is the current secant modulus (see, for example, Stowell, 1948; Hutchinson, 1974). It may be noted that the expression (4.10), which corresponds to a pure twisting mode, is also the limiting value of the critical load for a long rectangular plate subject to axial compression with one of the long edges free and the other three edges simply supported.

Experimental results for cruciform columns cited by Gerard and Becker (1957) are in good agreement with the deformation theory predictions which, as noted in Section 2.3, can be regarded as the bifurcation predictions of a flow theory of plasticity with a vertex at the current loading point.

Onat and Drucker (1953) argued that extremely small geometric imperfections could bring the maximum support loads predicted by a smooth yield surface flow theory in line with the experimentally observed values. More recently, Hutchinson and Budiansky (1976) have presented a detailed asymptotic and numerical analysis, which shows that an extremely small amount of pre-twist is sufficient to considerably reduce the maximum load carrying capacity predicted by J_2 flow theory. Other deviations from "perfect" conditions can have a similar effect. For example, Gjelsvik and Lin (1976) have shown that even for a perfectly flat plate prebifurcation stresses induced by friction at the loaded edges reduce the flow theory bifurcation stress considerably.

Here, the effect of a vertex on the yield surface will be analyzed based on J_2 corner theory as specified by (2.19)–(2.22). The transition function $f(\theta)$ used for $\theta_0 \leqslant \theta \leqslant \theta_c$ is one of those suggested by Christoffersen and Hutchinson (1979). This transition function was also used in an analysis by Hutchinson and Tvergaard (1980), where further details of the numerical implementation can be found. The angular measure θ_c of the cone angle is taken to be given by

$$\tan\theta_c = \sqrt{a}\tan\beta_c, \quad a = \left[\frac{E}{E_s} - 1\right]\left[\frac{E}{E_t} - 1\right]^{-1}, \tag{4.12}$$

where, in terms of the initial yield stress σ_y and the Mises effective stress σ_e,

$$\tan\beta_c = -\sigma_y/(\sigma_e^2 - \sigma_y^2)^{1/2} \quad \text{for} \quad \tfrac{1}{2}\pi \leqslant \beta_c \leqslant (\beta_c)_{\max}, \tag{4.13}$$

and otherwise $\beta_c = (\beta_c)_{\max}$.

The analysis is based on DMV theory with the approximations also used by Hutchinson and Budiansky (1976). Thus, each cross-section is assumed to rotate as a rigid body about the x_1-axis and to translate only in the x_1-direction. An initial imperfection is specified by the (x_1-independent) twisting angle $\bar{\phi}$ per unit length, ϕ is the additional twisting angle per unit length, and ε_0 is the strain of the column axis. Then, for the plate in the x_1–x_2 plane the displacement components are

$$u_1 = \varepsilon_0 x_1, \quad u_2 = -\tfrac{1}{2}(\phi^2 + 2\phi\bar{\phi})x_1^2 x_2, \quad w = \phi x_1 x_2, \tag{4.14}$$

and the only nonzero components of $\varepsilon_{\alpha\beta}$ and $\kappa_{\alpha\beta}$ are

$$\varepsilon_{11} = \varepsilon_0 + \tfrac{1}{2}(\phi^2 + 2\phi\bar{\phi})x_2^2, \quad \kappa_{12} = -\phi. \tag{4.15}$$

Furthermore, σ^{22} is taken to be zero, and thus the zero value of ε_{22} implied by (4.14) is not enforced in the constitutive relations.

A numerical solution is based on the incremental principle of virtual work (2.9). Here, the only nonzero components of the DMV stress measures are N^{11} and M^{12}, and in each increment corresponding values of $\dot{\varepsilon}_0$, $\dot{\phi}$, and $\dot{P}$ are determined, where P is the axial compressive force. The integrals in (2.8) and (2.9) are evaluated by 4-point Gaussian integration over the plate width and 5-point Simpson integration through the half-thickness.

In all the following examples the thickness to width ratio h/b is chosen such that the critical bifurcation load P_D for deformation theory equals half the elastic (or J_2 flow theory) bifurcation load P_E. The uniaxial stress–strain curve is taken to be given by (4.6), where $\sigma_y/E = 0.002$ is used in the following, together with Poisson's ratio $\nu = 0.3$. Also, as in Hutchinson and Budiansky (1976), the parameters $\bar{\xi} = \bar{\phi}b^2/h$ and $\xi = \phi b^2/h$ are introduced, where $\bar{\xi}h$ is the difference between the initial plate deflections at two points of the free plate edge a distance b apart.

The postbifurcation behavior of a perfect column shown in Fig. 4.5 for two different vertex descriptions is obtained numerically as the behavior of a column with a very small initial imperfection $\bar{\xi} = 10^{-5}$. The strain hardening exponent is $n = 4$. For a rather sharp vertex with $(\beta_c)_{\max} = \frac{3}{4}\pi$ in (4.4) and with a finite total loading range limited by $\theta_0 = \frac{1}{2}\theta_n$ a positive value of the initial postbifurcation slope λ_1 in (2.18) is found, as expected on the basis of the condition (2.23). On the other hand, for $\theta_0 = 0$, bifurcation takes place smoothly without a slope discontinuity in the P–ξ curve, as discussed in Section 2.3. In both cases bifurcation takes place under increasing load and subsequently a maximum is reached at a finite mode deflection, as also predicted by Hutchinson's theory (3.1) for a smooth yield surface. It may be noted that the postbifurcation curve predicted by deformation theory has zero initial slope, and that also J_2 flow theory predicts $\lambda_1 = 0$ for this particular structure.

Figure 4.6 shows the influence of an initial imperfection $\bar{\xi} = 0.01$. For the vertex

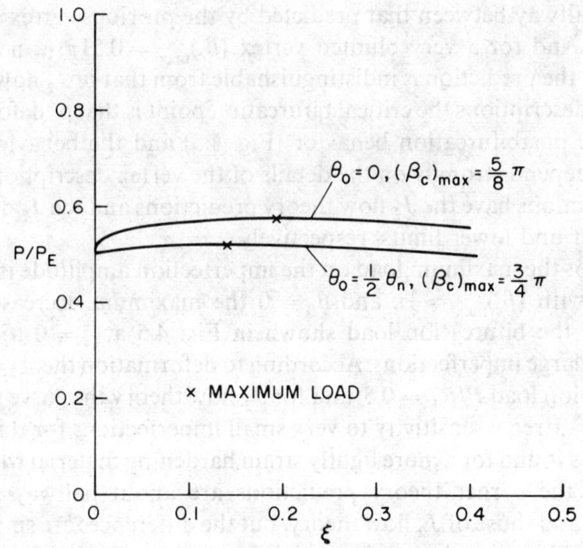

FIG. 4.5. Postbifurcation behavior of a perfect cruciform column according to two different vertex descriptions ($\sigma_y/E = 0.002$, $\nu = 0.3$, $n = 4$, $P_D/P_E = 0.5$).

characterized by $(\beta_c)_{max} = \frac{3}{4}\pi$ and $\theta_0 = \frac{1}{2}\theta_n$ the material is initially completely in the total loading range $\theta < \theta_0$ until, at $\xi \simeq 0.06$, a region in which the material response has entered the "nonlinear" range, $\theta_0 \leqslant \theta \leqslant \theta_c$ starts to spread into the material. In this case the maximum load is only slightly higher than that predicted by deformation theory. If the material response is totally nonlinear, $\theta_0 = 0$, with a less sharp vertex $(\beta_c)_{max} = \frac{5}{8}\pi$, the

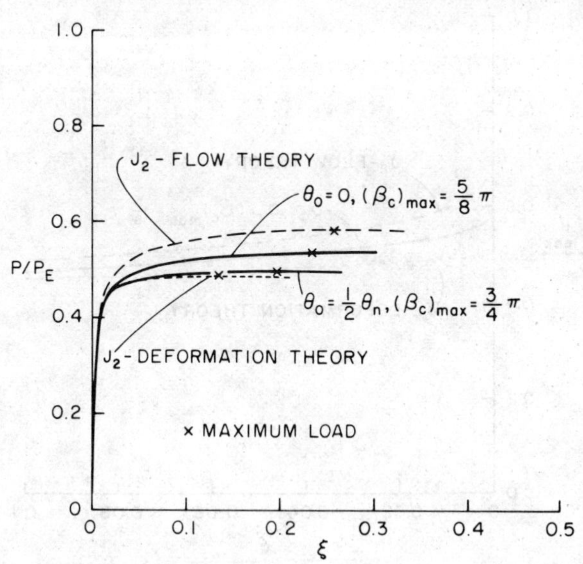

FIG. 4.6. Influence of an initial imperfection $\overline{\xi} = 0.001$ on a cruciform column for various constitutive descriptions ($\sigma_y/E = 0.002$, $\nu = 0.3$, $n = 4$, $P_D/P_E = 0.5$).

maximum occurs halfway between that predicted by the previous vertex and that predicted by J_2 flow theory. And for a very blunted vertex $(\beta_c)_{max} = 0.51\pi$ (a nearly smooth yield surface) with $\theta_0 = 0$ the prediction is indistinguishable from that of J_2 flow theory in Fig. 4.6. For all three vertex descriptions the critical bifurcation point is that of deformation theory, at $P/P_E = 0.5$, but the postbifurcation behavior (Fig. 4.5) and the behavior of an imperfect structure (Fig. 4.6) depend strongly on the details of the vertex description. As expected, the various vertex predictions have the J_2 flow theory predictions and the J_2 deformation theory predictions as upper and lower limits, respectively.

The dependence of the maximum load on the imperfection amplitude is shown in Fig. 4.7. For vertex theory with $(\beta_c)_{max} = \frac{5}{8}\pi$ and $\theta_0 = 0$ the maximum decreases from the value considerably above the bifurcation load shown in Fig. 4.5 at $\bar{\xi} = 0$ to values below the bifurcation load for large imperfections. According to deformation theory the decaying curve starts at the bifurcation load $P/P_E = 0.5$; and for J_2 flow theory the curve starts at $P/P_E = 1$, thus illustrating the extreme sensitivity to very small imperfections for this constitutive law.

Similar behavior is found for a more lightly strain hardening material with $n = 8$, as shown in Fig. 4.8. Again, the corner theory predictions are about halfway between those of deformation theory and those of J_2 flow theory, but the differences are smaller. In agreement with this trend, the differences found for a high hardening material ($n = 2$) in Fig. 4.9 are considerably larger. In this latter case no maxima are found for imperfections larger than about 0.04, due to the stiffening effect of postbuckling membrane forces in this plate made of a very high hardening material.

Even though the results presented in this Section are obtained for the special case of a cruciform column, it is expected that a yield surface vertex will have a similar effect in other buckling problems. When bifurcation takes place just above the initial yield stress, the deformation and flow theory predictions differ little, and thus no significant effect of a corner

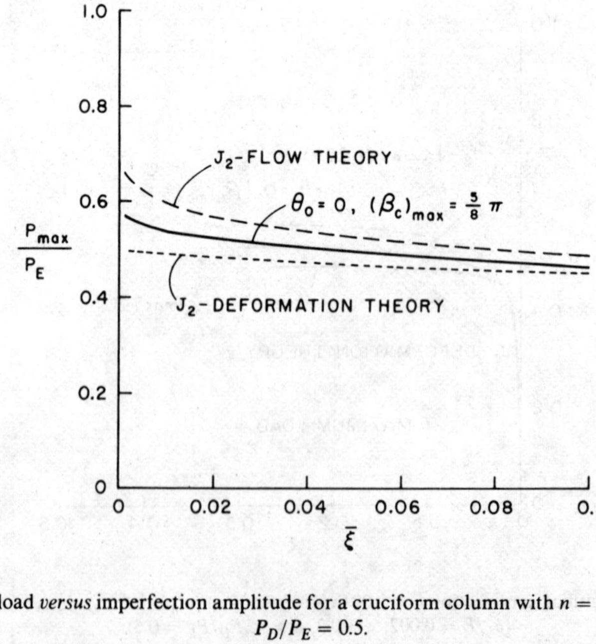

FIG. 4.7. Maximum load *versus* imperfection amplitude for a cruciform column with $n = 4$, $\sigma_y/E = 0.002$, $\nu = 0.3$, $P_D/P_E = 0.5$.

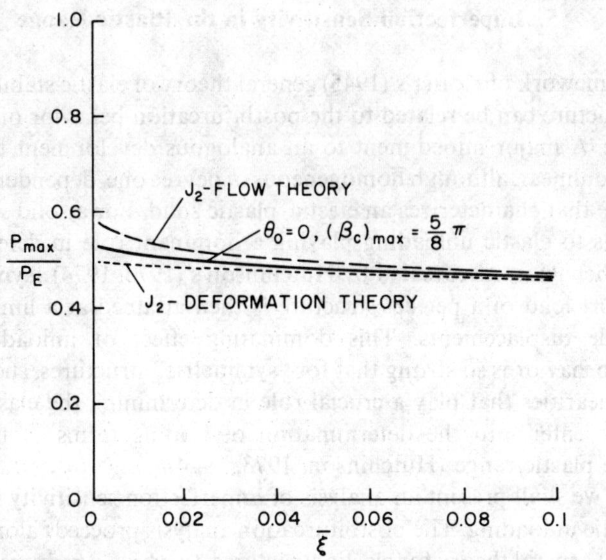

FIG. 4.8. Maximum load *versus* imperfection amplitude for a cruciform column with $n = 8$, $\sigma_y/E = 0.002$, $\nu = 0.3$, $P_D/P_E = 0.5$.

theory is expected. However, when bifurcation occurs more deeply into the plastic range, as for the cruciform columns considered here, the behavior will be more dependent on the details of a vertex description.

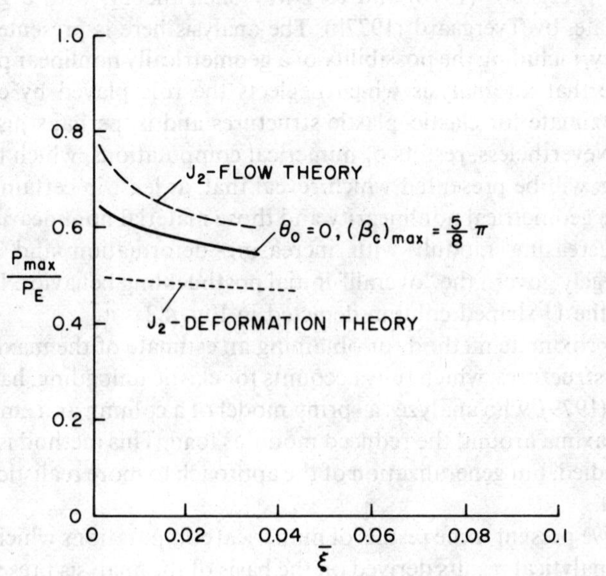

FIG. 4.9. Maximum load *versus* imperfection amplitude for a cruciform column with $n = 2$, $\sigma_y/E = 0.002$, $\nu = 0.3$, $P_D/P_E = 0.5$.

5. Imperfection Sensitivity in the Plastic Range

Within the framework of Koiter's (1945) general theory of elastic stability, the behavior of an imperfect structure can be related to the postbifurcation behavior of the corresponding perfect structure. A major impediment to an analogous development in the plastic range stems from the nonlinear, although homogeneous of degree one, dependence of the stress rate on the strain rate that characterizes an elastic–plastic solid. For a solid with a smooth yield surface this leads to elastic unloading playing a dominant role in determining the initial postbifurcation behavior. As revealed by Hutchinson's (1973, 1974) asymptotic analysis the maximum support load of a perfect structure is then attained at a limit point after finite bifurcation mode displacements. This dominating effect of unloading on the initial postbifurcation behavior is so strong that for "symmetric" structures, such as flat plates, the geometric nonlinearities that play a crucial role in determining the elastic postbifurcation behavior do not enter into the determination of leading terms of the postbifurcation expansion in the plastic range (Hutchinson, 1973, 1974).

In Section 5.1 we shall present an analysis of imperfection sensitivity in the plastic range that ignores elastic unloading. The postbifurcation analysis proceeds along the same lines as in Koiter's (1945) general theory for elastic structures, to which it reduces for an elastic solid. The maximum support load, if there is one in the vicinity of the bifurcation point, is the bifurcation load. Furthermore, an asymptotic expression can be determined which relates the behavior of an imperfect structure to the postbifurcation behavior of the corresponding perfect structure. The analysis to be presented here, approximate in general, derives from Hutchinson and Budiansky's (1976) exact (for certain parameter values) analysis of the imperfection sensitivity of a cruciform column in the plastic range. Hutchinson and Budiansky's (1976) analysis was generalized to the full von Kármán plate equations by Needleman and Tvergaard (1976) and to DMV shell theory, with a geometrically linear prebifurcation state, by Tvergaard (1977b). The analysis here is presented in the context of DMV shell theory, including the possibility of a geometrically nonlinear prebifurcation state.

We emphasize that an analysis which neglects the role played by elastic unloading is inherently approximate for elastic–plastic structures and is, perhaps justly, to be regarded with suspicion. Nevertheless, results of numerical computations, which take full account of elastic unloading, will be presented which reveal that, at least in certain circumstances, the interaction of the geometrical nonlinearity and those material nonlinearities included in this analysis—the decreasing moduli with increasing deformation and the material path dependence—largely govern the "overall" initial postbuckling behavior. In this regard, recall the behavior of the U-shaped column depicted in Fig. 4.2.

A different approximate method for obtaining an estimate of the maximum support load of elastic–plastic structures, which fully accounts for elastic unloading, has been proposed by van der Heijden (1979), who analyzes a spring model of a column and employs an expansion of the locus of maxima around the reduced modulus load. This method is found quite useful for the model studied, but generalization of the approach to more realistic structures remains to be carried out.

In Section 5.2 we present some results of numerical computations which have been directly compared with analytical results derived on the basis of the analysis presented in Section 5.1. Numerical solutions appear indispensable in plastic buckling problems due to the highly nonlinear nature of these problems. No attempt is made here to review the growing literature in this field. Tvergaard (1977a) surveys representative work in this area. Also, Hutchinson

(1974) and Budiansky and Hutchinson (1979) discuss aspects of plastic postbuckling behavior not touched on here.

5.1. *Asymptotic analysis of imperfection sensitivity*

Here, we consider the postbifurcation behavior and imperfection sensitivity of Hill's (1958, 1961) incrementally "linear comparison solid" for which the constitutive law takes the form

$$\dot{\sigma}_{\alpha\beta} = L^{\alpha\beta\gamma\delta}\dot{\eta}_{\gamma\delta}, \tag{5.1}$$

where the plastic loading branch of **L** is taken to be active everywhere in the current plastic zone regardless of the direction (in strain rate space) of $\dot{\eta}$. (Here and subsequently, for notational convenience, the plane stress moduli are denoted by **L** rather than $\hat{\mathbf{L}}$ as in Section 2.) In general, this incrementally linear comparison solid will be a path dependent hypoelastic solid.

We focus on problems for which the loading can be characterized by a single monotonically increasing loading parameter λ. Furthermore, attention is restricted to circumstances in which there is a unique eigenmode associated with the lowest critical load λ_c.

It is convenient to carry out the analysis in several steps. First, the initial postbifurcation behavior for a perfect structure is determined. Then for an imperfect structure a regular perturbation analysis valid for $\lambda < \lambda_c$ is carried out. The expansion determined in this part of the analysis breaks down as $\lambda \approx \lambda_c$. A singular perturbation analysis is then required for λ near λ_c. Unless the hypoelastic solid is actually a path independent elastic solid, this singular perturbation analysis results in an expression containing undetermined constants. Matching with the results of the regular perturbation analysis determines these constants.

The starting point of the analysis is the principle of virtual work. In much of the analysis it is convenient to use the total, rather than the incremental, form of this principle and to express the external virtual work in terms of quantities associated with the prebifurcation state.

$$\int_V [(\sigma^{\alpha\beta} - \sigma_0^{\alpha\beta})(\delta e_{\alpha\beta} + x_3\delta\kappa_{\alpha\beta}) + \sigma^{\alpha\beta}w_{,\alpha}\delta w_{,\beta} + \sigma^{\alpha\beta}\bar{w}_{,\alpha}\delta w_{,\beta} - \sigma_0^{\alpha\beta}w^0{}_{,\alpha}\delta w_{,\beta}]dV = 0, \tag{5.2}$$

where V is the volume occupied by the shell and in (5.2) and throughout this analysis a subscript or superscript $(\ )_0$ (placed where convenient) denotes field quantities associated with the prebifurcation state; the quantity $\bar{w}$ represents the initial geometric imperfection and w measures the additional lateral deflection. Furthermore, we shall, for convenience, work directly with the stress tensor $\boldsymbol{\sigma}$ rather than with the membrane force **N** and moment **M** separately.

(i) Initial postbifurcation behavior

The initial postbifurcation expansion takes the form

$$\lambda = \lambda_c + \lambda_1^{he}\xi + \lambda_2^{he}\xi^2 + \dots, \tag{5.3}$$

$$\begin{Bmatrix} u_\alpha \\ w \end{Bmatrix} = \begin{Bmatrix} u_\alpha^0 \\ w_0 \end{Bmatrix} + \xi\begin{Bmatrix} \overset{(1)}{u}_\alpha \\ \overset{(1)}{w} \end{Bmatrix} + \xi^2\begin{Bmatrix} \overset{(2)}{u}_\alpha \\ \overset{(2)}{w} \end{Bmatrix} + \dots, \tag{5.4}$$

where in (5.3) the notation λ_i^{he} has been employed to emphasize that this expansion is only valid for the hypoelastic linear comparison solid and not for an elastic–plastic solid.

Attention is confined to determining the leading term in (5.3) so that λ_2^{he} is only calculated if λ_1^{he} vanishes.

In order for the expansion parameter ξ to be well defined it is necessary to ensure that the functions $\overset{(i)}{w}$, $i > 1$, are orthogonal to $\overset{(1)}{w}$. We take this orthogonality condition to be

$$\int_V \sigma_c^{\alpha\beta} \overset{(1)}{w}_{,\alpha} \overset{(i)}{w}_{,\beta} dV = 0. \tag{5.5}$$

The expansion for the strains $\boldsymbol{\eta}$ is obtained by substituting (5.4) into (2.1) and expanding $w_0(\lambda)$ in a Taylor series about λ_c to obtain

$$\eta_{\alpha\beta} = \overset{0}{\eta}_{\alpha\beta} + \xi \overset{(1)}{\eta}_{\alpha\beta} + \xi^2 \overset{(2)}{\eta}_{\alpha\beta} + \xi^3 \overset{(3)}{\eta}_{\alpha\beta} + \dots, \tag{5.6}$$

where, with $\mathbf{e}$ denoting the linear part of the strain,

$$\eta_{\alpha\beta}^0 = e_{\alpha\beta}^0 + x_3 \kappa_{\alpha\beta}^0 + \tfrac{1}{2} w^0{}_{,\alpha}\, w^0{}_{,\beta}, \tag{5.7}$$

$$\overset{(1)}{\eta}_{\alpha\beta} = \overset{(1)}{e}_{\alpha\beta} + x_3 \overset{(1)}{\kappa}_{\alpha\beta} + \tfrac{1}{2} \overset{(1)}{w}_{,\alpha} w^c{}_{,\beta} + \tfrac{1}{2} w^c{}_{,\alpha} \overset{(1)}{w}_{,\beta}, \tag{5.8}$$

$$\overset{(2)}{\eta}_{\alpha\beta} = \overset{(2)}{e}_{\alpha\beta} + x_3 \overset{(2)}{\kappa}_{\alpha\beta} + \tfrac{1}{2} \overset{(2)}{w}_{,\alpha} w^c{}_{,\beta} + \tfrac{1}{2} w^c{}_{,\alpha} \overset{(2)}{w}_{,\beta} + \tfrac{1}{2} \overset{(1)}{w}_{,\alpha} \overset{(1)}{w}_{,\beta} + \tfrac{1}{2} \lambda_1^{he} w_{,\alpha}'^c \overset{(1)}{w}_{,\beta} + \tfrac{1}{2} \lambda_1^{he} \overset{(1)}{w}_{,\alpha} w_{,\beta}'^c, \tag{5.9}$$

$$\begin{aligned} \overset{(3)}{\eta}_{\alpha\beta} = {} & \overset{(3)}{e}_{\alpha\beta} + x_3 \overset{(3)}{\kappa}_{\alpha\beta} + \tfrac{1}{2} \overset{(3)}{w}_{,\alpha} w^c{}_{,\beta} + \tfrac{1}{2} w^c{}_{,\alpha} \overset{(3)}{w}_{,\beta} + \tfrac{1}{2} \overset{(1)}{w}_{,\alpha} \overset{(2)}{w}_{,\beta} + \tfrac{1}{2} \overset{(2)}{w}_{,\alpha} \overset{(1)}{w}_{,\beta} \\ & + \tfrac{1}{2} \lambda_2^{he} w_{,\alpha}'^c \overset{(1)}{w}_{,\beta} + \tfrac{1}{2} \lambda_2^{he} \overset{(1)}{w}_{,\alpha} w_{,\beta}'^c. \end{aligned} \tag{5.10}$$

Here, and subsequently, a subscript or superscript c denotes functions of λ evaluated at $\lambda = \lambda_c$ and $(\ )'^c \equiv d(\ )/d\lambda$ at $\lambda = \lambda_c$.

The expression (5.10) presumes $\lambda_1^{he} \equiv 0$, since only in this case will we find it necessary to calculate $\overset{(3)}{\eta}_{\alpha\beta}$.

The expansion for the stresses is obtained from the constitutive relation (5.1), with $(\dot{\ })$ identified as $d(\ )/d\xi$, and (5.6). The tensor of moduli is taken to be a function of the stress state and is expanded in a Taylor series about λ_c, so that

$$L^{\alpha\beta\gamma\delta} = L_c^{\alpha\beta\gamma\delta} + \frac{\partial L^{\alpha\beta\gamma\delta}}{\partial \sigma^{\mu\nu}} (\sigma^{\mu\nu} - \sigma_c^{\mu\nu}) + \tfrac{1}{2} \frac{\partial^2 L^{\alpha\beta\gamma\delta}}{\partial \sigma^{\mu\nu} \partial \sigma^{\rho\omega}} (\sigma^{\mu\nu} - \sigma_c^{\mu\nu})(\sigma^{\rho\omega} - \sigma_c^{\rho\omega}) + \dots. \tag{5.11}$$

The expansion for the stresses then takes the form

$$\sigma^{\alpha\beta} = \sigma_0^{\alpha\beta} + \xi \overset{(1)}{\sigma}{}^{\alpha\beta} + \xi^2 \overset{(2)}{\sigma}{}^{\alpha\beta} + \xi^3 \overset{(3)}{\sigma}{}^{\alpha\beta} + \dots, \tag{5.12}$$

where

$$\overset{(1)}{\sigma}{}^{\alpha\beta} = L_c^{\alpha\beta\gamma\delta}\overset{(1)}{\eta}_{\gamma\delta}, \quad \overset{(i)}{\sigma}{}^{\alpha\beta} = L_c^{\alpha\beta\gamma\delta}\overset{(i)}{\eta}_{\gamma\delta} + \overset{(i)}{D}{}^{\alpha\beta}, \quad i > 1. \tag{5.13}$$

If $\lambda_1^{he} \neq 0$,

$$\overset{(2)}{D}{}^{\alpha\beta} = \tfrac{1}{2}\overset{(1)}{\eta}_{\gamma\delta}\frac{\partial L^{\alpha\beta\gamma\delta}}{\partial\sigma^{\mu\nu}}\bigg|_c \overset{(1)}{\sigma}{}^{\mu\nu} + \tfrac{1}{2}\lambda_1^{he}\left[\overset{(1)}{\eta}_{\gamma\delta}\frac{\partial L^{\alpha\beta\gamma\delta}}{\partial\sigma^{\mu\nu}}\bigg|_c \sigma_c'^{\mu\nu} + \eta_{\gamma\delta}'^{c}\frac{\partial L^{\alpha\beta\gamma\delta}}{\partial\sigma^{\mu\nu}}\bigg|_c \overset{(1)}{\sigma}{}^{\mu\nu}\right] \tag{5.14}$$

and the expression for $\overset{(3)}{D}{}^{\alpha\beta}$ is not needed.

If $\lambda_1^{he} \equiv 0$, then $\overset{(2)}{D}{}^{\alpha\beta}$ is given by (5.14) with λ_1^{he} set zero and

$$\overset{(3)}{D}{}^{\alpha\beta} = \tfrac{1}{3}\overset{(1)}{\eta}_{\gamma\delta}\frac{\partial L^{\alpha\beta\gamma\delta}}{\partial\sigma^{\mu\nu}}\bigg|_c \overset{(2)}{\sigma}{}^{\mu\nu} + \tfrac{2}{3}\overset{(2)}{\eta}_{\gamma\delta}\frac{\partial L^{\alpha\beta\gamma\delta}}{\partial\sigma^{\mu\nu}}\bigg|_c \overset{(1)}{\sigma}{}^{\mu\nu} + \tfrac{2}{3}\lambda_2^{he}\eta_{\gamma\delta}'^{c}\frac{\partial L^{\alpha\beta\gamma\delta}}{\partial\sigma^{\mu\nu}}\bigg|_c \overset{(1)}{\sigma}{}^{\mu\nu}$$
$$+ \tfrac{1}{6}\overset{(1)}{\eta}_{\gamma\delta}\frac{\partial^2 L^{\alpha\beta\gamma\delta}}{\partial\sigma^{\mu\nu}\partial\sigma^{\rho\omega}}\bigg|_c \overset{(1)}{\sigma}{}^{\mu\nu}\overset{(1)}{\sigma}{}^{\rho\omega} + \tfrac{1}{3}\lambda_2^{he}\overset{(1)}{\eta}_{\gamma\delta}\frac{\partial L^{\alpha\beta\gamma\delta}}{\partial\sigma^{\mu\nu}}\bigg|_c \sigma_c'^{\mu\nu}. \tag{5.15}$$

Substituting (5.4), (5.6), and (5.12) into the principle of virtual work (5.2) with $\bar{w} \equiv 0$ and expanding $w_0(\lambda)$ in Taylor series about λ_c gives

$$\xi\int_V\left[\overset{(1)}{\sigma}{}^{\alpha\beta}(\delta e_{\alpha\beta} + x_3\delta\kappa_{\alpha\beta}) + \overset{(1)}{\sigma}{}^{\alpha\beta}w^c{}_{,\alpha}\delta w_{,\beta} + \sigma_c^{\alpha\beta}\overset{(1)}{w}{}_{,\alpha}\delta w_{,\beta}\right]dV$$
$$+\xi^2\int_V\left[\overset{(2)}{\sigma}{}^{\alpha\beta}(\delta e_{\alpha\beta} + x_3\delta\kappa_{\alpha\beta}) + \overset{(2)}{\sigma}{}^{\alpha\beta}w^c{}_{,\alpha}\delta w_{,\beta} + \sigma_c^{\alpha\beta}\overset{(2)}{w}{}_{,\alpha}\delta w_{,\beta}\right.$$
$$\left.+\overset{(1)}{\sigma}{}^{\alpha\beta}\overset{(1)}{w}{}_{,\alpha}\delta w_{,\beta} + \lambda_1^{he}(\overset{(1)}{\sigma}{}^{\alpha\beta}w_{,\alpha}'^{c}\delta w_{,\beta} + \sigma_c'^{\alpha\beta}\overset{(1)}{w}{}_{,\alpha}\delta w_{,\beta})\right]dV = 0. \tag{5.16}$$

The leading term in (5.16) is the variational equation for the eigenmodal quantities and thus vanishes identically. Furthermore, with $\delta(\;)$ replaced by $(^{(i)})$ in this leading term and making use of the orthogonality condition (5.5) we obtain the identity

$$\int_V\left[\overset{(1)}{\sigma}{}^{\alpha\beta}(\overset{(i)}{e}_{\alpha\beta} + x_3\overset{(i)}{\kappa}_{\alpha\beta}) + \overset{(1)}{\sigma}{}^{\alpha\beta}w^c_{,\alpha}\overset{(i)}{w}{}_{,\beta}\right]dV = 0 \tag{5.17}$$

for $i > 1$. Next, with $\delta(\;)$ replaced by $(^{(1)})$, the term of order ξ^2 in (5.16) yields

$$\int_V\overset{(2)}{\sigma}{}^{\alpha\beta}(\overset{(1)}{e}_{\alpha\beta} + x_3\overset{(1)}{\kappa}_{\alpha\beta} + w^c_{,\alpha}\overset{(1)}{w}{}_{,\beta})dV +$$
$$\int_V\left[\overset{(1)}{\sigma}{}^{\alpha\beta}\overset{(1)}{w}{}_{,\alpha}\overset{(1)}{w}{}_{,\beta} + \lambda_1^{he}(\overset{(1)}{\sigma}{}^{\alpha\beta}w_{,\alpha}'^{c}\overset{(1)}{w}{}_{,\beta} + \sigma_c'^{\alpha\beta}\overset{(1)}{w}{}_{,\alpha}\overset{(1)}{w}{}_{,\beta})\right]dV = 0. \tag{5.18}$$

From (5.9) and (5.7) with $i = 2$,

$$\int_V \overset{(2)}{\sigma}{}^{\alpha\beta}(\overset{(1)}{e}_{\alpha\beta} + x_3\overset{(1)}{\kappa}_{\alpha\beta} + w{,}^c_\alpha \overset{(1)}{w}{,}_\beta)dV = \int_V \left[\tfrac{1}{2}\overset{(1)}{\sigma}{}^{\alpha\beta}\overset{(1)}{w}{,}_\alpha\overset{(1)}{w}{,}_\beta + \lambda_1^{he}\overset{(1)}{\sigma}{}^{\alpha\beta}w{,}'^c_\alpha\overset{(1)}{w}{,}_\beta + \overset{(2)}{D}{}^{\alpha\beta}\overset{(1)}{\eta}_{\alpha\beta}\right]dV. \quad (5.19)$$

Employing (5.19) in (5.18) and solving for λ_1^{he} gives

$$\lambda_1^{he} = -\mathscr{A}/\mathscr{B}, \quad (5.20)$$

where

$$\mathscr{A} = \int_V \left[3\overset{(1)}{\sigma}{}^{\alpha\beta}\overset{(1)}{w}{,}_\alpha\overset{(1)}{w}{,}_\beta + \overset{(1)}{\eta}_{\alpha\beta}\overset{(1)}{\eta}_{\gamma\delta}\frac{\partial L^{\alpha\beta\gamma\delta}}{\partial\sigma^{\mu\nu}}\bigg|_c \overset{(1)}{\sigma}{}^{\mu\nu}\right]dV, \quad (5.21)$$

$$\mathscr{B} = \int_V \left[4\overset{(1)}{\sigma}{}^{\alpha\beta}w{,}'^c_\alpha\overset{(1)}{w}{,}_\beta + 2\sigma_c'^{\alpha\beta}\overset{(1)}{w}{,}_\alpha\overset{(1)}{w}{,}_\beta + \overset{(1)}{\eta}_{\alpha\beta}\overset{(1)}{\eta}_{\gamma\delta}\frac{\partial L^{\alpha\beta\gamma\delta}}{\partial\sigma^{\mu\nu}}\bigg|_c \sigma_c'^{\mu\nu} + \overset{(1)}{\eta}_{\alpha\beta}\eta'^c_{\gamma\delta}\frac{\partial L^{\alpha\beta\gamma\delta}}{\partial\sigma^{\mu\nu}}\bigg|_c \overset{(1)}{\sigma}{}^{\mu\nu}\right]dV. \quad (5.22)$$

This expression for λ_1^{he} is given by Hutchinson (1974).

For symmetric structures λ_1^{he} vanishes identically. Then, to calculate λ_2^{he}, the expansions (5.4), (5.6), and (5.12) are employed in the principle of virtual work, as in deriving (5.16). Now, to order ξ^3, with the term of order ξ vanishing identically,

$$\xi^2\int_V\left[\overset{(2)}{\sigma}{}^{\alpha\beta}(\delta e_{\alpha\beta} + x_3\delta\kappa_{\alpha\beta} + w{,}^c_\alpha\delta w{,}_\beta) + \sigma_c^{\alpha\beta}\overset{(2)}{w}{,}_\alpha\delta w{,}_\beta + \overset{(1)}{\sigma}{}^{\alpha\beta}\overset{(1)}{w}{,}_\alpha\delta w{,}_\beta\right]dV$$

$$+\xi^3\int_V\Big[\overset{(3)}{\sigma}{}^{\alpha\beta}(\delta e_{\alpha\beta} + x_3\delta\kappa_{\alpha\beta} + w{,}^c_\alpha\delta w{,}_\beta) + \sigma_c^{\alpha\beta}\overset{(3)}{w}{,}_\alpha\delta w{,}_\beta + \overset{(2)}{\sigma}{}^{\alpha\beta}\overset{(1)}{w}{,}_\alpha\delta w{,}_\beta$$

$$+\overset{(1)}{\sigma}{}^{\alpha\beta}\overset{(2)}{w}{,}_\alpha\delta w{,}_\beta + \lambda_2^{he}(\sigma_c'^{\alpha\beta}\overset{(1)}{w}{,}_\alpha\delta w{,}_\beta + \overset{(1)}{\sigma}{}^{\alpha\beta}w{,}'^c_\alpha\delta w{,}_\beta)\Big]dV + \ldots = 0. \quad (5.23)$$

The term of order ξ^2 constitutes an auxiliary boundary value problem for $\overset{(2)}{u}_\alpha$ and $\overset{(2)}{w}$. In variational form this auxiliary boundary value problem is

$$\int_V\left[\left(L_c^{\alpha\beta\gamma\delta}\overset{(2)}{\eta}_{\gamma\delta} + \overset{(2)}{D}{}^{\alpha\beta}\right)\left(\delta e_{\alpha\beta} + x_3\delta\kappa_{\alpha\beta} + w{,}^c_\alpha\delta w{,}_\beta\right) + \sigma_c^{\alpha\beta}\overset{(2)}{w}{,}_\alpha\delta w{,}_\beta + \overset{(1)}{\sigma}{}^{\alpha\beta}\overset{(1)}{w}{,}_\alpha\delta w{,}_\beta\right]dV = 0 \quad (5.24)$$

where $\overset{(2)}{\boldsymbol{\eta}}$ and $\overset{(3)}{\mathbf{D}}$ are given by (5.9) and (5.14) respectively, with λ_1^{he} set zero. The quantities $\overset{(2)}{\mathbf{u}}$ and $\overset{(2)}{w}$ satisfy appropriate homogeneous boundary conditions.

Once these field quantities are known the term of order ξ^3 in (5.23) is employed to give λ_2^{he}. Using the orthogonality condition (5.5), the identity (5.17) and the relations (5.10) and (5.15), the result is, as given in Tvergaard (1977b),

$$\lambda_2^{he} = -\mathscr{C}/\mathscr{D}, \quad (5.25)$$

where

$$\mathscr{C} = \int_V \Bigg[6\overset{(1)}{\sigma}{}^{\alpha\beta}\overset{(1)}{w}{}_{,\alpha}\overset{(2)}{w}{}_{,\beta} + 3\overset{(2)}{\sigma}{}^{\alpha\beta}\overset{(1)}{w}{}_{,\alpha}\overset{(1)}{w}{}_{,\beta} + \overset{(1)}{\eta}_{\alpha\beta}\overset{(1)}{\eta}_{\gamma\delta}\frac{\partial L^{\alpha\beta\gamma\delta}}{\partial\sigma^{\mu\nu}}\bigg|_c \overset{(2)}{\sigma}{}^{\mu\nu} + 2\overset{(1)}{\eta}_{\alpha\beta}\overset{(2)}{\eta}_{\gamma\delta}\frac{\partial L^{\alpha\beta\gamma\delta}}{\partial\sigma^{\mu\nu}}\bigg|_c \overset{(1)}{\sigma}{}^{\mu\nu}$$

$$+ \tfrac{1}{2}\overset{(1)}{\eta}_{\alpha\beta}\overset{(1)}{\eta}_{\gamma\delta}\frac{\partial^2 L^{\alpha\beta\gamma\delta}}{\partial\sigma^{\mu\nu}\partial\sigma^{\rho\omega}}\bigg|_c \overset{(1)}{\sigma}{}^{\mu\nu}\overset{(1)}{\sigma}{}^{\rho\omega}\Bigg] dV, \tag{5.26}$$

$$\mathscr{D} = \int_V \Bigg[6\overset{(1)}{\sigma}{}^{\alpha\beta} w'^{c}_{,\alpha}\overset{(1)}{w}{}_{,\beta} + 3\sigma'^{\alpha\beta}_c\overset{(1)}{w}{}_{,\alpha}\overset{(1)}{w}{}_{,\beta} + \overset{(1)}{\eta}_{\alpha\beta}\overset{(1)}{\eta}_{\gamma\delta}\frac{\partial L^{\alpha\beta\gamma\delta}}{\partial\sigma^{\mu\nu}}\bigg|_c \sigma'^{\mu\nu}_c$$

$$+ 2\overset{(1)}{\eta}_{\alpha\beta}\eta'^{c}_{\gamma\delta}\frac{\partial L^{\alpha\beta\gamma\delta}}{\partial\sigma^{\mu\nu}}\bigg|_c \overset{(1)}{\sigma}{}^{\mu\nu}\Bigg] dV. \tag{5.27}$$

(ii) Regular perturbation analysis for $\lambda < \lambda_c$

The initial imperfection $\bar{w}$ is written as $\bar{\xi}w^*$, where $\bar{\xi}$ is the amplitude of the imperfection and w^*, normalized in some convenient fashion, has unit amplitude. The additional displacements resulting from application of the load are written as

$$u_\alpha = u^0_\alpha + \xi\overset{(1)}{u}_\alpha + \hat{\xi}\hat{u}_\alpha, \quad w = w^0 + \xi\overset{(1)}{w} + \hat{\xi}\hat{w}, \tag{5.28}$$

where $\hat{u}_\alpha$ and $\hat{w}$ are orthogonal to the eigenmode in that (5.5) and (5.17) are taken to be satisfied with $(^{(i)})$ replaced by $(\hat{\ })$. Furthermore, the spatial forms of $\hat{u}_\alpha$ and $\hat{w}$ are permitted to vary during the deformation history as long as this spatial variation is such that orthogonality holds and that $\hat{u}_\alpha$ and $\hat{w}$ are consistently normalized.

During the initial stages of the loading history the load parameter λ increases monotonically with ξ and $\hat{\xi}$ and we seek the lowest order effect of a small imperfection ($\bar{\xi} \ll 1$) on the initial rising portion of the λ–ξ curve. Since in the regime under consideration λ is increasing, λ will be employed as the parameter characterizing the loading history and ξ and $\hat{\xi}$ will be considered as functions of λ. All equations will be linearized so that products of perturbation quantities and products of $\bar{\xi}$ with perturbation quantities are neglected.

The linearized principle of virtual work (5.2) takes the form

$$\int_V \Bigg[\Big(\sigma^{\alpha\beta} - \sigma^{\alpha\beta}_0\Big)\Big(\delta e_{\alpha\beta} + x_3\delta\kappa_{\alpha\beta} + w^0_{,\alpha}\delta w_{,\beta}\Big) + \xi\sigma^{\alpha\beta}_0\overset{(1)}{w}{}_{,\alpha}\delta w_{,\beta} + \hat{\xi}\sigma^{\alpha\beta}_0\hat{w}_{,\alpha}\delta w_{,\beta}$$

$$+ \bar{\xi}\sigma^{\alpha\beta}_0 w^*_{,\alpha}\delta w_{,\beta}\Bigg] dV = 0. \tag{5.29}$$

In order to obtain the linearized expression for $(\boldsymbol{\sigma} - \boldsymbol{\sigma}_0)$, we first substitute (5.28) into the strain displacement relation (2.1) and linearize to obtain

$$\eta_{\alpha\beta}=\eta^0_{\alpha\beta}+\xi\left[\overset{(1)}{\eta}_{\alpha\beta}+\tfrac{1}{2}\left(\lambda-\lambda_c\right)\left(V_\alpha\overset{(1)}{w}_{,\beta}+\overset{(1)}{w}_{,\alpha}V_\beta\right)\right]+\hat{\xi}\left[\hat{\eta}_{\alpha\beta}+\tfrac{1}{2}\left(\lambda-\lambda_c\right)\left(V_\alpha\hat{w}_{,\beta}+\hat{w}_{,\alpha}V_\beta\right)\right]$$

$$+\bar{\xi}\left[\tfrac{1}{2}\left(w^c_{,\alpha}w^*_{,\beta}+w^*_{,\alpha}w^c_{,\beta}\right)+\tfrac{1}{2}\left(\lambda-\lambda_c\right)\left(V_\alpha w^*_{,\beta}+w^*_{,\alpha}V_\beta\right)\right] \tag{5.30}$$

with

$$V_\alpha=\frac{w^0_{,\alpha}-w^c_{,\alpha}}{\lambda-\lambda_c} \tag{5.31}$$

and

$$\hat{\eta}_{\alpha\beta}=\hat{e}_{\alpha\beta}+x_3\hat{\kappa}_{\alpha\beta}+\tfrac{1}{2}(w^c_{,\alpha}\hat{w}_{,\beta}+\hat{w}_{,\alpha}w^c_{,\beta}). \tag{5.32}$$

Next expanding the tensor of moduli **L** in a Taylor series about the prebifurcation state gives to lowest order

$$L^{\alpha\beta\gamma\delta}=L_0^{\alpha\beta\gamma\delta}+\left.\frac{\partial L^{\alpha\beta\gamma\delta}}{\partial\sigma^{\mu\nu}}\right|_0(\sigma^{\mu\nu}-\sigma_0^{\mu\nu}), \tag{5.33}$$

where $|_0$ denotes derivatives evaluated at the prebifurcation state.

Noting that

$$\frac{d\sigma_0^{\alpha\beta}}{d\lambda}=L_0^{\alpha\beta\gamma\delta}\frac{d\eta^0_{\gamma\delta}}{d\lambda} \tag{5.34}$$

and substituting (5.30) into (5.1) gives

$$(\sigma^{\alpha\beta}-\sigma_0^{\alpha\beta})=L_0^{\alpha\beta\gamma\delta}(\eta_{\gamma\delta}-\eta^0_{\gamma\delta})+Q^{\alpha\beta}, \tag{5.35}$$

where $Q^{\alpha\beta}$ satisfies

$$\frac{dQ^{\alpha\beta}}{d\lambda}-R^{\alpha\beta}_{\cdot\cdot\gamma\delta}Q^{\gamma\delta}=S^{\alpha\beta\rho\omega}(\eta_{\rho\omega}-\eta^0_{\rho\omega}) \tag{5.36}$$

with

$$R^{\alpha\beta}_{\cdot\cdot\gamma\delta}=\eta'^0_{\mu\nu}\left.\frac{\partial L^{\alpha\beta\mu\nu}}{\partial\sigma^{\gamma\delta}}\right|_0, \tag{5.37}$$

$$S^{\alpha\beta\rho\omega}=\eta'^0_{\gamma\delta}\left.\frac{\partial L^{\alpha\beta\gamma\delta}}{\partial\sigma^{\mu\nu}}\right|_0 L_0^{\mu\nu\rho\omega}-\left.\frac{\partial L^{\alpha\beta\rho\omega}}{\partial\sigma^{\mu\nu}}\right|_0\sigma_0'^{\mu\nu}. \tag{5.38}$$

Here, and subsequently, $(\ )'$ denotes $d(\ )/d\lambda$.

The relation (5.36) constitutes a set of coupled first-order linear differential equations for **Q**. By superposition the solution to (5.36) can be written in the form

$$Q^{\alpha\beta}=\overset{(1)}{Q}{}^{\alpha\beta}+\hat{Q}^{\alpha\beta}+\bar{Q}^{\alpha\beta}, \tag{5.39}$$

where

$$\overset{(1)}{Q}{}^{\alpha\beta}=F^{\alpha\beta}_{\cdot\cdot\mu\nu}\int_0^\lambda(F^{\mu\nu}_{\cdot\cdot\gamma\delta})^{-1}S^{\gamma\delta\rho\omega}\left[\overset{(1)}{\eta}_{\rho\omega}+(\lambda-\lambda_c)V_\rho\overset{(1)}{w}_{,\omega}\right]\xi\,d\tau, \tag{5.40}$$

$$\hat{Q}^{\alpha\beta} = F^{\alpha\beta}_{\cdot\cdot\mu\nu} \int_0^{\lambda} (F^{\mu\nu}_{\cdot\cdot\gamma\delta})^{-1} S^{\gamma\delta\rho\omega}\left[\hat{\eta}_{\rho\omega} + (\lambda - \lambda_c) V_\rho \hat{w}_{,\omega} \right]\hat{\xi}\, d\tau, \tag{5.41}$$

$$\overline{Q}^{\alpha\beta} = F^{\alpha\beta}_{\cdot\cdot\mu\nu} \int_0^{\lambda} (F^{\mu\nu}_{\cdot\cdot\gamma\delta})^{-1} S^{\gamma\delta\rho\omega}\left[w^c_{,\rho} w^*_{,\omega} + (\lambda - \lambda_c) V_\rho w^*_{,\overline{\omega}} \right]\overline{\xi}\, d\tau. \tag{5.42}$$

The so-called fundamental matrix, **F** (see, for example, Bellman, 1960), is invertible as long as **R** is nonsingular. We assume the constitutive law and prebifurcation path are such that **R** is nonsingular for $0 \leqslant \lambda \leqslant \lambda_c$. Using (5.35), (5.39), and (5.30) in the linearized principle of virtual work (5.29) gives

$$\left[\xi(\lambda) I\left(\overset{(1)}{u}_\alpha, \overset{(1)}{w}, \delta u_\alpha, \delta w, \lambda \right) + q\left(\overset{(1)}{Q}{}^{\alpha\beta}, \delta u_\alpha, \delta w, \lambda \right) \right]$$

$$+ \left[\hat{\xi}(\lambda) I(\hat{u}_\alpha, \hat{w}, \delta u_\alpha, \delta w, \lambda) + q(\hat{Q}^{\alpha\beta}, \delta u_\alpha, \delta w, \lambda) \right]$$

$$+ \left[\overline{\xi} J(w^*, \delta u_\alpha, \delta w, \lambda) + q(\overline{Q}^{\alpha\beta}, \delta u_\alpha, \delta w, \lambda) \right] = 0 \tag{5.43}$$

In (5.43), we have introduced the notations

$$I(u_\alpha, w, \delta u_\alpha, \delta w, \lambda) = \int_V \left[L_0^{\alpha\beta\gamma\delta} \eta_{\gamma\delta} \delta\eta^c_{\alpha\beta} + \sigma_0^{\alpha\beta} w_{,\alpha} \delta w_{,\beta} \right] dV$$

$$+ (\lambda - \lambda_c) \int_V \left[L_0^{\alpha\beta\gamma\delta} V_\gamma w_{,\delta} \delta\eta^c_{\alpha\beta} + L_0^{\alpha\beta\gamma\delta} \eta_{\gamma\delta} V_\alpha \delta w_{,\beta} + (\lambda - \lambda_c) L_0^{\alpha\beta\gamma\delta} V_\delta w_{,\gamma} V_\alpha \delta w_{,\beta} \right] dV, \tag{5.44}$$

$$J(w^*, \delta u_\alpha, \delta w, \lambda) = \int_V \left[L_0^{\alpha\beta\gamma\delta} w^c_{,\gamma} w^*_{,\delta} \delta\eta^c_{\alpha\beta} + \sigma_0^{\alpha\beta} w^*_{,\alpha} \delta w_{,\beta} \right] dV$$

$$+ (\lambda - \lambda_c) \int_V \left[L_0^{\alpha\beta\gamma\delta} V_\gamma w^*_{,\delta} \delta\eta^c_{\alpha\beta} + L_0^{\alpha\beta\gamma\delta} w^c_{,\gamma} w^*_{,\delta} V_\alpha \delta w_{,\beta} + (\lambda - \lambda_c) L_0^{\alpha\beta\gamma\delta} V_\gamma w^*_{,\delta} V_\alpha \delta w_{,\beta} \right] dV, \tag{5.45}$$

$$q(Q^{\alpha\beta}, \delta u_\alpha, \delta w, \lambda) = \int_V \left[Q^{\alpha\beta} \delta\eta^c_{\alpha\beta} + (\lambda - \lambda_c) Q^{\alpha\beta} V_\alpha \delta w_{,\beta} \right] dV, \tag{5.46}$$

where

$$\delta\eta^c_{\alpha\beta} = \delta e_{\alpha\beta} + x_3 \delta\kappa_{\alpha\beta} + \tfrac{1}{2} w^c_{,\alpha} \delta w_{,\beta} + \tfrac{1}{2} w^c_{,\beta} \delta w_{,\alpha}. \tag{5.47}$$

In (5.43) note that

$$\lim_{\lambda \to \lambda_c} I\left(\overset{(1)}{u}_\alpha, \overset{(1)}{w}, \delta u_\alpha, \delta w, \lambda \right) = 0 \tag{5.48}$$

like $(\lambda_c - \lambda)$ for all δu_α and δw, since in this limit the bifurcation functional is obtained. On the other hand

$$\lim_{\lambda \to \lambda_c} I(\hat{u}_\alpha, \hat{w}, \delta u_\alpha, \delta w, \lambda) > 0 \tag{5.49}$$

for all δu_α and δw orthogonal to the bifurcation mode.

Now we write

$$\xi(\lambda) = \bar{\xi}\left(1 - \frac{\lambda}{\lambda_c}\right)^{-\psi} p(\lambda) \tag{5.50}$$

and from (5.40) and (5.41) observe that, as $\lambda \to \lambda_c$, $\overset{(1)}{Q}{}^{\alpha\beta}$ and $\hat{Q}^{\alpha\beta}$ behave like

$$\overset{(1)}{Q}{}^{\alpha\beta} \sim \Gamma_1^{\alpha\beta} \int_0^{\lambda_c} \xi(\tau)d\tau, \tag{5.51}$$

$$\hat{Q}^{\alpha\beta} \sim \Gamma_2^{\alpha\beta} \int_0^{\lambda_c} \hat{\xi}(\tau)d\tau, \tag{5.52}$$

where $\mathbf{\Gamma}_1$ and $\mathbf{\Gamma}_2$ are finite at $\lambda = \lambda_c$.

Set $\delta u_\alpha = \hat{u}_\alpha$ and $\delta w = \hat{w}$ in (5.43) and take the limit as $\lambda \to \lambda_c$, then from (5.48) through (5.52) and the fact that the term multiplying $\bar{\xi}$ in (5.43) remains finite in this limit

$$\lim_{\lambda \to \lambda_c} \left(1 - \frac{\lambda}{\lambda_c}\right)^{\psi} \hat{\xi}(\lambda) = 0. \tag{5.53}$$

The exponent ψ remains to be determined. Let $\delta u_\alpha = \overset{(1)}{u}_\alpha$ and $\delta w = \overset{(1)}{w}$ in (5.43), substitute (5.50) and differentiate with respect to λ to obtain

$$\begin{aligned}
&\frac{I\left(\overset{(1)}{u}_\alpha, \overset{(1)}{w}, \overset{(1)}{u}_\alpha, \overset{(1)}{w}, \lambda\right)}{1 - \lambda/\lambda_c}\frac{dp}{d\lambda} + \frac{p(\lambda)}{1 - \lambda/\lambda_c}\left[\frac{\psi I\left(\overset{(1)}{u}_\alpha, \overset{(1)}{w}, \overset{(1)}{u}_\alpha, \overset{(1)}{w}, \lambda\right)}{\lambda_c - \lambda} + \frac{dI\left(\overset{(1)}{u}_\alpha, \overset{(1)}{w}, \overset{(1)}{u}_\alpha, \overset{(1)}{w}, \lambda\right)}{d\lambda}\right] \\
&+\left(1 - \frac{\lambda}{\lambda_c}\right)^{\psi - 1} \frac{dq\left(\overset{(1)}{Q}{}^{\alpha\beta}, \overset{(1)}{u}_\alpha, \overset{(1)}{w}, \lambda\right)}{d\lambda} \\
&+\left(1 - \frac{\lambda}{\lambda_c}\right)^{\psi - 1} \frac{d}{d\lambda}\left[\hat{\xi}(\lambda) I\left(\hat{u}_\alpha, \hat{w}, \overset{(1)}{u}_\alpha, \overset{(1)}{w}, \lambda\right) + q\left(\hat{Q}^{\alpha\beta}, \overset{(1)}{u}_\alpha, \overset{(1)}{w}, \lambda\right)\right] \\
&+\left(1 - \frac{\lambda}{\lambda_c}\right)^{\psi - 1}\left[\bar{\xi} J\left(w_*^{*}, \overset{(1)}{u}_\alpha, \overset{(1)}{w}, \lambda\right) + q\left(\overline{Q}{}^{\alpha\beta}, \overset{(1)}{u}_\alpha, \overset{(1)}{w}, \lambda\right)\right] = 0.
\end{aligned} \tag{5.54}$$

The exponent ψ is determined from the condition that $p(\lambda)$ be regular at λ_c. Hence, from (5.54), (5.44) and (5.46),

$$\psi = 1 + \beta_1/\beta_2 \tag{5.55}$$

with

$$\left.\begin{aligned}\beta_1 &= \int_V \left[\eta'^{c}_{\gamma\delta} \frac{\partial L^{\alpha\beta\gamma\delta}}{\partial \sigma^{\mu\nu}}\bigg|_c \overset{(1)}{\sigma}{}^{\mu\nu} \overset{(1)}{\eta}_{\alpha\beta} - \overset{(1)}{\eta}_{\gamma\delta} \frac{\partial L^{\alpha\beta\gamma\delta}}{\partial \sigma^{\mu\nu}}\bigg|_c \sigma'^{\mu\nu}_c \overset{(1)}{\eta}_{\alpha\beta} \right] dV, \\ \beta_2 &= \int_V \left[2\overset{(1)}{\sigma}{}^{\alpha\beta} w'^{c}_{,\alpha} \overset{(1)}{w}_{,\beta} + \sigma'^{\alpha\beta}_c \overset{(1)}{w}_{,\alpha} \overset{(1)}{w}_{,\beta} + \overset{(1)}{\eta}_{\gamma\delta} \frac{\partial L^{\alpha\beta\gamma\delta}}{\partial \sigma^{\mu\nu}}\bigg|_c \sigma'^{\mu\nu}_c \overset{(1)}{\eta}_{\alpha\beta} \right] dV. \end{aligned}\right\} \tag{5.56}$$

For an elastic material, linear or nonlinear, $\psi = 1$, as is readily observed by noting that when the moduli are derivable from a strain energy function, β_1 vanishes.

As $\lambda/\lambda_c \to 1$, (5.50) can be rewritten as

$$\frac{\lambda}{\lambda_c} \sim 1 - [p(\lambda_c)]^{1/\psi} \left(\frac{\bar{\xi}}{\xi} \right)^{1/\psi}. \tag{5.57}$$

For a path dependent solid, $p(\lambda_c)$ can be determined from an incremental Galerkin solution to (5.54), which involves consideration of modes in addition to the bifurcation mode. If, as in Needleman and Tvergaard (1976) and Tvergaard (1977b) an extended orthogonality holds, then even for a path dependent solid the determination of $p(\lambda_c)$ involves only the bifurcation mode.

(iii) Singular perturbation analysis for λ near λ_c

We now seek an asymptotic expression valid for small $|\lambda - \lambda_c|$, ξ, and $\bar{\xi}$ that accounts for the lowest order effect of an imperfection. Following Budiansky (1974) the functional relation between λ, ξ, and $\bar{\xi}$ is constrained by the relation $\bar{\xi} = \alpha\xi^{\gamma}$, so that for properly chosen γ only integral powers of ξ are needed in the expansion about λ_c. Here, we assume $\lambda_1^{he} \neq 0$. The case $\lambda_1^{he} = 0$ will be commented on subsequently. The expansions about λ_c are assumed to be of the form

$$\lambda = \lambda_c + \bar{\lambda}_1^{he} \xi + \ldots, \tag{5.58}$$

$$\begin{Bmatrix} u_\alpha \\ w \end{Bmatrix} = \begin{Bmatrix} u_\alpha^0 \\ w^0 \end{Bmatrix} + \xi \begin{Bmatrix} \overset{(1)}{\bar{u}}_\alpha \\ \overset{(1)}{\bar{w}} \end{Bmatrix} + \xi^2 \begin{Bmatrix} \overset{(2)}{\bar{u}}_\alpha \\ \overset{(2)}{\bar{w}} \end{Bmatrix} + \ldots. \tag{5.59}$$

Barred quantities are functions of α and for a perfect structure reduce to their unbarred counterparts. By employing ξ as the expansion variable in (5.58) and (5.59), the result, to be verified subsequently, that $\overset{(1)}{\bar{w}} = \overset{(1)}{w}$ and $\overset{(1)}{\bar{u}}_\alpha = \overset{(1)}{u}_\alpha$ has been assumed. Furthermore, we require that $\overset{(i)}{\bar{w}}$, $i > 1$, be orthogonal to $\overset{(1)}{\bar{w}}\ (= \overset{(1)}{w})$ in the sense of (5.5). This gives rise to an identity analogous to (5.17) being satisfied by the barred quantities.

Substituting (5.59) into (2.1) gives the expansion for the strains $\boldsymbol{\eta}$,

$$\eta_{\alpha\beta} = \eta^0_{\alpha\beta} + \xi\overset{(1)}{\bar{\eta}}_{\alpha\beta} + \xi^2\overset{(2)}{\bar{\eta}}_{\alpha\beta} + \tfrac{1}{2}\alpha\xi^{\gamma}(w^*_{,\alpha}w^c_{,\beta} + w^c_{,\alpha}w^*_{,\beta}), \tag{5.60}$$

where $\boldsymbol{\eta}^0$ is given by (5.7) and

$$\overset{(1)}{\bar{\eta}}_{\alpha\beta} = \overset{(1)}{\bar{e}}_{\alpha\beta} + x_3\overset{(1)}{\bar{\kappa}}_{\alpha\beta} + \tfrac{1}{2}(w^c_{,\alpha}\overset{(1)}{\bar{w}}_{,\beta} + \overset{(1)}{\bar{w}}_{,\alpha}w^c_{,\beta}), \tag{5.61}$$

$$\overset{(2)}{\bar{\eta}}_{\alpha\beta} = \overset{(2)}{\bar{e}}_{\alpha\beta} + x_3\overset{(2)}{\bar{\kappa}}_{\alpha\beta} + \tfrac{1}{2}(w^c_{,\alpha}\overset{(2)}{\bar{w}}_{,\beta} + \overset{(2)}{\bar{w}}_{,\alpha}w^c_{,\beta}) + \tfrac{1}{2}\bar{\lambda}_1^{he}(w'^c_{,\alpha}\overset{(1)}{\bar{w}}_{,\beta} + \overset{(1)}{\bar{w}}_{,\alpha}w'^c_{,\beta}). \tag{5.62}$$

Assuming, as will be verified subsequently, that $\gamma \geqslant 2$, the expansion for the stresses can be written as

$$\sigma^{\alpha\beta} = \sigma^{\alpha\beta}_0 + \xi\overset{(1)}{\bar{\sigma}}{}^{\alpha\beta} + \xi^2\overset{(2)}{\bar{\sigma}}{}^{\alpha\beta} + \dots . \tag{5.63}$$

At any given stage of the deformation history the stress increments $\dot{\boldsymbol{\sigma}}$ are related to the strain increments $\dot{\boldsymbol{\eta}}$ via (5.1). In this aspect of the analysis the monotonically increasing parameter characterizing the loading history is taken to be ξ and $(\dot{\ })$ denotes the derivative with respect to ξ with $\bar{\xi}$ fixed. Hence,

$$\dot{\alpha} = -\alpha\gamma\xi^{-1}. \tag{5.64}$$

Thus, the expansions for the stress and strain increments are, from (5.58), (5.60), (5.63), and (5.64),

$$\dot{\eta}_{\alpha\beta} = L_1(\bar{\lambda}_1^{he}\eta'^c_{\alpha\beta} + \overset{(1)}{\bar{\eta}}_{\alpha\beta}) + \xi[L_2(\bar{\lambda}_2^{he}\eta'^c_{\alpha\beta} + \overset{(2)}{\bar{\eta}}_{\alpha\beta}) + \bar{\lambda}_1^{he}L_1(\bar{\lambda}_1^{he})\eta''^c_{\alpha\beta}] + \dots, \tag{5.65}$$

$$\dot{\sigma}^{\alpha\beta} = L_1(\bar{\lambda}_1^{he}\sigma'^{\alpha\beta}_c + \overset{(1)}{\bar{\sigma}}{}^{\alpha\beta}) + \xi[L_2(\bar{\lambda}_2^{he}\sigma'^{\alpha\beta}_c + \overset{(2)}{\bar{\sigma}}{}^{\alpha\beta}) + \bar{\lambda}_1^{he}L_1(\bar{\lambda}_1^{he})\sigma''^{\alpha\beta}_c] + \dots, \tag{5.66}$$

where $L_i(\ )$ is the linear differential operator introduced by Hutchinson and Budiansky (1976),

$$L_i(\ ) = i(\ ) - \alpha\gamma(\ )_{,\alpha}. \tag{5.67}$$

The moduli $\mathbf{L}$ are expanded in a Taylor series analogous to (5.11) which yields

$$L^{\alpha\beta\gamma\delta} = L^{\alpha\beta\gamma\delta}_c + \xi\bar{\lambda}_1^{he}\left.\frac{\partial L^{\alpha\beta\gamma\delta}}{\partial\sigma^{\omega\rho}}\right|_c \sigma'^{\omega\rho}_c + \dots . \tag{5.68}$$

Substituting (5.65), (5.66), and (5.68) into (5.1) and collecting like powers of ξ gives

$$L_1(\bar{\lambda}_1^{he}\sigma'^{\alpha\beta}_c + \overset{(1)}{\bar{\sigma}}{}^{\alpha\beta}) = L^{\alpha\beta\gamma\delta}_c L_1(\bar{\lambda}_1^{he}\eta'^c_{\gamma\delta} + \overset{(1)}{\bar{\eta}}_{\gamma\delta}), \tag{5.69}$$

$$\begin{aligned} L_2(\bar{\lambda}_2^{he}\sigma'^{\alpha\beta}_c + \overset{(2)}{\bar{\sigma}}{}^{\alpha\beta}) + \bar{\lambda}_1^{he}L_1(\bar{\lambda}_1^{he})\sigma''^{\alpha\beta}_c &= L^{\alpha\beta\gamma\delta}_c[L_2(\bar{\lambda}_2^{he}\eta'^c_{\gamma\delta} + \overset{(2)}{\bar{\eta}}_{\gamma\delta}) + \bar{\lambda}_1^{he}L_1(\bar{\lambda}_1^{he})\eta''^c_{\gamma\delta}] \\ &+ \bar{\lambda}_1^{he}[L_1(\bar{\lambda}_1^{he}\eta'^c_{\gamma\delta} + \overset{(1)}{\bar{\eta}}_{\gamma\delta})]\left.\frac{\partial L^{\alpha\beta\gamma\delta}}{\partial\sigma^{\omega\rho}}\right|_c \sigma'^{\omega\rho}_c + [L_1(\bar{\lambda}_1^{he}\eta'^c_{\gamma\delta} + \overset{(1)}{\bar{\eta}}_{\gamma\delta})]\left.\frac{\partial L^{\alpha\beta\gamma\delta}}{\partial\sigma^{\omega\rho}}\right|_c \overset{(1)}{\bar{\sigma}}{}^{\omega\rho}. \end{aligned} \tag{5.70}$$

Using the relations, derived from a Taylor expansion along the prebifurcation path,

$$\sigma_c'^{\alpha\beta} = L_c^{\alpha\beta\gamma\delta}\eta_{\gamma\delta}'^{c}, \tag{5.71}$$

$$\sigma_c''^{\alpha\beta} = L_c^{\alpha\beta\gamma\delta}\eta_{\gamma\delta}''^{c} + \eta_{\gamma\delta}'^{c}\left.\frac{\partial L^{\alpha\beta\gamma\delta}}{\partial\sigma^{\omega\rho}}\right|_c \sigma_c'^{\omega\rho}, \tag{5.72}$$

the relations (5.69) and (5.70) simplify to

$$L_1(\overset{(1)}{\bar\sigma}{}^{\alpha\beta}) = L_c^{\alpha\beta\gamma\delta}L_1(\overset{(1)}{\bar\eta}_{\gamma\delta}) \tag{5.73}$$

and

$$L_2(\overset{(2)}{\bar\sigma}{}^{\alpha\beta}) = L_c^{\alpha\beta\gamma\delta}L_2(\overset{(2)}{\bar\eta}_{\gamma\delta}) + \bar\lambda_1^{he}L_1(\overset{(1)}{\bar\eta}_{\gamma\delta})\left.\frac{\partial L^{\alpha\beta\gamma\delta}}{\partial\sigma^{\omega\rho}}\right|_c \sigma_c'^{\omega\rho}$$

$$+[L_1(\overset{(1)}{\bar\eta}_{\gamma\delta}) + L_1(\bar\lambda_1^{he})\eta_{\gamma\delta}'^{c}]\left.\frac{\partial L^{\alpha\beta\gamma\delta}}{\partial\sigma^{\omega\rho}}\right|_c \overset{(1)}{\bar\sigma}{}^{\omega\rho}. \tag{5.74}$$

The solution to (5.73) is

$$\overset{(1)}{\bar\sigma}{}^{\alpha\beta} = L_c^{\alpha\beta\gamma\delta}\overset{(1)}{\bar\eta}_{\gamma\delta} \tag{5.75}$$

plus the homogeneous solution to (5.73). This homogeneous solution gives rise to a stress field which does not vanish with ξ. Since there is no such stress field emanating from $\lambda = 0$, the arbitrary constant multiplying the homogeneous solution of (5.73) must be zero. Accordingly, such homogeneous solutions are discarded.

Substituting (5.59), (5.60), and (5.63) into the principle of virtual work while making use of (5.57) and (5.74) gives

$$\xi\int_V [\overset{(1)}{\bar\eta}_{\gamma\delta}L_c^{\alpha\beta\gamma\delta}(\delta e_{\alpha\beta} + x_3\delta\kappa_{\alpha\beta} + w^c_{,\alpha}\delta w_{,\beta}) + \sigma_c^{\alpha\beta}\overset{(1)}{\bar w}_{,\alpha}\delta w_{,\beta}]\,dV$$

$$+\xi^2\int_V [\overset{(2)}{\bar\sigma}{}^{\alpha\beta}(\delta e_{\alpha\beta} + x_3\delta\kappa_{\alpha\beta} + w^c_{,\alpha}\delta w_{,\beta}) + \sigma_c^{\alpha\beta}\overset{(2)}{\bar w}_{,\alpha}\delta w_{,\beta} + \bar\lambda_1^{he}\overset{(1)}{\bar\sigma}{}^{\alpha\beta}w'^{c}_{,\alpha}\delta w_{,\beta}$$

$$+\bar\lambda_1^{he}\sigma_c'^{\alpha\beta}\overset{(1)}{\bar w}_{,\alpha}\delta w_{,\beta}]\,dV + \alpha\xi^{\gamma}\int_V \sigma_c^{\alpha\beta}w^*_{,\alpha}\delta w_{,\beta}\,dV + \ldots = 0. \tag{5.76}$$

The lowest order terms in (5.76) give

$$\overset{(1)}{\bar u}_\alpha = \overset{(1)}{u}_\alpha, \quad \overset{(1)}{\bar w} = \overset{(1)}{w}, \tag{5.77}$$

with the supposition that the same normalization is employed for $\overset{(1)}{\bar{\mathbf{u}}}$ and $\overset{(1)}{\bar w}$ as for $\overset{(1)}{\mathbf{u}}$ and $\overset{(1)}{w}$.

Hence, the assertion that ξ is an appropriate expansion parameter is justified.

From (5.61), (5.75), and (5.77) it is seen that

$$\overset{(1)}{\bar{\eta}}_{\alpha\beta} = \overset{(1)}{\eta}_{\alpha\beta}, \quad \overset{(1)}{\bar{\sigma}}{}^{\alpha\beta} = \overset{(1)}{\sigma}{}^{\alpha\beta}. \tag{5.78}$$

Introducing $\Delta\bar{\lambda}_1^{he} = \bar{\lambda}_1^{he} - \lambda_1^{he}$, $\Delta\overset{(2)}{\bar{\sigma}}{}^{\alpha\beta} = \overset{(2)}{\bar{\sigma}}{}^{\alpha\beta} - \overset{(2)}{\sigma}{}^{\alpha\beta}$, etc. into (5.76) with $\delta(\ )$ replaced by $(^{(1)})$, we obtain

$$\xi^2 \int_V \left[\Delta\overset{(2)}{\bar{\sigma}}{}^{\alpha\beta} \overset{(1)}{\eta}_{\alpha\beta} + \Delta\bar{\lambda}_1^{he} \overset{(1)}{\sigma}{}^{\alpha\beta} w'^{c}_{,\alpha} w_{,\beta} + \Delta\bar{\lambda}_1^{he} \sigma'^{\alpha\beta}_c \overset{(1)}{w}_{,\alpha} \overset{(1)}{w}_{,\beta} \right] dV$$
$$+ \alpha\xi^\gamma \int_V \sigma_c^{\alpha\beta} w^*_{,\alpha} \overset{(1)}{w}_{,\beta}\, dV + \ldots = 0 \tag{5.79}$$

since the term of order ξ in (5.76) vanishes identically.

From (5.74), (5.75), and (5.13) and (5.14),

$$L_2(\Delta\overset{(2)}{\bar{\sigma}}{}^{\alpha\beta}) = L_c^{\alpha\beta\gamma\delta} L_2(\Delta\overset{(2)}{\bar{\eta}}_{\gamma\delta}) + \Delta\bar{\lambda}_1^{he}\, \overset{(1)}{\eta}_{\gamma\delta} \left.\frac{\partial L^{\alpha\beta\gamma\delta}}{\partial\sigma^{\omega\rho}}\right|_c \sigma'^{\omega\rho}_c + \Delta\bar{\lambda}_1^{he} \eta'^{c}_{\gamma\delta} \left.\frac{\partial L^{\alpha\beta\gamma\delta}}{\partial\sigma^{\omega\rho}}\right|_c \overset{(1)}{\sigma}{}^{\omega\rho}. \tag{5.80}$$

The solution to (5.80) is

$$\Delta\overset{(2)}{\bar{\sigma}}{}^{\alpha\beta} = L_c^{\alpha\beta\gamma\delta} \Delta\overset{(2)}{\bar{\eta}}_{\gamma\delta} + \tfrac{1}{2}\Delta\bar{\lambda}_1^{he} \left[\overset{(1)}{\eta}_{\gamma\delta} \left.\frac{\partial L^{\alpha\beta\gamma\delta}}{\partial\sigma^{\omega\rho}}\right|_c \sigma'^{\omega\rho}_c + \eta'^{c}_{\gamma\delta} \left.\frac{\partial L^{\alpha\beta\gamma\delta}}{\partial\sigma^{\omega\rho}}\right|_c \overset{(1)}{\sigma}{}^{\omega\rho} \right]$$
$$- \tfrac{1}{2}\alpha^{2/\gamma} \int^\alpha \alpha^{-2/\gamma} \frac{d(\Delta\bar{\lambda}_1^{he})}{d\alpha} d\alpha \left[\overset{(1)}{\eta}_{\gamma\delta} \left.\frac{\partial L^{\alpha\beta\gamma\delta}}{\partial\sigma^{\omega\rho}}\right|_c \sigma'^{\omega\rho}_c - \eta'^{c}_{\gamma\delta} \left.\frac{\partial L^{\alpha\beta\gamma\delta}}{\partial\sigma^{\omega\rho}}\right|_c \overset{(1)}{\sigma}{}^{\omega\rho} \right], \tag{5.81}$$

where the lower limit of the integral is left unspecified temporarily. From the identity (5.17), which arises from orthogonality,

$$\int_V \Delta\overset{(2)}{\bar{\eta}}_{\gamma\delta} L_c^{\alpha\beta\gamma\delta} \overset{(1)}{\eta}_{\alpha\beta}\, dV = \Delta\bar{\lambda}_1^{he} \int_V \overset{(1)}{\sigma}{}^{\alpha\beta} w'^{c}_{,\alpha} \overset{(1)}{w}_{,\beta}\, dV. \tag{5.82}$$

Substituting (5.81) and (5.82) into (5.79) gives

$$\frac{\beta_2}{2}(1+\psi)\xi^2 \left[\Delta\bar{\lambda}_1^{he} - \frac{1-\psi}{1+\psi}\alpha^{2/\gamma} \int^\alpha \alpha^{-2/\gamma} \frac{d(\Delta\bar{\lambda}_1^{he})}{d\alpha} d\alpha \right] + \alpha\xi^\gamma \int_V \sigma_c^{\alpha\beta} w^*_{,\alpha} \overset{(1)}{w}_{,\beta}\, dV + \ldots = 0, \tag{5.83}$$

where ψ is given by (5.55) and β_2 by $(5.56)_2$.

We shall proceed on the assumption that $\gamma > 2$. Then dividing by ξ^2 and taking the limit as $\xi \to 0$ in (5.83) gives an equation for $\Delta\bar{\lambda}_1^{he}$. The solution is

$$\Delta\bar{\lambda}_1^{he} = K\,\alpha^{(1+\psi)/\gamma\psi}. \tag{5.84}$$

The lower limit of the integral appearing in (5.81) and (5.82) can now be specified. Substituting (5.84) into (5.83) reveals that this lower limit must be ∞.

From (5.84), (5.58) and using $\alpha = \overline{\xi}\xi^{-\gamma}$, the expansion valid for $|\lambda - \lambda_c|$ small takes the form

$$\lambda = \lambda_c + \lambda_1^{he}\xi + K\overline{\xi}^{(1+\psi)/\gamma\psi}\xi^{-1/\psi}. \tag{5.85}$$

Let $\xi \to 0$, with $|\lambda_1^{he}\xi|$ small compared to the third term in (5.85), which itself must be small for (5.85) to be valid. Matching (5.85) and (5.57) gives

$$K = -\lambda_c[p(\lambda_c)]^{1/\psi}, \quad \gamma = 1 + \psi. \tag{5.86}$$

Since $\psi \geqslant 1$, with equality only holding for a path independent material, (5.86) gives $\gamma > 2$ if the solid is not elastic. If $\gamma = 2$, (5.83) is a linear inhomogeneous algebraic equation for $\Delta\overline{\lambda}_1^{he}$ and the result for an elastic structure is obtained (see, for example, Budiansky, 1974).

The expansions (5.84) with the matching condition pertain to an asymmetric structure for which $\lambda_1^{he} \neq 0$. For symmetric structures the result analogous to (5.85) is (Needleman and Tvergaard, 1976),

$$\lambda = \lambda_c + \lambda_2^{he}\xi^2 + K\overline{\xi}^{(2\psi+1)/\gamma\psi}\xi^{-1/\psi} \tag{5.87}$$

with

$$K = -\lambda_c[p(\lambda_c)]^{1/\psi}, \quad \gamma = 2\psi + 1, \tag{5.88}$$

where, here, ψ is given by (5.55) and $p(\lambda_c)$ is determined from (5.43).

For the asymmetric postbifurcation behavior, $\lambda_1^{he} \neq 0$, a maximum support load occurs for buckling in the direction that gives $\lambda_1^{he}\xi < 0$. According to (5.85) the maximum support load for an imperfect structure is given by

$$\frac{\lambda_{\max}}{\lambda_c} = 1 - \mu(|\overline{\xi}|)^{1/(\psi+1)}, \quad \lambda_1^{he}\overline{\xi} < 0, \tag{5.89}$$

where

$$\mu = (\psi+1)\psi^{-\psi/(\psi+1)}\left[\frac{|\lambda_1^{he}|}{\lambda_c}\right]^{1/(\psi+1)}[p(\lambda_c)]^{1/(\psi+1)}. \tag{5.90}$$

For symmetric postbifurcation behavior a maximum support load occurs if $\lambda_2^{he} < 0$, and from (5.87) the relation between the maximum support load and imperfection amplitude is

$$\frac{\lambda_{\max}}{\lambda_c} = 1 - \mu(\overline{\xi})^{2/(2\psi+1)}, \tag{5.91}$$

where

$$\mu = (2\psi+1)(2\psi)^{-2\psi/(2\psi+1)}\left[\frac{-\lambda_2^{he}}{\lambda_c}\right]^{1/(2\psi+1)}[p(\lambda_c)]^{2/(2\psi+1)}. \tag{5.92}$$

5.2. *Numerical studies of imperfection sensitivity*

Here, we discuss some results of numerical computations which have been compared with the predictions of the hypoelastic asymptotic analysis. These numerical analyses, as virtually all numerical postbuckling solutions carried out so far, employ the simplest flow theory of plasticity. However, we note that in all specific examples to be discussed here the plastic prestrains are small enough that the bifurcation predictions of flow theory and deformation theory do not differ significantly. Hence, the imperfection sensitivity exhibited in these cases is not the anomalous sensitivity associated with a smooth yield surface plasticity theory.

Our first example is a square plate simply supported on all four sides and subject to axial compression in one direction. In the elastic range the initial postbuckling behavior, as determined from Koiter's (1945) general theory, is stable symmetric, i.e. the load parameter λ and the buckling mode amplitude ξ are related by an expression of the form (5.3) with $\lambda_1^{he} = 0$ and $\lambda_2^{he} > 0$.

In the plastic range, a study of postbuckling behavior and imperfection sensitivity of such square plates has been carried out by Needleman and Tvergaard (1976). Two types of in-plane boundary conditions were investigated: one requiring all four edges of the plate to remain straight throughout the loading history, the other leaving the edges unconstrained. Numerically determined load-buckling deflection curves, from Needleman and Tvergaard (1976), for a rather high hardening material ($n = 3$ in (4.6)) are displayed in Fig. 5.1. Here the material and geometric parameters characterizing the plate were chosen to give a bifurcation stress 1.259 times the yield stress. The imperfections were taken in the form of the critical eigenmode and the amplitude $\bar{\xi}$ is normalized by the thickness. The curves labeled "perfect" are actually the results of computations with a very small initial imperfection.

In the plastic range the initial postbifurcation behavior, as determined from Hutchinson's (1973, 1974) asymptotic analysis, is independent of the in-plane boundary conditions. In accord with this analysis the load-buckling deflection curves for the perfect plates virtually coincide up to, and even somewhat beyond, the maximum load. Then the two curves deviate with that for the unconstrained edges monotonically decreasing while the plate with the edges

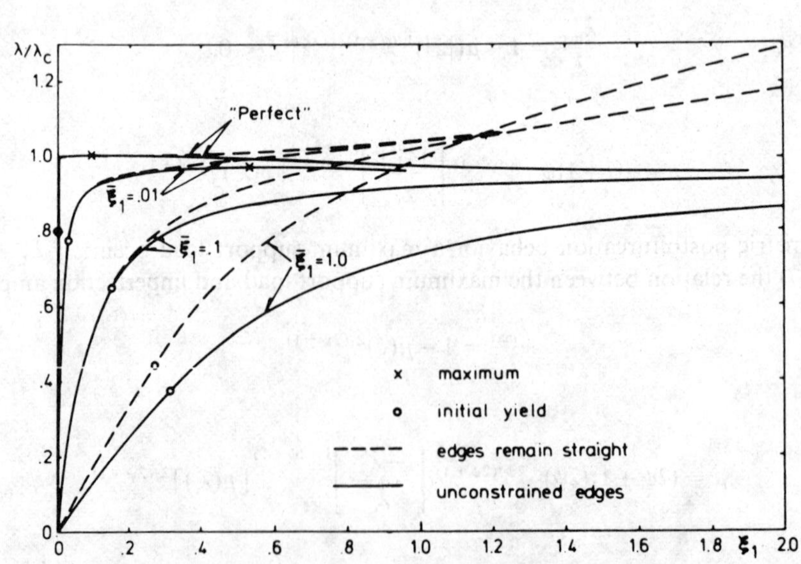

FIG. 5.1. Load *versus* buckling mode displacement for an axially compressed square plate made of a high hardening material for which bifurcation takes place in the plastic range.

constrained straight reaches a minimum load after which the load rises monotonically. An initial postbuckling analysis along the lines described in Section 5.1 was also carried out by Needleman and Tvergaard (1976). For the parameter values characterizing this case, $\lambda_2^{he} < 0$ with the unconstrained edge condition, while with the constrained edge condition $\lambda_2^{he} > 0$. Consistent with this result, the numerically determined overall postbuckling behavior of the plate with constrained edges is stable with the associated lack of imperfection sensitivity. On the other hand, the overall postbuckling behavior of the plate with the unconstrained edges is unstable and, as shown in Fig. 5.1, for this set of in-plane boundary conditions the plate exhibits a mild imperfection sensitivity. Needleman and Tvergaard (1976) also note that a plate made of a lightly hardening material ($n = 10$ in (4.6)) is imperfection sensitive for both of these in-plane boundary conditions, as also revealed by the asymptotic analysis of Section 5.1.

Another example is provided by Tvergaard's (1977b) investigation of the postbuckling behavior and imperfection sensitivity of an axially compressed elastic–plastic cylindrical panel of the type occurring in stiffened shells. The investigation carried out by Tvergaard (1977b) concentrates on the local buckling of such cylindrical panels between the stiffeners and assumes that the torsional rigidity of the stiffeners can be neglected. In the elastic range, Koiter (1956) demonstrated that sufficiently flat panels have a stable symmetric postbifurcation behavior, whereas more curved panels are imperfection sensitive. A comparison of numerical results and asymptotic hypoelastic predictions, taken from Tvergaard (1977b), is shown in Fig. 5.2. Here, the critical stresses are 1.104, 1.033, and 1.052 times the yield stress for the three cases shown. As can be seen in this Figure, the sensitivity to small imperfections agrees quite well with the predictions of the asymptotic analysis. However, larger imperfections give a smaller reduction in the maximum support load than predicted by the analysis. Additionally, Tvergaard (1977b) discusses panels for which the postbifurcation load for a perfect panel first attains a maximum, then decreases to a value slightly below the bifurcation load, and finally increases in the advanced postbuckling range. The initial decrease is consistent with the small negative value of λ_2^{he} found by Tvergaard (1977b). However, in this case, even for a small initial imperfection having a magnitude of 1% of the panel thickness, no maximum support load is found.

These examples show that the hypoelastic asymptotic analysis can account for the material nonlinearities—the decreasing moduli with increasing deformation and the material path dependence—that interact with the geometric nonlinearity to determine the "overall" initial postbuckling behavior. The major limitation of this approach appears, to us, to lie in its

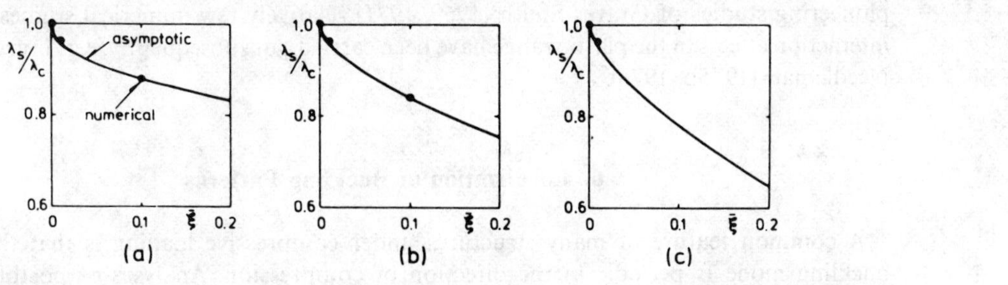

FIG. 5.2. Comparison of numerical results and asymptotic hypoelastic predictions for the imperfection sensitivity of cylindrical panels that bifurcate in the plastic range. (a) and (b) Panels made of a low hardening material. (c) Panel made of a high hardening material.

asymptotic nature rather than in the fact that elastic unloading is neglected. Due to the highly nonlinear nature of these problems, the range of validity of an asymptotic expansion about the bifurcation point can be quite limited. Indeed, this has already been revealed by the behavior of the nonlinear elastic U-shaped column discussed in Section 4.1. The asymptotic analysis reveals the asymmetric nature of the initial postbifurcation behavior. However, the numerical results (the dotted line in Fig. 4.2) show that, despite the initial rise of the postbifurcation load-buckling deflection curve for buckling towards the open end of the U, a load maximum is eventually reached at a finite buckling mode deflection. Furthermore, as also illustrated in Fig. 4.2, there is significant sensitivity to imperfections in this direction. This type of limitation on initial postbuckling asymptotic results is not unheard of in the elastic range. For example, oval cylindrical shells exhibit a highly unstable initial postbuckling behavior (Hutchinson, 1968), but numerical computations of the advanced postbuckling behavior (Kempner and Chen, 1966, 1969; Almroth *et al.*, 1971) have shown that load carrying capacities above the bifurcation load can be attained. In the plastic range, such instances, where the overall buckling behavior is not entirely disclosed by the immediate postbifurcation behavior, are apt to be far more prevalent.

For this reason, numerical solutions play an important role in elucidating plastic postbuckling behavior. One phenomenon that numerical solutions have played a significant role in elucidating is the imperfection sensitivity that arises when bifurcation of a perfect structure takes place just prior to elastic yielding. The U-shaped column depicted in Fig. 4.3 is one such example. Another is provided by Needleman's (1975) numerical solution for buckling of a simply supported circular plate subject to radial compression, where, for example, an initial imperfection with a magnitude of 6 % of the plate thickness results in a maximum support load less than 80 % of the bifurcation load, when bifurcation and initial yielding are nearly coincident. We note that for a highly imperfection sensitive structure such as an axially compressed monocoque cylindrical shell, plasticity can play a much less significant role when the bifurcation stress and yield stress nearly coincide (Hutchinson, 1974).

A number of numerical investigations have been directed toward determining the maximum support load of steel plating in the regime when bifurcation slightly precedes yielding, e.g. Crisfield (1975). Some of this work is directed toward understanding the complex postbuckling behavior of stiffened panels in the regime where plastic yielding effects play a significant role. For practical designs the postbuckling behavior of stiffened panels often involves the interaction of a column mode and a local buckling mode. Very little is known theoretically about mode interaction effects in the plastic range, and despite the early pioneering studies of Graves-Smith (1969, 1971) relatively few numerical studies of mode interaction effects in the plastic range have been carried out subsequently, e.g. Tvergaard and Needleman (1975b, 1976).

6. Localization of Buckling Patterns

A common feature of many structures under compressive loading is that the critical buckling mode is periodic in the direction of compression. Analyses of postbifurcation behavior and imperfection sensitivity for such structures, as discussed in Section 5, generally presume that the deflection pattern remains periodic as the modes grow and base the calculation of the maximum support load on this assumption. However, in some such structures, particularly when buckling occurs in the plastic range, it is observed that a transition stage is reached at which the periodic pattern is replaced rather abruptly by the localized growth of one or a few buckles. An axially compressed long mild steel plate simply supported at the edges is a rather well known example of a structure for which such localization occurs. Moxham (1971a, b) has observed this in over 100 tests and has used a model based on an assembly of single plates hinged together to argue that continued growth of all buckles is not stable beyond the maximum load.

In a recent paper Tvergaard and Needleman (1980) have treated buckling localization for a wider class of structures. A simple one-dimensional model was used to show that the basic mechanism of localization is a bifurcation at the maximum load point, analogous to the phenomenon of necking in tensile bars. Furthermore, this bifurcation away from a periodic buckling pattern was analyzed in detail for a column on a softening foundation and for a long rectangular plate made of a strain hardening elastic–plastic material.

6.1. *A one-dimensional model*

The one-dimensional model reflects the response of a buckled structure such as that of a nonlinear spring, which is on the average uniformly compressed as long as the buckling pattern in the structure remains periodic.

The model consists of a homogeneous axially compressed bar (Fig. 6.1) constrained to remain straight, but free to slide in the axial direction. The axial stress–strain curve is taken to be a non-linear function of the strain, as represented by the incremental relation

$$\dot{N} = C\dot{\varepsilon}, \quad \dot{\varepsilon} = \dot{u}_{,x}, \tag{6.1}$$

between the axial force N and the strain ε. Here, $C(\varepsilon)$ is the instantaneous modulus and u is the axial displacement. A small strain formulation is used, since the strains do remain small in the buckling problems of interest, and the effects of both material nonlinearities and geometric nonlinearities are incorporated in the stiffness C.

For a bar of length L, with the end displacements prescribed at $x = 0, L$, bifurcation away from a uniformly strained state is governed by the variational equation

$$\delta I = 0, \quad I = \int_0^L C\,\tilde{\varepsilon}\tilde{\varepsilon}\,dx, \tag{6.2}$$

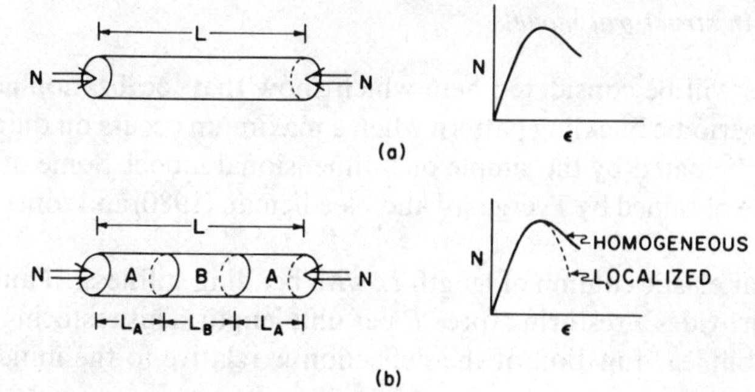

FIG. 6.1. Axially compressed bar model. (a) A homogeneous bar with its stress–strain curve. (b) A bar undergoing localization.

where $(\tilde{\ })$ denotes the difference between two solutions corresponding to the same prescribed end displacements. All bifurcation modes of the form

$$\tilde{u} = \sin\frac{m\pi x}{L}, \quad (m = 1, 2, 3, \ldots) \tag{6.3}$$

are critical simultaneously for $C = 0$, and thus bifurcation in this simple model occurs only if a maximum is reached.

We can also use the simple bar model to investigate whether a local region (B in Fig. 6.1 (b)) can undergo deformations different from those in the surrounding material (A in Fig. 6.1(b)). Incremental equilibrium requires

$$\dot{N}_A = \dot{N}_B \tag{6.4}$$

and the expressions for the increments of axial force are

$$\dot{N}_A = C_A\dot{\varepsilon}_A, \quad \dot{N}_B = C_B\dot{\varepsilon}_B. \tag{6.5}$$

Thus, as long as $\varepsilon_B = \varepsilon_A$, and consequently $C_B = C_A = C$, (6.4) and (6.5) result in

$$C(\dot{\varepsilon}_B - \dot{\varepsilon}_A) = 0, \tag{6.6}$$

which again shows that bifurcation is possible only for $C = 0$.

On the postbifurcation path or in the presence of initial imperfections the strains and stiffnesses are not identical in the two regions. Then the increment of the total strain of the bar is given by

$$\dot{\varepsilon} = (1-\rho)\dot{\varepsilon}_A + \rho\,\dot{\varepsilon}_B, \tag{6.7}$$

where $\rho = L_B/L$ and $L = 2L_A + L_B$. Using (6.4), (6.5), and (6.7) gives $\dot{\varepsilon}$ in terms of the axial force increment $\dot{N}$ as

$$\dot{\varepsilon} = \frac{\dot{N}}{C_B}\left[\rho + (1-\rho)\frac{C_B}{C_A}\right]. \tag{6.8}$$

It should be noted that the length L_B of the localized region is determined by factors not incorporated in the one-dimensional model. In a structure such as an axially compressed plate strip the localized deformation pattern will cover the length of at least one buckle in the fundamental periodic pattern.

6.2. *Localization in structural models*

A few examples will be considered here which show that localization actually occurs in structures with a periodic buckling pattern when a maximum occurs on the axial force *versus* strain curve, as anticipated by the simple one-dimensional model. Some of the results to be discussed are those obtained by Tvergaard and Needleman (1980) and some are extensions of these results.

Consider a linear elastic column of length L, with bending stiffness EI and supported by a foundation that provides a restoring force F per unit length. This restoring force is taken to be the following bilinear function of the deflection w relative to the initial column shape

$$F = \begin{cases} K_1 w, & \text{for } |w| \leqslant w_0, \\ \pm K_1 w_0 + K_2(w \mp w_0), & \text{for } |w| > w_0. \end{cases} \tag{6.9}$$

Here, the deflection w_0 is termed the yield point of the foundation, the upper sign is used for $w > 0$ and the lower for $w < 0$.

With a compressive force P and an initial imperfection $\bar{w}$ the governing differential equation takes the form

$$EIw_{,xxxx} + Pw_{,xx} + F = -P\bar{w}_{,xx}. \tag{6.10}$$

For a perfect column, $\bar{w} \equiv 0$, simply supported at the ends, the initial bifurcation mode is of the form

$$\tilde{w} = \sin\frac{m\pi x}{L}. \tag{6.11}$$

The length and the wavenumber shall be chosen here so that the wavelength is that critical for an infinite column, $L/m = \pi(EI/K_1)^{1/4}$, with a critical force $P_c = 2(K_1 EI)^{1/2}$.

For a softening foundation $K_2/K_1 < 1$ and an imperfection in the shape of the critical bifurcation mode (6.11), with amplitude $\delta_1 w_0$, Tvergaard and Needleman (1980) used closed form solutions of the linear differential equation (6.10) to determine the growth of the periodic buckling pattern and bifurcation away from this pattern. A maximum load is reached shortly after the wave-peaks have exceeded the deflection w_0, and bifurcation occurs with a little delay beyond the maximum. The fact that this delay is much larger for small m than for larger m represents a length effect analogous to that found for necking in tensile bars (see Needleman, 1972, and Hutchinson and Miles, 1974).

To determine whether this bifurcation actually leads to localization, as suggested by the simple one-dimensional model, or just to mode snapping into a mode with a number of half-waves different from m requires a postbifurcation analysis. Results of such an analysis will be shown here, based on a numerical solution for a column with an initial imperfection of the form

$$\bar{w} = w_0\left[\delta_1 \sin\frac{m\pi x}{L} - \delta_2 \sin\frac{(m-2)\pi x}{L}\right]. \tag{6.12}$$

Here, δ_1 is the amplitude of an imperfection in the shape of the initial bifurcation mode and δ_2 is the amplitude of a mode similar to the bifurcation mode occurring after the maximum load. As in the bifurcation analysis of Tvergaard and Needleman (1980), consideration is restricted to deflection patterns symmetric about the center $x = L/2$.

The numerical solution is obtained by a linear incremental method based on the incremental principle of virtual work

$$\int_0^L [\dot{N}\delta\dot{\varepsilon} + \dot{M}\delta\dot{\kappa} + \dot{F}\delta\dot{w} + N\dot{w}_{,x}\delta\dot{w}_{,x}]dx + \dot{P}[\delta\dot{u}]_0^L = 0. \tag{6.13}$$

Each half-wave is divided into six finite elements. Within an element the displacement components u and w are approximated by Hermitian cubics, and integrals along the column are evaluated by 4-point Gaussian quadrature.

The plot of load P *versus* average axial strain ε shown in Fig. 6.2 is obtained for $K_2/K_1 = 0.1$, $\delta_1 = 0.5$, and $m = 9$. Here, ε is the length change relative to initial length, and ε_y is the strain at which $|w| = |w_0|$ is reached at the wave peaks. For $\delta_2 = 0$ the previously obtained bifurcation just beyond the maximum load is confirmed. Even with a small additional imperfection, $\delta_2 = 0.025$, the buckling pattern remains nearly periodic up to around the bifurcation point and the maximum load is hardly changed at all. However, near the bifurcation point the central buckles start to grow rapidly while buckles away from the center start to decay, at the same time as the postbifurcation load drops below that for continued growth of a periodic pattern. Figure 6.3(b) shows the buckling pattern at the maximum load, and Fig. 6.3(c) shows the localized pattern considerably later, at $\varepsilon/\varepsilon_y = 1.72$.

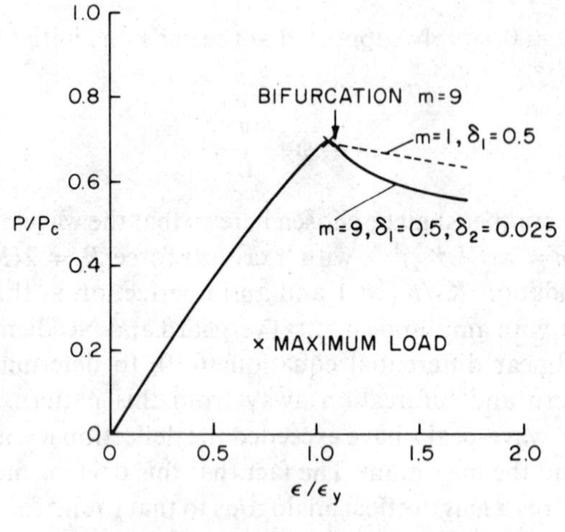

FIG. 6.2. Load *versus* average axial strain for an elastic column on an elastic–plastic foundation with $K_2/K_1 = 0.1$.

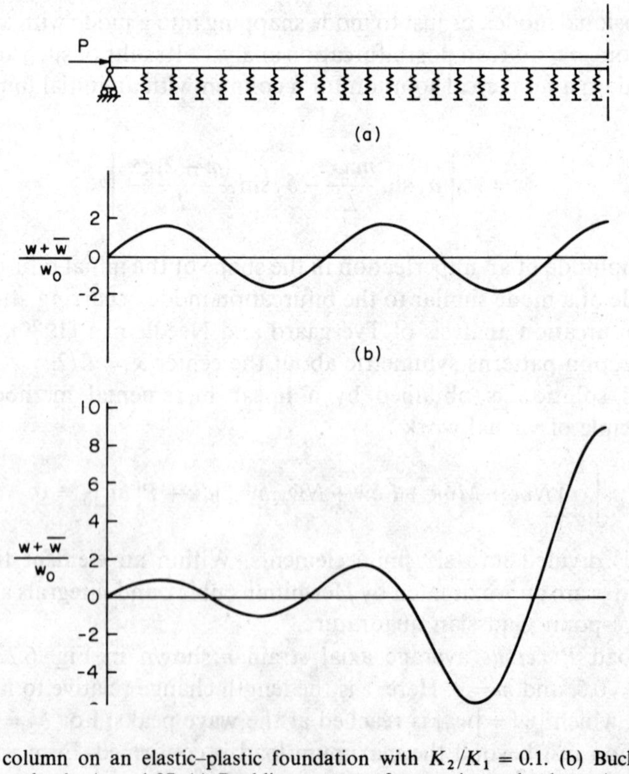

FIG. 6.3. (a) Elastic column on an elastic–plastic foundation with $K_2/K_1 = 0.1$. (b) Buckling pattern at the maximum load, $\varepsilon/\varepsilon_y = 1.07$. (c) Buckling pattern after maximum load, at $\varepsilon/\varepsilon_y = 1.72$.

The results in Figs. 6.2 and 6.3 are obtained for an "elastic–plastic foundation," in which the larger stiffness K_1 applies whenever $\dot{w}w < 0$. The computations have been repeated for an elastic foundation directly characterized by (6.9), without "elastic unloading." These purely elastic results differ little from those shown in Figs. 6.2 and 6.3 apart from the fact that the amplitudes away from the localized region are a little smaller in the elastic case. It may be noted that the column on an elastic–plastic foundation represents a fairly good model of a railway track, and in fact the localized deflection pattern shown in Fig. 6.3(b) is similar to that observed after thermal track buckling.

Another example of a structure for which such localization would be expected is that of an elastic–plastic column continuous over several equally spaced supports, which are free to move in the axial direction. For such a column a maximum load is reached due to the material nonlinearity. If L is the length of each bay, I is the area moment of inertia, and m is the number of bays, the critical bifurcation load for an initially straight column is $P_c = \pi^2 E_t I/L^2$, where E_t is the tangent modulus at the critical load. The uniaxial stress–strain curve is taken to be represented by the piecewise power law (4.6), with the initial yield stress σ_y and the strain hardening exponent n.

The column considered in Figs. 6.4 and 6.5 covers nine bays ($m = 9$), still assuming symmetry about the center $x = mL/2$, and the material parameters are $\sigma_y/E = 0.003$ and $n = 10$. The column has rectangular cross-section with a height to length ratio $h/L = 0.1$. The numerical scheme used to solve the problem is similar to that used for the column on a softening foundation. The necessary integration of current stresses and tangent moduli through the thickness are carried out by seven-point Simpson integration. As the imperfection (6.12) is not kinematically admissible in the present case, an initial deflection is assumed of the form

$$\bar{w} = h(\bar{\xi}_1 + \bar{\xi}_2 e^{-[(2x-mL)/L]^2}) \sin \frac{\pi x}{L}. \tag{6.14}$$

Here $\bar{\xi}_1$ is the amplitude of an imperfection in the shape of the initial bifurcation mode and $\bar{\xi}_2$ represents a localized imperfection.

The load *versus* average axial strain curves shown in Fig. 6.4 for $\bar{\xi}_1 = 0.02$ are rather similar to those in Fig. 6.2. With an initially periodic pattern ($\bar{\xi}_2 = 0$) bifurcation does occur a little beyond the maximum. Also in this case a small deviation, $\bar{\xi}_2 = 0.001$, from the periodic imperfection has negligible effect until near the bifurcation point, where deformations start to localize. Figures 6.5(b) and (c) show the nearly periodic pattern at the maximum and a localized pattern at a later stage, respectively.

The structure for which Moxham (1971a, b) observed and discussed localized buckling is a long, elastic–plastic, axially compressed plate strip supported at the edges. For such a plate made of a power hardening material Tvergaard and Needleman (1980) obtained results similar to those shown in Figs. 6.2–6.5. In the case of a low hardening material a point of bifurcation does occur somewhat beyond the maximum load and subsequently localization takes place. For a high hardening material, where no maximum is reached, localization is not found, as also implied by the simple one-dimensional model. Tvergaard and Needleman (1980) employ J_2 flow theory in their analysis of the plate. It is interesting to note that the choice of constitutive law is much less crucial in analyzing the bifurcation leading to localization than it is in determining the bifurcation point for the initial periodic mode.

Tvergaard and Needleman (1980) also found that small localized imperfections did not significantly affect the load carrying capacity of the plate, so that an analysis of the maximum

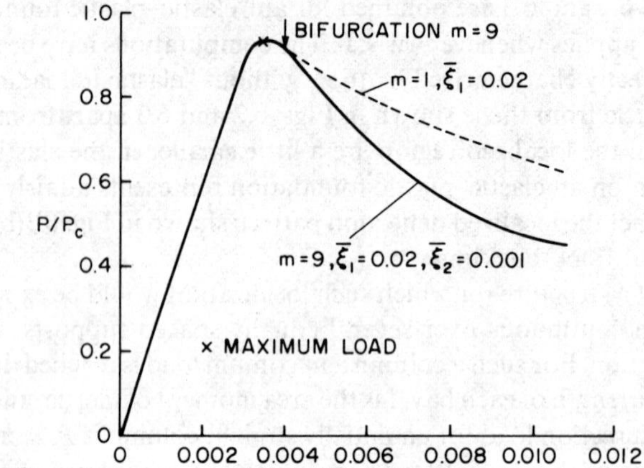

FIG. 6.4. Load *versus* average axial strain for a multiply supported elastic–plastic column with $\sigma_y/E = 0.003$, $n = 10$, $h/L = 0.1$.

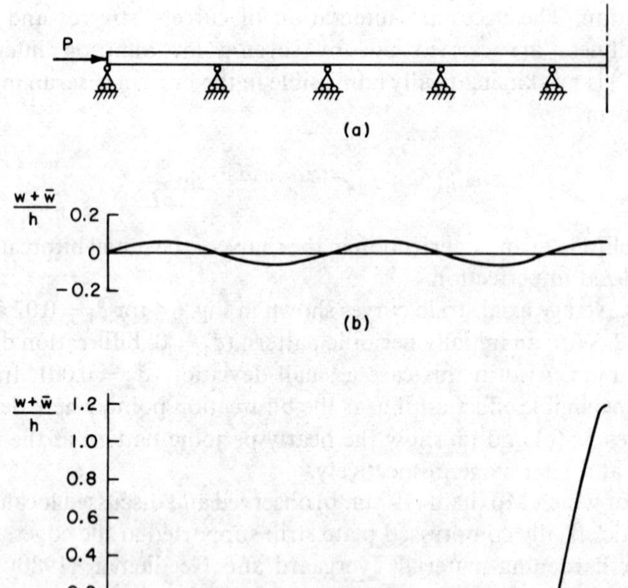

FIG. 6.5. (a) Multiply supported elastic–plastic column with $\sigma_y/E = 0.003$, $n = 10$, $h/L = 0.1$. (b) Buckling pattern at the maximum load, $\varepsilon = 0.0035$. (c) Buckling pattern after the maximum load, at $\varepsilon = 0.0108$.

support load which restricted attention to periodic modes and imperfections did give an accurate estimate of the load carrying capacity.

Acknowledgement

We are grateful to the Division of Engineering, Brown University, for supporting this work.

References

Almroth, B. O., Brogan, F. A., and Marlowe, M. B. (1971) Collapse analysis for elliptic cones, *AIAA Jl* **9,** 32–37.

Batdorf, S. B. (1949) Theories of plastic buckling, *J. Aeronaut. Sci.* **16,** 405–408.

Batdorf, S. B. and Budiansky, B. (1949) A mathematical theory of plasticity based on the concept of slip, Nat. Adv. Comm. Aeronaut. Tech. Note 1871.

Bellman, R. (1960) *Introduction to Matrix Analysis*, McGraw-Hill, New York.

Bijlaard, P. P. (1949) Theory and tests on the plastic stability of plates and shells, *J. Aeronaut. Sci.* **16,** 529–541.

Budiansky, B. (1974) Theory of buckling and post-buckling behavior of elastic structures, *Adv. Appl. Mech.* **14,** 1–65.

Budiansky, B. and Hutchinson, J. W. (1979) Buckling: progress and challenge. *Trends in Solid Mechanics*, edited by J. F. Besseling and A. M. A. van der Heijden, 93–116, Delft University Press, and Sijthoff and Noordhoff, Alphen aan den Rijn.

Christoffersen, J. and Hutchinson, J. W. (1979) A class of phenomenological corner theories of plasticity, *J. Mech. Phys. Solids* **27,** 465–487.

Crisfield, M. A. (1975) Full-range analysis of steel plates and stiffened plating under uniaxial compression, *Proc. Instn Civ. Engrs*, Part 2, **59,** 595–624.

Gerard, G. and Becker, H. (1957) *Handbook of Structural Stability*, Part I: *Buckling of Flat Plates*, Nat. Adv. Comm. Aeronaut. Tech. Note 3781.

Gjelsvik, A. and Lin, G.-S. (1976) Plastic buckling of plates with edge frictional shear effects. Columbia University, Department of Civil Eng. and Eng. Mech. Report N00014–75–C–0695/50.

Graves-Smith, T. R. (1969) The ultimate strength of locally buckled columns of arbitrary length, *Thin-walled Steel Structures* 35–60, Crosby Lockwood, London.

Graves-Smith, T. R. (1971) The effect of initial imperfections on the strength of thin-walled box columns, *Int. J. Mech. Sci.* **13,** 911–925.

Hecker, S. S. (1976) Experimental studies of yield phenomena in biaxially loaded metals, *Constitutive Equations in Viscoplasticity*, *AMD* **20,** 1–33, ASME, New York.

Hill, R. (1958) A general theory of uniqueness and stability in elastic–plastic solids, *J. Mech. Phys. Solids* **6,** 236–249.

Hill, R. (1961) Bifurcation and uniqueness in nonlinear mechanics of continua, *Problems of Continuum Mechanics*, 155–164, Soc. Ind. Appl. Math., Philadelphia.

Hill, R. (1966) Generalized constitutive relations for incremental deformation of metal crystals by multislip, *J. Mech. Phys. Solids* **14,** 95–102.

Hill, R. (1978) Aspects of invariance in solid mechanics, *Adv. Appl. Mech.* **18,** 1–75.

Hutchinson, J. W. (1968) Buckling and initial postbuckling behavior of oval cylindrical shells under axial compression, *J. Appl. Mech.* **35,** 66–72.

Hutchinson, J. W. (1970) Elastic–plastic behavior of polycrystalline metals and composites, *Proc. R. Soc. London* A **318,** 247–272.

Hutchinson, J. W. (1973) Postbifurcation behavior in the plastic range, *J. Mech. Phys. Solids* **21,** 163–190.

Hutchinson, J. W. (1974) Plastic buckling, *Adv. Appl. Mech.* **14,** 67–144.

Hutchinson, J. W. and Budiansky, B. (1976) Analytical and numerical study of the effects of initial imperfections on the inelastic buckling of a cruciform column, *Buckling of Structures*, edited by B. Budiansky, 98–105, Springer-Verlag, Berlin.

Hutchinson, J. W. and Miles, J. P. (1974) Bifurcation analysis of the onset of necking in an elastic–plastic cylinder under uniaxial tension, *J. Mech. Phys. Solids* **22,** 61–77.

Hutchinson, J. W. and Tvergaard, V. (1980) Surface instabilities on statically strained plastic solids, *Int. J. Mech. Sci.* **22,** 337–354.

Kempner, J. and Chen, Y.-N. (1966) Large deflections of an axially compressed oval cylindrical shell, *Proc. 11th Int. Cong. Appl. Mech.*, edited by H. Görtler, 299–306, Springer-Verlag, Berlin.

Kempner, J. and Chen, Y.-N. (1969) Postbuckling of an axially compressed oval cylindrical shell, *Proc. 12th Int. Congr. Appl. Mech.*, 246–256, Springer-Verlag, Berlin.

Koiter, W. T. (1945) Over de stabiliteit van het elastisch evenwicht, Thesis Delft, H. J. Paris, Amsterdam. English translations (a) NASA TT-F10, 833 (1967), (b) AFFDL-TR-70-25 (1970).

Koiter, W. T. (1956) Buckling and postbuckling behavior of a cylindrical panel under axial compression, Nationaal Luchtvaart Laboratorium **20,** Report S476.

Koiter, W. T. (1966) On the nonlinear theory of thin elastic shells, *Proc. Kon. Ned Akad. Wet.* Ser. B, **69,** 1–54.

Lin, T. H. (1971) Physical theory of plasticity, *Adv. Appl. Mech.* **11,** 255–311.

Moxham, K. E. (1971a) Buckling tests on individual welded steel plates in compression, Cambridge University Engineering Department Report CUED/C-Struct/Tr. 3.

Moxham, K. E. (1971b) Theoretical prediction of the strength of welded steel plates in compression, Cambridge University Engineering Department Report CUED/C-Struct/Tr. 2.

Needleman, A. (1972) A numerical study of necking in circular cylindrical bars, *J. Mech. Phys. Solids* **20,** 111–127.

Needleman, A. (1975) Postbifurcation behavior and imperfection sensitivity of elastic–plastic circular plates, *Int. J. Mech. Sci.* **17,** 1–13.

Needleman, A. and Tvergaard, V. (1976) An analysis of the imperfection-sensitivity of square elastic–plastic plates under axial compression, *Int. J. Solids Struct.* **12,** 185–201.

Onat, E. T. and Drucker, D. C. (1953) Inelastic instability and incremental theories of plasticity, *J. Aeronaut. Sci.* **20,** 181–186.

Sanders, J. L. (1954) Plastic stress–strain relation based on linear loading functions, *Proc. 2nd U.S. Nat. Congr. Appl. Mech.*, 455–460, University of Michigan, Ann Arbor.

Sewell, M. J. (1972) A survey of plastic buckling, *Stability*, edited by H. Leipholz, 85–197, University of Waterloo Press.

Sewell, M. J. (1974) A plastic flow rule at a yield vertex, *J. Mech. Phys. Solids* **22,** 469–490.

Shanley, F. R. (1947) Inelastic column theory, *J. Aeronaut. Sci.* **14,** 261–267.

Stowell, E. Z. (1948) A unified theory of plastic buckling of columns and plates, Nat. Adv. Comm. Aeronaut. Tech. Note 898.

Tvergaard, V. (1977a) Buckling behavior of plate and shell structures, *Theoretical and Applied Mechanics*, Proceedings 14th IUTAM Congress, edited by W. T. Koiter, 233–247, North-Holland, Amsterdam.

Tvergaard, V. (1977b) Buckling of elastic–plastic cylindrical panel under axial compression, *Int. J. Solids Struct.* **13,** 957–970.

Tvergaard, V. and Needleman, A. (1975a) On the buckling of elastic–plastic columns with asymmetric cross-sections, *Int. J. Mech. Sci.* **17,** 419–424.

Tvergaard, V. and Needleman, A. (1975b) Buckling of eccentrically stiffened elastic–plastic panels on two simple supports or multiply supported, *Int. J. Solids Struct.* **11,** 647–663.

Tvergaard, V. and Needleman, A. (1976) Mode interaction in an eccentrically stiffened elastic–plastic panel under compression, *Buckling of Structures*, edited by B. Budiansky, 160–171, Springer-Verlag, Berlin.

Tvergaard, V. and Needleman, A. (1980) On the localization of buckling patterns, *J. Appl. Mech.* **47,** 613–619.

van der Heijden, A. M. A. (1979) A study of Hutchinson's plastic buckling model, *J. Mech. Phys. Solids* **27,** 441–464.

Elastic Deformations of Rubberlike Solids

by R. W. Ogden

Department of Mathematics, Brunel University

Summary

The macroscopic elastic behaviour of rubberlike solids is reviewed on the basis of the general theory of isotropic elasticity. Attention is restricted to homogeneous biaxial deformations since these provide stress–deformation data sufficient to characterize the strain-energy function of rubberlike solids. The incompressible theory is described first and then experimental data are discussed and represented in graphical forms yielding immediate theoretical interpretation. The main stress–deformation characteristics of rubberlike solids are thereby exposed. Specific forms of strain-energy function are also reviewed, the role of the Valanis–Landel hypothesis being highlighted. Macroscopic implications of molecular statistical theories are assessed briefly. The compressible theory is then developed and details of the theory of "nearly incompressible" elastic solids examined with a view to characterizing volume changes accompanying biaxial deformation. Experimental data are discussed in relation to the theory which is also viewed *vis-à-vis* structurally based theories.

1. Introduction

In this article various aspects of the isothermal macroscopic elastic behaviour of rubberlike solids are reviewed. The standpoint is purely phenomenological and based on the general theory of isotropic elasticity and it is the mathematical representation of the macroscopic elastic properties which is the main interest rather than any explanation of those properties at the microscopic (structural) level. However, contact will be made with the molecular statistical theory, and the implications of the relation between microscopic and macroscopic theory discussed briefly.

The theoretical relation between stress and deformation is based on a general form of isotropic elastic strain-energy function. This generality is maintained throughout the mathematical development. Particular forms of strain-energy function are considered when the theory is assessed in relation to experimental data. The discussion is confined to quasi-static *homogeneous deformations* so the equilibrium equations are satisfied automatically and the mathematics is thereby kept as simple as is consistent with the use of a general isotropic elastic constitutive law.

In Section 2 the assumption of *incompressibility* is adopted in order to establish the main features of the elastic stress–deformation characteristics of rubberlike solids, and the theoretical stress–deformation relations are compared with experimental data for general homogeneous deformations. In this context a number of alternative general representations of the strain-energy function are discussed, and their relative advantages and disadvantages assessed. Since the principal extension ratios have immediate physical interpretation they are used as the basic deformation variables in this article. In terms of such variables the stress–deformation relations take on a concise and transparent mathematical form. The extensive agreement between the Valanis–Landel hypothesis and experimental data for general homogeneous biaxial deformations is examined in detail, following a discussion of the more important experimental contributions.

The strain-energy functions are also discussed briefly in relation to statistical theories of the elasticity of the molecular network structure with a view to helping in the interpretation and explanation of the statistical theories.

With forms of (incompressible) strain-energy function which characterize the macroscopic elastic response of the molecular network established, the next step is to develop the corresponding theory for compressible isotropic elastic solids. This is done in Section 3 and, in particular, the changes in the stress–deformation characteristics induced by the introduction of compressibility are examined.

General formulae relating the volume changes to the applied principal stretches for biaxial deformations are given and specialized to simple tension and other single-parameter deformations. The theory is related to experimental data, and calculation of the deformation-induced volume changes is a feature of the analysis. In this context the theory of *nearly isochoric deformations* is relevant since volume changes accompanying the deformation of rubberlike solids are "small" in the sense that the ratio of shear to bulk modulus is typically of order 10^{-4}. This theory is discussed in detail for homogeneous biaxial deformations, and specific forms of strain-energy function, modifications of those arising in the incompressible theory, are examined. In particular, their physical interpretations are assessed in relation to volume-change data.

Finally, the general theory of compressible isotropic elasticity is used to provide a broad perspective for certain microscopically (or structurally) based theories. In particular, the notion of *anisotropy of compressibility* is examined from a general standpoint and also for nearly isochoric deformations. The results of the theory are interpreted in terms of the underlying network structure and assessed in the light of experimental findings.

Overall the aim is to provide a fairly complete picture of the relation between theory and experiment for rubberlike solids in isothermal biaxial deformation. But the choice of the detailed contents of this review and, more particularly, its form of presentation and emphasis is necessarily coloured by the writer's personal preferences. Detailed references to both the experimental and the theoretical literature on which this article draws are included throughout, and for the most part the results described here have been published previously. However, a number of results are included here which have not before appeared in the literature.

2. Incompressible Rubberlike Solids

2.1. *Stress–deformation relations*

The homogeneous deformation of a cuboid-shaped specimen of material deformed parallel to its edges may be described by the three ratios of current to initial length of its sides.

These are the *principal stretches* of the deformation, denoted by $\lambda_1, \lambda_2, \lambda_3$. Here, we confine attention to deformations of this type.

For *incompressible* materials the deformation is necessarily *isochoric* (volume preserving) and the stretches satisfy the *incompressibility constraint*

$$\lambda_1 \lambda_2 \lambda_3 = 1 \tag{2.1}$$

(in general, $\lambda_1 \lambda_2 \lambda_3$ is the ratio of current to initial volume of the specimen). Thus, *only two of the stretches are independently assignable.*

The material is assumed to be *isotropic* relative to its undeformed and unstressed (or natural) configuration, and its elastic properties are to be characterized by a *strain-energy function* $W(\lambda_1, \lambda_2, \lambda_3)$ per unit volume which (by virtue of the isotropy) depends symmetrically on $\lambda_1, \lambda_2, \lambda_3$. Thus,

$$W(\lambda_1, \lambda_2, \lambda_3) = W(\lambda_1, \lambda_3, \lambda_2) = W(\lambda_3, \lambda_1, \lambda_2) \tag{2.2}$$

subject to (2.1).

That the elastic response of rubberlike solids can be regarded as isotropic is generally accepted. It is also generally accepted that the adoption of the incompressibility constraint has no significant effect on the determination of the primary deformation characteristics of rubberlike solids. We emphasize, however, that (2.1) is an *approximation* whose validity is put on a sound theoretical basis in Section 3.

The stress and deformation in the material are related by

$$\sigma_i = \lambda_i \frac{\partial W}{\partial \lambda_i} - p \quad (i = 1, 2, 3; \text{ no sum}), \tag{2.3}$$

where $\sigma_1, \sigma_2, \sigma_3$ are the principal components of the Cauchy (or true) stress tensor (i.e. they are the forces parallel to the cuboid edges per unit current area of cross-section) and p, a Lagrange multiplier arising from the constraint (2.1), is referred to as the "arbitrary" hydrostatic pressure. The corresponding *nominal* (or engineering) stresses are σ_i/λ_i $(i = 1, 2, 3)$ per unit *undeformed* area of cross-section.

Because of (2.1) it is convenient to work with just two of the stretches, λ_1 and λ_2, say, and to eliminate λ_3 from explicit consideration. Thus, we write

$$\hat{W}(\lambda_1, \lambda_2) = \hat{W}(\lambda_2, \lambda_1) = W(\lambda_1, \lambda_2, \lambda_1^{-1}\lambda_2^{-1}), \tag{2.4}$$

recalling (2.2), and from (2.3) we obtain

$$\sigma_1 - \sigma_3 = \lambda_1 \frac{\partial \hat{W}}{\partial \lambda_1}, \quad \sigma_2 - \sigma_3 = \lambda_2 \frac{\partial \hat{W}}{\partial \lambda_2}. \tag{2.5}$$

For consistency with the classical theory of incompressible isotropic elasticity we must have

$$\left.\begin{aligned} &\hat{W}(1, 1) = 0, && \frac{\partial \hat{W}}{\partial \lambda_\alpha}(1,1) = 0 \quad (\alpha = 1, 2), \\ &\frac{\partial^2 \hat{W}}{\partial \lambda_1 \partial \lambda_2}(1, 1) = 2\mu, && \frac{\partial^2 \hat{W}}{\partial \lambda_\alpha^2}(1, 1) = 4\mu \quad (\alpha = 1, 2), \end{aligned}\right\} \tag{2.6}$$

where μ is the classical shear modulus. In the present context we refer to μ as the *ground-state shear modulus.*

The two stress–deformation relations (2.5) provide the theoretical basis for relating experimentally determined values of the stress and the two independent deformation parameters λ_1 and λ_2. At this point we record some important special cases of (2.5) in which λ_1, λ_2 are not independent but depend on a single deformation parameter, λ say:

(a) *Simple tension*: $\lambda_1 = \lambda$, $\lambda_2 = \lambda^{-1/2}$, $\sigma_1 = \sigma$, $\sigma_2 = 0$ lead to

$$\sigma = \lambda \frac{d\tilde{W}}{d\lambda}, \quad \tilde{W}(\lambda) = \hat{W}(\lambda, \lambda^{-1/2}). \tag{2.7}$$

(b) *Pure shear*: $\lambda_1 = \lambda$, $\lambda_2 = \lambda^{-1}$, $\sigma_1 = \sigma$, $\sigma_2 = 0$ lead to

$$\sigma = \lambda \frac{d\tilde{W}}{d\lambda}, \quad \tilde{W}(\lambda) = \hat{W}(\lambda, \lambda^{-1}). \tag{2.8}$$

(c) *Equibiaxial tension*: $\lambda_1 = \lambda_2 = \lambda$, $\sigma_1 = \sigma_2 = \sigma$ lead to

$$\sigma = \tfrac{1}{2}\lambda \frac{d\tilde{W}}{d\lambda}, \quad \tilde{W}(\lambda) = \hat{W}(\lambda, \lambda). \tag{2.9}$$

In each case the nominal stress f, per unit undeformed area of cross-section, is given by $f = \sigma/\lambda$. Note that the hydrostatic pressure p does not occur explicitly in (2.5)–(2.9), but, if required, it can be found from the equation $\sigma_3 = 0$. The specializations of (2.6) for (a)–(c) are not made explicit here.

2.2. *Application to membranes*

The deformation of a thin-walled shell (or membrane) may be approximated by a homogeneous deformation of the type described in Section 2.1 under suitable loading conditions. Let λ_1, λ_2 be the principal stretches in the membrane surface and λ_3 that normal to the surface.

The current thickness h of the membrane is related to the undeformed thickness H by means of the *approximation*

$$h = \lambda_3 H \tag{2.10}$$

(which is exact for a plane membrane).

The membrane approximation also gives $\sigma_3 = 0$, in which case equations (2.5) become

$$\sigma_1 = \lambda_1 \frac{\partial \hat{W}}{\partial \lambda_1}, \quad \sigma_2 = \lambda_2 \frac{\partial \hat{W}}{\partial \lambda_2} \tag{2.11}$$

(these being exact for a plane membrane).

First we apply this membrane theory to the inflation of a spherical balloon whose undeformed wall thickness H is very much smaller than its undeformed radius R. Let r denote the current radius. (It is assumed that the balloon remains spherical during inflation. This assumption is not in general valid, however, and further discussion of this point is provided in Section 2.6.) Then the incompressibility constraint and (2.10) lead to the approximation

$$r = \lambda R, \tag{2.12}$$

where $\lambda_1 = \lambda_2 = \lambda$, the deformation in the skin being (approximately) equivalent to that of a uniform equibiaxial tension.

Let $T(=h\sigma_1=h\sigma_2)$ be the surface tension in the membrane and P the inflating pressure. Then, for equilibrium, we have $P=2T/r$, and it follows on use of (2.9) that

$$P^*=\lambda^{-2}\frac{d\tilde{W}}{d\lambda}, \tag{2.13}$$

where $P^*=PR/H$ and $\tilde{W}(\lambda)=\hat{W}(\lambda,\lambda)$.

Next we obtain the equations for the extension and inflation of a circular cylindrical membrane tube with closed ends. Let λ_1 be the stretch in the axial direction and λ_2 the circumferential stretch. Suppose (L,l), (R,r), and (H,h) are the (initial, current) lengths, radii, and thicknesses of the membrane. Then we have $l=\lambda_1 L$, which is exact, together with the approximations

$$r=\lambda_2 R,\quad h=\lambda_1^{-1}\lambda_2^{-1}H. \tag{2.14}$$

The inflating pressure P^* $(=PR/H)$ per unit current area and the axial force F^* $(=F/2\pi RH)$ per unit undeformed area of membrane cross-section are given by

$$P^*=\lambda_1^{-1}\lambda_2^{-1}\frac{\partial\hat{W}}{\partial\lambda_2},\quad F^*=\frac{\partial\hat{W}}{\partial\lambda_1}-\tfrac{1}{2}\lambda_2\lambda_1^{-1}\frac{\partial\hat{W}}{\partial\lambda_2}, \tag{2.15}$$

when the pressure on the end faces is taken into account. More detailed derivations of (2.15) have been given by Ogden (1979a) and Haughton and Ogden (1979). See also Alexander (1971a) and Vangerko and Treloar (1978).

It is convenient to rearrange (2.15) in the form

$$\sigma_1-\sigma_2=\lambda_1\frac{\partial\hat{W}}{\partial\lambda_1}-\lambda_2\frac{\partial\hat{W}}{\partial\lambda_2}=\lambda_1 F^*-\tfrac{1}{2}\lambda_2^2\lambda_1 P^* \tag{2.16}$$

in preparation for comparison with experimental data in Section 2.5. Since λ_1 and λ_2 can be varied independently in an extension–inflation experiment, (2.16) provides a theoretical framework for the experimentally determined relation between stress and general biaxial deformation.

Thus far we have maintained the strain-energy function in its most general form. We next examine alternative (general) representations for W and some specific forms of W as a prelude to comparing the above theory with experimental data.

2.3. *Strain–energy functions*

Since W depends symmetrically on λ_1, λ_2, λ_3 it may be regarded as a function of any two independent symmetrical functions of λ_1, λ_2, λ_3, recalling that only two of λ_1, λ_2, λ_3 are independent by (2.1). One such pair of independent functions is

$$I_1=\lambda_1^2+\lambda_2^2+\lambda_3^2,\quad I_2=\lambda_1^{-2}+\lambda_2^{-2}+\lambda_3^{-2}, \tag{2.17}$$

and it was on the basis of these that the theory of large elastic deformations was originally developed by Rivlin (1948). The majority of subsequent writers have also adopted (2.17).

A more general pair is

$$I_1=\lambda_1^{\alpha_1}+\lambda_2^{\alpha_1}+\lambda_3^{\alpha_1},\quad I_2=\lambda_1^{\alpha_2}+\lambda_2^{\alpha_2}+\lambda_3^{\alpha_2}, \tag{2.18}$$

where α_1 and $\alpha_2\neq\alpha_1$ are real numbers, not necessarily integers. The choice $\alpha_1=2$, $\alpha_2=-2$

gives (2.17) as a special case. To my knowledge, a general development of the theory using (2.18) rather than (2.17) has not appeared in the literature, although the possibility of using symmetrical functions of the type (2.18) was alluded to by Hill (1969); see also Hill (1978).

From (2.1) and (2.18) we have

$$I_1 = \lambda_1^{\alpha_1} + \lambda_2^{\alpha_1} + \lambda_1^{-\alpha_1}\lambda_2^{-\alpha_1}, \quad I_2 = \lambda_1^{\alpha_2} + \lambda_2^{\alpha_2} + \lambda_1^{-\alpha_2}\lambda_2^{-\alpha_2}, \tag{2.19}$$

and hence the alternative representations

$$W(\lambda_1, \lambda_2, \lambda_3) \equiv \hat{W}(\lambda_1, \lambda_2) \equiv \Phi(I_1, I_2) \tag{2.20}$$

for the strain-energy function. Thus, W may be represented equally generally and equivalently in terms of the principal stretches or an independent pair of symmetrical functions, such as (2.18). Indeed, for any real number α, λ_1^α, λ_2^α, λ_3^α are the principal values of a certain second-order symmetric tensor and as such are expressible as functions of an independent pair of invariants of that tensor. Such invariants are symmetrical functions of the form (2.18); in particular, $\alpha_1 = \alpha$ and $\alpha_2 = -\alpha$ give the principal invariants. Moreover, the principal values are themselves invariants of the tensor. An explicit formula for λ_1, λ_2, λ_3 in terms of the invariants (2.17) has been given by Rivlin and Sawyers (1976) and a suitable reference to the relevant theorems from linear algebra is Mirsky (1955). We emphasize the above because it is a quite common misconception that the theory as developed in terms of the principal stretches is different from that in terms of invariants.

Use of (2.19) and (2.20) in (2.5) gives

$$\sigma_\mu - \sigma_3 = \alpha_1(\lambda_\mu^{\alpha_1} - \lambda_3^{\alpha_1})\frac{\partial \Phi}{\partial I_1} + \alpha_2(\lambda_\mu^{\alpha_2} - \lambda_3^{\alpha_2})\frac{\partial \Phi}{\partial I_2} \quad (\mu = 1, 2). \tag{2.21}$$

More generally than (2.18), a pair of invariants of the form

$$\lambda_1^{\alpha_1}(\lambda_2^{\alpha_2} + \lambda_3^{\alpha_2}) + \lambda_2^{\alpha_1}(\lambda_3^{\alpha_2} + \lambda_1^{\alpha_2}) + \lambda_3^{\alpha_1}(\lambda_1^{\alpha_2} + \lambda_2^{\alpha_2}) \tag{2.22}$$

may be used, but it seems that little is to be gained by such a choice, nor indeed by any symmetrical function which is not "polynomial" in form. In fact any suitably well-behaved symmetrical function of λ_1, λ_2, λ_3 can be approximated as closely as required on any closed domain in (λ_1, λ_2)-space by polynomial expansions in polynomial-type invariants, such as (2.18).

The expansion

$$\Phi(I_1, I_2) = \sum_{m=0}^{\infty} \sum_{n=0}^{\infty} C_{mn}(I_1 - 3)^m (I_2 - 3)^n, \quad C_{00} = 0, \tag{2.23}$$

was used by Rivlin (1949), where I_1, I_2 are given by (2.17) and the C_{mn}'s are constants. This is equally valid in terms of the invariants (2.18). Surveys of particular forms of strain-energy function based on the invariants (2.17) have been given by Ogden (1972a) and Treloar (1975). Some of these are special cases of (2.23), while others are not polynomial in character (but can of course be approximated by (2.23) with suitable values of the constants C_{mn}). We do not detail these here.

Equally valid is a polynomial expansion for W in terms of λ_1, λ_2, λ_3, and if these are replaced by λ_1^α, λ_2^α, λ_3^α as the basic units, for any real number α, the expansion

$$W(\lambda_1, \lambda_2, \lambda_3) = \sum_{\alpha_1, \alpha_2} C(\alpha_1, \alpha_2)\{\lambda_1^{\alpha_1}(\lambda_2^{\alpha_2} + \lambda_3^{\alpha_2}) + \lambda_2^{\alpha_1}(\lambda_3^{\alpha_2} + \lambda_1^{\alpha_2}) + \lambda_3^{\alpha_1}(\lambda_1^{\alpha_2} + \lambda_2^{\alpha_2}) - 6\} \tag{2.24}$$

follows, where α_1 and α_2 are integral multiples of α and $C(\alpha_1, \alpha_2)$ are constants with $C(\alpha_2, \alpha_1) = C(\alpha_1, \alpha_2)$. This type of expansion was suggested by Ogden (1972a).

An alternative to (2.24), employing only two independent stretches, is

$$\hat{W}(\lambda_1, \lambda_2) = \sum_{\alpha_1, \alpha_2} A(\alpha_1, \alpha_2)(\lambda_1^{\alpha_1}\lambda_2^{\alpha_2} + \lambda_1^{\alpha_2}\lambda_2^{\alpha_1} - 2), \tag{2.25}$$

where the $A(\alpha_1, \alpha_2) = A(\alpha_2, \alpha_1)$ are constants. The expansion (2.24) can clearly be rewritten in the form (2.25) by use of (2.1), but (2.25) cannot be converted into (2.24) unless

$$A(\alpha_1, \alpha_2) = A(\alpha_1 - \alpha_2, -\alpha_2) = A(\alpha_2 - \alpha_1, -\alpha_1). \tag{2.26}$$

This latter requirement, however, is not essential for the validity of (2.25), but if (2.26) does not hold the form of $W(\lambda_1, \lambda_2, \lambda_3)$ corresponding to (2.25) is not expressible in simple symmetrical form since λ_1 and λ_2 then need to be replaced by their expressions in terms of symmetrical invariants. For further discussion see Ogden (1979a).

The expansions (2.23)–(2.25) are equally general. Strain-energy functions which are special cases of (2.24) have been considered previously by a number of authors. For example, Mooney (1940) used

$$W = \sum_{n=1}^{\infty} \{ a_n(\lambda_1^{2n} + \lambda_2^{2n} + \lambda_3^{2n} - 3) + b_n(\lambda_1^{-2n} + \lambda_2^{-2n} + \lambda_3^{-2n} - 3) \}, \tag{2.27}$$

where a_n and b_n are constants.

If just the first term in the expansion (2.27) is maintained then

$$W = C_1(\lambda_1^2 + \lambda_2^2 + \lambda_3^2 - 3) + C_2(\lambda_1^{-2} + \lambda_2^{-2} + \lambda_3^{-2} - 3) \tag{2.28}$$

and this is the *Mooney* (or Mooney–Rivlin) form of strain-energy function; it has played an important part in theoretical work on large elastic deformations. Note that this is linear in the invariants (2.17). The special case of (2.28) corresponding to $C_2 = 0$ is written

$$W = \tfrac{1}{2}\mu(\lambda_1^2 + \lambda_2^2 + \lambda_3^2 - 3), \tag{2.29}$$

where μ is the ground-state shear modulus (recall (2.26)), and this is referred to as the *neo-Hookean* strain-energy function.

Each of (2.27)–(2.29) is of the *separable form*

$$W = w(\lambda_1) + w(\lambda_2) + w(\lambda_3) \tag{2.30}$$

proposed by Valanis and Landel (1967). An equivalent of (2.30) had previously been mentioned by Carmichael and Holdaway (1961) as a generalization of (2.27). At this point it is useful to record that I_1, I_2, I_1^2, and I_2^2 are expressible in separable form by use of (2.1) but $I_1 I_2$ is not, and neither are the terms of third and higher order in (2.23).

An equivalent of the separation (2.30) was proposed by Ogden (1972a) in the form

$$W = \sum_{n=1}^{\infty} \mu_n I(\alpha_n), \tag{2.31}$$

where

$$I(\alpha) = (\lambda_1^\alpha + \lambda_2^\alpha + \lambda_3^\alpha - 3)/\alpha \tag{2.32}$$

and μ_n, α_n are constants. R. Hill (private communication, dated 1970) showed that a single-term strain-energy function of the form (2.32) could be fitted to experimental data by use of different values of α for different ranges of values of the stretches. Equation (2.31) is a special case of (2.24). The single-term form of (2.31) corresponding to $\alpha_1 = 1$ was employed extensively by Varga (1966).

Theoretical support for the separability is obtained from the expansion of $W(\lambda_1, \lambda_2, \lambda_3)$ in a neighbourhood of the undeformed configuration for "small" strains. It was shown implicitly by Ogden (1974) that, up to the *fifth* order in $\lambda_i - 1 (i = 1, 2, 3)$, W is necessarily separable and contains four independent constants. For consistency with this expansion the strain-energy function (2.31) must contain two terms with four independent constants $\mu_1, \mu_2, \alpha_1, \alpha_2$. It is noteworthy that the Mooney strain-energy function (2.28) contains only two independent constants.

Finally, we mention a specific form of strain-energy function proposed by Blatz *et al.* (1974) and based on the invariant (2.32). This may be written

$$W = \mu_1 I(\alpha_1) + \mu_2 \{I(\alpha_1)\}^{\alpha_2}, \tag{2.33}$$

where $\alpha_1 > 0$ and $\alpha_2 > 1$, but it is not separable unless $\alpha_2 = 2$.

As we have seen, there are many possible alternative representations for a general isotropic form of strain-energy function. With hindsight, therefore, it would seem that the previous preoccupation with the invariants I_1 and I_2 in (2.17) was unduly subjective. In particular, attempts at physical interpretation of $\partial\Phi/\partial I_1$ and $\partial\Phi/\partial I_2$ should be viewed with caution since the symmetrical invariants themselves have no immediate physical interpretation in general. However, use of I_1 from (2.17) has some physical justification in that the neo-Hookean strain-energy function (2.29) has been derived on the basis of the Gaussian statistical theory of the underlying molecular network; see Treloar (1975), for example. Deviation from the Gaussian theory is assessed by departure of $\partial\Phi/\partial I_1$ from constancy. There is no corresponding justification for use of I_2.

Since the stretches $\lambda_1, \lambda_2, \lambda_3$ have themselves simple interpretations as the ratios of deformed to undeformed lengths, it is natural to use these in preference to symmetrical invariants as independent variables. They have the added advantage that the corresponding stress–deformation relations (2.5) are simpler in form and interpretation than (2.21). For these reasons we use the stretches as independent variables in this article, subject to (2.1) for incompressible materials and with some modification for compressible materials.

Nevertheless, it is recognized that in some circumstances other variables may be more suitable. In general an objective overview, encompassing all possible variables and representations of W, should be maintained, having regard to physical interpretation and ease of manipulation for the problem at hand.

2.4. *Molecular basis of the theory*

The observed elastic behaviour of rubberlike solids is a macroscopic manifestation of the underlying molecular structure in which long-chain molecules are joined together by chemical cross-links to establish a three-dimensional network. Statistical arguments may be used to transmit the elastic properties of the individual long-chain molecules into the elastic properties of the bulk material. It is not our purpose to discusss the details of the molecular statistics here but we summarize the principal ideas and their macroscopic consequences which are relevant to the main aim of this article. Full details and references are given in Treloar (1975), for example.

On the basis of Gaussian statistical theory a strain-energy function of the form

$$W = \tfrac{1}{2}NkT\,(\lambda_1^2 + \lambda_2^2 + \lambda_3^2 - 3) \tag{2.34}$$

can be derived, where T is the absolute temperature, k is the Boltzmann constant, and N is the number of chains per unit volume. The principal stretches $\lambda_1, \lambda_2, \lambda_3$ are as defined in Section 2.1 and refer to the bulk material. (A chain is defined as the segment of a molecule between successive points of cross-linkage.) With $\mu = NkT$ the strain-energy function (2.34) is identified with the neo-Hookean form (2.29) discussed in Section 2.3.

In the derivation of (2.34) a number of important simplifying assumptions were adopted and we mention here four which have interpretations relevant in the phenomenological context:

(a) It is assumed that no volume change takes place during deformation and that there is no accompanying change in internal energy.
(b) It is assumed that the total entropy is the sum of the entropies of the individual chains or, equivalently in view of (a), that the total elastic strain energy is the sum of the elastic strain energies of the individual chains.
(c) The crucial assumption, which generates the transition from the microscopic to the macroscopic level, is the *affine deformation assumption.* This states that points of cross-linkage move in such a way that the components of the end-to-end length of each chain change in the same ratio as the corresponding dimensions of the bulk rubber during deformation.
(d) The *three-chain model* assumes that a network consisting of N chains per unit volume is equivalent (mathematically) to three independent (non-interacting) sets of $\tfrac{1}{3}N$ chains per unit volume parallel to the axes of the bulk rubber.

Because of these assumptions and the consequent simplicity of (2.34) it is not surprising that the observed stress–deformation data (which will be discussed in relation to the theory in Section 2.5) depart from the predictions of the Gaussian theory (2.34) for large deformations. This discrepancy has not been explained satisfactorily by any more refined structural model, and absence of such an explanation is a serious outstanding problem as has been emphasized by Treloar (1975, 1976).

The Mooney strain-energy function (2.28) has been used extensively as a platform for the *explanation* of deviations from the Gaussian theory. In particular, the influence on the constants C_1, C_2 of a variety of structural factors has been assessed mainly on the basis of data from simple tension tests. For example, Gumbrell *et al.* (1953) found that C_1 varies widely with the degree of cross-linking but is virtually independent of the amount of swelling while C_2 is closely connected with the degree of swelling and virtually independent of cross-linking. More recent work, reviewed by Mark (1975), has shown that the picture is not nearly so clear-cut. The Mooney strain-energy function provides a good fit to simple tension data for small and moderate deformations. But for large deformations in simple tension and, more generally, in biaxial deformation, the predictions of this theory deviate markedly from observation (details will be described in Section 2.5). Consequently, any deductions made about the molecular interpretation of the C_2-term on the basis of simple tension data must be viewed with caution (Treloar, 1975, p. 99). The predictions of the Mooney theory are, nevertheless, qualitatively if not quantitatively correct except for large deformations. However, no direct derivation of the $C_2(I_2 - 3)$-term by use of a structural model has been achieved and, moreover, no molecular explanation of this term has received general acceptance (Mark, 1975).

A fundamental limitation of the Gaussian statistical theory is that it is restricted to circumstances where the distance r between the ends of the chain is very much less than the fully extended length of the chain nl, n being the number of bonds in the chain and l the bond length. The theory is therefore applicable up to only moderate deformations. A (single-chain) statistical theory which does not suffer from this restriction involves the inverse Langevin function $\mathscr{L}^{-1}$, where

$$\mathscr{L}(x) = \coth x - x^{-1}.$$

Application of the assumptions (a)–(d) in this case leads to

$$\sigma_1 - \sigma_2 = \tfrac{1}{3}NkTn^{1/2}\{\lambda_1 \mathscr{L}^{-1}(\lambda_1 n^{-1/2}) - \lambda_2 \mathscr{L}^{-1}(\lambda_2 n^{-1/2})\}, \tag{2.35}$$

and the corresponding form of strain-energy function is

$$W = \tfrac{1}{3}NkTn \sum_{i=1}^{3} \{\lambda_i n^{-1/2} \mathscr{L}^{-1}(\lambda_i n^{-1/2}) - n^{-1/2} \mathscr{L}^{-1}(n^{-1/2}) + \ln[\mathscr{L}^{-1}(\lambda_i n^{-1/2}) \sinh \mathscr{L}^{-1}(n^{-1/2}) / \mathscr{L}^{-1}(n^{-1/2}) \sinh \mathscr{L}^{-1}(\lambda_i n^{-1/2})]\}. \tag{2.36}$$

In the derivation of (2.36) it is assumed that the chains have mean-square end-to-end distance in the undeformed configuration equal to the root-mean-square value for a free chain, that is, $ln^{1/2}$ (see Treloar, 1975, for details). This theory provides a very much better fit to simple tension data than the Gaussian theory for large deformations.

The first term in the series expansion of (2.36) for "small" r/nl yields the Gaussian approximation (2.34), which is independent of n, while in general we note that (2.36) involves essentially two independent material constants, namely $NkT\,(=\mu)$ and n. Clearly (2.36) is of the separable form (2.30) and exemplifies the close relationship between the Valanis–Landel hypothesis and the three-chain model. This raises the question of whether a statistical theory can be developed that accounts for the general separability (2.30) or, indeed, for a general form of isotropic elastic strain-energy function.

According to Treloar (1975, p. 245) one cannot expect the Valanis–Landel hypothesis to hold for very large strains. With this in mind Treloar and Riding (1980), employing inverse Langevin statistics, have taken chain orientation into account by rejecting the three-chain model while retaining the affine deformation assumption. This leads to a strain-energy function

$$W = \frac{NkTn}{4\pi} \int_\Omega \{\lambda n^{-1/2} \mathscr{L}^{-1}(\lambda n^{-1/2}) - n^{-1/2} \mathscr{L}^{-1}(n^{-1/2}) + \ln[\mathscr{L}^{-1}(\lambda n^{-1/2}) \sinh \mathscr{L}^{-1}(n^{-1/2}) / \mathscr{L}^{-1}(n^{-1/2}) \sinh \mathscr{L}^{-1}(\lambda n^{-1/2})]\} \sin\theta \, d\theta \, d\phi, \tag{2.37}$$

where

$$\lambda^2 = \lambda_1^2 \cos^2\theta + \lambda_2^2 \sin^2\theta \cos^2\phi + \lambda_3^2 \sin^2\theta \sin^2\phi$$

and integration is over the unit sphere Ω.

The corresponding stress–deformation relations are obtained from (2.5) with

$$\left.\begin{aligned} \lambda_1 \frac{\partial \hat{W}}{\partial \lambda_1} &= \frac{NkTn^{1/2}}{4\pi} \int_\Omega \lambda^{-1}(\lambda_1^2 \cos^2\theta - \lambda_3^2 \sin^2\theta \sin^2\phi) \mathscr{L}^{-1}(\lambda n^{-1/2}) \sin\theta \, d\theta \, d\phi, \\ \lambda_2 \frac{\partial \hat{W}}{\partial \lambda_2} &= \frac{NkTn^{1/2}}{4\pi} \int_\Omega \lambda^{-1}(\lambda_2^2 \cos^2\phi - \lambda_3^2 \sin^2\phi) \mathscr{L}^{-1}(\lambda n^{-1/2}) \sin^3\theta \, d\theta \, d\phi. \end{aligned}\right\} \tag{2.38}$$

The series expansion of (2.37) for "small" r/nl is of separable form up to only one term beyond the Gaussian approximation. Treloar and Riding (1980) have applied (2.38) to biaxial deformations and made numerical calculations which indicate a good agreement between theory and experiment, the departure from separability being noticeable only for very large biaxial deformations.

2.5. *Experimental results and their relation to the theory*

In the representation of experimental data many authors, following Rivlin and Saunders (1951), base their comparison with theory on the use of $\partial\Phi/\partial I_1$ and $\partial\Phi/\partial I_2$ from (2.20) with I_1 and I_2 given by (2.17). The usual practice is to plot values of $\partial\Phi/\partial I_1$ and $\partial\Phi/\partial I_2$ against I_1 or I_2 at fixed I_2 or I_1 calculated from measured values of $\lambda_1, \lambda_2, \sigma_1/\lambda_1$ and σ_2/λ_2 with $\sigma_3 = 0$. More generally, use of (2.18) leads to

$$\left.\begin{aligned} \frac{\partial\Phi}{\partial I_1} &= \frac{(\sigma_1-\sigma_3)(\lambda_2^{\alpha_2}-\lambda_3^{\alpha_2})-(\sigma_2-\sigma_3)(\lambda_1^{\alpha_2}-\lambda_3^{\alpha_2})}{\alpha_1[(\lambda_1^{\alpha_1}-\lambda_3^{\alpha_1})(\lambda_2^{\alpha_2}-\lambda_3^{\alpha_2})-(\lambda_2^{\alpha_1}-\lambda_3^{\alpha_1})(\lambda_1^{\alpha_2}-\lambda_3^{\alpha_2})]}, \\ \frac{\partial\Phi}{\partial I_2} &= \frac{(\sigma_1-\sigma_3)(\lambda_2^{\alpha_1}-\lambda_3^{\alpha_1})-(\sigma_2-\sigma_3)(\lambda_1^{\alpha_1}-\lambda_3^{\alpha_1})}{\alpha_2[(\lambda_1^{\alpha_2}-\lambda_3^{\alpha_2})(\lambda_2^{\alpha_1}-\lambda_3^{\alpha_1})-(\lambda_2^{\alpha_2}-\lambda_3^{\alpha_2})(\lambda_1^{\alpha_1}-\lambda_2^{\alpha_1})]}. \end{aligned}\right\} \tag{2.39}$$

This method has two main disadvantages. Firstly, for small strains experimental errors are compounded by the smallness of the terms in the denominators of (2.39) and $\partial\Phi/\partial I_1, \partial\Phi/\partial I_2$ are therefore not determined accurately, as Rivlin and Saunders themselves pointed out. Secondly, as mentioned in Section 2.3, $\partial\Phi/\partial I_1$ and $\partial\Phi/\partial I_2$ have no immediate physical interpretation although $\partial\Phi/\partial I_1$ (with I_1 given by (2.17)) is a measure of the departure of the data from the predictions of the Gaussian statistical theory.

In this paper we adopt a more natural approach and use the fundamental variables $\lambda_1, \lambda_2, \sigma_1, \sigma_2$ themselves with the stress–deformation relations (2.5), thereby circumventing both disadvantages mentioned above. In particular in the graphical representation of biaxial data we plot $\sigma_1 - \sigma_2$ against λ_1 at fixed values of λ_2 (the roles of λ_1 and λ_2 may be interchanged in this exercise). The predictions of the neo-Hookean and Mooney theories, (2.29) and (2.28), are also plotted on the same basis for comparison, primarily to illustrate the departure of the Gaussian theory from observed behaviour. First, however, we review briefly the more important experimental contributions.

The first systematic series of experiments on rubberlike solids was carried out by Treloar (1944). These experiments involved single-parameter tests, namely simple tension, pure shear, and equibiaxial tension, and were aimed at determining the extent of validity of the Gaussian statistical theory. The data from Treloar's experiments have been used subsequently by a number of writers as the basis for comparison of theory and experiment. In particular, Ogden (1972a), using a strain-energy function of the form (2.31) with corresponding specializations of (2.7)–(2.9), obtained very good agreement with Treloar's data (for λ up to about 7.0 in simple tension). For future reference we record that this strain-energy function has just three terms with values of the material constants given by

$$\left.\begin{aligned} \alpha_1 &= 1.3, & \alpha_2 &= 5.0, & \alpha_3 &= -2.0, \\ \mu_1^* &= 1.491, & \mu_2^* &= 0.003, & \mu_3^* &= -0.0237, \end{aligned}\right\} \tag{2.40}$$

where $\mu_n^* = \mu_n/\mu$. Justification for use of such forms of strain-energy function is provided on the basis of general biaxial data which are examined shortly.

In Figs. 2.1 (a), (b), and (c) respectively, we show Treloar's (1944) data on simple tension, pure shear, and equibiaxial tension with $f = \sigma/\lambda$ plotted against λ. The theoretical curves for the neo-Hookean strain-energy function and for (2.40) calculated on the basis of (2.7)–(2.9) are included for comparison, the units for f being kg cm^{-2}.

Treloar (1948), using a plane sheet of rubber with forces applied perpendicular to its edges and parallel to the plane, was the first to obtain biaxial data. However, the results were rather limited in the sense that the apparatus did not allow independent control of λ_1 and λ_2. For example, λ_2 could not be held fixed while λ_1 was varied.

An improved apparatus enabled Rivlin and Saunders (1951) to carry out experiments in which either I_1 or I_2 was held constant, thus determining values of $\partial\Phi/\partial I_1$ and $\partial\Phi/\partial I_2$. Rivlin and Saunders also obtained data for simple tension, pure shear, and equibiaxial tension. Their biaxial data, however, do not give a series of values of λ_1 (or λ_2) at each fixed value of λ_2 (or λ_1) and are therefore unsuitable for our present purposes, although their apparatus was capable of providing data in this form. The next systematic series of experiments was carried out by Becker (1967) who also represented data in terms of $\partial\Phi/\partial I_1$ and $\partial\Phi/\partial I_2$ and obtained results broadly similar to those of Rivlin and Saunders. He provided detailed results for the

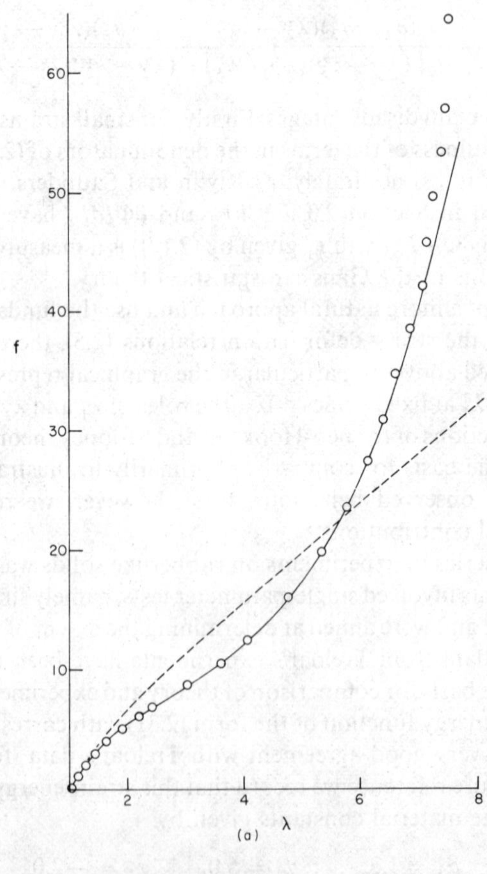

FIG. 2.1. (a) Nominal stress f plotted against the extension ratio λ for the neo-Hookean (-----) strain-energy function and the three-term strain-energy function (2.40) (——) compared with Treloar's (1944) data (○) in simple tension. The units of f are kg cm^{-2}.

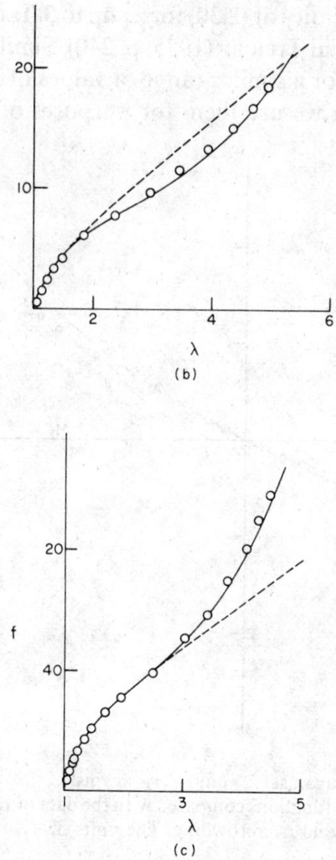

FIG. 2.1. (b) and (c) Theoretical curves and Treloar's (1944) data plotted as in (a), but in (b) pure shear, (c) equibiaxial tension. The units of f are kg cm^{-2}.

region of small deformations and concluded that the behaviour of $\partial\Phi/\partial I_1$ and $\partial\Phi/\partial I_2$ as functions of I_1 and I_2 is very complicated in this region. Similar complicated behaviour was reported by Kawabata and Kawai (1977).

It is not difficult to fit data from a single-parameter test, such as simple tension, and many different forms of W can be found in the literature which do this effectively. In general, however, it is more difficult to fit data from several different single-parameter tests and success in fitting data on simple tension, pure shear, and equibiaxial tension for a specific rubber is not common. The relevant literature has been reviewed by Ogden (1972a) who, as mentioned above, was notably successful in fitting Treloar's (1944) data. The fact that the strain-energy function used by Ogden is of the separable form (2.30) is the key to its success. However, it is not possible to assert the validity of the separability (i.e. of the Valanis–Landel hypothesis) (2.30) on the basis of single-parameter tests. General biaxial deformation data are required to assess the extent of the validity of (2.30).

Experiments specifically designed to test the Valanis–Landel hypothesis were performed by Obata *et al.* (1970). They considered general biaxial deformation of a rubber sheet in which one stretch was held fixed (at a series of different values) while the other stretch was varied, and they demonstrated the validity of (2.30) for λ_1 up to 3.0 and λ_2 up to about 2.0 jointly. The essential details are discussed in Treloar (1975, p. 240). Similar experiments were carried out by Jones and Treloar (1975) for a similar range of values of λ_1 and λ_2. Since their results are representative of biaxial data we use them for purposes of illustration.

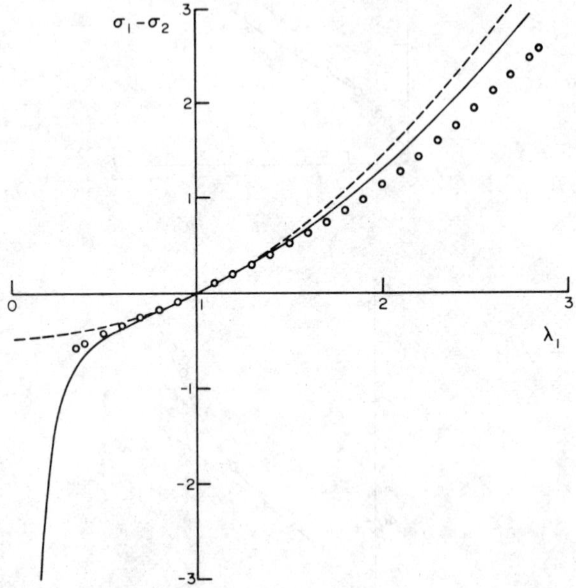

FIG. 2.2. (a)–(e) Plots of $\sigma_1 - \sigma_2$ against λ_1 (and $-\sigma_2$ against $\lambda_3; \sigma_3 = 0$) for the neo-Hookean (----) and Mooney–Rivlin (——) strain-energy functions compared with the data of Jones and Treloar (1975), (O). Some points have been omitted to avoid overcrowding. The units of $\sigma_1 - \sigma_2$ are N mm^{-2}. (a) $\lambda_2 = 1$.

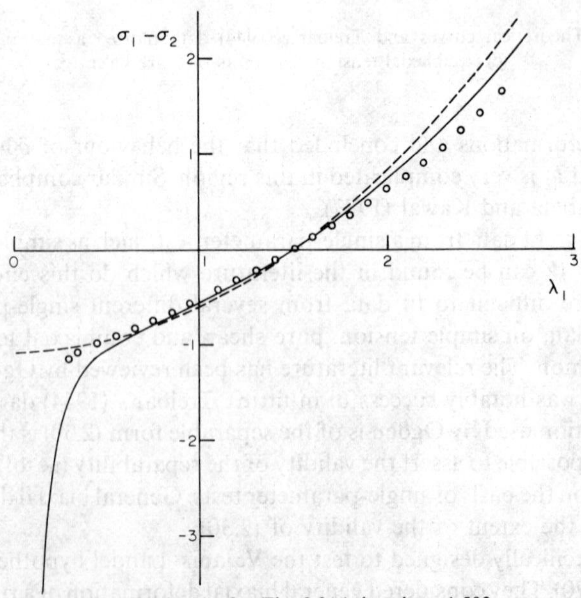

FIG. 2.2(b). As for Fig. 2.2(a), but $\lambda_2 = 1.502$.

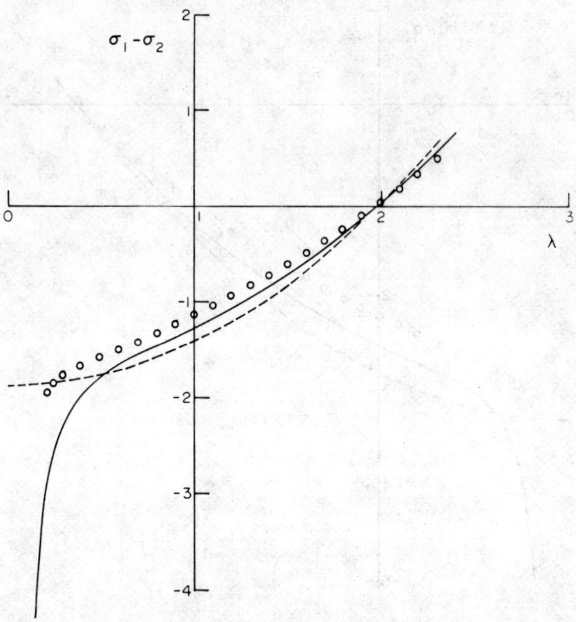

FIG. 2.2(c). As for Fig. 2.2(a), but $\lambda_2 = 1.984$.

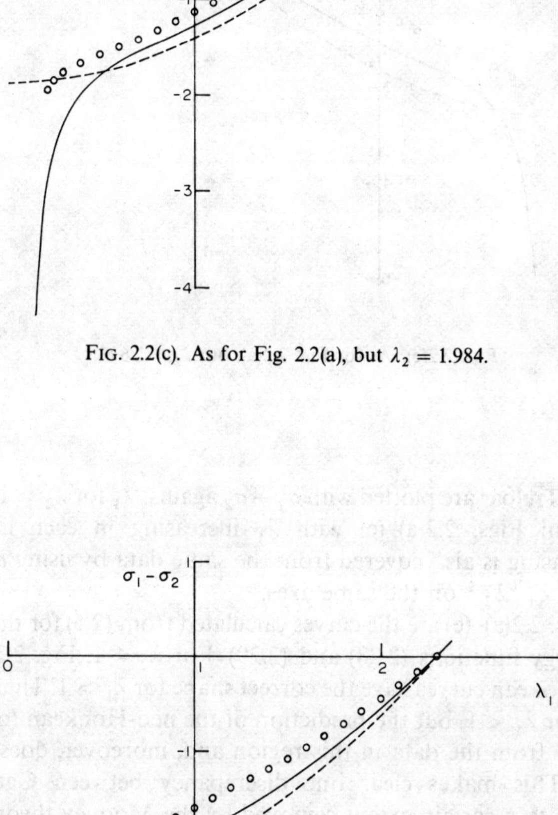

FIG. 2.2(d). As for Fig. 2.2(a), but $\lambda_2 = 2.295$.

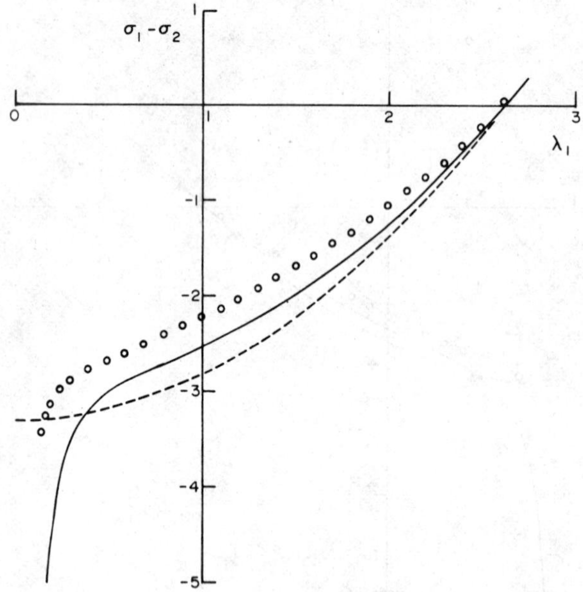

FIG. 2.2(e). As for Fig. 2.2(a), but $\lambda_2 = 2.623$.

Data of Jones and Treloar are plotted with $\sigma_1 - \sigma_2$ against λ_1 for $\lambda_2 = 1, 1.502, 1.984, 2.295, 2.623$, respectively, in Figs. 2.2(a)–(e) with λ_1 increasing in each case. The range of values with λ_1 decreasing is also covered from the same data by using $\sigma_3 - \sigma_2$ with $\sigma_3 = 0$ plotted against $\lambda_3 = \lambda_1^{-1}\lambda_2^{-1}$ on the same axes.

Also shown in Figs. 2.2(a)–(e) are the curves calculated from (2.5) for the Mooney and neo-Hookean strain-energy functions (2.28) and (2.29). For $\lambda_2 = 1$, Fig. 2.2(a) shows that the Mooney and neo-Hookean curves have the correct shape for $\lambda_1 > 1$. The same can be said of the Mooney curve for $\lambda_1 < 1$, but the prediction of the neo-Hookean (or Gaussian) theory diverges significantly from the data in this region and, moreover, does not conform with physical intuition. This makes clear one discrepancy between Gaussian theory and observation which is to a certain extent removed by the Mooney theory.

As Figs. 2.2(b)–(e) illustrate, the *shapes* of the Mooney and neo-Hookean curves are unaffected by a change in λ_2. Such a change merely shifts the curves vertically, and this is a consequence of the Mooney and neo-Hookean strain-energy functions being of the separable form (2.30). From (2.5) and (2.30) we have

$$\sigma_1 - \sigma_2 = \lambda_1 w'(\lambda_1) - w'(1) + [w'(1) - \lambda_2 w'(\lambda_2)]. \tag{2.41}$$

The shape of the curves is determined by the relation

$$\sigma_1 - \sigma_2 = \lambda_1 w'(\lambda_1) - w'(1) \tag{2.42}$$

which corresponds to $\lambda_2 = 1$ (pure shear), while the "vertical shift" for $\lambda_2 \neq 1$ is given by

$$w'(1) - \lambda_2 w'(\lambda_2). \tag{2.43}$$

The experimental data shown in Figs. 2.2(b)–(e) are similarly translated by a change in λ_2, and this is seen more clearly in Fig. 2.3. Superposition of the data onto the data curve for $\lambda_2 = 1$ by vertical transposition shows that the shapes of the data curves for different values of λ_2 are indistinguishable. This has been demonstrated by Jones and Treloar (1975) and we do not reproduce such a superposition here. The conclusion to be drawn from this is that biaxial data for the vulcanized rubber tested by Jones and Treloar can be represented by a separable form of strain-energy function (2.30), at least for the range of deformations examined. In other words these results confirm the findings of Obata *et al.* (1970) and provide further evidence in favour of the Valanis–Landel hypothesis.

As mentioned above, the Mooney and neo-Hookean strain-energy functions provide a reasonable representation of the data for moderate values of λ_1 when $\lambda_2 = 1$. But when $\lambda_2 > 1$ the representation is not good and deteriorates rapidly as λ_2 increases, as Figs. 2.2(b)–(e) demonstrate. This is because the *shift factor* calculated from (2.43) deviates from experimentally determined values. It is noticeable, however, that the Mooney theory provides a better correspondence with the data than the neo-Hookean theory.

Clearly, any separable form of strain-energy function must account for the shape of the data curve (using (2.42)) and the shift (using (2.43)). Jones and Treloar have fitted their data by a three-term strain-energy function of the form (2.31) with constants

$$\left.\begin{aligned} \alpha_1 &= 1.3, & \alpha_2 &= 4.0, & \alpha_3 &= -2.0, \\ \mu_1 &= 0.69, & \mu_2 &= 0.01, & \mu_3 &= -0.0122\ \mathrm{N\,mm^{-2}}. \end{aligned}\right\} \tag{2.44}$$

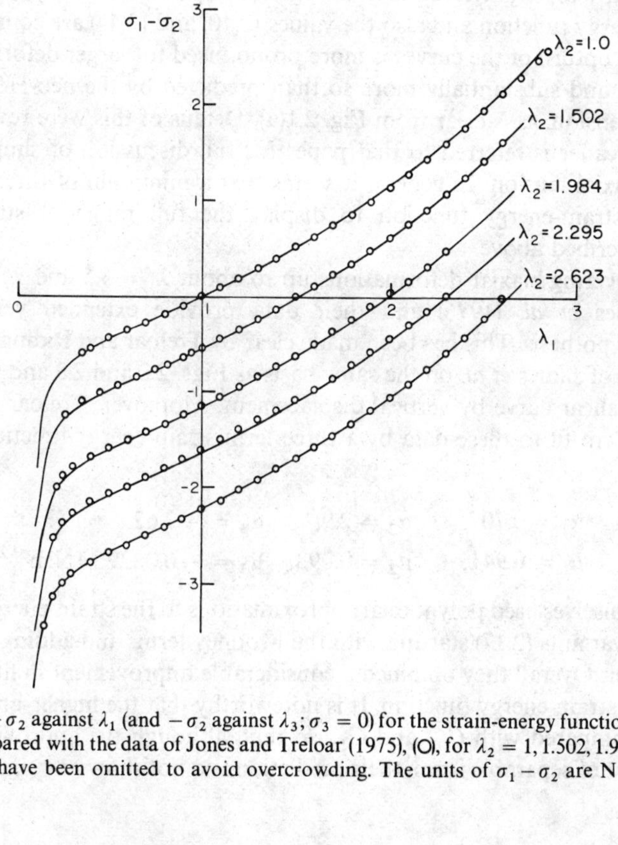

FIG. 2.3. Plot of $\sigma_1 - \sigma_2$ against λ_1 (and $-\sigma_2$ against λ_3; $\sigma_3 = 0$) for the strain-energy function (2.31) with material constants (2.44) compared with the data of Jones and Treloar (1975), (O), for $\lambda_2 = 1, 1.502, 1.984, 2.295, 2.623$. Some points have been omitted to avoid overcrowding. The units of $\sigma_1 - \sigma_2$ are N mm^{-2}.

Note the similarity between (2.40) and (2.44).

In Fig. 2.3 we replot the data of Jones and Treloar for all five values of λ_2 together with the theoretical curves calculated for (2.44). The agreement is excellent for both shape and shift. Apart from the vertical shift there are some features of the curves shown in Fig. 2.3 which require further comment. Without loss of generality, discussion of the shape can be based on $\lambda_2 = 1$. For definiteness we use the class of strain-energy functions (2.31) for which

$$\left.\begin{aligned} \sigma_1 - \sigma_2 &= \sum_{n=1}^{N} \mu_n(\lambda_1^{\alpha_n} - 1) \quad (\lambda_1 \geqslant 1,\ \lambda_2 = 1), \\ -\sigma_2 &= \sum_{n=1}^{N} \mu_n(\lambda_3^{\alpha_n} - 1) \quad (\lambda_3 \leqslant 1,\ \lambda_2 = 1), \end{aligned}\right\} \tag{2.45}$$

when only a finite number N of terms is retained in the summation. The corresponding shift factor is

$$\sum_{n=1}^{N} \mu_n(1 - \lambda_2^{\alpha_n}). \tag{2.46}$$

For small and moderate deformations it is easy to fit the data with $N = 1$ and a range of different values of α_1 (> 0) and μ_1 (> 0), as exemplified by the neo-Hookean strain-energy function ($\alpha_1 = 2$), the Varga strain-energy function ($\alpha_1 = 1$), and the strain-energy function ($\alpha_1 = 1.2$) used by Ogden (1972a). At larger values of λ_1 the upturn in the curves can only be accounted for by a positive exponent (α_2, say) with $\mu_2 > 0$ correspondingly. On the other hand, the downturn for $\lambda_3 < 1$ requires a negative exponent (α_3, say) with $\mu_3 < 0$. The Mooney strain-energy function and also the values (2.40) and (2.44) are consistent with these requirements. The upturn of the curves is more pronounced for larger deformations than are shown in Fig. 2.3, and substantially more so than predicted by the neo-Hookean exponent $\alpha_1 = 2$ in simple tension, as is clear from Fig. 2.1(a). Details of this were reviewed by Ogden (1972a), and the reader is referred to that paper for full discussion of simple tension, pure shear, and equibiaxial tension. In general it seems that a minimum of three (μ_n, α_n)-terms is required in the strain-energy function to display the full range of stress–deformation characteristics described above.

Experiments covering biaxial deformations up to about $\lambda_1 = 3.5$ and $\lambda_2 = 3.5$ have been reported by James *et al.* (1975) and their data provide extended verification of the Valanis–Landel hypothesis. This has been made clear by Treloar and Riding (1980) who have replotted the data of James *et al.* on the same basis as Figs. 2.2 and 2.3 and superposed these data on the pure shear curve by vertical displacement. Moreover, Treloar and Riding have obtained an excellent fit to these data by a three-term strain-energy function of form (2.31) with constants

$$\left.\begin{aligned} \alpha_1 &= 0.707, & \alpha_2 &= 2.9, & \alpha_3 &= -2.62, \\ \mu_1 &= 0.941, & \mu_2 &= 0.093, & \mu_3 &= -0.0029\ \mathrm{MN\,m^{-2}}. \end{aligned}\right\} \tag{2.47}$$

James *et al.* themselves used polynomial approximations to the strain-energy function (2.23) in terms of the invariants (2.17) starting with the Mooney terms and adding further terms up to the fourth order. Overall they obtained a considerable improvement in fit to that achieved with the Mooney strain-energy function. It is noteworthy that the higher-order constants are relatively small compared with C_{01} and C_{10} so that, although the approximations used by James *et al.* are not of separable form, their predictions do not show any significant departure from the Valanis–Landel hypothesis. Similar approximations have been derived recently by Haines and Wilson (1979) from simple tension, pure shear, and equibiaxial tension data, and these have been shown to give excellent agreement with the biaxial data of Jones and Treloar (1975).

Further confirmation of the Valanis–Landel hypothesis is provided in the review by Kawabata and Kawai (1977) for vulcanized natural rubber. Interestingly, although they do not give details, they indicate that the range of validity of the hypothesis may not be so extensive for other types of rubber.

It was mentioned in Section 2.2 that the deformation in the skin of a circular cylindrical membrane tube can be regarded as a general biaxial deformation. Experiments involving the inflation and extension of tubes with closed ends have been carried out by Alexander (1971a) and Charrier (1974). In these experiments the internal pressure and the axial loading were controlled separately to maintain the circular cylindrical shape. On the other hand, Vangerko and Treloar (1978), following Boonstra (1950), were able to fix the radius of the tube at selected values and extend the tube at each fixed radius by adjusting the internal pressure in order to maintain the shape. This latter method provides biaxial data in a form suitable for our present purposes, and Vangerko and Treloar achieved much larger biaxial deformations than did previous experimenters.

For two different rubbers Vangerko and Treloar's results again confirm the Valanis–Landel hypothesis for values of λ_1 and λ_2 up to about 3.0 and down to values of λ_3 lower than those achieved by Jones and Treloar (1975). But for λ_1 and λ_2 greater than about 3.0 Vangerko and Treloar observed some departure from the hypothesis.

In Fig. 2.4 (a) we have plotted Vangerko and Treloar's data for their 3% sulphur rubber compound on the same basis as used in Fig. 2.3. However, we have restricted attention to $\lambda_1 \geqslant 1$ since no departure from the Valanis–Landel hypothesis was observed in the region $\lambda_1 < 1$. In Fig. 2.4 (b) we have superposed these data on the pure shear curve by vertical displacement and this makes clear both the extent of validity of the Valanis–Landel

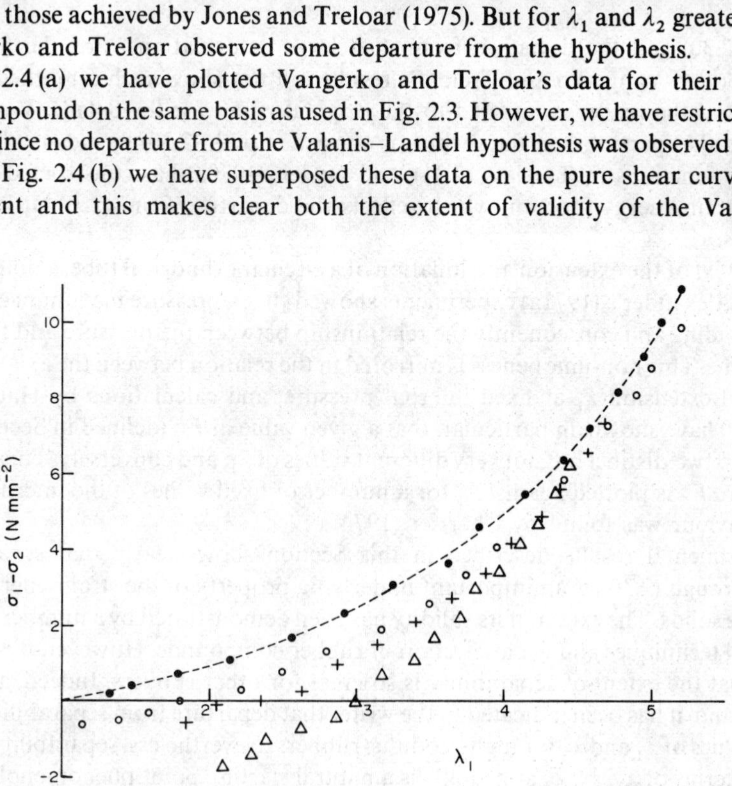

FIG. 2.4(a). Plot of $\sigma_1 - \sigma_2$ against λ_1 ($\geqslant 1$) in respect of the data of Vangerko and Treloar (1978) (3% sulphur rubber compound): $\lambda_2 = 1.02$ (●), 1.98 (○), 2.42 (+), 3.38 (Δ). The pure shear curve in represented by (-----).

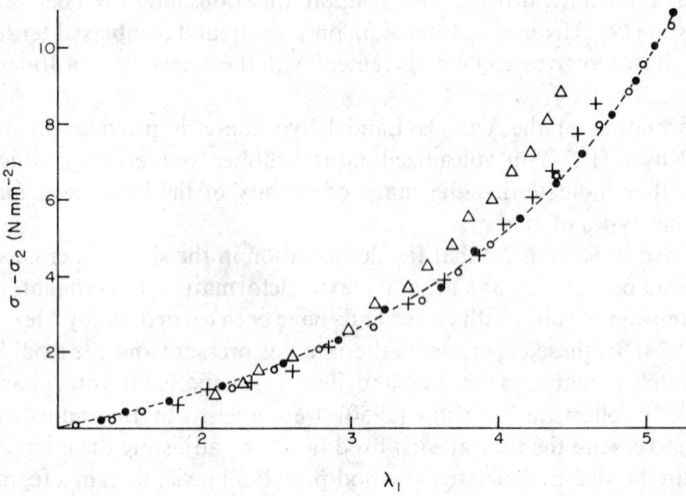

FIG. 2.4(b). Plot of $\sigma_1 - \sigma_2$ against $\lambda_1 (\geq 1)$ in respect of the data of Vangerko and Treloar (1978) (3% sulphur rubber compound): superposition of Fig. 2.4(a) by vertical shift onto the pure shear curve ($\lambda_2 = 1$). The pure shear curve is represented by (-----).

hypothesis (2.30) and the deviation from it at large deformations. For the 5% sulphur rubber compound Vangerko and Treloar's results did not show such a marked departure from (2.30) and the reader is referred to the original paper for the details. Vangerko and Treloar fitted that part of their data which is consistent with (2.30) by means of three-term strain-energy functions in the class (2.31), and Ogden (1979a) has constructed a modified three-term strain-energy function which reflects the departure from (2.30) at large biaxial deformations.

In the context of the extension and inflation of a circular cylindrical tube, a note of caution is called for. Alexander's (1971a) experiments showed that a pressure maximum can occur at fixed axial loading and consequently the relationship between the pressure and the radius is not one-to-one. This non-uniqueness is mirrored in the relation between the axial loading F^* and the axial extension λ_1 at fixed internal pressure, and calculations by Haughton and Ogden (1979) have shown, in particular, that a given value of F^* (defined in Section 2.2) can correspond to two distinct but not very different values of λ_1, and conversely. This is shown in Fig. 2.5 where F^* is plotted against λ_1 for a number of fixed values of the internal pressure. Similar behaviour was found by Charrier (1974).

The experimental results described in this Section show clearly that separability, as expressed through (2.30), is an important underlying property of the strain-energy function of rubberlike solids. The extent of its validity has been demonstrated by a number of different experimental techniques and for a selection of rubber compounds. However, it is premature to deduce that the extent of separability is so great for other rubbers. Indeed, although no data are to hand, it has been indicated to the writer that departure from separability is evident at smaller values of λ_1 and λ_2 for high-modulus rubbers. Nevertheless, separability, expressed concisely in terms of λ_1, λ_2, λ_3 and (2.30), is a natural starting point phenomenologically for the construction of strain-energy functions for rubberlike solids. At this point it is important to emphasize that separability is not in general expressible in simple closed algebraic form in

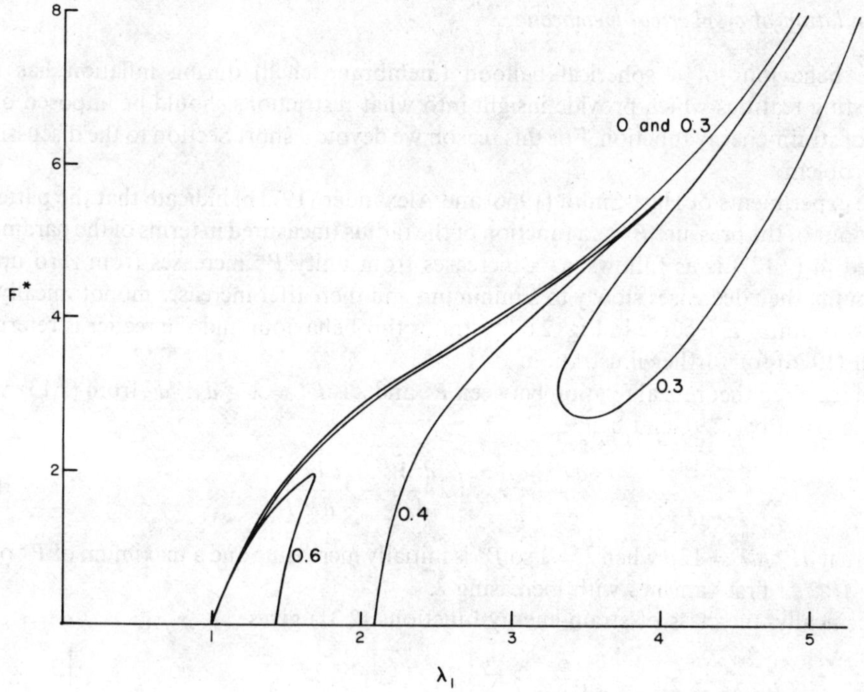

FIG. 2.5. Plot of the axial load F^* against the axial stretch λ_1 for an internally pressurized circular cylindrical tube with closed ends at fixed values $P^* = 0, 0.3, 0.4, 0.6$ of the internal pressure, calculated in respect of the three-term strain-energy function (2.40).

terms of any pair of symmetrical invariants I_1, I_2 as is clear from the work of Rivlin and Sawyers (1976). Indeed, the simple structure of (2.30) is abrogated by use of I_1 and I_2. However, as evidenced by the work of James *et al.* (1975) and Haines and Wilson (1979) forms of the W polynomial in I_1, I_2 can be fitted quite successfully to separable data if the non-separable terms give relatively small contributions to $\sigma_1 - \sigma_2$ compared with the Mooney terms except at very large deformations.

Separability, once established, provides justification for determining $w(\lambda)$ from data in single-parameter tests. This was done, without prior justification, by Ogden (1972a) and this approach has also been adopted implicitly by Chadwick *et al.* (1977) in respect of Treloar's (1944) data. With the Gaussian statistical theory in mind they used strain-energy functions of the form (2.31) with two or three terms and the leading exponent $\alpha_1 = 2$. Their material constants are consistent with the requirements laid down following (2.46).

Finally in this Section we emphasize the need for further detailed experimental work on biaxial deformations for a wide variety of rubberlike solids in order to obtain full information on the extent of separability. The observed separability is a reflection of the network structure at the microscopic level, and molecular statistical theories should be able to account for and explain separability. There is clearly a need for an improved structural theory which reflects not only the separability but also the departure from separability illustrated by Vangerko and Treloar (1978). The work of Treloar and Riding (1980) has provided a step in this direction.

2.6. *Inflation of a spherical membrane*

The behaviour of a spherical balloon (membrane shell) during inflation has some interesting features which provide insight into what restrictions should be imposed on the form of strain-energy function. For this reason we devote a short Section to the discussion of this problem.

The experiments of Hart-Smith (1966) and Alexander (1971b) indicate that the pattern of behaviour of the pressure P^* as a function of the radius (measured in terms of the parameter λ defined in (2.12)) is as follows: as λ increases from unity P^* increases from zero up to a maximum, then decreases slowly to a minimum and thereafter increases monotonically with λ up to rupture. The curve in Fig. 2.6 illustrates this behaviour and the reader is referred to Sewell (1976) for further illustration.

The required theoretical relation between P^* and λ is $P^* = \lambda^{-2}\, d\tilde{W}/d\lambda$ from (2.13), where $\tilde{W}(\lambda)$ is given by (2.9), and hence

$$\frac{dP^*}{d\lambda} = \lambda^{-3}\left(\lambda\frac{d^2\tilde{W}}{d\lambda^2} - 2\frac{d\tilde{W}}{d\lambda}\right). \tag{2.48}$$

Note that $dP^*/d\lambda = 12\mu$ when $\lambda = 1$ so P^* is initially increasing and a maximum of P^* occurs when $dP^*/d\lambda$ first vanishes with increasing λ.

Specifically, the class of strain-energy functions (2.31) gives

$$\left.\begin{aligned} P^* &= 2\sum_{n=1}^{N} \mu_n(\lambda^{\alpha_n-3} - \lambda^{-2\alpha_n-3}), \\ \frac{dP^*}{d\lambda} &= 2\sum_{n=1}^{N} \mu_n[(\alpha_n-3)\lambda^{\alpha_n-4} + (2\alpha_n+3)\lambda^{-2\alpha_n-4}]. \end{aligned}\right\} \tag{2.49}$$

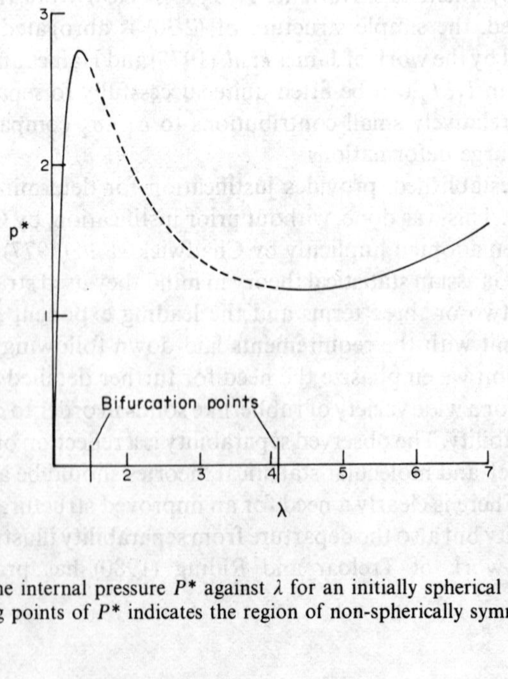

FIG. 2.6. Typical plot of the internal pressure P^* against λ for an initially spherical balloon. The broken curve between the turning points of P^* indicates the region of non-spherically symmetric deformation.

Clearly, for the existence of a maximum of P^* it is necessary that

$$-\tfrac{3}{2} < \alpha_n < 3 \quad (\text{some } n). \tag{2.50}$$

The neo-Hookean and Mooney strain-energy functions are consistent with this requirement, as are the three-term strain-energy functions with constants given by (2.40), (2.44), and (2.47). However, if (2.50) is satisfied for each n then P^* ultimately decreases monotonically as λ increases, and this is inconsistent with observation. Therefore, in order to reproduce theoretically the features shown in Fig. 2.6 it is necessary that $N \geqslant 2$ with one value of α_n satisfying (2.50) and another outside this range of values. This requirement rules out the neo-Hookean strain-energy function but not (2.28), (2.40), (2.44), or (2.47).

Thus far, the discussion has been based on the assumption that the spherical shape is maintained during inflation. In practice, however, this is not the case, and experiment shows that at some stage shortly after the maximum pressure the spherical symmetry is lost but is regained again near the pressure minimum and thereafter retained (Alexander, 1971b). This behaviour has been predicted theoretically by a number of authors at various levels of generality and on the assumption that the material is homogeneous. Recently Needleman (1977) has found points of bifurcation (i.e. values of λ at which departure from the spherical shape is possible) to occur between the maximum and minimum pressures in respect of certain strain-energy functions in the class (2.31). In Fig. 2.6 the broken curve indicates the non-spherical part of the deformation path. A more general analysis by Haughton and Ogden (1978) has confirmed Needleman's results and shown further that bifurcation occurs when the *surface tension* reaches a maximum, that is when

$$\lambda\frac{d^2\tilde{W}}{d\lambda^2} - \frac{d\tilde{W}}{d\lambda} = 0.$$

In respect of (2.31) a necessary condition for bifurcation to occur is

$$-1 < \alpha_n < 2 \quad (\text{some } n).$$

This rules out the Mooney strain-energy function, but not (2.40), (2.44), or (2.47).

The calculations referred to above assume that the balloon is perfectly homogeneous. However, similar conclusions are reached for balloons of non-uniform thickness, as shown by Haughton (1980). See also Needleman (1977). The incompatibility of the neo-Hookean and Mooney strain-energy functions with the observed behaviour of rubberlike balloons is consistent with the results described in Section 2.5.

3. Compressible Rubberlike Solids

3.1. *Stress–deformation relations*

For compressible materials the constraint (2.1) no longer holds in general, and we use the notation

$$J \equiv \lambda_1\lambda_2\lambda_3 \tag{3.1}$$

for the ratio of deformed to undeformed volume of the homogeneous and homogeneously deformed material. The strain-energy function is again denoted by $W(\lambda_1, \lambda_2, \lambda_3)$, but this is now per unit *initial* (undeformed) volume. The symmetries (2.2) remain intact.

The stress–deformation relations (2.3) are replaced by

$$J\sigma_i = \lambda_i \frac{\partial W}{\partial \lambda_i} \quad (i = 1, 2, 3), \tag{3.2}$$

where $\sigma_1, \sigma_2, \sigma_3$ are the principal components of Cauchy stress, as in Section 2. The corresponding nominal (or engineering) principal stresses f_1, f_2, f_3 are given by

$$f_i = \frac{\partial W}{\partial \lambda_i} \quad (i = 1, 2, 3). \tag{3.3}$$

Close contact with the incompressible theory is made if the deformation is separated into its dilatational and distortional parts by introduction of the *modified stretches* $\lambda_1^*, \lambda_2^*, \lambda_3^*$ defined by

$$\lambda_i^* = \lambda_i J^{-1/3} \tag{3.4}$$

following Flory (1961) and Ogden (1976). Then

$$\lambda_1^* \lambda_2^* \lambda_3^* = 1 \tag{3.5}$$

and, together with J, any pair of modified stretches may be used as independent variables. Note that $J = 1$ for an *isochoric* (volume preserving) or pure distortional deformation, while $\lambda_1^* = \lambda_2^* = \lambda_3^* = 1$ if the deformation is a pure dilatation.

Following Ogden (1976), we write

$$W(\lambda_1, \lambda_2, \lambda_3) \equiv W^*(\lambda_1^*, \lambda_2^*, \lambda_3^*, J) \tag{3.6}$$

per unit undeformed volume, subject to (3.5). The symmetry (2.2) of W with respect to $\lambda_1, \lambda_2, \lambda_3$ is reflected in that of W^* with respect to $\lambda_1^*, \lambda_2^*, \lambda_3^*$.

From (3.2) we obtain

$$J\sigma_i = \lambda_i^* \frac{\partial W^*}{\partial \lambda_i^*} - \left(\tfrac{1}{3} \sum_{j=1}^{3} \lambda_j^* \frac{\partial W^*}{\partial \lambda_j^*} - J \frac{\partial W^*}{\partial J} \right) \quad (i = 1, 2, 3), \tag{3.7}$$

and this may be compared with the corresponding equation from the incompressible theory, equation (2.3), to which it reduces in the limit $J \equiv 1$ (Ogden, 1976). It follows immediately from (3.7) that

$$\tfrac{1}{3}(\sigma_1 + \sigma_2 + \sigma_3) = \frac{\partial W^*}{\partial J}, \tag{3.8}$$

this being equivalent to a relationship first given by Penn (1970). Also,

$$J(\sigma_i - \sigma_j) = \lambda_i \frac{\partial W}{\partial \lambda_i} - \lambda_j \frac{\partial W}{\partial \lambda_j} = \lambda_i^* \frac{\partial W^*}{\partial \lambda_i^*} - \lambda_j^* \frac{\partial W^*}{\partial \lambda_j^*} \tag{3.9}$$

for each pair i, j.

By analogy with the incompressible theory, we write

$$\hat{W}^*(\lambda_1^*, \lambda_2^*, J) \equiv W^*(\lambda_1^*, \lambda_2^*, \lambda_1^{*-1} \lambda_2^{*-1}, J). \tag{3.10}$$

Then, from (3.8) and (3.9),

$$\tfrac{1}{3}(\sigma_1 + \sigma_2 + \sigma_3) = \frac{\partial \hat{W}^*}{\partial J}, \tag{3.11}$$

$$J(\sigma_1-\sigma_3)=\lambda_1^*\frac{\partial\hat{W}^*}{\partial\lambda_1^*},\quad J(\sigma_2-\sigma_3)=\lambda_2^*\frac{\partial\hat{W}^*}{\partial\lambda_2^*}. \tag{3.12}$$

Equation (3.11) in essence governs the dilatational response of the material, while equations (3.12) govern the distortional response (cf. (2.5)). We emphasize, however, that dilatational and distortional responses are in general coupled through $\hat{W}^*$.

For consistency with the classical theory of infinitesimal isotropic elasticity we must have

$$\hat{W}^*(1,1,1)=\frac{\partial\hat{W}^*}{\partial J}(1,1,1)=\frac{\partial\hat{W}^*}{\partial\lambda_1}(1,1,1)=\frac{\partial\hat{W}^*}{\partial\lambda_2}(1,1,1)=0, \tag{3.13}$$

$$\frac{\partial^2\hat{W}^*}{\partial\lambda_1^{*2}}(1,1,1)=\frac{\partial^2\hat{W}^*}{\partial\lambda_2^{*2}}(1,1,1)=2\frac{\partial^2\hat{W}^*}{\partial\lambda_1^*\partial\lambda_2^*}(1,1,1)=4\mu, \tag{3.14}$$

$$\frac{\partial^2\hat{W}^*}{\partial\lambda_1^*\partial J}(1,1,1)=\frac{\partial^2\hat{W}^*}{\partial\lambda_2^*\partial J}(1,1,1)=0,\quad \frac{\partial^2\hat{W}^*}{\partial J^2}(1,1,1)=\kappa, \tag{3.15}$$

where μ is the *ground-state shear modulus*, as in Section 2.1, and κ is the *ground-state bulk modulus*. Also, for a pure dilatation,

$$\frac{\partial\hat{W}^*}{\partial\lambda_1^*}(1,1,J)=\frac{\partial\hat{W}^*}{\partial\lambda_2^*}(1,1,J)=0 \tag{3.16}$$

from (3.12) and the corresponding hydrostatic stress σ is given by

$$\sigma=\frac{\partial\hat{W}^*}{\partial J}(1,1,J). \tag{3.17}$$

For a pure distortion, on the other hand,

$$\sigma_1-\sigma_3=\lambda_1^*\frac{\partial\hat{W}^*}{\partial\lambda_1^*}(\lambda_1^*,\lambda_2^*,1),\quad \sigma_2-\sigma_3=\lambda_2^*\frac{\partial\hat{W}^*}{\partial\lambda_2^*}(\lambda_1^*,\lambda_2^*,1) \tag{3.18}$$

but

$$\frac{\partial\hat{W}^*}{\partial J}(\lambda_1^*,\lambda_2^*,1)\neq 0 \tag{3.19}$$

in general (as distinct from in the classical theory).

On the basis of (3.18) the identification

$$\hat{W}^*(\lambda_1^*,\lambda_2^*,1)\equiv\hat{W}(\lambda_1,\lambda_2) \tag{3.20}$$

with $\lambda_1^*\equiv\lambda_1$, $\lambda_2^*\equiv\lambda_2$ can be made, where $\hat{W}(\lambda_1,\lambda_2)$ is the strain-energy function arising in the incompressible theory. Then (3.19) has the role of the arbitrary hydrostatic pressure in (2.3).

When $\sigma_3=0$ we have

$$J\sigma_1=\lambda_1^*\frac{\partial\hat{W}^*}{\partial\lambda_1^*},\quad J\sigma_2=\lambda_2^*\frac{\partial\hat{W}^*}{\partial\lambda_2^*}, \tag{3.21}$$

from (3.12), and (3.11) leads to

$$J\frac{\partial\hat{W}^*}{\partial J}=\tfrac{1}{3}\lambda_1^*\frac{\partial\hat{W}^*}{\partial\lambda_1^*}+\tfrac{1}{3}\lambda_2^*\frac{\partial\hat{W}^*}{\partial\lambda_2^*} \tag{3.22}$$

which generalizes a result given by Ogden (1976) for the case of simple tension. In the incompressible theory the equation $\sigma_3 = 0$ serves to determine the arbitrary hydrostatic pressure p as a function of λ_1 and λ_2, while here it provides an equation which in principle enables the volume ratio J to be determined as a function of λ_1^*, λ_2^* or, equivalently, of λ_1, λ_2.

As in the incompressible theory we consider three special deformations, the first of which was examined by Ogden (1976):

(a) *Simple tension*: $\lambda_1^* = \lambda^*, \lambda_2^* = \lambda^{*-1/2}, \sigma_1 = \sigma, \sigma_2 = 0$ lead to

$$\tfrac{1}{3} J\sigma = \tfrac{1}{3}\lambda^* \frac{\partial \tilde{W}^*}{\partial \lambda^*} = J\frac{\partial \tilde{W}^*}{\partial J}, \quad \tilde{W}^*(\lambda^*, J) = \hat{W}^*(\lambda^*, \lambda^{*-1/2}, J). \tag{3.23}$$

(b) *Pure shear*: $\lambda_1^* = \lambda^*, \lambda_2^* = \lambda^{*-1}, \sigma_1 = \sigma, \sigma_2 = 0$ lead to

$$\tfrac{1}{3} J\sigma = \tfrac{1}{3}\lambda^* \frac{\partial \tilde{W}^*}{\partial \lambda^*} = J\frac{\partial \tilde{W}^*}{\partial J}, \quad \tilde{W}^*(\lambda^*, J) = \hat{W}^*(\lambda^*, \lambda^{*-1}, J). \tag{3.24}$$

(c) *Equibiaxial tension*: $\lambda_1^* = \lambda_2^* = \lambda^*, \sigma_1 = \sigma_2 = \sigma$ lead to

$$\tfrac{2}{3} J\sigma = \tfrac{1}{3}\lambda^* \frac{\partial \tilde{W}^*}{\partial \lambda^*} = J\frac{\partial \tilde{W}^*}{\partial J}, \quad \tilde{W}^*(\lambda^*, J) = W^*(\lambda^*, \lambda^*, J). \tag{3.25}$$

In each case specializations of (3.13)–(3.15) arise but are not written down here. Also, in each of (a), (b), and (c) the nominal stress f is given by

$$f = J^{2/3}\sigma/\lambda^*. \tag{3.26}$$

For a membrane (discussed in Section 2.2) the approximation $\sigma_3 = 0$ gives (3.21) and (3.22). In particular, for an inflated spherical membrane

$$P^* = J^{-1}\lambda^{*-2}\frac{\partial \tilde{W}^*}{\partial \lambda^*} \tag{3.27}$$

analogously to (2.13), together with

$$\tfrac{1}{3}\lambda^* \frac{\partial \tilde{W}^*}{\partial \lambda^*} = J\frac{\partial \tilde{W}^*}{\partial J},$$

where λ^* and $\tilde{W}^*(\lambda^*, J)$ are as defined in (c) above, λ^* being the modified stretch in the membrane surface. For a circular cylindrical membrane tube under extension and inflation equations (2.15) are replaced by

$$P^* = J^{-1}\lambda_1^{*-1}\lambda_2^{*-1}\frac{\partial \hat{W}^*}{\partial \lambda_2^*}, \quad F^* = J^{-1/3}\frac{\partial \hat{W}^*}{\partial \lambda_1^*} - \tfrac{1}{2}J^{-1/3}\lambda_1^{*-1}\lambda_2^*\frac{\partial \hat{W}^*}{\partial \lambda_2^*}, \tag{3.28}$$

where F^* and P^* have the same meaning as in Section 2.2, and, in addition, (3.22) holds. Equations (3.27) and (3.28) have been included merely to complete the analogy with the incompressible theory in Section 2.2 and will not be used in subsequent Sections of the present article.

3.2. *Strain-energy functions*

As in the incompressible theory there are many possible general representations for the strain-energy function. For example, by analogy with (2.20) we may write

$$W(\lambda_1, \lambda_2, \lambda_3) = \Phi(I_1, I_2, I_3),$$

where I_1, I_2, I_3 are any three independent symmetric functions of $\lambda_1, \lambda_2, \lambda_3$. Those in most common use are

$$I_1 = \lambda_1^2 + \lambda_2^2 + \lambda_3^2, \quad I_2 = \lambda_2^2\lambda_3^2 + \lambda_3^2\lambda_1^2 + \lambda_1^2\lambda_2^2, \quad I_3 = \lambda_1^2\lambda_2^2\lambda_3^2. \tag{3.29}$$

Generalizations of (2.21) and (2.23) to the compressible context may be written down but will not be required here.

Several specific forms of strain-energy function (far fewer than in the incompressible theory) have been used. These have been based mainly on the neo-Hookean or Mooney functions from incompressible theory supplemented by an added function of the squared volume ratio I_3. For example,

$$\Phi = \tfrac{1}{2}\mu(I_1 - 3) + f(I_3) \tag{3.30}$$

has been adopted by Blatz and Ko (1962), Treloar (1969), Levinson (1972), Faulkner (1972), Faulkner and Haddow (1972), and others, in some cases $f(I_3)$ being specified from the outset. Blatz and Ko (1962), Price and Allen (1973), and Flory and Tatara (1975) have used the Mooney function as a starting point and allowed the material constants C_1 and/or C_2 in effect to depend on I_3. A noteworthy special case of (3.30) with

$$f(I_3) = \frac{\mu(1-2\nu)}{2\nu}\left[I_3^{\,-\nu/(1-2\nu)} - 1\right] \tag{3.31}$$

was proposed by Blatz and Ko (1962), the constant ν playing a role similar to that of Poisson's ratio in the classical theory.

Alternatives to the symmetrical invariants (3.29) have been used only in rare and isolated instances. For example, $\lambda_1 + \lambda_2 + \lambda_3, \lambda_2\lambda_3 + \lambda_3\lambda_1 + \lambda_1\lambda_2, \lambda_1\lambda_2\lambda_3$ were employed by Ogden (1970), while Sharda and Tschoegl (1976) have put (2.33) in a compressible framework by use of J.

Representation of W in terms of the stretches themselves has been examined by Ogden (1972b) who expresses W in the form

$$W = \sum_{\alpha_1,\alpha_2,\alpha_3} C(\alpha_1, \alpha_2, \alpha_3)\{[\lambda_1^{\alpha_1}(\lambda_2^{\alpha_2} + \lambda_3^{\alpha_2}) + \lambda_2^{\alpha_1}(\lambda_3^{\alpha_2} + \lambda_1^{\alpha_2}) + \lambda_3^{\alpha_1}(\lambda_1^{\alpha_2} + \lambda_2^{\alpha_2})](\lambda_1\lambda_2\lambda_3)^{\alpha_3} - 6\} \tag{3.32}$$

analogously to (2.24). A particular form of this discussed in detail by Ogden (1972b, 1976) is

$$W = \sum_{n=1}^{N} \mu_n I(\alpha_n) + F(J), \tag{3.33}$$

where $I(\alpha)$ is as defined in (2.32) except that $\lambda_1, \lambda_2, \lambda_3$ no longer satisfy (2.1). Forms equivalent to (3.33) have been used by Chadwick (1974) and Treloar (1978), while Hill (1978) has proposed (3.33), with $F(J) \equiv f(I_3)$ given by a generalized form of (3.31), for further study.

Penn (1970), with a different notation, introduced the invariants

$$I_1^* = I_1 I_3^{-1/3} \equiv \lambda_1^{*2} + \lambda_2^{*2} + \lambda_3^{*2}, \quad I_2^* = I_2 I_3^{\,-2/3} \equiv \lambda_1^{*\,-2} + \lambda_2^{*\,-2} + \lambda_3^{*\,-2} \tag{3.34}$$

(although he did not use the modified stretches (3.4) explicitly) for the express purpose of discussing volume changes accompanying deformation, and he used the representation $W = \Phi^*(I_1^*, I_2^*, I_3)$. In particular, from volume-change data on simple-tension experiments he deduced that Φ^* cannot be decoupled into dilatational and distortional parts, i.e. Φ^* cannot be written as the sum of a function of I_1^*, I_2^* and a function of I_3. The invariants (3.34) were used subsequently by Fong and Penn (1975) and Peng and Landel (1975), who constructed specific forms of strain-energy function designed to fit volume-change data on simple tension. The representation Φ^* is equivalent to the representation W^* we have adopted in Section 3.1.

If the modified stretches (3.4) along with J are used, as in Section 3.1, then (3.32) is replaced by

$$W^* = \sum_{\alpha_1,\alpha_2,\alpha_3} C(\alpha_1,\alpha_2,\alpha_3)\{[\lambda_1^{*\alpha_1}(\lambda_2^{*\alpha_2}+\lambda_3^{*\alpha_2})+\lambda_2^{*\alpha_1}(\lambda_3^{*\alpha_2}+\lambda_1^{*\alpha_2})+\lambda_3^{*\alpha_1}(\lambda_1^{*\alpha_2}+\lambda_2^{*\alpha_2})]J^{\alpha_3}-6\}. \tag{3.35}$$

A number of strain-energy functions, general and specific, based on this formalism have been examined by Ogden (1976, 1978, 1979b). They are special cases of the representation

$$W^* = \phi(\lambda_1^*, \lambda_2^*, \lambda_3^*) + \sum_{n=1}^{N} \chi_n(\lambda_1^*, \lambda_2^*, \lambda_3^*)h_n(J) + g(J) \tag{3.36}$$

which is equivalent to (3.35), the first and final terms in (3.36) corresponding to $\alpha_3 = 0$ and $\alpha_1 = \alpha_2 = 0$, respectively, in (3.35). In (3.36) ϕ and χ_n are symmetrical functions of $\lambda_1^*, \lambda_2^*, \lambda_3^*$ and g and h_n are functions of J which satisfy certain requirements for consistency with (3.13)–(3.15). For $N = 1$ particular cases of (3.36), with varying degrees of generality, have been studied by Fong and Penn (1975) and Peng and Landel (1975) in terms of the invariants I_1^*, I_2^*. We remark in passing that a form of W^* analogous to (2.25) may also be written down but we shall not require it here.

The main objective of the remainder of this article is the construction of strain-energy functions which (a) retain the essential stress–deformation characteristics described in Section 2, and (b) predict the observed features associated with change in volume for rubberlike solids.

3.3. *Nearly isochoric deformations*

In order to examine the magnitude of volume changes associated with biaxial deformation we set $\sigma_3 = 0$. Differentiation of the resulting (3.22) with respect to $\lambda_\alpha^*(\alpha = 1, 2)$ gives

$$\frac{\partial J}{\partial \lambda_\alpha^*} = \left[\tfrac{1}{3}\lambda_\beta^*\frac{\partial^2 \hat{W}^*}{\partial \lambda_\beta^* \partial \lambda_\alpha^*} + \tfrac{1}{3}\frac{\partial \hat{W}^*}{\partial \lambda_\alpha^*} - J\frac{\partial^2 \hat{W}^*}{\partial \lambda_\alpha^* \partial J}\right]\left[J\frac{\partial^2 \hat{W}^*}{\partial J^2} + \frac{\partial \hat{W}^*}{\partial J} - \tfrac{1}{3}\lambda_\gamma^*\frac{\partial^2 \hat{W}^*}{\partial \lambda_\gamma^* \partial J}\right]^{-1} \quad (\alpha = 1, 2) \tag{3.37}$$

with summation over β and γ from 1 to 2. Use of (3.13)–(3.15) shows that (3.37) reduces to

$$\frac{\partial J}{\partial \lambda_\alpha^*} = 2\mu/\kappa \quad (\alpha = 1, 2), \tag{3.38}$$

and for simple tension, with $\lambda_1^* = \lambda^*$, $\lambda_2^* = \lambda^{*-1/2}$, to

$$\frac{dJ}{d\lambda^*} = \mu/\kappa \tag{3.39}$$

in the undeformed configuration. For convenience we denote the dimensionless quantity μ/κ by η.

A *nearly incompressible* (or *slightly compressible*) material is defined as one in which $\eta \ll 1$ and, additionally, $\partial J/\partial\lambda_\alpha^* = O(\eta)$ in continued deformation ($\alpha = 1, 2$). It follows that the *dilatation* $(J-1)$ is itself of order η and the corresponding deformations are said to be *nearly isochoric.* We recall, however, that attention is restricted to $\sigma_3 = 0$ here. Thus, in simple tension, for example, the change δJ in J is of order η times the accompanying change $\delta\lambda^*$ in the deformation λ^*.

From (3.13) to (3.15) consideration of (3.37) indicates that for a nearly incompressible material the functions

$$\hat{W}^*, \quad \frac{\partial \hat{W}^*}{\partial J}, \quad \frac{\partial \hat{W}^*}{\partial \lambda_\alpha^*}, \quad \frac{\partial^2 \hat{W}^*}{\partial \lambda_\alpha^* \partial \lambda_\beta^*}, \quad \frac{\partial^2 \hat{W}^*}{\partial \lambda_\alpha^* \partial J}$$

are of order μ, while

$$\frac{\partial^2 \hat{W}^*}{\partial J^2}$$

is of order κ. We now adopt η as a "small" parameter and since $J-1 = O(\eta)$ we expand $\hat{W}^*$ as a Taylor series in the neighbourhood of $J=1$. Thus,

$$\hat{W}^*(\lambda_1^*, \lambda_2^*, J) = \hat{W}^*(\lambda_1^*, \lambda_2^*, 1) + (J-1)\hat{W}_J^*(\lambda_1^*, \lambda_2^*, 1) + \tfrac{1}{2}(J-1)^2 \hat{W}_{JJ}^*(\lambda_1^*, \lambda_2^*, 1) + \mu O(\eta^2), \tag{3.40}$$

terminating at the third term, where the subscript J denotes partial differentiation with respect to J.

In view of the above discussion $\hat{W}^*(\lambda_1^*, \lambda_2^*, 1)$ is typically $\mu O(1)$ while the second and third terms in (3.40) are $\mu O(\eta)$ and $\kappa O(\eta^2) \equiv \mu O(\eta)$ respectively. *It is assumed* that higher-order terms in the expansion are of order $\mu\eta^2$ and smaller. In other words $\hat{W}^*{}_{JJJ}/\hat{W}^*{}_{JJ} = O(1)$ and similarly for higher derivatives. This assumption is not essential if a series expansion method is not employed; indeed, for the Blatz–Ko material (3.30) with (3.31) it is not valid since essentially κ occurs through J^κ and then $\hat{W}^*{}_{JJJ}/\hat{W}^*{}_{JJ} = \kappa O(1)$ and we are not justified in terminating the expansion. The Blatz–Ko strain-energy function has nevertheless been used by some authors in circumstances where an expansion terminated at the third term, as in (3.40), is required (implicitly). See, for example, Faulkner and Haddow (1972). Discussion of the expansion has been provided by Ogden (1976) for simple tension and by Ogden (1978) for general deformations. An equivalent treatment has been given by Spencer (1964, 1970) in different variables.

Use of (3.40) in (3.21) and (3.22) leads to

$$\sigma_\alpha = \lambda_\alpha^* \frac{\partial \hat{W}^*}{\partial \lambda_\alpha^*}(\lambda_1^*, \lambda_2^*, 1) + \mu O(\eta) \quad (\alpha = 1, 2) \tag{3.41}$$

and

$$(J-1)\hat{W}^*{}_{JJ}(\lambda_1^*, \lambda_2^*, 1) = \tfrac{1}{3}(\sigma_1 + \sigma_2) - \hat{W}_J^*(\lambda_1^*, \lambda_2^*, 1) + \mu O(\eta) \tag{3.42}$$

with σ_1 and σ_2 given by (3.41). The latter equation determines the dilatation $(J-1)$ explicitly as a function of λ_1^* and λ_2^* with an error of order η^2. Note that, since $\lambda_\alpha^* = \lambda_\alpha[1 + O(\eta)]$ follows from (3.4), λ_α^* may be replaced by λ_α in (3.41) and (3.42) without affecting the order of approximation.

Comparison of (3.41) and the corresponding (2.11) in the incompressible theory shows that

$$\hat{W}^*(\lambda_1^*, \lambda_2^*, 1) = \hat{W}(\lambda_1, \lambda_2) + \mu O(\eta).$$

Hence, since $\lambda_\alpha^* \to \lambda_\alpha$ in the limit $J \to 1 (\eta \to 0)$ this becomes an identity

$$\hat{W}^*(\lambda_1, \lambda_2, 1) \equiv \hat{W}(\lambda_1, \lambda_2) \tag{3.43}$$

and therefore

$$\lambda_\alpha^* = \lambda_\alpha[1 + O(\eta)],$$

where λ_α, the stretches arising in the incompressible theory, equally may replace λ_α^* in (3.41) and (3.42).

Equation (3.41) therefore shows that the stresses associated with a given deformation in a nearly incompressible material differ only by a term $\mu O(\eta)$ from the stresses associated with the same deformation in the corresponding incompressible material, as is to be expected. Equation (3.42) shows that the dilatation is expressible in terms of the deformation λ_1, λ_2 and stress σ_1, σ_2 determined from the incompressible theory when the functions $\hat{W}_J^*(\lambda_1, \lambda_2, 1)$ and $\hat{W}_{JJ}^*(\lambda_1, \lambda_2, 1)$ are known.

For a nearly incompressible material the stress–deformation characteristics are clearly determined on the basis of the incompressible theory, in view of (3.41) and (3.43). The error $\mu O(\eta)$ is negligible for practical purposes. The additional functions $\hat{W}_J^*$ and $\hat{W}_{JJ}^*$ arise solely through the connection (3.42).

When the restriction $\sigma_3 = 0$ is not imposed σ_α is replaced by $\sigma_\alpha - \sigma_3$ in (3.41) while (3.42) is replaced by

$$(J-1)\hat{W}_{JJ}^*(\lambda_1^*, \lambda_2^*, 1) = \tfrac{1}{3}(\sigma_1 + \sigma_2 + \sigma_3) - \hat{W}_J^*(\lambda_1^*, \lambda_2^*, 1) + \mu O(\eta) \tag{3.44}$$

provided the applied stresses are such that the restriction $J - 1 = O(\eta)$ is maintained, that is, provided the stresses do not exceed $\mu O(1)$. The result (3.44) was given by Spencer (1964) in different variables, and in more general form by Ogden (1978).

3.4. *Experimental results and their relation to the theory*

For rubberlike solids experiment shows that η is typically of order 10^{-4} and the theory described in Section 3.3 is applicable. It is assumed that $\hat{W}(\lambda_1, \lambda_2)$ is known on the basis of the results described in Section 2. Then (3.42) enables $\hat{W}_J^*(\lambda_1, \lambda_2, 1)$ to be found from experiments which produce volume-change data on general biaxial deformation provided $\hat{W}_{JJ}^*(\lambda_1, \lambda_2, 1)$ is known. Information about this latter function is obtained from the bulk response of the material to a hydrostatic stress superposed on general biaxial stress.

For a general state of deformation the bulk modulus may be defined as

$$\tfrac{1}{3}\frac{\partial}{\partial J}(\sigma_1 + \sigma_2 + \sigma_3) \equiv \frac{\partial^2 \hat{W}^*}{\partial J^2}, \tag{3.45}$$

but from (3.40) this is just $\hat{W}_{JJ}^*(\lambda_1^*, \lambda_2^*, 1)$ since attention is confined to circumstances where $J - 1 = O(\eta)$. In pure dilatation $\lambda_1^* = \lambda_2^* = 1$ and $\hat{W}_{JJ}^*(1, 1, 1) = \kappa$, the *ground-state bulk modulus*, by $(3.15)_3$. More generally, it has been found in practice that the bulk modulus is "isotropic", i.e. effectively independent of λ_1^*, λ_2^*. Experimental verification of this is limited to simple tension and for stretches up to about 2.0. See Price *et al.* (1969), Christensen and

Hoeve (1970), and the review by Price (1976). In the earlier work of Gee (1946), for example, it was assumed that the bulk modulus is constant. For larger stretches in simple tension and biaxial deformation the bulk modulus may depend on λ_1^*, λ_2^*, but here we assume that it is constant and await further experimental results to support or negate this assumption.

Equation (3.44) now becomes

$$\kappa(J-1) = \tfrac{1}{3}(\sigma_1+\sigma_2+\sigma_3) - \hat{W}_J^*(\lambda_1^*, \lambda_2^*, 1) + \mu O(\eta). \tag{3.46}$$

This is the relationship on which attention is now focused, the purpose being to discover the nature of the function $\hat{W}_J^*(\lambda_1^*, \lambda_2^*, 1)$. In simple tension (3.46) reduces to

$$\kappa(J-1) = \tfrac{1}{3}\lambda \tilde{W}_\lambda^*(\lambda, 1) - \tilde{W}_J^*(\lambda, 1) + \mu O(\eta) \tag{3.47}$$

with

$$\sigma = \lambda \tilde{W}_\lambda^*(\lambda, 1) + \mu O(\eta) \tag{3.48}$$

and the notation as defined in (3.23) except that λ replaces λ^*. This case has been discussed in detail by Ogden (1976, 1979b). For pure shear and equibiaxial tension (3.47) still applies but with $\tilde{W}^*$ defined by (3.24) and (3.25) respectively.

For future reference we record that (3.40) simplifies to

$$\hat{W}^*(\lambda_1^*, \lambda_2^*, J) = \hat{W}^*(\lambda_1^*, \lambda_2^*, 1) + (J-1)\hat{W}_J^*(\lambda_1^*, \lambda_2^*, 1) + \tfrac{1}{2}\kappa(J-1)^2, \tag{3.49}$$

with appropriate specializations for simple tension, pure shear, and equibiaxial tension.

The only deformation for which data are available when the restriction $J-1 = O(\eta)$ is applicable is simple tension, so attention is now confined to this.

Typical volume-change data are represented in Fig. 3.1, where $\kappa(J-1)/\mu$ is plotted against λ. The continuous curve is equivalent to that found by Fong and Penn (1975) to fit the data of Penn (1970). Also shown for comparison (broken curve) is a plot of $\sigma/3\mu$ from Penn's data for

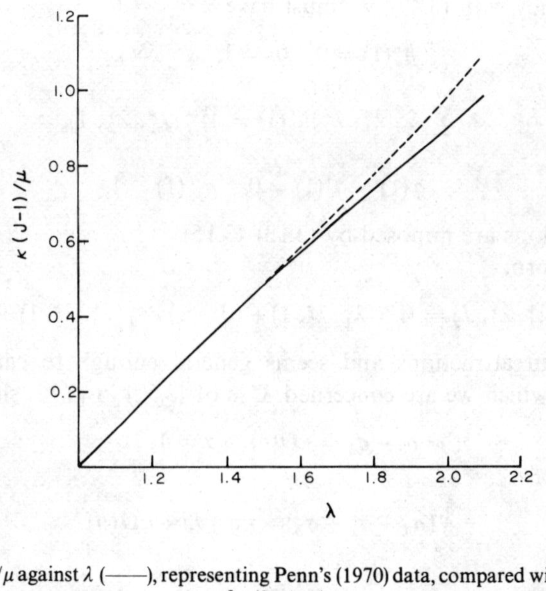

FIG. 3.1. Plot of $\kappa(J-1)/\mu$ against λ (——), representing Penn's (1970) data, compared with the corresponding plot of $\sigma/3\mu$ (----).

the same materials. This initial trend of volume changes has also been observed by Holt and McPherson (1936), Gee *et al.* (1950), Hewitt and Anthony (1958), Allen *et al.* (1963), and Christensen and Hoeve (1970) for stretches up to about 2.2 in simple tension.

The experimental results depicted in Fig. 3.1 should be viewed in relation to the theoretical result (3.47) with (3.48). In particular the result $3\kappa\varepsilon = \sigma$ does not hold except as an approximation for $1 \leqslant \lambda < 1.5$. For $\lambda > 1.5$ there is a clear departure from this relationship, and this led Penn (1970) to conclude the existence of a function equivalent to $\tilde{W}^*_J(\lambda, 1)$. This function must be positive in sign for $\lambda > 1.5$ and barely distinguishable from zero for $1 \leqslant \lambda < 1.5$. It is not unduly restrictive therefore to impose the inequality

$$\tilde{W}^*_J(\lambda, 1) \geqslant 0 \quad \text{for} \quad \lambda \geqslant 1 \tag{3.50}$$

bearing in mind that strict equality holds for $\lambda = 1$. The above experimental results have been discussed in relation to the theory by Fong and Penn (1975) and Ogden (1976) from different starting points.

For larger stretches the only data available are those of Holt and McPherson (1936). Their results indicate that $(J-1)$ reaches a maximum and thereafter decreases steadily, ultimately becoming negative. This trend can be accommodated without difficulty by an appropriate form of $\tilde{W}^*_J(\lambda, 1)$, as shown by Ogden (1976), but no further discussion is included here. We await more comprehensive data for large stretches in simple tension and for general biaxial deformations.

Attention is now focused on situations in which $(J-1)$ is not restricted to being of order η. The assumption that the bulk modulus is "isotropic" is retained, however, and this assumption is incorporated into the general representation (3.36) by substitution of $\kappa g(J)$ for $g(J)$. Elimination of λ_3^* then leads to

$$\hat{W}^*(\lambda_1^*, \lambda_2^*, J) = \hat{W}^*(\lambda_1^*, \lambda_2^*, 1) + \sum_{n=1}^{N} \hat{\chi}_n(\lambda_1^*, \lambda_2^*) h_n(J) + \kappa g(J), \tag{3.51}$$

where, for consistency with (3.49), we must have

$$h_n(1) = 0 \quad (n = 1, \ldots, N),$$

$$\sum_{n=1}^{N} \hat{\chi}_n(\lambda_1^*, \lambda_2^*) h_n'(1) = \hat{W}^*_J(\lambda_1^*, \lambda_2^*, 1),$$

$$g(1) = g'(1) = 0, \quad g''(1) = 1,$$

and further restrictions are imposed by (3.13)–(3.15).

The specialized form

$$\hat{W}^*(\lambda_1^*, \lambda_2^*, J) = \hat{W}^*(\lambda_1^*, \lambda_2^*, 1) + (J-1)\hat{W}^*_J(\lambda_1^*, \lambda_2^*, 1) + \kappa g(J) \tag{3.52}$$

of (3.51) has certain attractions and seems general enough to cater for the range of deformations with which we are concerned. Use of (3.52) in (3.12) shows that

$$\sigma_\alpha - \sigma_3 = \kappa O(\eta) \quad (\alpha = 1, 2),$$

and (3.11) leads to

$$\tfrac{1}{3}(\sigma_1 + \sigma_2 + \sigma_3) = \kappa g'(J) + \kappa O(\eta).$$

This shows that

$$\sigma_i = \kappa g'(J) + \kappa O(\eta) \quad (i = 1, 2, 3),$$

and hence, when $(J-1)$ exceeds $O(\eta)$, the stress is approximately hydrostatic, $\kappa g'(J)$ being the dominant term. For example, if $(J-1) = 0.01$ then the ratio $g'(J)/\eta$ is about 100. Clearly, large changes in volume can be achieved only if the stress is effectively hydrostatic.

Volume reductions of up to about 30% were obtained in hydrostatic compression tests by Adams and Gibson (1930), Bridgman (1945), and others (see Ogden, 1972b, for references). But in hydrostatic tension (Gent and Lindley, 1959) volume increases achieved have been relatively small. We therefore demonstrate the relation between theory and experiment by considering a hydrostatic pressure P, so that $P = -\kappa g'(J)$.

In Fig. 3.2 data from the experiments of Adams and Gibson (1930) and Bridgman (1945) are plotted in dimensionless form. The dimensionless pressure P^* is also plotted against $(1-J)$, where, following Ogden (1972b), we have chosen

$$P^* = 9P/\kappa = J^{-10} - J^{-1} \tag{3.53}$$

to illustrate the close agreement between theory and experiment. Full details are given by Ogden (1972b).

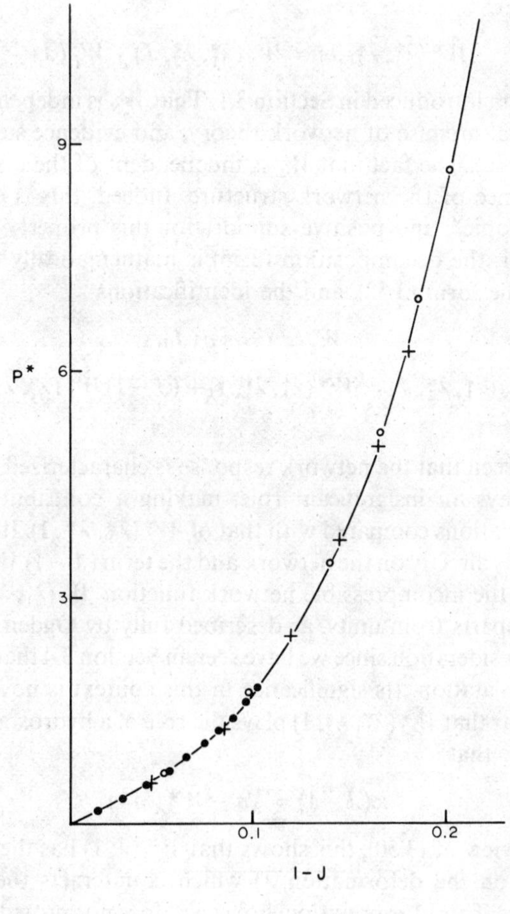

FIG. 3.2. Plot of the hydrostatic pressure P^* against $(1-J)$ based on the empirical function (3.53) compared with the data of Adams and Gibson (1930), (●), and Bridgman (1945), (○, +), following Ogden (1972b).

We conclude this Section by remarking that the form

$$g'(J) = J - 1$$

provides an adequate representation of the data in Fig. 3.2 up to $(1 - J) \simeq 0.1$, and hence the predictions of (3.49) are fully consistent with observation except in situations where a very large hydrostatic pressure is imposed. With $\hat{W}^*(\lambda_1^*, \lambda_2^*, 1)$ specialized to the Mooney form, similar conclusions were reached by Fong and Penn (1975).

3.5. *Structural interpretation of the phenomenological theory*

According to Flory (1961), the strain energy of a rubberlike solid arises from two independent sources. Firstly, there is the strain energy W_N associated with the elasticity of the long-chain molecular network and, secondly, the "liquid-like" strain-energy W_L associated with intermolecular forces. The total strain-energy is

$$W(\lambda_1, \lambda_2, \lambda_3) = W_N(\lambda_1, \lambda_2, \lambda_3) + W_L(J) \tag{3.54}$$

or, equivalently,

$$\hat{W}^*(\lambda_1^*, \lambda_2^*, J) = \hat{W}_N^*(\lambda_1^*, \lambda_2^*, J) + W_L(J) \tag{3.55}$$

in line with the notation introduced in Section 3.1. That W_N is independent of intermolecular forces is a fundamental premise of network theory, and evidence supporting this has been reviewed by Price (1976). The fact that W_L is independent of the distortional deformation reflects its independence of the network structure. Indeed, this is equivalent to the bulk modulus being "isotropic", and positive support for this property has been discussed in Section 3.4. Moreover, the decomposition (3.55) is mathematically consistent with strain-energy functions of the form (3.52), and the identifications

$$W_L(J) = \kappa g(J), \tag{3.56}$$

$$\hat{W}_N^*(\lambda_1^*, \lambda_2^*, J) = \hat{W}^*(\lambda_1^*, \lambda_2^*, 1) + (J - 1)\hat{W}_J^*(\lambda_1^*, \lambda_2^*, 1) \tag{3.57}$$

can be made.

However, we have seen that the network response is characterized by $\hat{W}^*(\lambda_1^*, \lambda_2^*, 1)$ alone and $\hat{W}_J^*(\lambda_1^*, \lambda_2^*, 1)$ plays an insignificant role, making a contribution of order η to the stress–deformation relations compared with that of $\hat{W}^*(\lambda_1^*, \lambda_2^*, 1)$. If (3.57) is accepted, then $\hat{W}_J^*(\lambda_1^*, \lambda_2^*, 1)$ depends directly on the network and the term $(J - 1)\hat{W}_J^*(\lambda_1^*, \lambda_2^*, 1)$ represents a small correction to the incompressible network function $W_N(\lambda_1, \lambda_2, \lambda_3)$ arising from the fact that $\lambda_1, \lambda_2, \lambda_3$ departs from unity, as described fully by Ogden (1979b). But this term must be taken into consideration since we have seen in Section 3.4 that it makes an important contribution to the dilatation. Its significance in this context is now examined.

From (3.46) it is clear that $\hat{W}_J^*(\lambda_1^*, \lambda_2^*, 1)$ plays the role of a hydrostatic stress. In particular, (3.47) and (3.48) show that

$$\kappa(J - 1) = \tfrac{1}{3}\sigma - \tilde{W}_J^*(\lambda, 1) \tag{3.58}$$

in simple tension. In view of (3.50), this shows that $\tilde{W}_J^*(\lambda, 1)$ has the effect of a hydrostatic pressure (dependent on the deformation λ) which counteracts the effect of the applied hydrostatic tension $\frac{1}{3}\sigma$. This interpretation, however, does not provide an explanation of the origin of $\tilde{W}_J^*(\lambda, 1)$ at the structural level. Flory's theory indicates that $\tilde{W}_J^*(\lambda, 1)$ is a function of the network and must therefore arise from molecular rearrangements caused by the deformation of the network, but the precise mechanism with which $\tilde{W}_J^*(\lambda, 1)$ is to be associated is unclear. Therefore, an alternative viewpoint is now discussed.

Mathematically, $\hat{W}_J^*(\lambda_1^*, \lambda_2^*, 1)$ is *independent* of $\hat{W}^*(\lambda_1^*, \lambda_2^*, 1)$ in general. This leads us to consider the consequences of it being independent of the network structure, although this is in conflict with Flory's (1961) "fundamental principle," equation (3.54). Since the network response, as characterized by $\hat{W}^*(\lambda_1^*, \lambda_2^*, 1)$ is essentially independent of intermolecular forces, an attractive explanation of $\hat{W}_J^*(\lambda_1^*, \lambda_2^*, 1)$ is that it is due to intermolecular forces which are redistributed by the deformation-induced molecular rearrangements. A more detailed discussion of this possible interpretation is given by Ogden (1979b). Equivalently, on the basis of (3.57), the network response can be regarded as a function of intermolecular forces through $\hat{W}_J^*(\lambda_1^*, \lambda_2^*, 1)$.

At present there is no positive evidence which enables us to discriminate in favour of one or other of the interpretations offered above. The main point of the above discussion is that it highlights the different roles of the functions $\hat{W}^*(\lambda_1^*, \lambda_2^*, 1)$ and $\hat{W}_J^*(\lambda_1^*, \lambda_2^*, 1)$, which may or may not be independent depending on the physical interpretation of the latter.

In the analysis of deformation-induced volume changes the notion of *linear compressibility* has a significant role, as exemplified by the work of Elliott and Lippmann (1945), Gee (1946), Flory (1961), and Price and Allen (1973). The relationship between linear compressibility and bulk compressibility (the inverse of the bulk modulus) has been discussed by Ogden (1979b). Justification for assuming that *the bulk modulus is isotropic* has been presented in Section 3.4, but there is little corresponding justification for assuming that the linear compressibility is isotropic. This is not surprising since the word "isotropic" has different connotations in the two cases.

An explicit formula for the *anisotropy of linear compressibility* was provided by Flory (1961), on the basis of the Gaussian theory, and more recently by Ogden (1979b) for a general form of strain-energy function. With the simple-tension equations (3.23) modified by a superposed hydrostatic pressure and the nominal stress f given by (3.26), Ogden's calculation gives

$$\tfrac{1}{3}J^{-1}\lambda^*(A-1) \equiv \left(\frac{\partial\lambda^*}{\partial J}\right)_f = \left\{\tfrac{1}{3}J^{-1}\frac{\partial\tilde{W}^*}{\partial\lambda^*} - \frac{\partial^2\tilde{W}^*}{\partial J\partial\lambda^*}\right\}\Big/\frac{\partial^2\tilde{W}^*}{\partial\lambda^{*2}}, \tag{3.59}$$

where A is the *anisotropy factor* introduced by Elliott and Lippmann (1945). The linear compressibility is said to be *isotropic* if $A \equiv 1$ or, equivalently, $(\partial\lambda^*/\partial J)_f = 0$. In essence this means that a change in hydrostatic pressure at fixed f does not affect the distortional part of the deformation, and in a more general context that the principal stretches $\lambda_1, \lambda_2, \lambda_3$ are changed in the same ratio.

In respect of (3.49), equation (3.59) leads to

$$\left(\frac{\partial\lambda^*}{\partial J}\right)_f = \left\{\tfrac{1}{3}\frac{\partial\tilde{W}^*}{\partial\lambda^*}(\lambda^*, 1) - \frac{\partial\tilde{W}_J^*}{\partial\lambda^*}(\lambda^*, 1)\right\}\Big/\frac{\partial^2\tilde{W}^*}{\partial\lambda^{*2}}(\lambda^*, 1) \tag{3.60}$$

apart from an error of order η. Except in the undeformed configuration ($\lambda^* = 1$) this is in general non-zero, but the (very specialized) condition of isotropy of linear compressibility can be imposed by setting

$$\tilde{W}_J^*(\lambda^*, 1) = \tfrac{1}{3}\tilde{W}^*(\lambda^*, 1), \tag{3.61}$$

in which case $\tilde{W}_J^*(\lambda^*, 1)$ cannot be treated as independent of the network. A case in favour of isotropy of linear compressibility has been argued by Price (1976).

For the Gaussian strain-energy function

$$W = \tfrac{1}{2}\mu(\lambda_1^2 + \lambda_2^2 + \lambda_3^2 - 3J^{2/3}) + \tfrac{1}{2}\kappa(J-1)^2, \tag{3.62}$$

with the J-dependence chosen for consistency with (3.13)–(3.15), we obtain

$$\tilde{W}_J^*(\lambda^*, 1) = \tfrac{2}{3}\tilde{W}^*(\lambda^*, 1) \tag{3.63}$$

and

$$A = 3/(\lambda^{*3} + 2), \tag{3.64}$$

the latter of which was obtained by Flory (1961) following Khasanovich (1959). In view of (3.63), equation (3.62) is consistent with (3.54), but the condition (3.61) is not met.

From (3.58) we obtain

$$\kappa(J-1) = \mu(1 - \lambda^{*-1}) \tag{3.65}$$

in respect of (3.62) specialized to simple tension. The prediction of (3.65) is plotted in Fig. 3.3 in dimensionless form and compared with the data of Penn (1970) reproduced from Fig. 3.1. Clearly, there is a wide discrepancy between the theory (3.65) and experimental data, and this has been discussed by Price (1976) and Treloar (1978).

The predictions of the Gaussian theory can be brought into closer correspondence with the experimental data by allowing μ to depend on J. For example, if μ is replaced by μJ^m in (3.62), where m is a disposable constant, then (3.63) and (3.65) respectively are generalized to

$$\tilde{W}_J^*(\lambda^*, 1) = (m + \tfrac{2}{3})\tilde{W}^*(\lambda^*, 1) \tag{3.66}$$

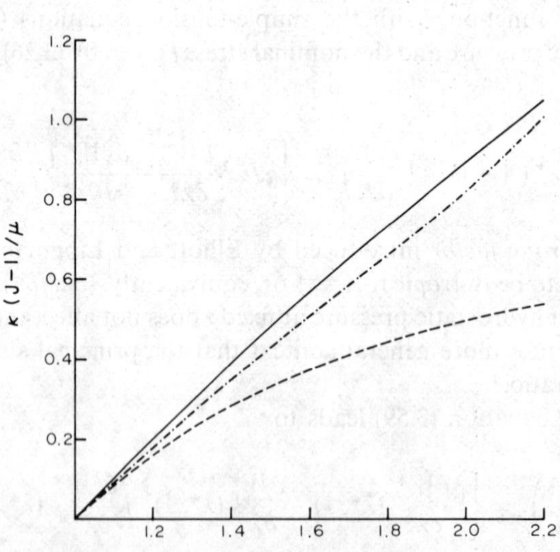

FIG. 3.3. Plot of $\kappa(J-1)/\mu$ against λ(——) in respect of Penn's data, as in Fig. 3.1, compared with the prediction of the Gaussian theory (-----) and with that of the Gaussian theory modified to fit the requirement of isotropy of linear compressibility (-·-·-·-).

and

$$\kappa(J-1)=\mu(1-\lambda^{*-1})-\tfrac{1}{2}m\mu(\lambda^{*2}+2\lambda^{*-1}-3). \qquad (3.67)$$

In particular, for $m=-\frac{1}{3}$ the linear compressibility becomes isotropic in accordance with (3.61). The corresponding prediction of (3.67) is plotted in Fig. 3.3, demonstrating a much closer correlation with the data than (3.65). However, it is dangerous to extrapolate beyond the range of values of λ^* included in Fig. 3.3 for we note, in particular, that (3.67) grows like λ^{*2} $(m<0)$.

Equation (3.66) shows that the above device of modifying μ is equivalent to introducing an additional term into $\tilde{W}_J^*(\lambda^*, 1)$, in which case $\tilde{W}_J^*(\lambda^*, 1)$ is directly related to the network response function. On the basis of the Mooney strain-energy function similar modifications have been employed by Price and Allen (1973) in order to make the linear compressibility isotropic, and by Flory and Tatara (1975) for a different purpose. Introduction of a term $\tilde{W}_J^*(\lambda^*, 1)$ which can be chosen independently of $\tilde{W}^*(\lambda^*, 1)$, on the other hand, provides a more general and flexible method of relating theory to experimental data, particularly for large stretches in simple tension and for general biaxial deformations. For simple tension the possibilities have been explored by Ogden (1976, 1979b). In particular, he has shown (1979b) that if the linear compressibility is assumed isotropic then, for quite a large class of strain-energy functions, the dilatation is necessarily positive, and the ultimate volume reduction observed by Holt and McPherson (1936) cannot be predicted. This argues against isotropy of linear compressibility and in favour of a term $\tilde{W}_J^*(\lambda, 1)$ independent of $\tilde{W}^*(\lambda, 1)$. However, there are clearly gaps in our knowledge of deformation-induced volume changes and much experimental work needs to be done before firm conclusions can be drawn on these matters.

Acknowledgements

The writer is grateful to Professor L. R. G. Treloar for supplying details of the experimental data from Treloar (1944), Jones and Treloar (1975), and Vangerko and Treloar (1978). This article was written whilst the author was a member of the School of Mathematics, University of Bath, Bath, England.

References

Adams, L. H. and Gibson, R. E. (1930) The compressibility of rubber, *J. Acad. Sci. Wash.* **20,** 213–223.

Alexander, H. (1971a) The tensile instability of an inflated cylindrical membrane as affected by an axial load, *Int. J. Mech. Sci.* **13,** 87–95.

Alexander, H. (1971b) Tensile instability of initially spherical balloons, *Int. J. Engng Sci.* **9,** 151–162.

Allen, G., Bianchi, U., and Price, C. (1963) Thermodynamics of elasticity of natural rubber, *Trans. Faraday Soc.* **59,** 2493–2502.

Becker, G. W. (1967) On the phenomenological description of the non-linear deformation behaviour of rubberlike high polymers, *J. Polymer Sci.* C **16,** 2893–2903.

Blatz, P. J. and Ko, W. L. (1962) Application of finite elastic theory to the deformation of rubbery materials, *Trans. Soc. Rheology* **6,** 223–251.

Blatz, P. J., Sharda, S. C., and Tschoegl, N. W. (1974) Strain-energy function for rubberlike materials based on a generalized measure of strain, *Trans. Soc. Rheology* **18,** 145–161.

Boonstra, B. B. S. T. (1950) Stress–strain properties of natural rubber in biaxial strain, *J. Appl. Phys.* **21,** 1098–1104.

Bridgman, P. W. (1945) The compression of 61 substances to 25,000 kg/cm² determined by a new rapid method, *Proc. Am. Acad. Arts Sci.* **76,** 9–24.

Carmichael, A. J. and Holdaway, H. W. (1961) Phenomenological elastomechanical behavior of rubbers over wide ranges of strain, *J. Appl. Phys.* **32,** 159–166.

Chadwick, P. (1974) Thermo-mechanics of rubberlike materials, *Phil. Trans. R. Soc. London* A **276,** 371–403.
Chadwick, P., Creasy, C. F. M., and Hart, V. G. (1977) The deformation of rubber tubes and cylinders by rotation, *J. Australian Math. Soc.* B **20,** 62–96.
Charrier, J.-M. (1974) Large elastic deformation theories and unstable deformations of rubber structures – thin tubes, *Rubber Chem. Technol.* **47,** 788–802.
Christensen, R. C. and Hoeve, J. A. C. (1970) Comparison between theoretical and experimental values of the volume changes accompanying rubber extension, *J. Polymer Sci.* (A–1) **8,** 1503–1512.
Elliott, D. A. and Lippmann, S. A. (1945) The thermodynamics of rubber at small extensions, *J. Appl. Phys.* **16,** 50–54.
Faulkner, M. G. (1972) Compressibility effects for nearly incompressible solids, *Appl. Sci. Res.* **25,** 328–336.
Faulkner, M. G. and Haddow, J. B. (1972) Nearly isochoric finite torsion of a compressible isotropic elastic circular cylinder, *Acta Mechanica* **13,** 245–253.
Flory, P. J. (1961) Thermodynamic relations for high elastic materials, *Trans. Faraday Soc.* **57,** 829–838.
Flory, P. J. and Tatara, Y.-I. (1975) The elastic free energy and the elastic equation of state: elongation and swelling of polydimethylsiloxane networks, *J. Polymer Sci.* (*Physics*) **13,** 683–702.
Fong, J. T. and Penn, R. W. (1975) Construction of a strain-energy function for an isotropic elastic material, *Trans. Soc. Rheology* **19,** 99–113.
Gee, G. (1946) The interaction between rubber and liquids, IX: The elastic behaviour of dry and swollen rubbers, *Trans. Faraday Soc.* **42,** 585–598.
Gee, G., Stern, J., and Treloar, L. R. G. (1950) Volume changes in the stretching of vulcanized natural rubber, *Trans. Faraday Soc.* **46,** 1101–1106.
Gent, A. N. and Lindley, P. B. (1959) Internal rupture of bonded rubber cylinders in tension, *Proc. R. Soc. London* A **249,** 195–205.
Gumbrell, S. M., Mullins, L., and Rivlin, R. S. (1953) Departures of the elastic behaviour of rubbers in simple tension from the kinetic theory, *Trans. Faraday Soc.* **49,** 1495–1505.
Haines, D. W. and Wilson, W. D. (1979) Strain-energy density function for rubberlike materials, *J. Mech. Phys. Solids* **27,** 345–360.
Hart-Smith, L. J. (1966) Elasticity parameters for finite deformations of rubberlike materials, *Z. angew. Math. Phys.* **17,** 608–626.
Haughton, D. M. (1980) Post-bifurcation of perfect and imperfect spherical elastic membranes, *Int. J. Solids Struct.* **16,** 1123–1133.
Haughton, D. M. and Ogden, R. W. (1978) On the incremental equations in non-linear elasticity, II: Bifurcation of pressurized spherical shells, *J. Mech. Phys. Solids* **26,** 111–138.
Haughton, D. M. and Ogden, R. W. (1979) Bifurcation of inflated circular cylinders of elastic material under axial loading, I. Membrane theory for thin-walled tubes, *J. Mech. Phys. Solids* **27,** 179–212.
Hewitt, F. G. and Anthony, R. L. (1958) Measurement of the isothermal volume dilation accompanying the unilateral extension of rubber, *J. Appl. Phys.* **29,** 1411–1414.
Hill, R. (1969) Some aspects of the incremental behaviour of isotropic elastic solids after finite strain, in *Problems in Mechanics: Deformation of Solid Bodies* (V. V. Novozhilov Anniversary Volume), 459–466, Leningrad.
Hill, R. (1978) Aspects of invariance in solid mechanics, *Adv. Appl. Mech.* **18,** 1–75.
Holt, W. L. and McPherson, A. T. (1936) Change of volume of rubber on stretching: effects of time, elongation and temperature, *J. Res. Nat. Bur. Stands* **17,** 657–678.
James, A. G., Green, A., and Simpson, G. M. (1975) Strain-energy functions of rubber, I: Characterization of gum vulcanizates, *J. Appl. Polymer Sci.* **19,** 2033–2058.
Jones, D. F. and Treloar, L. R. G. (1975) The properties of rubber in pure homogeneous strain, *J. Phys.* D, *Appl. Phys.* **8,** 1285–1304.
Kawabata, S. and Kawai, H. (1977) Strain-energy density functions of rubber vulcanizates from biaxial extension, *Adv. Polymer Sci.* **24,** 89–124.
Khasanovich, T. N. (1959) Note on the article by Hewitt and Anthony: "Measurement of the isothermal volume dilation accompanying the unilateral extension of rubber", *J. Appl. Phys.* **30,** 948–949.
Levinson, M. (1972) Finite torsion of slightly compressible rubberlike circular cylinders, *Int. J. Non-Linear Mech.* **7,** 445–463.
Mark, J. E. (1975) The constants $2C_1$ and $2C_2$ in phenomenological elasticity theory and their dependence on experimental variables, *Rubber Chem. Technol.* **48,** 495–512.
Mirsky, L. (1955) *An Introduction to Linear Algebra,* Clarendon Press, Oxford.
Mooney, M. (1940) A theory of large elastic deformation, *J. Appl. Phys.* **11,** 582–592.
Needleman., A. (1977) Inflation of spherical rubber balloons, *Int. J. Solids Struct.* **13,** 409–421.
Obata, Y., Kawabata, S. and Kawai, H. (1970) Mechanical properties of natural rubber vulcanizates in finite deformation, *J. Polymer Sci.* (A–2) **8,** 903–919.
Ogden, R. W. (1970) Waves in isotropic elastic materials of Hadamard, Green or harmonic type, *J. Mech. Phys. Solids* **18,** 149–164.
Ogden, R. W. (1972a) Large deformation isotropic elasticity: on the correlation of theory and experiment for incompressible rubberlike solids, *Proc. R. Soc. London* A **326,** 565–584.

Ogden, R. W. (1972b) Large deformation isotropic elasticity: on the correlation of theory and experiment for compressible rubberlike solids, *Proc. R. Soc. London* A **328,** 567–583.
Ogden, R. W. (1974) On isotropic tensors and elastic moduli, *Proc. Cambridge Phil. Soc.* **75,** 427–436.
Ogden, R. W. (1976) Volume changes associated with the deformation of rubberlike solids, *J. Mech. Phys. Solids* **24,** 323–338.
Ogden, R. W. (1978) Nearly isochoric elastic deformations: application to rubberlike solids, *J. Mech. Phys. Solids* **26,** 37–57.
Ogden, R. W. (1979a) Biaxial deformation of rubberlike solids: Comparison of theory and experiment, *J. Phys.* D, *Appl. Phys.* **12,** 1463–1472.
Ogden, R. W. (1979b) On the anisotropy of compressibility of rubberlike solids, *J. Phys.* D, *Appl. Phys.* **12,** 465–479.
Peng, S. T. J. and Landel, R. F. (1975) Stored-energy function and compressibility of compressible rubberlike materials under large strain, *J. Appl. Phys.* **46,** 2599–2604.
Penn, R. W. (1970) Volume changes accompanying the extension of rubber, *Trans. Soc. Rheology* **14,** 509–517.
Price, C. (1976) Thermodynamics of rubber elasticity, *Proc. R. Soc. London* A **351,** 331–350.
Price, C. and Allen, G. (1973) A study of dilation coefficients of cis-polybutadiene in simple tension, *Polymer* **14,** 576–578.
Price, C., Padget, J., Kirkham, M. J., and Allen, G. (1969) The effect of strain on the equilibrium properties of rubber networks, *Polymer* **10,** 495–499.
Rivlin, R. S. (1948) Large elastic deformations of isotropic materials, IV. Further developments of the general theory, *Phil. Trans. R. Soc. London* A **241,** 379–397.
Rivlin, R. S. (1949) A note on the torsion of an incompressible highly-elastic cylinder, *Proc. Cambridge Phil. Soc.* **45,** 485–487.
Rivlin, R. S. and Saunders, D. W. (1951) Large elastic deformations of isotropic materials, VII: Experiments on the deformation of rubber, *Phil. Trans. R. Soc. London* A **243,** 251–288.
Rivlin, R. S. and Sawyers, K. N. (1976) The strain-energy function for elastomers, *Trans. Soc. Rheology* **20,** 545–557.
Sewell, M. J. (1976) Some mechanical examples of catastrophe theory, *Bull. Inst. Maths Applics* **12,** 163–172.
Sharda, S. C. and Tschoegl, N. W. (1976) A strain-energy density function for compressible rubberlike materials, *Trans. Soc. Rheology* **20,** 361–372.
Spencer, A. J. M. (1964) Finite deformation of an almost incompressible elastic solid, *Second-order Effects in Elasticity, Plasticity and Fluid Dynamics* (*Proceedings of IUTAM Symposium, Haifa, 1962*), edited by M. Reiner and D. Abir, 200–216, Pergamon Press, Oxford.
Spencer, A. J. M. (1970) The static theory of finite elasticity, *J. Inst. Maths Applics* **6,** 164–200.
Treloar, L. R. G. (1944) Stress–strain data for vulcanized rubber under various types of deformation, *Trans. Faraday Soc.* **40,** 59–70.
Treloar, L. R. G. (1948) Stresses and birefringence in rubber subjected to general homogeneous strain, *Proc. Phys. Soc.* **60,** 135–144.
Treloar, L. R. G. (1969) Volume changes and mechanical anisotropy of strained rubbers, *Polymer* **10,** 279–289.
Treloar, L. R. G. (1975) *The Physics of Rubber Elasticity*, 3rd ed., Clarendon Press, Oxford.
Treloar, L. R. G. (1976) The mechanics of rubber elasticity, *Proc. R. Soc. London* A **351,** 301–330.
Treloar, L. R. G. (1978) Dilation of rubber on extension, *Polymer* **19,** 1414–1420.
Treloar, L. R. G. and Riding, G. (1980) A non-Gaussian theory for rubber in biaxial strain, I: Mechanical properties, *Proc. R. Soc. London* A **369,** 261–280.
Valanis, K. C. and Landel, R. F. (1967) The strain-energy function of a hyperelastic material in terms of the extension ratios, *J. Appl. Phys.* **38,** 2997–3002.
Vangerko, H. and Treloar, L. R. G. (1978) The inflation and extension of rubber tube for biaxial strain studies, *J. Phys.* D, *Appl. Phys.* **11,** 1969–1978.
Varga, O. H. (1966) *Stress–Strain Behavior of Elastic Materials*, Interscience, New York.

Elastic–Plastic Crack Growth

by J. R. Rice

Division of Applied Sciences, Harward University

Summary

The asymptotic structure of near-tip stress and deformation fields is analyzed for cracks growing in elastic–perfectly plastic solids. A general formulation is presented for materials of arbitrary yield condition and associated flow rule, including anisotropic response, although detailed results are presented only for isotropic materials of the Huber–Mises type. Centered fan sectors of singular straining at a crack tip are shown to be general, independent of details of material response, as are also the types of plastic strain singularities associated with stationary and growing cracks. Previously results for the Huber–Mises material are recovered by specialization from the general formulation, some corrections are made, and recent work on using results of such a near-tip analysis as a basis for predicting plane strain stable crack growth is reviewed briefly.

1. Introduction

This article describes the quasi-static analysis of stress–deformation fields very near the tips of growing cracks in materials which are idealized as being of the elastic–perfectly plastic type. Results are given for isotropic materials of the Huber–Mises type satisfying the Prandtl–Reuss flow rule, but much of the analysis is carried through for arbitrary forms of the yield condition and arbitrary anisotropy of material, restricted only by the requirements that the flow rule is of the associated type (normality) and that the material is homogeneous. Effects of the geometry of ongoing deformation on the formulation of stress (and stress rate) measures and on the form of the equilibrium equations are neglected.

The analysis is carried out for two-dimensional problems of plane strain, anti-plane strain, or general combinations of the two. Indeed, as regards a study of crack tip singularities, *three*-dimensional crack problems are included in the scope of the study as well, in the sense that the components of the strain tensor which can become infinite at a three-dimensional crack front are the plane and anti-plane components but not (generally) the component parallel to the crack front itself. Also, the analysis is formulated so that the generalized plane stress approximation for cracked thin sheets can be discussed as well as a special case.

With reference to Fig. 1.1 let cartesian axes x_1, x_2, x_3 be chosen so that x_3 is parallel to the crack front and x_1 points in the direction of crack growth. Let r, θ be polar coordinates in the

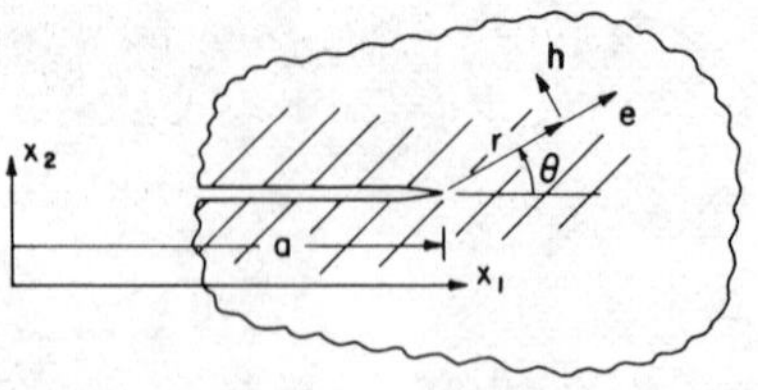

FIG. 1.1. x_1 and x_2 are fixed cartesian coordinates; polar coordinates r, θ are centered at the tip and move relative to the material as the crack grows.

x_1, x_2 plane, with $\theta = 0$ coinciding with the positive x_1 direction. These polar coordinates are always centered on the crack tip and move through the material with the tip; a is the measure of crack length. Also, as shown, **e** is a unit vector in the radial direction, **h** in the angular direction. Note that

$$\partial r/\partial x_\alpha = e_\alpha, \quad \partial\theta/\partial x_\alpha = h_\alpha/r, \tag{1.1}$$

and, for example, $e_1 = \cos\theta$, $e_2 = \sin\theta$. In the next four Sections we shall be concerned with an asymptotic analysis of the stress and deformation field as $r \to 0$ at the crack tip.

2. Governing Equations and Asymptotic Forms

2.1. *Equilibrium*

The equilibrium equations are

$$\partial\sigma_{\alpha\beta}/\partial x_\alpha + f_\beta = 0, \tag{2.1}$$

where $\sigma_{\alpha\beta} = \sigma_{\beta\alpha}$ is the stress tensor, f_β the body force, and where the letters $\alpha\beta\gamma\delta$ are used to index cartesian components of tensors, have range 1, 2, 3, and follow the summation convention. It is convenient to regard $\sigma_{\alpha\beta}$ as a function of r and θ so that with the help of (1.1) the previous equation becomes

$$(\partial\sigma_{\alpha\beta}/\partial\theta)(h_\alpha/r) + (\partial\sigma_{\alpha\beta}/\partial r)e_\alpha + f_\beta = 0. \tag{2.2}$$

Now since rf_β and $r\partial\sigma_{\alpha\beta}/\partial r \to 0$ as $r \to 0$ (the latter because stress is bounded in the type of perfectly plastic materials considered), the equilibrium equations (2.2) reduce, as $r \to 0$, to

$$h_\alpha \sigma'_{\alpha\beta} = 0, \quad \text{where} \quad \sigma'_{\alpha\beta} = [\partial\sigma_{\alpha\beta}(r,\theta)/\partial\theta]_{r=0}. \tag{2.3}$$

That is, the equations constrain only the variation of $\sigma_{\alpha\beta}$ with θ at $r = 0$. Equations (2.3) are, of course, just the equilibrium equations which one would write if it was assumed from the outset that $\sigma_{\alpha\beta} = \sigma_{\alpha\beta}(\theta)$. Equivalent forms, convenient for various purposes, follow either from recognizing that $(h_1, h_2) = (-e_2, e_1)$ so that, writing the equations out fully,

$$e_2\sigma'_{11} = e_1\sigma'_{21}, \quad e_2\sigma'_{12} = e_1\sigma'_{22}, \quad e_2\sigma'_{13} = e_1\sigma'_{23}, \tag{2.4}$$

or from resorting to polar coordinate representations of stress components in which case

$$\sigma_{rr} - \sigma_{\theta\theta} + \sigma'_{r\theta} = 0, \quad 2\sigma_{r\theta} + \sigma'_{\theta\theta} = 0, \quad \sigma_{r3} + \sigma'_{\theta 3} = 0. \tag{2.5}$$

Note that here $\sigma'_{r\theta}$, $\sigma'_{\theta\theta}$, and $\sigma'_{\theta 3}$ are derivatives of $\sigma_{r\theta}$, $\sigma_{\theta\theta}$, and $\sigma_{\theta 3}$ with respect to θ at $r = 0$ and *not* the $r\theta$, $\theta\theta$, and $\theta 3$ components of $\sigma'_{\alpha\beta}$.

A result important for some subsequent calculations follows at once from the equilibrium equations in the form (2.4). Let $H_{\alpha\beta}$ be an arbitrary tensor. Then by writing $\sigma'_{11} = \sigma'_{12}e_1/e_2$, $\sigma'_{22} = \sigma'_{12}e_2/e_1$, $\sigma'_{23} = \sigma'_{13}e_2/e_1$ one calculates that

$$\sigma'_{\alpha\beta}H_{\alpha\beta} = (\sigma'_{12}/e_1e_2)e_\alpha e_\beta H_{\alpha\beta} + \sigma'_{33}H_{33} + 2(\sigma'_{13}/e_1)e_\alpha H_{\alpha 3}.$$

But $e_\alpha e_\beta H_{\alpha\beta} = H_{rr}$, etc., where H_{rr} is the radial diagonal component of the tensor $H_{\alpha\beta}$, so that this may be rewritten as

$$\sigma'_{\alpha\beta}H_{\alpha\beta} = (\sigma'_{12}/e_1e_2)H_{rr} + \sigma'_{33}H_{33} + 2(\sigma'_{13}/e_1)H_{r3}. \tag{2.6}$$

It can be shown from (2.4) that $\sigma'_{12}/e_1e_2 = \sigma'_{11} + \sigma'_{22}$, and this latter form is sometimes convenient.

2.2. *Yield condition*

The yield condition is presumed to take the form $f(\sigma) = 0$, and it is supposed that f has been made to depend symmetrically on $\sigma_{\alpha\beta}$ and $\sigma_{\beta\alpha}$, and signed such that

$$P_{\alpha\beta} = \partial f(\sigma)/\partial\sigma_{\alpha\beta} \tag{2.7}$$

is a symmetric tensor which can be interpreted as the outer normal to the yield surface in stress space. The material is taken to be plastically incompressible, so that $P_{\alpha\alpha} = 0$. Of course, for the Huber–Mises criterion

$$f = \tfrac{1}{2}s_{\alpha\beta}s_{\alpha\beta} - \tau_0{}^2, \quad \text{and} \quad P_{\alpha\beta} = s_{\alpha\beta}, \tag{2.8}$$

where $s_{\alpha\beta}$ is the deviatoric stress tensor and τ_0 is the shear strength.

It is convenient to adopt the yield condition in the differential form

$$P_{\alpha\beta}\partial\sigma_{\alpha\beta}/\partial p = 0, \tag{2.9}$$

to be satisfied in plastic regions of the body, where the parameter p may represent either time or some spatial coordinate. In particular, as $r \to 0$, $\sigma'_{\alpha\beta}$ as defined in (2.3) becomes a suitable $\partial\sigma_{\alpha\beta}/\partial p$. Hence in ranges of the angle θ which do indeed correspond to plastic material response at the crack tip,

$$P_{\alpha\beta}\sigma'_{\alpha\beta} = 0. \tag{2.10}$$

From (2.6) we conclude therefore that in such plastic sectors

$$(\sigma'_{12}/e_1e_2)P_{rr} + \sigma'_{33}P_{33} + 2(\sigma'_{13}/e_1)P_{r3} = 0. \tag{2.11}$$

There are six unknown stress components $\sigma_{\alpha\beta}$ whose variation with θ is to be determined, but thus far we have assembled only four equations, namely (2.11) and any convenient form, such as (2.4) or (2.5), of the three equilibrium equations. Hence the general problem of deriving stress fields in plastic sectors at crack tips is *not* statically determined, although some special versions of the problem are. This issue is discussed further in Section 2.3.

The formulation in terms of $P_{\alpha\beta}$ is not adaptable directly to yield surfaces with vertices, although it may be assumed that such cases would emerge as a simple limit of the present formulation for smooth yield surfaces. Such expectations are reinforced by results for crack and analogous punch problems in anisotropic materials (Rice, 1967, 1973).

2.3. *Stress–strain relations*

These are expressed in rate form, with the rate of deformation defined by

$$D_{\alpha\beta} = \dot{\varepsilon}_{\alpha\beta} = \tfrac{1}{2}(\partial v_\alpha/\partial x_\beta + \partial v_\beta/\partial x_\alpha). \tag{2.12}$$

Here $\varepsilon_{\alpha\beta}$ is the (infinitesimal) strain, the superposed dot denotes time rate at a fixed material point, v_α is the velocity field, and $v_\alpha = \dot{u}_\alpha$ where u_α is the displacement field. The use of the notation $D_{\alpha\beta}$ avoids potential confusion with $\dot{\varepsilon}_{\alpha\beta}$ because, while $D_{22} = \dot{\varepsilon}_{22}$, for example, $D_{rr} \neq \dot{\varepsilon}_{rr}$ when the crack is moving (at least if $\dot{\varepsilon}_{rr}$ is thought of as the time rate of ε_{rr}; rather, $D_{rr} = e_\alpha e_\beta \dot{\varepsilon}_{\alpha\beta}$ and $\dot{\varepsilon}_{rr} = d(e_\alpha e_\beta \varepsilon_{\alpha\beta})/dt$).

For elastic response, whether in unyielded material or in material that has previously yielded but is not now plastically deforming,

$$D_{\alpha\beta} = D^e_{\alpha\beta} = M_{\alpha\beta\gamma\delta}\dot{\sigma}_{\gamma\delta}, \tag{2.13}$$

where M is the (constant) matrix of elastic compliances and has the usual symmetries. For example,

$$M_{\alpha\beta\gamma\delta} = \frac{1+v}{2E}(\delta_{\alpha\gamma}\delta_{\beta\delta} + \delta_{\alpha\delta}\delta_{\beta\gamma}) - \frac{v}{E}\delta_{\alpha\beta}\delta_{\gamma\delta} \tag{2.14}$$

in the case of isotropic elastic response, where E is the Young modulus and v the Poisson ratio. In the case of elastic–plastic material response, an associated flow rule is assumed so that

$$D_{\alpha\beta} = D^e_{\alpha\beta} + D^p_{\alpha\beta} = M_{\alpha\beta\gamma\delta}\dot{\sigma}_{\gamma\delta} + \dot{\Lambda}P_{\alpha\beta}, \tag{2.15}$$

where $\dot{\Lambda} \geqslant 0$ is an undetermined parameter.

To examine (2.15) near the crack tip, write $\sigma_{\gamma\delta} = \sigma_{\gamma\delta}(r, \theta, t)$ and observe that

$$\begin{aligned}\dot{\sigma}_{\gamma\delta} &= (\partial\sigma_{\gamma\delta}/\partial\theta)\dot{\theta} + (\partial\sigma_{\gamma\delta}/\partial r)\dot{r} + \partial\sigma_{\gamma\delta}/\partial t \\ &= (\partial\sigma_{\gamma\delta}/\partial\theta)(e_2\dot{a}/r) - (\partial\sigma_{\gamma\delta}/\partial r)(e_1\dot{a}) + \partial\sigma_{\gamma\delta}/\partial t,\end{aligned} \tag{2.16}$$

where use is made of $\dot{\theta} = e_2\dot{a}/r$, $\dot{r} = -e_1\dot{a}$. Hence it is seen that as the crack tip is approached,

$$r\dot{\sigma}_{\gamma\delta} \to \sigma'_{\gamma\delta}e_2\dot{a}. \tag{2.17}$$

This means that the stress rates, and hence the elastic strain rates $D^e_{\alpha\beta}$, are singular as $\dot{a}/r$ in angular sectors about the crack tip in which the cartesian stress components vary with θ (of course the elastic *strains* $\varepsilon^e_{\alpha\beta}$ remain finite). In general, one finds that such elastic strain rates, created in plastically yielding angular sectors, are not themselves compatible, in the sense of being derivable from a velocity field. Hence additional plastic strain rates are necessitated to meet compatibility, and these (as will be seen in subsequent examples) are more strongly singular than $\dot{a}/r$, leading to unbounded strains $\varepsilon^p_{\alpha\beta}$ at the crack tip.

Our goal at this point is to reduce (2.15) to tangible restrictions on the form of the near-tip deformation-rate field. This is done first by seeking consequences of (2.15) that do not explicitly involve the $P_{\alpha\beta}$ or $\dot{\Lambda}$. One such consequence follows from plastic incompressibility $P_{\alpha\alpha} = 0$:

$$D_{\alpha\alpha} = \partial v_\alpha/\partial x_\alpha = M_{\alpha\alpha\gamma\delta}\dot{\sigma}_{\gamma\delta}. \tag{2.18}$$

Multiplying by r and letting $r \to 0$ we therefore have

$$rD_{\alpha\alpha} = r\partial v_\alpha/\partial x_\alpha \to M_{\alpha\alpha\gamma\delta}\sigma'_{\gamma\delta}e_2\dot{a}, \tag{2.19}$$

which will be simplest to use with velocities referred to polar coordinate directions (note that $v_2 = \dot{u}_2$ but $v_r \neq \dot{u}_r$ when the crack is advancing, where $\dot{u}_r$ is the time derivative of $\dot{u}_r$). The result is

$$rD_{33} + \partial v_\theta/\partial\theta + \partial(rv_r)/\partial r \to M_{\alpha\alpha\gamma\delta}\sigma'_{\gamma\delta}e_2\dot{a}, \tag{2.20}$$

and we note that the first term, containing D_{33}, vanishes for arbitrary combinations of plane and anti-plane strain; for the plane stress approximation, D_{33} does not vanish, and the equation then simply serves to give D_{33} near the tip in terms of in-plane quantities.

Another significant restriction on the near-tip deformation rate field, following from (2.15), is obtained by multiplying both sides of that equation by $r\partial\sigma_{\alpha\beta}/\partial\theta$. Since $\dot{\Lambda}P_{\alpha\beta}\partial\sigma_{\alpha\beta}/\partial\theta = 0$ (in plastic regions because of (2.9), in elastic regions because $\dot{\Lambda} = 0$), we have

$$(\partial\sigma_{\alpha\beta}/\partial\theta)(rD_{\alpha\beta}) = (\partial\sigma_{\alpha\beta}/\partial\theta)M_{\alpha\beta\gamma\delta}(r\dot{\sigma}_{\gamma\delta}). \tag{2.21}$$

Hence, using the definition of $\sigma'_{\alpha\beta}$ in (2.3) and using (2.17) for the limit of $r\dot{\sigma}_{\gamma\delta}$, we have, as $r \to 0$,

$$\sigma'_{\alpha\beta}(rD_{\alpha\beta}) \to \sigma'_{\alpha\beta}M_{\alpha\beta\gamma\delta}\sigma'_{\gamma\delta}e_2\dot{a}. \tag{2.22}$$

Making use of (2.6), this becomes

$$\sigma'_{33}(rD_{33}) + (\sigma'_{12}/e_1e_2)(rD_{rr}) + (\sigma'_{13}/e_1)(2rD_{r3}) \to \sigma'_{\alpha\beta}M_{\alpha\beta\gamma\delta}\sigma'_{\gamma\delta}e_2\dot{a}.$$

Note that the first term on the left always vanishes (either $D_{33} = 0$ or $\sigma_{33} = 0$ for the class of problems considered). Thus, using the polar coordinate relations $D_{rr} = \partial v_r/\partial r$ and $2D_{r3} = \partial v_3/\partial r$ (note that $\partial v_r/\partial x_3$ vanishes for the problems considered), we have

$$(\sigma'_{12}/e_1e_2)(r\partial v_r/\partial r) + (\sigma'_{13}/e_1)(r\partial v_3/\partial r) \to \sigma'_{\alpha\beta}M_{\alpha\beta\gamma\delta}\sigma'_{\gamma\delta}e_2\dot{a}. \tag{2.23}$$

Equations (2.20) and (2.23) are two useful consequences of the flow rule for the present type of near-tip asymptotic analysis. They embody all such asymptotic restrictions on the velocity gradients within plastic singular sectors for states of plane and anti-plane strain (provided that one further, indirect consequence of the flow rule, namely $P_{33} = 0$—as will be discussed—is included also); the flow-rule serves further in such cases only to define $\dot{\Lambda}$, which must be checked as to its sign. Within angular sectors which currently respond elastically, the situation is different. Equations (2.20) and (2.23) are valid in such sectors also, but now (2.13) rather than (2.15) applies and, in cases such as plane or anti-plane strain, the content of the stress-rate–strain-rate relations not already embodied in (2.20) and (2.23) provides one additional restriction on the stress rates (rather than just a characterization of $\dot{\Lambda}$). Of course, while this one restriction is gained, the restriction set by (2.10) for plastic sectors is irrelevant in such elastic sectors, so that the total number of equations is the same in each type of sector.

As a closing comment on utilization of (2.20) and (2.23), note that the result (2.6) can be used to simplify the right-hand sides of each. Thus, for use in (2.20),

$$M_{\alpha\alpha\gamma\delta}\sigma'_{\gamma\delta} = M_{\alpha\alpha rr}(\sigma'_{12}/e_1e_2) + M_{\alpha\alpha 33}\sigma'_{33} + M_{\alpha\alpha r3}(2\sigma'_{13}/e_1) \tag{2.24}$$

and, for use in (2.23),

$$\sigma'_{\alpha\beta}M_{\alpha\beta\gamma\delta}\sigma'_{\gamma\delta} = M_{rrrr}(\sigma'_{12}/e_1e_2)^2 + M_{3333}(\sigma'_{33})^2 + M_{r3r3}(2\sigma'_{13}/e_1)^2 + 2M_{rr33}(\sigma'_{12}/e_1e_2)(\sigma'_{33}) + 2M_{33r3}(\sigma'_{33})(2\sigma'_{13}/e_1) + 2M_{rrr3}(\sigma'_{12}/e_1e_2)(2\sigma'_{13}/e_1). \tag{2.25}$$

Of course, in anisotropic materials compliances such as M_{rrrr}, M_{rr33}, M_{rrr3}, etc. generally depend on θ.

2.4. *Plan of further Sections*

In the following three Sections the general formulation given here is applied to the problem of determining near-tip stress and deformation states separately for the case of anti-plane strain and plane strain, and for plane stress. Naturally, in treating separately states of anti-plane and plane strain, it must be assumed that the anisotropy is not such that the two modes are inevitably coupled (e.g., one assumes $M_{rrr3} = 0$ and related constraints on the nature of plastic response). This proviso being made, it should be noted that there are, of course, cases of crack growth (in both isotropic and anisotropic materials) which involve significantly coupled anti-plane and in-plane plastic deformations. Slant fractures in ductile plates are an example. Such problems seem not yet to have been studied from the standpoint of elastic–plastic analysis. The formulation of the present Section provides a sufficient basis for addressing them, but a fuller analysis is deferred for possible future work.

3. Anti-plane Strain

The only non-vanishing displacement is u_3, and non-vanishing components of $\sigma_{\alpha\beta}, \varepsilon_{\alpha\beta}, P_{\alpha\beta}$, etc. are the 31 (= 13) and 32 (= 23) components only.

Hence equilibrium takes the form ((2.4) and (2.5))

$$e_2\sigma'_{13} = e_1\sigma'_{23} \quad \text{or} \quad \sigma_{r3} + \sigma'_{\theta 3} = 0, \tag{3.1}$$

whereas the meeting of the yield condition in plastic sectors requires that, from (2.11),

$$\sigma'_{13}P_{r3} = 0. \tag{3.2}$$

The flow rule reduces to

$$D_{\alpha 3} = 2M_{\alpha 3\beta 3}\dot{\sigma}_{\beta 3} + \dot{\Lambda}P_{\alpha 3}, \tag{3.3}$$

and its consequence in the form of (2.20) is irrelevant to the present case, whereas (2.23) is pertinent and reduces to

$$(\sigma'_{13}/e_1)(r\partial v_3/\partial r) \to 4M_{r3r3}(\sigma'_{13}/e_1)^2 e_2\dot{a}, \tag{3.4}$$

which can be rewritten using (3.1) in sectors in which $\sigma'_{\alpha 3} \neq 0$ as

$$r\partial v_3/\partial r \to 4M_{r3r3}\sigma'_{23}\dot{a}. \tag{3.5}$$

3.1. *Stress distribution in plastic sectors*

Observe that the anti-plane problem is statically determined in plastic sectors, at least in the sense that the equilibrium and yield equations, (3.1) and (3.2), are equal in number to the unknown stress components. It is obvious by inspection of these equations that plastically active sectors at the crack tip must have the form of either:

(i) *constant stress sectors* (i.e., σ_{13} = constant, σ_{23} = constant, where σ_{13} and σ_{23} meet the yield condition); *or*

(ii) *centered fan sectors* (i.e., $P_{r3} = 0$. The name "centered fan sectors" is applied because, in the corresponding rigid–plastic material model, directions along which the projection of $P_{\alpha 3}$ vanishes are characteristic lines. Hence $P_{r3} = 0$ corresponds to all radial lines from the crack tip being characteristics, thus forming a centered fan of characteristics.)

If we consider the isotropic Huber–Mises material, $P_{\alpha 3} = \sigma_{\alpha 3}$. In fact, $P_{\alpha 3}$ is proportional to $\sigma_{\alpha 3}$ for *any* isotropic perfectly plastic material in anti-plane strain. Thus, for such materials the condition $P_{r3} = 0$ in centered fan sectors reduces to $\sigma_{r3} = 0$ and, when (3.1) is used, the stress state is therefore

$$\sigma_{r3} = 0, \quad \sigma_{\theta 3} = \text{constant} = \pm\, \tau_0, \tag{3.6}$$

where τ_0 is the yield strength in shear. This field was found by Hult and McClintock (1956), and used by them, Hult (1957) and McClintock (1958), to discuss anti-plane problems of stationary and growing cracks. See also Rice (1968) and Chitaley and McClintock (1971).

3.2. *Deformation field in plastic sectors*

If (3.5) is regarded as an equality valid sufficiently close to the crack tip, then integration gives

$$v_3 = -4M_{r3r3}\sigma'_{23}\dot{a}\,\ln(\bar{R}/r) + f(\theta, t) \tag{3.7}$$

as the near-tip velocity field, where $\bar{R}$ is an arbitrary length to scale r in the logarithmic term and where $f(\theta, t)$ is a function which, within plastic sectors, is undetermined by the asymptotic analysis. The function can be related to the location and rate of movement of the elastic–plastic boundary, and Rice (1968) gives details for the isotropic case. Hence, a full solution is required to determine the near-tip field completely. The deformation-rate field in centered fan plastic sectors is therefore

$$\left.\begin{aligned} D_{r3} &= 2M_{r3r3}\sigma'_{23}\dot{a}/r, \\ D_{\theta 3} &= (1/2r)(\partial v_3/\partial\theta) \\ &= -2[d(M_{r3r3}\sigma'_{23})/d\theta](\dot{a}/r)\ln(\bar{R}/r) + (\partial f/\partial\theta)/2r. \end{aligned}\right\} \tag{3.8}$$

In this case the strain rate D_{r3} is entirely elastic (since $P_{r3} = 0$), whereas $D_{\theta 3}$ includes plastic contributions.

For centered fan sectors in isotropic materials, the above results reduce to

$$v_3 = (\tau_0/G)\dot{a}\sin\theta\,\ln(\bar{R}/r) + f(\theta, t), \tag{3.9}$$

$$\left.\begin{aligned} D_{r3} &= (\tau_0/2G)\dot{a}\sin\theta/r, \\ D_{\theta 3} &= (\tau_0/2G)(\dot{a}\cos\theta/r)\ln(\bar{R}/r) + (\partial f/\partial\theta)/2r. \end{aligned}\right\} \tag{3.10}$$

In fact, while derived here as asymptotically valid results, these actually apply throughout the plastic zone, within centered fan angular regions.

On the line $\theta = 0$ one has (e.g., Rice, 1968), for contained plasticity,

$$D_{23} = \dot{\varepsilon}_{23} = (D_{\theta 3})_{\theta = 0}$$
$$= (\tau_0/2G)\{(\dot{a}/r)[1 + \ln(R_0/r)] + \dot{R}_0/r\}, \tag{3.11}$$

where R_0 is the radius to the elastic-plastic boundary along $\theta = 0$. By integration, this result shows that for a stationary crack (i.e., $\dot{a} = 0$) under monotonic load increase,

$$\varepsilon_{23} = (\tau_0/2G)(R_0/r), \tag{3.12}$$

whereas for the opposite extreme of a crack which advances under steady conditions, in the sense that the loads are adjusted so that R_0 remains constant during growth,

$$\varepsilon_{23} = (\tau_0/2G)\{1 + \ln(R_0/r) + \tfrac{1}{2}[\ln(R_0/r)]^2\}. \tag{3.13}$$

The comparison of these last two results indicates at once the significant differences in patterns of strain accumulation at the tips of stationary versus growing cracks. Similar differences are known to exist for tensile modes too; e.g., see Sections 4 and 5. As McClintock (1958) and McClintock and Irwin (1965) emphasized, these differences are a source of extensive stable crack growth in ductile materials. For example, let us adopt the latter's simple fracture criterion that a critical plastic (engineering) shear strain $\eta\tau_0/G$ must be achieved at distance ρ ahead of the crack tip for growth. Then for the initiation of crack growth, the plastic zone must be enlarged to the size

$$(R_0)_{\text{init.}} = \rho(1 + \eta), \tag{3.14}$$

whereas for the maintenance of steady state crack growth as described above the plastic zone must be enlarged to the size

$$(R_0)_{\text{s.s.}} = \rho \exp[(1 + 2\eta)^{1/2} - 1]. \tag{3.15}$$

The ratio of $(R_0)_{\text{s.s.}}$ to $(R_0)_{\text{init.}}$ varies strongly with ductility, as expressed by η; that ratio is 1.04 when $\eta = 1$, 3 when $\eta = 10$, 18 when $\eta = 25$, 169 when $\eta = 50$, and 5100 when $\eta = 100$.

By comparison of the general results of (3.7, 8) to (3.9, 10) for isotropic materials, it is evident that similar results will occur for the character of strain singularities, and thus the distinctions noted between stationary and growing cracks are general in their application to all materials within the class considered here. Such distinctions are, of course, directly related to the path dependence and relative permanence of plastic deformations. Similar distinctions would not result, for example, in nonlinear elastic materials.

3.3. *Elastic sectors*

Elastic sectors at the crack tip may be of a constant stress form, $\sigma'_{\alpha 3} = 0$, which trivially satisfy the equilibrium equation. Such sectors result immediately adjacent to the crack surfaces for monotonic loading of a stationary crack (Fig. 3.1(a)), and the stresses within those sectors (occupying, for an isotropic material, $\frac{1}{2}\pi < |\theta| < \pi$) just meet the yield condition as r reaches zero. In fact σ_{23} vanishes in such sectors and σ_{13} is at yield.

However, a more general type of elastic sector, in which $\sigma'_{\alpha 3} \neq 0$, is possible for the growing crack. In such a sector, the equilibrium equation (3.1) must be met, and the constitutive relation has the form

$$D_{\alpha 3} = 2M_{\alpha 3 \beta 3}\dot{\sigma}_{\beta 3}. \tag{3.16}$$

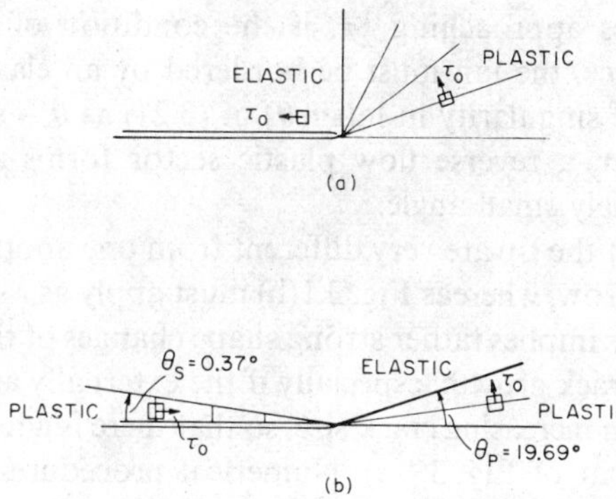

FIG. 3.1. Near-tip fields for anti-plane strain of isotropic material: (a) stationary crack; (b) growing crack.

One consequence of this constitutive relation is already embodied in (3.5), which applies whether plastic flow is occurring or not, namely,

$$2rD_{r3} = r\partial v_3/\partial r \to 4M_{r3r3}\sigma'_{23}\dot{a}. \tag{3.17}$$

Another consequence is readily shown to be

$$2rD_{\theta 3} = \partial v_3/\partial\theta = 4M_{\theta 3\beta 3}r\dot{\sigma}_{\beta 3} \to 4M_{\theta 3r3}\sigma'_{23}\dot{a}. \tag{3.18}$$

These last two results for derivatives of v_3 are consistent with one another only if

$$d(M_{r3r3}\sigma'_{23})/d\theta = 0, \tag{3.19}$$

and this (essentially, a deformation compatibility requirement), together with the equilibrium equation, determines the admissible forms of stress variation in elastic sectors.

For example, in the case of an elastically isotropic material, M_{r3r3} is independent of θ and (3.19) integrates to

$$\sigma_{23} = A + B\theta, \tag{3.20}$$

whereas (3.1) shows that

$$\sigma_{13} = C + B\ln|\sin\theta|; \tag{3.21}$$

here A, B, and C are integration constants.

3.4. *Assembly of results for the crack tip field*

The complete field as $r \to 0$ at the crack tip is given by assembling plastic and elastic sectors in such a way that appropriate continuity conditions are met at boundaries between sectors, and that the plastic flow rate amplitude Λ is nonnegative. The result as obtained by Hult and McClintock (1956) is shown in Fig. 3.1(a) for a stationary crack. Rice (1967) has shown the generalization of this construction for materials of arbitrary anisotropy.

The result for a growing crack in an isotropic material was obtained by Chitaley and McClintock (1971) and is shown in Fig. 3.1(b). One finds in this case that the centered fan in front of the crack cannot join onto a plastic sector of constant stress type, and the fan itself cannot continue to angles approaching $\frac{1}{2}\pi$, if the condition of nonnegative dissipation ($\dot{\Lambda} \geqslant 0$) is to be met. Hence, the fan must be bordered by an elastic sector as shown, but this sector (because of the singularity in $\ln(\sin\theta)$ of (3.21) as $\theta \to \pi$) cannot continue up to the crack surfaces. Rather, a reverse flow plastic sector forms as shown, although this sector occupies an extremely small angle.

The asymptotic states at the tip are very different from one another. Figure 3.1(a) applies until the crack begins to grow, whereas Fig. 3.1(b) must apply as $r \to 0$ even for the first few increments of growth. This implies rather strong shape changes of the plastic region near the tip in the early stages of crack growth, especially if the externally applied load is then being increased substantially with increasing crack size, so that there is a tendency continually to re-establish a pattern like that of Fig. 3.1(a). Numerical procedures are not well suited for documenting such efforts, but some finite element results, which suggest some details of the transition, have been given by Sorensen (1978). Also, Dean and Hutchinson (1980), in finite element work which solves the problem of steady state crack growth, conclude that the asymptotic field of Fig. 3.1(b), while presumed to be correct, becomes evident only on a finer scale than their numerical solution was able to resolve.

4. Plane Strain

For this case, the non-vanishing displacements are u_1, u_2, and non-vanishing components of $\sigma_{\alpha\beta}$, $\varepsilon_{\alpha\beta}$, $P_{\alpha\beta}$, etc. are the 11, 12 (= 21), 22, and 33 components. Hence the equilibrium equations (2.4) reduce to

$$e_2\sigma'_{12} = e_1\sigma'_{22}, \quad e_2\sigma'_{11} = e_1\sigma'_{21}, \tag{4.1}$$

or, equivalently, to

$$\sigma_{rr} - \sigma_{\theta\theta} + \sigma'_{r\theta} = 0, \quad 2\sigma_{r\theta} + \sigma'_{\theta\theta} = 0, \tag{4.2}$$

whereas the meeting of the yield condition in plastic sectors requires that, from (2.11),

$$(\sigma'_{11} + \sigma'_{22})P_{rr} + \sigma'_{33}P_{33} = 0, \tag{4.3}$$

where we have noted that $\sigma'_{12}/e_1e_2 = \sigma'_{11} + \sigma'_{22}$ by (4.1).

The flow rule has the form (2.15) and recognizing that $\varepsilon_{33} = 0$ for plane strain,

$$\varepsilon^e_{33} + \varepsilon^p_{33} = M_{33\alpha\beta}\sigma_{\alpha\beta} + \varepsilon^p_{33} = 0, \tag{4.4}$$

where $\dot{\varepsilon}^p_{33} = D^p_{33} = \dot{\Lambda}P_{33}$. The strain ε^p_{33} is therefore bounded at the crack tip, and by a development similar to that leading to (2.16), D^p_{33} can at most be singular of order $\dot{a}/r$ at the tip.

Now, if we consider what will be referred to as a *singular* plastically deforming sector at the crack tip, namely a sector in which one or more of the in-plane components of $D^p_{\alpha\beta}$ become infinite faster than $\dot{a}/r$ (so that in-plane components of $\varepsilon^p_{\alpha\beta}$ can develop singularities at the tip), it is evident that

$$D^p_{33}/[(D^p_{r\theta})^2 + (D^p_{rr} - D^p_{\theta\theta})^2/4]^{1/2} \to 0 \tag{4.5}$$

as $r \to 0$ in that sector. But $D^p_{\alpha\beta} = \dot{\Lambda}P_{\alpha\beta}$, and hence

$$P_{33}/[(P_{r\theta})^2 + (P_{rr} - P_{\theta\theta})^2/4]^{1/2} \to 0,$$

or, more simply,

$$P_{33} \to 0 \tag{4.6}$$

as $r \to 0$. Also, $(r/\dot{a})\dot{\Lambda}$ must become unbounded in such a sector as $r \to 0$, or as $\dot{a} \to 0$. Note that (4.6) would apply in singular plastic sectors even if (4.4) were replaced by the weaker condition that ε_{33} is bounded at the crack tip. This suggests a relevance to three-dimensional crack problems.

The consequences of the flow rule in the form of (2.23) and (2.20) reduce to

$$(\sigma'_{11}+\sigma'_{22})(r\partial v_r/\partial r) \to [M_{rrrr}(\sigma'_{11}+\sigma'_{22})^2+2M_{rr33}(\sigma'_{11}+\sigma'_{22})\sigma'_{33}+M_{3333}(\sigma'_{33})^2]e_2\dot{a} \tag{4.7}$$

and to

$$\partial v_\theta/\partial\theta+\partial(rv_r)/\partial r \to [M_{\alpha\alpha rr}(\sigma'_{11}+\sigma'_{22})+M_{\alpha\alpha 33}\sigma'_{33}]e_2\dot{a} \tag{4.8}$$

respectively.

4.1. *Stress distribution in plastic sectors*

Since there are four stress components to be determined at the crack tip (σ_{11}, $\sigma_{12}=\sigma_{21}$, σ_{22}, and σ_{33}), but only the two equilibrium equations and one requirement of the yield criterion (4.3), the stress state is not, in general, statically determined. The situation changes when *singular* plastic sectors are considered. In that case the consequence of the flow rule in the form (4.6) gives the additional condition that $P_{33}=0$. If we now adopt the condition $P_{33}=0$, which must be valid in singular sectors but the satisfaction of which does not, conversely, imply that a sector is singular, then (4.3) reduces to

$$(\sigma'_{11}+\sigma'_{22})P_{rr}=0. \tag{4.9}$$

Thus, remembering that $\sigma'_{11}+\sigma'_{22}=\sigma'_{12}/e_1e_2$ by (4.1), equations (4.1) and (4.3) (the latter is reduced to (4.9) when $P_{33}=0$) have solutions in the form of either:

(i) *constant stress sectors* (characterized by $\sigma_{11}=$ constant, $\sigma_{12}=\sigma_{21}=$ constant, and $\sigma_{22}=$ constant, where the constants are chosen to satisfy the yield condition); *or*
(ii) *centered fan sectors* (characterized by $P_{rr}=0$; again, the name derives from the corresponding rigid–plastic case in which characteristic directions are those along which extensional components of $P_{\alpha\beta}$ vanish, so that $P_{rr}=0$ corresponds to one of the two families of characteristics being a fan of radial lines).

In case of the isotropic Huber–Mises material, $P_{\alpha\beta}=s_{\alpha\beta}$. Hence, the conditions $P_{33}=0$ and $P_{rr}=0$, which are simultaneously met in centered fan sectors, require that

$$\sigma_{33}=\sigma_{rr}=\sigma_{\theta\theta}. \tag{4.10}$$

In this case, the equilibrium equations, now most simply expressed by (4.2), and the yield condition (2.8) lead to

$$\sigma_{r\theta}=\mp\tau_0, \quad \sigma_{33}=\sigma_{rr}=\sigma_{\theta\theta}=\text{constant}\pm 2\tau_0\theta \tag{4.11}$$

for the stress state in centered fan sectors.

The division of stress states into constant stress and centered fan sectors in the circumstances discussed above was first shown for the Huber–Mises material by Rice and

Tracey (1973) and for the Tresca material by Slepyan (1974); previous work (Rice, 1967, 1968) had either assumed directly that slipline theory was applicable in plastic sectors, which is equivalent to approximating the yield criterion by

$$(\sigma_{11}-\sigma_{22})^2/4+(\sigma_{12})^2 = \tau_0{}^2, \tag{4.12}$$

or had dealt with materials which were elastically as well as plastically incompressible (i.e., $\nu = \frac{1}{2}$, which implies that $s_{33} = 0$ everywhere.

Figures 4.1(a) and (b) show representations of the near-tip state by sliplines, and will be discussed subsequently.

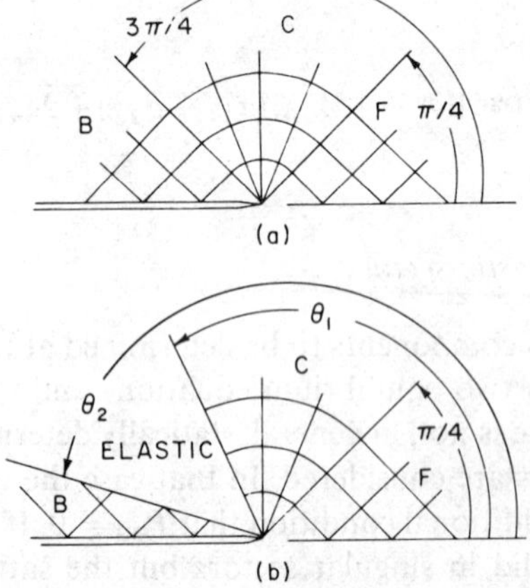

FIG. 4.1. Near-tip fields for plane strain of isotropic material; slipline representation of states in plastic regions: (a) stationary crack; (b) growing crack.

A further analysis of plastic sectors in which $P_{33} \neq 0$ is given in Section 4.4. These sectors cannot allow the accumulation of singular plastic strains $r = 0$, but their analysis is not of the statically determined type and is best deferred until after a consideration of deformations in singular plastic and in elastic sectors.

4.2. *Deformation states in singular plastic sectors*

Consider a centered fan singular plastic sector. The conditions $P_{33} = 0$, $P_{rr} = 0$, and the two equilibrium equations constrain the stress variations and enable the terms $\sigma'_{\alpha\beta}$ in (4.7) and (4.8) to be determined. Interpreting these as equations valid very near the crack tip, integration gives the velocity field in the centered fan sector as

$$\left.\begin{aligned} v_r &= Y(\theta)\dot{a}\ln(\bar{R}/r)+\partial f(\theta,t)/\partial\theta,\\ v_\theta &= \int_{\theta_0}^{\theta}[Z(\theta)+Y(\theta)]d\theta\dot{a}-\int_{\theta_0}^{\theta}Y(\theta)d\theta\dot{a}\ln(\bar{R}/r)-f(\theta,t)+g(r,t),\end{aligned}\right\} \tag{4.13}$$

where $f(\theta, t)$, $g(r, t)$ appear as functions of integration, the latter satisfying $g(0, t) = 0$, where θ_0 is the angle at which the fan begins (e.g., $\theta_0 = \frac{1}{4}\pi$ in Fig. 4.1), and where

$$\left.\begin{aligned} Y(\theta) &= -e_2[M_{rrrr}(\sigma'_{11}+\sigma'_{22})+2M_{rr33}\sigma'_{33}+M_{3333}(\sigma'_{33})^2/(\sigma'_{11}+\sigma'_{22})], \\ Z(\theta) &= e_2[M_{\alpha\alpha rr}(\sigma'_{11}+\sigma'_{22})+M_{\alpha\alpha 33}\sigma'_{33}]. \end{aligned}\right\} \tag{4.14}$$

In fact, since $g(r, t)$ does not contribute to the dominant singularities of strain rate, and since the formulae are valid asymptotically, it is appropriate to take $g(r, t)$ in the form $A(t)r$, where $A(t)$ is some undetermined function. This form contributes zero strain rates but allows the incorporation of a rigid rotation field, of spin rate $A(t)$, near the tip.

The strain rates within the centered fan are

$$\left.\begin{aligned} D_{rr} &= \partial v_r/\partial r = -\dot{a}Y(\theta)/r, \\ D_{\theta\theta} &= (1/r)(\partial v_\theta/\partial\theta + v_r) = \dot{a}[Z(\theta)+Y(\theta)]/r, \\ D_{r\theta} &= (1/2r)(\partial v_r/\partial\theta - v_\theta) + (1/2)(\partial v_\theta/\partial r) \\ &= (\dot{a}/2r)\ln(\bar{R}/r)[dY(\theta)/d\theta + \int_{\theta_0}^{\theta} Y(\theta)d\theta] \\ &\quad - (\dot{a}/2r)\int_{\theta_0}^{\theta} Z(\theta)d\theta + (1/2r)[\partial^2 f(\theta, t)/\partial\theta^2 + f(\theta, t)]. \end{aligned}\right\} \tag{4.15}$$

An interesting feature of these equations is the following: within the centered fan sector $P_{33} \to 0$ and $P_{rr} \to 0$ as $r \to 0$. Further, since $P_{\alpha\alpha} = 0$ for the materials considered, it is evident that

$$P_{rr} \to 0, \quad P_{\theta\theta} \to 0, \quad P_{33} \to 0 \quad \text{as} \quad r \to 0. \tag{4.16}$$

However, the plastic strain rates are given as

$$D^p_{\alpha\beta} = \dot{\Lambda} P_{\alpha\beta},$$

and it might therefore be expected that all of D^p_{rr}, $D^p_{\theta\theta}$, and $D^p_{33} \to 0$ as $r \to 0$. But such is not the case. Instead, when (4.15) for $D_{\alpha\beta}$ are compared to $D^e_{\alpha\beta} = M_{\alpha\beta\gamma\delta}\dot{\sigma}_{\gamma\delta}$, which reduces to

$$D^e_{\alpha\beta} = M_{\alpha\beta rr}(\sigma'_{11}+\sigma'_{22})e_2\dot{a}/r + M_{\alpha\beta 33}\sigma'_{33}e_2\dot{a}/r \tag{4.17}$$

near the tip, it becomes clear that all of $D^p_{rr} (= D_{rr} - D^e_{rr})$, $D^p_{\theta\theta}$, and D^p_{33} are singular of order $e_2\dot{a}/r$ at the tip. (This is the type of singular term that produces only bounded strains at the tip.) Of course, the dominant component of $D^p_{\alpha\beta}$, producing singular strain, is $D^p_{r\theta} (= D^p_{\theta r})$; this is seen from (4.15) to be singular with terms of order $(\dot{a}/r)\ln(\bar{R}/r)$, as well as with terms of order $1/r$ which remain when $\dot{a} = 0$. Hence, since $D^p_{r\theta} = \dot{\Lambda} P_{r\theta}$, $\dot{\Lambda}$ is evidently singular with terms of the same orders at $r = 0$, and although, for example, P_{rr} approaches 0 as $r \to 0$, it is evident that the *product* $\dot{\Lambda} P_{rr}$, which gives D^p_{rr}, does *not* approach zero as $r \to 0$ but rather becomes infinite of order $e_2\dot{a}/r$. The failure to realize this distinction is a source of a minor inaccuracy in previous work (Rice and Sorensen, 1978; Rice *et al.*, 1980) which generalized Rice's (1968) analysis of plane strain crack growth in an elastically incompressible, isotropic Huber–Mises material to the case $v \neq \frac{1}{2}$.

In fact, for the isotropic Huber–Mises material, the elastic compliances are given by (2.14) and the stresses in a centered fan plastic sector by (4.11). Accordingly, within such a plastic sector one computes from (4.14) that

$$Y(\theta) = (5-4v)(\tau_0/E)\sin\theta, \quad Z(\theta) = -6(1-2v)(\tau_0/E)\sin\theta. \tag{4.18}$$

These correspond to use of $\sigma'_{33} = (\sigma'_{11}+\sigma'_{22})/2 = -2\tau_0$ as appropriate for the fans shown in

Fig. 4.1 with θ positive (i.e., for the Mode I tensile crack). Thus the previous expressions for velocities and strain-rates in centered fan plastic sectors, beginning at $\theta_0 = \frac{1}{4}\pi$ (Fig. 4.1), become

$$\left.\begin{aligned} v_r &= (5-4\nu)(\tau_0/E)\dot{a}\sin\theta\ln(\bar{R}/r) + \partial f(\theta,t)/\partial\theta, \\ v_\theta &= -(5-4\nu)(\tau_0/E)\dot{a}[(1/\sqrt{2})-\cos\theta]\ln(\bar{R}/r) \\ &\quad + [(5-4\nu)-6(1-2\nu)](\tau_0/E)\dot{a}[(1/\sqrt{2})-\cos\theta] - f(\theta,t) \\ &\quad + A(t)r, \end{aligned}\right\} \tag{4.19}$$

and

$$\left.\begin{aligned} D_{rr} &= -(5-4\nu)(\tau_0/E)(\dot{a}/r)\sin\theta, \\ D_{\theta\theta} &= [(5-4\nu)-6(1-2\nu)](\tau_0/E)(\dot{a}/r)\sin\theta, \\ D_{r\theta} &= [(5-4\nu)/\sqrt{2}](\tau_0/E)(\dot{a}/2r)\ln(\bar{R}/r) + 6(1-2\nu)(\tau_0/E)(\dot{a}/2r)[1/\sqrt{2}) - \cos\theta] \\ &\quad + (1/2r)[\partial^2 f(\theta,t)/\partial\theta^2 + f(\theta,t)]. \end{aligned}\right\} \tag{4.20}$$

It may be observed that a deformation rate field of the type described by the term $f(\theta, t)$ is of the same form as that which describes the opening of a stationary crack (Rice, 1967, 1968). The additional terms in the velocity field, proportional to a, were first analyzed by Rice (1968) for the case of an elastically incompressible material ($\nu = \frac{1}{2}$) in steady state crack growth. As shown subsequently by Cherepanov (1974) and Rice (1974), the result for a moving crack in general unsteady conditions involves a superposition of these two types of fields, as in (4.19) and (4.20). Indeed, examination of (4.13) and (4.15) for the most general anisotropic material reveals that the same interpretation applies there also.

The results presented in this Section are the first fully correct analysis of the near-tip field for $\nu \neq \frac{1}{2}$. Previous discussions of this problem by Rice and Sorensen (1978), as modified further by Rice *et al.* (1980) to account for an elastic unloading zone, failed to notice the problem with limits discussed after (4.16). Instead they assumed that $D^p_{rr} = D^p_{\theta\theta} = 0$ near the tip, so that D_{rr} and $D_{\theta\theta}$ would be calculable directly as the elastic components D^e_{rr} and $D^e_{\theta\theta}$ as given by an equation like (4.17) as specialized to the isotropic case. This led to results for velocities and strain-rates which had the same form as (4.19) and (4.20) but in which $(5-4\nu)$ was replaced by $2(2-\nu)$ and in which the factor of 6 which precedes $(1-2\nu)$ was replaced by a factor of 4. The former is the more important, since it controls the amplitude of the strain singularity, which figures in the stable crack growth model developed in these references. Happily, the change is not large; $(5-4\nu)$ is equal to $2(2-\nu)$ when $\nu = \frac{1}{2}$, so that there is no effect on the results, and $(5-4\nu)$ exceeds $2(2-\nu)$ by approximately 12% when $\nu = 0.3$.

It may be verified directly that (4.17), when written for the isotropic case, gives the result $D^e_{r\theta} = 0$. Hence, the result for $D_{r\theta}$ in (4.20) is a result for $D^p_{r\theta}$ which is, in turn, expressible as $\dot{\Lambda}P_{r\theta}(=\dot{\Lambda}s_{r\theta} = \dot{\Lambda}\tau_0)$, so that the last member of (4.20) gives the asymptotic result for $\dot{\Lambda}$ at the tip.

4.3. *Elastic sectors*

Sectors which are instantaneously deforming elastically may be either of the constant stress form, which trivially satisfy all the requisite equations, or, if the crack is growing ($\dot{a} \neq 0$) may instead involve a variable state of stress of a form which is analyzed here. In such elastic sectors $D_{\alpha\beta} = D^e_{\alpha\beta}$, and is therefore given by (4.17). Since $D_{33} = 0$, we can eliminate σ_{33} by writing

$$\sigma'_{33} = -(M_{33rr}/M_{3333})(\sigma'_{11} + \sigma'_{22}) \tag{4.21}$$

so that (4.17) may be restated as

$$D_{\alpha\beta} = (M_{\alpha\beta rr} - M_{\alpha\beta 33}M_{33rr}/M_{3333})(\sigma'_{11} + \sigma'_{22})e_2\dot{a}/r. \tag{4.22}$$

Evidently, the in-plane stresses in elastic sectors are determined by the two equilibrium equations (4.1) and the condition that the deformation rate field of (4.22) be compatible with the existence of a velocity field. The structure of the equations is similar to those for non-singular plastic sectors (Section 4.4) and it is useful to have a general analysis. Accordingly, it is noted that for deformation rate fields of the type

$$D_{ij} = \dot{a}F_{ij}(\theta)/r \quad (i, j = 1, 2 \quad \text{or} \quad i, j = r, \theta) \tag{4.23}$$

to be compatible, it is necessary that

$$\partial^2 D_{22}/\partial x_1^2 + \partial^2 D_{11}/\partial x_2^2 - 2\partial^2 D_{12}/\partial x_1 \partial x_2 = (1/r^3)\dot{a}[d^2F_{rr}(\theta)/d\theta^2 + F_{rr}(\theta)] = 0. \tag{4.24}$$

(Details of the expression of the compatibility operation in terms of $F_{rr}(\theta)$ are straightforward but lengthy, and are omitted here.) Thus, for compatibility of (4.23) one must have

$$F_{rr}(\theta) \equiv F_{11}(\theta)\cos^2\theta + 2F_{12}(\theta)\sin\theta\cos\theta + F_{22}(\theta)\sin^2\theta = A_1\cos\theta + A_2\sin\theta, \tag{4.25}$$

where A_1 and A_2 are constants, and further analysis shows that the velocities have the form (to within an arbitrary rigid translation and rotation)

$$\left.\begin{aligned} v_1 &= \dot{a}A_1\ln(r\sin\theta/\bar{R}) - \dot{a}\int[F_{11}(\theta)/\sin\theta]d\theta, \\ v_2 &= \dot{a}A_2\ln(r\cos\theta/\bar{R}) + \dot{a}\int[F_{22}(\theta)/\cos\theta]d\theta. \end{aligned}\right\} \tag{4.26}$$

Thus, applying the results of (4.23)–(4.26) to the expression for $D_{\alpha\beta}$ in (4.22), one concludes that the stress-state in an instantaneously elastic sector at a growing crack tip must satisfy

$$[M_{rrrr} - (M_{rr33})^2/M_{3333}](\sigma'_{11} + \sigma'_{22})\sin\theta = A_1\cos\theta + A_2\sin\theta, \tag{4.27}$$

where A_1 and A_2 are constants, and this, together with (4.1) and (4.21), determines the stress state. Further, the velocity field is given by (4.26) when components of $F_{ij}(\theta)$ are chosen so that (4.23) coincides with (4.22).

For elastically isotropic materials (4.21) becomes

$$\sigma'_{33} = \nu(\sigma'_{11} + \sigma'_{22}), \tag{4.28}$$

and (4.25) implies that

$$\sigma'_{11} + \sigma'_{22} = [E/(1-\nu^2)](A_1\cos\theta/\sin\theta + A_2). \tag{4.29}$$

From these results and from (4.1), the full form of the stress field is

$$\left.\begin{aligned} 4(1-\nu^2)\sigma_{11}/E &= 4A_1\ln|\sin\theta| + A_1\cos 2\theta + A_2(2\theta + \sin 2\theta) + C_{11}, \\ 4(1-\nu^2)\sigma_{12}/E &= A_1(2\theta + \sin 2\theta) - A_2\cos 2\theta + C_{12}, \\ 4(1-\nu^2)\sigma_{22}/E &= -A_1\cos 2\theta + A_2(2\theta - \sin 2\theta) + C_{22}, \\ \sigma_{33} &= \nu(\sigma_{11} + \sigma_{22}) + d, \end{aligned}\right\} \tag{4.30}$$

where C_{11}, C_{12}, C_{22}, and d are also constants of integration. The corresponding velocity field is given by (4.26) with

$$\left.\begin{aligned} F_{11}(\theta) &= (\cos^2\theta - v\sin^2\theta - v^2)(A_1\cos\theta + A_2\sin\theta)/(1-v^2),\\ F_{22}(\theta) &= (\sin^2\theta - v\cos^2\theta - v^2)(A_1\cos\theta + A_2\sin\theta)/(1-v^2). \end{aligned}\right\} \quad (4.31)$$

The same or equivalent results for the stress and velocity fields in elastic sectors at a moving crack tip have been obtained first by Slepyan (1974) and later but independently by Gao (1980) (see also Gao and Hwang, 1981) and by Rice *et al.* (1980). Slepyan and Gao address the steady state crack growth problem only, whereas the latter reference considers general nonsteady conditions and gives estimates, from a numerical solution, for some parameters of the near-tip field which are left undetermined by the asymptotic analysis.

4.4. *Non-singular plastic sectors*

Non-singular plastic sectors in which $P_{33} \to 0$ as $r \to 0$ are necessarily constant stress sectors, as shown by the analysis of Section 4.2. Hence, to complete the analysis of all possible sectors at a crack tip, we need consider further only plastically active sectors in which $P_{33} \neq 0$ at $r = 0$. As remarked earlier, a sector with $P_{33} \neq 0$ cannot develop singular strains.

By using the flow rule (2.15) and noting that $D_{33} = 0$, one finds that

$$\dot{\Lambda} = -(1/P_{33})M_{33\alpha\beta}\dot{\sigma}_{\alpha\beta} \quad (4.32)$$

and that the strain rates are given by

$$D_{\alpha\beta} = [M_{\alpha\beta\gamma\delta} - (P_{\alpha\beta}/P_{33})M_{33\gamma\delta}]\dot{\sigma}_{\gamma\delta}. \quad (4.33)$$

(A constraint on any proposed solution is that (4.32) result in $\dot{\Lambda} \geqslant 0$.) Using the result that $\dot{\sigma}_{\gamma\delta} = \sigma'_{\gamma\delta}e_2\dot{a}/r$ sufficiently close to the tip, and then making use of (2.6), the asymptotic expression for $D_{\alpha\beta}$ is therefore

$$D_{\alpha\beta} = \{[M_{\alpha\beta rr} - (P_{\alpha\beta}/P_{33})M_{33rr}](\sigma'_{11} + \sigma'_{22}) + [M_{\alpha\beta 33} - (P_{\alpha\beta}/P_{33})M_{3333}]\sigma'_{33}\}e_2\dot{a}/r. \quad (4.34)$$

Now, the problem of determining stress distributions in plastic sectors with $P_{33} \neq 0$ is one of simultaneously solving the two equilibrium equations of (4.1) or (4.2), subject to continued satisfaction of the yield condition, as expressed by (4.3), and to the requirement that $D_{\alpha\beta}$ of (4.34) be compatible. The formalism of (4.23)–(4.26) may be applied to (4.34), so that the expression that $D_{\alpha\beta}$ be compatible is equivalent to writing (see (4.25))

$$[M_{rrrr} - (P_{rr}/P_{33})M_{33rr}](\sigma'_{11} + \sigma'_{22})\sin\theta + [M_{rr33} - (P_{rr}/P_{33})M_{3333}]\sigma'_{33}\sin\theta = A_1\cos\theta + A_2\sin\theta, \quad (4.35)$$

where A_1 and A_2 are constants, related to the velocity field in the sector as in (4.26).

Considering the isotropic Huber–Mises material, $P_{\alpha\beta} = s_{\alpha\beta} = \sigma_{\alpha\beta} - \sigma_{\gamma\gamma}\delta_{\alpha\beta}/3$, and (4.3) and (4.35) become

$$(\sigma'_{rr} + \sigma'_{\theta\theta})(2\sigma_{rr} - \sigma_{\theta\theta} - \sigma_{33}) + \sigma'_{33}(2\sigma_{33} - \sigma_{rr} - \sigma_{\theta\theta}) = 0 \quad (4.36)$$

and

$$\begin{aligned} &[1 + v(2\sigma_{rr} - \sigma_{\theta\theta} - \sigma_{33})/(2\sigma_{33} - \sigma_{rr} - \sigma_{\theta\theta})](\sigma'_{rr} + \sigma'_{\theta\theta})\sin\theta \\ &\quad - [v + (2\sigma_{rr} - \sigma_{\theta\theta} - \sigma_{33})/(2\sigma_{33} - \sigma_{rr} - \sigma_{\theta\theta})]\sigma'_{33}\sin\theta \\ &\qquad = E(A_1\cos\theta + A_2\sin\theta), \end{aligned} \quad (4.37)$$

respectively. These and (4.2) constitute a system of four equations for the stresses σ_{33}, σ_{rr}, $\sigma_{r\theta}$, and $\sigma_{\theta\theta}$ in non-singular plastic sectors; the formulation assumes $\sigma_{33} \neq (\sigma_{rr}+\sigma_{\theta\theta})/2$. The equations have the obvious solution $\sigma_{\alpha\beta} = \text{constant}$, although this requires that $A_1 = A_2 = 0$. The full nature of the solutions for (A_1, A_2) different from (0, 0) has not yet been determined.

4.5. *Assembly of sectors*

The previous subsections have categorized all possible crack tip sectors. These sectors must be assembled in a manner which is consistent with continuity of $\sigma_{\theta r}$, $\sigma_{\theta\theta}$, and u_θ along radial lines between them, and in which any discontinuities (e.g., of velocity) at the boundaries of sectors are consistent with the flow rule, when regarded as straining within a concentrated region.

It is well known that for the stationary crack the near-tip field is not unique. The Prandtl field of Fig. 4.1(a) is the only field appropriate to a stationary crack in an isotropic material, *if* it is assumed that the stress component σ_{rr} is fully continuous, that yield with $s_{33} = 0$ is met at all angles about the tip, and that the loading is Mode I (tension). This field was proposed by Rice (1967, 1968) for well-contained plastic yielding, and seems to be very accurately verified by numerical solutions (e.g., Rice and Tracey, 1973). Hutchinson (1968) gave the corresponding solution for Mode II (in-plane shear). From generalizations of slipline theory to the plane straining of materials of arbitrary anisotropy (Hill, 1950; Booker and Davis, 1972; Rice, 1973), it is known that slipline fields of identical form to the Prandtl field of Fig. 4.1(a) arise in solutions to the analogous problem of indentation by a smooth-surfaced punch. Hence, it seems likely that the construction of Fig. 4.1(a) is valid as a crack tip field for well-contained yielding in *all* materials of the type considered here, so long as there is symmetry of properties about the plane of the crack.

However, many solutions for stationary crack tip fields are known from plastic limit analyses (e.g., McClintock and Irwin, 1965; McClintock, 1971) which show less triaxial stress elevation than does the Prandtl field (which has $\sigma_{22} = (2+\pi)\tau_0$ ahead of the tip).

The first analyses of growing cracks in elastic–plastic materials (Rice, 1968, 1974; Cherepanov, 1974) solved for the logarithmically singular strain accumulation in centered fan sectors, but did not address the conditions between sectors. Slepyan (1974) was the first to demonstrate the necessity of the elastic unloading zone as shown in Fig. 4.1(b), as was discovered much later but independently by Gao (1980) and by Rice *et al.* (1980), the latter in a correction of the earlier analysis by Rice and Sorensen (1978) which had assumed validity of the Prandtl field of Fig. 4.1(a) for a growing crack. The necessity for the elastic unloading zone results from potential discontinuities of the logarithmically singular part of v_r, in (4.13) and (4.19), at boundaries of the centered fan region. The discontinuity at $\theta = \frac{1}{4}\pi$ is consistent with positive plastic work, but that which would result at $\theta = \frac{3}{4}\pi$ in Fig. 4.1(a) is not. In fact, it is found that any continuation of the fan sector into a constant stress plastic sector involves an inadmissible negative plastic work, and an elastic region must intervene. However, because material elements emerge from the fan with a strain ε^p_{11} which becomes negatively infinite as $r \to 0$, it is necessary that a further plastic sector form along the crack surfaces as shown in Fig. 4.1(b). By contrast to the anti-plane strain case (Fig. 3.1), this further plastic sector involves yield in the same sense (i.e., $D^p_{11} \geqslant 0$) as for the stationary crack.

For $\nu = \frac{1}{2}$, Slepyan (1974), Gao (1980), and Rice *et al.* (1980) all obtain $\theta_1 \simeq 112°$ and $\theta_2 \simeq 162°$. The problem was considered for the Huber–Mises material with $\nu \neq \frac{1}{2}$ only in the latter reference, and because of the difficulty of integrating the equations for nonsingular plastic sectors with $s_{33} \neq 0$, the approximation $s_{33} = 0$ was made in those sectors. The results, which may be subject to minor modifications for the various reasons mentioned, are $\theta_1 \simeq 115°$ and $\theta_2 \simeq 163°$ for $\nu = 0.3$.

Rice *et al.* (1980) also give the asymptotic form of the plastic strain field in the centered fan region. As is well known, for the stationary crack this has the form

$$\varepsilon^p_{\alpha\beta} = F_{\alpha\beta}(\theta)/r, \tag{4.38}$$

where $F_{\alpha\beta}(\theta)$ is undetermined by the asymptotic analysis, except that only the $r\theta(=\theta r)$ component of $F_{\alpha\beta}$ is nonzero (Rice and Tracey, 1973, give some numerical results for $F_{r\theta}$). For the growing crack, the plastic strain accumulates a step increase at the front boundary of the centered fan, due to the velocity discontinuity there. Their result for any point within the fan, when one incorporates the correction of replacing $2(2-\nu)$ by $(5-4\nu)$ as discussed in Section 4.2, is

$$\varepsilon^p_{\alpha\beta} = [(5-4\nu)/2\sqrt{2}](\tau_0/E)G_{\alpha\beta}(\theta)\ln(\bar{R}/r) + H_{\alpha\beta}(\theta), \tag{4.39}$$

where $H_{\alpha\beta}(\theta)$ is undetermined by the asymptotic analysis, and where

$$\left.\begin{aligned} G_{11}(\theta) &= -G_{22}(\theta) = -2\sin\theta, \\ G_{12}(\theta) &= G_{21}(\theta) = \ln[\tan(\theta/2)/\tan(\pi/8)] + 2[\cos\theta - (1/\sqrt{2})]. \end{aligned}\right\} \tag{4.40}$$

Also, an equivalent plastic shear strain γ^p may be defined by (Hill, 1950)

$$\dot{\gamma}^p = (2D^p_{\alpha\beta}D^p_{\alpha\beta})^{1/2},$$

and this has the form

$$\gamma^p = (5-4v)(\tau_0/E)\{1 + (1/\sqrt{2})\ln[\tan(\theta/2)/\tan(\pi/8)]\}\ln(\bar{R}/r) + h(\theta) \tag{4.41}$$

in the fan, where $h(\theta)$ is undetermined by the asymptotic analysis. [Gao (1980) gives results for plastic strains which seem to contain Poisson's ratio in a different manner than (4.39) and (4.41), but his analysis is valid only when $\nu = \frac{1}{2}$.] The strains at a steadily growing crack tip are less strongly singular for plane strain than for anti-plane strain (compare (3.13), which involves the square rather than first power of a logarithmic term) and for plane stress, although the singularity of plastic strain-rate is of the same form, $D^p_{\alpha\beta} \sim (\dot{a}/r)\ln(\bar{R}/r)$, in all cases. Rice (1974) explained the origin of this in the fact that the crack grows into its centered fan zones for anti-plane strain and plane stress, but not for plane strain tension.

The primary features of the stress field in Fig. 4.1(b) are verified by numerical results of Sham (1979), cited in Rice *et al.* (1980), and, less directly, of Dean and Hutchinson (1980) for crack growth under well-contained yielding. The Sham solution clearly shows elastic unloading of finite elements behind the tip, and was used to identify some parameters of the growing crack field which are left undetermined by the asymptotic analysis. The question of whether there may exist fields other than that of Fig. 4.1(b) (say, for crack growth in fully plastic geometries which are not well constrained) is not yet fully resolved.

Application of the growing crack analysis to the prediction of stable ductile crack growth is discussed in Section 6.

5. Plane Stress

5.1. *Stress distribution in plastic sectors*

For the two-dimensional plane stress theory, we set $\sigma_{33} = \sigma_{31} = \sigma_{32} = 0$. The asymptotic form of the equilibrium equations is the same as in (4.1) or (4.2), and the requirement (2.11) that the yield condition be met in plastic sectors simplifies to

$$(\sigma'_{11} + \sigma'_{22})P_{rr} = 0. \tag{5.1}$$

Hence, since there are three equations for the three non-zero stresses, $\sigma_{11}, \sigma_{12}(= \sigma_{21})$, and σ_{22}, one concludes at once that the stress field is of the statically determined type in plastic sectors. Further, all solutions of these equations define sectors which, similarly to results in Sections 3 and 4, are either:

(i) *constant stress sectors* (σ_{11} = constant, $\sigma_{12} = \sigma_{21}$ = constant, and σ_{22} = constant); *or*
(ii) *centered fan sectors* ($P_{rr} = 0$; the rigid–plastic plane stress equations have characteristics along directions for which the extensional components of $P_{\alpha\beta}$ vanish, if such exist, so that $P_{rr} = 0$ defines a fan of radial characteristics).

For isotropic Huber–Mises materials, the centered fan condition $P_{rr} = 0$ reduces to $2\sigma_{rr} = \sigma_{\theta\theta}$, and the equilibrium equations in the form (4.2) can be integrated at once to give

$$\left.\begin{aligned} \sigma_{\theta\theta} &= \pm 2\tau_0 \cos(\theta - \theta_1), \\ \sigma_{rr} &= \pm \tau_0 \cos(\theta - \theta_1), \\ \sigma_{r\theta} &= \pm \tau_0 \sin(\theta - \theta_1), \end{aligned}\right\} \tag{5.2}$$

where θ_1 is arbitrary but would, for example, be taken as zero to define the centered fan field ahead of a Mode I crack (Hutchinson, 1968).

5.2. *Deformation rate field in plastic sectors*

The consequence of the flow rule in the form of (2.20) is irrelevant for plane stress, except to define D_{33}. But (2.23), with (2.25), leads to

$$r\partial v_r/\partial r \to M_{rrrr}(\sigma'_{11} + \sigma'_{22})e_2\dot{a} \tag{5.3}$$

as $r \to 0$, and if this is interpreted as an equality very close to the crack tip, there results

$$v_r = -M_{rrrr}(\sigma'_{11} + \sigma'_{22})e_2\dot{a}\ln(\bar{R}/r) + f(\theta, t). \tag{5.4}$$

To constrain the other velocity component, we note that

$$\dot{\Lambda}P_{r\theta} = D_{r\theta} - M_{r\theta\alpha\beta}\dot{\sigma}_{\alpha\beta} \quad \text{and} \quad \dot{\Lambda}P_{\theta\theta} = D_{\theta\theta} - M_{\theta\theta\alpha\beta}\dot{\sigma}_{\alpha\beta},$$

so that

$$P_{r\theta}D_{\theta\theta} - P_{\theta\theta}D_{r\theta} = (P_{r\theta}M_{\theta\theta\alpha\beta} - P_{\theta\theta}M_{r\theta\alpha\beta})\dot{\sigma}_{\alpha\beta}. \tag{5.5}$$

Hence, when use is made of (2.17) and (2.6), there results

$$P_{r\theta}(rD_{\theta\theta})-P_{\theta\theta}(rD_{r\theta}) = P_{r\theta}(\partial v_\theta/\partial\theta+v_r)-(P_{\theta\theta}/2)(\partial v_r/\partial\theta-v_\theta+r\partial v_\theta/\partial r)$$
$$\rightarrow (P_{r\theta}M_{\theta\theta rr}-P_{\theta\theta}M_{r\theta rr})(\sigma'_{11}+\sigma'_{22})e_2\dot{a} \qquad (5.6)$$

as $r\rightarrow 0$. Interpreting this result as an equality very near the tip, it is seen that solutions for v_θ in the form

$$v_\theta = \dot{a}F(\theta)\ln(\bar{R}/r)+G(\theta,t)+A(t)r \qquad (5.7)$$

can be found.

For the isotropic Huber–Mises material, (5.4) gives in a centered fan ahead of a tensile crack (i.e., $\theta_1=0$)

$$v_r = 3(\tau_0/E)\dot{a}\sin^2\theta\ln(\bar{R}/r)+f(\theta,t), \qquad (5.8)$$

whereas (5.6) becomes

$$\sin\theta(\partial v_\theta/\partial\theta+v_r)-(1/2)(\cos\theta)(\partial v_r/\partial\theta-v_\theta+r\partial v_\theta/\partial r)$$
$$= 3v(\tau_0/E)\dot{a}\sin^2\theta. \qquad (5.9)$$

The expressions for $F(\theta)$ and $G(\theta,t)$ of (5.7) can be determined. In particular,

$$F(\theta) = 3(\tau_0/E)(\sin\theta)^{-1/2}\int_0^\theta(\sin\phi)^{1/2}(\cos^2\phi-\sin^2\phi)d\phi. \qquad (5.10)$$

In this case the strain-rate components $D_{\theta\theta}$ and $D_{r\theta}$ have terms which are singular of the orders $(\dot{a}/r)\ln(\bar{R}/r)$ and $(1/r)$, the latter remaining in the case of a stationary crack.

5.3. *Elastic sectors*

The analysis of elastic sectors closely parallels that of Section 4.3 for plane strain. Now, however, (4.22) is replaced by

$$D_{\alpha\beta} = M_{\alpha\beta rr}(\sigma'_{11}+\sigma'_{22})e_2\dot{a}/r \qquad (5.11)$$

and the compatibility condition (4.27) becomes

$$M_{rrrr}(\sigma'_{11}+\sigma'_{22})\sin\theta = A_1\cos\theta+A_2\sin\theta. \qquad (5.12)$$

Here A_1 and A_2 are related to the velocities as in (4.26), and the functions $F_{ij}(\theta)$ which further determine the velocity distribution are chosen so that (4.23) coincides with (5.11). The results for elastically isotropic materials are identical to what is presented in the first three of (4.30) and in (4.31), except that the term v^2 appearing in these equations should be deleted everywhere.

5.4. *Assembly of sectors*

Hutchinson (1968) assembled a candidate field for a stationary crack, consisting of a centered fan sector extending from $\theta=0°$ to $\theta\simeq 80°$, joined to two constant stress plastic sectors with a discontinuity in σ_{rr} at their border at $\theta\simeq 151°$. No solution seems yet to have been found for the growing crack, although it has been verified (K. K. Lo, Brown University, private communication, 1980) that just as for the plane strain and anti-plane strain cases, for non-negative plastic work at the trailing boundary of the fan sector, this sector must join onto an elastic unloading sector rather than a constant stress plastic sector.

6. Application to Prediction of Ductile Plane Strain Crack Growth

Rice *et al.* (1980), in an extension of an earlier study by Rice and Sorensen (1978), used asymptotic results of the type reviewed in Section 4, in combination with finite element numerical analysis, to determine approximately the form of near-tip crack openings for growing plane strain cracks. They observed that the opening rate $\dot{\delta}\ [= v_2(r, \pi) - v_2(r, -\pi)]$ across the crack faces could be written as

$$\dot{\delta} = \alpha \dot{J}/\sigma_0 + \beta(\sigma_0/E)\dot{a}\ \ln(R/r) \tag{6.1}$$

at small distances r behind the crack tip, where $\sigma_0 = \sqrt{3}\,\tau_0$ (ideally plastic tensile strength), J is some appropriately defined measure of the J integral, reducing to the classical line integral (e.g., Rice, 1968) taken in elastic material outside the plastically deformed region for crack growth in the small scale yielding range, and where β is defined by the asymptotic analysis whereas α and R are not.

Rice *et al.* (1980) give $\beta = 4.395$ for $v = 0.5$. They also report $\beta \simeq 5.08$ for $v = 0.3$, although a final analysis may show this value to increase by perhaps 10% or so due to the factors mentioned in Section 4.2. Comparison with finite element solutions for various combinations of load increase and crack growth suggest that for the small scale yielding range

$$\alpha \simeq 0.65, \quad R \simeq 0.2EJ/\sigma_0{}^2, \tag{6.2}$$

results which seem to be approximately independent of the ratio $\dot{J}/\dot{a}$ (if $\dot{a} \geqslant 0, \dot{J} \geqslant 0$) or of the amount of prior crack growth. The result for R is of the order of the plastic zone size (about 15% to 30% greater).

For monotonic loading of a stationary crack, (6.1) integrates to

$$(\delta)_{r=0} = \int \alpha \dot{J}/\sigma_0 = \bar{\alpha} J/\sigma_0 \quad (\simeq 0.65\ J/\sigma_0 \text{ for small scale yielding}), \tag{6.3}$$

which shows a characteristic feature of ideally plastic solutions that a discrete opening displacement develops at the crack tip. But when (6.1) is integrated for a continuously growing crack (dJ/da finite), there results very near the tip

$$\delta = \beta(\sigma_0/E) r \ln(\rho/r), \tag{6.4}$$

where

$$\rho \equiv R \exp[1 + (\alpha/\beta)(E/\sigma_0{}^2)(dJ/da)]. \tag{6.5}$$

Equation (6.4) for δ shows that there is no opening at the tip of the growing crack, although $d\delta/dr$ becomes unbounded there. Further, the result shows that the combination defined as ρ in (6.5) uniquely characterizes the intensity of the crack surface opening very near the tip.

A plausible candidate for a ductile crack growth criterion is therefore that $\rho =$ constant. This carries the implication that the near-tip deformation field remains approximately fixed in magnitude and distribution during growth. It may be appropriate for ductile metals of a type which are relatively unaffected by the preconditioning experienced in the crack stress concentration, before a material point is reached by the crack tip fracture process zone. Rice *et al.* (1980) suggest that such a growth criterion will be at least qualitatively similar to one based on the near-tip strain field (which is less well defined than δ due to difficulties of interpretation of numerical solutions near the crack tip).

If the criterion is adopted, then (6.5) enables solution for dJ/da in terms of ρ, presumed constant, and the length scale R. For small scale yielding this relation becomes, with (6.4),

$$dJ/da = (\beta/\alpha)(\sigma_0{}^2/E)\ln(\rho\sigma_0{}^2/0.2eEJ), \tag{6.6}$$

where e is the natural logarithm base, and the equation is to be integrated subject to $J = J_{Ic}$ at the onset of growth. The features of the integrals of this equation are that dJ/da is initially positive, at least if ρ is large enough, but dJ/da continually diminishes toward zero with increasing a, so that J increases from J_{Ic} and is asymptotic to a plateau level J_{ss} (for "steady state") at which no further increase in J is required to maintain growth. From (6.6),

$$J_{ss} = (1/0.2e)\rho\sigma_0{}^2/E = 1.6\rho\sigma_0{}^2/E. \tag{6.7}$$

Hermann and Rice (1980) have shown recently that the predictions of this model are in good agreement with observed "plane strain" crack growth, prior to general yielding, in specimens of a high-strength AISI-4140 steel ($\sigma_{\text{yield}} = 1173$ MPa, $\sigma_{\text{ult.}} = 1327$ MPa). They obtain amounts of growth ranging from approximately 5–10 mm prior to general yield, and define J as the value based on measurement procedures as would be appropriate to "deformation theory" for deeply cracked bend specimens. The parameters giving the best fit to the data were $J_{Ic} = 35$ kN/m and $\rho = 7.2$ mm, the latter leading to $J_{ss} = 80$ kN/m, which is the approximate level of a plateau in the J versus $a - a_0$ relation that was reasonably well defined by experimental points prior to general yielding.

The basic parameters of the fracture criterion are J_{Ic} and ρ, although the latter can be expressed instead in terms of J_{ss} by (6.7). Alternatively, following some of the parameters used by Rice *et al.* (1980), it is possible to express ρ in terms of the dimensionless "tearing modulus" T of Paris *et al.* (1977) at the onset of growth, defined by

$$T_0 = (E/\sigma_0{}^2)(dJ/da)_{a=a_0}. \tag{6.8}$$

From (6.6), with $(J)_{a=a_0} = J_{Ic}$, and (6.7),

$$T_0 = (\beta/\alpha)\ln(\rho\sigma_0{}^2/0.2eEJ_{Ic}) = (\beta/\alpha)\ln(J_{ss}/J_{Ic}), \tag{6.9}$$

which connects the alternative (and equivalent) growth parameters T_0, ρ, and J_{ss}. Figure 6.1 shows integrals of the crack growth criterion for small scale yielding, based on $\beta = 5.1$ and $\alpha = 0.65$. Values of T_0 and $j \equiv J_{ss}/J_{Ic}$ are indicated.

Finally, it is possible to rephrase the growth criterion as the achievement of a critical opening angle parameter δ_c/r_m at a small microstructural distance r_m behind the tip. This viewpoint enables some contact with microstructural fracture processes. For the reasons discussed by Rice and Sorensen (1978) it is plausible and convenient from the standpoint of interpretation to identify r_m as the discrete crack tip opening displacement $\delta_{Ic} = \alpha J_{Ic}/\sigma_0$ at the onset of growth, since this opening is a useful characterization of the fracture process zone size. Hence, from this viewpoint, $r_m (= \delta_{Ic})$ and the local opening angle parameter δ_c/r_m over the process zone size are taken as fundamental, and the macroscopic parameters are given in terms of them by

$$\left.\begin{aligned} J_{Ic} &= \sigma_0 r_m/\alpha, \\ \rho &= r_m \exp[(E/\beta\sigma_0)(\delta_c/r_m)]. \end{aligned}\right\} \tag{6.10}$$

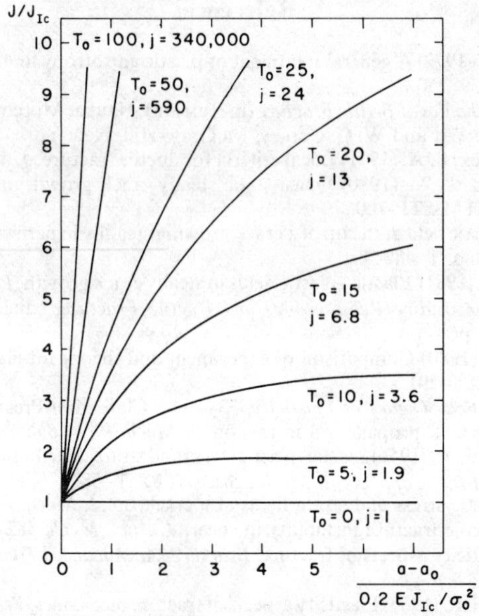

FIG. 6.1. Crack growth criterion for small scale yielding, based on fixed intensity of near-tip crack opening during growth. T_0 is the initial value of the Paris *et al.* (1977) tearing modulus; $j = J_{ss}/J_{Ic}$ where J_{ss} is the plateau level of J at which continued steady state growth occurs.

Thus, for example, with $\alpha = 0.65$,

$$J_{ss}/J_{Ic} = 1.15(\sigma_0/E)\exp[(E/\beta\sigma_0)(\delta_c/r_m)]. \tag{6.11}$$

As shown in various tabulations by Rice and Sorensen (1978), the above relations are such that the resistance to stable crack growth is strongly dependent on the ratio σ_0/E. For a given opening angle parameter (i.e., a given strain-like ductility), high σ_0/E materials are much less resistant to crack growth than are low σ_0/E materials. For example, taking $\beta = 5.7$, an opening angle parameter of 0.15 for δ_c/r_m results in J_{ss}/J_{Ic} values of approximately 30 000 for $\sigma_0/E = 0.0015$, 15 for $\sigma_0/E = 0.0030$, 1 for $\sigma_0/E = 0.0045$, and less than 1 for greater σ_0/E. Of course, values of J_{ss}/J_{Ic} less than unity imply that spontaneous failure accompanies initiation. On the other hand, such large values of J_{ss}/J_{Ic} as result for the lowest strength level in this example correspond to cases in which there is great resistance to ductile crack growth and in which unstable crack propagation is unable to occur, certainly within the well-contained yielding range, in ordinary size bodies.

The extension of the theoretical analysis of crack tip fields to the fully plastic range should provide a basis for analyzing ductile crack growth over the full range encountered in practical application.

Acknowledgements

The preparation of this paper was supported by the Materials Research Laboratory at Brown University, funded by the U.S. National Science Foundation. This article was written whilst the author was a member of the Division of Engineering, Brown University, Providence, Rhode Island, U.S.A.

References

Booker, J. R. and Davis, E. H. (1972) A general treatment of plastic anisotropy under conditions of plane strain, *J. Mech. Phys. Solids* **20,** 239–250.

Cherepanov, G. P. (1974) *Mechanics of Brittle Fracture* (in Russian), Nauka, Moscow. [English translation by A. L. Peabody, edited by R. de Wit and W. C. Cooley, McGraw-Hill, New York, 1979.]

Chitaley, A. D. and McClintock, F. A. (1971) Local criteria for ductile fracture, *J. Mech. Phys. Solids* **19,** 147–163.

Dean, R. H. and Hutchinson, J. W. (1980) Quasi-static steady crack growth in small scale yielding, *Fracture Mechanics*, 383–405, ASTM-STP 700.

Gao, Y.-C. (1980) Elastic–plastic field at the tip of a crack growing steadily in perfectly plastic medium (in Chinese), *Acta Mechanica Sinica*, No. 1, 48–56.

Gao, Y.-C. and Hwang, K.-C. (1981) Elastic–plastic fields in steady crack growth, *Proceedings IUTAM Symposium on Three-Dimensional Constitutive Relationships and Ductile Fracture*, edited by S. Nemat-Nasser, North Holland, Amsterdam (in press).

Hermann, L. and Rice, J. R. (1980) Comparison of experiment and theory for elastic–plastic plane strain crack growth, *Metal Science*, 285–291.

Hill, R. (1950) *The Mathematical Theory of Plasticity*, 334–340, Clarendon Press, Oxford.

Hult, J. A. H. (1957) Fatigue crack propagation in torsion, *J. Mech. Phys. Solids* **6,** 47–52.

Hult, J. A. H. and McClintock, F. A. (1956) Elastic–plastic stress and strain distributions around sharp notches under repeated shear, *Proc. 9th Int. Cong. Appl. Mech., Brussels* **8,** 51–58.

Hutchinson, J. W. (1968) Plastic stress and strain fields at a crack tip, *J. Mech. Phys. Solids* **16,** 337–347.

McClintock, F. A. (1958) Ductile fracture instability in shear, *J. Appl. Mech.* 582–588.

McClintock, F. A. (1971) Plasticity aspects of fracture, *Fracture: An Advanced Treatise*, edited by H. Liebowitz, **3,** 47–225, Academic Press, New York.

McClintock, F. A. and Irwin, G. R. (1965) Plasticity aspects of fracture mechanics, *Fracture Toughness Testing and Its Applications*, 84–113, ASTM-STP 381.

Paris, P. C., Tada, H., Zahoor, A., and Ernst, H. (1977) A treatment of the subject of tearing instability, U.S. Nuclear Regulatory Commission Report NUREG-0311/August 1977. (Available through NTIS, Springfield, Va., USA.)

Rice, J. R. (1967) Mechanics of crack tip deformation and extension by fatigue, *Fatigue Crack Propagation*, 247–311, ASTM-STP 415.

Rice, J. R. (1968) Mathematical analysis in the mechanics of fracture, *Fracture: An Advanced Treatise*, edited by H. Liebowitz, **2,** 191–311, Academic Press, New York.

Rice, J. R. (1973) Plane strain slip line theory for anisotropic rigid/plastic materials, *J. Mech. Phys. Solids* **21,** 63–74.

Rice, J. R. (1974) Elastic–plastic models for stable crack growth, *Mechanics and Mechanisms of Crack Growth*, edited by M. J. May, 14–39, British Steel Corp. Physical Metallurgy Centre Publication, Sheffield.

Rice, J. R., Drugan, W. J., and Sham, T. L. (1980) Elastic–plastic analysis of growing cracks, *Fracture Mechanics*, 189–219, ASTM-STP 700.

Rice, J. R. and Sorensen, E. P. (1978) Continuing crack tip deformation and fracture for plane strain crack growth in elastic–plastic solids, *J. Mech. Phys. Solids* **26,** 163–186.

Rice, J. R. and Tracey, D. M. (1973) Computational fracture mechanics, *Numerical and Computer Methods in Structural Mechanics*, edited by S. J. Fenves *et al.*, 585–623, Academic Press, New York.

Sham, T. L. (1979) A finite element analysis of quasi-static crack growth in an elastic–perfectly plastic solid, Sc.M. thesis, Brown University (Division of Engineering), Providence, R.I.

Slepyan, L. I. (1974) Growing crack during plane deformation of an elastic–plastic body, *Izv. Akad. Nauk SSSR, Mekh. Tela* **9,** 57–67.

Sorensen, E. P. (1978) A finite-element investigation of stable crack growth in anti-plane strain, *Int. J. Fract.* **14,** 485–500.

Legendre Transformations and Extremum Principles

by M. J. SEWELL
Department of Mathematics, University of Reading

Summary

A general study of both geometrical and analytical aspects of Legendre transformations is described, and particular attention is paid to how singularities of the transformation may be accommodated within a global description. Information from modern singularity theory is incorporated, and several examples from the mechanics of solids and fluids are treated in detail. An account is also given of a simple procedure, valid in many branches of applied mathematics, for deriving dual extremum principles from a single saddle functional. Examples of this are described which also reflect the role of Legendre transformations.

1. Introduction

The Legendre transformation has, in my view, a richness born of simplicity that makes it a very attractive mathematical device. It is both elegant and useful.

A clear and balanced appreciation of its role in applied mathematics is not widespread, however, in spite of the fact that the idea is at least two hundred years old. Its presence in classical discrete mechanics has not caused it to be recognized with equal facility in every branch of continuum mechanics, for example. Also, the twin aspects of geometrical and analytical duality inherent in a Legendre transformation do not always support each other in the literature with an emphasis which is sometimes possible.

The attribution may be a misnomer. According to Oravas and McLean (1966, p. 924) the transformation was used by Legendre in 1787, but it was discovered and used by Euler in 1770, and the first primitive instance of it has been found in Leibniz's researches. I shall not crusade for a lost cause against established usage in this respect, although one notes that this may not be the first injustice suffered by Euler. Apparently (op. cit., pp. 650 and 919) he used the so-called Lagrange multipliers in 1732 when Lagrange, even if subsequently precocious, was aged minus four.

The major part of this article is a connected account of some geometrical and analytical properties of the Legendre transformation treated in its own right, with especial attention paid to how singularities of the transformation may be accommodated within a global description. A review of some of this author's previous and recent work is incorporated. Selected examples from the mechanics of both solids and fluids are described in detail in Section 3. Some of these illustrations are chosen with the aim of achieving a ready integration with the second part of the article, whose theme is dual extremum principles. In Section 4 a very simple scheme proposed by the author for the generation of dual extremum principles from a single saddle functional is described. Examples from gas dynamics and plasticity are given in Section 5. These show how the particular Legendre transformations of Section 3 can be exploited. The article is broadly self-contained, and it includes some new material which arose during the research.

The existence of an extremum principle normally requires the idea of convexity to be present in some guise. In nonlinear problems of solid mechanics the first explicit recognition of this unifying role of convexity seems to be due to Hill (1956), who "saw what was needed and ferreted around to find out what it was called" (private communication). It was my good fortune, after listening to an undergraduate lecture by Rodney Hill in 1957, that this 1956 paper was the first research paper I read. It planted the seed from which the present article grew.

I have also tried to draw a judicious dose of stimulus from the rise of singularity theory, or catastrophe theory, since the Warwick Summer School of 1969. Singularity theory contains attractive mathematical and geometrical ideas, and it is unfortunate that interest in their application to solid mechanics may have been inhibited by exaggerated claims and neglect of pertinent references by some authors. Cusps were found in imperfection-sensitivity calculations by Koiter (1945), and the concept of the equilibrium surface was first introduced by Sewell (1966) with the objective of geometrizing bifurcation and stability analyses. At the time of writing (late 1979) catastrophe theory has contributed much pedagogic information about the shapes of certain equilibrium surfaces, but only one genuinely new result in the buckling of structures—that in multimode buckling the latent effect of imperfections reflecting modes higher than the first coincident ones has to be incorporated (see Fig. 13.38 of Poston and Stewart (1978)). Real progress in the application of singularity theory to the mechanics of solids has had so much leeway to make up, in terms of the detailed successes of conventional methods, that only now is one beginning to see serious new developments. Some of these are suggested by Golubitsky and Schaeffer (1979), and an account of underlying mathematical ideas oriented towards that aspect of bifurcation theory is given by Chillingworth (1981) elsewhere in this Volume. Uses of singularity theory in an interesting diversity of physical problems can be found in the volume edited by Güttinger and Eikemeier (1979).

There is not space here to treat bifurcation problems *per se*, but elastic/plastic bifurcations are discussed in this Volume by Needleman and Tvergaard (1981). Instead, singularity theory is used to focus on how to carry Legendre transformations through their singularities. It would be unnatural to confine the examples exclusively to solid mechanics, because compressible gas dynamics provides the concept of a "constitutive surface", introduced by Sewell and Porter (1980), whose explicit shape allows the knowledge of stationary and extremum principles for a general gas to be carried here further than can be found in the fluid mechanics literature; and this is of interest not only for its own sake in informing the calculation of flows, but also as a prototype for the study of Legendre transformations and extremum principles in more complicated "simple" materials, including solids, which have not yet yielded such explicit information.

2. General Properties of Legendre Transformations

2.1. *Geometrical duality*

We recall some particular properties of the classical duality between poles and polars with respect to convex surfaces. Let X, x_i ($i = 1, \ldots, n$) be cartesian coordinates spanning an $(n+1)$-dimensional space, and consider a paraboloid

$$2X = x_i^2 \tag{2.1}$$

drawn in it. (Throughout the article repeated suffixes, as implied here, will mean summation unless otherwise stated. Isolated suffixes will take their full range of values in turn.) Let Y, y_i denote a particular point in the space. The condition for a plane drawn through this point to meet the paraboloid tangentially is

$$X + Y = x_i y_i. \tag{2.2}$$

This is the equation of another plane, and all such tangency points must lie on it.

The point Y, y_i and the plane (2.2) are called pole and polar respectively, with respect to the paraboloid. They are apex and base of a cone touching the paraboloid. Given the pole we can write down the polar; given the polar equation (2.2) we can read off the coordinates of the pole. Some authors (e.g. Young, 1969, Section 21) regard this correspondence between points and planes as a Legendre transformation. Figure 2.1 shows the particular intercept made by the plane which contains both the pole and the axis of the paraboloid.

It is easy to show that the notion of pole and polar with respect to a sphere in n-dimensional space is conveyed by a second particular intercept, namely that by the plane $X = Y$ which is orthogonal to the axis of the paraboloid, remembering that at this stage Y is a given constant, say $K > 0$. For then the particular point y_i is the pole in x_i-space with respect to the sphere $2K = x_i^2$, and its polar plane is the appropriate section of (2.2), namely $2K = x_i y_i$. They are apex and base of a cone touching the sphere. This special type of duality is shown in Fig. 2.2 for the case $n = 2$. We find duality with respect to a sphere in classical dynamics, for instance (e.g. see Synge, 1960, Section 71), and we give another example in Fig. 3.6 below where a singularity is also accommodated.

Next we imagine the pole Y, y_i to move around in the $(n+1)$-dimensional space, say on the nominated smooth surface

$$Y = Y(y_i). \tag{2.3}$$

The polar plane (2.2) will then envelop another smooth surface

$$X = X(x_i). \tag{2.4}$$

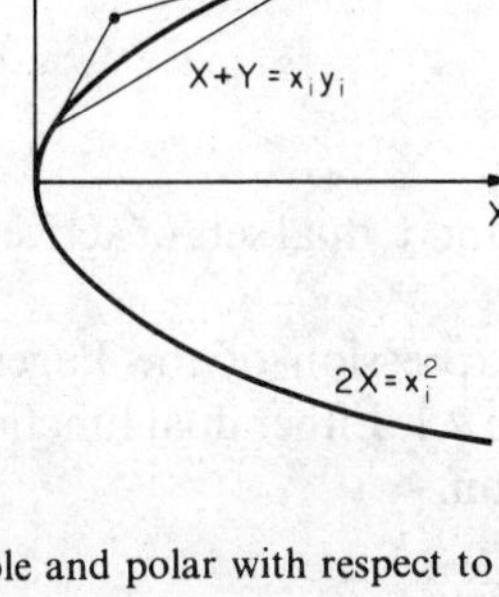

FIG. 2.1. Pole and polar with respect to paraboloid.

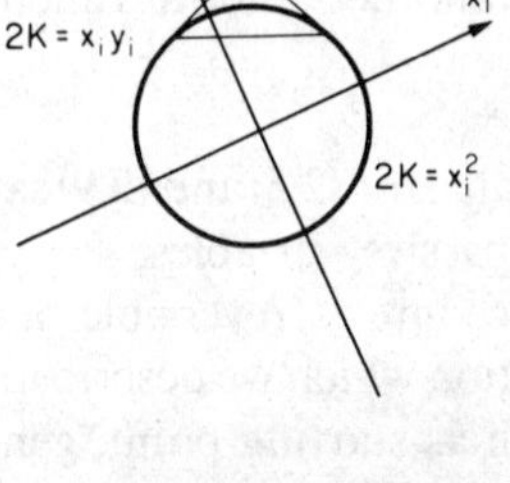

FIG. 2.2. Special duality with respect to sphere.

The symmetrical structure of the theory shows that the roles of these two surfaces can be interchanged, i.e. a pole moving around on (2.4) will define a polar enveloping (2.3). We shall say that either of (2.3) or (2.4) is a generating function of a Legendre dual transformation. This usage of the term "generating function" is slightly different from that sometimes found in generalized dynamics. The Legendre transformation now establishes a correspondence between two dual surfaces, and it might also be called a contact transformation because of its enveloping property.

The geometrical construction just described may be called "polar reciprocation". It is very useful for inferring the qualitative shape of one surface from the other when an analytical approach is difficult or impossible in closed form, before one resorts to the third approach of numerical plotting, e.g. of a possibly multi-valued dual function. Polar reciprocation is found in a special guise, with respect to a fixed sphere, when we need only a single cross-section of the dual of a given generating function, rather than the entire dual function. Suppose that a pole moves on a given surface $Y(y_i) = K$ in n-dimensional space, where K is the previous positive constant, and that we require the envelope $X(x_i) = K$ in x_i-space made by its polar with respect to the sphere $2K = x_i^2$. Such a construction can be based on Fig. 2.2 (when $n = 2$) or its generalization. It will be pertinent, for example, when (2.3) is a cone in the $(n+1)$-dimensional space, because then all its cross-sections perpendicular to the Y-axis will be similar. Exactly this circumstance arises in theories of wave propagation in anisotropic media, and we give an illustration of it in Section 3.5 below.

2.2. *Analytical duality*

Suppose that there exists a smooth function

$$Y = Y(y_i, u_\alpha) \tag{2.5}$$

of n variables y_i and m variables u_α ($\alpha = 1, \ldots, m$), having gradients which we denote by

$$x_i = \frac{\partial Y}{\partial y_i}, \quad v_\alpha = \frac{\partial Y}{\partial u_\alpha}. \tag{2.6}$$

Whenever the determinant

$$\left| \frac{\partial^2 Y}{\partial y_i \partial y_j} \right| \neq 0, \quad \pm\infty, \tag{2.7}$$

we can invert $(2.6)_1$ locally at fixed u_α, and express its inverse as the first of

$$y_i = \frac{\partial X}{\partial x_i}, \quad -v_\alpha = \frac{\partial X}{\partial u_\alpha}, \tag{2.8}$$

in terms of another smooth function

$$X = X(x_i, u_\alpha), \tag{2.9}$$

where the values of (2.5) and (2.9) satisfy (2.2). We call the x_i and y_i dual sets of active variables, and the u_α passive variables.

This procedure is reversible, and it is the analytical expression of the Legendre dual transformation which we described geometrically in Section 2.1. Either dual function (2.5) or (2.9), chosen as starting point, generates the transformation.

The smoothness hypothesis of C^3 functions will be adequate for the requirements of this article, and in places it will be more than sufficient. We have no real need of refinements such as Fenchel duality, or of other technicalities discussed, for example, by Rockafellar (1970). Instead one of our main concerns will be to review current understanding of how a Legendre transformation may be continued through its singularity, defined as a failure of hypothesis (2.7).

Straightforward differentiations employing implicit functions lead to the following relations between second derivatives of the dual generating functions (2.5) and (2.9). For brevity we denote second derivatives, with respect to either active or passive variables, by Latin or Greek suffixes. Then

$$X_{ik}Y_{kj} = \delta_{ij}, \tag{2.10}$$

$$X_{i\alpha} = -X_{ik}Y_{k\alpha}, \quad Y_{i\alpha} = -Y_{ik}X_{k\alpha}, \tag{2.11}$$

$$X_{\alpha\beta} + Y_{\alpha\beta} = -X_{\alpha k}Y_{k\beta} = -X_{\beta k}Y_{k\alpha}. \tag{2.12}$$

Consequently we have

$$X_{\alpha\beta} + Y_{\alpha\beta} = X_{ij}Y_{i\alpha}Y_{j\beta} = Y_{ij}X_{i\alpha}X_{j\beta}. \tag{2.13}$$

The formulae (2.11)–(2.13) are not customarily given in descriptions of the Legendre transformation. They are used in a discussion in Section 2.9 below of the shape of the dual functions.

A singularity of the Legendre transformation is thus defined by either determinantal equation

$$|X_{ik}| = 0 \quad \text{or} \quad |Y_{kj}| = 0. \tag{2.14}$$

2.3. *Response curves*

One may regard $(2.6)_1$ and $(2.8)_1$ as response equations like constitutive equations. They describe how the x_i and y_i are related to each other, at each fixed u_α, when these sets of variables have different *a priori* physical definitions, such as stress and strain. In the case $n = 1$ and with no passive variable, either equation

$$x = \frac{dY}{dy}, \quad y = \frac{dX}{dx}, \tag{2.15}$$

can be shown as a response curve in the x, y plane. This is a different viewpoint from that of Section 2.1, where the x_i and the y_i spaces were in effect superimposed. The dual functions satisfy (2.2) in the form

$$X(x) + Y(y) = xy. \tag{2.16}$$

This has the following interpretation, well known in the non-singular case. The rectangle between the axes, the origin, and the point x, y is divided by the response curve $x = x(y)$ or $y = y(x)$ into the two "complementary" areas $Y(y)$ and $X(x)$ lying between the curve and the axes, as indicated in Fig. 2.3(a).

This interpretation can be extended through a singularity, for example where $dy/dx = 0$ by (2.14), if we add the obvious convention (Sewell, 1978a) that areas subtended under a *falling* ordinate of the response curve, onto the ordinate axis, are counted as negative. Figure 2.3

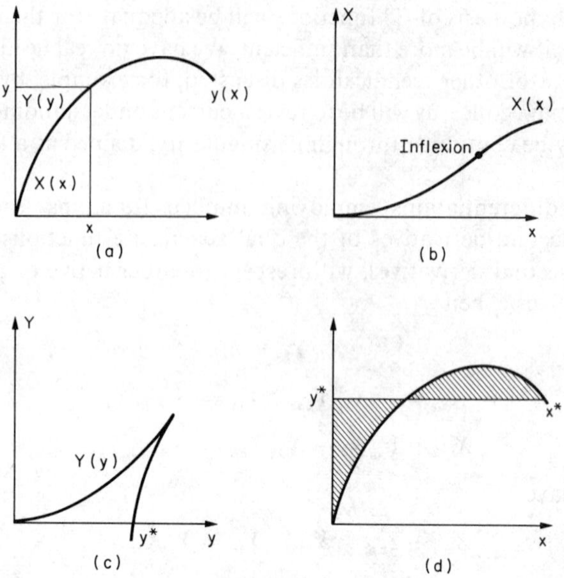

FIG. 2.3. Areas under a response curve with singularity.

illustrates a case in which a response curve begins at the origin but dy/dx decreases monotonically from positive through to negative values. At the maximum y the area $X(x)$ passes through an inflexion, and the complementary area $Y(y)$ passes through a cusp. The foregoing sign convention then comes into operation, allowing $Y(y)$ to start falling and even to pass through zero at a later point x^*, y^* (say) on the response curve. The two shaded areas in Fig. 2.3(d) are then equal.

A long-established example of Fig. 2.3(a) is found in the one-dimensional engineering theory of a bar in tension, where y is nominal stress (current load per unit initial cross-sectional area) and $x+1$ is uniform stretch (longitudinal after/before length ratio). Evidently the associated strain energy and complementary energy have the singularities displayed as inflexion and cusp respectively in Figs. 2.3(b) and (c), as remarked by Sewell (1978a). Hill (1957) exploited a nominal stress tensor in the first of his long series of papers on three-dimensional bifurcation theories in elasticity and plasticity, and Koiter (1976) and Ogden (1977) have given different modern discussions of complementary energy in nonlinear elasticity. In particular, Ogden claims that a certain "symmetrized Biot stress" measure is indispensable for the formulation of complementary stationary principles in that subject. How the inflexion and cusp of Fig. 2.3 are to be embedded in a geometrical description of energy singularities in three-dimensional theories is still an open question.

The usefulness of response curves in a plane diagram is by no means confined to one-dimensional theories, since they may relate invariants of vectors or tensors present in a three-dimensional theory. We give such an example from compressible inviscid fluid mechanics in Section 3.4 below, where the integrands in complementary stationary principles contain Legendre dual functions of the magnitude only of the fluid velocity and mass flow vectors, and these functions have inflexion and cusp at the sonic singularity.

In general the presence of a Legendre singularity (2.14) will be intimately associated with the occurrence of a bifurcation, which itself will often terminate the range of validity of an extremum principle. This is the point of connection between the two major themes in the title of this article.

The precise statement of such comparative results may depend crucially on the detailed choice of physical variables, e.g. on the particular measures used for stress and strain. In passing we can indicate this facet of singularity theory by an example of a plane response curve which expresses a singularity like that of Fig. 2.3 in a different way. Suppose the context contains two variables y and η of physical interest (e.g. nominal stress and true stress in the solid bar, or mass flow and particle speed in the fluid) which are related to a positive third variable $x+1 = v$ (say, for example, stretch in the solid, or reciprocal of density in the fluid) by response functions $y(v)$ and $\eta(v)$ satisfying

$$\eta(v) = vy(v). \tag{2.17}$$

Then

$$\frac{dy}{dv} = \frac{1}{v}\left(\frac{d\eta}{dv} - \frac{\eta}{v}\right). \tag{2.18}$$

This means that a singularity which appears as a stationary point on $y = y(v)$ or $y = y(x)$, like that of Fig. 2.3(a), can also be located on the response curve $\eta = \eta(v)$ by drawing a tangent from the origin $\eta = v = 0$ to the curve. This "tangent construction" in the stretched bar problem has been attributed to Considère (via an intermediate reference quoted by Nádai, 1950). For the fluid the construction was discussed by Calladine (1974) and by Sewell and Porter (1980).

One can recognize that elaborations of it will be present in other contexts. Certain formulae of Haughton and Ogden (1978) concerning the inflation of a spherical rubber membrane can be expressed in novel graphical form as the stack of pairs of tangent constructions in Fig. 2.4. Here $X(v)$ represents a strain energy function of v = final/initial radius ratio, and the quotient of each successive pair of curves is equal to v. In other words

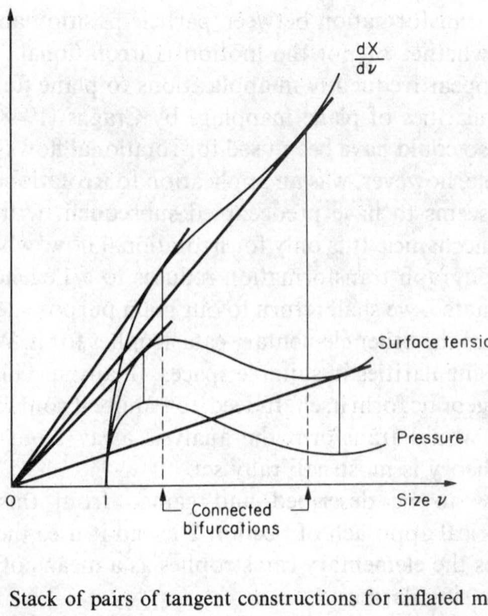

FIG. 2.4. Stack of pairs of tangent constructions for inflated membrane.

there are two formulae of type (2.17); for example, surface tension is proportional to v times inflation pressure. A pair of tangent constructions to the curve dX/dv locates singularities which are therefore also maximum and minimum values of surface tension, at which bifurcation into an aspherical mode can be shown to take place. Another pair of tangent constructions to the surface tension curve locates extrema of the internal pressure which do not coincide with but herald the bifurcation, in the sense that maximum (minimum) pressure precedes the bifurcation when v is imagined to be increasing (decreasing). Sewell (1976) suggested that the two aspherical bifurcated equilibrium paths will join up to form a closed loop. This was taken up and reinforced by Needleman (1977), who followed numerically the development of an aspherical mode in an imperfect spherical membrane, and has since been confirmed by Haughton (1980) working with a perfect initial sphere. It can be shown theoretically that a free soap bubble will have a film tension curve more like the upper than the middle curve of Fig. 2.4.

2.4. *Singularities*

In the next few subsections we use some of the more familiar elementary catastrophes of Thom (1969) to explore what one may expect to be the more common forms which a singularity in a Legendre transformation can take. The basis for this is that these catastrophes contain canonical expressions for generating functions which lead to diffeomorphisms, i.e. smooth changes of variable, whose singularities are stable with respect to certain mathematical perturbations.

For the sake of perspective it is well to be aware that the first three of these stable forms, namely the fold, cusp, and swallowtail, are the typical singularities of non-gradient mappings of $R^3 \to R^3$ (e.g. see Thorndike *et al.*, 1978). Likewise the fold and cusp are the only stable singularities of smooth plane mappings (Whitney, 1955). Therefore these three forms will appear more widely than as Legendre singularities, which can be viewed as singularities of gradient mappings. For example, in the three-dimensional kinematics of any continuous medium the hodograph transformation between particle position and velocity will exhibit the stated singularities whether or not the motion is irrotational. This is the underlying reason why cusps will appear frequently in applications to plane fluid flow. An early basic investigation of the singularities of plane mappings by Craggs (1948) was not restricted to gradient mappings, and so could have been used for rotational flows in spite of his misleading title. His only example, however, was an application to irrotational plane flows, and it is this special case which seems to have preoccupied subsequent writers on the hodograph transformation in fluid mechanics. It is only for irrotational flow, whether plane or in three dimensions, that the hodograph transformation reduces to a Legendre transformation.

With this perspective stated, we shall return to our main purpose described above and use fold, cusp, swallowtail, and the other elementary catastrophes for it. We seek insight into the qualitative shape of the singularities in suitable spaces. In any specific application one may not need the canonical algebraic form itself. Indeed, to require it could necessitate a search for diffeomorphisms which would transform the analysis away from the *ab initio* physical variables in which the theory is most naturally set.

The information now to be described and gained from this viewpoint therefore supplements the geometrical approach of Section 2.1, and it uses the notion of "structural stability" which underlies the elementary catastrophes as a means of focusing on the more likely forms of Legendre singularities.

2.5. *A singular example*

The following example of the geometrical and analytical procedures of Sections 2.1 and 2.2 is illustrated in Fig. 2.5. We seek the dual of the function

$$Y(y) = \tfrac{1}{3}y^3 + ay, \tag{2.19}$$

where a is some fixed value of a passive variable u. The gradient equation $(2.6)_1$ is then

$$x = y^2 + a \tag{2.20}$$

and this can be inverted locally except possibly through the singularity at $y = 0$. The inverse is

$$y = \pm (x - a)^{1/2}, \tag{2.21}$$

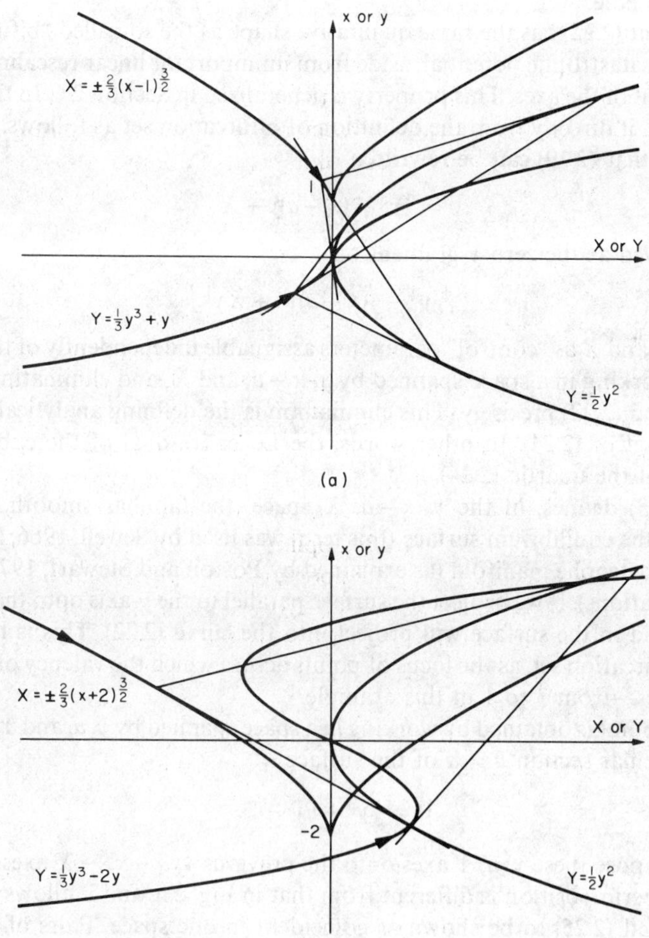

FIG. 2.5. (a) Legendre dual curves for $a = +1$. (b) Legendre dual curves for $a = -2$.

this being also the gradient equation $(2.8)_1$ derived from the dual function

$$X(x) = \pm\tfrac{2}{3}(x-a)^{3/2}. \tag{2.22}$$

The global inverse of (2.19) can now be defined simply as the union of the two branches of (2.22), together with the singular point itself where $X(a) = 0$. Thus it is possible to continue a Legendre transformation analytically through a singularity associated with a cubic-like inflexion, and the result has a cusp in the dual function. Figure 2.5 shows the functions (2.19) and (2.22), and also the parabola $2Y = y^2$ or $2X = x^2$ with respect to which they are the geometric duals, for the two cases $a = +1$ and $a = -2$. The curves are drawn to scale, and respective dual points can be traversed in the sense of the arrows. An example of the corresponding envelope properties in a general position is shown in Fig. 2.5(a), and in a special position in Fig. 2.5(b). A point on either curve which is outside the parabola can serve as a pole or as the tangency point of a polar. A point inside the parabola can serve as the latter but not as a real pole.

It is evident that (2.22) has the same qualitative shape as the so-called "bifurcation set" of the quartic cusp catastrophe potential, aside from unimportant linear rescaling and possible rigid displacement of the axes. This property is generalized in Section 2.6. In this example we can also establish it directly from the definition of bifurcation set as follows. We note from (2.2) with $n = 1$ that (2.19) can be rewritten

$$0 = \tfrac{1}{3}y^3 - (x-a)y + X. \tag{2.23}$$

This can be viewed as the zero y-gradient of

$$\tfrac{1}{12}y^4 - \tfrac{1}{2}(x-a)y^2 + Xy \tag{2.24}$$

by treating $x-a$ and X as "control" parameters assignable independently of the "behaviour" variable y. By working in a space spanned by y, $x-a$, and X, and eliminating y from (2.23) and (2.20), we find (2.22) precisely. This elimination is the defining analytical procedure for the bifurcation set of (2.24). In other words, the Legendre dual of the cubic (2.19) is the bifurcation set of the quartic (2.24).

Equation (2.23) defines, in the y, $x-a$, X space, the familiar smooth folded surface variously called the equilibrium surface (this term was used by Sewell, 1966, for equilibrium problems) or catastrophe manifold (as explained by Poston and Stewart, 1978, for example, in general applications). If we project the surface parallel to the y-axis onto the control plane, the perceived fold in the surface will project into the curve (2.22). This is the geometrical definition of bifurcation set, as the locus of points across which the valency of the projection mapping changes—from 3 to 1 in this example.

Another viewpoint is obtained by working in a space spanned by y, u, and Y, and regarding (2.19) as a particular section $u = a$ of the surface

$$0 = \tfrac{1}{3}y^3 + uy - Y. \tag{2.25}$$

Next we superimpose these y, u, Y axes onto the previous y, $a-x$, $-X$ axes respectively in Fig. 2.6. This superimposition is different from that in Fig. 2.5, and it allows the two folded surfaces (2.23) and (2.25) to be shown as coincident in one space. Pairs of Legendre dual functions can now be pictured as different aspects of the same three-dimensional diagram: the projection of the fold onto the $x-a$, X control plane is dual to any cubic section formed by a plane normal to the passive variable u-axis.

The surface in Fig. 2.6 can be regarded as made up of straight lines, on each of which

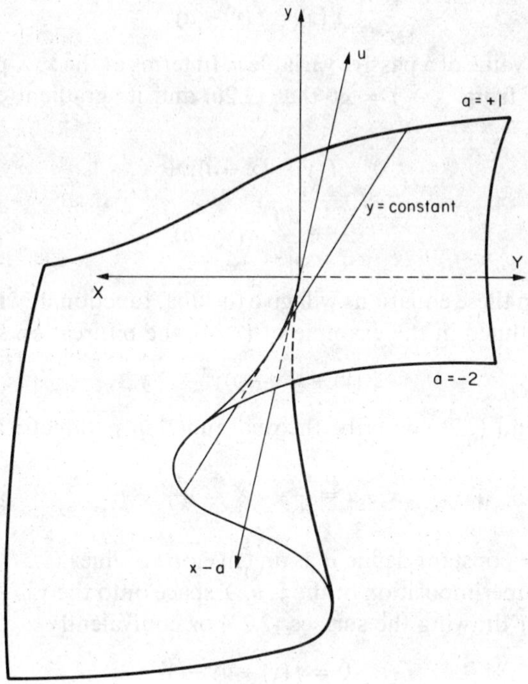

FIG. 2.6. Legendre dual functions in superimposed spaces.

$y =$ constant, and we show one of them. The reader may find it helpful to make his own string model of the surface, for example along the lines described in an article by Sewell (1978b), where a photograph is shown. In terms of this surface we can give the following reinterpretation of the defining geometrical relation between pole and polar, for this example. Take any particular point on the surface (2.25) as pole, for example with fixed values of y, $u = a$, and Y related by (2.19). Its polar is (2.2) with $n = 1$, i.e. $X + Y = xy$ in any x, X plane independently of the a-value, and is therefore precisely the line (2.23) in y, $x - a$, X space such as Fig. 2.6 illustrates. Hence we have the very simple result that, in this diagram, *the pole lies on its own polar*. We establish the same property more generally in Section 2.6.

The pedagogic value of Fig. 2.6 as an equilibrium surface describing the unimodal buckling of structures, some sensitive to imperfections and some not, is conveyed in some simple examples described by Sewell (1976, 1978c) and dating from unpublished 1969 work. Zeeman's diverse writings (1977) have been a strong source of stimulus in other contexts. In multimode buckling of compressed spheres (without the artificial restriction to rotationally symmetric modes) and cylinders, there are still outstanding problems (described in my 1976 paper) of fusing the singularity theory approach with results which have been obtained by conventional methods.

2.6. *Ladder for the cuspoids*

Let $F(y)$ be a given scalar function with gradient $f(y)$. We seek the Legendre dual of the function

$$Y(y) = f(y) + ay, \tag{2.26}$$

where a is some fixed value of a passive variable u. In terms of the x, X plane in which the dual function lies, we see from $X + Y = xy$ that (2.26) and its gradient equation $(2.6)_1$ can be rewritten

$$0 = f(y) - (x-a)y + X, \tag{2.27}$$

$$0 = \frac{df}{dy} - (x-a). \tag{2.28}$$

Elimination of y from these equations will give the dual function. It will also give, for exactly the same reasons outlined in the example of (2.24), the bifurcation set of the potential

$$F(y) - \tfrac{1}{2}(x-a)y^2 + Xy. \tag{2.29}$$

Equations (2.27) and (2.28) describe the dual function parametrically as

$$x - a = \frac{df}{dy}, \quad X = y\frac{df}{dy} - f. \tag{2.30}$$

Successive values $y =$ constant define it as an envelope of lines (2.27). This envelope can be found via the same superimposition of the y, u, Y space onto the y, $a-x$, $-X$ space that we used in Fig. 2.6, after drawing the surface (2.27) or equivalently

$$0 = f(y) + uy - Y \tag{2.31}$$

in this space. That is, we project onto the $x-a$, X plane any folds which there may be in the surface (2.27). The result is the dual function of any section $u = a$ of the same surface (2.31). We also see, more generally now than in the previous example $f(y) = \frac{1}{3}y^3$, that the pole of any point with given y, $u = a$, Y will lie on its own polar line (2.27) in y, $x-a$, X space.

These arguments apply in particular to the polynomial

$$F(y) = \frac{y^{n+1}}{n(n+1)} + \ldots + \tfrac{1}{6}by^3 \tag{2.32}$$

for any positive integer $n \geqslant 2$, where the coefficients $b, \ldots,$ are given fixed values of passive variables. The coefficient of the nth degree term can always be transformed to zero by a change of y-origin, and (2.26) is then often called a "cuspoid" potential. The conclusion reached at (2.29) then says that the Legendre dual of the cuspoid of degree n has the same shape as the bifurcation set of the cuspoid of degree $n+1$. This was proved by Sewell (1977a) using explicit scaling arguments, and expressed as a "ladder for the cuspoids". The first few steps of this ladder are indicated in Fig. 2.7. In the left column is the list of unfolded cuspoid potentials, typically of degree $n \geqslant 3$ in one behaviour variable and containing $n-2$ control parameters linearly. In the right column are surfaces lying in the $(n-2)$-dimensional control spaces but obtained in two different ways—as a Legendre transform by stepping up the ladder or as a bifurcation set by stepping down it. The second rung, showing that the Legendre transform of the fold potential has the same shape as the bifurcation set of the cusp potential, was analysed in detail in Section 2.5.

The significance of this ladder is that bifurcation sets have already been determined in catastrophe theory for the first few values of n (e.g. computer pictures for the cases from $n = 3$ to 8 were published by Woodcock and Poston (1974)). They are complicated surfaces, being typically composed of several branches, sometimes self-intersecting, but with continuous

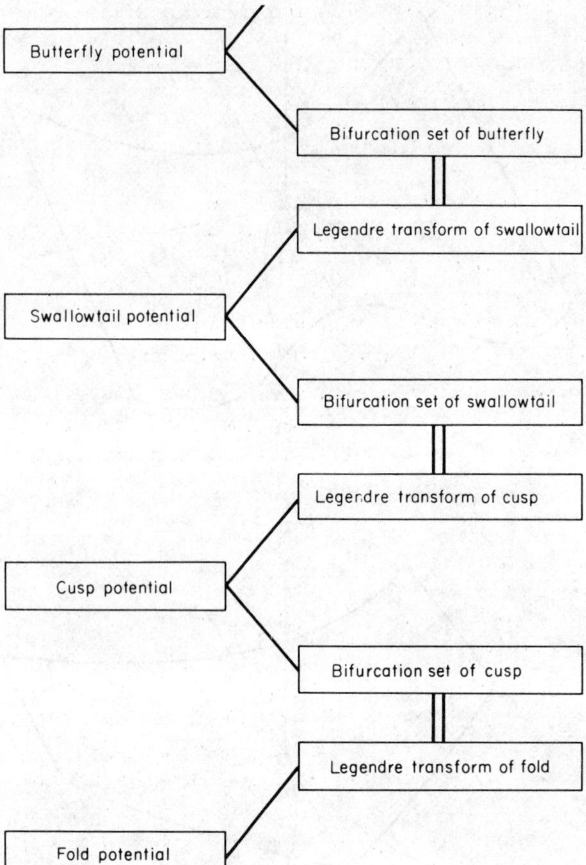

FIG. 2.7. Ladder for the cuspoids.

value and slope at cusped edges ("ribs") of regression. Nevertheless, they can be considered known, especially as regards their qualitative structure. In that sense the corresponding Legendre dual functions are therefore also known by reading them off the ladder.

Evidently the dual $X(x)$ of (2.26) is actually always of the form $X(x-a)$. It is sometimes also desirable to know its dependence on any other parameters which may be latent in $f(y)$, such as the $b, \ldots,$ in (2.32). The case $n = 4$ there makes (2.29) into the so-called swallowtail potential

$$\tfrac{1}{20}y^5 + \tfrac{1}{6}by^3 - \tfrac{1}{2}(x-a)y^2 + Xy. \tag{2.33}$$

Its bifurcation set lies in a three-dimensional space and is known. It is sketched in Fig. 2.8(b) in the space spanned by b, $x-a$, and X. By the arguments given this will also be the Legendre dual function $X(x-a, b)$ of

$$Y(y, a, b) = \tfrac{1}{4}y^4 + \tfrac{1}{2}by^2 + ay \tag{2.34}$$

given by (2.26), with passive b and a. The dual function $X(x-a, b)$ will be a surface ruled by lines along which $y =$ constant, and defined by the intersection of what are now pairs of

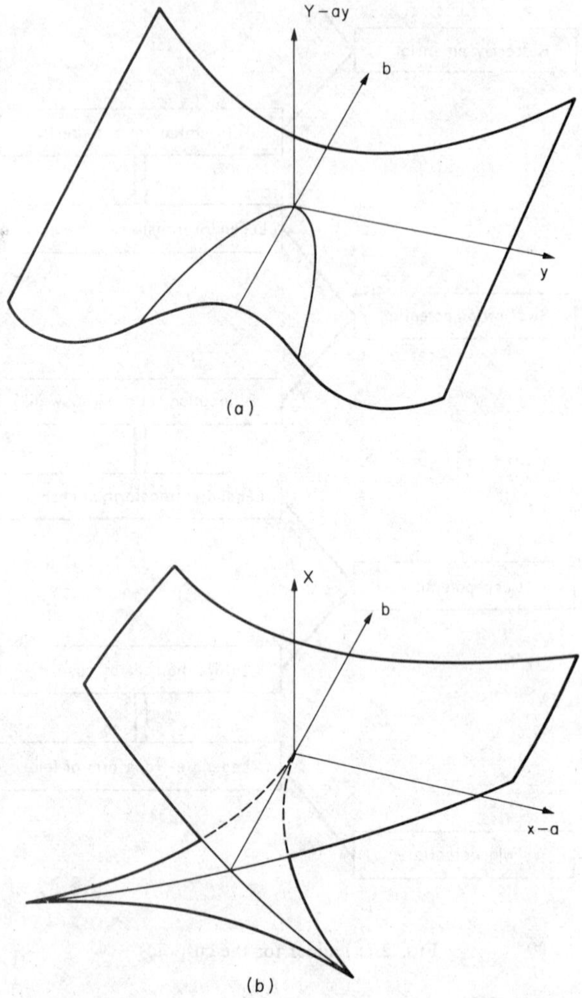

FIG. 2.8. (a) $Y - ay = \frac{1}{4}y^4 + \frac{1}{2}by^2$. (b) Swallowtail dual $X(x-a, b)$ of $Y(y, a, b)$.

planes (2.30) in the b, $x-a$, X space, by the linearity of b in (2.33). The details of this calculation, and a photograph of a string model of Fig. 2.8(b), were given by Sewell (1977a). In Fig. 2.8(a) we sketch (2.34) rotated about the b-axis into a symmetrical position to represent it in a three-dimensional space as $Y - ay = \frac{1}{4}y^4 + \frac{1}{2}by^2$. The b-axis lies in each surface of Fig. 2.8. For each fixed a, the singularity of the Legendre transformation appears as a locus of quartic inflexions on successive sections b = constant of Fig. 2.8(a), and as a cusped edge of regression in Fig. 2.8(b), this edge *itself* forming a cusp at $X = x - a = b = 0$ (called the "swallowtail point").

A Legendre dual function with this swallowtail shape has been deployed by Sewell and Porter (1980) in compressible gas dynamics and shallow water theory, and is described in Section 3.4.

2.7. *Umbilics*

The two variable polynomials

$$Y(y_1, y_2; u) = y_1^2 y_2 \pm y_2^3 + u y_1^2 \tag{2.35}$$

parametrized by u are, when linear terms are added, the universal unfoldings of the so-called hyperbolic (+ sign alternative) and elliptic (− sign) umbilics. To find the Legendre dual function $X(x_1, x_2; u)$ of either requires the simultaneous inversion of two equations $(2.6)_1$. The omitted linear terms will not affect the form of the result but, just as we found for the cuspoids, they will merely shift the x_i-origin. The singularity of the transformation is where the Hessian matrix

$$\left[\frac{\partial^2 Y}{\partial(y_1, y_2)}\right] = 2\begin{bmatrix} y_2 + u & y_1 \\ y_1 & \pm 3y_2 \end{bmatrix} \tag{2.36}$$

has zero determinant. When $u \neq 0$ this singularity traces a hyperbola or ellipse in the y_1, y_2 plane, which degenerates to a pair of asymptotes or a single point respectively if $u = 0$. On this conic the rank of (2.36) is one if $u \neq 0$, but zero if $u = 0$.

In this way either umbilic organizes the passage of a Legendre transformation through an isolated compound singularity of multiplicity two. This is not the only way in which a double zero can appear in the singularity of a Legendre transformation, but it is one of the simplest. Other typical ways are to be found via four more umbilics listed in the statement of Thom's Theorem (see Poston and Stewart, 1978, p. 122). In previous subsections we concentrated upon isolated simple singularities, i.e. of multiplicity one. The way to use these results, in any applied context where there are more active variables than the multiplicity of the isolated singularity (i.e. more than one or two as the case may be), is to seek smooth changes of variable which identifies those one or two contributing essentially to the singularity.

The detailed arguments needed to derive the Legendre transforms of the hyperbolic and elliptic umbilics are given in the paper of Sewell (1978d), where some information about their relation to the bifurcation set of the parabolic umbilic is also recorded. Only the results for the hyperbolic case having a constant $u < 0$ are sketched here.

The function (2.35+) shown in Fig. 2.9(a) is single-valued, continuous, and smooth and consists of three parts A, B, C separated by the trace of the singularity hyperbola $y_1^2 = 3y_2(y_2 + u)$ raised to the surface. C is saddle-shaped, and A and B are strictly jointly concave and convex respectively. The dual function shown in Fig. 2.9(b) also consists basically of three parts individually mapped from those of Fig. 2.9(a) and retaining the stated convexity and concavity properties. The parts join continuously, and smoothly but regressively, at a cusped edge itself having two branches (since it maps from the trace of the singularity hyperbola). This edge has a smooth-nosed branch A/C, and a branch B/C having a cusped nose at a swallowtail point; at the latter a self-intersection line in C also arrives and terminates. This swallowtail point is the map of the only place in Fig. 2.9(a) where the singularity trace is tangential to the eigenvector of (2.36) associated with the singularity (this property may help to guide investigation of other non-canonical two-variable examples). Elsewhere this eigenvector is transverse to the trace. The dual function is clearly multivalued, with multiplicity 0, 2, or 4. Both surfaces in Fig. 2.9 have a plane of symmetry ($y_1 = 0$ or $x_1 = 0$).

The Legendre dual of the elliptic umbilic contains three swallowtail points instead of one,

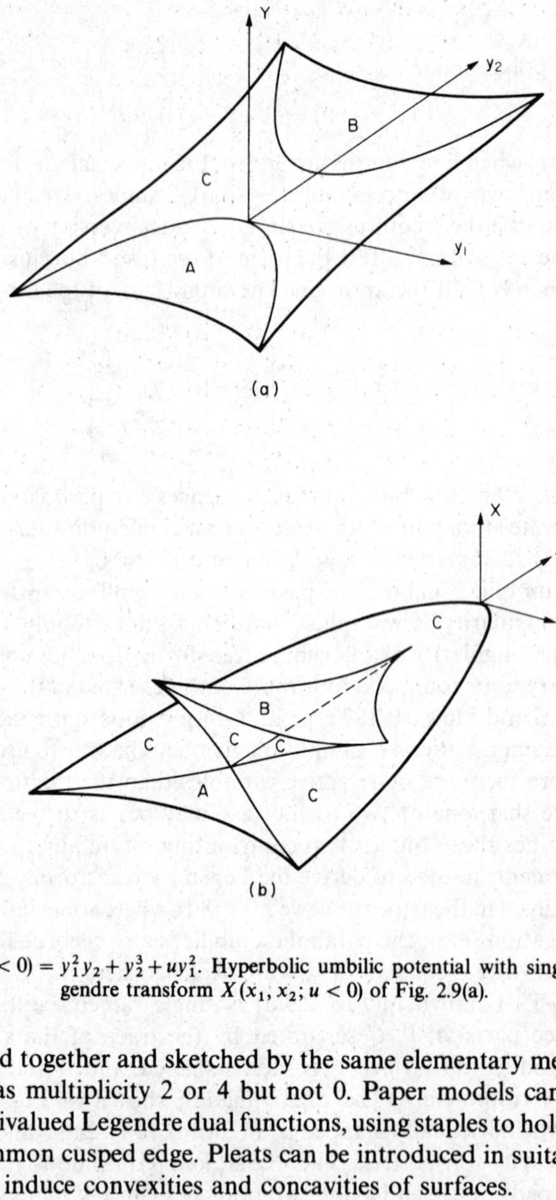

FIG. 2.9. (a) $Y(y_1, y_2;\, u < 0) = y_1^2 y_2 + y_2^3 + u y_1^2$. Hyperbolic umbilic potential with singular hyperbola. (b) Legendre transform $X(x_1, x_2;\, u < 0)$ of Fig. 2.9(a).

but can still be pieced together and sketched by the same elementary methods (see Fig. 9 of Sewell, 1978d). It has multiplicity 2 or 4 but not 0. Paper models can be made to assist visualization of multivalued Legendre dual functions, using staples to hold together branches meeting along a common cusped edge. Pleats can be introduced in suitable directions with scissors and tape to induce convexities and concavities of surfaces.

2.8. *Closed chain of Legendre transformations*

The information summarized in Sections 2.4–2.7 about isolated Legendre singularities provides a basis from which to address this difficulty in an informed way when it arises in applications, if not yet to surmount it in every case. In particular we can now expect sometimes to remove the restriction (2.7) from the analytical statement (2.5)–(2.9) of the Legendre transformation. The price of so doing, when isolated simple or double zeros occur in the Jacobian, is evidently that we must be ready to allow one or other of the continuous and smooth generating functions to be multivalued. In this sense we now regard (2.2) and (2.5)–(2.9) without (2.7) as a global, rather than local, statement of the transformation.

Next we explore the effect of interchanging the roles of active and passive variables. Thus we seek to invert $(2.6)_2$ rather than $(2.6)_1$. We shall arrive at another dual function, say

$$W = W(y_i, v_\alpha) \tag{2.37}$$

which satisfies

$$-x_i = \frac{\partial W}{\partial y_i}, \quad u_\alpha = \frac{\partial W}{\partial v_\alpha} \tag{2.38}$$

and

$$Y + W = u_\alpha v_\alpha. \tag{2.39}$$

Alternatively, we might invert $(2.8)_2$ instead of $(2.8)_1$, or $(2.38)_1$ instead of $(2.38)_2$. Either inversion leads to the same new dual function

$$Z = Z(x_i, v_\alpha) \tag{2.40}$$

say, satisfying

$$-y_i = \frac{\partial Z}{\partial x_i}, \quad -u_\alpha = \frac{\partial Z}{\partial v_\alpha}, \tag{2.41}$$

and

$$X + Y + Z + W = 0. \tag{2.42}$$

We can thereby construct a closed chain of four Legendre transformations, any of which may be global under the qualifications mentioned. A mnemonic summarizing these interconnections is the "diamond diagram" of Fig. 2.10. This was given by Noble and Sewell (1972), but effectively for the particular non-singular quartet indicated in Fig. 2.11. Now in Fig. 2.10 global generating functions are written at the corners of the diamond. They may be multivalued, and each link joining them represents a Legendre transformation admitting isolated singularities. The dual active variables in each link are written next to the arrows. This mnemonic carries more explicit information than the so-called "thermodynamic square".

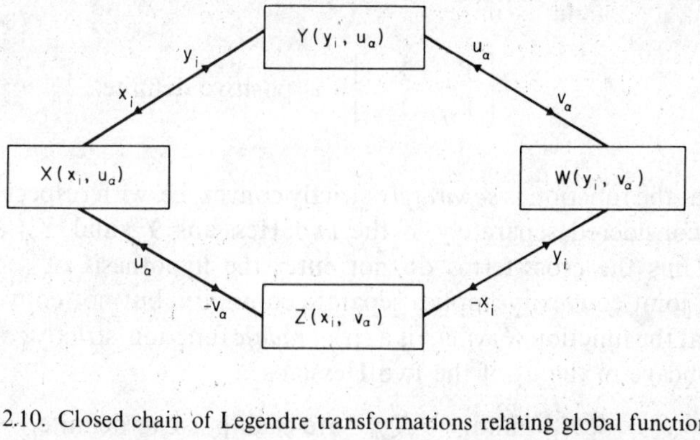

FIG. 2.10. Closed chain of Legendre transformations relating global functions.

We have now arrived at the symmetrical standpoint summarized by Fig. 2.10. Any corner of the diamond diagram may be chosen, according to convenience, as a starting point from which to develop certain aspects of a physical theory. For example, by differentiating any one generating function we obtain two sets of equations relating four sets of variables. Supplementary information will be required from other sources to complete a fully posed statement.

Sometimes there is not a real need for more than two of the four alternative starting points; but in other problems all four can be deployed to advantage. Any generating function might be global as described, and the Legendre transformation links are presumed to have been constructed even if they contain isolated singularities. We shall also see that non-isolated singularities can occur, for example when a generating function is linear in some variable, and in such cases one may have to accept that only part of the diamond diagram can survive in its full form because the associated Legendre inversions truly fail. One then uses whatever remains from Fig. 2.10.

2.9. *Convex and saddle functions*

Here we examine certain local relations between the shapes of a pair of generating functions connected by a non-singular Legendre transformation. When a whole circuit of Fig. 2.10 is non-singular it connects a particular quartet of branches of the global generating functions. We describe how the shapes of individual members of two of these quartets are related.

It is elementary that a sufficient condition for $Y(y_i, u_\alpha)$ to be a strictly convex function of the y_i, at any fixed set of the u_α, is that the Hessian Y_{ij} be positive definite everywhere, at least in a convex y_i-domain. Of course this is not a necessary prerequisite for convexity, but it is adequate for the present purpose. It is also elementary, from (2.10), that positive definite $Y_{ij} \Leftrightarrow$ positive definite X_{ij}: Hence strict convexity as assured in this way is transmitted through a non-singular Legendre transformation to the dual function. A particular example is the dual pair of segments labelled B in Fig. 2.9.

When the passive variables are no longer held fixed, the following rules relate the shapes of (for example) $Y(y_i, u_\alpha)$ and $X(x_i, u_\alpha)$.

(a) First consider any function whose arguments are split arbitrarily into two sets of variables. Taking $Y(y_i, u_\alpha)$ as an example, we say that it is *jointly* strictly convex with respect to both sets of variables taken together if the Hessian

$$\begin{bmatrix} Y_{ij} & Y_{i\beta} \\ Y_{\alpha j} & Y_{\alpha\beta} \end{bmatrix} \quad \text{is positive definite.} \tag{2.43}$$

We say that the function is *separately* strictly convex, i.e. with respect to the two sets of variables considered separately, if the two Hessians Y_{ij} and $Y_{\alpha\beta}$ are each positive definite. Thus the cross-terms do not enter the hypothesis of separate convexity. Therefore joint convexity implies separate convexity, but not conversely.

(b) We say that the function $X(x_i, u_\alpha)$ is a strict *saddle* function, strictly convex in the x_i and strictly concave in the u_α, if the two Hessians

$$X_{ij} \quad \text{and} \quad -X_{\alpha\beta} \quad \text{are each positive definite.} \tag{2.44}$$

It is straightforward to verify, using only (2.10)–(2.13), that for an arbitrary vector $[\lambda_i\ \lambda_\alpha]$ the quadratic form associated with the Hessian in (2.43) can be written

$$Y_{ij}\lambda_i\lambda_j + 2Y_{i\alpha}\lambda_i\lambda_\alpha + Y_{\alpha\beta}\lambda_\alpha\lambda_\beta = X_{ij}(Y_{ik}\lambda_k + Y_{i\alpha}\lambda_\alpha)(Y_{jl}\lambda_l + Y_{j\beta}\lambda_\beta) - X_{\alpha\beta}\lambda_\alpha\lambda_\beta. \tag{2.45}$$

This formula is valid without requiring any sign-definiteness hypotheses. However, we can base upon it a direct proof of, for example, the conclusion that

$$(2.43) \Leftrightarrow (2.44). \tag{2.46}$$

In other words, the Legendre transformation with passive u_α converts a function $Y(y_i, u_\alpha)$ which is jointly strictly convex in the sense of (2.43) into a function $X(x_i, u_\alpha)$ which is strictly saddle in the sense of (2.44); and conversely. This result was established by Noble and Sewell (1972) by different methods, namely mean value theorem arguments which can be applied to functionals over pairs of inner product spaces. Connections between definitions of jointly weakly convex functionals and weak saddle functionals were also given there.

(c) By reversing the roles of active and passive variables we see that either hypothesis (2.43) or (2.44) is sufficient for $Z(x_i, v_\alpha)$ to be a jointly strictly concave function, and for $W(y_i, v_\alpha)$ to be another strict saddle function but concave in the y_i and convex in the v_α. Hence we obtain the relations summarized in Fig. 2.11 between the shapes of a particular quartet of individual branches of the global and possibly multibranched functions contained in Fig. 2.10. It is convenient to have a name for the particular branches having the shapes of Fig. 2.11; some later examples indicate that "ground state branch" might be a suitably suggestive general name for each function branch there. Then the particular quartet of Fig. 2.11 would be the "ground state quartet", in the sense of being that special quartet (among those comprising Fig. 2.10) upon each branch of which the starting point of some physical or mathematical sequence of states is found.

(d) Suppose that (2.43) and (2.44) do not hold, but instead that $Y(y_i, u_\alpha)$ has the *qualified*

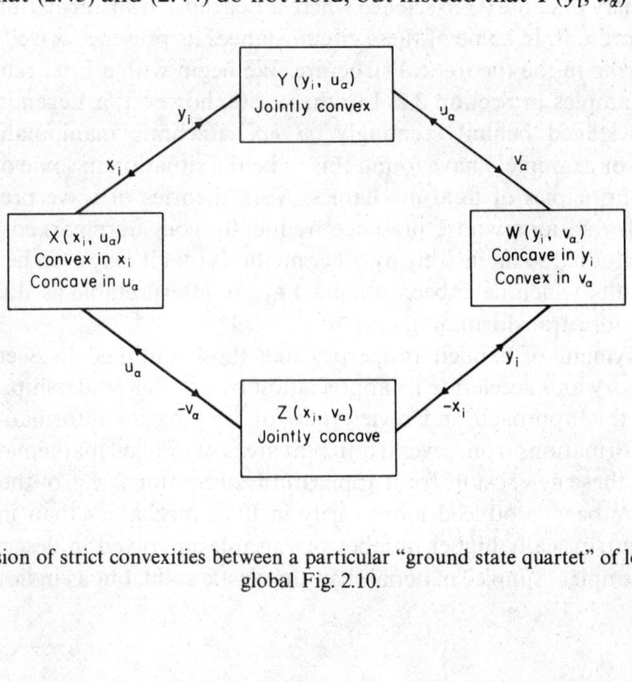

FIG. 2.11. Transmission of strict convexities between a particular "ground state quartet" of local branches of the global Fig. 2.10.

separate strict convexity assured by requiring that the three Hessians

$$\left.\begin{array}{r} Y_{ij}, \quad Y_{\alpha\beta} \\ \text{and} \qquad X_{\alpha\beta} = Y_{ij}^{-1} Y_{i\alpha} Y_{j\beta} - Y_{\alpha\beta} \end{array}\right\} \text{ are positive definite.} \tag{2.47}$$

Here (2.10) and (2.13) have been used, and $Y_{ij}^{-1} \equiv X_{ij}$ merely denotes the typical element of the matrix inverse of Y_{ij}. Then it follows that conditions of exactly the same structure as (2.47) apply to $X(x_i, u_\alpha)$. Therefore this qualified form of separate convexity is transmitted through the Legendre transformation. In plasticity examples both of this case and of Fig. 2.11 are described after (3.7) below.

With one more hypothesis added to (2.47), namely W_{ij} or Z_{ij} positive definite, we find that such qualified separate convexity is transmitted around another whole quartet of Fig. 2.10, different of course from the particular quartet of Fig. 2.11. The signs of all these Hessians could be changed, with negative definiteness in (2.47) being consistent with transmission of qualified separate concavity round a circuit. This can be illustrated in gas dynamics and in shallow water theory by quartets of branches round which the Mach number $M > 1$ or Froude number $F > 1$ there respectively. Each is smoothly joined to a ground state quartet satisfying Fig. 2.11 for which $M < 1$ or $F < 1$, and upon which fluid stagnation states lie as for example in Fig. 3.5.

We recall finally that a sufficient condition for strict convexity, expressed in terms of a Hessian, need not be invariant under nonlinear diffeomorphisms. Hill (1978, Section II A 3) remarked that a certain "relative" convexity *is* coordinate invariant.

3. Examples of Legendre Transformations

3.1. *Introduction*

There is a variety of contexts in science where a Legendre transformation, with or without singularities, is present. In some of these circumstances its presence is well recognized and it plays a pivotal role in the theoretical structure. We begin with a brief reminder of some of these known examples in Section 3.2. In other cases, however, a Legendre transformation lurks unacknowledged behind seemingly *ad hoc* algebraic manipulations or envelope constructions. For example, I have found this to be the situation in some of the literature on the variational principles of fluid mechanics. Also, theories of wave propagation contain Legendre transformations whose presence frequently goes unremarked. This is a subject which has been studied so intensively by other methods that it may even be fruitful to inquire whether the results which have been obtained are re-interpretable as throwing fresh light back onto Legendre transformations *per se*.

Timely deployment of explicit properties like those outlined in Section 2 above can streamline a theory and accelerate its appreciation by a wider readership. In Sections 3.3 *et seq.* we follow this approach and review new or very recent information about specific Legendre transformations from several different areas of applied mathematics. At the time of writing some of these new results from applications of Section 2, e.g. of the type described in Section 3.4, have been produced more easily in fluid mechanics than in solid mechanics because of the intrinsically higher number of variables involved in describing a solid. The perfect gas is a simpler "simple" material than the elastic solid, but as indicated in Section 1 it serves as a step along the road to the study of more complicated situations. It is useful again in Section 5.2 in our study of extremum principles.

3.2. *Classical examples*

The most elementary illustration of Fig. 2.11 is generated by any one of the canonical quartet of two variable "Morse functions"

$$\left.\begin{aligned} Y &= \tfrac{1}{2}(y^2+u^2), \quad W = \tfrac{1}{2}(-y^2+v^2), \\ X &= \tfrac{1}{2}(x^2-u^2), \quad Z = -\tfrac{1}{2}(x^2+v^2). \end{aligned}\right\} \tag{3.1}$$

This example was noted by Noble and Sewell (1972), and different aspects of its canonical nature are explained by Poston and Stewart (1976) and Sewell (1977b). From the viewpoint of applications, however, (3.1) may be too simple because the diffeomorphisms required to reach it might take us too far from natural physical variables, just as we remarked for the elementary catastrophes in Section 2.4.

The four thermodynamic potentials, namely specific internal energy ε, free energy ψ, enthalpy χ, and free enthalpy ζ, are probably one of the oldest examples of Fig. 2.10. Their general properties are conveniently described by, for example, Truesdell and Toupin (1960, Section 251). Perhaps the easiest non-singular case is the simple fluid having specific volume $v(=1/\text{density})$, entropy η, temperature θ, and pressure p. Under the standard restrictions $c_p > c_v > 0$ on the specific heats, and with $\theta > 0$ and real sound speed, each potential is a single-valued function of two variables possessing the interconnections and strict convexity properties of Fig. 3.1. This ground state quartet is one of several non-canonical illustrations of Fig. 2.11 recorded by Sewell (1974).

In any context where a phase transition can occur we can expect the presence of Legendre singularities. For example, Sewell (1977c) showed that a density jump in a van der Waals fluid can be described in terms of the particular free energy

$$\psi(\theta, v) = v - \frac{3}{v} - \frac{8}{3}\theta \log(3v-1) \tag{3.2}$$

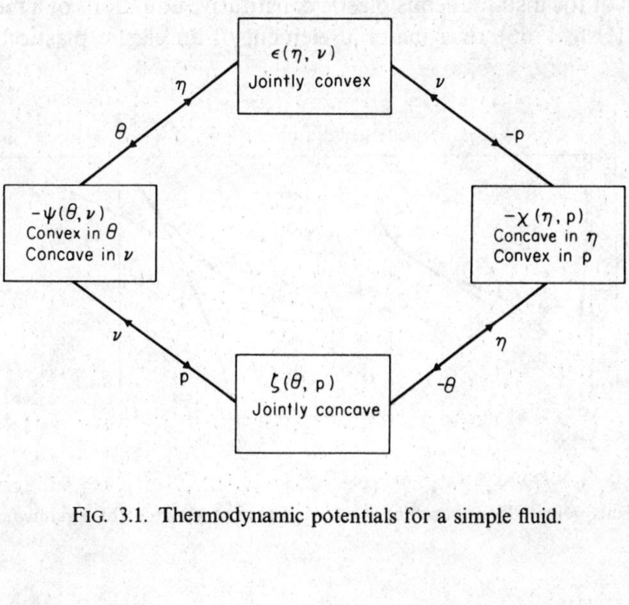

FIG. 3.1. Thermodynamic potentials for a simple fluid.

in variables normalized so that the liquid/gas critical point is $v = \theta = 1, p = 0$. The linearity of θ implies a non-isolated singularity so that only part of Fig. 2.10 can survive in full form. What remains has, for example at fixed $\theta < 1$, the two isolated singularities shown at the inflexions and cusps of Fig. 3.2. One may then propose as a convention for the suppression of superheating and supercooling that the central hump of the ψ-surface be removed and replaced by the dotted "supporting" tangent plane. Correspondingly the two-cusped "tail" of the swallowtail ζ-surface would be excised at the self-intersection. The jump in v at boiling or condensation would then be the jump in the v-abscissa of Fig. 3.2(a) or, since $v = \partial\zeta/\partial p$, the jump in slope at the self-intersection point of Fig. 3.2(b).

In metallurgy a two-phase alloy can have a free energy $\psi(v)$ per atom which sometimes has a central hump like that of Fig. 3.2(a), where v now would be the fractional composition of one phase (e.g. see Section 22 of Christian, 1975). "Spinodal" seems to be the name attached to the inflexion points, and spinodals and swallowtails have been commented upon by Kikuchi and de Fontaine (1976). There is not the space here to pursue all that could be said about the connection between Legendre singularities and modern theories of phase transitions, but we note the investigation of Dubois and Dufour (1977). Viewpoints different from those of the present article are described by Poston and Stewart (1978, Ch. 14).

The Lagrangian and Hamiltonian of classical mechanics generate a familiar example of a Legendre transformation which is often non-singular, as in the case of a holonomic scleronomic conservative system. This is a context in which, by contrast with Fig. 3.1, there is less physical interest in the full circuit of Fig. 2.10 even when so-called co-Lagrangians and co-Hamiltonians can be defined. Another such example, in the classical elastic solid, is generated by the elastic plus body force potential energy density, and its dual or complementary energy density. Here stress and strain are active variables, and displacement is a passive variable which appears linearly when the body force is gravity. Co-energies associated with the names of Millar, Brayton, and Moser receive some emphasis in electric circuit theory, and can also illustrate Fig. 2.11 (Sewell, 1974).

3.3. *Plastic constitutive equations*

Certain models of the instantaneous plastic constitutive equations of a metal single crystal capable of glide (Hill, 1966), or a material element of an elastic/plastic body (Hill, 1978,

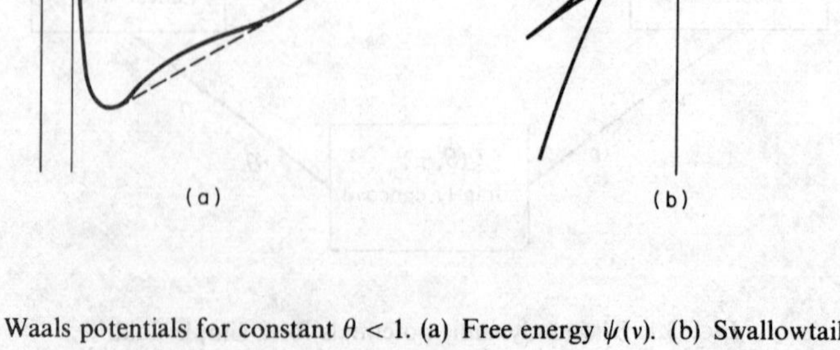

FIG. 3.2. Van der Waals potentials for constant $\theta < 1$. (a) Free energy $\psi(v)$. (b) Swallowtail free enthalpy $\zeta(p)$.

Ch. IIB) have required very careful analysis of the precise physical meaning and tensorial specification of the quasi-static rates of stress and strain contained therein, in particular before the sign of a hardening hypothesis can be given unambiguously. The basic formal structure of these equations has become accepted for many purposes, however, although it is not yet widely recognized that this can be succinctly described via another example of Fig. 2.10. This is shown in Fig. 3.3, and was described in detail by Sewell (1974).

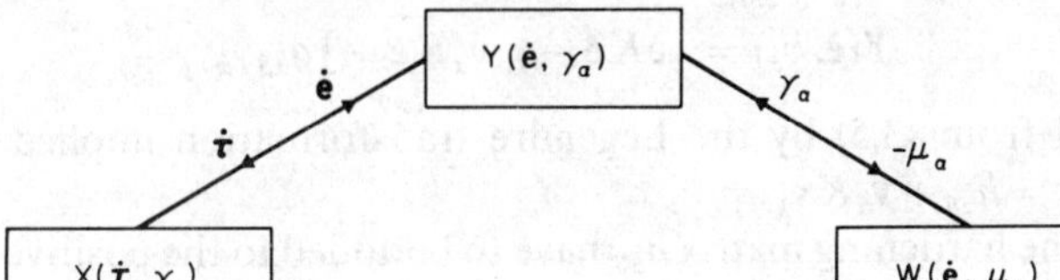

FIG. 3.3. Potentials generating plastic constitutive equations.

The chosen definitions of stress-rate $\dot{\boldsymbol{\tau}}$ and strain-rate $\dot{\mathbf{e}}$ belong to a space of second-order symmetric tensors, having inner product exemplified by $\dot{\boldsymbol{\tau}}\dot{\mathbf{e}} = \dot{\tau}_{ij}\dot{e}_{ij}$ in terms of tensor components on a common cartesian reference frame. A second inner product space consisting of m-tuples of scalar multipliers μ_α, γ_α, etc. is present, with typical inner product $\mu_\alpha\gamma_\alpha$ (repeated Greek suffixes are summed over the range $\alpha = 1, \ldots, m$). Up to four given global generating functions are assumed to exist and to satisfy the relationships implied by the links in Fig. 3.3 according to the conventions indicated by (2.6), (2.8), (2.38), and (2.41).

Adopting the symmetrical standpoint described towards the end of Section 2.8, we may generate two sets of equations relating four sets of variables from any convenient corner of the diamond diagram. Perhaps the most convenient starting point in this context is $X(\dot{\boldsymbol{\tau}}, \gamma_\alpha)$, which generates

$$\dot{\mathbf{e}} = \frac{\partial X}{\partial \dot{\boldsymbol{\tau}}}, \quad \mu_\alpha = \frac{\partial X}{\partial \gamma_\alpha}. \tag{3.3}$$

Whichever starting point is used, the supplementary information required to reach a fully posed statement of the elastic/plastic constitutive equations is the system of "flow conditions"

$$\mu_\alpha \leqslant 0, \quad \gamma_\alpha \geqslant 0, \quad \mu_\alpha\gamma_\alpha = 0, \tag{3.4}$$

on the multipliers. Locking conditions have the inequalities reversed.

The most familiar example is a consequence of

$$X(\dot{\boldsymbol{\tau}}, \gamma_\alpha) = \tfrac{1}{2}\dot{\boldsymbol{\tau}}\, M\, \dot{\boldsymbol{\tau}} + \gamma_\alpha \mathbf{v}_\alpha \dot{\boldsymbol{\tau}} - \tfrac{1}{2} h_{\alpha\beta}\gamma_\alpha\gamma_\beta. \tag{3.5}$$

Here M is a given positive definite fourth-order tensor of elastic moduli, the $\mathbf{v}_\alpha$ are m given second-order symmetric tensors representing local yield surface normals at a given stress state (the yield surface has a vertex there if $m > 1$, and is locally smooth if $m = 1$), and $h_{\alpha\beta}$ is the typical element of a given local $m \times m$ hardening matrix (not a tensor). From (3.3) and (3.5) we find

$$\dot{\mathbf{e}} = M\dot{\mathbf{t}} + \gamma_\alpha \mathbf{v}_\alpha, \quad \mu_\alpha = \mathbf{v}_\alpha \dot{\mathbf{t}} - h_{\alpha\beta}\gamma_\beta. \tag{3.6}$$

Here $M\dot{\mathbf{t}}$ symbolizes a second-order symmetric tensor with $M_{ijkl}\dot{\tau}_{kl}$ as its ij component. Then with (3.4) we can give interpretations of plastic strain-rate magnitudes to the γ_α, and "slack" stress-rate variables to the μ_α. Evidently (3.5) has the *values* of $X = \frac{1}{2}(\dot{\mathbf{t}}\dot{\mathbf{e}} + \mu_\alpha\gamma_\alpha)$. The alternative starting point

$$Y(\dot{\mathbf{e}}, \gamma_\alpha) = \tfrac{1}{2}\dot{\mathbf{e}}K\dot{\mathbf{e}} - \gamma_\alpha \mathbf{v}_\alpha K\dot{\mathbf{e}} + \tfrac{1}{2}g_{\alpha\beta}\gamma_\alpha\gamma_\beta \tag{3.7}$$

is easily generated from (3.5) by the Legendre transformation implied in Fig. 3.3. Here $K = M^{-1}$ and $g_{\alpha\beta} = h_{\alpha\beta} + \mathbf{v}_\alpha K \mathbf{v}_\beta$.

Hypotheses on the hardening matrix $h_{\alpha\beta}$ have to be added to the positive-definiteness of M before further details of the constitutive statements can be studied. If $h_{\alpha\beta}$ is positive definite, then (3.5) is a strict saddle function, and a single globally valid ground state quartet of generating functions with the shapes implied by Fig. 2.11 exists. If $h_{\alpha\beta}$ is negative definite while $g_{\alpha\beta}$ is still positive definite, we have instead the qualified separate convexity specified by (2.47) for both $X(\mathbf{t}, \gamma_\alpha)$ and $Y(\dot{\mathbf{e}}, \gamma_\alpha)$, as the explicit quadratics in (3.5) and (3.7) affirm. Whenever the determinant $|h_{\alpha\beta}| \neq 0$ we can seek to eliminate the "internal" variables μ_α and γ_α from the constitutive statements (3.4) and (3.6), so leaving relations between the "external" variables $\dot{\mathbf{t}}$ and $\dot{\mathbf{e}}$ alone. The latter can be multivalued whenever $h_{\alpha\beta}$ is not positive definite, as the easiest case $m = 1$ with $h < 0$ (strain softening) shows. However, if this elimination of internal variables is carried out *ab initio* or prematurely we can miss the sense of dealing with a simple mathematical structure which is familiar in other areas of applied mathematics, i.e. the structure of Fig. 3.3 supplemented by (3.4). When $|h_{\alpha\beta}| = 0$ we cannot construct an unambiguous function $Z(\mathbf{t}, \mu_\alpha)$ because its two sets of arguments are not independent, as a result of a non-isolated Legendre singularity; nor does there seem to be any physical guideline for converting this into an isolated singularity, e.g. by the introduction of higher order γ_α terms into (3.5) which preserve the "time" homogeneity.

Unless $|g_{\alpha\beta}| = 0$ we can construct from (3.7) the alternative starting point

$$W(\dot{\mathbf{e}}, \mu_\alpha) = -\tfrac{1}{2}\dot{\mathbf{e}}K\dot{\mathbf{e}} + \tfrac{1}{2}g^{-1}_{\alpha\beta}(\mathbf{v}_\alpha K\dot{\mathbf{e}} - \mu_\alpha)(\mathbf{v}_\beta K\dot{\mathbf{e}} - \mu_\beta), \tag{3.8}$$

where $g^{-1}_{\alpha\beta}$ is the typical element of the inverse of the matrix $g_{\alpha\beta}$. This function, when $g_{\alpha\beta}$ is positive definite, is convenient in elastic/plastic bifurcation theory for expressing the generalization of Hill's idea (1958 for $m = 1$) of a linear comparison solid to the case $m > 1$, as given by Sewell (1974, equation (77) and otherwise in 1972, equation (3.65)). Specifically, whereas the real elastic/plastic solid

$$-\dot{\mathbf{t}} = \frac{\partial W}{\partial \dot{\mathbf{e}}}, \quad \gamma_\alpha = -\frac{\partial W}{\partial \mu_\alpha}, \tag{3.9}$$

supplemented by the nonlinear conditions (3.4), is so generated by (3.8) from Fig. 3.3, the fictitious linear comparison solid replaces (3.4) by the different supplementary conditions that all $\mu_\alpha = 0$. These are inserted into (3.8) giving $W(\dot{\mathbf{e}}, 0) \equiv W_L(\dot{\mathbf{e}})$ say, so generating the stress-rates $-\partial W_L/\partial\dot{\mathbf{e}}$. As explained in the papers cited, and most recently by Hill (1978, equation (2.60)), the theorem emerges that a bifurcation in the elastic/plastic solid cannot precede one in the more easily analysed comparison solid, whose instantaneous moduli are the same for both loading and unloading.

3.4. *Constitutive surfaces in inviscid fluid dynamics*

Here we describe some recent new results obtained as a direct consequence of the realization that a global Legendre transformation containing isolated singularities could have a dual function with the swallowtail shape shown in Fig. 2.8(b). The results were given by Sewell and Porter (1980), who introduced the novel concept of a "constitutive surface" into compressible gas dynamics and shallow water theory.

Consider a perfect compressible fluid particle for which Fig. 3.1 applies with all convexities strict, and otherwise arbitrary equation of state. Then the enthalpy function $\chi(\eta, p)$ can be inverted to give a pressure function $p(\chi, \eta)$ since $v = \partial\chi/\partial p > 0$. If v is fluid speed and we introduce a new variable h, representing total energy, by the decomposition $\chi = h - \frac{1}{2}v^2$ we arrive at a new pressure function

$$p = p(v, h, \eta). \tag{3.10}$$

Sewell (1963) used the negative of this as generating function to construct a new Legendre transformation with v active, and h and η passive, leading to a dual generating function

$$P = P(Q, h, \eta) \tag{3.11}$$

such that

$$Q = -\frac{\partial p}{\partial v}, \quad v = \frac{\partial P}{\partial Q}. \tag{3.12}$$

The values of Q and P are those of mass flow ρv and "flow stress" $p + \rho v^2$, where density $\rho = 1/v$. Of course, a quantity with such *values* for P has long been familiar, e.g. at least since Bateman's time in the literature on variational principles in fluid mechanics. Nevertheless, the *function* (3.11) was not previously recognized as participating in an explicit Legendre transformation which controls the structure of the theory.

Higher derivatives of (3.10) and (3.11) can be calculated, for example

$$\frac{\partial^2 p}{\partial v^2} = \rho(M^2 - 1), \quad \frac{\partial^3 p}{\partial v^3} = \frac{M\rho}{c}\left[3 + M^2\left(\frac{\partial c^2}{\partial \chi} - 1\right)\right]. \tag{3.13}$$

Here $M = v/c$ is the local Mach number, and $c = c(\chi, \eta)$ is the local sound speed expressed as a function of χ and η for any equation of state which satisfies the convexity hypotheses of Fig. 3.1. (In the special case of the ideal gas η happens to be absent from $c^2 = (\gamma - 1)\chi$, where $\gamma = c_p/c_v$ is a fixed constant.) From $(3.13)_1$ we see that there is an isolated singularity of the Legendre transformation at sonic speed. With $\partial c/\partial\chi > 0$ it can be shown that, at each fixed h and η, $p(v)$ has a cubic-like inflexion and $P(Q)$ has a cusp at the unique sonic speed (Sewell, 1978d).

When h and η are allowed to vary as well, (3.10)–(3.12) each describe a three-dimensional hypersurface lying in a four-dimensional state space. Each surface can be determined once and for all in advance of the differential equations and jump conditions expressing balance of mass, momentum, and energy, and so each represents geometrically the constitutive hypotheses for the considered fluid particle. The history of a particle, as controlled by the balance conditions, defines a track which must always lie on any constitutive surface regardless of the type of motion (e.g. whether steady or unsteady, rotational or irrotational, in one, two, or three dimensions). When a shock is traversed the track may jump instantaneously from one part of the surface to be continued on another part.

The qualitative shape of such constitutive surfaces has been determined analytically by Sewell and Porter (1980), under detailed but light additional hypotheses (including $\partial^2 c/\partial\chi^2 < 0$, and postulated lower bound $k(\eta) < \chi(p, \eta)$ on that enthalpy function). Figure 3.4 sketches one of their results, namely the surface (3.11) in both of the two special cases when η is fixed (so that h is the effective variable along the third axis) and when h is fixed (with $-\eta$ increasing along the third axis). This surface is half of the tail of the swallowtail in Fig. 2.8(b), turned upside down. In other words $P \geqslant 0$ and $Q \geqslant 0$ are measured along the previous $-X$ and $x - a$ directions, and $h - k \geqslant 0$ or $-\eta \leqslant 0$ along the self-intersection line which is the previous $-b$ axis. The cusped edge of regression represents the sonic singularity, and joins the subsonic upper branch of the surface, on which either of $P(Q, h)$ and $P(Q, \eta)$ is jointly convex, to the supersonic lower branch which is separately but not jointly concave. The pressure constitutive surfaces $p(v, h)$ and $p(v, \eta)$ are like part of Fig. 2.8(a) over the quadrant $b \leqslant 0$, $y \geqslant 0$.

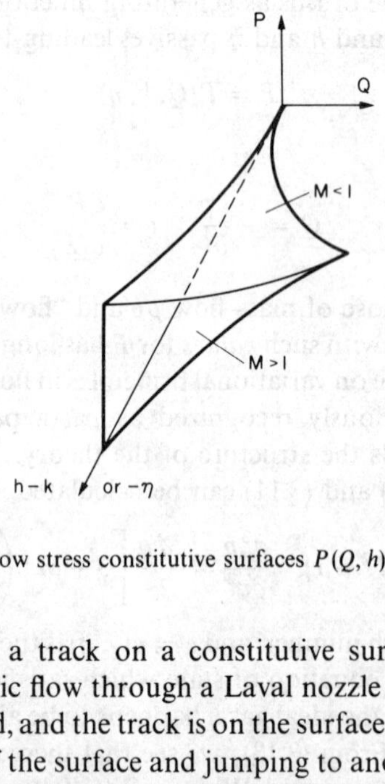

FIG. 3.4. Flow stress constitutive surfaces $P(Q, h)$ and $P(Q, \eta)$.

A very simple example of a track on a constitutive surface is provided by the one-dimensional steady homenergic flow through a Laval nozzle of a particle which traverses a shock normally. Thus h is fixed, and the track is on the surface $P = P(Q, \eta)$ beginning on one section $\eta = \eta_+$ (a constant) of the surface and jumping to another section $\eta = \eta_-$ (constant $\eta_- > \eta_+$) as the shock is traversed. Sewell and Porter (1980, Fig. 10) show that this track can be exhibited as a continuous curve in the P, Q plane shown in Fig. 3.5. By $(3.12)_2$ the sudden drop in speed at the shock is the drop in gradient of Fig. 3.5.

It is straightforward to verify that the shallow water approximation provides another illustration of constitutive surfaces connected by Fig. 2.10. Start with $-X(x_i, u_\alpha)$ as

$$\bar{p}(v, h) = \frac{1}{2g}(\tfrac{1}{2}v^2 - h)^2. \tag{3.14}$$

This is the height-averaged pressure function having gradients

$$\frac{\partial \bar{p}}{\partial v} = -Q, \quad \frac{\partial \bar{p}}{\partial h} = d. \tag{3.15}$$

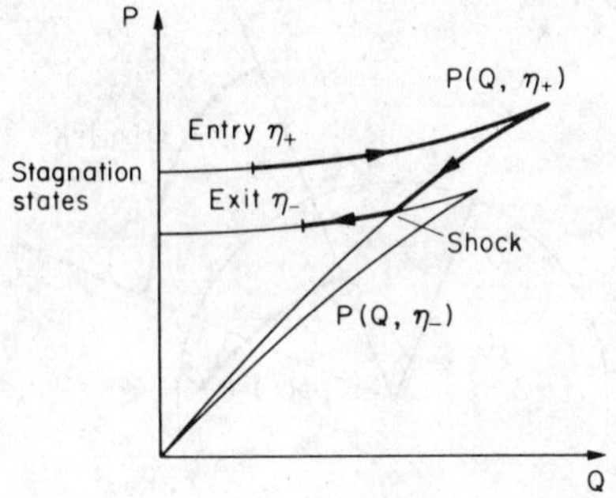

FIG. 3.5. Example of continuous particle history track with shock.

Here d is depth, v is horizontal fluid speed and g is the fixed acceleration due to gravity. There is a ground state quartet of branches upon which $F \equiv v/(gd)^{1/2} < 1$, joined smoothly to a second quartet of separately but not jointly concave branches upon which $F > 1$, so illustrating the negative-definite version of the property described after (2.47). This approach offers new viewpoints which are not found in standard texts such as that of Henderson (1966), for example for the discussion of hydraulic jumps and open-channel flows.

3.5. *Velocity and phase surfaces in wave propagation*

The search for plane wave solutions with direction ratios n_i begins, in various media, with the introduction of a phase ϕ which here will be taken to have the values of $\phi = n_i x_i - \omega t$, where x_i are the cartesian coordinates of position and t is time. This leads to a relation

$$\omega = \omega(n_i) \tag{3.16}$$

for wave speed ω/n, where $n = +(n_i^2)^{1/2}$. The function $\omega(n_i)$ is homogeneous of degree one because the wave speed depends only on the direction cosines n_i/n, but we do not wish to introduce constrained variables in direction space too early.

To illustrate the geometrical duality of Section 2.1 we construct two superimposed four-spaces spanned by ϕ and the x_i, and by ω and the n_i. Then at unit time, the polar of a point in either space, with respect to a paraboloid such as (2.1), is the plane

$$\omega + \phi = n_i x_i. \tag{3.17}$$

This has the structure of (2.2). We now regard (3.16) as a "velocity surface" lying in the four-space, even though actual velocity values are only to be found on its intercept with the cylinder $n = 1$. The velocity surface is conical, by homogeneity. As a pole moves around on the cone, its polar will envelop a phase surface

$$\phi = \phi(x_i) \tag{3.18}$$

lying in the dual four-space. Thus we reach another example of a global Legendre transformation, although it is not usually expressed as such. The literature emphasizes the geometrical rather than the analytical duality, evidently because the conical form of (3.16) means that the special type of polar reciprocation described at the end of Section 2.1 is applicable.

Figure 3.6 illustrates this for two spatial dimensions in a 45° segment, for a given constant

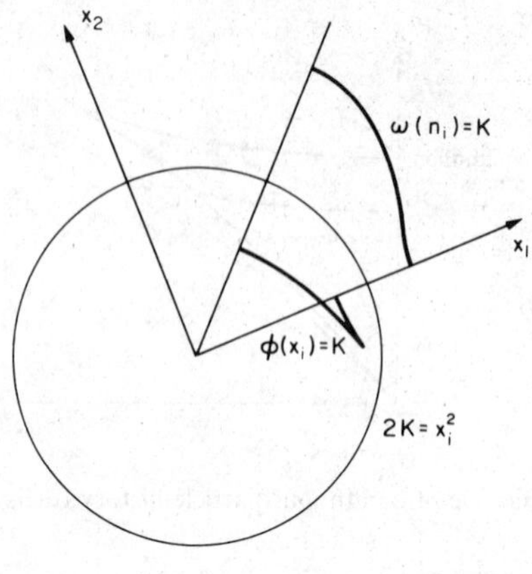

FIG. 3.6. Duality of cusp and inflexion with respect to circle.

$K > 0$. The section $\omega(n_i) = K$ might be called a "slowness surface" in direction space, and the envelope of its polar lines $2K = n_i x_i$ with respect to the circle is a "wave surface" $\phi(x_i) = K$ lying in physical space. The cusp has mapped from the hypothesized inflexion. When $\omega(n_i) = K$ can be continued by reflection in the n_1 or x_1 axis, the two matched segments of the wave surface combine to exhibit a swallowtail.

Wave surfaces containing swallowtails have been obtained before by numerical plotting, after discussions of aspects of polarity different from those emphasized here. Musgrave (1954), Miller and Musgrave (1956), and Buchwald (1959) give such results for anisotropic linear elasticity. The equation determining (3.16) for propagation in three dimensions is a cubic for ω^2 which itself can be represented geometrically as in Fig. 2.6.

Burridge (1967) gives cusped wave surfaces for several media, and he remarks that the region inside the swallowtail will be a "travelling zone of absolute quiet" or lacuna where the wave field is transitorily zero. Poston and Stewart (1978, Fig. 12.43) give a diagram which suggests that our Fig. 2.9(b) may be a typical wave surface, with u in (2.35) interpreted as a given time. None of these authors make explicit the connection with Legendre transformations described here. There is more to do in exploring the role of Legendre singularities in wave propagation in various media and, for example, their connection with singularities in the envelopes of rays and of characteristics. We note in particular the ideas discussed by Berry (1976).

3.6. *Infinite dimensional problems*

Let E be any finite or infinite dimensional vector space which has an inner product $(\cdot,\cdot)$ defined in it, and let $x, y, \ldots$ be typical elements of E. Let $X, Y, \ldots$ be real numbers. Then a straightforward generalization of (2.2) is

$$X + Y = (x, y). \tag{3.19}$$

Since in the infinite-dimensional case we can no longer draw the pictures, it may be less helpful than in some of the previous examples (e.g. Section 3.5) to emphasize the geometrical aspect of duality, although formally we could still speak of poles and polars with respect to the paraboloid $2X = \|x\|^2$ in $R \times E$.

Instead we seek workable analytical generalizations of some of the properties of global Legendre transformations described in Sections 2.2, 2.8, and 2.9, being guided by the structure of the theory described there. The right sides of (2.3) and (2.4) would be replaced by functionals $Y[y]$ and $X[x]$. We shall suppose that there is a second vector space F, usually different from E and having different inner product $\langle \cdot, \cdot \rangle$ and typical elements $u, v, \ldots$.

We then adopt as starting point a supposedly given functional

$$X = X[x, u] \tag{3.20}$$

on $E \times F$. We assume that partial gradients denoted by $\partial X/\partial x$ and $\partial X/\partial u$ can be defined, also as members of E and F respectively, for example in the Gateaux sense as illustrated by Noble and Sewell (1972). In place of (2.8) we then write

$$y = \frac{\partial X}{\partial x}, \quad -v = \frac{\partial X}{\partial u}. \tag{3.21}$$

We say that a global Legendre transformation is available with x and y active and u passive if there exists a second functional

$$Y = Y[y, u] \tag{3.22}$$

on $E \times F$ such that (3.19) holds and

$$x = \frac{\partial Y}{\partial y}, \quad v = \frac{\partial Y}{\partial u}. \tag{3.23}$$

However, it does not seem to be known how to characterize with precise hypotheses for infinite-dimensional spaces the types of singularity which may have to be accommodated in performing the implied inversion. In fact there is a need to explore what can happen with specific functionals, such as integrals.

Again we could say that another global Legendre transformation is available with x passive and u and v active if there exists a second functional

$$Z = Z[x, v] \tag{3.24}$$

on $E \times F$ such that

$$X + Z = -\langle u, v \rangle \tag{3.25}$$

and

$$-y = \frac{\partial Z}{\partial x}, \quad u = \frac{\partial Z}{\partial y}, \tag{3.26}$$

so generalizing (2.40) and (2.41). In these ways we can rewrite Fig. 2.10 formally in terms of functionals instead of functions. It is then a target for investigation to find how much of its global structure survives in infinite-dimensional problems.

The need for inversions such as those between $(3.21)_1$ and $(3.23)_1$, or $(3.21)_2$ and $(3.26)_2$, arises in seeking to satisfy side conditions of stationary principles. From what was said in Section 2.8 we expect that supplementary information will be required from other sources to add to, for example, the two sets of equations (3.21) before a fully posed statement is obtained. One type of supplementary information containing inequalities was illustrated in (3.4). Another type is the pair of equations

$$T^*u = y, \quad Tx = -v. \tag{3.27}$$

Here T and T^* are mutually adjoint linear operators satisfying

$$(x, T^*u) = \langle u, Tx \rangle \tag{3.28}$$

for all x and u in subspaces of E and F respectively.

By combining (3.21) and (3.27) we arrive at the generalized Hamiltonian equations

$$\left.\begin{aligned} T^*u &= \frac{\partial X}{\partial x}, \quad (\alpha) \\ Tx &= \frac{\partial X}{\partial u}, \quad (\beta) \end{aligned}\right\} \tag{3.29}$$

proposed for study by Noble (1964). They are the natural conditions of the free stationary principle

$$\delta\{(x, T^*u) - X[x, u]\} = 0, \tag{3.30}$$

where δ here implies that independent variations of both x and u are to be performed. If it is desired to eliminate x at the outset by imposing (3.29α) and inverting it as $(3.23)_1$ with $(3.27)_1$, the generalized Euler–Lagrange equation

$$T\left[\frac{\partial Y}{\partial T^*u}\right] + \frac{\partial Y}{\partial u} = 0 \tag{3.31}$$

emerges as the natural condition of

$$\delta\{Y[T^*u, u]\} = 0 \tag{3.32}$$

after free variations of u.

4. Dual Extremum Principles

4.1. *Introduction*

In Section 3.6 we remarked that the generalization of Fig. 2.10 to infinite-dimensional cases, i.e. as a structure relating global and possible multivalued functionals, is a target for future investigation. The same issue arises for the ground state quartet of branches described by Fig. 2.11. When the whole of that special quartet is either not available, or does not have equal physical interest, it may be appropriate to concentrate on the ground state branch of one particular member of that quartet, say $X(x_i, u_\alpha)$.

It is the generalization of this last situation to certain infinite dimensional problems which we examine in the remainder of this article. We explore, in particular, the consequences of a hypothesis that a functional $X[x, u]$, such as appears in (3.20), be saddle-shaped. Qualified and constrained saddle hypotheses are described, and problems such as (3.29) appear as particular examples in a wider setting.

The motivation is to provide a clean and symmetric theoretical structure for the generation of governing conditions and dual extremum principles in terms of the arguments and gradients of a single given saddle functional. The approach has some affinity with what might loosely be called "Lagrangian theory" as employed in different ways by various authors (e.g. see Luenberger, 1969, or Whittle, 1971). We really make a fresh start, however, in respect of the emphasis which we give to the central controlling role of a saddle functional.

4.2. *General governing conditions*

We use the inner product space notations of Section 3.6. We suppose, as in Sewell (1973, p. 9), that a functional $L[x, u]$ is given on some domain of $E \times F$. Sometimes three or more inner product spaces are needed in applications, as in the rigid/plastic yield point problem where the three spaces involve tensors (e.g. stress), vectors (e.g. displacement), and scalars (e.g. multipliers). We omit this extension from the general theory for brevity, since the manner of its incorporation into the following formalism will be straightforward and is illustrated in the example of Section 5.4. Let $\partial L/\partial x$ and $\partial L/\partial u$ denote partial gradients in the Gateaux sense, as in the familiar process of computing a "first variation" and picking out the coefficients of increments in the varied argument.

We wish to consider four different problems simultaneously. Each problem will be defined by two sets of conditions selected, in different ways which we shall describe, from the four sets of conditions which now follow.

$$\frac{\partial L}{\partial x} = 0; \quad (\alpha) \tag{4.1}$$

$$\frac{\partial L}{\partial u} = 0; \quad (\beta) \tag{4.2}$$

$$\left.\begin{aligned} \frac{\partial L}{\partial x} &\leqslant 0, \quad (\alpha) \\ x &\geqslant 0, \quad (\beta) \\ \left(x, \frac{\partial L}{\partial x}\right) &= 0; \end{aligned}\right\} \tag{4.3}$$

$$\left.\begin{aligned} \frac{\partial L}{\partial u} &\geqslant 0, \quad (\beta) \\ u &\geqslant 0, \quad (\alpha) \\ \left\langle u, \frac{\partial L}{\partial u} \right\rangle &= 0. \end{aligned}\right\} \tag{4.4}$$

The four individual problems are now posed by the following selected combinations of governing conditions.

PROBLEM I: (4.1) + (4.2). This statement corresponds to a free stationary principle. An illustration is provided by (3.29) and (3.30) in the special case in which the given $L[x, u]$ has the special form

$$L[x, u] = (x, T^*u) - X[x, u]. \tag{4.5}$$

By implication the given functional $X[x, u]$ is not expected to contain the linear operator T or T^*. This special form (4.5) is widespread, but we have no need to assume that it is the only possible form of $L[x, u]$.

PROBLEM II: (4.3) + (4.4). This problem has the structure of a so-called nonlinear programming problem, either finite or infinite-dimensional. It will be a linear programming problem if $L[x, u]$ is bilinear. Of course appropriate meanings have to be attached to the inequalities in (4.3) and (4.4). For finite-dimensional spaces inequalities might apply to every element of suitable matrices. For infinite-dimensional spaces ideas of polar cones (e.g. see Moreau, 1962, or Collins, 1976) may be required to achieve precision. The growing literature on unilateral constraints provides examples. Proofs of dual extremum principles for rather general initial motion problems in constrained continua, generalizing the cavitation examples of Moreau (1967), were given by Sewell (1969).

PROBLEM III: (4.1) + (4.4).

PROBLEM IV: (4.2) + (4.3). These last two pairs of mixed conditions are also of mathematical programming type, since they can only be stationary principles in at best a restricted sense.

Each problem is seen to contain a subset of conditions which we have labelled (α) and a second subset labelled (β). The role of these subsets is explained below. The special functional (4.5) can be inserted into the governing conditions, not only of problem I as already described, but also into those of problems II, III, or IV to give particular illustrations of these problems.

4.3. *Saddle functional*

At this point we need to replace hypotheses expressed in terms of Hessians, such as we employed, for example, in (2.44) and in proofs of transmission properties such as (2.45), by a more convenient infinite-dimensional statement. We make the hypothesis that $L[x, u]$ be a saddle functional, of one kind or another now to be specified.

We express this hypothesis in terms of the scalar quantity

$$S \equiv L_+ - L_- - \left(x_+ - x_-, \frac{\partial L}{\partial x_+}\right) - \left\langle u_+ - u_-, \frac{\partial L}{\partial u_-}\right\rangle \tag{4.6}$$

evaluated at certain pairs of points labelled x_+, u_+ and x_-, u_- in the domain of $L[x, u]$. Subscripts attached to L or its gradients will imply evaluation at either the "plus point" or the "minus point" (and $\partial L/\partial x_+$, for example, means $\partial L/\partial x$ evaluated at the plus point, and not merely a partial gradient with respect to x_+). The scalar quantity (4.6) was proposed by Sewell (1969, equation (2.50)) as a basis for describing saddle properties leading to the construction of dual extremum principles. Other ways of making saddle statements are familiar in the literature, but are less helpful for our purposes here. We now list some of the various levels of restriction on a saddle hypothesis which can be required in different circumstances.

HYPOTHESIS (i): $S > 0$ for every pair of distinct points in the domain of $L[x, u]$. The latter would then be called a strict saddle functional, strictly concave in x and strictly convex in u. The condition is quite strong.

A bilinear functional such as (x, T^*u) alone will make $S \equiv 0$ for every pair of points, as can be verified directly. Therefore in the particular example (4.5) the hypothesis is equivalent to requiring that $X[x, u]$ itself be a strict saddle functional of the opposite shape, i.e. strictly convex in x and strictly concave in u. When the functional $X[x, u]$ is an integral, its saddle property can be achieved *a fortiori* by requiring the integrand to satisfy (2.44) at every point of the field. This is proved by appropriate mean value theorems, as described by Noble and Sewell (1972) and Sewell and Noble (1978).

HYPOTHESIS (ii): $S \geqslant 0$ for every pair of distinct points in the domain of $L[x, u]$. The latter is then a weak saddle functional, concave in x and convex in u. Uniqueness of any existing solution for both x and u is no longer assured as it would be via (4.13) under (i) for any one of the problems I–IV, but dual extremum principles can still be proved as below. The condition is still rather strong, however, in the way that it invokes the whole domain of $L[x, u]$.

HYPOTHESIS (iii): $S \geqslant 0$ for every pair of distinct points which satisfy the (α)-subset of conditions in whichever of problems I–IV is being studied.

HYPOTHESIS (iv): $S \geqslant 0$ for every pair of distinct points which satisfy the (β)-subset of conditions. It is still possible to prove an extremum principle under either of these constrained weak saddle hypotheses.

Tighter extremum principles of this latter kind can be important in practice, since we must usually sail close to the wind with minimal hypotheses if we are to discover the most interesting effects. The price to be paid in going from (ii) to (iii) or (iv) will lie in the difficulty of solving the constraints to express the smaller number of competing pairs of points explicitly. In some cases this will require inversion by the Legendre transformations discussed earlier.

HYPOTHESES (v): Embedded in (ii), (iii), or (iv) may be cases of particular interest where, for example, $L[x, u]$ is strictly concave in x but only weakly convex (or even linear) in u. Then $S = 0$ for all distinct pairs with $x_+ = x_-$, but $S > 0$ for all distinct pairs with $x_+ \neq x_-$, regardless of whether or not $u_+ \neq u_-$. We do not describe all possible variants.

4.4. *General dual extremum principles*

The plus and minus points in the hypotheses (i) and (ii) of Section 4.3 are arbitrary distinct points in the domain of $L[x, u]$, and therefore are not yet constrained by any of the conditions (4.1)–(4.4); while in hypotheses (iii) and (iv) they are constrained as described. The proofs now to follow depend on the facility with which we can add additional interpretations to such pairs of points.

We adopt the notation that, whichever one of the problems I–IV is being studied, a suffix α is attached to any quantity evaluated at a point which satisfies the (α)-subset of governing conditions, but not necessarily the (β)-subset. In particular x_α, u_α denotes such a point itself. Likewise the suffix β denotes evaluation at any point x_β, u_β satisfying the (β)-subset but not necessarily the (α)-subset. A suffix zero is attached, for emphasis, to any quantity evaluated at a solution point x_0, u_0 (say, not necessarily unique) of the complete problem I, II, III, or IV.

We focus on problem II, since the other three problems will entail only obvious simplifications of the following procedure. Adopt first the hypothesis (ii). In that weak inequality on the scalar (4.6) choose first the particular interpretations

$$x_+, u_+ = x_\alpha, u_\alpha, \quad x_-, u_- = x_0, u_0. \tag{4.7}$$

This establishes the minimum principle

$$L_\alpha - \left(x_\alpha, \frac{\partial L}{\partial x_\alpha} \right) - L_0 \geqslant -\left(x_\beta, \frac{\partial L}{\partial x_\alpha} \right) + \left\langle u_\alpha, \frac{\partial L}{\partial u_0} \right\rangle \geqslant 0. \tag{4.8}$$

Second choose in hypothesis (ii) the fresh interpretation

$$x_+, u_+ = x_0, u_0, \quad x_-, u_- = x_\beta, u_\beta. \tag{4.9}$$

This choice establishes the maximum principle

$$L_0 - L_\beta + \left\langle u_\beta, \frac{\partial L}{\partial u_\beta} \right\rangle \geqslant -\left(x_\beta, \frac{\partial L}{\partial x_0} \right) + \left\langle u_0, \frac{\partial L}{\partial u_\beta} \right\rangle \geqslant 0. \tag{4.10}$$

By combining (4.8) and (4.10) we achieve the pair of dual extremum principles

$$L_\alpha - \left(x_\alpha, \frac{\partial L}{\partial x_\alpha} \right) \geqslant L_0 \geqslant L_\beta - \left\langle u_\beta, \frac{\partial L}{\partial u_\beta} \right\rangle. \tag{4.11}$$

In this generality they are due to Sewell (1973, Part I, Section IIc). Of course the pair is valid *a fortiori* under hypothesis (i).

The minimum principle in (4.8) is also valid more widely, namely under the constrained saddle hypothesis (iii). This is because, since a solution point x_0, u_0 satisfies the (α)-subset among other conditions both members of every pair of points (4.7) actually used in the proof of (4.8) satisfy the (α)-subset.

Likewise the maximum principle in (4.10) is also valid more widely under the different constrained saddle hypothesis (iv). For both members of every pair of points (4.9) actually used in the proof of (4.10) satisfy the (β)-subset. Since hypotheses (iii) and (iv) are not the same, however, neither of them individually leads to the continued inequality (4.11).

In problem III $\partial L/\partial x_\alpha = 0$, and in problem IV $\partial L/\partial u_\beta = 0$, so that the corresponding results (4.8) and (4.10) are simplified. In general, however, both extrema are not stationary in problems II, III, and IV. Only in problem I can both extrema be stationary, and then the dual extremum principles (4.11) become

$$L_\alpha \geqslant L_0 \geqslant L_\beta. \tag{4.12}$$

4.5. *Associated problems*

The viewpoint established here poses the question of what variety of circumstances allows the construction under precise hypotheses of a single saddle functional which generates governing conditions such as (4.1)–(4.4) and extremum principles such as (4.11). The development of this saddle functional viewpoint can be traced in the papers of Sewell (1969, 1973, 1977b) and Noble and Sewell (1972, 1978).

The original stimulus for this author's work on extremum principles came from the exploitation of convexity in the elastic/plastic bifurcation theories of Hill (1956, 1958, 1959, 1962, 1978). These lead to estimates which may coincide, in appropriate contexts, with particular points on a stability boundary or bifurcation set. In terms of saddle functionals, if we interchange the plus and minus labels in (4.6) and then add to the original S, we obtain

$$-\left(x_+ - x_-, \frac{\partial L}{\partial x_+} - \frac{\partial L}{\partial x_-} \right) + \left\langle u_+ - u_-, \frac{\partial L}{\partial u_+} - \frac{\partial L}{\partial u_-} \right\rangle. \tag{4.13}$$

This can be interpreted to express the contradiction needed for uniqueness proofs.

Detailed technicalities of functional analysis can be found in well-known texts such as Ekeland and Temam (1976) or Oden and Reddy (1976). Extremum principles for initial value problems have been developed rather recently (by comparison with the long history for boundary value problems), and approaches to them which are not radically different from that of the present article have been explored by Collins (1977), Herrera and Bielak (1976), and Herrera and Sewell (1978) for the heat and wave equations. A number of examples of complementary variational principles have been given by Robinson (e.g. 1978) in theoretical chemistry, Arthurs (1980) in mathematical physics, and their co-workers in recent years. We have not the space here, however, to make a detailed comparison of the differing viewpoints in the extensive literature, nor to record all the straightforward boundary value problems which can be reformulated in the saddle viewpoint along the lines of Section 5. The developing area of variational inequalities has been surveyed by Cryer (1979a, b) and applied in particular to plasticity; it will have connections with our inequality conditions (4.3) and (4.4). For example, it was shown by Noble (1979) that the Signorini membrane problem may be put into the setting described in this article.

5. Examples of Dual Extremum Principles

5.1. *Generalized Hamiltonian problems*

Complementary or dual stationary principles are obtained from (3.30) by imposing in turn either (3.29α) or (3.29β) at the outset as an "essential" condition. Then there emerges (3.29β) or (3.29α) respectively as the "natural" condition. These stationary principles can be strengthened to dual extremum principles (4.12) for problem I if $X[x, u]$ is a saddle functional, convex in x and concave in u, as described in Section 4.3.

In principle such dual extremum principles are available when this saddle hypothesis on $X[x, u]$ is either strict (as in (i)) or weak (as in (ii)), although they may be easier to apply when the convexity or concavity is strict with respect to at least one of the variables (as in (v)). In practice, one or other extremum principle, if not the dual pair, may be more explicitly stated when a constrained saddle hypothesis (iii) or (iv) applies.

For example, suppose (3.22) exists supplemented by $(3.27)_1$, in other words that (3.29α) is satisfied and can be inverted as

$$x = \frac{\partial Y}{\partial T^* u} \tag{5.1}$$

from $(3.23)_1$. Then there exists a functional

$$Y[T^*u, u] = J[u] \tag{5.2}$$

say, of the single variable u alone, such that the gradients of (4.5) are

$$\frac{\partial L}{\partial x} = 0, \quad \frac{\partial L}{\partial u} = T\left[\frac{\partial Y}{\partial T^* u}\right] + \frac{\partial Y}{\partial u} = \frac{dJ}{du}. \tag{5.3}$$

Then the constrained saddle inequality (iii), originally expressed via (4.6), can be rewritten using the special form (4.5) with (3.19) as

$$J_+ - J_- - \left\langle u_+ - u_-, \frac{dJ}{du_-} \right\rangle \geqslant 0 \tag{5.4}$$

for all points u_+ and u_- in the domain of $J[u]$, and therefore in that of T^*. In other words the constrained saddle inequality (iii) is equivalent to the convexity of $J[u]$ in this case (see inequality (3.23) of Sewell, 1977b). The corresponding minimum principle is

$$J[u] \geqslant L_0. \tag{5.5}$$

Likewise when (3.24) exists supplemented by $(3.27)_2$, there exists a functional $Z[x, -Tx] = K[x]$ say, of x alone such that $\partial L/\partial x = dK/dx$ and $\partial L/\partial u = 0$. Then the constrained saddle inequality (iv) is equivalent to the concavity of $K[x]$ with respect to all x in the domain of T. The corresponding maximum principle is

$$L_0 \geqslant K[x]. \tag{5.6}$$

Consider the example

$$X[x, u] = -(a, x) - \tfrac{1}{2}p(x, x) - \tfrac{1}{2}\langle u, u \rangle, \tag{5.7}$$

where a is a given vector in E and p is a given scalar. Here u has the role of an intermediate variable allowing decomposition of an *ab initio* problem

$$T^*Tx = px + a \tag{5.8}$$

into an example of problem I of type (3.29), namely

$$T^*u + px + a = 0 \quad (\alpha), \qquad Tx + u = 0 \quad (\beta). \tag{5.9}$$

If $p \neq 0$ we find

$$J[u] = -\frac{1}{2p}(T^*u + a, T^*u + a) + \tfrac{1}{2}\langle u, u \rangle \tag{5.10}$$

and for any p we find

$$K[x] = (a, x) - \tfrac{1}{2}\langle Tx, Tx \rangle + \tfrac{1}{2}p(x, x). \tag{5.11}$$

Obviously hypothesis (i) is satisfied if $p < 0$, and (ii) if $p \leqslant 0$. Associated dual extremum principles for this problem, and a nonlinear generalization of it, were studied by Noble and Sewell (1972), prompted by integral equations.

However, the constrained saddle hypotheses (iii) or (iv) applied to (5.7) permit positive p and allow either or both of the extremum principles (5.5) and (5.6) to hold for some $p > 0$. For example, the convexity of (5.10) and the concavity of (5.11) together require

$$\inf \frac{\langle Tx, Tx \rangle}{\langle x, x \rangle} \geqslant p \geqslant \sup \frac{(T^*u, T^*u)}{(u, u)}. \tag{5.12}$$

Alternatively we might require only one of these two inequalities. They reflect the frequent appearance in uniqueness theories, including those of Hill (1958, 1959), of Rayleigh quotients for determining a particular point on a bifurcation set. An example of (4.5) accommodating the ordinary differential equations of a heavy cantilever under compressive axial load p was given by Sewell (1977b, equation (4.14)). The light strut has unique deflection up to the positive Euler load, not merely for $p \leqslant 0$. Hill's theories apply to three-dimensional problems, as the illustration for an elastic/plastic strut by Hill and Sewell (1960a, b, 1962) shows.

The linear functional

$$X[x, u] = -(a, x) - \langle b, u \rangle \tag{5.13}$$

deployed via (4.5) in problem II instead of I is a symmetric starting point leading to both primal and dual programs (4.11) of an infinite-dimensional linear programming problem. Here b is a given member of F.

5.2. *Compressible inviscid fluid dynamics*

We use the context of Section 3.4 to give a precise and straightforward illustration, in a nonlinear problem, of the formal structure described in Section 3.6 *et seq.* for the development of dual extremum principles. We take the opportunity here to incorporate into their statement the explicit geometrical information about their integrands which is presented in, for example, Fig. 3.5.

A free stationary principle was given by Sewell (1963) for steady flow which was not necessarily homentropic nor homenergic. For a flow in which η and h are constant everywhere, and which is subsonic in an overall sense over the given flow region, dual extremum principles were proved. Consider such flow in a given three-dimensional region V having finite boundary consisting of two parts S_Q and S_ϕ. Denote the local velocity potential by ϕ and the mass flow vector by $\mathbf{Q}$, having magnitude Q. Let $\mathbf{l}$ be the unit normal to S_Q outward from the flow region, and assign boundary values C and b respectively to the normal mass flow and velocity potential on S_Q and S_ϕ. These boundary conditions can be incorporated with the governing differential equations as an illustration of (3.29) in the form

$$\begin{bmatrix} \nabla\phi \\ -\mathbf{l}\phi \\ 0 \end{bmatrix} = \begin{bmatrix} \dfrac{\partial F}{\partial \mathbf{Q}} & \text{in} & V \\ -\mathbf{l}b & \text{on} & S_\phi \\ 0 & \text{on} & S_Q \end{bmatrix}, \quad (\alpha)$$

$$\begin{bmatrix} -\nabla.\mathbf{Q} \\ 0 \\ \mathbf{l},\mathbf{Q} \end{bmatrix} = \begin{bmatrix} 0 & \text{in} & V \\ 0 & \text{on} & S_\phi \\ C & \text{on} & S_Q \end{bmatrix}. \quad (\beta) \tag{5.14}$$

Here the fluid velocity is written as the gradient of a function $F(\mathbf{Q}) = P(Q)$ obtained from that of (3.11) by inserting $Q = |\mathbf{Q}|$ and fixing h and η. (This F stands for "flow force", and has nothing to do with the space F now following.)

Precise definitions of the spaces E and F of Section 3.6 needed for this example were detailed by Noble and Sewell (1972, Section 12). They are unusual but very simple. The essence of them is as follows. They are both spaces of 3×1 column matrices. For E a typical trio of entries in the matrix are the values over V, S_ϕ, and S_Q of any integrable vector field. For F a typical trio are the values over V, S_ϕ, and S_Q of any integrable scalar field. The normal properties of a linear space (e.g. the sum of two members is another member) hold for such matrices. Each inner product $(\cdot,\cdot)$ and $\langle\cdot,\cdot\rangle$ is the sum of integrals of scalar products over V, S_ϕ, and S_Q. No continuity or smoothness is required for such members of E and F, and explicit examples are found in the column matrices on the *right*-hand sides of (5.14α) and (5.14β) respectively. However, smooth members of F and E are required to define the subspaces which are the domains of the linear operators T^* and T respectively. These operators range over E and F according to their composite definitions implied by the *left* sides of (5.14). These technicalities are all consistent, for example, with the fact that (3.28) expresses integration by parts in the form

$$\int \mathbf{Q}.\nabla\phi\, dV - \int \mathbf{Q}.\mathbf{l}\phi\, dS_\phi = -\int \phi\nabla.\mathbf{Q}\, dV + \int \phi\mathbf{l}.\mathbf{Q}\, dS_Q. \tag{5.15}$$

It follows that the current example of the single functional $L[x, u]$ of type (4.5) upon which the whole theory of problem I can be based is

$$L[\mathbf{Q}, \phi] = \int [\mathbf{Q}.\nabla\phi - F(\mathbf{Q})]\, dV - \int (\phi - b)\mathbf{l}.\mathbf{Q}\, dS_\phi - \int C\phi\, dS_Q. \tag{5.16}$$

(Symbols distinct from $\mathbf{Q}$ and ϕ should really be introduced on the left to distinguish the matrix members of E and F from the vector and scalar fields used to construct them, but we can manage without this extra notation in this subsection.) Since the linear and bilinear terms here contribute only identically zero terms to (4.6), the character of this functional is controlled by the volume integral of $F(\mathbf{Q})$, and therefore by a single cusped curve of Fig. 3.5 (say for η_+). In terms of cartesian coordinates of $\mathbf{Q}$ and of the fluid velocity vector $\mathbf{v}$ it may be shown that

$$\rho\frac{\partial^2 F}{\partial Q_i \partial Q_j} = \delta_{ij} + \frac{v_i v_j}{c^2 - v^2}. \tag{5.17}$$

This Hessian matrix is positive definite if $M < 1$ and indefinite if $M > 1$. Therefore if we are on the convex upper branch of $P(Q)$ in Fig. 3.5, $L[\mathbf{Q}, \phi]$ is a weak saddle functional whose concavity in $\mathbf{Q}$ is strict. Thus the dual extremum principles (4.12) follow, because an unconstrained saddle hypothesis (v) is satisfied. A limited supersonic region of flow could also be admitted without violating the weak saddle requirement of $L[\mathbf{Q}, \phi]$, since it would be enough for $\int F(\mathbf{Q})\,dV$ to be a convex functional overall rather than at every point. The extent of such a region may be unknown in advance, but shocks need not be present. The functionals appearing in (4.12) have the values of

$$L_\alpha = -\int p(v)\,dV - \int \phi C\,dS_Q \tag{5.18}$$

to be minimized among the fields satisfying (5.14α); and

$$L_\beta = -\int P(Q)\,dV + \int \mathbf{l}.\mathbf{Q}b\,dS_\phi \tag{5.19}$$

to be maximized among the fields satisfying (5.14β). In applications of (5.19) we can see from Fig. 3.5, quite explicitly for the first time, how trial values of Q can encounter multivaluedness in $P(Q)$ and therefore in (5.19), especially in a transonic regime. In other words, care will be needed to decide whether it is the subsonic or supersonic branch of $P(Q)$ (if either) which is intercepted by an ordinate erected at given Q to pass near the cusp. Finally we notice that the concavity of $P(Q)$ where $M > 1$ in Fig. 3.5 does not imply the concavity of $F(\mathbf{Q})$ in (5.16), because of the indefiniteness proved after (5.17). That is the reason (aside from the possible presence of shocks) why these dual extremum principles for continuous solutions do not extend to the supersonic case, except in the limited sense mentioned. The ancestry of associated principles in the previous fluids literature was described by Sewell (1963).

5.3. *Incremental elastic/plastic distortion*

Here we show how plastic constitutive equations having the structure of (3.3) supplemented by (3.4), and generated for example by (3.5), can be incorporated into a boundary value problem for incremental elastic/plastic distortion, superimposed on existing stress and strain, which itself has the structure represented by an elaboration of problem III in Section 4.2. An amplified account of this representation was given by Sewell (1973, Section VI), and is carried further here in the explicit links (5.34) and (5.36) with Fig..3.3. The original unambiguous formulation of such problems in *ab initio* style is due to Hill (1962, 1978).

The inner product spaces E and F again are spaces of 3×1 column matrices, as in the fluids problem of Section 5.2. Here, however, a typical trio of entries in a matrix of E consists of the fields, over a reference volume V and two parts S_u, S_s of its surface, of any unsymmetric second order tensor. And for F a typical trio of entries consists of the fields over V, S_u, and S_s of any vector. With obvious inner products, and smoothness hypotheses on subspaces of E and F, we introduce adjoint linear operators T and T^* by the definitions

$$T^*\dot{u} \equiv \begin{bmatrix} \dfrac{\partial \dot{u}_j}{\partial a_i} \\ -n_i\dot{u}_j \\ 0 \end{bmatrix}, \quad T\dot{s} \equiv \begin{bmatrix} -\dfrac{\partial \dot{s}_{ij}}{\partial a_i} \\ 0 \\ n_i\dot{s}_{ij} \end{bmatrix} \begin{matrix} \text{on} \quad V, \\ \text{on} \quad S_u, \\ \text{on} \quad S_s. \end{matrix} \tag{5.20}$$

Here the a_i are cartesian components of a material particle in the reference configuration, and $\dot{s}_{ij}$ and $\dot{u}_j$ are the cartesian components of the quasi-static rate of change of nominal stress and displacement. Adjointness (3.28) is thus expressed via integration by parts over the reference configuration. The formal symbols $\dot{s}$ and $\dot{u}$ designate matrices which are the actual members of appropriate subspaces of E and F, and are constructed out of smooth $\dot{s}_{ij}$ and $\dot{u}_j$. Rates are all evaluated in a supposedly known "current" configuration, differing from the reference configuration by a particle displacement u_j.

Convenient auxiliary variables describing the current finite strain from the reference configuration are defined by

$$d_{ij} \equiv \frac{\partial u_j}{\partial a_i}, \quad e_{ij} \equiv \tfrac{1}{2}(d_{ij} + d_{ji} + d_{ik}d_{jk}). \tag{5.21}$$

One invariant description of incipient constitutive response is in terms of the so-called Kirchhoff symmetric stress τ_{ij} and its rate $\dot{\tau}_{ij}$ satisfying

$$\dot{s}_{ij} = \dot{\tau}_{ij} + \dot{\tau}_{ik}d_{kj} + \tau_{ik}\dot{d}_{kj}. \tag{5.22}$$

We need a third inner product space G consisting of m-tuples $\gamma = [\gamma_1, \ldots, \gamma_m]$ of multipliers such as those which appear in (3.4), and having inner product $\{\gamma, \mu\} \equiv \int[\gamma_1\mu_1 + \ldots + \gamma_m\mu_m]dV$. Let g_j, b_j, and C_j denote given vectors on V, S_u, S_s respectively.

We are now in a position to write down a single functional, defined on a subspace of $E \times E \times F \times E \times G$, in terms of which all governing conditions, stationary principles, and extremum principles can be expressed. It is

$$L[\dot{s}, \dot{\tau}, \dot{u}, \dot{d}, \gamma] = \int\left[\dot{s}_{ij}\frac{\partial \dot{u}_j}{\partial a_i} - (\dot{s}_{ij} - \dot{\tau}_{ij} - \dot{\tau}_{ik}d_{kj})\dot{d}_{ij} + \tfrac{1}{2}\tau_{ik}\dot{d}_{ij}\dot{d}_{kj} - X[\dot{\tau}_{ij}, \gamma_\alpha] - g_j\dot{u}_j\right]dV$$
$$+ \int l_i\dot{s}_{ij}(b_j - \dot{u}_j)dS_u - \int C_j\dot{u}_j dS_s. \tag{5.23}$$

Here l_i denotes the given unit normal outwards from S_u, the current stress τ_{ik} and strains (5.21) are given, and the function $X[\dot{\tau}_{ij}, \gamma_\alpha]$ is given as in Fig. 3.3, with (3.5) as a particular example. Then the governing equations and inequalities of the problem are the following extension of problem III:

$$\frac{\partial L}{\partial \dot{s}} = 0, \quad (\alpha) \tag{5.24}$$

$$\frac{\partial L}{\partial \dot{\tau}} = 0, \quad (\alpha) \tag{5.25}$$

$$\frac{\partial L}{\partial \dot{u}} = 0, \quad (\beta) \tag{5.26}$$

$$\frac{\partial L}{\partial \dot{d}} = 0, \quad (\beta) \tag{5.27}$$

$$\frac{\partial L}{\partial \gamma} \geqslant 0, \quad (\beta) \tag{5.28}$$

$$\gamma \geqslant 0, \quad (\alpha) \tag{5.29}$$

$$\left\{\gamma, \frac{\partial L}{\partial \gamma}\right\} = 0. \tag{5.30}$$

Calculating the first variation of (5.23) will show that (5.24) gives the kinematic conditions

$$\frac{\partial \dot{u}_j}{\partial a_i} = \dot{d}_{ij} \quad \text{in} \quad V, \qquad \dot{u}_j = b_j \quad \text{on} \quad S_u; \tag{5.31}$$

and that (5.26) gives the conditions of continuing equilibrium

$$-\frac{\partial \dot{s}_{ij}}{\partial a_i} = g_j \quad \text{in} \quad V, \qquad l_i \dot{s}_{ij} = C_j \quad \text{on} \quad S_s; \tag{5.32}$$

that (5.27) gives (5.22) without extra boundary terms; that (5.25) gives $(3.3)_1$ when $\dot{e}_{ij}$ is defined as the rate of change of $(5.21)_2$; and that (5.28)–(5.30) become $(3.3)_2$ supplemented by (3.4).

As in Section 4.2 we have identified two subsets of governing conditions, labelled (α) and (β). We introduce now the saddle hypothesis that (5.23) is concave with respect to $\dot{s}$ and $\dot{\tau}$ and convex with respect to $\dot{u}$, $\dot{d}$, and γ. This can be expressed via a formula having the structure of (4.6), but with more terms (see inequality (142) of Sewell (1973)). Recall first that linear terms, and bilinear terms in oppositely paired variables, have only a minor role in such a hypothesis. Then supposing that $X[\dot{\tau}_{ij}, \gamma_\alpha]$ is convex in the $\dot{\tau}_{ij}$ and concave in the γ_α (assured for (3.5) when the elastic and hardening moduli are positive definite), the term which really controls the saddle hypothesis for (5.23) is seen to be the given stress τ_{ik}. If, for example, this matrix is non-negative definite we shall have an unconstrained saddle hypothesis.

By the obvious extension of (4.11) it then follows that any solution of the problem (5.24)–(5.30) which may exist will minimize

$$L - \left(\dot{s}, \frac{\partial L}{\partial \dot{s}}\right) - \left(\dot{\tau}, \frac{\partial L}{\partial \dot{\tau}}\right)$$

$$= \int\left[\dot{\tau}_{ij}\frac{\partial X}{\partial \dot{\tau}_{ij}} - X + \tfrac{1}{2}\tau_{ik}\dot{d}_{ij}\dot{d}_{kj} - g_j\dot{u}_j\right]dV - \int C_j\dot{u}_j\,dS_s \tag{5.33}$$

among the solutions of the (α)-subset (principle (144) of Sewell, 1973). The minimum need not be stationary with respect to the γ_α. In applications it will usually be convenient to use selected members of the (α)-subset at once to simplify this expression. For example, $(5.31)_1$ eliminates the $\dot{d}_{ij}$ in favour of $\dot{u}_j$. Also, $\dot{\tau}_{ij}$ can be eliminated if (5.25) in the form $(3.3)_1$ is invertible in the sense conveyed by Fig. 3.3, for the first two terms of the volume integrand in (5.33) are then the function $Y[\dot{e}_{ij}, \gamma_\alpha]$. This reduces (5.33) to a functional of $\dot{u}_j$ and the γ_α (and of the given u_j passively via (5.21)).

If (5.28) and (5.30) are imposed as *extra* constraints on the minimization, it may then be possible effectively to express the γ_α in terms of the $\dot{u}_j$ and express (5.33) as a functional of $\dot{u}_j$ alone. This will involve a composite strain-rate potential (Hill, 1962) which in terms of Fig. 3.3 will be given (with $m = 1$ for example) by

$$V[\dot{\mathbf{e}}] \equiv \begin{Bmatrix} Y[\dot{\mathbf{e}}, 0] & \text{if} & \mu < 0, \\ -W[\dot{\mathbf{e}}, 0] & \text{if} & \gamma > 0. \end{Bmatrix} \tag{5.34}$$

This takes the place of $\dot{\tau}\partial X/\partial\dot{\tau} - X$ in (5.33), and Hill shows that the principle becomes stationary under these extra constraints (1962, equation (22), where W is written in place of the present V and it is shown how more general boundary conditions can also be included. Hill adopts the current configuration as reference, effectively setting $u_j = 0$ after the differentiations.) For $m > 1$ a composite $V[\dot{\mathbf{e}}]$ can be built up from $\sum_{r=0}^{m} m!/(m-r)!\,r!$ potentials of $\dot{\mathbf{e}}$ and a mixture of $r\ \gamma_\alpha$ and $m - r\ \mu_\alpha$, obtainable by Legendre transformations, according to the number of active plastic modes.

Again from the extension of (4.11), any solution of (5.24)–(5.30) will maximize

$$L - \left\langle \dot{u}, \frac{\partial L}{\partial \dot{u}} \right\rangle - \left(d, \frac{\partial L}{\partial d} \right) - \left\{ \gamma, \frac{\partial L}{\partial \gamma} \right\}$$

$$= \int \left[\gamma_\alpha \frac{\partial X}{\partial \gamma_\alpha} - X - \tfrac{1}{2}\tau_{ik} \dot{d}_{ij} \dot{d}_{kj} \right] dV + \int l_i \dot{s}_{ij} b_j dS_u \tag{5.35}$$

among the solutions of the (β)-subset (principle (147) of Sewell, 1973). If (5.29) and (5.30) are imposed as extra constraints on the maximization here, the first two terms of the volume integrand may be written in terms of Fig. 3.3 as a composite stress-rate potential such as (for $m = 1$)

$$-V_c(\dot{\tau}) = \begin{Bmatrix} -X(\dot{\tau}, 0) & \text{if} & \mu < 0, \\ Z(\dot{\tau}, 0) & \text{if} & \gamma > 0. \end{Bmatrix} \tag{5.36}$$

The generalization to $m > 1$ again presents no difficulty of principle. Evidently $V + V_c = \dot{\tau}\dot{\mathbf{e}}$ in value, and attributes of a Legendre transformation attach to such composite potentials themselves.

Many associated details of various kinds, including connections with uniqueness investigations, are reviewed by Hill (1962, 1978), Hutchinson (1974), and Sewell (1972, 1973). In particular, either extremum principle (5.33) or (5.35) may hold under suitably constrained saddle hypotheses, such as allow, for example, some range of negative stress.

References

Arthurs, A. M. (1980) *Complementary Variational Principles*, 2nd edn., Clarendon Press, Oxford.

Berry, M. V. (1976) Waves and Thom's theorem, *Adv. Phys.* **25,** 1–26.

Buchwald, V. T. (1959) Elastic waves in anisotropic media, *Proc. R. Soc. London* A **253,** 563–580.

Burridge, R. (1967) Lacunas in two-dimensional wave propagation, *Proc. Camb. Phil. Soc.* **63,** 819–825.

Calladine, C. R. (1974) A new thermodynamic diagram for representing steady one-dimensional compressible fluid flow, *J. Mech. Engng Sci.* **16,** 192–195.

Chillingworth, D. R. J. (1981) Universal bifurcation problems (this Volume, pp. 101–133).

Christian, J. W. (1975) *The Theory of Transformations in Metals and Alloys*, 2nd edn., Pergamon Press, Oxford.

Collins, W. D. (1976) Upper and lower bounds for solutions of linear operator problems with unilateral constraints, *Proc. R. Soc. Edinburgh* **76A,** 95–105.
Collins, W. D. (1977) Dual extremum principles for the heat equation, *Proc. R. Soc. Edinburgh* **77A,** 273–292.
Craggs, J. W. (1948) The breakdown of the hodograph transformation for irrotational compressible fluid flow in two dimensions, *Proc. Camb. Phil. Soc.* **44,** 360–379.
Cryer, C. W. (1979a) The solution of the axisymmetric elastic–plastic torsion of a shaft using variational inequalities, Mathematics Research Center Report 1948, University of Wisconsin.
Cryer, C. W. (1979b) A survey of variational inequalities, Mathematics Research Center Report in preparation, University of Wisconsin.
Dubois, J.-G. and Dufour, J.-P. (1977) La théorie des catastrophes, V: Transformées de Legendre et thermodynamique, reprint, Departement de Mathématiques, Université du Québec, Montréal.
Ekeland, I. and Temam, R. (1976) *Convex Analysis and Variational Problems,* North-Holland, Amsterdam.
Golubitsky, M. and Schaeffer, D. (1979) A theory for imperfect bifurcation via singularity theory, *Communs Pure Appl. Math.* **32,** 21–98.
Güttinger, W. and Eikemeier, H. (eds.) (1979) *Structural Stability in Physics*, Springer-Verlag, Berlin.
Haughton, D. M. (1980) Post-bifurcation of perfect and imperfect spherical elastic membranes, *Int. J. Solids Struct.* **16,** 1123–1133.
Haughton, D. M. and Ogden, R. W. (1978) On the incremental equations in nonlinear elasticity, II: Bifurcation of pressurized spherical shells, *J. Mech. Phys. Solids* **26,** 111–138.
Henderson, F. M. (1966) *Open Channel Flow*, Macmillan, New York.
Herrera, I. and Bielak, J. (1976) Dual variational principles for diffusion equations, *Q. Appl. Math.* **34,** 85–102.
Herrera, I. and Sewell, M. J. (1978) Dual extremum principles for non-negative unsymmetric operators, *J. Inst. Math. Applics* **21,** 95–115.
Hill, R. (1956) New horizons in the mechanics of solids, *J. Mech. Phys. Solids* **5,** 66–74.
Hill, R. (1957) On uniqueness and stability in the theory of finite elastic strain, *J. Mech. Phys. Solids* **5,** 229–241.
Hill, R. (1958) A general theory of uniqueness and stability in elastic/plastic solids, *J. Mech. Phys. Solids* **6,** 236–249.
Hill, R. (1959) Some basic principles in the mechanics of solids without a natural time, *J. Mech. Phys. Solids* **7,** 209–225.
Hill, R. (1962) Uniqueness criteria and extremum principles in self-adjoint problems of continuum mechanics, *J. Mech. Phys. Solids* **10,** 185–194.
Hill, R. (1966) Generalized constitutive relations for incremental deformation of metal crystals by multislip, *J. Mech. Phys. Solids* **14,** 95–102.
Hill, R. (1978) Aspects of invariance in solid mechanics, *Adv. Appl. Mech.* **18,** 1–75.
Hill, R. and Sewell, M. J. (1960a) A general theory of inelastic column failure, I: *J. Mech. Phys. Solids* **8,** 105–111.
Hill, R. and Sewell, M. J. (1960b) A general theory of inelastic column failure, II: *J. Mech. Phys. Solids* **8,** 112–118.
Hill, R. and Sewell, M. J. (1962) A general theory of inelastic column failure, III: *J. Mech. Phys. Solids* **10,** 285–300.
Hutchinson, J. W. (1974) Plastic buckling, *Adv. Appl. Mech.* **14,** 67–144.
Kikuchi, R. and de Fontaine, D. (1976) On spinodals and swallowtails, *Scripta Metall.* **10,** 995–999.
Koiter, W. T. (1945) On the stability of elastic equilibrium, Delft Thesis, H. J. Paris, Amsterdam. English translation available as NASA TT F 10-833, 1967.
Koiter, W. T. (1976) Complementary energy, neutral equilibrium and buckling, *Proc. K. Ned. Akad. Wet.* B **76,** 183–200.
Luenberger, D. G. (1969) *Optimization by Vector Space Methods*, Wiley, New York.
Miller, G. F. and Musgrave, M. J. P. (1956) On the propagation of elastic waves in aeolotropic media, III: Media of cubic symmetry, *Proc. R. Soc. London* A **236,** 352–383.
Moreau, J. J. (1962) Décomposition orthogonale d'un espace hilbertien selon deux cônes mutuellement polaires, *C.R. Acad. Sci. Paris* **255,** 238–240.
Moreau, J. J. (1967) One-sided constraints in hydrodynamics, *Nonlinear Programming*, edited by J. Abadie, 261–279, North-Holland, Amsterdam.
Musgrave, M. J. P. (1954) On the propagation of elastic waves in aeolotropic media, I: General principles, *Proc. R. Soc. London* A **226,** 339–355; II: Media of hexagonal symmetry, *ibid.* A **226,** 356–366.
Nádai, A. (1950) *Theory of Fracture and Flow of Solids*, Vol. I, McGraw-Hill.
Needleman, A. (1977) Inflation of spherical rubber balloons, *Int. J. Solids Struct.* **13,** 409–421.
Needleman, A. and Tvergaard, V. (1981) Aspects of plastic postbuckling behavior (this Volume, pp. 453–498).
Noble, B. (1964) Complementary variational principles for boundary value problems, I: Basic principles, with an application to ordinary differential equations, Mathematics Research Center Report 473, University of Wisconsin.
Noble, B. (1979) An abstract approach to dual variational principles for physical problems, Lecture at University of Saskatchewan.
Noble, B. and Sewell, M. J. (1972) On dual extremum principles in applied mathematics, *J. Inst. Math. Applics* **9,** 123–193.

Oravas, G. A. and McLean, L. (1966) Historical development of energetical principles in elastomechanics, *Appl. Mech. Rev.* **19,** 647–658, 919–933.

Poston, T. and Stewart, I. N. (1976) *Taylor Expansions and Catastrophes*, Pitman, London.

Poston, T. and Stewart, I. N. (1978) *Catastrophe Theory and its Applications*, Pitman, London.

Robinson, P. D. (1978) New variational bounds on generalized polarizabilities, *J. Math. Phys.* **19,** 694–699.

Rockafellar, R. T. (1970) *Convex Analysis*, Princeton University Press.

Sewell, M. J. (1963) On reciprocal variational principles for perfect fluids, *J. Math. Mech.* **12,** 495–504.

Sewell, M. J. (1966) On the connexion between stability and the shape of the equilibrium surface, *J. Mech. Phys. Solids* **14,** 203–230.

Sewell, M. J. (1969) On dual approximation principles and optimization in continuum mechanics, *Phil. Trans. R. Soc. London* A **265,** 319–351.

Sewell, M. J. (1972) A survey of plastic buckling, *Stability*, edited by H. Leipholz, 85–197, University of Waterloo Press, Canada.

Sewell, M. J. (1973) The governing equations and extremum principles of elasticity and plasticity generated from a single functional, I and II, *J. Struct. Mech.* **2,** 1–32, 135–158.

Sewell, M. J. (1974) On applications of saddle-shaped and convex generating functionals, *Physical Structure in Systems Theory*, edited by J. J. van Dixhoorn and F. J. Evans, 219–245, Academic Press, London.

Sewell, M. J. (1976) Some mechanical examples of catastrophe theory, *Bull. Inst. Math. Applics* **12,** 163–172.

Sewell, M. J. (1977a) On Legendre transformations and elementary catastrophes, *Math. Proc. Camb. Phil. Soc.* **82,** 147–163.

Sewell, M. J. (1977b) Degenerate duality, catastrophes and saddle functionals, Lectures to the Summer School on Duality and Complementarity in the Mechanics of Solids, Polish Academy of Sciences, Warsaw, *Mechanics Today*, edited by S. Nemat-Nasser, **6,** 41–92, Pergamon Press, Oxford.

Sewell, M. J. (1977c) Elementary catastrophe theory, *Problem Analysis in Science and Engineering*, edited by F. H. Branin, Jr. and K. Huseyin, 391–426, Academic Press, London.

Sewell, M. J. (1978a) Complementary energy and catastrophes, *Variational Methods in the Mechanics of Solids*, edited by S. Nemat-Nasser, 163–168, Pergamon Press, Oxford.

Sewell, M. J. (1978b) A pure catastrophe machine, *Bull. Inst. Math. Applics* **14,** 212–214.

Sewell, M. J. (1978c) Some global equilibrium surfaces, *Int. J. Mech. Engng Educ.* **6,** 163–174.

Sewell, M. J. (1978d) On Legendre transformations and umbilic catastrophes, *Math. Proc. Camb. Phil. Soc.* **83,** 273–288.

Sewell, M. J. and Noble, B. (1978) General estimates for linear functionals in nonlinear problems, *Proc. R. Soc. London* A **361,** 293–324.

Sewell, M. J. and Porter, D. (1980) Constitutive surfaces in fluid mechanics, *Math Proc. Camb. Phil. Soc.* **88,** 517–546.

Synge, J. L. (1960) Classical dynamics, *Encyclopedia of Physics* III/1, 1–225, edited by S. Flügge, Springer-Verlag, Berlin.

Thom, R. (1969) Topological models in biology, *Topology* **8,** 313–335.

Thorndike, A. S., Cooley, C. R., and Nye, J. F. (1978) The structure and evolution of flow fields and other vector fields, *J. Phys.* A, *Math. and Gen.* **11,** 1445–1490.

Truesdell, C. and Toupin, R. A. (1960) The classical field theories, *Encyclopedia of Physics* III/1, 226–858, edited by S. Flügge, Springer-Verlag, Berlin.

Whitney, H. (1955) On singularities of mappings of Euclidean spaces, I: Mappings of the plane onto the plane, *Ann. Math.* **62,** 374–410.

Whittle, P. (1971) *Optimization under Constraints*, Wiley–Interscience, London.

Woodcock, A. E. R. and Poston, T. (1974) *A Geometrical Study of the Elementary Catastrophes*, Springer-Verlag, Berlin.

Young, L. C. (1969) *Lectures on the Calculus of Variations and Optimal Control Theory*, W. B. Saunders Co., Philadelphia.

Zeeman, E. C. (1977) *Catastrophe Theory, Selected Papers 1972–1977*, Addison-Wesley, Reading, Mass.

Deformation of Ideal Granular Materials

by A. J. M. SPENCER

Department of Theoretical Mechanics, University of Nottingham

Summary

The formulation of the equations which govern the deformation and flow of granular materials is an outstanding problem in continuum mechanics. This article concentrates on one model for the mechanical behaviour of granular materials, termed the "double-shearing model", in which deformation is assumed to occur by simultaneous shearing on the two families of characteristic curves of the stress equations. We review the formulation of the equations of the double-shearing model for plane deformations, and describe the known exact solutions of these equations. Further developments of the plane strain theory are then given in which the velocity equations are expressed in a new form from which some general results are derived. In the last part of the article the theory is extended to include problems in three dimensions. The three-dimensional theory is applied to some special cases of non-planar deformations, and in particular we formulate a new theory of axially symmetric deformations of granular materials, and apply this theory to the solution of some boundary value problems.

1. Introduction

The correct formulation of constitutive equations to describe the mechanical behaviour of granular materials on the macroscopic scale is one of the outstanding problems in continuum mechanics. If attention is confined to plane deformations, then the failure hypothesis originally proposed by Coulomb (1773) provides a basis for the determination of the stress in a granular material. Theories based on Coulomb's hypothesis have been widely accepted in soil mechanics, and in many cases seem to determine the stress field in a granular material in a satisfactory way. However, the formulation of equations which determine the flow and deformation of a granular material is a more difficult and controversial problem, even in the case of plane deformations. Many theories have been proposed, some of which will be discussed in Section 3, but none of them has found general acceptance.

Experimental evidence of the behaviour of granular materials is also often inconclusive. The behaviour of these materials is complex and varied and it is difficult to construct suitable experiments and to interpret their results. The soil mechanics literature abounds with accounts of tests on real and artificial soil samples under various conditions; for example, the book by Schofield and Wroth (1968) contains data on carefully conducted tests. There are also many accounts of observations of the flow of granular materials in hoppers and similar configurations: as recent examples we mention Lee *et al.* (1974) and Drescher *et al.* (1978). Other workers, such as Drescher and de Josselin de Jong (1972) and Drescher (1976) have worked with model granular materials. Nevertheless, it remains difficult to form a clear picture of the mechanics of the processes which take place in a deforming granular material. At the same time, the lack of a generally accepted theory hinders the construction of appropriate experiments.

The purpose of this article is not to undertake a comprehensive review of the many theories which have been proposed to describe the deformation and flow of granular materials, but rather to investigate a particular model in some detail. The theory we shall describe is based on work by Mandel (1947), de Josselin de Jong (1959, 1971, 1977), Spencer (1964), and Zagainov (1967), and we shall call it, for reasons which will appear later, the "double-shearing model" (de Josselin de Jong, 1971, 1977, uses the term "double-sliding", but we consider "double-shearing" to be a more suitable description). In addition to reviewing existing formulations and applications of this theory, we shall describe some new developments of it. In particular, we give new extensions of the plane deformation theory and we extend the theory to include three-dimensional deformations and states of stress, with particular applications to problems of axial symmetry.

As the formulation of constitutive equations for granular materials has on occasions been the subject of controversy, it is perhaps appropriate that the author should make clear his own point of view, which is that the question is an open one; none of the theories which has been proposed appears to be beyond criticism or capable of accounting for all the experimental facts. This is not surprising, for the behaviour of real granular materials is diverse and complicated and it is unlikely that any one model will be adequate to cover all circumstances under which granular materials might deform, and therefore it is probable that different models will be required not only for different materials but also for the same material under different conditions. Consequently the various proposed models are not necessarily in conflict. A possible way to make progress is to investigate the implications of the various sets of equations which have been proposed for the flow and deformation of granular materials, in the hope of deriving results which can be tested against the behaviour of real granular materials. When viewed in this context the double-shearing theory is attractive because it accounts for at least some aspects of the behaviour of granular materials, it is sufficiently simple to permit the solution of some non-trivial boundary value problems, it is self-consistent, and it possesses a certain amount of mathematical elegance. The extent to which it adequately models the behaviour of any real materials can only be decided by experiment.

The arrangement of this article is as follows. In Section 2 we formulate the equations which govern the stress in plane deformations of granular materials. In Section 3 we analyse some of the conditions which might apply to the velocity field in a deforming granular material, briefly review some of the theories which result from adopting various combinations of these conditions, and rehearse the arguments which support the adoption of the double-shearing model, again for plane deformations. In Section 4 we formulate the equations for plane deformations, and express them in various alternative but equivalent forms. In Section 5 we apply these equations to the solution of some boundary value problems. Section 6 contains a discussion of some further general properties of the plane deformation equations. In Section 7 we abandon the restriction to plane deformations and formulate a three-dimensional theory. Some particular classes of non-planar deformations are considered in Section 8. In Section 9 we develop and apply the theory of axially symmetric deformations.

Practical problems which involve the flow of granular materials are of two main types. The first type concerns the flow of dry granular materials such as powders, grain, and so on. These materials have insignificant moisture content, so that the bulk material is a two-phase composite composed of solid and gas. The second class of problems concerns the behaviour of soils. Natural soils are much more complicated materials, and usually contain a liquid phase as well as solid and gas. The liquid content of a soil is a very important variable which has a strong influence on its mechanical behaviour. To some extent this influence can be taken into account by introducing the effective stress, which is the difference between the true stress and the hydrostatic stress due to pore pressure in the fluid. The properties of the soil are then specified in terms of the effective stress. In this article we shall concentrate on the simpler case of dry granular materials, although we shall indicate in Section 2 how the effective stress may be introduced. The modifications to the theory which are needed when the true stress is replaced by effective stress are straightforward in principle, although they can have a large effect on the solutions of particular problems.

For brevity, in this article the theory is developed only for incompressible materials. A brief discussion of the significance of dilatancy and ways in which it may be incorporated in the model is given in Section 4.

2. The Stress Field for Plane Deformations

The equations which govern the stress field in plane flow of a granular material are well known. A comprehensive account is that by Sokolovskii (1965). The theory is summarized in this Section.

We introduce a rectangular cartesian coordinate system x, y, z, referred to which the matrix of components of the stress tensor σ is given by

$$\sigma = \begin{bmatrix} \sigma_{xx} & \sigma_{xy} & \sigma_{xz} \\ \sigma_{xy} & \sigma_{yy} & \sigma_{yz} \\ \sigma_{xz} & \sigma_{yz} & \sigma_{zz} \end{bmatrix}. \tag{2.1}$$

We follow the usual sign conventions of continuum mechanics in which tensile stress components are positive. It is a common convention in soil mechanics to take compressive stress components to be positive, but we do not adopt this.

The principal stress components are denoted by σ_{I}, σ_{II}, and σ_{III}, and are ordered so that

$$\sigma_{\mathrm{I}} \geqslant \sigma_{\mathrm{II}} \geqslant \sigma_{\mathrm{III}}. \tag{2.2}$$

We consider two-dimensional states of stress in which the stress components are independent of the coordinate z, and the principal axis of stress, which corresponds to the intermediate principal stress component σ_{II}, is directed in the z-direction. Then

$$\sigma_{zz} = \sigma_{\mathrm{II}}, \quad \sigma_{xz} = \sigma_{yz} = 0, \tag{2.3}$$

and the equations of equilibrium reduce to

$$\frac{\partial \sigma_{xx}}{\partial x} + \frac{\partial \sigma_{xy}}{\partial y} + X = 0, \quad \frac{\partial \sigma_{xy}}{\partial x} + \frac{\partial \sigma_{yy}}{\partial y} + Y = 0, \tag{2.4}$$

where X and Y are body forces, per unit volume, in the x- and y-directions respectively.

The principal stress axes which correspond to σ_{I} and σ_{III} lie in the (x, y)-plane. We denote by ψ the angle which the axis corresponding to σ_{I} makes with the x-direction, as illustrated in Fig. 2.1. We also denote

$$p = -\tfrac{1}{2}(\sigma_{\text{I}} + \sigma_{\text{III}}), \quad q = \tfrac{1}{2}(\sigma_{\text{I}} - \sigma_{\text{III}}) \geqslant 0. \tag{2.5}$$

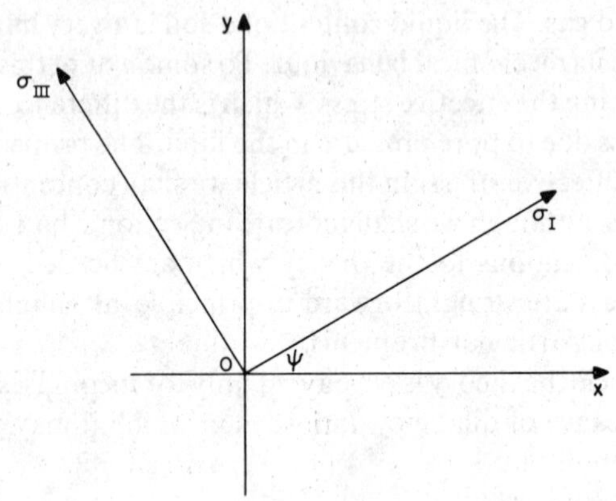

FIG. 2.1. Inclination of principal stress axes to x- and y-axes.

Then

$$\begin{aligned} \sigma_{xx} &= -p + q\cos 2\psi, \\ \sigma_{yy} &= -p - q\cos 2\psi, \\ \sigma_{xy} &= q\sin 2\psi, \end{aligned} \tag{2.6}$$

and

$$p = -\tfrac{1}{2}(\sigma_{xx} + \sigma_{yy}), \quad q = \tfrac{1}{2}\{(\sigma_{xx} - \sigma_{yy})^2 + 4\sigma_{xy}^2\}^{\frac{1}{2}}. \tag{2.7}$$

Consider an element of surface whose unit normal $\mathbf{n}$ is parallel to the (x, y)-plane, so that $\mathbf{n}$ has components $(n_x, n_y, 0) = (\cos\chi, \sin\chi, 0)$, where χ is the angle which $\mathbf{n}$ makes with the x-direction. We denote by σ_n the normal component of compressive traction, and by τ_n the magnitude of the shear component of traction, on the surface element. Thus

$$\begin{aligned} \sigma_n &= -(\sigma_{xx}n_x^2 + 2\sigma_{xy}n_xn_y + \sigma_{yy}n_y^2) = p - q\cos 2(\chi - \psi), \\ \tau_n &= \left|(\sigma_{xx} - \sigma_{yy})n_xn_y + \sigma_{xy}(n_x^2 - n_y^2)\right| = q\left|\sin 2(\chi - \psi)\right|. \end{aligned} \tag{2.8}$$

The Coulomb–Mohr hypothesis postulates, by analogy with the laws of friction between sliding surfaces, that slip can occur on the surface element with normal $\mathbf{n}$ if τ_n attains the critical value

$$\tau_n = c + \sigma_n \tan\phi, \tag{2.9}$$

where the parameters c and ϕ characterize the mechanical properties of the material. The parameter c is called the *cohesion* and ϕ is called the *angle of internal friction*. We assume that c and ϕ are constants. Thus at each point in the material

$$\tau_n - \sigma_n \tan\phi \leqslant c, \tag{2.10}$$

for all orientations of **n**. As χ varies, for given values of p, q, and ψ, the left-hand side of (2.10) attains its maximum value when

$$\chi = \psi \pm (\tfrac{1}{4}\pi - \tfrac{1}{2}\phi). \tag{2.11}$$

The corresponding values of σ_n and τ_n are denoted by σ and τ respectively, so that, from (2.8),

$$\sigma = p - q\sin\phi, \quad \tau = q\cos\phi, \tag{2.12}$$

and the condition (2.10) may be replaced by

$$q - p\sin\phi \leqslant c\cos\phi. \tag{2.13}$$

Thus, according to the Coulomb–Mohr hypothesis, the admissible states of stress are limited by the inequality (2.13). Slip may occur only if equality holds in (2.13), and then the orientations of the surfaces on which slip may take place are given by (2.11).

These results are often represented geometrically by the Mohr's circle diagram illustrated in Fig. 2.2. The line (2.9) is the envelope in the (σ_n, τ_n)-plane of the circles with centre $(p, 0)$ and radius q, where p and q are related by

$$q - p\sin\phi = c\cos\phi. \tag{2.14}$$

The circles touch their envelope at the points (σ, τ) given by (2.12).

The principal stress components σ_{I} and σ_{III} are invariants of the stress tensor $\boldsymbol{\sigma}$, and so p and q are invariants of $\boldsymbol{\sigma}$. Hence (2.13) is consistent with the property of material isotropy, and indeed we have made an implicit assumption that there are no preferred directions in the material. A natural generalization of (2.13) for isotropic materials is to abandon the linear relation between p and q and to postulate an inequality of the form

$$q \leqslant f(p), \tag{2.15}$$

where $f(p)$ is a specified function of p. The theory of materials whose behaviour is governed by

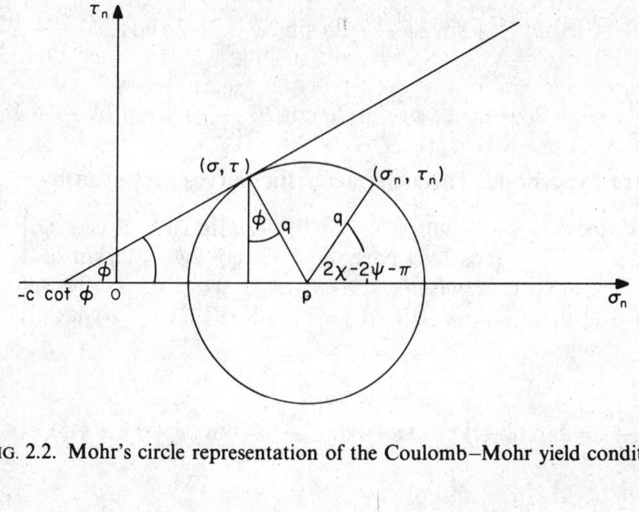

FIG. 2.2. Mohr's circle representation of the Coulomb–Mohr yield condition.

(2.15) has been developed by Mandel (1938a, b, 1962), Hill (1950), Sokolovskii (1965), and Kingston and Spencer (1970). In this article we shall, for simplicity, adopt the linear inequality (2.13), but much of the theory which follows can readily be extended to the more general form (2.15) if $\sin\phi$ is interpreted as $df(p)/dp$.

Coulomb originally formulated his hypothesis as a condition for slip along a surface separating two parts of a body of granular material, each of which moves without deformation as a rigid body. However, such a relative motion of rigid bodies (in two dimensions) can occur only if the slip surface intersects the (x, y)-plane in a straight line or a circle, and this seems to be unduly restrictive. A natural extension of the hypothesis is to admit the possibility of deformation by *shear* on surfaces on which τ_n attains the critical value given by (2.9). In this context shear may be regarded as the limit of a large number of small slipping motions on closely spaced parallel surfaces. This leads us to the formulation of theories of *soil plasticity*, in which deformation is not necessarily confined to isolated surfaces, but may extend throughout a region of a body. The relation (2.13) (or, more generally, (2.15)) may then be interpreted as a yield condition. When the strict inequality holds in (2.13), then the material may not deform (elastic deformations are not considered). In regions in which equality holds in (2.13), then deformation may occur by shear on the surfaces whose orientation is given by (2.11). Thus, in deforming regions, the stress is governed by (2.14) and the equilibrium equations (2.4). With (2.6), these give a system of three equations for the three unknowns p, q, and ψ or, equivalently, σ_{xx}, σ_{yy}, and σ_{xy}. The possibility of slip (i.e. a discontinuity in the component of velocity tangential to a slip surface) is not excluded because such a discontinuity may be regarded as the limiting case of a shear in which the velocity gradient becomes infinite. We note that, in the plane problem which is under discussion, the equations for the stress uncouple entirely from those for the velocity (which so far we have not even considered) and so the stress may be determined independently of the velocity field.

If $\phi = 0$, then the material is said to be *frictionless* or *purely cohesive*. The equations then reduce to those which determine the stress in the plane strain theory of metal plasticity. If $c = 0$, then the material is said to be *cohesionless*.

The system of (2.4), (2.6), and (2.14) has been studied extensively and applied to a wide range of problems. A comprehensive account is that by Sokolovskii (1965). If (2.6) are substituted into (2.4) and p is eliminated by means of (2.14), then the equilibrium equations (2.4) become the following two quasi-linear equations for q and ψ:

$$(\cos 2\psi - \operatorname{cosec}\phi)\frac{\partial q}{\partial x} + \sin 2\psi\frac{\partial q}{\partial y} - 2q\sin 2\psi\frac{\partial \psi}{\partial x} + 2q\cos 2\psi\frac{\partial \psi}{\partial y} + X = 0,$$

$$\sin 2\psi\frac{\partial q}{\partial x} - (\cos 2\psi + \operatorname{cosec}\phi)\frac{\partial q}{\partial y} + 2q\cos 2\psi\frac{\partial \psi}{\partial x} + 2q\sin 2\psi\frac{\partial \psi}{\partial y} + Y = 0.$$

These equations are hyperbolic. Their characteristic curves are given by

$$\begin{vmatrix} \cos 2\psi - \operatorname{cosec}\phi & \sin 2\psi & -2q\sin 2\psi & 2q\cos 2\psi \\ \sin 2\psi & -(\cos 2\psi + \operatorname{cosec}\phi) & 2q\cos 2\psi & 2q\sin 2\psi \\ 1 & dy/dx & 0 & 0 \\ 0 & 0 & 1 & dy/dx \end{vmatrix} = 0$$

which gives

$$\frac{dy}{dx} = \tan(\psi - \tfrac{1}{4}\pi - \tfrac{1}{2}\phi) \quad \text{or} \quad \frac{dy}{dx} = \tan(\psi + \tfrac{1}{4}\pi + \tfrac{1}{2}\phi). \tag{2.16}$$

The two families of curves represented by (2.16) are called α-lines and β-lines respectively. We see that the two directions $\mathbf{n}$ determined by (2.11) are the normals to the α- and β-lines, and so the α- and β-lines are the curves on which $\tau_n - \sigma_n \tan\phi$ attains the critical value c. The characteristic relations along the α- and β-lines are found by standard procedures to be

$$\begin{aligned} \cot\phi\, dq + 2q\, d\psi &= \{X \sin(\psi + \tfrac{1}{4}\pi + \tfrac{1}{2}\phi) - Y \cos(\psi + \tfrac{1}{4}\pi + \tfrac{1}{2}\phi)\} ds_\alpha \text{ on an } \alpha\text{-line}, \\ \cot\phi\, dq - 2q d\psi &= \{-X \sin(\psi - \tfrac{1}{4}\pi - \tfrac{1}{2}\phi) + Y \cos(\psi - \tfrac{1}{4}\pi - \tfrac{1}{2}\phi) ds_\beta \text{ on a } \beta\text{-line}, \end{aligned} \tag{2.17}$$

where ds_α and ds_β are elements of length along the α- and β-lines respectively. Equations (2.17) are known as Kötter's equations.

A number of exact solutions of these equations are known, and standard procedures exist for their numerical solution. Some examples will be given in Section 5.

If a pore pressure p_0 is present, then in (2.9) and subsequently σ_n, σ and p are replaced by the effective stresses and effective pressure $\bar{\sigma}_n = \sigma_n - p_0$, $\bar{\sigma} = \sigma - p_0$, and $\bar{p} = p - p_0$ respectively. The main effect is to introduce additional terms which involve the pore pressure gradient into the right-hand sides of (2.17).

Throughout this article we consider only quasi-static motions. The equations are all homogeneous in time, so that time t may be replaced by any parameter which orders the sequence of events. The extension of (2.17) to the dynamic case, so that it includes acceleration terms, has been given by Morrison and Richmond (1976).

3. Kinematic Theories for Plane Deformations

In this Section we discuss some of the assumptions which have been made in constructing theories of the kinematics of granular materials in plane deformations. As a preliminary to this it is convenient to mention some aspects of the behaviour of real granular materials in simple shearing flow.

Suppose a body of granular material is initially at rest in equilibrium. If it is then subjected to a shearing stress of sufficient magnitude to cause motion, the initial deformation is strongly dependent on the initial state of compaction and, hence, on the previous history of the body. When a well-compacted body of granular material undergoes a shearing deformation, it will usually expand in volume in the initial stages of the motion. However, if the shearing motion is maintained the material will quite soon settle to a fairly constant volume at which it can undergo a steady shear flow. Mandl and Fernandez Luque (1970) call this state fully developed plastic shear flow. The complicated initial motion is of importance in many problems, especially in soil mechanics, but in this article we concentrate on the simpler case of fully developed shear flow.

For simple shearing flows the Coulomb–Mohr hypothesis provides a means of determining the stress and the shear planes, as described in Section 2. However, to study inhomogeneous deformations it is necessary to formulate field equations for the velocity. These field equations must be based on some physical assumptions about the mechanical behaviour of granular materials. A number of plausible assumptions can be made; unfortunately these assumptions are not always consistent with one another, and this has proved to be a major obstacle to the development of the theory. We begin by stating and discussing some of the constitutive assumptions which have been employed in theories of the mechanics of granular material.

3.1. *Incompressibility*

As we have noted above, the initial motion of granular materials usually involves dilatation or, occasionally, compaction. However, in fully developed shear flow volume changes are often small, and it would then seem reasonable to adopt the approximation that the material behaves as an incompressible material. Experience in other branches of continuum mechanics suggests that if incompressibility is a valid approximation then it is a very convenient one, because the solution of boundary value problems is usually much simpler for incompressible materials than it is for compressible materials. Also from experience in other branches in continuum mechanics it is difficult to envisage a class of ideal materials which does not include an incompressible material as a special case.

3.2. *Isotropy*

Not all granular materials are isotropic on the macroscopic scale, but many are. Any theory for anisotropic materials will necessarily be more complicated than the corresponding theory for isotropic materials, and so it is natural to concentrate on the isotropic case in the first instance.

3.3. *Coaxiality of the stress and rate-of-deformation tensors*

This postulate states that (as in isotropic metal plasticity) the principal axes of the stress tensor and the rate-of-deformation tensor must coincide. This postulate has often been put forward as the condition for isotropy. In fact this is the case only in a restricted sense. If the rate-of-deformation depends only on the stress (or vice versa), and if the material is isotropic, then the rate-of-deformation is an isotropic tensor function of the stress. In three dimensions, a well-known representation theorem then states that the rate-of-deformation and the stress are coaxial. However, Mandl and Fernandez Luque (1970) have pointed out that this conclusion does not follow in two dimensions. Moreover, the argument that isotropy implies coaxiality is not valid if the rate-of-deformation depends not only on the stress but also on other tensors (e.g. a stress-rate tensor) or vectors.

3.4. *Associated flow rules*

In metal plasticity it is usually assumed that the yield function is a "plastic potential" for the rate-of-deformation tensor. Suppose that, referred to a system of rectangular cartesian coordinates x_i, the stress components are denoted by σ_{ij}, the velocity components by v_i, and the rate-of-deformation tensor components by d_{ij}, so that

$$d_{ij} = \tfrac{1}{2}\left(\frac{\partial v_i}{\partial x_j} + \frac{\partial v_j}{\partial x_i}\right) \quad (i, j = 1, 2, 3), \tag{3.1}$$

and that the yield condition is an expression of the form

$$F(\sigma_{ij}) \leqslant c, \tag{3.2}$$

where c is a yield stress. Then, according to the associated flow rule, d_{ij} are given by

$$d_{ij} = \dot{\Lambda}\partial F/\partial\sigma_{ij}, \tag{3.3}$$

where $\dot{\Lambda}$ is a scalar factor of proportionality (not a material constant).

The arguments usually advanced for the adoption of (3.3) are based on the material stability postulate of Drucker (1959, 1964). Drucker's postulate represents a classification of materials rather than a universal law of material behaviour. Among its consequences are that the surface $F(\sigma_{ij}) = c$ (regarded as a surface in six-dimensional σ_{ij} space) is convex, and the result (3.3) that d_{ij} (regarded as a vector in this space) is normal to the yield surface. The adoption of (3.3) has the great advantage that it makes possible the proof of some very useful variational theorems. If the material is isotropic, $F(\sigma_{ij})$ must be a function of the stress invariants and then (3.3) ensures also coaxiality of the stress and rate-of-deformation tensors. Thus the associated flow rule is closely related to the coaxiality assumption, though the two properties are not equivalent to one another.

3.5. *Coincidence of stress and velocity characteristics*

The proposed mechanism of deformation, based on Coulomb's hypothesis, that deformation in a granular material occurs by slip or shear on the surfaces on which τ_n reaches its critical value, is supported by substantial observational and experimental evidence. Curves which appear to correspond to the stress characteristics (2.16) are often observed in flowing and deformed granular materials. It might therefore be expected that these curves will also arise naturally in the solution of appropriate velocity equations, and in particular as the characteristic curves of a system of hyperbolic differential equations. It would seem to be a desirable property of the velocity equations that they be hyperbolic with characteristics which coincide with the stress characteristics.

We now review some of the theories which have been proposed in the light of the considerations listed above.

Hill (1950) proposed velocity equations based on incompressibility and on coaxiality of stress and rate-of-deformation tensors. These equations have orthogonal characteristics which do not coincide with the stress characteristics. A solution to the punch indentation problem by Cox (1963) using Hill's equations gives unrealistic results.

Drucker and Prager (1952) followed the procedure which has proved successful in metal plasticity (with the appropriate modifications for two-dimensional deformations) and assumed that the Coulomb–Mohr yield function

$$F = q \sec\phi - p\tan\phi \tag{3.4}$$

is the plastic potential. The standard theory of plane strain metal plasticity then corresponds to the special case $\phi = 0$. Their theory was subsequently extended and developed by several authors, including Shield (1953, 1955a) and Cox *et al.* (1961). It conforms to isotropy and coincidence of stress and velocity characteristics, and employs the associated flow rule. However, for $\phi \neq 0$ it is subject to a major objection because this theory predicts that a shearing deformation is always accompanied by a dilatation of a magnitude which is quite unrealistic for real granular materials. Although granular materials may expand on initial shearing, we do not expect this expansion to continue indefinitely or to be of the magnitude which the theory predicts for typical values of ϕ. Another difficulty with Drucker and

Prager's theory, which was pointed out by Mandl and Fernandez Luque (1970), is that the rate of energy dissipation is proportional to c, and so zero energy dissipation is predicted in a cohesionless material. It seems therefore that this theory, in its straightforward form, must be rejected.

Most of the theories which have been formulated subsequently fall into one of two categories. In the first category are theories which retain the associated flow rule, or some modification of it. For example, Drucker *et al.* (1957), Jenike and Shield (1959), and Jenike (1961) have proposed work-hardening models. Probably the most sophisticated model of this type is that described in the book by Schofield and Wroth (1968). These models incorporate isotropy. In general they require dilatancy, and the characteristics of stress and velocity equations do not coincide.

The second category of theories abandons the concept of the associated flow rule and returns directly to Coulomb's hypothesis, or a generalization of it. Such theories subdivide into two classes, which may be described as the single-shearing and the double-shearing models.

The single-shearing model seems to have been formulated first by Geniev (1958), and is described by Mandl and Fernandez Luque (1970). In it deformation is assumed to occur primarily by shear on one of the families of stress characteristics (the operative characteristics) so that two possible sets of velocity equations arise, depending on which family is the operative one. The model incorporates incompressibility, and the principal axes of stress and deformation-rate do not coincide but are inclined at the angle $\frac{1}{2}\phi$ to one another. Nevertheless, the model is not inconsistent with isotropy. One family of velocity characteristics coincides with the operative stress characteristics, and the second family is orthogonal to the first. A modified version of the model in which the material is not incompressible has been developed by Takagi (1962, 1979).

The double-shearing model admits deformation by simultaneous shearing on both families of stress characteristics, and this model forms the main subject of this article. Stress and deformation-rate are not, in general, coaxial; the model incorporates isotropy, and the stress and velocity characteristics are coincident. We shall develop the theory for incompressible materials, but at the end of Section 4 we mention some extensions to dilatant materials.

In general, theories such as the one described by Schofield and Wroth (1968), which employ an associated flow rule or some related concept, seem most likely to find applications to the study of initial motion, where dilatancy is usually important. These theories require dilatancy and have the possible defect that they do not, in general, predict coincidence of the stress and velocity characteristics. The single- and double-shearing theories are perhaps better suited to the description of fully developed flow; their main disadvantage is that the useful general results which follow from Drucker's material stability postulate do not apply to these models. The two categories of theory are not necessarily irreconcilable. The concept of material stability may well require modification for its application to granular materials. The relevance of Drucker's postulate to granular materials is discussed in detail by Mandl and Fernandez Luque (1970). On the other hand, it will be shown in Section 6 that in the double-shearing model a part of the deformation-rate (which in many cases is the predominant part) can be derived from a plastic potential. Hence the gap between the two types of theory may not be as wide as it appears to be. As to the choice between single-shearing and double-shearing models, we know of no experimental results which support one rather than the other. We have chosen to develop the double-shearing theory because it seems to us to possess a simpler mathematical structure and to be more in accord with the physics of the deformation of granular materials.

In the brief review given in this Section we have referred only to works which are directly relevant to the discussion. There are many other important papers which, for reasons of space, have not been described. Among the many significant contributions which are not cited above or elsewhere in this article we mention those by Jeske (1968), Dais (1970), Goodman and Cowin (1972), Nikolaevskii (1972), and Nemat-Nasser (1980). This list is by no means comprehensive. Further important contributions, in particular to the double-shearing model, have been reported by Mróz and Szymański (1979) and Mróz (1980) since this article was completed.

4. The Double-shearing Model for Plane Deformations

We shall formulate the equations for plane deformations using the double-shearing model in two distinct but equivalent ways.

4.1. *First formulation*

This approach is essentially the one which was followed by Spencer (1964). It is necessary first to specify carefully what is meant by a shearing deformation on an α-line or a β-line (or on any other curve), bearing in mind that in general the curves are not fixed in time, and that even if the stress does not depend on time t, the directions of the α- and β-lines through a given particle will change with t as the particle moves relative to the frame of reference. It is therefore convenient in the first instance to select a local moving frame of reference in which the particle at the origin is at rest and the α- and β-lines are fixed in direction. Since the slopes of the α- and β-lines are related by (2.16) to the directions of principal axes of stress, the last objective can be achieved by selecting a moving frame of reference which rotates with an angular speed which is equal to the rate-of-rotation of the principal axes of stress.

We therefore select a generic particle P whose velocity components relative to the fixed (x, y) coordinates are V_x and V_y. We take the position occupied by P as the origin of a new coordinate system which is translating with velocity (V_x, V_y) relative to the (x, y) system, so that P is a fixed point in the new coordinate system. The new coordinates are denoted by ξ, η, and for definiteness we select the ξ- and η-axes so that the ξ-axis coincides with the axis of the algebraically smaller principal stress component at P. The ξ- and η-axes are illustrated in Fig. 4.1. We also specify that the ξ- and η-axes rotate, relative to the x- and y-axes, with angular speed Ω, which is the rate-of-rotation (relative to (x, y) coordinates) of the principal axes of stress at the particle P. Thus Ω is the convected derivative $\dot{\psi}$ of ψ, so that

$$\Omega = \dot{\psi} = \frac{\partial \psi}{\partial t} + V_x \frac{\partial \psi}{\partial x} + V_y \frac{\partial \psi}{\partial y}, \tag{4.1}$$

where $\partial\psi/\partial t$ is the derivative at a point which is fixed relative to (x, y) coordinates. Let v_x, v_y be components of velocity relative to the (x, y) coordinates, and v_ξ, v_η components of velocity relative to the (ξ, η) coordinates. Then, with the above specification of the (ξ, η) system,

$$\begin{aligned} v_x &= V_x + (v_\xi - \Omega\eta)\cos\psi - (v_\eta + \Omega\xi)\sin\psi, \\ v_y &= V_y + (v_\xi - \Omega\eta)\sin\psi + (v_\eta + \Omega\xi)\cos\psi. \end{aligned} \tag{4.2}$$

From (2.16), the α- and β-lines make fixed angles $\mp(\frac{1}{4}\pi + \frac{1}{2}\phi)$ with the ξ- and η-axes at the

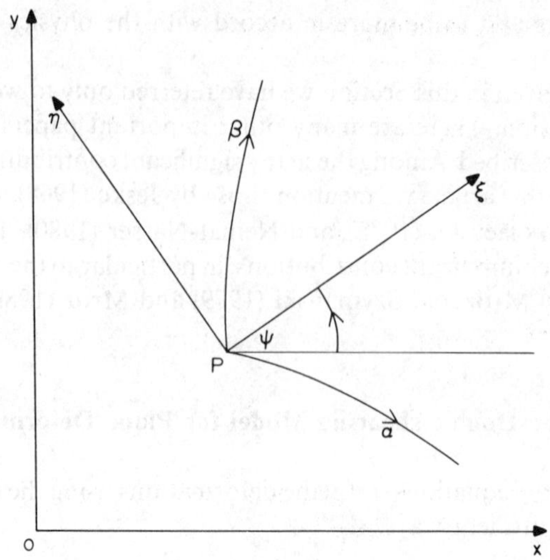

FIG. 4.1. Orientation of the moving (ξ, η) axes.

origin. Let v'_α and v'_β denote the components of velocity, relative to the (ξ, η) reference frame, in the directions of the α- and β-lines. Then, at P,

$$\begin{aligned} v_\xi &= (v'_\alpha + v'_\beta)\cos(\tfrac{1}{4}\pi + \tfrac{1}{2}\phi), \\ v_\eta &= (-v'_\alpha + v'_\beta)\sin(\tfrac{1}{4}\pi + \tfrac{1}{2}\phi). \end{aligned} \tag{4.3}$$

We note that v'_α and v'_β are not the projections of a velocity vector onto the α- and β-line directions, but components whose vector sum is the velocity vector.

We say that a motion consists of a shearing deformation along the α-lines if, in a coordinate system in which the α-line direction in the neighbourhood of a particle P is fixed, the relative velocity of two adjacent particles near P is directed in the direction of the α-line if the particles do not lie on the same α-line, and is zero if they do lie on the same α-line. Thus, in the (ξ, η) coordinate system, if $\mathbf{v}'$ denotes the particle velocity relative to the (ξ, η) coordinates, the motion in the neighbourhood of P is a shearing motion on the α-lines if

(i) $\partial \mathbf{v}'/\partial s_\alpha = \mathbf{0}$, (4.4)

and

(ii) $\partial \mathbf{v}'/\partial s_\beta$ is a vector which is directed in the direction of the α-line through P,

where $\partial/\partial s_\alpha$ and $\partial/\partial s_\beta$ denote directional derivatives in the α- and β-line directions respectively.

In general it is not possible to satisfy conditions (i) and (ii) throughout an extended region of an incompressible material (exceptions are the cases in which the α-lines are straight lines or concentric circular arcs). However, the incompressibility condition does not exclude the possibility of two simultaneous superposed shearing deformations on the α- and β-lines respectively, and the double-shearing model is based on the hypothesis that the deformation is of such a form. By analogy with (4.4), $\partial \mathbf{v}'/\partial s_\beta = \mathbf{0}$ in a shearing deformation on the β-lines. Hence, from (ii), in a superposed shearing deformation on both the α- and β-lines near P, the vector $\partial \mathbf{v}'/\partial s_\beta$ must be directed along the α-line through P. Therefore,

$$\frac{\partial v_\eta/\partial s_\beta}{\partial v_\xi/\partial s_\beta} = -\tan\left(\tfrac{1}{4}\pi + \tfrac{1}{2}\phi\right)$$

or

$$\sin\left(\tfrac{1}{4}\pi + \tfrac{1}{2}\phi\right)\frac{\partial v_\xi}{\partial s_\beta} + \cos\left(\tfrac{1}{4}\pi + \tfrac{1}{2}\phi\right)\frac{\partial v_\eta}{\partial s_\beta} = 0, \tag{4.5}$$

which, from (4.3), gives

$$\frac{\partial v'_\beta}{\partial s_\beta} = 0. \tag{4.6}$$

Similarly, in the superposed shearing deformation, the vector $\partial \mathbf{v}'/\partial s_\alpha$ is directed in the β-line direction, which gives

$$\sin\left(\tfrac{1}{4}\pi + \tfrac{1}{2}\phi\right)\frac{\partial v_\xi}{\partial s_\alpha} - \cos\left(\tfrac{1}{4}\pi + \tfrac{1}{2}\phi\right)\frac{\partial v_\eta}{\partial s_\alpha} = 0, \tag{4.7}$$

and hence

$$\frac{\partial v'_\alpha}{\partial s_\alpha} = 0. \tag{4.8}$$

Equations (4.6) and (4.8) state that, in the fictitious velocity field with velocity components v_ξ and v_η, the rate-of-extension of material line elements lying along the characteristic directions is zero.

It is also useful to express (4.5) and (4.7) in terms of ξ and η. Since

$$\begin{aligned}\frac{\partial}{\partial s_\alpha} &= \cos\left(\tfrac{1}{4}\pi + \tfrac{1}{2}\phi\right)\frac{\partial}{\partial \xi} - \sin\left(\tfrac{1}{4}\pi + \tfrac{1}{2}\phi\right)\frac{\partial}{\partial \eta},\\ \frac{\partial}{\partial s_\beta} &= \cos\left(\tfrac{1}{4}\pi + \tfrac{1}{2}\phi\right)\frac{\partial}{\partial \xi} + \sin\left(\tfrac{1}{4}\pi + \tfrac{1}{2}\phi\right)\frac{\partial}{\partial \eta},\end{aligned} \tag{4.9}$$

it follows that

$$\begin{aligned}\cos\phi\left(\frac{\partial v_\xi}{\partial \xi} + \frac{\partial v_\eta}{\partial \eta}\right) + \left(\frac{\partial v_\xi}{\partial \eta} + \frac{\partial v_\eta}{\partial \xi}\right) + \sin\phi\left(\frac{\partial v_\xi}{\partial \eta} - \frac{\partial v_\eta}{\partial \xi}\right) = 0,\\ -\cos\phi\left(\frac{\partial v_\xi}{\partial \xi} + \frac{\partial v_\eta}{\partial \eta}\right) + \left(\frac{\partial v_\xi}{\partial \eta} + \frac{\partial v_\eta}{\partial \xi}\right) + \sin\phi\left(\frac{\partial v_\xi}{\partial \eta} - \frac{\partial v_\eta}{\partial \xi}\right) = 0,\end{aligned} \tag{4.10}$$

and hence that

$$\frac{\partial v_\xi}{\partial \xi} + \frac{\partial v_\eta}{\partial \eta} = 0, \tag{4.11}$$

$$\left(\frac{\partial v_\xi}{\partial \eta} + \frac{\partial v_\eta}{\partial \xi}\right) + \sin\phi\left(\frac{\partial v_\xi}{\partial \eta} - \frac{\partial v_\eta}{\partial \xi}\right) = 0. \tag{4.12}$$

The components of the rate-of-deformation tensor and the spin tensor, relative to (ξ, η) coordinates, are defined by

$$d_{\xi\xi} = \frac{\partial v_\xi}{\partial \xi}, \quad d_{\eta\eta} = \frac{\partial v_\eta}{\partial \eta}, \quad d_{\xi\eta} = \tfrac{1}{2}\left(\frac{\partial v_\xi}{\partial \eta} + \frac{\partial v_\eta}{\partial \xi}\right), \tag{4.13}$$

$$\omega_{\xi\eta} = \tfrac{1}{2}\left(\frac{\partial v_\xi}{\partial \eta} - \frac{\partial v_\eta}{\partial \xi}\right), \tag{4.14}$$

so that (4.11) and (4.12) take the forms

$$d_{\xi\xi} + d_{\eta\eta} = 0, \qquad d_{\xi\eta} + \omega_{\xi\eta} \sin\phi = 0. \tag{4.15}$$

In terms of the velocity components, (4.15) may be written as

$$\frac{\partial v_\xi}{\partial \xi} + \frac{\partial v_\eta}{\partial \eta} = 0, \quad (1 + \sin\phi)\frac{\partial v_\xi}{\partial \eta} + (1 - \sin\phi)\frac{\partial v_\eta}{\partial \xi} = 0. \tag{4.16}$$

Equations (4.11), $(4.15)_1$, and $(4.16)_1$ are forms of the incompressibility condition, so that the double-shearing model is consistent with the assumption of material incompressibility. Equations (4.12), $(4.15)_2$, and $(4.16)_2$ each show that the principal axes of stress and rate-of-deformation do not coincide, except in the case $\phi = 0$ of a frictionless material, or when the spin component $\omega_{\xi\eta} = 0$. Thus the double-shearing model is one in which the stress and rate-of-deformation tensors are, in general, not coaxial.

We now express (4.15) and (4.16) in terms of the fixed (x, y) coordinate system. The components of deformation-rate and spin relative to the (x, y) coordinates are denoted by

$$d_{xx} = \frac{\partial v_x}{\partial x}, \quad d_{yy} = \frac{\partial v_y}{\partial y}, \quad d_{xy} = \tfrac{1}{2}\left(\frac{\partial v_x}{\partial y} + \frac{\partial v_y}{\partial x}\right),$$
$$\omega_{xy} = \tfrac{1}{2}\left(\frac{\partial v_x}{\partial y} - \frac{\partial v_y}{\partial x}\right). \tag{4.17}$$

Since the ξ- and η-axes are rotated through the angle ψ relative to the x- and y-axes, we have

$$\begin{pmatrix} d_{\xi\xi} & d_{\xi\eta} \\ d_{\xi\eta} & d_{\eta\eta} \end{pmatrix} = \begin{pmatrix} \cos\psi & \sin\psi \\ -\sin\psi & \cos\psi \end{pmatrix} \begin{pmatrix} d_{xx} & d_{xy} \\ d_{xy} & d_{yy} \end{pmatrix} \begin{pmatrix} \cos\psi & -\sin\psi \\ \sin\psi & \cos\psi \end{pmatrix} \tag{4.18}$$

and, since the (ξ, η) system is rotating with angular speed relative to the (x, y) system,

$$\omega_{\xi\eta} = \omega_{xy} + \Omega. \tag{4.19}$$

Hence (4.15) may be written as

$$d_{xx} + d_{yy} = 0,$$
$$d_{xy} \cos 2\psi - \tfrac{1}{2}(d_{xx} - d_{yy}) \sin 2\psi + \sin\phi\,(\omega_{xy} + \Omega) = 0, \tag{4.20}$$

or, in terms of x, y and the velocity components v_x and v_y, as

$$\frac{\partial v_x}{\partial x} + \frac{\partial v_y}{\partial y} = 0,$$
$$\left(\frac{\partial v_x}{\partial y} + \frac{\partial v_y}{\partial x}\right)\cos 2\psi - \left(\frac{\partial v_x}{\partial x} - \frac{\partial v_y}{\partial y}\right)\sin 2\psi + \sin\phi\left(\frac{\partial v_x}{\partial y} - \frac{\partial v_y}{\partial x} + 2\Omega\right) = 0. \tag{4.21}$$

Equations (4.21) satisfy the requirement that constitutive equations are independent of superposed rigid-body motions, for if such a motion is superposed, d_{xx}, d_{yy}, and d_{xy} remain unchanged, and ω_{xy} and Ω change by amounts which are equal in magnitude but opposite in sign.

It is also useful to express (4.21) as relations along the α- and β-lines. The components of velocity (relative to the fixed (x, y) coordinate system) in the directions of the α- and β-lines are denoted by v_α and v_β, so that

$$\begin{aligned} v_\alpha \cos\phi &= v_x \sin(\psi + \tfrac{1}{4}\pi + \tfrac{1}{2}\phi) - v_y \cos(\psi + \tfrac{1}{4}\pi + \tfrac{1}{2}\phi), \\ v_\beta \cos\phi &= -v_x \sin(\psi - \tfrac{1}{4}\pi - \tfrac{1}{2}\phi) + v_y \cos(\psi - \tfrac{1}{4}\pi - \tfrac{1}{2}\phi), \end{aligned} \tag{4.22}$$

and the directional derivatives along the α- and β-lines are denoted by $\partial/\partial s_\alpha$ and $\partial/\partial s_\beta$, so that

$$\begin{aligned} \frac{\partial}{\partial s_\alpha} &= \cos(\psi - \tfrac{1}{4}\pi - \tfrac{1}{2}\phi)\frac{\partial}{\partial x} + \sin(\psi - \tfrac{1}{4}\pi - \tfrac{1}{2}\phi)\frac{\partial}{\partial y}, \\ \frac{\partial}{\partial s_\beta} &= \cos(\psi + \tfrac{1}{4}\pi + \tfrac{1}{2}\phi)\frac{\partial}{\partial x} + \sin(\psi + \tfrac{1}{4}\pi + \tfrac{1}{2}\phi)\frac{\partial}{\partial y}. \end{aligned} \tag{4.23}$$

Then it follows after a little manipulation that (4.21) are equivalent to the relations

$$\begin{aligned} \cos\phi \frac{\partial v_\alpha}{\partial s_\alpha} - (v_\beta - v_\alpha \sin\phi)\frac{\partial \psi}{\partial s_\alpha} - \Omega \sin\phi &= 0, \\ \cos\phi \frac{\partial v_\beta}{\partial s_\beta} + (v_\alpha - v_\beta \sin\phi)\frac{\partial \psi}{\partial s_\beta} + \Omega \sin\phi &= 0. \end{aligned} \tag{4.24}$$

These equations reduce to Geiringer's velocity equations for a frictionless ideal plastic solid when $\phi = 0$. We observe that, from (4.1), Ω can be expressed in the form

$$\Omega = \frac{\partial \psi}{\partial t} + v_\alpha \frac{\partial \psi}{\partial s_\alpha} + v_\beta \frac{\partial \psi}{\partial s_\beta}. \tag{4.25}$$

Some further developments of equations (4.21) will be introduced in Section 5.

It has already been pointed out that in general the equations formulated in this Section imply non-coaxiality of the principal axes of stress and deformation-rate. This is not inconsistent with our assumption that the material is isotropic, because the equations are not simply relations between the stress (through the angle ψ) and the velocity gradients; they also involve $\Omega = \dot{\psi}$, which can be expressed in terms of stress and stress-rate components. Hence the constitutive equations involve a stress-rate tensor as well as stress and velocity gradient tensors.

4.2. *Second formulation*

This approach was initiated by de Josselin de Jong (1959) and has been developed by Zagainov (1967), de Josselin de Jong (1971, 1977), and Mandl and Fernandez Luque (1970). These authors employ a fixed coordinate system and concentrate on the case of steady motion. We depart slightly from this approach and first formulate the equations in the moving (ξ, η) reference frame.

Consider a motion which consists of a single shear on the α-lines with shear-rate of magnitude $2a$, so that $2a$ is the directional derivative of v_α in the direction normal to the α-line. The corresponding values of $d_{\xi\xi}$, $d_{\eta\eta}$, $d_{\xi\eta}$, $\omega_{\xi\eta}$ are

$$d^{(a)}_{\xi\xi} = -d^{(a)}_{\eta\eta} = -a\cos\phi, \quad d^{(a)}_{\xi\eta} = b\sin\phi, \quad \omega^{(a)}_{\xi\eta} = a. \tag{4.26}$$

Similarly, for a single shear of magnitude b on the β-lines, we have

$$d^{(b)}_{\xi\xi} = -d^{(b)}_{\eta\eta} = -b\cos\phi, \quad d^{(b)}_{\xi\eta} = b\sin\phi, \quad \omega^{(b)}_{\xi\eta} = -b. \tag{4.27}$$

Except for trivial cases, it is not possible to determine a so that the components (4.26) satisfy the compatibility conditions

$$\frac{\partial}{\partial\xi}\left(d^{(a)}_{\eta\eta}\right) = \frac{\partial}{\partial\eta}\left(d^{(a)}_{\xi\eta} + \omega^{(a)}_{\xi\eta}\right), \quad \frac{\partial}{\partial\eta}\left(d^{(a)}_{\xi\xi}\right) = \frac{\partial}{\partial\xi}\left(d^{(a)}_{\xi\eta} - \omega^{(a)}_{\xi\eta}\right),$$

and similarly for the components (4.27). Consequently a single shearing motion is possible only in a very restricted class of motions. However, it is possible to superpose the motions (4.26) and (4.27) to obtain motions of the form

$$d_{\xi\xi} = -d_{\eta\eta} = -(a+b)\cos\phi, \quad d_{\xi\eta} = -(a-b)\sin\phi, \quad \omega_{\xi\eta} = a-b, \tag{4.28}$$

and the compatibility conditions may then have non-trivial solutions for a and b. Then by eliminating a and b from (4.28), we have

$$d_{\xi\xi} + d_{\eta\eta} = 0, \quad d_{\xi\eta} + \omega_{\xi\eta}\sin\phi = 0, \tag{4.29}$$

which are identical to (4.16). The remainder of the formulation then proceeds as before.

de Josselin de Jong (1971, 1977) follows a similar argument to this, but he treats $2a$ and $2b$ as shear-rates relative to a fixed coordinate system, and admits an arbitrary rotation. Thus in de Josselin de Jong's theory Ω is not identified with the rotation of the principal stress axes; this theory results in equations of the same form as (4.21), but with Ω interpreted as a "free" undetermined rotation. Consequently de Josselin de Jong's equations form an indeterminate system, with an infinite number of solutions. de Josselin de Jong also imposes the requirement that the rate-of-working of the stress in each of the shears individually shall be non-negative; this requires that $a \leqslant 0$, $b \leqslant 0$. There seems to be no compelling physical reason why this restriction should be adopted, but it may be a useful way of defining a narrower class of materials. If $a \leqslant 0$, $b \leqslant 0$, then it follows from (4.28) that

$$(d_{\xi\xi} - d_{\eta\eta})\sin\phi \pm 2d_{\xi\eta}\cos\phi \geqslant 0, \tag{4.30}$$

which implies (since the (ξ, η) axes coincide with the principal axes of stress) that the angle i between the principal axes of maximum principal stress and maximum principal strain-rate satisfies the inequality

$$-\tfrac{1}{2}\phi \leqslant i \leqslant \tfrac{1}{2}\phi. \tag{4.31}$$

Zagainov (1967) and Mandl and Fernandez Luque (1970) consider steady motion with superposed shear-rates $2a$ and $2b$ relative to fixed local shear directions. Their argument leads to equations of the form (4.21) with $\Omega = 0$. Mandl and Fernandez Luque then allow for rotation of the principal stress axes by referring the equations to a frame which is rotating with angular speed $\partial\psi/\partial t$. This leads to equations of the form (4.21), except that the material derivative $\dot{\psi} = \Omega$ is replaced by the spatial derivative $\partial\psi/\partial t$. Such a system of equations is unacceptable because, as was shown by Spencer (1971), the equations are not form-invariant under superposed rigid translations. Physically, the discrepancy arises because, in steady motion, an element shearing on a curved slipline (say an α-line) undergoes a rotation-rate $v_\alpha\, \partial\psi/\partial s_\alpha$ due to its translation along the α-line, and this additional rotation is omitted by Mandl and Fernandez Luque. We have avoided this difficulty by using a frame of reference in which the particle velocity at the origin is zero. Equations (4.21) with $\Omega = 0$ are the equations proposed by Mandl and Fernandez Luque for steady motion.

A third (but historically the first) formulation is that of Mandel (1947) who postulated that the deformation is governed by equations which are equivalent to (4.6) and (4.8), but generalized to admit a superposed elastic strain-rate. Subsequently, Mandel (1966) has criticized the double-shearing model on the grounds that it does not allow a double-sliding (as opposed to a double-shearing) deformation mechanism, and that it does not conform to the associated flow rule.

Since the constitutive equations (4.21) involve the stress-rate (through Ω), it is also necessary in applications to ensure that equilibrium equations for the stress-rate are satisfied. For uniform constant body forces X and Y these may be expressed as

$$\frac{\partial}{\partial x}\left(\frac{\partial \sigma_{xx}}{\partial t}\right)+\frac{\partial}{\partial y}\left(\frac{\partial \sigma_{xy}}{\partial t}\right)=0, \quad \frac{\partial}{\partial x}\left(\frac{\partial \sigma_{xy}}{\partial t}\right)+\frac{\partial}{\partial y}\left(\frac{\partial \sigma_{yy}}{\partial t}\right)=0, \tag{4.32}$$

where, as before, $\partial/\partial t$ denotes the time derivative at a fixed point. There does not seem to be a convenient way of expressing these equations in terms of a properly invariant stress-rate tensor. From (2.6) and (2.14) it follows that

$$\begin{aligned}\frac{\partial \sigma_{xx}}{\partial t} &= -p_t + q_t \cos 2\psi - 2\psi_t q \sin 2\psi,\\ \frac{\partial \sigma_{yy}}{\partial t} &= -p_t - q_t \cos 2\psi + 2\psi_t q \sin 2\psi,\\ \frac{\partial \sigma_{xy}}{\partial t} &= q_t \sin 2\psi + 2\psi_t q \cos 2\psi,\end{aligned} \tag{4.33}$$

and

$$q_t = p_t \sin\phi,$$

where

$$p_t = \partial p/\partial t, \quad q_t = \partial q/\partial t, \quad \text{and} \quad \psi_t = \partial\psi/\partial t.$$

By substituting (4.33) into (4.32), and referring the equations to the α- and β-lines as oblique curvilinear coordinates, it can be shown that

$$\begin{aligned}\cot\phi\frac{\partial q_t}{\partial s_\alpha}+2q\frac{\partial \psi_t}{\partial s_\alpha}+2q_t\frac{\partial \psi}{\partial s_\alpha}+2\psi_t\left(\frac{\partial q}{\partial s_\alpha}+2q\tan\phi\frac{\partial\psi}{\partial s_\alpha}\right)+4\psi_t q\sec\phi\frac{\partial\psi}{\partial s_\beta}&=0,\\ \cot\phi\frac{\partial q_t}{\partial s_\beta}-2q\frac{\partial \psi_t}{\partial s_\beta}-2q_t\frac{\partial \psi}{\partial s_\beta}-2\psi_t\left(\frac{\partial q}{\partial s_\beta}-2q\tan\phi\frac{\partial\psi}{\partial s_\beta}\right)+4\psi_t q\sec\phi\frac{\partial\psi}{\partial s_\alpha}&=0.\end{aligned} \tag{4.34}$$

Alternative forms of (4.34) can be obtained by using (2.17). In particular, (4.34) can be simplified by the use of (2.17) when the body force components X and Y are zero.

Since inertia effects are neglected, the traction-rates must be continuous across any surface, and particularly across an α-line or a β-line. This gives the conditions

$$\begin{aligned} &\cot\phi\, q_t - 2q\psi_t \quad \text{is continuous across an } \alpha\text{-line}, \\ &\cot\phi\, q_t + 2q\psi_t \quad \text{is continuous across a } \beta\text{-line}. \end{aligned} \tag{4.35}$$

The stress-rate equations are trivially satisfied in problems of steady motion.

The dependence of the deformation on stress-rate as well as upon stress is not an entirely new phenomenon in the theory of rigid–plastic solids, for Hill (1956) has shown that in order to determine the velocity field in an ideal metal uniquely it is necessary to specify surface traction-rates. The stress-rate also appears in the constitutive equation for an elastic–plastic solid.

Throughout this Section it has been assumed that the material is incompressible. The extension of the theory to compressible and dilatant materials will not be discussed in detail, but we close this Section with a brief mention of some developments in this direction. Spencer and Kingston (1973) proposed a theory which admits compressibility by the superposition of a uniform expansion or contraction on the double-shearing deformation. An alternative theory which seems to be more satisfactory than that of Spencer and Kingston is based on a mechanism proposed by Butterfield and Harkness (1972) and Butterfield *et al.* (1973), in which shearing is accompanied by a stretch in the direction normal to the shear plane. The ratio of the normal stretch to the shear is governed by a parameter called the angle of dilatancy, which could be a function of deformation history. This theory has been developed by de Josselin de Jong (1977) for his free-rotating theory and by Mehrabadi and Cowin (1978) for the version of the double-shearing theory described in this article. Mehrabadi and Cowin show that the incompressible theory of this Section arises as the special case of their theory when the angle of dilatancy is zero, and the associated flow rule proposed by Drucker and Prager (1952) corresponds to the case in which the angle of dilatancy is equal to the angle of internal friction. There would seem to be no difficulty in principle in constructing a more general theory which combines the uniform expansion proposed by Spencer and Kingston (1973) with the normal stretch determined by the angle of dilatancy.

Recently, Mehrabadi and Cowin (1980) have discovered interesting connections between their theory and the yield-vertex model which was formulated by Rudnicki and Rice (1975) to explain deformation localization phenomena in the failure of geological materials. The relationship between these models is clearly a subject which merits further research.

5. Exact Solutions of the Velocity Equations

Many solutions of the stress equations described in Section 2 have been obtained. A number of these solutions are given in Sokolovskii (1965). However, only a few solutions of the velocity equations developed in Section 4 are known. As the equations are hyperbolic and have many features in common with the equations of plane strain metal plasticity, it is straightforward in principle to solve them numerically by integration along the characteristics. In such numerical solutions the requirement to satisfy stress-rate equilibrium equations is a complicating factor in problems of unsteady motion, but not in steady motion problems. However, we are aware of no published numerical solutions even for problems of steady motion, and numerical methods are not considered here. Instead we give brief descriptions of the known exact solutions. In all of these solutions body forces are neglected.

5.1. *Simple shear*

It is useful to confirm that the velocity equations (4.21) admit as a special case a single-shearing motion on a family of straight lines. If τ_n attains its critical value on the planes normal to the y-axis, then from (2.11) $\psi = \pm(\frac{1}{4}\pi + \frac{1}{2}\phi)$. Then $\Omega = \dot{\psi} = 0$, and $\cos 2\psi = -\sin\phi$, and so (4.21) have the solution

$$v_x = V(y), \quad v_y = 0,$$

which is a simple shearing deformation on the planes $y = $ constant.

5.2. *Uniform extensions*

The solution

$$\psi = 0, \quad v_x = dx, \quad v_y = -dy,$$

where d is constant, satisfies (4.21) and corresponds to a uniform stretching motion along the largest principal stress axis, with accompanying contraction in the transverse direction.

5.3. *Indentation by a flat smooth rigid punch*

The problem of punch indentation is a classical one in plasticity. Because it is a problem of initial motion, rather than of fully developed flow, it is not a very realistic application of the flow theory formulated in Section 4. However, it has some interesting features which justify its consideration.

A possible plastic stress distribution for a material which satisfies the Coulomb–Mohr condition was given by Prandtl (1920); an alternative plastic stress field was determined by Shield (1953). The latter is the analogue of Hill's (1949) solution for an ideal metal. The corresponding velocity field, which we give below, was obtained by Spencer (1964). A similar solution can be obtained for the velocity field associated with Prandtl's stress field.

The configuration is illustrated in Fig. 5.1. The material occupies the region $y \leqslant 0$ and the punch is in contact with the surface $y = 0$ over the segment $A'A$, of width $2a$. Since the configuration is symmetrical about $x = 0$, only the region $x \geqslant 0$ is considered. The boundary conditions are

$$\begin{aligned} \sigma_{xy} &= 0, \quad \dot{\sigma}_{xy} = 0 \quad &&\text{on} \quad y = 0, \quad 0 \leqslant x < \infty, \\ \sigma_{yy} &= 0, \quad \dot{\sigma}_{yy} = 0 \quad &&\text{on} \quad y = 0, \quad a < x < \infty, \\ v_y &= -V \quad &&\text{on} \quad y = 0, \quad 0 \leqslant x < a, \end{aligned} \tag{5.1}$$

where V is the downward speed of the punch and superposed dots denote material derivatives. In Fig. 5.1, $ODCB$ is an α-line, AD and AC are β-lines. In OAD and ABC the α- and β-lines are parallel straight lines. In ADC the β-lines are straight lines through A and the α-lines are logarithmic spirals $r = r_0 \exp(\theta \tan\phi)$, where r_0 is constant on an α-line and r, θ are the polar coordinates illustrated in Fig. 5.1. Then the stress field is given by

$$\left.\begin{aligned} p &= c\cot\phi\{(1-\sin\phi)^{-1}\exp(\pi\tan\phi) - 1\}, \\ \psi &= 0, \end{aligned}\right\} \text{in } OAD, \tag{5.2}$$

MOS - U*

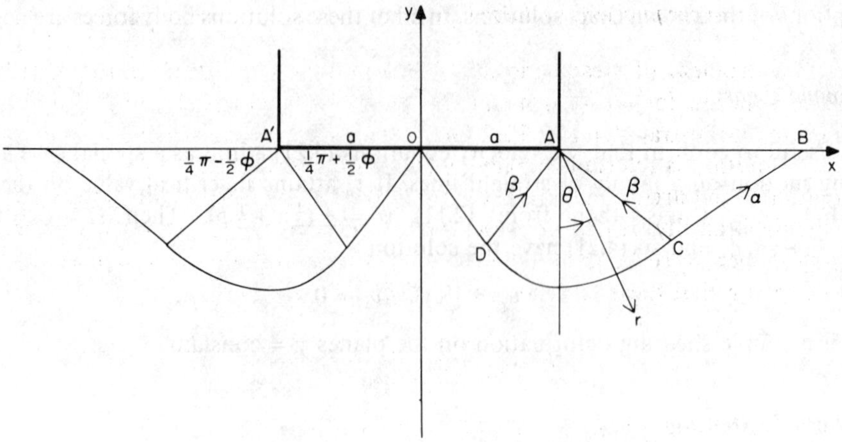

FIG. 5.1. Slipline field for indentation by a flat smooth rigid punch.

$$\left.\begin{aligned} p &= c\cot\phi[(1-\sin\phi)^{-1}\exp\{(\tfrac{1}{2}\pi+\phi-2\theta)\tan\phi\}-1], \\ \psi &= \theta+\tfrac{1}{4}\pi-\tfrac{1}{2}\phi, \end{aligned}\right\}\text{in } ACD, \tag{5.3}$$

$$\left.\begin{aligned} p &= c(1-\sin\phi)^{-1}\cos\phi, \\ \psi &= \tfrac{1}{2}\pi, \end{aligned}\right\}\text{in } ABC. \tag{5.4}$$

The indentation pressure on OA is of magnitude

$$c\cot\phi\{\tan^2(\tfrac{1}{4}\pi+\tfrac{1}{2}\phi)\exp(\pi\tan\phi)-1\}.$$

We seek solutions in which $v_\beta = 0$ throughout $OABCD$. This value satisfies the continuity condition on $ODCB$ and is suggested by the solution of the analogous problem for an ideal metal (Hill, 1949). Then the velocity equation $(4.24)_2$ gives

$$\Omega = \dot{\psi} = 0, \quad \text{in } OABCD, \tag{5.5}$$

and the solution of $(4.24)_1$, which satisfies the boundary and continuity conditions, is

$$v_\alpha = \begin{cases} V\sec(\tfrac{1}{4}\pi-\tfrac{1}{2}\phi), & \text{in } OAD, \\ V\sec(\tfrac{1}{4}\pi-\tfrac{1}{2}\phi)\exp\{-(\theta+\tfrac{1}{4}\pi-\tfrac{1}{2}\phi)\tan\phi\}, & \text{in } ACD, \\ V\sec(\tfrac{1}{4}\pi-\tfrac{1}{2}\phi)\exp(-\tfrac{1}{2}\pi\tan\phi), & \text{in } ABC. \end{cases} \tag{5.6}$$

It remains to consider the stress-rate equilibrium equations (4.34). It can be verified that when $\Omega = 0$, v_α is given by (5.6), and the body forces are zero, then equations (4.34) and the traction-rate boundary conditions are satisfied by

$$\dot{p} = p_t + v_\alpha \partial p/\partial s_\alpha + v_\beta \partial p/\partial s_\beta = 0. \tag{5.7}$$

The condition $(4.35)_2$ is also satisfied across the β-lines AC and AD, and so the solution in $OABCD$ satisfies all the requirements. The solution is incomplete in the sense that it has not been shown that it can be extended into the rigid region outside $OABCD$.

5.4. *Flow in a converging wedge-shaped channel*

This is a problem of steady motion. The solution is due to Pemberton (1965). The analogous solution for an ideal metal was given by Nádai (1924) and Hill (1950). The configuration is illustrated in Fig. 5.2. Material flows in a wedge-shaped channel with rough walls. The solution is referred to plane polar coordinates r, θ.

A solution of the stress equations is given in Sokolovskii (1965). We denote by ψ_1 the angle which the algebraically greatest principal axis of stress makes with the radial direction, so that $\psi = \psi_1 + \theta$. Then, by seeking solutions of the stress equations in which ψ_1 depends only on θ, it can be shown that these equations are satisfied if

$$\theta = \int_0^{\psi_1} \frac{\cos 2\psi_1 + \sin\phi}{C - \cos 2\psi_1}\, d\psi_1,$$

where C is a constant and $C > 1$. It then follows that

$$\left.\begin{matrix}\sigma_r \\ \sigma_\theta\end{matrix}\right\} = -Ar^n\left(\frac{1 \mp \sin\phi\cos 2\psi_1}{\sin\phi}\right)(C - \cos 2\psi_1)^{\frac{1}{2}n} + c\cot\phi,$$

$$\tau_{r\theta} = Ar^n \sin 2\psi_1 (C - \cos 2\psi_1)^{\frac{1}{2}n},$$

where

$$n = 2\sin\phi\,(C + \sin\phi)\sec^2\phi,$$

A is a constant of integration, and σ_r, σ_θ, $\tau_{r\theta}$ denote stress components referred to (r, θ) coordinates. The constant A can be expressed in terms of C, the channel geometry, the material constants c and ϕ, and the pressure across a specified section of the channel. Then C is determined by the conditions at the channel wall. Pemberton (1965) considers the cases of (a) a cohesionless material with Coulomb sliding friction at the walls, and (b) a cohesive or cohesionless material with perfectly rough walls.

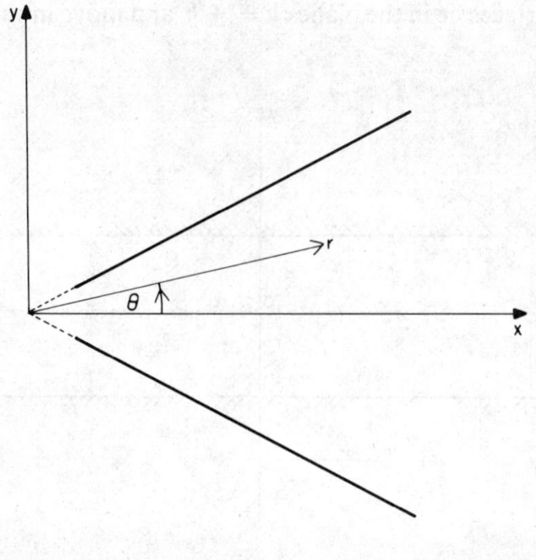

FIG. 5.2. Flow in a converging wedge-shaped channel.

The velocity equations (4.21) can be expressed in terms of (r, θ) coordinates as

$$\frac{\partial v_r}{\partial r}+\frac{v_r}{r}+\frac{1}{r}\frac{\partial v_\theta}{\partial \theta}=0,$$

$$\left(\frac{1}{r}\frac{\partial v_r}{\partial \theta}+\frac{\partial v_\theta}{\partial r}-\frac{v_\theta}{r}\right)\cos 2\psi_1-\left(\frac{\partial v_r}{\partial r}-\frac{v_r}{r}-\frac{1}{r}\frac{\partial v_\theta}{\partial \theta}\right)\sin 2\psi_1 \tag{5.9}$$

$$+\sin\phi\left(\frac{1}{r}\frac{\partial v_r}{\partial \theta}-\frac{\partial v_\theta}{\partial r}-\frac{v_\theta}{r}+2\Omega_1\right)=0,$$

where $\Omega_1 = \dot{\psi}_1$. When ψ_1 is given by (5.8) and $v_\theta = 0$, then $\Omega_1 = 0$. We then seek solutions of (5.9) with $v_\theta = 0$. This gives the velocity solution

$$v_r = -\frac{B}{r(C-\cos 2\psi_1)}, \tag{5.10}$$

where $B > 0$ is a constant of integration which can be related to the flow rate through the channel.

5.5. *Compression between rough parallel plates*

This solution is due to Marshall (1967). The problem is not one of steady motion but is one in which geometrical similarity is preserved (for a general discussion of such problems in metal plasticity, see Hill, 1950, Chapter 8). The solution is the analogue of Prandtl's "cycloid" solution in metal plasticity, for which the stress solution was given by Prandtl (1923) and the velocity solution by Nádai (see Hill, 1950, pp. 233–234). In metal plasticity this solution gives, to a good approximation, the stress in a block compressed between rough plates, except near the ends of the block. A numerical solution which incorporates end effects was obtained by Hill *et al.* (1951).

The configuration is illustrated in Fig. 5.3. A block of material is compressed between rough plates whose surfaces lie in the planes $y = \pm h$, and move inwards with speed U. For a

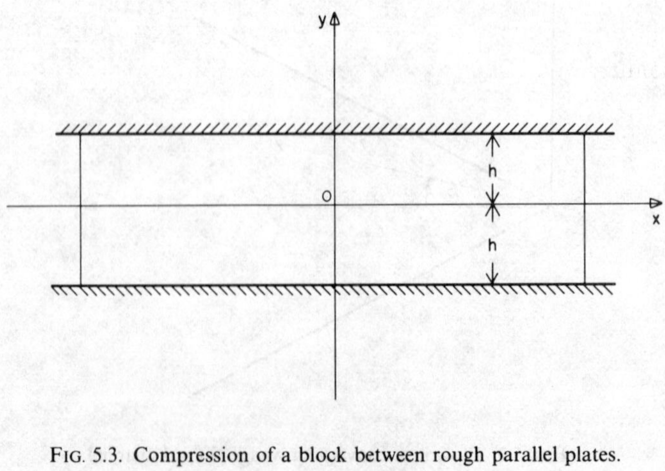

FIG. 5.3. Compression of a block between rough parallel plates.

material which satisfies the Coulomb–Mohr condition a stress solution was given by Hartmann (1925). By seeking solutions of the stress equations in which ψ depends only on y, one obtains the solutions

$$\frac{y}{h} = \frac{\sin 2\psi + 2\psi \sin \phi}{C},$$
$$q = A \exp\left\{\frac{\sin \phi}{\cos^2 \phi}\left(\frac{Cx}{h} - \cos 2\psi\right)\right\}, \tag{5.11}$$

where A and C are constants of integration. The α- and β-lines are the two families of curves whose parametric equations are

$$\begin{aligned} Cy &= h(\sin 2\psi + 2\psi \sin \phi), \\ Cx &= h(\cos 2\psi \mp 2\psi \cos \phi). \end{aligned} \tag{5.12}$$

The constant A is determined by the traction in the x-direction on the ends of the block. The value of C is then found by using the friction conditions at the interfaces between the block and the plates. Marshall (1967) determines C in the two cases of (a) a cohesive material compressed between perfectly rough plates, and (b) a cohesionless material with Coulomb sliding friction at the interfaces.

When ψ is given implicitly by (5.11), Marshall showed that the velocity equations (4.21) have the solution

$$\frac{v_x}{U} = \frac{x}{h} - \frac{2}{C}\cos 2\psi + B, \quad \frac{v_y}{U} = -\frac{y}{h}, \tag{5.13}$$

where $2h$ denotes the current thickness of the block and $U = -dh/dt$. B is a constant of integration which is related to the flow rate in the x-direction at a given section (say $x = 0$). Marshall also observed that $\Omega = 0$ in this solution.

5.6. *Discussion*

It is noteworthy, and possibly significant, that in all the known solutions based on the theory of Section 4 the rate of rotation Ω of the principal axes of stress at a particle is zero. It would be of interest to find a solution in which this is not the case.

All of the solutions described above are analogues of well-known solutions in metal plasticity. It is also of interest to note that there are other well-known solutions in metal plasticity which do not appear to have analogous solutions in the theory developed in Section 4. One such problem is the geometrically similar problem of indentation of a half-space by a smooth rigid wedge. Dais (1967a) has presented a solution which satisfies the stress equations of Section 2 and the velocity equations (4.24), but (Dais, 1967b) this solution is not compatible with the stress-rate equations (4.34). Another problem (to which my attention has been drawn by Drescher, 1978) in which difficulty arises is that of extrusion through a smooth square die. For certain special reductions of area the stress equations are satisfied by a slipline field which in the deforming region consists of straight lines and logarithmic spirals. When $\phi = 0$ the spirals become circular arcs and there exists an associated velocity field which satisfies the boundary conditions. However, when $\phi \neq 0$ the slipline field does not seem to possess an associated velocity field for which the velocity boundary conditions are all satisfied. There may be some other slipline field for which both stress and velocity equations and boundary conditions are obeyed, but no such field is known.

6. Further Development of the Plane Strain Theory

In this Section we give alternative forms for the velocity equations (4.20), and use these to derive some new general results.

From (2.6) we have

$$\tan 2\psi = \frac{2\sigma_{xy}}{\sigma_{xx}-\sigma_{yy}}, \tag{6.1}$$

and it follows from (2.6) and (2.7) that

$$\Omega = \dot{\psi} = \frac{(\sigma_{xx}-\sigma_{yy})\dot{\sigma}_{xy}-\sigma_{xy}(\dot{\sigma}_{xx}-\dot{\sigma}_{yy})}{4q^2}, \tag{6.2}$$

where, as before, the superposed dots denote material times derivatives. Also, from (2.6), equation $(4.20)_2$ may be written as

$$(\sigma_{xx}-\sigma_{yy})d_{xy}-\sigma_{xy}(d_{xx}-d_{yy})+2q(\omega_{xy}+\Omega)\sin\phi = 0. \tag{6.3}$$

Hence, from (6.2) and (6.3),

$$(\sigma_{xx}-\sigma_{yy})\{2qd_{xy}+\dot{\sigma}_{xy}\sin\phi\}-\sigma_{xy}\{2q(d_{xx}-d_{yy})+(\dot{\sigma}_{xx}-\dot{\sigma}_{yy})\sin\phi\}+4q^2\omega_{xy}\sin\phi = 0,$$

or, using (2.7),

$$(\sigma_{xx}-\sigma_{yy})[2qd_{xy}+\{\dot{\sigma}_{xy}+(\sigma_{xx}-\sigma_{yy})\omega_{xy}\}\sin\phi]-\sigma_{xy}[2q(d_{xx}-d_{yy})+\{(\dot{\sigma}_{xx}-\dot{\sigma}_{yy}) - 4\sigma_{xy}\omega_{xy}\}\sin\phi] = 0.$$

This equation is rearranged in the form

$$\frac{d_{xx}-d_{yy}+\frac{1}{2}q^{-1}\sin\phi\{(\dot{\sigma}_{xx}-\dot{\sigma}_{yy})-4\sigma_{xy}\omega_{xy}\}}{\sigma_{xx}-\sigma_{yy}} = \frac{d_{xy}+\frac{1}{2}q^{-1}\sin\phi\{\dot{\sigma}_{xy}+(\sigma_{xx}-\sigma_{yy})\omega_{xy}\}}{\sigma_{xy}} \tag{6.4}$$

If we now set each side of (6.4) equal to a multiplier λ, we obtain

$$\begin{aligned} d_{xx}-d_{yy} &= \dot{\lambda}(\sigma_{xx}-\sigma_{yy})-\tfrac{1}{2}q^{-1}\sin\phi\{(\dot{\sigma}_{xx}-\dot{\sigma}_{yy})-4\sigma_{xy}\omega_{xy}\}, \\ d_{xy} &= \dot{\lambda}\sigma_{xy}-\tfrac{1}{2}q^{-1}\sin\phi\{\dot{\sigma}_{xy}+(\sigma_{xx}-\sigma_{yy})\omega_{xy}\}. \end{aligned} \tag{6.5}$$

Introducing the incompressibility equation $(4.20)_1$ then gives

$$\begin{aligned} d_{xx} &= \tfrac{1}{2}\dot{\lambda}(\sigma_{xx}-\sigma_{yy})-\tfrac{1}{4}q^{-1}\sin\phi\{(\dot{\sigma}_{xx}-\dot{\sigma}_{yy})+4\sigma_{xy}\omega_{xy}\}, \\ d_{yy} &= \tfrac{1}{2}\dot{\lambda}(\sigma_{yy}-\sigma_{xx})-\tfrac{1}{4}q^{-1}\sin\phi\{(\dot{\sigma}_{yy}-\dot{\sigma}_{xx})-4\sigma_{xy}\omega_{xy}\}, \\ d_{xy} &= \dot{\lambda}\sigma_{xy}-\tfrac{1}{2}q^{-1}\sin\phi\{\dot{\sigma}_{xy}+(\sigma_{xx}-\sigma_{yy})\omega_{xy}\}. \end{aligned} \tag{6.6}$$

Equation (6.6) can be expressed concisely using indicial notation. We now denote

$$d_{xx} = d_{11}, \quad d_{yy} = d_{22}, \quad d_{xy} = d_{12},$$
$$\omega_{xy} = \omega_{12} = -\omega_{21}, \tag{6.7}$$
$$\sigma_{xx} = \sigma_{11}, \quad \sigma_{yy} = \sigma_{22}, \quad \sigma_{xy} = \sigma_{12}.$$

It is also convenient to introduce the deviatoric stress components s_{ij}, defined by

$$s_{ij} = \sigma_{ij} - \tfrac{1}{2}\sigma_{kk}\delta_{ij}, \tag{6.8}$$

where δ_{ij} denotes the Kronecker delta, and the usual convention regarding summation on repeated indices is assumed, with indices here taking the values 1, 2. Then

$$s_{ii} = 0, \quad q^2 = \tfrac{1}{2}s_{ij}s_{ij}, \tag{6.9}$$

and

$$s_{11} = -s_{22} = \tfrac{1}{2}(\sigma_{11} - \sigma_{22}) = \tfrac{1}{2}(\sigma_{xx} - \sigma_{yy}), \quad s_{12} = \sigma_{12} = \sigma_{xy}. \tag{6.10}$$

Equations (6.6) can then be written as

$$d_{ij} = \dot{\lambda}s_{ij} - \tfrac{1}{2}q^{-1}\sin\phi\{\dot{s}_{ij} + s_{ip}\omega_{pj} + s_{jp}\omega_{pi}\}. \tag{6.11}$$

The derivative

$$\frac{\mathscr{D}s_{ij}}{\mathscr{D}t} = \dot{s}_{ij} + s_{ip}\omega_{pj} + s_{jp}\omega_{pi}, \tag{6.12}$$

is the "Jaumann" or "co-rotational" time derivative of s_{ij}, which (see, for example, Green and Adkins, 1960) has the property that it is independent of superposed rigid-body rotations, and so is suitable for use in properly invariant constitutive equations which involve the stress-rate. Hence (6.11) becomes

$$d_{ij} = \dot{\lambda}s_{ij} - \frac{\sin\phi}{2q}\frac{\mathscr{D}s_{ij}}{\mathscr{D}t}. \tag{6.13}$$

This is an alternative form of the velocity equations (4.20). It reduces to the flow rule for plane flow of a frictionless perfectly plastic isotropic material, namely

$$d_{ij} = \dot{\lambda}s_{ij}$$

when the angle of internal friction ϕ is zero.

It can be verified that

$$\frac{\mathscr{D}}{\mathscr{D}t}\left(\frac{s_{ij}}{q}\right) = \frac{1}{q}\frac{\mathscr{D}s_{ij}}{\mathscr{D}t} - \frac{\dot{q}}{q^2}s_{ij}. \tag{6.14}$$

Hence (6.13) may be expressed in the alternative form

$$d_{ij} = \dot{\mu}\left(\frac{s_{ij}}{q}\right) - \tfrac{1}{2}\sin\phi\frac{\mathscr{D}}{\mathscr{D}t}\left(\frac{s_{ij}}{q}\right), \tag{6.15}$$

where $\dot{\mu} = q\dot{\lambda} - \tfrac{1}{2}\dot{q}q^{-1}\sin\phi$ is also an arbitrary multiplier.

When the velocity equations are expressed in either of the forms (6.13) or (6.15) it is readily verified that they satisfy the condition for isotropy and the requirement of invariance under rigid-body rotations.

6.1. *Decomposition of* d_{ij}

Equation (6.15) decomposes the rate-of-deformation into two parts. We have

$$d_{ij} = d_{ij}^{(1)} + d_{ij}^{(2)}, \tag{6.16}$$

where

$$d_{ij}^{(1)} = \dot{\mu}\left(\frac{s_{ij}}{q}\right), \qquad d_{ij}^{(2)} = -\tfrac{1}{2}\sin\phi\,\frac{\mathscr{D}}{\mathscr{D}t}\left(\frac{s_{ij}}{q}\right). \tag{6.17}$$

Then

$$\begin{aligned} d_{ij}^{(1)} d_{ij}^{(2)} &= -\tfrac{1}{2}\dot{\mu}\sin\phi\frac{s_{ij}}{q}\frac{\mathscr{D}}{\mathscr{D}t}\left(\frac{s_{ij}}{q}\right) \\ &= -\tfrac{1}{2}\dot{\mu}\sin\phi\frac{s_{ij}}{q}\left\{\frac{D}{Dt}\left(\frac{s_{ij}}{q}\right) + s_{ip}\omega_{pj} + s_{jp}\omega_{pi}\right\}, \end{aligned}$$

where D/Dt denotes the convected derivative. However, since $s_{ij} = s_{ji}$, $\omega_{ij} = -\omega_{ji}$, it follows that

$$s_{ij}s_{ip}\omega_{pj} = 0, \quad s_{ij}s_{jp}\omega_{pi} = 0,$$

and hence

$$d_{ij}^{(1)} d_{ij}^{(2)} = -\tfrac{1}{2}\dot{\mu}\sin\phi\frac{s_{ij}}{q}\frac{D}{Dt}\left(\frac{s_{ij}}{q}\right) = -\tfrac{1}{4}\dot{\mu}\sin\phi\frac{D}{Dt}\left(\frac{s_{ij}s_{ij}}{q^2}\right) = 0, \tag{6.18}$$

because $s_{ij}s_{ij} = 2q^2$. Hence (6.16) decomposes d_{ij} into two parts, which are orthogonal in the sense that the trace of their tensor product is zero.

The deformation-rate components $d_{ij}^{(1)}$ and $d_{ij}^{(2)}$ can be interpreted geometrically in the following way. Suppose that $p = -\frac{1}{2}(\sigma_{11}+\sigma_{22})$, $\frac{1}{2}(s_{11}-s_{22}) = \frac{1}{2}(\sigma_{11}-\sigma_{22})$, and $s_{12} = \sigma_{12}$ are chosen as axes of a three-dimensional rectangular cartesian coordinate system. Then the Coulomb–Mohr yield condition (2.14) is represented by a cone in this space, as illustrated in Fig. 6.1. Since $d_{11}+d_{22} = 0$, the vector $\{-\frac{1}{2}(d_{11}+d_{22}), \frac{1}{2}(d_{11}-d_{22}), d_{12}\}$ always lies in a

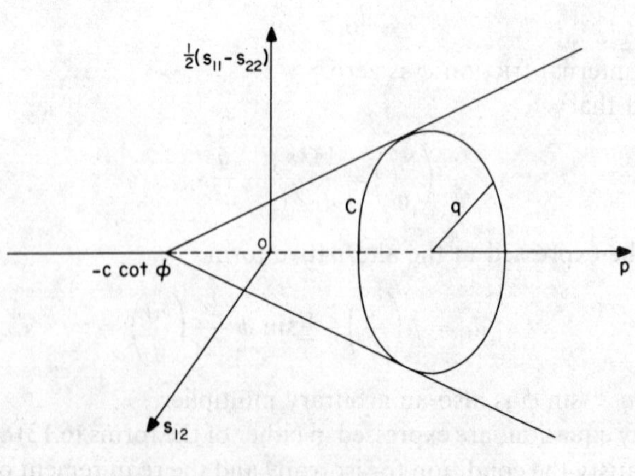

FIG. 6.1. Conical yield surface in (s_{ij}, p)-space.

plane $p =$ constant in this space. Such a plane is shown in Fig. 6.2; it intersects the Coulomb–Mohr surface in a circle C of radius q. Because $d_{ij}^{(1)}$ is proportional to s_{ij}, the vector $\{-\frac{1}{2}(d_{11}^{(1)}+d_{22}^{(1)}),\ \frac{1}{2}(d_{11}^{(1)}-d_{22}^{(1)}),\ d_{12}^{(1)}\}$ is directed along the radius to the point $\{\frac{1}{2}(s_{11}-s_{22}),\ s_{12}\}$ in this plane, i.e. along the normal to C. The vector $\{-\frac{1}{2}(d_{11}^{(2)}+d_{22}^{(2)}),\ \frac{1}{2}(d_{11}^{(2)}-d_{22}^{(2)}),\ d_{12}^{(2)}\}$ is directed along the tangent to C at the point $\{\frac{1}{2}(s_{11}-s_{22}),\ s_{12}\}$.

The part $d_{ij}^{(1)}$ of the deformation-rate is of the same form as the deformation-rate which occurs in isotropic plane strain metal plasticity. Since

$$d_{ij}^{(1)} = 2\dot{\mu}\frac{\partial q}{\partial \sigma_{ij}},$$

the stress invariant q may be regarded as plastic potential for $d_{ij}^{(1)}$, and $d_{ij}^{(1)}$ is given by a relation resembling an associated flow rule. We note, however, that $d_{ij}^{(1)}$ is not directed along the normal to the yield surface, but is constrained to lie in a plane $p =$ constant. The property that $d_{ij}^{(1)}$ can be derived from a potential suggests the possibility of various generalizations to anisotropic and dilatant materials, but such ideas are speculative and we shall not pursue them.

The part $d_{ij}^{(2)}$ of the deformation-rate has no equivalent in metal plasticity. It is completely determined by the current stress and stress-rate, and contains no arbitrary multiplier.

6.2. *Plastic work-rate*

The rate-of-working $\dot{W}$ of the stress, per unit volume, in an incompressible material is $s_{ij}d_{ij}$. From (6.18) it follows that

$$s_{ij}d_{ij}^{(2)} = 0. \tag{6.19}$$

Thus only $d_{ij}^{(1)}$ contributes to the plastic work-rate, and, from (6.9) and (6.17),

$$\dot{W} = s_{ij}d_{ij}^{(1)} = \dot{\mu}q^{-1}s_{ij}s_{ij} = 2\dot{\mu}q. \tag{6.20}$$

Since q is positive, it follows that if the rate-of-plastic working is to be positive, then $\dot{\mu} > 0$. Henceforth we assume this to be the case.

The natural decomposition of d_{ij} into two parts, one of which makes no contribution to the

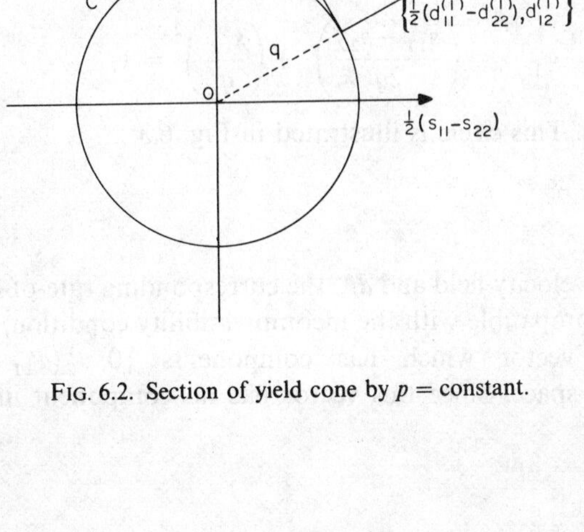

FIG. 6.2. Section of yield cone by $p =$ constant.

energy dissipation, seems to be an unusual situation in continuum mechanics. However, several authors (Brulin and Hjalmars, 1970, 1972; Edelen, 1977) have considered problems in which the stress is the sum of two parts, one of which does no work in any deformation.

6.3. *Material stability*

Various criteria for material stability have been proposed. One such criterion proposes that in any deformation of an incompressible material

$$d_{ij}\frac{\mathscr{D}s_{ij}}{\mathscr{D}t} \geqslant 0. \tag{6.21}$$

From (6.12) it can be shown that, since $d_{ij} = d_{ji}$, $s_{ij} = s_{ji}$ and $\omega_{ij} = -\omega_{ji}$, then $d_{ij}\mathscr{D}s_{ij}/\mathscr{D}t = d_{ij}\dot{s}_{ij}$, and so (6.21) is equivalent to $d_{ij}\dot{s}_{ij} \geqslant 0$. Now, from (6.14) and (6.15),

$$d_{ij}\frac{\mathscr{D}s_{ij}}{\mathscr{D}t} = \left\{\dot{\mu}\frac{s_{ij}}{q} - \tfrac{1}{2}\sin\phi\,\frac{\mathscr{D}}{\mathscr{D}t}\left(\frac{s_{ij}}{q}\right)\right\}\left\{q\frac{\mathscr{D}}{\mathscr{D}t}\left(\frac{s_{ij}}{q}\right) + \frac{\dot{q}}{q}s_{ij}\right\}.$$

However, it was shown above that $s_{ij}\,\mathscr{D}(s_{ij}/q)/\mathscr{D}t = 0$, and so

$$d_{ij}\frac{\mathscr{D}s_{ij}}{\mathscr{D}t} = 2\dot{\mu}\dot{q} - \tfrac{1}{2}q\sin\phi\,\frac{\mathscr{D}}{\mathscr{D}t}\left(\frac{s_{ij}}{q}\right)\frac{\mathscr{D}}{\mathscr{D}t}\left(\frac{s_{ij}}{q}\right) \leqslant 2\dot{\mu}\dot{q} = 2\dot{\mu}\dot{p}\sin\phi.$$

The inequality (6.21) may or may not be satisfied if $\dot{p} > 0$, in which case deformation occurs under increasing pressure, and is not satisfied if $\dot{p} < 0$. It may not be unreasonable that a granular material should be unstable when deforming under decreasing pressure.

6.4. *Inequalities*

In ideal metal plasticity, there exist two fundamental inequalities which form the basis of a number of extremum and limit analysis theorems. We formulate the corresponding inequalities for the theory developed here, although in the present case the inequalities do not seem to have such far-reaching consequences.

We first recall that a section of the Coulomb–Mohr yield surface (Fig. 6.2) by a plane p = constant is a circle of radius q, where $q = c + p\sin\phi$. It is convenient to introduce scaled coordinates $\tfrac{1}{2}(s_{11} - s_{22})/q$ and s_{12}/q in this plane, so that points on the yield surface are represented by points on the unit circle

$$\left(\frac{s_{11} - s_{22}}{2q}\right)^2 + \left(\frac{s_{12}}{q}\right)^2 = 1, \tag{6.22}$$

where $q = c + p\sin\phi$. This circle is illustrated in Fig. 6.3.

6.5. *First inequality*

Let $\mathbf{v}^*$ be a virtual velocity field and d^*_{ij} the corresponding rate-of-deformation. Suppose this deformation is compatible with the incompressibility condition, so that $d^*_{11} + d^*_{22} = 0$. We introduce the vector which has components $\{0,\ \tfrac{1}{2}(d^*_{11} - d^*_{22}),\ d^*_{12}\}$ in the $\{p,\ \tfrac{1}{2}(s_{11} - s_{22}),\ s_{12}\}$-space. Since this vector has no component in the p-direction, it is

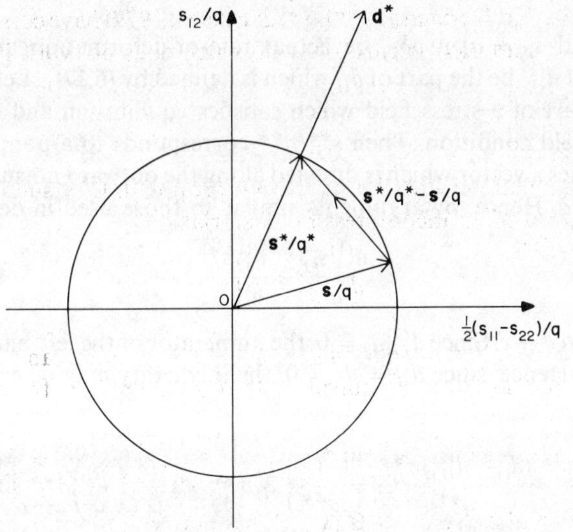

FIG. 6.3. Unit circle illustrating proof of the first inequality.

nowhere (for $\phi > 0$) normal to the yield surface, and so the usual normality and convexity arguments cannot be applied. However, in two dimensions, the vector $\mathbf{d}^*$ with components $\{\frac{1}{2}(d^*_{11} - d^*_{22}), d^*_{12}\}$ is directed along the outward normal to (6.22) at a unique point with position vector $\mathbf{s}^*/q^*$, where $\mathbf{s}^*$ has components $\{\frac{1}{2}(s^*_{11} - s^*_{22}), s^*_{12}\}$. Thus if $\mathbf{s}$ is the actual deviatoric stress, it follows from the convexity of the circle (6.22) (Fig. 6.3) that

$$\left(\frac{\mathbf{s}^*}{q^*} - \frac{\mathbf{s}}{q}\right).\mathbf{d}^* \geqslant 0, \tag{6.23}$$

where $q^* = c + p^* \sin \phi$, $q = p + c \sin \phi$. In terms of components, (6.23) may be written as

$$\frac{s^*_{ij} d^*_{ij}}{q^*} \geqslant \frac{s_{ij} d^*_{ij}}{q}, \tag{6.24}$$

or, since $s_{ij} = \sigma_{ij} + p\delta_{ij}$, $s^*_{ij} = \sigma^*_{ij} + p^*\delta_{ij}$, and $d^*_{ii} = 0$, as

$$\frac{\sigma^*_{ij} d^*_{ij}}{q^*} \geqslant \frac{\sigma_{ij} d^*_{ij}}{q}. \tag{6.25}$$

In the case of frictionless material, $\phi = 0$, and $q^* = q = c$, so that (6.25) reduces to

$$\sigma^*_{ij} d^*_{ij} \geqslant \sigma_{ij} d^*_{ij}. \tag{6.26}$$

It is then possible to integrate (6.26) through a volume, transform by the divergence theorem, and so derive a useful global inequality. However, for (6.25) the presence of q^* and q in the denominators means that application of the divergence theorem fails to produce useful results, unless we make additional assumptions which are difficult to justify.

6.6. *Second inequality*

Again let d_{ij} and $s_{ij} = \sigma_{ij} + p\delta_{ij}$ be actual rate-of-deformation and deviatoric stress components, and let $d_{ij}^{(1)}$ be the part of d_{ij} which is defined by $(6.17)_1$. Let $s_{ij}^{**} = \sigma_{ij}^{**} + p^{**}\delta_{ij}$ be the deviatoric part of a stress field which satisfies equilibrium and does not violate the Coulomb–Mohr yield condition. Then s_{ij}^{**}/q^{**} corresponds to a point in or on the circle (6.22), and $d_{ij}^{(1)}$ defines a vector which is directed along the outward normal at the point which corresponds to s_{ij}/q. Hence, by arguments similar to those used in deriving (6.24),

$$\frac{d_{ij}^{(1)} s_{ij}}{q} \geqslant \frac{d_{ij}^{(1)} s_{ij}^{**}}{q^{**}}, \tag{6.27}$$

where $q^{**} = p^{**} \sin\phi + c$. Since $d_{ij}^{(2)} s_{ij} = 0$, the numerator of the left-hand side of (6.27) may be written as $d_{ij}s_{ij}$. Hence, since $d_{ii}^{(1)} = d_{ii} = 0$, the inequality may be expressed in the form

$$\frac{d_{ij}\sigma_{ij}}{q} \geqslant \frac{(d_{ij} - d_{ij}^{(2)})\sigma_{ij}^{**}}{q^{**}} = \frac{1}{q^{**}}\left\{d_{ij}\sigma_{ij}^{**} + \tfrac{1}{2}\frac{\mathscr{D}}{\mathscr{D}t}\left(\frac{s_{ij}}{q}\right)\sigma_{ij}^{**} \sin\phi\right\}. \tag{6.28}$$

When $\phi = 0$, (6.28) reduces to $d_{ij}\sigma_{ij} \geqslant d_{ij}\sigma_{ij}^{**}$, and the usual transformations can be applied. For $\phi \neq 0$, it is not clear how (6.28) may be exploited.

Limit analysis theorems for materials with non-associated flow rules have been described by Palmer (1966), de Josselin de Jong (1964), and Collins (1973). However, the lower bound theorems in particular are considerably weaker than the corresponding theorems for materials with an associated flow rule.

7. Three-dimensional Theory for the Double-shearing Model

No generalization of the double-shearing model for plane deformations to three-dimensional states of stress and deformation has previously been published. Shield (1955a) has correctly formulated the yield condition in three dimensions and the appropriate associated flow rule. Cox *et al.* (1961) applied Shield's theory in an exhaustive study of axially symmetric deformations. In this Section we follow Shield's approach but discard the associated flow rule in favour of the double-shearing model.

7.1. *Yield condition*

As in Section 2 we denote the principal components of stress by σ_{I}, σ_{II}, and σ_{III}, and order them so that

$$\sigma_{\mathrm{I}} \geqslant \sigma_{\mathrm{II}} \geqslant \sigma_{\mathrm{III}}. \tag{7.1}$$

Then Shield's formulation of Coulomb's yield condition states that flow may occur when the shearing stress τ_n on any surface attains the critical value $c + \sigma_n \tan\phi$. In terms of the principal stress components the condition is

$$\sigma_{\mathrm{I}} - \sigma_{\mathrm{III}} \leqslant 2c\cos\phi - (\sigma_{\mathrm{I}} + \sigma_{\mathrm{III}})\sin\phi, \tag{7.2}$$

so that yielding is not influenced by the intermediate principal stress component. If $\sigma_1, \sigma_2, \sigma_3$ are the principal stress components in arbitrary order, then in principal stress space the yield surface is a right hexagonal pyramid equally inclined to the σ_1-, σ_2-, σ_3-axes, with its vertex at the point $\sigma_1 = \sigma_2 = \sigma_3 = c \cot \phi$, as illustrated in Fig. 7.1. The yield condition (7.2) is a generalization of Tresca's yield condition for an ideal metal, and reduces to Tresca's condition when $\phi = 0$.

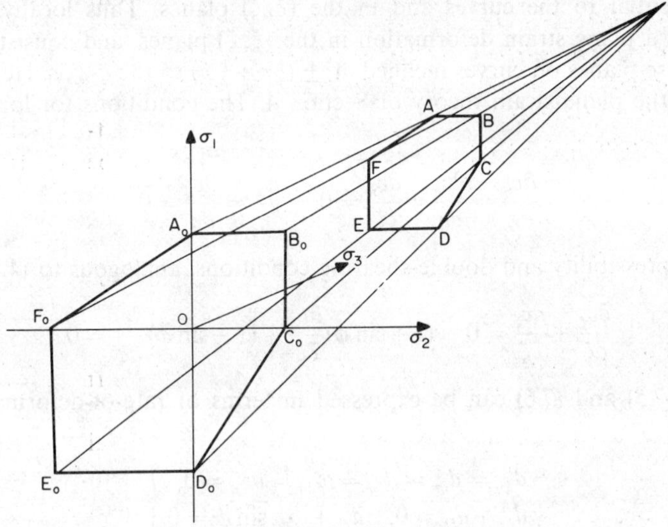

FIG. 7.1. Yield surface in principal stress space.

7.2. *Flow rule*

Following and extending the procedure described in Section 4, we select a generic particle P and introduce a new rectangular cartesian coordinate system ξ, η, ζ whose origin translates with the particle P. The directions of the ξ-, η-, and ζ-axes coincide with the principal axes of stress at P, and the axes rotate with the principal stress axes. If two of the principal stress components are equal, so that the principal axes of stress are not uniquely defined, the coordinate axes are chosen to coincide with any orthogonal set of principal stress axes.

We now consider the motion relative to the (ξ, η, ζ) coordinate system and refer all quantities to this system. Velocity components relative to the moving (ξ, η, ζ) frame of reference are denoted by v_ξ, v_η, v_ζ. The rate-of-deformation components $d_{\xi\xi}, d_{\xi\eta}$, etc., and the spin components $\omega_{\xi\eta}$, etc., relative to the (ξ, η, ζ) frame, are defined by

$$d_{\xi\xi} = \frac{\partial v_\xi}{\partial \xi}, \quad d_{\xi\eta} = \tfrac{1}{2}\left(\frac{\partial v_\xi}{\partial \eta} + \frac{\partial v_\eta}{\partial \xi}\right), \quad \text{etc.,} \quad \omega_{\xi\eta} = \tfrac{1}{2}\left(\frac{\partial v_\xi}{\partial \eta} - \frac{\partial v_\eta}{\partial \xi}\right), \quad \text{etc.} \tag{7.3}$$

Case I

We first consider the case in which the strict inequalities

$$\sigma_{\mathrm{I}} > \sigma_{\mathrm{II}} > \sigma_{\mathrm{III}} \tag{7.4}$$

hold. This corresponds to a stress point on a face of the yield pyramid of Fig. 7.1. In the neighbourhood of P the maximum and minimum principal stress directions coincide with the directions of the ξ- and ζ-axes, and the intermediate principal stress direction coincides with the direction of the η-axis. The maximum value of $\tau_n - \sigma_n \tan\phi$ is generated in the (ξ, ζ) planes on curves which are inclined at $\pm(\frac{1}{4}\pi + \frac{1}{2}\phi)$ to the ξ-axis. As in Sections 2 and 4, we interpret Coulomb's theory to imply that deformation occurs by double-shearing on these curves. The direction of shear is the direction of the shear traction on the curves, which is the direction tangential to the curves and in the (ξ, ζ) planes. Thus locally the proposed deformation is a plane strain deformation in the (ξ, ζ) planes, and consists of a double-shearing in these planes on curves inclined at $\pm(\frac{1}{4}\pi + \frac{1}{2}\phi)$ to the ξ-axis. Hence, locally, we can take over the plane strain theory of Section 4. The conditions for local plane strain give

$$\frac{\partial v_\eta}{\partial \xi} = \frac{\partial v_\eta}{\partial \eta} = \frac{\partial v_\eta}{\partial \zeta} = 0, \quad \frac{\partial v_\xi}{\partial \eta} = \frac{\partial v_\zeta}{\partial \eta} = 0, \tag{7.5}$$

and the incompressibility and double-shearing conditions, analogous to (4.16), are

$$\frac{\partial v_\xi}{\partial \xi} + \frac{\partial v_\zeta}{\partial \zeta} = 0, \quad (1 + \sin\phi)\frac{\partial v_\xi}{\partial \zeta} + (1 - \sin\phi)\frac{\partial v_\zeta}{\partial \xi} = 0. \tag{7.6}$$

Alternatively, (7.5) and (7.6) can be expressed in terms of rate-of-deformation and spin components as

$$\left.\begin{aligned} d_{\xi\eta} = d_{\eta\eta} = d_{\zeta\eta} = \omega_{\xi\eta} = \omega_{\zeta\eta} = 0, \\ d_{\xi\xi} + d_{\zeta\zeta} = 0, \quad d_{\xi\zeta} + \omega_{\xi\zeta} \sin\phi = 0. \end{aligned}\right\} \tag{7.7}$$

Although the case (7.4) in which the principal stress components are distinct may appear to be the most likely case to arise, it is in fact rather restrictive because it compels the local deformation to be plane strain. Therefore we also expect importance to attach to stress states in which two of the principal stress components are equal. These will now be considered.

Case II

We now consider that

$$\sigma_{\text{I}} = \sigma_{\text{II}} > \sigma_{\text{III}}, \tag{7.8}$$

which corresponds to a stress point on an edge of the pyramidal yield surface.

In metal plasticity (and in soil plasticity when the associated flow rule is adopted) it is usually assumed that the deformation-rate at an edge of the yield surface is a linear combination of the deformation-rates associated with the two faces which meet at that edge, with scalar multipliers whose values are usually restricted by normality and convexity requirements, but are otherwise arbitrary. We adopt a similar procedure and propose that when (7.8) applies, the motion relative to the (ξ, η, ζ) coordinates consists of a superposition of two double-shearing motions. Since the ζ-direction is the direction of the axis of the minimum principal stress component, these double shears take place in the (ξ, ζ) and (η, ζ) planes respectively.

For the double-shearing in the (ξ, ζ) planes we denote the rate-of-deformation and spin components by single primes ($d'_{\xi\xi}$, $\omega'_{\xi\eta}$, etc.) and corresponding quantities for double-shearing in the (η, ζ) planes are denoted by use of double primes ($d''_{\xi\xi}$, $\omega''_{\xi\eta}$, etc.). The resultant rate-of-deformation and spin are thus

$$d_{\xi\xi} = d'_{\xi\xi} + d''_{\xi\xi}, \quad \omega_{\xi\eta} = \omega'_{\xi\eta} + \omega''_{\xi\eta}, \quad \text{etc.} \tag{7.9}$$

For double-shearing in (ξ, ζ) planes we have, from (7.7),

$$\begin{gathered} d'_{\xi\eta} = d'_{\eta\eta} = d'_{\zeta\eta} = \omega'_{\xi\eta} = \omega'_{\zeta\eta} = 0, \\ d'_{\xi\xi} + d'_{\zeta\zeta} = 0, \quad d'_{\xi\zeta} + \omega'_{\xi\zeta} \sin\phi = 0. \end{gathered} \tag{7.10}$$

Similarly, for the double-shearing in (η, ζ) planes,

$$\begin{gathered} d''_{\xi\xi} = d''_{\xi\eta} = d''_{\xi\zeta} = \omega''_{\xi\eta} = \omega''_{\xi\zeta} = 0, \\ d''_{\eta\eta} + d''_{\zeta\zeta} = 0, \quad d''_{\eta\zeta} + \omega''_{\eta\zeta} \sin\phi = 0. \end{gathered} \tag{7.11}$$

From (7.9), (7.10), and (7.11) it follows that

$$\begin{gathered} d_{\xi\eta} = 0, \quad \omega_{\xi\eta} = 0, \quad d_{\xi\xi} + d_{\eta\eta} + d_{\zeta\zeta} = 0, \\ d_{\xi\zeta} + \omega_{\xi\zeta} \sin\phi = 0, \quad d_{\eta\zeta} + \omega_{\eta\zeta} \sin\phi = 0. \end{gathered} \tag{7.12}$$

In terms of components, these equations take the forms

$$\begin{gathered} \frac{\partial v_\xi}{\partial \eta} = 0, \quad \frac{\partial v_\eta}{\partial \xi} = 0, \quad \frac{\partial v_\xi}{\partial \xi} + \frac{\partial v_\eta}{\partial \eta} + \frac{\partial v_\zeta}{\partial \zeta} = 0, \\ (1+\sin\phi)\frac{\partial v_\xi}{\partial \zeta} + (1-\sin\phi)\frac{\partial v_\zeta}{\partial \xi} = 0, \quad (1+\sin\phi)\frac{\partial v_\eta}{\partial \zeta} + (1-\sin\phi)\frac{\partial v_\zeta}{\partial \eta} = 0. \end{gathered} \tag{7.13}$$

We observe that (7.12) and (7.13) are form-invariant under rotations of the coordinate system about the ζ-axis, and so the arbitrary choice of the (ξ, η) axes imposes no real restriction.

Case III

The analysis for the case

$$\sigma_{\text{I}} > \sigma_{\text{II}} = \sigma_{\text{III}} \tag{7.14}$$

is similar to Case II. The equations which correspond to (7.12) and (7.13) are

$$\begin{gathered} d_{\eta\zeta} = 0, \quad \omega_{\eta\zeta} = 0, \quad d_{\xi\xi} + d_{\eta\eta} + d_{\zeta\zeta} = 0, \\ d_{\xi\eta} + \omega_{\xi\eta} \sin\phi = 0, \quad d_{\xi\zeta} + \omega_{\xi\zeta} \sin\phi = 0, \end{gathered} \tag{7.15}$$

$$\begin{gathered} \frac{\partial v_\zeta}{\partial \eta} = 0, \quad \frac{\partial v_\eta}{\partial \zeta} = 0, \quad \frac{\partial v_\xi}{\partial \xi} + \frac{\partial v_\eta}{\partial \eta} + \frac{\partial v_\zeta}{\partial \zeta} = 0, \\ (1-\sin\phi)\frac{\partial v_\eta}{\partial \xi} + (1+\sin\phi)\frac{\partial v_\xi}{\partial \eta} = 0, \quad (1-\sin\phi)\frac{\partial v_\zeta}{\partial \xi} + (1+\sin\phi)\frac{\partial v_\xi}{\partial \zeta} = 0. \end{gathered} \tag{7.16}$$

In all three cases the equations are expressed in terms of the velocity relative to the moving (ξ, η, ζ) frame of reference. They may be expressed in terms of the velocity relative to the fixed (x, y, z) frame by procedures similar to those used in Section 4. The resulting equations will involve the stress (because the (ξ, η, ζ) directions are principal stress directions) and the stress-rate (because the (ξ, η, ζ) frame rotates with the principal stress axes). In each case the equations have to be solved in conjunction with the equations of equilibrium and the yield condition.

We shall not pursue the general theory further at this stage, but in the following two Sections we investigate some special cases of non-planar deformations.

8. Some Non-planar Deformations

8.1. *Uniform compression of a cylinder*

The simplest example of a deformation which involves all three velocity components is the problem of uniform compression (or extension) of a right cylinder. We take the z-axis as parallel to the axis of the cylinder, and x- and y-axes to lie in a normal cross-section. Then if no traction is applied to the lateral surfaces, and body forces are neglected, the stress field which satisfies the yield condition and equilibrium equations is

$$\sigma_{xx} = \sigma_{yy} = \sigma_{xy} = \sigma_{xz} = \sigma_{yz} = 0, \quad \sigma_{zz} = -2c\cos\phi/(1-\sin\phi). \tag{8.1}$$

The x-, y-, and z-axes are principal axes of stress, so this is an example of Case II of Section 7, with

$$\sigma_{\mathrm{I}} = \sigma_{xx} = \sigma_{\mathrm{II}} = \sigma_{yy}, \quad \sigma_{\mathrm{III}} = \sigma_{zz}. \tag{8.2}$$

Hence the theory for Case II applies, with the ξ-, η-, and ζ-axes identified with the x-, y-, and z-axes respectively. Therefore the velocity equations (7.13) become

$$\frac{\partial v_x}{\partial y} = 0, \quad \frac{\partial v_y}{\partial x} = 0, \quad \frac{\partial v_x}{\partial x} + \frac{\partial v_y}{\partial y} + \frac{\partial v_z}{\partial z} = 0,$$

$$(1+\sin\phi)\frac{\partial v_x}{\partial z} + (1-\sin\phi)\frac{\partial v_z}{\partial x} = 0,$$

$$(1+\sin\phi)\frac{\partial v_y}{\partial z} + (1-\sin\phi)\frac{\partial v_z}{\partial y} = 0.$$

These equations are satisfied by the uniform compression solution

$$v_x = \tfrac{1}{2}Dx, \quad v_y = \tfrac{1}{2}Dy, \quad v_z = -Dz, \tag{8.3}$$

where D is constant. The homogeneous deformation (8.3) can be regarded as a superposition of two double-shearing deformations in the (x, z) and (y, z) planes respectively.

8.2. *Radial expansion of a sphere or spherical cavity*

Another simple example is that in which a hollow sphere or a spherical cavity expands (or contracts) uniformly in the radial direction. Referred to spherical polar coordinates r, θ, Φ, the non-zero stress components are σ_r and $\sigma_\theta = \sigma_\Phi$. These are therefore principal stress components. For expansion, $\sigma_\theta = \sigma_\Phi > \sigma_r$, and so again the stress corresponds to Case II, with

$$\sigma_{\mathrm{I}} = \sigma_\theta = \sigma_{\mathrm{II}} = \sigma_\Phi, \quad \sigma_{\mathrm{III}} = \sigma_r, \tag{8.4}$$

and the directions of increasing θ, Φ, r are to be identified with the directions of ξ, η, ζ respectively. The yield condition (7.2) gives

$$\sigma_\theta = \sigma_\Phi = \{2c\cos\phi + \sigma_r(1-\sin\phi)\}/(1+\sin\phi). \tag{8.5}$$

The equation of radial equilibrium is

$$\frac{d\sigma_r}{dr} + \frac{2\sigma_r - \sigma_\theta - \sigma_\Phi}{r} = 0, \tag{8.6}$$

and it readily follows from (8.5) and (8.6) that

$$\sigma_r = Ar^{-n} + c\cot\phi, \quad \sigma_\theta = \sigma_\Phi = A\frac{(1-\sin\phi)}{(1+\sin\phi)}r^{-n} + c\cot\phi, \tag{8.7}$$

where A is arbitrary and $n = 4\sin\phi/(1+\sin\phi)$.

With the appropriate substitutions, (7.12) become

$$\begin{aligned} &d_{\theta\Phi} = 0, \quad \omega_{\theta\Phi} = 0, \quad d_{rr} + d_{\theta\theta} + d_{\Phi\Phi} = 0, \\ &d_{\theta r} + \omega_{\theta r}\sin\phi = 0, \quad d_{\Phi r} + \omega_{\Phi r}\sin\phi = 0, \end{aligned} \tag{8.8}$$

where d_{rr}, etc. denote components of rate-of-deformation, and $\omega_{\theta\Phi}$, etc. denote components of spin, all referred to (r, θ, Φ) coordinates.

For the radially symmetric deformations under consideration, the only non-zero velocity component is the radial component v_r, and this depends only on r. Hence

$$d_{rr} = \frac{dv_r}{dr}, \quad d_{\theta\theta} = d_{\Phi\Phi} = \frac{v_r}{r},$$

$$d_{r\theta} = d_{\theta\Phi} = d_{\Phi r} = \omega_{r\theta} = \omega_{\theta\Phi} = \omega_{\Phi r} = 0,$$

and all of (8.8) are immediately satisfied except the incompressibility condition $(8.8)_3$, which has the solution

$$v_r = Br^{-2},$$

where B is constant. Radial expansion can be regarded as a superposition of two double-shearing motions in the surfaces θ = constant and Φ = constant.

Radially symmetric deformations of ideal soils have been considered, under various constitutive assumptions, by several authors, including Chadwick (1959) and Chadwick *et al.* (1964).

8.3. *Rectilinear flow*

We consider rectilinear flow in which each particle moves in the direction of the x-axis, with speed which depends on y and z. Thus

$$v_x = v(y, z), \quad v_y = 0, \quad v_z = 0, \tag{8.9}$$

$$\begin{aligned} &d_{xx} = d_{yy} = d_{zz} = d_{yz} = 0, \quad d_{xy} = \omega_{xy} = \tfrac{1}{2}\partial v/\partial y, \\ &d_{xz} = \omega_{xz} = \tfrac{1}{2}\partial v/\partial z, \quad \omega_{yz} = 0. \end{aligned} \tag{8.10}$$

These flows are of interest for problems of slope stability, and in the flow of granular materials in pipes and down chutes. The theory of rectilinear flow is developed from first principles in a forthcoming paper (O'Mahony and Spencer, 1982). We show here that it can be derived as a special case of the theory of Section 7.

Consider a surface on which v_x is constant. Since $v_x = v(y, z)$, this surface is cylindrical with its generators parallel to the x-axis. Suppose the surface intersects a plane x = constant in a curve C, as illustrated in Fig. 8.1(a), and let P be a point on C at which the tangent to C makes the angle χ with the y-axis. Then in the neighbourhood of P the motion is a shearing deformation, with the tangent to C defining the shear plane and with the x-direction as the direction of shear. There is no velocity component in the direction of the tangent to C and the directional derivative of v_x in this tangential direction is zero. Also, (8.9) satisfies the incompressibility condition. Therefore, if we choose the η-direction to be the direction of the tangent to C, all of (7.7) are satisfied immediately except for the equation

$$d_{\xi\zeta} + \omega_{\xi\zeta} \sin \phi = 0. \tag{8.11}$$

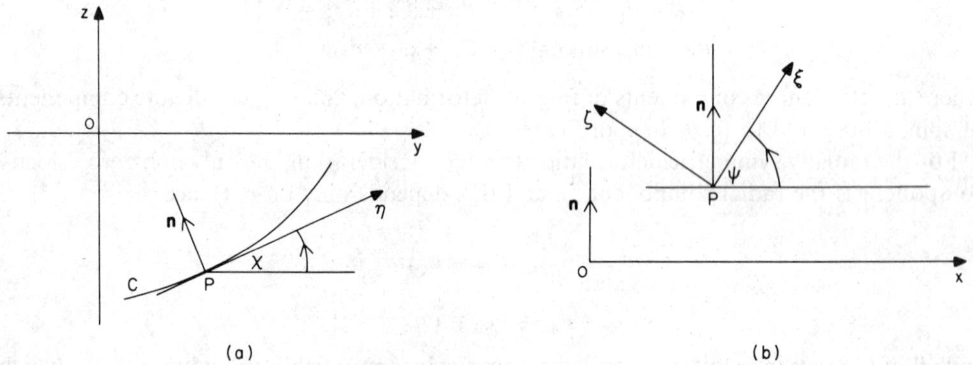

FIG. 8.1. Principal axes of stress in rectilinear flow.

We recall that in Case I of Section 7 (which is the appropriate case for this class of deformations) the η-axis is the principal stress axis which corresponds to the intermediate principal stress component. It is now necessary to select the ξ- and ζ-axes (which are principal stress axes corresponding to the maximum and minimum principal stress components) so that (8.11) is satisfied. The ξ- and ζ-axes are orthogonal to the η-axis, and so they lie in the plane through P which contains the normal **n** to C and is parallel to the x-axis, as shown in Fig. 8.1(b). Suppose the ξ-axis makes the angle ψ with the x-axis. Then

$$v_\xi = v \cos \psi, \quad v_\eta = -v \sin \psi, \tag{8.12}$$

$$\frac{\partial}{\partial \xi} = \cos \psi \frac{\partial}{\partial x} + \sin \psi \frac{\partial}{\partial n}, \quad \frac{\partial}{\partial \zeta} = -\sin \psi \frac{\partial}{\partial x} + \cos \psi \frac{\partial}{\partial n}, \tag{8.13}$$

where $\partial/\partial n$ denotes the directional derivative in the **n**-direction. Equation (8.11) may be written as

$$(1 + \sin \phi)\frac{\partial v_\xi}{\partial \zeta} + (1 - \sin \phi)\frac{\partial v_\zeta}{\partial \xi} = 0,$$

which, from (8.12) and (8.13), and since v is independent of x, gives

$$\{(1+\sin\phi)\cos^2\psi-(1-\sin\phi)\sin^2\psi\}\frac{\partial v}{\partial n}=0.$$

Hence, for $\partial v/\partial n \neq 0$,

$$\psi=\tfrac{1}{4}\pi+\tfrac{1}{2}\phi. \tag{8.14}$$

Thus, as in the plane strain theory, the shear plane makes the angle $\frac{1}{4}\pi+\frac{1}{2}\phi$ with the maximum principal stress axis.

The direction cosines of the principal axes of stress relative to the (x, y, z) coordinates have thus been determined as follows:

$$\begin{aligned}&\xi\text{-axis}\quad (\cos\psi,\ -\sin\psi\sin\chi,\ \sin\psi\cos\chi),\\ &\eta\text{-axis}\quad (0,\ \cos\chi,\ \sin\chi),\\ &\zeta\text{-axis}\quad (-\sin\psi,\ -\cos\psi\sin\chi,\ \cos\psi\cos\chi),\end{aligned} \tag{8.15}$$

where ψ is given by (8.14). Hence the stress components referred to (x, y, z) coordinates may be written down in terms of the principal stress components σ_{I}, σ_{II}, σ_{III}, and the angle χ. These quantities are determined by the three equilibrium equations and the yield condition (7.2). A particular solution of interest is that in which there is no shear stress acting in the (y, z) planes. Then $\sigma_{\mathrm{II}}=\sigma_{yy}=\sigma_{zz}$, $\sigma_{yz}=0$, and the stress referred to (x, y, z) coordinates takes the form

$$\sigma=\begin{bmatrix}-\sigma-2\tau\tan\phi & -\tau\sin\chi & \tau\cos\chi\\ -\tau\sin\chi & -\sigma & 0\\ \tau\cos\chi & 0 & -\sigma\end{bmatrix}, \tag{8.16}$$

where $\tau=c+\sigma\tan\phi$, and τ, σ, and χ are functions of y and z. If the body force is constant, then σ can be chosen so that the equations for equilibrium in the y- and z-directions are satisfied. Then the equation for equilibrium in the x-direction yields a first-order equation for χ. Further developments and applications of this theory are given by O'Mahony and Spencer (1982).

9. Axially Symmetric Deformations

The theory of axially symmetric plastic deformations of soils was formulated by Cox *et al.* (1961) using the Coulomb–Mohr yield condition and associated flow rule. Their formulation followed that used by Shield (1955b) for the theory of axially symmetric deformation of plastic materials which follow Tresca's yield condition and its associated flow rule. We follow the approach used by Cox *et al.*, and as far as possible conform to their notation and terminology, but discard the associated flow rule in favour of the theory developed in Section 7.

We employ cylindrical polar coordinates r, θ, z. For axially symmetric states of stress, the non-zero stress components are denoted by σ_r, σ_θ, σ_z, and σ_{rz}. For axially symmetric deformations, the tangential velocity component $v_\theta=0$, and the radial and axial velocity components are denoted by v_r and v_z respectively. The stress and velocity components depend only on the coordinates r and z. We neglect body forces, so that the equations of equilibrium reduce to

$$\frac{\partial \sigma_r}{\partial r}+\frac{\partial \sigma_{rz}}{\partial z}+\frac{\sigma_r-\sigma_\theta}{r}=0, \quad \frac{\partial \sigma_{rz}}{\partial r}+\frac{\partial \sigma_z}{\partial z}+\frac{\sigma_{rz}}{r}=0. \tag{9.1}$$

The stress component σ_θ is always a principal stress component and the θ-direction a principal axis of stress. The other two principal components of stress are denoted by σ_1 and σ_2, and are ordered so that $\sigma_1 \geqslant \sigma_2$. Hence

$$\left.\begin{matrix}\sigma_1\\ \sigma_2\end{matrix}\right\} = \tfrac{1}{2}(\sigma_r+\sigma_z) \pm \{\tfrac{1}{4}(\sigma_r-\sigma_z)^2+\sigma_{rz}^2\}^{\frac{1}{2}}. \tag{9.2}$$

The principal axes of stress corresponding to σ_1 and σ_2 lie in the (r, z) planes. We denote by ψ the angle which the axis corresponding to σ_1 makes with the radial direction.

The principal stress components σ_θ, σ_1, and σ_2 are some permutation of $\sigma_{\rm I}$, $\sigma_{\rm II}$, and $\sigma_{\rm III}$, where $\sigma_{\rm I} \geqslant \sigma_{\rm II} \geqslant \sigma_{\rm III}$. Various possibilities arise depending on the relative magnitudes of σ_θ, σ_1, and σ_2. In Fig. 9.1 we illustrate a section of the yield surface of Fig. 7.1 by a plane $\sigma_3 = $ constant, where σ_3 is identified as σ_θ. Since we have specified $\sigma_1 > \sigma_2$, it is necessary to consider only the segment $BAFE$. The state of stress in the various plastic regimes is summarized in Table 9.1, which also gives the case (from Section 7) to which each stress state belongs.

The non-zero rate-of-deformation and spin components are given by

$$\begin{gathered} d_{rr}=\frac{\partial v_r}{\partial r}, \quad d_{\theta\theta}=\frac{v_r}{r}, \quad d_{zz}=\frac{\partial v_z}{\partial z}, \\ d_{rz}=\tfrac{1}{2}\left(\frac{\partial v_r}{\partial z}+\frac{\partial v_z}{\partial r}\right), \quad \omega_{rz}=\tfrac{1}{2}\left(\frac{\partial v_r}{\partial z}-\frac{\partial v_z}{\partial r}\right). \end{gathered} \tag{9.3}$$

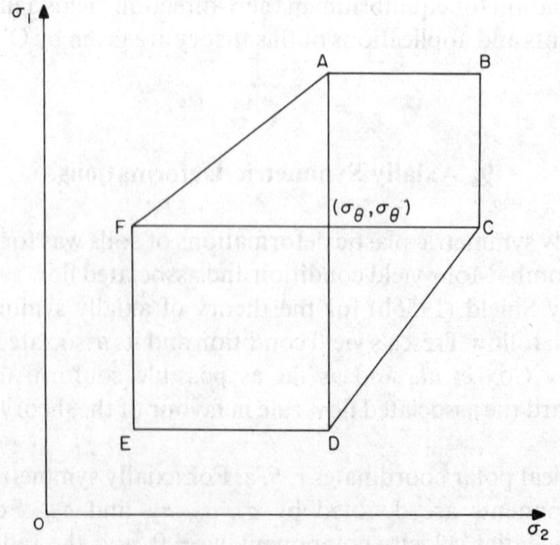

FIG. 9.1. Yield hexagon for axially symmetric deformations.

TABLE 9.1. Plastic regimes for axially symmetric deformations

Plastic regime	σ_{I}	σ_{II}	σ_{III}	Stress state	Case
B	σ_1	σ_2	σ_θ	$\sigma_1=\sigma_2$, $\sigma_1(1+\sin\phi)=2c\cos\phi+\sigma_\theta(1-\sin\phi)$	II
E	σ_θ	σ_1	σ_2	$\sigma_\theta(1+\sin\phi)=2c\cos\phi+\sigma_1(1-\sin\phi)$, $\sigma_1=\sigma_2$	III
AB	σ_1	σ_2	σ_θ	$\sigma_1>\sigma_2>\sigma_\theta$, $\sigma_1(1+\sin\phi)=2c\cos\phi+\sigma_\theta(1-\sin\phi)$	I
EF	σ_θ	σ_1	σ_2	$\sigma_\theta>\sigma_1>\sigma_2$, $\sigma_\theta(1+\sin\phi)=2c\cos\phi+\sigma_1(1-\sin\phi)$	I
A	σ_1	σ_2	σ_θ	$\sigma_1(1+\sin\phi)=2c\cos\phi+\sigma_\theta(1-\sin\phi)$, $\sigma_2=\sigma_\theta$	III
F	σ_θ	σ_1	σ_2	$\sigma_\theta(1+\sin\phi)=2c\cos\phi+\sigma_2(1-\sin\phi)$, $\sigma_\theta=\sigma_1$	II
AF	σ_1	σ_θ	σ_2	$\sigma_1(1+\sin\phi)=\sigma_2(1-\sin\phi)+2c\cos\phi$	I

We now consider the velocity equations for each of the plastic regimes.

9.1. *Regime B*

In this regime $\sigma_1=\sigma_2$, and so we have $\sigma_r=\sigma_z$, $\sigma_{rz}=0$. Then the yield condition and equilibrium equations admit only the solution (Cox *et al.*, 1961)

$$\sigma_r=\sigma_z=c\cot\phi-Dr^m,\quad \sigma_\theta=c\cot\phi-D(m+1)r^m, \tag{9.4}$$

where $m=2\sin\phi/(1-\sin\phi)$ and D is an arbitrary constant. The principal axes of stress in the (r,z) plane may be taken to be directed in the r- and z-directions. Thus, in Case II of Section 7 the ξ-, η-, and ζ-directions may be identified with the directions of increasing r, z, and θ respectively. Hence, with the appropriate substitutions, (7.12) reduce to

$$\frac{\partial v_r}{\partial z}=0,\quad \frac{\partial v_z}{\partial r}=0,\quad \frac{\partial v_r}{\partial r}+\frac{v_r}{r}+\frac{\partial v_z}{\partial z}=0, \tag{9.5}$$

which admit only the solutions

$$v_r=\tfrac{1}{2}Ar+Br^{-1},\quad v_z=-Az+C, \tag{9.6}$$

where A, B, and C are arbitrary constants. The values of A, B, and D are restricted by the requirement that the plastic work is positive, and possibly by requiring that the rate of working in each of the superposed double-shearing deformations is positive.

9.2. *Regime E*

This is similar to regime B and we omit the details.

9.3. *Regime AB*

Here Case I of Section 7 applies. The ξ- and η-axes are therefore chosen to coincide with the principal stress axes which correspond to the principal stress components σ_1 and σ_2, and the ζ-axis is in the direction of increasing θ. Then from $(7.7)_1$ we have $d_{\xi\eta} = 0$, and so the principal axes of stress and deformation-rate coincide. It follows that $d_{\eta\eta}$ is a principal component of the deformation-rate, and so $(7.7)_2$ gives

$$\left(\frac{\partial v_r}{\partial r}+\frac{\partial v_z}{\partial z}\right)-\left\{\left(\frac{\partial v_r}{\partial r}-\frac{\partial v_z}{\partial z}\right)^2+\left(\frac{\partial v_r}{\partial z}+\frac{\partial v_z}{\partial r}\right)^2\right\}^{\frac{1}{2}}=0. \tag{9.7}$$

Also $(7.7)_2$ and $(7.7)_6$ give the incompressibility condition

$$\frac{\partial v_r}{\partial r}+\frac{v_r}{r}+\frac{\partial v_z}{\partial z}=0. \tag{9.8}$$

The remainder of (7.7) are automatically satisfied by the symmetry conditions, except for $(7.7)_4$, namely $\omega_{\xi\eta} = 0$. But $(7.7)_4$ gives no information, because it only specifies that the spin component is zero relative to the (ξ, η)-axes, which themselves rotate with an unknown angular speed.

Equations (9.7) and (9.8) are the same as the equations obtained by Shield (1955b) for the corresponding regime for an ideal metal with Tresca's yield condition and associated flow rule, and they are also a special case of the equations obtained by Cox *et al.* (1961) for this regime using the Coulomb–Mohr condition and associated flow rule. The properties of the equations are discussed by these authors. When a solution of (9.7) and (9.8) is known, the stress components are determined by the equilibrium equations, yield condition, and the result that the principal axes of stress and deformation-rate coincide in this regime.

9.4. *Regime EF*

This is similar to regime *AB* and we omit details.

9.5. *Regimes A and F*

These are the regimes in which the hypothesis of Haar and von Kármán (1909) that $\sigma_\theta = \sigma_1$ or $\sigma_\theta = \sigma_2$ is satisfied. The yield condition gives

$$\begin{aligned}\tfrac{1}{2}(\sigma_r+\sigma_z)\sin\phi+\{\tfrac{1}{4}(\sigma_r-\sigma_z)^2+\sigma_{rz}^2\}^{\frac{1}{2}}&=c\cos\phi,\\ \sigma_\theta=\tfrac{1}{2}(\sigma_r+\sigma_z)-\varepsilon\{\tfrac{1}{4}(\sigma_r-\sigma_z)^2+\sigma_{rz}^2\}^{\frac{1}{2}}&,\end{aligned} \tag{9.9}$$

where $\varepsilon = +1$ for regime *A*, and $\varepsilon = -1$ for regime *F*. Equations (9.9) do not depend on the choice of flow rule, and so the analysis of Cox *et al.* (1961) applies directly. We may write

$$\sigma_r = -p+q\cos 2\psi, \quad \sigma_z = -p-q\cos 2\psi, \quad \sigma_{rz} = q\sin 2\psi, \tag{9.10}$$

and (9.9) then give

$$\sigma_\theta = -p-\varepsilon q, \quad q = p\sin\phi+c\cos\phi. \tag{9.11}$$

It is apparent that the equations closely resemble the plane strain equations (2.4), (2.6), and (2.14) which were discussed in Section 2. By substituting (9.10) and (9.11) into the equilibrium equations (9.1), we obtain a pair of hyperbolic quasi-linear first-order differential equations, with characteristic curves

$$\frac{dy}{dx} = \tan(\psi - \tfrac{1}{4}\pi - \tfrac{1}{2}\phi), \quad \frac{dy}{dx} = \tan(\psi + \tfrac{1}{4}\pi + \tfrac{1}{2}\phi), \tag{9.12}$$

which are termed (as in plane strain theory) α-lines and β-lines respectively. If body forces are neglected, the characteristic relations are found to be

$$\begin{aligned} \cos\phi\, dp + 2q\, d\psi &= r^{-1}q\{\varepsilon \cos\phi\, dr - (1 + \varepsilon\cos\phi)dz\} \quad \text{on an } \alpha\text{-line}, \\ \cos\phi\, dp - 2q\, d\psi &= r^{-1}q\{\varepsilon \cos\phi\, dr + (1 + \varepsilon\cos\phi)dz\} \quad \text{on a } \beta\text{-line}. \end{aligned} \tag{9.13}$$

Our formulation of the velocity equations differs from that of Cox *et al.* In regime A, Case III of Section 7 applies. The (ξ, η) axes become the directions of the principal stress axes in the (r, z) plane, and the ζ-axis coincides with the tangential (θ) direction. For axially symmetric deformations, (7.15) reduce to

$$\begin{aligned} d_{\xi\xi} + d_{\eta\eta} + \frac{v_r}{r} &= 0, \\ d_{\xi\eta} + \omega_{\xi\eta} \sin\phi &= 0. \end{aligned} \tag{9.14}$$

Except for the term v_r/r in $(9.14)_1$, these equations are identical in form to the plane strain equations (4.15). They are of course referred to the (ξ, η) coordinate axes which are translating and rotating relative to the (r, z) coordinate axes. However, (9.14) may be referred to fixed axes r, θ, z in a manner which is precisely analogous to that which was used to transform (4.15) into the forms (4.20) and (4.21). When this is done, (9.14) take the forms

$$\begin{aligned} d_{rr} + d_{zz} + d_{\theta\theta} &= 0, \\ d_{rz}\cos 2\psi - \tfrac{1}{2}(d_{rr} - d_{zz})\sin 2\psi + \sin\phi\,(\omega_{rz} + \Omega) &= 0, \end{aligned} \tag{9.15}$$

or equivalently

$$\begin{aligned} \frac{\partial v_r}{\partial r} + \frac{\partial v_z}{\partial z} + \frac{v_r}{r} &= 0, \\ \left(\frac{\partial v_r}{\partial z} + \frac{\partial v_z}{\partial r}\right)\cos 2\psi - \left(\frac{\partial v_r}{\partial r} - \frac{\partial v_z}{\partial z}\right)\sin 2\psi + \sin\phi\left(\frac{\partial v_r}{\partial z} - \frac{\partial v_z}{\partial r} + 2\Omega\right) &= 0, \end{aligned} \tag{9.16}$$

where $\Omega = \dot{\psi}$.

Precisely the same equations (9.15) and (9.16) obtain for regime *F*.

The system of equations (9.16) is hyperbolic with the α- and β-lines (9.12) as characteristics. The characteristic relations are

$$\begin{aligned} \cos\phi \frac{\partial v_\alpha}{\partial s_\alpha} - (v_\beta - v_\alpha \sin\phi)\frac{\partial \psi}{\partial s_\alpha} + \frac{v_r}{2r}\cos\phi - \Omega\sin\phi &= 0 \quad (\alpha\text{-lines}), \\ \cos\phi \frac{\partial v_\beta}{\partial s_\beta} + (v_\alpha - v_\beta \sin\phi)\frac{\partial \psi}{\partial s_\beta} + \frac{v_r}{2r}\cos\phi + \Omega\sin\phi &= 0 \quad (\beta\text{-lines}), \end{aligned} \tag{9.17}$$

where v_α and v_β are velocity components in the directions of the α- and β-lines. The analogy with (4.24) is apparent. When $\phi = 0$, equations (9.17) reduce to the equations given by Shield (1955b) for frictionless materials in the Haar–von Kármán regime.

The velocity equations (9.15), (9.16), and (9.17) were proposed by Pemberton (1969) on the basis of the analogy with the plane strain equations. Physically, these equations imply that the deformation consists of a double-shearing deformation on the α- and β-lines in the (r, z) planes, on which is superposed a circumferential stretch or contraction which is necessary to preserve the volume of an element if the element undergoes radial motion.

Equations (9.15) may be subjected to analysis similar to that which was used in Section 6. In particular, they may be expressed in the following form, which is analogous to (6.13):

$$\begin{aligned} d_{rr} + \tfrac{1}{2} d_{\theta\theta} &= \dot{\lambda} s_{rr} - \frac{\sin\phi}{2q'} \frac{\mathscr{D} s_{rr}}{\mathscr{D} t}, \\ d_{zz} + \tfrac{1}{2} d_{\theta\theta} &= \dot{\lambda} s_{zz} - \frac{\sin\phi}{2q'} \frac{\mathscr{D} s_{zz}}{\mathscr{D} t}, \\ d_{rz} &= \dot{\lambda} s_{rz} - \frac{\sin\phi}{2q'} \frac{\mathscr{D} s_{rz}}{\mathscr{D} t}, \end{aligned} \tag{9.18}$$

where

$$s_{rr} = \tfrac{1}{2}(\sigma_r - \sigma_z), \quad s_{zz} = \tfrac{1}{2}(\sigma_z - \sigma_r), \quad s_{rz} = \sigma_{rz},$$
$$q' = \{\tfrac{1}{4}(\sigma_r - \sigma_z)^2 + \sigma_{rz}^2\}^{\frac{1}{2}},$$

and

$$\frac{\mathscr{D} s_{rr}}{\mathscr{D} t} = \dot{s}_{rr} - 2\omega_{rz} s_{rz}, \qquad \frac{\mathscr{D} s_{zz}}{\mathscr{D} t} = \dot{s}_{zz} + 2\omega_{rz} s_{rz},$$

$$\frac{\mathscr{D} s_{rz}}{\mathscr{D} t} = \dot{s}_{rz} + \omega_{rz}(s_{rr} - s_{zz})$$

are components, referred to (r, θ, z) coordinates, of a stress-rate tensor which is independent of superposed rotations in the (r, z) planes.

9.6. *Regime EF*

In regime EF, Case I of Section 7 applies. The intermediate principal stress is σ_θ, and so the η-direction coincides with the θ-direction. Hence $d_{\theta\theta} = 0$ and so $v_r = 0$. The incompressibility condition then gives $v_z = w(r)$, so that the only possible deformation is a telescopic shear. This is a special case of the rectilinear flow motions discussed in Section 8.

9.7. *Flow under uniaxial compressive stress*

The Haar–von Kármán regimes A and F seem to be ones of most interest, and the examples we give relate to these regimes. The regime F admits the homogeneous stress state, with $\psi = 0$,

$$\begin{aligned} \sigma_r = \sigma_\theta &= -p(1 - \sin\phi) + c\cos\phi, \\ \sigma_z &= -p(1 + \sin\phi) - c\cos\phi, \quad \sigma_{rz} = 0, \end{aligned} \tag{9.19}$$

where p is constant. There is a corresponding solution for the stress point A, but for definiteness we discuss (9.19). We consider velocity solutions which can be associated with (9.19). For a material which obeys Tresca's yield condition and its associated flow rule, three such velocity solutions were given by Shield (1955b). Analogous solutions for a material which conforms to the Coulomb–Mohr yield condition and its associated flow rule were obtained by Cox *et al.* (1961). The solutions which follow are due to Pemberton (1969) and are the analogues for regime A and the double-shearing model of the solutions obtained by Shield and by Cox *et al.*

When $\psi = 0$, it follows that $\Omega = 0$, and (9.16) become

$$\frac{\partial v_r}{\partial r} + \frac{v_r}{r} + \frac{\partial v_z}{\partial z} = 0,$$
$$(1+\sin\phi)\frac{\partial v_r}{\partial z} + (1-\sin\phi)\frac{\partial v_z}{\partial r} = 0. \tag{9.20}$$

These equations have (among others) the following solutions:

(a) $v_r = \frac{1}{2}Dr$, $v_z = -Dz$, where D is constant.
This is simply the uniform compression solution (8.3).

(b) $$\frac{v_r}{V} = \frac{2}{\pi\gamma}\left(1 - \frac{z^2}{r^2\gamma^2}\right)^{\frac{1}{2}}, \quad \frac{v_z}{V} = \frac{2}{\pi}\cos^{-1}\left(\frac{z}{r\gamma}\right). \tag{9.21}$$

(c) $$\frac{v_r}{V} = \frac{1}{2\pi a\gamma}\left\{r\cos^{-1}\left(\frac{z}{r\gamma}\right) - \frac{z}{\gamma}\left(1 - \frac{z^2}{r^2\gamma^2}\right)^{\frac{1}{2}}\right\},$$
$$\frac{v_z}{V} = \frac{1}{\pi a}\left\{r\left(1 - \frac{z^2}{r^2\gamma^2}\right)^{\frac{1}{2}} - \frac{z}{\gamma}\cos^{-1}\left(\frac{z}{r\gamma}\right)\right\}, \tag{9.22}$$

where $\gamma^2 = (1+\sin\phi)/(1-\sin\phi)$, and V and a are constants. These solutions are of interest for the interpretation of the triaxial test for cylindrical soil specimens, which is discussed in this connection by Haythornthwaite (1960).

9.8. *Flow through a converging conical channel*

The solution for conical channel flow is also due to Pemberton (1969). It is the analogue of the solution for flow in a converging wedge-shaped channel which was given in Section 5. There is a corresponding solution for diverging flow, which is not discussed here.

We seek solutions which correspond to the Haar–von Kármán regime A, and introduce spherical polar coordinates s, Θ, θ (the notation differs from that used in Section 8) such that

$$s^2 = r^2 + z^2, \quad \tan\Theta = r/z,$$

and whose origin is at the virtual apex of the channel. Then Pemberton (1969) showed that (9.9) and the equilibrium equations have solutions of the form

$$\psi_2 = \psi_2(\Theta), \quad q = s^n g(\Theta),$$

where $\psi_2 \equiv \frac{1}{2}\pi - \psi - \Theta$ is the angle the σ_1 principal stress direction makes with the direction of s, and n is a constant to be determined, provided that $\psi_2(\Theta)$ and $g(\Theta)$ satisfy the equations

$$\frac{d\psi_2}{d\Theta}+1=\frac{n\cos^2\phi-(1+\sin\phi)\sin\phi\left\{\dfrac{\sin(2\psi_2+\Theta)}{\sin\Theta}+1\right\}}{2(\cos 2\psi_2+\sin\phi)\sin\phi},$$

$$\frac{dg}{d\Theta}=\frac{2g\sin\psi_2\left\{n\cos\psi_2-\dfrac{\sin\phi\sin(\psi_2+\Theta)}{\sin\Theta}\right\}}{\cos 2\psi_2+\sin\phi}. \tag{9.23}$$

The velocity components in the s- and Θ-directions are denoted by v_s and v_Θ. We seek solutions in which $v_\Theta = 0$, and then $\Omega = 0$. Then the velocity equations (9.16) have solution of the form

$$v_s = -Vr^{-2}h(\Theta),$$

where $h(\Theta)$ is given by

$$\frac{dh}{d\Theta}=-\frac{3h(\Theta)\sin 2\psi_2}{\cos 2\psi_2+\sin\phi}. \tag{9.24}$$

Unlike the corresponding equations for flow in a wedge-shaped channel, (9.23) and (9.24) do not seem to have analytical solutions. However, their numerical integration is straightforward. Pemberton (1969) gave numerical solutions and compared them with the corresponding solutions for flow in a wedge-shaped channel. The boundary conditions for which solutions of this form can be obtained are (a) Coulomb friction at the wall, for a cohesionless material, and (b) a perfectly rough wall, for a cohesionless or cohesive material.

References

Brulin, O. and Hjalmars, S. (1970) Rate-independent non-viscous potential stress, *Int. J. Engng Sci.* **8,** 925–933.

Brulin, O. and Hjalmars, S. (1972) Rate-independent non-viscous stress, *Int. J. Engng Sci.* **10,** 643–648.

Butterfield, R. and Harkness, R. M. (1972) The kinematics of Mohr–Coulomb materials, *Proceedings of the Roscoe Memorial Symposium, "Stress–Strain Behaviour of Soils"*, edited by R. H. G. Parry, 220–233, G. T. Foulis, Henley.

Butterfield, R., Harkness, R. M., and Andrawes, K. Z. (1973) Idealised granular materials, *Proceedings of the Symposium on the Role of Plasticity in Soil Mechanics*, edited by A. C. Palmer, 174–187, Cambridge University Engineering Department.

Chadwick, P. (1959) The quasi-static expansion of a spherical cavity in metals and ideal soils, *Q. Jl Mech. Appl. Math.* **12,** 52–71.

Chadwick, P., Cox, A. D., and Hopkins, H. G. (1964) Mechanics of deep underground explosions in soils, *Phil. Trans. R. Soc. London* A **256,** 235–300.

Collins, I. F. (1973) A note on the interpretation of Coulomb's analysis of the thrust on a retaining wall in terms of the limit theorems of plasticity theory, *Géotechnique* **23,** 442–447.

Coulomb, C. A. (1773) Essai sur une application des régles de maximis et minimis à quelques problèmes de statique, relatifs à l'architecture, *Mém. de Math. de l'Acad. Roy. des Sci. Paris* **7,** 343–382. In English translation: Heyman, J. (1972) *Coulomb's Memoir on Statics: An Essay in the History of Civil Engineering*, Cambridge University Press.

Cox, A. D. (1963) The use of non-associated flow rules in soil plasticity, unpublished RARDE Report.

Cox, A. D., Eason, G., and Hopkins, H. G. (1961) Axially symmetric plastic deformations in soils, *Phil. Trans. R. Soc. London* A **254,** 1–45.

Dais, J. L. (1967a) An isotropic frictional theory for a granular medium with or without cohesion, Brow University Division of Engineering Report NSF-GK 1013/6.

Dais, J. L. (1967b) Private communication.

Dais, J. L. (1970) An isotropic frictional theory for a granular medium with or without cohesion, *Int. J. Solids Struct.* **6,** 1185–1191.

de Josselin de Jong, G. (1959) *Statics and Kinematics of the Failable Zone of a Granular Material*, Uitgeverij Waltman, Delft.

de Josselin de Jong, G. (1964) Lower bound collapse theorem and lack of normality of strain rate to yield surface for soils, *Proc. IUTAM Symposium, Grenoble, "Rhéologie et Mécanique des Sols"*, edited by J. Kravtchenko and P. M. Sirieys, 69–74, Springer-Verlag, Berlin.

de Josselin de Jong, G. (1971) The double-sliding, free-rotating model for granular assemblies, *Géotechnique* **21,** 155–163.

de Josselin de Jong, G. (1977) Mathematical elaboration of the double-sliding, free-rotating model, *Archs Mech.* **29,** 561–591.

Drescher, A. (1976) An experimental investigation of flow rules for granular materials using optically sensitive glass particles, *Géotechnique* **26,** 591–601.

Drescher, A. (1978) Private communication.

Drescher, A. and de Josselin de Jong, G. (1972) Photoelastic verification of a mechanical model for the flow of a granular material, *J. Mech. Phys. Solids* **20,** 337–351.

Drescher, A., Cousens, T. W., and Bransby, P. L. (1978) Kinematics of the mass flow of granular material through a plane hopper, *Géotechnique* **28,** 27–42.

Drucker, D. C. (1959) A definition of stable inelastic material, *J. Appl. Mech.* **26,** 101–106.

Drucker, D. C. (1964) Concept of path independence and material stability for soils, *Proc. IUTAM Symposium, Grenoble, "Rhéologie et Mécanique des Sols"*, edited by J. Kravtchenko and P. M. Sirieys, 23–46, Springer-Verlag, Berlin.

Drucker, D. C. and Prager, W. (1952) Soil mechanics and plastic analysis or limit design, *Q. Appl. Math.* **10,** 157–165.

Drucker, D. C., Gibson, R. E., and Henkel, D. J. (1957) Soil mechanics and work-hardening theories of plasticity, *Trans. ASCE* **122,** 338–346.

Edelen, D. B. (1977) Properties of an elementary class of fluids with nondissipative viscous stresses, *Int. J. Engng Sci.* **15,** 727–731.

Geniev, G. A. (1958) *Problems of the Dynamics of a Granular Medium* (in Russian), Akad. Stroit. Archit. SSSR, Moscow.

Goodman, M. A. and Cowin, S. C. (1972) A continuum theory for granular materials, *Archs Ration. Mech. Anal.* **44,** 249–266.

Green, A. E. and Adkins, J. E. (1960) *Large Elastic Deformations*, Clarendon Press, Oxford.

Haar, A. and von Kármán, Th. (1909) Zur Theorie der Spannungszustände in plastischer und sandartigen Medien, *Nachr. Ges. Wiss. Göttingen, Math-Phys. Kl.* 204–218.

Hartmann, W. (1925) Über die Integration der Differentialgleichungen des ebenen Gleichgewichtszustandes für den allgemein-plastischen Körper, Thesis, Göttingen.

Haythornthwaite, R. M. (1960) Mechanics of the triaxial test for soils, *J. Soil Mech. Fdns Div., Proc. ASCE* **86,** 35–62.

Hill, R. (1949) The plastic yielding of notched bars under tension, *Q. Jl Mech. Appl. Math.* **2,** 40–52.

Hill, R. (1950) *The Mathematical Theory of Plasticity*, Clarendon Press, Oxford.

Hill, R. (1956) On the problem of uniqueness in the theory of a rigid–plastic solid, I, *J. Mech. Phys. Solids* **4,** 247–255.

Hill, R., Lee, E. H., and Tupper, S. J. (1951) A method of numerical analysis of plastic flow in plane strain and its application to the compression of a ductile material between rough plates, *J. Appl. Mech.* **18,** 46–52.

Jenike, A. W. (1961) Gravity flow of bulk solids, *Bull. Univ. Utah* **52,** No. 29.

Jenike, A. W. and Shield, R. T. (1959) On the plastic flow of Coulomb solids beyond original failure, *J. Appl. Mech.* **26,** 599–602.

Jeske, T. (1968) On the kinematics of a granular medium in the case of limit equilibrium state under plane strain conditions, *Arch. Mech. Stosow.* **20,** 211–224.

Kingston, M. R. and Spencer, A. J. M. (1970) General yield conditions in plane deformations of granular media, *J. Mech. Phys. Solids* **18,** 233–243.

Lee, J., Cowin, S. C., and Templeton, J. S. (1974) An experimental study of the kinematics of flow through hoppers, *Trans. Soc. Rheol.* **18,** 247–269.

Mandel, J. (1938a) Sur les équilibres par tranches planes parallèles des milieux plastiques à la limite d'écoulement et en particulier des terres et des métaux ductiles, *C. r. Acad. Sci. Paris* **206,** 317–318.

Mandel, J. (1938b) Sur les équilibres par tranches parallèles des terres et des milieux plastiques à la limite d'écoulement, *C. r. Acad. Sci. Paris* **206,** 583–585.

Mandel, J. (1947) Sur les lignes de glissement et le calcul des déplacements dans la déformation plastique, *C. r. Acad. Sci. Paris* **225,** 1272–1273.

Mandel, J. (1962) *Séminaire de Plasticité (Ecole Polytechnique, 1971)*, Chapitre 4, Publications Scientifiques et Techniques de Ministère de l'Air, Paris.

Mandel, J. (1966) Sur les équations à écoulement des sols idéaux en déformation plane et le concept de double glissement, *J. Mech. Phys. Solids* **14,** 303–308.
Mandl, G. and Fernandez Luque, R. (1970) Fully developed plastic shear flow of granular materials, *Géotechnique* **20,** 277–307.
Marshall, E. A. (1967) The compression of a slab of ideal soil between rough plates, *Acta Mech.* **3,** 82–92.
Mehrabadi, M. M. and Cowin, S. C. (1978) Initial planar deformation of dilatant granular materials, *J. Mech. Phys. Solids* **26,** 269–284.
Mehrabadi, M. M. and Cowin, S. C. (1980) Pre-failure and post-failure soil plasticity models, *J. Engng Mech. Div., Proc. ACSE* **106,** No. EM5, 991–1003.
Morrison, H. L. and Richmond, O. (1976) Application of Spencer's ideal soil model to granular materials flow, *J. Appl. Mech.* **43,** 49–53.
Mróz, Z. (1980) Deformation and flow of granular materials, *Proc. 15th International Congress of Theoretical and Applied Mechanics, Toronto, "Theoretical and Applied Mechanics"*, edited by F. P. J. Rimrott and B. Tabarrok, 119–132, North-Holland, Amsterdam.
Mróz, Z. and Szymański, Cz. (1979) Non-associated flow rules in description of plastic flow of granular materials, *CISM Course 217 (1974), Udine, "Limit Analysis and Rheological Approach in Soil Mechanics"*, edited by W. Olszak and L. Suklje, 51–94, Springer-Verlag, Vienna and New York.
Nádai, A. (1924) Uber die Gleit- und Verzweigungsflächen einiger gleichgewichtzuständer bildsamer Massen, *Z. Phys.* **30,** 106–138.
Nemat-Nasser, S. (1980) On behavior of granular materials in simple shear, *Soils and Foundations* **20,** 59–73.
Nikolaevskii, V. N. (1972) *Mechanical Properties of Soils and the Theory of Plasticity* (in Russian), Itogi Naukii Teknici, Moscow.
O'Mahony, T. C. and Spencer, A. J. M. (1982) The theory of the stability of long slopes of finite width (to be published).
Palmer, A. C. (1966) A limit theorem for materials with non-associated flow laws, *J. de Mécanique* **5,** 219–222.
Pemberton, C. S. (1965) Flow of imponderable granular materials in wedge-shaped channels, *J. Mech. Phys. Solids* **13,** 351–360.
Pemberton, C. S. (1969) Some stress and velocity solutions in plasticity theory of soils, Ph.D. Thesis, University of Nottingham.
Prandtl, L. (1920) Über die Härte plastischer Körper, *Nachr. Ges. Wiss. Göttingen, Math-Phys. Kl.* 74–85.
Prandtl, L. (1923) Anwendungsbeispiele zu einem Henckyschen Satz über das plastische Gleichgewicht, *Z. angew. Math. Mech.* **3,** 401–406.
Rudnicki, J. W. and Rice, J. R. (1975) Conditions for the localization of deformation in pressure-sensitive dilatant materials, *J. Mech. Phys. Solids* **23,** 371–394.
Schofield, A. and Wroth, P. (1968) *Critical State Soil Mechanics*, McGraw-Hill, London.
Shield, R. T. (1953) Mixed boundary-value problems in soil mechanics, *Q. Appl. Math.* **11,** 61–75.
Shield, R. T. (1955a) On Coulomb's law of failure in soils, *J. Mech. Phys. Solids* **4,** 10–16.
Shield, R. T. (1955b) On the plastic flow of metals under conditions of axial symmetry, *Proc. R. Soc. London* A **233,** 267–287.
Sokolovskii, V. V. (1965) *Statics of Granular Media*, Pergamon Press, Oxford.
Spencer, A. J. M. (1964) A theory of the kinematics of ideal soils under plane strain conditions, *J. Mech. Phys. Solids* **12,** 337–351.
Spencer, A. J. M. (1971) Discussion of "Fully developed shear flow of granular materials" by G. Mandl and R. Fernandez Luque, *Géotechnique* **21,** 190–192.
Spencer, A. J. M. and Kingston, M. R. (1973) Plane mechanics and kinematics of compressible ideal granular materials, *Rheol. Acta* **12,** 194–199.
Takagi, S. (1962) Plane plastic deformation of soils, *J. Engng Mech. Div., Proc. ASCE* **88,** No. EM3, 107–151.
Takagi, S. (1979) Steady in-plane deformation of noncoaxial plastic soil, *Int. J. Engng Sci.* **17,** 1048–1072.
Zagainov, L. S. (1967) Equations of plane steady-state motion of a granular medium (in Russian), *Inzh. Zh. Mekh. Tverdova Tela* **2,** 188–196. Translated as *Mech. Solids* **2** (1970) 130–134.

Elasticity Theory of Composites

by J. R. Willis

School of Mathematics, University of Bath

Summary

The determination of the overall elastic moduli of random composites is approached using variational principles. Information on the configuration of a composite is never complete, and strict estimates can do no more than take the form of bounds that use whatever statistical information is available. The classical Hashin–Shtrikman bounds are deduced as special cases of bounds that allow explicitly for any two-point correlations, and it is shown that these general bounds take a simple form not given before for any two-phase composite. Self-consistent estimates are related to the variational formulation. Bounds of higher order are discussed and equations that define a new optimum bound of third order are given explicitly. The concluding Sections treat wave propagation problems from the standpoint of a new variational principle which reduces to the Hashin–Shtrikman principle in the static limit.

1. Introduction

The subject of this article is the theoretical determination of the overall properties of inhomogeneous materials such as fibre-reinforced plastics. The distinguishing feature of these materials, to which the generic term "composite" is applied, is that they are strongly inhomogeneous relative to a small length scale (the microscale) and yet, relative to a larger scale (the macroscale) over which variations in applied loads are significant, they appear either homogeneous or else their "average" properties appear to vary smoothly. Generally, a composite may comprise *n* different phases, distributed either deterministically or at random, and it will be assumed that the microscale defined by the local phase geometry is sufficiently large for each phase to be treated as a continuum: this is certainly true of fibre-reinforced plastics and it is true also of polycrystalline aggregates. The latter represent a limiting case in which the number of phases is indefinitely large, each crystal orientation being regarded as defining a different phase.

The article will be concerned with the determination of the overall elastic response of a composite, since this particular aspect of material behaviour can be treated with reasonable precision. Some of the methods to be described have wider application (when suitably generalized), e.g. to the elastoplastic response of a polycrystal, but study of the simpler elastic problem permits some assessment of limitations inherent in these methods by comparison with results obtained from methods that allow greater precision but are less easily generalized.

The elastic properties of composites have been studied quite intensively during the past 20 years and many of the simpler problems are at a definitive state of development. A reasonably comprehensive review is contained in the recent book by Christensen (1979). The major gaps in Christensen's presentation concern allowance for random phase geometry when this is not isotropic, and wave propagation in randomly inhomogeneous composites. Progress has, in fact, been made in both of these areas which happen, too, to be of particular interest to the writer. Correspondingly, attention is focused in this article upon problems for random media, which need not be isotropic. The presentation is self-contained: the longer-established results are deducible as special cases and some of the more important of these are discussed explicitly, though not in the detail given by Christensen. Apart from introductory material, the present review is largely complementary to one recently prepared by the writer (Willis, 1981). This covered a somewhat broader range of topics and emphasized more strongly composites consisting of a matrix reinforced by inclusions. Distinctive features of the present work are the discussion of bounds, including a new simple formula which allows for arbitrary two-point correlations in a two-phase composite, new optimum bounds of third order, and a variational treatment of wave propagation problems.

It will soon become apparent to the reader that, as with so many areas in solid mechanics, Rodney Hill's work on the elasticity of composites has been of fundamental importance. The pages that follow are unlikely to maintain the "Hill standard" and will perhaps indicate how much the subject could benefit from his renewed attention.

2. Overall Properties

Underlying the "overall modulus" concept is the belief that, in a specimen subjected to some system of loading, the stress and strain fields vary in a complicated fashion from point to point but, if some form of "running average" is taken which, in effect, replaces the fields at **x** by weighted means evaluated over some neighbourhood centred on **x**, whose dimensions are small relative to overall specimen dimensions and yet large relative to the microscale, then the "averaged" stress and strain fields will vary smoothly relative to the macroscale. It is difficult to define the averaging procedure precisely for a general composite but it corresponds, hopefully, to what might be measured by a strain gauge or a transducer. The details of the fields in the neighbourhood of **x** are then expected to be determined just by their mean values at **x** and the local configuration of the material. These expectations are indeed borne out for composites with a periodic structure, for which the stress and strain fields may be expressed, asymptotically at least, as products of periodic functions with the period of the structure, and amplitudes that vary smoothly on the macroscale. For static problems, these ideas were enunciated by Sanchez-Palencia (1974). The Floquet theory approach of Kohn *et al.* (1972) for dispersion relations for periodic composites is essentially similar, except that it does not even require the wavelength associated with the "amplitude" function to be large.

Overall moduli are defined so that they relate "average" stress to "average" strain. Symbolically, if stress and strain tensors are denoted $\boldsymbol{\sigma}$, $\mathbf{e}$, and their local averages are correspondingly represented as $\bar{\boldsymbol{\sigma}}$, $\bar{\mathbf{e}}$, then the tensor of overall moduli $\bar{\mathbf{L}}$ is defined so that

$$\bar{\boldsymbol{\sigma}} = \bar{\mathbf{L}}\,\bar{\mathbf{e}}. \tag{2.1}$$

Strictly, if the relation (2.1) is to apply when $\bar{\boldsymbol{\sigma}}$, $\bar{\mathbf{e}}$ are generated from arbitrary boundary conditions, then $\bar{\mathbf{L}}$ should be a non-local operator (Kröner, 1977) but, asymptotically, when boundary conditions are such that $\bar{\boldsymbol{\sigma}}, \bar{\mathbf{e}}$ vary smoothly relative to the microscale (and so hardly at all in a neighbourhood of $\mathbf{x}$), $\bar{\mathbf{L}}$ reduces to a tensor of moduli except, of course, in a thin layer close to the boundary of the specimen which is excluded from detailed consideration.

If the validity of (2.1) is accepted, the overall moduli $\bar{\mathbf{L}}$ may be determined experimentally by studying the response of a macroscopically uniform specimen when it is subjected to loads that generate uniform average stress and strain fields. In fact, the tensor of moduli $\bar{\mathbf{L}}$ may be defined in an unambiguous manner by subjecting a specimen that occupies a region V with boundary ∂V to surface displacements that have the form

$$u_i = \bar{e}_{ij} x_j \tag{2.2}$$

relative to a Cartesian basis, the summation convention being implied for repeated suffixes. Application of Gauss' theorem then shows that the volume average of the strain actually produced throughout V is $\bar{\mathbf{e}}$ precisely ($\bar{\mathbf{e}}$ having components $\bar{e}_{ij}$). If the average over V of the stress $\boldsymbol{\sigma}$ is denoted by $\bar{\boldsymbol{\sigma}}$, then $\bar{\boldsymbol{\sigma}}$ depends linearly upon $\bar{\mathbf{e}}$ and defines the tensor $\bar{\mathbf{L}}$ in (2.1). The above discussion assumes, of course, that the phases of the composite are linearly elastic and that there is perfect adhesion across phase boundaries. It may be noted that the components $\bar{\sigma}_{ij}$ of the mean stress may be determined from measured values of surface traction through the relation

$$\bar{\sigma}_{ij} = \frac{1}{V} \int_{\partial V} x_i t_j \, dS, \tag{2.3}$$

where V has also been used to denote the volume of the region V to avoid further symbols, and $t_j = \sigma_{jk} n_k$, the normal to ∂V having components n_k. Equation (2.3) follows from Gauss' theorem, together with the requirement that the stress $\boldsymbol{\sigma}$ should be self-equilibrated; that is,

$$\sigma_{ij,j} = 0 \tag{2.4}$$

except at interphase boundaries, across which $\sigma_{ij} n_j$ is continuous. The latter proviso is included in (2.4) if the derivatives are interpreted in the sense of generalized functions (Gel'fand and Shilov, 1964): this interpretation will be used throughout the sequel to avoid the need for explicit discussion of jump conditions and it should not be assumed that sharp changes in properties are thereby excluded.

An alternative definition for $\bar{\mathbf{L}}$ is obtained by applying to the specimen the traction boundary condition

$$\sigma_{ij} n_j = \bar{\sigma}_{ij} n_j. \tag{2.5}$$

Equation (2.3) then ensures that the average of the actual stress over V is $\bar{\boldsymbol{\sigma}}$ and $\bar{\mathbf{L}}$ follows from (2.1), this time with $\bar{\mathbf{e}}$ calculated as the average of the strain $\mathbf{e}$ over V. Again, $\bar{\mathbf{e}}$ can be determined from surface values of displacement alone, by use of the identity

$$\bar{e}_{ij} = \frac{1}{2V} \int_{\partial V} (u_i n_j + u_j n_i) \, dS. \tag{2.6}$$

Thus, either experiment determines $\bar{\mathbf{L}}$ from values that are accessible to measurement, in principle at least. They were both proposed, as precise mathematical definitions for $\bar{\mathbf{L}}$, by Hill (1963a). The two definitions are not equivalent to one another for arbitrarily inhomogeneous bodies, and the extent to which they agree in their estimates for $\overline{\mathbf{L}}$ provides a partial check on the validity of the "overall modulus" concept.

The definitions are interesting for another reason, also developed by Hill (1963a), that, when either of the boundary conditions (2.2) or (2.5) is applied, then the average energy density $\overline{U}$ in the specimen is given by

$$2\overline{U} = \overline{\boldsymbol{\sigma}}\,\overline{\mathbf{e}} = \overline{\mathbf{e}}\overline{\mathbf{L}}\overline{\mathbf{e}} \tag{2.7}$$

precisely, again by application of Gauss' theorem combined with one of (2.3), (2.6). The shorthand notation implicit in (2.1) has been extended in (2.7) which, in suffixes, would read

$$2\overline{U} = \bar{\sigma}_{ij}\bar{e}_{ij} = \bar{e}_{ij}\bar{L}_{ijkl}\bar{e}_{kl}. \tag{2.8}$$

In the sequel, suffixes will be avoided wherever possible.

The derivation of (2.7) has assumed the usual symmetry of the tensor of moduli $\mathbf{L}$ which applies to the actual inhomogeneous specimen. Corresponding symmetry of $\overline{\mathbf{L}}$ may be demonstrated by calculating the increment $\delta\overline{U}$ in mean energy corresponding to an increment $\delta\overline{\mathbf{e}}$ in mean strain: the argument that led to (2.7) gives

$$\delta\overline{U} = \overline{\boldsymbol{\sigma}}\,\delta\overline{\mathbf{e}} = \delta\overline{\mathbf{e}}\overline{\mathbf{L}}\overline{\mathbf{e}} \tag{2.9}$$

which, when compared with $\delta\overline{U}$ calculated by differentiating (2.7), gives the desired result.

Bounds upon components of the tensor of overall moduli $\overline{\mathbf{L}}$ may now be developed from (2.7) and the classical variational principles for elasticity. For the boundary condition (2.2), the minimum energy principle asserts that

$$2\overline{U} \leqslant \frac{1}{V}\int_V \mathbf{e}^*\,\mathbf{L}\mathbf{e}^*\,d\mathbf{x} \tag{2.10}$$

for a strain field $\mathbf{e}^*$ derived from any displacement field $\mathbf{u}^*$ that conforms with (2.2) on ∂V. The simplest field that can be substituted into (2.10) is $\mathbf{e}^* = \overline{\mathbf{e}}$; this gives

$$\overline{\mathbf{L}} \leqslant \mathbf{L}_v, \tag{2.11}$$

where $\mathbf{L}_v$ denotes the average of $\mathbf{L}$ over V; it was introduced by Voigt (1889). The symbol $\leqslant$ in (2.11) defines an ordering in the sense that the quadratic form $\overline{\mathbf{e}}(\mathbf{L}_v - \overline{\mathbf{L}})\overline{\mathbf{e}}$ is positive semidefinite for symmetric second-order tensors $\overline{\mathbf{e}}$. Dually, for the boundary condition (2.2), the complementary energy principle asserts that

$$\frac{2}{V}\int_{\partial V} \mathbf{t}^*\,\mathbf{u}\,dS - \frac{1}{V}\int_V \boldsymbol{\sigma}^*\,\mathbf{M}\boldsymbol{\sigma}^*\,d\mathbf{x} \leqslant 2\overline{U}, \tag{2.12}$$

where $t_i^* = \sigma_{ij}^* n_j$, $\boldsymbol{\sigma}^*$ is any self-equilibrated stress field, $u_i = \bar{e}_{ij}x_j$, and $\mathbf{M}$ is the tensor of compliances inverse to $\mathbf{L}$. The simplest stress field $\boldsymbol{\sigma}^*$ to substitute into (2.12) is a constant; this gives, upon use of Gauss' theorem,

$$2\boldsymbol{\sigma}^*\overline{\mathbf{e}} - \boldsymbol{\sigma}^*\mathbf{M}_R\boldsymbol{\sigma}^* \leqslant 2\overline{U}, \tag{2.13}$$

where $\mathbf{M}_R$ is the average of $\mathbf{M}$ over V, introduced by Reuss (1929). The left side of (2.13) is maximized by taking

$$\boldsymbol{\sigma}^* = \mathbf{L}_R\overline{\mathbf{e}}, \tag{2.14}$$

where $\mathbf{L}_R$ is inverse to $\mathbf{M}_R$. It follows then that

$$\mathbf{L}_R \leqslant \overline{\mathbf{L}} \leqslant \mathbf{L}_v. \tag{2.15}$$

The result (2.15) is due to Hill (1952). It can also be deduced in a corresponding fashion if $\mathbf{L}$ is defined by the boundary condition (2.5). The bounds $\mathbf{L}_R$, $\mathbf{L}_v$ depend only upon the moduli and concentrations of the phases and take no account of how the phases are distributed. The construction of better bounds, which make more allowance for the structure of the composite, is considered in later Sections. Before proceeding, however, it may be remarked that corresponding mean value theorems can be developed for a body subjected to finite deformation. The relation (2.3) is valid in any continuum but it is useful to have a description that employs Lagrangian variables based upon a fixed reference state. Relative to such a state, Hill (1972) has shown that an equation of the form (2.3) applies if the Cauchy stress $\boldsymbol{\sigma}$ is replaced by nominal stress and x_i is interpreted as a Lagrangian coordinate. Alternatively, $\boldsymbol{\sigma}$ may be replaced by Kirchhoff stress and x_i is interpreted as the current (Eulerian) coordinate, expressed as a function of the Lagrangian coordinates. An equation like (2.6) (not symmetrized) applies to the deformation gradient, and mean Kirchhoff stress, nominal stress, and deformation gradient are related exactly as though the fields were uniform. Similar relations apply to rates which allow, for elastic materials, generalizations of (2.7) and (2.9); some implications for nonlinearly elastic composites have been explored by Ogden (1978).

3. Stress Polarization and Variational Principles

Bounds better than (2.15) require the construction of stress and strain fields that make some explicit allowance for the microscopic arrangement of the composite. For this purpose, it proves useful to introduce a homogeneous "comparison" material with moduli $\mathbf{L}_0$ and to set

$$\mathbf{L} = \mathbf{L}_0 + \delta\mathbf{L}. \tag{3.1}$$

Substitution of (3.1) into the constitutive relation

$$\boldsymbol{\sigma} = \mathbf{L}\mathbf{e} \tag{3.2}$$

and thence into the equilibrium equation (2.4) gives

$$\operatorname{div}(\mathbf{L}_0\,\mathbf{e}) + \operatorname{div}(\delta\mathbf{L}\mathbf{e}) = 0, \quad \mathbf{x} \in V, \tag{3.3}$$

which is to be solved together with whatever boundary condition is given. The discussion to follow will be phrased in terms of the displacement condition

$$\mathbf{u} = \mathbf{u}_0, \quad \mathbf{x} \in \partial V, \tag{3.4}$$

with a mind to subsequent specialization to (2.2). If Green's function $\mathbf{G}$ for the homogeneous comparison material is known, (3.3) and (3.4) may be expressed together in the form

$$\mathbf{u}(\mathbf{x}) = \mathbf{u}_0(\mathbf{x}) + \int_V \mathbf{G}(\mathbf{x}, \mathbf{x}')\operatorname{div}(\delta\mathbf{L}\mathbf{e})(\mathbf{x}')\,d\mathbf{x}', \tag{3.5}$$

where $\mathbf{u}_0(\mathbf{x})$ is the field that the boundary condition (3.4) would generate by itself in the homogeneous comparison body. An integral equation for the strain $\mathbf{e}(\mathbf{x})$ is now deduced by integration by parts, followed by differentiating with respect to $\mathbf{x}$. This gives

$$\mathbf{e}(\mathbf{x}) = \mathbf{e}_0(\mathbf{x}) - \int_V \boldsymbol{\Gamma}(\mathbf{x}, \mathbf{x}')\delta\mathbf{L}(\mathbf{x}')\mathbf{e}(\mathbf{x}')\,d\mathbf{x}', \tag{3.6}$$

where $\mathbf{e}_0$ is the strain associated with $\mathbf{u}_0$ and $\mathbf{\Gamma}(\mathbf{x}, \mathbf{x}')$ has components

$$\Gamma_{ijkl}(\mathbf{x}, \mathbf{x}') = \partial^2 G_{jk}(\mathbf{x}, \mathbf{x}')/\partial x_i \partial x'_l \big|_{(ij)(kl)}. \tag{3.7}$$

It has a singularity of order $|\mathbf{x}-\mathbf{x}'|^{-3}$, which is to be interpreted in the sense of generalized functions (Gel'fand and Shilov, 1964). In the sequel, (3.6) will be written even more briefly as

$$\mathbf{e} = \mathbf{e}_0 - \mathbf{\Gamma}\delta\mathbf{L}\mathbf{e}. \tag{3.8}$$

If the components of $\delta\mathbf{L}$ are small, (3.8) can be solved by iteration, and the resulting series, suitably truncated, can be substituted into the energy and complementary energy principles. This has been developed in detail by Dederichs and Zeller (1973); the validity of the bounds is, of course, independent of convergence of the perturbation series.

It is possible, however, to make rather better use of the work implicit in the construction of the bounds of Dederichs and Zeller (1973), by considering a form for $\mathbf{e}$ that contains some parameters which may be chosen optimally. It is helpful for this purpose (as well as for other, subsequent, developments) to introduce the stress polarization $\boldsymbol{\tau}$ so that

$$\boldsymbol{\tau} = \delta\mathbf{L}\mathbf{e} = (\mathbf{L}-\mathbf{L}_0)\mathbf{e} \tag{3.9}$$

and

$$\boldsymbol{\sigma} = \mathbf{L}_0\mathbf{e}+\boldsymbol{\tau}. \tag{3.10}$$

Equation (3.8) gives

$$\mathbf{e} = \mathbf{e}_0 - \mathbf{\Gamma}\boldsymbol{\tau} \tag{3.11}$$

and, for any choice $\boldsymbol{\tau}^*$ of $\boldsymbol{\tau}$, (3.11) and (3.10) generate strain and stress fields $\mathbf{e}^*$ and $\boldsymbol{\sigma}^*$ that are, respectively, compatible with boundary conditions and self-equilibrated. If the mean energy density in the actual body is called $\bar{U}$, the minimum energy principle gives

$$2\bar{U} \leqslant (\mathbf{e}^*, \mathbf{L}_0\mathbf{e}^*)+(\mathbf{e}^*, \delta\mathbf{L}\mathbf{e}^*), \tag{3.12}$$

where the "inner product" is defined so that

$$(\mathbf{f}, \mathbf{g}) = \frac{1}{V}\int_V \mathbf{f}\mathbf{g}\,d\mathbf{x}. \tag{3.13}$$

The first term on the right side of (3.12) is simplified by employing (3.10) and (3.11) for $\boldsymbol{\sigma}^*$, $\mathbf{e}^*$, and $\boldsymbol{\tau}^*$ and using the "virtual work" equality

$$(\boldsymbol{\sigma}_1, \mathbf{e}_2) = 0 \tag{3.14}$$

for any self-equilibrated stress field $\boldsymbol{\sigma}_1$ and strain field $\mathbf{e}_2$ compatible with zero boundary displacements. Then, in terms of $\boldsymbol{\tau}^*$, (3.12) becomes

$$2(\bar{U}-\bar{U}_0) \leqslant (\boldsymbol{\tau}^*, \mathbf{\Gamma}\boldsymbol{\tau}^*)+(\mathbf{e}_0, \delta\mathbf{L}\mathbf{e}_0)-2(\mathbf{e}_0, \delta\mathbf{L}\mathbf{\Gamma}\boldsymbol{\tau}^*)+(\mathbf{\Gamma}\boldsymbol{\tau}^*, \delta\mathbf{L}\mathbf{\Gamma}\boldsymbol{\tau}^*), \tag{3.15}$$

where $2\bar{U}_0 = (\mathbf{e}_0, \mathbf{L}_0\mathbf{e}_0)$. The inequality (3.15) will be used in full later. Clearly, the last term is the one whose evaluation is the most complicated. A variational principle in which it does not appear can, in fact, be generated by noting that the exact polarization $\boldsymbol{\tau}$ must satisfy the equation

$$[(\delta\mathbf{L})^{-1}+\mathbf{\Gamma}]\boldsymbol{\tau} = \mathbf{e}_0 \tag{3.16}$$

from (3.9) and (3.11). This equation is self-adjoint and hence the actual field $\boldsymbol{\tau}$ extremizes the functional

$$\mathscr{F}(\boldsymbol{\tau}^*) = (\boldsymbol{\tau}^*, (\delta\mathbf{L})^{-1}\boldsymbol{\tau}^*) + (\boldsymbol{\tau}^*, \boldsymbol{\Gamma}\boldsymbol{\tau}^*) - 2(\boldsymbol{\tau}^*, \mathbf{e}_0). \tag{3.17}$$

This principle was deduced from the field equations by Hashin and Shtrikman (1962a) and directly from (3.16) by Willis (1977). Here, however, it seems particularly apt to give a derivation which corresponds to that of Hill (1963b). This shows how the Hashin–Shtrikman principle is related to the classical energy principles. Reverting to (3.15), therefore, the offending last term is rewritten

$$\begin{aligned}(\boldsymbol{\Gamma}\boldsymbol{\tau}^*, \delta\mathbf{L}\boldsymbol{\Gamma}\boldsymbol{\tau}^*) = {} & ([(\delta\mathbf{L})^{-1} + \boldsymbol{\Gamma}]\boldsymbol{\tau}^* - \mathbf{e}_0, \delta\mathbf{L}\{[(\delta\mathbf{L})^{-1} + \boldsymbol{\Gamma}]\boldsymbol{\tau}^* - \mathbf{e}_0\}) \\ & - (\boldsymbol{\tau}^*, (\delta\mathbf{L})^{-1}\boldsymbol{\tau}^*) - 2(\boldsymbol{\tau}^*, \boldsymbol{\Gamma}\boldsymbol{\tau}^*) + 2(\mathbf{e}_0, \boldsymbol{\tau}^* + \delta\mathbf{L}\boldsymbol{\Gamma}\boldsymbol{\tau}^*) - (\mathbf{e}_0, \delta\mathbf{L}\mathbf{e}_0);\end{aligned} \tag{3.18}$$

since the exact $\boldsymbol{\tau}$ satisfies (3.16), the first term on the right of (3.18) is zero when $\boldsymbol{\tau}^* = \boldsymbol{\tau}$ and is small when $\boldsymbol{\tau}^*$ is a good approximation to $\boldsymbol{\tau}$. Hence, (3.15) gives

$$2(\bar{U} - \bar{U}_0) \leqslant -(\boldsymbol{\tau}^*, (\delta\mathbf{L})^{-1}\boldsymbol{\tau}^*) - (\boldsymbol{\tau}^*, \boldsymbol{\Gamma}\boldsymbol{\tau}^*) + 2(\boldsymbol{\tau}^*, \mathbf{e}_0) + (\boldsymbol{\varepsilon}^*, \delta\mathbf{L}\boldsymbol{\varepsilon}^*), \tag{3.19}$$

where $\boldsymbol{\varepsilon}^*$ has been written for the "error" $[(\delta\mathbf{L})^{-1} + \boldsymbol{\Gamma}]\boldsymbol{\tau}^* - \mathbf{e}_0$. If, now, $\mathbf{L}_0$ is chosen in such a way that $\delta\mathbf{L}$ is negative definite at each point of V, discarding the last term in (3.19) gives

$$2(\bar{U} - \bar{U}_0) \leqslant -\mathscr{F}(\boldsymbol{\tau}^*). \tag{3.20}$$

The extreme value of $\mathscr{F}(\boldsymbol{\tau}^*)$ is $2(\bar{U}_0 - \bar{U})$, whatever $\mathbf{L}_0$ is chosen. The inequality (3.19) shows that the Hashin–Shtrikman principle embodied in (3.20) is weaker than the corresponding classical principle through neglect of the term quadratic in $\boldsymbol{\varepsilon}^*$. Hill (1963b) made no use of the operator equation (3.16) but instead gave $\boldsymbol{\varepsilon}^*$ in the form

$$\boldsymbol{\varepsilon}^* = (\delta\mathbf{L})^{-1}(\boldsymbol{\sigma}^* - \mathbf{L}\mathbf{e}^*). \tag{3.21}$$

The complementary energy principle

$$2(\boldsymbol{\sigma}^*, \mathbf{e}_0) - (\boldsymbol{\sigma}^*, \mathbf{M}\boldsymbol{\sigma}^*) \leqslant 2\bar{U} \tag{3.22}$$

may be expanded similarly: with the notation

$$\mathbf{M} = \mathbf{M}_0 + \delta\mathbf{M} \tag{3.23}$$

so that $\mathbf{M}_0$ is inverse to $\mathbf{L}_0$, and some use of the identity

$$\mathbf{L}_0\delta\mathbf{M} + \delta\mathbf{L}(\mathbf{M}_0 + \delta\mathbf{M}) = \mathbf{0}, \tag{3.24}$$

it follows that

$$-\mathscr{F}(\boldsymbol{\tau}^*) - (\mathbf{L}_0\boldsymbol{\varepsilon}^*, \delta\mathbf{M}\mathbf{L}_0\boldsymbol{\varepsilon}^*) \leqslant 2(\bar{U} - \bar{U}_0). \tag{3.25}$$

If $\mathbf{L}_0$ is chosen so that $\delta\mathbf{M}$ is negative definite (or, equivalently, so that $\delta\mathbf{L}$ is positive definite), (3.25) implies

$$-\mathscr{F}(\boldsymbol{\tau}^*) \leqslant 2(\bar{U} - \bar{U}_0). \tag{3.26}$$

The traction boundary value problem can be treated in a similar way. Its duality with the displacement problem is best displayed by introducing a strain polarization $\boldsymbol{\eta}$, so that

$$\boldsymbol{\eta} = \mathbf{M}_0\boldsymbol{\tau} \quad \text{and} \quad \mathbf{e} = \mathbf{M}_0\boldsymbol{\sigma} - \boldsymbol{\eta}. \tag{3.27}$$

If tractions are specified that would generate the stress field $\boldsymbol{\sigma}_0$ in the comparison material, an operator $\boldsymbol{\Delta}_1$ may be defined so that

$$\boldsymbol{\sigma} = \boldsymbol{\sigma}_0 + \boldsymbol{\Delta}_1 \boldsymbol{\eta}, \tag{3.28}$$

which is analogous to (3.11). The corresponding analogue of (3.16) is

$$[\boldsymbol{\Delta}_1 + (\delta \mathbf{M})^{-1}]\boldsymbol{\eta} = -\boldsymbol{\sigma}_0, \tag{3.29}$$

with which the functional

$$\mathscr{G}_1(\boldsymbol{\eta}^*) = -(\boldsymbol{\eta}^*, [\boldsymbol{\Delta}_1 + (\delta \mathbf{M})^{-1}]\boldsymbol{\eta}^*) - 2(\boldsymbol{\eta}^*, \boldsymbol{\sigma}_0) \tag{3.30}$$

may be associated. The minimum energy principle and the complementary energy principle now in effect change roles. The analogue of (3.12), from the complementary principle, is

$$2\overline{U} \leqslant (\boldsymbol{\sigma}^*, M_0 \boldsymbol{\sigma}^*) + (\boldsymbol{\sigma}^*, \delta \mathbf{M} \boldsymbol{\sigma}^*) \tag{3.31}$$

which, expanded in terms of $\boldsymbol{\eta}^*$, is

$$2(\overline{U} - \overline{U}_0) \leqslant -(\boldsymbol{\eta}^*, \boldsymbol{\Delta}_1 \boldsymbol{\eta}^*) + (\boldsymbol{\sigma}_0, \delta \mathbf{M} \boldsymbol{\sigma}_0) + 2(\boldsymbol{\sigma}_0, \delta \mathbf{M} \boldsymbol{\Delta}_1 \boldsymbol{\eta}^*) + (\boldsymbol{\Delta}_1 \boldsymbol{\eta}^*, \delta \mathbf{M} \boldsymbol{\Delta}_1 \boldsymbol{\eta}^*). \tag{3.32}$$

This, finally, can be subjected to the reasoning that produced (3.19), to give

$$2(\overline{U} - \overline{U}_0) \leqslant -\mathscr{G}_1(\boldsymbol{\eta}^*) + (\mathbf{L}_0 \boldsymbol{\varepsilon}^*, \delta \mathbf{M} \mathbf{L}_0 \boldsymbol{\varepsilon}^*), \tag{3.33}$$

so that

$$2(\overline{U} - \overline{U}_0) \leqslant -\mathscr{G}_1(\boldsymbol{\eta}^*) \tag{3.34}$$

whenever $\mathbf{M}_0$ is chosen so that $\delta \mathbf{M}$ is negative definite (or, equivalently, so that $\delta \mathbf{L}$ is positive definite). The "error" $\boldsymbol{\varepsilon}^*$ in (3.33) still conforms to (3.21); in terms of $\boldsymbol{\eta}^*$,

$$\mathbf{L}_0 \boldsymbol{\varepsilon}^* = -[\boldsymbol{\Delta}_1 + (\delta \mathbf{M})^{-1}]\boldsymbol{\eta}^* - \boldsymbol{\sigma}_0. \tag{3.35}$$

It is, of course, possible to express the functional $\mathscr{F}$ in terms of $\boldsymbol{\eta}^*$ and $\mathscr{G}_1$ in terms of $\boldsymbol{\tau}^*$, with $\boldsymbol{\tau}^*$ and $\boldsymbol{\eta}^*$ related by $(3.27)_1$. For example, (3.16) can be written

$$[\mathbf{L}_0 (\delta \mathbf{L})^{-1} \mathbf{L}_0 + \mathbf{L}_0 \boldsymbol{\Gamma} \mathbf{L}_0]\boldsymbol{\eta} = \boldsymbol{\sigma}_0 \tag{3.36}$$

from which follows, using a rearrangement of (3.24), an equation of the form of (3.29), with the operator $\boldsymbol{\Delta}_1$ replaced by $\boldsymbol{\Delta}$, where

$$\boldsymbol{\Delta} = \mathbf{L}_0 - \mathbf{L}_0 \boldsymbol{\Gamma} \mathbf{L}_0. \tag{3.37}$$

Correspondingly, $\mathscr{F}(\boldsymbol{\tau}^*) = \mathscr{G}(\boldsymbol{\eta}^*)$, where $\mathscr{G}(\boldsymbol{\eta}^*)$ is defined by (3.30) but with $\boldsymbol{\Delta}_1$ replaced by $\boldsymbol{\Delta}$. Dually, $\mathscr{G}_1(\boldsymbol{\eta}^*) = \mathscr{F}_1(\boldsymbol{\tau}^*)$, where $\mathscr{F}_1(\boldsymbol{\tau}^*)$ is defined by (3.17), with $\boldsymbol{\Gamma}$ replaced by $\boldsymbol{\Gamma}_1$, where

$$\boldsymbol{\Gamma}_1 = \mathbf{M}_0 - \mathbf{M}_0 \boldsymbol{\Delta}_1 \mathbf{M}_0. \tag{3.38}$$

Finally, there are duals of (3.25) and (3.26), obtained from the minimum energy principle for the traction problem:

$$-\mathscr{G}_1(\boldsymbol{\eta}^*) - (\boldsymbol{\varepsilon}^*, \delta \mathbf{L} \boldsymbol{\varepsilon}^*) \leqslant 2(\overline{U} - \overline{U}_0) \tag{3.39}$$

and

$$-\mathscr{G}_1(\boldsymbol{\eta}^*) \leqslant 2(\overline{U} - \overline{U}_0) \tag{3.40}$$

whenever $\delta \mathbf{L}$ is negative definite everywhere in V.

4. Inclusion Problems

For any composite that comprises a matrix reinforced (or weakened) by inclusions, a fundamental problem is to determine the way in which any particular inclusion perturbs the field that the matrix would support in its absence. Interaction effects between inclusions are significant, but the problem of finding the effect of a single inclusion in an infinite homogeneous matrix provides an essential building block for considering a general composite. This basic problem is discussed now. The integral equations (3.8) or (3.16) are relevant, and it is natural to identify the comparison material with the matrix. With a slight distortion of earlier notation, therefore, the matrix is taken to have moduli $\mathbf{L}_0$ while the inclusion, which occupies a region D, is taken to be uniform, with moduli $\mathbf{L}$. The field $\mathbf{e}_0$ is now the strain field that would be present if the matrix contained no inclusion and the term involving $\mathbf{\Gamma}$ represents the perturbation of this field produced by the inclusion. $\delta\mathbf{L}$ is zero except over D and (3.8) is now an integral equation only for $\mathbf{x} \in D$; elsewhere, it provides a representation for the strain field $\mathbf{e}$. Equivalently, $\boldsymbol{\tau}$ is non-zero only over D and (3.16) applies only for $\mathbf{x} \in D$. Either of these equations could be studied equally well but there is some later advantage in concentrating upon (3.16).

The region D occupied by the inclusion will be assumed to be bounded. The three-dimensional problem will be considered first but afterwards a short discussion will be given of the corresponding two-dimensional problem for which the inclusion may be an infinite cylinder. The matrix is infinite in extent and so the operator $\mathbf{\Gamma}$ that appears in (3.16) is obtained, via (3.7), from the infinite-body Green's function $\mathbf{G}^\infty$. This satisfies, in matrix notation, the differential equation

$$\mathbf{L}_0(\nabla)\mathbf{G}^\infty + \mathbf{I}\,\delta(\mathbf{x}) = \mathbf{0}, \tag{4.1}$$

where $\mathbf{L}_0(\nabla)$ is a differential operator whose symbol $\mathbf{L}_0(\boldsymbol{\xi})$ has components

$$[\mathbf{L}_0(\boldsymbol{\xi})]_{ik} = (\mathbf{L}_0)_{ijkl}\xi_j\xi_l. \tag{4.2}$$

Equation (4.1) can be solved by noting the plane-wave decomposition

$$\delta(\mathbf{x}) = -\frac{1}{8\pi^2}\int_{|\boldsymbol{\xi}|=1} dS\,\delta''(\boldsymbol{\xi}\,.\,\mathbf{x}) \tag{4.3}$$

given by Gel'fand and Shilov (1964). Solving the differential equation (4.1) with $\delta(\mathbf{x})$ replaced by $\delta''(\boldsymbol{\xi}\,.\,\mathbf{x})$ and then performing the superposition implied by (4.3) gives

$$\mathbf{G}^\infty(\mathbf{x}) = \frac{1}{8\pi^2}\int_{|\boldsymbol{\xi}|=1} dS[\mathbf{L}_0(\boldsymbol{\xi})]^{-1}\delta(\boldsymbol{\xi}\,.\,\mathbf{x}). \tag{4.4}$$

The components of $\mathbf{\Gamma}^\infty$ (so called to correspond with $\mathbf{G}^\infty$) now follow using (3.7):

$$\Gamma^\infty_{ijkl}(\mathbf{x},\mathbf{x}') = -\frac{1}{8\pi^2}\int_{|\boldsymbol{\xi}|=1} dS\xi_i[\mathbf{L}_0(\boldsymbol{\xi})]^{-1}_{jk}\xi_l\delta''[\boldsymbol{\xi}\,.\,(\mathbf{x}-\mathbf{x}')]\Big|_{(ij)(kl)} \tag{4.5}$$

and substitution of (4.5) into (3.16) gives

$$(\delta\mathbf{L})^{-1}\boldsymbol{\tau}(\mathbf{x}) - \frac{1}{8\pi^2}\int_{|\boldsymbol{\xi}|=1} dS\,\tilde{\mathbf{\Gamma}}^\infty(\boldsymbol{\xi})\check{\boldsymbol{\tau}}''(\boldsymbol{\xi}\,.\,\mathbf{x},\boldsymbol{\xi}) = \mathbf{e}_0(\mathbf{x}), \quad \mathbf{x} \in D, \tag{4.6}$$

where

$$\check{\boldsymbol{\tau}}''(p, \boldsymbol{\xi}) = \frac{\partial^2}{\partial p^2}\check{\boldsymbol{\tau}}(p, \boldsymbol{\xi}) = \frac{\partial^2}{\partial p^2}\int \boldsymbol{\tau}(\mathbf{x}')\delta(p - \boldsymbol{\xi}\,.\,\mathbf{x}')\,d\mathbf{x}'. \tag{4.7}$$

$\check{\boldsymbol{\tau}}(p, \boldsymbol{\xi})$ is called the Radon transform of $\boldsymbol{\tau}(\mathbf{x})$. In (4.6), the term $\tilde{\boldsymbol{\Gamma}}^{\infty}(\boldsymbol{\xi})$ represents the part of the integrand of (4.5) that depends upon $\boldsymbol{\xi}$ only; it is actually the Fourier transform of $\boldsymbol{\Gamma}^{\infty}$.

It is not clear whether (4.6) is superior to (3.16) in general but it is particularly useful when D is an ellipsoid. To illustrate this, consider first the case for which D represents a sphere of radius a, centred at the origin. If $\boldsymbol{\tau}$ is taken constant over D, its Radon transform is obtained as $\boldsymbol{\tau}$ times the area of intersection of the plane $\boldsymbol{\xi}\,.\,\mathbf{x} = p$ with D; that is,

$$\check{\boldsymbol{\tau}}(p, \boldsymbol{\xi}) = \boldsymbol{\tau}\pi(a^2 - p^2)H(a^2 - p^2). \tag{4.8}$$

The second derivative that appears in (4.6) is then constant when $|\boldsymbol{\xi}| = 1$ and $\mathbf{x} \in D$, so that the left side of (4.6) is constant over D. If $\mathbf{e}_0(\mathbf{x})$ is taken constant, therefore, $\boldsymbol{\tau}$ may be taken constant and must satisfy the algebraic equation

$$[(\delta\mathbf{L})^{-1} + \mathbf{P}]\boldsymbol{\tau} = \mathbf{e}_0, \tag{4.9}$$

where the constant tensor $\mathbf{P}$ is given by

$$\mathbf{P} = \frac{1}{4\pi}\int_{|\boldsymbol{\xi}|=1} dS\,\tilde{\boldsymbol{\Gamma}}^{\infty}(\boldsymbol{\xi}). \tag{4.10}$$

This result was first given by Kneer (1965). The case for which D is the ellipsoid

$$D = \{\mathbf{x} : \mathbf{x}^T\mathbf{A}^T\mathbf{A}\mathbf{x} < 1\} \tag{4.11}$$

can be reduced to the one just considered by the dual transformations

$$\mathbf{y} = \mathbf{A}\mathbf{x}, \quad \boldsymbol{\xi} = \mathbf{A}^T\boldsymbol{\zeta}; \tag{4.12}$$

this reproduces equation (4.9) except that $\mathbf{P}$ is now given as

$$\mathbf{P} = \frac{1}{4\pi|\mathbf{A}|}\int_{|\boldsymbol{\xi}|=1} dS\,\tilde{\boldsymbol{\Gamma}}^{\infty}(\boldsymbol{\xi})[\boldsymbol{\xi}^T(\mathbf{A}^T\mathbf{A})^{-1}\boldsymbol{\xi}]^{-3/2}. \tag{4.13}$$

This result was given first by Khatchaturyan (1967) for a generally anisotropic matrix. For an isotropic matrix, the result was derived using methods of potential theory by Eshelby (1957), who expressed $\mathbf{P}$ in terms of elliptic integrals. Eshelby (1961) showed further that, for an inclusion in an isotropic matrix, a polynomial field $\mathbf{e}_0(\mathbf{x})$ would generate a polynomial field within the inclusion: the corresponding result for a generally anisotropic matrix was given by Asaro and Barnett (1975) and Willis (1975), essentially by demonstrating that, if $\boldsymbol{\tau}(\mathbf{x})$ is a polynomial of degree n in $\mathbf{x}$, then $\check{\boldsymbol{\tau}}(p, \boldsymbol{\xi})$ is a polynomial of degree $n+2$ in p.

The tensor $\mathbf{P}$ given by (4.13) has generally to be computed. However, for illustration, it will be evaluated for a spherical inclusion in an isotropic matrix. Thus, $\mathbf{L}_0$ is taken to have components

$$(\mathbf{L}_0)_{ijkl} = \kappa\delta_{ij}\delta_{kl} + \mu(\delta_{ik}\delta_{jl} + \delta_{il}\delta_{jk} - \tfrac{2}{3}\delta_{ij}\delta_{kl}), \tag{4.14}$$

so that it has bulk modulus κ and shear modulus μ. It is helpful to use the symbolic notation

$$\mathbf{L}_0 = (3\kappa, 2\mu) \tag{4.15}$$

introduced by Hill (1965a), in which the product of two isotropic tensors $\mathbf{A} = (3\kappa_A, 2\mu_A)$ and $\mathbf{B} = (3\kappa_B, 2\mu_B)$ is given by

$$\mathbf{AB} = [(3\kappa_A)(3\kappa_B), (2\mu_A)(2\mu_B)]. \tag{4.16}$$

It is easily deduced from (4.14) that

$$[\mathbf{L}_0(\boldsymbol{\xi})]_{ik}^{-1} = \left[|\boldsymbol{\xi}|^2 \delta_{ik} - \left(\frac{3\kappa+\mu}{3\kappa+4\mu}\right)\xi_i\xi_k\right] \Big/ \mu|\boldsymbol{\xi}|^4 \tag{4.17}$$

and the components of $\tilde{\mathbf{\Gamma}}^\infty$ follow correspondingly. The tensor $\mathbf{P}$ is isotropic and its "moduli" κ_p, μ_p may be obtained from the scalars

$$P_{iikk} = 9\kappa_p, \quad P_{ijij} = 3\kappa_p + 10\mu_p, \tag{4.18}$$

for each of which the integrand of (4.10) reduces to a constant. Completing the details gives

$$(3\kappa_p, 2\mu_p) = \left(\frac{1}{3\kappa+4\mu}, \frac{3(\kappa+2\mu)}{5\mu(3\kappa+4\mu)}\right). \tag{4.19}$$

No other special case is particularly simple. However, some reduction is possible when the inclusion is long and thin, so that it approximates to a cylinder with elliptic cross-section. If its principal axes are aligned with coordinate axes and the 3-axis is the one of greatest length, the matrix $\mathbf{A}$ may be taken diagonal, with entries $(1/a_1, 1/a_2, \varepsilon)$, where ε is small. The integral in (4.13) may now be transformed by projection onto the circular cylinder of unit cross-section defined by $\xi_1^2 + \xi_2^2 = 1$, $-\infty < \xi_3 < \infty$. The element of area on this cylinder is $ds\, d\xi_3$, s representing arc length around a cross-section, and the element of solid angle dS is given by

$$dS = ds\, d\xi_3\, (1+\xi_3^2)^{-3/2}. \tag{4.20}$$

The function $\tilde{\mathbf{\Gamma}}^\infty(\boldsymbol{\xi})$ is unchanged by the projection because it is homogeneous of degree zero. The variable ξ_3 is now replaced by $\xi_3' = \xi_3/\varepsilon$ and (4.13) takes the form

$$\mathbf{P} = \frac{a_1 a_2}{4\pi}\int_{\xi_1^2+\xi_2^2=1} ds \int_{-\infty}^{\infty} d\xi_3'\, \tilde{\mathbf{\Gamma}}^\infty(\boldsymbol{\xi})[a_1^2\xi_1^2 + a_2^2\xi_2^2 + \xi_3'^2]^{-3/2}(1+\varepsilon^2\xi_3'^2)^{-3/2}. \tag{4.21}$$

The integral with respect to ξ_3' may be estimated asymptotically by setting $\varepsilon = 0$ directly; performing the resulting elementary integration with respect to ξ_3' then gives

$$\mathbf{P} \sim \frac{a_1 a_2}{2\pi}\int_{\xi_1^2+\xi_2^2=1} ds\, \tilde{\mathbf{\Gamma}}^\infty(\xi_1, \xi_2, 0)[a_1^2\xi_1^2 + a_2^2\xi_2^2]^{-1}. \tag{4.22}$$

Equation (4.22) offers a solution to the cylindrical inclusion problem that is an alternative to the usual one involving complex variables.

Reverting now to composites, suppose that a composite comprises a matrix with moduli $\mathbf{L}_0$, which contains n different types of ellipsoidal inclusion, with moduli $\mathbf{L}_r$, $r = 1, 2, \ldots, n$, distributed at volume concentrations c_r, where each c_r is small. The overall moduli of the composite may be estimated by imposing a uniform strain $\bar{\mathbf{e}}$ at infinity. The mean stress $\bar{\boldsymbol{\sigma}}$ in the composite is given by the formula

$$\bar{\boldsymbol{\sigma}} = \mathbf{L}_0\bar{\mathbf{e}} + \sum_{r=1}^{n} c_r \boldsymbol{\tau}_r, \tag{4.23}$$

where $\boldsymbol{\tau}_r$ represents the mean polarization in the rth phase. To lowest order in c_r, the inclusions do not interact and $\boldsymbol{\tau}$ may be estimated as the solution of (4.9), with $\mathbf{e}_0 = \bar{\mathbf{e}}$, $\delta\mathbf{L} = \mathbf{L}_r - \mathbf{L}_0$, and $\mathbf{P} = \mathbf{P}_r$, the value appropriate to an inclusion of rth type. The low-concentration estimate of the tensor of overall moduli then follows, using (2.1), as

$$\bar{\mathbf{L}} \sim \mathbf{L}_0 + \sum_{r=1}^{n} \mathbf{c}_r[(\mathbf{L}_r - \mathbf{L}_0)^{-1} + \mathbf{P}_r]^{-1}. \tag{4.24}$$

This result is (asymptotically) exact. It is affected by the symmetry of the individual inclusions but not by their relative positions because they are by definition far apart. Estimates that apply at any concentration are considered in the following Sections.

5. The Self-consistent Method

Consider now a composite comprising a matrix containing n different types of inclusions, not necessarily ellipsoidal, with tensors of moduli $\mathbf{L}_r$ and volume concentrations c_r, $r = 1, 2, \ldots, n$, not necessarily small. A comparison material different from the matrix will be used, so the matrix is now regarded as phase $n+1$, with moduli $\mathbf{L}_{n+1}$ and volume concentration c_{n+1}. Let the average values of the stress and strain over the rth phase be denoted $\boldsymbol{\sigma}_r$, $\mathbf{e}_r$. Then

$$\boldsymbol{\sigma}_r = \mathbf{L}_r \mathbf{e}_r, \tag{5.1}$$

$$\bar{\mathbf{e}} = \sum_{r=1}^{n+1} c_r \mathbf{e}_r, \tag{5.2}$$

and

$$\bar{\boldsymbol{\sigma}} = \sum_{r=1}^{n+1} c_r \boldsymbol{\sigma}_r = \sum_{r=1}^{n+1} c_r \mathbf{L}_r \mathbf{e}_r. \tag{5.3}$$

Eliminating $c_{n+1}\mathbf{e}_{n+1}$ between (5.2) and $(5.3)_2$ gives

$$\bar{\boldsymbol{\sigma}} = \mathbf{L}_{n+1}\bar{\mathbf{e}} + \sum_{r=1}^{n} c_r(\mathbf{L}_r - \mathbf{L}_{n+1})\mathbf{e}_r \tag{5.4}$$

so that, if $\bar{\mathbf{e}}$ is prescribed and

$$\mathbf{e}_r = \mathbf{A}_r \bar{\mathbf{e}}, \tag{5.5}$$

the tensor of overall moduli $\bar{\mathbf{L}}$ is given by

$$\bar{\mathbf{L}} = \mathbf{L}_{n+1} + \sum_{r=1}^{n} c_r(\mathbf{L}_r - \mathbf{L}_{n+1})\mathbf{A}_r. \tag{5.6}$$

The problem now is to estimate the tensor $\mathbf{A}_r$. An approximate method for performing this estimation is to embed a single inclusion of rth type in an infinite matrix composed of the overall material. If the inclusion is an ellipsoid, $\mathbf{A}_r$ follows from the results of the preceding Section, with $\mathbf{L}_0 = \bar{\mathbf{L}}$ and $\delta\mathbf{L} = \mathbf{L}_r - \mathbf{L}_0$. The polarization $\boldsymbol{\tau}_r$ is found from (4.9) with $\mathbf{e}_0 = \bar{\mathbf{e}}$ and then (3.9) gives

$$\mathbf{A}_r = (\mathbf{L}_r - \bar{\mathbf{L}})^{-1}[(\mathbf{L}_r - \bar{\mathbf{L}})^{-1} + \mathbf{P}_r]^{-1} = [\mathbf{I} + \mathbf{P}_r(\mathbf{L}_r - \bar{\mathbf{L}})]^{-1}. \tag{5.7}$$

If the rth inclusion is not an ellipsoid, the solution of the inclusion problem is less straightforward. It is convenient for the discussion to follow, however, to treat (5.7) as defining $\mathbf{P}_r$ in such cases. Substituting (5.7) into (5.6) now gives an algebraic equation for $\bar{\mathbf{L}}$, since it appears on the right side both directly and implicitly through the tensor $\mathbf{P}_r$. The estimate for $\bar{\mathbf{L}}$ that results from solving this equation is usually termed the self-consistent estimate. The prescription was developed in essentially this form by Hill (1965c) and Budiansky (1965). It allows for inclusion shape and also approximately for interactions because the inclusion is embedded in the "overall" material. It also reduces correctly to (4.24) in the limit of low concentrations. A possible objection is that the inclusions actually interface with "overall" material rather than with the matrix. One method of overcoming this is to surround the inclusion by a "shell" of matrix material and then to embed this new compound inclusion in an infinite matrix of "overall" material. For spherical inclusions, it is reasonable to take the shell as spherical and to choose its radius so that the inclusion concerned occupies exactly its share of the matrix. Kerner (1956) solved this problem for isotropic phases, exactly for an imposed dilatation and approximately for an imposed shear. The exact solution for an imposed shear has been provided by Christensen and Lo (1979).

Suppose now that the composite has no clearly defined matrix phase or, equivalently, that $c_{n+1} = 0$. The prescription (5.6) fails in this case but, if the shapes of the phases can still be distinguished, it remains possible to estimate the tensors $\mathbf{A}_r$ by solving inclusion problems. Three possible prescriptions for estimating $\bar{\mathbf{L}}$ are now available. Equation (5.2) requires that

$$\sum_{r=1}^{n} c_r[\mathbf{I} + \mathbf{P}_r(\mathbf{L}_r - \bar{\mathbf{L}})]^{-1} = \mathbf{I}, \tag{5.8}$$

$(5.3)_2$ requires

$$\sum_{r=1}^{n} c_r\mathbf{L}_r[\mathbf{I} + \mathbf{P}_r(\mathbf{L}_r - \bar{\mathbf{L}})]^{-1} = \bar{\mathbf{L}} \tag{5.9}$$

and, finally, $\bar{\boldsymbol{\sigma}} = \bar{\mathbf{L}}\bar{\mathbf{e}} + \Sigma c_r\boldsymbol{\tau}_r$ implies

$$\sum_{r=1}^{n} c_r[(\mathbf{L}_r - \bar{\mathbf{L}})^{-1} + \mathbf{P}_r]^{-1} = \mathbf{0}. \tag{5.10}$$

Equations (5.8)–(5.10) are not all equivalent, for (5.8) can be expressed in the form

$$\sum_{r=1}^{n} c_r\mathbf{P}_r[(\mathbf{L}_r - \bar{\mathbf{L}}) + \mathbf{P}_r]^{-1} = \mathbf{0} \tag{5.11}$$

and (5.10) is equivalent to

$$\left\{\sum_{r=1}^{n} c_r\mathbf{L}_r[\mathbf{I} + \mathbf{P}_r(\mathbf{L}_r - \bar{\mathbf{L}})]^{-1}\right\}\left\{\sum_{s=1}^{n} c_s[\mathbf{I} + \mathbf{P}_s(\mathbf{L}_s - \bar{\mathbf{L}})]^{-1}\right\}^{-1} = \bar{\mathbf{L}}. \tag{5.12}$$

It is clear from these relations that (5.8)–(5.10) become equivalent when $\mathbf{P}_r = \mathbf{P}$ for all r, i.e. when all phases can be modelled by similar ellipsoids. The prescriptions (5.8)–(5.10) were proposed variously by Hershey (1954), Kröner (1958), and Eshelby (1961) for estimating the overall moduli of a polycrystal by treating each grain as a sphere. Their equivalance was proved, essentially as above, by Hill (1965a). It is true also that, when $\mathbf{P}_r = \mathbf{P}$ for all r, (5.6) can be manipulated into any of the forms (5.8)–(5.10), with n replaced by $n+1$. This was noted for spheres in an isotropic matrix by Budiansky (1965) and generally by Hill (1965c).

The great advantage of the self-consistent method is its simplicity. Furthermore, it will be seen when examples are discussed later that it often yields results of acceptable accuracy. There are, however, some situations with which it cannot be expected to cope. If, for example, an isotropic matrix contained spherical isotropic inclusions, arranged on a lattice, the simple self-consistent method would estimate the overall moduli as isotropic even though they should plainly reflect the symmetry of the lattice. Although exact methods are available for periodic composites, real composites are usually not perfectly periodic, though they may well display some directional bias "on average"; the inadequacy of the self-consistent method would remain in this case also. The Section that follows presents a method that makes systematic allowance for such phase geometry.

6. Bounds for Overall Moduli

If every detail of the geometry of a composite were known, overall moduli could, in principle, be calculated exactly by solving the relevant boundary value problem. In practice, however, except in cases such as those displaying periodicity, a complete solution could not even be computed because of the detail that it would contain. Also, in practice, only certain "average" features of the microstructure will be known. If, for example, only the moduli and mean concentrations of the phases are known, the only completely accurate statement that can be made is embodied in the inequalities (2.15); that is, that the Reuss and Voigt estimates bound the overall moduli. If more information is available, however, it should be possible to provide better bounds. This information is likely to be statistical in character, and it is necessary now to make a slight digression to discuss random media and introduce thereby the quantities that will be employed.

6.1. *Random media*

A random medium is one of a family, any member of which may be characterized by a label α that belongs to a sample space $\mathscr{S}$. The value of α is taken as defining the medium completely. It has associated with it a probability density $p(\alpha)$ which is defined over $\mathscr{S}$. The sample space $\mathscr{S}$ could be of low dimension, e.g. if the random medium were known to be perfectly periodic, except for uncertainty over the position of any one representative cell, or it could be of infinite dimension, with α representing the values of r at each individual point of the medium. For an n-phase material, it is convenient to introduce the indicator function $f_r(\mathbf{x})$, that takes the value 1 if $\mathbf{x}$ lies in phase r and zero otherwise; it depends, of course, on α. The mean value, or ensemble average, of $f_r(\mathbf{x})$ defines the probability $P_r(\mathbf{x})$ of finding phase r at $\mathbf{x}$. Thus,

$$P_r(\mathbf{x}) = \langle f_r(\mathbf{x}) \rangle = \int_{\mathscr{S}} f_r(\mathbf{x}, \alpha) p(\alpha) \, d\alpha. \tag{6.1}$$

Likewise, the probability $P_{rs}(\mathbf{x}, \mathbf{x}')$ of finding simultaneously phase r at $\mathbf{x}$ and phase s at $\mathbf{x}'$ is

$$P_{rs}(\mathbf{x}, \mathbf{x}') = \langle f_r(\mathbf{x}) f_s(\mathbf{x}') \rangle = \int_{\mathscr{S}} f_r(\mathbf{x}, \alpha) f_s(\mathbf{x}', \alpha) p(\alpha) \, d\alpha. \tag{6.2}$$

Probabilities involving more points are defined similarly. If, now, phase r has moduli $\mathbf{L}_r$,

$r = 1, 2, \ldots, n$, the value $\mathbf{L}(\mathbf{x})$ of the tensor of moduli at $\mathbf{x}$ can be written

$$\mathbf{L}(\mathbf{x}) = \sum_{r=1}^{n} \mathbf{L}_r f_r(\mathbf{x}) \tag{6.3}$$

and this has, using (6.1), mean value

$$\langle \mathbf{L}(\mathbf{x}) \rangle = \sum_{r=1}^{n} \mathbf{L}_r P_r(\mathbf{x}). \tag{6.4}$$

Similarly,

$$\langle \mathbf{L}(\mathbf{x}) \otimes \mathbf{L}(\mathbf{x}') \rangle = \sum_{r=1}^{n} \sum_{s=1}^{n} \mathbf{L}_r \otimes \mathbf{L}_s P_{rs}(\mathbf{x}, \mathbf{x}'). \tag{6.5}$$

The individual components in (6.5) represent two-point correlation functions. Correlation functions for more points follow a similar pattern; clearly, they are expressible in terms of $P_r(\mathbf{x})$, $P_{rs}(\mathbf{x}, \mathbf{x}')$, and so on, for any physical quantities whose values are given for each phase.

The foregoing description is useful for developing statements "on average", taken over many specimens produced, perhaps, by some manufacturing process with which the probability $p(\alpha)$ is associated. A statistically uniform material is one for which the probabilities $P_r(\mathbf{x})$, $P_{rs}(\mathbf{x}, \mathbf{x}')$, etc. are insensitive to translations; thus $P_r(\mathbf{x})$ reduces to a constant, $P_{rs}(\mathbf{x}, \mathbf{x}')$ becomes a function of $(\mathbf{x} - \mathbf{x}')$ only, and so on. For such a medium it is usual to make an assumption of ergodic type, that local configurations occur over any one specimen with the frequency with which they occur over a single neighbourhood in an ensemble of specimens. In this case, the probability P_r is identified with the volume average of the function $f_r(\mathbf{x})$ and so represents the volume concentration c_r of phase r, while $P_{rs}(\mathbf{x}, \mathbf{x}')$ becomes the average of $f_r(\mathbf{x} + \mathbf{x}'') f_s(\mathbf{x}' + \mathbf{x}'')$ as $\mathbf{x}''$ ranges over a large volume. Strictly, these notions require the medium to be infinite but they provide useful empirical definitions in practice. Clearly, it will be possible to determine multipoint probabilities only up to some relatively low order. Correspondingly, Kröner (1977) proposed classifying materials as uniform "of grade k" if probabilities involving up to k points were known to be translation-invariant. He classified other properties (such as isotropy) similarly.

For future use, the two-point probabilities $P_{rs}(\mathbf{x}, \mathbf{x}')$ are now discussed for a two-phase material. They satisfy the relations

$$\begin{aligned} &P_1(\mathbf{x}) + P_2(\mathbf{x}) = 1, \\ &P_{11}(\mathbf{x}, \mathbf{x}') + P_{12}(\mathbf{x}, \mathbf{x}') = P_1(\mathbf{x}) = P_{11}(\mathbf{x}', \mathbf{x}) + P_{21}(\mathbf{x}', \mathbf{x}), \\ &P_{12}(\mathbf{x}', \mathbf{x}) + P_{22}(\mathbf{x}', \mathbf{x}) = P_2(\mathbf{x}) = P_{21}(\mathbf{x}, \mathbf{x}') + P_{22}(\mathbf{x}, \mathbf{x}') \end{aligned} \tag{6.6}$$

always. Therefore, when the medium is statistically uniform (up to grade two),

$$P_{11} = c_1^2 + c_1 c_2 h, \quad P_{12} = c_1 c_2 (1 - h), \quad P_{22} = c_2^2 + c_1 c_2 h, \tag{6.7}$$

where h is some function that depends upon $(\mathbf{x} - \mathbf{x}')$ only which, furthermore, is an even function of its argument. If, in addition, the medium has no long-range order, $P_{rs} \sim c_r c_s$ at large separations of $\mathbf{x}$ and $\mathbf{x}'$ and h tends to zero as $|\mathbf{x} - \mathbf{x}'|$ tends to infinity. If the distribution of phases is isotropic (up to grade 2), then h is a function of $|\mathbf{x} - \mathbf{x}'|$ only; it should be noted that this statement makes no reference to the elastic symmetries of the phases themselves so that the composite could still be elastically anisotropic.

Suppose now that the composite comprises a matrix in which are embedded inclusions of known shapes and orientations. All mean values for this composite follow from knowledge of the probability of finding any specified number of inclusions in any specified region. This will not be discussed in detail but, for purposes of later application, if the matrix contains just a single population of inclusions, centred at $\mathbf{x}_A$, $A = 1, 2, \ldots, N$, P_A will be defined as the probability density for finding an inclusion centred at $\mathbf{x}_A$, P_{AB} will denote the joint probability density for finding different inclusions centred at $\mathbf{x}_A$, $\mathbf{x}_B$, and so on. Statistical uniformity, isotropy, etc. for such a composite is expressed in terms of P_A, P_{AB}, etc. in a similar manner to that described above for P_r, P_{rs}, etc.

6.2. *Hashin–Shtrikman bounds*

In most applications, it is unlikely that information of grade higher than two (in Kröner's terminology) will be available. Therefore, although formal extensions are possible (and will be outlined later), attention will be focused now on the best bounds that can be obtained using two-point correlations only. Inequalities such as (3.15) which are derived from the classical energy principles are excluded immediately for, written explicitly, the last term in (3.15) is

$$V^{-1}\int d\mathbf{x} \int d\mathbf{x}'\, \boldsymbol{\Gamma}(\mathbf{x}, \mathbf{x}')\boldsymbol{\tau}^*(\mathbf{x}')\,\delta\mathbf{L}(\mathbf{x}) \int d\mathbf{x}''\, \boldsymbol{\Gamma}(\mathbf{x}, \mathbf{x}'')\boldsymbol{\tau}^*(\mathbf{x}''),$$

which involves points taken three at a time. The Hashin–Shtrikman inequalities (3.20), (3.26) were constructed to contain no such term and bounds can be obtained from them by considering trial fields of the form

$$\boldsymbol{\tau}^*(\mathbf{x}) = \sum_{r=1}^{n} \boldsymbol{\tau}_r(\mathbf{x}) f_r(\mathbf{x}), \tag{6.8}$$

for an n-phase material, where the $\boldsymbol{\tau}_r$ are functions of $\mathbf{x}$ alone, independent of the configuration: any relaxation of this restriction would introduce terms into $\mathcal{F}(\boldsymbol{\tau}^*)$ that would involve more than two points, so that (6.8) is the most general trial field that can be allowed.

Substituting (6.8) into (3.17) now gives

$$\mathcal{F}(\boldsymbol{\tau}^*) = V^{-1}\Bigg\{ \sum_{r=1}^{n} \int d\mathbf{x}\, f_r(\mathbf{x})\boldsymbol{\tau}_r(\mathbf{x})\big[(\mathbf{L}_r - \mathbf{L}_0)^{-1}\boldsymbol{\tau}_r(\mathbf{x}) - 2\,\mathbf{e}_0(\mathbf{x})\big]$$

$$+ \sum_{r=1}^{n}\sum_{s=1}^{n} \int d\mathbf{x}\, f_r(\mathbf{x})\boldsymbol{\tau}_r(\mathbf{x}) \int d\mathbf{x}'\, \boldsymbol{\Gamma}(\mathbf{x}, \mathbf{x}')\boldsymbol{\tau}_s(\mathbf{x}') f_s(\mathbf{x}')\Bigg\}. \tag{6.9}$$

This depends upon the sample α, as it must. Further simplification is possible subject to the assumption that the functions f_r vary rapidly over regions that are still sufficiently small relative to the macroscale for the functions $\boldsymbol{\tau}_r(\mathbf{x})$, $\boldsymbol{\Gamma}(\mathbf{x}, \mathbf{x}')$ to be treated as effectively constant. It is credible (though not rigorously proven) in this case that the value of $\mathcal{F}(\boldsymbol{\tau}^*)$ is independent of α and so identical with its mean value, or ensemble average. Then (6.9) reduces to

$$\mathcal{F}(\boldsymbol{\tau}^*) = V^{-1}\Bigg\{ \sum_{r=1}^{n} \int d\mathbf{x}\, P_r(\mathbf{x})\boldsymbol{\tau}_r(\mathbf{x})\big[(\mathbf{L}_r - \mathbf{L}_0)^{-1}\boldsymbol{\tau}_r(\mathbf{x}) - 2\,\mathbf{e}_0(\mathbf{x})\big]$$

$$+ \sum_{r=1}^{n}\sum_{s=1}^{n} \int d\mathbf{x}\, \boldsymbol{\tau}_r(\mathbf{x}) \int d\mathbf{x}'\, \boldsymbol{\Gamma}(\mathbf{x}, \mathbf{x}')\boldsymbol{\tau}_s(\mathbf{x}') P_{rs}(\mathbf{x}, \mathbf{x}')\Bigg\}. \tag{6.10}$$

This is extremized when

$$(\mathbf{L}_r - \mathbf{L}_0)^{-1}\boldsymbol{\tau}_r(\mathbf{x})P_r(\mathbf{x}) + \sum_{s=1}^{n}\int d\mathbf{x}'\,\boldsymbol{\Gamma}(\mathbf{x},\mathbf{x}')\boldsymbol{\tau}_s(\mathbf{x}')P_{rs}(\mathbf{x},\mathbf{x}') = \mathbf{e}_0(\mathbf{x})P_r(\mathbf{x}),\qquad r = 1, 2, \ldots, n, \tag{6.11}$$

to give the extreme value $-(\langle\boldsymbol{\tau}\rangle, \mathbf{e}_0)$, $\langle\boldsymbol{\tau}\rangle$ denoting the ensemble average of $\boldsymbol{\tau}(\mathbf{x})$.

It may be noted that, if the integral equation (3.16) is averaged conditionally upon phase r being present at $\mathbf{x}$, it gives

$$(\mathbf{L}_r - \mathbf{L}_0)^{-1}\langle\boldsymbol{\tau}(\mathbf{x})\rangle_r^r + \sum_{s=1}^{n}\int d\mathbf{x}'\,\boldsymbol{\Gamma}(\mathbf{x},\mathbf{x}')\langle\boldsymbol{\tau}(\mathbf{x}')\rangle_{rs}^{s}P_{s|r}(\mathbf{x}',\mathbf{x}) = \mathbf{e}_0(\mathbf{x}), \tag{6.12}$$

where $\langle\boldsymbol{\tau}(\mathbf{x})\rangle_r^r$ denotes the expectation value of $\boldsymbol{\tau}(\mathbf{x})$, conditional upon phase r being present at $\mathbf{x}$, $\langle\boldsymbol{\tau}(\mathbf{x}')\rangle_{rs}^{s}$ denotes the expectation value of $\boldsymbol{\tau}(\mathbf{x}')$, conditional upon phases r and s being present at $\mathbf{x}$, $\mathbf{x}'$ respectively and $P_{s|r}(\mathbf{x}', \mathbf{x})$ is the corresponding conditional probability for finding phase s at $\mathbf{x}'$, so that

$$P_{rs}(\mathbf{x},\mathbf{x}') = P_{s|r}(\mathbf{x}',\mathbf{x})P_r(\mathbf{x}). \tag{6.13}$$

Equation (6.12) reduces to (6.11) if $\langle\boldsymbol{\tau}\rangle_r^r$ is identified with $\boldsymbol{\tau}_r$ and, in addition,

$$\langle\boldsymbol{\tau}(\mathbf{x}')\rangle_{rs}^{s} = \langle\boldsymbol{\tau}(\mathbf{x}')\rangle_{s}^{s}. \tag{6.14}$$

Equation (6.14) is not, of course, strictly correct unless the composite has some special structure such as periodicity. It is an example of the "quasicrystalline approximation", introduced as an *ad hoc* approximation by Lax (1952); its appearance from the variational principle at least partially explains its success.

The discussion so far has been rather general; it is desirable now to specialize to the case of a statistically uniform medium and to the boundary condition (2.2), so that $\mathbf{e}_0 = \bar{\mathbf{e}}$, to obtain bounds upon overall moduli. The body has already been assumed to be large relative to the microstructure, and it might be expected that $\boldsymbol{\Gamma}(\mathbf{x}, \mathbf{x}')$ could be replaced by its infinite-body form (4.5). This is not quite correct because this is of order $|\mathbf{x}|^{-3}$ at large $|\mathbf{x}|$ so that the integral in (6.11) would converge only conditionally. It may be noted, however, that the mean value of the exact $\boldsymbol{\Gamma}$ over V is zero, since it is associated with zero boundary displacement. Consequently, for any field $\boldsymbol{\tau}$ that oscillates rapidly about a mean $\bar{\boldsymbol{\tau}}$, $\boldsymbol{\Gamma}\boldsymbol{\tau} = \boldsymbol{\Gamma}(\boldsymbol{\tau} - \bar{\boldsymbol{\tau}})$ exactly and, in this latter form, a convergent integral is obtained if $\boldsymbol{\Gamma}$ is now replaced by $\boldsymbol{\Gamma}^{\infty}$. This does not prove that this prescription is correct, but further arguments for its plausibility have been advanced by Korringa (1973), Willis and Acton (1976), and Willis (1977), which are also developed slightly differently in the review of Willis (1981). These essentially hinge upon representing $\boldsymbol{\Gamma}$ in terms of the infinite-body operator as above, plus an additional term whose influence will be felt only in a layer close to the boundary ∂V. In this approximation, the operator $\boldsymbol{\Gamma}$ is insensitive to translations and the functions $\boldsymbol{\tau}_r(\mathbf{x})$ reduce to constants $\boldsymbol{\tau}_r$ that satisfy the algebraic equations

$$(\mathbf{L}_r - \mathbf{L}_0)^{-1}\boldsymbol{\tau}_r + \sum_{s=1}^{n}\int d\mathbf{x}'\,\boldsymbol{\Gamma}^{\infty}(\mathbf{x}')[P_{s|r}(\mathbf{x}') - c_s]\boldsymbol{\tau}_s = \bar{\mathbf{e}},\quad r = 1, 2, \ldots, n, \tag{6.15}$$

where $P_{s|r}(\mathbf{x}')$ represents the probability of finding phase s at $\mathbf{x}'$, given phase r at the origin. The concentration c_s is its mean value. Having solved the system (6.15) to give

$$\boldsymbol{\tau}_r = \mathbf{S}_r\bar{\mathbf{e}}, \tag{6.16}$$

say, the extreme value of $\mathscr{F}(\boldsymbol{\tau}^*)$ is expressible in the form

$$\mathscr{F}(\boldsymbol{\tau}^*) = \bar{\mathbf{e}}(\mathbf{L}_0 - \tilde{\mathbf{L}})\bar{\mathbf{e}}, \tag{6.17}$$

where the estimate

$$\tilde{\mathbf{L}} = \mathbf{L}_0 + \sum_{r=1}^{n} c_r \mathbf{S}_r \tag{6.18}$$

for the overall moduli bounds $\bar{\mathbf{L}}$ from below if $\mathbf{L}_r - \mathbf{L}_0$ is positive definite for each r, and from above if $\mathbf{L}_r - \mathbf{L}_0$ is negative definite for each r.

The solution (6.18) takes a particularly simple form if all of the functions $P_{s|r}(\mathbf{x}')$ are isotropic and so depend upon $|\mathbf{x}'|$ only. The result given in Section 4, that the integral of $\boldsymbol{\Gamma}^\infty$ over any sphere is $\mathbf{P}$, independently of its radius, implies that the integral of $\boldsymbol{\Gamma}^\infty$ over any shell $|\mathbf{x}'| = r > 0$ is zero. The integrals in (6.15) may therefore be replaced by integrals over a small sphere centred at the origin, within which $P_{s|r}$ differs negligibly from its value at the origin, namely, δ_{rs}. Evaluating the integral then gives

$$[(\mathbf{L}_r - \mathbf{L}_0)^{-1} + \mathbf{P}]\boldsymbol{\tau}_r - \mathbf{P}\langle \boldsymbol{\tau} \rangle = \bar{\mathbf{e}}, \tag{6.19}$$

where $\langle \boldsymbol{\tau} \rangle = \sum c_s \boldsymbol{\tau}_s$. It follows now that

$$\langle \boldsymbol{\tau} \rangle = \{\textstyle\sum c_r[\mathbf{I} + (\mathbf{L}_r - \mathbf{L}_0)\mathbf{P}]^{-1}\}^{-1} \textstyle\sum c_s[\mathbf{I} + (\mathbf{L}_s - \mathbf{L}_0)\mathbf{P}]^{-1}(\mathbf{L}_s - \mathbf{L}_0) \tag{6.20}$$

and, correspondingly,

$$\tilde{\mathbf{L}} = \{\textstyle\sum c_r[\mathbf{I} + (\mathbf{L}_r - \mathbf{L}_0)\mathbf{P}]^{-1}\}^{-1} \textstyle\sum c_s[\mathbf{I} + (\mathbf{L}_s - \mathbf{L}_0)\mathbf{P}]^{-1}\mathbf{L}_s. \tag{6.21}$$

The estimate (6.21) was originally produced by Hashin and Shtrikman (1962a, b, 1963) from their variational principle and by a different but equivalent method by Walpole (1966a, b). Both derivations relied upon the isotropy of the phase geometry from an early stage, however, and the equations (6.15), which allow for any two-point probabilities, were derived much more recently, by Willis (1977). It was also observed by Willis (1977) that the same formal simplification occurs if the function has "ellipsoidal" symmetry, such as would be obtained if an isotropic distribution were subjected to an affine transformation. Overall moduli are still bounded by (6.21) except that the tensor $\mathbf{P}$ is now defined by (4.13).

Equations (6.15) can also be solved for a two-phase composite with arbitrary geometry. In terms of the function h introduced in equations (6.7), let

$$\mathbf{P}' = \int dx'\, \boldsymbol{\Gamma}^\infty(\mathbf{x}')h(\mathbf{x}'). \tag{6.22}$$

Equations (6.15) now take the form of (6.19), with $n = 2$ and $\mathbf{P}$ replaced by $\mathbf{P}'$ and their solution yields the estimate (6.21), similarly modified, for $\tilde{\mathbf{L}}$.

Willis (1977) suggested that the estimates (6.21), with $\mathbf{P}$ taking "ellipsoidal" form, might be appropriate to discuss a composite containing aligned ellipsoids including, as limiting cases, thin needles and flat cracks. Justification was provided by Willis (1978), through explicit study of a composite containing inclusions, whose distribution was described by the probability densities P_A, $P_{B|A}$, etc. for the positions of their centres. If $P_{B|A}$ was "ellipsoidal", (6.21) was reproduced. Willis (1980a) also generated a result rather like the one above with $\mathbf{P}$ replaced by $\mathbf{P}'$ for a matrix containing a single population of aligned inclusions of any shape, distributed arbitrarily. Only the "ellipsoidal" example has been worked out in detail, however.

6.3. *Bounds for compliances*

Bounds for compliances are obtained by inversion of $\tilde{\mathbf{L}}$ to produce corresponding estimates $\tilde{\mathbf{M}}$ for $\overline{\mathbf{M}}$. These can also be obtained directly, however, by considering the traction boundary value problem, with boundary condition (2.5). The bounds follow by extremizing the functional $\mathscr{G}_1(\boldsymbol{\eta}^*)$ given by (3.30): their structure is exactly like that for $\overline{\mathbf{L}}$, except that $(\mathbf{L}_r - \mathbf{L}_0)$ is replaced by $(\mathbf{M}_0 - \mathbf{M}_r)$ and $\boldsymbol{\Gamma}$ is replaced by $-\Delta_1$. The infinite-body form for the operator $\boldsymbol{\Delta}_1$ can be deduced by noting that, when V is large, (3.16) and (3.29) must be consistent, with $\boldsymbol{\tau} = \mathbf{L}_0\boldsymbol{\eta}$, $\mathbf{e}_0 = \bar{\mathbf{e}}$, and $\boldsymbol{\sigma}_0 = \bar{\boldsymbol{\sigma}} = \mathbf{L}_0\bar{\mathbf{e}} + \langle \boldsymbol{\tau} \rangle$. The result is that, for large V, $\boldsymbol{\Delta}_1$ is defined by

$$\boldsymbol{\Delta}_1 : \boldsymbol{\eta} \rightarrow (\mathbf{L}_0 - \mathbf{L}_0\boldsymbol{\Gamma}^\infty\mathbf{L}_0)(\boldsymbol{\eta} - \bar{\boldsymbol{\eta}}). \tag{6.23}$$

Then, for a composite with "ellipsoidal" symmetry, for example,

$$\tilde{\mathbf{M}} = \{\sum c_r[\mathbf{I} + (\mathbf{M}_0 - \mathbf{M}_r)\mathbf{Q}]^{-1}\}^{-1}\sum c_s[\mathbf{I} + (\mathbf{M}_0 - \mathbf{M}_s)\mathbf{Q}]^{-1}\mathbf{M}_s, \tag{6.24}$$

where

$$\mathbf{Q} = \mathbf{L}_0\mathbf{P}\mathbf{L}_0 - \mathbf{L}_0. \tag{6.25}$$

It can be verified directly, using (3.24), that $\tilde{\mathbf{L}}$ and $\tilde{\mathbf{M}}$, as given by (6.21) and (6.24), are inverses.

6.4. *Bounds of higher order*

The bounds that are given by (6.18) are not the best that are obtainable from the trial field (6.8) for $\boldsymbol{\tau}^*$ since they were derived from the Hashin–Shtrikman principle rather than the stronger classical principles. The penalty for using the latter is that three-point probabilities will also be involved. Still, assuming that these are known, (6.8) is now substituted into the minimum energy principle (3.15), to give

$$2(\bar{U} - \bar{U}_0) \leqslant V^{-1}\int d\mathbf{x}\{\mathbf{e}_0(\mathbf{x})\sum_r(\mathbf{L}_r - \mathbf{L}_0)f_r(\mathbf{x})\mathbf{e}_0(\mathbf{x}) + \sum_r\sum_s[\boldsymbol{\tau}_r(\mathbf{x})\mathbf{A}_{rs}(\boldsymbol{\tau}_s - 2(\mathbf{L}_s - \mathbf{L}_0)\mathbf{e}_0) + \mathbf{B}_{rs}(\boldsymbol{\tau}_r, \boldsymbol{\tau}_s)]\}, \tag{6.26}$$

where

$$\mathbf{A}_{rs}: \boldsymbol{\tau}_s \rightarrow f_r(\mathbf{x})\int d\mathbf{x}'\,\boldsymbol{\Gamma}(\mathbf{x}, \mathbf{x}')f_s(\mathbf{x}')\boldsymbol{\tau}_s(\mathbf{x}') \tag{6.27}$$

and

$$\mathbf{B}_{rs}: (\boldsymbol{\tau}_r, \boldsymbol{\tau}_s) \rightarrow \sum_t \int d\mathbf{x}'\,\boldsymbol{\tau}_r(\mathbf{x}')f_r(\mathbf{x}')\boldsymbol{\Gamma}(\mathbf{x}', \mathbf{x})(\mathbf{L}_t - \mathbf{L}_0)f_t(\mathbf{x})\int d\mathbf{x}''\,\boldsymbol{\Gamma}(\mathbf{x}, \mathbf{x}'')f_s(\mathbf{x}'')\boldsymbol{\tau}_s(\mathbf{x}''). \tag{6.28}$$

Specializing now to a statistically uniform medium, the integrals that appear in (6.27) and (6.28) may be evaluated as in the preceding Section: for (6.27), $f_s(\mathbf{x}')$ is replaced by $f_s(\mathbf{x}') - c_s$ and $\boldsymbol{\Gamma}$ is replaced by $\boldsymbol{\Gamma}^\infty$, and similar replacements are made in (6.28). The expectation values of the operators $\mathbf{A}_{rs}$, $\mathbf{B}_{rs}$ then become translation-invariant and, if $\mathbf{e}_0(\mathbf{x})$ is taken equal to $\bar{\mathbf{e}}$, the functions $\boldsymbol{\tau}_r$ may be taken constant. If the right-hand side of (6.26) is now replaced by its expectation value, it follows by extremizing that

$$\bar{\mathbf{e}}\bar{\mathbf{L}}\bar{\mathbf{e}} \leqslant \bar{\mathbf{e}}\mathbf{L}_v\bar{\mathbf{e}} - \sum_r \sum_s \boldsymbol{\tau}_r \langle \mathbf{A}_{rs} \rangle (\mathbf{L}_s - \mathbf{L}_0)\bar{\mathbf{e}}, \tag{6.29}$$

where $\mathbf{L}_v$ is the Voigt estimate $\sum c_r \mathbf{L}_r$ and $\boldsymbol{\tau}_r$, $r = 1, 2, \ldots, n$, satisfy the equations

$$\sum_s (\langle \mathbf{A}_{rs} \rangle + \langle \mathbf{B}_{rs} \rangle)\boldsymbol{\tau}_s = \sum_s \langle \mathbf{A}_{rs} \rangle (\mathbf{L}_s - \mathbf{L}_0)\bar{\mathbf{e}}, \tag{6.30}$$

with

$$\langle \mathbf{A}_{rs} \rangle = \int d\mathbf{x}' \boldsymbol{\Gamma}^\infty(\mathbf{x}')[P_{rs}(\mathbf{0}, \mathbf{x}') - c_r c_s] \tag{6.31}$$

and

$$\langle \mathbf{B}_{rs} \rangle = \sum_t \int d\mathbf{x}' \boldsymbol{\Gamma}^\infty(\mathbf{x}')(\mathbf{L}_t - \mathbf{L}_0) \int d\mathbf{x}'' \boldsymbol{\Gamma}^\infty(\mathbf{x}'')$$
$$\times [P_{rst}(\mathbf{x}', \mathbf{x}'', \mathbf{0}) - c_r P_{st}(\mathbf{x}'', \mathbf{0}) - c_s P_{tr}(\mathbf{0}, \mathbf{x}') - c_t P_{rs}(\mathbf{x}', \mathbf{x}'') + 2c_r c_s c_t]. \tag{6.32}$$

Following Beran and Molyneux (1966) and Dederichs and Zeller (1973), Kröner (1977) generated an "optimized" third-order bound, essentially by substituting into (3.15) the trial field

$$\boldsymbol{\tau}^* = \delta \mathbf{L} \mathbf{e}_1, \tag{6.33}$$

in which the constant tensor $\mathbf{e}_1$ was arbitrary. This possibility is included within (6.8) and the bound (6.29) is correspondingly tighter. The work required to produce the Beran–Molyneux bound is only marginally less than is needed to produce (6.29), since it requires evaluation of $\langle \mathbf{A}_{rs} \rangle$ and $\langle \mathbf{B}_{rs} \rangle$.

Kröner (1977) also produced a tighter bound for a special type of material that he termed "disordered". Its distinctive feature (which he describes in terms of the moduli and topology being uncorrelated) is that it permits the replacement of the awkward function $\boldsymbol{\Gamma}(\mathbf{x})$ in (6.31) and (6.32) by $\mathbf{P}\delta(\mathbf{x})$, where the constant tensor $\mathbf{P}$ is defined by (4.10). For (6.31), this is legitimate so long as P_{rs} is isotropic, but for (6.32) it requires more justification: Kröner (1977) considers the replacement justified so long as $\mathbf{L}_0$ is identified with $\mathbf{L}_v$, when $\langle \delta \mathbf{L} \rangle = \mathbf{0}$. Without arguing for or against this prescription, it is simply remarked that its adoption yields simple forms for $\langle \mathbf{A}_{rs} \rangle$, $\langle \mathbf{B}_{rs} \rangle$ for which (6.29) and (6.30) lead to the upper bound (6.21), previously derived from the Hashin–Shtrikman principle, except that now $\mathbf{L}_0$ is identified with $\mathbf{L}_v$.

Analogous lower bounds of third order can be obtained from (3.22); no explicit development will be given.

Bounds that involve still more statistical information can be found by allowing trial fields more complex than (6.8). For example, the trial field

$$\boldsymbol{\tau}^*(\mathbf{x}) = \sum_r f_r(\mathbf{x})[\boldsymbol{\tau}_r(\mathbf{x}) + \sum_s \int \boldsymbol{\phi}_{rs}(\mathbf{x}, \mathbf{x}') f_s(\mathbf{x}') d\mathbf{x}'], \tag{6.34}$$

in which the functions $\boldsymbol{\tau}_r(\mathbf{x})$ and $\boldsymbol{\phi}_{rs}(\mathbf{x}, \mathbf{x}')$ are to be determined optimally, introduces correlations involving up to four points in the Hashin–Shtrikman principle, and up to five points in the classical principles. These are not considered further, except to remark that Willis (1978) employed a polarization similar to (6.34) for a matrix containing a single set of inclusions and obtained, by considering only the terms of low order, a variational estimate for the c^2-coefficient in the low-concentration expansion for $\bar{\mathbf{L}}$, c being the volume concentration of the inclusions.

6.5. *Further comments on the self-consistent method*

The Hashin-Shtrikman estimate (6.18) for $\bar{\mathbf{L}}$ provides a lower bound when $\mathbf{L}_0$ is sufficiently small and an upper bound when $\mathbf{L}_0$ is sufficiently large. At intermediate values of $\mathbf{L}_0$, it does not yield a bound, but it is still a variational estimate and it might be postulated that this estimate will be closest to $\bar{\mathbf{L}}$ when $\mathbf{L}_0 = \bar{\mathbf{L}}$, for then the mean polarization will be zero. An estimate of self-consistent type is therefore obtained by postulating that $\tilde{\mathbf{L}}$ is identical to $\bar{\mathbf{L}}$ when $\mathbf{L}_0 = \bar{\mathbf{L}}$. That is,

$$\tilde{\mathbf{L}}(\bar{\mathbf{L}}) = \bar{\mathbf{L}} \tag{6.35}$$

or, equivalently,

$$\sum c_r \mathbf{S}_r = \mathbf{0} \quad (\mathbf{L}_0 = \bar{\mathbf{L}}). \tag{6.36}$$

The prescription (6.35) of course is not exact because $\tilde{\mathbf{L}}$ is estimated from a piecewise-constant polarization field. If the composite has isotropic (or, more generally, "ellipsoidal") phase geometry, $\tilde{\mathbf{L}}$ is given by (6.21) and it is easily seen that the self-consistent prescription (6.35) yields any of (5.8), (5.9), or (5.10), with $\mathbf{P}_r = \mathbf{P}$ for all r. The interesting point is that the prescription (6.35) allows for two-point statistics in a general way and makes no reference to particular inclusion shapes. Thus, it provides some justification for the use of (5.8)–(5.10) for composites such as isotropic polycrystals, in which the grains are certainly not all spheres, and, in addition, offers an extension of the self-consistent method to composites in which the microstructure displays spatial bias.

7. Examples

The application of the preceding theory will now be illustrated for some simple examples.

7.1. *An isotropic mixture of isotropic phases*

In the notation of Hill (1965a), introduced at (4.15), let a composite have n phases with moduli $\mathbf{L}_r = (3\kappa_r, 2\mu_r)$, $r = 1, 2, \ldots, n$, distributed statistically uniformly and isotropically. It is reasonable to take the comparison material as isotropic so that $\mathbf{L}_0 = (3\kappa, 2\mu)$, say. The Hashin–Shtrikman bounds for this composite then follow from (6.21), with the tensor $\mathbf{P}$ given by (4.19). Explicitly,

$$\tilde{\kappa} = \left\{\sum_r c_r \frac{(3\kappa+4\mu)}{(3\kappa_r+4\mu)}\right\}^{-1} \sum_s c_s \frac{(3\kappa+4\mu)\kappa_s}{(3\kappa_s+4\mu)},$$
$$\tilde{\mu} = \left\{\sum_r c_r \frac{5\mu(3\kappa+4\mu)}{6\mu_r(\kappa+2\mu)+\mu(9\kappa+8\mu)}\right\}^{-1} \sum_s c_s \frac{5\mu(3\kappa+4\mu)\mu_s}{6\mu_s(\kappa+2\mu)+\mu(9\kappa+8\mu)}. \tag{7.1}$$

Equations (7.1) yield upper bounds for $\bar{\kappa}$, $\bar{\mu}$ whenever κ, μ are chosen larger than κ_r, μ_r for each r and the smallest upper bounds are obtained by setting $\kappa = \kappa_g = \max(\kappa_r)$, $\mu = \mu_g = \max(\mu_r)$. Similarly, the greatest lower bounds are found by setting $\kappa = \kappa_l = \min(\kappa_r)$, $\mu = \mu_l = \min(\mu_r)$. These bounds were given by Hashin and Shtrikman (1963), with the implicit restriction that κ_g, μ_g had both to be obtained from the same phase, and κ_l, μ_l similarly. This restriction was removed by Walpole (1966a).

Equations (7.1) also provide self-consistent equations for the determination of $\bar{\kappa}$, $\bar{\mu}$, if $\tilde{\kappa}$ and κ are identified with $\bar{\kappa}$, and $\tilde{\mu}$ and μ are identified with $\bar{\mu}$. For a two-phase material, elimination of $\bar{\kappa}$ from (7.1) yields a quartic equation for $\bar{\mu}$. This will not be given but a difficulty with the self-consistent method will be highlighted by considering the special case $\kappa_1 = \mu_1 = 0$, so that the composite represents a porous medium whose matrix phase has moduli κ_2, μ_2. It is easy to show in this case that

$$\bar{\kappa} = \frac{4(1-c_1)\kappa_2\bar{\mu}}{3c_1\kappa_2 + 4\bar{\mu}}, \tag{7.2}$$

while $\bar{\mu}$ satisfies the quadratic equation

$$16\bar{\mu}^2 + \bar{\mu}[\kappa_2(3-c_1) - 4\mu_2(4-5c_1)] - 3\mu_2\kappa_2(1-2c_1) = 0. \tag{7.3}$$

If, now, $\kappa_2 \to \infty$, (7.2) and (7.3) have solutions

$$\bar{\kappa} = \frac{4(1-c_1)(1-2c_1)\mu_2}{c_1(3-c_1)},$$

$$\bar{\mu} = \frac{3(1-2c_1)\mu_2}{(3-c_1)}. \tag{7.4}$$

The corresponding Hashin–Shtrikman upper bounds are found from (7.1) with $\kappa = \kappa_2 \sim \infty$, $\mu = \mu_2$:

$$\bar{\kappa} = 4\mu_2(1-c_1)/3c_1, \quad \tilde{\mu} = 3\mu_2(1-c_1)/(3+2c_1). \tag{7.5}$$

The Hashin–Shtrikman lower bounds, obtained with $\kappa = \mu = 0$, are zero.

Equations (7.4) and (7.5) agree at low values of c_1 but the self-consistent estimates (7.4) reduce to zero at a pore concentration equal to 1/2. This is unrealistic and demonstrates the need for caution in applying the self-consistent method in extreme cases. Hill (1965c) and Budiansky (1965) both discussed the example presented in this Section with $n = 2$, in the context of a matrix containing spherical inclusions but, with the interpretation of Section 6.5, any phase geometry is allowed, so long as it is isotropic.

For a two-phase composite for which κ_g, μ_g are both derived from one phase and κ_l, μ_l are both derived from the other, the bounds obtained from (7.1) agree with the estimates of overall moduli given by Kerner (1956), which were obtained by self-consistently embedding into a matrix of "overall" material a composite inclusion consisting of a sphere of one material encased in a spherical shell of the other, the radii being chosen to give the correct volume ratios of each phase. Either bound can be reproduced, by interchanging the two phases. Christensen and Lo (1979) noted that Kerner's solution for an imposed shear violated a continuity condition, though his solution for imposed dilatation or compression was correct. They provided the true solution for the shear case and demonstrated that it estimates the shear modulus as lying between the Hashin–Shtrikman bounds. For exteme differences between phase moduli, it may be remarked that the model of Kerner and Christensen and Lo is not necessarily more reliable than the simple self-consistent model. In surrounding an inclusion by a uniform shell of matrix, it is likely to underestimate interactions between inclusions while, to judge from (7.4), the simple self-consistent model appears to overestimate them.

7.2. *Polycrystals*

A polycrystal may be regarded as a limiting case of an n-phase material in which $n \to \infty$, each phase being associated with a different crystal orientation. Relative to crystallographic axes, any one crystal has moduli $\mathbf{L}_c$, say, and the moduli $\mathbf{L}_r$ of phase r are obtained from $\mathbf{L}_c$ by a rotation.

Consider now an isotropic polycrystal, in which each crystal orientation is equally likely and all of the functions P_{rs} are isotropic. Estimates for the overall moduli $\bar{\mathbf{L}}$ are obtained from (6.21) with $\mathbf{P}$ taken isotropic. The manipulations are easier to perform if (6.21) is written in the alternative form

$$\tilde{\mathbf{L}} = \mathbf{L}_0 + [\mathbf{I} - \langle \mathbf{T} \rangle \mathbf{P}]^{-1} \langle \mathbf{T} \rangle, \tag{7.6}$$

where $\langle \mathbf{T} \rangle = \sum c_r \mathbf{T}_r$

and
$$\mathbf{T}_r = [(\mathbf{L}_r - \mathbf{L}_0)^{-1} + \mathbf{P}]^{-1}. \tag{7.7}$$

Since all orientations are equally likely, the tensor $\langle \mathbf{T} \rangle$ is isotropic, so that $\langle \mathbf{T} \rangle = (3\kappa_T, 2\mu_T)$. The bulk and shear "moduli" associated with $\langle \mathbf{T} \rangle$ can be found from its scalar invariants, as given by (4.18) for the isotropic tensor $\mathbf{P}$. The corresponding scalars calculated for $\mathbf{T}_r$ are independent of rotation and so may be calculated for any one orientation, say that for which $\mathbf{L}_r = \mathbf{L}_c$. Once κ_T and μ_T have been found in this way, (7.16) gives the bulk and shear moduli $\tilde{\kappa}$, $\tilde{\mu}$ in the form

$$\begin{aligned} \tilde{\kappa} &= \kappa + \kappa_T/(1 - 9\kappa_p \kappa_T), \\ \tilde{\mu} &= \mu + \mu_T/(1 - 4\mu_p \mu_T), \end{aligned} \tag{7.8}$$

where $\mathbf{L}_0 = (3\kappa, 2\mu)$. If $\mathbf{L}_c$ is cubic, so that

$$\sigma_{kk} = 3\kappa_c e_{kk}, \quad \sigma_{11} - \sigma_{22} = 2\mu_c(e_{11} - e_{22}), \quad \sigma_{12} = 2\mu_c' e_{12}, \tag{7.9}$$

with equations for other components obtained by cyclic permutation of the suffixes, it follows that

$$(\mathbf{L}_c)_{iikk} = 9\kappa_c, \quad (\mathbf{L}_c)_{ijij} = 3\kappa_c + 4\mu_c + 6\mu_c'. \tag{7.10}$$

Then, if the notation $\mathbf{L}_c = (3\kappa_c, 2\mu_c, 2\mu_c')$ introduced by Walpole (1966b) is employed, products and inverses can be worked out directly and, if $\mathbf{T}_c$ represents $\mathbf{T}_r$ referred to the cube axes,

$$\mathbf{T}_c = \left\{ \left[\frac{1}{3(\kappa_c - \kappa)} + 3\kappa_p \right]^{-1}, \ \left[\frac{1}{2(\mu_c - \mu)} + 2\mu_p \right]^{-1}, \ \left[\frac{1}{2(\mu_c' - \mu)} + 2\mu_p \right]^{-1} \right\}. \tag{7.11}$$

Then, from (7.10) and (4.18),

$$\begin{aligned} 3\kappa_T &= \left[\frac{1}{3(\kappa_c - \kappa)} + 3\kappa_p \right]^{-1}, \\ 2\mu_T &= \frac{2}{5} \left[\frac{1}{2(\mu_c - \mu)} + 2\mu_p \right]^{-1} + \frac{3}{5} \left[\frac{1}{2(\mu_c' - \mu)} + 2\mu_p \right]^{-1}. \end{aligned} \tag{7.12}$$

Bounds for $\bar{\kappa}$, $\bar{\mu}$ follow by specifying the κ, μ appropriately. The unique choice for κ is κ_c, which gives $\kappa_T = 0$ and $\tilde{\kappa} = \kappa_c$, exactly. μ, on the other hand, may take either of the values μ_c, μ_c', the greater giving an upper bound and the lower a lower bound for $\bar{\mu}$. When $\mu = \mu_c$, $\tilde{\mu} = \tilde{\mu}_1$, where

$$\tilde{\mu}_1 = \mu_c + 3\left(\frac{5}{\mu_c' - \mu_c} + 8\mu_p\right)^{-1} \tag{7.13}$$

and, when $\mu = \mu_c'$, $\tilde{\mu} = \tilde{\mu}_2$, where

$$\tilde{\mu}_2 = \mu_c' + 2\left(\frac{5}{\mu_c - \mu_c'} + 12\mu_p'\right)^{-1}. \tag{7.14}$$

In these equations, μ_p and μ_p' are obtained from $\mathbf{P}$, evaluated with $\mu = \mu_c$ and μ_c', respectively. The bounds (7.13), (7.14) were first given by Hashin and Shtrikman (1962b).

A self-consistent estimate may be obtained from (6.23), which implies that $\kappa_T = \mu_T = 0$. The first of these gives $\bar{\kappa} = \kappa_c$, exactly, while the second can be manipulated to give

$$8\bar{\mu}^3 + (9\kappa_c + 4\mu_c)\bar{\mu}^2 - 3\mu_c'(\kappa_c + 4\mu_c)\bar{\mu} - 6\kappa_c\mu_c\mu_c' = 0. \tag{7.15}$$

This result was first given by Hershey (1954) in the form of a quartic equation with factors (7.15) and $(8\bar{\mu} + 9\kappa_c)$. He assumed spherical grains and applied the prescription (5.9).

Anisotropic polycrystals require a computation. Kneer (1965) studied textured polycrystals, in which grains were modelled as spheres but not all crystal orientations were equally likely. This required the evaluation of $\mathbf{P}$ for an anisotropic comparison material, using (4.10). Examples for which the actual phase geometry is anisotropic do not yet appear to have been studied, though the development of Section 6 now makes this possible.

7.3. *A composite containing long aligned fibres*

The prototype of the composite now to be considered consists of a matrix reinforced by long parallel fibres, distributed randomly on any cross-section. Such a composite is transversely isotropic and was first studied systematically by Hill (1964) and by Hashin and Rosen (1964). For present purposes, there is nothing to be gained from immediate specialization to this extent and so an n-phase composite will be considered, with each phase transversely isotropic with symmetry axis parallel to the 3-axis, and with two-point probabilities $P_{rs}(\mathbf{x} - \mathbf{x}')$ that depend only upon $(x_\alpha - x_\alpha')(x_\alpha - x_\alpha')$, Greek suffixes taking the values 1, 2 only. The composite may thus be treated as two-dimensional, or else may be regarded as the limit of a composite with "spheroidal" symmetry. In either case, estimates of overall moduli are obtained from (6.21), with the tensor $\mathbf{P}$ taking the form (4.16) with $a_1 = a_2$.

The tensor of moduli $\mathbf{L}$ is most usefully expressed, in the notation of Hill (1964, 1965b) and Walpole (1969), as

$$\mathbf{L} = (2k, l, l', n, 2m, 2p) \tag{7.16}$$

when $\boldsymbol{\sigma} = \mathbf{L}\mathbf{e}$ implies

$$\begin{aligned} \tfrac{1}{2}(\sigma_{11} + \sigma_{22}) &= k(e_{11} + e_{22}) + le_{33}, \\ \sigma_{33} &= l'(e_{11} + e_{22}) + ne_{33}, \\ (\sigma_{11} - \sigma_{22}) &= 2m(e_{11} - e_{22}), \quad \sigma_{12} = 2me_{12}, \\ \sigma_{23} &= 2pe_{23}, \quad \sigma_{31} = 2pe_{31}. \end{aligned} \tag{7.17}$$

In this notation, the product $\mathbf{L}_1\mathbf{L}_2$ of two such tensors is given by

$$\mathbf{L}_1\mathbf{L}_2 = (4k_1k_2 + 2l_1l'_2,\ 2k_1l_2 + l_1n_2,\ 2l'_1k_2 + n_1l'_2,\ 2l'_1l_2 + n_1n_2,\ 4m_1m_2,\ 4p_1p_2) \tag{7.18}$$

and

$$\mathbf{L}^{-1} = (n/2\Delta,\ -l/2\Delta,\ -l'/2\Delta,\ k/\Delta,\ 1/2m,\ 1/2p), \tag{7.19}$$

where

$$\Delta = kn - ll'. \tag{7.20}$$

When $\mathbf{L}$ is a tensor of elastic moduli, it has the symmetry $l = l'$ but products do not share this property. The moduli k, m are the plane-strain bulk and shear moduli. Other useful quantities are the plane-strain Poisson's ratio ν and the longitudinal Young's modulus E. These are

$$\nu = l/2k, \quad E = n - l^2/k. \tag{7.21}$$

Positive definiteness of $\mathbf{L}$ requires that all of k, m, E, and p are positive.

Taking the comparison material as transversely isotropic now, with components as in (7.16) with $l = l'$, it is fairly easily derived from (4.16) that

$$\mathbf{P} = 1/2(k+m),\ 0,\ 0,\ 0,\ (2m+k)/4m(k+m),\ 1/4p); \tag{7.22}$$

this was given by Walpole (1969) and Laws and McLaughlin (1979). All that is required now is routine substitution into (6.21). Walpole (1969) gives the results in the form

$$\begin{aligned}
\tilde{k} &= [\textstyle\sum c_r/(k_r+m)]^{-1} - m, \quad \tilde{l} = [\sum c_r l_r/(k_r+m)]/[\sum c_r/(k_r+m)],\\
\tilde{n} &= \textstyle\sum c_r n_r + \{\sum c_r l_r[\sum c_s(l_s - l_r)/(k_s+m)]/(k_r+m)\}/[\sum c_r/(k_r+m)],\\
\tilde{m} &= \{\textstyle\sum c_r/[m_r + mk/(k+2m)]\}^{-1} - mk/(k+2m),\\
\tilde{p} &= [\textstyle\sum c_r/(p_r+p)]^{-1} - p,
\end{aligned} \tag{7.23}$$

from which both bounds and self-consistent estimates follow. Equivalent formulae, together with useful variants involving E and ν, were given by Hill (1964, 1965b) for a two-phase composite, except that he obtained no bound for $\tilde{m}$. Hashin (1979) recently reconsidered the two-phase problem and derived estimates of moduli from a "composite cylinder" model. It generated as estimates the bounds that are obtained from (7.23).

7.4. *Composite containing short fibres*

Fibre-reinforced composites frequently contain fibres whose length to thickness ratio is of the order of 10 to 100. The analysis of Section 7.3 therefore applies only approximately and some assessment of its range of validity is called for. Such an assessment can be made if the fibres are modelled as aligned spheroids. This allows a fairly straightforward self-consistent analysis and, if the distribution of the fibres is taken as "spheroidal", bounds can also be constructed using (6.21). The self-consistent approach requires the evaluation of $\mathbf{P}$ for a transversely isotropic matrix. Although the integrals can be worked out analytically, their form is still sufficiently complicated to require numerical evaluation by computer except, of course, that for long fibres, $\mathbf{P}$ approaches the value (7.22). Laws and McLaughin (1979) have carried out the required computations for a glass-reinforced polyester resin. Their results will be discussed later. A complete set of bounds has not been found but, in the course of a study of waves in composites, Willis (1980b, c) obtained as a by-product the estimate (6.21) for "spheroidal" **P**, when both phases were isotropic and the comparison material was identified with one of them. Again, general results could be obtained only with a computation, but some general feeling for the influence of fibre length can be gained by considering the asymptotic form taken by **P** for a spheroid whose thickness to length ratio is ε, where $\varepsilon \ll 1$. Willis (1980b) gave the formula

$$\mathbf{P} \sim \left(\frac{3}{2(3\kappa+4\mu)}, 0, 0, 0, \frac{3\kappa+7\mu}{4\mu(3\kappa+4\mu)}, \frac{1}{4\mu}\right)$$
$$+\frac{\varepsilon^2 \ln(2/\varepsilon)}{2\mu(3\kappa+4\mu)}[3\kappa+2\mu, -(3\kappa+\mu), -(3\kappa+\mu), 2(3\kappa+4\mu), -3\mu, -3\kappa/2]$$
$$+\frac{3\varepsilon^2}{4\mu(3\kappa+4\mu)}[-(3\kappa-\mu), 3\kappa+\mu, 3\kappa+\mu, -6(\kappa+\mu), (3\kappa-11\mu)/2, 5\kappa+2\mu/3]. \quad (7.24)$$

Russel (1973) did not give **P** but derived an equivalent result calculated from general formulae of Eshelby (1957). If $\mathbf{L}_0$ is identified with one of the phases, say $\mathbf{L}_0 = \mathbf{L}_1$, it is advantageous to use (6.21) in the equivalent form

$$\tilde{\mathbf{L}} = \mathbf{L}_1 + c_2[(\mathbf{L}_2 - \mathbf{L}_1)^{-1} + c_1\mathbf{P}]^{-1}, \quad (7.25)$$

whose similarity with (4.24) may be noted. Clearly, now, the error involved in disregarding terms of order ε^2 is small so long as $(\mathbf{L}_2 - \mathbf{L}_1)^{-1}$ is of the same order as μ^{-1}, for example. Within this range, fibres with an aspect ratio of 10 or more may be regarded as long. If, however, $\mathbf{L}_2$ is much stiffer than $\mathbf{L}_1$ so that $(\mathbf{L}_2 - \mathbf{L}_1)^{-1}$ is small, terms of order ε^2 make a significant contribution because the $O(1)$ part of **P** is singular. The extreme case is that of phase 2 being rigid, when (7.25) gives

$$\tilde{\mathbf{L}} = \mathbf{L}_1 + c_2\mathbf{P}^{-1}/c_1, \quad (7.26)$$

in which a singular result is obtained if the terms of order ε^2 are ignored. Explicitly,

$$\tilde{\mathbf{L}} = \mathbf{L}_1 + \frac{c_2\mu}{c_1\varepsilon^2[\ln(2/\varepsilon) - 9(\kappa+\mu)/2(3\kappa+4\mu)]}(0, 0, 0, 1, 0, 0). \quad (7.27)$$

The example treated by Laws and McLaughlin (1979) falls in the intermediate range, the ratio of Young's moduli for *E* glass and their polyester resin being around 20, and the results of their calculation differed appreciably from the infinite-fibre predictions, for aspect ratios up to about 100. Clearly a similar situation would occur for many other composites.

Real composites contain imperfectly aligned fibres of variable length. Laws and McLaughin (1979) made some approximate allowance for these features but no rigorous study has yet been made. One possibility would be to obtain both bounds and self-consistent estimates from (7.25), but with **P** replaced by the tensor **P**′ defined by (6.22). The difficult task, of course, would be that of realistically estimating the function *h* which defines the two-point probabilities.

7.5. *A body containing aligned cracks*

A body containing aligned penny-shaped cracks, whose centres are distributed uniformly at number density n_2, can reasonably be treated as a limiting case of a composite with "spheroidal" symmetry, the thickness to diameter ratio of the spheroids being $\varepsilon \ll 1$. Willis (1980b) gave the result

$$\mathbf{P} \sim [0, 0, 0, 3/(3\kappa+4\mu), 0, 1/2\mu] + \frac{\pi\varepsilon}{\mu(3\kappa+4\mu)}[(3\kappa+11\mu)/8, \\ -3(\kappa+\mu)/8, -3(\kappa+\mu)/8, (3\kappa-\mu)/4, 9(\kappa+3\mu)/16, 9(\kappa+2\mu)/8] \qquad (7.28)$$

so that, if $\mathbf{L}_0$ is identified with $\mathbf{L}_1$ ($\mathbf{L}_2 = \mathbf{0}$ for the cracks), substitution of (7.28) into (7.25) yields an upper-bound estimate for $\mathbf{L}$. Again, the presence of the term of order ε in (7.28) is crucial, since without it the inverse in (7.25) does not exist. The result was given by Willis (1980c). It is complicated when expressed in terms of $\tilde{\mathbf{L}}$ but takes a much simpler form when expressed in terms of the inverse $\tilde{\mathbf{M}}$:

$$\tilde{\mathbf{M}} = \mathbf{M}_1 + \frac{4(n_2 a^3)(3\kappa+4\mu)}{3\mu}[0, 0, 0, 1/(3\kappa+\mu), 0, 2/9(\kappa+2\mu)], \qquad (7.29)$$

where the relation $c_2 = 4\pi n_2 a^3 \varepsilon/3$ between volume concentration and number density has been employed and the limit $\varepsilon \to 0$ has been taken with number density fixed. Equation (7.29) confirms that the cracks interact only with the stress components σ_{13}, σ_{23}, σ_{33}, as they should. Hoenig (1979) has performed self-consistent calculations for the same configuration. His estimates agree with (7.29) at low concentrations; more generally, (7.29) bounds the compliances from below.

8. Wave Propagation

Problems involving wave propagation necessarily involve spatial variation and effective properties cannot be defined through global averages in the way outlined for static problems in Section 2. Instead, at least for random composites, equations which govern the ensemble average $\langle \mathbf{u} \rangle$ of the solution to any particular boundary value problem are sought or, alternatively, $\langle \mathbf{u} \rangle$ is constructed directly. Work published to date has relied on one of two methods. The first involves solving the dynamical analogue of (3.8) by perturbation theory. If this is done in the most elementary manner, it leads to a series some of whose individual terms contain integrals that diverge, which are then grouped together in a way that eliminates the divergence by a technique known as renormalization. This is a well-known technique in many branches of physics; its application to composites has recently been discussed by McCoy (1979). Alternatively, the equations can be solved by the method of smoothing introduced by Karal and Keller (1964), in which an essential intermediate step is to represent $\mathbf{u}' = \mathbf{u} - \langle \mathbf{u} \rangle$ in terms of $\langle \mathbf{u} \rangle$. The statistically fluctuating term $\mathbf{u}'$ can then be eliminated from the starting equation to give an equation for $\langle \mathbf{u} \rangle$ itself. Karal and Keller (1964) truncated their formal iterative solution after one iteration to obtain results valid for weakly inhomogeneous media. McCoy (1973), on the other hand, considered the series to arbitrary order, to obtain formal expressions containing correlation functions of all orders. Since these are not generally known, however, the method is most useful for weakly inhomogeneous media. An entirely different method has been applied to composites consisting of a matrix with distributed inclusions. Each inclusion is treated as a scatterer, which scatters the field incident upon it. The complication is that the field incident upon any one inclusion has contributions from the scattered fields of every other inclusion. The scattered fields have always been represented explicitly as eigenfunction expansions (typically involving Hankel functions), so that the method appears to be restricted to an isotropic matrix, at least in the form presented. After some complex manipulations, equations for the coefficients defining the fields scattered from the inclusions are derived. These are ensemble averaged (essentially as in (6.12)) and the system is closed by making the quasi-crystalline approximation of Lax (1952) (essentially as in (6.14)). Particular examples of this approach are provided in the papers of Bose and Mal (1973, 1974), Datta (1977, 1978), Twersky (1977), Varadan *et al.* (1978), and Varadan and Varadan (1979).

A compressed and preliminary account will now be given of a variational approach. It is a natural extension of that already given for static problems and offers a unified treatment for materials such as polycrystals (for which only perturbation theory has been available) and matrix-inclusion composites (for which only the multiple scattering method has been available), without restriction to an isotropic matrix.

8.1. *Variational formulation*

The equation of motion for any continuum may be given, in the absence of body forces, in the form

$$\operatorname{div} \boldsymbol{\sigma} = \dot{\mathbf{p}}, \tag{8.1}$$

where $\mathbf{p}$ denotes momentum density and the superposed dot represents differentiation with respect to time t. Equation (8.1) is supplemented by the constitutive relations

$$\boldsymbol{\sigma} = \mathbf{L}\mathbf{e}, \quad \mathbf{p} = \rho\dot{\mathbf{u}}, \tag{8.2}$$

where $\mathbf{L}$, ρ are the tensors of moduli and the density respectively, together with suitable boundary and initial conditions. The similarity between $(8.2)_{1,2}$ suggests introducing a comparison material with moduli $\mathbf{L}_0$ and density ρ_0 and defining a momentum polarization $\boldsymbol{\pi}$ so that

$$\boldsymbol{\pi} = (\rho - \rho_0)\dot{\mathbf{u}}, \quad \mathbf{p} = \rho_0\dot{\mathbf{u}} + \boldsymbol{\pi}, \tag{8.3}$$

to parallel (3.9), (3.10) for $\boldsymbol{\tau}$. Substituting (3.10) and $(8.3)_2$ into (8.1) then gives an equation for the displacement that would be generated in the comparison material by the body force $\operatorname{div}\boldsymbol{\tau} - \dot{\mathbf{p}}$. Willis (1980b) showed that its solution could be expressed, in terms of Green's function $\mathbf{G}$ for the comparison body, in the form

$$\mathbf{u} = \mathbf{u}_0 - \mathbf{S}\boldsymbol{\tau} - \mathbf{M}\boldsymbol{\pi}, \tag{8.4}$$

generalizing (3.5), where the operators $\mathbf{S}$, $\mathbf{M}$ are defined by

$$\mathbf{S}\colon \tau_{ij}(\mathbf{x}, t) \to \int dt' \int_V d\mathbf{x}'\, S_{pij}(\mathbf{x}, t, \mathbf{x}', t')\,\tau_{ij}(\mathbf{x}', t'), \tag{8.5}$$

$$\mathbf{M}\colon \pi_i(\mathbf{x}, t) \to \int dt' \int_V d\mathbf{x}'\, M_{pi}(\mathbf{x}, t, \mathbf{x}', t')\,\pi_i(\mathbf{x}', t') \tag{8.6}$$

and

$$S_{pij}(\mathbf{x}, t, \mathbf{x}', t') = \frac{\partial G_{pi}}{\partial x'_j}(\mathbf{x}, t, \mathbf{x}', t')\bigg|_{(ij)}, \tag{8.7}$$

$$M_{pi}(\mathbf{x}, t, \mathbf{x}', t') = -\frac{\partial G_{pi}}{\partial t'}(\mathbf{x}, t, \mathbf{x}', t'). \tag{8.8}$$

The displacement $\mathbf{u}_0$ that appears in (8.4) satisfies the elastic wave equation (without body forces) in the comparison material, together with the prescribed boundary and initial conditions. The initial conditions, incidentally, must be prescribed displacement and momentum density, not velocity: the distinction is usually irrelevant but (8.4) relies upon it.

Combination of (8.4) with the definitions (3.9), $(8.3)_1$ of $\boldsymbol{\tau}$, $\boldsymbol{\pi}$ now generates the operator equations

$$\begin{aligned}(\mathbf{L}-\mathbf{L}_0)^{-1}\boldsymbol{\tau}+\mathbf{S}_x\boldsymbol{\tau}+\mathbf{M}_x\boldsymbol{\pi}&=\mathbf{e}_0,\\(\rho-\rho_0)^{-1}\boldsymbol{\pi}+\mathbf{S}_t\boldsymbol{\tau}+\mathbf{M}_t\boldsymbol{\pi}&=\dot{\mathbf{u}}_0,\end{aligned}\tag{8.9}$$

where $\mathbf{e}_0$ is the strain associated with $\mathbf{u}_0$; $\mathbf{S}_x$, $\mathbf{M}_x$ have kernels

$$\begin{aligned}(\mathbf{S}_x)_{pqij}&=\left.\frac{\partial^2 G_{pi}}{\partial x_q\partial x'_j}\right|_{(pq)(ij)},\\(\mathbf{M}_x)_{pqi}&=-\left.\frac{\partial^2 G_{pi}}{\partial x_q\partial t'}\right|_{(pq)},\end{aligned}\tag{8.10}$$

and $\mathbf{S}_t$, $\mathbf{M}_t$ are the derivatives with respect to t of $\mathbf{S}$, $\mathbf{M}$.

Equations $(8.9)_1$ and $(8.10)_1$ generalize (3.16) and (3.7). The object now is to produce a variational principle, analogous to the Hashin–Shtrikman principle, that is equivalent to equations (8.9). One possibility is to construct an analogue of Hamilton's principle using a Lagrangian that is quadratic in $\boldsymbol{\tau}$, $\boldsymbol{\pi}$. Such a principle was displayed by Willis (1981) but an alternative, which appears to be better, is to obtain a principle, similar to that of Gurtin (1964), which is tailored to the type of initial conditions that are actually imposed. To develop this, let $\mathbf{u}_1$, $\mathbf{p}_1$, $\boldsymbol{\sigma}_1$, $\mathbf{e}_1$ be displacement, momentum, stress, and strain fields generated by polarizations $\boldsymbol{\tau}_1$, $\boldsymbol{\pi}_1$ with homogeneous initial and boundary conditions, so that they satisfy (8.1), (3.10), $(8.3)_2$ (but not necessarily (8.2)) and define fields with a suffix 2 similarly. It is now easy to show that the bilinear form

$$\{(\boldsymbol{\tau}_1,\boldsymbol{\pi}_1),(\boldsymbol{\tau}_2,\boldsymbol{\pi}_2)\}=\frac{1}{V}\int_V d\mathbf{x}\,(\boldsymbol{\tau}_1*\mathbf{e}_2+\boldsymbol{\pi}_1*\dot{\mathbf{u}}_2),\tag{8.11}$$

where $*$ denotes the ordinary operation of convolution with respect to time, is symmetric and it follows immediately that equations (8.9) are equivalent to the variational principle

$$\delta\mathscr{F}(\boldsymbol{\tau},\boldsymbol{\pi})=0,\tag{8.12}$$

where

$$\mathscr{F}(\boldsymbol{\tau},\boldsymbol{\pi})=\frac{1}{V}\int d\mathbf{x}\,\{\boldsymbol{\tau}*[(\mathbf{L}-\mathbf{L}_0)^{-1}\boldsymbol{\tau}-\mathbf{e}-\mathbf{e}_0]+\boldsymbol{\pi}*[(\rho-\rho_0)^{-1}\boldsymbol{\pi}-\dot{\mathbf{u}}-\dot{\mathbf{u}}_0]\},\tag{8.13}$$

$\mathbf{e}$ and $\dot{\mathbf{u}}$ being expressed in terms of $\boldsymbol{\tau}$ and $\boldsymbol{\pi}$ via (8.4). The operator (8.13) is a generalization of (3.17).

The principle (8.12) may now be employed to generate approximate solutions by the procedure given in Section 6. If $\mathscr{F}$ is replaced by its expectation value in (8.12), this equation generates exactly "hierarchy equations" such as (6.12), by suitable choice of variations. These are of no use as they stand but the observation that they can be obtained in this way encourages the production of approximations generated from trial fields $\boldsymbol{\tau}^*, \boldsymbol{\pi}^*$. The natural extension of (6.8) is

$$\boldsymbol{\tau}^*(\mathbf{x}, t) = \sum_{r=1}^{n} \boldsymbol{\tau}_r(\mathbf{x}, t) f_r(\mathbf{x}), \quad \boldsymbol{\pi}^*(\mathbf{x}, t) = \sum_{r=1}^{n} \boldsymbol{\pi}_r(\mathbf{x}, t) f_r(\mathbf{x}); \tag{8.14}$$

the extremum of $\langle \mathscr{F}(\boldsymbol{\tau}^*, \boldsymbol{\pi}^*) \rangle$ is obtained when $\boldsymbol{\tau}_r, \boldsymbol{\pi}_r$ satisfy

$$\begin{aligned} (\mathbf{L}_r - \mathbf{L}_0)^{-1}\boldsymbol{\tau}_r + \sum_s (\mathbf{S}_x P_{s|r}\boldsymbol{\tau}_s + \mathbf{M}_x P_{s|r}\boldsymbol{\pi}_s) &= \mathbf{e}_0, \\ (\rho_r - \rho_0)^{-1}\boldsymbol{\pi}_r + \sum_s (\mathbf{S}_t P_{s|r}\boldsymbol{\tau}_s + \mathbf{M}_t P_{s|r}\boldsymbol{\pi}_s) &= \dot{\mathbf{u}}_0. \end{aligned} \tag{8.15}$$

These equations generalize (6.11) and are the same as would be produced from the exact "hierarchy equations" by making the quasi-crystalline assumption (6.14) with regard to both $\boldsymbol{\tau}$ and $\boldsymbol{\pi}$. Similar equations were produced by Willis (1980c) for a matrix containing inclusions. In that work, however, they were derived making the quasi-crystalline assumption explicitly, without reference to a variational principle. More general approximate equations could also be obtained from (8.12), whose full implications are still under investigation. The article is concluded, however, with a short discussion of the properties of (8.15).

8.2. *Wave speeds*

The polarizations (8.14) generate a mean wave

$$\langle \mathbf{u} \rangle = \mathbf{u}_0 - \sum_s (\mathbf{S} P_s \boldsymbol{\tau}_s + \mathbf{M} P_s \boldsymbol{\pi}_s), \tag{8.16}$$

obtained by taking the mean of (8.4). Correspondingly, using (8.16), equations (8.15) can be recast in the form

$$\begin{aligned} (\mathbf{L}_r - \mathbf{L}_0)^{-1}\boldsymbol{\tau}_r + \sum_s [\mathbf{S}_x (P_{s|r} - P_s)\boldsymbol{\tau}_s + \mathbf{M}_x (P_{s|r} - P_s)\boldsymbol{\pi}_s] &= \langle \mathbf{e} \rangle, \\ (\rho_r - \rho_0)^{-1}\boldsymbol{\pi}_r + \sum_s [\mathbf{S}_t (P_{s|r} - P_s)\boldsymbol{\tau}_s + \mathbf{M}_t (P_{s|r} - P_s)\boldsymbol{\pi}_s] &= \langle \dot{\mathbf{u}} \rangle. \end{aligned} \tag{8.17}$$

An elementary and important problem for the ordinary elastodynamic equations is to determine the speeds with which plane waves may travel in an infinite, homogeneous medium. The corresponding problem for (8.15) is to determine possible plane-wave solutions, for which

$$\begin{aligned} \boldsymbol{\tau}_r(\mathbf{x}, t) &= \boldsymbol{\tau}_r \exp[-i(k\mathbf{n}\,.\,\mathbf{x} + \omega t)], \\ \boldsymbol{\pi}_r(\mathbf{x}, t) &= \boldsymbol{\pi}_r \exp[-i(k\mathbf{n}\,.\,\mathbf{x} + \omega t)], \end{aligned} \tag{8.18}$$

with $\boldsymbol{\tau}_r, \boldsymbol{\pi}_r$ constants, in an infinite, statistically uniform medium. Thus, plane-wave solutions of (8.15) are required, when $\mathbf{u}_0 = \mathbf{0}$, the probabilities $P_r, P_{s|r}$ are translation-invariant and the operators $\mathbf{S}$, $\mathbf{M}$ take limiting forms corresponding to infinite V. The limit is easily taken for the equivalent equations (8.16), (8.17). Equation (8.16) gives (with $\mathbf{u}_0 = \mathbf{0}$)

$$\langle \mathbf{u} \rangle = -\tilde{\mathbf{S}}(k\mathbf{n}, \omega)\langle \boldsymbol{\tau} \rangle - \tilde{\mathbf{M}}(k\mathbf{n}, \omega)\langle \boldsymbol{\pi} \rangle, \tag{8.19}$$

where

$$\langle \boldsymbol{\tau} \rangle = \sum_s c_s \boldsymbol{\tau}_s \exp[-i(k\mathbf{n}.\mathbf{x}+\omega t)],$$
$$\langle \boldsymbol{\pi} \rangle = \sum_s c_s \boldsymbol{\pi}_s \exp[-i(k\mathbf{n}.\mathbf{x}+\omega t)] \tag{8.20}$$

and $\tilde{\mathbf{S}}$, $\tilde{\mathbf{M}}$ are the Fourier transforms of the operators **S**, **M**. Equation $(8.17)_1$ implies, when V is large,

$$(\mathbf{L}_r-\mathbf{L}_0)^{-1}\boldsymbol{\tau}_r+\sum_s \int d\mathbf{x}'\overline{\mathbf{S}}_x^{\infty}(\mathbf{x}',\omega)[P_{s|r}(\mathbf{x}')-c_s]\exp[-ik\mathbf{n}.\mathbf{x}']\boldsymbol{\tau}_s$$
$$+\sum_s \int d\mathbf{x}'\overline{\mathbf{M}}_x^{\infty}(\mathbf{x}',\omega)[P_{s|r}(\mathbf{x}')-c_s]\exp[-ik\mathbf{n}.\mathbf{x}']\boldsymbol{\pi}_s$$
$$=-\tilde{\mathbf{S}}_x(k\mathbf{n},\omega)\sum_s c_s\boldsymbol{\tau}_s-\tilde{\mathbf{M}}_x(k\mathbf{n},\omega)\sum_s c_s\boldsymbol{\pi}_s, \tag{8.21}$$

where $\overline{\mathbf{S}}_x^{\infty}(\mathbf{x},\omega)$, $\overline{\mathbf{M}}_x^{\infty}(\mathbf{x},\omega)$ are time-reduced operators obtained from the time-reduced infinite-body Green's function $\overline{\mathbf{G}}^{\infty}(\mathbf{x},\omega)$. Equation $(8.17)_2$ can be expanded similarly. The choice $\mathbf{x}=\mathbf{0}$ could be made in (8.21) because of the translation-invariance of $P_{s|r}$.

Further progress now depends upon finding a suitable representation for Green's function $\overline{\mathbf{G}}^{\infty}$. This satisfies the equation

$$\mathbf{L}_0(\nabla)\overline{\mathbf{G}}^{\infty}+\rho_0\omega^2\overline{\mathbf{G}}^{\infty}+\mathbf{I}\delta(\mathbf{x})=\mathbf{0}, \tag{8.22}$$

which may be solved by the method that was explained for the special case (4.1): Willis (1980b) showed in detail that

$$\overline{\mathbf{G}}^{\infty}(\mathbf{x},\omega)=\frac{1}{8\pi^2}\sum_{r=1}^{3}\int_{|\xi|=1} dS\,\frac{\mathbf{u}^r(\xi)[\mathbf{u}^r(\xi)]^T}{\rho_0 c_r^2}\left\{\delta(\xi.\mathbf{x})+\frac{i\omega}{2c_r}\exp[i\omega|\xi.\mathbf{x}|/c_r]\right\}, \tag{8.23}$$

where $c_r(\xi)$, $\mathbf{u}^r(\xi)$ satisfy

$$[\mathbf{L}_0(\xi)-\rho_0 c_r^2\mathbf{I}]\mathbf{u}^r=\mathbf{0}, \tag{8.24}$$

so that they represent the speeds and polarizations of plane waves that can propagate in the direction ξ in the comparison material. In (8.23), the eigenvectors $\mathbf{u}^r$ are taken as normalized. The limiting case $\omega=0$ reproduces the static formula (4.4).

Attention will now be restricted to long waves or, equivalently, to small ω, and a lowest-order approximate solution of (8.21) and the corresponding equation for $\boldsymbol{\pi}$ will be constructed by retaining only terms of order zero in k or ω. Amongst the operators $\overline{\mathbf{S}}_x^{\infty}$, $\overline{\mathbf{M}}_x^{\infty}$, $\overline{\mathbf{S}}_t^{\infty}$, $\overline{\mathbf{M}}_t^{\infty}$, only the first has a term of order zero in ω, since the other three involve at least one time derivative. Furthermore, $\overline{\mathbf{S}}_x^{\infty}\sim\boldsymbol{\Gamma}^{\infty}$, defined by (4.5). The Fourier transforms $\tilde{\mathbf{S}}_x^{\infty}$, etc., on the other hand, are all homogeneous of degree zero and are retained exactly. Equation (8.21) now reduces to

$$(\mathbf{L}_r-\mathbf{L}_0)^{-1}\boldsymbol{\tau}_r+\sum_s \int d\mathbf{x}'\boldsymbol{\Gamma}^{\infty}(\mathbf{x}')(P_{s|r}(\mathbf{x}')-c_s)\boldsymbol{\tau}_s=\langle\mathbf{e}\rangle \tag{8.25}$$

while the corresponding equation for $\boldsymbol{\pi}$ becomes

$$(\rho_r - \rho_0)^{-1}\boldsymbol{\pi}_r = -i\omega\langle \mathbf{u} \rangle, \tag{8.26}$$

where $\langle \mathbf{e} \rangle$ is the strain associated with $\langle \mathbf{u} \rangle$ and both are evaluated at $\mathbf{x} = \mathbf{0}$, $t = 0$. Equations (8.25) are identical with (6.15). Therefore, by definition of $\tilde{\mathbf{L}}$, they imply

$$\langle \boldsymbol{\tau} \rangle = (\tilde{\mathbf{L}} - \mathbf{L}_0)\langle \mathbf{e} \rangle \tag{8.27}$$

(each side having the exponential dependence restored). Also, from (8.26),

$$\langle \boldsymbol{\pi} \rangle = (\bar{\rho} - \rho_0)\langle \dot{\mathbf{u}} \rangle, \tag{8.28}$$

where $\bar{\rho} = \sum c_r \rho_r$ is the mean density of the composite. Equations (8.27) and (8.28) show that the mean polarizations $\langle \boldsymbol{\tau} \rangle$, $\langle \boldsymbol{\pi} \rangle$ are those that would be generated relative to $\mathbf{L}_0, \rho_0$, if the mean wave $\langle \mathbf{u} \rangle$ propagated in a homogeneous medium with moduli $\tilde{\mathbf{L}}$ and density $\bar{\rho}$. This conclusion can also be reached by algebraic manipulation. Retention of terms of higher order in the equations produces both dispersion and attenuation. Attenuation in a matrix containing inclusions has been studied by Willis (1980c). A more complete study of the present system is being conducted and will be reported elsewhere.

Acknowledgements

This article was written while the author was a guest in the Division of Applied Mechanics, Department of Mechanical Engineering, Stanford University. Thanks are due to Professor G. Herrmann for generous provision of facilities. Partial financial support from the United States Department of the Navy, Office of Naval Research under Contract No. N00014-76-C-0054, and from the Electric Power Research Institute, Palo Alto, California under Contract No. RP609-1 is gratefully acknowledged.

References

Asaro, R. J. and Barnett, D. M. (1975) The non-uniform transformation strain problem for an anisotropic ellipsoidal inclusion, *J. Mech. Phys. Solids* **23,** 77–83.

Beran, M. J. and Molyneux, J. (1966) Use of classical variational principles to determine bounds for the effective bulk modulus in heterogeneous media, *Q. Appl. Math.* **24,** 107–118.

Bose, S. K. and Mal, A. K. (1973) Longitudinal shear waves in a fiber-reinforced composite, *Int. J. Solids Struct.* **9,** 1075–1085.

Bose, S. K. and Mal, A. K. (1974) Elastic waves in a fiber-reinforced composite, *J. Mech. Phys. Solids* **22,** 217–229.

Budiansky, B. (1965) On the elastic moduli of some heterogeneous materials, *J. Mech. Phys. Solids* **13,** 223–227.

Christensen, R. M. (1979) *Mechanics of Composite Materials*, Wiley-Interscience, New York.

Christensen, R. M. and Lo, K. H. (1979) Solutions for effective shear properties in three phase sphere and cylinder models, *J. Mech. Phys. Solids* **27,** 315–330.

Datta, S. K. (1977) A self-consistent approach to multiple scattering by elastic ellipsoidal inclusions, *J. Appl. Mech.* **44,** 657–662.

Datta, S. K. (1978) Scattering by a random distribution of inclusions and effective elastic properties, *Continuum Models of Discrete Systems*, edited by J. W. Provan, 111–127, University of Waterloo Press.

Dederichs, P. H. and Zeller, R. (1973) Variational treatment of the elastic constants of disordered materials, *Z. Phys.* **259,** 103–116.

Eshelby, J. D. (1957) The determination of the elastic field of an ellipsoidal inclusion and related problems, *Proc. R. Soc. London* A **241,** 376–396.

Eshelby, J. D. (1961) Elastic inclusions and inhomogeneities, *Progress in Solid Mechanics*, Vol. II, edited by I. N. Sneddon and R. Hill, 87–140, North-Holland, Amsterdam.

Gel'fand, I. M. and Shilov, G. E. (1964) *Generalized Functions*, Vol. I, *Properties and Operations*, Academic Press, New York.

Gurtin, M. E. (1964) Variational principles in linear elastodynamics, *Arch. Ration. Mech. Anal.* **16,** 34–50.
Hashin, Z. (1979) Analysis of properties of fiber composites with anisotropic constituents, *J. Appl. Mech.* **46,** 543–550.
Hashin, Z. and Rosen, B. W. (1964) The elastic moduli of fiber-reinforced materials, *J. Appl. Mech.* **31,** 223–232.
Hashin, Z. and Shtrikman, S. (1962a) On some variational principles in anisotropic and nonhomogeneous elasticity, *J. Mech. Phys. Solids* **10,** 335–342.
Hashin, Z. and Shtrikman, S. (1962b) A variational approach to the theory of the elastic behaviour of polycrystals, *J. Mech. Phys. Solids* **10,** 343–352.
Hashin, Z. and Shtrikman, S. (1963) A variational approach to the theory of the elastic behaviour of multiphase materials, *J. Mech. Phys. Solids* **11,** 127–140.
Hershey, A. V. (1954) The elasticity of an isotropic aggregate of anisotropic cubic crystals, *J. Appl. Mech.* **21,** 236–240.
Hill, R. (1952) The elastic behaviour of a crystalline aggregate, *Proc. Phys. Soc.* A **65,** 349–354.
Hill, R. (1963a) Elastic properties of reinforced solids: some theoretical principles, *J. Mech. Phys. Solids* **11,** 357–372.
Hill, R. (1963b) New derivations of some elastic extremum principles, *Progress in Applied Mechanics, The Prager Anniversary Volume*, 99–106, Macmillan, New York.
Hill, R. (1964) Theory of mechanical properties of fibre-strengthened materials: I, Elastic behaviour, *J. Mech. Phys. Solids* **12,** 199–212.
Hill, R. (1965a) Continuum micromechanics of elastoplastic polycrystals, *J. Mech. Phys. Solids* **13,** 89–101.
Hill, R. (1965b) Theory of mechanical properties of fibre-strengthened materials: Self-consistent model, *J. Mech. Phys. Solids* **13,** 189–198.
Hill, R. (1965c) A self-consistent mechanics of composite materials, *J. Mech. Phys. Solids* **13,** 213–222.
Hill, R. (1972) On constitutive macro-variables for heterogeneous solids at finite strain, *Proc. R. Soc. London* A **326,** 131–147.
Hoenig, A. (1979) Elastic moduli of a non-randomly cracked body, *Int. J. Solids Struct.* **15,** 137–154.
Karal, F. C. and Keller, J. B. (1964) Elastic, electromagnetic and other waves in a random medium, *J. Math. Phys.* **5,** 537–547.
Kerner, E. H. (1956) The elastic and thermoelastic properties of composite media, *Proc. Phys. Soc.* B **69,** 808–813.
Khatchaturyan, A. G. (1967) Some questions concerning the theory of phase transformations in solids, *Soviet Phys., Solid State.* (English trans.) **8,** 2163–2168.
Kneer, G. (1965) Calculation of elastic moduli of polycrystalline aggregates with texture, *Phys. Stat. Solidi* **9,** 825–838.
Kohn, W., Krumhansl, J. A., and Lee, E. H. (1972) Variational methods for dispersion relations and elastic properties of composite materials, *J. Appl. Mech.* **94,** 327–336.
Korringa, J. (1973) Theory of elastic constants of heterogeneous media, *J. Math. Phys.* **14,** 509–513.
Kröner, E. (1958) Berechnung der elastichen Konstanten des Vielkristalls aus den Konstanten des Einkristalls, *Z. Phys.* **151,** 504–518.
Kröner, E. (1977) Bounds for effective elastic moduli of disordered materials, *J. Mech. Phys. Solids* **21,** 9–17.
Laws, N. and McLaughlin, R. (1979) The effect of fibre length on the overall moduli of composite materials, *J. Mech. Phys. Solids* **27,** 1–13.
Lax, M. (1952) Multiple scattering of waves: II, The effective field in dense systems, *Phys. Rev.* **85,** 621–629.
McCoy, J. J. (1973) On the dynamic response of disordered composites, *J. Appl. Mech.* **40,** 511–517.
McCoy, J. J. (1979) On the calculation of bulk properties of heterogeneous materials, *Q. Appl. Math.* **37,** 137–149.
Ogden, R. W. (1978) Extremum principles in non-linear elasticity and their application to composites, *Int. J. Solids Struct.* **14,** 265–282.
Reuss, A. (1929) Calculation of the flow limits of mixed crystals on the basis of the plasticity of mono-crystals, *Z. angew. Math. Mech.* **9,** 49–58.
Russel, W. B. (1973) On the effective moduli of composite materials: effect of fiber length and geometry at dilute concentrations, *Z. angew. Math. Phys.* **24,** 581–600.
Sanchez-Palencia, E. (1974) Comportements local et macroscopique d'un type de milieux physiques hétérogènes, *Int. J. Engng Sci.* **12,** 331–351.
Twersky, V. (1977) Coherent scalar field in pair-correlated distributions of aligned scatterers, *J. Math. Phys.* **18,** 2468–2486.
Varadan, V. K. and Varadan, V. V. (1979) Frequency dependence of elastic (SH) wave velocity and attenuation in anisotropic two-phase media, *Wave Motion* **1,** 53–63.
Varadan, V. K., Varadan, V. V., and Pao, Y. H. (1978) Multiple scattering of elastic waves by cylinders of arbitrary cross-section: I, SH waves, *J. Acoust. Soc. Am.* **63,** 1310–1319.
Voigt, W. (1889) Ueber die Beziehung zwischen den beiden Elasticitäts-constanten isotroper Körper, *Ann. Phys.* **38,** 573–587.
Walpole, L. J. (1966a) On bounds for the overall elastic moduli of inhomogeneous systems: I, *J. Mech. Phys. Solids* **14,** 151–162.

Walpole, L. J. (1966b) On bounds for the overall elastic moduli of inhomogeneous systems: II, *J. Mech. Phys. Solids* **14,** 289–301.

Walpole, L. J. (1969) On the overall elastic moduli of composite materials, *J. Mech. Phys. Solids* **17,** 235–251.

Willis, J. R. (1975) The interaction of gas bubbles in an anisotropic elastic solid, *J. Mech. Phys. Solids* **23,** 129–138.

Willis, J. R. (1977) Bounds and self-consistent estimates for the overall moduli of anisotropic composites, *J. Mech. Phys. Solids* **25,** 185–202.

Willis, J. R. (1978) Variational principles and bounds for the overall properties of composites, *Continuum Models of Discrete Systems*, edited by J. W. Provan, 185–215, University of Waterloo Press.

Willis, J. R. (1980a) Relationships between derivations of the overall properties of composites by perturbation expansions and variational principles, *Variational Methods in Mechanics of Solids*, edited by S. Nemat-Nasser, 59–66, Pergamon Press, Oxford.

Willis, J. R. (1980b) A polarization approach to the scattering of elastic waves: I, Scattering by a single inclusion, *J. Mech. Phys. Solids* **28,** 287–305.

Willis, J. R. (1980c) A polarization approach to the scattering of elastic waves: II, Multiple scattering from inclusions, *J. Mech. Phys. Solids* **28,** 307–327.

Willis, J. R. (1981) Variational and related methods for the overall properties of composites, to be published in *Advances in Applied Mechanics*, **20,** Academic Press, New York.

Willis, J. R. and Acton, J. R. (1976) The overall elastic moduli of a dilute suspension of spheres, *Q. J. Mech. Appl. Math.* **29,** 163–177.

Subject Index